KV-620-508

Strasburger's Textbook of Botany

New English Edition

Strasburger's Textbook of Botany

Rewritten by
Dietrich von Denffer, *Giessen*
Walter Schumacher, *Bonn*
Karl Mägdefrau, *Tübingen*
Friedrich Ehrendorfer, *Wien*

New English Edition

Translated from the Thirtieth German Edition by
Peter Bell, *University College, London*
David Coombe, *Christ's College, Cambridge*

Longman
London and New York

Longman Group Limited London

Associated companies, branches and representatives throughout the world

Published in the United States of America by Longman Inc., New York

Translated from *Lehrbuch der Botanik* (30th edition) published by Gustav Fischer Verlag, Stuttgart, 1971

Library of Congress Cataloging in Publication Data

Strasburger, Eduard, 1844–1912
Strasburger's textbook of botany

Translation of Lehrbuch der Botanik
Bibliography: p.
Includes index
1. Botany. I. Denffer, Dietrich von. II. Bell, Peter Robert.
III. Coombe, David. IV. Title. V. Title: Textbook of botany.
QK47.S8913 581 75–11731
ISBN 0 582 44169 2

Set in IBM Journal, 10 on 11pt
and printed in Great Britain by
Lowe & Brydone (Printers) Ltd, Thetford, Norfolk

This textbook was founded by Eduard Strasburger, Fritz Noll, Heinrich Schenck and A. F. Wilhelm Schimper and first appeared in 1894 when all four were colleagues at Bonn. Subsequent German editions have been revised as shown below. Although the entire work has been rewritten with close collaboration between the various authors, each has been particularly responsible for one of the principal sections.

Introduction and Morphology

Eds.		
Eds.	1–11 (1894–1911)	**Eduard Strasburger**
	12–26 (1913–1954)	**Hans Fitting**
	27–30 (1958–1971)	**Dietrich von Denffer**

Physiology

Eds.		
Eds.	1– 9 (1894–1908)	**Fritz Noll**
	10–16 (1909–1923)	**Ludwig Jost**
	17–21 (1928–1939)	**Hermann Sierp**
	22–30 (1944–1971)	**Walter Schumacher**

Lower plants

Eds.		
Eds.	1–16 (1894–1923)	**Heinrich Schenck**
	17–30 (1928–1971)	**Richard Harder**

Seed plants

Eds.		
Eds.	1– 5 (1894–1901)	**A. F. W. Schimper**
	6–19 (1904–1936)	**George Karsten**
	20–29 (1939–1967)	**Franz Firbas**
	30 (1971)	**Friedrich Ehrendorfer**

Plant geography

Eds.		
Eds.	20–29 (1939–1967)	**Franz Firbas**
	30 (1971)	**Friedrich Ehrendorfer**

Translators' Preface

Our 1965 English edition of 'Strasburger' has been followed by two more German editions; two new authors have joined the team of what has long been known as the 'four man book', the text has been substantially re-written throughout (with the exception of Part Four, Plant Geography, by the late Professor F. Firbas); numerous original illustrations have been incorporated; and an entirely new section on the general principles of systematics and evolution has been written by Professor F. Ehrendorfer. Details of medicinal plants which can be found in the German and British pharmacopeias are now omitted; nevertheless the 30th German edition, of which this is a translation, is 15 per cent longer than the 28th.

This extensive revision and the inclusion of much new material maintain the position of 'Strasburger' as the most comprehensive, balanced, and up-to-date textbook of Botany in a single volume; excellent as are many of the specialized works on the numerous plant sciences, they tend to overwhelm the student by their numbers, their diversity, and their tendency to fragment the interrelated aspects of form, function, ecology and history of plants.

'Strasburger' has always been successful in combining a synoptic view with what Agnes Arber has called (in *The Mind and the Eye*) 'the delightful detail of the factual multiplicity of living things', in being both a general textbook and a reference work of lasting value. It is, furthermore, the product of the accumulated wisdom and experience of fourteen acknowledged experts familiar with the whole range of botanical learning; hence it carries a degree and breadth of authority rare in a general text.

As before, we have regarded ourselves primarily as translators rather than as editors or commentators, although we are well aware that the process of translation inevitably leads to slight changes of meaning, either when a German word has no unique English equivalent in all contexts (e.g. 'Sippe') or when the German has overtones (e.g. 'Staubblatt' and 'Fruchtblatt') absent from the English equivalents ('stamen' and 'carpel'). We have corrected a few small errors—and no doubt created some of our own; more significantly we have attempted to bring the references up-to-date, especially to recent works in English. We make no apology for retaining most of the German references: the serious student of Botany is still well advised to learn at least to read German.

D. E. Coombe
Christ's College, Cambridge

P. R. Bell
University College, London

March 1975

Contents

Part one: **Morphology**

Part two: **Physiology**

Part three: **Systematics and evolution**

Part four: **Plant geography**

Chronology

c. 300 B.C. Theophrastus (371–286 B.C.) *Enquiry into Plants*

1530 First volume of Otto Brunfels's (1488–1534) *Herbarium Vivae Eicones* published in Strasbourg.

1539 Herbal of Hieronymus Bock, known as Tragus (1498–1554).

1542 Herbal of Leonhard Fuchs (1501–66).

1590 Invention of the microscope by Johann and Zacharias Jansen.

1665 Discovery of the cellular structure of organisms: Robert Hooke (1635–1703).

1675 *Anatomia plantarum*: Marcello Malpighi (1628–94).

1682 *The Anatomy of Plants*: Nehemiah Grew (1628–1711).

1683 First illustration of bacteria: Antonius van Leeuwenhoek (1632–1723).

1694 *De sexu plantarum epistola*. Discovery of plant sexuality: Rudolph Jacob Camerarius (1665–1721).

1753 *Species plantarum*: Carl von Linné (Linnaeus) (1707–78).
The 'priority rule' in taxonomy applies from the date of this publication (1 May).

1774 Demonstration of oxygen: Joseph Priestley (1733–1804).

1779 Discovery of photosynthesis: Jan Ingenhousz (1730–99).

1790 *Metamorphose der Pflanzen*, concept of plant metamorphosis: Johann Wolfgang von Goethe (1749–1832).

1793 First studies of the biology of flowers: Christian Konrad Sprengel (1750–1816).

1804 *Recherches chimiques sur la végétation*, discovery of the gaseous exchange of plants: Nicolas Théodore de Saussure (1767–1845).

1809 *Philosophie zoologique*, evolutionary theory: Jean Baptiste de Lamarck (1744–1829).

1822 Discovery of endosmosis: Joachim Dutrochet (1776–1847).

1833 Discovery of the cell nucleus: Robert Brown (1773–1858).

1838 Formulation of the cell theory: Matthias Jacob Schleiden (1804–81) together with the zoologist Theodor Schwann (1810–82).

1840 Mineral nutrition of plants established, overthrow of the humus theory: Justus von Liebig (1803–73).

1842 Law of the conservation of energy: Julius Robert von Mayer (1814–78).

1846 Introduction of the concept of 'protoplasm' into botanical science: Hugo von Mohl (1805–72).

1851 Discovery of the homologies in plant reproduction: Wilhelm Hofmeister (1824–77).

1855 *Omnis cellula e cellula*: Rudolf Virchow (1821–1902).

1858 Formulation of the micellar theory: Carl von Nägeli (1817–91).

1859 *Origin of Species*: Charles Darwin (1809–82).

1860 Plant growth in nutrient solutions: Julius Sachs (1832–97).

1860 Refutation of the doctrine of special creation: Hermann Hoffman (1819–1891) and Louis Pasteur (1822–95).

1866 Laws of inheritance: Gregor Mendel (1822–84).

1866 Biogenetic law (*Generelle Morphologie*): Ernst Haeckel (1834–1919).

1866 Foundation of experimental plant physiology: Julius Sachs, *Handbuch der Experimental-Physiologie der Pflanzen*.

1876 Identification of the anthrax bacillus as a pathogen: Robert Koch (1843–1910).

1877	Wilhelm Pfeffer (1845–1920): *Osmotische Untersuchungen*.
1879	Discovery of nuclear division in plants: Eduard Strasburger (1844–1912).
1882	Gottlieb Haberlandt (1854–1945): *Physiologische Pflanzenanatomie*.
1884	*Vergleichende Morphologie und Biologie der Pilze und Bakterien*, a comparative treatment of the lower plants: Heinrich Anton de Bary (1831–88).
1888	Discovery of fertilization in flowering plants by Eduard Strasburger.
1900	Rediscovery of Mendel's laws: Erich Tschermak-Seysenegg (1871–1962), Carl Correns (1864–1933), Hugo de Vries (1848–1935).
1901	Hugo de Vries: *Die Mutationstheorie*.
1913	Elucidation of the structure of chlorophyll: Richard Willstätter (1872–1942) and co-workers.
1933	H. Wieland (1877–1957): *Über den Verlauf der Oxydationsvorgänge* (theory of respiration).
1935	P. Boysen-Jensen (1883–1959): *Die Wuchsstofftheorie* (theory of growth-regulating substances).
1935	Crystallization of the tobacco mosaic virus: W. M. Stanley.
1937	Citric acid cycle: H. A. Krebs and co-workers.
1937	Photolysis of water by means of isolated chloroplasts: R. Hill.
1939	First evidence of significance of nucleic acid in transmission of heredity: E. Knapp and H. Schreiber.
1940	Invention of the electron microscope: E. Ruska and H. Mahl.
1943	First evidence of genetic transformation by means of DNA: O. T. Avery, C. M. McLeod and M. McCarty.
1953	Molecular model of DNA: J. D. Watson and F. H. C. Crick.
1957	Sugar cycle in carbon assimilation: M. Calvin and co-workers.

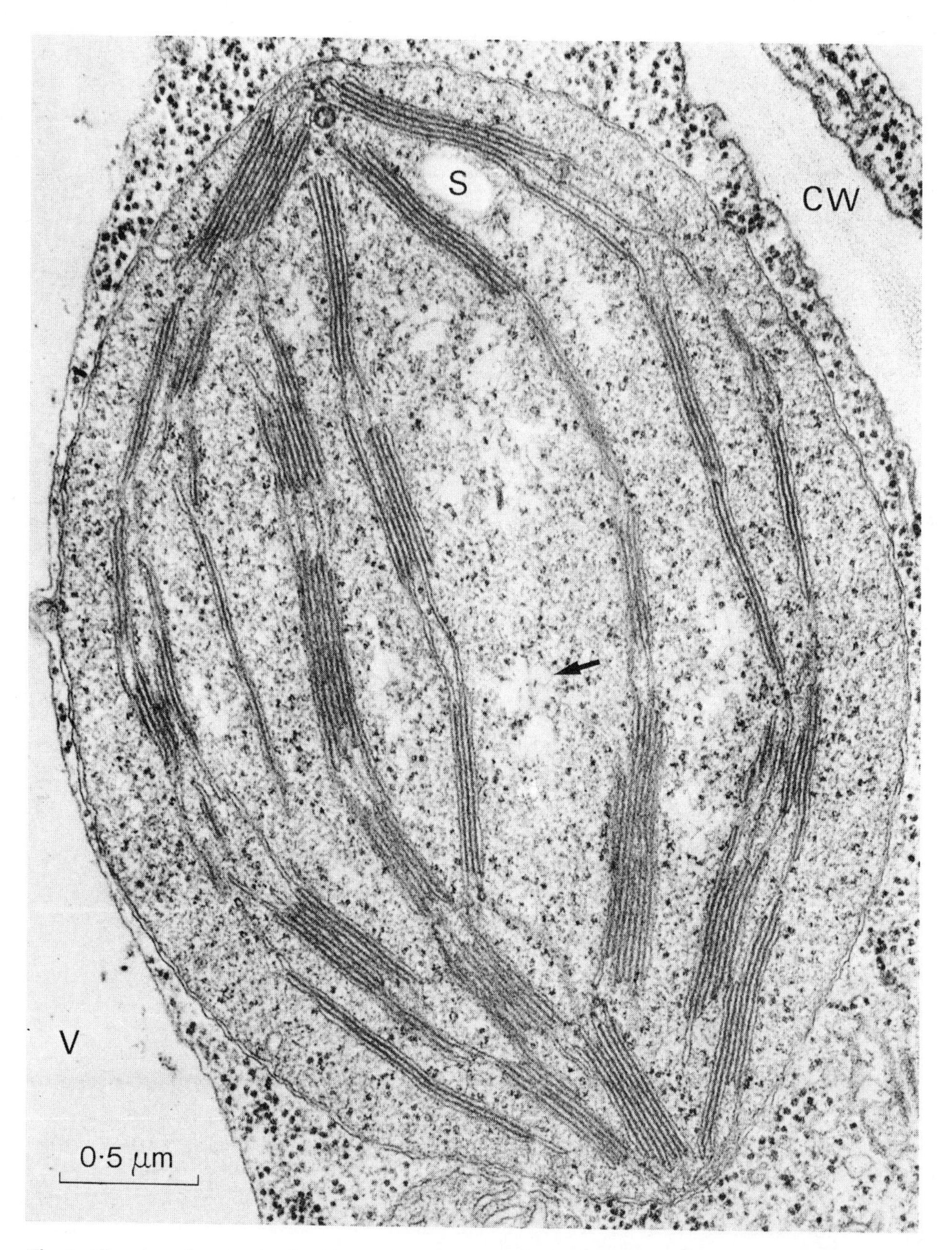

Fig. 1. Electron micrograph of a thin section of a chloroplast in the palisade tissue of spinach (*Spinacea oleracea*), fixed in glutaraldehyde and stained with lead and uranyl acetate. The envelope clearly consists of two membranes, the inner of which is continuous with peripheral vesicles (top left), and perhaps with the stroma lamellae (shown indistinctly, bottom right). The membranes forming the sacs ('thylakoids') of the grana are clearly in continuity with the stroma lamellae. The staining treatment reveals the spaces containing the DNA fibrils (an example is arrowed), and the ribosomes. Those in the stroma of the chloroplast can be seen to be slightly smaller than those in the ground cytoplasm. S, starch grain; CW, cell wall; V, vacuole. Micrograph supplied by A. D. Greenwood.

Introduction

Botany is the natural history of the Plant Kingdom. Along with zoology, the natural history of the Animal Kingdom, and anthropology, the natural history of Mankind, botany forms part of biology, the science of life itself.

General observations on the living state.[1] In contradistinction to non-living materials, living forms—plants, animals and Man—constitute a 'new category' in the sense of the philosopher Nicolai Hartmann. When oxygen and hydrogen, two gaseous substances, combine in a certain way they give rise to an entirely new property—fluidity—possessed by the water produced. Again, other atoms, themselves colourless, brought together chemically in a certain combination, suddenly generate colour. In the same way life can be regarded as a novel property of matter, arising only when certain molecules, which, if isolated or in simple chemical compounds, would remain dead for all time, occur together in a certain definite arrangement. A living system is thus more than the sum of its parts.

This specific organization of living material, revealing the fundamentally material character of life, has morphological and dynamic consequences. Morphologically it expresses itself in the building up of individuals clearly set off from their environment, and usually distinguished by a well-defined form.

The dynamic consequences are seen as three new properties, foreign to non-living material in general: metabolism, productivity and irritability. In metabolism, a constant activity, dead material is taken up from the environment and built into the living system (anabolism, assimilation), while at the same time dead refuse materials, produced by processes of breaking down, are constantly being returned to the inanimate realm (catabolism, dissimilation). The organization of the living material itself remains unchanged; it exists in a state of dynamic equilibrium. Productivity manifests itself in growth and reprodution. Growth usually occurs when the processes of building up exceed those of breaking down. In reproduction or self-multiplication (idiosynthesis) an individual gives rise to descendants which resemble in characters and properties the initial form (autoreduplication, identical reduplication, heredity). Finally, by the phenomenon of irritability is understood the capacity of an organism to react to changes in the outer and inner environments in a manner which cannot be accounted for in terms of the energy involved in the stimulus, but in which energy is supplied from the reserves of the organism itself (excitation mechanisms).

Although some of these properties may occasionally occur among inanimate objects (e.g. the form of a crystal, the metabolism of a candle flame, the autocatalytic reproduction of certain chemicals and the 'irritability' of a mousetrap when set), their occurrence together at one and the same time is exclusively a property of living organisms.

In both plants and animals the basic material of all the vital phenomena is the protoplasm, a highly organized system of numerous different, partly simple, partly very complex, chemical compounds. Among these compounds the macromolecules of the proteins and nucleic acids are of outstanding importance, since they are carriers of specific organizers, and are passed on unchanged from generation to generation. The investigation of the protoplasm and its sub-microscopic fine structure is thus one of the most important tasks of biology. It demands the closest cooperation between biologists, biochemists and biophysicists.

The viruses, sub-microscopically small, filterable agents of disease in plants and animals, also always contain protein and nucleic acid. Since their discovery they have figured prominently in discussions of the origin of life. Like organisms, viruses are capable of self-reproduction. Since, however, they lack a metabolism of their own, and can consequently multiply only with the help and at the expense of a more highly organized organism, the crystallizable viruses cannot be counted as proper living forms (see p. 299).

Origin of life.[2] We can only make conjectures about the origin of life on our earth. We have good reason to believe that living forms have been present on it for several hundred million years (see Fig. A1, p. 794). It is also clear that the first living forms were very much more simply organized than the great majority of those in existence today.

In ancient time, and even up to the nineteenth century, the view was widespread that life was capable of being spontaneously and sporadically created from inanimate matter. According to Aristotle (384–322 B.C.) not only plants but also worms, larvae of flies and other insects were said to proceed from dew, mud, rotting manure and excrement by a process of spontaneous generation (*generatio spontanea*). Even in the seventeenth century, one of the most famous scholars of that time, van Helmont, maintained that mice could arise from wheat bran and the effluvia of old worn shirts! The speculative nature philosophy at the end of the eighteenth and the beginning of the nineteenth century interpreted petrifactions as the 'awakening' of organisms to a living state.

Scientific evidence that all life as we know it today is always derived from life was not obtained until the careful experiments of Hermann Hoffmann (1819–91) and Louis Pasteur (1822–95), although this view had already been put forward very much earlier by a number of careful observers. The famous English anatomist, William Harvey (1578–1657), for example, maintained that all animals arose from eggs: *omnia animalia ex ovo*. The physiologist W. Preyer later put this axiom into a more general form: *omne vivum e vivo*. Many biologists still regard this today as one of the most fundamental principles of their science.

In more recent time the view has grown, especially among biochemists, that it is not only very probable, but also logically inevitable, that life must have arisen spontaneously in the particular conditions which must have prevailed on our planet many hundreds of millions of years ago (see Fig. A1). The idea, going back as far as Anaxagoras (*c*. 500–428 B.C.), that life is eternal and has arrived on the Earth in the form of 'ethereal seeds' was later elaborated by the Swedish physicist Svante Arrhenius as the hypothesis of panspermy. This hypothesis has lost much of its credibility now that cosmic radiation is known to have such lethal effects. On the other hand, the American chemist Stanley L. Miller has shown that organic compounds important for life (among which the amino acids glycine and alanine have been demonstrated with certainty) can be produced by passing electrical discharges through an oxygen-free atmosphere consisting of hydrogen, steam, ammonia and simple hydrocarbons, such as might have existed very early in the Earth's history. The gap between amino acids and protein is still, of course, very considerable. Amino acids are far from proteins, and pure proteins are still a long way from protoplasm.

Animals and plants. The maintaining of the specific organization which characterizes life is a process utilizing energy. Life is consequently an energy-consuming process, and it is dependent upon the possibility of setting up sufficient reserves of energy for the maintenance of the organized structures involved. The moment the supply of energy ceases, active life comes to an end, and the organization begins to break down.

According to the second law of thermodynamics, all changes in inorganic systems are accompanied by a loss of organization (or increase of entropy). Living forms, on the contrary, possess the capacity of taking up energy uninterruptedly from natural energy gradients, and using it for the maintenance of their organization. Thus they reverse the natural tendency for increase in entropy, and in this way actively escape the constant progression towards atomic disorganization to which all inorganic matter is subject.

Animals obtain the energy necessary for the maintenance of life solely from their nutrition, but green plants obtain it directly from the sun. Animals and green plants are thus fundamentally different in the nature of their energy supply. The plants, with the exception of a few families and specialized forms (see p. 4), are autotrophic; animals are heterotrophic. All the other differences between the organizations of animals and plants rest in the last analysis on this primary and fundamental distinction.

Green plants owe their autotrophic nutrition and habit to their possession of the pigment chlorophyll. With its help the radiant energy absorbed from the sun is made available for the synthesis of organic molecules of very high energy values. The energy required for the maintenance of life and for reproduction is drawn from the reserves created in this way. Only green plants possess chlorophyll, and consequently only they are able to provide directly, i.e. by the exclusive use of the sun's radiation, the supply of energy necessary for the synthesis of organic carbon compounds. Ultimately therefore all animal life is dependent upon that of plants, and were there no green plants in existence animal life on earth would be inconceivable.

The most important element for the synthesis of organic substances, carbon, is distributed everywhere in the air as carbon dioxide (about 0.03 per cent by volume). Less self-evident is the presence of water, which yields the second indispensable element of all organic materials, hydrogen. In addition, natural water usually contains a number of other important elements in solution, such as nitrogen, sulphur, phosphorus, iron, potassium, calcium and magnesium, all essential for the building up of the protoplasm. Plants will grow anywhere where light, air and water are available. With the exception of completely dry desert regions, and the summits of high mountains and polar regions perpetually covered with ice and snow, there is hardly any place on Earth's surface which could not be colonized by plants.

Since the green plants extract the energy required for the assimilation of carbon and of other elements from sunlight, they need not, like the animals, which must forage for their food, be mobile. The typical higher plant is consequently firmly rooted to its habitat.

There are, however, a number of exceptions to this rule. There is, for example, a whole group of aquatic plants, mostly microscopic, which so long as they are alive swim freely with the help of special locomotory organs. Again, numerous lower plants develop motile reproductive stages or zoospores from time to time (Fig. 2). At first this was regarded as a remarkable instance of plants becoming animals, so firmly was the view held that free locomotion was a property found only in animals. On the other hand, animals with an adequate nutrient supply may become firmly attached to the substratum like plants and engulf the necessary food materials with the help of special mechanisms. Examples are provided by the sponges and several coelenterates, classified by Linnaeus with the plants.

As a consequence of the different modes of nutrition of plants and animals, they have fundamentally different methods of growth. Both animals and plants require large surfaces for the functioning of their metabolism. While, however, following an early

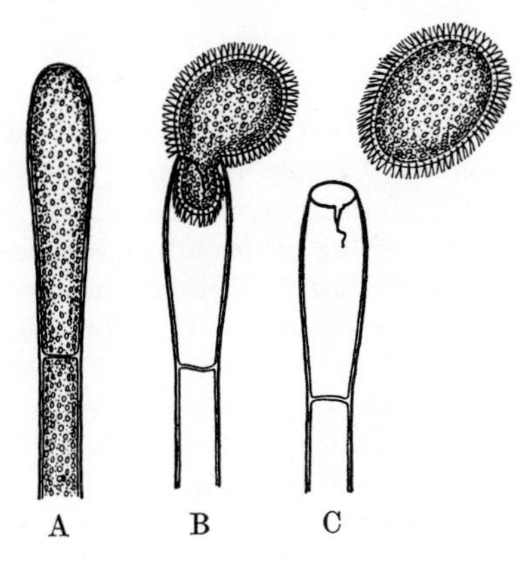

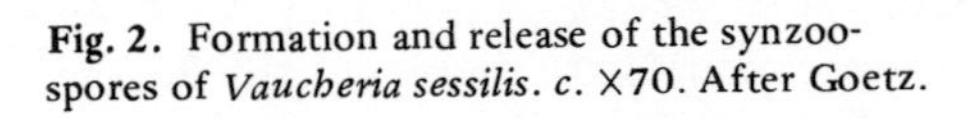

Fig. 2. Formation and release of the synzoospores of *Vaucheria sessilis*. *c*. ×70. After Goetz.

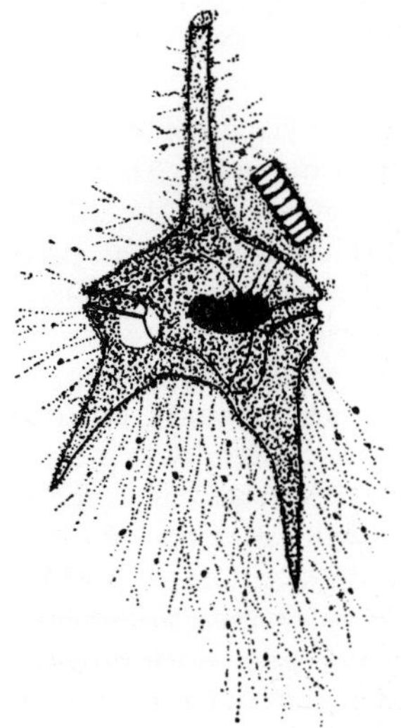

Fig. 3. (Right) *Ceratium hirundella*. This unicellular organism, capable of photosynthesis, swims freely in the water. It is surrounded by delicate cytoplasmic threads which project from its membrane, and which catch and digest small planktonic organisms. ×500. After Hofeneder.

invagination, the development of animals is directed towards the creation of the largest possible internal surface for the uptake and absorption of nutrients, that of plants is in the reverse direction and aims at the largest possible external assimilating surface accessible to light. Since plants have no need to move, no external limits are set to their growth. In contradistinction to the animals, therefore, which cease to grow after a certain period of juvenility and adolescence, plants continue to grow and produce new organs until they die. This distinction is usefully expressed by saying that animals have a 'closed' and plants an 'open' form.

Plants are usually richly branched, and at the extreme tips of the branches, especially in the higher plants, are growing points, i.e. zones where the embryonic condition is never lost. These perpetually embryonic cells make it possible in certain circumstances, e.g. by the separation of individual shoots at an appropriate time (propagation by cuttings, p. 203), to continue almost indefinitely the life of a 'plant' even after the death of the initial individual.

Finally, the attachment of plants to their habitat and their immobility are also bound up with an organizational and structural difference at the microscopic level. With few exceptions both plants and animals are seen under the microscope to consist of millions upon millions of minute units, the cells. Almost without exception, the cells of plants are enclosed within a special cellulose membrane, the cell wall. (The cell walls of bacteria, however, and of many lower and all higher fungi usually consist of other substances. In the bacteria various highly complex proteinaceous compounds (mucopeptides) predominate, while in the fungi the principal wall material is chitin, a polymer of acetylglucosamine.) Animal cells only rarely show a cell wall, and cellulose is more or less confined to a single phylum, the tunicates.

To summarize, the most important characters distinguishing plants and animals are as follows:

Plants	*Animals*
Chlorophyll present	Chlorophyll absent
Autotrophs	Heterotrophs
Capable of immediately utilizing solar energy	Energy obtained indirectly by the taking up of energy-rich organic substances as food
Metabolism directed towards building up of complex materials	Metabolism directed towards breaking down of complex materials
Growth and form indeterminate	Growth and form determinate
Non-mobile, often firmly attached to substratum	Mobile, not bound to one site
Firm cell wall	Naked cells

Both animals and plants, especially the lower, provide exceptions to many of the characteristics given. Thus in the course of evolution many plants, e.g. fungi, probably a very old group, have lost the capacity for autotrophic life. Exactly as the animals, they must accordingly feed as parasites or saprophytes upon living or dead organic material, initially created by other plants (see p. 291 seq.). It is not usually difficult, however, to assign such organisms to their systematic position in the Plant Kingdom (see, for example, Fig. 218, p. 198; Fig. 728, p. 731).

There are, however, other groups of organisms which it is extremely difficult or even impossible to classify as plants or animals. Such are the microscopically small, unicellular, Protista, which appear in the textbooks of both botanists and zoologists. They combine typically animal properties (e.g. motility and the uptake of complex nutrients) with the possession of chlorophyll and the capacity for autotrophic life (Fig. 3). In the context of the theory of evolution, according to which the more complex forms have descended from the simpler, such mixotrophic organisms (which obtain their energy partly directly as autotrophs, and partly indirectly as heterotrophs) are regarded as the starting-point of two great developmental series which have led on the one side to the 'typical' plants and on the other to the 'typical' animals. At the lowest grade of organization there is still no sharp distinction between the two possible kinds of development. It is, however, generally agreed that all organisms characterized by the possession of chlorophyll be considered as plants, even though they may at the same time be capable of an animal-like nutrition.

The subdivisions of botany

The investigation of the Plant Kingdom can proceed from a number of very different points of view. The manner of the investigation may, for example, be determined by the level of organization to which it relates, and this may range from the order of molecules and cells, through tissues and organs to the whole plant, and even beyond to populations and plant communities. Yet other approaches are seen in the study of floras, both of the geological past and present, of the interrelationships of plants, of the uses of plants by Man, and so on. No matter what kind of investigation is being made, however, the method of working is first to analyse individual phenomena in detail, then to proceed by way of comparison to generalizations, and thence to the identification of unifying principles. In any such investigation it is essential that both the static and dynamic aspects of the problem are considered, for they complement each other. On the one hand, therefore, there is the recognition and interpretation of structures and forms, on the other the analysis of living processes, function and development. The ultimate goal of both disciplines, however, must always remain the same, namely to try to discover how form and function have a continual and reciprocal relationship.

The different points of view we have discussed, and the use of techniques appropriate to them, have led to the development of numerous, interrelated, specialist fields in botany. Here we begin with morphology. In the widest sense this is the general study of the structure and form of plants, and includes cytology, concerned with the fine structure of the cell (which, in the realm of the molecular, cuts across aspects of molecular biology), and histology, the study of tissues. Both cytology and histology concern the inner structure, or the anatomy, of plants, and are distinct from organography (morphology in the narrow sense), which is the study of the outer structure. In treating of adaptive structures morphology borders on plant ecology, the study of the relations between plants and their environment. Here autecology devotes its attention to the environmental characteristics of an individual species, while synecology attempts to fathom the interrelationships of whole plant communities with climate and soil, a study referred to again under the headings physiology and geobotany.

Plant physiology follows in the second part of the book. This concerns itself with the general course of metabolism, with morphogenesis (including growth and development) and with plant movements, all questions involving ultimately aspects of molecular biology. A branch of biological research today largely independent is genetics, the study of inheritance, concerned with the identical reproduction of organisms and with hereditary changes (mutations). Genetics can be considered only very briefly here, and principally in relation to reproduction, morphogenesis and plant evolution.

The third part of the book deals with the systematics of plants. As the study of the interrelatedness of plants it draws on the results from all other disciplines. Besides external morphology, important sources of information are cytology, anatomy, palynology (structure of pollen and spores), embryology (development of the gametophyte generation and of the embryo), phytochemistry (chemical constituents of plants), genetics and geobotany (plant geography). A subdivision of systematics is taxonomy, which has the task of describing, naming and arranging in a systematic way the over 500,000 plants living today. Another aspect of systematics is the elucidation of the lines of descent in the Plant Kingdom (phylogeny), particularly helped by palaeobotany, the science of plants of earlier geological time, and also by studies of evolution, which seek to discover the laws and causes of affinity between organisms. It is possible to allude here only briefly to fields which deal intensively with individual groups of organisms, as, for example, microbiology (microorganisms), bacteriology (bacteria) and mycology (fungi), and to those applied disciplines which investigate the plants of economic significance to Man and which are consequently allied to agriculture and forestry, pharmacy, and so on.

Sufficient be it to mention horticulture, plant breeding, phytopathology (plant diseases and their prevention) and pharmacognosy (plant medicines and drugs).

The last section of the book is devoted exclusively to geobotany (plant geography). This discipline, encompassing the study of plant distribution, of plant assemblages (including plant sociology) and habitats, together with that of the history of floras and plant communities, seeks to explain the principles and causes of the spread and association of plants on the surface of the Earth in space and time.

Until the middle of the preceding century interest was predominantly in the static aspects of biology, and the approach consequently descriptive and comparative. It was then justifiable to contrast botany and zoology as descriptive sciences with the so-called 'exact' sciences of physics and chemistry. Latterly, however, the dynamic aspects, and with them the causal-analytical approach, have come ever more into the foreground. Intervention on the part of the investigator into the normal course of development is one of the most important techniques in this kind of approach; experiment replaces mere observation. Advances in equipment and apparatus have at the same time facilitated an expansion of research from the level of whole plants and organs into that of the cellular, and ultimately of the molecular. These developments, together with our continually growing recognition of the causal interrelationships of all phenomena in the biosphere, are gradually leading to an ever-increasing interaction of the different aspects of botany, and even of biology as a whole and the natural science generally.

Part one: **Morphology**

Tasks and methods. The Plant Kingdom confronts us with overwhelming diversity. The task of morphology is to trace out the general laws which underlie the individual forms. This end can be achieved in two different ways. Firstly, the type of a group of forms can be found by a comparative examination, i.e. by looking at a succession of individual forms. Secondly, naturally and artificially produced modifications of the normal structure reveal the potentialities lying dormant in the different forms and their organs, and allow relationships to come to light which would otherwise lie hidden.

The first way constitutes comparative morphology. It starts from the assumption that a variety of forms are merely developments from a type. It considers its goal to be the recognition and the abstraction of these types from among the individual forms. The best known exponent of this method is Goethe, who in his *Metamorphose der Pflanzen* (1790) attempted to formulate the type of the *Urpflanze* (primordial plant). The second way is that of experimental or analytical morphology, from which modern developmental physiology has arisen. Finally, the phylogenetic approach successfully utilizes both methods. It attempts to reconstruct phylogenetic sequences by tracing arrays of present-day forms back to common starting points.

Every plant should perhaps be regarded as having been developed in two ways, and without taking into account this twofold development it cannot be fully comprehended. On the one hand, it has arisen from a fertilized egg cell, and in the course of its individual development or ontogeny it has passed in an orderly manner through a succession of developmental stages. On the other hand, the species to which it belongs has usually in the course of geological time evolved from simpler ancestors. Consequently it is quite impossible to understand many of a plant's characteristics without at least some idea of its evolutionary history or phylogeny.

The actual development of a species in time, which we can regard as a sequence of countless ontogenies, is referred to as its hologeny. The phylogeny or evolution of an organism therefore encompasses the successive modifications of the ontogenies which have contributed to its hologeny (see p. 389 seq.). The ultimate form of an organism or of a plant tissue can only be understood from its developmental history. On the other hand, the present-day form of a species can be effectively understood only with a knowledge of its evolutionary history.

Organs which have arisen from a common basic form, i.e. which have an identical ground plan, are said to be homologous, i.e. identical in origin (see p. 178). On the other hand, adaptations externally very similar can sometimes be traced back to very dissimilar initial forms. Forms which have arisen in this way by convergence are analogous, i.e. identical in function.

Organizational characters, adaptations, indifferent characters. To arrive at a true understanding of the different forms of plants, it is not sufficient to study merely the causal factors governing their development. An attempt must also be made to assess their significance. The ultimate goal of this last approach is to comprehend the organism and its organs as a contingent expression of its adaptation to its particular habitat. In the course of its phylogeny every plant has accumulated structural peculiarities which make it possible for it to live in certain environmental conditions, but by no means everywhere on the Earth's surface. Aquatic conditions are, for example, very different from those of the desert, and water plants and desert plants are accordingly very differently constructed. They are adapted to their habitats and are capable of thriving only in the conditions to which they are accustomed, or in closely similar ones. Characters intelligible from the point of view of evolution are termed organizational, whereas structural peculiarities acquired as functional modifications in relation to the environment are termed adaptations. Close examination shows that besides the organizational characters and adaptations there are numerous other characters which so far it is impossible to interpret either historically or in relation to any function (indifferent characters).

The number of the various parts of the flower of a species can usually be taken as an organizational character. Their arrangement (radial or bilateral symmetry, or asymmetry), on the other hand, generally shows the influence of adaptation. Lipped and papilionate flowers, for example, are adapted to particular forms of pollination. Flower colour, which frequently serves to attract the pollinator from afar, can also in most instances be regarded as a similar adaptational character. On the other hand, many patterns of pigmentation which differ from one another only in detail, such as the patterns in the lips of the flowers of many Labiatae and Orchidaceae, are probably generally of little significance to the species concerned, and should be regarded as indifferent characters. In many species of orchids, e.g. those of *Ophrys*, numerous varieties are found growing together in the same habitat, clearly distinguished from one another by the hereditary patterns of pigmentation of their flowers, but none of the varieties seems to have any advantage or disadvantage in relation to the others. It is possible, of course, for characters at first indifferent rapidly to become adaptations. Thus conspicuously coloured areas of sap (honey-guides) may indicate to pollinators the way to the nectar.

In general, detailed investigation of the wealth of forms encountered in nature shows that the diversity in growth and development exceeds that of the environmental conditions able to support life.

Since we must consider plants predominantly as living organisms, the first aspect of them to engage our attention is the site of the vital activity, namely the protoplasm. We usually encounter this substance in the form of microscopically small units known as cells. These cells possess the capacity to reproduce themselves, a property which we have come to regard as one of the most important and basic phenomena of life. Consequently we must study the structure of cells in detail before anything else. This study is known as cytology.

Large tracts of similarly differentiated cells are referred to as tissues. An aggregation of tissues with a definite and characteristic form, and certain specialized functions, is called an organ. The study of tissues is the task of histology. Cytology and histology together form anatomy. In this respect the botanical terminology differs from that customary in zoology and medicine. What is referred to in these two disciplines as anatomy corresponds in botany to organography, which is concerned with the investigation of the external morphology of the basic organs of the plant. Plant anatomy and organography cannot be clearly distinguished from one another, since there is very considerable interaction between inner organization and outer form. It is only to facilitate the presentation of the information that in the following account a short consideration of the most important kinds of tissues precedes the study of organs and their differentiation.

Section one: **Cytology**[3]

I. The cell as the unit of life

In his *Micrographia* of 1665, Robert Hooke first reported his observation, sensational at the time, that the apparently uniform and firm vegetable matter of bottle corks was in reality composed of innumerable tiny 'boxes' or 'cells'. The term 'cell' which he then introduced came later into general use. It was coined, like the term 'cytology', on the false assumption that the cells were filled only with air or water. Since Hooke made his discovery on dead material, it is not surprising that he saw only the non-living framework of cell walls and cell membranes. The true living matter, contained within the lumen or inner part of the cell, and known as the protoplast, was not discovered until much later. The concept of protoplasm (first used by the Czech physiologist Purkinje in 1829) was brought into botanical science by Hugo von Mohl in 1846.

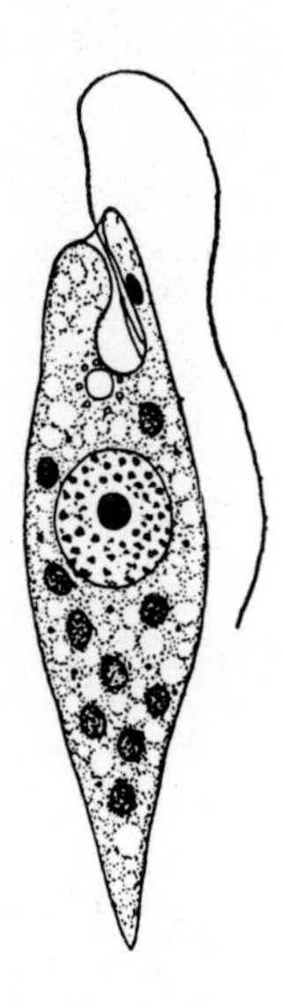

Fig. 4. *Euglena* sp. The flagellum at the anterior end is inserted within a throat-like vacuole. There is also a pulsating vacuole and an eye-spot. In the centre is the nucleus with its nucleolus. The chromatophores are indicated as dark bodies. The cell is naked and can therefore change its form. ×750. After Haupt.

A. Form and size of cells

The form and size of cells can vary very markedly, corresponding to their different functions. The simplest form—the sphere (Fig. 442E, p. 482; Fig. 445, p. 486)—occurs only rarely. Cells occurring in tissues are usually isodiametric (cubical or polyhedral), together with others more or less elongated, and prismatic or filamentous in form. The isodiametric cells occur predominantly in ground tissue or parenchyma, and are termed parenchymatous, while elongated cells in general are termed prosenchymatous. Many Protista consist throughout life of a single cell, often showing a strikingly complex differentiation (Figs. 3; 4).

The smallest cells are found in those bacteria which exist as isolated cells. Micrococcal cells have a diameter of less than 0.2 μm. Assuming that the approximately spherical spores of such micrococci have a diameter of only 0.12 μm the construction of such an 'atom of life' nevertheless involves the vast number of some 100 million chemical atoms. A truly living organism of smaller size is not known. The gulf between the first artificially synthesized constituent of organic life and a genuine living organism remains enormous.

The transverse diameter of plant cells commonly lies between 10 and 100 μm. The upper epidermal cells of many leaves, the cells of many plant hairs and the cells of the flesh of a ripe mealy apple or of boiled potatoes can even be seen with the naked eye. Prosenchymatous cells only a few microns in diameter may reach several millimetres or centimetres in length (see the following table). Multinucleate latex ducts (see p. 119) may even extend several metres.

Orders of size 1 μm = 1/1000 mm = 10^{-3} mm; 1 nm = 1/1000 μm = 10^{-6} mm
(*L*: length *D*: diameter)

Cells of higher plants	
Boehmeria nivea—fibre (*L*)	250–550 mm
Urtica dioica—fibre (*L*)	50–75 mm
Linum usitatissimum—fibre (*L*)	40–65 mm
Musa tracheid (*L*)	8–10 mm
Vinca minor—fibre (*L*)	1–2 mm
Pinus sylvestris—tracheid (*L*)	1–2 mm
Sambucus—pith parenchyma (*D*)	0.2 mm
Rosa—epidermal cells (*D*)	0.04 mm = 40 μm

Algal cells	
Chara—internode cells (*L*)	40–80 mm
Acetabularia (*L*)	50–60 mm
Chlamydomonas (*D*)	0.02 mm = 20 μm

Bacterial cells	
Thiospirillum jenense (*L*)	0.08 mm = 80 μm
Escherichia coli (*L*)	0.003 mm = 3.0 μm
Micrococcus (*D*)	0.0002 mm = 0.2 μm

Intracellular structures	
Chloroplasts of land plants (*D*)	4.0–8.0 μm
Golgi bodies (dictyosomes) (*L*)	0.2–5.5 μm
Mitochondria (*D*)	0.5–0.8 μm

Limit of resolution of light microscope	
Visible light	*c.* 0.40 μm = 400 nm
Ultra-violet light	*c.* 0.24 μm = 240 nm

Viruses	
Tobacco mosaic virus (*L*)	0.28 μm = 280 nm
Influenza virus (*D*)	0.12 μm = 120 nm
Tz *E. coli* bacteriophage (*L*)	95 nm
Foot-and-mouth disease virus (*D*)	10 nm

Intracellular structures	
Interlamellar space of endoplasmic reticulum (*D*)	25–30 nm
Ribosomes (*D*)	10–15 nm
Unit membrane profile (*D*)	6–8 nm
Limit of resolution of electron microscope	*c.* 0.8 nm
Organic molecules	
Haemoglobin	6.4 nm
Chlorophyll (*L*)	3.5 nm
DNA helix (*D*)	2.5 nm
Glucose (*L*)	0.7 nm
Hydrogen atom (*D*)	0.1 nm

B. Significance of cellular organization for organisms[4]

Cells are not only the units out of which organisms are built, they are also the sites of growth and development. Indeed all organisms develop initially from a single cell.

One of the most important discoveries in biology was the clarification towards the end of the nineteenth century, through the work of Bütschli, Strasburger, Fleming and others, of the complicated process of cell division. Eventually Virchow's famous dictum *omnis cellula e cellula* became substantiated by exact scientific observations. At first, of course, the idea that cells multiply themselves by division was very distant. At the beginning of the nineteenth century it was still generally believed that cells differentiated themselves like crystals out of unorganized living material. Today we know that the cells are in fact the sites of identical reproduction, which in greater units would probably be impossible. Cellular organization was therefore the necessary starting-point for any higher development. Even in so small a unit as a single cell, the exact passing on of the specific structures which characterize the living form concerned is still an extremely difficult and complicated process. Not only must it be possible to pass on the mere capacity for life, but also the specific kind of life, such, for example, as distinguishes the tulip from the apple tree. Not only do the reproductive cells, newly differentiated each time, show themselves capable of repeating the developmental sequence in a subsequent generation but also, it is clear, any somatic cell of the plant may in certain conditions give rise to a complete new plant. Such plants are quite normal with respect to growth and function (Figs. 273; 274). These facts force us to look within the cell itself for those permanent features which make this remarkable behaviour possible.

C. The structure of a typical plant cell[5]

Figure 5 is a highly diagrammatic representation of a differentiated cell from the assimilatory tissue of a green leaf. Figure 6 shows the fine structure of a very young (meristematic) cell, as revealed by the very much higher resolving power of the electron microscope.

The typical plant cell is surrounded by a cell wall (W). The space within, the cell lumen, is occupied by a transparent, mucilaginous and viscous mass, the protoplasm. Even superficial observation reveals that the protoplasm is not uniform in structure. Within a ground cytoplasm, which is more or less homogeneous and optically empty as seen in the light microscope, various components can be recognized, varying in size, but of well defined morphology. These components are delimited by special membranes, and fulfil

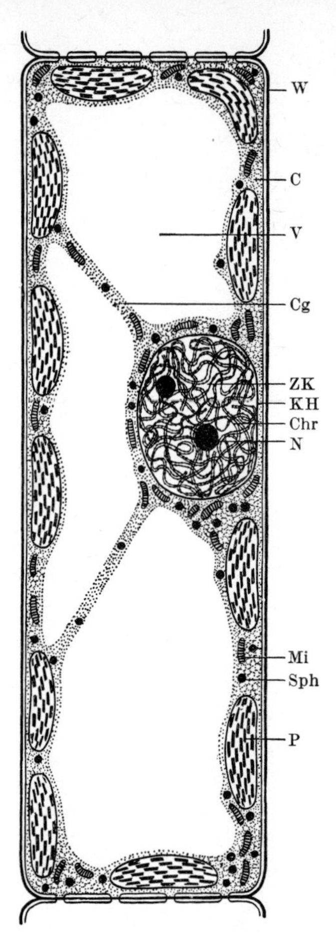

Fig. 5. Diagram of a mature cell from the assimilatory parenchyma of a leaf: W, cell wall; C, cytoplasm; V, vacuole, traversed by strands of cytoplasm (Cg); ZK, cell nucleus; KH, nuclear envelope; Chr, chromatin (extended chromosomes); N, nucleolus; P, plastid (chloroplast); Mi, mitochondrion, Sph, spherosome.

quite definite functions in the life of the cell. They can thus be justifiably referred to as 'organs of the cell', or cell organelles.

The most important structure differentiated within the protoplasm is the nucleus (ZK). This is a globular or spindle-shaped body, enclosed within a special membrane. In embryonic cells it is usually at the centre, but in older cells it often lies at one side close against the cell wall. The protoplasm of the nucleus itself is called the karyoplasm (or nucleoplasm), the remainder, outside the nucleus, the cytoplasm. The karyoplasm is distinguished from the cytoplasm by the presence in the former of a special protein (nucleoprotein), forming (after fixation) a fibrous network with special staining properties, the chromatin (Chr). Also in the nucleus are found one or more highly refractive, spherical bodies, the nucleoli (N). The nucleus is surrounded by its own envelope of two membranes, referred to as the nuclear membrane by light microscope cytologists, and the nuclear envelope by electron microscopists (KH). The envelope is perforated by pores about 25 nm in diameter.

All green cells of land plants contain lens-shaped chromatophores, the chloroplasts (Figs. 5, P; 38C, p. 45). They are usually numerous, the number in a cell reaching or exceeding 100. They are the bearers of the photosynthetic pigments, and the place where the assimilation of carbon dioxide of the air is accomplished. In illuminated cells the chloroplasts often contain starch, the first visible product of photosynthesis (Fig. 38G, p. 45).

Chloroplasts are already represented in colourless embryonic and meristematic cells as very small and unpigmented, but clearly recognizable, proplastids (Fig. 6, P). As growth takes place, the proplastids also develop, accompanied by a gradual accumulation of the pigments, until finally they become the typically constructed chromatophores (Fig. 5, P). Ontogenetically, morphologically, and functionally, the chloroplasts can be assigned to a multifarious class of organelles, occurring in both green and colourless cells, the plastids.

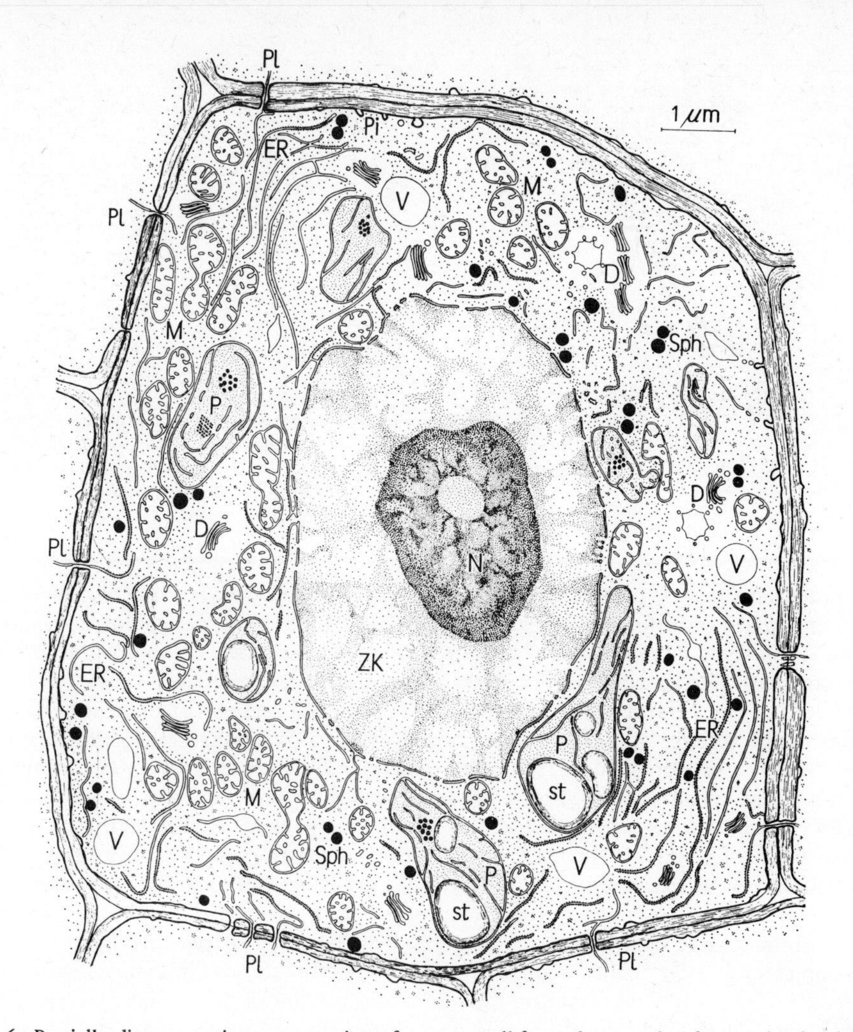

Fig. 6. Partially diagrammatic representation of a young cell from the root tip of a germinating pea as it appears in the electron microscope. ZK, nucleus; N, nucleolus; P, proplastid with starch; M, mitochondrion; D, Golgi body (dictyosome); V, vacuole; Sph, spherosome or lipid droplet; ER, endoplasmic reticulum; Pl, plasmodesma; Pi, site of pinocytosis. The ground plasm contains numerous minute ribosomes. Much of the endoplasmic reticulum is also thickly beset with ribosomes ('rough e.r.'), but some remain quite free of them ('smooth e.r.'). ×10,000. From an original drawing by E. Perner.

The particular cytoplasm contained within plastids is sometimes referred to as the plastidoplasma.

Besides the large, and consequently easily detectable plastids, every plant cell contains very numerous colourless, mitochondria (chondriosomes) (Mi). These are of the order of 1–2 μm in length, and their shape, which is capable of change, may be spherical, cylindrical, rod-like or thread-like. The mitochondria, like the plastids, are also surrounded by an envelope of two membranes. Their plasma is referred to as chondrioplasma.

Cell nuclei reproduce solely by division. Plastids and mitochondria also probably reproduce principally by division, and correspondingly contain their own charge of information in chemical form (plastid and mitochondrial DNA). These three large organelles thus distinguish themselves from the remaining, usually very much smaller, components of the protoplasm. They are consequently considered in detail in a later section.

With some objects it is possible to detect, in certain conditions even with the light microscope, small, droplet-like, spheres, about 1 μm in diameter, referred to as spherosomes (Sph), and also more or less disc-like structures known as Golgi bodies, or dictyosomes. In the electron microscope these are found to consist of stacks of flattened vesicles, each surrounded by a single membrane (Fig. 6, D), the stack itself lacking any particular sheath. In many cells the electron microscope has revealed extremely delicate

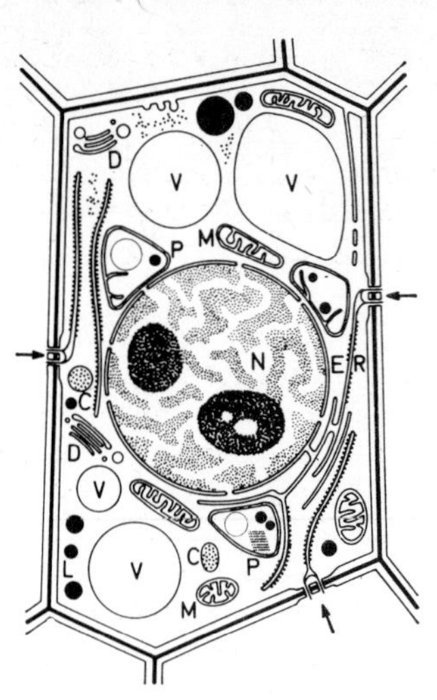

Fig. 7. Diagrammatic representation of a young plant cell. N, nucleus with two nucleoli; V, vacuole; M, mitochondrion; P, proplastid; C, lysosome; ER, endoplasmic reticulum; D, Golgi body (dictyosome); L, lipid droplet. The arrow indicates primary pit fields with plasmodesmata. After Sitte.

membranous structures, present in the form of sheets or tubules, referred to collectively as the endoplasmic reticulum (Figs. 6 and 7, ER). In life this reticulum is probably constantly changing its extent and distribution. In certain specially favourable instances it can be seen with the light microscope if phase-contrast optics are used.

The cells of the heterotrophic animals differ principally in the lack of a firm cell wall, and also in the absence of plastids, which, however, are also lacking in the cells of heterotrophic bacteria, blue-green algae, fungi and slime moulds. Furthermore, in the cytoplasm of mature plant cells there are always regions occupied by a watery cell sap, the vacuoles. Their formation depends upon the fact that in cell enlargement, facilitated principally by the absorption of water, the increase in the amount of protoplasm does not keep in step with the increase in cell volume (Figs. 5; 6; 7, V). Cells not infrequently expand to 1000 times their initial volume, and during this process of growth the vacuoles often amalgamate to form a large central vacuole (Fig. 5, V). The nucleus, plastids, mitochondria, spherosomes and golgi bodies remain throughout within the cytoplasm, which in the mature cell forms only a thin layer on the inside of the cell wall. It may become so thin as to be not immediately visible. Removing water from the cell with, for example, a strong solution of salt or sugar causes the layer of cytoplasm to become apparent by detaching it and drawing it away from the wall (plasmolysis; Fig. 228, p. 219).

When this occurs it can often be seen, even with the light microscope, that the cytoplasm is delimited, both on the outside, adjacent to the cell wall, and on the inside, adjacent to the vacuole, by a hyaline plasma membrane, clearly set off from the intervening granular cytoplasm. The inner plasma membrane surrounding the vacuole is known as the tonoplast, and that adjacent to the cell wall as the plasmalemma. Both membranes are important because of the special part they play in regulating the permeability of the cell to water and solutes (see p. 233).

Valuable insight into the fine structure of the protoplast, as discernible with the light microscope, is given by various staining methods. Their value derives from the fact that the different components of the fixed protoplasm take up dyes differentially from solutions, and retain them when the specimen is washed with solvent alone. It is also possible to stain components of the living protoplasm with certain dyes (Chrysoidin, Neutral Red, Malachite Green, Acridine Orange), a technique known as vital staining. Besides bright-field stains (diachromes), fluorescent stains (fluorochromes) play an important role in cytology. The absorption of only a minute quantity of the stain causes the cell component concerned to fluoresce brightly in ultra-violet light (e.g. Fluorescein, Acridine Orange). Dilute caustic soda, chloral hydrate and Eau de Javelle cause general dissolution of the protoplast, and are consequently used for the clearing of microscopic preparations.

With the help of dark ground illumination, and especially with the use of phase-contrast optics, it is possible to make structures visible which cannot normally be seen in ordinary transmitted illumination. This purely physical technique, involving no chemical change in the object, can be used to observe, and even to film, the movement of components of the living cytoplasm. Dark ground illumination involves directing light obliquely on to the cell. Structures with refractive indices different from that of the ground cytoplasm diffract light at their boundaries, and consequently appear brilliantly illuminated against a black background. In phase-contrast microscopy phase differences in the light emerging from the specimen, caused by differences in the refractive indices of the components, are transformed into amplitude differences by interference in the image plane, and thus made visible to the human eye.

The most valuable information about the morphology of the protoplasmic structures lying at the limit of this resolution of the light microscope has however been obtained by means of the electron microscope. This instrument enables the resolution of fine structure, as compared with the light microscope, to be advanced by the order of 1000-fold, and thus takes direct observation into the realm of the macromolecular. The endoplasmic reticulum at the magnification possible in an electron microscope is revealed as an intricate and continuous mesh of tubes or flattened sacs (cisternae). Parts of the reticulum run through the pores between adjacent cells and so connect the protoplasts together into a unified symplast. All the dead materials excluded from the cellular metabolism (cell walls, crystals, starch, etc.) are said to constitute, as apposed to the symplast, the apoplast.

Because of the artifacts which inevitably arise in the fixing and embedding of the material concerned, electron micrographs must always be interpreted with caution. The less damaging method of freeze-etching, however, yields three-dimensional pictures which reveal the naturally-occurring structures almost unchanged. In this method the material is very rapidly cooled to −160°C, and so fixed. It has been shown that yeast cells frozen in this way are capable of resuming all their living functions when they are returned to room temperature, and it can thus be inferred that fixation by rapid cooling ensures authentic preservation of all the structures essential for life. The reflecting electron microscope ('Stereoscan'), although an instrument of low resolving power, yields impressive, three-dimensional, pictures of superficial structures (Figs. 57E; 98).

D. Cells and energids[6]

The cells of the bacteria and of the Cyanophyceae are very much more simply organized than those of most plants and animals. Although they contain the nucleoproteins characteristic of the cell nucleus, their quantity is considerably less and their arrangement more primitive. A special nuclear membrane is wholly absent. Bacteria and Cyanophyceae are accordingly distinguished as proto- or procaryotes from those organisms, referred to as eucaryotes, whose cells contain distinct nuclei. The organization of the procaryotes, which also contain no clearly demarcated chromatophores or mitochondria, is evidently that of a primitive and elementary cell.

Occasionally eucaryotes are encountered consisting of a single, uncompartmented protoplast containing numerous true nuclei. For example, in the plasmodia of the slime-fungi (Myxomycetes), which may become several hundred square centimetres in area, many millions of nuclei, dividing more or less synchronously, occur in a slimy naked mass of protoplasm (Fig. 8). The cells of the siphonaceous algae and phycomycete fungi are also almost always multinucleate (Fig. 479, p. 507), as are the 'internode' cells of the Characeae (Fig. 73B, p. 76), each of which may reach a length of 10 cm. In the flowering plants the multinucleate condition is encountered in phloem fibres, latex ducts and endosperm cells.

In all these instances we are concerned with protoplasts without morphologically defined boundaries, but there is no reason to doubt that the reciprocal physiological relationships between the nuclei and the cytoplasm are similar to those in clearly defined cells. To every nucleus belongs a corresponding volume of cytoplasm. This functional unit, lacking a recognizable morphological boundary, is referred to as an energid, and accordingly multinucleate 'cells' such as those described above are best described as polyenergids. Not infrequently the polyenergid condition is transformed, as a consequence of wall formation, into the monoenergid. In these instances the polyenergid is

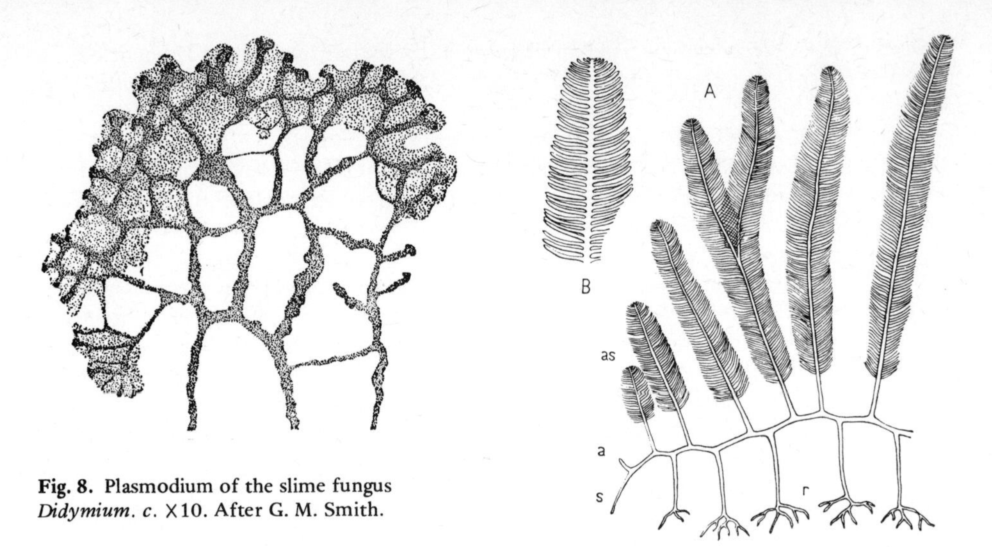

Fig. 8. Plasmodium of the slime fungus *Didymium. c.* X10. After G. M. Smith.

Fig. 9. (Right) *Caulerpa sertularioides.* **A.** The thallus consists of a creeping axis (s) bearing rhizoids (r), and pinnately branched assimilators (as). Both the axis (s) and the young assimilators (a) grow at their tips. Despite the complex morphology, the interior is filled with a single mass of protoplasm. **B.** Terminal segment of an assimilator. Natural size. Original.

revealed to be actually a latent form of multicellular organization (see, for example, Fig. 36, p. 44). In certain fungi (Asco- and Basidiomycetes), at particular stages of their development, the cells regularly possess two nuclei which share a common cytoplasm (Figs. 506, p. 523; 513, p. 528).

II. The protoplasm

A. Chemical components of the protoplasm

Protoplasm is not chemically uniform, but a mixture highly organized into organelles and other sub-structures. It consists for the greater part of organic compounds, but there are also accompanying inorganic substances. The molecules are partly in a more or less colloidal dispersion, and partly in solid form, and in living protoplasm they are all for the most part in continual transformation ('turn over').

a. Fundamental components. Most of the dry matter of the protoplast consists of proteins (40–50 per cent). Protoplasm thus coagulates on heating (denaturation). Proteins are thrown out of solution by trichloracetic acid and other protein precipitants, in particular the salts of the heavy metals (e.g. uranyl acetate), and this property can be used to purify them. Millon's reagent (which yields a red coloration) is used to demonstrate proteins in the cell. The nucleoproteins (associated with deoxyribo- and ribonucleic acids, and by no means confined to nuclei) are also classed as proteins in the wide sense. Deoxyribonucleic acid (DNA) occurs in chloroplasts and mitochondria as well as in nuclei, while ribonucleic acid (RNA) is the principal component of nucleoli and of the ribosomes.

A second important class of compounds, always encountered in protoplasm are the lipids (2–3 per cent of the dry weight; of the plastids up to 50 per cent). They are frequently loosely combined with proteins as lipoproteins. As a consequence of the polar nature of their molecules, which enables them to associate with both hydrophilic and hydrophobic substances, the lipids play an important part in the organization of membranes within, and at the surface of the cytoplasm. They are able to form monomolecular or bimolecular films which facilitate the segregation of individual reaction spaces (compartments) within the cell. The plasmalemma, tonoplast, and the chromatophores are especially rich in lipid. Carbohydrates account for 15–20 per cent of the dry matter of the protoplast. The simpler molecular forms (sugars) are conspicuous as sources of metabolic energy, while the polymerized forms (polysaccharides) serve as reserve substances. Fats (10–20 per cent of the dry matter) represent the most concentrated form of energy reserve. It is not unexpected therefore that they are the most frequent storage product encountered in the various forms of reproductive bodies capable of prolonged dormancy (e.g. spores and seeds).

b. Electrolytes. Salts of various organic and inorganic acids are found in the cytoplasm and in the vacuoles. Amongst other functions, the electrolytes are of considerable importance in the regulation of the water content of the cell, and in determining the properties of bounding membranes.

c. Enzymes (Biocatalysts). Enzymes (or ferments, in older terminology) are concerned with the metabolism of the cell. They are simple or complex, large protein molecules which, as with inorganic catalysts, merely by their presence accelerate or sometimes actually initiate reactions or sequences of reactions, without themselves appearing in the end product. They have appropriately been compared with lubricants which facilitate transformations by reducing friction (in the case of enzymes, by reduction of the activation energy required by the reaction). Corresponding to the multiplicity of metabolic reactions, the number of enzymes, mostly highly specific, is also great. Their actual quantity, however, is fairly small, since they are not themselves consumed. In all, over 1000 of these biocatalysts have now been identified. Over 100 of them have already been isolated in chemically pure form and investigated in greater detail (see p. 259).

Energy carriers. Energy may become stored in many chemical compounds. In all living cells, however, the temporary storage of energy and its rapid transport from one metabolic process to another is undertaken by special phosphate compounds, the energy-rich phosphates, such as, for example, acetyl phosphate and especially adenosine triphosphate (ATP). Just as an electric motor can be driven only with electrical energy, and all other forms of energy (e.g. atomic, light, hydraulic, coal and oil) must first be converted into electrical, so in the living cell energy (e.g. light energy, chemical energy) can flow into the system only if it is first incorporated into energy-rich bonds which then, as a kind of battery, keep the metabolism of the cell in operation (see p. 266 seq.).

Intermediary and secondary products of metabolism. In the chemical transformations which characterize life large numbers of substances are formed which are immediately subjected to further metabolism. These substances are termed intermediary metabolic products. Secondary metabolic products accumulate in cells in large quantities, and since they are not of primary importance for life, they are either laid down in vacuoles or even excreted right out of the cell. Their significance for the plant is often little known. Many such secondary metabolic products (e.g. resin, wax, dyes, tannins, essential oils) have industrial uses.

Many of the substances mentioned above (e.g. the enzymes and the growth-regulating substances, as well as many of the intermediary and secondary metabolic products) very probably arise initially in definite organelles and from there enter the ground plasm secondarily. Some, on the other hand, may remain localized at the sites of their formation.

B. Physico-chemical properties of the cytoplasm

The consistency of the cytoplasm is peculiarly intermediate between that of the solid and the liquid states. It was for a long time interpreted as consisting of a colloidal solution in the physico-chemical sense, i.e. a solution whose characteristic properties (such as viscosity and elasticity) depend predominantly on the size of the dissolved molecules.

Actually the cytoplasm behaves in many respects wholly as a fluid. This is particularly clearly seen in the streaming of the cytoplasm, which can be observed both in the free,

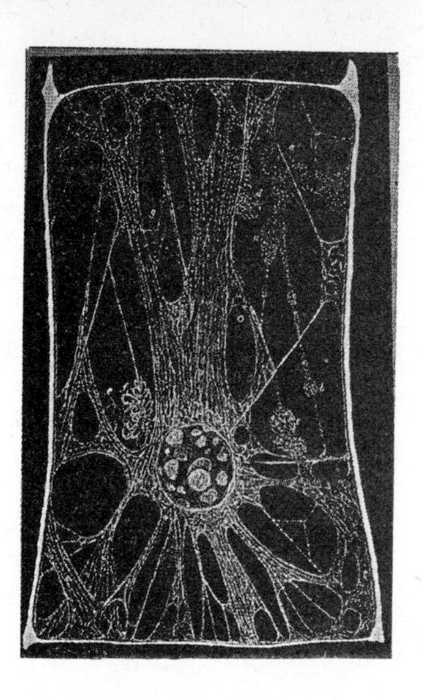

Fig. 10. Cell from a hair of *Cucurbita*. Dark field illumination. The lumen is traversed by numerous strands of cytoplasm, and the nucleus is suspended at the centre. *c.* ×120. After Heidenhain.

naked plasmodium of the slime fungi (Fig. 8) and in the enclosed cells of any of the more highly organized plants (see, for example, Fig. 10; also p. 384). Three types of movement can be distinguished, namely rotation, circulation, and a flowing movement. In plasmarotation (as, for example, in the cells of several water plants, such as *Chara, Nitella, Elodea, Vallisneria,* especially after initial stimulation) the cytoplasm, confined to a thin layer next to the wall, contains a broad stream which moves continuously in one direction between other areas at rest, forming an endless band flowing back on itself. Portions of protoplasm escaping from cut cells usually round themselves off like drops of fluid, and the streaming may persist in these droplets for a considerable time. Plasmacirculation is seen especially beautifully in the cells of the hairs of several land plants, as well as in those of the root hairs of water plants, where strands or lamellae of the cytoplasm traverse the central vacuole (Fig. 10). Movement occurs both in the strands and in the layer of cytoplasm against the wall. The currents may be narrow or broad; they frequently branch, and in both the strands and peripheral layer adjacent currents may flow in different directions. The nucleus and the other organelles may be carried along by both rotating and circulating cytoplasm, the speed of movement depending upon the size of the organelle. Flowing protoplasmic movement is seen occasionally in the long, growing, tube-like filaments of fungi and algae. Included here are the often rhythmical movements in one direction of the whole cell mass. This behaviour, as well as the amoeboid changes of form of which many single-celled organisms are capable, points to the freedom of movement of the individual particles of the protoplasm, a condition which is characteristic of the fluid state.

In spite of these observations, purely theoretical considerations have led many investigators to ascribe to the protoplasm an inner structure. It has justifiably appeared to them quite unthinkable that so many living processes going on simultaneously could be based on a fluid substrate. In favour of such a submicroscopic structure, which would invest protoplasm with a certain internal coherence, are its insolubility in water, its capacity for swelling, its plasticity and its tendency to form long, thin threads (Fig. 10), as well as its frequently observable flow birefringence.

It is, of course, possible to distinguish fluid and firm areas within the cytoplasm by careful examination with the light microscope. The differentiation into the hyaline plasma membranes and the granular cytoplasm has already been mentioned (p. 16). The plasma membranes consist of a relatively firm plasmagel, also referred to as the ectoplasm, whereas the granular cytoplasm consists of a fluid plasmasol, also known as the endoplasm. The plasmasol and the plasmagel are not, however, simply other names for the inner and outer cytoplasm; the distinction between them is frequently much more complicated. Moreover, the plasmagel and the plasmasol can in no way be regarded as permanent states in living cells: the one can be seen to change into the other with relative rapidity. If a cell be wounded the complexity of the cytoplasm at the exposed surface is reduced very rapidly, and the wound is closed by a new skin of plasmagel.

C. Molecular structure of the cytoplasm[8]

The cytoplasm in which the larger cell organelles are embedded is shown by the electron microscope to contain yet other differentiated structures (Fig. 6, p. 15), no doubt responsible in part for its physico-chemical properties. One of these, already mentioned (p. 16), is the endoplasmic reticulum, hardly resolvable in the light microscope. That which remains 'empty', even in the electron microscope, is a finely granular matrix, referred to as the ground cytoplasm or ground plasm.

All assertions about the nature of this ground plasm depend upon indirect methods of investigation. It is highly hydrated and certainly contains freely soluble proteins of linear and globular form, not structurally bound. A knowledge of the fundamentals of protein chemistry is consequently essential for an understanding of its structure. Simple and compound proteins (proteins and proteids) consist of long chains, partly folded or coiled up, of some hundreds to a thousand amino acids (Fig. 259, p. 279). In the proteids the amino-acid chains are coupled with a non-protein component (the prosthetic group). An almost limitless range of possible compounds can be obtained by varying the sequence (permutation) in which the twenty known amino acids are joined together in the chain, as well as by varying the prosthetic group. The number of possible isomers increases rapidly with the number of amino acids involved. If the twenty amino acids occurring most commonly in proteins are linked together, each amino acid being represented only once, then theoretically 2.4×10^{18} different icosapeptides are possible. These will have molecular weights of approximately 2400. Even ten different amino acids, each taken only once, can give rise to 3.6×10^{6} different decapeptides. These figures already vastly exceed the total number of kinds of existing organisms (about 380,000 plants and 2,000,000 animals). In actual fact the natural proteins, as can be seen from their molecular weights (which reach into millions) consist of polypeptide chains of considerable length. The number of theoretically possible proteins is thus altogether unimaginable (see also p. 278 seq.).

The side chains of the amino acids projecting from the polypeptide chain of the protein can be either positively or negatively charged. According to their chemical nature they may also be hydrophilic or hydrophobic. Certain highly active groups may also tend to cross-link with similarly active groups in side chains of adjacent polypeptides (e.g. sulphur bridges may form as a consequence of the oxidation of neighbouring sulphhydryl groups: $R . SH\ HS . R \longrightarrow R . S . S . R$).

In this way a great number of different combinations are possible, involving loose or firm links between adjacent molecules. Further, the ability to form bonds of this kind allows the structural proteins and proteids of the ground plasm to come together to form chains and fibrils, or flat sheets, or even to become packed in three-dimensional crystalline arrays. All such structures must, however, be regarded as labile. Experiments with radioactive carbon and nitrogen have shown that in protein molecules there is a constant interchange of the amino acids, the 'bricks' of the protein. Thus the order and arrangement of the materials of the cell are certainly not static, but present a dynamic picture. Chemically the cytoplasm can be regarded as an entity in a state of constant flux.

The theory of junctions makes many of the properties of the cytoplasm explicable. Water, containing dissolved salts, could be loosely held by the hydrophilic main and side chains of the protein molecules. Moreover, perhaps unattached water could be retained by capillary attraction in the interstices of the protein framework which the theory envisages. The process of water uptake is referred to as imbibition (see p. 213), and the extent of the imbibition by the cytoplasm determines its hydration. A slight change in the affinity of the side chains for water can lead to loss of hydration and turgescence. Hydration causes a loosening of the junctions, dehydration favours their establishment. The turgescence of the cytoplasm is of critical importance in the phenomena of life. In protoplasm deprived of water, characteristic of resting cells, life is latent. Absolute alcohol, high temperatures and, because of their strong charges which block the side chains, heavy metals bring about irreversible loss of turgescence (coagulation). These agents harden the structure of the cell and result in death, a process known as fixation. Other fixatives are formaldehyde, osmic acid, chromic acid, mercuric chloride and many other substances which collectively cause the irreversible formation of inner bonds and entanglement of the side chains.

Besides the structural proteins and numerous 'soluble' enzymes (i.e. those that are not bound to structural components and are consequently easily extractable), the ground plasm is undoubtedly well furnished with lipids. Without further treatment it is not miscible with water, and it can be vitally stained with the lipid stain Rhodamin B. These lipids are responsible for a further property of the ground plasm, namely its capacity to form membranes, and in this way to cut off even smaller reaction spaces or compartments.

D. Structure of cytoplasmic membranes

a. Simple membranes. As already mentioned, many lipids, especially the phospholipids such as lecithin (Fig. 11A), are furnished at one end with so-called 'polar' hydrophilic groups (e.g. -OH, $-NH_2$) and at the other with 'apolar' hydrophobic (or lipophilic) groups (e.g. $-CH_3$, the methyl groups of the fatty acid chain). This means that they are able to lie, as emulsifying agents and detergents, at interfaces between hydrophilic and hydrophobic media. They do this by forming monomolecular films or lamellae, in which every single lipid molecule stands with its hydrophilic 'head' immersed in the aqueous phase, while the hydrophobic 'tail' either emerges into the air or is associated with other hydrophobic molecules, which may form a distinct lipophilic phase. In a purely aqueous environment such polar (or 'structural') lipids are therefore capable of forming bimolecular membranes, consisting of two monolayers with the apolar groups of one turned towards those of the other (Fig. 11B, C, D).

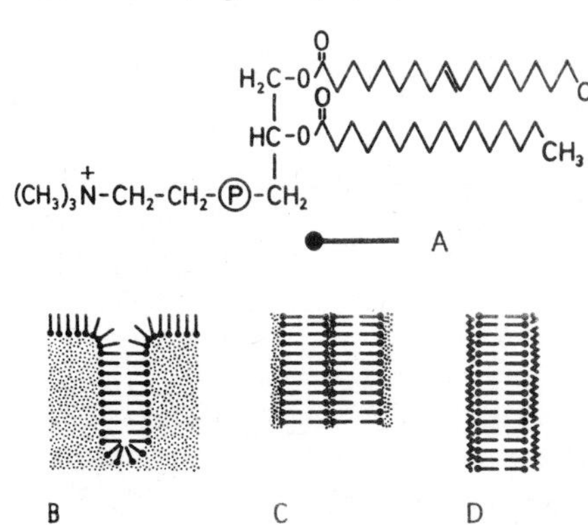

Fig. 11. A. Structural formula of lecithin, often denoted by the symbol beneath. The small black circle represents the hydrophilic end of the molecule. **B.** Monolayer of phospholipid on the surface of water. The monolayer is under compression and a fold is developing. **C.** Part of a myelin figure showing its molecular structure. **D.** Molecular model of a 'unit membrane'. The bimolecular lipid leaflet is coated with protein on each hydrophilic face. After Panicki, Dawson and Robertson, from Sitte.

Such bimolecular membranes or 'leaflets' can easily be made experimentally in the form of so-called 'myelin figures' (for example, by placing the flesh of the seed of *Ginkgo biloba*, very rich in lipids, in water). These figures consist of numerous bimolecular leaflets stacked one upon the other. Because of this regular symmetry (referred to as a paracrystalline structure), myelin figures display birefringence in the polarizing microscope (and can be regarded as 'fluid crystals').

In the living cell, membranes of lipid and protein, in which the lipid has a similar bimolecular conformation, play an important role in dividing up the ground plasm, consisting principally of hydrophilic proteins, into discrete reaction spaces. These lipoprotein membranes, some 7 to 10 nm thick, give the fluid ground plasm structure and an internal consistency, so they can be regarded as providing the cytoplasm with a kind of framework embedded in the matrix of ground plasm.

In the electron microscope these elementary membranes appear to consist of three layers, two dark lines (which can be approximately identified with the outer faces of protein) being separated by a clear space (the bimolecular leaflet of lipid). This characteristic tripartite appearance of a simple membrane in section is referred to as the 'unit membrane profile' (Fig. 11D). Sometimes profiles are seen which suggest that the two lipid layers of the membrane are coming apart, possibly a consequence of the chemical fixation of the cell. Freeze-etching shows that most membranes have some kind of granular structure, and it is not clear how this is to be reconciled with the unit membrane profile. It should be stressed that the concept of the unit membrane, arising from the study of fixed and sectioned material, is largely descriptive and may be given by a number of different kinds of molecular structure. Membranes differing in chemical composition, thickness and permeability probably occur together in every living cell.

The presence of such membranes effectively suppresses the free diffusion of the larger organic molecules, and the result is the formation of numerous separate reaction spaces. These are often distinguished by their containing quite definite enzymes, and in consequence are associated with certain sequences of reactions and metabolic products which occur here and nowhere else. This spatial subdivision and differentiation of the protoplast is referred to as the compartmentation of the cell, and the individual reaction spaces as compartments. Current investigations indicate that biochemical compartmentation may go even further than it is at present possible to demonstrate cytologically.

Even the larger cell organelles described on p. 27 are nothing more than such cell compartments with typical form and function. Unfortunately the concept of 'organelle' is used in botany, zoology and medical science in a number of quite different senses. Originally coined for the relatively large components of unicellular Protista recognizable in the light microscope (e.g. the flagella) by analogy with the multicellular organs of the Metazoa, the term later became used indiscriminately for all structures within the cell visible in the light or electron microscopes which appeared to have specific form or function. Even the centromeres of chromosomes are regarded by many investigators as organelles.

b. Endoplasmic reticulum. Besides the stable membrane systems, such as those which delimit the various larger cell organelles (see p. 27) from the ground cytoplasm, there is another system of membranes, probably fully mobile and to some extent labile, in the

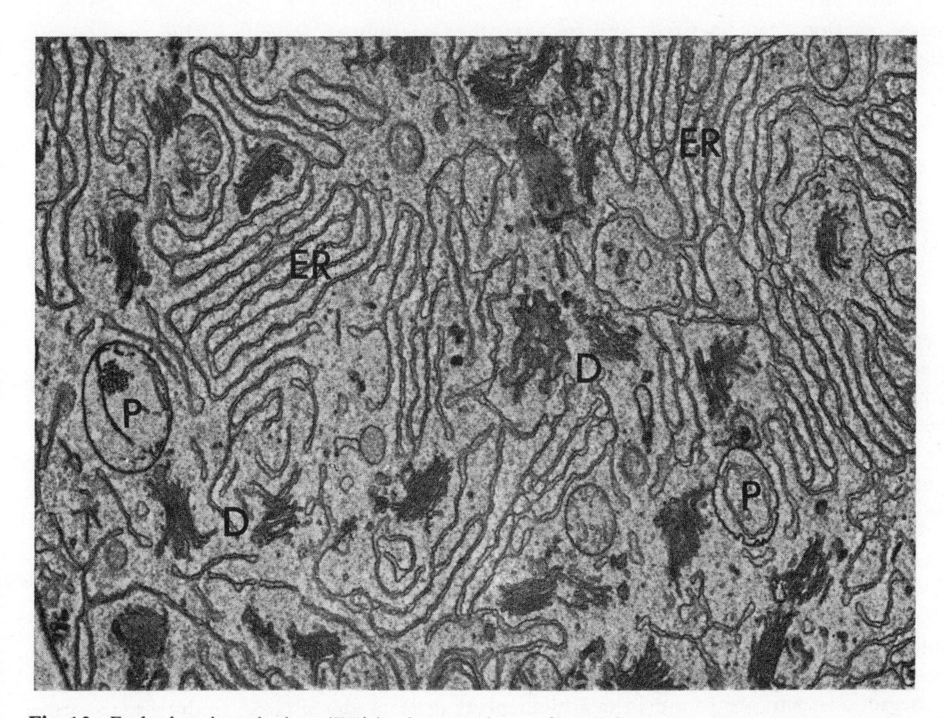

Fig. 12. Endoplasmic reticulum (ER) in the cytoplasm of a cell from an axillary scale (squamula) of *Elodea canadensis*. The cell also contains conspicuous Golgi bodies (dictyosomes) (D), and several more or less circular profiles of proplastids with very little internal differentiation (P). ×10,000. After W. Menke.

cytoplasm. Because of the net-like appearance often displayed by this system of membranes when seen in ultrathin sections in the electron microscope, it is referred to as the endoplasmic reticulum (often abbreviated to ER) (Figs. 6; 7, ER; Fig. 12). The reticulum is especially conspicuous in meristematic cells, and also in older cells in which there is marked metabolic activity.

Time-lapse photography with the phase-contrast microscope has clearly shown that the static picture of plant and animal cells, derived from the study of fixed and stained preparations with the light microscope, must be discarded in favour of a dynamic concept. This is particularly true of young meristematic cells, and of those which are actively metabolizing.

The endoplasmic reticulum is a system of paired membranes. The interlamellar space occasionally expands, forming cisternae. All parts of the reticulum are intricately connected, so that it has the form of a highly perforated, but continuous, system of sheets and tubes. Some of these may pass through pores in the cell wall so that adjacent protoplasts are connected, and may be considered to constitute a symplast. The interlamellar material (except sometimes within cisternae) is apparently fluid and notably less dense than the ground cytoplasm. All cell organelles and other protoplasmic inclusions, such as plastids, mitochondria, golgi bodies, ribosomes and spherosomes lie in the ground cytoplasm. Only the spherosomes and vacuoles can be related to the interlamellar space. Many investigators hold the view that they arise directly from local expansions (vesicles) of the reticulum. Their contents would then have been in continuity with the material of the interlamellar space (sometimes referred to as the intra-cisternal phase).

The interlamellar space forms a continuous system, certainly connected with the space between the two layers of the nuclear envelope (the perinuclear space) and possibly also with the exterior at the plasmalemma. Since during nuclear division the nuclear envelope regularly breaks down, and subsequently reassembles around the daughter nuclei, the envelope is today no longer regarded as a truly functional element of the nucleus itself, but as a sheath which shields the chromosomes, the essential components of the nucleus, from the ground cytoplasm during a particular phase of nuclear activity. The endoplasmic reticulum serves on the one hand to expand considerably the intracellular surfaces at which reactions can be sited, and on the other possibly to provide within the aqueous interlamellar channels a means of transport of soluble materials within the cell and also (by way of the pores in the wall which are penetrated by the endoplasmic reticulum, see Fig. 6, Pl) between neighbouring cells.

E. Ribosomes and polysomes

The faces of the endoplasmic reticulum directed towards the ground cytoplasm are frequently beset with small, approximately spherical, particles, from 15 to 25 nm in diameter and consisting of protein and ribonucleic acid. These particles are termed ribosomes, and the endoplasmic reticulum bearing them is frequently referred to as 'rough' or 'granular' in contradistinction to the 'smooth' reticulum which is free of ribosomes. Similar ribosomes are also found free in the cytoplasm, frequently in great numbers (of the order of 10^5 or more) (Figs. 6; 14). Frequently they are grouped in spirals or helices, the so-called polysomes (Fig. 13A, B). Ribosomes and polysomes are the sites of protein synthesis (cf. p. 282 seq.). They are especially numerous in cells which are growing vigorously or are otherwise metabolically active.

F. Golgi bodies (dictyosomes)

Golgi bodies consist of stacks of flattened, disc-like, cisternae, each cisterna being bounded by a single membrane. They are believed to be the sites of synthesis of a number of secretions, and are found in nearly all plant cells. A single Golgi body is made up of anywhere from 3 to 15 cisternae, each approximately circular and from 0.5 to 2 μm in diameter (in rare instances, e.g. in the alga *Micrasterias*, up to 5.5 μm in diameter) arranged in a stack which, when seen from the side, is often slightly hour-glass shaped. The

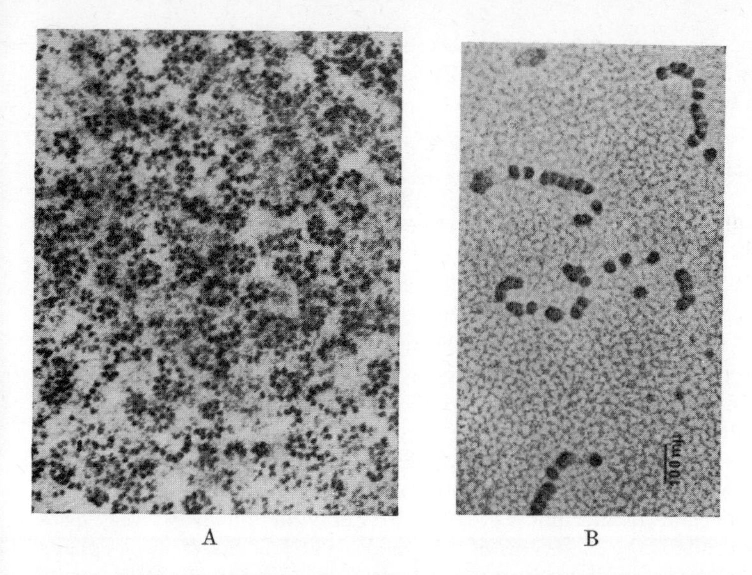

Fig. 13. Polysomes. **A.** In the cytoplasm of a cell from the cotyledon of *Vicia faba*. **B.** In an extract from the leaves of tobacco. **A.** ×35,000. From Opik. **B.** ×80,000. From Milne.

Golgi bodies (also referred to as dictyosomes) lie free in the cytoplasm without any special sheathing membrane.

The number of Golgi bodies varies, according to the function and physiological condition of the cell, from a few to several hundred. It is greatest in cells or parts of cells in which there is notable synthetic activity, e.g. in gland cells, or in meristematic cells at the site of the formation of the new cell wall. At the margins the individual cisternae are frequently perforated in a reticulate fashion, and, when the Golgi body is actively

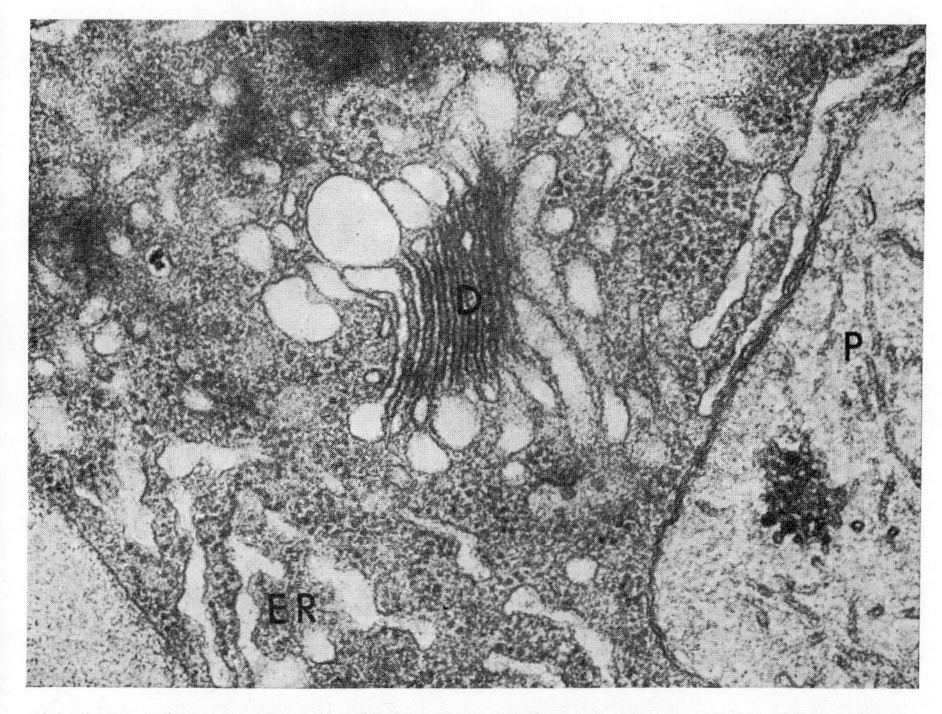

Fig. 14. Part of the cytoplasm of a glandular hair cell of *Mentha piperita*. D, Golgi bodies (dictyosomes), the cisternae dilated with the glandular secretion, and some of the vesicles apparently breaking down. ER, endoplasmic reticulum with clearly expanded interlamellar spaces. The ground cytoplasm is filled with ribosomes. Part of a proplastid (P) is visible on the right. ×55,000. After F. Amelunxen.

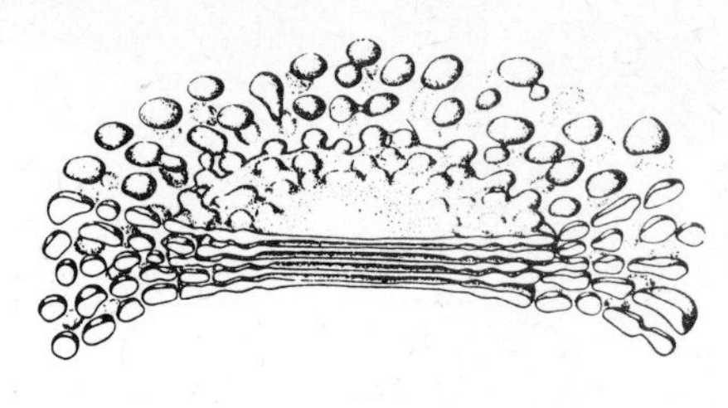

Fig. 15. Active Golgi body (dictyosome) from *Micrasterias rotata*. Numerous vesicles are being pinched off from the periphery of the cisternae. ×10,000. From Drawert and Mix.

secreting, small vesicles become pinched off in this region. These so-called Golgi vesicles, each bounded by a single membrane are filled with secretion, then drift away into the cytoplasm (Figs. 14; 15). Recent investigations have shown that many Golgi bodies are polar in structure. Very flat cisternae become apposed to one side which is convex and is regarded as proximal. Each cisterna then appears to move up through the stack, becoming progressively more filled with secretion on the way. As it approaches the distal (or production) side, it becomes gradually dispersed in the form of vesicles.

The Golgi bodies are concerned principally with the synthesis of wall substances: acid polysaccharides, pectins and hemicelluloses. Other products are mucilages (e.g. in a number of brown algae, the slimy secretions of the glands of various insectivorous plants (*Drosera, Drosophyllum*), and the lubricatory mucilage on the outside of the root cap) and ethereal oils (as in the glandular hairs of many Labiatae). In the stalked glands of *Drosophyllum* every Golgi body yields about three vesicles every minute, and after two to three minutes these have discharged at the surface. The membrane of the vesicle fuses with the plasmalemma, which then opens at the point of contact, allowing the contents of the vesicle, which now flattens and becomes indistinguishable from the plasmalemma, to be extruded from the protoplast.

The secretory activity of the Golgi apparatus is to a large extent dependent on an adequate supply of energy from respiration. In dormant seeds the cells of the embryo are almost completely free of Golgi bodies. Not until two to three days after germination are they present in great numbers in the cytoplasm, apparently having been newly formed. In this 'mobilization phase' of germination the endoplasmic reticulum also becomes conspicuously greater in quantity.

G. Microsomes, cytosomes, spherosomes, microtubules

Formerly all cytoplasmic inclusions which lay at the limit of resolution of the light microscope, and which were seen with dark-field illumination as bright spots (often showing active Brownian motion) between the larger cell organelles were designated microsomes. Today, however, the term is used solely for that fraction of mechanically homogenized protoplasm in which there is a concentration of fragments of endoplasmic reticulum and ribosomes.

The electron microscope has shown that the bodies formerly known as 'microsomes' differ widely in size and structure, and also in their biochemical nature. Some are simple inorganic particles, such as minute crystals. Others are spherical bodies, about 1 μm in diameter, rich in lipid and surrounded by a protein sheath, known as spherosomes. Of similar form and size are cytosomes, consisting of a single membrane enclosing a dense granular matrix. Although the metabolic significance of the cytosomes is at the moment no clearer than that of the spherosomes, many investigators suspect that cytosomes may be analogous to the lysosomes of animal cells. These are believed to be the sites of formation of hydrolytic enzymes.

Microtubules are tubular structures, about 25 nm in diameter and of indefinite length. Their walls are made up of globules about 7 nm in diameter, probably arranged in helices, consisting entirely of protein. Microtubules are preserved only in certain conditions of fixation, and were not discovered until 1957. The mitotic spindle (see p. 33) is formed of such microtubules, but they are also found just beneath the plasmalemma. They are possibly concerned in the generation and arrangement of the cellulose microfibrils of the cell wall (see p. 69). The more detailed investigation of all these structures is currently proceeding.

III. Structure of the larger cell organelles

A. The nucleus and cell division[10]

1. Significance of the nucleus

The nucleus is usually spherical or lens-shaped; rarely is it spindle-shaped and elongated, or does it continually change its form in an elastic manner. It is the most important carrier of information in the cell. In embryonic cells (Fig. 6) it often takes up more than half the width of the lumen. In fully developed cells (Fig. 5) it is relatively smaller, since as a rule its size is little changed in the expansion of the cell. New cell nuclei are derived solely by the division of pre-existing nuclei: *omnis nucleus e nucleo*.

Especially large nuclei (up to 0.6 mm diameter) occur in the egg cells of cycads and conifers. The smallest nuclei, with a diameter of less than 0.5 μm, are found in fungi.

Enucleate cells are not capable of prolonged life. If uninucleate elongated cells, as occur, for example, in the alga *Spirogyra* and the hairs of various plants, are plasmolysed (see p. 219) their protoplasts are frequently broken up into several parts (Fig. 16). Only those parts which have retained nuclei (Fig. 16A, m) or which have remained in connection with nucleate parts by means of protoplasmic threads (Fig. 16B) are capable of remaining alive and forming a new cell wall.

In *Acetabularia* (an alga which, although consisting of a single cell with only one nucleus, may reach up to 6 cm in height) transplantation experiments between two young plants of different species have shown that the nucleus determines the specific characters expressed in the fully grown plant (Fig. 16C).

The nucleus must contain, therefore, the most important genetic factors or genes responsible for the development of the specific characteristics. The investigation of the changes in the form and structure of the nucleus during cell division and the formation of the germ cells consequently provides a most important foundation for the understanding of genetical phenomena (see p. 319 seq.).

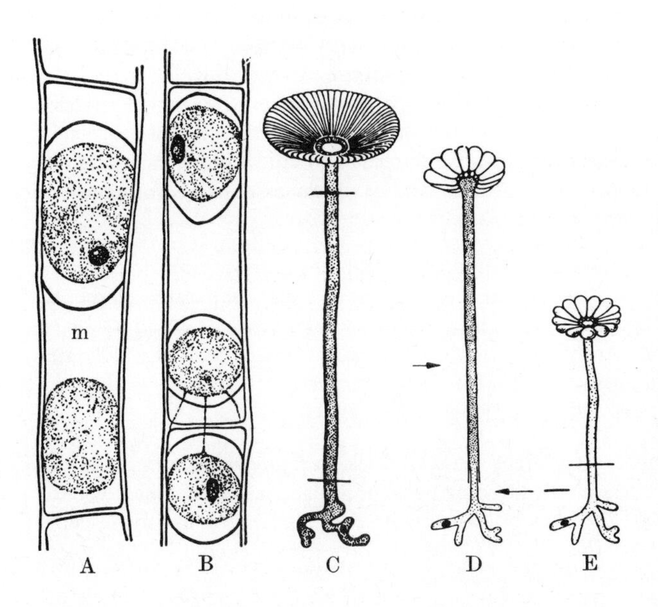

Fig. 16. Evidence for morphogenetic control by the nucleus. **A.** and **B.** Hair cells of *Cucurbita pepo*. **A.** Protoplast divided into two as a result of plasmolysis. Only that portion containing the nucleus has regenerated a membrane (m). **B.** Here the portion lacking a nucleus has developed a membrane because it is connected by strands of cytoplasm with a fragment containing a nucleus in the neighbouring cell. **C., D.** and **E.** *Acetabularia*, showing the importance of the nucleus in determining the specific features of the cap-like reproductive region. **C.** *A. mediterranea*. **E.** *A. wettsteinii*. **D.** The result of grafting a young thallus of *mediterranea* on to a rhizoid of *wettsteinii* containing a nucleus. The cap produced by the *mediterranea* thallus has the features of that of *wettsteinii*, owing to the presence of the *wettsteinii* nucleus. **A.** and **B.** ×50. After Townsend. **C., D.** and **E.** ×1.5. After Hämmerling.

2. Changes in the form of the nucleus

The nutrients necessary for the growth of the nucleus are absorbed from the cytoplasm. All its fine structures responsible for heredity are present in duplicate. In a region of a plant where active growth is occurring, nuclear division is frequently seen, in the course of which the appearance of the nucleus is fundamentally changed. In all, three different conditions of the nucleus can be distinguished: the interphase nucleus, the mitotic nucleus, and the working nucleus. 1. Formerly the nucleus between two successive divisions was falsely thought to be in a condition of rest because its appearance in the light microscope at this time is one of relative quiescence. Actually the activity of such nuclei is very considerable, and includes, for example, the identical reproduction of the structures bearing the genetic information. The nucleus between divisions is consequently better referred to as the interphase nucleus. 2. The mitotic nucleus is the nucleus undergoing division, in the course of which the structures duplicated during interphase are separated. 3. The working nucleus is a term recently used to denote the nucleus of a fully differentiated cell. Such a nucleus is no longer preparing to divide, but plays a definite controlling role within the general organization of the plant.

The study of the cell nucleus and its different structures is facilitated by the sectioning and staining of fixed material or by the preparation of squashes. Recently, with the aid of phase-contrast equipment, it has been possible to observe the behaviour of nuclei directly in living unstained material.

a. The interphase nucleus. Between cell divisions the nucleus is separated from the cytoplasm by a special sheath, the nuclear membrane. This membrane is in fact an envelope of two membranes, not individually resolvable in the light microscope, perforated by pores. It is thus very similar to the endoplasmic reticulum, and according to recent interpretations is derived from it. Connections between the nuclear envelope and the endoplasmic reticulum are indeed frequently seen in electron micrographs. Within the nuclear membrane it is possible to distinguish a sol- or gel-like humour, the karyolymph, and one or more highly refractive nucleoli (Fig. 22.1 nl), together with a gelatinous and threadlike framework (Fig. 22.2 ch), which, since it has an affinity for the basic nuclear stains (such as haematoxylin, methyl green and carmine in acid solution), is referred to as the chromatin. The fact that the chromatin takes up only certain specific stains depends upon its content of nucleoproteids. The nucleic acids of the nucleus form the non-protein component (prosthetic group) of these proteids.

The Feulgen reaction has come to occupy a particularly important place for the identification of the specific nucleoproteid of nuclei. Deoxyribonucleic acid (previously known as thymonucleic acid), detached from its union with the protein by preliminary hydrolysis with hydrochloric acid, gives a red-violet colour with fuchsin–sulphurous acid. Deoxyribonucleic acid (DNA) is confined to the chromatin of the nucleus and to the nucleus-like structures in bacteria and blue-green algae, and also to some extent in plastids and mitochondria. Ribonucleic acid (RNA), on the other hand, although closely related chemically, is Feulgen-negative and occurs in both nucleus and cytoplasm. The nucleolus is composed predominantly of ribonucleic acid, and it is an essential component of ribosomes.

b. Chromosomes.[11] The most important structures within the nucleus are the chromosomes, the carriers of the genetic information. They appear as clearly separated elements, readily visible in the light microscope, only in nuclei undergoing mitosis (Fig. 22.3–8). In interphase nuclei they are in an unravelled, functional form, and individual chromosomes can only very rarely be distinguished with the light microscope. Instead we see the network of delicate threads mentioned earlier (Figs. 6 ZK; 22.1).

α. Individuality of chromosomes. If a larger series of division figures is examined in, for example, fixed and stained sections of the growing parts of a plant it is seen that not only the number but also the form of the chromosomes reappears unchanged in all the cells. Within an individual cell, however, the chromosomes can differ from each other considerably in form and size (the law of chromosome individuality).

All higher plants begin their development from the fusion of two germ cells or gametes of different sex (cf. p. 206). In this fusion not only do the protoplasts of the germ cells intermingle but also the two nuclei become united. If the gametes are of the same species two identical sets of chromosomes (referred to as homologous chromosomes) are brought together as a consequence of this fusion. The nucleus of the gamete contains only one set of chromosomes and is said to be haploid (n). This set is composed of single representatives of each kind of chromosome. The haploid nucleus contains, nevertheless, the complete assemblage of genetic factors or genes (referred to as the genome) which determine the characteristics of the species. Consequently, two homologous sets of chromosomes and genomes are incorporated into the zygotic nucleus formed as a result of fertilization. The individuality of the chromosomes is retained, so the new membrane of the zygotic nucleus encloses two chromosome sets of different origin, one descended from the paternal germ cells and the other from the maternal. The zygotic nucleus is said to be diploid (2n). All the nuclei formed by simple mitosis in subsequent development from the zygote retain this diploid condition. The smallest chromosome number so far observed in higher plants is n = 2, 2n = 4 (*Haplopappus gracilis* (Compositae)). Usually, however, the numbers are higher (e.g. *Crepis* spp. n = 3, *Vicia faba* n = 6; *Zea mays* n = 10; *Triticum aestivum* n = 21). The highest chromosome numbers, difficult to count, are found in the Pteridophyta (e.g. *Dryopteris filix-mas* n = 82, *Equisetum* n = 108, *Ophioglossum reticulatum* n = about 630), and in the algae (e.g. in the brown alga *Nectrium digitus* n = about 600).

The chromosome at the instant of nuclear division is typically condensed and highly chromatic, and is broadly cylindrical or spherical in shape. Frequently it is divided into two arms of different length by a narrow segment, or commissure, which is Feulgen-negative (achromatic). The centromere (or kinetochore), the position from which the movement of the chromosome is governed, is situated in this achromatic segment. During nuclear division the spindle fibres attach themselves to the chromosomes at the centromeres.

In addition to the primary constriction, secondary constrictions may also be present, and the positions of these can be used to identify individual chromosomes.

β. Nucleoli. Of special interest are the satellite- or SAT-chromosomes. In these one arm bears at its end a usually minute satellite or trabant, set off from the body of the chromosome by a threadlike filament, achromatic and free from nucleic acid (SAT = sine acido thymonucleinico). According to newer views, however, this achromatic segment contains deoxyribonucleic acid, but in such small quantity that it cannot be demonstrated by the Feulgen reaction.

The filament of the SAT-chromosome is also referred to as the nucleolar thread or the nucleolar organizer, since while the nucleus is leaving the dividing and entering the resting state the nucleolus usually appears at this position (Fig. 24B, C, D). The number of nucleoli in a cell thus usually corresponds to the number of SAT-chromosomes. It sometimes happens, however, that several nucleoli arising in different positions fuse together.

Nucleoli contain about 40 per cent protein, but consist principally of ribonucleic acid (up to 60 per cent of their dry weight). They are regarded today as the most important collecting site for the ribonucleic acid formed in the nucleus. They disappear at mitosis, and the ribonucleic acid may at that time be released into the cytoplasm.

γ. Structure of the chromosome. The chromosomes at mitosis, when fixed and stained and observed with the light microscope, appear to be made up of two different substances, namely a strongly chromatic axial body or filament and a very much more weakly chromatic substance which ensheathes it. This latter is the matrix or kalymma; its chemical nature has hardly been investigated, and many investigators have doubted its authenticity. In many objects it can be seen that the axial body of the highly contracted mitotic chromosomes consists of a series of narrow coils packed together in a close helix (Fig. 17A, B). In a few especially favourable cases it has been possible to demonstrate that these easily visible major spirals are formed in turn by filaments wound up in even finer minor spirals. The chromosome thus consists of a coiled coil, like the filaments in many electric lamps (Fig. 18A).

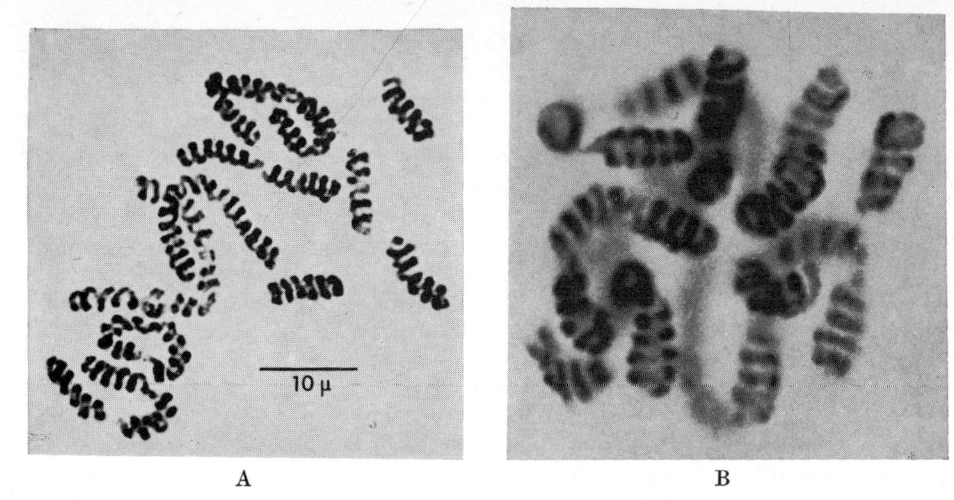

Fig. 17. Spirally contracted meiotic chromosomes from *Tradescantia virginiana* (metaphase of meiosis I). **A.** About ×1300, from Vosa. **B.** About ×4000, from Darlington and La Cour.

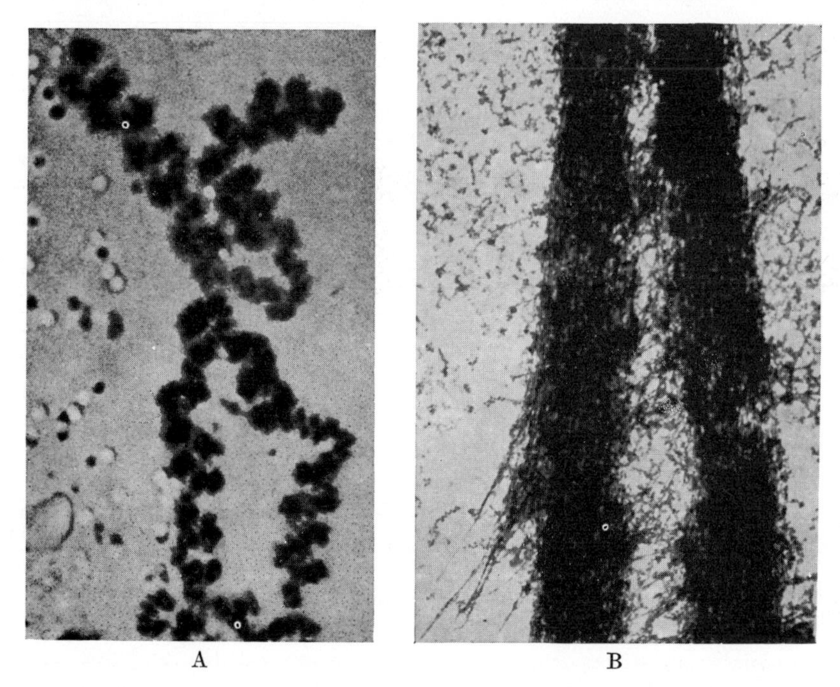

Fig. 18. Electron-micrographs of plant chromosomes. **A.** Terminal segment, consisting of two intertwined chromatids, of a chromosome of *Lilium candidum*. Each chromatid is supercoiled and forms a double helix (cf. Fig. 25E). **B.** Highly magnified segment of a chromosome of *Vicia faba* which has been treated with trypsin. The two chromatids are made up of numerous fibrillar elements about 15 nm in diameter. **A.** About ×10,000. From Taylor. **B.** ×32,500. From Wolfe and Martin.

It can be readily seen, even at a lower magnification, that before nuclear division the chromatic axial portion of the chromosome is divided longitudinally into two elements, known as chromatids, wound spirally around each other as the strands of a rope (Fig. 22.2–5).

In contrast to their strongly condensed form at mitosis, the chromonemata in interphase nuclei lose their spiral coiling and become extended. At this stage the kalymma is no longer visible; it either becomes dispersed or at least strongly swollen (Fig. 24E, F). In favourable objects at certain stages immediately before the onset of cell division, careful preparation and staining reveals that even extended chromonemata still display a fine structure (Fig. 29). They appear to consist of a more or less achromatic thread on which is arranged—at least before the prophase of the first meiotic division (see p. 37)—at varying intervals a series of strongly chromatic granules of different sizes, the chromomeres.

Sometimes in the transition from the mitotic to the interphase nucleus even whole chromosomes or large segments of chromosomes may remain in the spirally contracted condition. They consequently remain visible in the resting nucleus and are referred to as chromocentres. The corresponding segments of the chromosomes are often already visible in the mitotic chromosomes as strongly stained heterochromatic zones, distinguishable from the more weakly staining euchromatic zones. The chromomeres of the heterochromatic chromosomes or regions of chromosomes are usually larger than elsewhere, but probably genetically inactive.

The electron microscope has not yet made possible any firm conclusions about the submicroscopic structure of the chromosomes (Fig. 18A). Although both light and electron micrographs not infrequently appear to demonstrate a multi-strandedness of the chromosome (Fig. 18B), especially in the region of the centromere, a single sequence of genetic factors is to be inferred from most breeding experiments. Since, however, the deoxyribonucleic acid in a *Vicia faba* chromosome if fully extended would reach over a metre in length, the linear replication of such a molecule would present immense difficulties. Observations made on the incorporation of tritiated thymidine into chromosomes by means of autoradiography indicate no clear sequential uptake of the thymidine. The replication of the deoxyribonucleic acid seems in fact to begin more or less simultaneously over the whole chromosome. This has led to the view that the chromosome can be considered as consisting of a large number of replicating segments, or replicons. A replicon is possibly of the same order of size as the strand of deoxyribonucleic acid in procaryote (referred to as the chromoneme or genophore). Some investigators go so far as to suggest that the large chromosomes of eucaryotes can be regarded as each consisting of a chain of several hundred of such chromonemes bound together by histone and a sheath of acid protamine to form a superstructure, the details of which are not yet understood.

δ. *Giant chromosomes.* In rare cases repeated multiplication of chromonemata leads to multistranded (polytenic), cable-like, giant chromosomes.

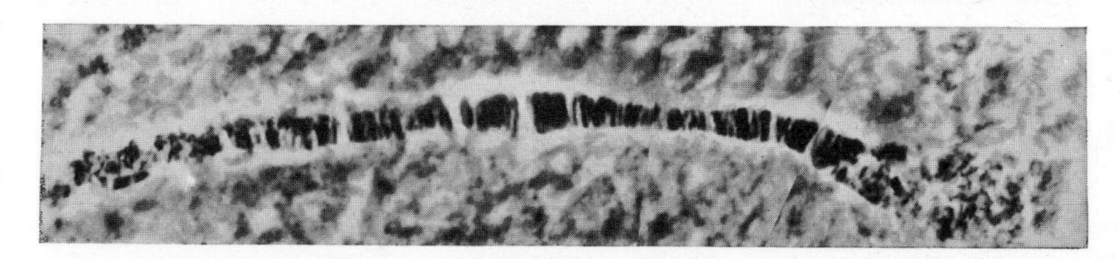

Fig. 19. Giant chromosome, polytene and consisting of about 2048 strands, from the suspensor of the embryo of *Phaseolus vulgaris*. Unstained, phase-contrast. ×1500. From W. Nagl.

Giant chromosomes were first described from the salivary glands of various species of insects. Subsequently, however, they have also been found in plants, as, for example, in the haustorial cells of the suspensor of *Phaseolus coccineus*, *P. vulgaris*, *Gagea lutea* and *Loasa* (Fig. 19). Since the chromonemata (the individual strands) of these giant chromosomes do not contract, and may number as many as 4096, the chromosomes are some thirty to forty times longer and thicker than normal metaphase chromosomes. It was with such chromosomes (first in animals, subsequently in plants) that it was possible to demonstrate cytologically the localized activity of specific chromosome segments ('puffing', see p. 330).

c. Typical nuclear division or mitosis. The identical reduplication of the chromonemata thus takes place while the chromosomes are still in the extended condition in the interphase nucleus (Figs. 24E, F; 25B, C, D). In the course of a normal mitosis these duplicate

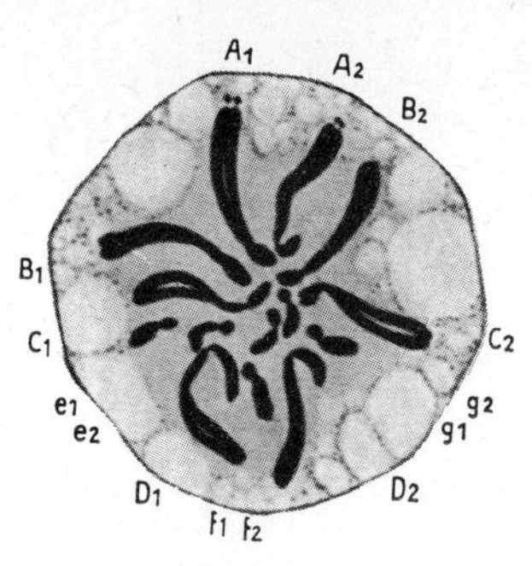

Fig. 20. Metaphase in polar view (equatorial plate) in *Aloe thraskii* (2n = 14, cf. Fig. 22.6). All the chromosomes are divided longitudinally into two. The homologous chromosomes are indicated by the same letter. *c.* ×1000. After Schaffstein.

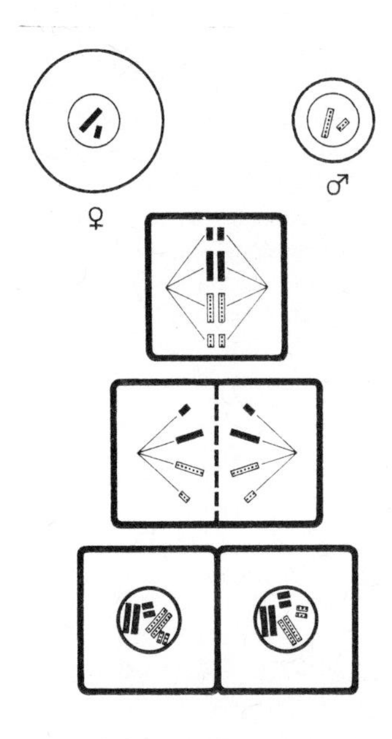

Fig. 21. Diagrammatic representation of mitosis. Top line, two germ cells. Female gamete (♀) with two black chromosomes, male (♂) with two white chromosomes. Below (framed in black), diploid somatic cell in the course of mitosis. The chromosomes, both maternal (black) and paternal (white), have each split into two chromatids. The daughter chromosomes then separate (anaphase), and finally two daughter cells are formed, each nucleus containing one maternal and one paternal genome.

structures are separated from each other, one member of each pair passing to a daughter nucleus.

The course of mitosis is identical in fundamentals in plants and animals. The sequence of events in the division of a nucleus and a cell is shown in Fig. 22. The drawings are based on a series of longitudinal sections through the meristematic region of a root, fixed and stained. The nucleus contains fourteen chromosomes. The same sequence is shown diagrammatically in Fig. 21, where four chromosomes (n = 2) are assumed to be present. It is customary to divide the whole process of mitosis, the mitotic cycle, into a series of phases which are referred to by special terms.

Prophase is that in which the chromatin reticulum of the nucleus begins gradually to disentangle itself. Fine chromosome threads become progressively more clearly visible as a consequence of the contraction of their chromonemata and the increasing chromaticity. These chromosomes are always present in a number characteristic of the organism.

Metaphase. The chromosomes become strongly contracted as a result on the internal spiral coiling of the chromonemata. That each chromosome consists of two chromatids is clearly visible from this time onwards. At the same time the chromosomes collect themselves together in the middle plane of the cell, often forming in this plane a star-like figure, the equatorial plate. The constancy in the number of the chromosomes and their

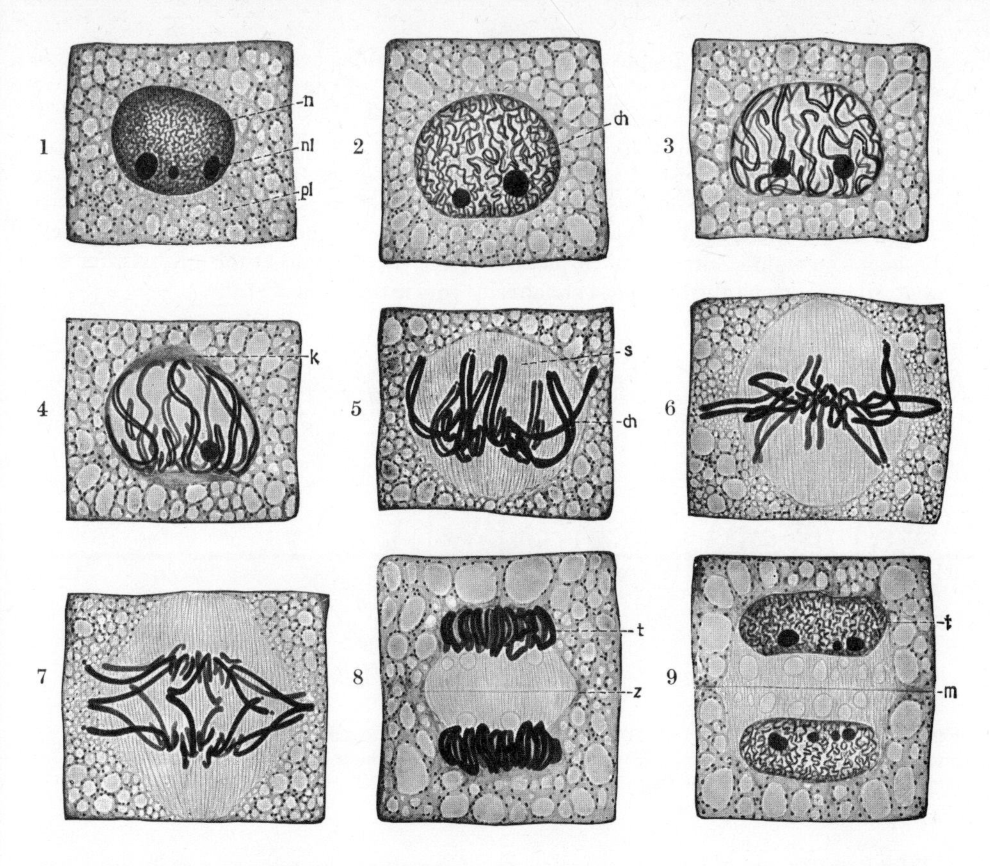

Fig. 22. Mitosis and cell division in an embryonic cell from the root tip of *Aloe thraskii*: n, nucleus; nl, nucleolus; ch, chromosomes; pl, cytoplasm; s, spindle; k, polar cap; t, daughter nucleus; z, beginning of dividing wall, still traversed by fibrils (phragmoplast); m, cell plate. **1.** Interphase nucleus. **2.** to **4.** Prophase. **5.** Transition to metaphase. **6.** Metaphase (see also Fig. 21). **7.** Anaphase. **8.** Early telophase. **9.** Late telophase. ×1000 approx. After Schaffstein.

separateness from one another can best be seen at this stage. Consequently, in diploid cells this is the stage at which the presence of two homologous sets of chromosomes can most easily be established (Figs. 20; 22.6). The nucleolus disperses, and the nuclear envelope falls into fragments which are indistinguishable from elements of the endoplasmic reticulum.

A loose pairing of homologous chromosomes resembling, but not proceeding to the extent characteristic of meiosis (see p. 37), is occasionally seen at metaphase.

Anaphase. At this stage the chromosomes contract to their maximum extent. Independently of the contraction of the chromosomes, a fibrous system, the nuclear spindle, differentiates in the cytoplasm. This is preceded by the formation on each side of the equatorial plate of two saucer-shaped masses of fibres known as the polar caps (Fig. 22.4, k). Where the dividing cells form flagella as a means of locomotion, a phenomenon still occurring sporadically amongst plants as highly evolved as the gymnosperms (e.g. the formation of male gametes in the cycads and in *Ginkgo*), the spindle fibres originate from a small granule called the centriole (Fig. 23A–D). The centrioles are self-reproducing. They are lacking in the cells of higher plants, but present in those of animals. A further relationship between centrioles and flagella, demonstrated by the electron microscope, is that the structure of the base of the flagellum (basal body) is almost indistinguishable from that of a centriole.

Individual fibres of the spindle apparatus push themselves from the polar caps up to the chromosomes, and attach themselves to the already existing centromeres. Subsequently, by a mechanism still not elucidated in detail, the two chromatids, from now on referred to as daughter chromosomes, are drawn to the poles of the cell (Fig. 22.6–8). In the process those longer daughter chromosomes with more or less central centromeres may assume the shape of a V, because of their ends trailing behind and approaching one another (Figs. 20.7; 24B, C). In the electron microscope the spindle fibres are seen to have the same structure as microtubules.

In the last phases of this stage it is generally possible to see that the daughter chromosomes are already again divided into two longitudinally, these chromatids corresponding to half chromatids of the original mother chromosome. This subdivision clearly indicates that not only has the split enabling the division of the chromosome at the immediately subsequent metaphase already taken place, but that the divisions at the mitosis after that and possibly even still further ahead are already anticipated (Fig. 25).

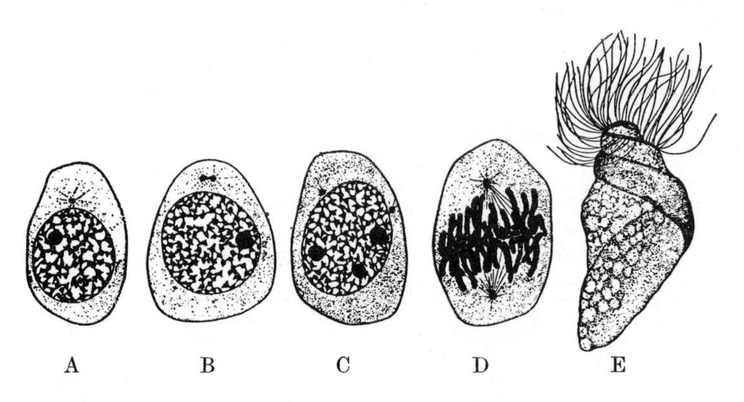

Fig. 23. A., B., C., D. Appearance of the centrosomes in spermatogenesis of *Equisetum*. When the multiflagellate spermatozoids (**E.**) mature the centrosomes become transformed into ribbon-like blepharoplasts, from which the flagella emerge. After Sharp.

Telophase. In telophase the two groups of daughter chromosomes each form new daughter nuclei. This is accomplished by the chromosomes aggregating and, following dissolution or swelling of the kalymma, losing their spiral coiling. The whole aggregate becomes surrounded by a new nuclear membrane. In resting nuclei the chromatin generally appears more or less diffusely distributed, or it may be no longer visible at all. In only a few species of plants possessing chromosomes with heterochromatic segments (see section b) are strongly staining chromocentres still visible in resting nuclei (examples are provided by *Antirrhinum, Beta,* and some Cruciferae.

During the mitotic cycle the nucleoli go through a striking change of form. They gradually disappear during prophase and early metaphase (Fig. 22.2–4), remain undetectable during anaphase and first appear again in telophase, to be regenerated to their original size in the interphase nuclei (Fig. 24D–F). In some instances (e.g. *Cucurbita, Canna*) the nucleoli remain intact. They are then ejected from the nucleus during division, and new nucleoli are formed in the subsequent interphase nuclei.

d. Nuclear division cycle. The copying of the hereditary information in the chromosomes, an essential preliminary to nuclear division, implies in macromolecular terms the replication of the double helix of deoxyribonucleic acid. Cytochemical investigations (using microspectrophotometric and autoradiographic methods) have shown that this replication takes place in the interphase nucleus.

The synthesis of both ribonucleic acid and deoxyribonucleic acid is confined to interphase. While, however, the synthesis of ribonucleic acid is continuous and more or less uniform throughout interphase, that of the deoxyribonucleic acid is limited to a relatively short period (1–2 hours), known as the S period. Usually the S period is preceded by a post-mitotic quiescent period (G_1 period) lasting several hours, and followed by a second resting period (G_2 period), preceding the next mitosis.

Interphase can thus be regarded as falling into three sub-phases:

1. Restitution of the quantitative relationships between nucleus and cytoplasm, disturbed at the time of mitosis (G_1 period);

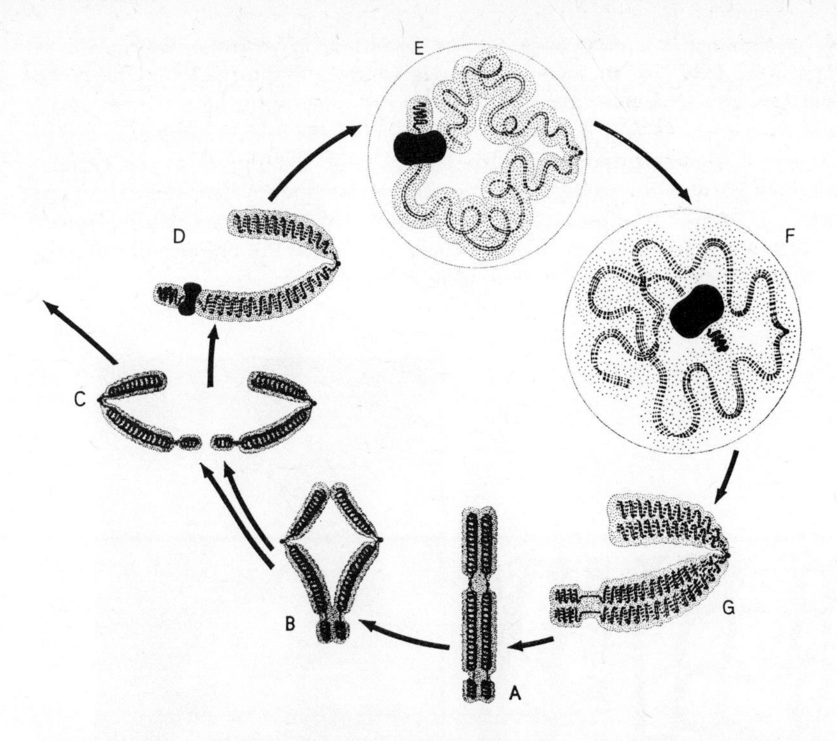

Fig. 24. Diagrammatic representation of the changes in form of a single chromosome in the course of the mitotic cycle. **A.** Metaphase chromosome split along its length into two chromatids. **B.** Anaphase; the two chromatids, now daughter chromosomes, separate from each other and begin to move apart. **C.** Early telophase. **D.** Late telophase and transition to interphase. **E., F.** Weakening of the spiral contraction of the chromosome sheath (kalymma), and reformation of the nucleolus. Interphase, the period during which the DNA of the nucleus (and the hereditary information stored therein) replicates itself, has now begun (see p. 34). **G.** Prophase of the next mitosis. Each chromosome now consists of two spirally contracted, identical, nucleoprotein strands (chromatids). These lie parallel, each spirally twisted about the other. They separate in the subsequent mitosis. After Kühn, expanded and modified.

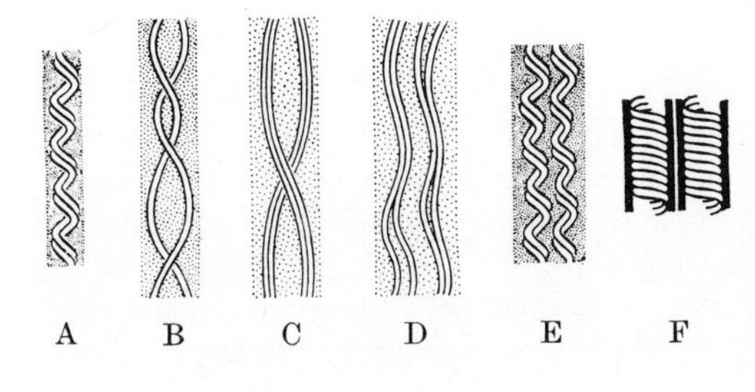

Fig. 25. Diagrammatic representation of the changes in chromosome structure during the mitotic cycle. **A.** Anaphase. **B.** Telophase (loosening of the spiral). **C.** and **D.** Interphase (identical reproduction). **E.** Prophase (renewed spiral contraction). **F.** Replicated metaphase chromosome with two chromatids. After Kühn.

2. Replication of the deoxyribonucleic acid, and consequently of the genetic information in the nucleus (S period);
3. Preparation for mitosis (G_2 period).

Mitosis (M) itself, involving contraction of the chromosomes, formation of the spindle and the separation of the daughter chromosomes, takes about one-tenth of the time of the whole cycle, a period shorter than that of any of the sub-phases of interphase. The whole nuclear cycle can be conveniently indicated by the symbols G_1, S, G_2, M.

In the formation of polytenic giant chromosomes, the G_2 period and mitosis are absent; instead successive replications of the deoxyribonucleic acid occur at short intervals, with the geometrical consequence that the newly generated chromonemata accumulate to form a cable-like structure continually increasing in thickness.

e. Polyploidy. The normal nuclear cycle can be disturbed in yet other ways. Mitosis consists essentially of two processes which, although admittedly normally coupled, are entirely independent of each other, namely: 1. the longitudinal splitting and consequent doubling of the chromosomes (for which the replication of the deoxyribonucleic acid in the previous S period is an essential prerequisite); and 2. the elaboration of the spindle, the apparatus which governs the passage of the separated daughter chromosomes into the daughter nuclei. If the chromosomes divide without the daughter chromosomes separating to form daughter nuclei, a single nucleus with twice the number of chromosomes results. Such a nucleus is termed polyploid.

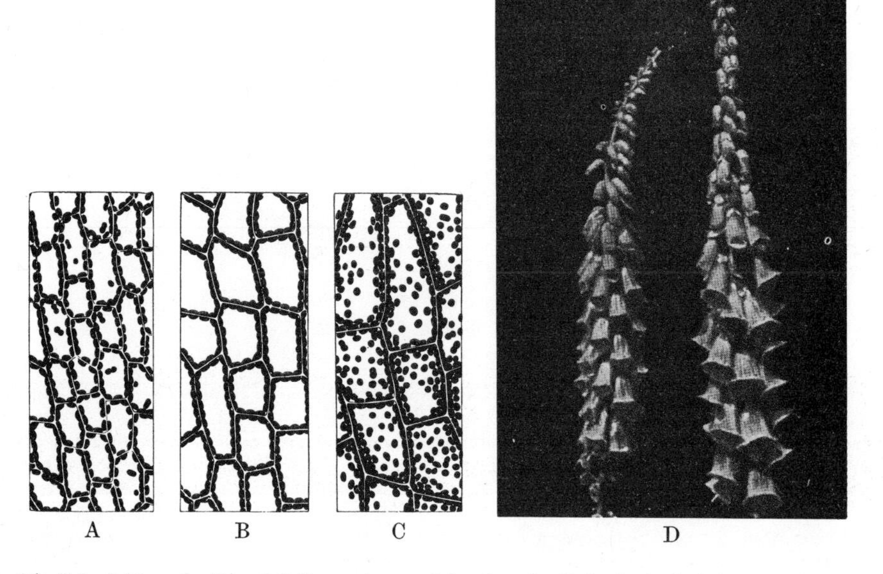

Fig. 26. Polyploidy. **A., B.** and **C.** Dependence of the size of cells in the leaf of the moss *Funaria hygrometrica* on the number of chromosome sets in the nuclei. **A.** Haploid. **B.** Diploid. **C.** Tetraploid. Black dots represent chloroplasts. **D.** Inflorescences of diploid (left) and tetraploid (right) forms of *Digitalis purpurea*. **A., B.** and **C.** X200. After F. von Wettstein, redrawn. **D.** After Schwanitz.

α. Autopolyploidy. Chromosome duplication can in fact occur quite independently of any spindle formation, as can be shown by the use of colchicine which prevents the formation of this apparatus (p. 332). The consequence is a doubling of the chromosome number from 2n to 4n. If this process of polyploidization is repeated in descendants of this cell, yet higher even multiples of the original genome result, so that we obtain a series 2n (diploid), 4n (tetraploid), 8n (octaploid), etc. Since there are close reciprocal relationships between the nucleus and cytoplasm (the nucleus being responsible for the synthesis and nature of the plasma proteins), the increase in nuclear volume consequent upon polyploidization inevitably has effects upon cell size, and not infrequently on the size of the whole organism as well (Fig. 26A–D). However, this increase in size, although often a desirable quality from the point of view of plant breeding, is almost always associated in autopolyploid plants with reduced vitality, and always with reduced fertility.

The formation of polyploid nuclei is to some extent a feature of normal development. For example, the formation of restitution nuclei (the aggregation of already divided chromosomes as a single new nucleus) is characteristic of the tapetal cells of many angiosperm anthers (see p. 648).

β. Endopolyploidy. Still more widespread as an aspect of normal tissue differentiation is the phenomenon of endopolyploidy. Many especially large cells, or cells especially active metabolically, and sometimes whole tracts of tissue, are distinguished by their nuclei having become polyploid by endomitosis. In this kind of polyploidization daughter chromosomes are formed without the appearance of a spindle and without dissolution of the nuclear membrane. In *Tropaeolum majus*, for example, the degree of polyploidization of the nuclei in the ground parenchyma of the petiole reaches 32n, in the pith of the stem 128n and in the integument 1024n. In the haustorial cells of the endosperm of *Arum maculatum* (n = 14), polyploidization to the extent of 24,576n has been demonstrated, corresponding to 344,064 chromosomes, naturally no longer possible to count individually. Finally, the unusually large primary nucleus of the alga *Acetabularia* (Fig. 16C–E) has been shown to reach a high grade of endopolyploidy.

γ. *Allopolyploidy.* The forms of polyploidy so far considered concern the doubling (or further multiplication) of like genomes. They may arise either spontaneously or be induced artificially. Occasionally, in the course of sexual reproduction, a doubling of the unlike genomes of hybrids may occur. This kind of polyploidy is called, in contradistinction to autopolyploidy, allopolyploidy. The hybrid *Nicotiana paniculata* (2n = 24) x *N. undulata* (2n = 24) yields, for example, the fertile and true-breeding allotetraploid *N. rustica* (2n = 48). In a similar manner *Brassica napus* (2n = 38) has arisen from a hybrid between *B. oleracea* (2n = 18) and *B. campestris* (2n = 20).

f. Meiosis. In sexual reproduction, as we have seen, the haploid chromosome number n is doubled. A continuous series of sexual fusions would thus soon cause the chromosome number to become immense, and ordered mitosis impossible. The development of sexuality was therefore of necessity accompanied by the simultaneous development of a mechanism by which the diploid chromosome number of the zygote, or of the cells derived from it, could be reduced to the original haploid number (chromosome reduction). This mechanism is known as meiosis. It likewise depends, as does the contrary process of polyploidization, on the fundamental independence of chromosome multiplication and the mechanism of nuclear division. In meiosis a single division of the chromosomes is accompanied by two consecutive formations of the spindle system. In the first meiotic or reduction division it is not the chromatids which become partitioned between the daughter nuclei, as in mitosis, but entire chromosomes, though these may already be split longitudinally (Fig. 28). In this way the homologous chromosomes derived in fertilization from the parents are again separated. Their distribution among the daughter nuclei is, however, independent of whether they came originally from the maternal or the paternal genome (rearrangement of the genome, Fig. 28B). Not until the second reduction division, in practice following without pause upon the first, are the halves of the split chromosomes pulled apart, exactly as in normal mitosis. The result is a group of four cells (referred to as gonospores or meiospores) in which the haploid condition is restored. Each contains a single genome. The four cells resulting from a reduction division are said to form a tetrad. Frequently the four cells resulting from meiosis separate from one another and become unicellular, more or less thick-walled, meiospores, providing a means of distributing and propagating the species (see p. 205).

The significance of meiosis lies not only in the re-establishment of the haploid chromosome number. Closer investigation has revealed that not only does a reassortment of the homologous chromosomes from the two preceding parents take place in reduction division but also that the genes lying within homologous chromosomes can be exchanged and new genetic combinations formed (structural modification of the chromosome). The reduction of the chromosome number in the process of meiosis, and the recombination of the genetic factors derived from the parents are made possible by a pairing of the homologous chromosomes which takes place in prophase. This phase accordingly merits special consideration.

We will now follow the course of a typical meiosis phase by phase. The prophase of meiosis, unlike that of mitosis, is subdivided into a number of stages, each of which is distinguished by a name.

Prophase of the first division of meiosis. Before the onset of the division proper, the nuclei about to undergo meiosis are rendered conspicuous by their unusually large size and a correlated dispersal of the chromatin. At the beginning of the meiotic prophase,

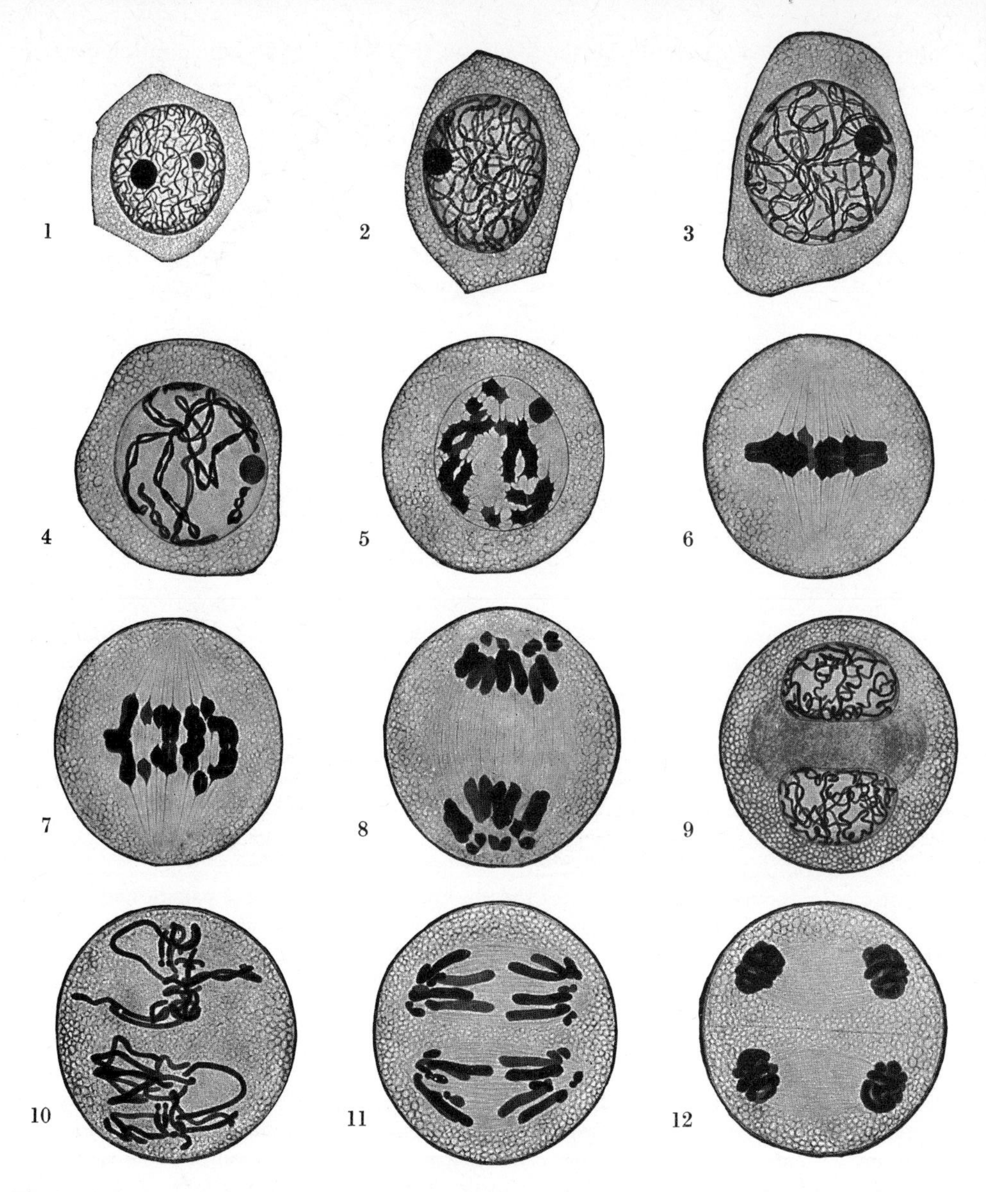

Fig. 27. Meiosis in a pollen mother cell of *Alöe thraskii* (cf. Fig. 22 showing mitosis in the root tip of the same plant). **1.** to **9.** First division of meiosis. **10.** to **12.** Second division. **1.** to **5.** Prophase of the first division. **1.** Leptotene. **2.** Zygotene. **3.** Pachytene. **4.** Diplotene. **5.** Diakinesis. **6.** Metaphase of the first division. **7.** Anaphase. **8.** Telophase. **9.** Interkinesis. **10.** Metaphase of the second division. **11.** Anaphase. **12.** Formation of the four haploid nuclei of the pollen grains. *c.* ×1000. After Schaffstein.

just as at the beginning of normal mitosis, there become visible greatly extended chromosomes, apparently entangled and coiled up with one another, and split into their two chromatids (Fig. 27.1 and 2). But whereas in mitotic prophase the spiralling and the accompanying contraction of the chromosomes begins after a few hours, meiotic prophase, in which the pairing of the homologous chromosomes takes place, can extend over weeks, sometimes even months.

α. Leptotene. As a result of shrinkage of the kalymma and a general spiral contraction, the chromosomes become visible as single, fine threads (Figs. 27.1; 29A). The longitudinal splitting is not usually visible at this stage with the light microscope.

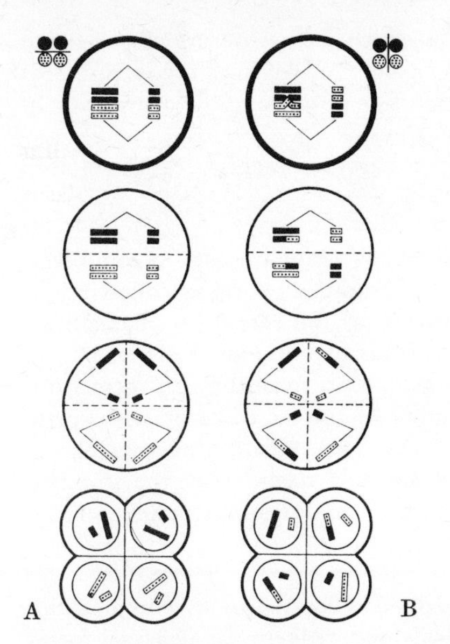

Fig. 28. Diagrammatic representation of meiosis in a pollen mother cell with the chromosome number 2n = 4 (maternal chromosomes black, paternal white). **A.** Pre-reduction. The four chromatids of the bivalent separate reductionally. **B.** Mixed reduction. As a result of a chiasma in the left-hand bivalent there is an interchange between two of the four chromatids. Distal to the chiasma the separation is thus equational, and the maternal and paternal portions remain together until the second division of meiosis. There is thus a mixture of pre-reduction (proximal to the chiasma) and post-reduction (distal to the chiasma).

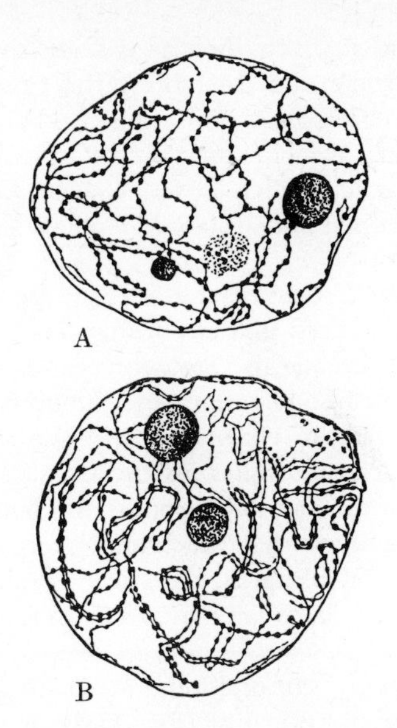

Fig. 29. Prophase of the first meiotic divisions of the nucleus of a pollen mother cell of *Trillium erectum* (Liliaceae). **A.** Leptotene. **B.** Zygotene. The pairing of the homologous chromosomes, and the approximating of similar chromomeres is already far advanced. The double nature of the chromosomes however is not yet visible. After Huskins and Smith.

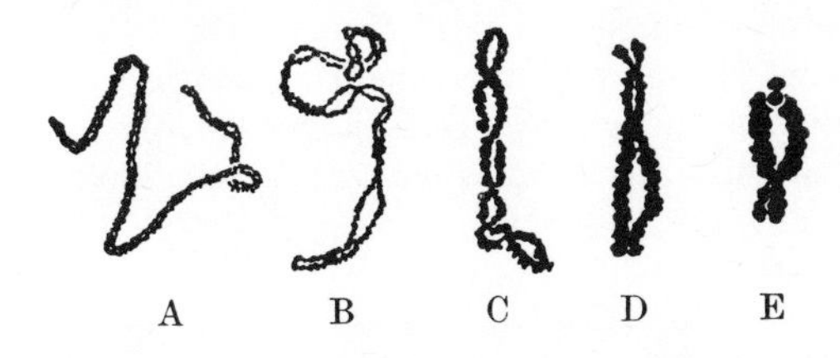

Fig. 30. *Anemone baicalensis*. Reduction in the number of chiasmata (terminalization) from pachytene (**A.**) to metaphase (**E.**) of the first meiotic division. After Moffet.

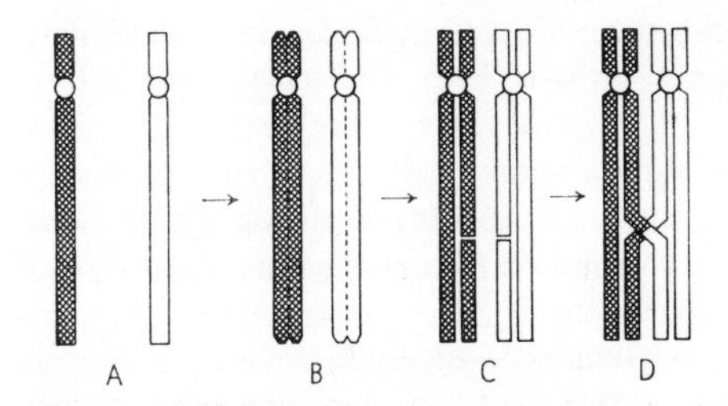

Fig. 31. Formation of a chiasma according to the chiasma-type hypothesis. **A.** and **B.** Pairing of the homologous chromosomes. **C.** Fracture of two adjacent chromatids at the same relative positions. **D.** Cross-wise rejoining of the chromatids. The consequence is that the parts of the chromosomes near the centromere undergo pre-reduction, those on the other side of the chiasma post-reduction. After Rieger and Michaelis.

β. Zygotene. The pairing of the homologous chromosomes begins. Closer observation shows that pairing begins by segments of the chromosomes, recognizable as homologous from the arrangement of their chromomeres, coming here and there to lie parallel with one another (conjugation, Fig. 27.2). Homologous chromomeres evidently have a particular attraction for each other, but the first coming together of the chromosomes seems to take place more or less fortuitously. Once pairing has begun, it proceeds rapidly, extending in both directions from the positions in which it first occurs (Fig. 29B), a process somewhat resembling the closing of a zip-fastener. Sometimes the intrusion of another chromosome between two homologues presents an obstacle to the completion of the pairing. We will come back to this situation in diakinesis.

Similar, but much looser, association of homologous chromosomes is occasionally seen during normal mitosis in many plants and animals (see p. 33).

γ. Pachytene. In pachytene the pairing of homologous chromosomes is completed (Fig. 27.3). Each pair of chromosomes consists of four chromatids, forming collectively a tetrad of chromatids. Two chromatids (sister chromatids) of this tetrad have come from the genome of one parent, and the other two from the other parent (Figs. 30B, C; 31B, C). These tetrads of chromatids are also referred to as gemini or, to distinguish them from unpaired chromosomes, as bivalents. The total number of tetrads of chromatids corresponds to the haploid chromosome number (pseudoreduction). Electron micrographs of this stage show that the homologous chromosomes are held together by the so-called synaptinemal complex, seen in longitudinal section as a tri-axial structure maintaining an even distance between the two chromosomes.

δ. Diplotene. In diplotene the pairing of the still extended chromosomes, completed in pachytene, is again lost (Fig. 27.4). A continually widening 'reductional cleavage' forms between the two homologues, while at the same time a so-called 'equational cleavage' separates the sister chromatids. Nevertheless, any two non-sister chromatids which have been particularly intimately associated will almost always be seen to remain joined together at certain points (Fig. 30B, C, D), often predictable in position.

According to the chiasmatype theory, two homologous chromatids of different origin (that of one being maternal, of the other paternal) break at this time at exactly corresponding positions, and almost immediately rejoin diagonally (Fig. 31C, D), forming a so-called chiasma. In the following segregation these two reciprocally related chromatids are separated and pass to different daughter nuclei. The consequence is that the genetic information in the chromatids concerned (which eventually become daughter chromosomes) becomes recombined, and new groupings of the genes emerge (recombination).

Thus there occurs in meiosis not only a reassortment of the entire chromosomes, so that each daughter cell contains descendants of those coming from both parents (Fig. 28A), but also the chromosomes themselves undergo structural changes (Fig. 28B). The chiasmata in diplotene are the morphological aspect of the genetic exchange frequently encountered in breeding experiments. Although the chiasmatype theory is currently widely accepted, it cannot explain all phenomena concerned with crossing-over in an entirely satisfactory manner.

ε. Diakinesis. In diakinesis, the last stage before the dissolution of the nuclear membrane, the fully extended, euchromatic, chromonemal portions of the bivalents begin once more to become spirally coiled. The coiling, however, is much wider and looser than in mitosis, and the eventual shortening of the chromosome much greater, the final length being a sixth to a tenth of that at a similar stage in mitosis (Figs. 27.5; 33E). For a short time they are distributed more or less symmetrically around the nuclear wall. It can often be seen at this time that as a consequence of one chromosome lying on the top of another during the pairing in zygotene, as mentioned previously, the bivalents are interlocked. This is without effect on the course of meiosis. The prophase of meiosis is concluded with diakinesis.

Metaphase and anaphase of the first division of meiosis. In metaphase the bivalents, as the univalents in mitotic metaphase, arrange themselves on the equatorial plate (Fig. 27.6), a delicate kalymma becomes visible, the nuclear membrane and nucleoli disappear and the nuclear spindle is differentiated.

During anaphase the partners of the bivalents are separated (Fig. 27.7). These are entire chromosomes. Before the chromosomes finally detach themselves from each other, the chiasmata are frequently displaced towards the ends of the chromosomes, or terminalized (Fig. 30D, E). While the homologous chromosomes, once again separated, are migrating towards opposite poles of the spindle, their two chromatids are already separated from each other, except at the centromere.

Telophase, interkinesis and the second meiotic division. The chromosomes assembled at the poles of the cell after the first meiotic division thus each consist of two chromatids almost wholly separated from each other. Consequently, they are quite different from the telophase chromosomes of normal mitosis; they resemble in fact those of the prophase of mitosis. The two daughter nuclei lack the capacity to turn themselves into normal resting nuclei. Admittedly there is some loosening of the spiral coiling of the chromosomes, and the formation of a delicate nuclear membrane (Fig. 27.9), but the stage of interkinesis lasts only a short time. The chromonemata contract once again, and with the formation of the spindle the nuclei move into the second meiotic division. In this, as in normal mitosis, the chromatids become completely separated from each other. Thus the partitioning of the four chromatids of the pachytene tetrads among the tetrad of nuclei formed as a consequence of meiosis is now completed (Fig. 27.10–12).

Usually new cell walls are laid down after each of the two nuclear divisions of meiosis, and the formation of the four cells of the tetrad is said to be successive. Occasionally no cell walls are laid down until after the second meiotic division and the mother cell becomes transiently four nucleate. The formation of the tetrad is then said to be simultaneous.

Pre-reduction, post-reduction, mixed reduction. The four chromatids of each chromatid tetrad at pachytene are, as a consequence of the preceding pairing of the chromosomes, only superficially similar. Actually two are from one parent and two from the other. The two cleavages which separate the four chromatids are therefore of different nature. Certainly, however, one of the two divisions of meiosis must again separate the homologous chromosomes or segments of chromosomes which have been paired in zygotene. For those segments adjacent to the centromeres (kinetochores) of the chromosomes, this 'reduction' or separation of the homologues usually takes place at the first division of meiosis ('pre-reduction'). This is because the spindle fibres attach themselves to the still undivided centromeres and draw the homologous chromosomes apart. With parts of chromosomes further from the centromere, however, it can often happen, as a consequence of chiasma formation, that a chromatid of maternal origin becomes paired in part with a segment of a chromatid of paternal origin, and reciprocally in the homologous chromosome. In these regions separation of the maternal and paternal material does not take place until the second division of meiosis (post-reduction). Such instances were first observed in fungal genetics, but it is now known that post-reduction of individual chromosome segments is a widely distributed phenomenon (mixed reduction).

In the Ascomycetes, where meiosis leads to the formation of spores, known as ascospores, the way in which certain genetic factors are distributed among the spores can be immediately seen from the colours of the individual spores. A peculiarity in the formation of ascospores is that each nucleus of the tetrad produced as a consequence of meiosis divides again mitotically, resulting in eight, instead of four, haploid spores. These spores lie in a row one behind the other in the tube-like sporangium, known as the ascus. If a genetic factor determining spore colour has been different in the two parents, then the colour of the spore will depend upon the source of its colour-determining factor. In some cases it is possible to tell immediately from the distribution of the two colours among the

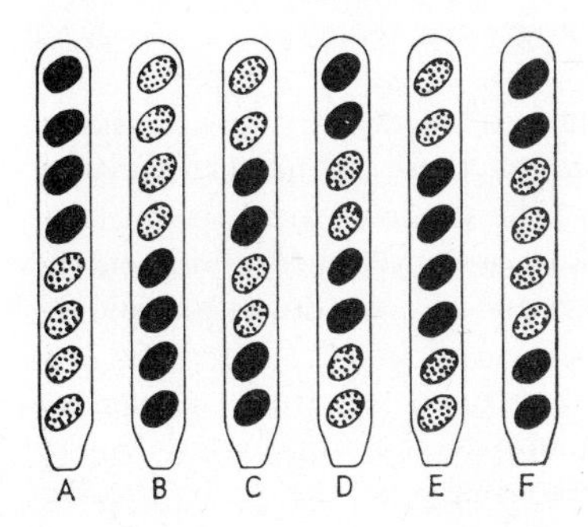

Fig. 32. Diagrammatic representation of factor segregation during spore-formation in *Neurospora sitophila* (Ascomycetes). **A.** and **B.** Pre-reduction. **C.–F.** Various possible arrangements of the two sorts of ascospores (depending upon the presence of the maternal or paternal genome) with post-reduction. The independent segregation of another pair of factors was demonstrated in the same experiment (mixed reduction). After Dodge.

spores which of the two divisions of meiosis has been reductional and which equational (Fig. 32).

The more or less random combination of the different ways in which chromatids and chromatid segments can be distributed amongst the four nuclei resulting from meiosis, described in the foregoing, leads to a thorough mixing and recombination of the totality of the genetic information inherited from the parents. The four nuclei of the tetrad receive in fact different genomes (Fig. 28B).

Time and place of meiosis. The interval in time, and also in the amount of growth, between fertilization (karyogamy) and meiosis has changed in the course of evolution. In the simplest case, found in numerous unicellular and primitive algae, meiosis sets in immediately on germination of the zygote and the change of phase of the nucleus is said to be zygotic. In such organisms all the stages of development, except for the zygote itself, are haploid and the organism is said to be a pure haplont (Fig. 33A). In many of the more highly developed algae, however, the diploid zygote grows first into a multicellular diploid plant, the sporophyte. After the development of numerous diploid spore-mother cells meiosis takes place freely. Thus from the original nuclear fusion, not one tetrad, but a multitude of haploid cells (gonospores or meiospores) is produced. These organisms are termed diplohaplonts, and are said to show an antithetic or heterophasic alternation of generations (Fig. 33B). Each generation is commonly independent of the other, and the germ cells from which they arise are different, the zygote yielding the sporophyte and the gonospore or meiospore the gametophyte. The change of phase of the nucleus in diplohaplonts is said to be intermediate. Finally, meiosis can be deferred right up to the formation of gametes. In this case, rare in plants, all the cells other than the gametes are diploid and the organisms are termed pure diplonts. The change of phase of the nucleus is said to be gametic (Fig. 33C; examples are *Codium, Acetabularia, Fucus* and the diatoms).

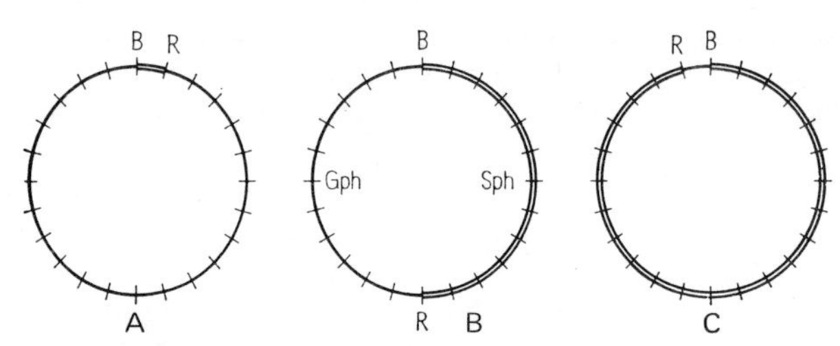

Fig. 33. Diagrammatic representation of the various possible alternations between the haplophase (single line, followed clockwise) and diplophase (double line). B, fertilization; R, meiosis; Gph, gametophyte; Sph, sporophyte. **A.** Zygotic change of nuclear phase (only the zygote is diploid: purely haploid organism). **B.** Intermediate change of nuclear phase (alternation of a haploid and a diploid phase of development: heterophasic alternation of generations). **C.** Gametic change of nuclear phase (only the gametes are haploid: purely diploid organism).

With zygotic change of nuclear phase following each sexual fusion, there can be only one opportunity for recombination of the genetic information coming from the parents. When the change of nuclear phase is intermediate or gametic, numerous different new combinations can arise following a single nuclear fusion. In the course of evolution there can be seen a clear tendency for the diploid phases of development to become more elaborated, while conversely the haploid phases suffer a progressive reduction.

g. The working nucleus. The nuclei of differentiated and no longer meristematic tissue, are often marked by morphological peculiarities, by which they can be distinguished clearly from resting or interphase nuclei which still have the capacity for division. As a result of endomitosis, the nuclei of differentiated tissue are often polyploid and in consequence considerably larger than the nuclei of embryonic cells. Grades of polyploidy of 2048n and 4096n are not unknown.

The grade of polyploidy can in some cases be inferred approximately from the number of chromocentres present. In others it is possible to stimulate the working nucleus to undergo mitosis by experimental treatment, such as wounding, and the grade of polyploidy can then be discovered. Nuclei containing multistranded (polytenic) giant chromosomes may also be present (see p. 31).

While in mitotic and interphase nuclei only one identical reproduction of the karyoplasm occurs in the mitotic cycle, in working nuclei an over-production of certain materials often takes place. These become transferred to the cytoplasm and then from here enter into the general metabolism of the organism. Note also the production of morphogenetic substances by the nucleus (p. 27). In relation to this activity, working nuclei are not uncommonly marked by particularly large nucleoli. In *Acetabularia* the volume of the working nucleus is more than 100 times that of the nucleus of the zygote, an increase partly attributable to polyploidy, but principally to an enormous growth of the nucleolus.

3. Division of the protoplast

Generally, at the conclusion of nuclear division the division of the protoplast begins. The individual components of the cell, which have reproduced themselves in the preceding growth phase, are now distributed between the two daughter cells.

The division of the cytoplasm of those Flagellatae with naked cells begins with the appearance of a furrow. This gets progressively deeper, until finally the cell is pinched into two parts (Fig. 403). This mode of division, which might well be primitive, is referred to as simple cleavage (schizotomy).

In those Flagellatae where the cell is furnished with a wall, the protoplast first becomes naked by detaching itself from the wall, and then, corresponding to the number of nuclear divisions, divides itself up by fission to form two or more naked uninucleate cells. Not until this has occurred do the cells become surrounded by a new wall, a process which usually occurs simultaneously throughout the group of cells (Fig. 408B) (schizogeny). In many Flagellatae (Dinoflagellatae) and other unicellular algae (diatoms, desmids), where daughter cells are produced in pairs, each daughter cell retains half of the original parental wall and forms anew the missing half (Figs. 388D; 400).

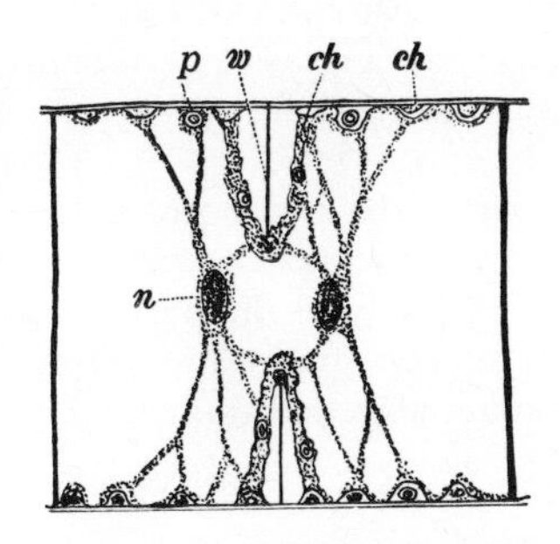

Fig. 34. Centripetal formation of the new cell wall after nuclear division in a cell of *Spirogyra*. The wall (w) gradually grows out from the margins towards the centre, bisecting the tubular cell like an iris diaphragm: n, nucleus; ch, band-like chromatophore in section; p, pyrenoid. After Strasburger.

In the elongated cells of many filamentous algae the new cell walls arise centripetally. A ledge of thickening attached on the inner side of the wall gradually closes itself like an iris diaphragm and eventually the protoplast is divided (e.g. *Anabaena* (Fig. 68B), *Spirogyra* (Fig. 34), *Cladophora*).

In most of the more highly organized plants, on the other hand, the differentiation of the dividing wall is centrifugal. While the daughter nuclei are forming, some new fibres become visible between them from about late anaphase on, the so-called connecting fibrils (Fig. 22.7). Like the spindle fibres, they are at first directed towards the nuclei, but often do not reach them, and frequently pass far to the side of the nuclear region. Ultimately they form a barrel-shaped body, the phragmoplast (Fig. 22.8 and 9). Golgi vesicles (see p. 24), containing the acid hemicelluloses of the middle lamella and primary wall, collect at its equator. Beginning at the centre they fuse together and so give rise to the optically isotropic, cellulose-free, middle lamella of the dividing wall. The membranes of the vesicles form the new plasmalemma on each side of the wall. Here and there the cytoplasms of the two daughter cells remain in connection through narrow pores, the plasmodesmata. The formation of the wall is thus centrifugal, proceeding from the centre to the periphery (Fig. 35A–C). When long narrow cells divide longitudinally (e.g. of the

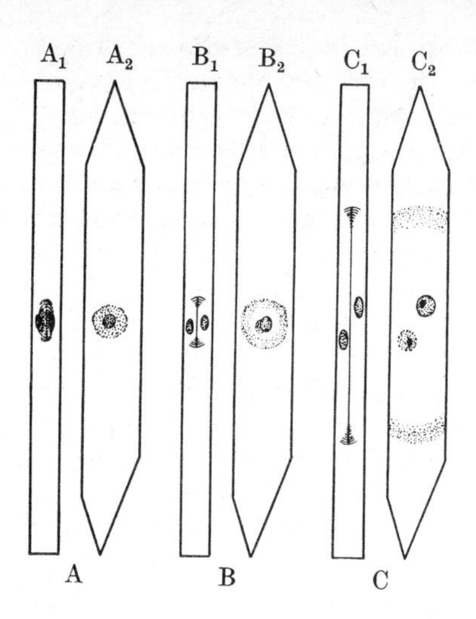

Fig. 35. Diagrammatic representation of centrifugal wall formation preceded by a phragmoplast in a cambial cell (cf. Figs. 157, 158 and 159). **A.** Telophase of the nuclear division and the development of the phragmoplast. **B.** and **C.** The phragmoplast grows out centrifugally all around its periphery and forms a more or less disc-like dividing wall freely suspended in the cytoplasm. **C.** The phragmoplast has reached the lateral walls of the elongated prosenchymatous cell, but not yet the two ends of the cell which are consequently still undivided. (A_1, B_1, C_1; viewed tangentially: A_2, B_2, C_2; viewed radially.)

cambium), they can consequently be fully divided in the middle while still undivided at the ends (Fig. 35C).

Departures from typical cell division. Free nuclear divisions, that is, divisions not accompanied by cell division, occur in those Thallophyta showing the polyenergid condition (see p. 17). In the higher plants free nuclear divisions occur only in a few specialized kinds of cells (e.g. schlerenchyma fibres, latex ducts), as well as in the embryo sacs of angiosperms following the double fertilization. Here, not uncommonly, thousands of nuclei arise by repeated divisions from the triploid secondary endosperm nucleus. They are distributed more or less equidistantly in the plasmatic layer lining the wall of the embryo sac. Eventually the divisions cease and there follows an active formation of phragmoplasts throughout the cytoplasm, either in all parts simultaneously or progressively from the base to the micropylar end. In this way the cytoplasm is divided up into as many cells as nuclei (Fig. 36). The propagules of many algae (e.g. Figs. 415; 438) and fungi are formed in a similar manner, free nuclear division being followed by the simultaneous formation of cell walls (Figs. 469; 509B).

In free cell formation, as occurs for example in the formation of the ascospores of the Ascomycetes (see Fig. 493E), the daughter nuclei produced as a consequence of the free nuclear division become converted into integumented cells lying freely in the plasma of the sporangium. Only a part of the plasma of the mother cell is taken up into the spores, and the remainder serves to prevent the spores coming into contact with one another.

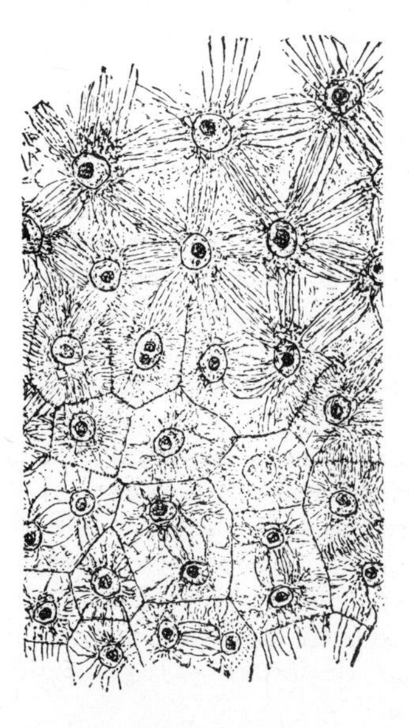

Fig. 36. Portion of the protoplasmatic polyenergid forming the wall layer of an embryo-sac of *Reseda*. Generalized wall-formation is proceeding upwards. *c.* ×240. After Strasburger.

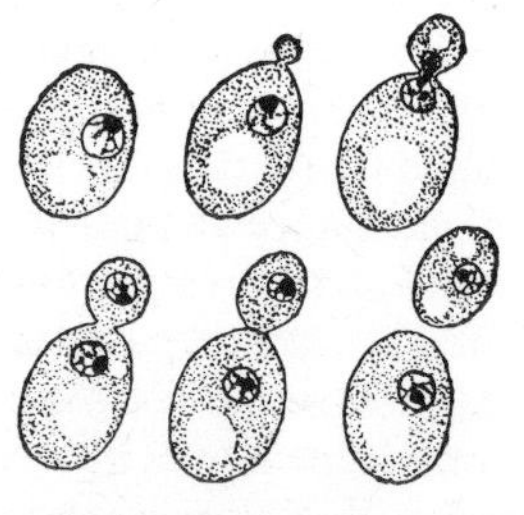

Fig. 37. *Saccharomyces cerevisiae*. Budding. ×1100. After Guilliermond, modified.

Another departure from normal cell division is the process of cell 'budding' encountered in the yeasts. In this kind of division the mother cell does not become halved. Instead, before the division of the nucleus, the mother cell puts forth a process, which after the minute daughter nucleus has passed into it becomes completely pinched off by a dividing wall (Fig. 37). Those propagules of numerous other fungi known as *conidia* (e.g. Fig. 475), as well as the basidiospores of the Basidiomycetes (Fig. 499), are formed in a similar manner.

B. The plastids[13]

The typical organelle of plant cells, absent only in the bacteria, fungi and blue-green algae, is the plastid. We are concerned here with a reaction space rich in lipid, separated from the ground cytoplasm by a delicate envelope of two membranes. Plastids are frequently made conspicuous by fat-soluble pigments (lipochromes). They participate in the anabolic metabolism of the cell, since they are the sites of the photosynthetic assimilation of carbon, and of the synthesis of starch.

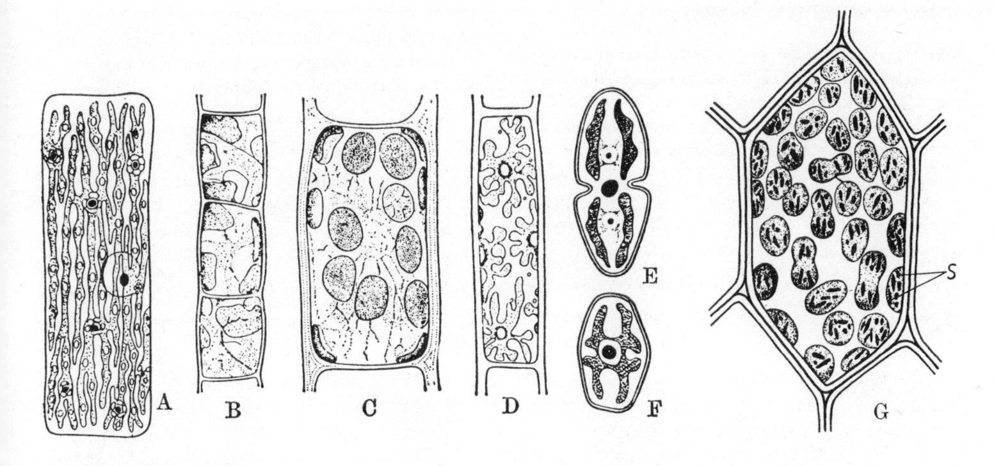

Fig. 38. Chloroplasts of different algae and a moss. **A.** Reticulate chromatophore of *Oedogonium* with pyrenoids (indicated by black spots). **B.** *Leptonema fasciculatum*. **C.** *Pilayella varia*. **D.** *Rhodochorton floridulum*. **E.** and **F.** *Euastrum dubium* (**E.** side view; **F.** transverse section). **G.** Cell from the leaf of *Mnium* sp. (consisting of a single layer of cells). Chloroplasts in various stages of division. All the plastids contain assimilatory starch (S). **A.** to **F.** ×800. After Sharp. **G.** ×700. After Molisch, modified.

All plastids reproduce by fission. They contain deoxyribonucleic acid, in their juvenile form (proplastids) amounting to 3 per cent and in their fully developed condition to 1 per cent of their dry weight. In many algal plastids the deoxyribonucleic acid is evident as fine fibrils in otherwise empty circumscribed areas of the stroma. The proplastids of meristematic cells are relatively small and often show amoeboid movements (Figs. 6, P; 7, P; 41A). They grow with the cell to several times their original size, and by infolding of the inner membrane (Fig. 41B, C) of the envelope acquire a comparatively large internal surface (Fig. 41D, E), on which the photosynthetic pigments are laid down in an ordered array. Ageing plastids often contain spheroidal droplets of lipid, the so-called plastoglobuli.

In sexual reproduction the plastids are believed to be transmitted (as proplastids) from one generation to the next in the gametes. The male gamete, however, frequently has little cytoplasm and lacks plastids. When the plastids are confined in this way to the abundant cytoplasm of the egg, any pigment mutations carried by the plastids are inherited solely through the female (plastid inheritance; see p. 333). It is significant in relation to plastid inheritance that plastids contain a deoxyribonucleic acid different from that of the nucleus.

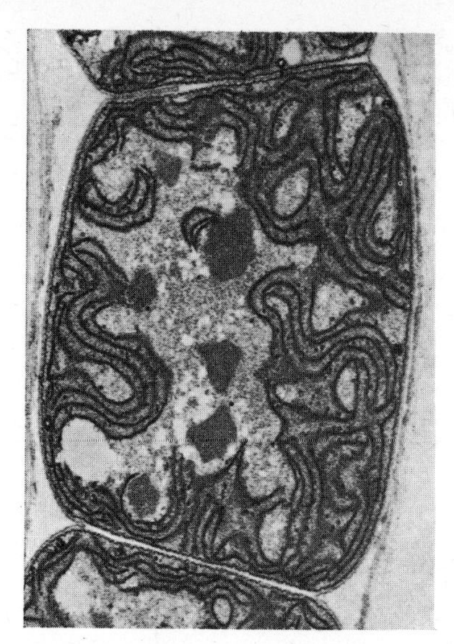

Fig. 39. A cell of the procaryotic blue-green alga *Nostoc muscorum*. The 'chromatoplasma' with numerous thylakoid membranes surrounds, but is not clearly delimited from, the membrane-less 'centroplasma'. X18,000. After Menke.

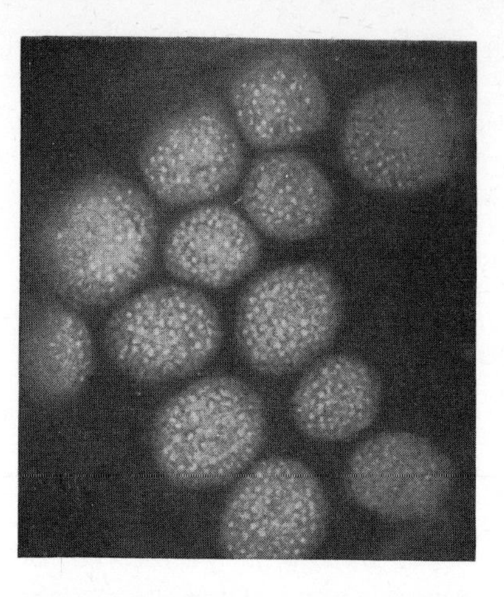

Fig. 40. Chloroplasts of *Arabidopsis thaliana* as seen in the fluorescence microscope. The strongly fluorescing grana are seen against a background of a very much weaker fluorescing stroma. The thylakoid membranes are many fewer in the stroma. X2000. After Röbbelen.

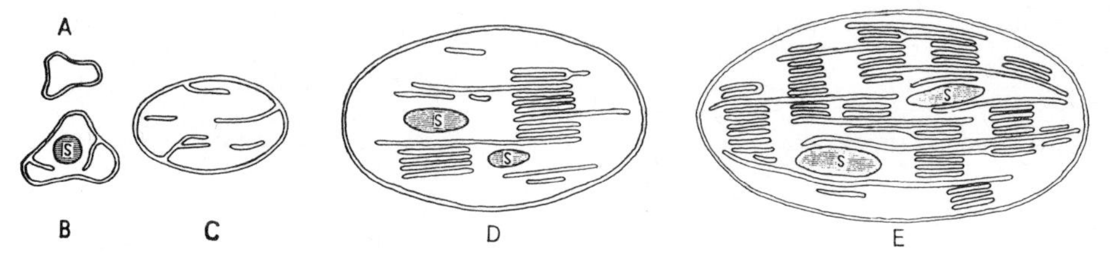

Fig. 41. Development of chloroplasts from proplastids (**A.**). The intermediate stages (**B.** and **C.**) show thylakoid membranes growing into the stroma from the inner membrane of the envelope. These eventually separate themselves and form stacks (**D.**), leading ultimately to the fully developed, green **chloroplast (E.)** with starch grains (S). X8000. After Frey-Wyssling and Mühlethaler.

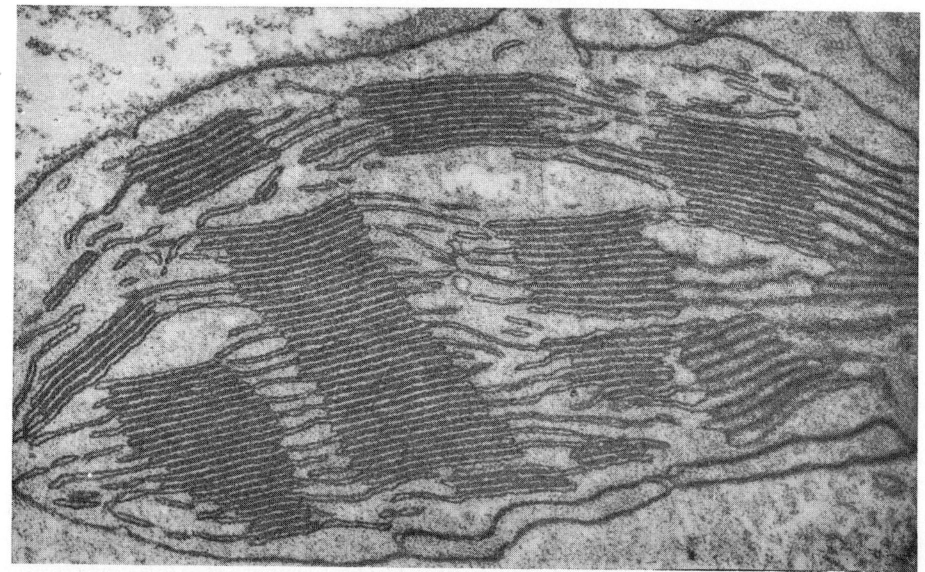

Fig. 42. Part of a median section, perpendicular to the grana thylakoids, through a differentiated chloroplast of *Antirrhinum majus*. X40,000. From Menke.

In the photosynthetic bacteria and in the blue-green algae the photosynthetic pigments are found not in special organelles, but in a peripheral chromatoplasma. The electron microscope shows that this pigmented region has a lamellated structure otherwise found only in chromatophores.

In the more highly organized algae the form of the chromatophores can show considerable diversity. Besides the plate-like chromatophores of *Mougeotia* (Fig. 331), and the spirally banded of *Spirogyra* (Fig. 425A), are found reticulate (*Oedogonium*; Fig. 38A), stellate (*Euastrum* and other Desmidiacese; Figs. 38E, F), cup-shaped (Volvocales; Figs. 408A, K) and irregularly lobed forms (*Rhodochorton*; Fig. 38D). Evidently the lens-shaped chloroplast of some 4–8 μm diameter did not become the predominant form until the evolution of the most highly developed algae and the land plants, in which it occurs with astonishing uniformity (Fig. 38C, G). All chloroplasts are capable of changing their form and volume in relation to light intensity (usually contracting in the dark).

The fully developed plastids of mature cells can be of three forms: 1. photosynthetically active chromatophores (chloroplasts, phaeoplasts, rhodoplasts); 2. photosynthetically inactive chromatophores (chromoplasts); 3. leucoplasts, both photosynthetically inactive and colourless.

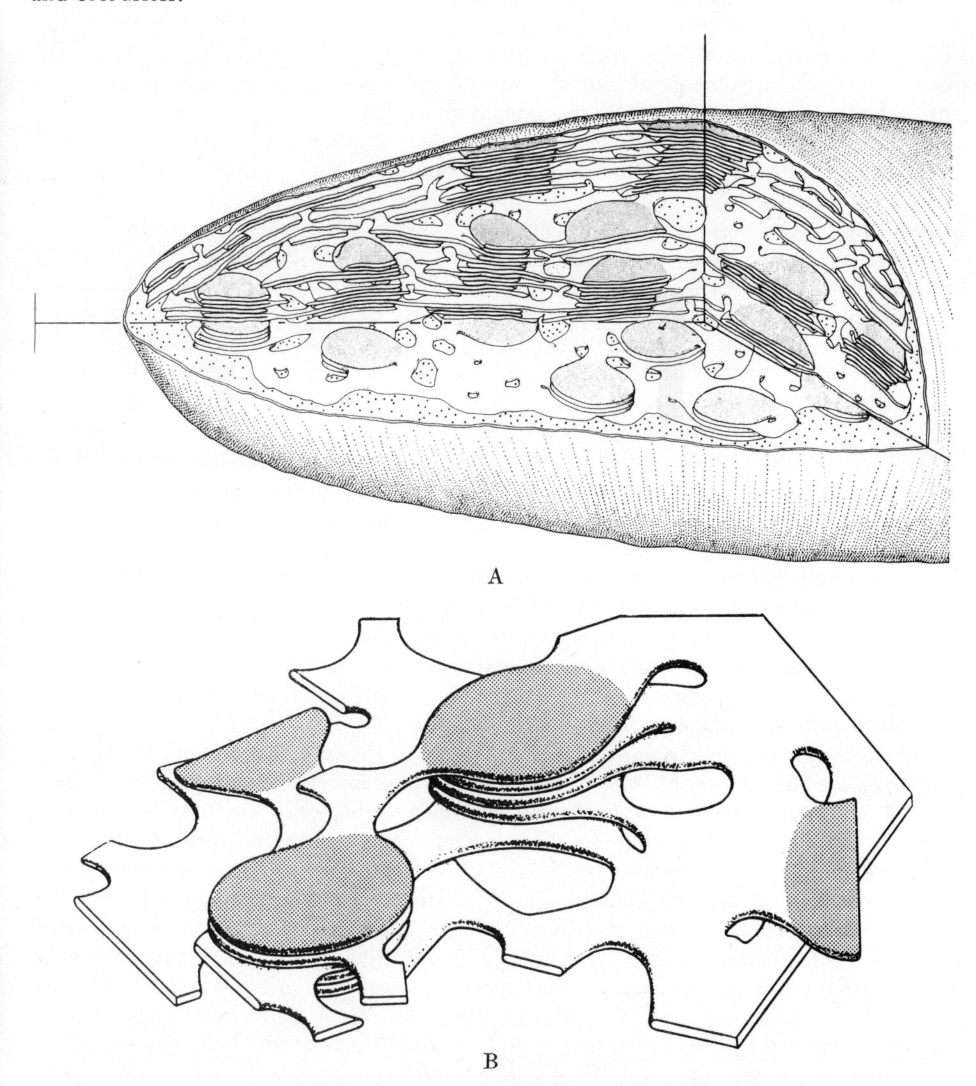

Fig. 43. A. Model of a differentiated chloroplast of a higher plant cut open to show the numerous grana. These are formed by the stacking of overlapping portions of stroma thylakoids. **B.** Detail of grana and stroma thylakoids. **A.** ×20,000. **B.** ×100,000. From original drawings by Wehrmeyer.

1. Photosynthetically active chromatophores

The most important plant pigment, and that making possible photosynthesis, is chlorophyll. This substance occurs in all photosynthetically active chromatophores, embedded in layers of lipoprotein in a colourless ground substance, referred to as the stroma. Sometimes the colour of the chlorophyll is masked by other pigments present, so that the chromatophore appears no longer green, but brown or red. Those plastids which from their high content of chlorophyll are green in colour are called chloroplasts. In the phaeoplasts of the brown algae (Phaeophyceae) the chlorophyll is masked by brown carotenoid pigments (of which fucoxanthin is one). In the rhodoplasts of the red algae (Rhodophyceae), the predominant pigments include not only the reddish carotenoids but also the red pigment phycoerythrin (related to the gall pigments) and the blue pigment phycocyanin.

Chlorophyll (insoluble in water and strongly lipophilic) occurs in most higher plants in two forms, namely chlorophyll a and chlorophyll b, their proportions usually being of the order 3:1. A whole series of other, but still related, chlorophylls has now been isolated from the algae. In addition to chlorophyll a, for example, the brown algae, dinoflagellates and diatoms contain chlorophyll c, and the red algae chlorophyll d.

The chlorophylls are chemically similar to haemoglobin, the red pigment of blood. They have a porphyrin nucleus built of four pyrrole rings, but instead of enclosing an atom of iron, as in haemoglobin, in chlorophyll there is an atom of magnesium in the centre of the molecule (Fig. 243a). The porphyrin nucleus is hydrophilic, but the phytol 'tail' attached at one corner is rich in methyl groups, and is consequently hydrophobic and lipophilic. The chlorophyll molecules are thus polar, and are well suited to lie as monolayers between hydrophilic protein and hydrophobic lipid lamellae. In short-wave ultra-violet light chlorophyll fluoresces a brilliant red (Fig. 40).

Besides the green chlorophyll, orange-red and yellow carotenoids are also present, although to a much smaller extent. The carotenoids, which are soluble in fats, are unsaturated hydrocarbons and can be regarded chemically as tetraterpenes. The oxygen-free carotenes, which have the general formula $C_{40}H_{56}$ (Fig. 243b) and which are usually red or orange in colour, can be distinguished from the oxygen-containing xanthophylls, which are usually yellowish or brownish. The yellow pigment lutein occurring in leaves is a xanthophyll and has the composition $C_{40}H_{56}O_2$.

The chlorophylls and carotenoids are insoluble in water. They can, however, easily be extracted from the green parts of plants with aqueous 80 per cent acetone or 80–90 per cent alcohol at boiling point. These pigments account for 0.5 to at the most 1.0 per cent of the dry weight of the green parts (8–10 per cent of the dry matter of the chloroplast). The chlorophylls probably occur in the plastids in the form of chromoproteids (compounds of chlorophyll and protein). The smallest physiological unit capable of photosynthesis is believed to contain some 250 chlorophyll molecules (see p. 246).

Even with the light microscope it is possible to see in many chloroplasts that the pigments are confined to small lens-shaped grana, 0.3–0.7 μm in diameter, embedded in the colourless stroma (Fig. 40A, B).

The electron microscope permits closer inspection of the fine structure of plastids. It is possible to see that the proplastids of meristematic and embryonic cells are already surrounded by an envelope of two membranes. In the development of the functional chloroplast, the inner of these two membranes invaginates here and there, so forming tongue-like projections into the interior (Fig. 41B, C), which continue to flatten as they extend. The invaginations occasionally bend back on themselves, and also fork, the lobes overlapping and sometimes fusing laterally with other invaginations thus forming anastomoses. In consequence a multilayered and interconnected system of chambers develops (Fig. 43A, B), which in transverse section appears to consist of numerous flattened sacs, the so-called thylakoids (Figs. 41, 42, 43). The plastids of plants reared in the dark contain, in place of photosynthetic lamellae, a crystalline prolamellar body, from which a typical thylakoid system is generated on illumination.

In many flagellates and red algae the whole chromatophore is more or less uniformly traversed by such thylakoids, an arrangement probably representative of the most primitive kind of plastid. With a few exceptions, the arrangement of the thylakoids in small, interconnected, stacks, recognizable in the light microscope as grana (Figs. 41–3) is not encountered until the higher plants.

The investigation of the molecular architecture of the thylakoid membrane, about 4 nm thick and consisting of about equal parts of protein and lipid, has become one of the main problems of current research. All the photosynthetic pigments, and all the enzymes concerned with the electron transport involved in the conversion of light into chemical energy are lodged in this membrane. Biophysical and biochemical methods are being used to attack this problem, in addition to electron microscopy. The technique of freeze-etching, in which the membrane is split, has yielded information about its internal structure. Although relatively smooth on the outside, the membrane evidently contains particles, of which two classes can be recognized. Large particles, each consisting of four sub-units about 6 nm diameter, are interspersed with smaller particles, also about 6 nm diameter. These two kinds of particles apparently fit between each other in a sufficiently regular order for the intact membrane to have a definite crystalline pattern (Fig. 44A, B).

Fig. 44. Shadowed preparations of frozen-dried thylakoid membranes from spinach chloroplasts (*Spinacia oleracea*). The 'tetra-particles', each consisting of a regular, almost crystalline, arrangement of four sub-units, can be clearly seen. **A.** ×100,000 approx. From Park and Biggins. **B.** ×300,000 approx. From Park and Healy.

This crystallinity of the membrane can be demonstrated by means of X-ray diffraction. Direct observation and inference from physical properties thus complement each other in a very satisfactory manner.

Pyrenoids. In the chloroplasts (Fig. 38A, D, F), phaeoplasts and rhodoplasts of the algae there are often special areas referred to as pyrenoids. These act as centres for the formation of starch or, as for example in the diatoms, fat. The pyrenoids usually reproduce themselves by simple fission, but in some cases, as in *Chlamydomonas*, they disappear at the time of cell division and are newly formed in the daughter cells.

2. Photosynthetically inactive chromatophores

The yellow and orange colours of many flowers (e.g. *Cytisus, Forsythia, Viola, Tropaeolum*, Fig. 45), as well as the bright red of many fruits (e.g. hips, haws, peppers, and tomatoes), are caused, at least partly, by photosynthetically inactive chromatophores (chromoplasts). These are developed either directly from the colourless proplastids or arise from normal green chloroplasts which lose their chlorophyll (as, for example, in *Trollius* spp., where the green flower buds become transformed into yellow flowers, and in the tomato, where the fruit changes from green to red in ripening). Chromoplasts may occur in organs which are morphologically roots (as, for example, in the tap roots of *Daucus carota*).

The colour of the chromatophores depends upon the presence of red carotenes and usually yellow xanthophylls, related to and partly identical with the carotenoids of the green chloroplasts. In all, over seventy different carotenoids are known. Of these we need refer here only to the carotene of the carrot, *Daucus carota*, which consists of three closely related components and from which the name of this whole class of compounds has been taken, to the lycopene of the tomato and other red fruits, and to the yellow violaxanthine whose derivatives are responsible for the yellow colour of the plastids of the flowers of *Viola, Arnica* and *Narcissus*. These highly unsaturated, brightly coloured, isoprene derivatives are present in the colourless stroma either as numerous, usually small, granules, or in regular orientation within a fibrillar stroma matrix.

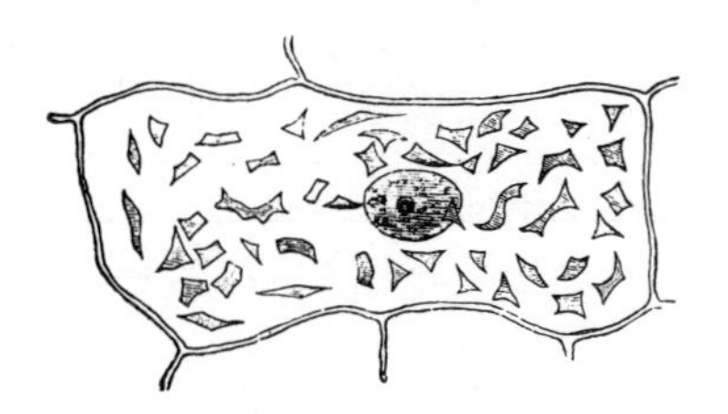

Fig. 45. Cell with nucleus and chromoplasts from the yellow calyx of *Tropaeolum majus*. X540. After Strasburger.

The yellow coloration of the leaves of many woody plants in autumn is a consequence of the fact that the green chlorophylls are the first pigments to be destroyed and removed from the leaves by way of the vascular tissue, the carotenoids alone remaining. If the vascular tissue is blocked by an insect gall or by artificial means, then the yellowing of the affected parts is delayed. The yellow or orange-red colour of oranges and lemons depends upon an enrichment of the carotenoid pigments, accompanied by a disappearance of the chlorophyll. The red coloration of leaves in the autumn is a different phenomenon; it results from the cell sap becoming coloured with anthocyanin (see p. 53). The browning of dying leaves is caused by the production following death of brown, water-soluble, pigments.

White flowers usually contain no pigment. The whiteness is caused instead by the reflection of the white daylight by intercellular layers of air. If the air is removed by bruising, then the bruised place becomes transparent.

The eye-spots of many flagellates (Fig. 4, at the base of the flagellum) and swarm spores of algae are usually considered as chromatophores. They contain carotenoids and are usually orange or bright red in colour. They are believed to be concerned with the receiving of light stimuli (see p. 383). Not infrequently starch grains are formed in them, clearly revealing their plastid nature.

3. Leucoplasts

Leucoplasts, which are colourless, are also closely related to chloroplasts. They are found in all those heterotrophic, saprophytic (see p. 291) and parasitic flowering plants which lack chlorophyll, and also in the whitish-yellow leaves and shoots of numerous variegated plants. In green plants the leucoplasts are confined as a rule to the colourless organs, especially to the subterranean parts such as roots and storage organs, and to the primary epidermis of leaves and stems. They often possess the capacity to develop chlorophyll in the light (as, for example, in the potato). In storage organs (e.g. tubers and root-stocks) and tissues (e.g. pith and endosperm) they are concerned with the conversion of sugar into starch, and in these instances are accordingly termed amyloplasts. Other metabolic products often found within leucoplasts are protein crystals (Fig. 49) and lipid droplets.

C. Mitochondria

In contradistinction to the plastids, in which the reactions are predominantly anabolic, the mitochondria (chondriosomes) are concerned principally with the supply of energy by the breaking down of carbon compounds. These minute organelles are present in every plant and animal cell. They contain all the essential enzymes of the citric acid (Krebs) cycle, of the respiratory chain, and of fatty acid metabolism, together with those related to the associated phosphorylation systems. The mitochondria can thus be considered not only as the source of respiratory energy, but also as the chief suppliers of ATP (see p. 262 seq.).

Mitochondria show selective staining with Janus green B, and can thus be seen in the light microscope (Figs. 5 Mi; 6 and 7 M). They are a few microns in length and 0.5–0.8 μm in breadth, and from oval to thread-like in form. In contrast to the proplastids and leucoplasts, which often move by amoeboid activity, mitochondria remain largely passive,

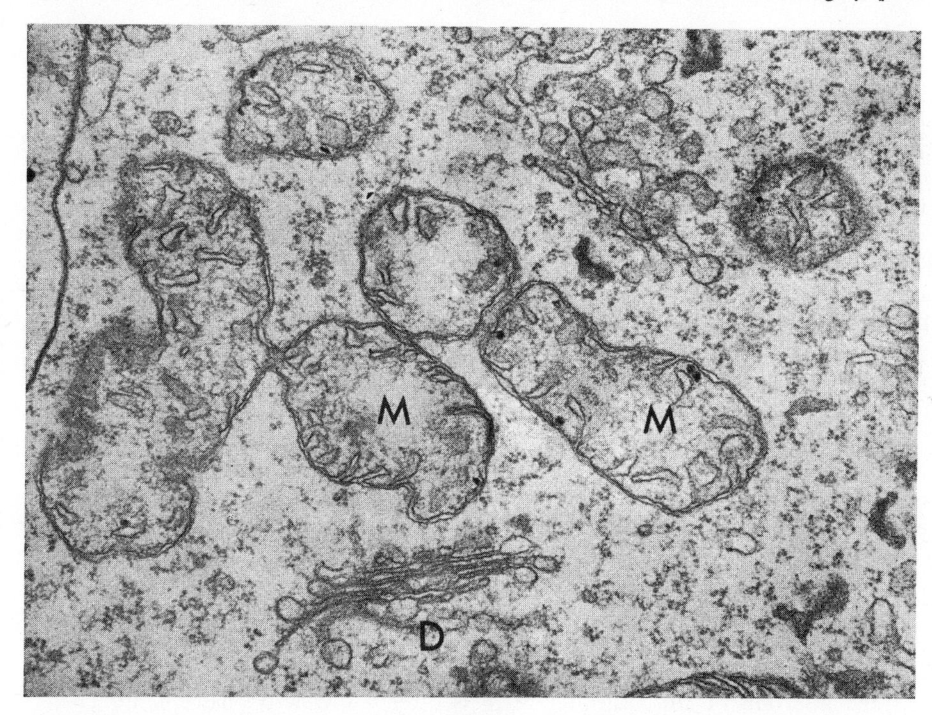

Fig. 46. Mitochondria (M) in a root meristem of *Pisum*. A Golgi body (dictyosome) (D) is also present. ×30,000. From Amelunxen.

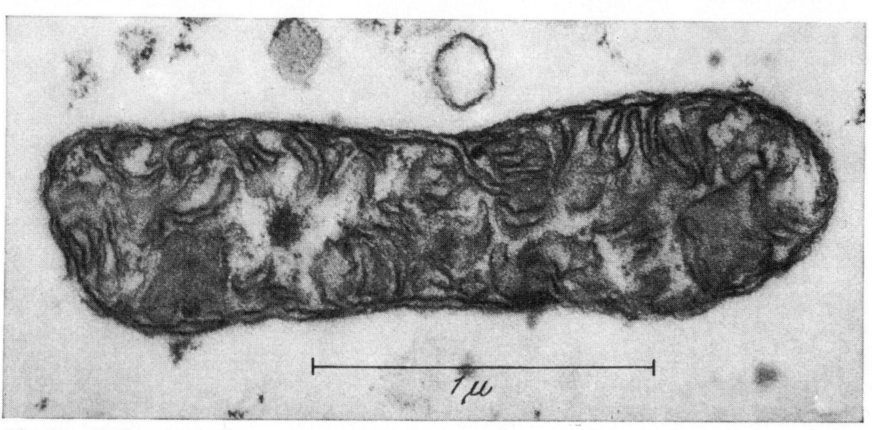

Fig. 47. Single mitochondrion from a meristem cell of *Elodea callitrichoides*. ×45,000. From Falk and Sitte.

merely showing elastic, flexing movements during cytoplasmic streaming. The fine structure of mitochondria is basically the same in animals and plants; they are bounded by a two-layered envelope, the inner layer of which forms folds projecting into the interior, where they end blindly (*sacculi mitochondrialis*), so forming a system of interconnected chambers. If the invaginations are in the form of pouches or discs, they are referred to as cristae, if tubular, as villi. Chemically the mitochondria are distinguished (amongst other features) by their high content of lipid.

Observations with the phase-contrast microscope on living cells show that mitochondria are capable of multiplication by division or budding, in relation to which the presence of a specific mitochondrial deoxyribonucleic acid is significant. In the mitochondria of many brown algae less dense regions occur in which threads, identified with deoxyribonucleic acid, have been demonstrated. This recalls the situation in plastids (see p. 45).

D. Flagella

Flagella (where very numerous sometimes referred to as cilia (Figs. 23E; 421, z)) are the locomotory organelles produced in varying number by many bacteria (Fig. 375) and unicellular algae (Figs. 4; 388; 389), as well as by the motile stages (zoospores and gametes) of numerous, more highly developed algae (Figs. 416; 428; 430; 435) and fungi (Figs. 453; 456; 466; 468; 470). They are also characteristic of the male gametes (spermatozoids) of certain land plants, such as the mosses (Fig. 530F), ferns (Figs. 566F; 573N; 577B; 602F), and even of some gymnosperms (cycads (Fig. 623G–J) and *Ginkgo*). They may be attached to the anterior end of the cell and generate a pulling force (acrokont or subacrokont attachment), or at the posterior end and generate a pushing force (opisthokont attachment) (Fig. 456).

Flagella are generally extremely delicate objects (their diameter may be less than 10 nm), and they can be seen with the light microscope only with dark field illumination, or with direct illumination after special pretreatment and staining. The electron microscope has facilitated more detailed investigation, and the cilia and flagella of all organisms investigated so far (with the exception of the bacteria, which in this respect as in others show a more primitive condition) appear to be constructed in a very uniform manner out of eleven fibrils. Two of the fibrils form an elastic axial thread, about which nine double fibrils are symmetrically arranged (Fig. 385). The fibrils consist of protein, the outer double fibrils probably of a contractile protein. In the electron microscope they have the form of microtubules (see p. 26). In many organisms the flagella bear minute hairs (Figs. 386B; 456; 457) (flimmer), or even scales.

At the insertion of the flagellum there is often a distinctive basal body. There is evidently a close relationship between this structure and the centriole (see p. 33). In certain instances (e.g. in the myxomycetes and bryophytes) the centriole may lie close to the origin of the spindle fibres (as normally) and subsequently become a basal body. Further, the electron microscope has shown that basal bodies and centrioles are structurally almost indistinguishable.

IV. Secretory products of the protoplast

It is today customary to refer to all those products which accumulate as a consequence of the metabolic activity of the organelles and cytoplasm generally as ergastic substances. They are often concentrated within spaces surrounded by special membranes, and may even form crystals. Eventually they may either be discharged from the active metabolism altogether, or set aside temporarily as food reserves, to be drawn on at some later stage, perhaps after a resting period, as sources of structural and energy-providing materials.

A. Vacuoles

Almost all differentiated plant cells contain one or more vacuoles, a feature in which they differ from most animal cells. There is as yet, however, no general agreement about how the vacuoles of a cell (referred to collectively as the vacuome) arise. Some maintain that they originate from cisternae of the endoplasmic reticulum which become detached, and subsequently form an independent system. This would adequately account for their being surrounded by a single membrane, the tonoplast. Others point out that vacuoles containing, for example, ethereal oils might well arise from Golgi vesicles which swell after their release (Fig. 14). A further possibility is that pockets of fluid accumulate here and there in the cytoplasm, each becoming surrounded by a membrane. Some freshwater flagellates possess contractile or pulsating vacuoles. These beat in a regular rhythm, first filling gradually with cell sap (diastole), and then discharging into the surrounding water (systole).

The composition of the cell sap naturally differs from species to species. It also varies with the organ, the nature of the tissue, and even from cell to cell. In certain specialized differentiated cells the vacuoles may contain practically only one metabolic product (e.g. tannin, mucilage, protein, oil). Many such inclusions are reserve materials, and they may accumulate in large amounts in the cells of storage organs (e.g. tubers, bulbs, seeds). Others, the so-called secondary plant metabolites, have an excretory character, and are without any evident significance for life. Their presence or subsequent change confers neither advantages nor disadvantages on the plant.

1. Carbohydrates. Soluble carbohydrates, especially the disaccharides sucrose and fructose, are the most frequent components of the cell sap. Sucrose is often stored as a food reserve, as, for example, in sugar beet and sugar cane, the sources of the 'cane sugar' of commerce. Glucose and fructose (as its name implies) are widely distributed in fruits.

Plants belonging to the Compositae accumulate the soluble polysaccharide inulin, composed of fructose molecules. It is precipitated by alcohol in the form of beautiful sphaerocrystals (see p. 58). In contradistinction to potato starch, built up from glucose molecules, inulin is not dangerous to diabetics, so that artichokes (the tubers of *Helianthus tuberosus*) can take the place of potatoes in their diets. Glycogen, a reserve carbohydrate frequently encountered in the Animal Kingdom and, like starch, also a polysaccharide composed of α-glucose units, occurs instead of starch or cane sugar in fungi, bacteria, slime fungi and blue-green algae. A mucilage consisting of carbohydrates is found, for example, in the vacuoles of the cells of many bulbs and of orchid tubers. Among aerial organs, mucilage is a conspicuous component of the stems and leaves of succulents. Mucilage can, however, also arise through the disintegration of components of the cell wall (see p. 69).

2. Vacuolar pigments (chymochromes). The blue, violet and reddish-purple colours of many flowers and fruits are usually attributable to the water-soluble anthocyanins. Yellow colours, however, may arise in two ways, either from the presence of water-soluble anthoxanthin, or from fat-soluble pigments in specialized chromatophores (see p. 49, chromoplasts). These fat-soluble pigments are commonly referred to as lipochromes to distinguish them from the water-soluble chymochromes. In many instances complex combinations of both kinds of pigmentation occur, and microscopic or biochemical analysis is essential to determine the extent of each.

The blue, violet, or reddish-purple anthocyanins always consist of a sugar-free component, the anthocyan-aglucone or anthocyanidin, and a sugar (glucose, galactose or rhamnose). The latter is responsible for the ready solubility of these chromosaccharides in water. So far eight different anthocyanidins have been isolated. The basic unit of their structure is flavone. The most important anthocyanidins result from the introduction of hydroxyl and methyl groups into the side ring (B-ring); examples are the salmon-coloured pelargonidin (in *Pelargonium, Dahlia, Papaver rhoeas*), red cyanidin (in *Rosa, Pulmonaria, Centaurea* and the leaves of red cabbage), paeonidin (in *Paeonia* and red species *Impatiens*), blue delphinidin (in *Delphinium, Malva*), blue petunidin (in species of *Petunia* and *Primula*), violet oenidin (in the skins of grapes), and the reddish-purple malvidin (also in species of *Petunia* and *Malva*). Several anthocyanidins may occur together.

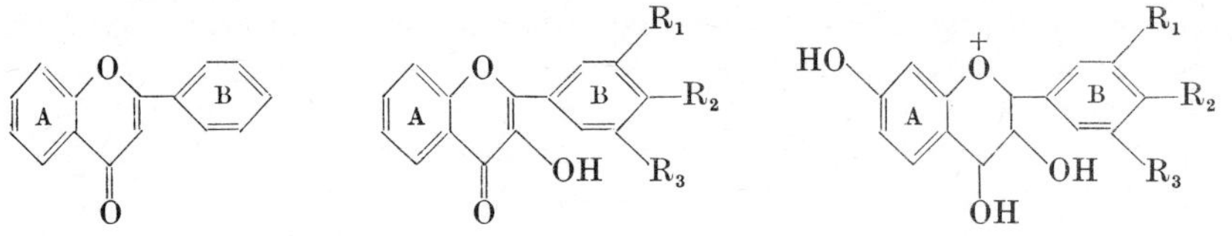

Flavone　　Flavonol　　Cyanidin

The final colour is not only, however, determined by the accompanying or co-pigments. As a consequence of their phenolic-OH groups and the positively charged oxygen atom (see formula), many anthocyanins can form salts of various colours. The actual colour in a living cell may, for example, come about through the presence of a mixture of the monomolecular anthocyanins and the same anthocyanins chelated with trivalent metals (e.g. ferric iron, aluminium), If these complexes are destroyed with acid the colour changes as the anthocyanins are freed (thus the blue cornflower chelate changes to the red of the cyanin). Conversely the rose-red flowers of many hydrangeas, for example, can be made to turn blue (as a consequence of chelate formation) if the plants are given liberal doses of aluminium salts (e.g. ammonia alum). Not infrequently the colours of flowers change as they fade. This often results from a change in the number of sugar molecules, with which the anthocyanidins are linked; examples are provided by *Pulmonaria officinalis, Lathyrus vernus* and *Myosotis*, in which there is a transition from red, through violet, to blue, as the flowers age.

Anthoxanthins are also flavonoids, and closely related chemically to the anthocyanins. Like the anthocyanins, they also form glycosides. The flavonols, oxidation products of the anthocyanidins, are the most widely distributed aglucone of the anthoxanthins. Evidence that a yellow flower colour is due to anthoxanthin is a striking deepening of the colour in the presence of ammonia vapour. Where pale yellow and blue or red flowers occur in the same genus (e.g. *Primula, Aconitum, Digitalis*), closely related anthoxanthins and anthocyanins are usually involved. The bright butter-yellow flowers of pansies, however, owe their colour to carotenoids in chromatophores.

Blood-red coloration, seen, for example, in the reddish-brown leaves of varieties of the beech and hazel, arises from the interaction of the red of an anthocyanin in the epidermis with the green of the assimilatory tissue beneath. In the red beet (*Beta vulgaris* var. *esculenta*) the vacuoles of almost all the cells contain anthocyanin. The bright red coloration of many leaves in the autumn also derives from anthocyanins, the formation of which is often promoted by an increase in the amount of sugar present in the cells. Thus coiled-up or broken branches of vines or of dogwood (*Cornus sanguinea*) show a premature 'autumn coloration' as a consequence of the interrupted flow of sugar away from the leaves.

3. **Glycosides.** Besides the pigments already discussed, a whole series of widely distributed secondary metabolites belong to this class of sugar compounds. Some are partly aromatic in character; examples are the coumarin glycosides (as in the woodruff (*Asperula odorata*)), and the sulphur-containing mustard-oil glycosides (e.g. sinigrin, the glycoside of black mustard (*Brassica nigra*) and horseradish). Often the glycosides are to a greater or lesser extent poisonous, such as the bitter glycoside present in members of the Gentianaceae (species of *Centaurium* and *Gentiana*), the cyanide-containing glycosides of almonds and other seeds of the stone fruits, the glycoside saponin, present in *Saponaria* and *Primula*, and the glycosides of *Digitalis* and *Strophanthus*. Each of these glycosides can be broken down into its components by a specific enzyme; for example, the enzyme emulsin decomposes amygdalin, the glycoside of bitter almonds, into glucose, benzaldehyde and hydrocyanic acid. The fragrant component of coumarin does not become detectable until, when the plant is withering, it is freed from the sugar residue. Consequently fresh material of *Asperula odorata* and *Anthoxanthum odoratum* is almost scentless.

4. Alkaloids. This class of metabolite (of which over 1200 are known), containing further important plant poisons, is more easily defined biologically than chemically. As the name implies, they are basic and form 'salts' with acids. Chemically they are very various, but commonly have a heterocyclic ring structure, usually containing nitrogen. They are typically very stable, and can consequently accumulate in large quantities as end products of metabolism. Here belong, for example, the active principles of coffee (caffeine), tea (caffeine and theophylline), cocoa (caffeine and theobromine), cinchona (quinine), poppy (morphine, codeine, narkotine and others), as well as the dangerous poisons cocaine, strychnine, coniine and aconitine. Plants considered closely related on other grounds frequently possess closely related alkaloids. In the Solanaceae, for example, there occur solanine, atropine, hyoscyamine and scopolamine, and also (especially in species of *Nicotiana*) the strongly poisonous nicotine, but this is not chemically similar to the preceding group. Ergot (*Claviceps purpurea*) yields not only ergotamine and ergobasine, widely used in gynaecology, but also lysergic acid derivatives (first isolated from the seeds of tropical species of *Ipomoea*) which are effective in the treatment of nervous disorders. Lysergic diethylamide (LSD), the most frequently used drug, is not, however, a natural product. In men and animals lysergic acid derivatives cause hallucinations. Other hallucinogens are psilocybine, contained in certain Mexican 'sacred' mushrooms, and convalline and the other alkaloids of *Cannabis*, the use of which is becoming increasingly popular in the West.

Despite its alkaloid-like constitution, colchicine, which contains the nitrogen in a side chain, is not usually counted amongst the alkaloids. This important mitotic poison (see p. 36) occurs not only in the autumn crocus (*Colchicum autumnale*), but also in various other members of the Liliaceae (e.g. *Gloriosa*).

5. Tannins. Vacuoles filled with tannin are present particularly in bark (brown phlobaphene, an oxidized form of tannin) and fruits (accounting for the astringency of the fruits of *Vaccinium vitisidaea* and *V. myrtillus*). The leaves of tea and the seeds of coffee are also rich in tannin. Tannins are mixtures of quite different aromatic substances, partly glycosidic in nature. Of frequent occurrence in them are gallic acid, gallotannic acid, ellagic acid (present in gall apples) and chlorogenic acid. Tannic acids in a tissue will often protect it from attack by micro-organisms.

6. Oils and fats. In the flowering plants these energy-rich foodstuffs are commonly found in the storage tissues of the seeds. In seeds especially rich in fats they may amount to some 70 per cent of the dry weight. Also among lower plants, e.g. some single-celled algae (*Chlorella*, diatoms), up to 65 per cent of the dry weight may, in certain conditions, consist of fat. In cytoplasm containing little water oil is present as an emulsion of minute droplets, and oil globules do not separate themselves out until the cytoplasm is hydrated. In cytoplasm containing abundant water, as for example in germinating seeds, large highly refractive fat vacuoles appear (Fig. 49A). The most frequently occurring fatty acids in plant fats are the unsaturated oleic ($C_{18}H_{34}O_2$), linoleic ($C_{18}H_{32}O_2$), and linolenic ($C_{18}H_{30}O_2$) acids, as well as the saturated palmitic ($C_{16}H_{32}O_2$) and stearic ($C_{18}H_{36}O_2$) acids. The oil fruits and oil seeds most important for human nutrition are, in order of increasing economic significance, soya beans (*Glycine max*, the seeds of which contain, in addition to 40 per cent protein, 20 per cent soya oil), cotton seeds (*Gossypium*, with 30 to 40 per cent cotton seed oil), ground nuts (*Arachis hypogaea*, with 40 to 50 per cent oil), olives (*Olea europaea*, with 40 to 60 per cent oil), coconuts (*Cocos nucifera*, the dried endosperm of which (copra) contains 60 to 70 per cent coconut oil), and oil palm (*Elaeis guineensis*, with some 60 to 70 per cent of oil in the fleshy part of the fruit). Many of these oils are a starting point in the manufacture of margarine.

7. Terpene derivatives. A whole range of secondary metabolic products is found in plants which has as its common feature the isoprene molecule, C_5H_8. Isoprene molecules may themselves be linked together, or more frequently oxidized derivatives in the form of alcohols, aldehydes, ketones, phenols and carboxylic acids, so forming polyterpenes. Isoprene itself does not occur in plants.

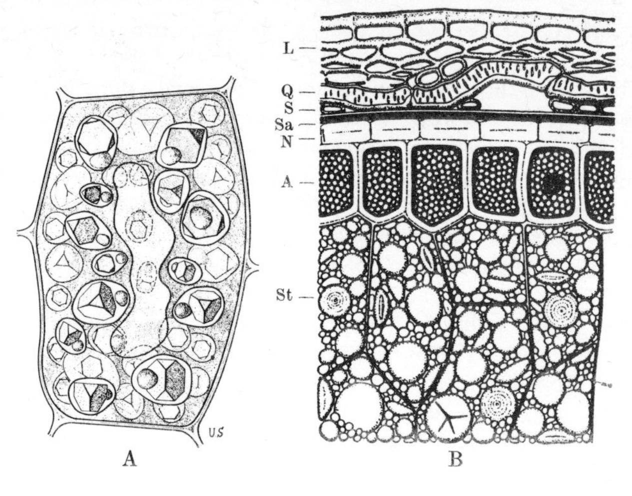

Fig. 48. Aleurone tissue. **A.** Storage cell from the endosperm of *Ricinus communis* with a large central oil vacuole and numerous aleurone grains of various sizes. Each aleurone grain contains a tetrahedral crystalloid and an amorphous spherical component. **B.** Transverse section through the surface layers of a wheat grain. The fruit is bounded by an epidermis, below which lie elongated and crushed parenchymatous cells (L), cross cells (elongated transversely to the axis of the fruit) (Q), and tube cells (S). The crushed epidermis of the seed (Sa) encloses the remains of the nucellus (N). Adjacent to this is the aleurone layer (A) (the outermost layer of the endosperm). The remaining cells of the endosperm are rich in starch (St). **A.** ×1000 approx. Original. **B.** ×200. After Gassner.

Ethereal oils are mixtures of hydrocarbons, alcohols, aldehydes and ketones of various mono- and sesquiterpenes, as well as phenylpropane derivatives. They are present, for example, as highly refractive droplets in the cells of roots and rhizomes (e.g. *Acorus calamus*, ginger (*Zingiber*)), barks (e.g. cinnamon (*Cinnamomum*)), leaves (e.g. laurel (*Laurus nobilis*), *Dictamnus albus*), skins of fruits and in seeds (e.g. pepper). Oil-filled cells are often suberized, and their protoplasts dead. Some living cells excrete the oil as it is formed (e.g. glandular hairs (Fig. 127) and gland cells (Fig. 129A)). Others disintegrate, the oil collecting together in a large drop in the space so created. Oil cavities of this kind are said to be lysigenous in origin (Fig. 129C). The cells of many petals contain ethereal oils transiently. The droplets become extruded through the outer wall of the epidermis and there evaporate, accounting for the characteristic scents of many flowers.

Ethereal oils, balsams and resins are not exact scientific concepts, but terms of commerce. The ethereal oils contain highly volatile terpene derivatives; the balsams, however, are semi-fluid, and the resins viscous fluids or solid mixtures of ethereal oils (which can be distilled off) and non-volative resin acids and other components. The proportions of these substances can show considerable variations according to the stage of development, and also to the time of year. In the conifers the resin is formed in special gland cells, and later secreted into the resin canals (Fig. 130). The resins of various pines yield, on distillation, turpentine, the remaining substance, a hard resin, enters commerce as colophonium. Amber is derived from fossil species of *Pinus* and *Picea*. Species of *Boswellia* yield incense. The biological significance of the resins and balsams perhaps lies in their preservative and antiseptic properties.

Guttapercha and caoutchouc (unvulcanized rubber) are macromolecular, polyisoprenes, consisting of about 100 isoprene residues in *cis*-configuration (guttapercha) or from 500 to 1000 residues in *trans*-configuration (caoutchouc). Large quantities of caoutchouc occur in the latex of *Hevea brasiliensis* (Euphorbiaceae), extensively cultivated as a source of caoutchouc for the rubber industry. Caoutchouc is also obtained from the latex of various species of *Ficus* (Moraceae), as well as from the herbaceous Compositae *Taraxacum bicorne* (kok-saghyz) and *Scorzonera tau-saghyz* (tau-saghyz).

B. Protein crystals and aleurone granules

Food vacuoles often contain proteins dissolved in the cell sap. These can be demonstrated by irrigation with protein precipitants (e.g. formaldehyde, tannic acid, osmic acid and

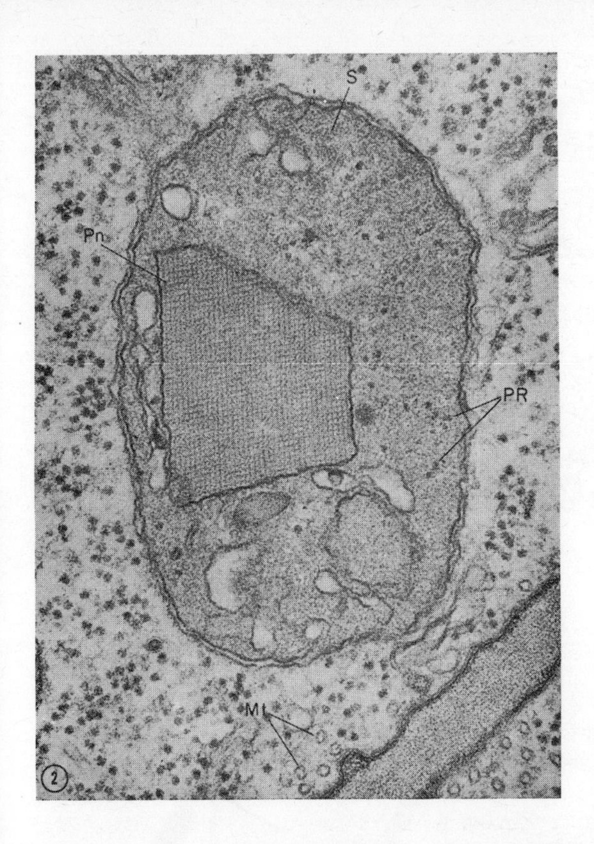

Fig. 49. Protein crystal (Pn) in a plastid in a cell of a root meristem of *Phaseolus vulgaris*. The crystal lies in the stroma (S) which also contains many plastid ribosomes (PR). Microtubules (Mt) are visible in the cytoplasm. ×60,000. From E. H. Newcomb.

other fixatives). In relatively dry storage tissues, found, for example, in many seeds, protein granules (aleurone grains) (Fig. 48A) occur in the cells. These are formed by the dehydration of numerous small vacuoles, filled with protein, distributed throughout the cytoplasm. The globulins, being the least soluble proteins, are the first to precipitate, and they can often be seen as beautiful crystals embedded within the albuminous matrix of the grain.

Protein crystals are distinguished from true crystals by their ability to imbibe water. Aleurone grains frequently also contain one to several globoids, consisting of amorphous phytin, a calcium-magnesium salt of inositol hexaphosphate. In wheat the cells of the outermost layer of the endosperm are densely packed with aleurone grains (Fig. 48B). In the manufacture of white, finely milled flour this layer, particularly rich in proteins and fat, remains with the bran, and is thus lost from the flour and the white bread baked from it. It is, however, present in wholemeal flour. Protein crystals occasionally occur in leucoplasts (e.g. in the root tips of *Phaseolus vulgaris* (Fig. 49)).

C. Other crystals

Calcium oxalate. The sparingly soluble crystals of calcium oxalate are a widely distributed metabolic end-product. They may consist of either the monohydrate ($Ca(C_2O_4).H_2O$) or the dihydrate ($Ca(C_2O_4).2H_2O$), depending upon the conditions in which they are formed. The monohydrate forms monoclinic crystals and the dihydrate tetragonal. The crystals of the monohydrate may occur solitarily, or they may be deposited as raphides (Fig. 50A, B), a sand-like mass (Fig. 50D), or as druses (numerous individual crystals radiating irregularly from a central point, Fig. 50C), usually in specialized cells of the ground parenchyma (idioblasts). The tetragonal crystals of the dihydrate occur likewise individually (Fig. 50E) or in druses. The walls of the calcium oxalate idioblasts are usually suberized. Calcium oxalate crystals are soluble in hydrochloric acid, but not in acetic acid. Sulphuric acid transforms them into crystals of gypsum. In this way they can be distinguished microchemically from other crystals occurring in cells.

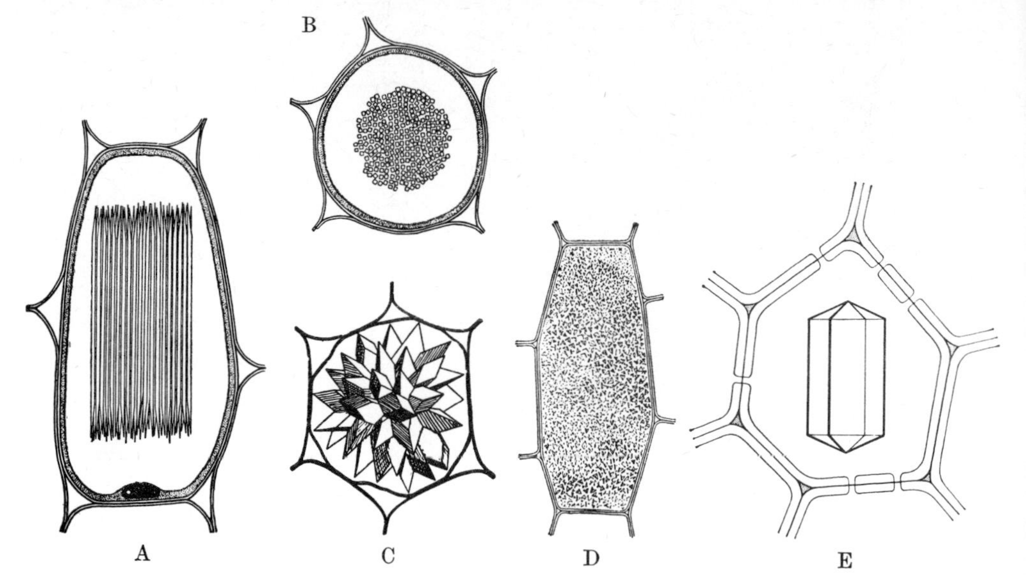

Fig. 50. Forms of crystallization of calcium oxalate monohydrate. **A.** and **B.** Raphides in longitudinal and transverse view respectively (*Impatiens*). **C.** Druse (*Opuntia*). **D.** 'Crystalline sand' (*Solanum*). ×500.

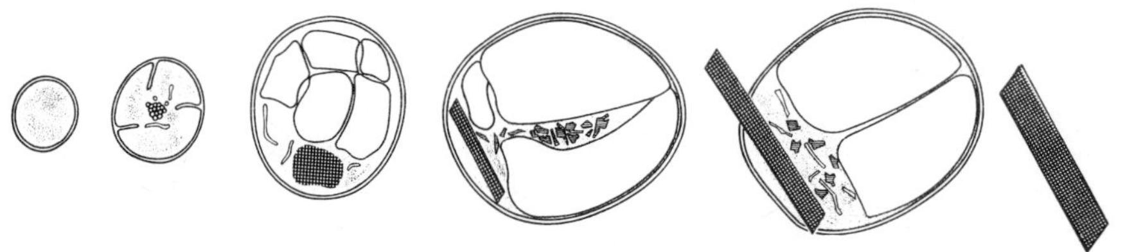

Fig. 51. Ontogeny of a chromoplast in the root parenchyma of *Daucus carota*, and the development of a carotene crystal. ×10,000. After Susumu Toyama.

Carotene. Carotene (or carotenoids) sometimes accumulate to such an extent in the lipid sacs of the chromoplasts that they crystallize, and eventually become extruded as naked crystals in the cytoplasm. Naked carotene crystals of this kind are found regularly in the swollen tap root of the domestic carrot (*Daucus carota*), hence the name of this whole class of pigments (Figs. 38G; 51).

D. Starch

After prolonged illumination chloroplasts usually contain few to several very small, lenticular, starch grains (Figs. 1; 38G; 41, S). This primary or assimilatory starch always appears in the colourless stroma between the thylakoids. At night, or in artificial darkening, it is again converted into sugar, which is then translocated to the sites of storage. Here starch is again formed in association with the colourless amyloplasts. This, the so-called reserve starch, consists of numerous large grains and provides the 'starch' of commerce. The amount of starch in storage organs and tissues is often very great. The starch in a potato tuber, for example, amounts to 20–30 per cent of the fresh weight, and that in the grain of wheat may even reach 70 per cent of the fresh weight. The reserve starch of plants is the most important basic foodstuff of man, although it may reach him from a number of different sources, such as bread, potatoes, rice, millet, sweet potatoes or bananas.

Chemically starch, being a polysaccharide, is classified as a carbohydrate (see p. 257 seq.). In the formation of the starch grains, the macromolecules, which are long, chain-like polymers of a high order, radiate from a central point, forming a so-called sphaerocrystal. When placed between crossed Nicol prisms or polaroids, such sphaerocrystals

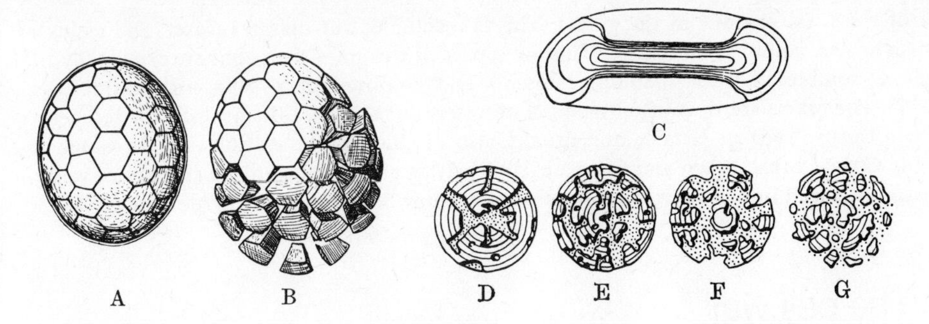

Fig. 52. **A.** and **B.** Large compound starch grains from oat. **C.** Dumb-bell shaped starch grain from the latex of *Euphorbia splendens* (the leucoplast is visible as a thin swollen outline). **D.** to **G.** Corrosion of a starch grain from the endosperm of a germinating wheat caryopsis. **D.** and **G.** From Noll, modified.

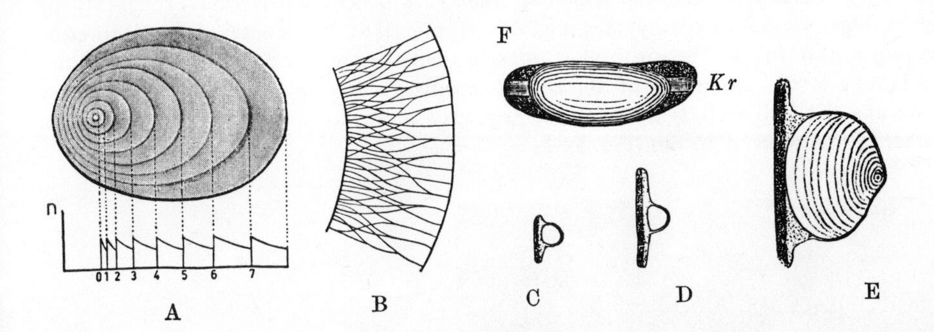

Fig. 53. **A.** Eccentrically layered starch grain of potato, with (below) a graph of the corresponding refractive indices (n). Each layer begins with a high refractive index, which then falls centrifugally. **B.** Submicroscopic structure of one layer. Initially there is close packing of the branched starch molecules (left), but then they become more widely spaced. The intervening spaces are filled with water. **C.** to **F.** Growth of the starch grains in the leucoplasts of the root tubers of *Phajus* (Orchidaceae). **C.**, **D.** and **E.** Side view. **F.** Viewed from above. Kr, protein crystalloid. **A.** ×400. **C.** to **F.** ×540. After Strasburger.

show a 'polarization cross', the arms of which intersect at the centre of the grain. This is because those macromolecules whose axes coincide with the planes of polarization of the prisms remain dark while the remainder are brightly illuminated.

Many starch grains appear clearly layered under the microscope. Within each layer, the density diminishes steadily towards the periphery, only to increase again sharply at the boundary of the next layer (Fig. 53A). According to recent investigations, the deposition of these layers seems to be independent of external factors, such as light and temperature, but instead to reflect an endogenous rhythm. The starch grains of the potato, for example, enlarge by two to three layers of this kind in each day's growth. In the starch grains of the Gramineae and Leguminosae the layers are generally symmetrical and concentrically arranged, whereas in the grains of other plants, especially those with very large starch grains, the layers are not infrequently eccentrically arranged (Fig. 53A, E). This situation comes about when the hilum of the grain is not exactly at the centre of the amyloplast. More starch is laid down on that side of the origin where the amyloplast is thicker, and in consequence the layers become unequally thickened. If the amyloplast contains more than one hilum, compound starch grains arise. Double and triple grains are often found together with the simple grains in potato tubers. In some other species very complex compound grains regularly occur. In oats, for example, large grains consisting of up to 200 units are found (Fig. 52A, B), and in rice similar grains of up to 300 units, while in *Spinacia* and other Chenopodiaceae complexes of up to 30,000 grains have been reported.

Many starch grains are bounded externally by a sheath of amylopectin (some 80 per cent of the total amount in the grain). This is insoluble, but imbibes water, and yields a reddish-brown colour with iodine. The inner part of the grain then consists principally of amylose, soluble in water ('soluble starch') and coloured deep blue with iodine (see p. 257). The grains are thus insoluble in cold water, but swell readily at ordinary temperatures in caustic soda or potash, and usually also in chloral hydrate solution. In water at 60–90°C they break down and form a paste. Heating and melting in the absence of water transforms starch into a water-soluble substance, known in commerce as 'dextrin'.

V. The cell wall[16]

Animal cells are generally naked, but the typical plant cell possesses a special cell wall or membrane. The most important constituent of the cell wall is cellulose. Only in the fungi, particularly the higher fungi, is the cellulose replaced by chitin.

The only naked plant cells are some flagellates (Fig. 389) and lowermost fungi (Myxomycetes, Figs. 8; 453; Phycomycetes, Fig. 462B), as well as the gametes and zoospores of many algae and fungi. The protoplast of these cells is surrounded by a thin, elastic, membranous layer, consisting of thickened ectoplasm. In the bacterium *Micrococcus radiodurans* the cell wall is built up of regularly arranged globular particles, with a diameter of the order of 15 nm (Fig. 54).

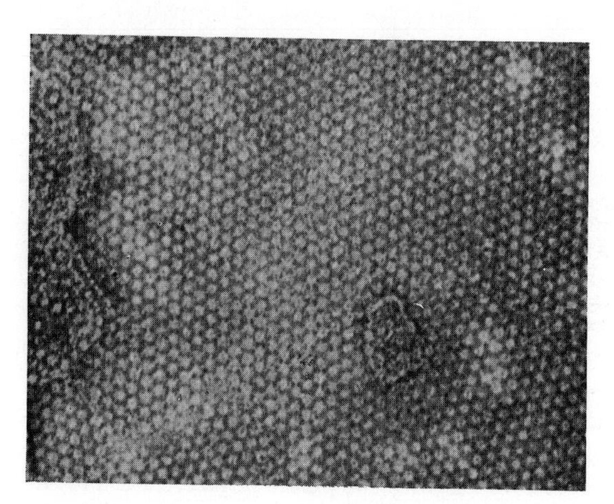

Fig. 54. Isolated wall of a bacterial cell (*Micrococcus radiodurans*). The wall is composed of globular particles in a regular array. ×100,000. From A. M. Glauert.

A. Ontogeny of the cell wall

The sexual cells of the higher plants have no cellulose membrane, but up to the time of fertilization they are enveloped in the parental cytoplasm. Not until after fertilization does a thin membrane appear on the surface of the egg. At each subsequent division of nucleus and cytoplasm dividing walls are formed, so that the whole of the embryonic tissue is divided up by delicate partitions into a reticulum of cells. At this stage the cell walls are still plastic.

The comparatively small and more or less isodiametric cells of the embryo grow to acquire their ultimate size and special forms, and this growth is accompanied by an increase in the surface area of the cells. Usually this surface growth is distributed uniformly over the cell, and is in register with that of the neighbouring cells (symplastic growth). Occasionally, however, in the instance of elongated cells, it may be limited to the tip (interpositional growth; see p. 151). Surface growth is brought about either by irreversible stretching of the primordial wall, which consists predominantly of protopectin, and the deposition of new wall material (by this stage containing cellulose) within the interstices of that already present (intussusception), or by the deposition of cellulosic lamellae on to the extended primordial wall (apposition; 'multi-net' growth, see p. 302 and Fig. 64).

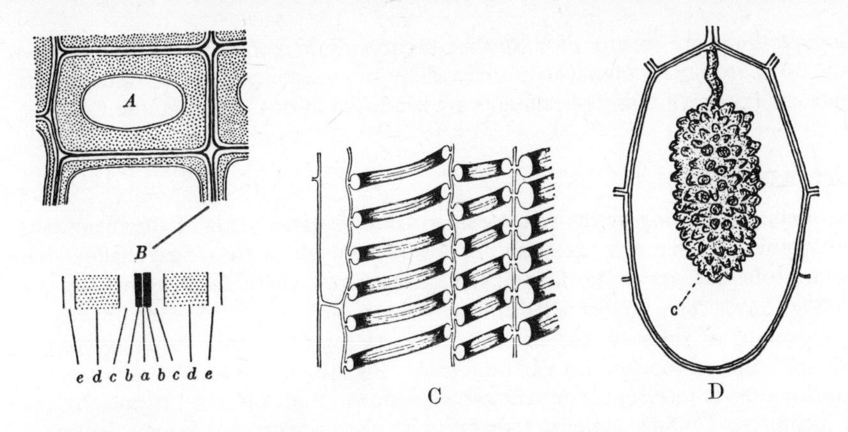

Fig. 55. Secondary wall thickenings. **A.** and **B.** Fine structure of the wall of a conifer tracheid. **A.** General view. **B.** Structure at greater magnification: a, middle lamella (or primary membrane) consisting of protopectin; b, primary walls of two adjacent tracheids; c, d, e, layers of the secondary wall (c, transitional lamella; d, central layer; e, terminal layer). **C.** Part of a section through a column of parenchyma cells and three spirally thickened vessels of a member of the Cucurbitaceae. **D.** Cell from the leaf of *Ficus elastica* showing a cystolith (c). **A.** ×800. **B.** ×4000. After Bailey. **C.** ×370. After Rothert. **D.** ×240. After Fitting.

The cell wall serves to protect and, above all, to provide a rigid container for the protoplast. The latter is achieved on the one hand by the tensing of the wall as a consequence of cell turgor (see p. 219), and on the other by its growth in thickness. Growth in thickness begins during cell expansion, but (more important) is able to continue after expansion has finished. Growth in thickness follows almost exclusively from the uniform laying down of wall material (apposition).

Since this laying down of wall material within a group of differentiating cells generally takes place simultaneously in adjacent cells, the developing dividing walls between even quite young cells are usually already three-layered. Each cell forms its own cellulose wall (primary wall; Fig. 55B, b), which abuts on the semi-fluid primordial wall or middle lamella (Fig. 55B, a). The primary wall also consists for the most part of protopectin and non-cellulosic polysaccharides. Its cellulose content lies between 8 and 14 per cent.

The cellulose content of the secondary membrane laid down after the conclusion of surface growth (Fig. 55B, c, d, e) is considerably higher. Wall material laid down by apposition is generally in the form of scale-like lamellae. As a result, thick, dense, lamellae may alternate with thinner lamellae which are less dense, contain more water and are often chemically different. The denser lamellae are more highly refractive, so that under the microscope the wall appears layered, the denser lamellae appearing brighter than the remainder (Fig. 56A; see also Fig. 53A). Walls which appear homogeneous under the microscope will often clearly show a layered structure after they have been made to swell by treatment with strong acids or alkalis. The outermost layers of the secondary wall adjacent to the primary wall are referred to as transitional lamellae (Fig. 55B, c). Then follows the middle or central layer of the secondary wall (in itself also layered). The innermost, so-called terminal lamellae (Fig. 55B, e), frequently have a different composition and structure.

Not infrequently the secondary thickening is limited to certain regions of the wall. It is then deposited either on the inside (centripetally) or on the outside of the cell (centrifugally). The strengthening of the water-conducting and water-storing cells (vessels and tracheids, Figs. 55C; 119) is brought about by local centripetal thickenings laid down in the form of rings, spirals or reticulately fused bars. Many centripetal thickenings can assume quite bizarre forms, such as the remarkable cystoliths in the leaves of various members of the Moraceae, Acanthaceae and Cucurbitaceae. The cystolith of *Ficus elastica*, reproduced in Fig. 55D, consists of a silicified base, from which hangs a mulberry-shaped body consisting of cellulose and callose impregnated with calcium carbonate (>90 per cent).

Centrifugal thickening of the wall can be brought about by intussusception, or can be laid on from without by means of a special periplasma or periplasmodium (see, for example, pp. 547, 562). The particularly rich variety of thickenings found on the outer walls of spores and pollen grains, for example, are produced in this latter manner.

B. Pits and plasmodesmata

In the centripetal thickening of the wall by apposition, isolated areas of the membrane remain unthickened. In this way little cavities are formed, which are at first shallow, but which become tube-like canals as the thickening increases. These gaps in the wall are called pits (Fig. 56A). The pits of adjacent cells meet each other (Fig. 56B, m). With continued thickening of the wall, the courses of two or more pits may fuse towards the centre of the cell and a branched pit can be formed (Fig. 56A). The canal of a pit shared by neighbouring cells is interrupted by the pit membrane (Fig. 56B, sch), formed by the primordial membrane (middle lamella) together with the primary membranes deposited by the two cells.

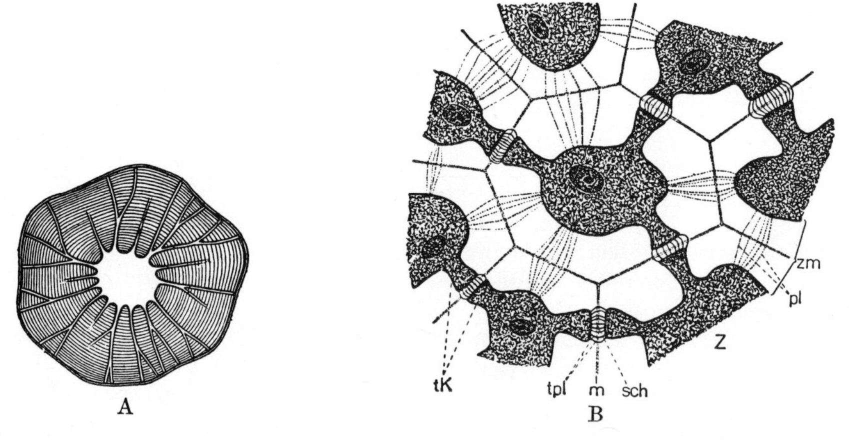

Fig. 56. Secondary wall thickening and pit formation. **A.** Stone cell from the endocarp of a walnut showing branched pit canals and layering of the wall. The incomplete canals are running obliquely upwards and downwards. **B.** Section through the swollen and pigmented endosperm of vegetable ivory (*Phytelephas*). Z, cell lumen with protoplast; tK, pit canal; m, middle-lamella; sch, closing membrane of the pit canal with plasmodesmata; zm, cell wall; pl, plasmodesmata external to a pit canal. **A.** ×1000. After Rothert and Reinke. **B.** ×350. After Halbsguth.

The pit membranes are perforated like sieves (forming the primary pit field; Fig. 62B), and are penetrated by extremely fine threads of protoplasm or plasmodesmata. As a result a group of cells, despite the dividing walls, behaves physiologically as a unit. The electron microscope shows that plasmodesmata are frequently traversed by strands of endoplasmic reticulum, so that the interlamellar spaces of the endoplasmic reticulum of adjacent cells may thus be in continuity (Figs. 6 and 7).

Similar plasmodesmata may also penetrate the cell walls outside the pits (Fig. 56B, pl). A single, more or less isodiametric, cell from the growing tissue of an onion root is connected by some 10,000 to 20,000 of such plasmodesmata with its neighbours. The walls of all fully-grown cells are also traversed by thousands of tenuous strands of protoplasm. Their existence is revealed when the protoplast is caused by plasmolysis (see p. 219) to draw away from the wall, and the process is watched with dark-field illumination. In many instances it can then be clearly seen how the protoplast remains connected in numerous places with the cell wall, so that the body of the contracted protoplast eventually appears suspended in a network of fine threads. These threads, observed before their nature was understood, were formerly known as 'Hecht's bars'. These interconnections show how the protoplasts of the individual cells form a living unity, referred to as the symplast.

In the dead water-conducting elements of the higher plants (p. 109 seq.) many pits are

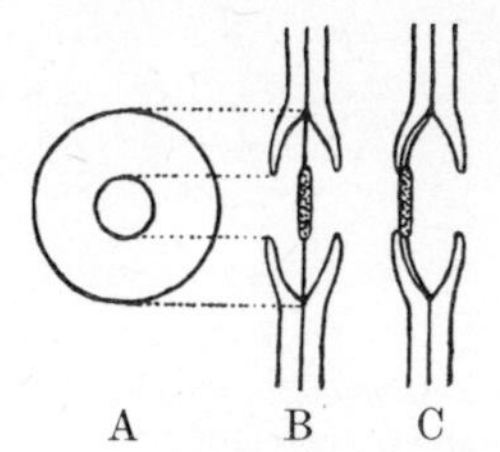

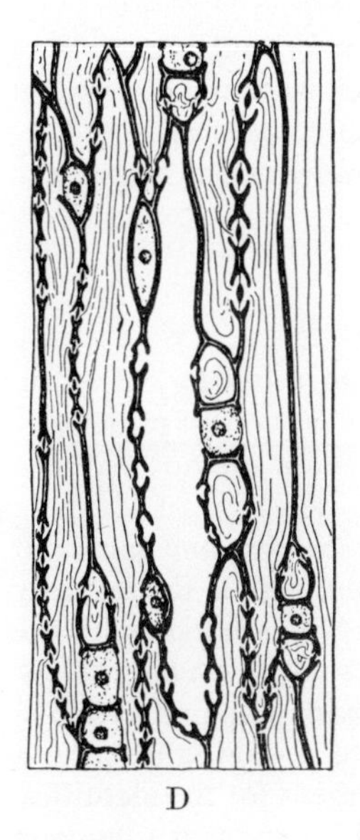

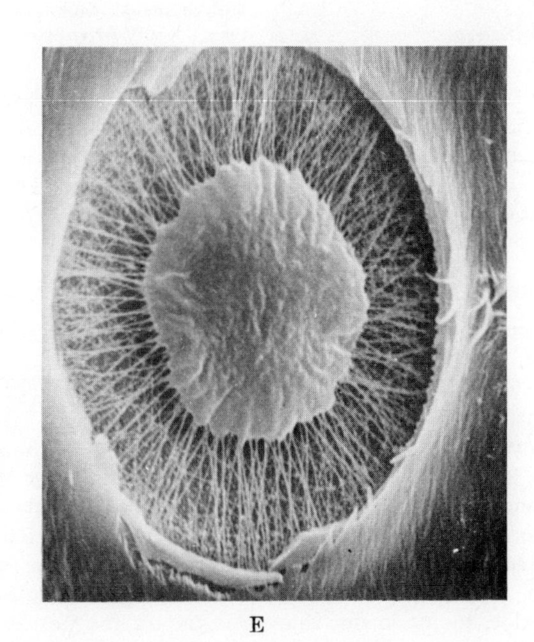

Fig. 57. Bordered pits of a conifer, diagrammatic. **A.** Surface view. **B.** and **C.** Longitudinal section showing the valve-like action of a pit bordered on both sides (**B.** Normal position. **C.** Movement in response to a pressure difference). **D.** Diagrammatic illustration of the valve-like function. Air has entered the central tracheid. The water in the adjacent cells (indicated by the fine longitudinal lines) is under negative pressure, and has drawn the tori of all the bordered pits communicating with the air-filled cell tightly against the pores on the water-filled side. **E.** *Pinus silvestris*. View into a bordered pit from the lumen of the tracheid, the nearer border having been torn away during splitting. Encrusting materials have been removed by solvents and the torus is seen supported by the strands of the margo (probably largely cellulose), corresponding to the diagrammatic section **B. A., B.** and **C.** ×1200. **D.** ×300. All after W. D. James. **E.** Scanning electron micrograph (×5000) supplied by J. Levy (photograph by W. B. Banks).

found whose canals broaden towards the pit membrane like funnels (Fig. 57B). Because of this such pits, known as bordered pits, appear in surface view as two concentric circles (Fig. 57A). The inner ring shows the boundary of the opening of the canal into the cell lumen (the pore), while the outer ring (the border) shows the boundary of the canal at its greatest diameter adjacent to the pit membrane. Such bordered pits are widely distributed in the conifers (Fig. 104). Where the pit fields are oval or elongated, the bordered pits take on a similar shape, a typical situation in many broad-leaved trees (Fig. 161I).

Bordered pits, especially amongst the Coniferae, are often furnished with a thickening in the middle of the pit membrane, known as the torus, from which cellulose strands, forming the margo, extend radially to the cell wall (Fig. 57E). So long as the torus lies in the middle between the two pores (Fig. 57B), water can circulate without impediment between these strands, from cell to cell. If the torus is displaced to one side, however, it can close the pore like a valve (Fig. 57C, E). Where the water-conducting elements adjoin living cells, their pits are usually bordered only on the water conducting side.

Somewhat surprisingly, pit-like structures are found not only in the walls between adjacent cells, but occasionally also in the outer walls of the upper epidermal cells, in contact with the air ('sensory pits' of certain tendril-bearing plants, see p. 363).

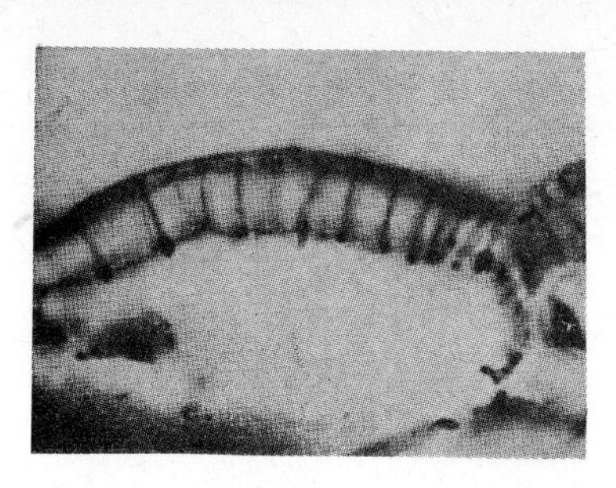

Fig. 58. Epidermal cell of *Passiflora coerulea* with plasmodesmata in the outer wall (ectodesmata). ×500. After Schumacher.

Apart from these specialized pits, filled with protoplasm, the outer epidermal walls of many species of plants are penetrated by pit-like perforations reaching to the cuticle, but probably quite devoid of protoplasmic content. They are termed ectodesmata, and are probably concerned with the uptake of water and dissolved nutrients by leaves and shoots, as well as with the secretion of cutin and waxes (Fig. 58).

C. Chemical composition of the cell wall

In living cells the cell wall is always saturated with water and swollen (see p. 213). Consequently, when water is withdrawn it tends to shrivel up. Its lamellae consist principally of carbohydrates, in particular cellulose, although hemicelluloses and pentosans may also be present, usually occurring together.

Plant walls never consist solely of cellulose, even though for brevity one speaks of 'cellulose lamellae'. Cellulose occurs in the walls of most plants, although in the majority of the fungi and bacteria it is absent. Like starch, cellulose is a polysaccharide with the general formula $(C_6H_{10}O_5)_n$ (see p. 261), but unlike starch it does not give a blue colour with an aqueous solution of idodine unless previously treated with sulphuric acid, or subjected to solutions of iodine containing high concentrations of certain salts, such as zinc chloride. Iodine in zinc chloride is, in fact, the most useful reagent for the identification of cellulose. Cellulose is insoluble in water, even when boiling, nor is it soluble in dilute acids, in alkalis or even in concentrated potassium hydroxide. It dissolves, however, when subjected to gentle hydrolysis, for example in a solution of copper oxide in ammonia (Schweizer's reagent). Concentrated sulphuric acid converts cellulose to glucose. The enzyme cellulase also brings about its conversion to sugar.

Many hemicelluloses also give a weak blue coloration with iodine in zinc chloride. The hemicelluloses are a series of carbohydrates which stand very close to cellulose, but which are broken down by weak acids and certain enzymes into sugars other than glucose, such as mannose and galactose. Mucilages and 'celluloses' serving as food reserves are especially rich in hemicelluloses. Such reserve celluloses form hard but easily soluble deposits on the cell walls in many fruits and seeds. The stony endosperm of the palm *Phytelephas macrocarpa* is used in commerce and is known as 'vegetable ivory' (Fig. 56B). Also the seeds (kernels) of the date palm consist largely of reserve cellulose. Just as cellulose and the hemicelluloses are polysaccharides built up from hexoses, and having in consequence the general formula $(C_6H_{10}O_5)_n$, so are the pentosans, with the general formula $(C_5H_8O_4)_n$, high polymers of the pentoses $(C_5H_{10}O_5)$, such as arabinose and xylose.

Primordial walls (which become later the middle lamellae) and primary walls consist principally of protopectin. In contradistinction to cellulose, protopectin gives a yellow-brown colour with iodine in zinc chloride, and stains intensely with methylene blue, saffranin, and especially ruthenium red. After preliminary hydrolysis with weak acids, protopectin can be readily removed with alkalis from tissue sections, a procedure sometimes used in histochemistry. Occasionally in the ripening of fruits a natural, enzymatic, dissolution of the protopectin occurs, with the result that the cells of the flesh lose their cohesion and separate to form a mealy mass (*Symphoricarpus*, certain kinds of cultivated fruits). The same effect is produced in the maceration of tissues by chemical means (e.g. by boiling in potassium chlorate or sodium nitrate solutions) and in the retting of flax by digestion with bacterial enzymes. Protopectin can therefore be regarded as a cementing substance holding the cells of a tissue together.

Chemically the pectins are derivatives of pectic acid, a linear macromolecule consisting of a chain of galacturonic acid units. Its structure is thus similar to that of cellulose, but the molecules are not packed, as are those of cellulose, in a crystalline array. Partial esterification of the carboxyl groups projecting laterally from the chain with methyl alcohol leads to the water-soluble pectins. These are found in the cell sap of many (especially young) fruits. Subsequent polymerization of these pectins leads to the 'setting' of sugary decoctions of fruits, such as jams and jellies. In the presence of bivalent ions (e.g. Ca^{++} and Mg^{++}) the pectins can form 'salts' (pectates), in which the cations cross-link adjacent molecules. The consequence is a reticulum of macromolecules which comes out of solution. The protopectin of the middle lamella is formed in this way.

Chitin, present in the cell walls of most fungi, is also a linear macromolecule, similar to cellulose. The structural unit here, however, is an acetylated and aminated sugar, acetylglucosamine ($C_8H_{14}O_6N$).

D. Physical properties and submicroscopic structure of the cell wall

Investigations with the electron microscope have shown that the plant cell wall consists of a porous ground mass, saturated with water, in which the cellulose is embedded in the form of a network of microfibrils (Figs. 59B; 60; 62). The ground mass is composed of

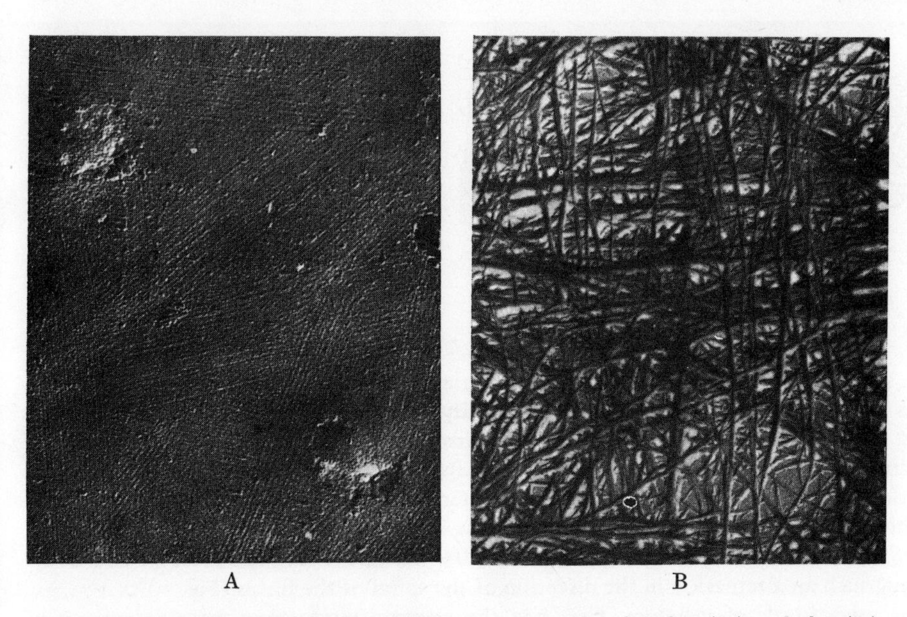

Fig. 59. Cell wall of the unicellular desmid *Pleurotaenium trabecula* before (A.), and after (B.), treatment with chromic acid. Removal of the interfibrillar incrusting material reveals the cellulose framework. The structure of the wall is thus analogous to that of reinforced concrete. ×22,000. After Drawert and Mix.

pectins and hemicelluloses, and its chemical composition largely coincides with that of the primordial wall. While the fibrils in the primary wall, which is still capable of growth, are intertwined with one another and run in all directions irregularly (giving the disperse texture shown in Fig. 60A), in the secondary wall the fibrils are much closer together and frequently arranged parallel to one another (Fig. 60B). Secondary walls often show a layering corresponding to a diurnal rhythm ('day rings', Fig. 56A). This comes about because during the day principally cellulose is laid down, and during the night principally hemicellulose.

If the fibrils of the secondary wall are orientated parallel to the longitudinal axis of the cell (as in ramie, hemp, flax and nettle fibres), they are said to have a fibre texture; if they run transversely round the cell, as in tracheids with annular thickening, they are said to have a ring texture; and if spirally, as in the tracheids of conifers and the xylem fibres of angiosperms, a spiral texture. If the spiral described by the fibres is not very

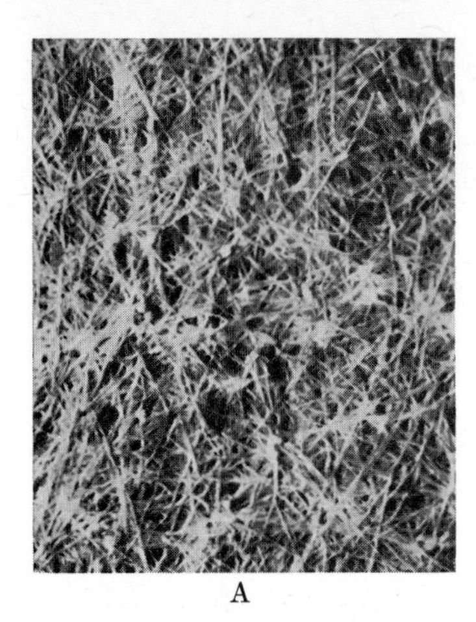

Fig. 60. **A.** Primary wall of a young cell of the alga *Valonia*. The cellulose fibrils are irregularly intertwined ('disperse texture'). **B.** Secondary wall of a cell of *Valonia* with parallel orientated cellulose fibrils ('spiral texture'). The direction of the fibrils changes in successive layers, the fibrils of one layer making a definite angle with those of the next. Both ×5000. From Steward and Mühlethaler.

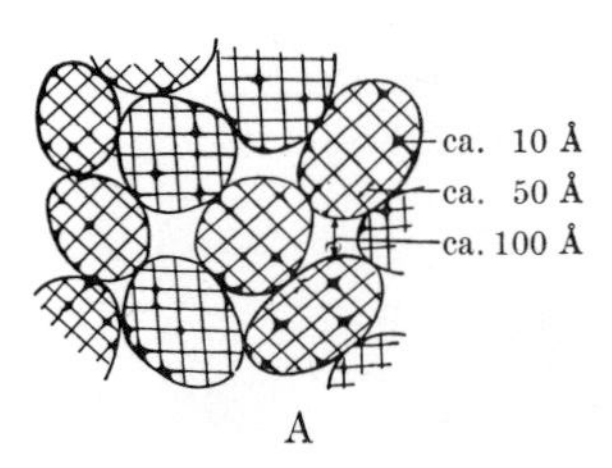

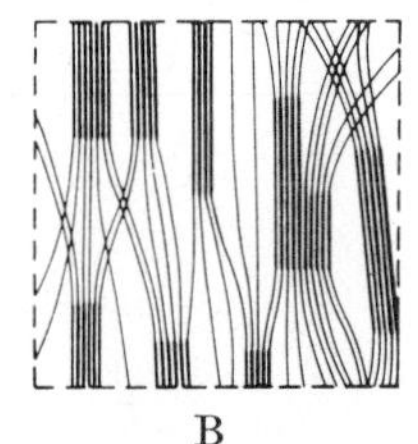

Fig. 61. Submicroscopic structure of a cell wall in transverse section. **A.** *c.* ×25,000. White areas indicate the microfibrils in transverse section; black, the interfibrillar spaces. **B.** Portion of **A.** at a greater magnification. Each microfibril is made up of about twenty micellar strands. Between the micellar strands are intermicellar spaces of diameter from 5 to 10 Å (10 Å = 1 nm). After Frey-Wyssling.

steep (as in the fibres of *Cocos*, where the angle made with the longitudinal axis of the cell is about 45°), the fibres are often markedly elastic, since the fibrils act as a helical spring capable of extension. Cells with the fibrils in a steep spiral are clearly well adapted to stand up to longitudinal stress, since the individual fibrils will press against each other like the strands in a twisted rope. The resistance to stretching becomes considerably enhanced if the directions of the spirals in the successive lamellae of the wall alternate (Fig. 65). In the alga *Valonia* the wall is only 0.04 mm thick, yet there are over 700 layers showing such an alternation in the direction of the spiral of the fibrils (Fig. 60B).

The microfibrils of cellulose, visible in the electron microscope, have a diameter of 20–30 nm. They consist of fifteen to twenty elementary fibrils or micellar strands (diameter 6–7 nm) which can be separated by ultrasonic disintegration. Each micellar strand consists in turn of individual micelles, each of these being composed of about 100 cellulose macromolecules arranged in a not very sharply defined bundle. Some of these macromolecules appear to cross to neighbouring micellar strands, and it is probable that the individual micelles are interlinked in this way (Fig. 61B). Very small molecules, such as those of iodine, water and alkalis, can enter the intermicellar spaces (Fig. 61A, the small black squares representing about 1 nm), but larger molecules, such as those of lignin, cutin, suberin and various pigments, can enter only into the inter-fibrillar capillary spaces (Fig. 61A, the large quadrangular spaces representing about 10 nm). That the submicroscopic structure of the wall should consist of a more or less parallel arrangement of cellulose micelles renders comprehensible the optical properties of the wall, and its laminated appearance when made to swell. Penetration by water molecules has scarcely any effect on the cohesion of the fibrils in the longitudinal direction; for example, if dry flax fibres are made to swell their diameter increases by 20 per cent, but their length by only 0.01 per cent (see p. 377).

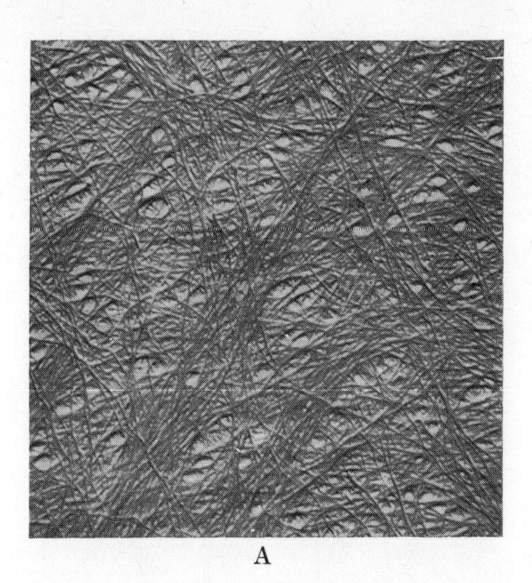

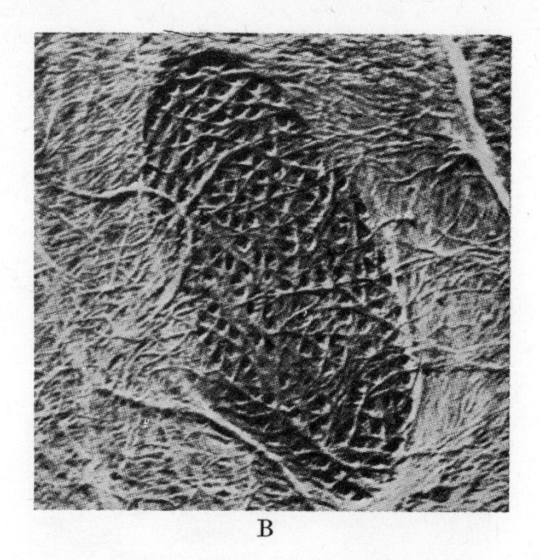

A B

Fig. 62. Formation of pits in the primary wall. **A.** Transverse wall of a meristematic cell from a young petiole of *Ricinus communis*. Disperse texture with numerous pores which remain and become occupied by plasmodesmata. **B.** Primary pit field in the primary wall of a cell from the root of *Zea mays*. The holes which become occupied by plasmodesmata are clearly visible. **A.** ×120,000. From Mühlethaler. **B.** ×100,000. From Wyckoff and Mühlethaler.

E. Secondary changes in the wall

During the life of a cell the cell wall often acquires new chemical and physical properties. The chemical nature of the layers present may become changed, or new and chemically distinct substances may be embedded in, or deposited on, those already present. These changes are often closely related to the demands made upon the cell. The turgid cellulose lamellae are plastic and markedly elastic. Purely cellulose walls offer very little resistance not only to the passage of water, but also of solutes. The most important secondary changes in the wall are lignification, suberization and cutinization. Lignification diminishes the extensibility of the cell wall and its capacity for swelling very considerably, but increases its rigidity, without destroying the permeability of the wall to water and dissolved substances. Suberized and cutinized walls, on the other hand, are little permeable to water or gases, and consequently strongly reduce evaporation.

a. Lignification. This consists of the deposition of lignin within the cellulose framework of the lamellae, often leading to a swelling of the wall. In this way complexes are formed owing their tensile strength to cellulose and their rigidity to lignin. In structure and mechanical properties they are analogous to reinforced concrete, except of course that they differ in being permeable to water and solutes. The innermost lamellae of many lignified walls often consist of pure cellulose. Lignins are complex polymers of various derivatives of phenyl propane (e.g. coniferyl alcohol among others), and in contrast to the linear molecules of cellulose they are richly branched. Lignified cell walls stain yellow with aniline sulphate, and with iodine in zinc chloride, but with alcoholic phloroglucin and hydrochloric acid produce a red colour. In microscopic sections lignin can be dissolved out of cell walls by Eau de Javelle, leaving only the carbohydrate lamellae. To remove lignin in the manufacture of cellulose and paper, the wood pulp is boiled for a long time under pressure with calcium bisulphite or sodium hydroxide.

b. Suberization. The deposition of suberin is generally limited to the secondary thickening of the wall, so that suberin lamellae, impervious to water, are laid directly upon unsuberized lamellae. While the cells are alive, small pores are left in the suberin lamellae, allowing the access of nutrients and continued metabolism. As suberization becomes completed, however, even these become occluded with suberin deposits.

Cutin is closely related to suberin, and cutin is excreted into or on to walls in contact with the external air. Suberins and cutins are highly polymerized esters of saturated and

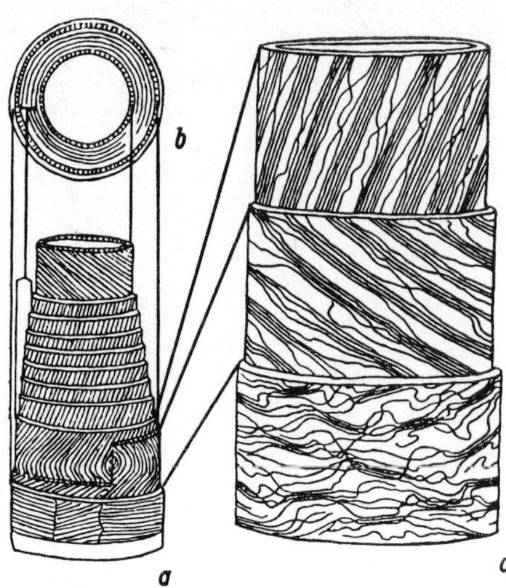

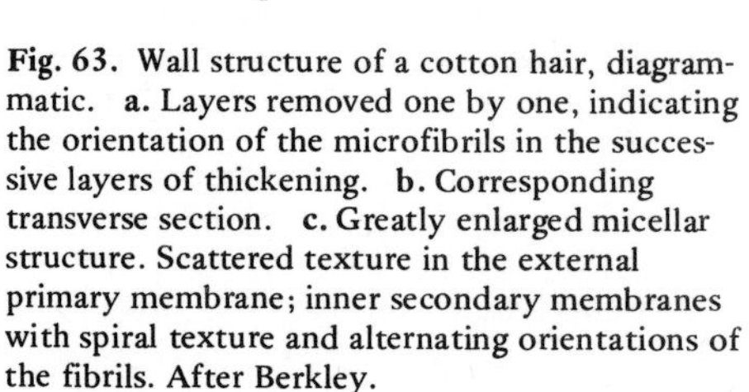

Fig. 63. Wall structure of a cotton hair, diagrammatic. **a.** Layers removed one by one, indicating the orientation of the microfibrils in the successive layers of thickening. **b.** Corresponding transverse section. **c.** Greatly enlarged micellar structure. Scattered texture in the external primary membrane; inner secondary membranes with spiral texture and alternating orientations of the fibrils. After Berkley.

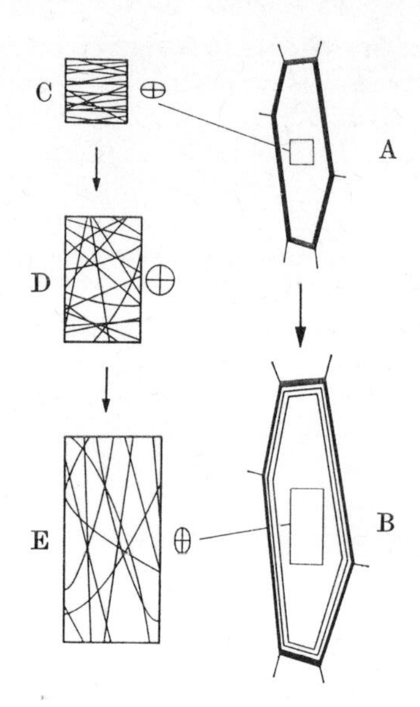

Fig. 64. Diagrammatic representation of the surface growth of the cell wall according to the multi-net theory. The initial situation, with the microfibrils oriented predominantly transversely, is shown in **A.** and **C.**, the later in **B.**, **D.** and **E.** The primary wall (shown by the heavy line) increases its surface, but decreases in thickness. This is accompanied by a loosening of the texture and a re-arrangement of the fibrils (**D.** and **E.**). After Sitte.

unsaturated fatty acids and oxyacids (frequently C_{15} chains), the suberins, for example, consisting of phellonic acid, suberinic acid and similar acids of very high molecular weight. In contradistinction to suberin, cutin contains only minute amounts of the unsaturated fatty acids. Both substances are hydrophobic and readily form polymers and molecular lattices, but the degree of polymerization is probably less in suberin than in cutin. The substance showing the greatest degree of polymerization is probably sporopollenin, chemically related to cutin and suberin, and found specifically in the walls of fungal spores and pollen grains. Sporopollenin is so resistant to saponification and decay that cell walls impregnated with it have persisted unchanged in peat deposits for millions of years. Fat dyes, such as Sudan red in glycerin, or a solution of chlorophyll, impart their colour to both cutin and suberin, and can be used for the identification of these substances. Neither cutin nor suberin is attacked by Schweizer's reagent (ammoniacal copper oxide), or by concentrated sulphuric acid.

c. Mineralization. Not only organic, but also inorganic substances may be laid down as incrustations in cell walls. Silica, for example, is present in the epidermal cell walls of grasses, sedges, and species of *Equisetum* rendering them very hard. It is also found in other plants such as diatoms and certain flagellates. The tips of the stinging hairs of various nettles (Urticaceae, Loasaceae, Fig. 107A, B, C) owe their glass-like fragility to deposits of silica. They break off at the least touch at a predetermined place. Extensive secretion of calcium carbonate on to the surface of the wall makes many plants (e.g. certain Characeae of our lakes and ponds) stiff and brittle. Also many red algae lay down so much calcium carbonate in and on their walls that they form rocks (corals). Whole masses of rock in the Alps, for example, are of this nature. In flowering plants calcium carbonate is often found in the cell walls of hairs (e.g. in the Cucurbitaceae and Boraginaceae). The lower parts of the stinging hairs of nettles are also usually calcified. A further example of calcification is provided by the walls of the fruits of *Lithospermum*, to which they owe their characteristic hardness. The deposition of calcium carbonate in and on cystoliths has already been mentioned (p. 61). Calcium oxalate (see p. 57) also becomes laid down in crystalline form within cell walls (e.g. in the phloem of many Cupressaceae).

d. Other incrustations. Frequently, cell walls ultimately become dark coloured, and resistant to decay, as a consequence of their becoming permeated by tannin derivatives. Examples are seen in the testas of seeds, in heart wood and in bark. In those cases where commercial dyes are extracted from wood (see p. 157) the pigments concerned, all flavones, are present in the cell wall. Red, yellow, green and violet pigments occur in the cell walls of the fruiting body of numerous agarics and give them their characteristic appearance.

Firm cell walls may become converted to mucilage; a notable example is seen in the 'gummosis' that affects certain trees. In species of *Prunus* the individual layers of the cell wall become converted successively to mucilage, and ultimately the cell contents themselves become added to it. The 'gum' resembles other plant mucilages chemically (see p. 53). In many bacteria, blue-green algae and higher algae (e.g. *Spirogyra, Batrachospermum*) the outer layers of the wall regularly become mucilaginous. As a consequence of this property bacteria are able to form zoogloea (p. 441) and the blue-green algae gelatinous masses (as in *Nostoc*), and the algae *Batrachospermum* owes its name to its resemblance, both in appearance and consistency, to frogs' spawn (see p. 487). Also, on moistening, the cell walls of the testas of many higher plants (e.g. linseed) swell to form a mucilaginous jelly.

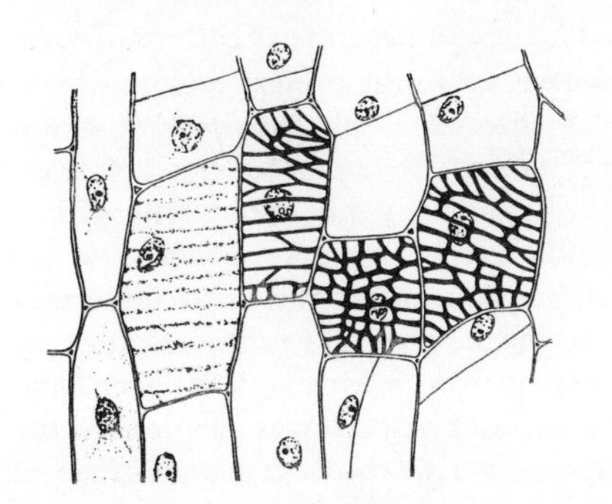

Fig. 65. Regeneration of xylem elements in the pith of a wounded shoot of *Coleus*. Local concentrations of the cytoplasm can be seen in those places where the secondary layers will be deposited, indicating cytoplasmic activity preceding the deposition of the thickening material. After Sinnot and Bloch.

Numerous observations lead us to the view that not only the cellulose itself but also the specific, fibrillar arrangement of its molecules in the cell wall is a product of the cytoplasm. Among the evidence in support of this is the fact that before conspicuous formations of the wall (such as ridges of tertiary thickening (Fig. 65) begin to be differentiated, local aggregations of cytoplasm are already visible in the places in which they will ultimately be formed. We can only conclude that at the sites of growth of the wall there is an intimate interrelation between it and the cytoplasm. Moreover, the fully formed cell wall is, as we have seen, freely penetrated by plasmodesmata. The possible significance of dictyosomes and microtubules in the formation of cell walls has already been mentioned (see p. 26).

Section two: **The grades of morphological organization**

As has already been stressed in the introduction, the plant forms of the present day can only be properly understood against the background of their historical development. The only records which we possess of the course of evolution are those plant remains which have become fossilized, the study of which is known as palaeobotany.

The comparative examination of plant remains found hitherto, particularly those of the earlier epochs in the history of the Earth, shows that these plants were always more simply organized than the more recent. The number of different types of specialized cells, and of tissues constructed from these cells, has considerably increased in the course of evolution. While the oldest plant remains which can be identified with certainty consist of single cells or thread-like aggregations of cells all similar to each other (see Fig. A1), some seventy to eighty different kinds of cells, varying in form and specialized function, are involved in the structure of many species of recent flowering plants. This relationship between evolutionary age and structure is referred to as the **principle of progressive differentiation.**

Although many species still living have retained even up to the present day numerous primitive characteristics (and serve as examples of antecedent plant forms), it is not possible to reconstruct phylogenetic trees from them alone. The reason for this is that we cannot say with certainty of any recent form that it has actually inherited unchanged all the primitive characteristics of a morphologically similar earlier form. Nevertheless we are well able to deduce from an examination of living species what the most important advances in plant organization must have been in the course of evolution. Such an examination indicates a progressive development leading from microscopically small and aquatic unicellular plants to multicellular and highly differentiated forms adapted for life on land. It becomes clear that each transition from a lower to the next higher grade of organization is accompanied by the acquisition of some important new property or capacity which results in a fundamental change in structure and way of life.

It is possible to recognize three major grades of morphological complexity in the Plant Kingdom, grades which are of course connected by transitional forms. These are the Protophyta, Thallophyta and Cormophyta. The Protophyta include all unicellular plants, and also all those plants which consist of loose aggregations of cells showing little or no functional differentiation and which can therefore easily break up at any time to give rise again to a unicellular condition. From this primitive stage has evolved the grade at which the plant is composed of a functionally differentiated thallus, consisting of numerous cells, or a polyenergid. Examples are seen in the more highly organized algae and fungi. This grade of morphological organization has been achieved by different routes in different lines of evolution.

The opportunity for the development of multicellular thalli came with the evolution of the rigid cell wall composed of highly polymerized substances such as cellulose and chitin. This made it possible for the two daughter cells resulting from a division to be separated by a common wall perforated only by the pits, which, although it certainly parted the two protoplasts, did not completely isolate them from each other. The presence of fine protoplasmic connections from cell to cell (the plasmodesmata (see p. 62)) allowed the development of true multicellular individuals in which the component cells, losing their individuality, became subordinated to a new and common unity and acquired specialized functions.

Like the Protophyta, the Thallophyta are again mostly confined to life in water or at least to an environment continuously saturated with water vapour, since they are not provided with wall materials capable of preventing evaporation, such as cutin and suberin. Their water economy is consequently not stabilized in relation to the atmospheric conditions. In dry conditions they can if necessary persist in a state of latent life (see p. 21). These plants thus display a relationship between their hydration and that of the environment analogous to that between the poikilothermic animals and the ambient temperature; they can accordingly be referred to as poikilohydric.

The typical thallophyte does not usually have special strengthening elements at its disposal. Consequently, if a thallophyte leaves the water, the medium that normally supports it, and takes up life on land, it will be a more or less prostrate plant. Only a few thallophytes, such as the higher fungi and lichens, have succeeded in becoming fully adapted to terrestrial life.

The Cormophyta, distinguished by a series of specific adaptations to life on land, have reached the highest level of differentiation. They have also stabilized their water economy, and are consequently capable of growing in atmospheres with high saturation deficits (homoiohydric organization). Anatomically they are characterized by the development of morphologically and functionally differentiated tissues, such as epidermal, water-conducting and strengthening tissues, as well as those concerned with water uptake and transpiration. The most conspicuous of the external features distinguishing the Cormophyta from the Thallophyta is the morphological differentiation of the former into roots, stems and leaves (Fig. 131D).

The only plants included in the Cormophyta are the ferns and fern-allies (Pteridophyta) and the seed plants (Spermatophyta). The mosses and liverworts (Bryophyta) occupy an intermediate position in that, although they possess small stems and leaves, they do not have true roots. Also their stems and leaves retain a very primitive kind of construction, and are not to be regarded as equivalent to those of the true Cormophyta.

I. Protophyta[17]

A. Unicellular plants. The simplest plants consist of a single, naked, cell (e.g. *Euglena* (Fig. 4) and other flagellates (Figs. 389; 408)). In the larger unicellular plants, however, the organization may reach an astonishing level (Figs. 3; 386; 388). The simplest geometrical form in which these single cells can occur is the sphere; this is frequently adopted in the resting stage, when a wall may be present (Figs. 411; 412). Many of the permanently integumented bacteria and blue-green algae (Figs. 66; 384B, E) pass their lives in this form, as well as the most primitive of the permanently integumented algae

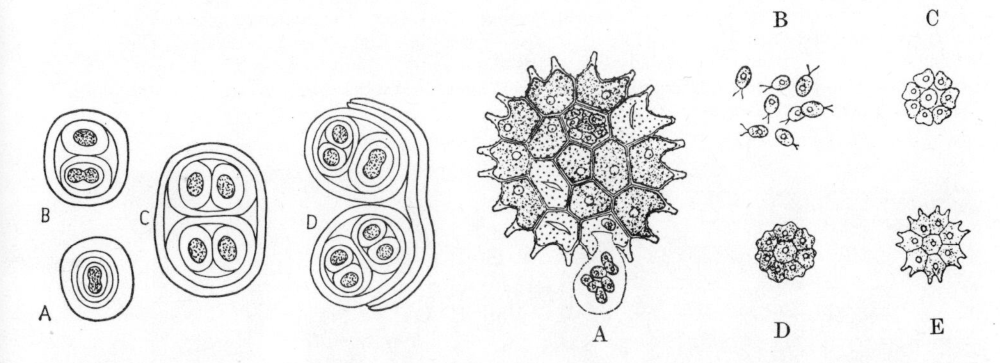

Fig. 66. *Gloeocapsa* sp. **A.** Single cell in division. **B.** and **C.** Continued cell division and formation of a coenobium. **D.** Break-up of the coenobium following rupture of the first-formed, swollen cell walls. After Strasburger and Wille.

Fig. 67. *Pediastrum boryanum.* Vegetative reproduction by means of zoospores. **A.** Escape of the zoospores. **B.** Zoospores in the swarming condition. **C.** to **E.** Three stages in the aggregation of the zoospores to form daughter colonies. ×300. After G. M. Smith, simplified. Stages **B.** to **E.** actually take place within a vesicle as shown in **A.** (cf. Fig. 414).

proper (Fig. 411). Many unicellular plants already show a clear polarization, often accompanied by an elongation of the cell. In the flagellates, for example, this is made apparent in the development of the flagellum and eye-spot at different points of the cell (Figs. 4; 408A, K).

B. Coenobia. Each cell division of a unicellular plant leads to the complete partitioning of the protoplast (Fig. 403) and the formation of two new independent individuals. At this level of organization cell division implies at the same time vegetative reproduction. If the daughter cells remain together in a commonly excreted jelly, or are held together within the wall of the original mother cell, a loose association of cells, or *coenobium*, will be formed.

Many bacteria form mucilaginous jellies (e.g. the zoogloea which form on fruit juices) or films, such as will form on an infusion of hay. Bluish-green coatings of slime found on the damp walls of greenhouses are formed by various species of Cyanophyceae (such as *Gloeocapsa* (Fig. 66), and *Nostoc* (Fig. 384E), among others). Many diatoms and Chrysophyceae (e.g. *Hydrurus foetidus*) form quite rigid coenobia in flowing water, often even irregularly branched and thallus-like, consisting of cells held together in a common jelly (Fig. 396).

In *Gloeocapsa* (Fig. 66) the daughter cells are retained after division within the expanding wall of the mother cell. Not until after several further divisions does this wall become mucilaginous and ultimately wholly dissolved. If the planes of division within a

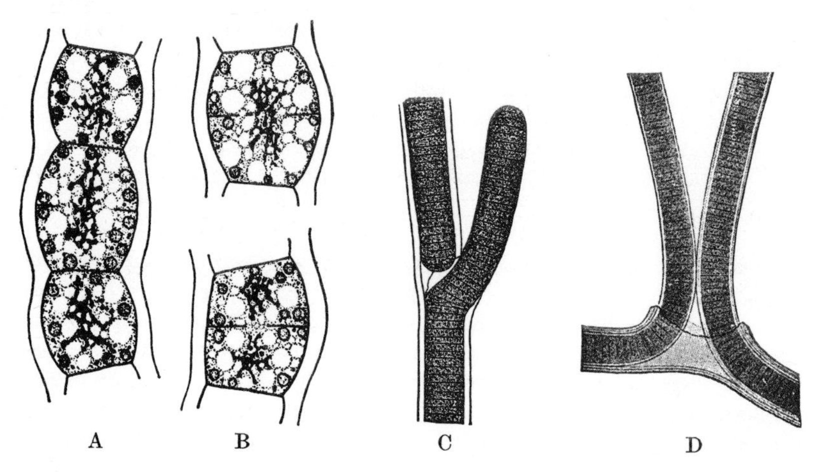

Fig. 68. Coenobia of Cyanophyceae. **A.** and **B.** *Anabaena circinalis*. The nucleoids (equivalent to nuclei) form irregular central groups of Feulgen positive granules. Right, two cells dividing. The dividing wall grows in from the periphery like an iris diaphragm. **C.** and **D.** False branching of filamentous Cyanophyceae. **C.** *Plectonema wollei*. **D.** *Plectonema mirabile*. **A.** and **B.** ×500. After Haupt. **C.** ×200. After Kirchner. **D.** ×200. After Bornet.

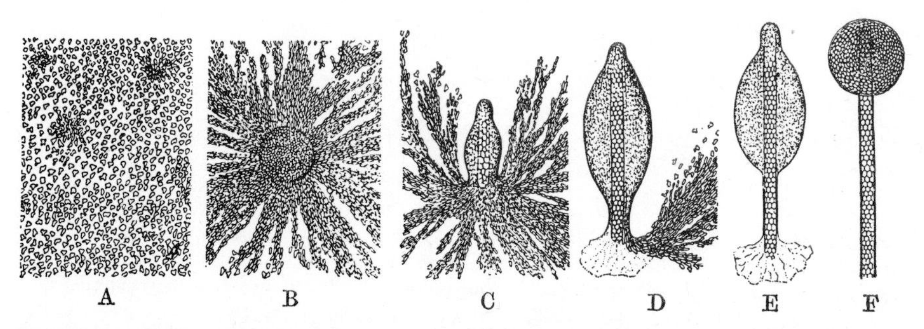

Fig. 69. Development of a sporangium of the slime fungus *Dictyostelium mucoroides*, diagrammatic. **A.** Mobile phase with numerous individual myxamoebae. **B.** Appearance of an aggregation centre. **C.** to **E.** The myxamoebae creep together to form a stalk and eventually a sporangium. **F.** The mature sporangium. ×40. After Kühn.

coenobium lie more or less parallel to each other, then a thread-like assemblage of cells results. Often the only factor causing the individual cells of such a thread to hold together is a continuous sheath of mucilage, and each cell within the thread preserves its complete independence. Owing to breaks developing here and there in the sheath, as in *Hydrurus*, the thread may appear to branch (e.g. *Plectonema* (Fig. 69C, D), *Sphaerotilus* (Fig. 377)). For a consideration of true branching see p. 129 seq. Breaking up of the thread, sometimes caused by the death of individual cells (e.g. in *Spirogyra*), brings about vegetative reproduction.

C. Plasmodia. These are multinucleate masses of protoplasm, found in the Myxomycetes or slime fungi (Figs. 8; 454). In extreme cases they are capable of slow flowing and creeping movements, by means of which they may cover considerable distances, even up to several decimetres.

The pseudoplasmodia of the Acrasieae (e.g. *Dictyostelium*, Fig. 69) are particularly interesting. They consist of several thousand independent cells living in close association. At maturity, after a sexual process, the individual cells, capable of amoeboid movement, creep together to form a fruiting body some several millimetres in height. Similar fruiting bodies are found also in the Myxobacteria, and they are again formed by the coming together of previously freely moving individual cells (Fig. 379).

II. Thallophyta[18]

The true thallus is always multicellular, or is a polyenergid (e.g. *Caulerpa* (Fig. 9)). The cell walls consist of cellulose, of chitin or of both substances. The external morphology varies from simple, spherical or thread-like assemblages of cells at one extreme, to highly complex forms at the other (e.g. Figs. 437; 440). These never, however, achieve the internal anatomical complexity characteristic of the Cormophyta.

The alga *Acetabularia* occupies an anomalous position. Its thallus, which reaches several centimetres in length, remains for a long time uninucleate (Figs. 16C–E; 423D). This nucleus does not divide until immediately before the formation of the reproductive cells. At this stage the nucleus divides into numerous daughter nuclei, which ascend into the umbrella-like cap. Each then becomes associated with a portion of cytoplasm and the units so formed differentiate as motile reproductive cells.

True thalli can be formed in two ways, namely, 1. by aggregation of previously free individual cells, and 2. by incomplete separation of the daughter cells after cell division. True multicellular thalli can arise only by the latter process.

A. Cell aggregates. These are found in many green algae, such as *Pediastrum* (Fig. 67) and *Hydrodictyon* (Fig. 415).

In those species forming cell aggregates it is usual for a large number of freely motile individual cells to be formed in vegetative reproduction. Later these cease to be motile, and come together to form a new single organism (Fig. 67B–E).

While in those species forming cell aggregates, the cohesion between the initially free individual cell is established after some delay (post-genital), the cohesion of the individual cells of truly multicellular species is present from the beginning (congenital). Only in truly multicellular forms, by virtue of the plasmodesmata which traverse the pits in the dividing walls, do the cells form a genuine physiological unit.

B. Cell colonies. The freely motile cell colonies of the Volvocales provide a bridge between those primitive forms in which the cells, all of which are of equal form and function, are aggregated together and the functionally differentiated thalli of the next grade of organization. The most advanced members of the genus *Volvox* (Fig. 410) do, in fact, already provide examples of true multicellular individuals. The different kinds of cells in these species, and the fact that some cells are non-reproductive and die, indicate functional differentiation within the colony (see p. 446 seq.). The individual cells of primitive colonies (*Pandorina, Eudorina*) are, exactly as in coenobia, all of equal standing. Nevertheless, and in contradistinction to coenobia, the colony already has a definite form and behaves as a functional unit. It can no longer be arbitrarily dissected without losing its individual character.

The organisms possessing the grades of organization considered hitherto may be either attached with mucilage or gelatinous material to the substratum, or float freely in the water. Those freely floating forms are known as plankton, and the attached as benthos. Planktonic organisms often possess special contrivances to increase their buoyancy, such as oil vacuoles in the cytoplasm or protuberances of the wall which increase the surface area and consequently the stability of the organism in the water as, for example, in *Ceratium*.

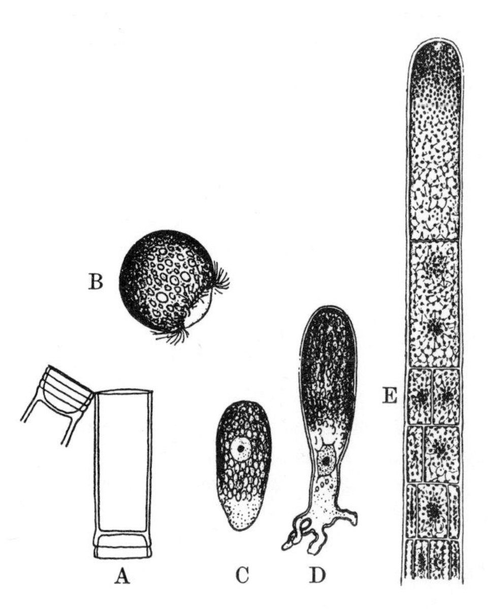

Fig. 70. Polar differentiation. Escape and germination of a zoospore of *Oedogonium concatenatum*. The zoospore (**B.**), released from the cell **A.**, swims around by means of a fringe of cilia. Eventually it becomes attached to the substratum by means of its apical pole (**C.**), which immediately becomes transformed into the rhizoid (**D.**). **E.** *Sphacelaria racemosa*. Tip of a branch of the thallus with the apical cell. **A.** to **D.** ×300. After Hirn. **E.** ×40. After Reinke.

C. Filamentous thalli. a. Cell chains. If in the formation of a true multicellular thallus all the division spindles remain orientated in one direction as a consequence of some strong polarity, a unidimensional filament (Fig. 424O) results, such as we have already seen in *Anabaena* and *Plectonema* (p. 72). True filamentous thalli are, however, distinguished from thread-like aggregations such as these by the firmness with which the individual cells are attached to one another. This results from their being congenitally united and possessing a common cellulose wall (Figs. 405; 416). In the simplest forms all the cells of the thread are alike, and any cell may divide and contribute to the growth of the thread.

b. Polarized differentiation. We have already seen that even single cells are often polarized (Fig. 4). A similar polarization is also seen in the swarm spores (zoospores) of the green (Figs. 70A; 416; 420) and brown (Fig. 430) algae, as well as of the fungi (Figs. 462; 466). It is usually at the apical pole that these unicellular motile spores attach themselves, subsequently to grow out to form new filamentous thalli. The basal pole generally develops into a holdfast or rhizoid (Fig. 70C).

The spindle of the first division following on the attachment is usually perpendicular to the plane of the substratum. In this way a filamentous sporeling develops attached at one end. In the simplest forms any of the cells may divide.

When polarity is especially strong in the sporeling the capacity for division remains more or less confined to the apical region of the filament (apical growth). In extreme cases the meristematic activity may be confined to the apical cell itself, then referred to as the apical initial.

We can envisage that in each cell certain substances or structures are distributed unequally in a polar manner, so that a concentration gradient comes into existence. Each

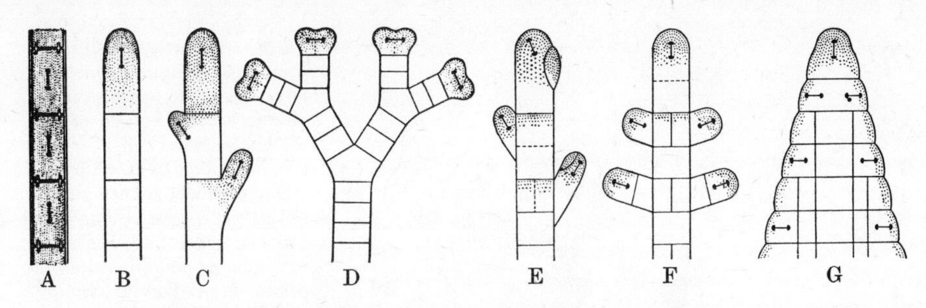

Fig. 71. Types of growth and branching of algal thalli growing as filaments or two-dimensionally as flattened sheets. **A.** Filament with generally distributed intercalary growth. **B.** Filament growing from an apical cell. **C.** The same, but showing in addition apical-polar branching. **D.** Uniform dichotomous branching resulting from the periodic interpolation of a longitudinal division in the apical cell. **E.** Sub-apical lateral branching of the apical cell. **F.** Lateral branching from the uppermost segments cut off from the apical cell. **G.** Congenital fusion of the lateral branches (giving rise to a marginal meristem and a flattened sheet-like thallus).

dividing wall formed transversely to this gradient inevitably divides the cell into two daughter cells of unequal potentiality. The cell cut off towards the apex retains the greater part of the meristematic activity, while that towards the base is capable of undergoing only a limited number of divisions. Frequently the basal cell is capable of dividing only after a short or a prolonged resting period, or after a regeneration phase. This is an example of the phenomenon of unequal division (Fig. 80).

In the brown algae belonging to the Sphacelariaceae, which have uninucleate cells, the polarized differentiation of the protoplasm in the large apical cell is made evident by the difference in colour between the apical and the basal regions (Fig. 70E). The deposition of a brown, sometimes almost black, substance ('fusosan', a resinous compound) towards the tip makes the apical region of the cell very much darker than the basal (the blackened tips of these plants are responsible for the name 'burnt algae'). Unequal division results in new cells being added basipetally. These cells can divide further, but of prime importance is the first transverse division, since this produces an upper nodal cell and a lower internodal cell. The further division of the latter is not resumed until after a resting period, during which the nodal cell becomes divided longitudinally. Renewed division of the internodal cell then leads to extension growth accompanied by branching at the node (Figs. 71E, F; 72 and 73).

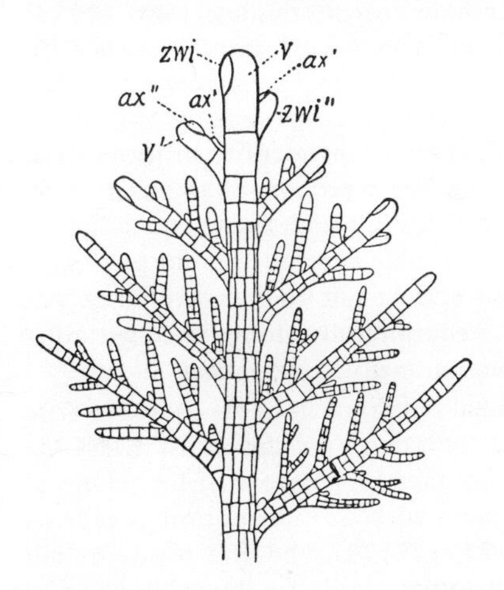

Fig. 72. *Halopteris filicina*. Termination of a long shoot. The large apical cells (V and V') cut off segments posteriorly in a regular rhythm and by unequal division. In addition, obliquely placed convex walls are formed alternately to the right and left cutting off branch initials (zwi). Similar walls in the apical cells of the side branches cut off the pseudo-auxiliary primordia (ax' and ax'') of the branches of the second order. ×40. After Goebel.

c. Branching. Every branching of the filament of cells reflects a change in the orientation of the mitotic spindle. In the simplest situation the direction of division changes at regular intervals in the apical cell itself. From time to time an equal longitudinal division follows upon a series of unequal transverse divisions, leading to the formation of two equivalent apical cells (Fig. 71D; see also Fig. 80B). Growth now takes place from two

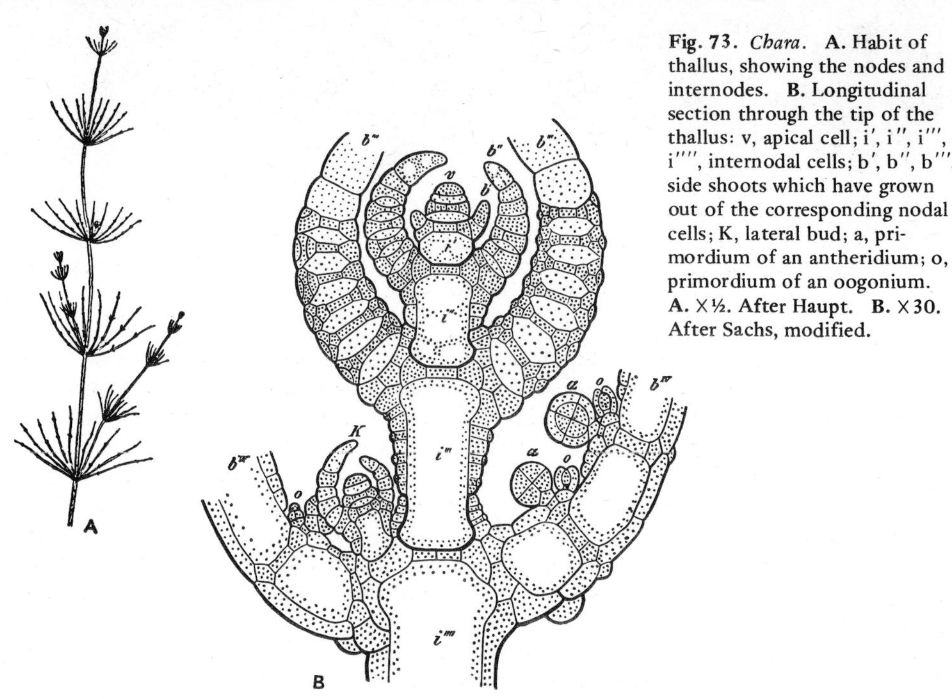

Fig. 73. *Chara*. **A.** Habit of thallus, showing the nodes and internodes. **B.** Longitudinal section through the tip of the thallus: v, apical cell; i′, i″, i‴, i⁗, internodal cells; b′, b″, b‴, side shoots which have grown out of the corresponding nodal cells; K, lateral bud; a, primordium of an antheridium; o, primordium of an oogonium. **A.** × ½. After Haupt. **B.** × 30. After Sachs, modified.

apical cells instead of one, and the resulting filaments diverge. This is referred to as forking, or dichotomous branching, or simply as dichotomy.

As a result of a rhythmical variation in the direction of division, brought about by unknown internal factors (an example of an endogenous rhythm), a regular arrangement of branches is formed. If the deflection of the spindle occurs not in the apical cell itself, but further down among the products of the apical cell, after growth has recommenced in the nodal cells, the result is lateral branching (Fig. 71C, E, F). Because of the polarity present in the cells of the filament, discussed in the preceding section, often only those new cells produced towards the apex can give rise to lateral apical cells of the first order. Consequently, the filament comes to display a very striking morphological differentiation into nodes at which branching takes place, and unbranched internodes (Figs. 72; 73). Moreover, growth of this kind does not lead to an irregular tangle of lateral branches, but to a fixed pattern of branching which repeats itself symmetrically.

d. Principles of symmetry. Symmetry in its literal sense means identity of proportion. Both in those forms in which a rhythmical change in the direction of the axis of the spindle leads to dichotomy (Figs. 71D; 79; 80) and in those in which the unequal division of the sub-apical cells leads to lateral branching of the filament and the morphological differentiation of the thallus into nodes and internodes, the plant body is divided up into a number of equal segments which repeat themselves rhythmically along the longitudinal axis of the filament. It can therefore be said to be longitudinally symmetrical.

Since new side branches, because of the longitudinal polarization of the axis, can arise only towards its tip, those nearest the apex are at the same time the shortest and the youngest. The sequence in which they arise is thus acropetal. A system of branching of this kind, in which the growth of the main axis continues without interruption, is called a monopodium, and the branching monopodial (Figs. 71F; 72; 73). The similar production of side shoots of the second, third and even higher orders leads to the production of intricate racemose systems of branching. The side branches may, like the main axis, be capable of unlimited growth (long shoots), or their growth may soon cease (short shoots).

Moreover, the arrangement of the side branches follows definite rules of symmetry which, to distinguish them from the longitudinal symmetry of the main axis, are referred to as lateral symmetry. A symmetry like that of mirror images (where symmetry is used in its strict sense) is especially striking.

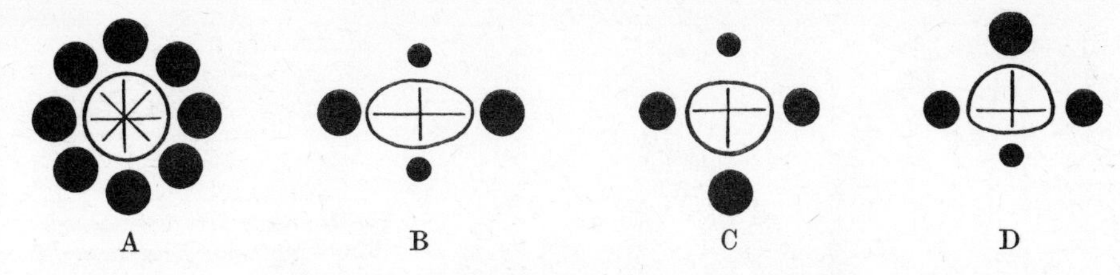

Fig. 74. Various possible developments of lateral symmetry. **A.** Radial or multilateral symmetry. **B.** Bilateral symmetry (amphitony). **C.** Dorsiventral symmetry with promotion of the lower side (hypotony). **D.** The same with promotion of the upper side (epitony). After Rauh.

If branching takes place equally all round the main axis the planes of symmetry are innumerable, and the symmetry is then said to be radial or multilateral (Figs. 73A; 74A). If, on the other hand, the lateral branching is confined to two planes intersecting at right angles (Fig. 73B), the symmetry is said to be bilateral. Branching may sometimes be confined to one plane (Fig. 71F, G), so that the system is then only two-dimensional. Related to bilateral symmetry are instances where one side of the system is different from the other (frequently seen, for example, in prostrate systems); only one plane of symmetry is then present and the system is said to be dorsiventral. Promotion of the lower side in a dorsiventral system is termed hypotony (Fig. 74C), that of the upper side epitony (Fig. 74D).

Lateral symmetry can be either determined by internal factors (autonomously) or induced by certain stimuli, particularly light and gravity.

e. Tissues consisting of interwoven cells (plectenchyma) and pseudoparenchyma. Richly branched and dense filamentous systems may, as a result of the intertwining of the cells and, in certain circumstances, the post-genital fusion of the filaments, form more highly organized structures which at a superficial glance may simulate the advanced algae and some of the Cormophyta, all constructed of true tissues. Especially well-developed examples of cell-associations of this kind occur in the red algae and in the fruiting bodies of the higher fungi.

The simplest way in which such a tissue-like association of cells can be formed is by the swelling up of the cell walls to form a water-insoluble jelly. This envelops the whole system of filaments, forming what is called plectenchyma (Fig. 78A, B). Such pseudoparenchyma is very difficult to distinguish from true tissue which has developed from a single growing point, even when examined under the microscope (Fig. 78C).

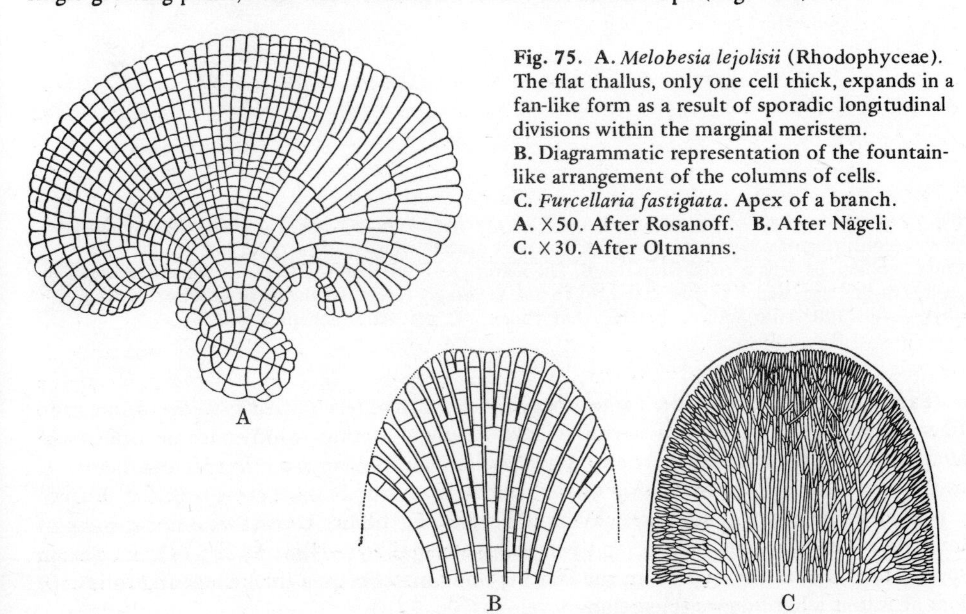

Fig. 75. **A.** *Melobesia lejolisii* (Rhodophyceae). The flat thallus, only one cell thick, expands in a fan-like form as a result of sporadic longitudinal divisions within the marginal meristem. **B.** Diagrammatic representation of the fountain-like arrangement of the columns of cells. **C.** *Furcellaria fastigiata*. Apex of a branch. **A.** ×50. After Rosanoff. **B.** After Nägeli. **C.** ×30. After Oltmanns.

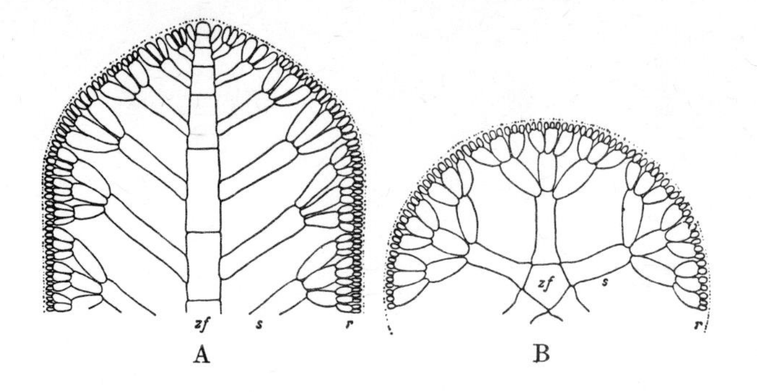

Fig. 76. *Chondria tenuissima* (Rhodophyceae). Diagrammatic longitudinal (**A.**) and transverse (**B.**) sections: zf, central filament; s, lateral branch; r, cortex. After Falkenberg, from Stocker.

In those members of the Rhodophyceae possessing plectenchymatous thalli of the kind described above it is possible to distinguish, on the basis of their differing longitudinal symmetry, two fundamentally distinct types of organization. The first ('fountain type') consists of a two- (Fig. 75A) or three- (Fig. 75C) dimensional dichotomous system, with apical growth of the individual filaments. The second ('central filament type') rests upon a monopodial system with sub-apical branching (Fig. 76). If the side branches become fused together at their initiation (Fig. 77A) circumscribed leaf-like thalli may be formed, very similar to the highly differentiated true leaves of the Cormophyta in external appearance (Fig. 77B, C). The identity of the component filaments is made apparent, however, even in the fully developed thallus by the fact that pit connections are formed only between the cells of one and the same branch (Fig. 77A).

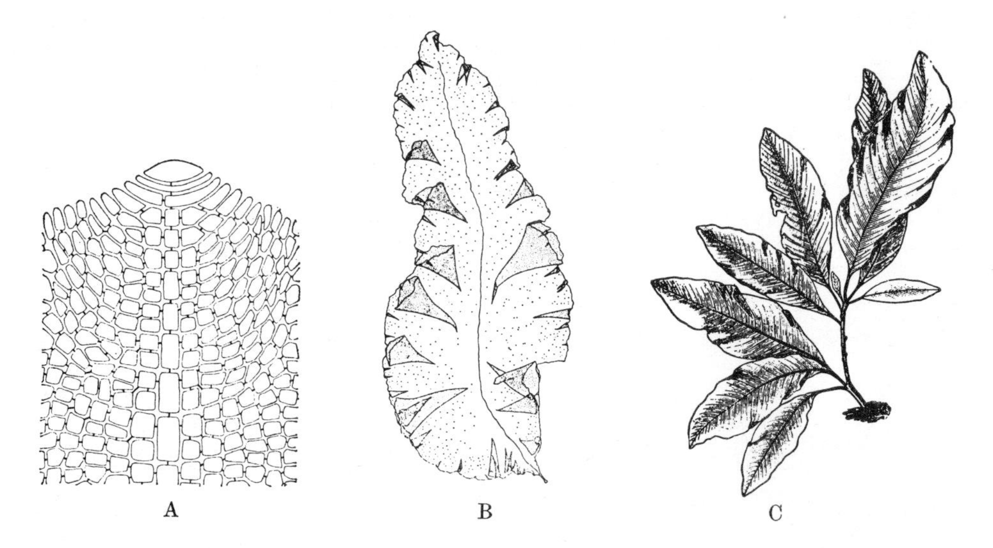

Fig. 77. Leaf-like thalli of red algae. **A.** and **B.** *Grinnellia americana*. **A.** Tip of the two-dimensional thallus, consisting of a single layer of cells. The large apical cell is clearly visible, and behind it the central column of cells derived from it. **B.** The mature leaf-like thallus. **C.** *Delesseria sanguinea*. The thallus of this alga, which occurs in the North Sea, strikingly resembles the leafy shoot of a cormophyte. **A.** X300. After Smith. **B.** X½. After Tilden. **C.** X¼. After Schenk.

The vegetative thalli of the higher fungi consist of highly ramified systems of finer and finer branches (hyphae) which run widely through rotting wood or some other substratum, and which absorb nutrients for growth. The hyphae are referred to collectively as the mycelium. In the formation of the fruiting body the hyphae grow together to form a plectenchymatous pseudoparenchyma. The fruiting bodies consist of a dense mass of interwoven hyphae, held together by mucilage (Figs. 79A; 498; 512; 514). In certain species the walls of the hyphae in the fruiting body may become thickened, and following loss of water, a hard dense sclerotium develops (Fig. 77B).

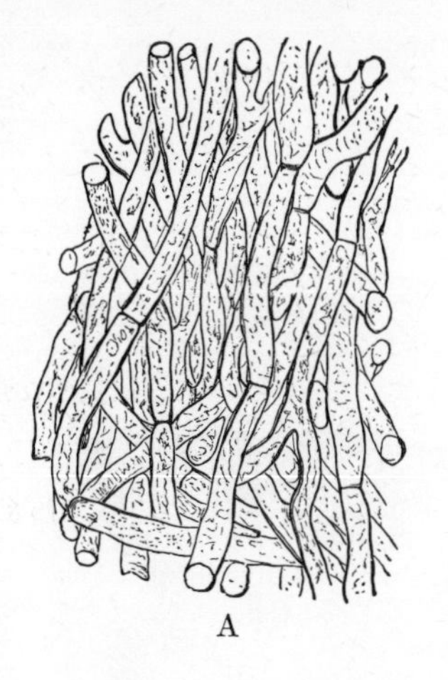

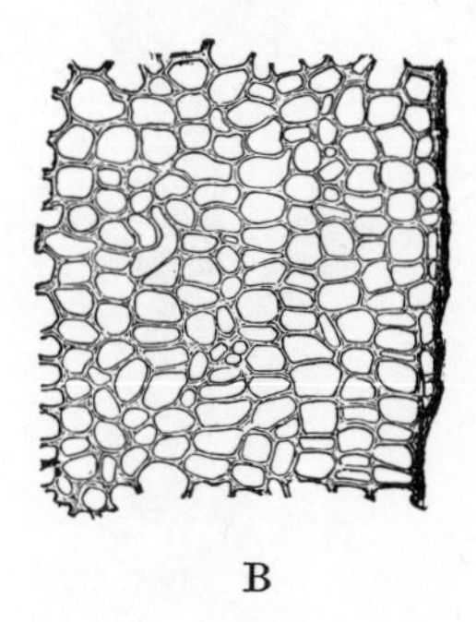

Fig. 78. A. Plectenchyma. Longitudinal section through the stalk of the fruit body of a lichen. **B.** Pseudoparenchyma. Section through the sclerotium of *Claviceps purpurea*. Both ×300. After Schenck.

D. Thalli composed of true tissue. In the most highly developed marine brown algae a true tissue is already found, the origin of which can be traced back to the activity of a single vegetative apex. In the simplest examples this apex is occupied by a single large apical cell whose descendants all remain joined together beneath one another (Fig. 79B–E). If the apical cell, as in the filamentous thalli, can cut off new cells only posteriorly (i.e. the apical cell has only one cutting face), and these are eventually transformed into a three-dimensional structure by growth and further division, cord-like thalli are formed like that of *Chorda filum*. Such thalli may also be flattened like leaves, as, for example, in *Dictyota* (Fig. 79A). Apical cells with several cutting faces are often found, as in *Fucus* (Fig. 440D), where the apical cell has five cutting faces. Such an apical cell cuts off new cells both posteriorly and laterally. These undergo further development and differentiation, leading to three different kinds of tissue, namely a central rib, ground and cortical tissue.

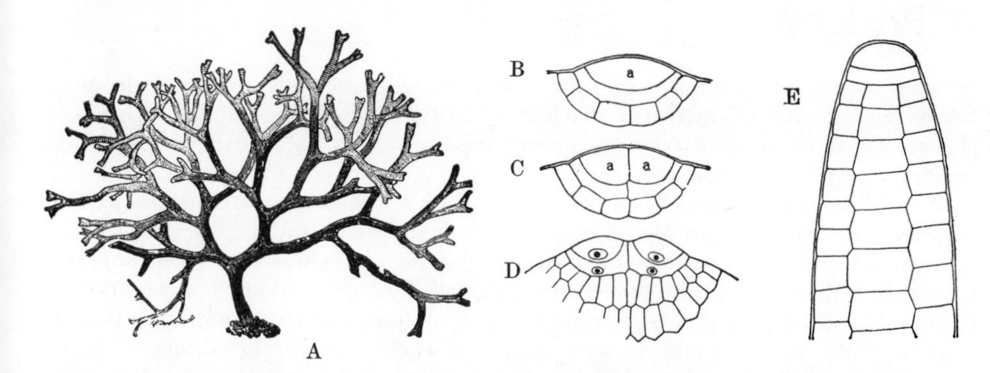

Fig. 79. *Dictyota dichotoma* (Phaeophyceae). **A.** Habit (*c.* ×½). **B.** The lens-shaped apical cell. **C.** and **D.** Longitudinal division of the apical cell leading to dichotomous branching. **E.** Longitudinal section through the thallus which becomes secondarily three-layered (×250). **A.** After Schenck. **E.** After de Wildeman.

The flat band-like thalli of the brown alga *Dictyopteris* develop not from a single apical cell, but from several initials forming a short apical border (Fig. 81B, C). This is an example of convergent evolution, since such an apical group represents a grade of organization displayed by most of the Cormophyta. Only in Pteridophyta is a single apical cell still frequently found, elsewhere in the Cormophyta a large group of initial cells is present at the growing point (see p. 89 seq.).

In the most highly developed Phaeophyceae the external differentiation is also remarkably far advanced, and has led to the production of leaf-like assimilatory organs, stem-like supporting organs and root-like organs for attachment (Fig. 437A, B, D, E). To distinguish them from the true leaves, stems and roots of the Cormophyta, all showing a very

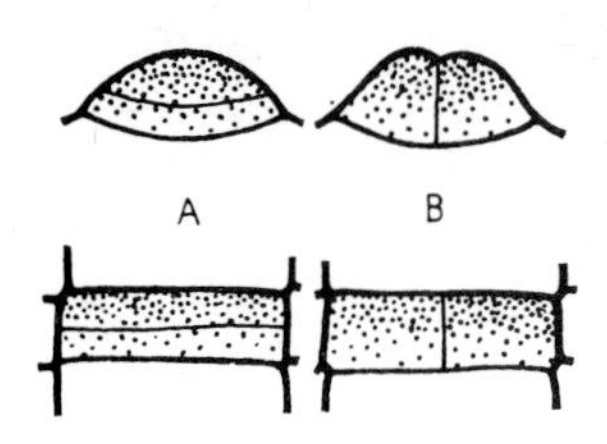

Fig. 80. The importance of polarity and the orientation of the spindle for the equality or inequality of the products of the division of the apical cell (above) or of any other cell (below). **A.** Unequal division perpendicular to the polarity gradient. **B.** Equal division in the direction of the polarity gradient. Diagrammatic, after Bünning.

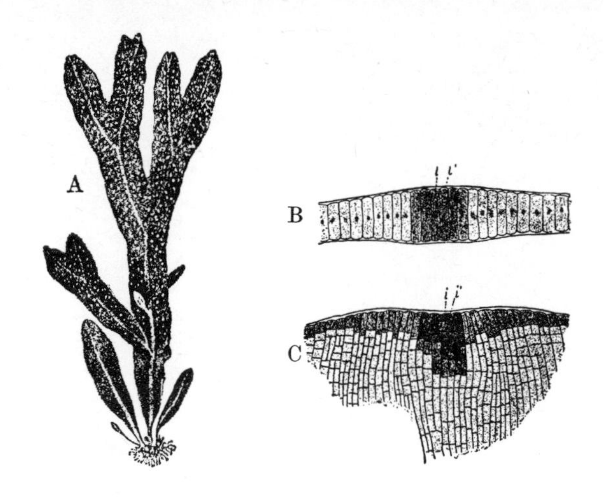

Fig. 81. *Dictyopteris polypodioides*. **A.** Habit. **B.** Apex from above, with group of apical initials (i, i′). **C.** The same from the side. The dark central group yield the mid-rib. **A.** *c.* ×½. **B.** and **C.** ×50. After Reinke.

much more complex anatomy, these organs are referred to respectively as phylloids, cauloids and rhizoids (but note the very much finer and more numerous rhizoids of the Bryophyta (Fig. 85) and of fern prothalli (Fig. 599)). Many species with attachment discs or lobes, or root-like rhizoids, are strikingly reminiscent of true Cormophyta. Some may become very large. The brown alga *Macrocystis*, for example, may reach over 120 m in length and is among the largest of plants.

III. Bryophyta

Between the parenchymatous thalli of the aquatic algae, developing from a single apical cell or from a marginal meristem, and the well-differentiated thalli of the higher land plants come those of the Bryophyta (mosses and liverworts). Their adaptation to life on land is in many respects still imperfect.

Many bryophytes are consequently limited in their distribution to conspicuously damp situations (e.g. the liverwort *Marchantia*, often around springs, and the rich moss-flora of the mist forest of tropical mountains). Nevertheless, as a consequence of their primitive organization they are capable of persisting through long dry periods in a condition of latent life, and because their epidermal tissue is for the most part a not very effective boundary layer, they can take up water over their whole surface from a saturated atmosphere and can utilize any dew which forms upon them. Consequently, many mosses are found even on places normally dry, such as rocks and walls, provided these are occasionally moistened by some form of precipitation.

In the thalli of bryophytes, which are usually several cells thick, the growing point is always apical and is often occupied by a single apical cell. In ribbon-like thalli (e.g. *Aneura* and *Metzgeria*), as in morphologically similar algae, the apical cell is wedge-shaped (Fig. 82B, C) and has two cutting faces. Cells are cut off from these faces alternately to the right and the left and are arranged obliquely in front of one another. They undergo further divisions and build up the thallus. The apparently purely dichotomous branching of liverworts with such growing points is, however, no longer dependent upon an equal division of the apical cell, as in *Dictyota* (Fig. 79C, D; 80B), but can be traced back to the development of a new apical cell which precedes the actual branching. This is derived from the abaxial portion of a cell recently cut off from the main apex (Fig. 82C, b).

The thallus of *Blasia pusilla* (Fig. 533) is lobed at the margins, as if leaf-like structures were beginning to emerge. The liverworts most complex morphologically (Jungermaniales

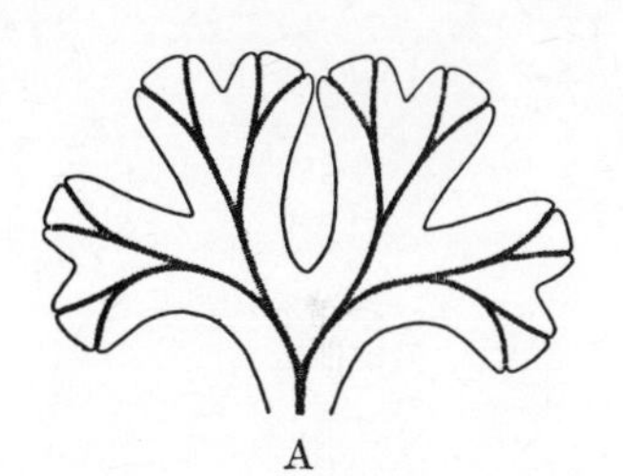
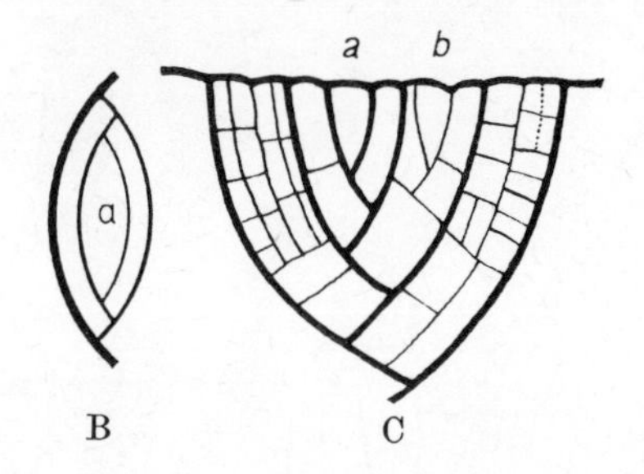

Fig. 82. A. *Riccia rhenana* (a liverwort). Dichotomously branched thallus of the land form. **B.** and **C.** Diagram of the growing point of *Metzgeria furcata* (a liverwort). **A.** View from the front of the apical cell (a) with its two cutting faces, and two segments cut from it. **B.** View from above at the initiation of a branch: a, apical cell of maternal axis; b, apical cell of the daughter axis arising secondarily in a segment cut from a. **A.** ×3. After Klingmüller. **B.** ×370. After Fitting. **C.** ×370. After Kny.

Fig. 533A–G), have branching thread-like stems bearing extremely delicate, nerveless leaves usually only one cell thick. Both stem and leaves are, however, only analogous to the corresponding organs in the Cormophyta, and they show no appreciable internal differentiation. In both mosses and liverworts the side branches originate beneath the leaflets.

Some liverworts and many mosses have acquired an upright habit of growth. Such upright thalli usually possess at their apices tetrahedral apical cells. The three cutting faces are directed posteriorly and they cut off segments basipetally in regular sequence (Fig. 84A, B). Each segment develops and undergoes further divisions by walls both periclinal (orientated parallel to the surface) and anticlinal (orientated perpendicular to the surface). These divisions not only contribute cells to the ground and cortical tissue but also lead to the formation of the primordia of a leaflet and lateral shoot. The leaflets of mosses are for the most part only one cell thick, and they generally grow at first from an apical cell with two cutting faces (Fig. 83A), but this eventually ceases to be active. Continued growth in area of the leaf follows from further divisions among the cells cut off from the apical cell. In the bog moss *Sphagnum* it has been shown that the typical unequal division of the cells in the developing leaflet into two chlorophyllose cells towards the apex and a hyaline cell towards the base (Fig. 83B, H; see also Fig. 542G) cannot take place while the apical cell of the leaflet is still active. Not until after the extinction of the activity of the apical cell do the rest of the cells become once again, and for a brief period, meristematic.

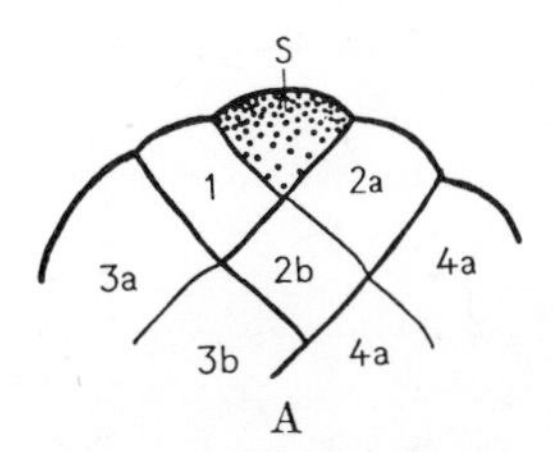

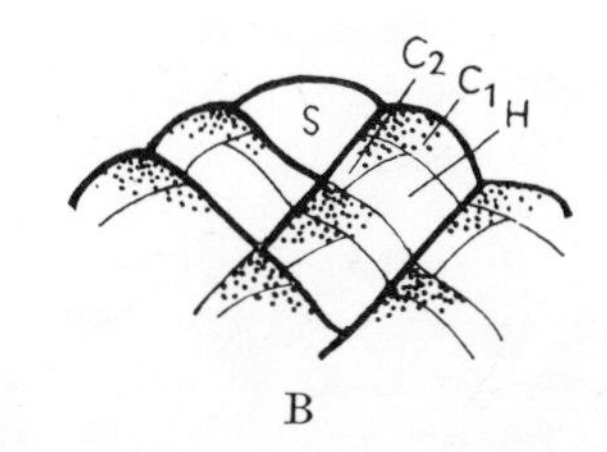

Fig. 83. Diagrammatic representation of the sequence of division in a leaf of *Sphagnum*. **A.** The apical cell (S) with its two cutting faces cuts off segments to the right and left (1, 2, 3, 4). Each of these cells divides again to give rise to two more or less equally sized rhomboidal cells (2a, 2b; 3a, 3b; 4a, 4b). **B.** After the meristematic activity in the apical cell ceases each rhomboidal cell undergoes two unequal divisions yielding two chlorophyllose cells (C_1 and C_2) and one hyaline cell (H). (Cf. Fig. 542.) ×150. After Bünning.

Since the cells are cut off from the three faces of the apical cell in regular rotation, the leaf primordia are arranged in a spiral along the stem. It would also be expected that each fourth leaf would stand directly above the first (Fig. 84B, D), and in some mosses (e.g. *Fontinalis, Barbula paludosa*) this arrangement is actually found. Usually however, each newly formed segmenting wall moves a little forward in the direction of the leaf spiral

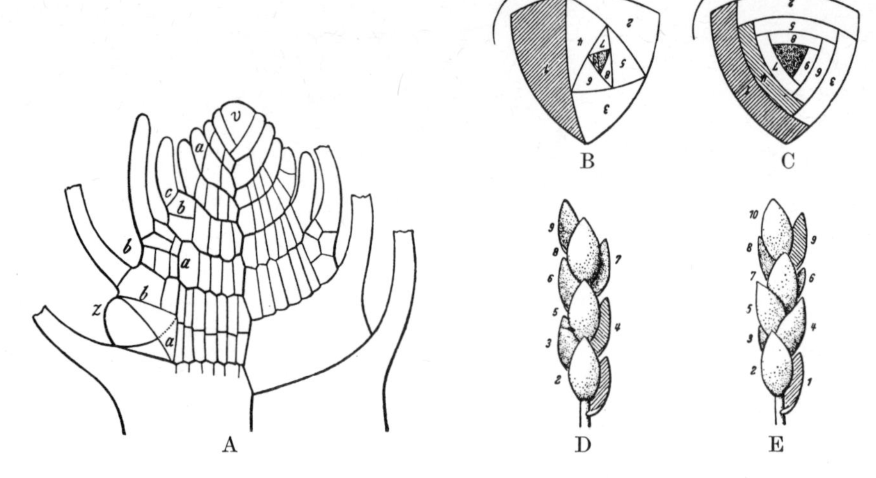

Fig. 84. **A.** Longitudinal action through the apical region of a stem of *Fontinalis antipyretica* (a moss): v, apical cell with three cutting faces. Each segment becomes divided by a periclinal wall (a) into an inner and an outer (cortical) cell. The latter yields cortical tissue and a leaf. Side shoots arise sporadically beneath the leaves as a result of the development of lateral apical cells, also with three cutting faces (z). **B.** to **E.** Leaf arrangement in mosses (diagrammatic). **B.** Apical regions of *Fontinalis antipyretica* from above. The cutting off of segments in regular rotation by the apical cell leads to a regular three-ranked arrangement of the leaves (**D.**). **C.** The same of *Eurhynchium rusciforme*. The segments are cut off similarly by the apical cell, but the more rapid differentiation leads to a spiral arrangement of the leaves (**D.**). (Cf. p. 123 seq.) **A.** ×100. After Leitgeb. **B.** to **E.** After Stocker.

(owing to asymmetrical division), so that a more complex spiral phyllotaxy results (Fig. 84C, E).

Roots are lacking in the Bryophyta, but in their place most possess delicate rhizoids. In the liverworts they consist of simple hairs, cut off by a single transverse wall and confined to the lower surface of the thallus. In the mosses they consist of filaments, one cell broad and usually freely branched. Since the uptake of water takes place over the whole surface of the thallus, the rhizoids principally serve to anchor the plant to the soil.

Primitive conducting strands occur in the Bryophyta. They consist of more or less elongated cells arranged centrally. They are found in some liverworts, forming a kind of midrib (Fig. 82A), and in the most highly developed mosses they occur in the centre of the stem (Fig. 85 l) and often also in the several-layered nerve present in the leaf (which is otherwise only one cell thick). The leaf strands join on to the strand in the stem. The

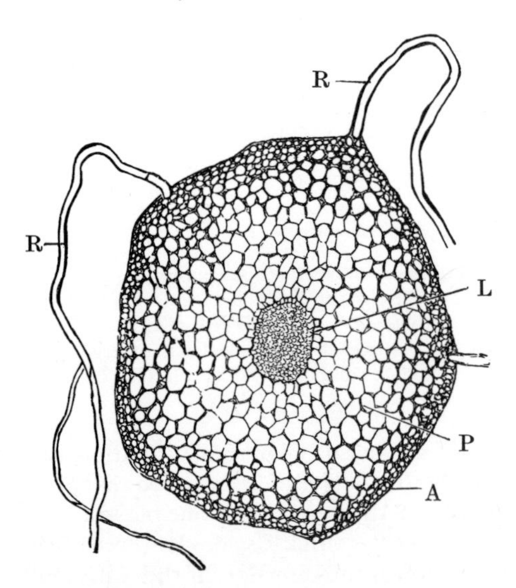

Fig. 85. Transverse section of the stem of *Mnium*: L, conducting strand; P, cortical tissue; A, thick-walled outer layer of cells; R, rhizoid. ×90. After Strasburger.

latter reaches its highest development in the Polytrichaceae, where the centre of the stem is occupied by thin-walled, prosenchymatous elements with oblique end walls (responsible for up to 60 per cent of the water conduction) and thick-walled elements which have a structural function. Adjacent to this central core are cells containing protein and carbohydrate, and these possibly serve to transport the products of assimilation. An undoubted epidermal tissue with pores to facilitate the exchange of gases is only rarely developed in the Bryophyta (e.g. in *Marchantia* (Fig. 528) and the sporogonia of many mosses (Fig. 541)).

IV. Cormophyta

A. Thallus and cormus.[19] Since, as we have seen, the vital processes can go on only when adequate water is retained in the protoplasm, the development of large aerial plant bodies was bound up with a number of far-reaching modifications. The fundamental and most important property required in this respect was the improved stabilizing of the amount of water in the plant. This was achieved by the impregnation of the permeable cellulose membranes of the outer layers of cells with the water-repelling substances cutin and suberin. The outer layer of cells was thus transformed into a barrier limiting transpiration, and this specialized layer, the cells of which frequently have the outer walls heavily thickened and cutinized, is referred to as the epidermis (Fig. 97A, B, C). Nevertheless, in order not to prevent completely the gas exchange essential for life, this primary insulating tissue had to be furnished with closable pores or stomata. Furthermore, the requirement of a stabilized water supply to the tissues necessitated the development of specialized systems for the uptake and conduction of water.

As a matter of fact, the step from aquatic to terrestrial habit has, as we have seen, already been taken by several members of the Protophyta and Thallophyta. The success of this transition must, however, remain limited, both with regard to the range of colonizable habitats and the size of life forms developed. This is inevitable from the nature of a thallus, which, having originated in and being adapted to life in water, permits the free exchange of water and water vapour over its entire surface. Thus colonies of *Nostoc*, the thalli of lichens, and bryophytes can rapidly take up rain and dew, and even mist, but they cannot hold on to the absorbed water when they are later in contact with incompletely saturated air. They are completely dependent upon the environmental conditions from one moment to the next, and consequently their existence passively fluctuates between periods of active life when water is adequate and periods of latent life when it is not (poikilohydric organization, cf. p. 71).

Finally, since on land the supporting effect of the water is no longer available, special structural elements had to be developed, which would at the same time facilitate the effective and ordered distribution of the photosynthetic organs. For this purpose the differentiation of the plant body into a stem, functioning as a support, and leaves, serving principally for gas exchange and photosynthesis, proved very satisfactory (Fig. 131D), a development facilitated by the hardening of the elastic cellulose of cell walls by the deposition of lignin (see p. 67).

Thus the transition from an aquatic to a terrestrial habit, and the accompanying stabilizing of the water content, brought with it an extensive modification of the plant body of the Thallophyta. Only very gradually, of course, and in small steps did this lead to the level of organization present today. Consequently, between the cormus of the higher plant, typically divided up into root, stem and leaf, and the thallus of the lower there are transitional forms of every kind.

B. The telome theory.[20] The earliest land plants (Psilophytatae, see p. 562 seq.) still resemble quite closely in their external form the dichotomously branched thalli of many of the more highly organized brown algae. In place of a single apical cell, still found at the summit of the apparently dichotomizing axes of the living Psilotales, the Psilophytales often grew from apical meristems consisting of numerous and more or less identical cells.

Their aerial parts were not yet differentiated into shoots and leaves, but they consisted, as many marine brown algae still do today, of quite uniformly forking shoot systems, not yet displaying any localization of function (Fig. 553A). This does not, of course, mean to say that the land plants were derived from the brown algae; of the immediate ancestors of the first land plants as good as nothing is yet known.

Since the morphological terminology used for the differentiated organs of the higher plants cannot be used for the uniform branchlets of the Psilophyta, they are referred to as *telomes*, or more strictly as primordial telomes. Each telome consists of a central strand of prosenchymatous lignified cells, serving as a water-conducting and structural tissue (primordial stele or protostele; Fig. 582B), and a parenchymatous sheath of cortical tissue, delimited on the outside by a cutinized epidermis.

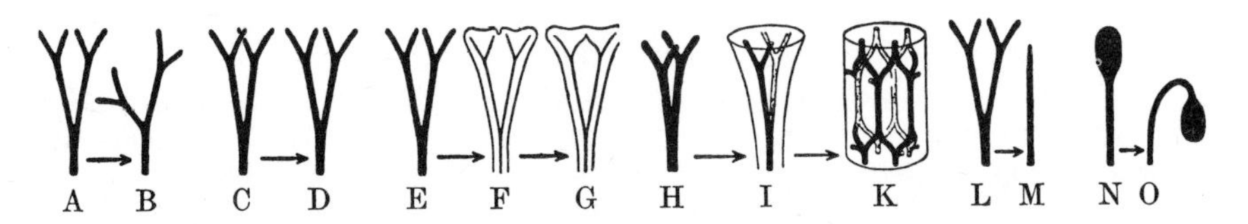

Fig. 86. Diagrammatic representation of the five basic processes which, according to the telome theory, have led to the structure of plants as we see them today. After Zimmermann. Explanation in text.

According to the telome theory, the typical organs of the highly differentiated cormophyte—reproductive organs, leaves and possibly also the roots—are said to be derived by a series of 'elementary processes' from the primordial telomes. According to this view, the most important 'elementary processes' which would have led to the external form of the typical cormophyte are: overtopping, planation, fusion, reduction and involution (Fig. 86).

a. Overtopping (Figs. 86A, B; 87A, B) might have been the means by which morphological differentiation and division of functions was introduced into a system of telomes originally all exactly alike. A system of equal branching would thus have led to a main axis, with a supporting function, bearing lateral photosynthetic axes, or systems of axes. This differentiation might in turn have led to stems and leaves. The overtopping main shoot retains a greater growth potential than the overtopped sister shoot, and it consequently becomes the leader, while the overtopped shoot (and any system developing from it) comes to be a laterally placed appendage (Fig. 86A, B).

The process of overtopping can be readily envisaged by comparing the shoot systems of the recent club mosses *Lycopodium selago* and *L. cernuum* (Fig. 87C, D). Both species grow from groups of apical initials, and from time to time the apices dichotomize. In *L. selago* both arms of the dichotomy grow with more or less equal vigour, so that an apparently ancient form of growth results (isotomy; Fig. 87A, C). In *L. cernuum*, on the other hand, one arm regularly overtops the sister arm, so that an apparently monopodial (see p. 129) habit of growth results (anisotomy; Fig. 87B, D).

b. Planation. The holders of the telome theory regard the process of planation as the second important step in the development of leaves. It envisages the telomes of the side branches coming to lie in one plane (Fig. 86B).

c. Fusion. The theory holds that telomes in a three-dimensional arrangement may fuse together (Fig. 86I, K), as well as those which have come into one plane as a result of planation (Fig. 86F, G). In the first case an appreciable 'stem' is produced, no longer enclosing a single central protostele, but two or more bundles of conducting and mechanical tissue (vascular bundles). As a consequence, the rigidity of the plant body is considerably enhanced. On the other hand, if the telomes previously brought into one plane by planation subsequently join up among themselves, we arrive at a typical 'leaf' (macrophyll) with dichotomously branched 'venation', between which is stretched the assimilatory tissue (Figs. 86K; 177; 583; 584). The integuments of ovules are considered to have arisen in the same way (Fig. 617).

d. Reduction. The reduction of individual telomes may lead to extreme simplification or even to complete suppression (Fig. 86L, M). It may be that the small one-nerved leaflets (microphylls) of the Lycopodiatae and species of *Asteroxylon* (Figs. 87C, D; 556B; 563) arose by a strong reduction of this kind, but the derivation of the microphyll is very controversial. Another view is that they should be regarded as organs *sui generis* which have arisen in early times as emergences from the cortical tissue of the still leafless stems of land plants resembling *Rhynia* (Fig. 553). The microphyll, because of its phylogenetically earlier appearance, is regarded as the authentic antecedent of the macrophyll.

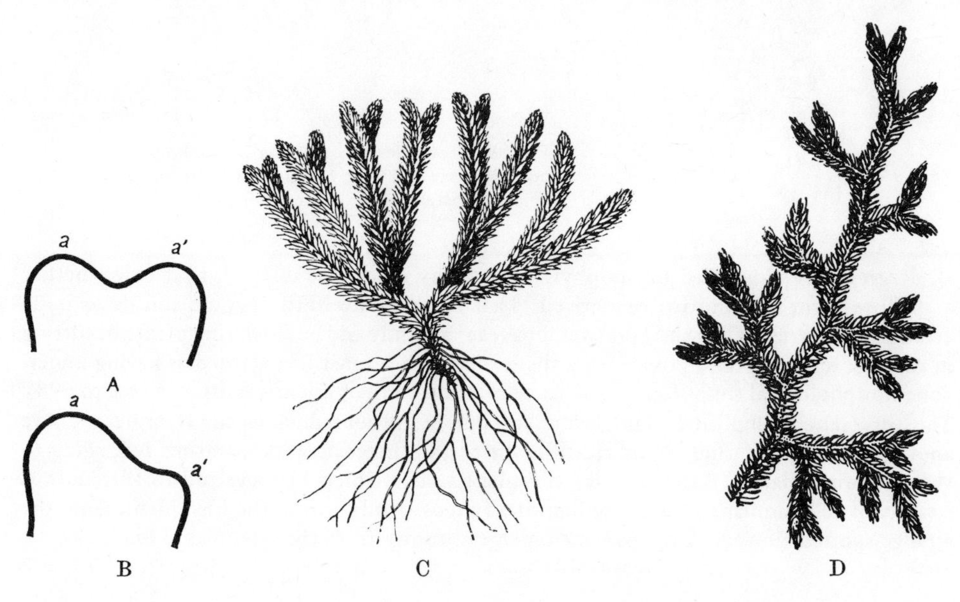

Fig. 87. **A.** and **B.** Isotomous (**A.**) and anisotomous (**B.**) branching of the growing point (diagrammatic): x, termination of parent axis; a and a', the primordia of the two branches. In **B.** a is promoted in relation to a'. **C.** *Lycopodium* (*Huperzia*) *selago*. The branch system shows cruciate dichotomy. Some of the branches bear sporophylls. **D.** *Lycopodium cernuum*. Anisotomous branching ('overtopping'), leading to a pseudo-monopodial growth form. **A.** and **B.** After Troll. **C.** $\times ^1/_3$. **D.** $\times ^1/_3$. Both after Troll.

e. Involution. This is the consequence of unequal growth on opposite sides of a developing telome, and plays a large part in the morphogenesis of the reproductive organs (Figs. 86N, O; 585B, C). This 'elementary process' is the outcome of the two sides of the shoot, in ancestral forms of equal status, having undergone unequal development. Usually it is the upper and lower sides that are affected, leading to a structure with dorsiventral symmetry.

According to the telome theory, all the 'elementary processes' here described have taken place several times over and independently of one another in the course of evolution. Particularly striking are the Primofilices (see p. 580) of the Middle and Upper Devonian. They show a range of forms which, interpreted according to the telome theory, lie almost without discontinuity between the early land plants and the morphologically complex plants of today. Despite these facts, however, the theory, as all concerned with phylogeny, is necessarily largely speculative in character.

C. The gametophyte of the Cormophyta. In the Cormophyta the life-cycle is heterophasic (see Fig. 33B), and the haploid sexual generation remains thallose. In the Pteridophyta the thallus of the gametophyte, here termed a prothallus, is usually very simple in structure, and is filamentous (Figs. 600; 601A), or flattened (Figs. 88; 599; 601B) and attached to the soil with simple rhizoids, or freely branched (Fig. 577A). The prothallus contains chlorophyll and is completely independent. In the Spermatophyta the thallose gametophyte is very much simplified and, being immersed in the tissue of the sporophyte, is invisible from without. It is also colourless and depends upon the sporophyte for its nutrition (see p. 654 seq.).

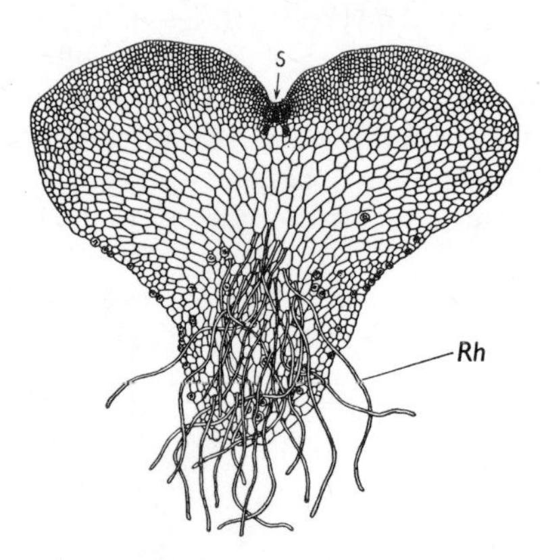

Fig. 88. Prothallus of a member of the Polypodiaceae viewed from beneath: Rh, rhizoid; S, apical meristem. ×20. After Haupt.

D. Regressively simplified Cormophyta. In many cases it is difficult to decide whether a thallose form is primitive or derived. That is, there are both upgrade and downgrade evolutionary series. Thus the Podostemonaceae, a family of the flowering plants occurring in tropical waterfalls and possessing a thallus-like habit, must be regarded as having undergone morphological simplification as an adaptation to their aquatic existence (see p. 698). The very much simplified plant bodies of our indigenous duck-weeds (Lemnaceae) are another example of such simplification. In certain heterotrophic parasites belonging to the flowering plants (Rafflesiaceae) the plant body, which has reverted to filamentous strands of cells similar to the mycelium of a fungus, grows within the host plant. Only the strange spotted flowers of the parasite emerge abruptly from the host plant (Fig. 217).

Section three: **The characters of tissues**[21]

(The histology of the Cormophyta)

Only in the most primitive of the multicellular thallophytes does the whole plant body consist of cells identical in form and function and with no division of labour, so that every cell still performs all the vital functions.

A striking example of this primitive condition is provided by the large leaf-like thallus of the sea-lettuce (*Ulva stenophylla*, Fig. 89). Although this reaches 30–50 cm in length, almost all the cells (with the exception of those of the holdfast) can give rise to zoospores or gametes at reproduction, so that ultimately only an empty skeleton of cell walls remains.

However, in the more highly organized thalli there is already, as we have seen, a localization of function. An embryonic growing zone (consisting of an apical meristem or a line of initial cells), reproductive organs (sporangia and gametangia), and regions of photosynthesis and food storage are present. In the most highly developed brown algae even special conducting tracts may arise, leading from the places adapted for photosynthesis (such as phylloids and assimilators) to the 'stems' (cauloids). The latter persist for several years, show a differentiation into 'pith' and 'cortex' and serve as regions of food storage.

As a rule similar cells are grouped together, and an association of similar cells is called a tissue. The different tissues are distinguished from one another by the form and content of the cells composing them, and by the nature of the cell walls. The more highly organized a plant is, the greater the number of different kinds of tissues it contains. There is thus an objective criterion for the assessment of the level of organization of an organism, namely, the degree of diversity of its parts, that is, the number of the different kinds of cells and tissues participating in its construction. This criterion naturally fails where evolution has led to a secondary simplification of the plant body (see p. 86).

The differentiation of a range of new kinds of tissues was necessarily bound up with the three new tasks which the land plants had to undertake, namely, 1. the restriction of the water loss, 2. the uptake, conduction and excretion of water from the soil or from rain, and 3. the mechanical strengthening of the plant body, which on land would lack the support of the water. The new tissues developed to meet these needs are either completely lacking in the Thallophyta, or only the first suggestions of them are present in a few particularly highly developed forms. In addition to the tissues already present at the second grade of morphological complexity, namely, 1. embryonic tissue, 2. ground tissue and 3. reproductive tissue (in the sporangia and gametangia), the following new tissues had therefore to be developed by the Cormophyta, 4. boundary tissue, 5. tissue concerned with water uptake, 6. conducting tissue, 7. mechanical tissue and 8. excretory tissue.

In connection with the new demands upon them and the large number of specialized cells required, the tissues of the Cormophyta are not infrequently composed of several kinds of cells. Thus in most of the epidermal tissue of the aerial parts of the plant there must necessarily be openings, called stomata, which facilitate gas exchange. These are formed by specially differentiated cells (see p. 97 seq.). Similarly, in the conducting tissue there occurs differentiation into those elements which conduct water and those which transport organic materials. In these cases, therefore, we are no longer concerned with uniform tissues, but with tissue systems, constructed of several different types of cells, and of both morphological and physiological significance. Functional units in the plant

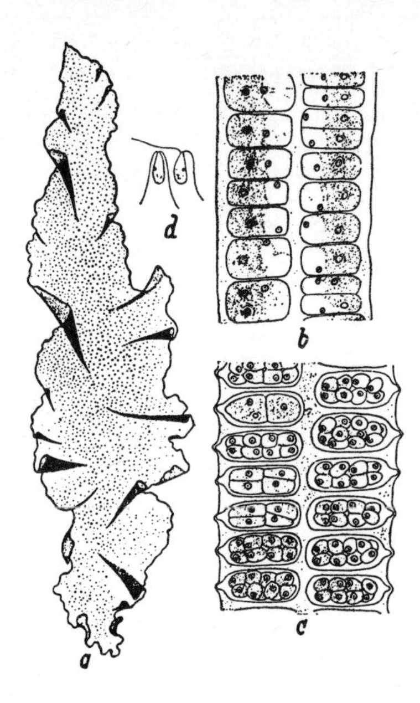

Fig. 89. **a.** Thallus of *Ulva stenophylla* (Chlorophyceae). **b.** Transverse section of the vegetative thallus. **c.** The same of the gametangial region. **d.** The biflagellate isogametes. **a.** ×½. **b.** and **c.** ×120. After G. M. Smith.

thus become characterized by their structural peculiarities. On the other hand, individual cells with specialised functions may now and again occur scattered in a large tract of tissue composed of quite different cells. Such isolated cells are referred to as *idioblasts*.

In the Cormophyta it is possible to distinguish on ontogenetic grounds two main groups of tissues; these are the embryonic tissues and the mature permanent tissues.

I. The embryonic tissue or meristem[22]

The fertilized egg cell, or zygote, of the higher plants develops into an embryo which consists at first exclusively of cells that are embryonic, i.e. capable of division. These cells are thin-walled, rich in cytoplasm and to all appearances very similar to each other (Fig. 90). The future axis of polarity is established very early, usually with the first division of

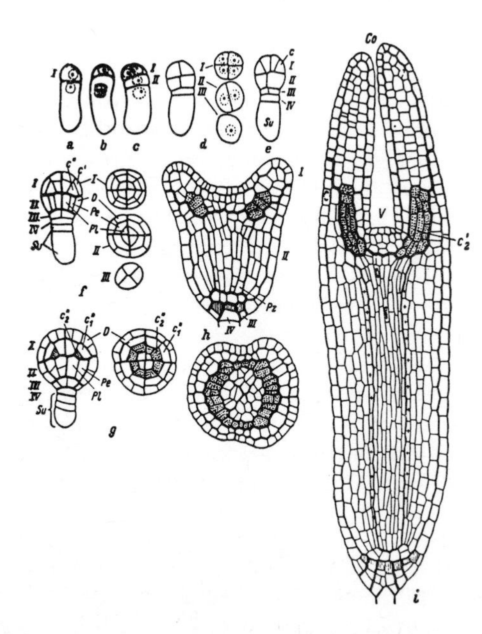

Fig. 90. Embryogeny of *Biophytum dendroides*. **a.** to **e.**, The first regular divisions leading to the suspensor (Su) and the four tiers of embryonal cells (I to IV). **f.** and **g.**, Differentiation of the protoderm (D) and division of the second tier of embryonal cells into an outer region (Pe) and an inner region (Pl) by means of periclinal walls. **h.** Development of the primary root and cotyledons. **i.** Mature embryo derived from the four tiers of embryonal cells. V, apical meristem of shoot. ×100. After Noll, from Kühn.

the zygote. At a later stage the apical pole gives rise to the shoot apex with the primordia of the first leaves, while the basal pole yields the first root and frequently in addition an haustorial organ concerned with the nutrition of the embryo, the suspensor (and, in the pteridophytes, the foot).

The development of the embryo in *Equisetum* and the ferns, which still, like the Bryophyta, grow from a single apical cell, follows a path different in detail from that seen in the Lycopodiatae and the seed plants, but consideration of this is deferred until the systematic section.

As soon as the embryo has become a little larger, cell divisions become confined to the extreme tips of the two poles, and apical growth ensues at both the lower root pole and the upper shoot pole. Thus at a very early stage we arrive at the differentiation of the plant body, characteristic of the growth of the higher plant, into embryonic tissue remaining capable of cell division, and mature tissue serving specialized functions, in which the cells have lost temporarily or permanently their capacity to divide. Primary embryonic tissues are those which are derived ontogenetically directly from the tissue of the embryo, and they are referred to as primordia or promeristems. Within a region which has already become transformed into permanent tissue, large complexes, plates or strands of cells may persist in a promeristematic condition. These can conveniently be termed residual meristems to distinguish them from the apical meristems.

The cells of many permanent tissues sooner or later again become capable of division, and in this way secondary meristems can arise. If very small groups of cells or even individual cells retain their initial activity, or after a temporary pause revert to renewed meristematic activity, they are referred to as meristemoids.

In addition, even the cells of tissues, which in the normal way remain unchanged throughout life, can often be stimulated to divide by experimental means, such as by isolating them from the other cells of the tissue. This makes it clear that they have remained potentially meristematic. Only very thick-walled mature cells and, of course, all those elements which do not become functional until the death of the cell (such as mechanical elements, those concerned with the conduction of water, and cork) permanently lose their capacity to divide.

A. Apical meristems. The cells of the promeristems lying at the tips of the shoot and root are always, like the embryonic cells from which they are directly descended, relatively small and nearly isodiametric, i.e. of a more or less uniform diameter whatever the direction of measurement. Their walls, mostly set at right angles to each other, are very delicate and poor in cellulose, consisting principally of protopectins. All the cells are closely applied to each other, leaving no spaces. The lumen is filled with dense protoplasm, with a relatively large nucleus (Fig. 6).

Microscopic examination shows the apical meristems of both the shoot and the root to vary in shape from more or less conical to flattened domes (Figs. 91A, C, D; 92; 132A, B), meristems of the first kind often being referred to as apical cones. The group of cells, situated at the very tip in the shoot, but in the root somewhat immersed, from which the mitotic activity of the apex takes its origin, are termed the apical initials (and are sometimes referred to collectively as initial complexes).

It can be readily seen that, within the apex as a whole, the mitotic activity of the initial cells is by no means the most conspicuous. On the contrary, the descendants of the initial cells must themselves undergo numerous divisions before a new division of the initial cells can occur. It is possible to demonstrate by feeding apices with tritiated thymidine (which becomes incorporated into deoxyribonucleic acid) and subsequently preparing autoradiographs that in both shoot and root apices few divisions actually take place where one might expect the greatest mitotic activity. The existence of this 'quiescent centre' in apices has given rise to much misunderstanding.

The further development and differentiation of the descendants of the initial cells is strongly influenced by their position. In the lower plants it is commonly found that an unequal division, either of an initial cell or of one of its descendants, determines the fate of the cells concerned (Figs. 71; 72; 73). In the higher plants, however, differentiation is more usually of groups of cells, and is dependent upon influences from adjacent tissues

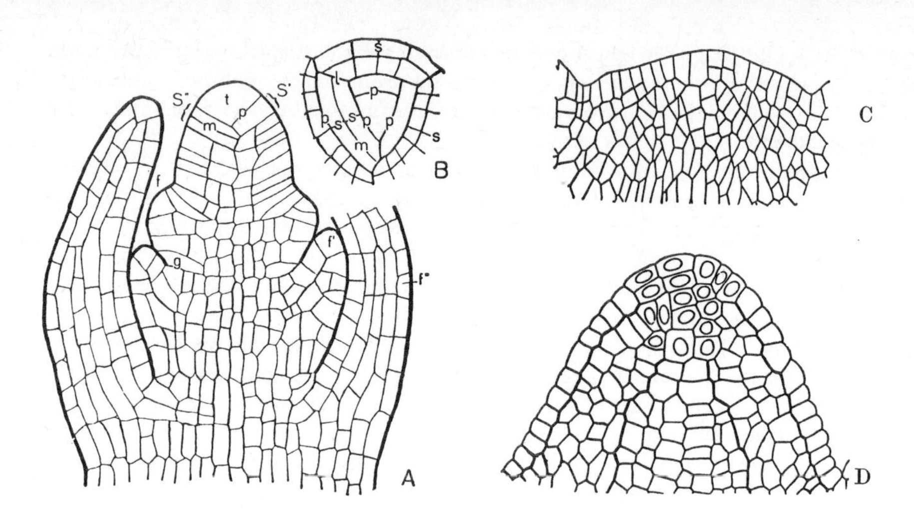

Fig. 91. A. and **B.** Vegetative apex of *Equisetum*. **A.** Longitudinal section. **B.** View from above. t, apical cell which cuts off segments in rotation (S′ and S″) to the posterior by an oblique wall (p). (These segments later become subdivided by anticlinal walls (m)); f, f′ f″, leaf primordia; g, initial cell of a lateral branch; l, radial wall separating segments. **C.** and **D.** Vegetative apices of *Lycopodium selago* (**C.**) and *Sequoia sempervirens* (**D.**). The superficial cells (in **D.** shown with prominent nuclei) divide by both periclinal and anticlinal walls. There is as yet no separation into tunica and corpus. **A.** and **B.** ×180. After Strasburger. **C.** ×120. After Haertel. **D.** ×140. After Cross.

(dependent differentiation). Hence differentiation into organs usually precedes differentiation of cells and tissues (Figs. 132B; 150).

The apices of the stem and the root must necessarily be considered separately. The shoot apex produces lateral outgrowths immediately beneath the growing point (Fig, 132B), which subsequently grow out to form the leaves and side shoots. The root apex by contrast is enclosed in a special sheath, the root cap, and remains unbranched.

a. The shoot apex. As in most of the marine algae and the Bryophyta, the promeristem in *Equisetum* and most of the ferns still originates from a special apical cell lying at the growing point. This cell often has the form of a tetrahedron (a three-sided pyramid), with the convex basal wall on the outside, so that the three cutting faces are directed backwards (Fig. 91A, t). Seen from above, such a single initial cell appears as an equilateral triangle (Fig. 95B). Apical cells with four and five cutting faces occur sporadically. The apical cell cuts off descendants, termed segments, basipetally in spiral sequence. Subsequently these divide further, the first divisions often following a regular pattern (Fig. 91A, m).

In those Pteridophyta with a single apical-initial the leaf primordia also usually develop from a single initial, in this case wedge-shaped and with two cutting faces (Fig. 91A, f, f′). The primordia of the lateral buds (Fig. 91A, g) each originate from a cell which becomes a new apical initial.

In many Pteridophyta, especially in the Lycopodiales, as well as in the majority of the gymnosperms, the promeristem possesses no clear apical cell. Instead there is a group of initial cells at the growing point, capable of dividing both anticlinically (perpendicularly to the surface) and periclinally (parallel to surface) (Fig. 91C, D). Finally, in the angiosperms, and in some of the more advanced gymnosperms, the initial cells are arranged in several storeys. Only the innermost group divides both periclinally and anticlinally, thereby building up the ground mass of the apical meristem, referred to as the corpus. The upper layers divide only anticlinally and contribute principally to surface growth, forming a 'tunica' which ensheaths the 'corpus' (Fig. 132B).

By definition the concepts 'tunica' and 'corpus' have merely descriptive significance, and they imply nothing in relation to the nature of the tissues into which these cells and their descendants ultimately differentiate. The tunica-corpus concept is thus directly opposed to the earlier generally accepted histogen concept. This was based on the false assumption that the fate of individual cells was already determined as they were formed in the apex. The histogen concept is in fact only applicable to the outermost layer of the tunica which eventually becomes the primary boundary layer, and can accordingly be referred to as the protoderm or dermatogen.

A single-layered tunica is found in some gymnosperms (e.g. in *Thujopsis* and *Phyllocladus*), in most monocotyledons (oat, wheat, maize) and in many species of Cactaceae. Over half of the dicotyledons investigated have two-layered tunicas. Multilayered tunicas have been reported in the Compositae and Hippuridaceae, and occasionally elsewhere. A nine-layered tunica is said to occur in *Oxycoccus macrocarpus*. The number of layers in the tunica can not only vary within one and the same species (e.g. in *Silene maritima* between one and four), but it can also change during ontogeny, possibly in relation to the change in the width of the apex in the transition to the flowering condition.

Experiments in which polyploidy has been induced by colchicine (see p. 36) have shown that where the tunica is several layers thick each layer possesses its own group of initial cells. In the thorn apple (*Datura*), for example, polyploidy can arise independently in each of the two layers of the tunica or in the corpus. In this way a so-called chimaera (see p. 311) is formed, in which cells derived from the tunica have nuclei of a size and number of chromosomes different from those of the cells derived from the corpus (Fig. 92). Such chimaeras are referred to as periclinal chimaeras. Artificially produced poly-

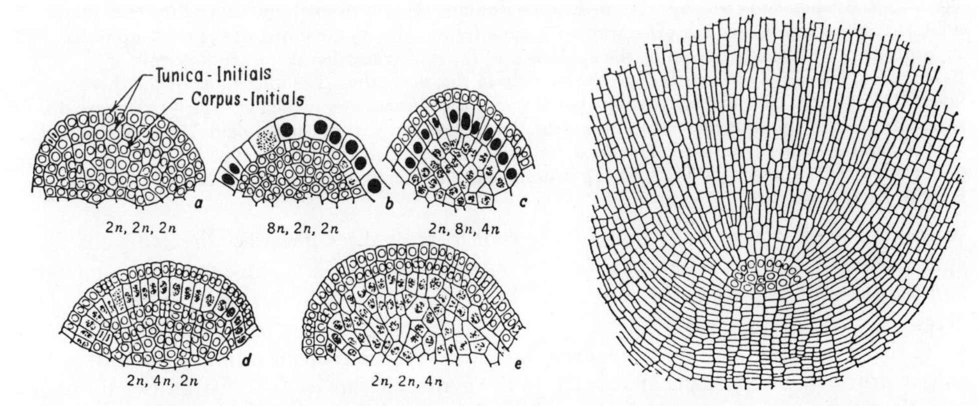

Fig. 92. Vegetative apices of *Datura*. **a.** Normal diploid plant. **b.** to **e.** Various cytochimaeras produced by treatment with colchicine, in **b.** the outer layer of the tunica (protoderm) is 8n, in **c.** the second layer of the tunica is 8n, in **d.** the second layer of the tunica is 4n, in **e.** the corpus is 4n. ×80. After Satina, Blakeslee and Avery.

Fig. 93. Longitudinal section through the root tip of *Pinus cembroides* var. *edulis*. The body of the root and the root cap are generated from the same group of initial cells (distinguished in the drawing by the insertion of the nuclei). ×100. After Chamberlain.

ploidy has shown, too, that more than one initial cell occurs in each layer. Thus, besides the periclinal chimaeras there always appear as well the so-called sectorial chimaeras, in which the polyploidy remains confined to individual sectors of the shoot. This indicates that the different sectors arise from their own initial cells. If there is a single initial cell in each layer, and the polyploidy arises, not in the initial cell itself, but in a descendant, then the polyploid sector is left behind in the longitudinal growth of the shoot. It becomes replaced towards the apex by the normal diploid descendants of the initial cell of the layer concerned (mericlinal chimaeras).

b. The root apex. The embryonic cells of the root apex need special protection, since the root is forcing its way into the earth. This is provided by the root cap or calyptra, which encloses the actual apical cone like a finger-stall. The outermost (and oldest) cells of the cap become mucilaginous and eventually detached, but new cells are simultaneously being added to the cap by the root meristem.

In most of the Pteridophyta a terminal apical cell (Fig. 94A, t), tetrahedral in form, occupies the summit of the growing point of the root as of the shoot. Besides the

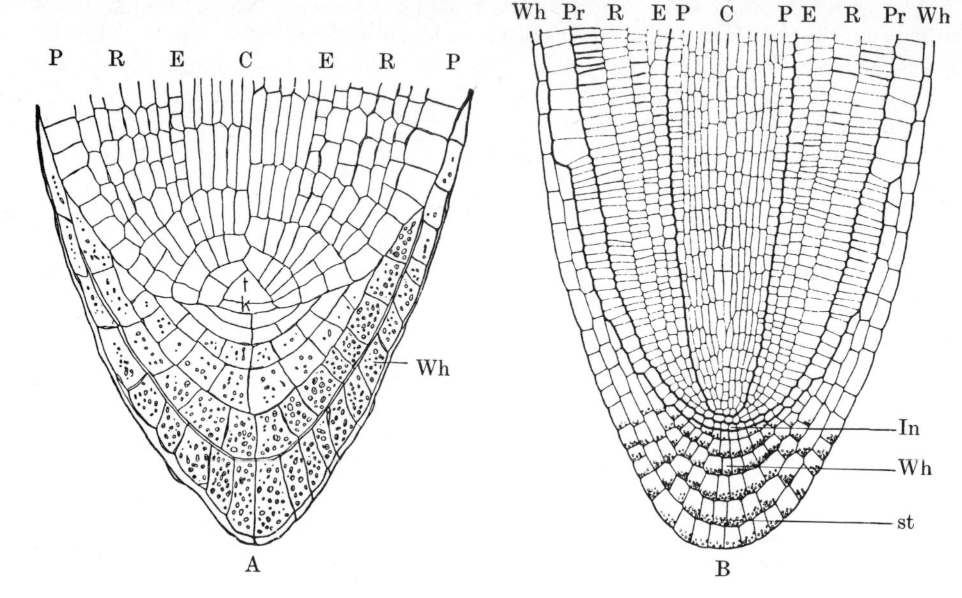

Fig. 94. Root apex and root cap. **A.** Median longitudinal section through the root of the fern *Pteris cretica*: t, apical cell; k, cell cut off towards the apex functioning as the initial of the root cap (Wh); Pr, protoderm, eventually becoming the epidermis of the root (rhizodermis); E, endodermis.
B. Median longitudinal section through the root tip of *Brassica napus*. The outermost of the three layers (In) of initial cells (dermato-calyptrogen) yields the dermatogen (which eventually becomes the rhizodermis), and the root cap or calyptra (Wh). The layer immediately above yields the cortex of the root (R) and the endodermis (E). The third, innermost layer yields the central cylinder (C) and pericambium (P). The cells of the root cap contain statolith starch (st). **A.** ×100. After Strasburger. **B.** ×50. After Kny.

segments cut off towards the body of the root from its three inwardly directed walls, the apical cell also cuts off others towards the outside (Fig. 94A, k). These segments undergo further divisions and form the root cap. Thus the apical cell in this case has four cutting faces.

In the gymnosperms and the angiosperms, however, the growing point of the root is no longer provided with a single apical cell. In the gymnosperms we find instead two more or less clearly defined groups of initial cells adjacent to one another (Fig. 93). The inner group divides by both anticlinal and periclinal walls, and contributes to the body of the root. The outer meanwhile, by at first predominantly periclinal divisions, gives rise to the cortical tissue and the root cap, in this case not clearly delimited at the growing point from the cortical tissue. Finally, in the angiosperms, at the tip of the root, as at that of shoot, we frequently find a stratified growth centre consisting of several independent groups of initial cells. The various permanent tissues (root cap, epidermis, cortex and central cylinder; see p. 171 seq.) develop by anticlinal and periclinal divisions from this centre, the manner of their development differing in the different systematic groups. For example, in the roots of grasses, the outermost layer of the promeristem (protoderm), which gives rise to the epidermis of the root (rhizodermis), coalesces at the apex with the meristematic layer beneath, from which the cortex is derived, forming a single group of initials. External to this lies the calyptrogen, the meristematic layer giving rise to the root cap. In most of the dicotyledons, however, the latter is derived by periclinal divisions from the same group of initial cells as that giving rise to the protoderm (this meristem being consequently referred to as the dermato-calyptrogen; see Fig. 94B, *Brassica*). Beneath this meristem lies a second storey of initial cells which yield the cortex and its inner delimiting layer, the endodermis (Fig. 94, R and E). Finally, a third storey of initial cells forms the central cylinder C, and the pericambium P. These three storeys of initial cells, which correspond to the tangential partitioning seen in the quadrants of the original embryo (see Fig. 90, f, g), constitute the initial complex. Accompanying such 'closed' root apices, whose layers of initial cells can be regarded as true histogens, are also 'open' types in which the originally well-defined histogens soon become disrupted by irregular divisions within the initial cells. In this way the clear distinction between the layers is lost, and apices can arise similar (but secondarily so) to those of the gymnosperms.

B. Residual meristems. Already closely behind the growing point the cells of the promeristem begin to assort themselves into strands and layers of differently shaped cells, at first still capable of division. At an early stage the cell walls here and there come apart from each other and give rise to intercellular spaces (see p. 94). This zone may be referred to as the zone of determination, since it is here that is determined the manner in which the cells will become transformed into permanent tissue in the adjacent zone of differentiation. In general, this transformation involves various kinds of extension growth, thickening and chemical change of the cell wall, until finally the capacity for division is wholly lost and the permanent condition is reached.

In the course of this differentiation of tissues portions of the promeristem are left behind as complete layers of cells, or as large groups or strands of cells, which retain their embryonic properties and a latent capacity for division. Thus in many monocotyledons, for example, the basal segments of the internodes remain meristematic for a long time, and form an interpolated (intercalary) growing zone. In the vascular bundles of the dicotyledons the strand-like fascicular cambium is the site at which later the secondary tissues of the shoot are generated (see p. 147 seq.). In roots the pericycle is capable of renewed meristematic activity, and is the site of origin of the lateral roots (see p. 175).

C. Secondary meristems. In contradistinction to the primary and residual meristems, which are descended directly from the promeristem of the embryo, the secondary meristems arise anew from the cells of permanent tissues which regain their capacity to divide. The cells of the secondary may resemble those of the primary meristems, but more usually they are prosenchymatous and have the form of elongated, plate-like prisms (Figs. 156; 157). They also contain large vacuoles, which are never present in the cells of the apical meristem. Examples of secondary meristems are the cork cambium, or phellogen (p. 159), and the interfascicular cambium (p. 150), which is formed in the parenchyma between the vascular bundles.

D. Meristemoids.[23] Small, few-celled, secondary meristems are referred to as meristemoids. Even in regions where general cell division is still going on, meristemoids may become apparent by their particularly high activity, causing the differentiation of, for example, stomata and hairs, and also the primordia of leaves and pith rays. Not infrequently they distinguish themselves by their capacity to stimulate division in immediately adjacent cells, although in the vicinity generally such division is characteristically suppressed. Thus fields or zones of *inhibition* are formed around meristemoids. The net result of such fields is that the differentiated products of meristemoids come to be arranged in a regular pattern. Although the physiology of the inhibitory effects of meristemoids is not understood, they probably form, together with unequal cell division (see p. 75) an important and fundamental factor in the differentiation of plants (see, for example, p. 123).

II. Permanent tissues[24]

In permanent tissues further cell divisions do not normally occur. The fully differentiated cells of permanent tissues are almost always considerably larger than the cells of the meristems, relatively poor in cytoplasm and not infrequently even dead, in which case they are often filled with water or air.

We can distinguish between primary and secondary permanent tissues according to their origin. Primary permanent tissues are formed from the promeristems, while secondary permanent tissues are derived from the activity of secondary meristems.

Considered from a purely histological point of view, however, the primary and secondary permanent tissues frequently consist of similar elements. Consequently, in a discussion of

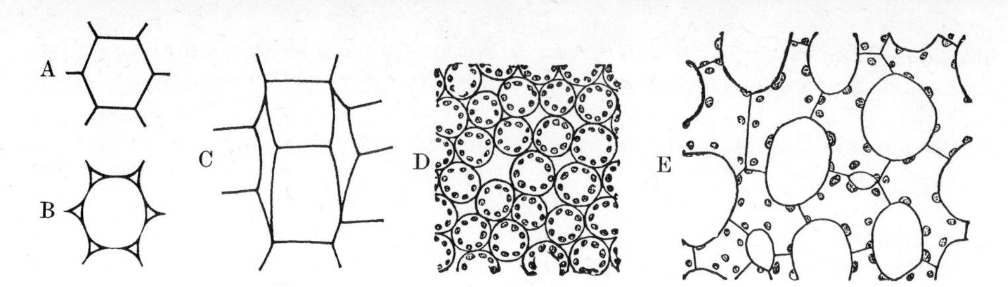

Fig. 95. Diagrams to show the schizogenous formation of intercellular spaces. The cell walls separate at the angles. **A.** and **B.** Transverse sections. **C.** Longitudinal section. **D.** and **E.** Horizontal sections through the mesophyll of the leaf of *Helleborus foetidus* showing the intercellular spaces. **D.** Palisade tissue. **E.** Spongy mesophyll. **A.** and **B.** After Rothert. **D.** and **E.** ×360. After Fitting.

the different kinds of tissues it is usually unnecessary to give attention to this histogenetic difference. On the other hand, when the structure and development of individual organs is being considered the order of the appearance of its tissues assumes great importance. The primary and secondary conditions of the organ in question must therefore be considered separately.

A. The formation of intercellular spaces and the aeration of the permanent tissues

When embryonic cells become transformed into their permanent condition the middle lamellae of the walls, which are underoing thickening, almost always break down at the edges and angles of the cells. As a consequence, the cell walls become split at these places, and the secondary layers of adjacent cells can draw away from each other. In this way, even at a very early stage of the stems and roots, air-filled intercellular spaces arise, and penetrate as far as the most distant apices. Spaces formed by such local separations of the cell walls are said to arise schizogenously.

In transverse section the intercellular spaces appear tri- or quadrangular (Fig. 95B, D), while in longitudinal section they appear as long narrow clefts (Figs. 95C; 132B) and form an interconnecting system of fine freely branching canals (intercellular system). These run along the margins of the cells, penetrate the tissue in all directions and are connected with the external air by way of special pores in the epidermis (the stomata; see p. 97). They are important for the aeration of the more deeply lying living cells.

The enhanced growth of certain portions of the cell walls may eventually cause neighbouring cells to become completely detached. Further divisions in these cells may then cause the intercellular spaces to enlarge and form large chambers or passages of a more or

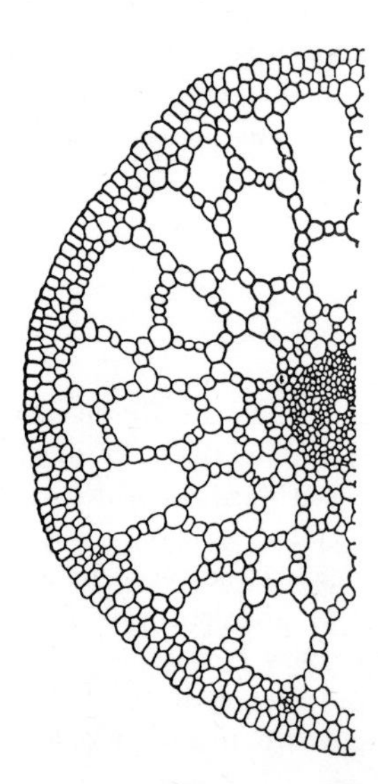

Fig. 96. *Elodea canadensis*. Transverse section of a shoot. The strongly reduced conducting strand in the centre lacks vessels and is surrounded by chlorophyll-containing aerenchyma with wide intercellular spaces. ×80. After Weaver and Clements.

less regular shape (Figs. 95E; 96; 130). Hollows in tissues may also arise through cells disintegrating; this may be caused either by the cells being pulled apart by unequally distributed growth (rhexigenously) (Fig. 126G, H), or by the breaking down of the cell walls (lysigenously) (Fig. 129B, C). Many stems, for example, become hollow in a rhexigenous manner.

B. Ground tissues

The main mass of the plant body of herbaceous plants consists of ground tissue (parenchyma). Normal living parenchyma cells usually have only a lightly thickened, elastic, cellulose wall; rarely is this lignified. The cells are often relatively large, and the cytoplasm surrounds a large vacuole, which—especially in storage parenchyma—may contain numerous foodstuffs. Besides leucoplasts and chloroplasts, chromoplasts may also occur (e.g. in the cytoplasm of the carrot). The leucoplasts and chloroplasts frequently contain starch. The cells of the ground parenchyma are frequently filled with sap under pressure, and because of this turgor the ground tissue contributes largely to the firmness of the plant. Wilting can always be traced back in the first place to loss of water by the parenchyma.

Corresponding to their different functions, we can distinguish assimilatory parenchyma, food-storage, parenchyma, transfusion parenchyma and parenchyma specialized for ventilation (aerenchyma).

The assimilatory tissue of the leaves (the mesophyll), which occupies the space between the upper and lower epidermises (both of which lack chlorophyll) (Figs. 95D, E; 178), may be classified as assimilatory parenchyma, as may also the green cortical tissue of many shoots. Assimilatory parenchyma is always richly provided with intercellular spaces (Fig. 95E), facilitating the exchange of gases, essential for the assimilation of carbon dioxide. Those of the spongy parenchyma on the lower side of the leaf are especially large (Fig. 95E), and these serve not only for the exchange of gases but also for the giving off of water vapour into the atmosphere (*transpiration*).

Typical food-storage parenchyma is found in the pith and in the cortex of shoots and roots, but especially in food-storage organs, such as tubers and swollen tap roots, as well as in the endosperm of seeds (Fig. 48). In these latter the parenchyma is brought into a condition of latent life by the almost complete dehydration of the tissue (see p. 21), and this is not terminated until the renewed uptake of water on germination. The most important food-storage tissue of the woody plants is the wood parenchyma. This forms an interconnecting network of living cells within the mass of xylem, much of which is non-living (Fig. 167A, p). The parenchymatous rays (Figs. 164; 166; 167, m) serve for both food-storage and conduction. The water-storage tissue of the fleshy desert plants (succulents) and many epiphytes (see pp. 184 and 194) would also be classified as food-storage parenchyma in the wide sense.

In many aquatic and swamp plants ventilation tissue (Fig. 96) furnished with broad intercellular spaces facilitates the gas exchange of the submerged organs. This intercellular system connects with the external air by way of the stomata in the leaves and shoots projecting above the water or floating on its surface.

C. Boundary tissues

Here it is not possible to ignore the distinction between primary and secondary tissues, since they differ so markedly. The primary boundary tissues are in the form of external or internal sheets of cells. The most important of the external boundary tissues is the single-layered epidermis, which is derived from the protoderm (see p. 91). In many shoots and roots, however, the outermost layers of the subepidermal cortical parenchyma are functionally involved in the formation of this tissue and form a hypodermis. When this occurs, the epidermis and the hypodermis together form a multi-layered primary boundary tissue.

There are also, of course, true multi-layered epidermises, produced as a result of periclinal divisions in the protoderm (Fig. 196A).

Primary inner (i.e. internal, among other tissues) partitioning layers are referred to as endodermises. They occur here and there in shoots, regularly in roots, and separate the conducting tissue, aggregated in the central cylinder, from the parenchymatous primary cortex.

Secondary boundary tissues occur in the form of layers or flakes of cork. They arise from secondary meristems—the cork cambia.

1. The epidermis

The epidermis develops from the outer layer of the promeristem, the protoderm. Although it serves as the protecting envelope on the outside of the plant, it permits at the same time the exchange of metabolites with the external environment. Typically the epidermis consists of one layer of living cells (Fig. 97A, B), joined together to form an unbroken skin. In surface view they are polygonal or elongated, often with sinuate (Fig. 99A) or dentate (Fig. 106C) outlines, so that adjacent cells are firmly interlocked. In transverse section the cells are rectangular or elliptical (Figs. 97A, B; 178, ep). Their protoplasts consist usually of only a thin layer next to the wall, and their large vacuoles contain colourless or pigmented sap. The plastids are in the form of small leucoplasts, or they are completely lacking.

Only in the ferns, and in many of the phanerogamous shade- and submerged water-plants (e.g. Fig. 193) are chloroplasts commonly found in the epidermis of the green parts.

The outer walls of the epidermal cells of most aerial parts of the plant, with the exception of the ephemeral perianth members, are to some extent thickened (Fig. 97A, B, C), and usually penetrated by ectodesmata (Fig. 58). This thickening, brought about by the development of secondary layers of cellulose, becomes very considerable in many plants of dry places. Much cutin may be deposited, particularly in the outer cellulose lamellae. The outer walls of the epidermal cells of submerged regions, or of plants growing in very damp air, as well as of the subterranean parts of plants, are thin and lack cutin. This is particularly significant in the case of the root, where the surface layer has a special function, namely the absorption of water and salts (see p. 172, the rhizodermis).

The outer walls of the epidermal cells, whether thickened or unthickened, cutinized or uncutinized, are always, except for those of the roots, covered on the outside by a delicate but firmly attached film of free cutin, the cuticle (Fig. 97A, B, C). This cutin is excreted by the protoplast through the cellulose wall and forms an unbroken film over the outside of the epidermis.

The cuticle is frequently somewhat folded, the folds bearing no relation to the cell boundaries beneath (Figs. 98; 106C). In surface view it then appears irregularly striate. The cutin in the cuticle and the cutinized layers make them much less permeable to water

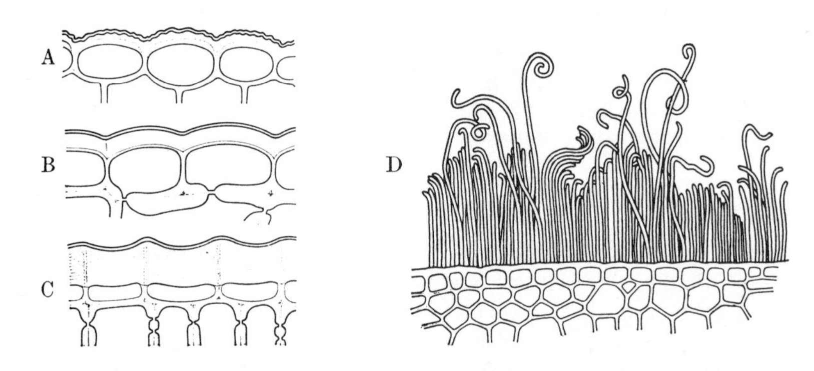

Fig. 97. Epidermis and epidermal secretions. **A.** Transverse section of the epidermis of the leaf of *Dianthus plumarius*. The outer wall is thickened and bears a corrugated cuticle. **B.** and **C.** Heavily thickened and cutinized epidermis from the stem of a cactus (*Cereus*). **D.** Rod-like secretions of wax on the epidermis of a node of the stem of sugar cane (*Saccharum officinarum*). **A.**, **B.** and **C.** ×220. After Mohl. **D.** ×140. After de Bary.

and gases than membranes consisting of cellulose alone. This prevents the tissues of the plant losing water to a dangerous extent through excessive evaporation. The thickening of the outer walls of the epidermal cells also inhibits evaporation, and at the same time increases the mechanical strength of these cells.

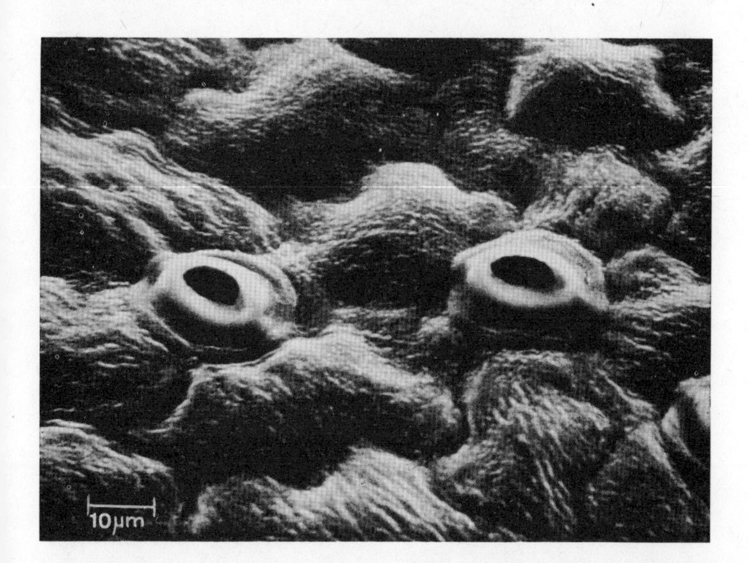

Fig. 98. *Helleborus niger*. Surface view of the lower epidermis with two stomata. All that is seen is the vestibule, bordered by the lip of cuticle. The stoma itself is partly immersed and therefore hidden. The bulging out of the individual epidermal cells is clearly seen, and also the numerous minute folds in the cuticle. ×900. Stereoscan picture by F. Amelunxen.

Wax is often deposited in the cutinized layers of the wall, and it then becomes particularly impermeable to water. The wax is sometimes extended beyond the cuticle, forming a light-grey bloom which can be easily wiped off. Such waxy surfaces are found on many leaves (e.g. red cabbage), fruits (especially striking on plums and grapes) and other organs. The wax may be in the form of granules (as on plums and grapes), or little rods of various lengths (Fig. 97D) or scales. A waxy surface makes the leaf unwettable (e.g. the leaves of the lotus flower, *Nelumbo*).

In certain plants the rigidity of the outer walls of the epidermal cells is increased by the deposition of chalk or silica (as, for example, in the Gramineae and Cyperaceae). In some species of *Equisetum* so much silica is present in the epidermis that they have been used for polishing pewter vessels.

In many leaves the epidermis serves as a water reservoir. Such epidermises often consist of especially large cells rich in sap, and sometimes, as a result of periclinal divisions in the cells of the protoderm, they are even multi-layered.

Frequently in the epidermal cells of fruits, and even more so in those of seeds, we find a striking diversity in the kinds of thickening and the form of the thickened layers. In these structures the epidermis is not usually functioning merely as a bounding layer. By bearing hairs, for example, it may assist in distribution, or by becoming mucilaginous ensure the attachment of the fruit and seeds to the soil. It may also contribute to the formation of a hard shell.

a. The stomatal apparatus. A very characteristic feature of the aerial green parts of the higher plants are conspicuous pairs of cells, usually bean-shaped and with a gap (the cleft or pore) left free between them (Figs. 99B; 100A). These cells are termed guard cells, and a pair of guard cells and the opening between them forms a stoma. The stomata facilitate gas exchange, and also the passage of water vapour from the tissues of the leaf (transpiration). Occasionally the cells adjoining the guard cells are distinct from those of the rest of the epidermis and take part in the functioning of the stoma; these cells are referred to as

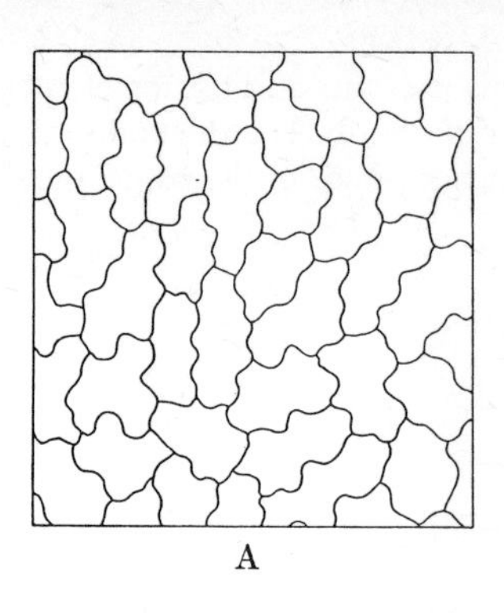

A

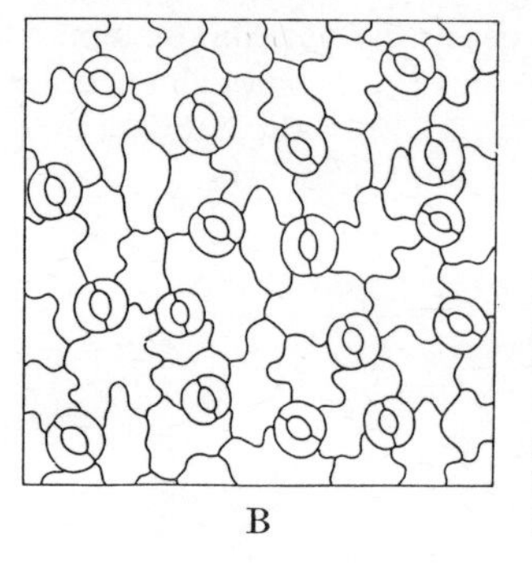

B

Fig. 99. Epidermis from the upper (**A.**) and lower (**B.**) sides of the leaf of *Helleborus niger*. On the lower side are numerous stomatal pores, each bounded by two bean-shaped guard cells. X100.

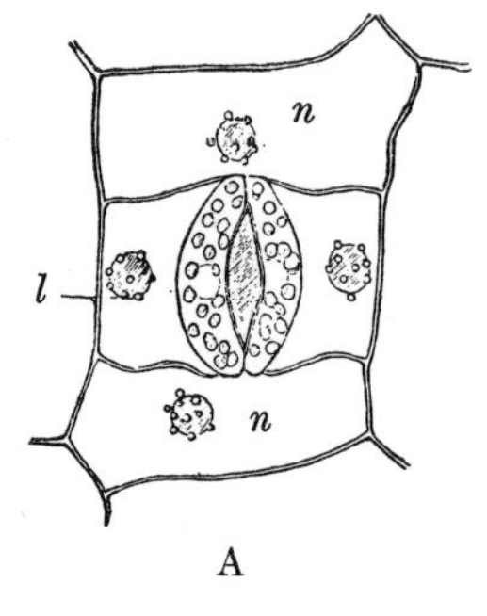

A

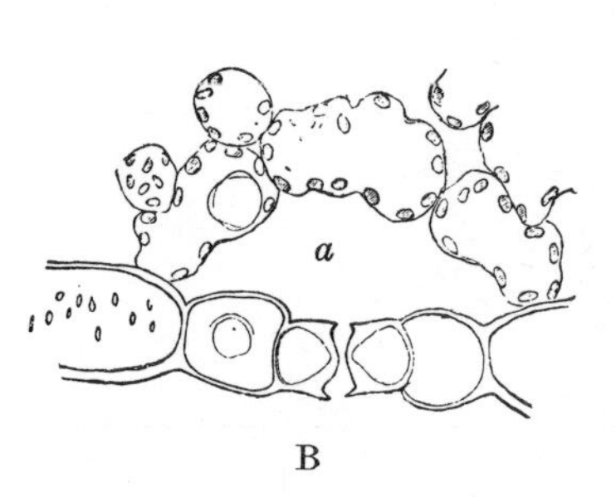

B

Fig. 100. Epidermis from the lower side of the leaf of *Tradescantia*. **A.** Surface view of a stoma; n, subsidiary cells with leucoplasts (l) around their nuclei. **B.** Transverse section; sp, stomatal pore; a, respiratory cavity. X240. After Strasburger.

subsidiary cells (Fig. 100A, n). The stomata and the subsidiary cells together form the stomatal apparatus. The epidermis of the perianth segments contains only few stomata, and they are generally absent from that of the root.

The pores of the stomata interrupt the otherwise unbroken layer of epidermal cells and thus provide a connection between the external air and the system of intercellular spaces of the subepidermal tissue. The space directly below the stoma itself is generally particularly large and is referred to as the substomatal cavity (and sometimes as the respiratory cavity, although it has little to do with respiration).

The guard cells always contain chloroplasts (Fig. 100A), usually with numerous starch inclusions. Their cell walls are almost always unequally thickened, quite clearly visible in transverse section (Figs. 100B; 102B). Usually on the ventral wall of each cell there are two ledges of thickening which project towards the pore, while the middle part of the ventral wall and the whole of the dorsal wall remain thin (Fig. 100B). These cells are called guard cells because, on account of their peculiar structure, variations in turgor (see p. 371 seq.) cause their shape to change in such a way that the pore between them is opened or closed. The stomata are thus regulators of both gas exchange and transpiration.

As the transverse section shown in Fig. 102B demonstrates, the pore is narrowest at its central part and widens towards the so-called vestibule or anterior cavity on the outside, and towards the posterior cavity on the inside. The thickened outer walls and inner walls of the subsidiary cells become attenuated, often abruptly, at the insertion of the guard cells, so that membranous hinges and joints are formed. These facilitate the changes in the shape of the guard cells.

The structure of the guard cells (extending even to the micellar structure of their cell walls) is very diverse, and in consequence so is the direction of movement of their walls, since this is dependent upon the structure. Three main types (omitting the aberrant forms seen in the conifers) can be distinguished. In the first type, probably the most primitive, the walls of the guard cells move principally in a direction perpendicular to the upper surface of the epidermis (*Mnium*-type; Fig. 102A). In the second the movement is almost entirely parallel to the surface (as in the Gramineae; Fig. 103). Most widely distributed, however, are numerous transitional forms in which the movement parallel to the surface is combined with a more or less well-defined vertical component (*Helleborus*-type; Figs. 102B; 104).

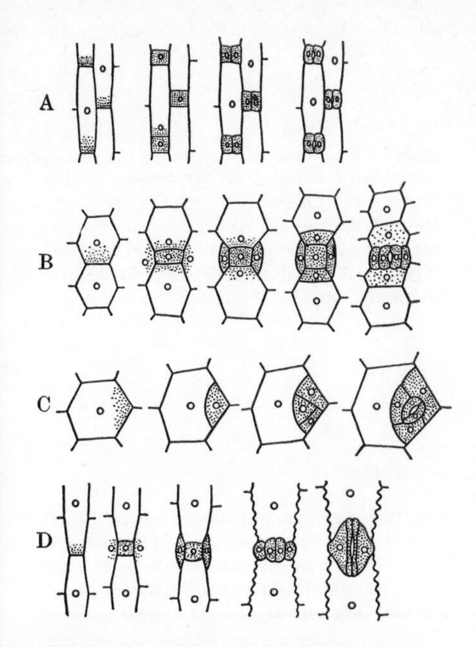

Fig. 101. Ontogeny of a number of different kinds of stomata, diagrammatic. **A.** *Iris*. **B.** *Tradescantia*. **C.** *Sedum*. **D.** *Zea mays*. Freely adapted from Popham.

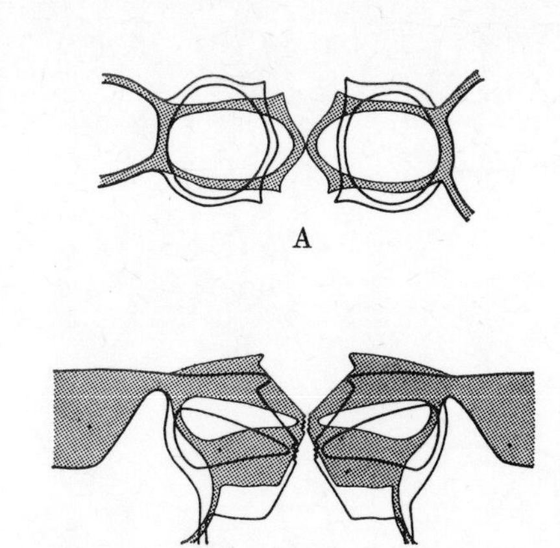

Fig. 102. **A.** Stoma of *Adiantum capillus-veneris* (*Mnium* type). Shaded: closed; unshaded, turgid and consequently open. **B.** Stoma of *Helleborus niger* (*Helleborus* type) in transverse section. Shaded: closed; unshaded, turgid and consequently open. **A.** ×1000. After Kraus. **B.** ×1500. Original.

In the *Mnium* type, which occurs principally in mosses and ferns, the ventral walls of the bean-shaped guard cells are always thin, while the dorsal, inner and outer walls are either thickened or unthickened. As the turgor of the guard cell increases, so the outer and the inner walls move away from each other. This causes the convexity of the ventral wall towards the pore, visible in transverse section, to be diminished, and the pore to be widened, the dorsal wall remaining in more or less the same position throughout (Fig. 102A). In the Gramineae type, widely distributed in the grasses and sedges, the guard cells are approximately dumb-bell-shaped (Fig. 103B, C). The widened ends are thin-walled (Fig. 103A, b), but the narrow intervening part has very strongly thickened upper and lower walls (Fig. 103A, c). An increase in turgor of the guard cells causes the thin-walled ends to expand elastically, and consequently the thickened middle portions to move slightly apart (Fig. 103D). The *Helleborus* type is found in many mono- and dicotyledons. The guard cells are again bean-shaped (Figs. 102B; 104), as in *Mnium*, but in contrast to the condition there, the ventral wall of the guard cell has a pronounced ledge of thickening on the upper and lower side. The dorsal wall remains thin and elastic. With increase in turgor the elastic dorsal wall moves back against the subsidiary cells, drawing the thickened and inelastic ventral wall with it, so causing the pore to open. The junction of the guard cells and the outer walls of the epidermis is usually marked by an attenuated 'joint', so that when the stoma opens it moves either slightly up or down out of the epidermal surface.

Besides the normal stomata, described in the foregoing, in many plants (e.g. *Alchemilla, Fuchsia, Tropaeolum* and the Gramineae) water stomata, or hydathodes, are found on the tips and serrations of the leaves, or singly at the ends of the main veins (Fig. 105). These, derived phylogenetically from the normal stomata, serve for the active excretion of droplets of water, a phenomenon referred to as guttation, and are seen for example in the leaves of *Tropaeolum* (the garden nasturtium) (Fig. 232). Only occasionally do the guard cells of hydathodes remain alive, and capable of opening and closing the pore as in normal stomata. In most species they lose their living contents at an early stage, and the pore then remains widely open all the time. The excreted water is not infrequently rich in calcium carbonate, and after some time this collects around the pores as minute and conspicuously white scales, such as cover the hydathodes on the leaf

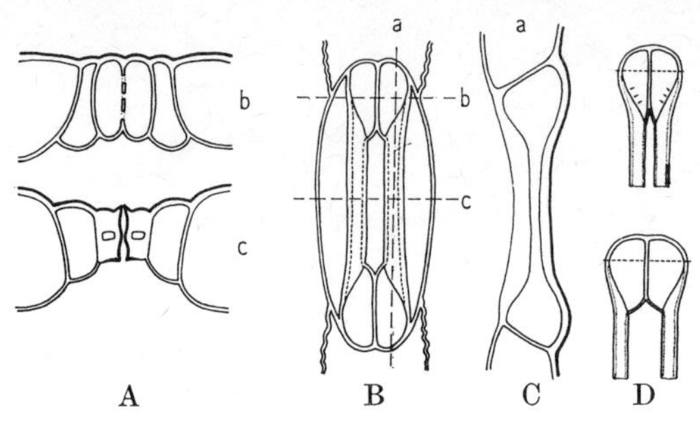

Fig. 103. Stomata of the Gramineae type (*Zea mays*). **B.** Surface view showing the three planes of section a, b and c. **A.** Transverse sections through **B.** at b and c. **C.** Longitudinal section through **B.** at a. **D.** Above, guard cells relaxed, slit closed; below, guard cells turgid, slit open. ×1000. **A.**, **B.** and **C.** Original. **D.** After Schwendener.

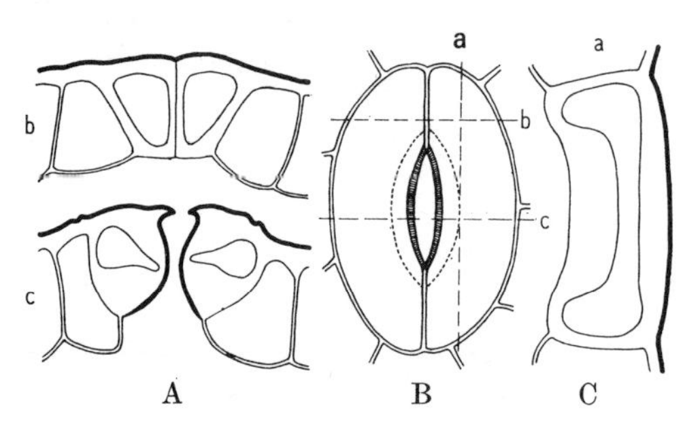

Fig. 104. Stomata of *Prunus cerasus*. **B.** Surface view showing the three planes of section a, b, c. **A.** Transverse section through **B.** at b and c. **C.** Longitudinal section through **B.** at a. ×1000. Original.

margins of many species of *Saxifraga*. Nectar, the sugary solution excreted from nectaries, found both in flowers and elsewhere in plants, issues through pores comparable in structure with hydathodes.

The guard cells arise by unequal division of the epidermal meristemoids (Fig. 101). In the simplest case (seen in many Liliaceae) the meristemoid divides into a smaller cell rich in contents, which immediately becomes the mother cell of the guard cells, and a larger cell which differentiates as an epidermal cell or as a subsidiary cell of the stoma. The mother cell rounds itself off to an ellipsoid and divides by a longitudinal wall to form the two guard cells. The pore is then formed schizogenously in the longitudinal wall (Fig. 101A). Where the stomatal apparatus has several subsidiary cells either several divisions take place consecutively in various directions within a young epidermal cell before the mother cell of the guard cells originates (e.g. in the Crassulaceae; Fig. 101C), or the subsidiary cells arise by division of the young epidermal cells which adjoin the stoma (e.g. in *Tradescantia* (Fig. 101B) and *Zea mays* (Fig. 101D)).

Fig. 105. Water stoma at the margin of the leaf of *Tropaeolum* together with the adjacent epidermal cells. ×160. After Strasburger.

b. Hairs. Of frequent occurrence are epidermal appendages, consisting of from one to many cells, and known as hairs or trichomes. They originate exclusively from the epidermal meristemoids (see p. 93), and generally from a single cell, the initial cell. This cell can, however, by vigorous extension growth, followed by cell divisions, give rise to a large multicellular structure. The protoplasm of the cells of hairs, so long as they remain alive, often shows strikingly active circulation (e.g. in *Cucurbita* (Fig. 10), and in the root hairs of various water plants). Dead hair cells are usually filled with air and consequently, by reflecting most of the incident light, appear white.

Single-celled hairs may be shaped like papillae (Fig. 106C, D), or tubular (Fig. 106B) or awl-shaped (Fig. 107A), and occasionally they may also be branched (Fig. 107D, E).

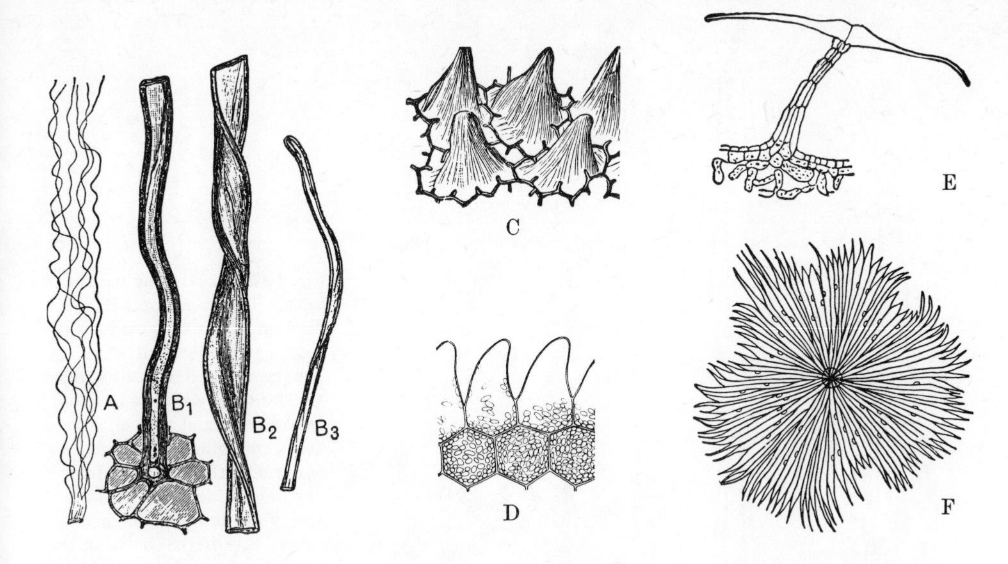

Fig. 106. **A.** and **B.** Hairs from cotton seeds (*Gossypium herbaceum*). **A.** Hairs with a portion of the seed coat. **B_1**, **B_2**, **B_3**. A cotton hair at its insertion, in the middle region, and at its tip respectively. **C.** and **D.** Epidermal papillae. **C.** Surface view of a petal of *Viola tricolor*. **D.** Transverse section through the epidermis of the petal of *Lupinus luteus* (numerous chromatophores present in the sub-epidermal layer). **E.** and **F.** Scale-like hairs from the lower side of the leaf of *Shepherdia canadensis* (Elaeagnaceae). **E.** Longitudinal section. **F.** In surface view. **A.** and **B.** ×225. After Strasburger. **C.** ×250. After Schenck. **D.** ×100. After Frank. **E.** and **F.** ×180. After Fitting.

They have either thin and delicate walls, or the walls may be heavily thickened and rigid, perhaps lignified, calcified or silicified as well, and sometimes furnished with a piercing tip (as, for example, in bristles and some stinging hairs). The part of the hair inserted into the epidermis, the foot, is always distinct from the projecting body of the hair. The epidermal cells surrounding the foot are often arranged in a ring or appear to radiate from the insertion; these cells are referred to as the subsidiary cells of the hair. The hairs of the seed of cotton, which exceed 4 cm in length, and out of which cotton wool and cloth are made, become laterally compressed after death (Fig. 106B_2).

The single-celled bristle-like stinging hairs of species of *Urtica* and *Loasa* (Fig. 107) have a very peculiar construction. The foot of the hair, very thin-walled and distended with cell sap, is surrounded by a cup-like upgrowth of the adjacent epidermal cells. At the same time cell divisions in the adjoining tissue provide the hair with a columnar base. The elongated cell of the hair narrows sharply towards the tip and ends in a small obliquely placed head, beneath which the wall remains unthickened (Fig. 107B). The tip of the hair is silicified, and the remaining parts of the wall, except the swelling below, are calcified. If the head of the brittle glass-like hair is lightly touched it breaks off at the position shown by a–b in Fig. 107B. The end of the hair then resembles a minute canula (Fig. 107C). It penetrates the skin, into which the contents of the hair are then discharged. The fluid within the hair contains, besides sodium formate, both acetyl choline and histamine. These substances cause inflammation and stinging pain (particularly severe in the case of certain tropical nettles) at the site of the puncture.

The simplest form of multicellular hair is that in which the cells are arranged in a single unbranches row (Fig. 127A). A woolly flocculent tomentum, as for example in *Verbascum*, is formed by the regular and repeated branching of the hairs. If, on the other hand, the branching is confined to the uppermost cell of the hair, then there arises a layer of cells one cell thick, which may be stalked or unstalked and resemble in form a disc or a minute leaf (examples are seen in scale-like hairs (Fig. 106E, F) and the chaff-like scales of ferns).

Often the hairs assist the epidermis in its specific functions. Thus the silky down of living hairs on a freshly unfolded leaf can, by its increasing the surface area, promote transpiration. More frequently, however, dense felts of dead hairs, resembling white felt,

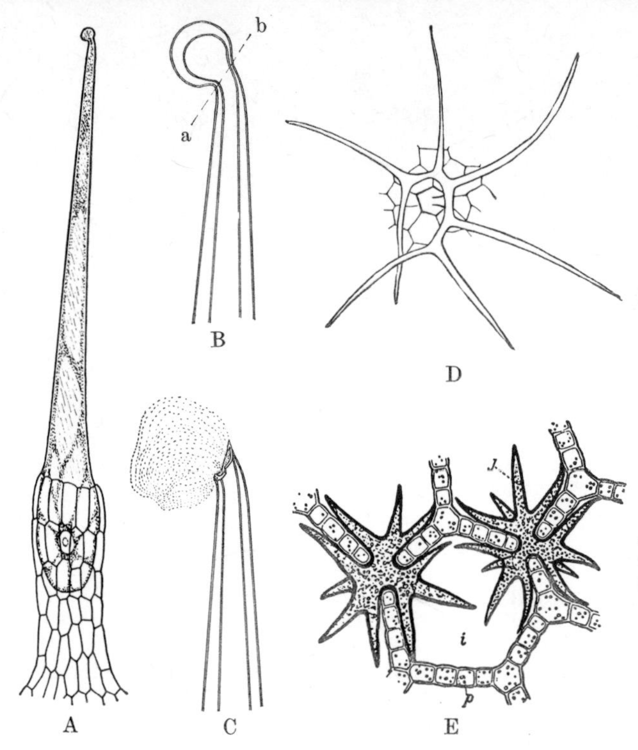

Fig. 107. A. Stinging hair of the nettle, *Urtica dioica*. **B.** Silicified tip of the stinging hair at a higher magnification, showing the pre-formed site of breakage. **C.** The same with the head broken off, and the cell contents emerging. **D.** Single stellate, antler-like, hair from the lower side of the leaf of *Matthiola annua* in surface view. **E.** 'Internal hairs' in the aerenchyma of the petiole of *Nuphar*: p, parenchyma; i, intercellular space; J, internal hair. **A.** ×60. **B.** and **C.** ×400. All original. **D.** ×90. After Strasburger. **E.** ×150. After Dalitzsch.

Fig. 108. A secretory papilla with glandular epithelium (cf. p. 120) from the stipule of *Viola tricolor*. Beside it a unicellular hair. ×160. After Strasburger.

reduce transpiration and at the same time act as a protection against direct sunlight. In many climbing plants (e.g. *Humulus, Galium aparine*) the slipping of the twining or climbing shoots is prevented by hook-like climbing hairs.

Absorption hairs (see p. 106) serve for the uptake of metabolites, especially of water (e.g. root hairs, Fig. 110). Glandular hairs (Fig. 127) excrete metabolites of various kinds. Where parts of plants respond to contact, shock or injury, the perception of the stimulus is often facilitated by papillae, hairs or bristles, peculiar in construction, and with cells rich in protoplasm (as, for example, in *Dionaea* (Fig. 221C, D, E) and the filaments of many *Centaurea* spp.). 'Internal hairs' occur as scattered cells of conspicuously unusual form (idioblasts) in, for example, the aerenchyma of water plants (Nymphaeaceae (Fig. 107E), Gentianaceae), and the aerial roots of *Monstera*. They are also found in *Dryopteris filix-mas*.

c. Emergences. These are outgrowths in the formation of which not only the epidermis but also fairly deeply lying parts of the subjacent tissue participate. Often they bear glands, as in the glandular teeth of the stipules of the pansy (Fig. 108) and the tentacles of *Drosera* (Fig. 221B), or they serve as attachment organs, as, for example, the thorns of the rose and the blackberry. The basal column of the stinging hairs of *Urtica* (Fig. 107A) is an emergence. The flesh of citrus fruits (oranges, lemons) is formed by internal emergences full of sap which grow into the loculi of the ovary.

2. Boundary tissues composed of suberized cells

Where the epidermis does not persist throughout the whole life of the organ it encloses, the bounding function is often even more effectively undertaken by layers of suberized cells. Occasionally layers of such cells may form inner boundaries and separate one mass of living tissue from another. They may be of primary or secondary origin.

The deposition of suberin within a cell usually takes place as follows. Upon the primary wall, at first wholly or almost wholly unsuberized, is deposited a layer of pure amorphous suberin, perforated only by very fine pits. This is often covered on the inside by a further layer of cellulose. Finally, the cells die, the pits at the same time also becoming occluded with suberin (Fig. 109A).

We can distinguish three kinds of suberized boundary tissues: 1. the outer skin; 2. the endodermis; and 3. the cork.

a. The outer skin originates as a consequence of the eventual suberization of either the epidermis or of the sub-epidermal parenchyma, which usually lacks discontinuities and may be of varying thickness. Topographically these layers are referred to as the hypodermis. Many old roots, for example, where the epidermis (rhizodermis) has died at an early stage, are bounded externally by a sheath consisting of several layers of cells which were originally hypodermal. Such a sheath is referred to as an exodermis (Fig. 187, ex). The cells of the outer skin usually retain their living contents.

b. Endodermis. An endodermis is a sheet of cells which separates tissues from one another within the plant (Figs. 187; 189; 190). It usually consists of a single layer of cells and is normally primary in origin. An endodermis occurs regularly in roots, where it separates the central vascular bundle from the cortex, but more rarely in shoots and homologous organs. Endodermal cells are elongated and prismatic in shape, living and joined together to form a continuous sheet (Fig. 190B). In the young (or primary) condition their cellulose walls are still elastic and unsuberized. In the radial walls, however, there is inserted a very narrow or broader strip of a substance of still unknown chemical composition, being both fatty, resembling suberin, and containing lignin. This material forms an inelastic band running round the cell (Fig. 109C). Should the endodermal walls reversibly contract, following perhaps a falling off in turgor, then this strip, permeable with difficulty by many substances in aqueous solution, appears in radial longitudinal sections through the endodermis (Fig. 109C, in direction b–c) as a dark band, in tangential sections (Fig. 109C, in direction a–b) as a wavy line, and in transverse sections as a dark spindle-shaped thickening of the wall (the Casparian strip, see Fig. 109B), although it is in fact no thicker than the rest of the wall. In many older endodermal cells a suberin lamella is deposited all over the cell wall, just as in the cells of the outer skin (the secondary condition). To this may be added thick, often strongly lignified, layers of cellulose (the tertiary condition, Fig. 187), but these may be laid down principally on one side of the cell, giving it a U-shaped appearance in transverse section. In such a secondarily thickened endodermis it is usual for isolated 'passage cells' to escape the thickening (Figs. 111; 121A; 187B, d).

c. Cork. This is always a secondary tissue, owing its origin to a secondary meristem, the cork cambium (see p. 159). Like the epidermis, the cork usually forms a peripheral sheath, although always consisting of many layers of cells. It may appear as a thin grey or a smooth brown skin, or as thicker crusts cracked on the outside. The cells are regularly arranged in radial rows and are rectangular in transverse section (Figs. 109A; 170 K; 171 K). Cork is formed in the older aerial and subterranean parts of plants where the epidermis has died, or where living parenchyma has been exposed by wounding. After death the cork cells are usually coloured brown by tannin derivatives (phlobaphene), and are joined together without intervening spaces. Often (e.g. in *Betula*) layers of very thin-walled and thick-walled cells alternate (laminated cork).

A tertiary layer of thickening, consisting of cellulose and often thicker on one side of the cell than the other, is often deposited upon the secondary suberin lamella of the cell wall (Fig. 109A). This cellulose layer and the middle lamella may be lignified, giving rise to so-called 'rock cork'. Because of the complete suberization of the cells, even thin skins

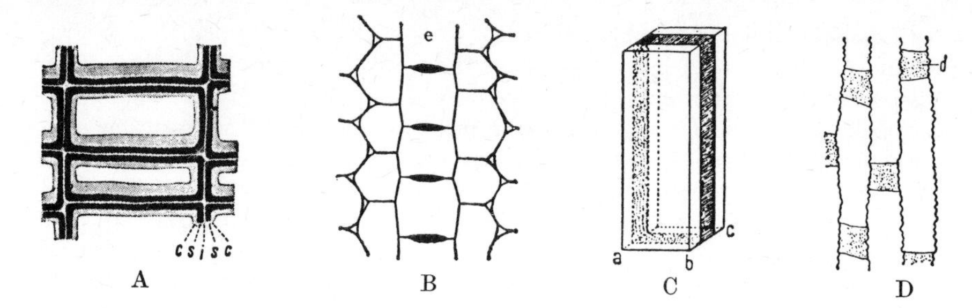

Fig. 109. A. Diagram of the structure of the wall of a cork cell: i, middle lamella (white); s, suberin lamella (black); c, cellulose layer (grey). **B.** Typical endodermis of root (e) in transverse section. **C.** Diagrammatic three-dimensional reconstruction of a single endodermal cell with the Casparian strips. **D.** Endodermis in surface view with passage cells rich in cytoplasm (d). **A.** After Rothert. **B.** and **C.** ×200. After Fitting. **D.** ×100. After Schwendener, modified.

of cork, consisting of only a few layers of cells (Fig. 166 K), diminish the loss of water by evaporation much more effectively than the epidermis. A thick crust of cork also prevents the entrance of parasites, and, because it is such a slow conductor of heat, cork tissue may additionally act as an effective protection against short periods of excessive heat.

Skins of cork cover the outsides of many mature stems (*Viscum*, switch plants and cacti providing some of the few exceptions) and roots, and of tubers (e.g. the jacket of the potato), many bud scales and fruits (e.g. *Mespilus, Pinus* spp.). Occasionally thick skins of cork peel off in numerous paper-like layers ('flake cork'). This occurs especially strikingly on the trunks of birch, and also of the sweet cherry (*Prunus avium*). Where the layers form concentric cylinders, the tree is said to possess a 'ring bark' (see p. 159). The walls of the mature cork cells either remain relatively thin (Fig. 170A) or become more or less heavily thickened (Fig. 109A). Although most cork cambia persist for only a relatively short time, that of the cork oak (*Quercus suber*) remains active for many years. Each season the cambium produces successive layers of wide, thin-walled, cork cells, gradually changing to narrower layers of flatter cells as the year's production ends. This leads to the formation of crusts of cork several centimetres in thickness. In the commercial cultivation of cork the first jacket of cork, together with the cork cambium, is artificially removed from the trunk when the tree is about fifteen years old. A new jacket of cork then forms from a new and much more rapidly dividing cork cambium which arises a few layers of cells deeper within the cortex. It is this cork which is commercially valuable, and is the source of bottle corks. It is peeled off after nine to ten years, whereupon another jacket of cork is regenerated in the same manner. As the stem itself is increasing in thickness during their formation, such thick jackets of cork eventually develop longitudinal fissures on the outside. Our native cork elm (*Ulmus carpinifolia* var. *suberosa*) and cork maple (*Acer campestre* var. *suberosa*) also have long-lived cork cambia producing wing-like ridges of cork on twigs and branches.

In older stems and roots woody tissue replaces the cork, forming an even more effective boundary layer. This is the bark, consisting of a complex mass of tissues (see pp. 161 seq.).

D. Absorptive tissues

1. Rhizodermis. In contradistinction to the surface of the shoot and leaves, the epidermis of the typical young root, the so-called *rhizodermis*, is not cutinized and is not covered by a cuticle. We are concerned here in fact with a tissue which is directly specialized for permeability to and the uptake of water. At some distance from the root tip, more or less where extension growth has ceased, the epidermis of the root (except in aerial roots) gives rise to root hairs (Figs. 110; 185; 186). These important appendages, which serve to increase the surface area of the root, can be seen particularly well in seeds which have been allowed to germinate in damp air. They are very numerous (in *Zea mays*, for example, numbering about 420 per mm^2) and appear to the naked eye as a delicate down on the surface of the root.

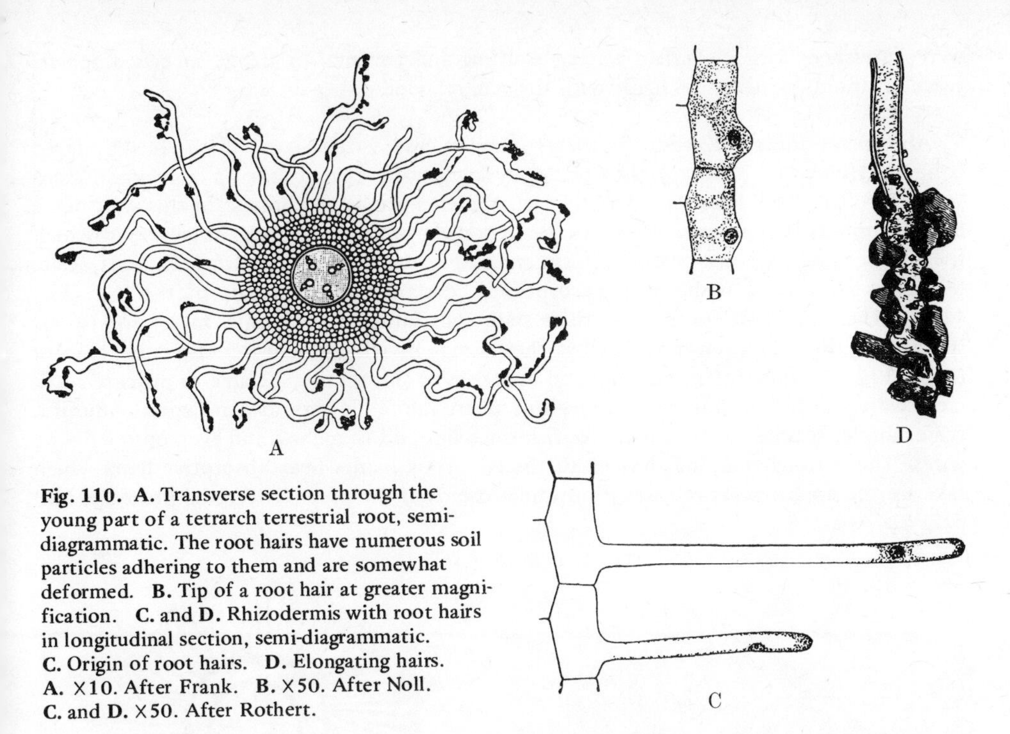

Fig. 110. A. Transverse section through the young part of a tetrarch terrestrial root, semi-diagrammatic. The root hairs have numerous soil particles adhering to them and are somewhat deformed. **B.** Tip of a root hair at greater magnification. **C.** and **D.** Rhizodermis with root hairs in longitudinal section, semi-diagrammatic. **C.** Origin of root hairs. **D.** Elongating hairs. **A.** ×10. After Frank. **B.** ×50. After Noll. **C.** and **D.** ×50. After Rothert.

The root hairs themselves (Fig. 110A, C) are very thin-walled tubular outgrowths of the cells, and may be from 0.1 to 8.0 mm in length. The outer lamellae of the walls become mucilaginous, so that each hair is enveloped in a thin film of mucilage. The root hairs very effectively increase the surface available for water absorption; in the pea, for example, by a factor of about twelve. Usually they remain alive for only a few days. Many aquatic plants and those of swampy places, however, completely lack root hairs.

In many species each cell of the rhizodermis bears its own root hair, and where this is so the hair does not usually arise from the middle of the cell, but from towards the end nearer the root tip (Fig. 110B). In other cases the root hairs of the rhizodermis originate from the uniform meristematic protoderm by unequal cell division, a process corresponding exactly with that seen in the epidermis in the differentiation of hairs and stomata. The inequality first becomes evident in the cells which will give rise to hairs (trichoblasts) being smaller, but containing a cytoplasm denser than that of the remaining cells of the rhizodermis. Their nuclei stain more intensely and are polyploid as a result of endomitosis. In *Hydromystria* there is the interesting situation that the trichoblast apparently undergoes exactly the same number of endomitoses as the sister cell normal mitoses. Thus, with each thirty-two-ploid trichoblast are linked developmentally approximately sixteen normal diploid cells. In some other plants (e.g. *Lycopodium, Zea, Najas, Nymphaea*) trichoblasts and cells without hairs alternate in the rhizodermis in a regular pattern. Occasionally the trichoblasts reveal their meristemoid character by undergoing repeated divisions, so that the root hairs arise not singly but bunched together.

Even before the rhizodermis dies, a zone of suberized cells, one or several cells thick, arises in the sub-epidermal cortex. This is the exodermis, already mentioned as a limiting tissue of suberized cells (Fig. 184). The fact that the cells of this tissue remain alive shows that the extent of their suberization is relatively small. Single cells or groups of cells of the exodermis may remain quite free of suberin for a considerable time and form a secondary absorption system. These passage cells often originate with an unequal division in the same way as the root hairs.

2. Ligules of the Selaginellales, Lepidodendrales and Isoetales. The leaves of the recent *Selaginella* and Isoetales bear on the upper side close to the base a small membranous scale lacking chlorophyll (Figs. 564C, li; 573C, li). This *ligule* is capable of taking up very rapidly any moisture deposited on the leaves. The leaves of the extinct Lepidodendrales were furnished with a similar water-absorbing mechanism. In many species a special vascular strand connects the ligule with the general conducting system.

3. Absorptive hairs. Except for water plants, and a few ferns of especially damp habitats (Hymenophyllaceae), the Cormophyta generally cannot take up large amounts of water through their leaves (and even with the exceptions the uptake in this way never exceeds about 10 per cent of the total). In some tropical epiphytes, however, in which the leaves come together below to form an urn-like water reservoir (Bromeliaceae), scale-like hairs concerned with water absorption occur on the upper side of the leaf (Fig. 112A). These hairs are inserted by their stalk cells into the epidermis. In dry conditions the heavily thickened outer walls close the passage by which the water enters like a valve (Fig. 112B). In this way even completely rootless epiphytes may be able to procure sufficient water for life in those tropical regions where rainfall is periodic. In Central America, for example, species of *Tillandsia* flourish on twigs and branches, and even on telephone wires. The various trapping devices of insectivorous plants bear absorptive hairs which take up the amino acids released from the proteins of the prey by secreted proteases (see p. 296).

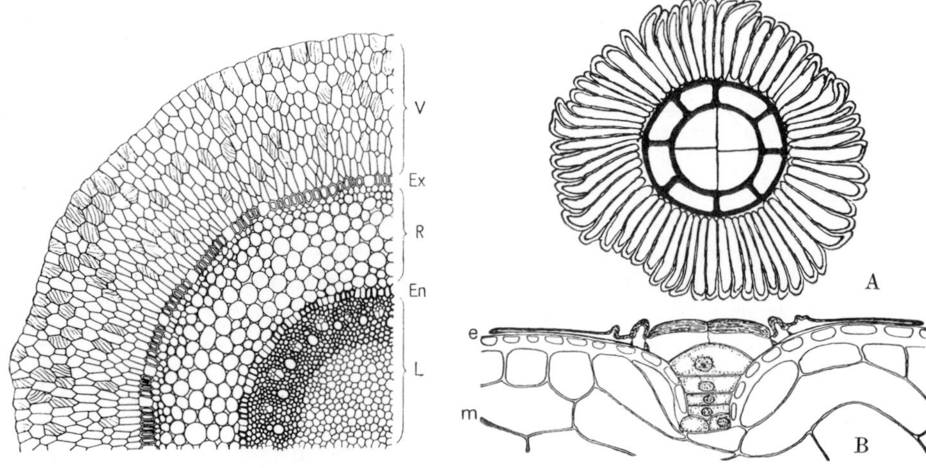

Fig. 111. Transverse section of an aerial root of the epiphytic tropical orchid *Dendrobium nobile*: V, velamen radicum; Ex, exodermis with passage cells; R, cortex; En, endodermis with groups of passage cells; L, central, radially symmetrical, vascular bundle. ×30. Original.

Fig. 112. Peltate scale of the epiphyte *Vriesea splendens* (Bromeliaceae). **A.** View from above. **B.** Transverse section through the scale, closely appressed to the epidermis (e). The outer walls of the epidermal cells are thin, the inner thick. The sub-epidermal tissue (m) is thin-walled. ×330. After Fitting.

4. Glandular emergences of water plants. Glandular structures are not infrequently found on the submerged leaves of water plants. For a long time they were regarded as being concerned with excretion, but has now been demonstrated that they also are concerned with absorption. Apparently they act as 'ion traps', and take up minerals from the water.

5. Velamen. The aerial roots of many monocotyledons, especially the epiphytic members of the Orchidaceae and Araceae (see p. 194), possess a so-called velamen. This consists of several layers of cells and is formed from the protoderm by repeated periclinal divisions. The cells of the velamen quickly die, and being furnished with large pores, soak up rain by capillary attraction like a sponge (Fig. 111). In the dry condition the cells are filled with air and the velamen is dirty white, but when saturated the green colour of the inner cortical tissue appears through it. The water taken up into the velamen is gradually absorbed through the passage cells of the exodermis into the cortex of the root, and from there through the endodermis into the interior (Fig. 111, Ex). The cells of the velamen are often reinforced by spiral or reticulate thickening of the walls.

E. Conducting tissues

The larger a plant's mass becomes, the greater the number of cells of which it is composed. Even more important, the more parts it bears aerially, outside the soil or water, the greater becomes the necessity to replace the water lost by evaporation, and to transport metabolites rapidly from one organ to another, such as from the root to the leaves and vice versa. Diffusion through elongated parenchyma cells is almost never adequate for this purpose, even if the movement is facilitated by the development of pits in their transverse walls. Thus special conducting tissues composed of very striking cell elements have been developed. These are customarily tubular and elongated in the direction in which conduction mainly occurs, and to facilitate the passage of metabolites the end walls are often steeply inclined so that their area is significantly increased (Figs. 113A, B; 118). Later in development the end walls often develop pores, and the individual cells more or less coalesce to form characteristic ducts. These are always joined up to form a connected system running through the whole plant.

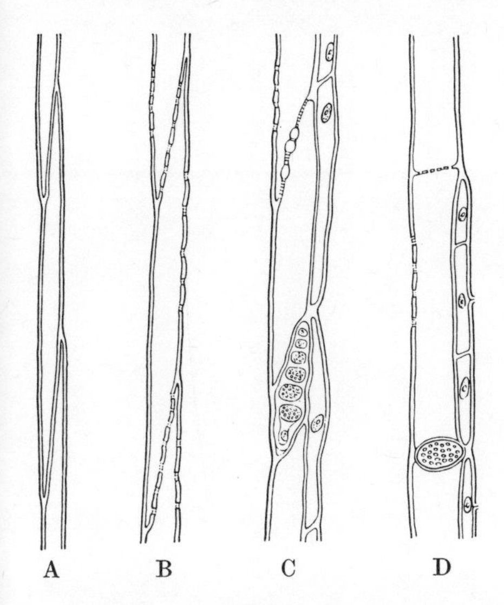

Fig. 113. Phylogeny of the sieve element. **A.** Prosenchymatous early form without special structure of the wall (e.g. Rhyniaceae). **B.** Development of rudimentary sieve areas (e.g. sieve cell of the Lycopodiaceae). **C.** Sieve tube with sieve plates (e.g. Cucurbitaceae). After Zimmermann.

1. Sieve cells and sieve tubes. In ferns and gymnosperms the transport of organic materials is brought about by elongated (prosenchymatous) sieve cells, tapering at each end. In most angiosperms this primitive conducting system is further developed to continuous sieve tube system, consisting of elongated cells with perforated, sieve-like, transverse walls, the sieve tube elements (Fig. 113C, D). Sieve cells and sieve tube elements contain living protoplasts with mitochondria and plastids, often with starch (Fig. 114). During differentiation the nucleus and tonoplast soon disintegrate. The protoplasm eventually fills the whole lumen of the cell as a loose meshwork of highly hydrated proteinaceous fibrils. The sieve cells and sieve tubes owe their name to localized perforations in their walls (consisting of cellulose and always unlignified), forming sieve-like connections between adjacent members of the conducting system. As with the primary pit fields of parenchyma cells (Fig. 62B), the fine pores in the walls of sieve cells and sieve tubes are grouped into large sieve fields (Fig. 114, sf).

Many Cucurbitaceae and other climbing plants have unusually broad sieve tubes (Fig. 115G) with correspondingly broad pores in the transverse walls. They are always filled with dense protoplasm, and traversed by strands of endoplasmic reticulum. This close plasmatic connection between the individual members of the whole sieve tube system, as well as its high osmotic pressure (a consequence of the substantial concentration of sugar in its sap) has the consequence that the system is extremely sensitive to wounding and similar damage. In dicotyledons with secondary thickening, and the consequent yearly increment of phloem, the life of the sieve tubes is not infrequently limited to a single season. At the end of the year they collapse, and their function is taken over by the newly-forming phloem. On the other hand, in long-lived species where secondary thickening is absent, as in many arborescent monocotyledons (e.g. the palms) sieve tubes may function for 50 and possibly even for as long as 100 years.

In many species the entire transverse or slightly oblique end wall between consecutive sieve-tube elements is a single sieve plate (Fig. 114, spl, e). If, on the other hand, the end walls of the elements are inclined to any extent, then they usually contain several sieve fields, sometimes a large number. They are arranged in a row, one above the other, and separated by unperforated parts of the wall, forming a compound sieve plate (Figs. 114, spl, z; 169). Large (Fig. 114, spl, l) or small sieve plates with very fine pores may also be differentiated in circumscribed areas of varying extent in the longitudinal walls between adjacent sieve tubes.

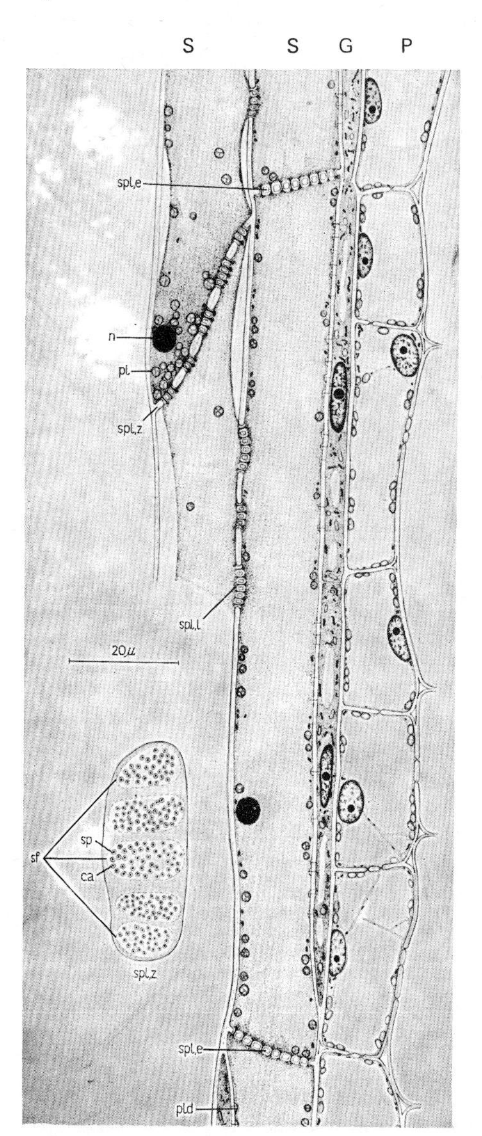

Fig. 114. Partially diagrammatic representation of sieve tube members (S), with companion cells (G) and phloem parenchyma (P), in *Passiflora coerulea*. The lumen of the sieve tube members is filled with a uniform, highly dispersed, cytoplasm. Also present are plastids (pl), and the nucleolus (n) of the otherwise dispersed nucleus. The sieve plates are inclined (spl, z), or transverse (spl, e); some are lateral (spl, l). Protoplasmic connections (pld) are present between the sieve tube and the companion cell. Lower left: sieve plate with five sieve fields (sf) in surface view, showing the sieve pores (sp), each ringed by callose. *c.* ×750. After Kollmann.

Towards the end of the growing period the sieve plates become choked with a thick layer of hyaline callose, thus interrupting the exchange of materials between the sieve-tube elements. The same substance is in fact present from an early stage lining the pores within the plate (Fig. 114, ca). Callose, a polysaccharide insoluble in water, is soluble in cold 1 per cent potassium hydroxide, but insoluble in Schweizer's reagent. It stains red-brown with iodine in zinc chloride, red with corallin and blue with resorcin blue (a specific reaction).

Companion cells (Figs. 113C, D; 114G; 115G) are found only in angiosperms. They arise by a process of unequal division from the same mother cells as the sieve-tube elements (Fig. 115A–F). Subsequently they divide transversely, frequently several times. Their lumina are much narrower than those of the sieve-tube elements; they have dense cytoplasm and a large nucleus. Numerous fine pits and of the sieve tubes; they contain a dense cytoplasm, have large (often polyploid) nuclei, lack plastids, but possess many mitochondria. They remain in intimate contact with sieve tubes by way of numerous small pits and plasmodesmata.

2. Tracheids and vessels. Specifically concerned with the conduction of water are dead, more or less elongated, cells with lignified walls, richly furnished with pits. In their functional condition they are always filled with water. Usually this contains mineral salts derived from the soil, and in woody plants the system also serves for the rapid transport of organic materials in the spring from storage regions to the growing points.

The adjoining end walls of tracheids are often steeply inclined, and, although always well pitted, never authentically perforated (Fig. 166B, t). The vessels, on the other hand, consist of columns of members or segments which individually are only a little elongated, and are predominantly barrel-shaped. These members represent the original cells which, as a consequence of extensive or complete dissolution of their transverse walls, fuse to form closed, tubular systems, often of considerable length (Figs. 116C, D; 120, h, i). During their differentiation the nuclei of vessel segments often become many times polyploid (16–64n) as a consequence of endomitosis, and, particularly in angiosperms, the segments may grow considerably in breadth (Fig. 116A–D) before their final differentiation. In many lianes the diameter of the vessels may reach 0.7 mm so that uninterrupted passages can be clearly seen if a piece of stem several centimetres long is directed towards the light. Even in our native species of *Quercus* the vessels of the spring wood reach a diameter of 0.3 mm and are clearly recognizable to the naked eye ('ring porous' wood, see p. 154). In *Tilia*, however, the average diameter of the vessels amounts to only 0.06 mm.

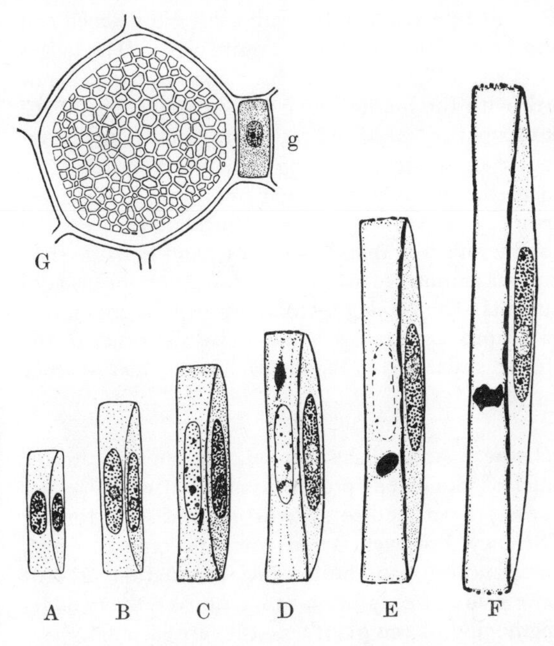

Fig. 115. Sieve tubes and companion cells. **A.** to **F.** *Vicia faba*. Development of a sieve-tube element with associated companion cell, semi-diagrammatic. **A.** Immediately after the unequal division of the mother cell. **B.** and **C.** Beginning of the growth of the cell and nucleus. **D.** and **E.** Dissolution of the nucleus of the sieve-tube element. **F.** Final stage showing the perforation of the end wall of the sieve-tube element and the cytoplasm reduced to a thin layer lining the wall. **G.** *Cucurbita pepo*. Sieve tube and companion cell (g) in transverse section. The sieve plate is in surface view. **A.** to **F.** After Resch. **G.** ×600. After Fitting.

When stems are cut air usually enters immediately into the vessels. Consequently, when they were first discovered they were thought to constitute an aerating system, similar to the tracheal tubes of insects, after which they were named. The term tracheid, implying a reduced form of trachea, or vessel, was not coined until later.

The walls of the conducting elements are usually strengthened by striking lignified

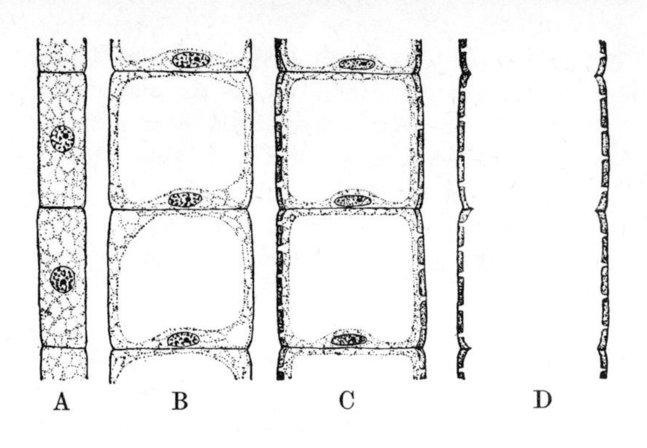

Fig. 116. Development of a vessel from a column of originally independent cells. ×150. After Sinnot.

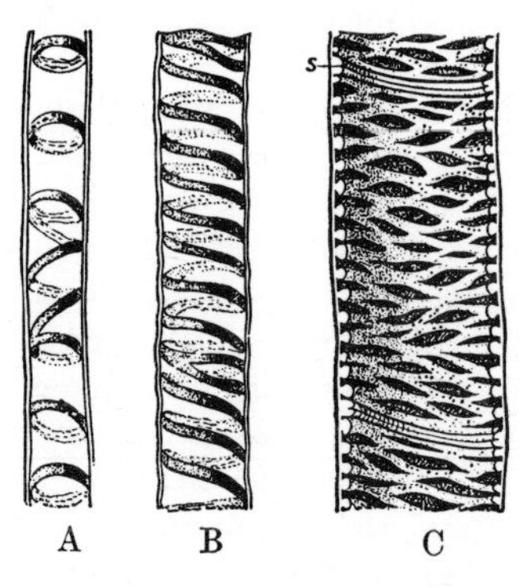

Fig. 117. A. Portion of a tracheid with spiral and annular thickening. **B.** The same with entirely spiral thickening. **C.** A vessel with reticulate thickening cut in halves longitudinally: s, one of the two perforated cross walls. ×240. After Schenck (C. modified).

thickenings of various kinds (Fig. 117). These prevent the dead water-conducting cells from collapsing if, owing to vigorous transpiration of the plant, there is sustained tension within them (see p. 224). At the same time the tracheary elements are so constructed that they can easily release water laterally to the surrounding tissue or take up water from it.

In many conducting elements the thickenings of the wall are confined to narrow rigid lignified bands or ledges (Fig. 117A, B). The remainder of the wall is little thickened and therefore easily permeable to water, and being unlignified is also quite elastic. The ledges of thickening may form isolated rings (Figs. 119A; 151C, a, b), complex spiral bands (Figs. 117B; 151C, c, d, e) or a network with the meshes lying transversely (Figs. 117C; 151C, f, g). Accordingly, we can speak of annular, spiral and reticulate thickening respectively. Only those elements with annular or spiral thickening remain extensible (Fig. 119A, B, C) and only such are found in the growing parts of plants (protoxylem; see p. 111 seq.). In pitted tracheary elements, however, the lignified thickening covers the greater part of the wall. Nevertheless, numerous pits remain within it, and in shape they may be circular or polygonal with a certain amount of transverse elongation, or broadly or narrowly elliptic, or little more than slits. The closing membranes remain unlignified (see Fig. 117C). If transversely elongated pits are arranged one above the other in the lateral walls the arrangement is said to be scalariform (Fig. 118). Intermediates occur between all the forms of thickening described.

3. Phylogeny of the water-conducting tissue. As we have seen (p. 82), there is already present in the Bryophyta a central strand of elongated (prosenchymatous) cells, lacking contents and with thickened walls. These are referred to as hydroids and serve for the internal conduction of water. Their efficiency, however, is very low, and only at a high humidity, with correspondingly low evaporation, can the internal conduction prevent desiccation of the plant. At this primitive grade of adaptation to life on land the conduction of water and the mechanical strengthening of the plant are still carried out by one and the same tissue. This is because the paths of the specialization of cells for these two functions have the same beginnings, namely, 1. marked axial extension of both the water-conducting and mechanical elements, and 2. a strengthening of the cell wall by the laying down of lamellae of secondary thickening which are subsequently made rigid by the deposition of lignin.

In the ferns and gymnosperms the resistance to conduction is diminished by an expansion of the cross-sectional area of the conducting elements, as well as by a particularly free pitting of their transverse walls. In some especially highly developed Pteridophyta and gymnosperms (e.g. *Pteridium aquilinum* (Fig. 118); *Ephedra*) the pit membranes of the pits in the transverse walls eventually completely disintegrate, so that the resistance to movement in a longitudinal direction is reduced to a minimum.

The distinction between exclusively water-conducting and exclusively mechanical elements (wood or libriform fibres; Fig. 161) was not achieved until relatively late in the course of evolution. In the gymnosperms the load-bearing part of the stem is still made up of essentially a single kind of cell with a tracheidal character, namely the tracheid with bordered pits, serving both for water conduction and strengthening. The only specialization seen is that the tracheids formed in the spring have somewhat thinner walls and a more frequent pitting than those of the 'late wood' formed in the summer; the former serve principally for the conduction of water, and the latter for strengthening the stem.

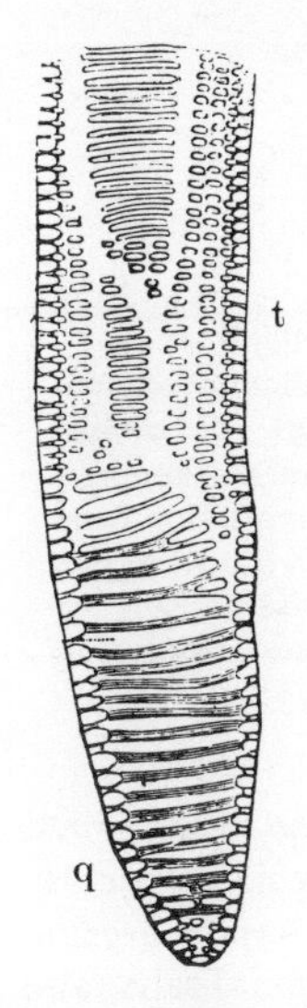

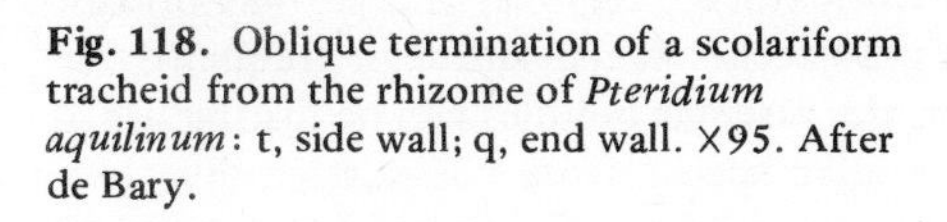

Fig. 118. Oblique termination of a scolariform tracheid from the rhizome of *Pteridium aquilinum*: t, side wall; q, end wall. ×95. After de Bary.

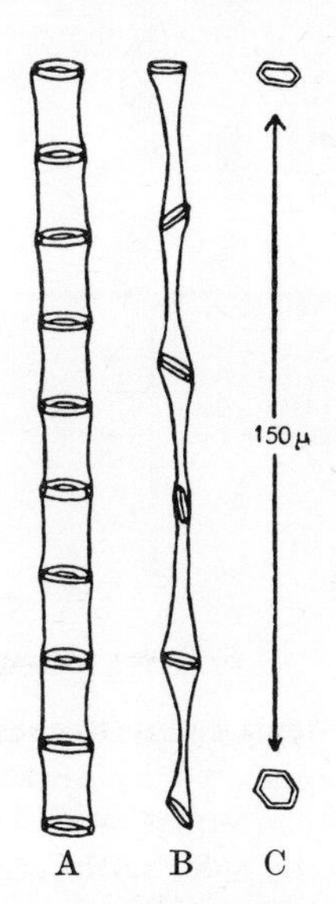

Fig. 119. Passive extension and rupturing of tracheidal protoxylem elements. **A.** Tracheid of *Aristolochia durior* with annular thickening showing a ten-fold extension. **B.** The same, showing a twenty-five-fold extension. **C.** Tracheid of *Cucurbita pepo* with annular thickening extended 120 times and ruptured. After Frey-Wyssling.

The division of the two functions between two different kinds of tissue element was not accomplished, and the development of systems of long ducts lacking cross walls, the vessels, did not reach its highest level until the advent of the angiosperms in more recent geological time. According to these considerations, in many circles of affinity the primitive types will still possess scalariform perforation plates, while in the advanced genera the cross walls will have completely disintegrated. Within the angiosperms, however, even highly specialized vessels are regularly accompanied by some tracheidal elements (see p. 153).

At intervals, even in the most highly developed conducting systems, the transverse walls remain intact. Consequently the vessels have a definite length. It amounts in general to less than 1 m, usually to only a few centimetres. On the other hand, in many lianes and in 'diffuse porous' woods (see p. 154), vessels may exceed 10 m in length.

4. Vascular bundles. In the primary tissues of the higher plants the strands of sieve tubes and of tracheids and vessels only rarely occur in isolation. Usually they are aggregated

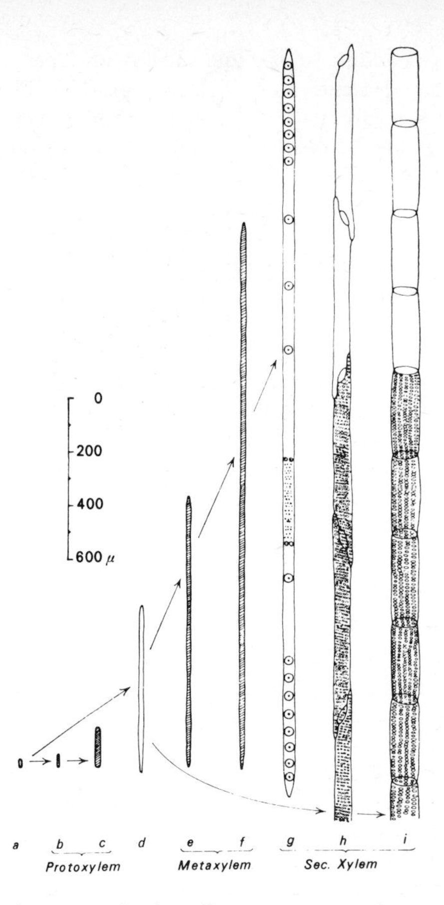

Fig. 120. Forms of conducting elements and their development. **a.** Meristematic cell. **b.** and **c.** Short conducting elements of the protoxylem (tracheids with annular thickening). **d.** Cambial cell. **e.** and **f.** Tracheids with spiral thickening developed by extension and differentiation from the cambial cell. **g.** Tracheid with bordered pits similarly developed. **h.** and **i.** Vessels arising by expansion of the vessel elements and dissolution of the end walls (cell fusion). *c.* ×35. After Frey-Wyssling.

into vascular bundles. In stems and roots these frequently run parallel to the longitudinal axis, but they are interconnected by anastomoses, giving rise to a reticulate vascular system. The elements concerned with the conduction of water and organic materials normally lie side by side, so that the water with its dissolved mineral salts (and in early spring also organic materials released from storage regions) and surplus organic assimilates produced in the leaves move along adjacent tracts, although frequently in opposite directions. Even at a low magnification the vascular bundles can be distinguished by their narrow elements and lack of intercellular spaces from the surrounding less-compact tissue. Often they are visible as strands even to the naked eye, e.g. in leaves and in the translucent stems of species of *Impatiens*, and their more or less circular outlines are clearly visible in transverse sections of stems and roots.

The vascular bundles in the stem are round, or broadly or narrowly elliptic in transverse section. The strand of sieve-tube elements constitutes the phloem, and that of the tracheids or vessels the xylem of the bundle.

The phloem (apart from any accompanying mechanical tissue) is also referred to as the leptome, and the xylem (also excluding any mechanical tissue) as the hadrome.

It is possible to distinguish radial, concentric and collateral bundles, according to the arrangement and development of the phloem and xylem. The radial vascular bundle (Fig. 121A) is circular in section, and each of the several strands of xylem and phloem in it lies on a radius, the strands of different nature alternating with one another. The xylem strands (Fig. 121A, X) are usually flattened laterally and form a star-like outline in transverse section. The phloem strands (Fig. 121A, Ph) lie in the intervals between the arms of xylem, from which they remain separated by one or several layers of parenchyma. In such bundles differentiation of both xylem and phloem begins near the periphery and proceeds towards the centre of the bundle. Radial bundles typically occur in roots; in shoots they are found only rarely, and where present are always single (e.g. in the stems of many species of *Lycopodium*).

In concentric bundles, which are also circular or elliptic in section, a xylem or phloem

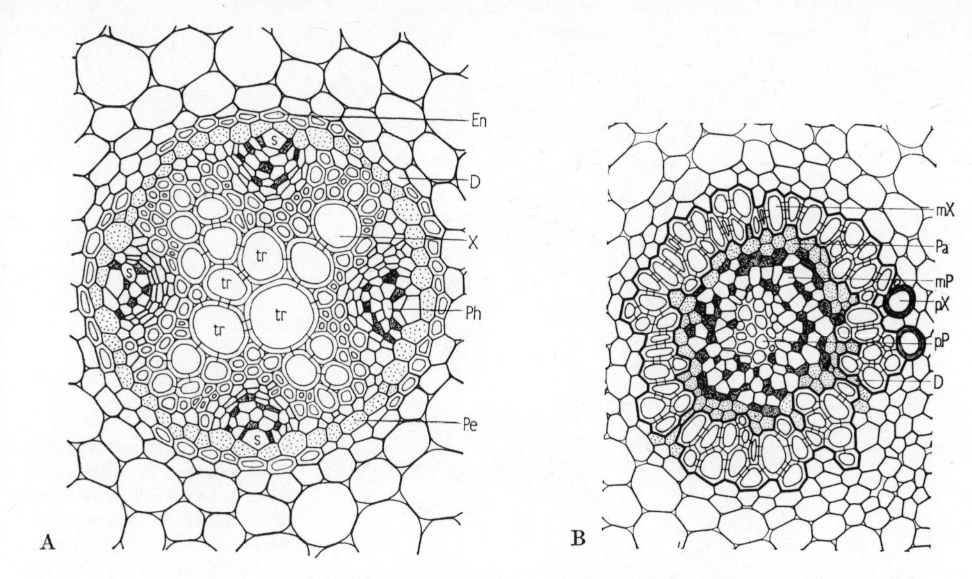

Fig. 121. A. Transverse section through the tetrarch, central, radially symmetrical vascular bundle of the root of *Ranunculus acris*. En, endodermis; D, passage cells; X, xylem; tr, tracheid; Ph, phloem with sieve tubes and (stippled) companion cells; Pe, pericycle. ×150. **B.** Transverse section through a concentric vascular bundle with external xylem in the rhizome of *Convallaria majalis*. pX, protoxylem; mX, metaxylem; pP, protophloem; mP, metaphloem; Pa, vascular parenchyma. ×200. Both original.

strand is completely enclosed by a sheath-like strand of phloem or xylem. If the xylem lies within and the phloem without, the bundle is said to be concentric with included xylem, and with the reverse arrangement, concentric with included phloem. In most of the ferns the bundles are concentric with included xylem. In certain rhizomes, however, and in the stems of many monocotyledons (Fig. 121B) concentric bundles with included phloem are found. The differentiation of the xylem and phloem in concentric bundles is not completed according to a set pattern; the places at which the differentiation of the xylem and phloem begins are variable.

Collateral vascular bundles, which may be circular, elliptic or narrow, almost linear, in transverse section (Figs. 122; 123; 124), are widely distributed in the gymnosperms and angiosperms. In such bundles there is only one strand of xylem and usually only one of phloem. The xylem lies adaxial to the phloem.

In collateral vascular bundles generally the xylem is directed towards the middle of the stem, and the phloem outwards. Such bundles are characteristic of the shoots of seed plants and species of *Equisetum*. There are also, however, bicollateral bundles, where phloem is not only on the outside but also on the inside. Such are found in the Solanaceae and Cucurbitaceae. In the monocotyledons the collateral bundles are usually closed, i.e. the whole bundle consists entirely of permanent tissue, and the xylem abuts directly on to the phloem (Fig. 122). In gymnosperms and dicotyledons, on the other hand, they are usually open, i.e. the phloem and xylem remain permanently separated by a layer of meristematic tissue, the fascicular cambium (Figs. 123, Kf; 124, K). This consists of radial rows of cells, whose characteristic arrangement is the result of continued periclinal divisions.

The differentiation of the phloem in a collateral bundle begins from the margin adjacent to the periphery of the stem and proceeds towards the centre of the bundle. The differentiation of the xylem begins from a more or less opposite point and also proceeds to the centre. Consequently, the first-formed tracheary elements (the protoxylem (Figs. 122; 123; 124, pX)) of the collateral bundle lie at the adaxial margin of the xylem (as seen in transverse section), and the first-formed sieve-tube elements (the protophloem (Figs. 122; 123, pP)) at the abaxial margin of the phloem. If all the meristematic tissue is used up in the formation of the bundle, then a closed collateral bundle is formed; if some remains, then an open bundle results, and the undifferentiated cells ultimately become part of the cambium (see p. 147).

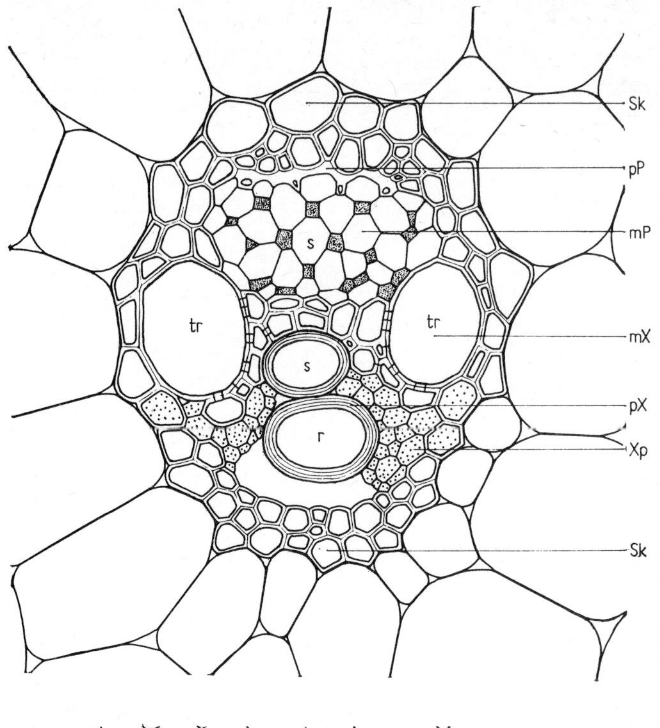

Fig. 122. Transverse section through a closed, collateral vascular bundle from the stem of *Zea mays*. Sk, sclerenchymatous sheath; pX, protoxylem (r, annular, s, spirally thickened tracheids); mX, metaxylem (tr, tracheids); Xp, xylem parenchyma; pP, protophloem; mP, metaphloem; S, sieve tubes, companion cells stippled. ×200. Original.

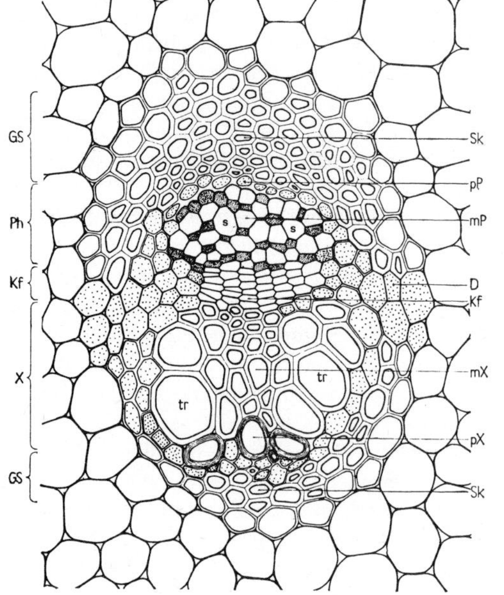

Fig. 123. Transverse section through an open, collateral vascular bundle from the stem of *Ranunculus repens*. GS, bundle sheath; Sk, sclerenchyma fibres; Ph, phloem; pP, protophloem; mP, metaphloem; Kf, dormant cambium; D, transfusion tissue; X, xylem; pX, protoxylem; mX, metaxylem; tr, tracheids. ×200. Original.

In all the forms of vascular bundle the xylem consists principally of narrow or wide lignified elements whose function is to conduct water. In angiosperms the tracheary elements are generally tracheids and vessels (Figs. 122; 123, tr), in Pteridophyta and gymnosperms usually tracheids alone. The tracheary elements are embedded in xylem parenchyma (Fig. 122, Xp), consisting of living, elongated and often unlignified parenchyma cells with no intercellular spaces. The elements formed after the protoxylem are usually reticulately thickened or uniformly thickened and pitted, and are capable of only little extension. In the Pteridophyta, however, only scalariformly thickened tracheids occur (Fig. 118).

The phloem of the vascular bundle (Figs. 121B; 122; 123, mP; 124, P) contains the sieve tubes. They are always accompanied by other living cells, and are either surrounded (as in many monocotyledons) entirely by companion cells (Figs. 121; 122; 123, grey tone), rich in cytoplasm but always free from starch, or by companion cells and other elongated parenchyma cells (phloem parenchyma), or by the latter alone. In the pteridophytes and gymnosperms there are still no sieve tubes, but merely strands of sieve cells without companion cells.

Each bundle is enclosed by a sheath, which may consist of parenchyma free from intercellular spaces (and called, if rich in starch, a starch sheath), of mechanical tissue or of a layer of endodermal cells. The sheath is not considered part of the bundle. In collateral bundles sheaths of mechanical elements are especially frequent on the outside of the phloem, and also on the adaxial side adjacent to the xylem, appearing in transverse section as sclerenchymatous caps of semi-lunar form (Figs. 122; 123; 124, SK). At each side of these bundles, i.e. opposite the junction of the xylem and phloem, the bundle sheath is interrupted by bands of less heavily thickened parenchymatous cells which facilitate the exchange of water and nutrients between the bundle and the surrounding parenchyma (Fig. 123, D).

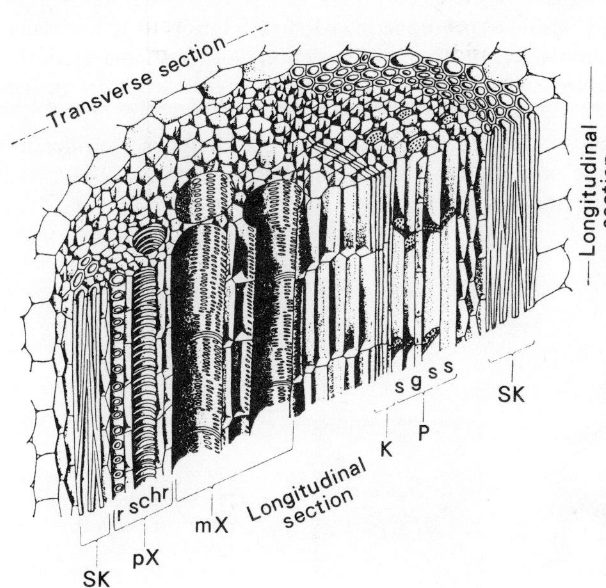

Fig. 124. Three-dimensional representation of a collateral bundle torn open. SK, sclerenchyma; pX, protoxylem; r, annular tracheids; schr, spirally thickened tracheids; mX, metaxylem; K, cambium; P, phloem; s, sieve tubes; g, companion cells. ×200. After Mägdefrau.

F. Mechanical tissues

Without a certain rigidity the plant could not maintain its proper shape, usually essential for the continuance of its life. Considerable rigidity is produced in thin-walled cells solely by the tension in their walls generated by the turgor of the cell (p. 220). The rigidity of growing tissues is produced entirely in this way (see p. 221 seq.). Since, however, such rigidity disappears in wilting, it is not adequate to provide land plants with permanent support. That solutions have been found to this problem becomes evident from, for example, the fact that a hollow stem of rye, reaching a height of 1.5 m, has a diameter at its base of hardly 3 mm, and yet bears at its tip the weight of the ear. Similarly, the slender stems of palms support not only the heavy fruits but also withstand the strain of the great, sail-like leaves. In species of *Raphia* these leaves may reach 15 m in length and a considerable width.

Besides rigidity, however, the plant body also requires high elasticity. The long stem of rye, for example, yields to gusts of wind, and the tip can be momentarily bent over without the stem breaking. The mechanical qualities of the plant body thus show remarkable perfection. Indeed, man makes very considerable use of the structural materials provided by plants. Wood, for example, is used for joists and beams, and the fibres for string, ropes and textiles (e.g. linen).

In most plants rigidity and elasticity depend upon the presence of specialized cells, aggregated together to form mechanical tissues. The walls of these cells are heavily thickened, either partly (as in collenchyma) or generally (as in sclerenchyma). In sclerenchyma the thickening may lead to the almost total occlusion of the lumen of the cell, and the protoplast usually sooner or later dies (Fig. 125C).

1. Collenchyma. In those parts of plants which are still actively growing collenchyma is the only mechanical tissue present. It is always primary in origin, and the walls are only partly thickened.

The primary collenchyma cells are usually prosenchymatous and may reach 2 mm in length, the ends running out to awl-like points. The cells often undergo secondary transverse divisions, so that a strand of cells (collenchyma fibre) results. After maceration the fibre, surrounded by the original primary wall, remains intact. In many respects they resemble parenchyma cells and like these may contain chloroplasts, but collenchyma cells are distinguished by the fact that their cellulose walls are unequally thickened. The thickening is especially marked at the angles of the cells (angular collenchyma) (Fig. 125A, B) or on the tangential walls. It consists to only a small extent of cellulose, the remainder consisting of strongly swollen protopectin, so that fixation in dehydrating fixatives causes considerable shrinkage. In transverse sections of fresh material the thickened parts of the wall glisten conspicuously. Intercellular spaces are more or less lacking or very small.

As a result of the thickening of the cell walls, collenchyma possesses considerable resistance to tearing. At the same time the long unthickened areas of the wall allow an almost unimpeded exchange of metabolites.

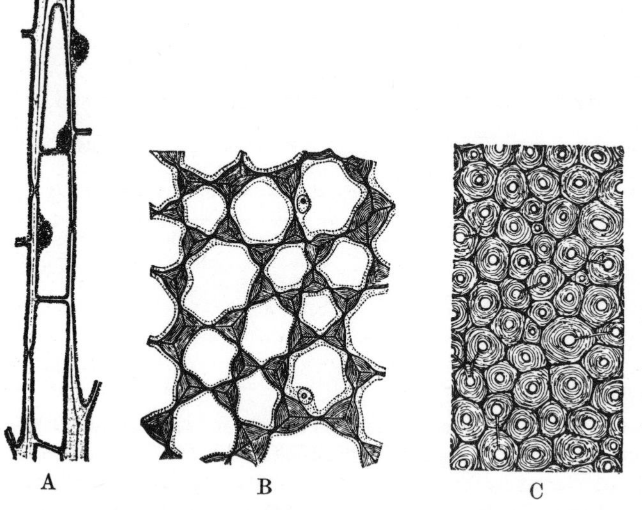

Fig. 125. Strengthening tissue. **A.** A column of collenchymatous cells in *Salvia*, side view. **B.** Transverse section through the angular collenchyma of a stem of *Cucurbita*. **C.** Transverse section through the sclerenchyma in the leaf of *Phormium tenax* (Liliaceae). **A.** ×160. After Haberlandt, freely adapted. **B.** and **C.** ×240. Both after Fitting.

2. Sclerenchyma. In the fully developed parts of the plant dead sclerenchyma takes over the task performed in the growing parts by the living collenchyma. According to whether this mechanical tissue is called upon to withstand pressure or tension, so it consists either of thick-walled stone cells, or of more or less thick-walled, elongated, sclerenchyma fibres. The stone cells of sclereids (Fig. 56A) are more or less polyhedral, and their walls, almost always very heavily lignified and consequently rigid, contain numerous round, tubular, branched or unbranched pits. The sclerenchyma fibres, on the other hand, are prosenchymatous, spindle-shaped and often become, as a consequence of tip growth, extremely elongated cells (Figs. 125C; 161D, E). The ends are sharply pointed, and the pits are few, obliquely ascending and slit-like, or are absent. All sclerenchymatous elements are polygonal in transverse section and have conspicuously narrow lumina, often appearing only as a dot (Fig. 125C). Their walls are either almost unlignified (as, for example, in *Linum*) and are then very elastic, or they are lignified to a varying extent and correspondingly rigid (as, for example, in *Cannabis*). Sclerenchyma fibres, although of small diameter, average 1–2 mm in length, unusually long for plant cells. Indeed, in flax they reach 0.4–6.5 cm, in *Urtica dioica* 7.5 cm and in *Boehmeria* (the source of ramie fibre) they extend to 30 cm and sometimes even to 55 cm. They often become multinucleate (polyenergid, see p. 17).

Because of their length and mechanical properties, such fibres are well suited for the production of textile yarns. They do not become completely differentiated until after extension growth of the organ has been completed. Their development frequently involves active apical growth (see Fig. 162).

The resistance to rupture of a sclerenchyma fibre is considerably enhanced by the spiral arrangements, often in converse senses, of the microfibrils in the individual layers of the wall (Fig. 63). With this arrangement, tension, as in a rope, causes the microfibrils to be pressed even more closely together. The cohesion of a tissue of sclerenchyma fibres depends upon the uninterrupted interlocking of the individual fibres.

Within the limit of elasticity, the load-bearing capacity of sclerenchyma fibres taken from the living plant is, generally speaking, equal to that of the best wrought iron, and in some plants even reaches that of steel. Moreover, the elasticity itself is some ten to fifty times greater than that of wrought iron, which is also inferior to the fibres in its much greater weight. When the limit of elasticity is overstepped the fibres rupture immediately, whereas this does not occur with iron until the load exceeds three times the limit.

The hardness of, for example, the rigid and brittle shells of nuts and the stones of drupes usually arises from the presence of stone cells. The resistance to bending of stems and many leaves, the columnar rigidity of the trunks and the resistance to stretching of roots depend, on the other hand, on the presence of tissues composed of sclerenchyma fibres. Both kinds of mechanical tissue contribute to the resistance which many organs offer to incision and other kinds of penetration.

Sclerenchyma cells and fibres are partly primary and partly secondary in origin, and are either scattered singly or arranged in groups. They are usually in strands, bands, sheaths, or shell-like layers, without intercellular spaces, and so arranged that they produce the required firmness in the organ with the minimum amount of strengthening material.

3. The principles governing the arrangement of the mechanical tissue. If a beam supported from beneath at each end, be mechanically bent (Fig. 126A, B) its convex side becomes longer, thus setting up tensile stress, while its concave side becomes conversely shortened and compressed. Only the middle zone, the so-called neutral plane, remains scarcely affected mechanically (apart from slight curvature). The danger of the structural elements tearing apart is therefore greater the further they lie above or below the neutral plane. Thus, whenever resistance to bending is required it is desirable that the strengthening elements be arranged peripherally.

In buildings of reinforced concrete the mechanical elements—provided by the framework of steel girders—are held in places of greatest stress by the less mechanically strong filling of ground material—the concrete (Fig. 126D). The arrangement of mechanical tissues, both collenchymatous and parenchymatous, in shoots is comparable with this kind of structure. The ground tissue or parenchyma represents the concrete, and is in fact superior to it by virtue of its elastic properties. Since the effectiveness of the reinforcement increases with its distance from the neutral axis, it is especially advantageous if it is at or protrudes from the periphery. Such an arrangement is seen in the four projecting wing-like ridges of many stems of Labiatae (Fig. 126H).

Completely different arrangements are found in those plants and plant organs which are subjected principally to tension. Such are the ascending shoots of water plants, and the roots and root-stocks of land plants strained by the stem moving in the wind. Only if the tension is quite uniform and exactly in the direction of the longitudinal axis are all the fibres in the transverse section of such an organ engaged simultaneously. If, as is more usual in nature, the tension shifts, being first greater on one side and then on the other, strands of strengthening elements distributed uniformly over the whole transverse section of the organ would be ruptured individually one after the other. Accordingly, in those

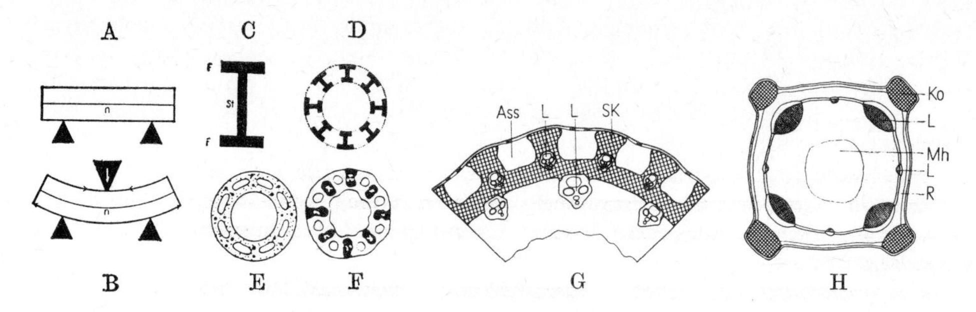

Fig. 126. Principles of mechanical strengthening. **A.** and **B.** Stresses in a beam subjected to bending: extension of the convex side; shortening (compression) of the concave side, the 'neutral plane' merely becoming bent and undergoing no change in length. **C.** Double T-beam (St, web; F, flange). **D.** Hollow cylinder, the walls strengthened with double T-beams. **E.** A factory chimney cast in concrete; the reinforcement is limited to eight pairs of opposed iron rods, making for a great saving in constructional materials. **F.** Transverse section through the stem of *Trichophorum germanicum*: the arrangement of the tracts of strengthening tissue and the hollow centre is exactly as in **E.** **G.** Transverse section through the stem of *Molinia coerulea* (Poaceae); SK, sclerenchyma; L, vascular bundle; Ass, assimilatory parenchyma. **H.** Transverse section of a shoot of *Lamium album*. The four-angled stem is supported by four columns of collenchyma (Ko). L, vascular bundle; R, cortical parenchyma; Mh, pith cavity. **E.** and **F.** After Rasdorski. **G.** ×25. **H.** ×10. Original.

roots and shoots subject to tension the strengthening elements are usually fused together into a central cable-like strand, of which the vessels with their thickened walls form a conspicuous component (Figs. 187A; 189; 190).

G. Excretory and secretory tissues

In plants, unlike animals, there is no expulsion of waste products or excrement, if we exclude the fall of leaves and the scaling off of old cortical tissue in the form of bark. On the other hand, in all land plants there is a continuous, albeit inconspicuous, giving off of gaseous end-products (particularly oxygen, carbon dioxide and water vapour). The tissues through which this release of metabolites into the atmosphere is accomplished can justifiably be referred to collectively as excretory systems.

Thus the green spongy mesophyll of leaves, already mentioned in the section dealing with parenchymatous tissues, with its well-developed intercellular spaces connected with

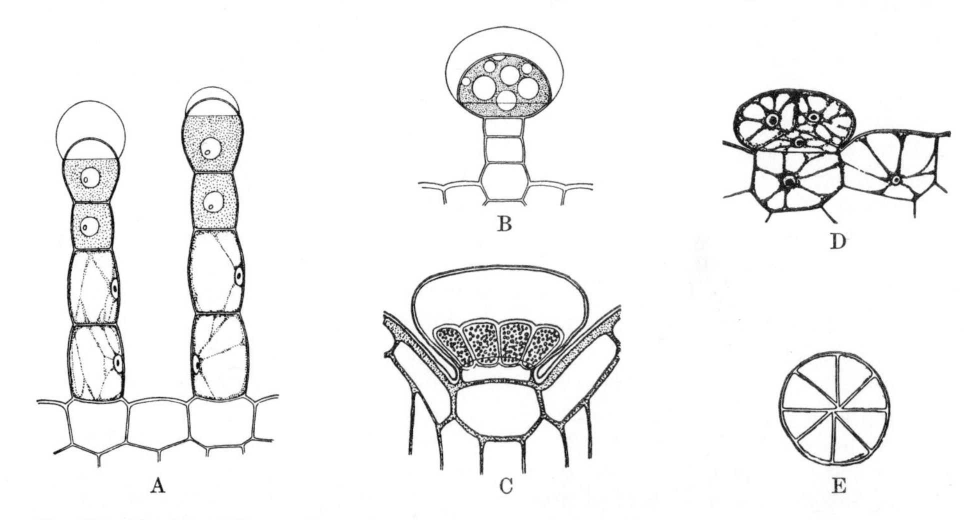

Fig. 127. Glandular hairs. **A.** Hairs from the petiole of *Primula obconica*. The secretion, which collects beneath the cuticle, can cause an itching dermatitis. **B.** Hair from the peduncle of *Pelargonium*. **C.** Orbicular glandular scale in the epidermis of the upper side of the leaf of *Thymus vulgaris* in vertical section. The cuticle forced up by the secretion. **D.** and **E.** Sessile digestive gland on the upper side of the leaf of *Pinguicula vulgaris*. **D.** Longitudinal section. **E.** View from above. **A.** and **B.** ×100. Original. **C.** ×200. After de Bary. **D.** and **E.** ×200. After Fenner.

the external air by way of numerous stomata, can be classed as an excretory tissue in the widest sense. The green primary cortical parenchyma, also well aerated by way of intercellular spaces and lenticels, can be similarly classified. At the same time by facilitating gas exchange these tissues fulfil the important function of absorption.

In addition, there are also specific unicellular (e.g. idioblasts, hairs) or multicellular internal and external systems whose metabolism is directed particularly to the production of certain secretions or excretions (see p. 289 seq.). If the extruded materials remain within the lumina of the cells they are then referred to as excretory (or secretory) cells, or excretory (or secretory) tissues. If, however, they are expelled through the wall of the cell we are then concerned with glandular cells or glandular tissues.

1. Excretory cells and tissues. These are secondary tissues scattered sporadically in the various primary tissues. Excretory cells are isodiametric to tubular (e.g. Fig. 128B). They occur as idioblasts (e.g. the cells with the crystalline druses in Fig. 171), or they form a complete tissue complex (Fig. 129B) or columns (Fig. 130) or layers (Fig. 169A and B, Kr) of cells. Their excretion products accumulate in the protoplasts as numerous small or fewer larger vacuoles frequently eventually occupying the whole lumen of the cell. The cell itself often becomes suberized and finally may even die. Excretory products of widespread occurrence are: mucilages, gums, gum-resins, resins, ethereal oils, tannins, alkaloids and crystals of calcium oxalate (see p. 57 seq.).

Simple latex ducts are also forms of excretory or secretory cells, the latex being the extruded material. These ducts, although usually richly branched, lack cross walls. They are tubes round in section, with a smooth elastic wall consisting of cellulose. There is a layer of living cytoplasm on the wall, possessing numerous nuclei, and often also starch grains. In place of normal cell sap they contain a milky, white or occasionally otherwise coloured, watery emulsion which flows out of damaged tubes and rapidly coagulates in the air. Multinucleate, branched, yet still simple (i.e. each consisting of a single cell) latex ducts are found in *Euphorbia* (Euphorbiaceae), *Asclepias* (Asclepiadaceae), *Nerium* (Apocyanaceae) and *Ficus* (Moraceae). Unbranched simple ducts occur in *Vinca* (Apocyanaceae), *Cannabis* (Moraceae) and *Urtica* (Urticaceae). Latex ducts usually develop from meristematic cells of the young embryo which already in the seedling have become multinucleate and cylindrical. They continue to grow with the whole plant, growth being principally more or less parallel to the longitudinal axes of its organs. They branch continuously, penetrate into all parts of the plant and in this way may become many metres in length. They are the largest plant cells known lacking cross walls.

Sometimes dissolution of the dividing walls may occur and several secretory cells fuse together. If only the transverse walls disappear, then a tubular secretory reservoir is formed, examples of which are the compound latex ducts. They appear quite similar to, and their contents resemble those of the simple ducts, but they are distinguished from these by their having arisen from a row of cells whose cross walls have eventually disintegrated. Compound latex ducts are thus segmented. In contradistinction to the simple

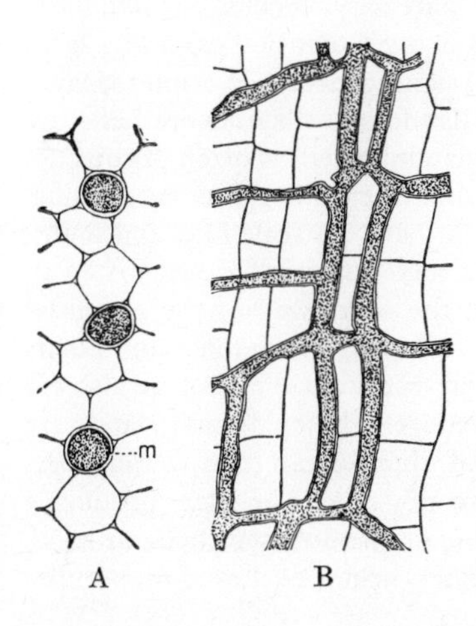

Fig. 128. **A.** Transverse sections of latex ducts (m) in parenchyma. **B.** Tangential longitudinal section from the periphery of the root of *Taraxacum* showing the compound latex ducts fused to form a reticulum. **A.** ×450. After Fitting. **B.** ×160. After Fitting.

latex ducts, the compound are often joined together to form a reticulum (Fig. 128). Segmented latex ducts are also confined to certain families. Segmented, but not anastomosing, ducts are found in the Musaceae (*Musa*, banana), Apocyanaceae (*Nerium*), Convolvulaceae (*Convolvulus*, *Ipomoea*), and the Papaveraceae (*Chelidonium*, with its orange-red latex). Anastomosing latex systems occur in *Papaver*, in the Euphorbiaceae (e.g. in the rubber tree, *Hevea*), and in the liguliflorous Compositae (e.g. *Taraxacum*, *Lactuca*, *Tragopogon*).

Dissolved in the latex of both simple and compound ducts occur the following: sugars, tannins, glycosides, many poisonous alkaloids (e.g. morphine and other alkaloids in the latex of *Papaver*, together constituting opium), and especially calcium malate. In *Ficus carica* and *Carica papaya* peptase enzymes occur as well. Other substances occurring as droplets in emulsion are ethereal oils, waxes, mixtures of gums and resins, and the polyterpenes caoutchouc and guttapercha. Solid components are starch, numerous protein granules and here and there albuminous crystals. The significance of the latex ducts for the plant is still unknown. Latex may serve to discourage grazing by vertebrates, but the caterpillars of certain moths seem in no way repelled.

Lysigenous secretory reservoirs arise from groups of cells rich in secretory products whose walls and protoplasts gradually disintegrate. Examples among others of these structures are the lysigenous secretory sacs or cavities, filled with ethereal oil, of oranges, lemons and other members of the Rutaceae (Fig. 129), and of many Myrtaceae.

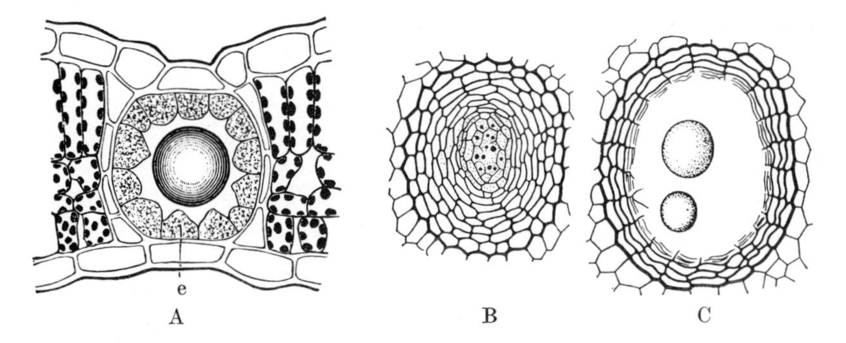

Fig. 129. A. Schizogenous oil sac in transverse section in the leaf of *Hypericum perforatum*: e, grandular epithelium. **B.** and **C.** Lysigenous oil sac in the peel of *Citrus vulgaris*, before (**B.**) and after (**C.**) release of the oil from the cells. **A.** ×80. After Haberlandt. **B.** and **C.** ×40. After Tschirch.

2. Glandular cells and glandular tissue. Besides secretory cells, and like them either singly or in groups, we often find glandular cells in the epidermis, parenchyma or in other kinds of tissue. These cells may occur as primary or secondary elements, and, in contrast to secretory cells, release their secretions outside the plant body or into the intercellular spaces. Glandular cells, which are always living, resemble parenchyma cells, but like meristematic cells often retain much cytoplasm (containing many Golgi bodies) and a large nucleus. The extruded materials often have an ecological significance. A continuous layer of glandular cells is termed a glandular epithelium. Glandular cells and epithelia are frequently developed in epidermal tissue. The cuticle of glandular cells is often porous.

In the epidermis occur glandular hairs, e.g. capitate hairs, where the glandular terminal cell is developed as a head (Fig. 127A, B). Other glandular hairs are scale-like, and multicellular glandular papillae are also found (Fig. 108). The secretion itself is very often an ethereal oil or resin. In such cases it arises first between the outer wall of the glandular cell and the cuticle, so that the latter is raised up (Fig. 127A, B, C), and in many plants (e.g. *Primula*, *Pelargonium*, and many Labiatae) ultimately ruptured. This is true also for the extrusion of many other viscous materials and mucilage. The epidermal glands are classified according to the nature of the excreted material, thus we can refer to mucilage, resin, salt, oil and digestive glands (Fig. 127D, E), and to nectaries. The latter produce a sugary secretion with attracts insects. Nectaries, consisting of glandular epithelia or hairs, occur principally within flowers (floral nectaries). Extrafloral nectaries, found on petioles

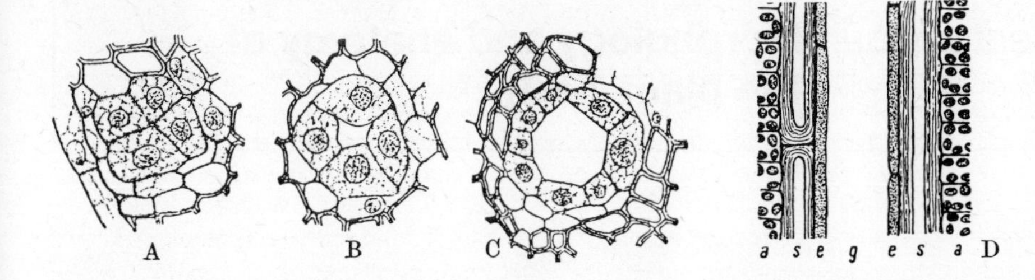

Fig. 130. A. to **C.** Schizogenous origin of a resin canal in the wood of *Pinus*. **D.** Resin canal (g) in the leaf of *Pinus* in longitudinal section (transverse section is shown in Fig. 179, H): e, epithelial cell; s, sclerenchymatous sheath; a, assimilatory parenchyma. **A.** to **C.** ×250. After W. H. Brown. **D.** ×240. After Haberlandt.

(e.g. *Prunus*, *Acacia*), stipules (e.g. *Vicia*) or in the angles between the nerves of the leaf (e.g. *Catalpa*), have a quite different construction (see also p. 100).

Schizogenous secretory reservoirs. The glandular cells or epithelia enclosed in parenchyma or other tissue always border on to irregular or more or less spherical intercellular spaces (Fig. 129A), or on to branched or unbranched intercellular canals, often forming an intercommunicating system penetrating the whole plant (Fig. 130B, C, D). These intercellular spaces are formed by the glandular cells moving apart from each other, and their secretion accumulates in them. These reservoirs are thus schizogenous in origin. According to the nature of the contents we can again distinguish oil, resin, gum or mucilage canals or sacs.

Resin canals formed in this way occur, for example, in many conifers (Figs. 163, h; 164); oil canals (with ethereal oil) in, for example, the Umbelliferae; mucilage and gum canals in the Cycadaceae. Spherical or elongated schizogenous sacs filled with ethereal oils are found in species of *Hypericum* (Fig. 129).

Hydathodes. In many mono- and dicotyledons, especially at the tips of the leaves (e.g. Gramineae), or on the teeth of the leaf margin (*Alchemilla, Fuchsia*), or singly at the ends of the main veins of the leaf (e.g. *Tropaeolum*), are found groups of small parenchymatous cells, lacking chlorophyll, which excrete droplets of water. These groups of cells, which occur beneath special water stomata (Fig. 105), are referred to as epithems, and a group of cells with its water stoma forms an epithemal hydathode.

As well as from these parenchymatous excretory tissues, liquid water can also be expelled from epidermal hydathodes. These are formed either by groups of metamorphosed epidermal cells or by multicellular hairs, the trichome hydathodes. Glands concerned specifically with secreting gas are found in the air bladders of many brown algae (e.g. *Nereocystis*). Islands of tissue in many flowers, formerly misinterpreted as providing 'fodder' for visiting insects, are now known to constitute distinct scent glands (osmophores) releasing volatile substances which attract pollinators.

Section four: **Morphology and anatomy of the plant body**

The basic pattern of structure seen in all plants, from the simplest multicellular algae to the most highly developed Cormophyta, is related to the autotrophic manner of life, already discussed in the Introduction. This pattern follows from the polarized differentiation of, on the one hand, a shoot directed towards the light and furnished with flattened assimilators of the largest possible size and, on the other, an extended organ serving as an anchor in or on the substrate, and in land plants as a water-absorbing organ as well. Consequently, the fundamental units of function of all higher land plants are a shoot furnished with leaves, and a more or less richly branched root system.

I. Construction of the typical plant body

Even the very young embryo (see Figs. 90H, I; 131A, B) is already differentiated into a root (radicle) and a stem (plumule). In the seed plants this differentiation has usually occurred long before the seed is fully formed. The plumule bears one to several leaves (cotyledons; Fig. 131C, D). In seed plants the lowest part of the stem of the seedling, from the boundary between the root and the stem up to the cotyledons, is called the hypocotyl (Figs. 131C, D, Hy; 172, Hy; 173, Hy). The succeeding portion of stem generated by the growing point of the seedling, extending from the cotyledons to the next leaf, is termed the epicotyl (Fig. 172E).

Later the surface area of the plant body almost always becomes considerably increased by branching. Both the shoot and root axes produce laterals, thereby giving rise to a shoot system and a root system. Later in development a shoot system may be generated afresh by suckers arising from the roots, and a root system from adventitious roots arising from the stem (Fig. 131D, w).

A. The shoot

The shoot consists of the axis, called the stem and normally developed as a cylindrical rod-like body, bearing the leaves, which are lateral appendages of the axis, usually of limited growth (see p. 162). The leaves are usually more or less flattened, and in the aerial shoot they are present mainly as green foliage leaves.

Thus the shoot axis bears the leaves. Additionally the axis is responsible for the transport of metabolites between the leaves and the roots, and reserve materials are stored within it. The foliage leaves, usually green, like the flattened assimilators of the Thallophyta, serve for photosynthesis, but at the same time they are also the most important organs of transpiration. The structure of the stems and leaves, both externally and internally, is such as to fit them for these diverse functions.

In very many perennial herbaceous plants, besides the assimilating aerial shoots, there also occur shoots running below the level of the soil and furnished with reduced leaves and adventitious roots. These serve as food-storage or resting organs, and provide a means of tiding over periods unfavourable for growth (Fig. 205A, B).

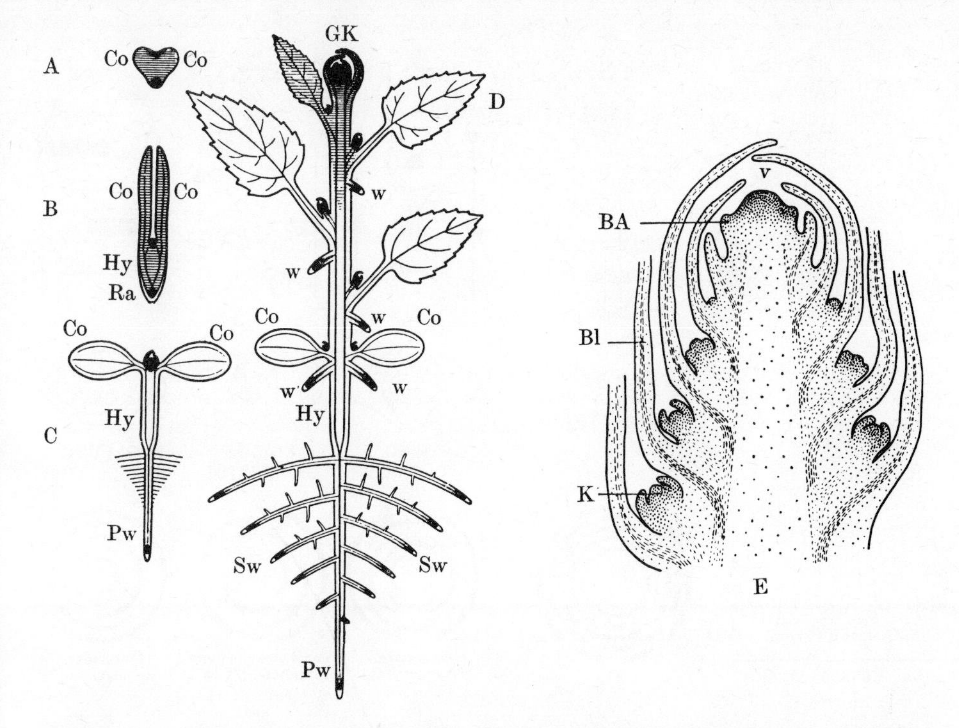

Fig. 131. Diagrammatic representation of a dicotyledonous plant. **A.** Young embryo. **B.** Mature embryo. **C.** Seedling with the two cotyledons (Co). **D.** Juvenile plant in a purely vegetative state. Ra, radicle; Pw, primary root; Sw, lateral root; Hy, hypocotyl; w, adventitious root; GK, apical bud. **E.** Longitudinal section of the shoot apex of a seed plant: v, the growing point; BA, leaf primordium. The primordia of the lateral shoots (K) arise in the axils of the young leaves (Bl) which envelop the growing point. **A.** to **D.** After Sachs, with additions from Troll. **E.** ×60. After Strasburger, modified.

1. The bud

Since the microscopically small growing point is usually enclosed in older leaves, it is not visible unless these are carefully removed, or unless a longitudinal section is examined with the lens (Figs. 131E; 132A, B). Leaf primordia can then be seen arising laterally close beneath the tip, and a little lower down between these leaf primordia those of the lateral branches also become evident (Fig. 131E).

The increase in diameter below the summit of the apex, as well as the internal and external development of this region, follow from the ontogenetic processes by which new lateral members are organized from the embryonic tissue of the tip. This growth usually begins with a pronounced extension of the leaf primordia. These soon extend above the stem tip and, as a consequence of greater growth on the lower side, the older leaf primordia come together to form a more or less dome-like covering above the growing point (Fig. 131D, E) and the younger primordia. Thus the larger and older leaf primordia provide an effective protection for the delicate apex and youngest primordia (e.g. against desiccation), and together with the growing point form a bud. The bud is therefore nothing more than the juvenile and incompletely developed end of a leafy shoot.

2. Principles of leaf arrangement[26]

The primordia of the leaves arise at the apex from the outer layers of cells (exogenously) and in sequence towards the tip (acropetal sequence) as a result of periclinal divisions in the cells of the tunica (Fig. 132B).

In the dicotyledons these divisions do not usually take place in the outermost layers of cells, but in fairly narrowly confined places in one or more of the layers lying beneath.

After the eventual elongation of the shoot axis the places at which the leaves are inserted on the shoot often appear thickened (as, for example, in the Gramineae). Accordingly, these positions are referred to as nodes, and the intervening portions of shoot free of leaves are distinguished as internodes.

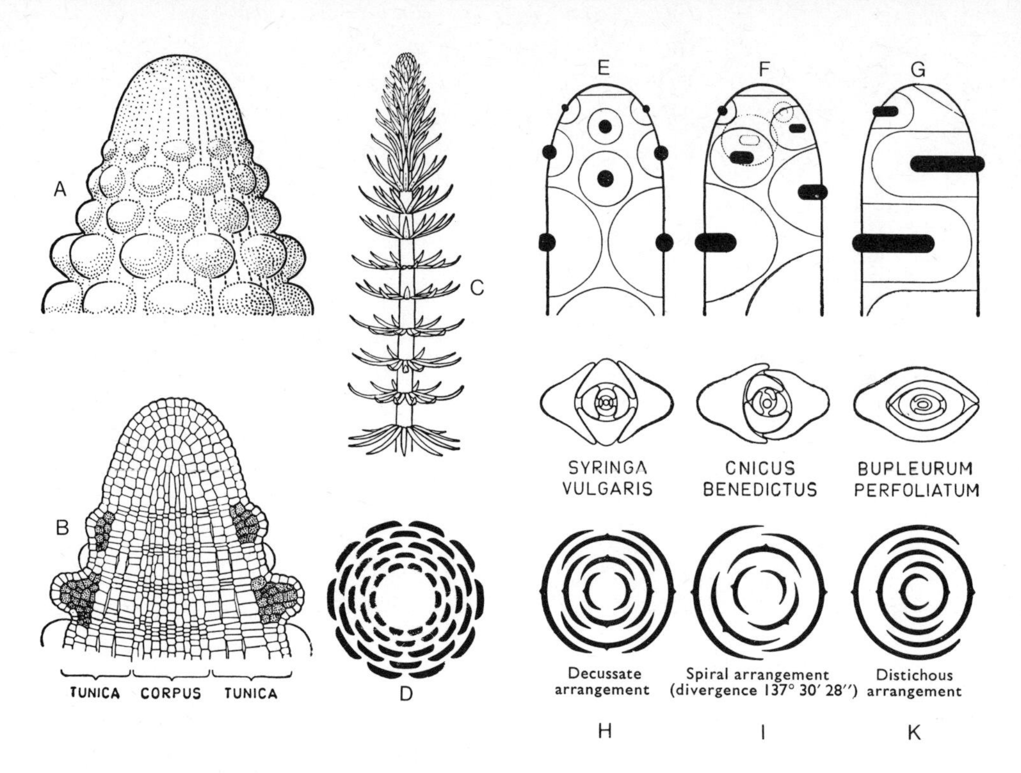

Fig. 132. **A.** to **D.** *Hippuris vulgaris.* **A.** General view of shoot apex. **B.** Apex in longitudinal section. The leaf primordia arise as a result of periclinal divisions in the second layer of the tunica. **C.** Side view of the shoot showing the whorled arrangement of the leaves. **D.** Diagram of the leaf arrangement. **E.** to **G.** Diagrammatic representation of the three most important types of leaf arrangement. **E.** and **H.** Decussate arrangement (two-membered whorls). **F.** and **I.** Spiral arrangement, the angle of divergence being the limiting angle. **G.** and **K.** Distichous arrangement. In the upper row the interference fields (see p. 93) are indicated about each leaf primordium (black). These are smallest in **E.** and largest in **G.** The transition from distichous to spiral phyllotaxy may lead to either a right-handed or a left-handed spiral. **A.** and **B.** ×30. **B.** After Strasburger, modified. **C.** ×½.

The number of leaf primordia which arise at one time, and therefore at one and the same node, is determined by the sizes of the individual primordia and the spaces available for them. This in turn depends upon the circumference of the growing point. If the leaf primordia are relatively small, then the greater is the number of available places on the perimeter. Thus there arises a many-membered (polymerous) whorl or verticil of leaves (as, for example, in *Equisetum*, *Ceratophyllum*, *Casuarina*, *Hippuris*; Fig. 132C, D).

The relative positions of the leaves can be seen most clearly if they are indicated in a diagrammatic ground plan. At the centre lies the median point of the apex of the shoot. The leaves lying closest to the centre are then the youngest and uppermost primordia, those further from it are successively older and more developed. Each node can be conveniently indicated by a circle. The leaves of a whorl will thus be shown on one circle (Fig. 132D, H, I, K). The larger the individual leaf primordia, the fewer find a place on the same circle. The line running from the centre of the axis through the middle of a leaf primordium is referred to as the median line of that leaf.

The intervals between successive leaf primordia are usually equal, so that the leaves are distributed uniformly about the shoot (principle of equidistance). The leaves of successive nodes are usually inserted with their median lines lying exactly in the middle of the space between the leaves of the preceding whorl (principle of alternation; Fig. 132A, D, E, H). In the interval of time between the origin of the leaves at two consecutive nodes, referred to as the plastochrone, the growing point changes its form in a characteristic manner (see, for example, Fig. 134A, B).

Three-membered whorls occur, for example, in *Elodea* and *Juniperus*. An arrangement

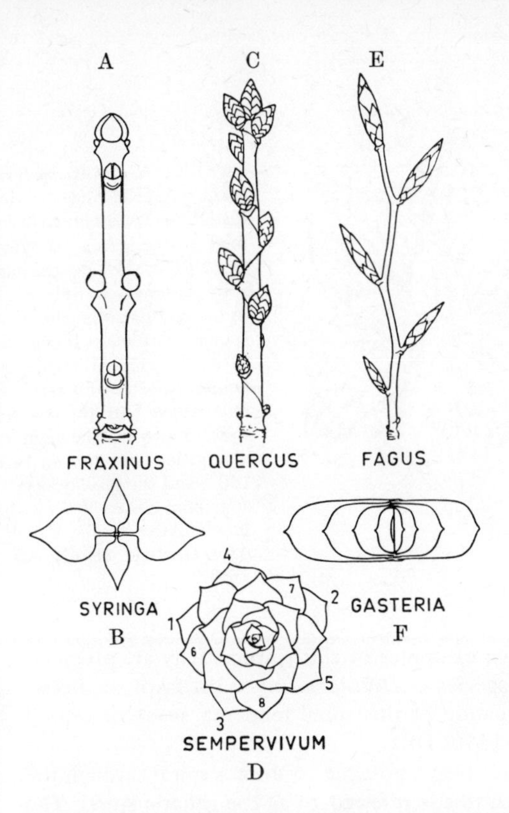

Fig. 133. The three most important types of leaf arrangement. **A.** and **B.** Decussate arrangement. **C.** and **D.** Spiral arrangement. **E.** and **F.** Distichous arrangement. (The branches of *Fraxinus, Quercus* and *Fagus* are in the winter condition. The positions at which the leaves were borne are indicated by the axillary buds.)

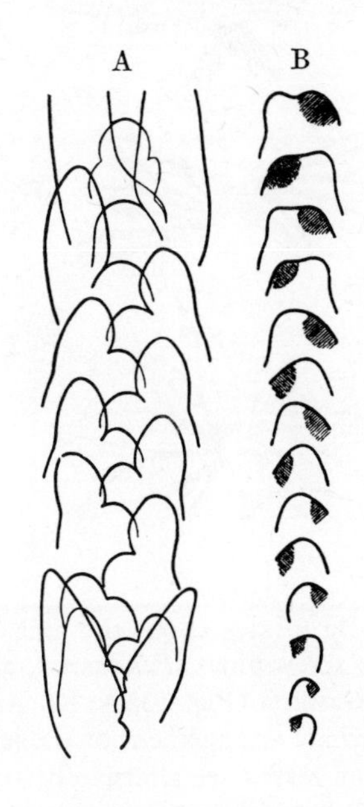

Fig. 134. *Juncus bufonius.* From below upwards; changes in form of the vegetative apex in successive stages of development. The growing point is displaced alternately towards the right and the left by the swelling leaf primordia. In **B.** the meristem of the main shoot is cross-hatched. ×20. After Gliem-Riebesel.

of the leaves in two-membered (dimerous) whorls is very frequently met with. Such a decussate arrangement of the leaves is found in *Acer, Fraxinus, Aesculus* and in the Labiatae. In a decussate arrangement the leaves are arranged on the shoot in four vertical lines or orthostichies. The number of the orthostichies is therefore, as is evident from the diagram, twice the number of leaves at each node (Figs. 132E, H; 133A, B).

If the leaf primordia are attached to the shoot axis by broad bases, surrounding a half or more of the apical cone (Fig. 132G, K; also in many monocotyledons, e.g. Fig. 133F), only one primordium can arise at each node. In this case the succeeding leaf primordium generally originates on the immediately opposite side of the apical cone. In the outgrown shoot the leaves then stand in two vertical ranks on opposite sides of the stem (Figs. 133F; 135A), the insertions of the leaves following each other as the swings of a pendulum (Figs. 133; 134A, B). This two-ranked or distichous leaf arrangement is found, for example, in the Gramineae and in many other monocotyledons (*Iris, Gasteria*; Fig. 135A); but occurs also in other alliances (*Pisum, Phaseolus, Aristolochia*).

A special case of distichy is to be observed in the more or less horizontally growing side branches of many broad-leaved trees (e.g. *Corylus, Ulmus, Fagus*). The upright (orthotropic) primary axes of these species display, in contrast to the lateral branches, a spiral arrangement of the leaves. Already by the third year, however, the main axis becomes inclined and its growth habit correspondingly plagiotropic rather than orthotropic (see p. 355 seq.). Simultaneously the no longer appropriate spiral arrangement of the leaves (and branches) gives place to a distichous arrangement, thus making optimal use of the incident illumination.

In many distichous monocotyledons, on the other hand, there is a striking tendency towards a spiral arrangement of the leaves. It is as if the plane of the pendulum's swing

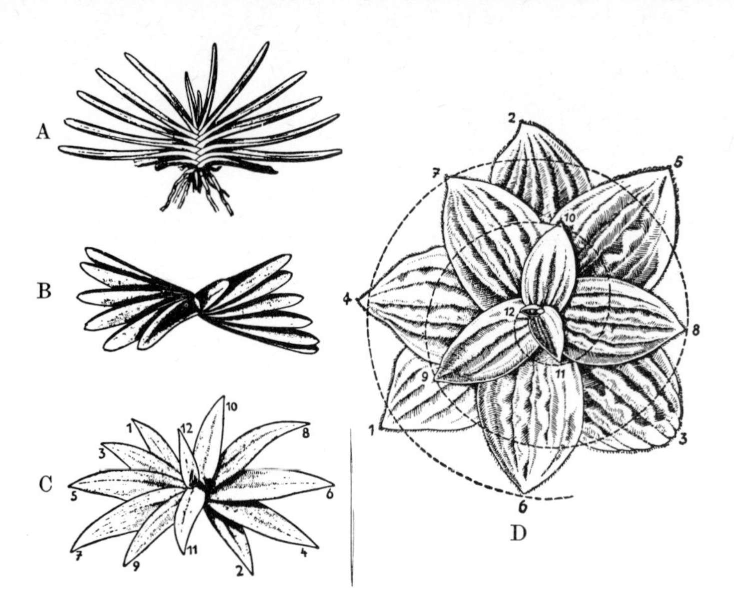

Fig. 135. **A.** and **B.** *Gasteria picta.* **A.** Side view showing t distichous (two-ranked) arran ment of the leaves. **B.** View from above showing the slight spirodistichy. **C.** Spirodistich in *Crinum powellii*, the sequence of leaves being indicated (1–11). **D.** *Plantago media*. Rosette of leaves seen from above with the genetic spiral drawn in. The angle of divergence lies between 140 an 150°, and the phyllotaxy approximates to $^3/_8$. **A.** and **B.** ×¼. After Troll. **C.** ×$^1/_{10}$. After Goebel. **D.** ×½. After Troll.

gradually rotates about the axis. Well-known examples of this spirodistichy are provided by the screw-pines (*Pandanus*), and many species of *Dracaena* and of the favourite house plant *Gasteria* (Fig. 135A, B). An intensification of this spiral tendency leads to a spiral or alternate arrangement of the leaves (Fig. 135C, D).

When leaves are alternately arranged it is always possible to draw a spiral through the median lines of successive leaves (Fig. 135); this is referred to as the genetic spiral. The sense of this genetic spiral varies from seedling to seedling and from branch to branch, and appears in most cases to be determined accidentally. Left- and right-handed spirals occur with approximately equal frequency (see Fig. 137D, as well as cones of *Pinus* and *Picea*).

The principle of equidistance is still valid for spirally arranged leaves. The angle between the median lines of successive leaves, the so-called angle of divergence, is constant. Usually, however, the divergence is not expressed by the size of this angle, but as a fraction of the circumference of the stem. The numerator of this fraction is the number of revolutions of the stem which are necessary to reach the next leaf of the same orthostichy; the denominator the number of plastochrones passed over until this leaf is reached. The most frequently observed divergences are those of the so-called main series of Schimper and Braun, namely, $\frac{1}{2}$(180°) (e.g. Gramineae); $\frac{1}{3}$(120°) (e.g. Cyperaceae); $\frac{2}{5}$(144°) (e.g. *Rosa, Corylus, Betula*); $\frac{3}{8}$(135°) (e.g. *Aster, Brassica, Plantago* (Fig. 135D), cones of *Tsuga* and *Pseudotsuga*); $\frac{5}{13}$(138° 27′) (e.g. *Sempervivum*, cones of *Pinus strobus*); $\frac{8}{21}$(137° 8′) (e.g. the cones of *Pinus sylvestris, P. nigra* and *Picea abies* (Fig. 137A)); $\frac{13}{34}$(137° 38′), $\frac{21}{55}$(137° 27′), $\frac{34}{89}$(137° 31′) (e.g. the capitula of various Compositae (Fig. 136)). These empirically discovered divergences have been a puzzling feature in the study of leaf arrangement, or phyllotaxy. As can be seen all these divergences form a series in which the numerator and denominator of each fraction are given by the sum of the numerators and denominators respectively of the two preceding fractions. Thus both the numerators and the denominators follow the so-called *Fibonacci series*. As the divergence increases the angular equivalents of the divergence fractions approach a limiting value, the limiting divergence angle, approximately 137° 30′. This particular angle divides the circle in the ratio of the 'golden mean'.

A stricter analysis of the angular displacement of the leaf primordia at the growing point has now shown that the higher divergences of the Schimper–Braun series are in fact by no means exactly followed. With a phyllotaxy of $\frac{2}{5}$, for example, the sixth leaf should stand exactly above the first. Accordingly, in plants with this divergence we should find five orthostichies (the denominator of the fraction gives the number of orthostichies). In fact, an unambiguous orthostichy is never seen at the growing point, and all higher divergences are found to approach the limiting divergence. In many plants it can be seen

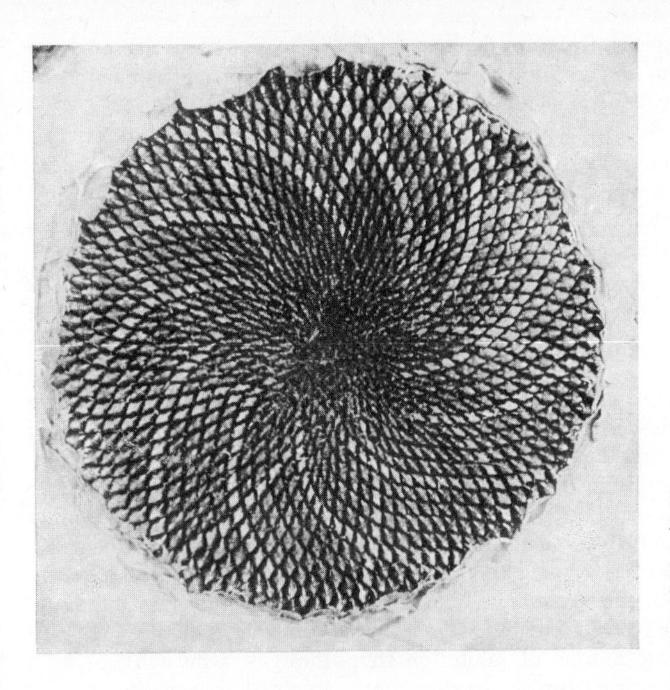

Fig. 136. Capitulum of *Helianthus annuus* in fruiting condition. The fifty-five left-handed and eighty-nine right-handed parastichies can be clearly seen. $\times 1/3$. After O. Habermann.

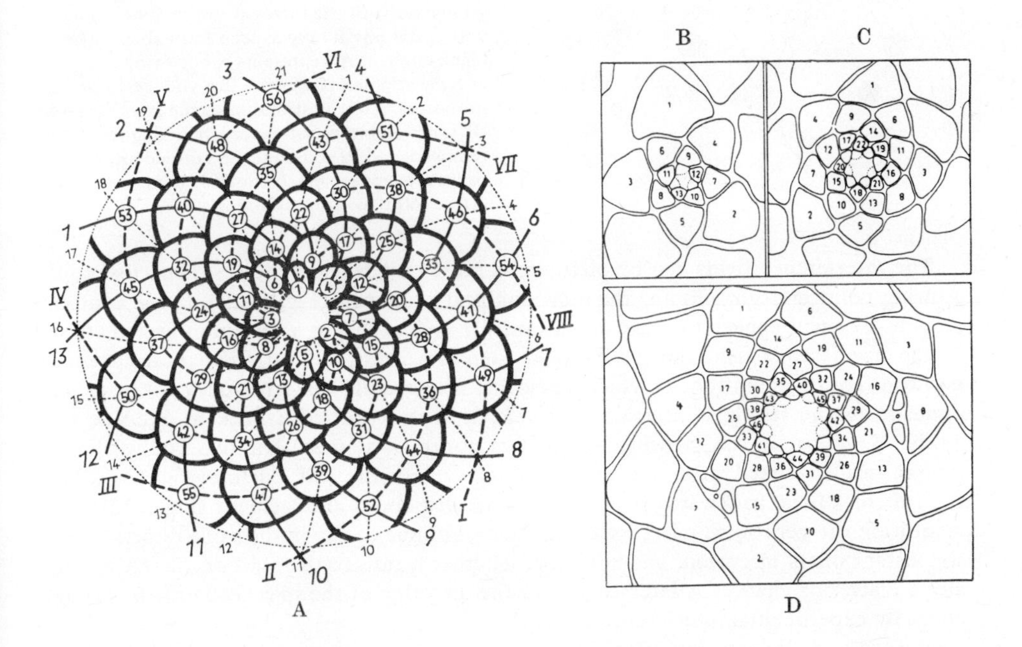

Fig. 137. **A.** Semi-diagrammatic view of a pine cone from below. The cone scales are arranged with the limiting divergence (successive numbering from 1 to 56). The dashed lines I–VIII and the continuous lines 1–13 are drawn in a clockwise and in an anticlockwise sense respectively through the scales with extensive mutual contact. (Contact parastichies with the contact numbers 1–9 and 1–14 respectively.) The dotted lines 1–21 are drawn through those scales apparently inserted above one another in vertical rows (orthostichies). In fact, these are also slightly spiral in a clockwise sense. **B.** to **D.** Transverse sections through three apices of different size of the conifer *Araucaria excelsa*. The size relationships between the apical cone and the leaf primordia determine the position of the latter in respect to each other and consequently the divergence. **B.** Divergence $3/8$ (genetic spiral right-handed). **C.** Divergence $5/13$ (left-handed). **D.** Divergence $13/21$ (left-handed). **A.** Original. **B.**, **C.** and **D.** $\times 25$. After Church, diagrammatic.

immediately how during ontogeny different divergences of the Schimper–Braun series can succeed one another without interruption (Fig. 137B, C, D). The divergence which we observe is clearly ultimately dependent upon the kind of spatial relationship which the young leaf primordium enters into with the preceding leaves of the genetic spiral. This spatial relationship, which determines the number of oblique rows, or parastichies (especially clearly seen in the cones of conifers (Fig. 137A) and the capitula of the Compositae (Fig. 136)), is itself a resultant of the relative sizes of the leaf primordium and the circumference of the growing point (Fig. 137B, C, D).

Since the Fibonacci angle divides the circle irrationally, the arrangement of the leaves with the limiting divergence of 137° 30′ ensures that each leaf receives the best possible illumination. Theoretically no newly formed leaf can ever come to stand directly above a preceding leaf.

To what are the initial departure from distichy and the transition to a spiral arrangement to be attributed? Each leaf primordium has the character of a broad meristemoid (see p. 93), and various observations and experiments have shown that during development every existing primordium inhibits the simultaneous appearance of further primordia in its neighbourhood. The apex of the shoot is itself surrounded by a similar interference field (Fig. 138A). If the apex be removed, then the inhibition it generated vanishes, and the leaf primordia are able to arise much closer to the tip than formerly (Fig. 138B). The nature of these interference fields is still obscure. They may be caused by the young primordium excreting chemical agents which impede growth in the surrounding area and so generate a zone of inhibition (hypothesis of repulsion), or be a consequence of each leaf requiring for its development a minimal area of the growing point.

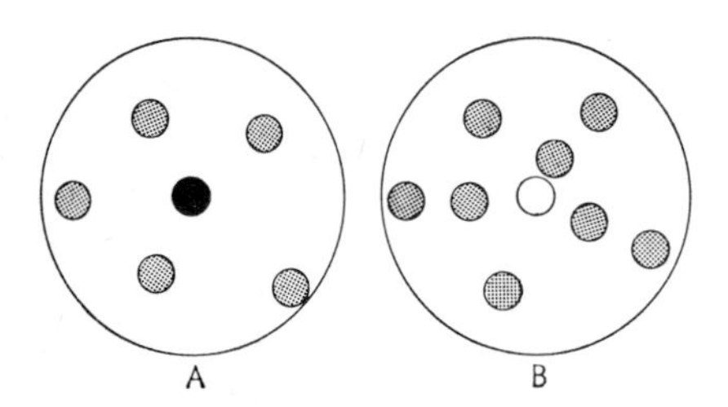

Fig. 138. Diagrammatic representation of the arrangement of the leaves at the apex of a plant with spiral phyllotaxy as seen from above. The black circle in **A.** represents the growing point, and the stippled circles the five youngest leaf primordia. After destruction of the growing point (**B.**) the inhibitory influence emanating from it is removed, and in consequence subsequent leaf primordia appear much closer to the centre. After Wardlaw, from Bünning.

The interference fields can be pictured as more or less circular discs lying close to the growing point and surrounding the individual leaf primordia (Fig. 132E). In the course of growth they will appear to move downwards. It is clear that at a certain intermediate size of the interference field the lowest (and therefore ontogenetically the most advanced) position free of inhibition, at which a new leaf primordium can arise, does not lie exactly opposite to the last leaf insertion (Fig. 132G), but is shifted a little to one side (Fig. 132F). Looking at the growing point from above, it is seen that two equivalent and symmetrically related positions are possible, namely the leaf can move either a little fowards or a little backwards from the original distichous arrangement (Fig. 132I), thus instituting the genetic spiral. This spiral must continue to run in the same direction from the instant of its inception, and therefore whether it runs to the right or the left is generally a matter of chance. A later change in the direction of the spiral can only be brought about by experimental interference.

Not only distichous but also decussate leaf arrangements can be transformed into spirals. In fact, the seedlings of alternate-leaved plants frequently show leaf arrangements which are different from those of the mature plants. Often an arrangement in whorls, usually restricted to the two cotyledons, is superseded by a transient distichy which sooner or later goes over to an alternate arrangement (e.g. *Alliaria petiolata*, *Aloe* (Fig. 139A)). In *Isoetes lacustris* the first eight to twelve leaves of the sporeling are strictly distichous. The divergence then gradually changes to $\frac{1}{3}$, $\frac{2}{5}$, $\frac{3}{8}$, $\frac{5}{13}$ and so on until a spiral arrangement with the limiting divergence is reached. However, the opposite can also occur: on the side shoots of the large Phyllocacti, whose areoli correspond to side shoots in the axils of reduced leaves (see Fig. 202), it can occasionally be seen that an initially

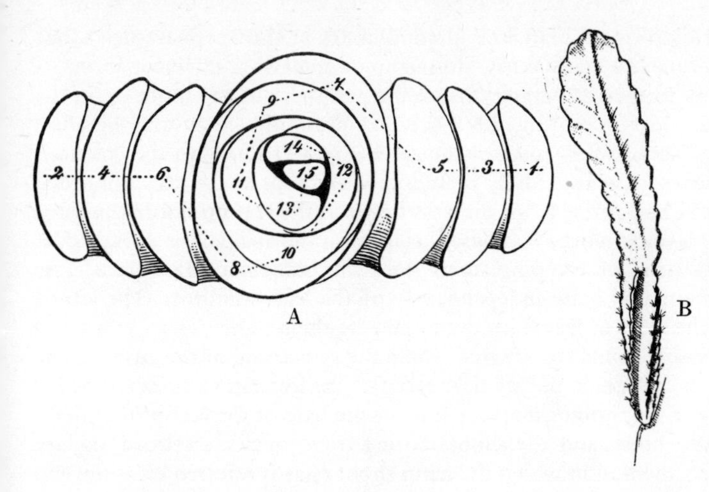

Fig. 139. A. Transverse section through the shoot of *Aloe serra*, showing a transition from a distichous (1–6) to a spiral (7–15) arrangement of the leaves. **B.** *Nopalxochia* (*Phyllocactus*) *phyllanthoides*. A side shoot with five ribs, and showing a spiral tendency, has grown out to form a flattened two-ribbed distichous shoot. **A.** Natural size. After Sachs. **B.** ×½. After Troll.

spiral phyllotaxy changes over into a perfect two-ranked arrangement (Fig. 139B). The leaf-like habit of the flattened side shoots (platycladodes) is dependent upon this change.

The divergences of the Schimper–Braun series are accordingly not determined by heredity. What is inherited is merely, as always, a pattern of behaviour, capable of tolerating a certain amount of variation. The phyllotaxy which is eventually adopted is dependent upon inner and outer factors (such as the size relationships between the apical meristem and leaf primordia, influenced in turn by nutrition, illumination and gravity), which influence each other in an intimate fashion.

3. Longitudinal symmetry

The internodes begin to elongate only a few millimetres behind the growing point. At the same time the young leaves unfold as a result of pronounced extension growth of their upper surfaces, and expand.

The elongation of internodes may, as the result of the activity of residual intercalary meristems (see p. 93), persist for a long time (intercalary growth). Individual leaves are then separated from each other by long sections of bare stem. In contrast to the often unbroken mitotic activity of the apical meristem, that of intercalary meristems is usually exhausted after some time. In the rosette plants all the internodes remain shortened throughout life, and consequently the leaves form a close rosette (e.g. *Plantago*, Fig. 135D); when this occurs the inflorescences arise on a side shoot. In the semi-rosette plants development again begins with a close rosette of leaves. In these, however, as the plant enters the flowering phase, the youngest internodes begin to elongate, so that a stem with longer internodes and usually simplified leaves emerges from the basal rosette. This stem terminates in a flower or an inflorescence (e.g. *Raphanus, Beta* and all grasses).

4. Branching of the shoot

As in the Thallophyta, the shoots of the Cormophyta can also branch in two different ways, namely, either by forking of the initial axis into two daughter axes (dichotomy, found almost only in the lycopods and in some of the Pteridophyta related to them), or by the initiation of daughter axes laterally on an axis which continues to grow without deflection, referred to as lateral branching (found in the ferns, *Equisetum* and in all seed plants).

a. Dichotomous branching. In the lycopods the apex of the shoot, independently of the production of leaves divides itself into two new apices of either equal or unequal size and activity (Fig. 87A, B). This results from growth being shifted laterally towards two diametrically opposite points on its broadening summit. Of these growing points the smaller (Fig. 87B, a′) is very soon displaced by the larger (a) as a result of its more vigorous growth, so that lateral branching is simulated (anisotomy, see Fig. 87D). If those branches which continue the branching all lie in more or less the one direction, and the others are at an angle to them, then a branch system arises which can so closely resemble racemose branching as to be mistaken for it.

b. Lateral branching. This form of branching predominates in the Spermatophyta, but it is also present in *Equisetum* and many ferns, although in a somewhat different form.

Lateral buds are usually formed as outgrowths on the periphery of the shoot or on the lowermost portions of the leaf primordia. Like the leaf primordia, they are therefore exogenous and are formed successively towards the apex (Fig. 131 K). Thus the positions in which the lateral branches arise are closely bound up with those of the leaf primordia.

In *Equisetum* the lateral buds arise from the axis between the leaf primordia. In ferns with upright shoots they develop on the leaf bases, usually on the abaxial (or dorsal) side. Only in ferns with dorsiventral or creeping stems does the branching take place close behind the apex of the main axis quite independently of the leaf primordia. The lateral branches may then be at the sides of leaves, or even opposite them.

In all seed plants the lateral buds are situated where the upper side of the cushion-like leaf primordium adjoins the tissue of the stem, the axil of the leaf. Sometimes the bud is more towards the stem, and sometimes more or less on the base of the leaf primordium. They are therefore axillary buds, and the shoots arising from them are termed axillary shoots. To avoid confusion the bud in which the main shoot ends is referred to as the end or terminal bud.

The leaf in whose axil a bud stands is spoken of as the subtending leaf or bract. In general, the axillary bud stands directly above the median line of the leaf (Fig. 140I and II); only rarely is it displaced laterally.

While in the angiosperms each leaf normally subtends an axillary bud (which does not necessarily of course, begin immediately to develop; see Section 5, Systems of Branching), in many gymnosperms (*Taxus, Picea*) bud are produced in only a few leaf axils.

A less common occurrence is the formation of further lateral buds by the first axillary bud. These either arise one above the other (serial buds), as for example in *Lonicera, Robinia, Forsythia* (Fig. 140III, upper) or beside one another (collateral buds), as for example in many Liliaceae, such as species of *Allium* and *Muscari*, and in the inflorescence of the banana (Fig. 140III, lower).

Besides this normal branching, the shoot system is occasionally augmented by adventitious shoots, such as those which emerge from the roots of herbs (e.g. *Coronilla varia*

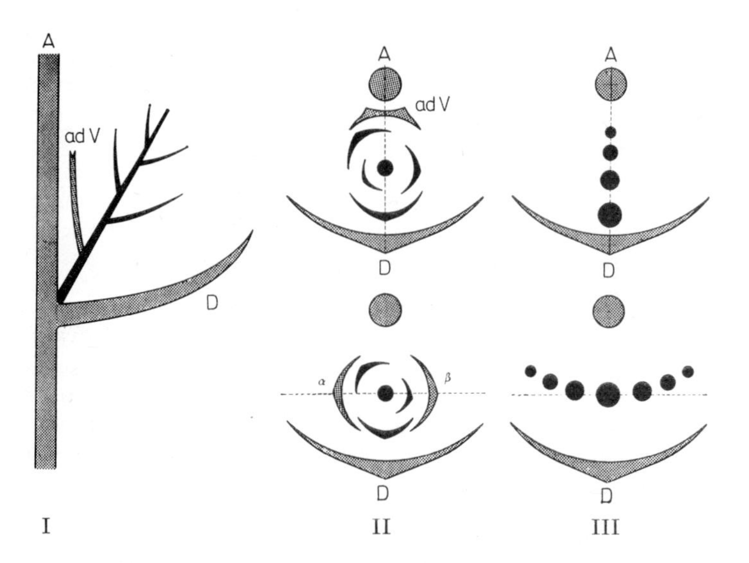

Fig. 140. Different kinds of insertion of axillary buds in angiospermous cormophytes. **I.** Side view. **II.** and **III.** Arrangement as seen in transverse section. A, main (parent) axis; D, subtending leaf or bract; ad V, adaxial prophyll (in monocotyledons); α and β, lateral prophylls (in dicotyledons). (The single prophyll of the monocotyledons has perhaps been derived by the fusion of two lateral prophylls as still seen in the dicotyledons.) **III.** Above, serial axillary buds; below, collateral axillary buds. In **II.** and **III.** above, the median plane is indicated (dotted), and the single prophyll of the monocotyledons, and serial axillary buds, are seen to lie in this plane. By contrast the two prophylls of the dicotyledons (α and β), and collateral axillary buds, lie in a tangential plane (**II.** and **III.**, below, dotted).

(Fig. 192), *Convolvulus arvensis, Rumex acetosella*), shrubs (e.g. *Rosa, Rubus, Corylus*) and trees (e.g. *Salix, Populus, Robinia*), and which in the instance of woody plants are commonly referred to as 'suckers'. Although much less common, adventitious shoots can also arise from leaves, taking their origin from a single or small group of epidermal cells (e.g. in many ferns; *Bryophyllum calycinum*, Fig. 223B).

In many plants the formation of adventitious buds does not occur until the plant is damaged. The shoots arising from the stumps of felled trees are an example of this. Gardeners frequently propagate plants by means of the adventitious buds which appear on cuttings, particularly those of portions of stems, rootstocks and leaves.

c. The arrangement of the members of the side shoot relative to the main axis. If it is desired to investigate the positional relationships of the organs on a side branch of any order, then it is always orientated so that the subtending leaf is towards the observer (Fig. 140II and III), and the parent axis away from him. The median line of the subtending leaf then lies in the median plane of the axillary shoot (Fig. 140II and III, upper). The plane containing the longitudinal axis of the side shoot and at right angles to the median plane is known as the transverse plane of the side shoot (Fig. 140II and III, lower). Those organs on the side shoot which fall in the median plane are said to be median, and those in the tranverse plane lateral; all other positions between the median and transverse planes are said to be diagonal.

Usually, the lowermost leaves on the side shoot, which follow directly after the subtending leaf, assume a definite position in relation to this leaf and the main axis, independently of the positions of the leaves higher up the side shoot. These first leaves of the side shoot, which are frequently of simple form (see p. 169), are termed prophylls. In the monocotyledons there is only one such prophyll (Fig. 140I and II, upper), in the dicotyledons two, arranged in a characteristic way. In the monocotyledons the prophyll is either lateral or more usually median and adjacent to the parent axis. In this instance the prophyll is said to be ventral. Two lateral veins, forming marginal keels, often run into this single prophyll, but the middle vein is wanting. In many cases it may be that the prophyll has arisen in the course of evolution by the fusion of two lateral prophylls. In the dicotyledons the two prophylls (often signified by α and β) are usually arranged laterally, one to the right and the other to the left of the axillary bud. They may be opposite or alternate, and above them the other leaves follow in an unconnected whorled or alternate arrangement. The arrangement of the leaves on the axillary branch may resemble or differ from that of the leaves on the main axis.

Intercalary growth in the tissue at the base of the axillary bud may bring about displacement so that the original relationships between the subtending leaf and the axillary bud are apparently changed. Thus buds may be removed from the axils of their

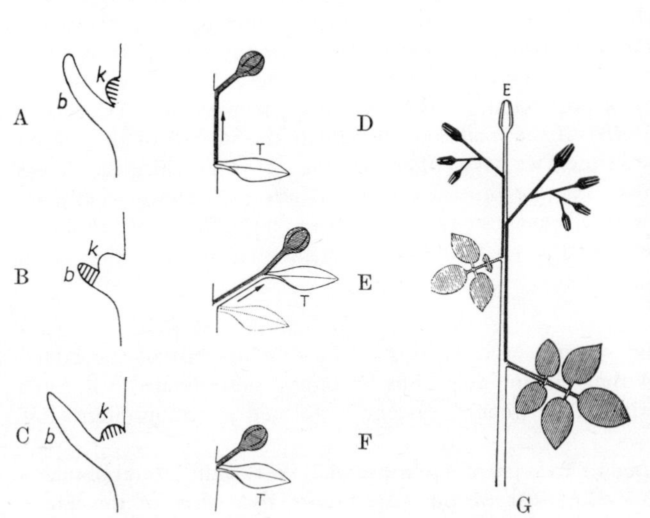

Fig. 141. Examples of metatopy. The various possible relationships between an axillary bud and its subtending leaf are shown in longitudinal section. **A.**, **B.** and **C.** Show that part of the complex which appears last as shaded: b, subtending leaf or bract; k, bud. **D.** Shows an example of concaulescence. **E.** Shows an example of recaulescence (T, subtending leaf or bract). **G.** Inflorescence of *Solanum tuberosum*, showing concaulescence of the two uppermost lateral branches (each terminating in a partial inflorescence) with the main axis, which itself ends in a flower (E.). **A.**, **B.** and **C.** After Goebel. **D.** to **G.** After Troll.

subtending leaves as a result of their becoming congenitally fused with the main axis. The individual buds are then situated much higher on the stem than their subtending leaves (concaulescence, e.g. *Symphytum*, *Solanum tuberosum* (Fig. 141B, G)). The subtending leaf may also, as a result of growth of its base beneath the axillary bud, carry the bud out with it, so that the axillary shoot is sited on the subtending leaf (recaulescence, e.g. *Thesium*, *Tilia* inflorescence). Especially complicated relationships are found in the Solanaceae (e.g. *Atropa belladona*). Here each node bears two leaves, each subtending an axillary shoot. One shoot is congenitally fused with the petiole of its subtending leaf and continues the shoot system, while the other remains suppressed. The main axis itself terminates at each node in a flower. Spatial displacements of this kind are termed metatopic.

5. Systems of branching

Each system of shoots attains its appearance and habit as a result of 1. the number of orders of side branches developed; 2. the symmetry and the direction of growth of the main axis and its lateral axes; 3. the position of these lateral branches on the parent axes; and 4. the extent of the development and the orientation of the side branches in relation, on the one hand, to others of the same order, and on the other, to their parent axes.

a. Sequence of shoots. If the growing point of the main axis is already capable of producing the reproductive organs, then the plant has only one axis and is said to be haplocaulous. An example of such a plant is *Papaver rhoeas*, in which the first shoot arising with germination may terminate in a flower. More usually the ability to produce flowers is acquired only by axes of the second, third or higher orders, i.e. by the lateral branches. Then the plant may have two (diplocaulous), three (triplocaulous) or an indefinite number of axes. An example of a triplocaulous plant is *Plantago major*. Its first axis bears only a rosette of leaves at ground level. The lateral branches of the first order (the axes of the inflorescences) bear only bracts; in the axils of these are the branches of the second order, and these terminate with the flowers. In many perennial plants, e.g. our native trees, the ability to flower is confined to those shoots produced after extensive ramification. But in most branch systems many lateral branches cease to branch before the order of the flowering shoots is reached.

b. Symmetry and direction of growth of the shoot system. The shoot system acquires its character as much if not more from the symmetry and direction of growth of the main axis as from the arrangement of the side branches and their subsidiary branches.

If the main axis is vertical it is said to be orthotropous and the plant upright (Fig. 144A). In this case the radially symmetrical main axis, if the system is growing freely, produces its more or less dorsiventral and plagiotropic side branches uniformly around the stem. The further branching of these side branches either takes place on the upper side (epitony) (Fig. 74D), as in many shrubs, or on the lower side (hypotony) (Fig. 74C), as in many trees, or laterally, as, for example, in *Abies, Picea* and *Taxus* (amphitony). If the main axis grows obliquely or horizontal, i.e. it is itself plagiotropic (Fig. 205), then it is usually dorsiventral (Fig. 205B). If the main axes remain at the surface of the soil, or grow horizontally beneath the surface, we have either creeping plants, in which the shoots usually bear roots on the lower side, or prostrate plants where such roots usually are lacking. In creeping plants the side branches customarily arise on the flanks of the shoot. If the side branches take up a vertical position their branching is similar to that of an upright plant.

c. Monopodial and sympodial systems of branching. The development of the lateral axes may fall behind that of their parent axis, thus becoming subordinate to it. Such growth is termed monopodial, and the branch system so formed a monopodium (Fig. 142A).

If the development of the main axis is promoted more than that of the lateral branches of the first order, and the development of the latter more than that of the lateral

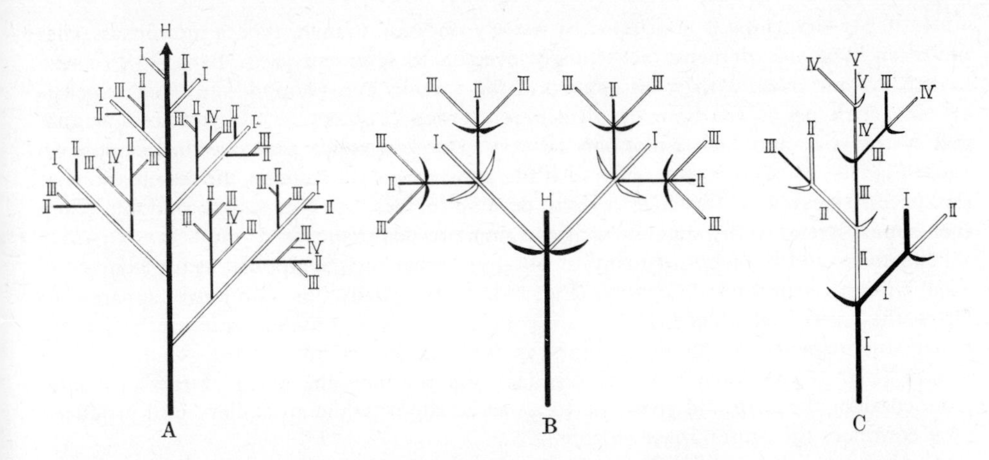

Fig. 142. A. Diagrammatic representation of a monopodial shoot system with lateral (racemose) branching. H, main axis; I, II, III, IV, lateral axes of the first, second, third, and fourth orders respectively. **B.** and **C.** Sympodial branching. **B.** Dichasium. H, primary axis; I, II, III, lateral axes of the first, second and third orders. **C.** Monochasium. I, primary axis; II, III, IV, V, lateral axes of the second, third, fourth and fifth orders.

branches of the second order, and so on, then the branching is said to be racemose (Figs. 142A; 144A). In this instance a single true main axis runs through the whole branch system. Such monopodial branching occurs, for example, in *Quercus*, *Fraxinus* and *Acer*, although here the continued growth of the main axis ceases after a time. Monopodial growth is especially typical of *Abies, Picea* and other conifers with pyramidal or cone-shaped outlines. The radially symmetrical main shoot, influenced by gravity, continues to grow vertically upwards. The side branches of the first order are usually dorsiventral, and they radiate from all sides of the main axis in a horizontal or oblique direction. The uppermost are the youngest and their age increases basipetally.

In other instances the lateral branches may be promoted relative to the main axis. Sometimes the main axis may cease to develop altogether after forming the side branches, its terminal bud either becoming dormant, or forming an inflorescence, or even dying. Growth, and the production of further side branches, is then continued by one or more of the original side branches, usually the uppermost. Repetition of this manner of growth (referred to as cymose branching) leads to a sympodial shoot system (Fig. 142B and C).

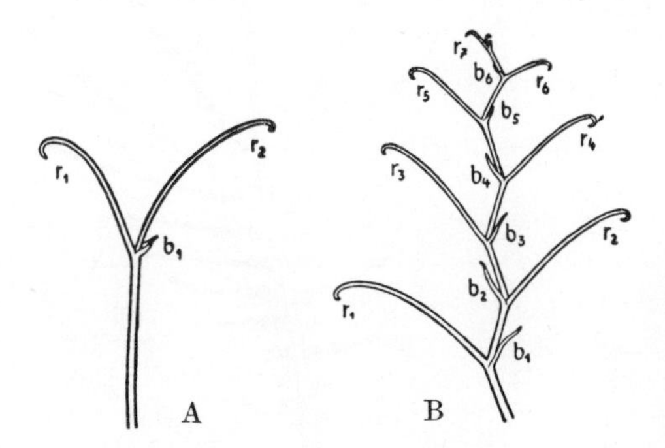

Fig. 143. Sympodial tendril system of *Vitis vinifera* (**A.**) and *Parthenocissus quinquefolia* (**B.**). The branching forms a rhipidium (cf. Fig. 146J): r_1–r_7 indicate the tendrilar axes, and b_1–b_6 the scale-like bracts subtending the lateral branches. ×½. After Troll.

I. If the branching is continued on each occasion by two side branches of the same order (which stand more or less exactly opposite each other) the system becomes a dichasium. This is shown schematically in Figs. 142B; 146D. Such a system of branching, which occurs, for example, in *Viscum*, can recall dichotomy in appearance. The lateral branches are not usually confined to one plane, as shown in the diagram, but extend in several directions. This often results from the planes of the successive branchings not being parallel, but at right angles to each other. Consequently, only the ground plan (Fig. 146E) can give a true picture of the arrangement of the branches of the system.

II. If the branching is continued by merely one side branch, then a monochasium is produced. The side branches accordingly overtop their parent axes (Fig. 146B), and a branch system arises with what appears to be a main axis (sympodium), but which is actually made up of side branches of different orders (Figs. 142C; 146B). Such a sympodial system may approach a monopodial very closely in appearance (as, for example, in *Vitis*; Fig. 143A). This is especially so if the apparent axis is straight, the terminations of the successive parent axes, which cease to develop further, being pushed to one side. They then appear lateral to the apparent axis and simulate side branches.

The stems and branches of many of our trees form such sympodia, as for example in *Tilia, Ulmus, Castanea* and *Carpinus* (Fig. 144B), but usually this is no longer apparent so far as the trunk and older branches are concerned. The sympodial structure frequently remains permanently visible in rhizomes, as, for example, in those of *Polygonatum multiflorum* (Fig. 205B). Each year the terminal bud for the time being of the horizontal monochasium turns up and gives rise to an aerial shoot, while an axillary bud produced by it continues the subterranean rhizome.

Not infrequently several different kinds of branching are present simultaneously in one shoot system. Thus cymosely branched lateral shoots may be borne upon a racemosely branched seedling axis.

d. Position of the active buds on their parent axes. Except for many herbs, only a small number of the buds on a parent axis expand into shoots. Those expanding in the year of their formation (as in some trees) give rise to so-called 'Lammas shoots', and those in the subsequent spring to 'innovations'. In the richly branched shoot systems of most trees the peripheral buds, i.e. those at the tips of the branches, are preferentially stimulated into growth (acrotony; Fig. 144). Here there is the best chance of the new leaves getting

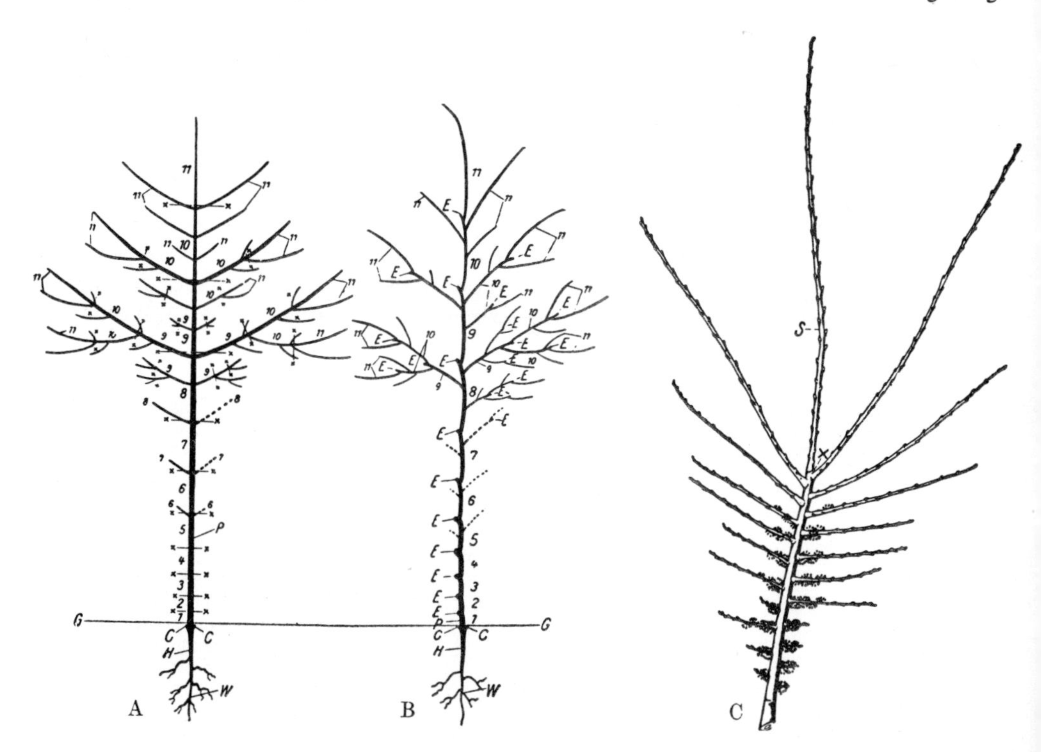

Fig. 144. A. and **B.** Growth form and branching of a monopodial (**A.**) and a sympodial (**B.**) tree. The numbers 1–11 refer to successive years' growth, the yearly increments being demarcated by x–x. E, aborted apex. W, main root. H, hypocotyl. C, cotyledonary node. G, soil surface. The root system is merely schematic. **C.** *Ulmus carpinifolia*. Two-year-old shoot with one-year-old side branches showing acrotonal promotion of growth. The aborted terminal bud of the parent axis is at x, and the uppermost side shoot (S) has become a leading shoot. The basal regions of the lower and middle shoots bear flowers. **A.** and **B.** Both after Rauh. **C.** $\times\frac{1}{10}$. After Troll.

adequate light. In shrubs and perennial herbs, on the other hand, the buds at the lower end or in the middle of the parent axis tend to develop rather than those elsewhere (basitony and mesotony respectively; Fig. 145).

Almost all our native trees restrict themselves during a season's growth to the elongation of the branches coming from the expanded winter buds, and the development of buds on these branches. These buds do not usually become active until the beginning of a new period of growth, and then it is principally the uppermost that give rise to further branches. These new branches may be arranged in a true or apparent whorl (*Araucaria*, *Abies*, *Picea*), or more usually the uppermost buds give rise to long shoots and those beneath short shoots, becoming shorter towards the base (e.g. *Ulmus* (Fig. 144C; see also below)).

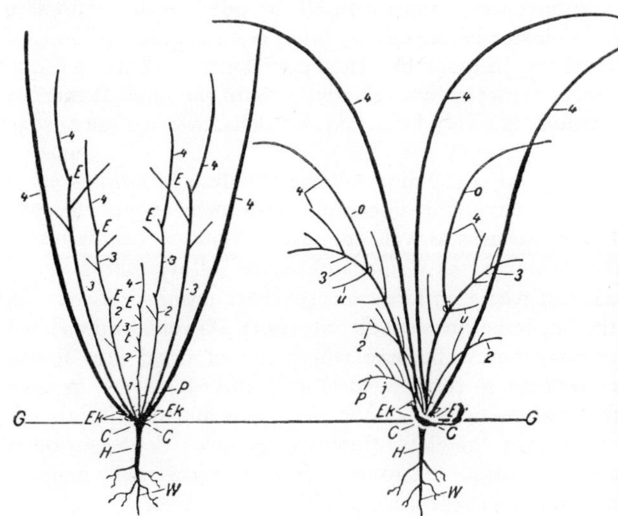

Fig. 145. Growth form and branching of two shrubs (diagrammatic). **A.** *Corylus avellana*. **B.** *Sambucus nigra*. The numbers 1–4 indicate successive years' growth. P, primary shoot; Ek, renewal buds; H, hypocotyl; C, cotyledonary node; W, main root; o, promoted upper branch; u, retarded lower branch. Root system merely schematic. After Rauh.

Expansion of all the buds would be a useless and even dangerous waste of the plant's resources, for many branches would so strongly shade each other that as a result they would die. Consequently, most of the lateral buds remain resting as 'dormant eyes', or soon degenerate. Certain pathogenic fungi and other parasites can, however, stimulate the outgrowth of all the buds in an affected area, forming a mass of shoots, usually remaining short and stubby ('witches' brooms'; see p. 315 seq.).

e. Long shoots and short shoots. The extent of the development of the side branches on a parent axis may vary considerably. Thus only a few may grow into long shoots with long internodes, while the remainder become rosette-like, contracted, short shoots. This difference is often an expression of a division of the function of the shoot as a whole. The short shoots often have only a limited life, and since they usually undergo no, or only a little, further branching, they consequently contribute little to the formation of the permanent framework of branches. In many woody plants the bearing of leaves is restricted, at least at maturity, to the short shoots (e.g. *Pinus*, *Ginkgo*). In *Larix* long shoots with spirally arranged needles are found only at the ends of the branches (they are therefore acrotonous), whereas at the bases of the older and leafless long shoots only short shoots occur, each giving rise to a tuft of needles (Fig. 637). Flowering frequently occurs exclusively on short shoots (e.g. *Prunus cerasus*, *Malus*, *Ginkgo biloba*).

f. Resting buds and cauliflory. As a consequence of the acrotonic opening of the buds, most trees possess resting buds in the lower parts of each year's shoot. These may remain viable for an indefinite time, and often show development only after many years. In *Quercus* and *Fagus*, for example, resting buds may persist for as much as 100 years. It is probably buds of this kind which give rise to shoots from old stems or stumps after injury to the crown, or destruction of apical dominance (e.g. *Tilia*, *Salix*). It appears very strange when resting buds, after prolonged secondary thickening of their parent branches or stems, yield flowers or inflorescences (cauliflory; Judas tree, *Cercis siliquastrum*).

g. Inflorescences. The inflorescences of the seed plants provide particularly beautiful examples of branch systems. The inflorescence is the entire branch system which gives rise to flowers, without undergoing any further vegetative development, and which is more or less clearly set off from the non-reproductive part of the plant. The bracts of the flower-bearing side branches may retain their leaf-like character even to well within the inflorescence (frondose inflorescences). More frequently, however, they become much less conspicuous (bracteose inflorescences), or even wholly reduced (naked inflorescences).

Simple and complex inflorescences can be distinguished according to the degree of their branching. In the simple inflorescences, or racemes, all the side branches terminate in a single flower. These simple inflorescences are usually interpreted as reduced forms of originally more complex inflorescences. In these, branched partial inflorescences, bearing few or many branches and flowers of higher orders, take the place of the single flowers of the simple inflorescences. Extraordinarily richly branched and dense systems can arise in this way.

Of special significance for the description and morphological analysis of inflorescences is the behaviour of the apex of the main axis and—in complex inflorescences—of the apex of the main axes of the partial inflorescences as well. In closed inflorescences these axes end in flowers (apical or terminal flower in the instance of simple inflorescences, lateral terminal flower of the central axes of partial inflorescences). These terminal flowers can be recognized by the fact that they open before the adjacent lateral flowers. In the closed inflorescences all the apical growing points become wholly transformed into flower primordia, and all the branches become so terminated (Figs. 146A–J; 147A). In open inflorescences, however, the terminal meristems of the main axis and of the primary laterals eventually cease activity without thereby differentiating into a flower; in other words they remain indeterminate (Fig. 146K–P, broken arrow). It occasionally happens that such open inflorescences show renewed vegetative growth.

1. Closed inflorescences. The phylogenetically oldest kind of inflorescence is probably the closed panicle. Disperse closed panicles (e.g. *Hydrangea paniculata*, Fig. 146A) and

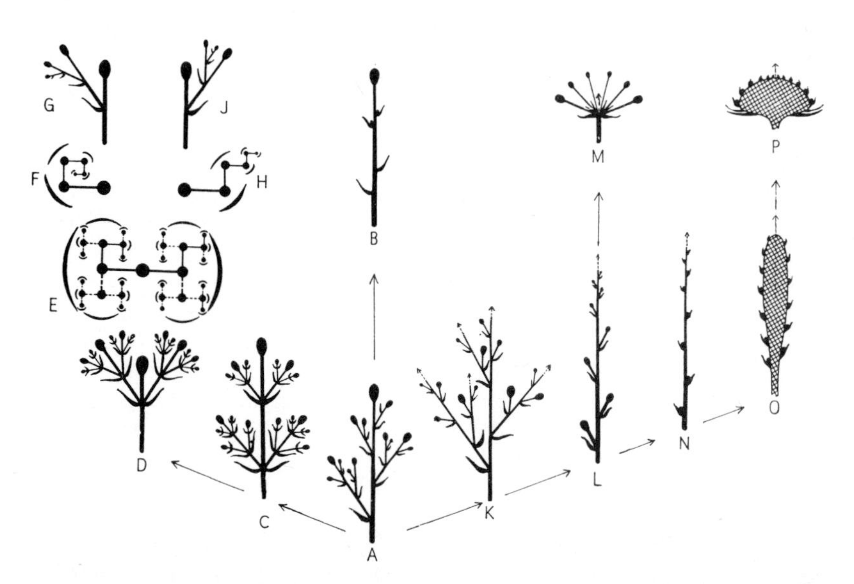

Fig. 146. Derivation of the most important types of inflorescence from the panicle (A.). **B.** to **J.** Closed inflorescences. **K.** to **P.** Open inflorescences. **B.** Single flower. **C.** Thyrsus. **D.** and **E.** Dichasium in side view and ground plan. **F.** Bostryx (helicoid cyme). **G.** Drepanium. **H.** Cincinnus (scorpioid cyme). **J.** Rhipidium. **K.** Open panicle. **L.** Raceme. **M.** Umbel. **N.** Spike. **O.** Spadix. **P.** Capitulum. From Zimmerman, expanded and modified.

decussate closed panicles (e.g. *Syringa*, Fig. 146C) can be distinguished according to the arrangement of the bracts. In the corymb (e.g. *Viburnum opulus*) there is basitonal promotion of the lower branches so that all the flowers move up to form an umbrella-like surface. In *Filipendula ulmaria* the terminal flower is even overtopped as a consequence of the marked extension of the lowermost laterals of the inflorescence. These richly-branched, closed systems can be transformed and simplified in many ways as a consequence of reduction. In the extreme case there is reduction of all the branches to single flowers (Fig. 146B, e.g. *Papaver*, *Paeonia*, *Trollius*, *Nigella*). Frequently, as a consequence of strictly acrotonous promotion of the branching, only the laterals at the upper nodes extend. If at the same time extension of the internodes of the main axis is wholly suppressed, then a pleiochasium results (e.g. *Sedum*, *Sempervivum*. The cyathia of many *Euphorbia* spp. are also pleiochasially arranged.) Where the leaf arrangement is decussate, often it is only the two laterals of the uppermost node beneath the terminal flower which grow out (cymoid inflorescence, Fig. 146D; many Caryophyllaceae, e.g. *Cerastium*). Within the partial inflorescences, either both bracts can be fertile (dichasium), or on each occasion only one (monochasium). Four kinds of monochasia can be distinguished according to their three-dimensional symmetry (Fig. 146F–I; for their derivation refer to D and E). These are termed the rhipidium (Fig. 146I; many species of *I is*), the drepanium (Fig. 146G; various rushes, e.g. *Juncus bufonius*), the bostryx or helicoid cyme (Fig. 146F; *Hypericum*, *Hemerocallis*), and the cincinnus or scorpioid cyme (Fig. 146H; *Saxifraga cymbalaria*, *Silene pendula*, the partial inflorescences of many Boraginaceae (e.g. *Myosotis*, *Symphytum*) and Crassulaceae (e.g. *Sempervivum*, *Sedum*)). A complex closed inflorescence whose partial inflorescences display a dichasial or monochasial character is termed a thyrsus (e.g. *Echium*, *Cynoglossum*, *Aesculus*).

2. Open inflorescences. Whole families, e.g. the Cruciferae, Labiatae and Scrophulariaceae, are characterized by open inflorescences. The archetype of all open inflorescences is the open panicle. As in closed panicles so also in open panicles the branching below tends to be more extensive than above, so that the whole panicle has a pyramidal form (Fig. 146K). Open inflorescences are distinguished from closed by the fact that their main axes, and the primary laterals never terminate in flowers, but end their growth in an undifferentiated condition (as, for example, in the paniculate grasses, and inflorescences of *Vitis*). In the raceme the lateral partial inflorescences are reduced to stalked single flowers, and a simple inflorescence is thus derived from a complex (Fig. 146L, e.g. *Lupinus*). In the spike the lateral pedicels are absent, and the flowers are sessile (Fig. 146N, e.g. *Orobanche*, *Plantago*, *Oenothera*). The spadix is distinguished from the spike by its thickened axis (Fig. 146O, e.g. *Acorus calamus*, female inflorescence of *Zea mais*). In capitula the individual flowers are unstalked and arranged on the expanded receptacle (Fig. 146P, e.g. *Knautia*, and Compositae). In the umbel (Fig. 146M, e.g. *Hedera*, *Primula*) the individual flowers, each with a relatively long pedicel, appear to spring from one level as a consequence of compression of the internodes. The open thyrsus is distinguished from the closed merely by the absence of a flower at the end of the main axis (e.g. many Labiatae and Scrophulariaceae, *Aesculus*, *Dictamnus*).

3. Typology of inflorescences. In many instances the fertile, flower-bearing, region is in fact much more complex in its structure than the classical concepts of descriptive morphology, detailed in the foregoing, would suggest. It has, therefore, been found necessary to introduce a number of typological concepts which make it possible to consider inflorescences comparatively in relation to their position in the plant as a whole.

Complex inflorescences are often encountered which are themselves composed of complex inflorescences of lower orders (Fig. 147B; polytelic inflorescences or synflorescences). The tip of the main axis bears the primary or principal florescence, beneath which a zone of branching yields lateral axes terminating in coflorescences, thus completing the whole inflorescence. The laterals repeat in large measure the symmetry of the main axis, and are termed paraclades. The principal florescence and the coflorescences may in turn be simple or complex. If partial florescences stand in the place of single flowers, then these partial florescences are always closed systems.

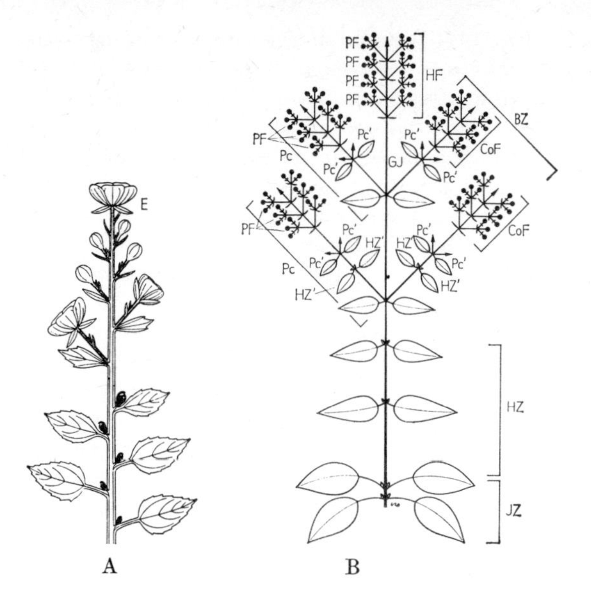

Fig. 147. Diagrammatic representation of a closed inflorescence with a terminal flower (E), and acropetal opening of the laterals (botryoid inflorescence). **B.** Structure of a complex synflorescence. JZ, vegetative zone; HZ, HZ^1, zones of inhibition; BZ, 'enrichment zone' in which branches terminating in coflorescences complete the inflorescence; HF, principal florescence; PF, coflorescence; Pc, paraclade; Pc^1, paraclade of the second order; GJ, basal internode. After Troll and Weberling.

Immediately adjoining the zone of branching is commonly a zone in which the axillary buds develop in only very unusual conditions, while in the basal, purely vegetative, region axillary buds may serve as the source of innovations in the following growing season. Purely vegetative zones, inhibitory zones, and zones of branching collectively form the infrastructure of the inflorescence, often clearly set off from the principal inflorescence by a conspicuously elongated basal internode (Fig. 147B). The typological distinction between monotelic and polytelic inflorescences, which largely coincides with that between the closed and open inflorescences of purely descriptive morphology, must be left to more specialist treatment.

6. The primary tissues of the stem[27]

There is a close relationship between the outer form of the shoot and its internal structure, or anatomy. The products of assimilation, coming from the leaves as they become capable of photosynthesis, together with the water and salts taken up by the roots from the soil, must be conducted to the growing tip of the shoot and to the leaf primordia. This means that conducting tissue must be developed for the transport of both assimilation products and water. Furthermore, mechanical tissue must be differentiated to support the embryonic tissue of the growing point, at first weak and upheld solely by the turgor of the cells.

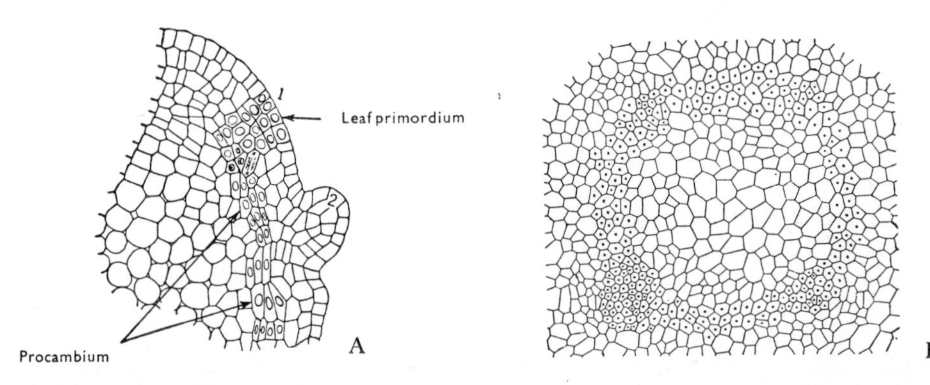

Fig. 148. A. Longitudinal section through the apex of *Linum*. Leaf primordia occur at 1 and 2. The procambial strands differentiate themselves deeper in the apex below the leaf primordia. **B.** Transverse section through the apex of *Ranunculus acris* close to the summit. The ring of meristematic cells is indicated by dots. The differentiation of the procambial strands is beginning at four sites. **A.** ×120. After Esau. **B.** ×100. After Helm.

The differentiation of these various tissues takes place successively and at different distances from the growing point (see p. 89 seq.). Moving back from the tip a series of different zones can be distinguished corresponding to the maturity of the region examined. These zones, of course, run smoothly into each other, and the states of maturity of the different tissues may overlap. The extreme tip of the growing point (that is, a zone of hardly more than 0.01–0.05 mm in length) is, as we have seen (Figs. 91; 92), composed of primordial meristematic cells with extremely thin walls. They can be referred to as the initial cells of the shoot. Immediately adjoining is the zone of determination (0.02–0.08 mm), which extends from the zone of initial cells without any visible boundary. It is in this zone that the ground plan of the future shoot—its pattern of leaves and side shoots—is laid down, accompanied by the first segregation of the future cortical, vascular and pith tissue.

In the dicotyledons, at a certain distance from the surface of the growing point and in the determination zone, a group of cells can be distinguished from the remainder with the help of special staining methods and specific enzyme reactions. In transverse section these cells are arranged approximately in a circle, and three-dimensionally as a cylinder or cone (Fig. 148). In certain conditions they are conspicuous without special preparation because of their dense contents. At the same time the centrally placed protopith and peripheral protocortex are beginning to mature by progressive vacuolation of the cells, thus beginning to approach in appearance the typical, little specialized, ground tissue or parenchyma. The zone of conspicuous cells can thus be regarded as being a remnant of the primordial meristem which remains behind in the maturing and ageing segment, and which retains in full the mitotic activity of the embryonic tissue. Because of its circular appearance in transverse section, this zone is also referred to as the ring meristem.

The determination zone gives way smoothly to the differentiation zone, where the different kinds of development prepared for in the determination zone now become visible anatomically. The cells of the protocortex and protopith display some meristematic activity, and by divisions principally transverse to the longitudinal axis build up the mass of the isodiametric parenchymatous ground tissue. Meanwhile, in the region of the ring meristem, and characteristically in relation to the leaf primordia which are arising exogenously at this stage (see p. 122 seq.), the divisions are principally longitudinal. In longitudinal section it is seen that these divisions give rise to groups of prosenchymatous elements. These mark the positions of the future vascular bundles destined to supply the leaves, and in this primary condition they are referred to as procambial strands (Figs. 148; 149).

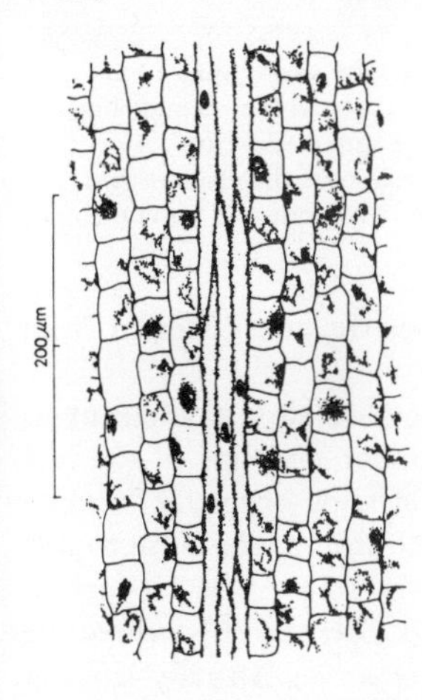

Fig. 149. Initiation of a fibre bundle in the ground tissue of *Sansevieria* (Liliaceae). The extension growth proceeds in contact with the surrounding ground tissue, still partly meristematic. ×150. After Meeuse.

The cell divisions which lead to the strengthening of the shoot (primary growth in thickness) also occur in this sub-apical differentiation zone. In contrast to the differential divisions (which determine the nature of subsequent cell lineages) occurring in the determination zone, the cell divisions in the differentiation zone are concerned with the multiplication of like cells. Consequently the response of the cell divisions in this zone to

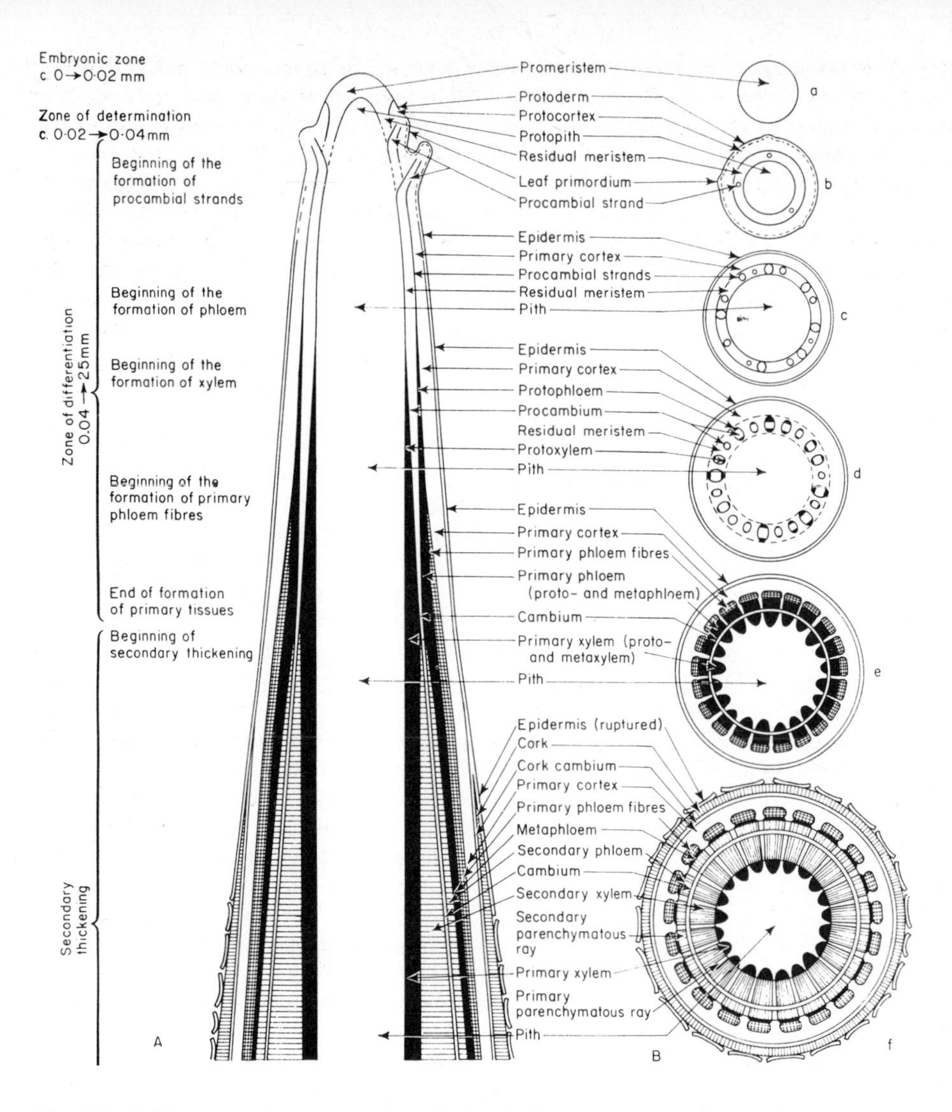

Fig. 150. A. Diagrammatic representation of a longitudinal section through the shoot apex of a woody dicotyledon. **B.**, a to f. The corresponding transverse sections. The procambial strands begin to form from the meristematic ring at the same time as the leaf primordia begin to develop exogenously at the surface. After some time, accompanying the differentiation of the strands into protoxylem and metaxylem, protophloem and metaphloem, they become confluent and remain separated only by narrow primary pith rays. Further explanation in text.

growth-regulating substances (e.g. gibberellins; see p. 303) is specific, and different from that of the divisions in the determination zone.

The distances between the procambial strands, whether they lie closely adjacent to each other or distant from each other, depends upon the number and size of the leaf primordia produced at the apex, as well as the number of the strands normally forming the leaf trace, a feature varying from species to species. Where the strands are close to one another they may, by coming into contact laterally at an early stage, fuse to form a closed procambial cylinder. This cylinder is perforated for a short distance only where the future leaf traces bend out and depart towards their appropriate leaves. Usually the leaf gaps arising in this way are closed shortly above the departing leaf trace because the procambium bordering the gap generates additional procambial tissue which rapidly fills it (Fig. 152D).

The differentiation of the procambium proceeds continuously towards the apex in such a way that the newly arising elements, at first procambial and later vascular, join on to the vascular system already present. This holds particularly for the first sieve tubes, the protophloem, arising in the adjoining maturation zone. The protophloem is differentiated in an acropetal direction out into the protruding leaf primordia, and they are thus provided with the metabolites necessary for further development. Then usually somewhat later the procambial elements adjacent to the developing pith differentiate to form the first water-conducting tracts, the protoxylem (Fig. 150B, a–d). Not infrequently the first-formed protoxylem is not in contact with the vascular system proper of the shoot. When this occurs, differentiation of protoxylem proceeds not only acropetally into the leaf primordia but also basipetally towards the vascular system already present. The joining up of the water-conducting system is thus brought about secondarily.

At the same time as the maturation and the eventual completion of the formation of the tissues there usually occurs both the principal extension growth and the primary growth in thickness of the young shoot. The differentiation zone and the principal extension zone of the young stem thus largely coincide.

7. Arrangement of the primary permanent tissues[27]

The primary conformation of the axis is determined by the processes of differentiation described in the foregoing. In the dicotyledons in general the primary permanent tissues at this stage of development are arranged concentrically in the following way (Fig. 150B, e). In the centre of the stem lies a more or less extensive parenchymatous pith. This can serve as a storage tissue, or alternatively a hollow pith may arise by a secondary pulling apart of the cells (schizogenously) or by tearing (rhexigenously). The pith is surrounded by the tracts of vascular tissue, the xylem of which is nearly always on the inside, while the phloem, separated from the xylem by a residue of undifferentiated procambium, lies on the outside. If the adjacent procambial strands fuse at an early stage the vascular tissue forms an almost closed hollow cylinder. Where fusion does not occur the bundles anastomose and form a cylinder perforated by small or large leaf gaps. In extreme cases the vascular tissue may be dispersed, and the cylinder consist of many or few individual vascular strands (Fig. 152E, F). The tissue, usually parenchymatous, separating the bundles and forming a connection between the pith and cortical parenchyma is said to form the primary pith rays. Closed vascular cylinders (Figs. 151A; 152C) are found in most trees (with the exception of many lianes), while the separation of the vascular tissue into anastomosing strands is encountered in the first stage of development of most herbs and perennial plants with annual aerial systems (Figs. 152E, F; 153A).

The very first elements of the protoxylem have to be laid down in a position where subsequently vigorous extension growth will take place. Accordingly, these elements have annular or spiral thickenings (Figs. 117A; 119; 151C, a–d), and the unthickened parts are able to endure an extension of more than 120 times. The protophloem has similar properties; here again the elements first produced are at the same time the longest and the thinnest. Their functioning is limited to the period during which the axis is extending. At the conclusion of extension growth the tracts of protoxylem are usually ruptured, the protophloem likewise. Meanwhile their roles are taken over by the elements of the metaxylem and metaphloem, differentiated somewhat later. Generally the metaxylem consists at least partly of spirally thickened elements which are still able to undergo a small amount of extension (Fig. 151C, e). In the angiosperms the metaphloem can frequently be distinguished from the protophloem by the presence in the former of companion cells, usually not yet present in the latter.

Sclerenchyma fibres are frequently differentiated on the outside of the vascular bundle, thereby more or less clearly setting off the inner part of the axis, the so-called central cylinder or stele, from the primary cortex. In many species (e.g. *Linum*) these phloem fibres arise directly in the protophloem so that there is no doubt of their belonging to the central cylinder.

In other species (e.g. in *Aristolochia*) the sclerenchymatous elements arise further towards the periphery. They then occasionally form a complete cylinder, previously regarded as part of the primary cortex. However, since even in these species the mantle of sclerenchyma can be interpreted as being composed of the outer sheaths of the vascular bundles fused together, it is reasonable to continue to attribute the primary phloem fibres

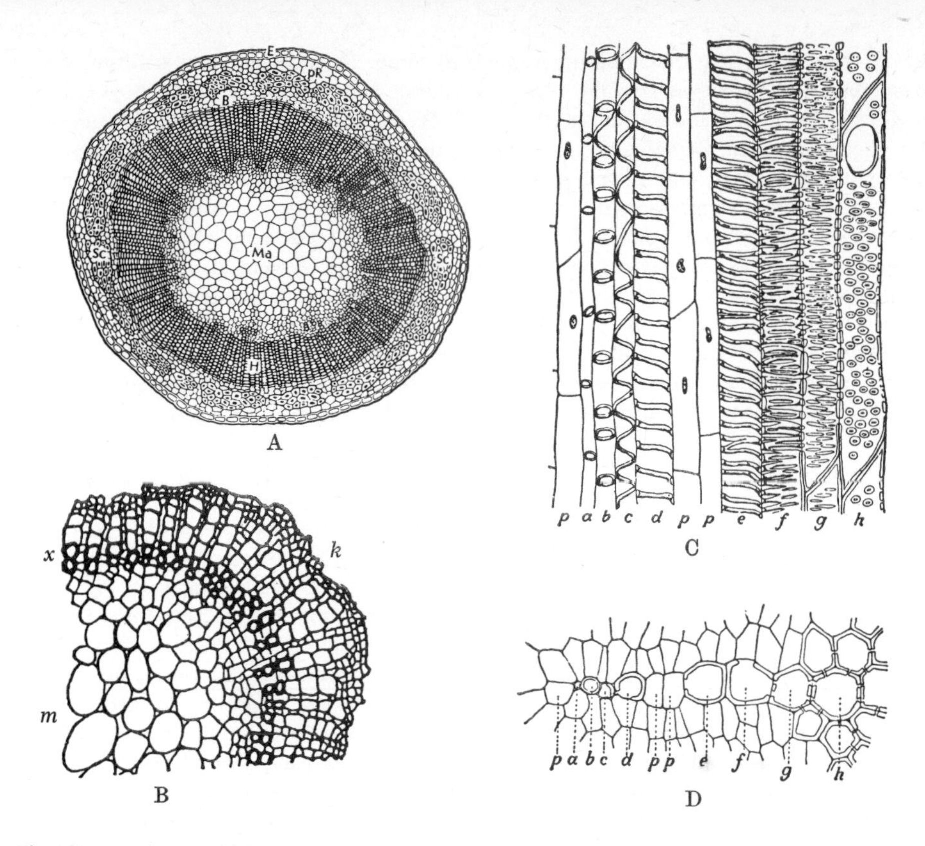

Fig. 151. **A.** *Linum usitatissimum*. Transverse section through a secondarily thickened stem: E, epidermis; pR, primary cortex; Sc, strands of sclerenchyma in the primary cortex; B, phloem; H, secondary xylem; Ma, pith. **B.** *Galium mollugo*. Inner part of a young stem, cut transversely: m, pith; x, xylem; k, cambium. **C.** and **D.** *Lobelia inflata*. Longitudinal section (**C.**) and transverse section (**D.**) through the protoxylem (a to d) and the metaxylem (e to h): p, parenchyma cells; a and b, tracheids with spiral thickening; f, conducting element with scalariform thickening; g, element with reticulate thickening; h, element with bordered pits. **A.** ×25. After Frank and Tschirsch. **B.** ×80. After Kostytchew. **C.** and **D.** ×200. After Eames and McDaniels.

to the central cylinder. Additional strands of phloem fibres, secondary in origin, may develop later in the cortical parenchyma.

In many species the innermost layer of the cortex contains large and easily movable starch grains, and is distinguished as a starch sheath. This always arises and lies outside the mantle of phloem fibres. In rhizomes and in the stems of water plants the innermost layer of the cortex may differentiate as a typical, one-layered, endodermis.

The mass of the peripheral cortical tissue normally consists of green assimilatory and storage parenchyma. The outermost sub-epidermal layers are not infrequently collenchymatous, or form a hypodermis one or several layers of cells in thickness and almost free of intercellular spaces. The outer limiting layer forms the epidermis, usually compact, lacking chlorophyll and perforated by stomata.

In many monocotyledons no clear distinction can be made between the primary cortex and central cylinder. The outermost of the scattered vascular bundles, each with its sheath of sclerenchyma, occur embedded in and adjacent to the sclerenchymatous hypodermis (e.g. the stems of many grasses; Fig. 153C).

Corresponding to their function, the vascular bundles form unbroken strands which can be followed from the tips of the roots to those of the shoots and leaves.

In some plants (e.g. many Pteridophyta) the bundles may not run out into the leaves, but ascend to the apex of the stem. Such bundles, because they belong to the stem, are termed cauline. Foliar bundles, by contrast, are those which run into the stem from the leaves. Immediately after their entry, however, they fuse with cauline bundles.

The group of bundles which departs from the stem to the leaf is said to form its trace, and the individual bundles of the trace are termed strands. The leaf trace may consist of a single (Fig. 152B–F) or several (Fig. 153B) strands.

8. The stelar theory[28]

The shoot of the land plant demands not so much tensile strength as resistance to bending (see p. 117). Towards this end the water-conducting and mechanical elements, strengthened by the deposition of lignin, become arranged at the periphery of the axis. According to the stelar theory, the vascular bundles of an axial organ should, in the light of their phylogenetic history, be regarded as forming a functional unit, which can be traced back to the centrally placed column of tracheids (the protostele) of the primitive land plants (Fig. 152A–F). The latter was surrounded by a mantle of little differentiated prosenchymatous cells, which can be regarded as constituting a primitive phloem. Even today the juvenile stages of many ferns are furnished with a typical protostele.

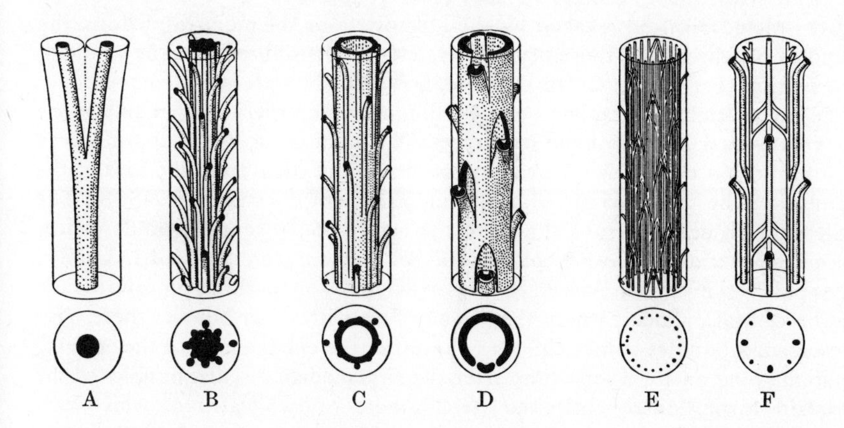

Fig. 152. Types of arrangement of the conducting tissue in the stem. **A.** Protostele (dichotomous branching). **B.** Actinostele with the leaf traces branching off at the sides. **C.** Siphonostele, the leaf traces departing as in **B. D.** Vascular cylinder with reticulate perforations (dictyostele). Each leaf trace departs from a broad base, leaving above it a parenchymatous leaf gap (seen in transverse section as a pith ray). **E.** and **F.** The stele divided up into discrete bundles (eustele): **E.**, *Linum*; **F.**, shoot with decussate leaves. **E.** and **F.** Derived from Esau and A. Braun.

In many large living seaweeds we find already central strands of prosenchymatous elements, mechanical (or possibly also sometimes conducting) in function, forming a kind of midrib to the phyllode (Figs. 77 and 81). They show us how the protostele may have arisen in the extinct migrant forms which colonized the land.

We can consider the transformation of the protostele into the stele of the angiosperms, which is broken up into numerous individual vascular bundles (Fig. 152E, F), as having taken place in a number of stages. Almost all the transitional forms are found in fact in the Pteridophyta, so that it is quite possible to demonstrate the phylogenetic sequence with a morphological series of existing types of stele.

Widely distributed, especially in the microphyllous Pteridophyta, the Lycopodiatae, is the actinostele, which in transverse section resembles a star (Figs. 152B; 557B). It originates if the lateral appendages, reduced by 'overtopping', as envisaged by the telome theory, fuse congenitally over some distance with the main axis. The process can also be interpreted from the opposite point of view, and be regarded as a basipetal displacement of the site of branching. If such a fissuring and lobing of the stele be imagined as continued to an extreme, we come by way of the deeply fissured plectostele (Fig. 556C) (found in various species of *Lycopodium*) to the polystele, consisting of a system of individual vascular bundles, distributed over the whole of the transverse section (Fig. 596).

In the siphonostele (Fig. 152C) the more or less centrally arranged vascular strands form a cylinder enclosing a parenchymatous pith (seen, for example, in the Gleicheniaceae, Schizaeaceae and in *Lepidodendron*). The pith, like the cortex, serves as food-storage tissue. In many ferns (e.g. *Osmunda* spp.) the branching off of leaf traces leaves gaps in the vascular cylinder occupied by parenchyma (Fig. 152D). The parenchyma which connects the pith and the cortex is said to form a pith ray. When the vascular cylinder is perforated in a reticulate manner it is called a dictyostele.

The ultimate stage reached is the dissolution of the original vascular cylinder into a network of individual bundles, the eustele. This is found in most herbaceous dicotyledons (Fig. 152E, F). The leaves are now supplied by one or several collateral vascular bundles. These can be traced basipetally in the axis for some distance as a leaf trace.

According to the telome theory (see p. 83), the eustele can be thought of as having been derived in another manner. A tendency for the stem to become thicker in the course of evolution would have required the displacement of the individual bundles of the polystele to the periphery to give mechanical strength. These could then have fused with one another to give the siphonostele, or transverse anastomoses may have developed leading to the widely distributed eustele of today (cf. Fig. 152E, F).

A somewhat isolated position is taken by the atactostele of the monocotyledons, the individual bundles of which, like those of the polystele, are distributed over the whole of the transverse section (Fig. 153B, C). In this case, however, the stele is not composed, as is the polystele, of a number of cauline bundles running more or less parallel, but of the very numerous individual strands of the leaf traces. This kind of stele is found in a very pronounced condition in the Palmae; here numerous bundles extend from the base of the leaf, which embraces the axis, into the whole circumference of the stem (Fig. 153B). The median bundle on each occasion runs almost to the centre of the central cylinder, while the lateral bundles penetrate the less deeply the further they are from the median bundle. As the bundles pass downwards, however, they all slowly pass outwards again towards the periphery of the central cylinder, where they finally fuse with other bundles there. This course of the vascular bundles comes about as a result of the enlargement of the growing point of the stem going on for a long time after the first median vascular bundles of the leaf traces are laid down. Consequently, the lateral bundles of each leaf trace, which later arise in succession towards the margin of the leaf, become increasingly less deeply immersed in the stem.

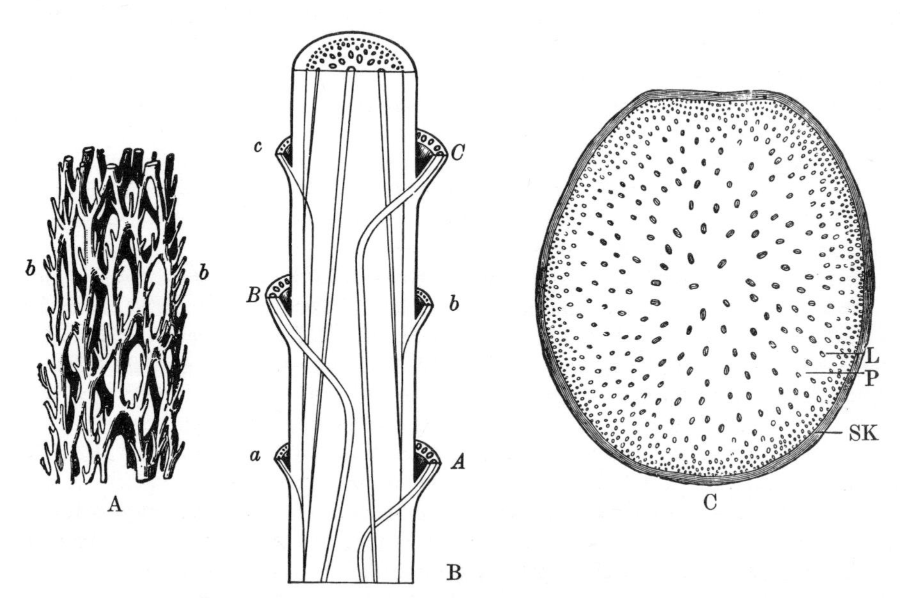

Fig. 153. Conducting systems. **A.** The cylinder of vascular bundles, together with the attachments of the leaf traces (b), of the fern *Dryopteris filix-mas*, obtained by maceration. **B.** Diagrammatic representation of the course of vascular bundles in the stem of a palm. The leaves are distichously arranged and amplexicaul, and the longitudinal section lies in the median plane of the leaves. The leaves Aa, Bb and Cc are cut off close to their insertions; the capital letters indicate the mid veins of the leaves. Above, half the stem is shown in section. **C.** Transverse section through an internode of the stem of *Zea mays*: L, vascular bundle; P, ground parenchyma; SK, hypodermal sclerenchyma. **A.** $\times\frac{4}{3}$. After Reinke. **B.** From Rothert, freely adapted from Rostafinski. **C.** $\times 2$. After Schenck.

9. The increase in thickness of the stem[27]

The seedling, at first small and with few leaves, may in time as a result of growth, usually accompanied by branching of the axis as well, yield a relatively enormous plant with innumerable leaves. *Sequoiadendron*, for example, may exceed 100 m, and *Eucalyptus* 120 m in height. The increase in size of the aerial shoot, and the concomitant increase in the number of leaves, demand the absorption of increasing amounts of water by the roots. These demands are met by the growth and branching of the roots, and often by the development of adventitious roots as well. Such increase in the root system can take place only if adequate amounts of organic food materials are formed in the leaves and conducted to the roots. Thus the development of the leaves and that of the roots are closely interdependent. The increase in the shoot and root systems can again take place only if an adequate number of conducting elements, both for water and for organic materials, are differentiated in the axes. These must also be sufficiently strong to support the increasing weight of leaves and branches, even in violent winds.

a. Growth forms. Many long-lived Cormophyta consequently increase the rigidity of their perennial axes (and often of their roots also) by the formation of masses of heavily lignified tissue. Such tissue is hard and mechanically strong. Woody plants behaving in this way are called, according to the symmetry of their branching (see p. 132 seq.), trees or shrubs (Figs. 144; 145). Two kinds of trees can be distinguished, those producing a crown of branches (Fig. 144) and those in which the leaves (or fronds) are confined to a comal tuft surmounting a slender, woody stem. In such trees the leaves are usually large. Examples are seen in the tree ferns (Figs. 592; 595), the Cycadales (Fig. 645A) and the Palmae. Only occasionally do the stems of such trees branch. Shrubs which remain less than 0.5 m in height are termed dwarf shrubs (e.g. *Vaccinium myrtillus, V. vitis-idaea, Calluna vulgaris* and other Ericaceae). Herbs are those Cormophyta whose aerial shoots remain more or less weak and sappy. At most they produce a little woody tissue at the base of the main axis, and herbs usually, in contrast to trees and shrubs, either die completely at the end of the growing period (annual plants) or die back to their perennating subterranean parts (Fig. 211). Shrubby herbs are intermediate forms. In these only the lower parts of the shoots are woody, the upper parts remain herbaceous. At the close of the growing period only the herbaceous shoots die away (e.g. *Salvia officinalis, Lavandula, Ruta graveolens*).

Some perennial herbs, such as *Veratrum* and particularly the tree-like *Musa* (banana) and some Zingiberaceae of the Tropical Rain Forest, form, besides the leafy flowering shoots, false stems which consist only of parenchymatous overlapping leaf sheaths.

The life span of a plant and the nature of its aerial shoots are generally indicated by a sign in systematic works. Woody plants are denoted either by ♄ (trees) or ♄ (shrubs). Herbs by ⊙ (annuals), ⊙ (biennials) and ♃ (perennials).

The increase in the number of conducting elements and in the amount of strengthening material, essential for the increase in size of a plant, is brought about in a number of different ways. Two very different growth processes, in the dicotyledons one subsequent to the other, contribute to the growth in thickness of the axis, namely primary and secondary thickening. Primary thickening is confined to the vicinity of the apical meristem and is effective for only a limited time. Secondary thickening, however, begins only after the completion of the primary thickening, and then persists until the death of the plant. Plants with exclusively primary thickening (as the palms) thus necessarily remain slender, but plants with secondary thickening, which may also reach a great age (e.g. the 3000 years or more of many specimens of *Sequoiadendron*) can in the course of time build up enormous trunks, sometimes exceeding 12 m in diameter.

b. Primary thickening of the monocotyledons. The stems of most monocotyledons, even of the tallest palms (e.g. the oil palm (30 m), the coconut palm (35 m), the South American wax palm, *Ceroxylon* (reputed to reach 60 m)), reach their ultimate diameter already after a few years entirely as a result of primary thickening. In these plants a mantle-like meristem lies between the tunica and corpus (Fig. 154A). This meristem, as a consequence of continued periclinal divisions, widens to form a collar lying beneath the

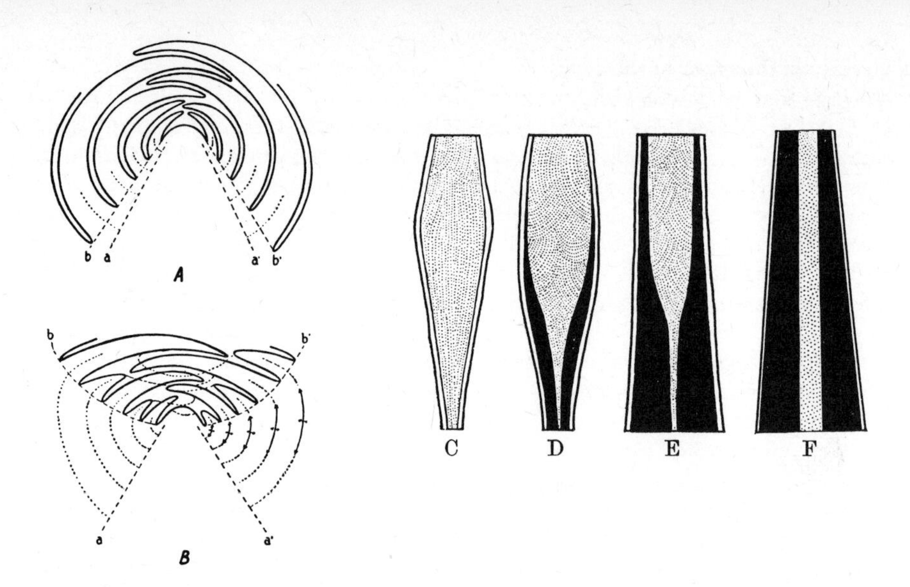

Fig. 154. Primary thickening and strengthening. **A.** and **B.** Primary growth in thickness in the apical region of a palm. **A.** Initial condition (a—b and a'—b', meristematic cone). **B.** Formation of a depressed apex as a result of cambial activity of the conical meristem. **C.** to **F.** Diagrammatic representation of strengthening growth in dicotyledons. **C.** Without being masked by secondary thickening. **D.** Partial masking by secondary thickening (black) in the lower region. **E.** Complete masking. **F.** Exclusively secondary growth in thickness in a woody plant. Pith dotted; cortex clear. **A.** and **B.** After Helm, from Troll, diagrammatic. **C.** to **F.** After Rauh.

young leaf primordia (Fig. 154B). As a result the top of the stem eventually acquires the form of a crater several decimetres broad. While the leaves are developing from their primordia the continued growth of the apex causes them to be displaced laterally over the margin of this crater on to the outside of the massive apical cone. This kind of growth persists until the apex reaches its ultimate diameter, at which time longitudinal growth begins. During this extension the diameter of the apex remains unchanged, so that a tall, slender, uniformly columnar trunk, characteristic of the palms, is generated.

The slender stems of other monocotyledons (e.g. of the grasses and bamboos) are formed in an exactly similar manner, but in a much shorter time. Only a very few monocotyledons (e.g. *Dracaena*) have at their disposal a special form of secondary thickening (see section **e**).

c. Primary thickening of dicotyledons. In the dicotyledons the primary thickening is completed without the participation of a special meristem. It rests upon irregular cell multiplication either within the pith parenchyma ('medullary' thickening, e.g. kohlrabi (Fig. 206), celeriac, potato tubers (Fig. 207)), or within the parenchyma of the primary cortex ('cortical' thickening, e.g. Cactaceae (Fig. 202)). Both kinds of thickening may occur together. Primary thickening may go to such lengths in some dicotyledons that even here the apices become immersed in grooves (e.g. many species of *Cactus*, potato tubers, *Plantago*). Adequate nutrition of the expanded stem is assured by the development of additional vascular systems in the pith and cortex. If these secondary bundles are furnished with lignified, sclerenchymatous sheaths the stem may become very 'woody' (e.g. kohlrabi).

d. Reinforcement of the axis. The extent of the primary thickening is closely related to the age and stage of development of the plant. It is usually only weakly expressed in the young stages, and then increases gradually as the plant becomes established, leading to the formation of a top-shaped axis (Fig. 154C). This conical form often becomes masked later by active secondary thickening of the slender lower part of the seedling (Fig. 154D, E), thus increasing the stability of previously more vigorous development of the superstructure. In the transition to the reproductive phase the diameter of the apical cone

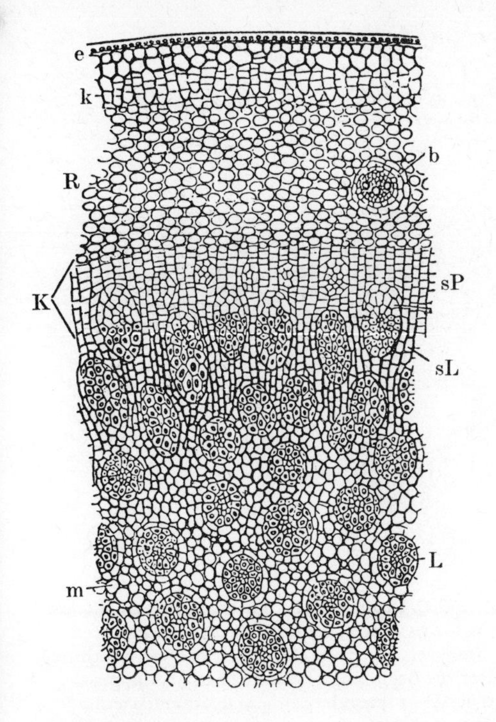

Fig. 155. Transverse section of the stem of *Dracaena* sp. after the onset of secondary thickening: e, epidermis; k, cork; R, cortex (containing a leaf trace (b)); L, primary, concentric, vascular bundle of the stem; m, primary parenchyma; K, cambium; sL, secondary vascular bundle; sP, secondary parenchyma. ×20. After Sachs.

frequently diminishes again, so giving the axis the form of a spindle, the narrow base and apex being separated by a more massive central region.

e. Secondary thickening. Usually in gymnosperms and dicotyledons the primary thickening soon gives way to secondary. In woody plants the necessary increase in conducting and strengthening elements in the stem depends, in fact, entirely upon this form of growth. It is provided by a peripheral meristem, the cambium, which remains active until the death of the plant.

The cambium is derived from the remains of the procambium of the apical region. When active it forms a closed cylindrical layer, separating the central mass of wood from the surrounding mantle of phloem and cortex (Fig. 150). The cambial initials have thin walls, and intercellular air spaces are notably absent. They cut off daughter cells both centripetally and centrifugally, and these clearly show their cambial origin by their alignment in radial rows (Figs. 151A, G; 165; 169B). This regular arrangement of the cells, very conspicuous in transverse sections of the stem, is consequently a reliable sign of secondary tissue.

Secondary thickening first took place in certain extinct Pteridophyta, but not until the gymnosperms and dicotyledons did it become of general occurrence.

Secondary thickening of monocotyledonous stems is found only in certain dendroid Liliiflorae, e.g. *Dracaena* (Fig. 155), *Cordyline*, *Yucca* and *Aloë*. Here the cambium arises outside the primary vascular bundles, scattered, as in all monocotyledons, within the central cylinder. It originates in the inner layer of the adjacent cortex either from primary meristematic cells, seen in transverse section to form a circular zone, or from parenchymatous cells of the cortex which begin to undergo renewed division by tangential walls (in *Dracaena* and *Cordyline* taking place only at a considerable distance from the apex). The cambium gives rise to a cylinder consisting of several layers of embryonic cells, prismatic in longitudinal section, arranged in radial rows and lacking intercellular spaces. After some time they cut off cells towards the inside and later also towards the outside. The relatively few cells cut off towards the outside go to form secondary parenchymatous cortical cells, while the more numerous cells cut off towards the centre give rise partly to vascular bundles (Fig. 155, sL) and partly to parenchyma (sP), the cell walls of which become thickened and lignified.

Since the cambium cuts off cells towards the centre (Fig. 157), it is, as a result of its

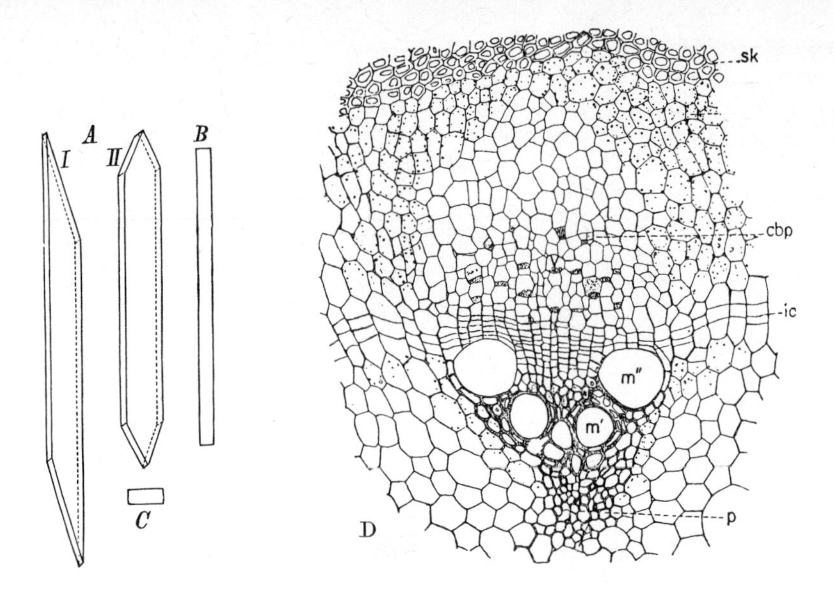

Fig. 156. Cambial cells and cambial activity. **A.**, **B.** and **C.** Diagrammatic representation of the form of cambial cells. **A.**, I and II. The two most frequent forms viewed tangentially and shown three-dimensionally. **B.** Radial longitudinal section. **C.** Transverse section. **D.** Transverse section through a vascular bundle of a twig of *Aristolochia durior* after the beginning of cambial activity: p, protoxylem; m′, pitted tracheid in metaxylem; m″, pitted vessel of secondary origin; ic, interfascicular cambium, continuous on both sides with the fascicular cambium; cbp, protophloem; sk, ring of sclerenchyma. **A.** to **C.** After Rothert. **D.** ×80. After Strasburger.

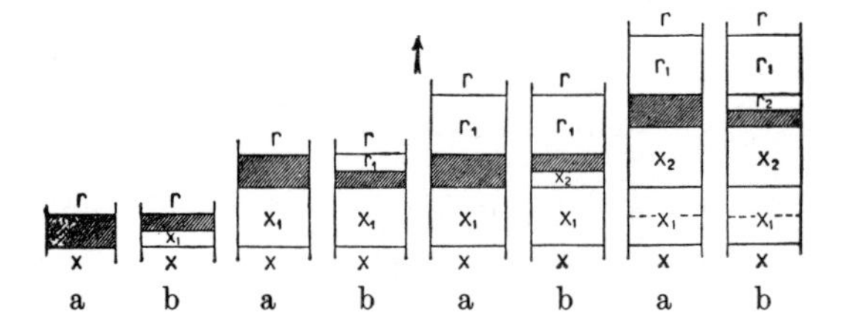

Fig. 157. Diagrammatic representation of the sequence of divisions of a cambial cell in transverse section, the cambial initials cross-hatched and the direction of the periphery of the organ indicated by the arrow: x, x_1, x_2, young xylem cells; r, r_1, r_2, young phloem cells. The stages indicated by **a.** are immediately before, and those indicated by **b.** immediately after a tangential division of the cross-hatched initial cell. Note how the initial cells are continually displaced in the direction of the periphery. After Jost.

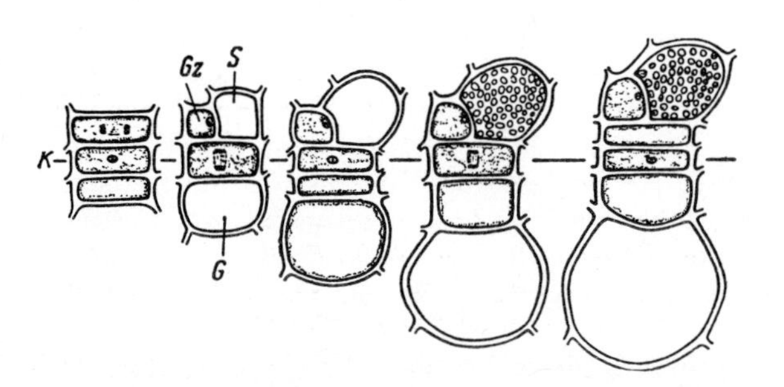

Fig. 158. Diagrammatic representation of the differentiation of a number of different secondary elements from the same cambial initial (K) in transverse section: G, vessel element; S, sieve tube element; Gz, companion cell. After Holman and Robbins.

own activity, continually moving outwards (Fig. 150). Accordingly, the circumference of the cambial cylinder must continually increase, a process known as dilatation.

This is only possible because growth, and cell multiplication as well, takes place in a tangential as well as a radial direction. Such tangential growth is brought about either by occasionally inserted vertical radial walls (as in storeyed cambia, e.g. *Robinia pseudoacacia* (Fig. 159)) or by the occurrence of divisions primarily transverse, but followed by the transverse walls coming into an oblique position (symplastic growth), so that eventually the originally superimposed cells now lie adjacent to each other. In this way a fusiform cambium is formed in which the precise sites of tangential growth can no longer be perceived.

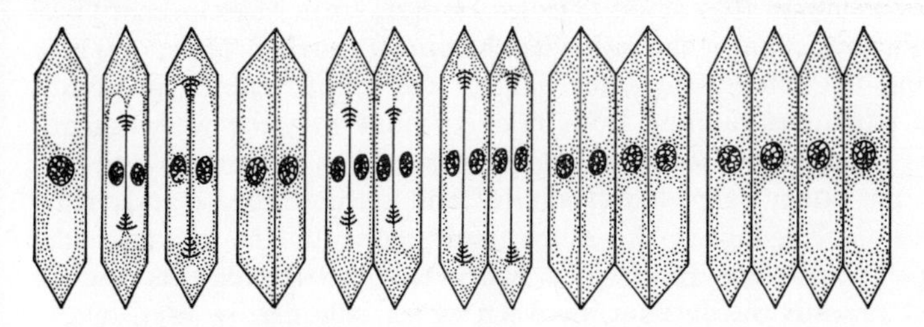

Fig. 159. Tangential division of a cambial cell by the insertion of a radial longitudinal wall. In this way a storeyed cambium is formed. Tangential sections, diagrammatic. After Bailey.

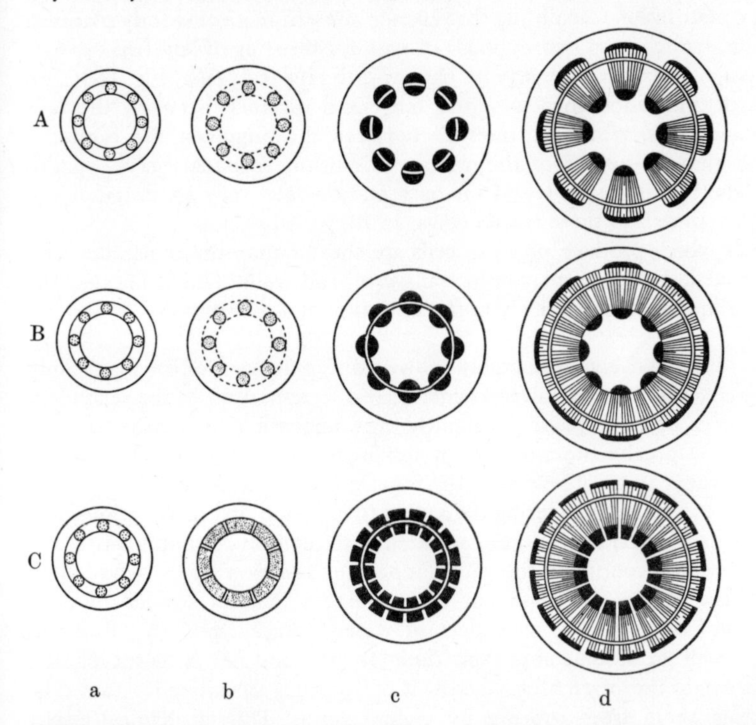

Fig. 160. Diagrams showing the three main types of secondary thickening in dicotyledons. **A.** *Aristolochia*. **B.** *Ricinus*. **C.** *Tilia*. The stippled areas represent the procambium. Further explanation in text.

In all those woody plants in which the procambium, and thus the primary xylem and phloem, are from the first in the form of closed cylinders (Figs. 160C, d; 151A, B), the primary cambium is similarly complete and its activity immediately leads to uniform growth around the whole circumference of the stem. In many herbs and lianes, however, the fascicular cambia of the individual vascular bundles are at first separated from one another by broad rays of parenchyma (Fig. 160A). Here as well, before secondary thickening begins, the cambium comes to form a complete cylinder as a result of arcs of cells in the intervening pith rays becoming meristematic, thus bridging the gaps (Fig.

156D, ic). This interfascicular cambium is by definition a secondary meristem (see p. 93).

All the permanent secondary tissue produced towards the centre of the stem as a result of the activity of the cambium is called wood, whether it consists of cells with lignified, and consequently hard and rigid, walls or, more rarely, of cells with weak unlignified walls. The secondary tissue produced on the outside of the cambium, mostly composed of unlignified cells, and which contributes to the phloem or cortex, is sometimes referred to as bast.

The secondary tissue produced on the inside of the fascicular cambium resembles the primary xylem, and that produced on the outside resembles the primary phloem. The activity of the interfascicular cambium causes the pith rays to be continuously extended on both the bast and wood sides of the cambium (Figs. 150B, f; 160). These primary pith rays of the wood and bast thus connect the pith with the primary cortex. These rays may extend unchanged during the subsequent thickening of the stem, resulting in simple, broad, parenchymatous partitions running from node to node between wedge-shaped segments of xylem. This arrangement, which should be regarded as exceptional, yields a stem with little rigidity, but rather a rope-like flexibility, and is accordingly characteristic of many lianes (e.g. *Aristolochia*, Fig. 160A). Much more frequently the interfascicular cambium cuts off from the beginning, and over its whole face, xylem elements towards the interior and phloem towards the exterior, between which only narrow rays, one to two cells broad, and of limited height, are laid down (e.g. *Ricinus*, Fig. 160B). In this way in many herbs and biennials with large and massive stems the axis eventually attains a mechanically strong construction resembling that already present in most woody plants in the primary condition, namely a closed cylinder of wood, consisting of lignified vascular and strengthening tissue, interrupted merely by narrow pith rays (e.g. *Tilia*, Fig. 160C, c).

As the diameter of the wood increases, the tangential distance between the rays inevitably gradually increases. When the intervals between the origins at the cambium become about double that at the end of the primary condition, secondary rays begin to be produced. These end blindly in the wood and bast, and the later they are initiated, the less deeply do they penetrate into these tissues (Figs. 150B, f; 160 a).

Those cambial cells which produce pith ray cells are shorter than the remainder, and have less steeply inclined, indeed in part even transverse, end walls. This is because the cambial cells divide obliquely or transversely at the initiation of a new pith ray.

f. Secondary wood. Progressively more highly evolved differentiation of the plant body is accompanied by increasing anatomical and functional differentiation in the secondary xylem. Consequently it is often difficult, frequently even impossible, to ascribe to a particular kind of xylem one precise function. In the less highly evolved forms one and the same kind of xylem tissue may fulfil several of the functions associated with the wood. It is therefore more satisfactory in considering the xylem to examine first its functions, and then all the particular histological conditions associated with each function in turn.

The conduction of water counts as the most important function of the wood. This hydraulic aspect of the xylem is dependent upon the longitudinal strands of dead tracheids or vessels, and occasionally, in respect of water storage, upon the dead and living wood fibres as well. A second important function of wood lies in its mechanical properties, making possible the often massive crowns of branches ascending far above the ground, and protecting them from rupture by violent winds. This mechanical aspect depends, in highly differentiated woods, upon longitudinal strands of dead sclerenchymatous cells, filled with air (wood fibres). In the more primitive gymnospermous woods, however, wood fibres are lacking and the tracheids serve both as conducting, and as strengthening, elements. A third important function of wood is the storage of assimilates during resting periods. This storage and conducting system for organic materials consists of the longitudinal and radial strands of living parenchyma cells (wood parenchyma and ray parenchyma), as well as the living, longitudinally orientated substitution fibres.

The individual elements of the wood can be separated from one another by maceration (brought about by boiling with acid (e.g. concentrated nitric acid and potassium chlorate), or by enzymic dissolution of the middle lamellae with the aid of bacteria), and examined individually in the microscope (Fig. 161).

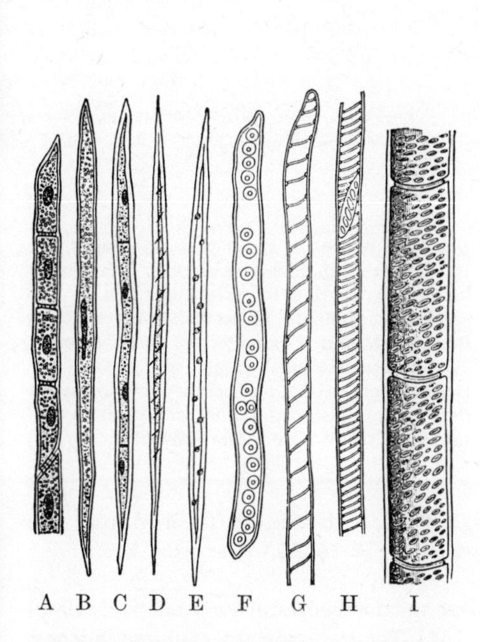

Fig. 161. Elements occurring in macerated secondary wood of a dicotyledon. **A.** Wood parenchyma. **B.** Undivided, and **C.** divided substitution fibres. **D.** Wood fibre. **E.** Fibre tracheid. **F.** Tracheid with bordered pits. **G.** Tracheid with spiral thickening. **H.** Vessel segments with scalariform perforation plate. **I.** Vessel segments with transverse vestigial end walls.

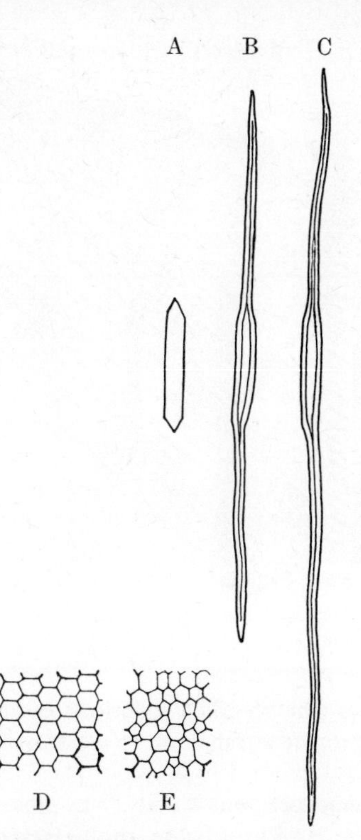

Fig. 162. Diagrammatic representation of the tip growth of wood fibres. **A.** Cambial initial of *Robinia pseudoacacia*. **B.** and **C.** Wood fibres developed from the initial by tip growth at each end of the cell. **D.** and **E.** Transverse section through a bundle of wood fibres before (**D.**) and after (**E.**) tip growth and interdigitation of the cells. **A.**, **B.** and **C.** After Eames and McDaniels. **D.** and **E.** After Rothert.

The tracheids and wood fibres, often 0.1–5.0 mm in length, are essentially longer than the cambial cells from which they are derived. The increase in length is brought about, just as is the outstanding size of most vessels, by growth at the tips (Fig. 162B, C), accompanied by dissolution of the middle lamellae and the formation of new wall material, so that the tips of the cells insinuate themselves between the adjacent cells (interpositional growth). The way in which the walls of the elongating sclerenchymatous cells force themselves between the yielding neighbouring cells thus resembles the impression that a 'caterpillar' track makes on soft earth, and the expression 'gliding growth', often used in relation to the extension of these cells is misleading. In the formation of the wood parenchyma the cambial initials repeatedly divide transversely. Consequently, this parenchyma consists of columns of cells running longitudinally, i.e. parallel to the conducting elements and fibres. The origin of these columns from cambial cells is indicated by their ending above and below in attenuated cells. (Fig. 161A).

Where conducting elements adjoin one another the pits formed in the dividing wall are bordered on each side. Where dead conducting elements abut on to living wood or pith ray parenchyma cells, however, the pits are bordered on one side only, namely on the side of the element. The longitudinal walls between the conducting elements and the fibres as well as those between the fibres and the parenchyma cells, are usually wholly unpitted.

Not uncommonly transitions between the typical elements are encountered in woody tissues. Narrow vessels (Fig. 161H) lead to relatively wide tracheids (Fig. 161F, G) and narrow sharply pointed tracheids with minute, reduced bordered pits (fibre tracheids; Fig. 161E), probably serving principally a mechanical function, lead to wood fibres (Fig. 161D). Wood fibres with only slightly thickened walls and which retain their living

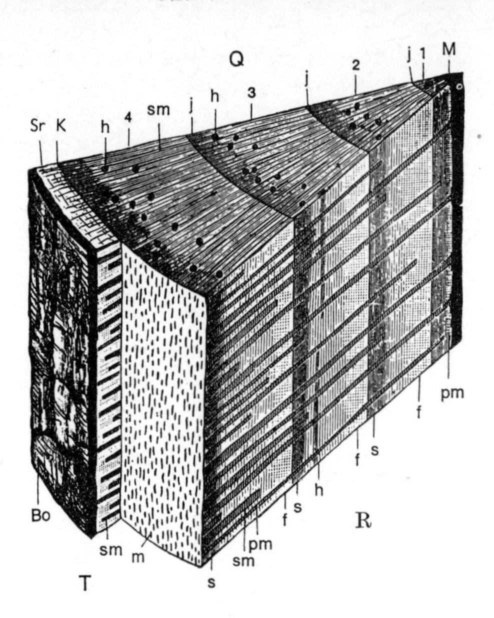

Fig. 163. Portion of a four-year-old stem of pine, cut in winter: Q, transverse section; T, tangential longitudinal section; R, radial longitudinal section. K, cambium; Sr, secondary cortex (bast); Bo, bark; M, pith; j, boundary of a year's growth; 1–4, successive annual rings; f, spring wood; s, late (summer) wood; pm, primary ray; sm, secondary ray in radial longitudinal section; m, rays in tangential section. ×6. After Schenck.

contents (the so-called substitution fibres; Fig. 161B), with (Fig. 161C) or without cross walls, provide a transition to wood parenchyma cells (Fig. 161A).

1. Gymnospermous woods. In most gymnosperms the wood still has a relatively simple structure, and vessels are wholly lacking (Fig. 164). The tracheids, 0.5–5.0 mm in length, are arranged in regular radial rows, corresponding to their origin. This is because the tracheids grow almost only in a radial and hardly at all in a tangential direction; longitudinal growth is also only slight. Consequently, the tracheids resemble the cambial cells in form and length, and they are often pitted only in their radial walls. They are best seen therefore in radial section.

Numerous primary and secondary pith rays, ribbon-like and usually only one layer of cells broad, run in a radial direction between the tracheids. Their cells are usually elongated radially, rich in starch and accompanied by intercellular spaces. The rays serve principally to transport the assimilation products, formed in the leaves and flowing downwards in the phloem, in a radial direction into the wood of the stem and roots, there to be stored. In the opposite direction they conduct water away from the wood. The rays can fulfil these functions, since, as we have seen, they penetrate equally into the phloem and xylem. The intercellular spaces of the pith rays connect with the system of intercellular spaces in the cortex.

In some conifers, e.g. *Pinus*, in the rays of the wood the cells of certain rows, especially those of the upper and lower margins, are longer, dead and differentiated as tracheids. These cells are connected with one another and with the tracheids running longitudinally by bordered pits. They facilitate the exchange of water radially. The parenchymatous cells of the wood rays are also closely connected with the tracheids by large pits, bordered on one side only.

In *Pinus, Picea* and *Larix* the wood parenchyma is found solely around the schizogenous resin canals. These penetrate the wood longitudinally (Fig. 164), and join with others which run radially in some of the broad pith rays, thus forming a network. Consequently, large quantities of resin often flow from wounded stems. In some other conifers the wood parenchyma is confined to simple columns of cells between the tracheids; later their lumina become filled with resin. In many conifers (e.g. *Taxus*) wood parenchyma (apart from the pith rays) is completely lacking.

In many transverse sections of wood annual rings can be seen with the naked eye (Figs. 163; 166A). At higher magnification it can be seen that at each ring the older inner elements (Figs. 163, f; 164) of each radial row of tracheids have wide lumina and thin walls, while the younger outer elements have narrow lumina and thick walls (Figs. 163, s;

164). Within an annual ring the transition from the broad to the narrow elements is quite gradual, but that from the narrow to the broad of the next annual ring is abrupt (Fig. 166A). The annual rings in the wood come about as a result of a periodicity in the activity of the cambium. With us this is in phase with the alternation of the seasons. When the new shoots develop in spring, particularly broad conducting elements are formed, giving rise to a porous, and relatively thin-walled, spring wood, which serves for the conduction of water to the sites of utilization and dispersal. Later the less porous summer wood is formed, and this serves principally to increase the rigidity of the stem.

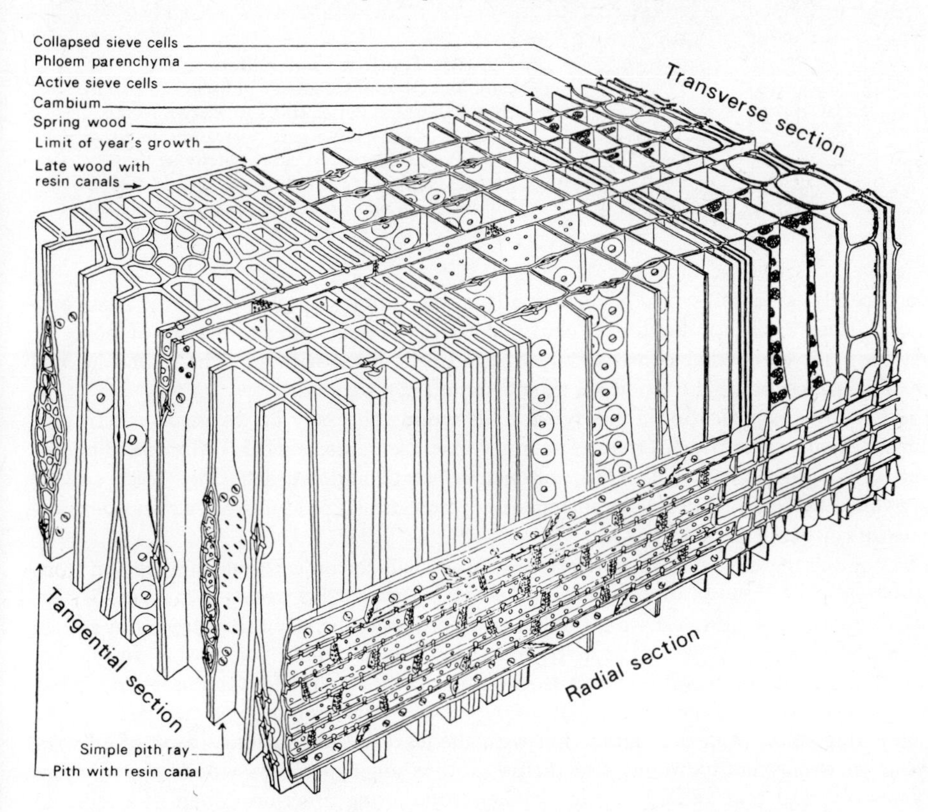

Fig. 164. Fragment from the xylem mass of a conifer in three-dimensional representation. The block is rotated through 180° relative to that shown in Fig. 163, the centre of the stem lying to the left, and the cambium and secondary phloem to the right. The bordered pits of the tracheids, and the sieve plates of the sieve tubes are confined to the radial walls. A pith ray is shown in longitudinal section below. In the xylem portion it consists of four rows of ray parenchyma cells, bordered above and below by a row of ray tracheids. ×200 approx. After Mägdefrau.

2. Dicotyledonous woods. While at the organizational level of the gymnosperms the uniformly tracheidal wood serves both a conducting and a mechanical function (tracheidal level of organization), at the level of the dicotyledons a differentiation between these two functions can be recognized. Corresponding to the emergence of vessels in dicotyledonous woods, the conduction of water by the tracheidal component becomes progressively restricted. Finally, the tracheids, as a consequence of the interpolation of living cells (substitution fibres and wood parenchyma), become wholly isolated from the vessels, and the vessels alone then continue the conducting function.

In some dicotyledonous woods (e.g. *Castanea*) both tracheids and vessels are equally concerned in the conduction of water. Another kind of organization is seen in the woods of *Quercus*, *Ulmus* and *Juglans* where a spatial and functional differentiation is found between water-conducting regions; consisting of tracheids and vessels, and complexes of wood or substitution fibres concerned with strengthening the stem or the storage of starch. A higher grade of organization is seen in the woods of *Vaccinium* and *Aesculus* where the conduction of water is confined to the vessels, the tracheidal component

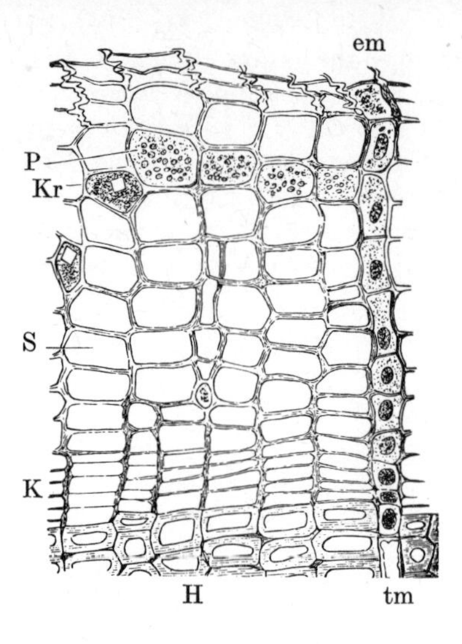

Fig. 165. Transverse section through the cambium (K) and secondary phloem and parenchyma (bast) of *Pinus*. H, outer margin of the xylem; S, sieve cells, regularly arranged in radial rows; P, secondary cortical parenchyma (bast parenchyma); Kr, cell with crystalline inclusion; em, cells rich in protein in ray; tm, ray tracheid. X240. After Schenck.

serving merely as a reservoir. A further advance is shown by the woods of *Acer* and *Fraxinus*. Here again conduction is confined to the vessels, and the matrix of fibres in which they are embedded consists only of dead cells filled with air. The hydraulic and mechanical functions of the wood are thus completely separated.

Even at low magnification a dicotyledonous wood can easily be distinguished from a gymnospermous, since the different elements (vessel elements, tracheids, fibres and parenchyma cells) grow to very different extents in length and breadth (Fig. 158), and in consequence the original radial arrangement of the cambial products can suffer considerable distortions (Figs. 166A, 167A).

The conducting elements form continuous tracts which run without interruption from the roots into the thinnest branches. While in the diffuse-porous woods (e.g. *Acer, Betula, Fagus, Populus, Tilia*) vessels are distributed more or less uniformly throughout the annual ring, in the ring-porous woods (e.g. *Quercus, Fraxinus, Ulmus*) very broad vessels are confined to the spring wood, whereas in the summer wood fibre-like tracheids and wood fibres with narrow lumina predominate. Only those water-conducting elements of the youngest ring stand in direct connection with the leaves of the current season of growth.

While in the gymnospermous and diffuse-porous angiospermous woods annual rings ten years old, and even older, may retain their conducting function (albeit at a reduced rate, some 1.2 to 1.4 m/h in the gymnospermous, 2 to 6 m/h in the diffuse-porous angiospermous woods), in the ring-porous angiospermous woods the functional life of the vessels is generally shorter. In these species conduction in the annual rings of preceding years is usually confined to the narrow tracheids of the late wood, the broad vessels of the spring wood having become filled with air and non-functional. In *Quercus* even the late wood has ceased to conduct water by the third year. This shorter functional life of the conducting elements in the ring-porous wood is, however, correlated with higher rates of conduction (4–44 m/h).

In many woody plants the number of leaves does not increase during the summer. Accordingly, in summer wood the cambium forms predominantly those elements providing mechanical support, namely the wood or libriform fibres (Figs. 166B, f; 167, f). In hardwoods they account for 50–66 per cent of the woody tissue.

In the wood of most dicotyledons the wood parenchyma (Figs. 166B, p; 167, p) is abundantly present, likewise forming strands or tangential layers running longitudinally. In the wood these sooner or later end blindly both above and below. Here and there they always adjoin conducting elements, either surrounding them or running alongside, and they also become associated with pith rays, often fusing with them transversely.

The pith rays (Figs. 166A, m; 167A, m; 167B, sm and tm) are, as in the gymnosperms, ribbons of cells of variable height, one or more layers in thickness, and running radially. They also abut here and there on to the vascular strands, from which they take up water, giving it up when required to other living cells. Together with the wood parenchyma they

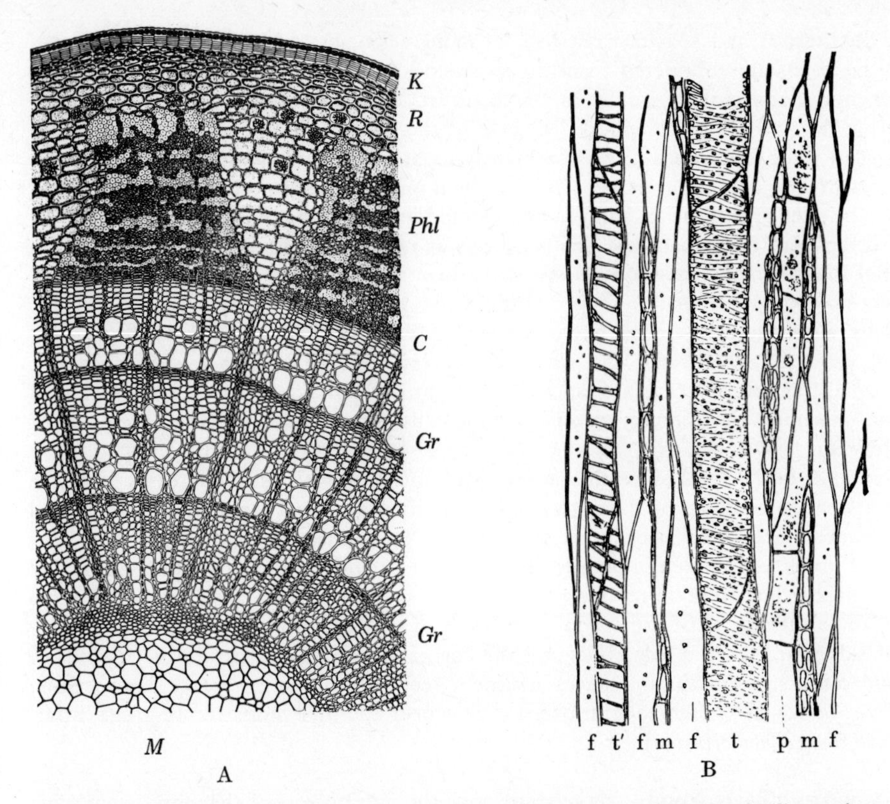

Fig. 166. *Tilia cordata.* **A.** Transverse section through a three-year-old twig of *Tilia*. At the outside (above) a several-layered jacket of cork (K) still covered by the epidermis; R, primary cortex; Phl, phloem showing the primary pith rays very much broadened (dilated) towards the periphery; C, cambium; M, pith; Gr, boundary of a year's growth. Between C and M the wood shows three annual rings and many pith rays one and two cells thick. **B.** Tangential longitudinal section of the wood: f, wood fibre; t, broad pitted vessel in spring wood; t′, tracheid of summer wood with narrow lumen; p, wood parenchyma; m, pith ray. **A.** ×20. After Kny. **B.** ×100. After Schenck.

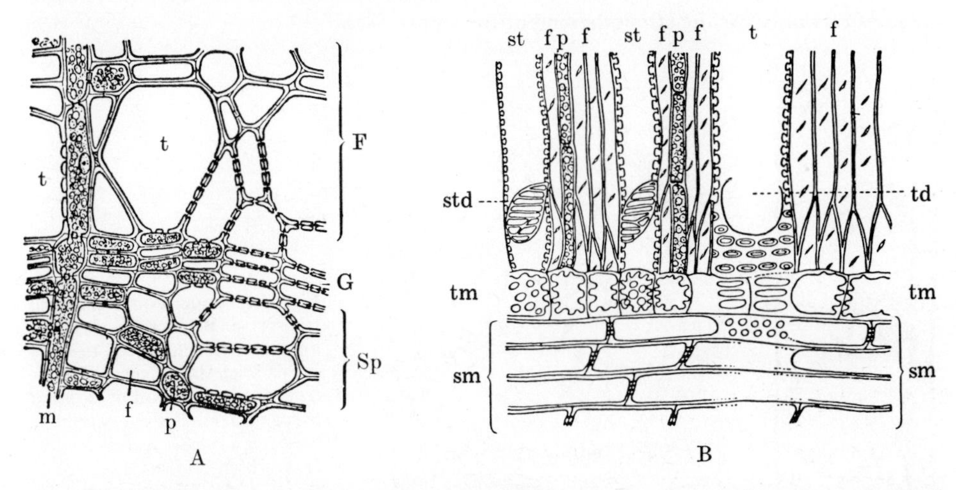

Fig. 167. **A.** Transverse section through the wood of *Tilia cordata* at the boundary of a year's growth (a portion of the stem illustrated in Fig. 166 at a higher magnification): F, spring wood; G, boundary; Sp, summer wood; t, tracheid; f, wood fibre; p, wood parenchyma. **B.** Radial longitudinal section through the wood of *Fagus sylvatica*: t, broad vessel of spring wood with simple perforation plates (td); st, narrower vessel of the late wood with scalariform perforation plates (std); m, pith ray (somewhat displaced from its initial plane by the broad vessel t) with the marginal cells (tm) higher and more conspicuously pitted (especially adjacent to the vessels) than the remainder (sm); p, wood parenchyma; f, wood fibres. The longitudinal walls of the vessels (mostly cut away) have elliptical bordered pits. **A.** ×360. After Strasburger. **B.** ×160. After Huber.

form a continuous and spacious network of living cells, aerated by intercellular spaces. Organic materials are conducted from those passing through the phloem into this system of parenchyma and stored, usually as starch or fat. In late winter and spring, preceding and during the unfolding of the buds, the cells of the ray and wood parenchyma give up a large part of their food reserves, predominantly sugar and a small quantity of nitrogenous substances, to the adjoining vessels. They are then rapidly transported with the ascending water to the sites of utilization and can be detected in the rising sap (see p. 228).

The height and breadth of the parenchymatous rays can be more easily determined in tangential than in radial longitudinal sections, since in the former they are seen in transverse section and appear spindle shaped (Fig. 166B). In most woods the height of the rays varies only within narrow limits, but in some (e.g. *Quercus* and *Fagus*), in which some of the rays are strikingly broad, the variation is very considerable. In many lianes, e.g. *Aristolochia*, the primary rays are especially high and broad, extending over the length of an entire internode. As in the gymnosperms, the cells of rays of dicotyledons are often of two forms (Fig. 167B, sm, tm).

'Grain' and 'bird's-eye'. Since the number of annual rings steadily decreases from the base of a stem towards its tip, both rings and stem have a conical form. Individual annual rings, however, are often broader at the tip of a stem than at its base. In tangential sections of stems, e.g. as seen in boards, the annual rings will not usually appear as parallel lines, but—according to the direction of the cut—as steep-sided conical sections. These are referred to as the 'grain' of the wood. As a result of crooked growth, a crowded development of side branches and side roots and the formation of wound tissue, the 'figure' of the wood can in certain circumstances assume a very irregular and obscure course. Such 'bird's-eye' woods are much sought after in the veneer and pipe industries (e.g. the 'briar' pipe made from *Erica arborea*).

3. Sap wood and heart wood. In general, only the outer layers of the wood, consisting of the younger annual rings and referred to as the sap wood, retain living cells. Consequently, here alone is there storage of food materials. The conduction of water is also confined to the sap wood, frequently (e.g. in ring-porous woods) even to its outermost ring. Only the vessels of this ring stand in direct connection with the leaves and with the younger side roots. The older part of the wood, the so-called heart wood, where the vessels, as a consequence of the penetration of air, no longer form part of the hydraulic system, serves only for the strengthening of the stem.

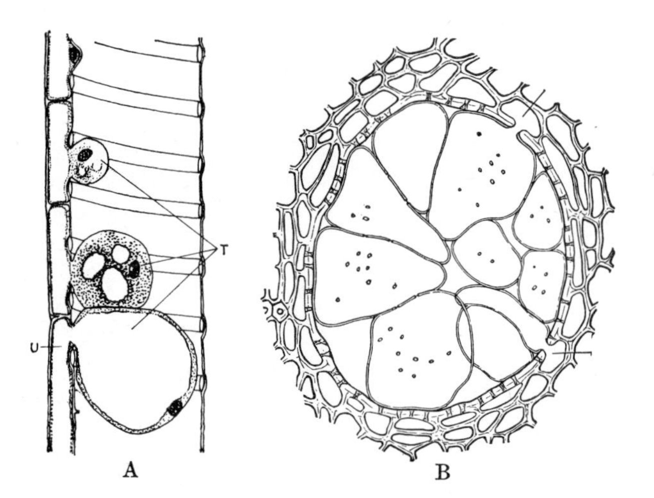

Fig. 168. Longitudinal (**A.**) and transverse (**B.**) sections through a vessel with tyloses (r), together with the adjacent cells, from the heart wood of *Robinia pseudoacacia*. The connection between a tylosis and the cell from which it arises is seen at u. **A.** ×150. Diagrammatic after Holman and Robbins. **B.** ×300. After Schenck.

4. Secondary changes in heart wood. In many trees the vessels in the heart wood become blocked as a result of living wood parenchyma cells penetrating adjacent vessels. The parenchyma cells cause enlargement of the pit closing membranes and bladder-like distensions enter the vessels and more or less occlude their lumina. These bladder-like intrusions, known as tyloses (Fig. 168), often obstruct the lumina of damaged vessels.

The heart wood can often be distinguished by its dark colour from the lighter outer sap wood. Frequently it is also denser, harder and heavier, as well as less liable to rot as a result of various substances deposited in it. The living parenchymatous elements of the wood die as it becomes heart wood. Before death, however, the reserve substances within them are liberated and various chemical compounds are formed, among them tannins, which infiltrate the walls of the surrounding elements. In other cases dyestuffs are also formed, such as haematoxylin in the log wood (*Haematoxylon campechianum*). Other cells accumulate resinous and gummy substances which fill their vacuoles and often also the lumina of the vessels. Exceeding all others, however, are the oxidation products of the tannins, referred to as phlobaphene, which give the heart wood its dark pigmentation. The tannins are responsible for protecting the dead heart wood from decay. Inorganic materials are not infrequently deposited as well in heart wood, e.g. calcium carbonate occurs in the vessels of *Ulmus carpinifolia* and *Fagus sylvatica*, and amorphous silica in the vessels of the extraordinarily durable wood of teak. The heart wood is consequently usually the most valuable part of the wood commercially.

Typical sources of heart wood are *Pinus, Larix, Juniperus, Taxus, Quercus, Robinia, Ulmus, Juglans, Prunus*. The darker the heart wood, the more durable it usually is. Especially valuable heart woods from abroad are mahogany (*Swietenia mahagoni*), rosewood (*Dalbergia*), teak (*Tectona grandis*) and the jet-black ebony (*Diospyros* spp.).

In the 'sap wood' trees (e.g. *Acer pseudoplatanus* and *A. platanoides, Alnus, Betula, Carpinus, Populus*) there is no pigmented or otherwise distinguishable heart wood. The whole mass of wood consists of cells with the same high water content, and the stems in section are uniformly coloured. In some other species (e.g. *Abies, Acer campestre, Fagus, Picea, Tilia*) the heart wood is completely dead, but it merely dries out without secondary changes, or impregnation by preservative substances. Consequently the central part of the trunk, like that of 'sap wood' trees, often suffers attack by fungi and becomes hollow.

5. Dendrochronology. The age of a tree can often, and with considerable accuracy, be determined from the number of annual rings. Thus it has been shown that the ages of specimens of the North American *Sequoiadendron*, with trunks reaching a diameter of 5–6 m at the base, are of the order of 3500 years, and that those of some specimens of *Pinus aristata* in California may even exceed 4600 years.

The annual rings also give information about climatic changes during the life of the tree. In dry periods the rings are narrow, with much summer wood, in wet periods they are broader. In secondarily thickened shoots or roots the closer towards the tips transverse sections are taken, the fewer the annual rings found. This follows inevitably from the apical growth of these organs and the fact that their age diminishes acropetally. Not infrequently the number of the rings in the wood exceeds the age of the shoot. The principal cause of this is a loss of leaves through frost, caterpillar attack or other destructive agency, stimulating a precocious opening of the buds laid down for the growing period of the succeeding year. The renewed production of leaves leads to the renewed formation of wood resembling spring wood. On the other hand, in trees in which the production of annual rings is normally strictly regular, the number of recognizable rings may sometimes appear to fall short of the age of the individual in question, because of some of them failing to be sharply set off from each other. In the wood of roots the boundaries of the rings are usually obscure. In those trees of the humid tropics which grow uninterruptedly annual rings may be absent from the wood (e.g. *Guajacum* and *Quassia*). However, even here the wood of many trees shows evidence of rhythmical growth, indicating a periodicity in the activity of the cambium corresponding to that with which the leaves are renewed, not in this case an annual rhythm.

g. Secondary phloem and cortex (bast). As with the phloem of the vascular bundles, whose function it takes over only a short distance behind the apex, the principal task of the secondary phloem is also the transport of assimilates. In addition, however, it may provide storage regions. In both gymnosperms and angiosperms the secondary phloem and cortex are composed of three kinds of tissues, namely 1. the sieve cells and sieve tubes, forming the most important system for transporting assimilates and organic reserve materials; 2. the ray parenchyma, providing a connection with the storage tissue in the wood, and, together with the parenchyma of the secondary cortex, itself also able to act as a storage region (Fig. 169A and B, M); and, 3. plates of sclerenchyma and layers of cork which together form a protective system, the hardness and elasticity of which averts mechanical damage, the insulating properties injury by the sun's heat and fire, and the antiseptic tannins infection by bacteria and fungi. The latter may also explain the significance of the latex ducts, mucilage cells and resin canals which frequently occur in the secondary cortex, and those contents serve to seal wounds after damage. Not infrequently cells containing crystals are also present (Fig. 169A and B, kr), but their function is unknown.

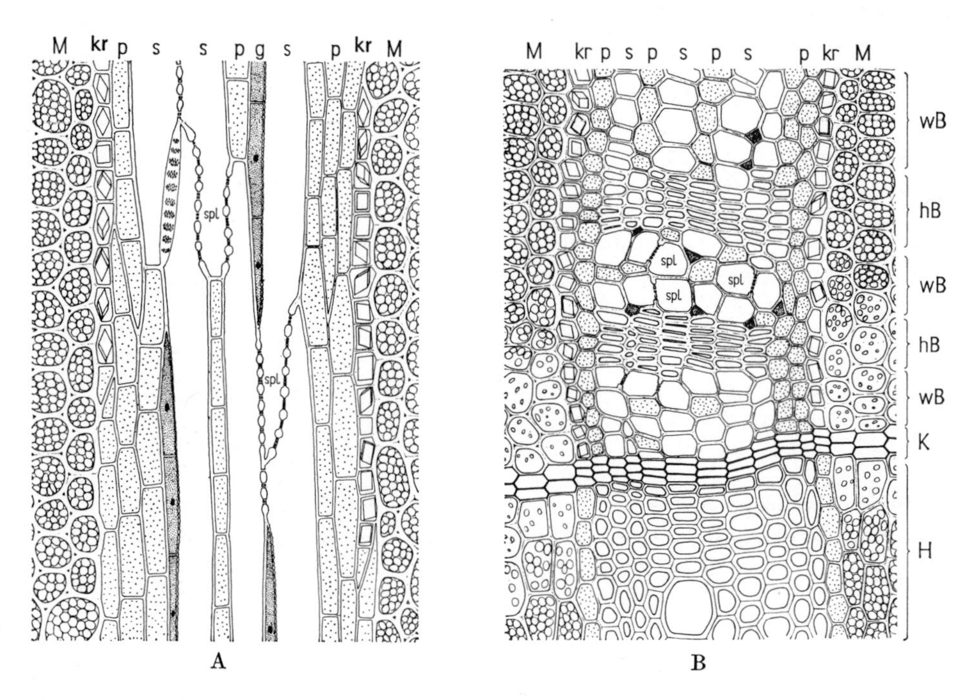

Fig. 169. *Vitis vinifera.* Vine. Secondary phloem. **A.** Tangential longitudinal section. **B.** Transverse section. s, sieve tube with sieve plates (spl); g, companion cell; p, parenchyma; M, ray parenchyma; kr, cell containing crystal; wB, 'soft bast' (sieve tubes and companion cells); hB, 'hard bast' (fibres); K, cambium; H, edge of secondary xylem. ×200. Original.

The sieve cells (Fig. 164, H) and sieve tubes (Fig. 169, s) of the phloem appear in tangential view attenuated at each end like chisels. The end walls are thus radially arranged, and usually very oblique, and are perforated by a variable number of sieve plates lying one above the other (Fig. 169A and B, spl). The sieve elements are thin-walled, unlignified and usually active for only a short time, after which they collapse or become compressed. The phloem or bast fibres are often very long and with narrow lumina, and their walls, usually lignified, are heavily thickened (Fig. 169B, hB) and sparingly furnished with oblique slit-like pits. The parenchyma cells (Fig. 169A and B, p) are similar to those of the wood; they have living contents, are elongated in the direction of the strand, rich in food reserves, thin walled and usually unlignified. The cells of the rays (Fig. 169A and B, M) usually contain abundant starch.

1. Arrangement of the tissues in the phloem. In the dicotyledons the individual tissues of the phloem and those of the wood are similarly arranged. Thus the sieve tubes form usually unbroken, branched, longitudinal pathways from the root up to the crown of leaves. In addition, the sieve tubes come into contact here and there with the ribbon-like rays (Figs. 164; 165) also present in the phloem, or at least are connected with them by living parenchyma cells. The rays in the secondary phloem are a radial continuation of those in the wood (Figs. 164; 166A). Thus the assimilation products from the crown of leaves, streaming down in the secondary phloem towards the roots, can also pass radially through the rays into the living cells of the wood or of the bast.

The tissues of the secondary phloem (sieve tubes, companion cells and parenchyma, on the one hand, and phloem fibres, on the other) often form regular layers, penetrated only by the rays. These layers, however, have nothing to do with the formation of annual rings. Often they will alternate with one another several times in one growing season (Figs. 166A, Sr; 169B, hB and wB), a result of some endogenous rhythm still not investigated in detail. Plates of hard fibres in the phloem of *Tilia* and *Salix* spp. yield the 'bast' of gardeners, although this has been increasingly superseded in recent times by phloem fibres from the leaves of the Raffia palm (*Raphia ruffia*), a native of western Madagascar. The secondary phloem is particularly regular in construction in many conifers. In the Cupressaceae, for example, single-layered tangential bands of fibres, sieve cells, parenchyma, sieve cells and once again fibres, sieve cells and so on alternate with one another continuously. In other conifers, e.g. *Larix*, the phloem formed first in the spring consists exclusively of sieve cells (ten or more rows), and not until later is there a regular alternation between sieve cells and parenchyma cells. Thus in larch there are genuine annual rings in the phloem, otherwise a rare occurrence.

h. The consequences of secondary thickening for the tissues external to the cambium

1. Dilatation. Since the cambium continues to produce new wood towards the inside and new phloem towards the outside, the stem or root undergoes a secondary increase in size. In consequence, the permanent tissues lying outside the cambium become tangentially stretched or ruptured and radially compressed, so far as they do not keep up with the expansion of the periphery by tangential growth (dilatation, see p. 149).

In tangential growth the living parenchymatous cells of the primary cortex, of the primary phloem and in some woody plants even of the epidermis divide by anticlinal (radial) walls. In the phloem this growth is often especially striking in the primary rays. In *Tilia*, for example, there is actually the formation of a secondary meristem which cuts off rows of parenchymatous cells on both sides by radial walls. Since the older outer zones undergo this tangential growth for a longer time than the younger inner zones, the parenchymatous rays are broadened towards the outside like wedges (Fig. 166A). On the other hand, the outer columns of phloem fibres, consisting of dead sclerenchyma laid down at a time when the branch was thinner, are narrower than the inner columns. Frequently they are ruptured tangentially as a result of the expansion. This is especially true of the primary strengthening zone of phloem fibres present in many species (e.g. *Aristolochia*), which separates the central cylinder from the primary cortex (p. 141) and is the first region to succumb to the dilatation. Frequently numerous parenchymatous cells from the adjacent living tissues (the parenchyma of the secondary phloem or cortex) grow into the gaps left by rupture, and they may become transformed wholly or partly into thick-walled stone cells (Fig. 171, S), thus closing the breaks once again. Complex cylinders of mechanical tissue arise in this way (Fig. 171). The secondary sieve tubes and companion cells, both of which are active for only a short time, and which then die and collapse, are crushed with any secretory cells present and their contents resorbed.

In various species of *Acer, Citrus, Cornus, Ilex, Rosa, Viscum* and Cactaceae, tangential growth of the epidermis occurs for several years. Such epidermal cells usually have considerably thickened outer walls. As the outer walls are split with the increasing surface, these species are able to repair them from within by the deposition of new layers of thickening.

2. Formation of periderm. Generally the epidermis takes no part in the process of dilatation, it becomes passively stretched and ultimately ruptured. Before this occurs a secondary boundary layer is differentiated, the *cork* (see p. 103). It replaces the epi- or exodermis and protects those parts which are in the process of expansion from desiccation. The cork is generated by a special secondary meristem, the cork cambium (or phellogen), which is formed at the periphery of the expanding organ (Fig. 170A, KK).

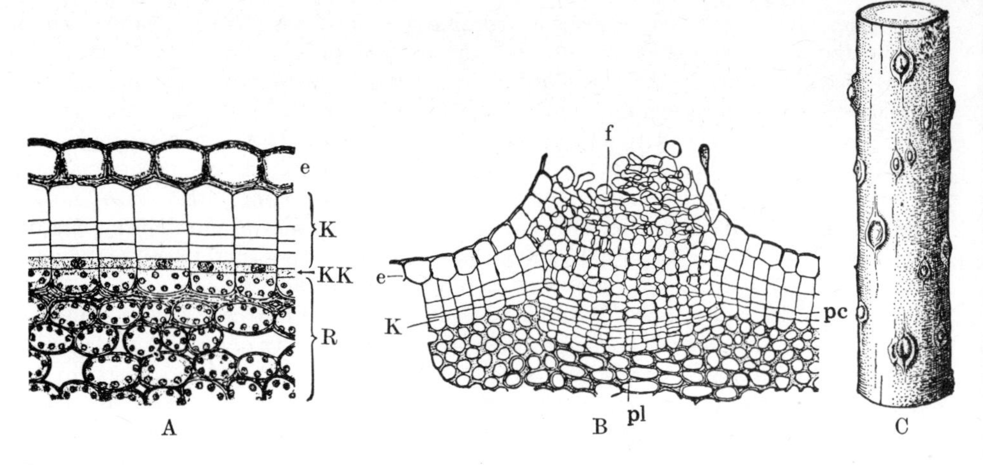

Fig. 170. Secondary insulating tissue and lenticels in *Sambucus nigra*. **A.** Transverse section through the outer part of a one-year-old twig of *Sambucus nigra* (elder) in summer. Beginning of the formation of periderm: R, cortical parenchyma; e, epidermis; K, cork; KK, cork cambium. **B.** and **C.** Older twig with large lenticels, several years old, in the cork. **B.** Transverse section through a small lenticel in the first year: e, epidermis; pc, cork cambium; pl, cork cambium of the lenticel; f, loosened cells filling the lenticel. **A.** ×500. After Fitting. **B.** ×60. After Strasburger. **C.** Natural size. After Fitting.

The formation of this cork cambium usually begins in the first period of growth, with us in June or July, soon after the beginning of the other secondary growth or even before. It may arise from the epidermis by tangential (periclinal) division of its cells (as in apple and pear). Usually, however, it arises in a similar way from the layer beneath the epidermis (Fig. 170A, B), or from both this layer and the epidermis (e.g. in potato tubers), or more rarely from deeper layers of the cortex (e.g. *Ribes*, *Thuja*). In roots it usually arises from the pericycle (Fig. 189). The cork cambium produces radial rows of cells, principally towards the outside. These constitute, whether they be suberized or not, the cork. The few cells cut off by the cork cambium inwards form the phelloderm. These are unsuberized and become chlorophyllose cortical cells. The cork cambium, along with the cork and the phelloderm, are referred to collectively as the periderm. As the formation of the periderm begins, so the outside of the stem, previously green, becomes brown or grey. Eventually periderm formation usually takes place even in those species where the epidermis initially undergoes tangential growth (e.g. *Acer, Ilex, Rosa*). It is absent in species of *Viscum*.

3. Lenticels. The replacement of the epidermis by a jacket of cork lacking any intercellular spaces would prevent the exchange of gases between the external air and the interior of the organ were nothing provided to take the place of the stomata. The function of the stomata is taken over, at least to a certain extent, by the lenticels. These are visible to the naked eye, and on the skin of cork enclosing the branches of our woody plants they appear as wart-like areas (Fig. 170B, C). The lenticels provide an example of a variation in the development of the secondary limiting tissue becoming progressively improved in the course of evolution. They consist of more or less clearly defined regions of the cork conspicuous for the high activity of their meristematic layers. The presence of numerous intercellular spaces at these sites makes the lenticels permeable to gases. The spaces connect with the external air, and in the other direction through the cork cambium to the intercellular system of the living tissue.

In bottle corks the lenticels appear as canals filled with a dark-brown powder, consisting of dead cells. These canals run radially through the whole thickness of cork. The lenticels usually arise beneath stomata and at the same time often above multicellular medullary rays; indeed, they often appear before the rest of the cork. Beneath the stomata, the cork cambium, which has radially arranged intercellular spaces, produces numerous complementary cells towards the outside (Fig. 170B, f), so that the lenticels eventually force up the epidermis and break through it. Besides the very loose and usually unsuberized com-

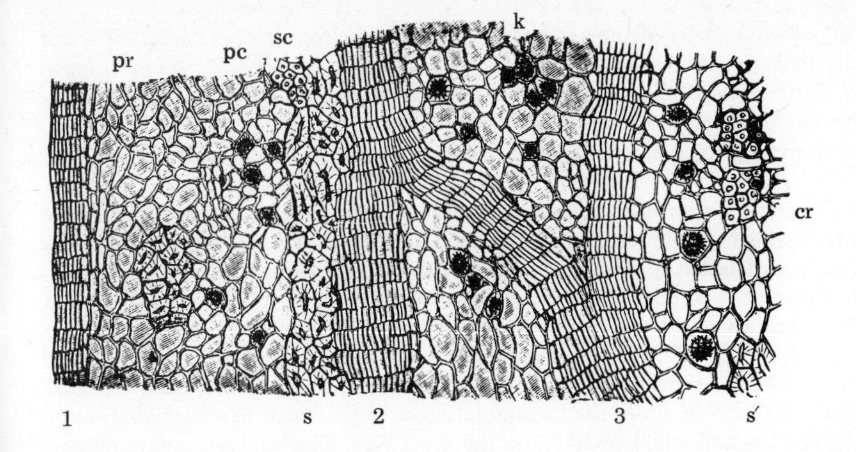

Fig. 171. Transverse section through the bark of *Quercus petraea*: 1, 2 and 3, successive layers of cork; pr, primary cortex modified by dilation; pc, outer boundary of the central cylinder identified by the sclerenchyma fibres; sc, remains of a primary zone of strengthening tissue, originally forming an almost complete cylinder (cf. Fig. 150B, e) but ruptured as a result of dilatation (Fig. 150B, f); s, secondarily differentiated stone cells; s′, further stone cells arising during secondary growth; cr, phloem fibres of secondary origin; k, excretory cells containing druses of calcium oxalate. All the tissues outside the innermost layer of cork are brown and necrotic. ×255. After Schenck.

plementary tissue, the cork cambium can also produce, especially towards the end of the growing period, layers of more firmly united suberized cells with only narrow intercellular canals. These layers, however, are always ruptured by the complementary cells which arise subsequently and push against them. In roots generally, and in the shoots of many gymnosperms, lenticels occur without the initial presence of stomata.

i. Formation of bark. Those tissues which lie outside the cork cambium, e.g. the epidermis, and in most roots the entire cortex (Fig. 189), are cut off from supplies of water and metabolites, and die. Usually the first cork cambium in stems and roots loses its activity (but in *Betula* not for many years, and in *Carpinus* and *Fagus*, as well as in the cork oak *Quercus suber*, never). Its cells then become a layer of the periderm, and a new cambium arises deeper within the cortex. After some time it similarly loses its activity, and a third cambium arises further in, and so on (as, for example, in *Quercus*; Fig. 171). Eventually the primary tissues are exhausted and the new cork cambium arises within secondary tissues, especially the living parenchyma of the secondary cortex. Thus in an old stem the living and dead tissues external to the wood are all of secondary origin.

The tissue segregated in this way, together with the intervening layers of periderm, are referred to as bark. Naturally, since it is cut off from supplies of water and nutrients, the bark is no longer able to take part in the further expansion of the stem or root. In time it is either shed or, as in most old trees, it becomes longitudinally fissured (Fig. 163, Bo).

Thus eventually the outer oldest part of the phloem is shed with the bark. The consequence is that in the trunks of old trees the layer of phloem always remains thin. Mechanically strengthening elements can therefore only become a permanent component of the stem if they arise on the inside of the cambium, i.e. in the wood. If the individual layers of cork form complete cylinders, then the masses of tissue encircling the stem give rise to ring bark as in birch (*Betula*) and cherry (*Prunus cerasus*). (This should not be confused with flake cork, p. 104.) If the cylindrical layers of cork are interrupted by columns of parenchyma, the bark then comes away in long strips, as in *Vitis, Lonicera* and *Clematis*. If the layers of cork and the cambium from which they have arisen encompass, as is usually the case, only a limited part of the stem surface, the margins of the younger layers curving and adjoining older layers lying towards the periphery, then scale-like portions of tissue are disjoined from the stem. Bark produced in this way is called scale bark (as in *Platanus*, *Quercus* (Fig. 171) and many conifers). Where the bark exfoliates, as in *Platanus* and *Pinus*, this results from the presence of abscission layers consisting of parenchyma cells or thin-walled cork cells inserted between the often heavily thickened cork layers of the periderm. These become ruptured as a result of transverse tensions arising in the bark hygroscopically. The brown or red colour of most barks is caused by derivatives of the tannins similar to those causing the colouring of many heart woods. These strongly antiseptic materials account for the resistance of bark to decay. The white colour of the bark of birch results from the laying down of betulin.

j. Wound healing. If a wound exposes a tissue still young a callus is generally formed at first, i.e. all the living cells bordering on the wound proliferate. Then in most cases a cork cambium is differentiated at the periphery of this callus tissue, giving rise to cork towards the outside. Extensive wounds in the older parts of the stems of conifers and dicotyledons which reach to the wood, and also the transverse wounds caused by pruning, become covered by bark. The vascular cambium bordering on the margins of the wound grows out in the form of a ring and produces a callus. The callus insulates itself with cork on the outside, while on its inner side a cambial layer arises which becomes continuous with that of the stem and which is active in the same way. Thus the callus extends over the surface of the wound and gradually covers it with new layers of wood. If the callus succeeds in bridging over the surface of the wound completely the cambium eventually again forms a uniform meristematic layer. The wood covering the wound does not fuse with that laid bare in the wounding; this has become brown and its parenchyma has died. Thus in stems which have been cut into to the depth of the xylem, the marks can still be seen in the wood for very many years after the callusing over. In the same way wires bound round stems gradually 'grow into' the wood and are eventually completely callused over. The wood formed above wounds is at first different from normal wood in structure and is referred to as wound wood. It consists of almost isodiametric cells, following which elongated elements gradually appear. In *Prunus avium* after wounding the cambium produces instead of normal xylem tissue nests of thin-walled parenchyma cells. These produce gum which protects the wound (p. 69).

B. The leaves[29]

We see the leaf primordia arise exogenously at the growing point as lateral protuberances of transverse ridges, initially undifferentiated (Figs. 131E; 132A; 137; 138). In contrast to the shoot axis, which generally continues to grow uninterruptedly at its tip, the growth of leaf primordia is, with rare exceptions, limited and is apical usually only for a short time (acroplastic growth). The activity of the apical meristem generally dies away very rapidly, and it is replaced by a basal or by one or more intercalary meristematic zones (basiplastic growth).

In the conifers and many monocotyledons the transition from acroplastic to basiplastic growth takes place when the leaf primordium is barely 0.3 mm in length. Only in a few pteridophytes (whose leaves often continue to grow from an apical cell or an apical marginal meristem) and in the ancient cycads is the whole of the leaf expanse generated acroplastically. In these plants the young leaf is coiled up at the tip of the shoot. Among the dicotyledons acroplastic growth persists for a long time in some of the plants specializing in carnivorous nutrition (*Drosophyllum, Utricularia, Pinguicula*).

Those monocotyledons furnished with narrow ribbon-like leaves (e.g. Gramineae, *Clivia*) show particularly persistent basiplastic growth. *Welwitschia mirabilis* (Fig. 648A) is very peculiar and differs from all other cormophytes. In this species, other than the two cotyledons, only a single pair of decussately arranged leaves is produced. These increase annually by a little basiplastic growth, while their ends gradually weather away.

Growth in breadth of the lamina takes place in most cases from sub-epidermal marginal cells—a so-called marginal meristem. In only the ferns, the delicate leaves of many water plants, the grasses, and in many scale leaves, are the cells of the marginal meristem at the surface. They then behave as edge-like apices. Marginal growth of the lamina is often complemented by surface growth, also involving cell multiplication.

The main axis of the seedling bears first of all the cotyledons, one in the monocotyledons, two in most of the dicotyledons (Figs. 172; 173), two or more in many gymnosperms. Following upon the cotyledons come first scale leaves (Fig. 173), or

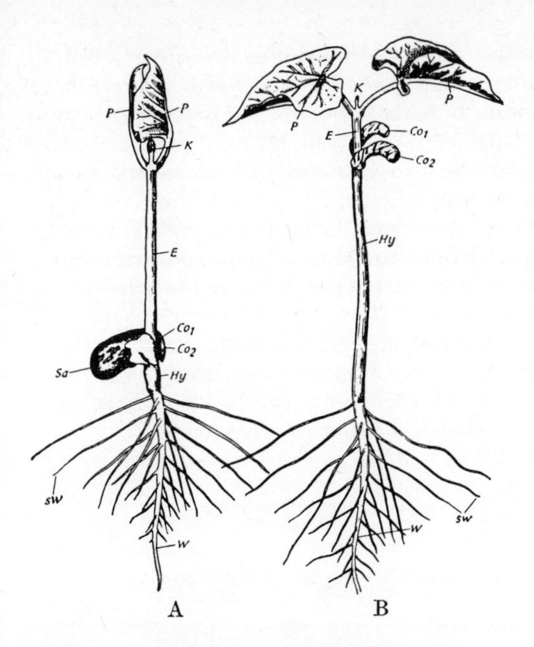

Fig. 172. Hypogeal (**A.**) and epigeal (**B.**) germination shown by seedlings of *Phaseolus coccineus* and *P. vulgaris* respectively: Sa, seed coat; Co, cotyledon; E, epicotyl; P, primary leaf; K, shoot apex; Hy, hypocotyl; W, main root; Sw, lateral roots. $\times\frac{1}{2}$. After Rauh.

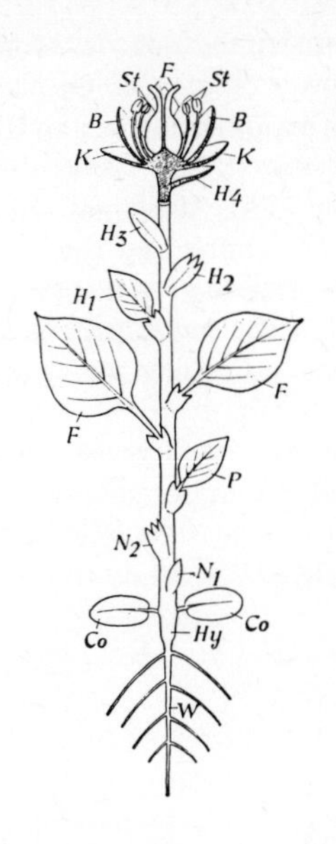

Fig. 173. Diagrammatic representation of the leaf sequence on the stem of a dicotyledonous plant. W, root; Hy, hypocotyl; Co, cotyledons (seed leaves); N_1, and N_2, scale leaves; P, transitional leaf; F, normal foliage leaves; H_1 to H_4, bracts; K, calyx members (sepals); B, perianth members (petals); St, stamens; F, carpels (together forming the ovary). After Rauh.

directly normal foliage leaves (Figs. 172A, B; 173, F). In many flowering plants bracts of simple form (Fig. 173, H_1–H_4) succeed the foliage leaves towards the reproductive part of the aerial shoot. The components of the calyx and corolla, and the reproductive organs are discussed in the systematic section (p. 601 seq.).

1. The cotyledons

The cotyledons, the life of which is usually only short, are almost always less complex in form than the normal leaves.

In the germination of the seed they may remain enclosed within the testa and concealed beneath the surface of the soil (hypogeal germination), e.g. in *Quercus, Aesculus, Pisum, Vicia faba, Phaseolus coccineus* (Fig. 172A). In these cases the cotyledons are generally fleshy reservoirs of food materials, consisting principally of food-storage parenchyma. In epigeal germination the cotyledons burst through the testa and appear above the surface of the earth. Here they become green, and for a variable period assimilate carbon dioxide just as foliage leaves, which they more or less resemble in external form and inner structure (e.g. *Phaseolus vulgaris* (Fig. 172B), *Tilia, Fagus, Carpinus* (Fig. 186), *Sinapis* (Fig. 185)).

In the remarkable tropical genus *Monophyllaea* and some species of *Streptocarpus* (both Gesneriaceae) no leaves are produced other than the two cotyledons. At first these are equal in size, but later one of them enlarges very considerably and forms the single long-lived 'leaf'. Eventually many inflorescences arise in its axil.

2. The foliage leaves

The foliage leaf usually consists of a number of parts. Thus we can recognize the flat, green, frequently very thin, expanse or lamina of the leaf, the stem-like petiole, and the leaf base. This may be differentiated as a leaf sheath, or may bear stipules (Figs. 174, s; 183, st). In many leaves, however, the leaf base is not specially developed, but goes over gradually into the petiole.

Usually the leaf primordium, at first ridge-like, begins at a very early stage to become weakly constricted near its base, thus dividing the primordium into upper (Fig. 174B, C, D) and lower (Fig. 174, U) parts. The former undergoes vigorous growth in breadth, at first from marginal initials and later by means of surface growth, and possibly growth in thickness as well, and yields the lamina of the leaf (Figs. 228; 229) and, where present, the petiole. This does not arise until relatively late as a result of intercalary growth between the lamina and the lower part of the leaf, initiated in some cases from the upper part of the leaf (e.g. *Malus* (Fig. 183)) and in others from the lower (e.g. *Helleborus* (Fig. 184)). From the lower part of the leaf develops the leaf base, often of course remaining quite small, and arising therefrom, again where present, the stipules (Fig. 174D and E).

a. The lamina. *1. External form.* The lamina of the leaf is usually markedly dorsiventral and the upper side darker green than the lower. It may be undivided (Fig. 173, F) or divided into leaflets (Figs. 174E; 181B; 182A). The monocotyledons show predominantly simple leaves, while those of the dicotyledons are frequently compound.

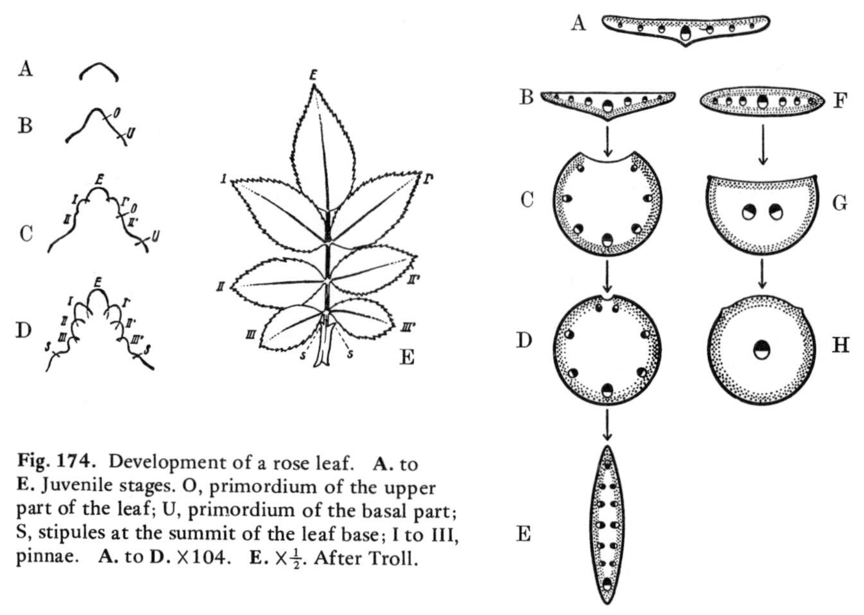

Fig. 174. Development of a rose leaf. **A.** to **E.** Juvenile stages. O, primordium of the upper part of the leaf; U, primordium of the basal part; S, stipules at the summit of the leaf base; I to III, pinnae. **A.** to **D.** ×104. **E.** ×$\frac{1}{2}$. After Troll.

Fig. 175. (Right) Diagrammatic transverse sections through different types of leaves. **A.** Normal bifacial flattened leaf. **B.** Inverted bifacial leaf (e.g. *Allium ursinum*). **C.** and **D.** Derivation of the unifacial cylindrical leaf (e.g. *Allium sativum, Juncus effusus*). **E.** Unifacial flattened leaf (e.g. *Iris*). **F.** Equifacial flattened leaf (e.g. *Lactuca serriola*). **G.** Equifacial needle-like leaf (e.g. *Pinus*). **H.** Equifacial cylindrical leaf (e.g. *Sedum album*). The lower side of the leaf is indicated by a heavy line, the palisade parenchyma by stippling and the xylem part of the vascular bundle is made to appear black. Based on Troll and Rauh.

Where the lamina is undivided the margin is either entire or only slightly toothed or lobed (Figs. 181A; 182B; 867). If there are deeper incisions reaching as far as the midline of the leaf, then the leaf is said to be digitate, pedate (Fig. 184E) or pinnate, according to whether the incisions run towards the base of the lamina, there approaching one another or whether they are directed towards the midrib, which in this case becomes the rachis of a pinnate leaf.

Developmentally, pinnation is brought about by the marginal growth of certain sections of the lamina being retarded while that of the others is promoted (Fig. 174D and E). The deeply lobed, and often in addition perforated, laminae of *Monstera*, a member of the Araceae frequent in cultivation, arise as a result of islands of tissue between the veins in the very young leaves dying and disintegrating. The segments of the leaves of the 'fan-leaved' and 'feather-leaved' palms also arise through the subsequent partitioning of the initially undivided lamina. The young lamina is folded backwards and forwards like a closed paper fan. Before the unfolding of the leaf either strips of tissue (usually on the upper edges of the folds, more rarely on the lower) die or the cell walls in one of these positions become mucilaginous and separate. The latter process is seen in *Cocos*.

Usually the upper and lower sides of the lamina are derived from the corresponding sides of the leaf primordium (bifacial leaves; Fig. 175A). In the unifacial leaves of certain monocotyledons, however, the whole surface of the lamina develops only from the lower surface of its primordium. Thus are formed the cylindrical or filamentous leaves of many species of *Allium* and *Juncus*, as well as the sword-like leaves of *Iris* (Fig. 175C, D and E).

The laminae of unpetioled (sessile) leaves usually adjoin the stem with a broad leaf base. If this embraces the stem the leaf is said to be *amplexicaul*.

The lamina is penetrated by veins or nerves, often bright green in colour, forming a richly branched network (Figs. 176A; 177).

The thicker nerves, also referred to as ribs, usually project on the lower surface of the leaf like a framework, while there may be furrows corresponding to them on the upper

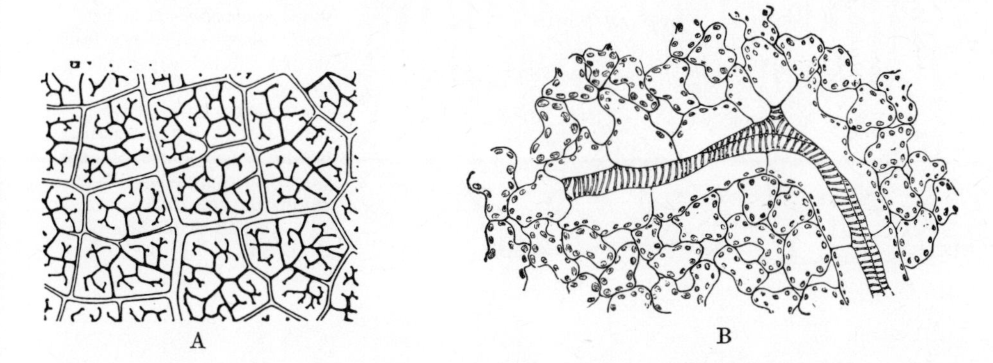

Fig. 176. A. *Morus alba*. Part of the leaf lamina showing the reticulate venation. Note how there is a more or less uniform degree of ramification between the principal veins. **B.** Termination of a vascular bundle in the leaf of *Impatiens parviflora*. **A.** ×10. After Wylie. **B.** ×240. After Schenck.

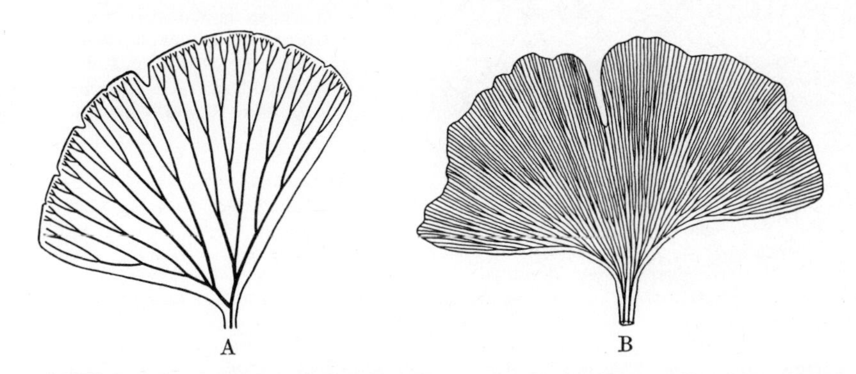

Fig. 177. A. *Adiantum tenerum* var. *farlayense* (a maiden hair fern), showing the dichotomous venation of a pinnule. Diagrammatic. **B.** *Ginkgo biloba*. Lamina with dichotomous venation. **A.** After Goebel. **B.** ×$\frac{2}{3}$. After Haupt.

surface. Their finer branches cannot, however, be seen unless the leaf is held up to the light. Often a nerve running in the midline of the leaf is especially well developed, and is then referred to as the principal vein or midrib. The lateral veins arise from it (Fig. 173, F).

Not infrequently in spring leaves are found on the ground which have rotted away all but for the filigree network of their veins. Such 'leaf skeletons' show especially clearly that the veins not only have the function of supplying the leaf with water and removing from it assimilation products but also that at the same time their lignified elements contribute physical support to the paper-like lamina. While most dicotyledons and ferns show reticulate venation, that of the monocotyledons is generally simpler and parallel. The needle-like leaves of many conifers are single veined. The primitive dichotomous venation, in which no midrib was developed, persists, except for some ferns (Fig. 177A), only in *Ginkgo* (Fig. 177B), and in certain primitive Ranunculaceae (*Kingdonia*).

2. *Inner structure*. The differentiation of that part of the leaf bounded above and below by epidermis, the mesophyll, corresponds to the two main functions of leaves, namely the photosynthesis of carbohydrates and the transpiration of water. The assimilatory parenchyma, rich in chloroplasts, occurs in most leaves beneath the upper epidermis. It consists of one (Fig. 178) or several (Fig. 196A) layers of cylindrical cells arranged with their long axes perpendicular to the surface of the leaf. Because of its characteristic appearance in

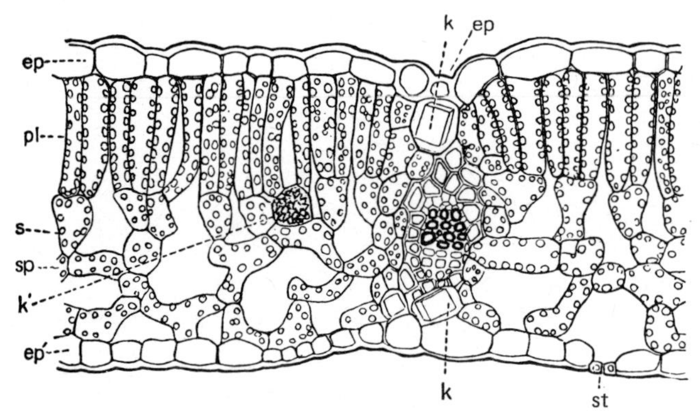

Fig. 178. Transverse section through the bifacial leaf of a beech (*Fagus sylvatica*): ep, upper epidermis; ep′, lower epidermis; pl, palisade; s, funnel-shaped cells below palisade; sp, spongy mesophyll; k, cells containing crystals between which lies a collateral vascular bundle (xylem above, phloem below) surrounded by a scleren-chymatous sheath; k′, druse; st, stoma. ×360. After Strasburger, modified.

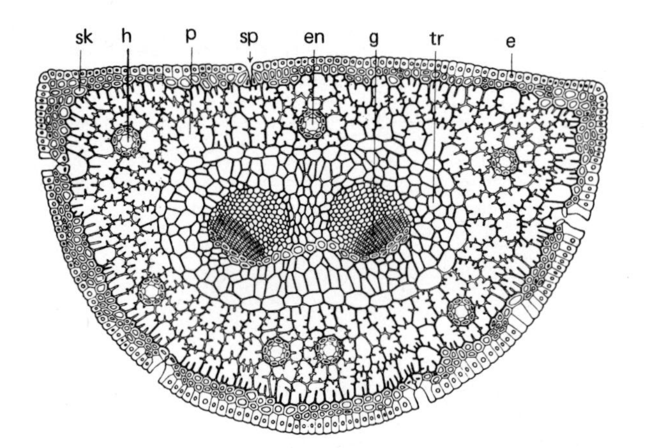

Fig. 179. Transverse section through the equifacial needle-like leaf of *Pinus nigra*; g, vascular bundle; tr, transfusion tissue; en, endodermis; p, assimilatory parenchyma; h, resin canal; e, epidermis; sp, stoma; sk, hypodermal sclerenchyma. ×30. After R. von Wettstein.

transverse section this tissue is known as the palisade parenchyma. As a surface section will show (Fig. 95D), the individual palisade cells are more or less completely separated from each other by intercellular spaces. Beneath the palisade parenchyma, towards the lower side of the leaf, is the spongy mesophyll, consisting principally of irregularly shaped cells (Figs. 95E; 178, s). The intercellular spaces are usually strikingly wide, and are directly connected with the numerous stomata always present in the lower side of the leaf (Fig. 178, st). Thus there is little impediment to the movement of water vapour, and to the uptake or giving off of carbon dioxide and oxygen when the stomata are open (see p. 225). The typical foliage leaf is thus dorsiventral or bifacial in construction (Figs. 175A; 178).

It has been estimated that in the leaf of *Ricinus communis* some 82 per cent of the chloroplasts belong to the palisade tissue and only 18 per cent to the spongy mesophyll. Stomata are frequently completely absent from the upper side of the leaf (e.g. in many broad-leaved trees); such leaves are said to be hypostomatic, in contradistinction to amphistomatic leaves in which the stomata occur on both sides. The number of stomata per square mm of surface varies between 20 and over 800 (e.g. *Sedum acre* 18; in maize, upper surface 95, lower 160; in pea 100/220; apple 0/250; beech 0/340; olive 0/550; sycamore 0/860. The stomata of floating leaves are usually exclusively on the upper side, and such leaves are therefore epistomatic.

Many leaves, especially those of species occupying extremely sunny and relatively dry places (Fig. 195A), together with others such as the needle-like leaves of *Pinus* (Fig. 179), and also those of many submerged water plants (Fig. 194A) are similar in structure above and below. Such leaves are said to be equifacial (Fig. 175F, G and H), and there is little or no distinction between the palisade and spongy mesophyll.

The structure of the vascular bundles in the lamina (Fig. 178, between k and k) corresponds to that of the bundles in the stem. In the seed plants the bundles of the leaf, which consist almost entirely of primary tissues, are generally collateral and closed. Those of the midrib in gymnosperms and dicotyledons may, however, be open. Since they are continuations of the stem bundles, bending outwards, the xylem is orientated towards the upper side of the leaf and the phloem towards the lower. As the bundles ramify in the lamina and become smaller (Fig. 176) so does their structure become simpler. Only reticulately or spirally thickened tracheids remain in the xylem, and the phloem correspondingly diminishes. The sieve tubes diminish in width and eventually vanish, while the xylem remains represented solely by short spirally thickened tracheids, with which it ultimately ends blindly. The vascular bundles of leaves are surrounded by a parenchymatous sheath. These are single- or double-layered, and invest uninterruptedly even the finest ramifications (Fig. 176B).

In the needle-like leaves of the conifers usually only one or two bundles occur running longitudinally (Fig. 179). Adjacent to the outer margin of the xylem is a border of dead tracheidal cells with bordered pits, while adjacent to the phloem is a corresponding border of cells rich in cytoplasm (albuminous cells). This transfusion tissue serves for the exchange of substances between the nerve and the mesophyll (Fig. 179, tr).

Frequently strands of sclerenchyma fibres lie along one or two sides of the larger vascular bundles. In transverse section they appear as crescents and they cause, often together with sub-epidermal strands of collenchyma, the projection of the ribs on the lower side of the lamina, so rendering it resistant to bending. In many leaves strands of sclerenchyma also occur sub-epidermally between the vascular bundles, and again at the leaf margin. The strengthening of the margin in this way with sclerenchyma and collenchyma will protect it against shearing forces tending to tear the lamina. Large laminae lacking such protection become torn into strips by the wind (e.g. the leaves of the banana and of the coconut palm).

b. The petiole. Apart from the conduction of water to and of assimilatory products from the lamina, the petiole serves as a support and, coupled with growth movements of the shoot axis, to extend the lamina into space and towards the light.

Usually, assisted by correlated growth movements of the stem, the lamina becomes so placed that it is perpendicular to the incident light. Frequently these movements of the leaves are brought about by special localized swellings at the base or at the tip of the petiole, or of the stalk of the pinnae (pulvini), which function as joints (as, for example, in *Mimosa* (Figs. 314; 315) and *Rhynchosia* (Fig. 319)). In various species of *Acacia* adapted to dry habitats the pinnate lamina is reduced, and the function of photosynthesis taken on by the flattened leaf-like petioles (phyllodes (Fig. 180A, 7–9)). In a few instances the juvenile leaves reveal the morphological nature of these peculiar formations. (Biogenetic law; see p. 393.)

Developmentally the petiole owes its origin to an earlier limitation of marginal growth in this region. It is usually bifacial in construction, more rarely unifacial, as in peltate leaves (Fig. 232). In many angiosperms the numerous vascular bundles in the petiole are arranged in transverse section in a ring, or at least in an arc open above. In many leaves, especially the broadly inserted leaves of many monocotyledons as well as the needle-like leaves of conifers, the petiole is virtually absent. In these instances the leaf base is in direct continuity with the more or less linear lamina.

c. The leaf base. In contrast to the petiole the leaf base is usually broadened and flattened. As a leaf sheath it frequently envelops and protects the axillary bud subtended by the leaf, as well as the young terminal region of the shoot and the intercalary growing zone of the internode above its insertion.

Outgrowths of the leaf base, the stipules are almost entirely confined to the dicotyledons. Where the petiole is bifacial, they always arise in pairs and are inserted laterally (lateral stipules (Figs. 174E; 183F; 213A and B, n)). Where unifacial, the stipules are fused to form a tongue- or hood-like structure in the median position (axillary or median stipule). In the Polygonaceae (e.g. rhubarb) the tip of the shoot is enclosed in the bud by median stipules which are developed to an extraordinary extent. These stipules are not perforated by the tip of the shoot until leaf expansion begins, and they then remain on the stem as dry frilly collars, each known as an ochrea. Similar enveloping median stipules are seen in the India-rubber plant (*Ficus elastica*), frequently cultivated as a house plant, in which the ochrea eventually becomes slit and is shed.

Where leaves are decussately arranged it is often found that on each side between the two leaves of the whorl there is only one stipule. This is referred to as an interpetiolar stipule (e.g. *Galium cruciata*), and is usually regarded as being formed by the fusion of two normal stipules. Such interpetiolar stipules are often difficult to distinguish from the leaves. In other Rubiaceae (e.g. *Asperula*) an increase in the number of interpetiolar stipules leads to the formation of several-membered 'false whorls'.

d. Juvenile and mature foliage. The earliest leaves of the young plant, i.e. the so-called juvenile leaves following upon the cotyledons, are often very different in shape from the leaves of the mature plant. Nevertheless, the two extremes may be interconnected by intermediate forms (Figs. 173; 180).

In many species (e.g. *Vicia, Helleborus*) the juvenile leaves are divided to a much smaller extent than the mature. Young trees of *Eucalyptus* produce for a long time only sessile, decussately inserted, oval, bifacial juvenile leaves. Later they produce instead stalked, spirally arranged, falcate, equifacial, mature leaves. The juvenile leaves of the ivy are lobed, while the mature leaves, which do not appear until the reproductive phase, are ovate-lanceolate and entire.

e. Anisophylly and heterophylly. Different forms of leaves by the side of each other can even occur after the juvenile phase. In dorsiventral shoots, for example, adjacent

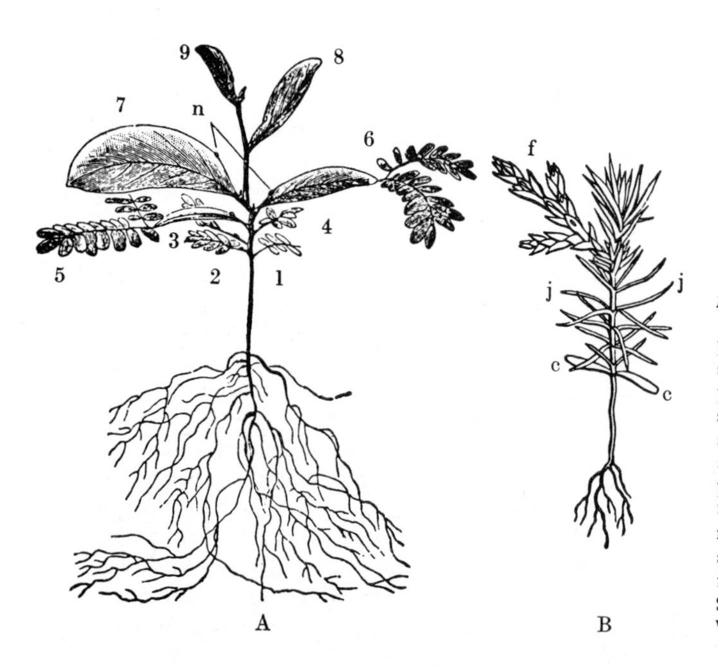

Fig. 180. Juvenile and mature leaves. **A.** Seedling of *Acacia pycnantha*, the cotyledons already shed: 1–6, 'typical' pinnate juvenile leaves (1–4 simply pinnate; 5 and 6 doubly pinnate). Leaves 5 and 6 already show the winged petiole; in 7, 8 and 9 the petiole is developed as a phyllode: n, nectaries on the phyllode. **B.** *Thuja occidentalis*. Seedling with cotyledons (c), needle-like juvenile leaves (n) and typical scale-like mature foliage (s). **A.** ×½. After Schenck. **B.** Natural size. After Warming.

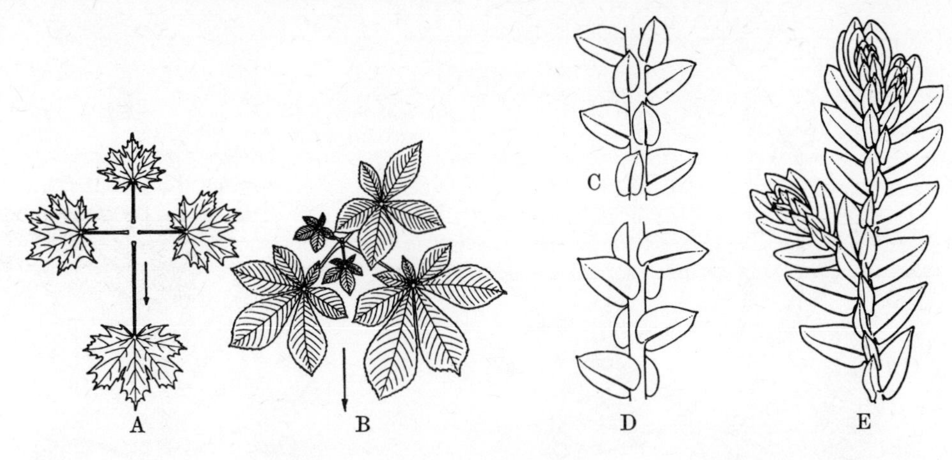

Fig. 181. Anisophylly. **A.** *Acer platanoides*. Leaves of two adjacent two-membered whorls of a plagiotropic branch. **B.** *Aesculus hippocastanum*. View from the anterior of an obliquely orientated branch. The direction of the earth's gravitational field is indicated by the arrow. The growth of the downwardly directed leaves and leaflets is promoted. **C.**, **D.** and **E.** Anisophylly in *Selaginella*. **C.** and **D.** *S. uncinata*, portion of shoot from above (**C.**) and below (**D.**). **E.** *S. douglasii*, tip of shoot from above. A large ventral and a small dorsal leaf occur at each node. **A.** After Troll. **B.** After Nordhausen. **C.**, **D.** and **E.** After Goebel.

leaves on the upper and lower sides are frequently different in size (see Fig. 74C and D), even at the same node (anisophylly). Occasionally, however, the difference is not merely one of size, but of form (heterophylly).

Anisophylly can be induced by gravity (facultative anisophylly of lateral shoots, as in *Acer* and *Aesculus* (Fig. 181A and B), or it can be genetically determined and present throughout the whole shoot system (obligate anisophylly, e.g. *Selaginella* (Fig. 181C, D, E)).

In a few instances the leaves on the upper and lower surfaces of regularly dorsiventral shoots are not only of different size, but also of completely different form (obligate heterophylly, e.g. *Salvinia* (Fig. 608A, B)). More frequent is the situation in which heterophylly occurs as a result of progressive change in the internal or external environment, such as the change in the form of the leaves along the stem as the plant matures. In many water plants the submerged leaves are highly dissected, but the later leaves, reaching and floating on the surface, are merely lobed (e.g. *Ranunculus peltatus*, Fig. 182A). The stag's horn fern (*Platycerium*) produces alternately more or less cordate nest-leaves (which soon die and retain humus and moisture) and deeply divided assimilatory and fertile leaves (Fig. 613). In many flowering plants the leaves on the long shoots are reduced and very different from the normal (e.g. the 'thorn leaves' of *Berberis* (Fig. 199)). In *Cucurbita* (Fig. 212) the leaves on the short shoots are in the form of tendrils. For other instances of heterophylly see pp. 179 and 194.

3. Scale leaves and bracts

Although scale leaves and bracts are indistinguishable in their primordia from leaves, when fully developed they contain many fewer parts (Fig. 182B) than the leaves, with which they are frequently connected by intermediate forms (Figs. 183; 184). Their ontogeny is marked by an early promotion of the lower part of the leaf primordium in relation to the upper part, after which all development ceases.

Scale leaves, colourless or green, precede the formation of the leaves proper in many seedlings (e.g. *Pisum, Vicia faba*), and on the innovations of perennial herbs (see p. 186), and occur as bud-scales on the annual shoots of most trees (Figs. 173; 183). Furthermore, scale leaves are the only leaves produced by rhizomes (Fig. 205), and they may be large (Fig. 238), or often so small as to be hardly visible, and usually short-lived.

Bracts, although similar in structure and nature to scale leaves, and often connected with normal leaves by transitional forms (Figs. 173; 184F–J), are often coloured differently from, and larger than, scale leaves. They often follow the leaves on the vegetative

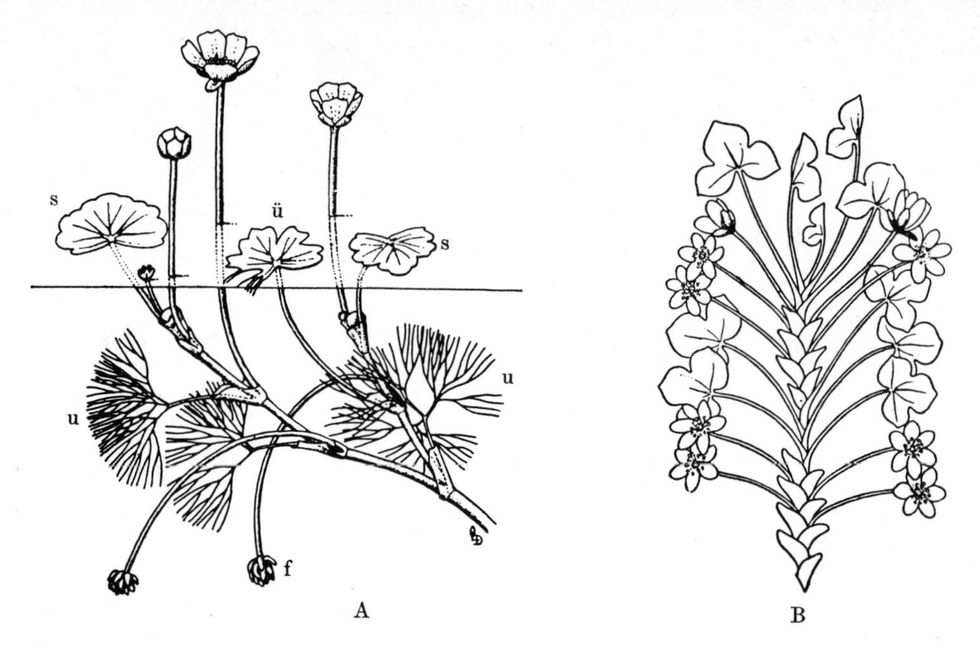

Fig. 182. Heterophylly. **A.** *Ranunculus peltatus*. Sympodially branched flowering shoot showing floating (s) and submerged (u) leaves: ü, transitional leaf with a few linear segments; f, mature submerged fruits. **B.** *Anemone hepatica*. Diagrammatic representation of two years' growth of the shoot. Growth begins with a series of scale-like bracts bearing flowers in their axils. The rosette of normal three-lobed leaves does not develop until later. **A.** Original. **B.** After A. Braun, diagrammatic.

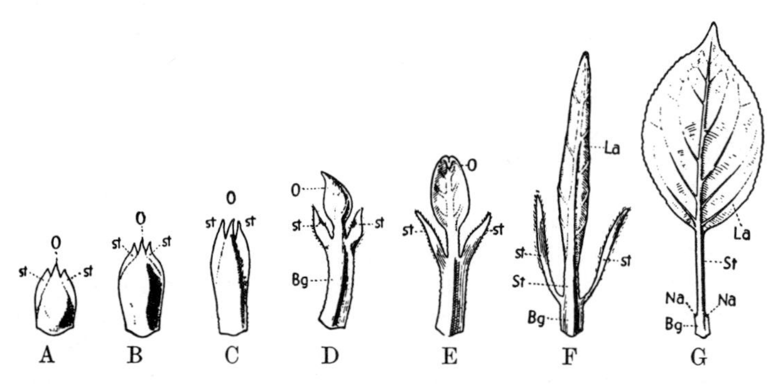

Fig. 183. *Malus baccata*. **A.** to **C.** Bud scales. **D.** and **E.** Transitional leaves. **F.** Leaf before expansion. **G.** The same in expanded condition; st, stipules; Na, scars of fallen stipules; St, petiole; La, lamina; O, primordium of the upper part of the leaf. **A.** to **F.** Approx. natural size. **G.** $\times\frac{1}{3}$. After Troll.

stem, and are associated with the flowers and branches of the inflorescence. In many cases they are succeeded without any sharp distinction by perianth segments (e.g. in species of *Helleborus* (Fig. 184H, J, K)).

The frequently mottled spathe of the Araceae is also a bract, as again is the prophyll which is partly fused with the peduncle in the inflorescence of *Tilia*. That scale leaves and bracts are reductions or transformation of leaves is inferred not only from their ontogeny but also from the fact that normal leaves sometimes develop from their primordia. Thus if the shoot is decapitated or defoliated many primordia which would normally have developed into scale leaves generate foliage leaves instead. Also many rhizomes, if forced to develop in daylight, produce foliage leaves from primordia which beneath the soil would have yielded scale leaves.

4. The life span of leaves

In almost all woody plants the leaves have a much shorter life than the axes, and they are

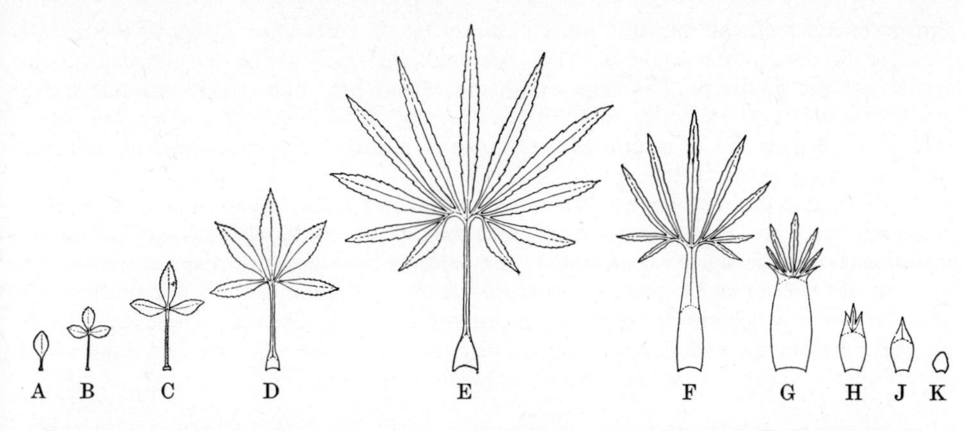

Fig. 184. *Helleborus foetidus*. Leaf sequence. **A.** Cotyledon. **B.** and **C.** Juvenile leaves. **D.** Foliage leaf of the first year of growth. **E.** Foliage leaf of the second year of growth. **F.** Leaf immediately preceding flowering stage. **G., H., J.** Bracts of the third year of growth. **K.** Perianth member. Original.

sooner or later shed, leaving leaf scars on the stem showing their former positions. Plants which retain their leaves for several growing periods are said to be evergreen, as opposed to deciduous plants, in which the leaves function for only one growing period.

Even 'evergreen' leaves, however, have a limited life-span. In *Ilex, Olea europaea* and *Pinus silvestris*, for example, it amounts to about two years, in *Laurus nobilis* and *Picea abies* to five or six.

The fall of leaves is brought about by a layer of small parenchyma cells, rich in cytoplasm and starch, which become differentiated by cell division across the leaf base, often not until shortly before the fall of the leaf. At this region all the mechanical tissues are reduced, the only lignified tissue being the vascular elements. While the leaf is still alive it detaches itself at this abscission layer as a result of the middle lamellae becoming mucilaginous and, consequent upon turgor changes, the cells rounding themselves off and rupturing the vascular tracts. The resulting leaf scars are sealed by the cells of the outermost layers of the exposed surface becoming lignified. Beneath this, usually before the leaf falls, a layer of cork is formed, becoming continuous with the cork of the stem. (See also p. 303; abscisin.)

The peculiar leaves of *Welwitschia mirabilis* behave very differently from those of other cormophytes. The plants may become 600 or more years old, but during their life produce only one pair of foliage leaves, placed transversely in relation to the cotyledons. These foliage leaves grow continuously from a basal meristem, their ends being worn away by weathering and attrition.

C. The root[30]

Roots, usually radially symmetrical and initially thread-like, are generally confined to the soil, but occasionally occur above it (aerial roots). They always lack leaves. Roots anchor the plant in the soil, and take from it water and minerals and conduct them to the shoot. They may also serve for the storage of reserve materials. Usually a root carries out several of these functions simultaneously.

In only a few seed plants does the root remain suppressed in the embryo. To these 'rootless' species belong the water-plants *Ceratophyllum* and *Utricularia*, the saprophytic orchids *Epipogium* and *Corallorhiza*, and the epiphytic bromeliad *Tillandsia usneoides*.

1. Growing point

The growth of the root, like that of the stem, takes place from a blunt, conical, apical meristem. The delicate embryonic cells of this meristem require protection, and this is provided by some persistent parenchymatous cells, forming the root cap or calyptra (Figs. 93; 94). Consequently, the growing point lies within the tissues of the root tip. The oldest and outermost cells of the root cap constantly become detached owing to the middle lamellae becoming mucilaginous. They are replaced from within by the continuous activity of the meristem. The cells within the cap often contain freely movable starch grains (Fig. 94B). These are possibly displaced under the influence of gravity, and play a role in the orientation of the root in relation to the earth's gravitational field (statolith hypothesis of graviperception; see p. 359).

The root cap can usually be seen only in a median longitudinal section of the root apex, but there are some thick roots in which the cap is immediately apparent as a brown investment over the tip (e.g. *Pandanus*). The conspicuous caps at the ends of the aquatic roots of our species of *Lemma* and *Hydrocharis* are, however, not root caps, but remains of shoot tissue which enveloped the root primordia in a cap-like way. These caps, which are derived from the endodermis and pericycle, no doubt take over the function of root caps proper, and can thus be regarded as analogous formations.

2. External morphology of the root

The extension of the root, usually indefinite in amount, involves the transformation into permanent tissue of the embryonic cells at the base of the apical cone. In subterranean roots this extension, in contrast to that of aerial shoots and many aerial roots, is limited to a very short zone, of the order of 5–10 mm in length, close behind the growing point (Fig. 272). The shortness of the extension zone prevents the resistance of the soil causing the root to buckle.

A few millimetres behind the root tip, about where extension growth ceases, the subterranean root produces root hairs (Figs. 110; 185, Wh; 186, wh). Since they increase the surface area of the root very considerably (Fig. 110A), they facilitate the uptake by the root of water and salts, but they remain alive only for a few days. Consequently, only a limited length of the root, of the order of a centimetre or less, is covered by them at any one time (Fig. 186).

Many plants lack root hairs. This is especially true of many water and marsh plants which can take up water without difficulty, and also of plants whose nutrition is facilitated by mycorrhizal fungi (see p. 294).

3. Primary internal structure of the root

The differentiated root consists at first, as does the stem, exclusively of primary tissue, usually arranged in radial symmetry (Fig. 187A). The younger parts are protected by a one-layered rhizodermis, which, together with the root hairs has an absorptive function. Corresponding to this function is the insignificant thickening of the outer walls of the rhizodermis in contrast to those of the epidermis of the shoot, as well as the lack of a cuticle and stomata. The rhizodermis dies with the root hairs, and its place is taken by a typical secondary boundary tissue, the exodermis (Fig. 187A, ex). This arises from one or several of the sub-epidermal layers of the cortex, the cells of which, not infrequently remaining alive, lay down suberin lamellae on their cellulose walls, thus forming a barrier. Small unsuberized cells, absorption or passage cells, may occur at more or less regular intervals in the exodermis.

The remaining tissue, as in the stem, can be regarded as consisting of cortex and central cylinder. The cortex of subterranean roots consists of colourless, parenchymatous, storage tissue. In the cortex of aerial roots chlorophyll is often present. The innermost layer of the cortex is usually differentiated as an endodermis (see p. 103) (Figs. 187B, en; 189), providing a sharp boundary between the cortex and the central cylinder. In the older parts of roots the endodermal cells are suberized and often, especially in many monocotyledons, additionally thickened by tertiary cellulose lamellae, usually lignified. This tertiary thickening is often unequal, being confined to the side of the cell adjacent to the central cylinder (Fig. 187B, en). Certain endodermal cells, however, usually in front of the xylem strands of the vascular bundle, remain unchanged (passage cells; Figs. 109D; 187B, d).

The outermost parenchymatous layer of the central cylinder of the root, i.e. the layer immediately beneath the endodermis, is referred to as the pericycle (Figs. 187B; 190, Per). The pericycle is of one or several layers, and is of great significance for the

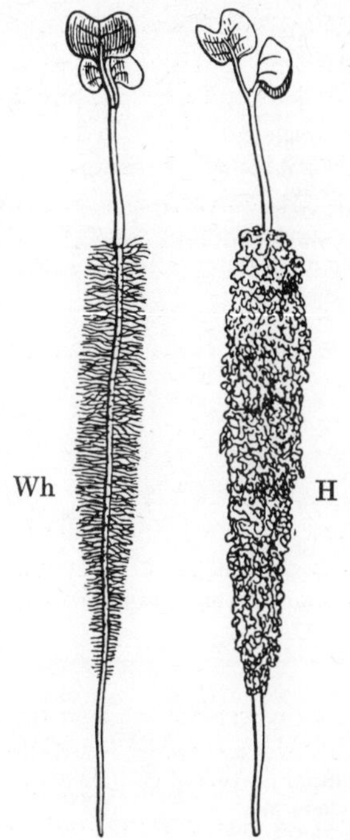

Fig. 185. Seedling of *Sinapis alba*. Wh, root hairs. H shows the sheath of soil particles normally adhering to the root (cf. Fig 110). Natural size. After Sachs.

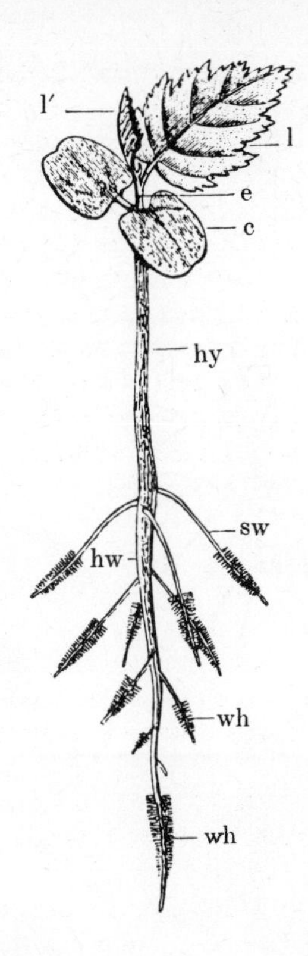

Fig. 186. Seedling of *Carpinus*: hy, hypocotyl; e, epicotyl; c, cotyledons; l, l′, normal foliage leaves; hw, main root; sw, lateral root; wh, root hairs. Natural size. Freely adapted from Noll.

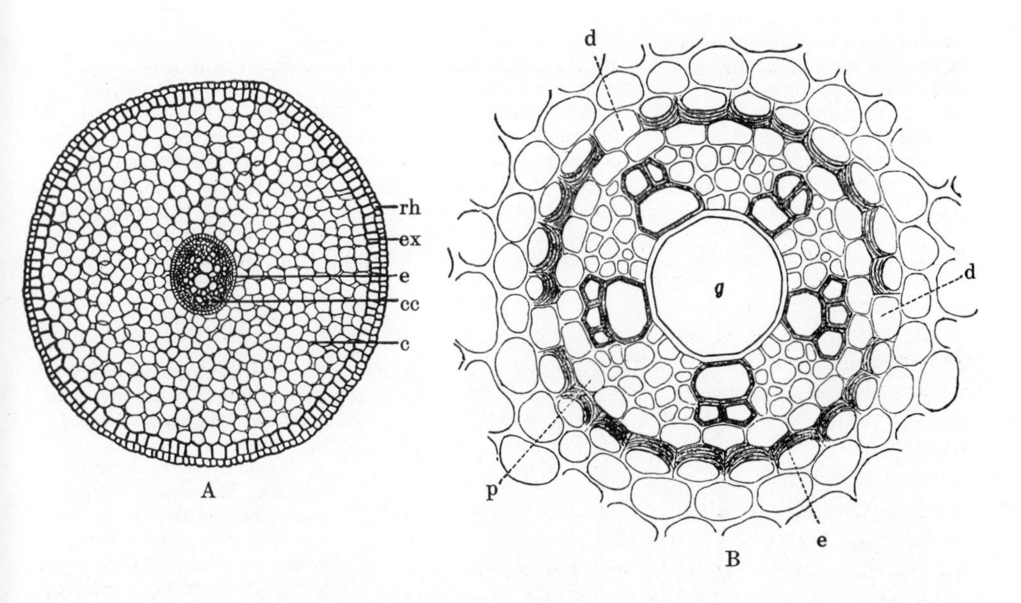

Fig. 187. A. Transverse section through a root of the domestic onion (*Allium cepa*): rh, rhizodermis; ex, exodermis; c, cortex; e, endodermis; cc, central cylinder with central, radially symmetrical, vascular bundle. **B.** Pentarch radially symmetrical vascular bundle of the root of *Allium ascalonicum*: e, endodermis, the inner tangential walls thickened; d, passage cells; p, pericycle; g, large central vessel. **A.** ×45. After M. Koernicke, modified. **B.** ×40. After Haberlandt.

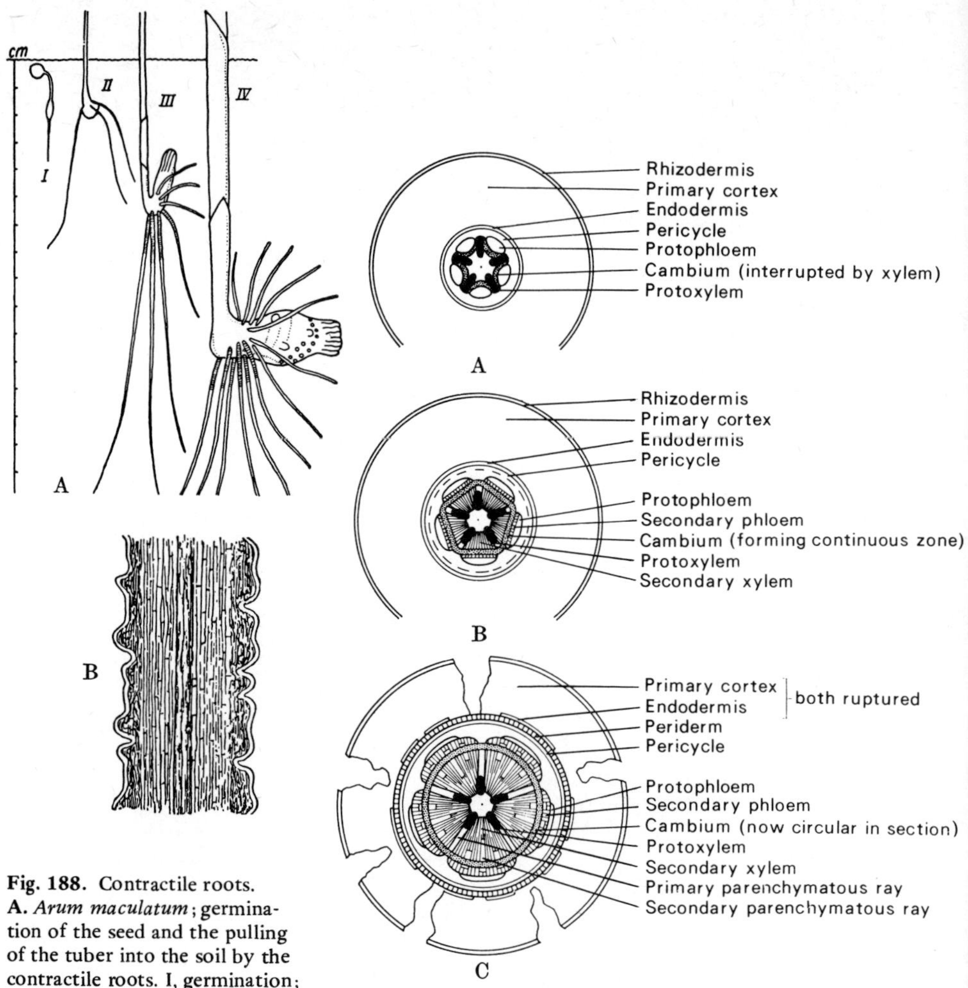

Fig. 188. Contractile roots. **A.** *Arum maculatum*; germination of the seed and the pulling of the tuber into the soil by the contractile roots. I, germination; II, depth of the young tuber at the beginning of the second year; III, further sinking during the second year; IV, tuber of the mature plant at its normal depth. **B.** Part of a longitudinal section through a contracted root of *Lilium martagon*. **A.** ×½. **B.** ×6. After Rimbach.

Fig. 189. Secondary growth in a root, diagrammatic. **A.** Primary condition with a central, pentarch, radially symmetrical vascular bundle. **B.** The cambium (dotted), star-shaped in outline in **A.**, has resumed activity between the primary xylem groups and pushed the primary phloem towards the outside. **C.** Activity is now uniform right round the cambium. The primary cortex is becoming fissured as a result of dilatation. The pericycle forms, within the similarly ruptured endodermis, an endogenous secondary insulating layer, the periderm.

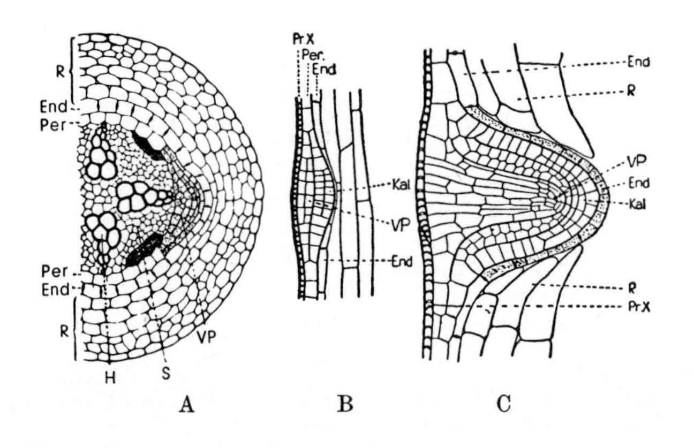

Fig. 190. Endogenous origin of lateral roots from the pericycle (Per) and their penetration of the cortex (R). **A.** Root of *Vicia faba* in transverse section. **B.** and **C.** The same of *Reseda* in longitudinal section. H, xylem; S, phloem of radially symmetrical bundle of the root; VP, growing point of the lateral root; Kal, calyptra; End, endodermis; Per, pericycle; PrX, protoxylem. **A.** ×40. After Fitting. **B.** and **C.** ×100. After van Tieghem.

subsequent formation of secondary cortical tissue, and especially for the initiation of lateral roots. In all roots the conducting tracts form a central, radially symmetrical, vascular bundle.

According to whether there are one, two, three, four or several strands of xylem present in the bundle, so the root is referred to as monarch, diarch, triarch, tetrarch (Figs. 121A; 190A), or polyarch respectively. Pentarch (Figs. 187B; 189) and hexarch roots also occur. The xylem elements either come together in the centre of the bundle (Figs. 121A; 187B) or there may be a central strand of parenchyma (Fig. 190A) or of sclerenchyma, or frequently of both together. The sclerenchyma increases the tensile strength of the root (see pp. 117 seq.). In prop roots above the surface of the soil a peripheral zone of strengthening tissue is frequently differentiated in the outer cortex, making such roots resistant to both bending and pressure.

4. Secondary growth in thickness in the root

As we have seen (Fig. 121A), in radial vascular bundles the strands of xylem and phloem are separated from each other laterally by radially arranged plates of parenchyma. When secondary thickening begins in the root of a gymnosperm or dicotyledon lines of cells within these plates divide approximately radially and give rise to cambial strips between the xylem and phloem strands. On both sides of each xylem strand the margins of the two cambial strips meet the pericycle, the cells of which on the outside of the xylem strand also give rise to a cambium. Thus arises a complete girdle of cambium, stellate in transverse section and purely secondary in origin (shown diagrammatically in Fig. 189A). This cambium gives rise to xylem on the inside and phloem on the outside, exactly as does that of the shoot.

Particularly intense production of xylem on the adaxial side of the phloem causes the concavities of the cambium to be smoothed out, so that it becomes circular in transverse section (Fig. 189B, C). As in the shoot (p. 149), the cortex of the root becomes tangentially extended and ultimately fissured in consequence of dilatation. Before this occurs, however, a tertiary insulating layer has already formed. In contrast to the situation in the shoot, this root periderm does not arise peripherally, but deep in the pericycle. Since the xylem and phloem of the root have the same structure as in the stem, the transverse section of a root with several years' secondary thickening is hardly to be distinguished from that of a stem of comparable age. However, in the root the radial primary xylem strands can still be seen at the centre of the axis (Fig. 189C). The cambium forms conspicuous parenchymatous rays, referred to as primary rays, in front of these primary xylem strands.

Older naked parts of roots often show a clear transverse wrinkling. This is probably caused principally by a passive contraction of the relatively thick cortical parenchyma, occasioned by the secondary thickening (Fig. 190B). This dilatation causes initially an extension of the thin-walled cells in a radial direction, followed by a secondary shortening in the longitudinal. In this way the whole root, including its central cylinder, experiences a contraction, and so anchors the shoot firmly in the ground (contractile roots; Fig. 188A).

5. Branching of the root

The acropetal branching of the main (or primary) root, and the subsequent similar branching of the laterals so formed, leads to the development of a root system which may occasionally exceed in length and extent that of the aerial shoots. The total length of all the roots can reach unexpectedly high values. Those of a free-standing plant of wheat (Fig. 191B), for example, extend in the aggregate some 80 km.

The initials of the lateral roots are formed endogenously, i.e. within the parent root (Fig. 190). In the seed plants these initials arise in the outer layer of cells of the central cylinder, i.e. in the pericycle. In the pteridophytes, however, they arise in the innermost layer of the cortex. The developing lateral roots must therefore always penetrate the entire cortex of their parent root. During this time either the endodermis is immediately ruptured (Fig. 190B, C), or it grows for a limited period (accompanied by a few cell divisions) so that a further hood is formed over the root cap. Consequently lateral roots are not infrequently fringed at the point of emergence by a collar of torn cortical tissue.

The lateral roots arise either from in front of the longitudinally running xylem strands of the parent root (Fig. 190) or, less commonly, from the parenchymatous plates lying between the strands of xylem and phloem. Consequently, they always arise in clearly defined ranks, the number of these ranks being either equal to the number of the xylem strands in the radial bundle of the root, or twice the number of these strands.

Many dicotyledons (e.g. *Lupinus, Quercus*) and gymnosperms (e.g. *Pinus*) acquire, since the secondary thickening extends down into the root, a radially symmetrical tap root. It is the elongated root of the seedling, continuing the main axis vertically downwards into the earth. Such tap roots may reach impressive depths. In the building of the Suez Canal, for example, roots of *Tamarix* were still encountered at depths of 30 m. Lateral roots of the first order arise on the primary root and spread out either horizontally or

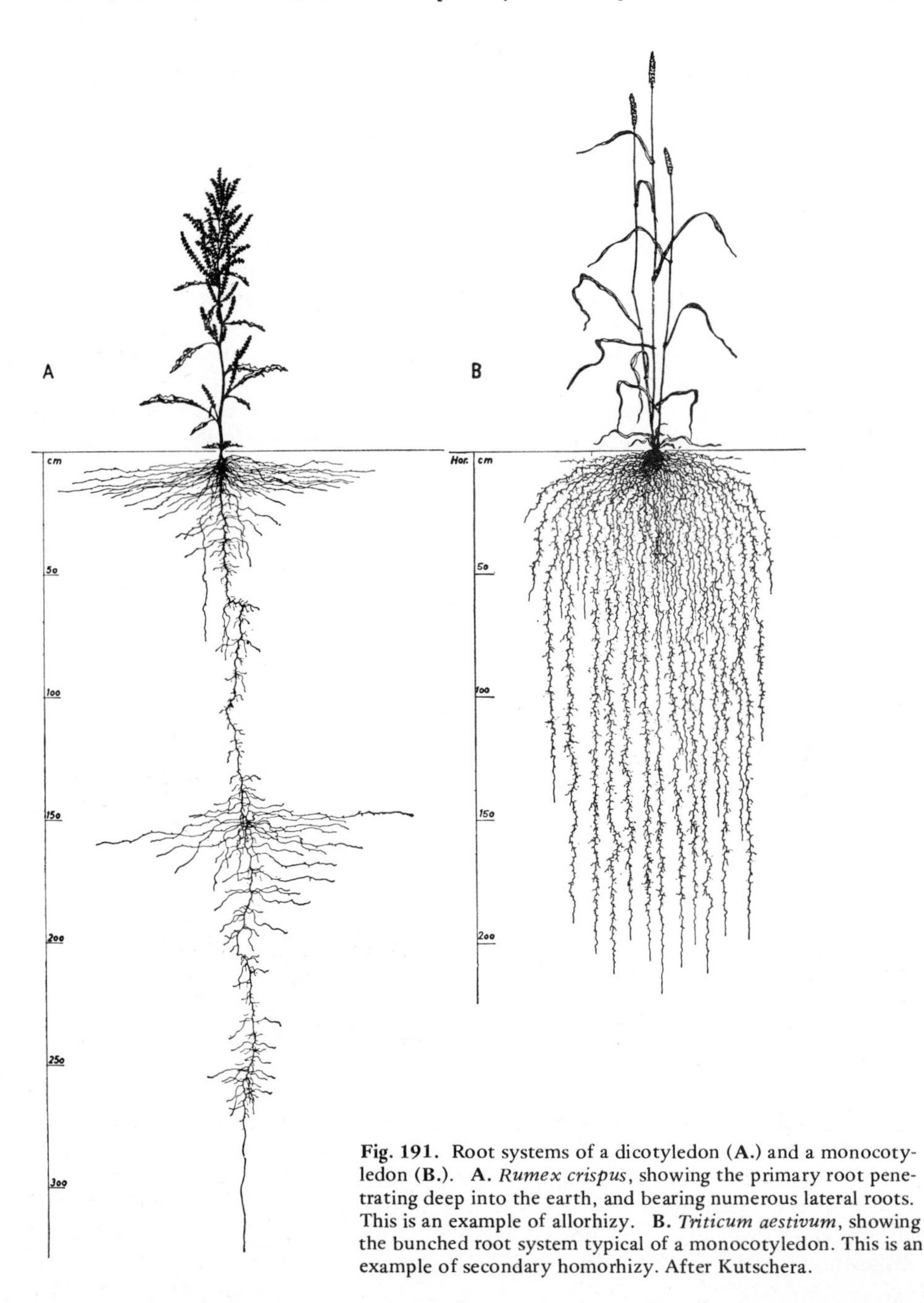

Fig. 191. Root systems of a dicotyledon (**A.**) and a monocotyledon (**B.**). **A.** *Rumex crispus*, showing the primary root penetrating deep into the earth, and bearing numerous lateral roots. This is an example of allorhizy. **B.** *Triticum aestivum*, showing the bunched root system typical of a monocotyledon. This is an example of secondary homorhizy. After Kutschera.

obliquely at a definite angle (plagiotropically) into the earth (Figs. 131D, Sw; 172, Sw). On these arise here and there the lateral roots of the second order, and these usually grow out in all directions into the soil. Thus the branches of the root system penetrate the soil in all possible directions, and their extensive ramifications leave no areas of the soil unexploited. The terminal branchlets usually remain short, and it is mainly through these that the soil water and its dissolved salts are absorbed. A particularly well-developed root system can form in regions of the soil favoured by dampness or abundance of mineral nutrients. Such proliferation may also be more circumscribed, the localized branching giving rise to 'nests' of rootlets. In certain dicotyledons and gymnosperms the root system may remain more superficial (e.g. *Picea* and the root system of the potato and many species of cacti).

6. Adventitious roots and suckers

Roots may also be formed, again usually endogenously, on the shoot system, and they may arise from both stems and leaves. Such roots are termed adventitious roots. They are produced in the normal course of development from the subterranean and creeping stems of perennial plants (e.g. *Polygonatum* (Fig. 205B) and the stolons of *Fragaria, Jussieua* (Fig. 192D) and *Ranunculus repens*). Adventitious roots may also be produced in response to injury (e.g. in cuttings of *Begonia* or treatment with hormones (Fig. 275). Strictly speaking, the term 'adventitious' (which means arising not in the customary place, or at an unusual time) should be restricted to this latter class of root.

In the Pteridophyta all the roots are produced on the stem. The first root primordium of the embryo is already lateral (Fig. 577C), and the embryo in consequence exhibits a unipolar structure. This kind of root system, where all the roots are of the same nature, can be referred to as homorhizal, as opposed to the allorhizal systems of dicotyledons and gymnosperms, where there is a main root bearing laterals.

In the monocotyledons the embryo is bipolar, and the primary root is always laid down at the pole opposite to the shoot. This primary root, which usually lacks secondary thickening, dies (except in many palms) at an early stage, together with the lowest part of the stem, and it is replaced by a crown of adventitious roots, arising especially at the

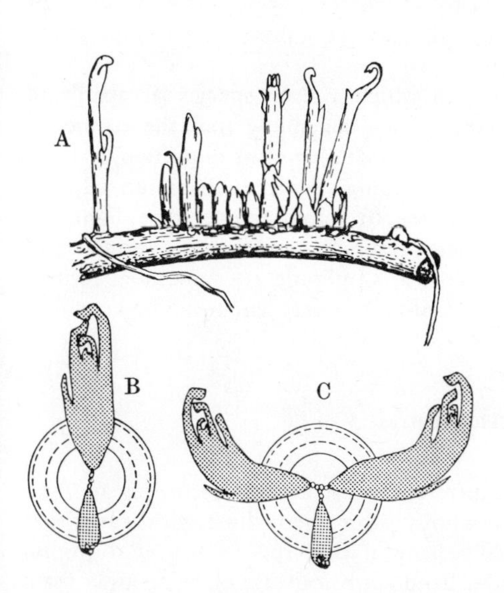

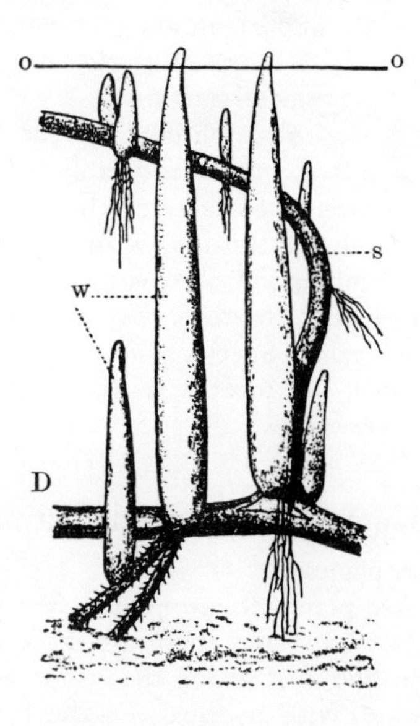

Fig. 192. Shoots arising from roots, and adventitious roots. **A., B.** and **C.** Shoots arising from roots of *Coronilla varia*. **A.** Lateral root with numerous shoot buds arising endogenously. **B.** Transverse section through a diarch root at greater magnification. **C.** Similar section through triarch root. The shoots arise in the pericycle opposite the xylem groups of the vascular bundle, and the growth of those on the upper side is promoted. **D.** *Jussieua repens* (Onagraceae). Portion of a shoot s with adventitious roots and pneumatophores (w). The water level is indicated by o–o. **A.** ×2. **B.** and **C.** ×10. All after Rauh. **D.** Natural size. After Giesenhagen.

nodes (secondary homorhizy). This process results in a large number of arching roots of more or less equivalent development running out in all directions, and forming an intricate and close mat in the upper layers of the soil. The grasses provide particularly good examples (Fig. 191B).

Root shoots (suckers), where present, arise endogenously in the pericycle of older roots (Fig. 192). They serve, as runners and stolons, as a means of vegetative propagation.

II. Modifications of the plant body: adaptations to manner of growth and environment[31]

The vegetative organs are not always constructed in the typical manner described in the foregoing, but may be modified in various ways. Their outer form and inner structure are usually conspicuously adapted in some degree to their manner of growth and environment. Thus the plants of one locality, and of different localities with similar climates, will show a number of features in common (i.e. the physiognomy is more or less uniform), and conversely, the vegetation of places of very different climates will show considerable differences in physiognomy.

Organs of a cormophyte appearing to be abnormal are almost always only modifications of one of the three basic organs, namely root, stem and leaf. To demonstrate this, and to recognize of which they are homologues, nevertheless often requires a detailed investigation.

The outer form and the function of the completely differentiated organ can particularly easily mislead. Thus not infrequently one fundamental organ, e.g. a stem, can assume the form and function of another, e.g. a leaf, and specialized organs with very similar forms and a common function may in different species be modifications of a different fundamental organ. Such specialized organs would be merely analogous and not homologous (see p. 9). Nevertheless, when all the morphological properties of an organ are taken into account there is usually no doubt about its derivation.

The most important environmental factors to which the cormophytes must adapt themselves, both morphologically and physiologically, are the water supply, temperature, light, and the mineral nutrition.

The ecological amplitude, i.e. the limits within which a given species is capable of surviving, differs very considerably with different species, depending upon the manner of their evolution and selection. Thus, each species, corresponding to its morphological and physiological constitution, is suited for a particular ecological niche, characterized by a specific combination of environmental factors related to water, temperature, light and mineral supply. For every species each one of these factors has a minimum, optimal, and maximum value. Species with a very broad ecological amplitude are sometimes termed euryoecious, and those with narrow (and hence confined to very circumscribed environments) stenoecious.

A. Adaptations to water and atmospheric humidity

1. Water plants

The water plants (hydrophytes) show very different constructions according to their manner of life. Thus there are those that are wholly submerged, those with leaves that float on the surface of the water and finally 'amphibious' species (e.g. *Polygonum amphibium*) with an aquatic 'water form' and a 'land form' capable of living away from the water. Such 'amphibious' species provide a transition to the marsh plants (helophytes) which stand with only their roots and the lowermost parts of their shoots in the water. Submerged stems and leaves show a specialized structure (hydromorphy). The hydrophytes are capable of taking up carbon dioxide, oxygen and mineral salts directly from the water. Many floating plants are in consequence rootless throughout life (e.g. *Ceratophyllum, Utricularia, Lemna trisulca, Wolffia arrhiza*).

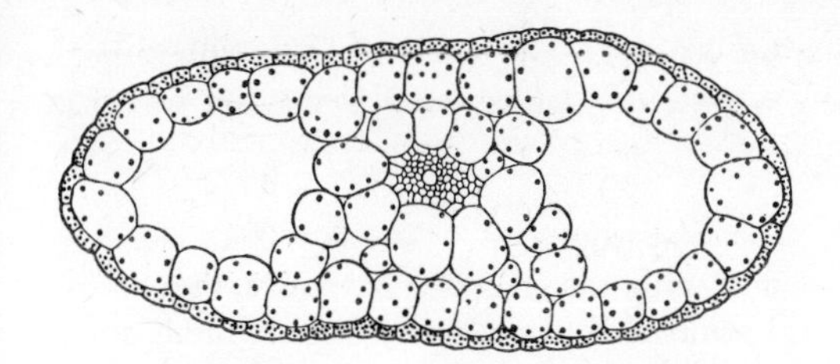

Fig. 193. Transverse section through the equifacial leaf of the submerged water plant *Zanichellia palustris*. ×150. After Schenck.

One litre of air contains about 210 cm^3 oxygen and 0.3 cm^3 carbon dioxide. A litre of water when saturated with air at 20° C contains only about 6.36 cm^3 oxygen, but again about 0.3 cm^3 carbon dioxide. The water organisms thus have no less carbon dioxide, but very much less oxygen at their disposal than in the air, assuming that the water is not very shallow or in very vigorous movement. The amount of carbon dioxide in water is usually, when the carbonic acid loosely bound in the dissolved bicarbonate is considered, even greater than that in the air.

The outer walls of the epidermis of submerged shoots are always very thin, and the very delicate cuticle offers hardly any impediment to the entry of dissolved gases, water and salts. The submerged leaves of water plants are usually very delicate, thin, rich in sap and often divided into filaments. This amplification of the surface area is closely related to the slowness with which gases diffuse in water, and to the relative poorness of natural waters in salts. The leaves floating on the surface, however, and the aerial leaves are usually similar in form to those of land plants (hetero- and anisophylly; see Fig. 182A).

The epidermis of the submerged leaves contains chlorophyll and usually lacks stomata (Fig. 193), and generally also hairs. The mesophyll, where it is not wholly lacking (as in *Elodea*), consists of more or less uniform, large-celled parenchyma, with numerous large intercellular spaces, and it is thus not usually differentiated into palisade and spongy mesophyll. Water-conducting elements become rudimentary or are wholly missing. The upthrust of the water also makes mechanical tissue unnecessary in stems and leaves. At most is the laying down of a vascular bundle at the centre of the stem, providing the necessary tensile strength in rapidly flowing water (see p. 117).

Even in submerged water plants there is evidence of at least a weak transpiration stream. It is often bound up with the excretion of water from special glands, or from the tips of the leaves, where there are not infrequently water pores. These pores arise as a result of the tips of the leaves dying at an early stage, and through the apical pit thus arising water and dissolved substances can be excreted (guttation; see p. 227).

A conspicuous feature in all water and march plants is the enormous development of intercellular spaces (e.g. Figs. 96; 193). The wide air canals are also gas chambers, on the one hand increasing the buoyancy of the plant, and on the other making possible an activc diffusion of gases into the interior of the tissue. Such aeration tissue is referred to as aerenchyma (see p. 95).

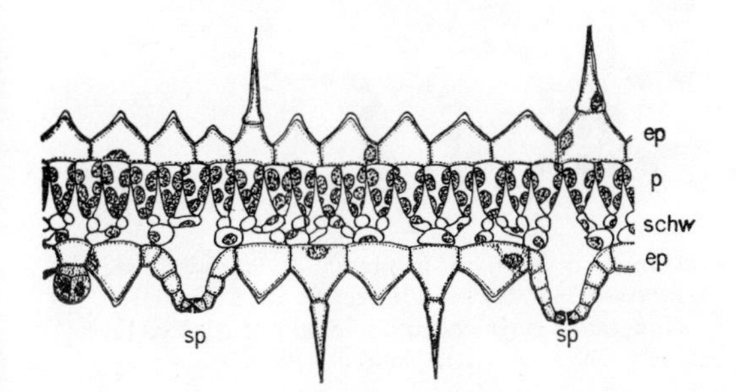

Fig. 194. Transverse section through the leaf of the tropical shade plant *Ruellia portellae* (Acanthaceae): ep, epidermis (the lower containing raised stomata (sp)); p, funnel-shaped palisade cells; schw, spongy parenchyma. ×100. After Fitting.

In many of the species of mangrove of tropical saline swamps and coastlines, growing in mud very deficient in oxygen, the normal roots, spreading out horizontally in the mud, give rise to branch roots which ascend vertically. These branch roots, usually markedly woody and unbranched, come above the mud and extend into the air. These are known as pneumatophores, since it is assumed that oxygen is able to enter them through lenticel-like openings (pneumatothodes), and reach by way of the aerenchyma the subterranean parts. Smaller pneumatophores, containing a glistening white aerenchyma, are found in *Jussieua repens* (Onagraceae) (Fig. 192D).

2. Plants of damp habitats (hygrophytes)

Land plants which have to contend habitually with the necessity to move water from permanently damp earth into a very humid atmosphere (e.g. many hygrophilous shade plants and plants of the tropical rain forest) have, like those hydrophytes which come above the surface of the water, peculiarities of structure which promote transpiration (hygromorphy). Consequently, in dry air such plants wilt rapidly and dry out. The damper the habitats in which they grow, the more they come to resemble in outer form and anatomical structure the aquatic and swamp plants.

In the hygrophytes we find large, thin, delicate laminae, rich in sap. They are glabrous, or possess a velvet-like surface as a result of the development of numerous hairs or papillae. In this way the transpiring surface is significantly increased. The epidermis contains chlorophyll and may consist of large cells; these perhaps serve as water reservoirs and act as a protection against too large a loss of water during dry periods of the day. The outer walls of the epidermal cells are very thin and covered by a delicate cuticle. The stomata are not immersed, but frequently are actually raised above the epidermis (Fig. 194, sp). The mesophyll consists of only a few layers of large, thin-walled cells, with very wide intercellular spaces. Many hygrophytes possess glandular hairs or epithemal hydathodes (p. 121) from which water can be actively expressed into saturated air (as in tropical rain forest). The large leaf surface promotes photosynthesis in the weak light of shaded habitats. The root and vascular systems are usually developed only weakly, corresponding to the small amount of transpiration in permanently damp and shaded conditions.

3. Plants of dry habitats (xerophytes)

Plants which at least from time to time can tolerate great dryness of habitat, especially of the soil, are known as xerophytes. They are xeromorphic, i.e. they possess features which tend permanently, or at least temporarily, to limit water loss. Xerophytes usually have very long roots in order to ensure an adequate uptake of water from the desiccated soil. Naturally many plants from regions of extreme aridity, such as deserts and steppes, are particularly xeromorphic, as are also plants from dry rocks, and many epiphytes (see p. 194 seq.). Some of the features which reduce transpiration may at the same time be regarded as a protection against too intense insolation and heating.

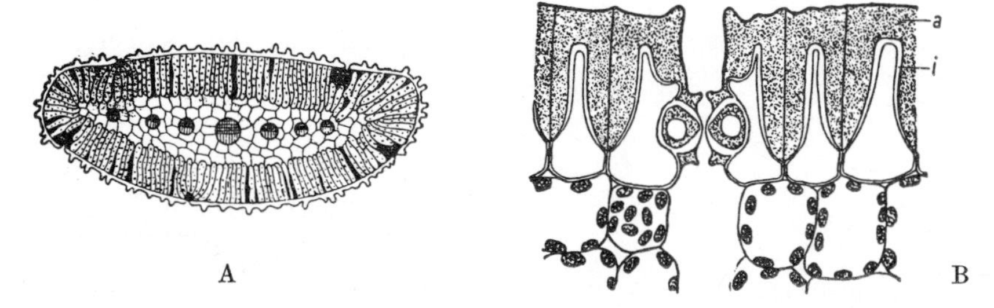

Fig. 195. Anatomical adaptations of plants of extremely dry and hot habitats. **A.** Transverse section through the equifacial leaf of the desert plant *Reaumuria hirtella* (Tamaricaceae). **B.** Transverse section through the epidermis of *Gasteria nigricans*: a, outer cutinized and i, inner uncutinized layers of thickening of the outer walls of the epidermal cells. **A.** ×30. After Volkens. **B.** ×180. After Strasburger, modified.

The usually small evergreen leaves of the xerophytes are frequently coriaceous and poor in sap (sclerophylls: e.g. *Laurus*, *Myrtus*, *Olea*). The small-celled and thick-walled mesophyll is often reinforced by special sclerenchymatous elements (sclereids). It is not usually differentiated into palisade and spongy mesophyll, and it contains only a few intercellular spaces; the leaves are thus equifacial (Fig. 195A). The contrivances which serve to limit transpiration are as follows: thickening of the outer walls of the epidermal cells, in particular the thickening and toughening of the cuticle and the cutinized layers (Fig. 195B); an epidermis consisting of several layers of cells (Fig. 196A); sub-epidermal layers of sclerenchyma; development of investments of wax, resin and calcium carbonate. Spaces containing still, humid air are produced by the narrowing and sinking of the stomata (Figs. 195B; 196A), or by the overarching of the stomata with a tomentum of hairs. In this way the saturation deficit of the internal air is reduced and evaporation is restricted. However, the number of the stomata—as in 'sun' leaves—is frequently particularly high. This peculiarity favours photosynthesis, provided the stomata are opened when it is not too dry.

Excessive heating of the lamina is frequently prevented by the leaves being placed vertically. In the *Eucalyptus* trees of Australia (see p. 792), for example, the stalked falcate mature leaves hang vertically downwards ('shadeless woods'). The native 'compass plant' (*Lactuca serriola*), not infrequent on sunny roadsides, bears all its spirally arranged leaves vertically and orientated more or less in a N–S direction, the movement being brought about by torsion of the leaf base. The result is that the sun at its highest merely strikes the narrow edges of the leaves (thermotropism; see p. 365). In many species of *Acacia* the laminae are obsolete, and they are functionally replaced by a lamina-like development of the petiole (phyllodes; Fig. 180A).

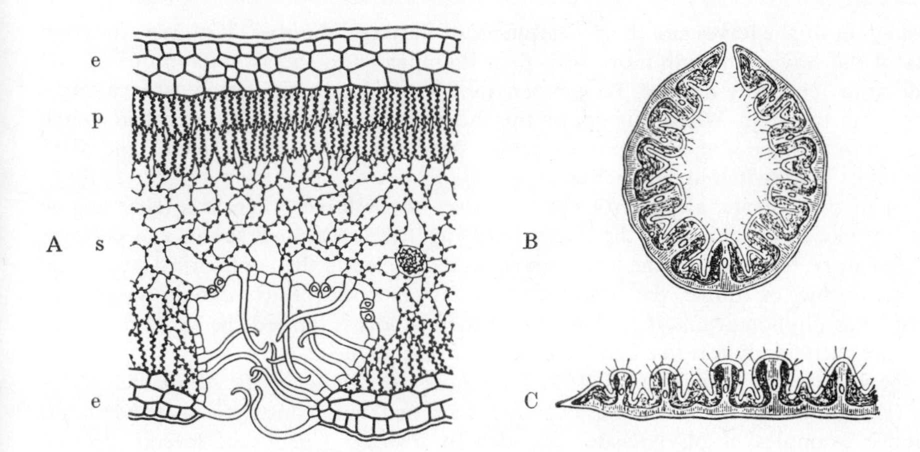

Fig. 196. Anatomical adaptations of plants of habitats occasionally very hot and dry. **A.** *Nerium oleander*. Transverse section of the leaf: e, multi-layered epidermis, the outermost walls thickened; p, palisade layer, two cells deep; s, spongy mesophyll. The stomata are in special chambers formed by depressions in the lower side of the leaf lined with hairs. **B.** and **C.** Transverse sections of the leaf of *Stipa capillata*. **B.** In rolled up, dry condition. **C.** Opened out after adequate rainfall. **A.** ×60. Original. **B.** and **C.** ×20. Both after Kerner von Marilaun.

The most effective and frequent protection against excessive transpiration is a drastic reduction of the transpiring surface in relation to the total volume, such as the fall of leaves frequently encountered at the beginning of the dry period. In evergreen plants other devices met with are dwarfing of the whole plant (*nanism*), brought about by more circumscribed branching and reduction in the amount of foliage, together with a shortening of the axes and a reduction in the size of the lamina.

In certain Ericaceae the exposed surface of the lamina is permanently and strongly reduced by the rolling up of the leaf. The same process occurs temporarily in some native dune and steppe grasses during the dry period. In species of *Genista*, in the Cupressaceae and in certain species of *Veronica* of similar habit from New Zealand the leaves are reduced to small scales. Similar leaves are found in *Salicornia herbacea*, a plant occurring on strongly saline soils, such as those of salt marshes. Plants of such localities are termed halophytes. In many Cactaceae (Figs. 197B; 202B, C; 203A) the leaves are represented by thorns. In other cacti, tree-like species of *Euphorbia*, and some Asclepiadaceae the leaves either quickly perish or are very reduced.

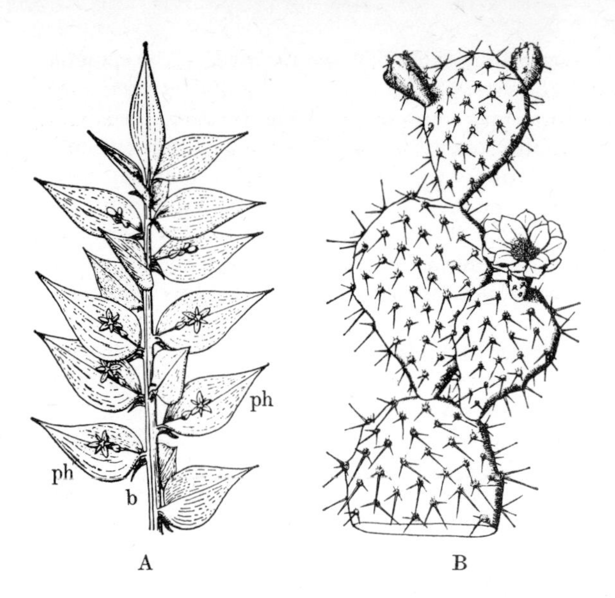

Fig. 197. Flattened shoots and phylloclades. A. Shoot of *Ruscus aculeatus* (Liliaceae): b, scale leaf; ph, leaf-like phylloclade. B. *Opuntia* sp. Flattened shoot with thorn-like leaves, flowers and fruits. A. Natural size. After Schenck, modified. B. $\times\frac{1}{5}$. After Schumann, modified.

a. Reduction of the leaves and the development of flattened shoots. With the reduction in area of the leaves, and still more with their disappearance, the assimilation of carbon dioxide must inevitably decline. To compensate for this loss, assimilation parenchyma is developed in the stem. When this occurs the shoots are green, as in the so-called switch plants. *Sarothamnus scoparius*, for example, produces only small, green, lanceolate leaflets on its long switch-like branches.

Often in such plants, along with the reduction of the leaves, there is a flattening or even a leaf-like development of the shoot axis (Fig. 197A, B). Such shoots can perform the assimilatory functions of the leaves much more efficiently than cylindrical stems, but at the same time, of course, the transpiration is also increased. Such leaf-like shoots are referred to as phyllomorphs. If the growth of such shoots is limited the short shoots so formed are particularly leaf-like and are then known as phylloclades.

In the well-known *Opuntia* (Fig. 197B), *Phyllocactus*, and also *Muehlenbeckia platyclados* (Polygonaceae) the whole fleshy stem is flattened, with constrictions at the nodes. Instructive examples of phylloclades are seen in *Ruscus*, a genus of several shrubby species. The branches of *Ruscus aculeatus*, for example, bear broad, dark-green phylloclades, excurrent into a sharp point, and arising in the axils of reduced scale-like leaves (Fig. 197A). The general appearance is that of leaves, but about half-way along the

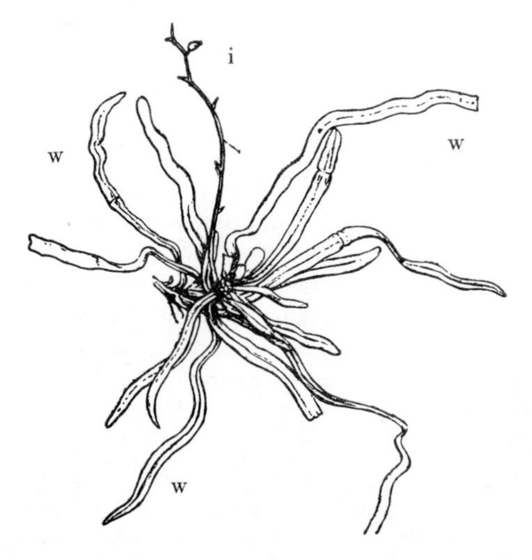

Fig. 198. *Taeniophyllum* sp. An epiphytic orchid with a very reduced shoot, and green widely spreading band-like aerial roots (w) which take over the function of assimilation. i, inflorescence. $\times\frac{1}{2}$. After Goebel.

midline of the phylloclade there arise one or more flowers from the axil of a minute, scale-like bract, thereby revealing the axial nature of the parent 'leaf'.

Particularly remarkable are some epiphytic Orchidaceae in which not only the leaves but the whole vegetative shoot is dispensed with. In these plants the green aerial roots, besides anchoring the plant on the substratum and absorbing water, also take on the function of leaves (Fig. 198).

b. Thorns. The great abundance of sclerenchyma, characteristic of the shoots of many xerophytes, maintains their rigidity during periods of water loss of both small and large extent. The development of sclerenchyma is often accompanied by the production of pungent thorns. Thorns do, of course, occur in some plants of other climates that are not xeromorphic, and in many climbing plants, but it is characteristic of the thorny shoots of many xerophytes of deserts, steppes, dry woodlands and thorn-bush communities that they act as a particularly effective, and in these regions valuable, protection against grazing animals.

Thorns are pointed, awl-shaped, branched or unbranched structures, richly provided with strengthening materials, and therefore rigid. They are derived from leaves or leaflets (e.g. Cactaceae, Fig. 197B; *Berberis*, Fig. 199F, G), and from shoots (Fig. 200A, B), or in rare cases from roots. Their formation often involves the production of much lignified tissue.

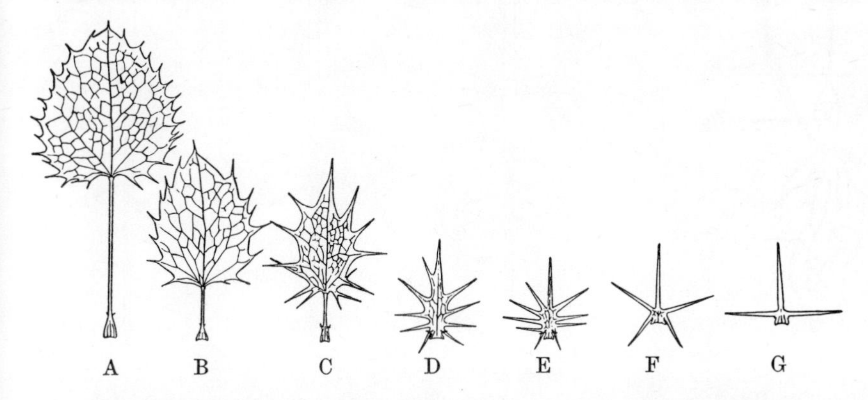

Fig. 199. *Berberis vulgaris.* **A.** Normal foliage leaf. **F.** Five- and **G.** three-rayed thorn-like leaves. **B.** to **E.** Transitional forms. $\times\frac{2}{3}$. After Troll.

In *Berberis vulgaris* the leaves of the main shoots are each represented by a thorn, usually three-rayed. The lateral short shoots, which produce exclusively normal foliage leaves, stand in the axils of these thorns (Fig. 199). In *Robinia, Acacia* and in many succulent species of *Euphorbia* (see below) with ephemeral leaves, the two stipules develop into thorns. Thorns derived from shoots occur, for example, in the hawthorn

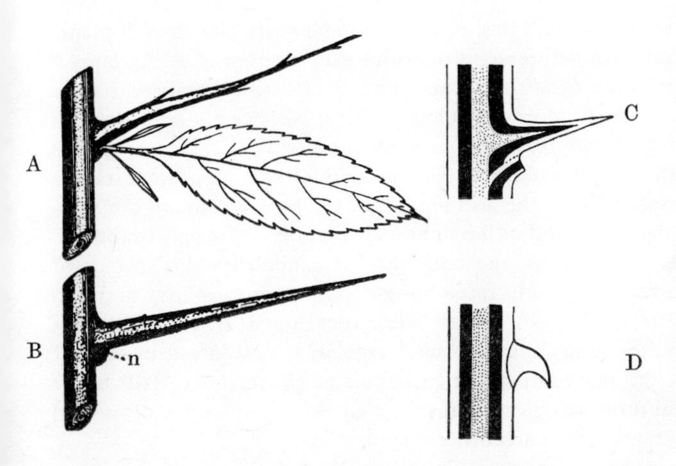

Fig. 200. A. and **B.** Shoot thorns of *Crataegus* sp. **A.** The shoot shortly after its formation in the axil of a leaf furnished with narrow stipules. Note how the young thorn-like shoot bears reduced leaves. **B.** The hard lignified thorn, showing the scar of the subtending leaf (n). **C.** Diagrammatic longitudinal section of the thorn. This contrasts with, for example, the prickle of a rose (**D.**), which is formed exclusively from the cortical tissue and is correspondingly easily broken off. **A.** and **B.** Natural size. After Fitting.

(*Crataegus*, Fig. 200A, B) and blackthorn (*Prunus spinosa*). Such thorns may sometimes be branched (e.g. in *Gleditschia tricanthos*). Thorns derived from roots occur, among the monocotyledons, on the stems of some palms (e.g. *Acanthorhiza*), and, among the dicotyledons, in the epiphytic *Myrmecodia* (Rubiaceae).

c. Succulence. Many xerophytes not only effectively diminish the output of water but in addition store water during the short wet season in special water-storage tissue for utilization in the subsequent, often prolonged, dry season. Sometimes the epidermal (as, for example, in the leaves of various Piperaceae, and species of *Ficus* and *Tillandsia*) or sub-epidermal cells are differentiated as a one- or several-layered, superficial, water-storage tissue (p. 97). At other times the water-storage tissue is more centrally placed, such as the inner parenchymatous water-storage tissue of the leaves of *Aloë*, *Mesembryanthemum* and *Lithops* (Fig. 201). Organs in which water-storage tissue is extensively

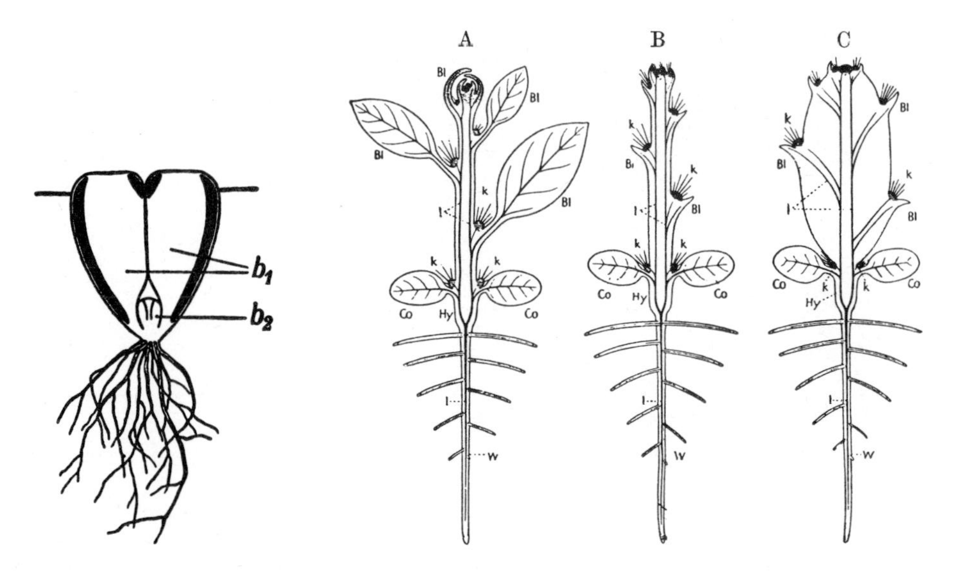

Fig. 201. *Lithops* sp. Diagrammatic longitudinal section. A primordial whorl of leaves (b_2) is enclosed between the two fleshy leaves (b_1). The upper horizontal line represents the soil surface, and black the assimilatory tissue. After Geitler.

Fig. 202. Diagrams illustrating the possible derivation of the 'cactus' habit. **A.** Initial form (e.g. *Peireskia*). The axillary buds (k) are transformed into areoles each bearing a tuft of thorns representing leaves. **B.** Intermediate form (e.g. *Opuntia* sp.) in which the leaves (Bl) are reduced to insignificant rudiments. **C.** The cortical tissue expanded to serve for water-storage. W, root; Hy, hypocotyl; Co, cotyledon; l, vascular trace. After Troll.

developed become as a result very thick and fleshy, with abundant sap. Hence such plants are called succulents. In certain Umbelliferae, Cucurbitaceae, Compositae, Asclepiadaceae, and species of *Pelargonium* and *Oxalis* of steppes and deserts, thickened roots serve as water reservoirs (root succulents). More frequent are leaf succulents (e.g. *Sedum, Sempervivum, Agave, Aloë, Mesembryanthemum*) or stem succulents (Cactaceae, species of *Euphorbia, Stapelia* and other Asclepiadaceae, *Kleinia* among the Compositae, *Cissus cactiformis*). Particularly characteristic of the arid regions of the New World are the club-shaped or columnar cacti, lacking leaves and either unbranched or only sparingly branched. In the Old World their place is taken by the columnar or candelabra-like species of *Euphorbia* and the Asclepiadaceae, so closely resembling certain Cactaceae that they can be mistaken for them (Fig. 203B, C). The 'cactus' habit, developed in quite different circles of affinity as an adaptation to dry climates with regular, if only brief, periods of precipitation, provides one of the most impressive examples of convergence. This is the phenomenon whereby identical morphological forms arise under the influence of natural selection in species widely separated from each other systematically.

As Goethe originally demonstrated, the seedlings of cacti are often hardly distinguishable from those of other seed plants. The later morphological transformation depends essentially upon the subsequent development deviating from the normal in three ways (Fig. 202), namely, the differentiation of the water reservoir (originating from the cortical parenchyma), the transformation of the leaves into thorns and the reduction of the lateral branches to clusters of spines (the areolae). In extreme cases the axis is condensed to a spherical form, thereby achieving the smallest possible surface subject to transpiration. The surface of the plant body may also become folded, the orthostichies of areolae standing out as ribs. This allows the multi-layered hypodermis, itself impermeable to water, to undergo concertina-like movements corresponding to changes in the hydration of the water storage lying within. The shade cast by these folds also possibly serves to prevent any portion of the surface from being too long in direct sunlight.

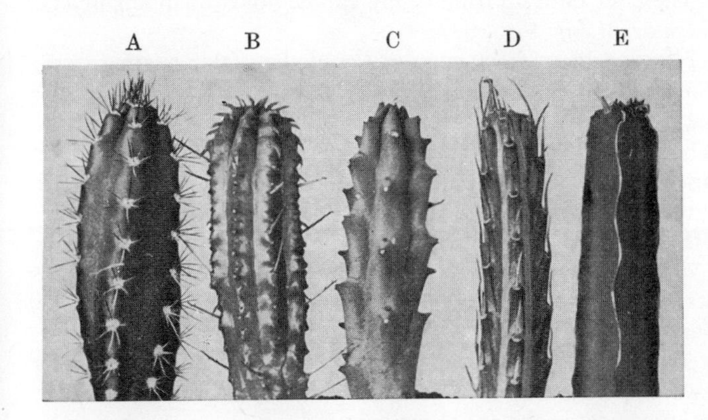

Fig. 203. Convergent evolution of stem succulence under the influence of a dry climate with short periods of heavy rainfall. **A.** *Cereus iquiquensis* (Cactaceae). **B.** *Euphorbia fimbriata* (Euphorbiaceae). **C.** *Heurnia verekeri* (Asclepiadaceae). **D.** *Senecio stapeliiformis*. **E.** *Cissus cactiformis* (Vitaceae). All $\times\frac{1}{2}$.

B. Adaptations to temperature

The effects of temperature are inevitably closely dependent on the nature of the water supply. On the one hand, where there is no lack of water in the substratum, the potentially damaging effects of high temperatures are allayed by enhanced transpiration, and the accompanying cooling effect of the evaporation at the leaf surfaces. Consequently, only in extremely dry desert regions do high temperatures make plant life impossible. On the other hand, prolonged temperatures below 0°C, and the consequent freezing and immobilization of the soil water (a form of drought), may cause the death of highly organized cormophytes, although more lowly organisms may survive and even multiply without difficulty in these conditions.

1. Plants of hot zones

In general growing plants are unable to withstand temperatures greater than 55°C. The only exceptions are a few protophytes, such as thermophilous Cyanophyceae and bacteria, able to exist in temperatures up to 80°C (see p. 335). However, in desert regions and dry scrubland ground temperatures of 60° and more are no rarity. In dry scrubland near Santa Maria (Colombia), for example, temperatures of almost 70°C were recorded at midday. Convection, however, ensures that in these conditions the air near the soil surface continually rises, its temperature concurrently falling, much lower values already being reached within a few decimetres of the soil surface. At these levels it is thus not so much the warmth of the soil as the infra-red radiation which can cause the leaf temperature to be considerably greater (up to 10°C) than that of the surrounding air.

With the tree-like phyllocactus *Pereskia colombiana*, for example, leaf temperatures of 42°C have been recorded in air temperatures of 30°C. The development of very small and narrow leaflets, favouring the dissipation of heat by convection and the nyctinastic drooping of leaves at high light intensities (day-time 'sleep movements'; e.g. *Acacia* spp.), of thick reflecting felts of white hairs (e.g. *Leucadendron*), and of thick-walled epidermises with a lacquer-like surface (e.g. *Laurus nobilis*), can all be interpreted as protections against heating by radiation. All these devices naturally also have the effect of limiting transpiration.

2. Plants adapted to seasonal climates (tropophytes)
How closely soil and climatic factors work together in shaping the morphological and physiological adaptation to the environment is particularly well shown by the tropophytes. These are plants whose external aspect and endogenous physiological rhythms are always in exact harmony with the yearly cycle of the seasons. Often it is difficult to decide whether the rhythmical change is habit and metabolism shown by these plants is determined more by the water relationships or more by temperature.

a. Woody plants. Most woody plants in regions of strongly seasonal climate (periodic change of temperature in temperate zones, periodic rainy seasons in many parts of the Tropics) protect their delicate apical meristems in the period unsuitable for growth (i.e. the cold period in temperate regions, and the dry in tropical) by the formation of resting buds (Fig. 133A, C and E). These are built up from firmly apposed or overlapping bracts, the bud-scales (e.g. *Acer, Aesculus, Malus*: Fig. 183A–C).

In other cases the bud-scales originate from stipules (as in the catkin-bearing trees, *Quercus, Carpinus, Fagus, Corylus,* and also in *Tilia*).

The outer bud-scales are coriaceous and usually brown. They are made effective barriers to desiccation by suberin and tomentose coverings, by excretions of resin, gum or mucilage, and also by enclosed air. Peculiarly formed glandular hairs which occur on the bud-scales of many of our trees (e.g. *Aesculus*) secrete a mixture of gum and resin which is freed by the bursting of the cuticle of the gland, and which spreads between the scales and sticks them together. When the buds break open in the spring the bud-scales are usually shed. In the shoots of trees (e.g. *Fagus*) the lowermost internodes, which lie between the bud-scales, usually remain conspicuously short. The positions of the bud-scales are thus indicated by crowded scars, rendering it possible to distinguish one year's growth from the next. In water plants (e.g. *Utricularia*) tightly closed winter buds are formed by shortening of the internodes in the apical region.

b. Perennial herbs. In seasonal climates the perennial herbs sacrifice at least those parts of their shoots which extend higher into the air than the rest, and which are therefore especially subject to drying. Some overwinter in a leafless condition, while others, such as the biennial herbs, retain some leaves during the winter. The resting buds arise from aerial or subterranean stems and are placed at ground level or just above it. Some perennial herbs sacrifice the whole aerial system and die back completely to subterranean stems. These stems persist through the winter and bear the subterranean resting buds.

Resting buds above the surface of the soil, on creeping or prostrate aerial shoots, are found in numerous native woody plants. During the winter when there is danger of desic-

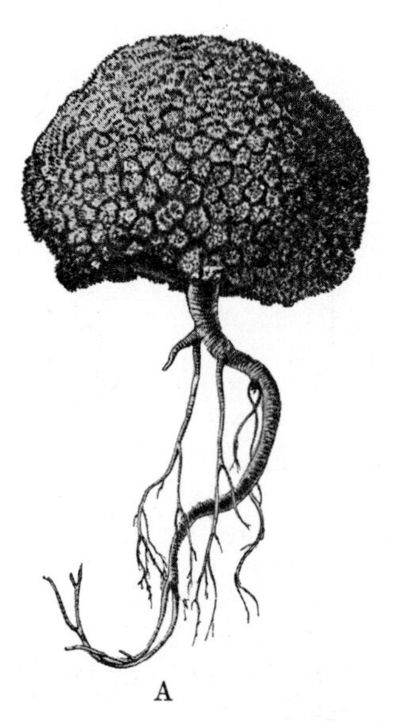

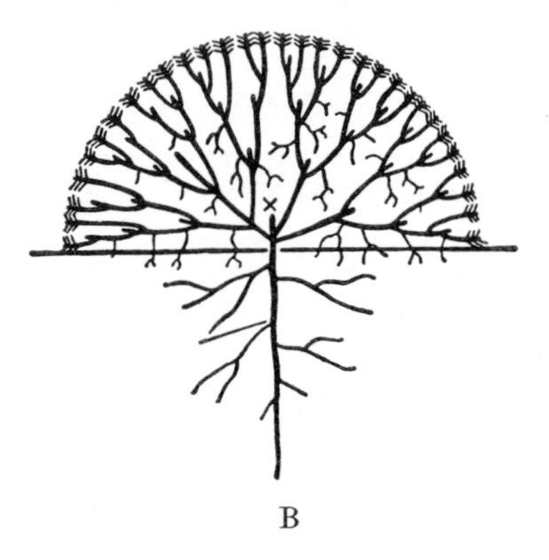

Fig. 204. A. *Azorella selago* (Umbelliferae). A cushion plant from Kerguelen. **B.** Diagram showing the sympodial structure of a cushion plant. **A.** $\times\frac{1}{4}$. After Schimper. **B.** After Rauh.

cation they are usually protected by a layer of snow. In the 'cushion plants', characteristic of high Alpine and Arctic regions, the shoots, which are richly branched and contracted into rosettes, are packed together to form firm cushions, often astonishingly large (Fig. 204). These cushions remain close to the ground or become hemispherical in form. All the terminal branchlets, which alone bear the small persistent leaves in close sequence, are found at the surface of the cushion.

Supplies of metabolites for the synthesis of structural materials, and for providing energy, are required in the spring to facilitate the outgrowth of the overwintered resting buds. To this end energy-rich materials have been laid down in most of the subterranean food-storage organs during the preceding year. Because of their richness in valuable organic materials, such food-storage organs are, in addition to fruits and seeds, among the most valuable sources of vegetable foodstuffs for animals and man. All three fundamental organs of the plant body can serve as food reservoirs. The following different kinds are distinguished: 1. root-stocks or rhizomes–subterranean stems capable of unlimited growth; 2. stem tubers of limited growth, formed either above or below the surface of the soil; 3. bulbs, in which the food-storage function is essentially relegated to the fleshy thickened leaves; 4. root tubers, which are derived from adventitious roots of limited growth; and 5. tap roots, which are formed by thickening of the principal root.

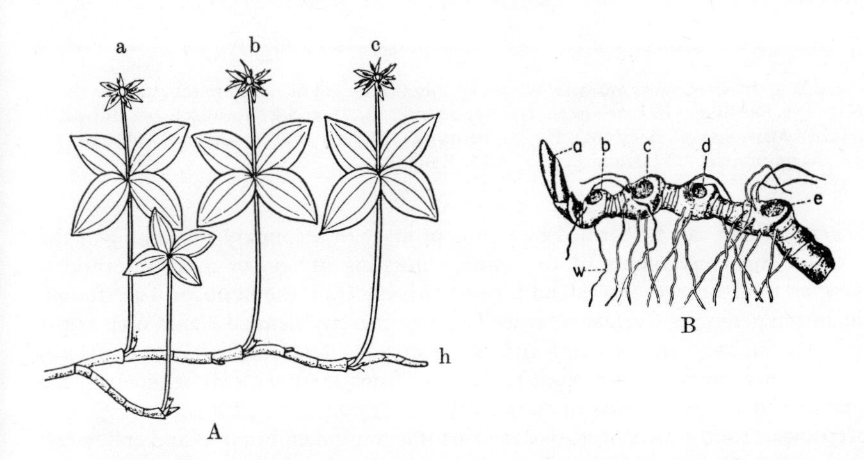

Fig. 205. A. *Paris quadrifolia*. Diagram of a monopodial rhizomatous plant: a, b and c are the flowering shoots of three successive years arising as lateral branches. h, the main axis, which continues to grow monopodially. **B.** Sympodial rhizome of *Polygonatum multiflorum*: a, bud surrounded by scales which will form the next year's aerial shoot; b, scar of the current year's aerial shoot; c, d and e, scars of the acrial shoots of the three preceding years; w, adventitious roots. **A.** $\times\frac{1}{10}$. After A. Braun, modified by Troll. **B.** $\times\frac{1}{2}$. After Schenck.

1. Rootstocks and rhizomes. These are usually sympodially, rarely monopodially, branched subterranean shoots, with more or less thickened axes and usually short internodes. They are usually the products of several seasons' growth (Fig. 205A, B), and they grow in the soil with either a little or no branching, some being horizontal and others erect. The older parts die during the year. Since, however, the rhizome grows continuously from the apical region, in the course of years rhizomatous plants can cover large areas, and probably become long established (e.g. stands of *Convallaria majalis* (lily-of-the-valley), and of *Mercurialis perennis* (dog's mercury)).

The rhizome produces adventitious roots continuously, either all round or confined to the lower side, and colourless, membranous, scale leaves. Rhizomes can be distinguished from roots by the presence of these scales or their scars, by the production of buds, by the absence of a root cap and additionally by their anatomical structure. The same holds for most subterranean stem tubers, which, as with bulbs, are related to rhizomes by a number of transitional forms.

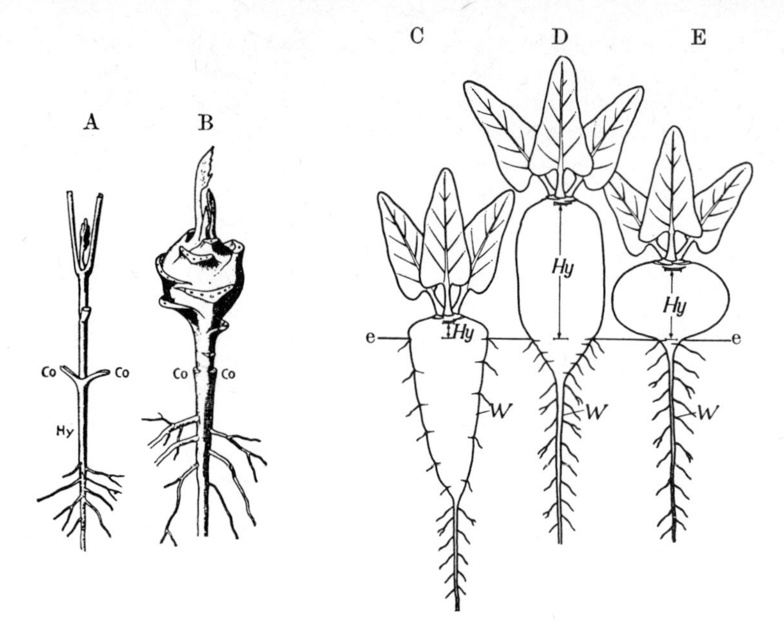

Fig. 206. **A.** and **B.** *Brassica oleracea* var. *gongylodes* (Kohlrabi). Origin of the stem tuber above the cotyledons (Co). **A.** Seedling. **B.** Later stage. Hy, hypocotyl. **C.**, **D.** and **E.** Various forms of beet (**C.** Sugar beet. **D.** Mangold. **E.** Beetroot). Hy, hypocotyl; W, primary root; e, soil level. **A.** and **B.** After Rauh, diagrammatic. **C.**, **D.** and **E.** $\times\frac{1}{10}$. After Rauh.

2. Stem tubers. These can be formed by strong primary or secondary thickening of the hypocotyl, or of higher segments of the shoot consisting of one or more internodes. Tubers serving as food-storage organs and formed entirely from the hypocotyl are found, for example, in the perennial *Cyclamen*, as well as in numerous biennial plants such as the radish (*Raphanus sativus* var. *radicula*) and the beetroot (*Beta vulgaris* var. *esculenta*) (Fig. 206E). A typical stem tuber derived exclusively from an upper leafy segment of the stem is shown by the kohlrabi (*Brassica oleracea* var. *gongylodes*; Fig. 206B).

The subterranean stem tubers bear, like the rhizome, only membranous and ephemeral scales, or their scars. They differ from rhizomes in their greater thickness, limited growth and usually in their lack of roots. Apart from those tubers derived from the hypocotyl, stem tubers again differ from rhizomes in persisting only from one growing season to the next. In perennials they are replaced in the next growing season by a new tuber in a variable manner.

Potato tubers (Fig. 207) arise at the ends of subterranean, plagiotropic, side branches (stolons) as the result of the swelling of several internodes. Besides serving as food reservoirs, they bring about the vegetative propagation of the parent plant. The regularly distributed pits ('eyes') visible on every potato tuber conceal lateral buds. The small scale-like leaves, in whose axils the eyes arise, are visible in only very young tubers, merely their scars being present on mature tubers. The parent plant dies after the production of the tubers.

In those perennial plants with annual subterranean tubers (e.g. *Ranunculus bulbosus, Colchicum, Crocus*) the subterranean base of the orthotropous main shoot itself swells to form the overwintering stem tuber, or *corm*. In the following spring a lateral bud grows out to form the new shoot, and during the growing period the base of this shoot forms the next corm. In *Crocus* the new corm arises near the summit of the old and appears to sit upon it; in *Colchicum*, on the other hand, the new corm is lateral.

3. Bulbs. A bulb (e.g. of the onion, tulip and hyacinth) is usually a subterranean and very strongly contracted shoot with thickened, fleshy scale leaves (Fig. 208, S_1–S_4) which serve as food-storage organs.

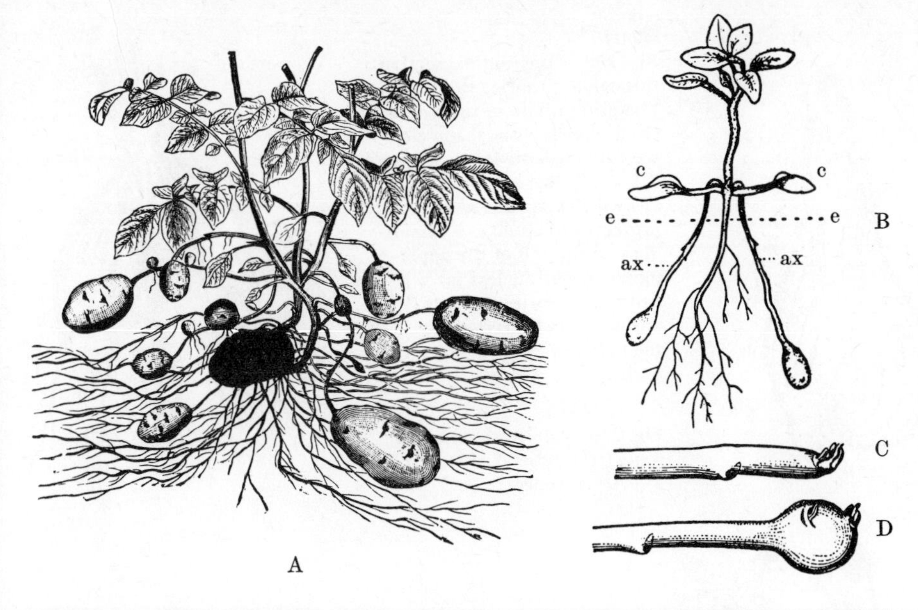

Fig. 207. *Solanum tuberosum*. **A.** Lower part of a mature plant. The central black tuber is the parent tuber from which the plant has developed. **B.** Seedling: e, soil level; c, cotyledon; ax, axillary shoots from the seedling which are positively geotropic and terminate in young tubers. **C.** and **D.** Stolons with developing tubers. **A.** $\times\frac{1}{4}$. After Schenck.

The scales of the bulb are either entire scale leaves (e.g in *Tulipa, Fritillaria* and *Lilium martagon*) or scale-like and closed leaf sheaths surrounding the stem, derived from the bases of dead leaves (e.g. in *Allium*). Both kinds of scale may alternate regularly in the bulb (e.g. *Lilium candidum, Hyacinthus*). The scales are set on a strongly contracted disc-like axis (Fig. 208), from which the growing point of the aerial shoot later arises. In many cases the bulb of the current year dies, and a new daughter bulb is formed annually from a bud in the axil of a scale.

4. Root tubers. These are formed by a number of perennial herbs and biennials from adventitious roots. They often resemble stem tubers, to which they are analogous, but their homology with adventitious roots is made apparent by the presence of a root cap, the absence of leaf primordia and their anatomy.

Root tubers occur, for example, in *Dahlia* (Fig. 209) and *Ranunculus ficaria* (Fig. 210A, B). The root tubers of many terrestrial orchids have a peculiar construction. They are ovate (Fig. 210C, D) or palmately divided (as in *Gymnadenia*), and always arise from a single adventitious root, the palmate tuber being formed as a result of its dichotomous branching. An older and a younger tuber always occur together. In Fig. 210C, D the tuber of the preceding year (K_1) has already produced a flowering shoot (B); the tuber is wrinkled and on the point of death. At the base of this shoot a bud in the axil of a lowermost bract (N) has developed an adventitious root. This breaks through the bract and swells to form the root tuber (K_2) of the current year The shoot bud Kn will continue the sympodial system in the coming year by producing a new inflorescence, and so on indefinitely. In all root tubers storage occurs in heavily thickened cortical parenchyma. This thickening is exclusively primary.

5. Tap roots. Carrot-like tubers consist wholly or at least partly of swollen tap roots. Accordingly, they are found only in the allorhizic dicotyledons. Since what is essentially part of the hypocotyl also takes part in the structure of the swollen region, these, despite their outer uniformity, are morphologically heterogeneous and very variable in their anatomy.

In the carrot and sugar beet (*Beta vulgaris* var. *altissima*, Fig. 206A) the tap root itself forms the major part of the swollen region. In the mangold (*Beta vulgaris* var. *rapa* f. *crassa*; Fig. 206B) and radish portions of the hypocotyl are also included in the swollen

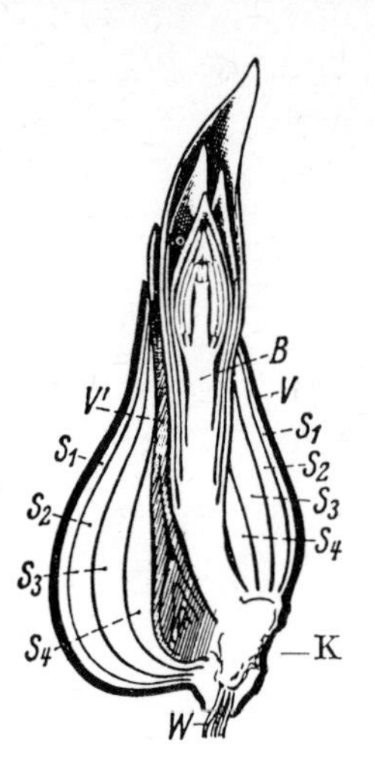

(Left)
Fig. 208. Longitudinal section through a sprouting tulip bulb. The primordium of the flower (B) is already visible, and also seen are the scale leaves (S_1 to S_4), the brown scarious sheath (V) which surrounded and protected the bulb as a primordium in the axil of the uppermost scale of the leaf (as V′), and the adventitious roots (W) arising laterally from the base of the bulb (K). Natural size. After Sachs.

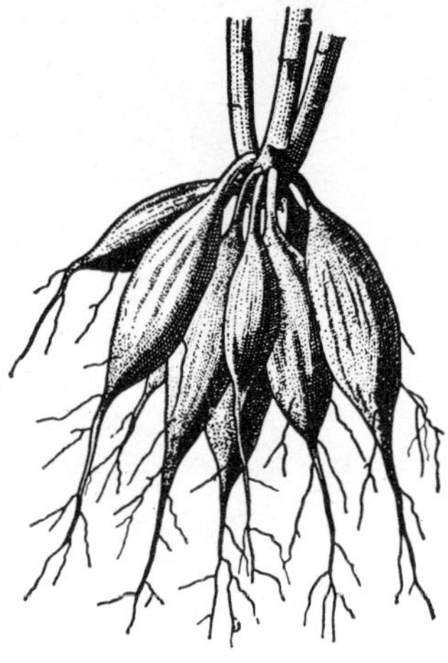

Fig. 209. Adventitious storage roots (root tubers) of *Dahlia*. $\times\frac{1}{5}$. After Weber.

Fig. 210. A. and **B.** *Ranunculus ficaria*. **A.** Bulbils (B) arising in the axils of the leaves, and root tubers (nK) appearing adventitiously at the base of the shoot; aK, old tuber from which a plant has developed; N, adventitious root. **B.** Single bulbil in longitudinal section. The tuber W is formed by the thickening of an adventitious root arising at the base of axillary bud K. **C.** and **D.** *Orchis militaris*: root tubers. K_1, tuber of the previous year from which the flowering shoot of the current year (B) has developed. An adventitious root arising in the axil of the lowermost scale leaf (N) gives rise to the tuber of the current year K_2. Normal adventitious roots (W) arise at the same node as the tuber. Kn, primordium of the shoot of the following year. **A.** and **B.** After Troll. **C.** and **D.** Natural size. After R. Wettstein.

region. In the swede (*Brassica napus* var. *napobrassica*) and celeriac (*Apium graveolens*) the swollen region extends even to the basal portion of the stem, together with its leaves, above the hypocotyl. These latter therefore show a transition to those tubers formed purely from the hypocotyl or from the stem, found as we have already seen under Section 2 in the beetroot and kohlrabi respectively.

Thus the main mass of the edible tissue of the carrot (*Daucus carota*) is derived from cortical parenchyma, but in the black radish (*Raphanus sativus* var. *niger*) from secondary xylem. Finally, in the sugar beet (*Beta vulgaris* var. *altissima*) the thickening comes about as a result of the activity of a succession of cambial zones arising in the cortical parenchyma. These are clearly indicated in transverse section by the secondary xylem elements derived from them.

Many bulbs, tubers and rhizomes are specific in the depth at which they will grow, although this depth may vary with the nature of the soil. Thus for example, the rhizomes of *Paris* are found at a depth of 2–5 cm, the tubers of *Arum* (Fig. 188A) at 6–12 cm, of *Colchicum* at 10–16 cm and the rhizomes of *Asparagus officinalis* at 20–40 cm. The seeds, however, germinate on or only a little under the soil. The underground stems of these plants must therefore descend deeper into the soil as the plant becomes mature. This is brought about partly by the growth movements of their plagiotropic axes and partly by contractile roots (Fig. 188B). In *Lilium* all the roots are contractile; in other cases the ability to contract is limited to a few or even to a single root (heterorhizy; e.g. *Crocus, Gladiolus*). In the various rosette and semi-rosette plants even the hypo- and epicotyls undergo persistent contraction during their secondary thickening. By this means the growing point of each year is drawn downwards to the same extent that it is elevated by apical growth, so that the rosette always stays at the level of the soil (e.g. *Gentiana lutea*).

3. Life forms

The Cormophyta can be classified in five life forms (Fig. 211) according to the length of life of the shoots, together with the position of the resting buds and the way in which they are protected during the unfavourable season (i.e. winter, or a summer dry period). The percentage representation of the life forms in the vegetation of any region depends upon its climate and location.

1. Phanerophytes. These bear their resting buds more than 50 cm above the soil surface (Fig. 211C). They include all the larger woody plants such as evergreen and deciduous trees and shrubs, most climbing plants (see p. 192), as well as many of the large upright herbs and the epiphytes (see p. 194) of the humid Tropics.

2. Chamaephytes. These bear their resting buds not far removed (within 10–50 cm) from the soil surface (Fig. 211A, B). Thus in climates where snow is frequent they enjoy an effective protection from frost, since the layer of snow covering them is a particularly bad conductor of heat. Many of the prostrate and creeping woody plants of the northern tundra and Alpine regions belong here. However, many of the Ericaceae of oceanic heaths (e.g. *Calluna, Erica tetralix*) and the cushion plants (Fig. 204) are also chamaephytes. Many of the woody plants and colonizers of extreme desert regions are again chamaephytes, often showing pronounced xeromorphy.

3. Hemicryptophytes. The resting buds of the hemicryptophytes lie at or very close to the soil surface. They include tufted plants, such as many of the Gramineae, the biennial and perennial rosette plants (e.g. *Taraxacum* (Fig. 211D), *Plantago* (Fig. 135D), *Beta* (Fig. 206C)), those plants whose renewal buds lie at the bases of the dead stems which bore leaves the preceding summer (e.g. *Artemisia, Urtica, Lysimachia vulgaris* (Fig. 211F)), and perennial stoloniferous herbs such as *Fragaria vesca, Potentilla reptans* and *Ranunculus repens* (Fig. 211E).

4. Cryptophytes (or *geophytes*). Here the resting buds are still better protected. They lie beneath the surface of the soil, on rhizomes (Fig. 211G) or in bulbs (Fig. 211H).

5. Therophytes. Finally, there are the therophytes or annuals which tide over the unfavourable season as embryos protected within resistant seeds regularly sacrificing the plant body built up during the year (Fig. 211J). To them belong many of our most important field crops (e.g. the summer cereals) as well as their weeds. The so-called winter annuals (e.g. the autumn-sown cereals), which germinate in the autumn and overwinter as seedlings, are included in the hemicryptophytes above.

In the Tropics, with their warm and damp, equable climate, the phanerophytes lianes and epiphytes are by far the most abundant components of the vegetation. Here occur numerous meso- and hygromorphic perennial herbs, many with enormous leaves (e.g. *Musa*, Araceae, ferns). The more extreme the periodicity of the climate, particularly in

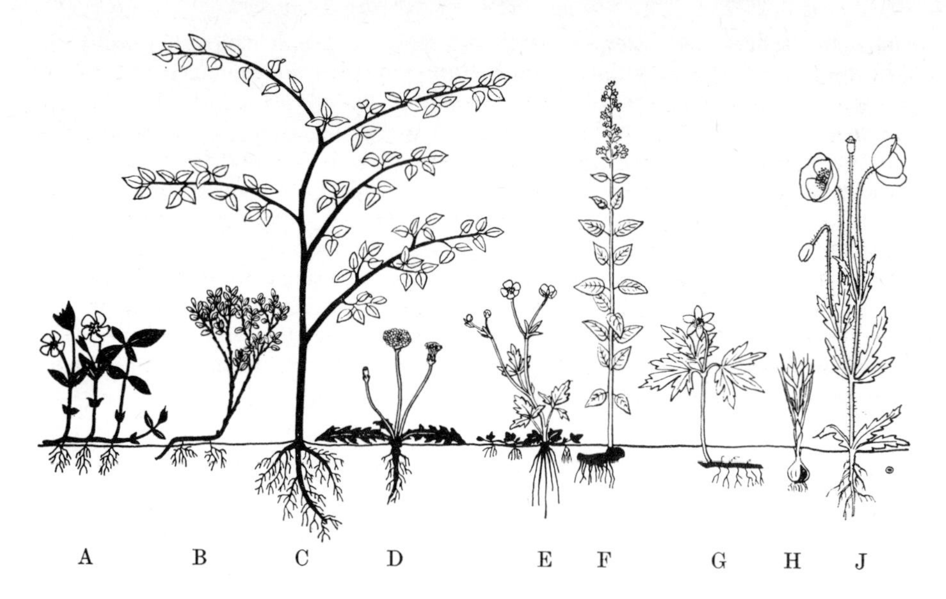

Fig. 211. Diagrammatic representation of the most important life forms (the black parts overwinter, the remainder die in the autumn). **A.** and **B.** Chamaephytes. **C.** Phanerophyte. **D.** to **F.** Hemicryptophytes. **D.** Rosette plant. **E.** Plant overwintering with creeping shoots. **F.** Plant overwintering with bud at base of dead summer shoot. **G.** and **H.** Cryptophytes. **G.** Rhizomatous geophyte. **H.** Geophyte with bulb or tuber. **J.** Therophyte (an annual). After Walter.

the dry regions of the earth where the arid period is long and the transition of climate abrupt (steppes, semi-deserts and deserts), the greater becomes the percentage of the deciduous tropophytes. Eventually the vegetation comes to consist of small-leaved or leafless xerophytes with efficient adaptations against desiccation, together with geophytes and annuals. The temperate zone can be considered as providing a hemicryptophyte climate, since over half the species occurring here belong to this group. In the northern tundra and high mountains the percentage rises to 60 and 70 per cent. Finally, in the high alpine and polar regions the chamaephytes predominate. Their cushion-like forms are protected by snow for at least some of the time during the period unfavourable for growth, while during the short time they are free from snow assimilation can immediately begin again. At the extreme limits of plant life, such as at heights above 3000 m, the proportion of cushion plants rises to 80 per cent.

C. Adaptations for gaining light

In the struggle for light and space two groups of the Cormophyta have arisen with peculiar growth forms, namely the climbing plants (including lianes) and the epiphytes.

1. Climbing plants (lianes)

The thin stems of these plants climb up on other plants, and also on rocks and walls. In this way, and without developing any strong supporting stems, their foliage is rapidly brought up from the shaded regions of the forest floor into the sunlight.

Climbing is accomplished in many different ways. In some plants (e.g. *Solanum dulcamara*) it is brought about by spreading, hook-like side shoots in others by stiff hairs (e.g. *Galium aparine, Humulus lupulus*), in others by prickles (*Rosa* spp., *Rubus* spp.) or thorns (*Lycium, Bougainvillea*). In root climbers there are adventitious roots which are often negatively phototropic (p. 349), but positively haptotropic (p. 361 seq.), acting as attachment organs (e.g. *Hedera*, many Araceae). In twining plants the stem ascends through winding movements of the stem, the internodes of which are very long (*Humulus, Phaseolus*, etc.; Fig. 316).

The tendril climbers have developed special attachment organs of quite a different morphological nature, and remarkable for their extreme sensitivity to contact. These again provide an impressive example of convergent evolution. Typical tendrils are branched or unbranched thread-like organs which wrap themselves around foreign objects and thus anchor the shoot in the complex of branches (see p. 361 seq.). In the Vitaceae (*Vitis vinifera, Parthenocissus*) the tendrils are metamorphosed shoot axes, and, although terminal, the sympodial structure of the shoot system causes them to become displaced to a lateral position (Fig. 143). In *Passiflora* the main shoot is monopodial and the thread-like tendrils arise as unbranched side shoots from the leaf axils. In the Cucurbitaceae (e.g. *Bryonia*; Fig. 309) the tendrils are derived from leaves reduced to midribs alone (Fig. 212). Leaf tendrils occur also in the Leguminosae, but here frequently only part of a lamina is transformed into a tendril, and this is often branched. In the pea and vetch, for example, it is the upper leaflets of the pinnate leaf that are so modified (Fig. 213A). In *Lathyrus aphaca* the lamina is reduced to a simple tendril, and the function of assimilation is taken over by the two large stipules (Fig. 213B).

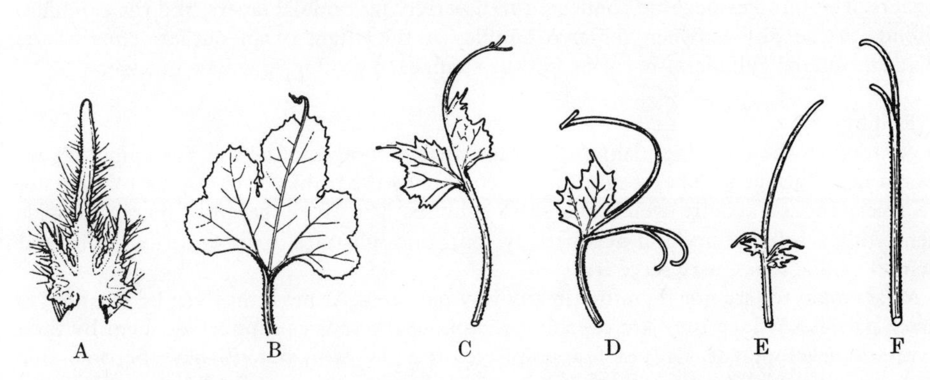

Fig. 212. *Cucurbita pepo*. Leaf tendril (**F.**), an almost normal foliage leaf (**B.**), and transitional forms (**C.**, **D.** and **E.**). **A.** Juvenile stage of a normal leaf showing the precocious extension of the midrib and principal lateral veins.

In certain species of the wild vine (*Parthenocissus*; Fig. 213C) the branches of the tendril are capable of developing terminal discs which adhere to flat surfaces such as walls. There are, however, climbing plants which have no special organs for this purpose. Here climbing takes place by various other organs embracing thin objects; thus internodes of elongated lateral branches, or petioles (e.g. *Tropaeolum* spp., *Nepenthes*) may have the same irritability as tendrils. The main and secondary rachises of the leaf (e.g. *Clematis*), or

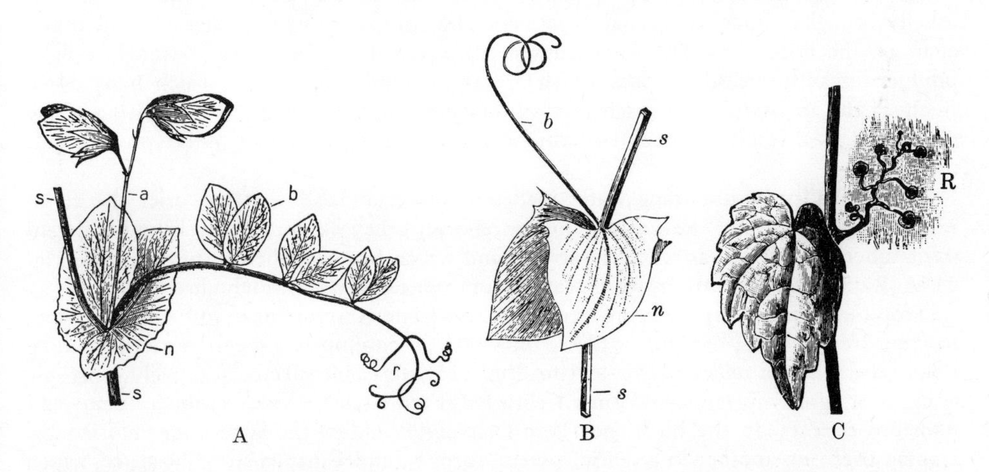

Fig. 213. Types of tendrils. **A.** Portion of the stem and leaf of *Pisum*: s, stem; n, stipules; b, leaflet of the simply pinnate leaf; r, the leaflets transformed into tendrils; a, the flower-bearing axillary shoot. **B.** Portion of the stem of *Lathyrus aphaca*: s, stem; b, leaf tendril; n, the leaf-like stipules. **C.** *Parthenocissus tricuspidata*: R, shoot tendril with adhesive discs. **A.** ×½. After Schenck, modified. **B.** ×½. After Schenck. **C.** ×¼. After Noll.

even the laminae of the pinnae (certain species of *Fumaria* and *Corydalis*) or the elongated tip of the leaf (*Gloriosa*) may behave similarly. In some tropical climbers (e.g. *Vanilla*) the aerial roots are irritable and grasp adjacent vegetation.

The stems of almost all climbing plants have unusually broad conducting elements and sieve tubes. In the tropical climbing plants anomalous secondary thickening is widely distributed, and the mass of xylem produced is banded, furrowed or fissured, or even divided up. As a result, the long and flexible stems, resembling in appearance ribbons, cables or ropes, are capable of bending and torsion. Furrowed masses of xylem are produced in many lianes belonging to the Bignoniaceae.

The existence of several cylinders of xylem in the stem is a condition peculiar to a number of tropical lianes belonging to the genera *Serjania* and *Paullinia* (Sapindaceae). They are able to arise, since the vascular bundles in the primary part of the stem are not arranged in transverse section in a closed circle, but in a ring with more or less extensive outliers. The bundles become connected transversely by cambial layers, and the cambium extends in the pith between the two bundles at the origin of an outlier. Thus several adjacent cambial cylinders are formed in the stem, each producing a mass of xylem.

2. Epiphytes

In contrast to the climbing plants, which are always rooted in the soil, the epiphytes are always raised above it. They begin life by colonizing the trunks or the upper branches of the trees, and so acquire well-illuminated habitats. The trees therefore merely provide them with a substratum, and occasionally inorganic substrata, such as rocks, roofs, and even telephone wires, may serve well.

Most epiphytes are not therefore in any way parasites. At most they can be regarded as 'space parasites', since they are capable of choking the species supporting them by their luxuriant development. Only a few epiphytes (e.g. *Viscum*, p. 199) have become true parasites, tapping and exploiting their host plants by special haustorial organs.

Naturally, only those plants are capable of the epiphytic habit whose seeds (or spores) continue to be dispersed either by air currents or by animals on to the trunks or branches of the trees. In cool temperate regions the epiphytic habit is confined principally to the corticolous algae, lichens and mosses, many species of which tolerate temporary desiccation.

For the larger epiphytes of the Cormophyta, the obtaining of the water and minerals essential for life poses an even more critical problem. Accordingly, the most favourable conditions for epiphytes are found in regions where appreciable rainfall occurs almost daily and the humidity is constantly high, conditions found particularly in the rain forests of the tropical lowlands and mountains. The attachment of epiphytes to the host plant is usually brought about by special negatively phototropic roots which often branch and embrace the host stem. The so-called semi-epiphytes (e.g. *Ficus* spp.; 'strangling figs') produce in addition other roots which are at first unbranched and which hang down freely in the air. After they reach the soil they contribute to the mineral nutrition of the plant, and as a result of secondary thickening may also become stem-like prop or stilt roots.

The difficulty of obtaining water makes it understandable that the drier the air in which epiphytes grow, the greater the xeromorphy they exhibit. Many have developed stem tubers which act as water reservoirs and which become filled during rainfall (Fig. 214A, B). Special adaptations which catch rainwater rapidly are widely distributed.

Tropical orchids epiphytic on trees, and also some Araceae, have roots often green, hanging freely in the air (Fig. 214A), on which is developed a special water-absorbing tissue, the *velamen radicum* (see p. 106; Fig. 111). In other species negatively geotropic aerial roots grow upwards to form a richly branched lattice, between which humus and moisture collects. In the birds' nest fern (*Asplenium nidus*) the large undivided fronds, lapping over one another in a dense rosette, form a funnel-shaped nest-like space, which becomes filled with humus above the enclosed growing point. In species of *Polypodium* and *Platycerium* special leaves are developed in a regular rhythm behind which humus and water can be collected (heterophylly, Fig. 613). In *Dischidia rafflesiana* (Asclepiadaceae) some of the leaves undergo even greater modification. Considerable surface growth and a

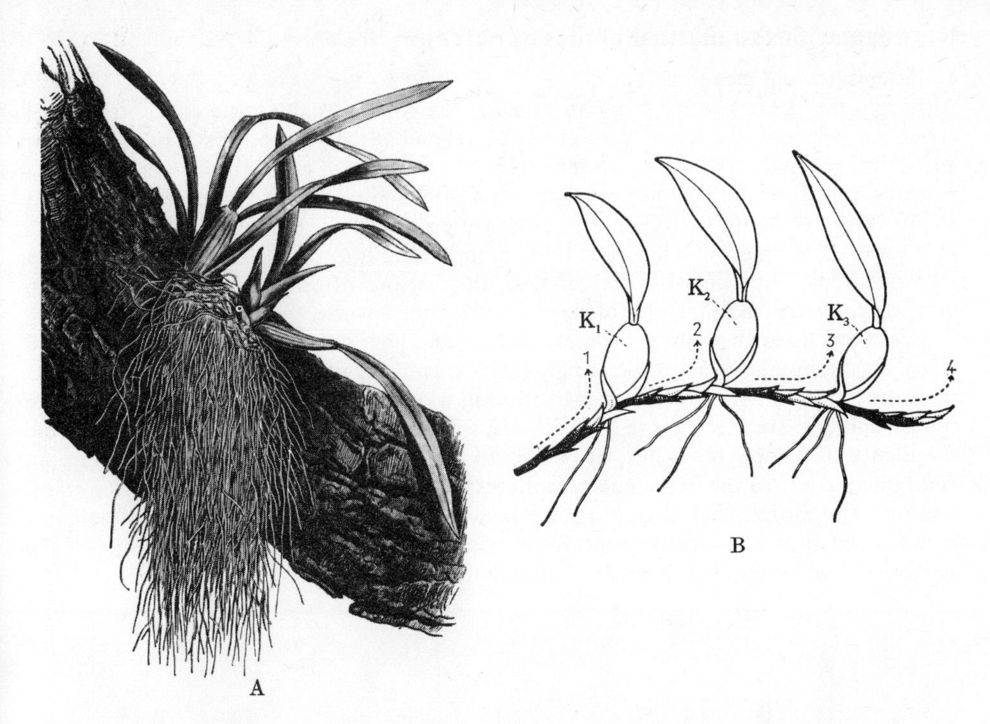

Fig. 214. Epiphytic orchids. **A.** *Oncidium* sp., an epiphytic orchid of tropical regions showing tuberous shoots adapted for water storage, and aerial roots. **B.** *Coelogyne* sp. Note the sympodial succession of three generations of shoots (K_1, K_2 and K_3), each ending in a tuber. The shoot currently advancing (4) will shortly give rise to a new tuber. **A.** $\times\frac{1}{10}$. After Kerner von Marilaun. **B.** $\times\frac{1}{5}$. After Troll, diagrammatic.

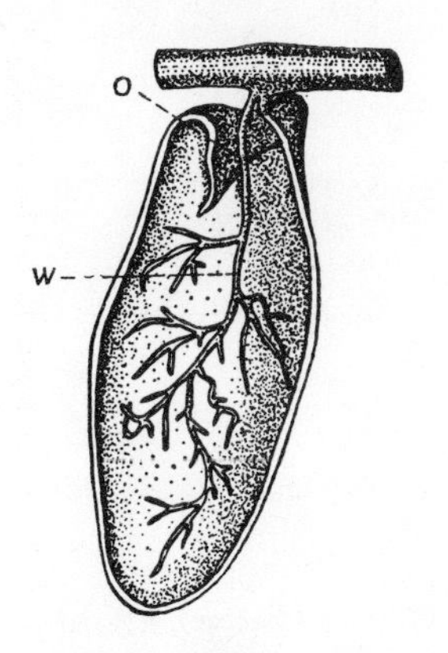

Fig. 215. *Dischidia rafflesiana.* Longitudinal section through an ascidiform leaf: o, opening to the interior; w, adventitious root which penetrates the cavity and ramifies within. $\times$½. After Fitting.

simultaneous restriction of marginal growth causes the leaves to develop into narrow-mouthed pouches or urns (Fig. 215), the inner surface of which corresponds to the lower side of the leaf. These cavities usually contain colonies of ants which drag in earth, and also leave inside corpses and faeces. Moisture can also collect within; for example, by condensation of water vapour on the inner walls of the urn which are cooled by transpiration. An adventitious root arising at the same node as the leaf grows into each urn, so the species creates as it were its own flower pot.

In the Bromeliaceae the roots are represented only by short wire-like attachment organs. In many species, such as species of *Tillandsia* which grow even on telephone wires and some of which resemble the barbate lichens, roots may be wholly lacking. In these plants water is absorbed entirely through absorption hairs inserted into the leaves (see p. 106; Fig. 112). In other species the shoot forms a rosette, and the leaf bases lying closely above one another create 'cisterns' in which considerable quantities of rainwater can collect and later be gradually used.

D. Adaptations to unusual modes of nutrition

1. Halophytes and mangroves

Although the great oceans show an average (and little variable) salt concentration of about 3.5 per cent, the halophytes of sea coasts and at the margins of salt pans in steppe and desert regions may have to endure, as a consequence of evaporation, concentrations reaching and exceeding 10 per cent (the concentration of salt in a saturated solution is about 38 per cent). In such habitats plants are presented not only with the problem of a soil water containing much larger than normal amounts of inessential, and possibly injurious, ions, but also with that of wide fluctuations of osmotic pressure, heavy rain bringing a substantial fall, and prolonged dryness the converse.

Coast and desert halophytes compensate for the high concentration of salt in the soil water by the uptake of correspondingly larger quantities of salt into the cell sap, so that the concentration here exceeds that of the soil water. In addition, many halophytes, like the xerophytes, are succulent (e.g. *Salicornia*, Fig. 216). The ecological significance of the succulence of halophytes is not, of course, to be sought in the storage of water since, in contradistinction to the succulent xerophytes, there are no modifications limiting transpiration. The morphological convergence between halophytes and xerophytes should not therefore be allowed to lead to spurious conclusions about physiological similarities (such as the putative 'physiological dryness' of such habitats).

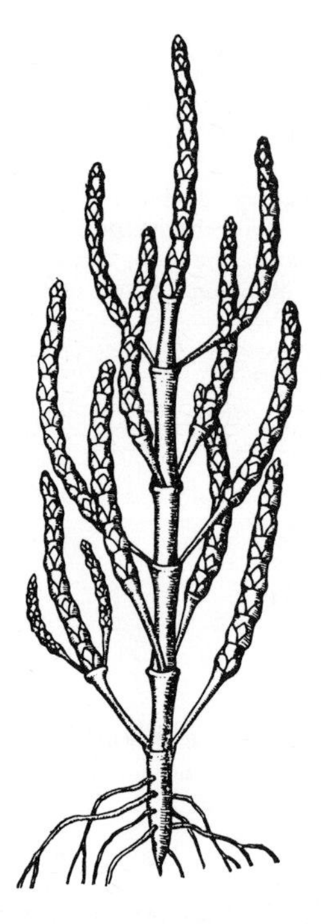

Fig. 216. *Salicornia europaea* (Chenopodiaceae). X½. After Fitting.

Many halophytes are provided with special glands which excrete salt (in the form of a highly concentrated solution). Tamarisk, and other plants of salt deserts (e.g. *Statice, Reaumurea*) have a glaucous appearance by day as a consequence of the dusting of salt crystals on the surface, but by night, when the issuing salt attracts moisture hygroscopically from the atmosphere, they appear green and bedewed.

In the Tropics a typical woody plant community (the mangrove swamp) is found in tidal estuaries and salt marshes which has adapted itself not only to the high salt concentration of the soil, but also to the poor aeration of the root system and the regular mechanical action of the tides. The salt concentration in the pools left behind at low tide rises to very high levels as evaporation proceeds, and correspondingly the cell sap of the mangroves shows some of the highest osmotic pressures known in plants (see p. 222). Aerial roots compensate for the lack of oxygen in the soil, and prop- (or stilt-) roots, adventitious in origin, provide extra anchorage resisting the action of waves and tides.

2. Partly or completely heterotrophic cormophytes

a. Parasites and semi-parasites. As in the Thallophyta, so also in the Cormophyta there are plants which have specialized nutritionally. They have wholly or partly lost the capacity for autotrophic nutrition, and in consequence obtain their organic nutrition at the expense of other plants—their hosts. They are referred to as parasites. While the semi-parasites can often at first glance hardly be distinguished from their green, wholly autotrophic relatives, the total or holo-parasites display a complete or almost complete loss of chlorophyll. Additionally they often show a very peculiar morphology, deviating considerably from that of their green relatives.

With the reduction in chlorophyll the leaves of parasites become superfluous and reduced to inconspicuous yellowish scales, or they may be lacking altogether. Even the shoot axes may become simplified to some extent, and contain only a little or no chlorophyll. Since, as a result of the reduction in the leaves, transpiration is curtailed, and also since many parasites tap the water-conducting elements of their hosts, even roots have frequently disappeared. The conducting elements of the vascular bundle also remain weakly developed, and the formation of woody tissue takes place to only a trifling extent. The atrophy of the assimilatory functions has led, however to the development of new structural features. Such are the special absorptive organs (haustoria) which allow the parasite to penetrate into the body of the attacked plant as far as the conducting tracts, and to nourish itself upon their contents (see pp. 292 seq.).

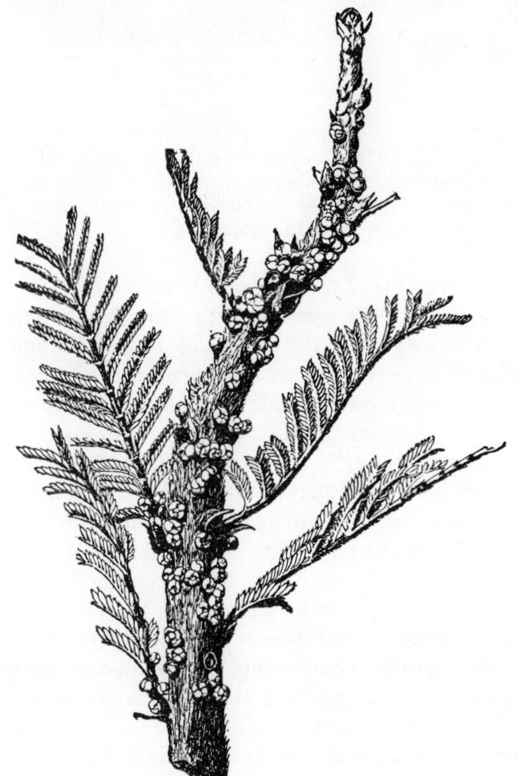

Fig. 217. Branch of a leguminous tree bearing flowers of the parasite *Pilostyles ulei* (Rafflesiaceae). *c.* $\times \frac{1}{3}$. After Goebel.

Many exotic parasites (e.g. the Rafflesiaceae) have become so completely adapted to the parasitic life that their vegetative organs are usually no longer visible externally, and they have ceased to resemble those of a normal cormophyte in structure. Instead they grow as thread-like strands of cells, similar to the hyphae of fungi, within the host plant, from which eventually the strange flower of the parasite breaks forth (e.g. *Pilostyles*, Fig. 217). The flower of *Rafflesia arnoldii*, reaching a diameter of 1 m and the largest of all flowers, emerges in this way from its host, a species of *Cissus* (Vitaceae) native to Sumatra.

A European parasite is *Cuscuta europaea* (Fig. 218). Although its pale-yellow shoots which are furnished only with minute scale leaves, contain merely a trace of chlorophyll, the plant still gives evidence of its evolutionary relationship to normal plants assimilating carbon dioxide. Its much reduced root dies at an early stage, but the stem of the seedling immediately elongates to a long thin thread. Its free end circumnutates continuously and it can thus find a host plant growing in the vicinity. If no host plant can be reached from the place of germination the seedling may creep along the ground for a short distance.

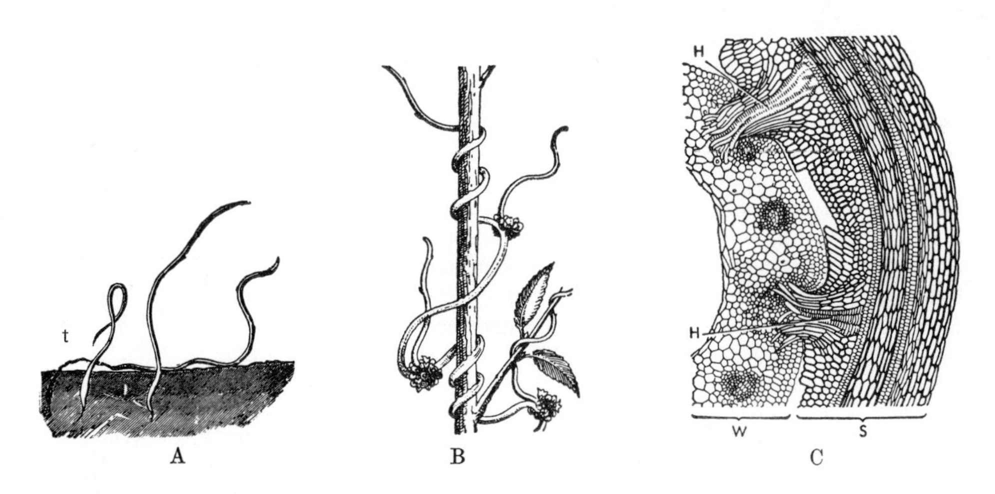

Fig. 218. *Cuscuta europaea.* **A.** Seedlings: the longest is growing along on the soil to the right and the lower end (t) to the left is dying away. **B.** A willow twig encircled by the parasite. **C.** Transverse section through an encircled stem of the host (W) with a portion of the stem of the parasite (S) in longitudinal section: H, haustorium. **A.** and **B.** $\times\frac{1}{2}$. After Noll. **C.** $\times 30$. After King.

This is made possible by the posterior end dying (Fig. 218A, h), and releasing nutrients which are utilized by the elongating anterior end. If the free end of the thread-like stem eventually encounters an appropriate host (e.g. a stem of *Salix* or *Urtica*) it winds around it in the same way as the stem of a twining plant. After a short time emergences, at first papillose, develop from the epidermis of the parasite on the side appressed to the host but they do not penetrate its tissues. If these processes find favourable conditions they are then very rapidly replaced by special absorptive organs the haustoria (Fig. 218C, H). These emerge from the interior of the parasite, perforate the cell walls of the host and, by causing the dissolution of the middle lamellae, are able to penetrate deeply into its tissues, eventually reaching the vascular bundles. At the same time individual files of cells, resembling fungal hyphae, grow out freely from the body of the haustorium and force their way into the delicate parenchyma of the host. Where an haustorium has forced its way as far as a vascular bundle conducting elements differentiate in it and join themselves on to the xylem and phloem of the host (Fig. 218C). The parasite is then able to draw from the host both water for transpiration and all the organic and inorganic nutrients it requires.

The seeds of the parasitic *Orobanche*, also native with us, do not usually germinate until they are in contact with the roots of the host. The only part of the plant to appear above the soil is the flowering shoot bearing bright-yellow, reddish-brown or amethyst-blue scale leaves; the shoot resembles that of asparagus. They emerge from the interior of a root tuber which is connected underground by an haustorium with a root of the host. The species of *Orobanche* also still contain a few plastids with chlorophyll. Both *Cuscuta* (dodder) and *Orobanche* (broom-rape) are dreaded by the agriculturalist, since they damage cultivated plants and are difficult to eradicate.

The flowering shoots of some of the orchids (e.g. *Neottia, Corallorhiza, Epipogium*) and of *Monotropa* (Pyrolaceae), which grow in humus of woodland soils, have a habit

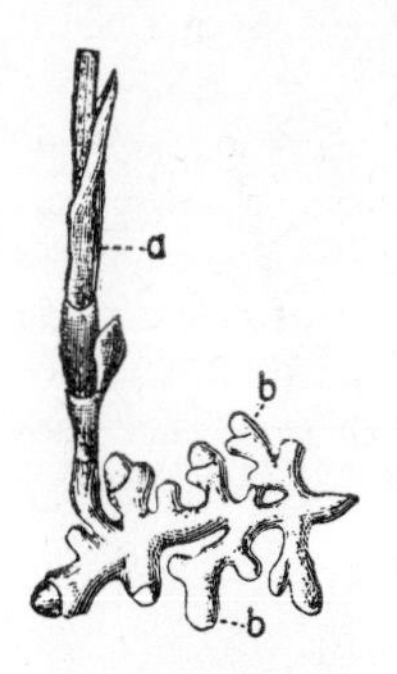

Fig. 219. *Corallorhiza trifida* (Orchidaceae): a, flowering shoot; b, primordia of new branches on rhizome. ×⅔. After Schacht.

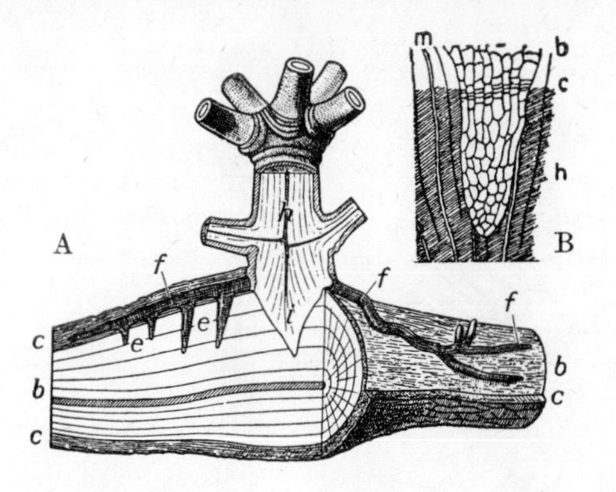

Fig. 220. A. Base of the stem of mistletoe (*Viscum album*), its xylem (h) inserted into that of the host. For further details see text. **B.** Longitudinal section of the tip of the mistletoe stem base in the host xylem h, (cross-hatched): b, phloem; c, cambium and meristematic zone of the stem base; m, pith ray. ×30. After Solms-Laubach.

similar to that of *Orobanche*. The scarcity or complete lack of chlorophyll and the reduction of leaves to scales indicates that these forms also obtain organic materials from without. In this case it is from fungal hyphae, upon which they are parasitic (see mycorrhiza, p. 294). In *Corallorhiza* and *Epipogium* even roots are lacking (Fig. 219).

A well-known native parasite is the epiphytic mistletoe (*Viscum*; Fig. 220), rendered particularly conspicuous in winter on its deciduous host trees (lime, poplar apple, pear etc.) by its winterhardy green leaves. The main stem (Fig. 220A, a) gives rise at its base to green cortical roots (cortical absorbing strands; Fig. 220A, f), which proliferate in the cortex of the host. Both the main stem and these cortical roots send out peg-like processes (Fig. 220A, i, e) perpendicular to the xylem of the host. These processes take up water and minerals. They also possess an intercalary meristem (Fig. 220B, c) at the same level as the cambium of the host, so that they elongate in step with, and become enclosed within, the secondary thickening of the stem.

Some herbs rooting in soil are also semi-parasites. Their roots are connected by small haustoria with the roots of their green hosts (often species of Gramineae). Examples are *Thesium* (Santalaceae) and, among the Scrophulariaceae, *Euphrasia, Rhinanthus, Pedicularis* and *Melampyrum*.

b. Carnivorous plants. Plants with specialized nutrition also occur on soils poor in minerals, especially those poor in nitrogen (e.g. the peat of raised bogs and volcanic ash). They assimilate carbon dioxide with their normal green leaves and can live completely autotrophically, but are in addition provided with adaptations enabling them to capture and imprison small animals, predominantly insects, which they digest and utilize as a complementary source of nutrition (see p. 296). These are the carnivorous or insectivorous plants. The adaptations for the capture of the animals are many and various.

The leaf of *Drosera* (Fig. 221) bears emergences (the tentacles) resembling the horns of a snail, each one penetrated by a vascular bundle. Their glandular heads excrete a glistening droplet of a sticky liquid, with a fragrance somewhat resembling that of honey, which attracts small animals. These remain attached to the glands and in their attempts to escape they come into contact with still more glands and so become even more firmly held. As a result of the stimulation so caused, the tentacles bend themselves towards the middle of the leaf (Fig. 221A), whereby the leaf becomes somewhat concave and the insect is enclosed. In *Pinguicula*, also a native plant, small animals are firmly caught on the sticky heads of glandular hairs on the upper side of the leaf.

In European species of *Utricularia*, occurring submerged in standing water, the leaves are finely dissected and some of the tips are transformed into small green bladders (Fig. 222). The upper side of the tip forms the inner surface of the bladder. The bladders are water-filled animal traps. Each possesses a small mouth, closed by a valve-like lid, and at first impassable by water. The lid opens elastically, but only inwards (Fig. 222B). If small

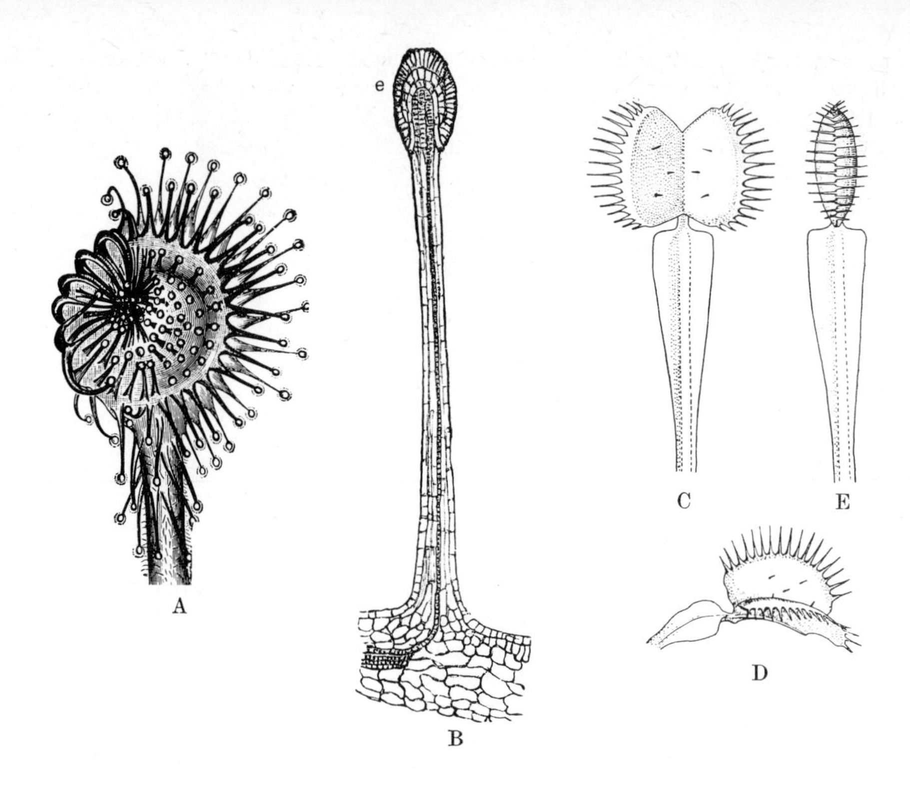

Fig. 221. Insect-catching leaves of the Droseraceae. **A.** and **B.** *Drosera rotundifolia*. **A.** Leaf from above, some of the tentacles incurved over a trapped insect. **B.** Tentacle, showing the glandular epithelium (e). **C.**, **D.** and **E.** *Dionaea muscipula*. Leaf trap open (C.) and closed (E.). **D.** Side view. **A.** ×4. After Darwin. **B.** ×60. After Strasburger. **C.**, **D.** and **E.** Natural size. Original.

Fig. 222. **A.** to **F.** *Utricularia vulgaris*. **A.** Portion of a leaf with five bladders. **B.** Bladder in longitudinal section. **C.** Bladder seen from the front. **D.**, **E.** and **F.** Development of a bladder from a leaf tooth. **G.** *Nepenthes rafflesiana*. Pitcher leaf. **A.** Natural size. **B.** and **C.** ×10. **D.**, **E.** and **F.** ×20. All original. **G.** ×½. From Wettstein.

animals push against the bristles projecting from the outside of the lid they act as levers and the lid is opened. The animals are then sucked by a stream of water into the bladder. The engulfing process is brought about by the release of tension in the walls of the bladder, which are initially under a negative pressure, and in consequence strongly indented, as the result of a cohesion mechanism (Fig. 328). The lid now springs back into its original position and again firmly closes the trap, thus preventing loss of the prey. At this stage hairs on the inner side of the walls of the bladder excrete a digestive fluid, and in turn resorb the digested materials from solution into the bladder. As a result of this resorption, and also of the excretion of water through the walls of the bladder to the outside, a negative pressure is regenerated within the cavity and in about fifteen minutes the trap is reset.

The trapping mechanisms of the exotic carnivorous plants are even more impressive and effective. The two halves of the lamina of *Dionaea* (the Venus fly trap), native on damp, sandy, grassy areas in the Carolinas, come together with amazing rapidity when the 'sensitive hairs' (see p. 371) are touched, and an insect settling thereon is immediately imprisoned. The lamina is toothed at its margins like a gin trap and is densely covered on its upper side by digestive glands (see Fig. 221C, where the leaf of *Dionaea* is shown in the open, 'sprung', condition). The very rare, submerged, rootless water plant *Aldrovanda* (also Droseraceae), native in Europe, possesses leaves similar in construction to those of *Dionaea*, but very much smaller.

In *Nepenthes, Cephalotus, Sarracenia* and the related *Darlingtonia* hollow jug- or urn-shaped (ascidiform) leaves serve as traps for animals. The ascidiform leaf of *Nepenthes* is usually surmounted by an immovable flap (Fig. 222G). The urn is derived from the lamina of a leaf, while the leaf base is flattened and broadened. The petiole lying between these two parts may, when stimulated by contact, react like a tendril, and in this way the heavy urn is held firmly among the branches. Lying within the urn of *Nepenthes*, whose inner side corresponds to the upper side of the leaf, is a watery fluid excreted by glands on the wall. Animals are attracted by the mottled colour of the urn and by the excretion of nectar from glands on its very smooth margin. When they alight on this margin they slide and fall into the fluid within, and they are prevented from climbing out by a waxy layer on the upper part of the inner wall. They subsequently drown and are digested.

Section five: **Reproduction**

A fundamental property of all life is its productivity. It expresses itself in growth, cell multiplication, and in the propagation of the species. While growth, by the accumulation of assimilated organic substances, leads principally to an enlargement of the individual, reproduction ensures the maintenance of the species over and above that of the individuals composing it.

Theoretically the maintenance of the species would require only that each individual leave a single descendant. Usually, however, each individual generates numerous progeny during its life-span, so that the representation of a species can sometimes increase enormously. It has been estimated, for example, that a giant puff-ball (*Calvatia gigantea*) produces more than 7×10^{12} spores, each of which can yield a new mycelium. The parent mycelium can bear several of these football-size fructifications in the course of a year (see p. 532).

I. Forms of reproduction

A prerequisite of all reproduction is either that the organism is uniformly divided into physiologically equivalent parts, or that the division is into unequal parts, the smaller of which then show special properties of regeneration. In the extreme, and frequently occurring, form of the latter situation, the smaller parts are single cells.

Such single-celled propagules may behave in two ways Either they grow immediately into new individuals genetically identical with the parent (agametic reproduction), or, before growth is possible, they must fuse (gametic reproduction, syngamy). In the latter case meiosis in the parent frequently leads to two kinds of gametes, termed either + or −, or male (♂) and female (♀), and only in rare cases can these develop immediately into new plants (apogamy). Their further development is normally dependent upon a male and a female cell fusing (copulation, plasmogamy). Sooner or later the two haploid nuclei also fuse, forming the diploid zygote nucleus (fertilization, karyogamy). Only the diploid zygotic cell is in a condition to give rise to the new generation. In the unicellular Protophyta the initial stage of sexual reproduction thus does not involve a multiplication of individuals, but rather a reduction. Sooner or later however (in the pure haplonts already in the germination of the zygote) this is compensated for by the formation of the haploid gonospores (or meiospores) at reduction division. The cycle recommences with their gametic fusion. In species in which both the haploid and diploid phases show somatic development or where somatic development is confined to the diploid, the number of meiospores formed as a consequence of a single fertilization may amount to many millions (e.g. the puff-ball *Calvatia gigantea*, and the massive production of pollen by a pine tree).

In summary, three different forms of reproduction can be distinguished

A. Vegetative reproduction by partition and fragmentation
B. Asexual reproduction by means of special cells
C. Sexual reproduction by means of syngamy and meiosis

A. Vegetative reproduction by partition and fragmentation

1. Vegetative reproduction of unicellular organisms

Numerous unicellular Protophyta show the simplest possible method of reproduction, namely binary fission (schizotomy: e.g. *Ankylonoton*, Fig. 403). Repetitive division can also occur (schizogony: e.g. *Chlamydomonas*, Fig. 408B; *Chlorococcum*, Fig. 411). Here the protoplast detaches itself from the cell wall of the mother cell and divides into a large number of daughter individuals. The mother cell then breaks open, and all the daughter cells, which rapidly grow to the size of the parent, are released at the same time. Also in reproduction by budding (e.g. *Saccharomyces*, Fig. 37) the daughter cells although at first much smaller than the parent, soon reach the same size.

The interval between the formation of a set of daughter individuals, and those individuals themselves reaching the condition in which they are about to divide is termed the generation time. The smaller an organism, the shorter its generation time. In many bacteria, in optimal culture conditions, it amounts to less than twenty minutes. The reproductive potential of such minute organisms is clearly enormous. Theoretically, such a bacterium in the course of a single day could produce 2^{72} (over 4×10^{21}) descendants. In a few days the volume of the biomass so generated would surpass that of the Earth itself.

The totality of the descendants of a single individual (produced vegetatively) is termed a clone. All the individuals of a clone, assuming no induced or spontaneous mutations, are genetically identical.

2. Vegetative reproduction of multicellular organisms

Simply organized coenobia reproduce themselves in suitable conditions by fragmentation (e.g. *Spirogyra, Plectonema, Oscillatoria*). In more complex coenobia fragmentation may occur adjacent to special cells (e.g. the heterocysts of *Nostoc*, Fig. 384E). Even the thalli of many highly organized marine algae and lichens (e.g. *Caulerpa, Fucus, Cladonia* spp.) reproduce themselves freely by simple fragmentation. This typically plant-like method of reproduction is also found not uncommonly amongst land plants, from the bryophytes to the flowering plants.

The liverworts *Scapania nimbosa* and *Sc. ornithopodioides*, in which sex organs have never been recorded, have probably propagated themselves by fragmentation for millions of years. In this respect, therefore, they resemble some of the oldest forms of life.

Many perennial herbs multiply themselves by the older parts of their branch systems rotting away, the younger branches thus becoming separated from each other and independent (e.g. *Convallaria, Anemone, Fragaria*). Alexander Braun observed with justice that in respect of such plants the concept of an 'individual' is hardly appropriate; they should instead be regarded as 'dividuals'.

In Europe, for example, *Elodea canadensis* has multiplied exclusively by fragmentation since its introduction in the middle of the last century, since only the female plant has so far reached here. Nevertheless, so effective has been its spread that it has occasionally seriously interfered with navigation and fisheries.

Vegetative propagation by means of grafting, runners, layering or cuttings is frequently utilized in the cultivation of economic and garden plants (e.g. roses stone fruits, potatoes, strawberries, pineapples, bananas, sisal (*Agave*), Lombardy poplars).

3. Vegetative reproduction by means of propagules

Several- or many-celled propagules (gemmae, bulbils, leaf-buds) are already encountered in the brown and red algae (e.g. *Sphacelaria* and *Plumaria* spp.). They are quite common in the liverworts (e.g. *Marchantia*, Fig. 529) and mosses (e.g. *Tetraphis pellucida*, Fig. 549J″), and occasionally occur in ferns and seed plants.

Frequently segments of shoots or side shoots, recognizable from their structure as reproductive bodies (propagules), serve for the tiding over of periods unfavourable for growth (e.g. potato tubers, turions of *Utricularia, Hydrocharis, Stratiotes*, and other water plants). In other instances swollen condensed shoots (bulbils) arise on the aerial part of the plant in leaf axils (e.g. *Cardamine bulbifera*, Figs. 223A, 224B; *Lilium bulbiferum*), in

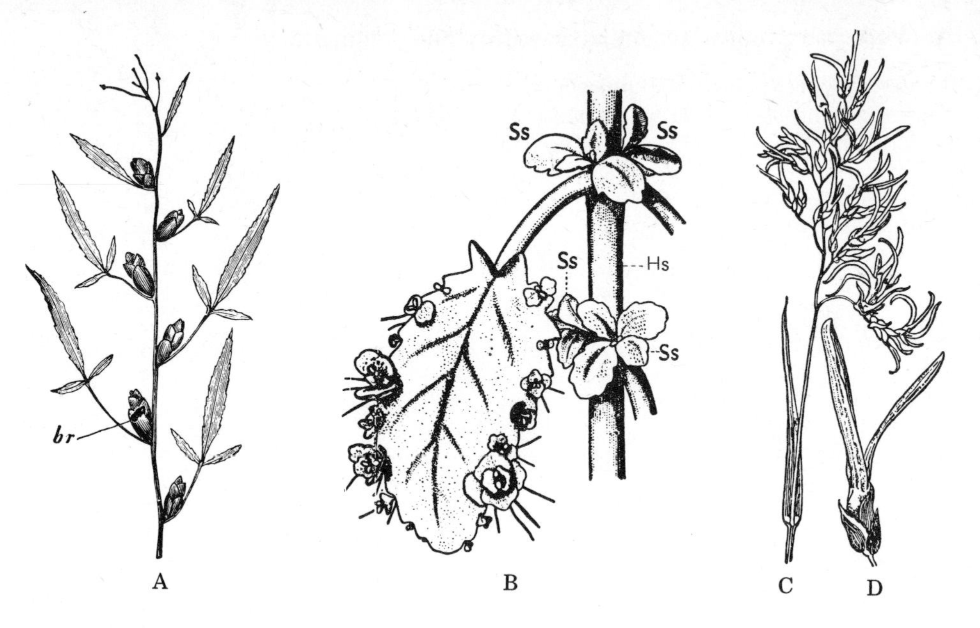

Fig. 223. Vegetative multiplication by means of bulbils and offsets. **A.** *Dentaria bulbifera*: br, bulbils. (Cf. Fig. 224B). **B.** *Bryophyllum calycinum* (Crassulaceae). Development of vegetative offsets from the notches of the leaf margin. Hs, main shoot; Ss, lateral shoots. **C.** and **D.** *Poa bulbosa* var. *vivipara*. **C.** Inflorescence with bulbils (pseudovivipary). **D.** Single bulbil developed from a proliferated spikelet. **A.** After Schenck. **B.** After Troll. **C.** and **D.** After Dalitsch. All natural size.

the inflorescence (e.g. *Allium* spp.), above the veins of the leaf (e.g. *Asplenium bulbiferum, A. viviparum*), or at the leaf margin (e.g. *Ceratopteris thalictroides*, and the plant made famous by Goethe *Bryophyllum calycinum* (Fig. 223B)). While still attached to the parent plant these propagules throw out adventitious roots and develop into small plantlets. Subsequently, separated from parent plant by an abscission layer, they fall to the ground and are able to continue growing without pause.

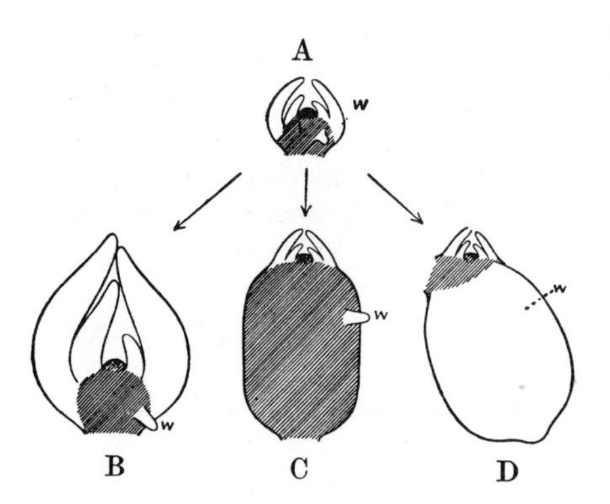

Fig. 224. Different types of bulbils and their derivation (diagrammatic). **A.** Axillary bud with primordium of an adventitious root (w). The axial portion is cross-hatched and the growing point black. **B., C.** and **D.** Bulbils formed by development of different portions of the bud. Type **B.** is shown by *Lilium bulbiferum* and *Cardamine bulbifera*, **C.** by *Polygonum viviparum* and **D.** by *Ranunculus ficaria*. *c.* ×3. After Troll.

In many grasses (e.g. *Poa bulbosa* var. *vivipara*, Fig. 223C) the glumes of the spikelets develop as small leaves, and the whole inflorescence is transformed into a great number of minute shoots. As soon as these come into contact with the soil adventitious roots arise, and they are thus an effective means of vegetative reproduction. Numerous other grasses under severe conditions (especially in the extreme north) adopt vegetative multiplication by bulbils. This spurious form of 'vivipary' should not be confused with the true vivipary of certain mangroves, in which the seed germinates on the mother plant and forms a relatively large seedling before being shed.

B. Asexual reproduction by means of special cells

Numerous algae and fungi actively reproduce themselves by special cells formed as a consequence of normal mitosis. For historical reasons these have been described as various kinds of spores or conidia. Unfortunately these two terms have been applied to reproductive cells which, in relation to their positions in the life-cycle, are by no means equivalent. As we shall see in a moment in considering sexual reproduction, quite similar unicellular spores can also arise as products of meiosis. In all instances in which the chromosomal relationships have been clearly established, it is therefore advisable to refer to those spores which owe their origin to mitosis as mitospores, thus distinguishing them from the meiospores formed by a process of meiosis.

Mitospores can be formed in both the haploid and diploid phases of the life-cycle. They may be furnished with flagella and motile (e.g. *Botrydium*, Fig. 406B; *Saprolegnia*, Fig. 468A, B), or they may be passively distributed by water (e.g. carpospores of *Ceramium, Chlorella* (Fig. 412C)) or by wind (e.g. *Aspergillus*, Fig. 482). They may be cut off externally (e.g. *Cunninghamella*, Fig. 475), but more frequently they are formed within special chambers (mitosporangia), as in *Mucor* (Fig. 473). Mitospores are usually uninucleate, but occasionally binucleate (e.g. uredospores of *Phragmidium*, Fig. 506), or even multinucleate (e.g. *Vaucheria*, Fig. 407B; *Sporodinia*, Fig. 474).

In a few rare instances (as an adaptation to the transition from an aquatic to a terrestrial existence) whole multinucleate mitosporangia become detached and distributed by wind (e.g. *Plasmopara*, Fig. 470B).

Mitospores of land plants are usually surrounded by thick walls, and thus particularly resistant to desiccation. They are thus well suited for enduring conditions unfavourable to life, and also for distribution by wind and animals (anemochory and zoochory).

C. Sexual reproduction

While in the vegetative form of reproduction considered so far the formation of all the reproductive cells (propagules) involves simple mitosis, so that they are necessarily genetically identical with their parents, in sexual reproduction the new generation arises from the fusion of two or more haploid gametes (syngamy). Thus a daughter or filial generation (F_1) produced sexually regularly contains in its diploid set of chromosomes two sets of inherited information from its genetically different parents. Before sexual reproduction can recur, the initial haploid condition must be restored (change in phase of the nucleus, see p. 42). In the four meio- or gonospores produced in this process the parental genes are commonly recombined, so that any chance variations (mutations) become reassorted, as playing cards in shuffling. The development of sexuality, with its conjoined phenomena of copulation and meiosis, thus had great significance in relation to evolution. On the basis of a relatively few mutations, as a consequence of the presence of homologous maternal and paternal genes and the possibility of extensive re-assortment, a wide range of genetically different individuals can arise and be subject to natural selection.

1. Meiospores

Since syngamy and meiosis are interdependent processes, it is inappropriate to refer only to the gametophytic generation as the sexual, and to regard the diploid, bearing the meiospores, as the asexual. The term 'asexual reproduction' should be reserved for vegetative reproduction, as described in sections A and B, irrespective of whether it concerns the haploid or diploid phase of the life cycle.

It is frequently difficult to distinguish meiospores from mitospores. Often both are produced in equally large numbers, and both may serve in an apparently identical manner to propagate and distribute the species. The distinction can most readily be made, without prolonged cytological investigation, where the meiospores arise in tetrads so revealing their meiotic origin (see, for example, Figs. 445; 565B).

The meiospores of aquatic species may, like the mitospores, be flagellate and freely notile (e.g. *Oedogonium*, Fig. 411G). In the Myxomycetes the meiospores are thick-walled and distributed by wind, but on settling each yields a flagellate cell, and this in turn sheds its flagella and becomes amoeboid. Meiospores may also, like mitospores, be cut off externally (e.g. in the Basidiomycetes, Fig. 499C). More frequently, however, they arise within special containers, which, analogous to the mitosporangia are termed meiosporangia (e.g. the so-called tetrasporangia of many Phaeophyceae and Rhodophyceae; Figs. 436D; 445A). In the Ascomycetes the four meiospores usually each divide yet again, so that the mature, so-called ascus comes to contain not four, but eight meiospores, often in a linear arrangement (polymeiospores, Fig. 493E). In some instances the divisions may proceed even further, and the number of meiospores become correspondingly greater (e.g. *Chorda filum*, Fig. 438D; *Polyphagus euglenae*, Fig. 462F, G). In the heterosporous archegoniate plants we encounter two kinds of meiosporangia containing different kinds of meiospores, namely the megasporangia containing a few large megaspores, and the microsporangia with many small microspores (e.g. *Selaginella*, Fig. 565). In the seed plants three of the four megaspores arising from meiosis customarily degenerate, so that the mature megasporangium contains a single large megaspore (Fig. 608F). For historical reasons the megasporangium of the seed plants is termed the micellus.

2. Gametes

In only very few plants are the gametes formed directly in meiosis (meiogametes), although this is commonly the situation in animals (gametic meiosis, gametic change of nuclear phase (see p. 42)). Examples are provided by the diatoms, *Fucus*, and some green algae (*Acetabularia, Codium*, among others). Much more frequent, and typical for plants is the interpolation of a few- or many-celled gametophyte generation between meiosis and gametogenesis (Fig. 33). The gametes are then formed by simple mitosis and can be referred to as mitogametes to distinguish them from meiogametes.

In the simplest case, represented in many algae and lower fungi, the gametes are naked, and of equal form and size. They do, nevertheless, differ physiologically. Since no morphological distinction as male and female is possible they are designated as + (the male or donor gamete) and – (the female or receptor gamete); because of the similarity of form, however, this kind of syngamy is referred to as isogamy (Fig. 225A).

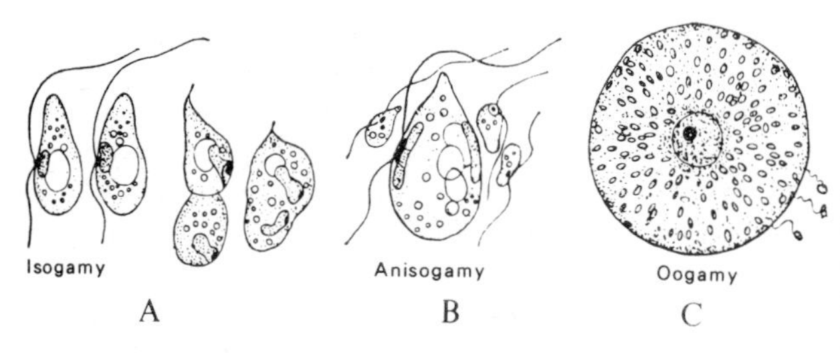

Fig. 225. Developmental series from isogamy (**A.**) through anisogamy (**B.**) to oogamy (**C.**) in the Phaeophyceae. **A.** Isogamy, as in *Sphacelaria bipinnata* and *Cladostephus spongiosus*. **B.** Anisogamy, as in *Nemoderma tingitana*. **C.** Oogamy, as in *Dictyota dichotoma*. **A.** and **B.** ×400. After Kuckuck. **C.** ×120. After Williams.

The naked, flagellate, freely motile isogametes of the lowest forms are often morphologically indistinguishable from the haploid mitospores or meiospores of the same species (e.g. *Olpidium*). In many species these two kinds of motile propagules are not fundamentally different, so that a haploid mitospore can behave in certain conditions as a gamete (facultative sexuality; e.g. *Chlamydomonas eugametos*, many lower fungi).

As distinct from the haploid mitospores, which can develop immediately to new haploid plants, the gametes, in response to a reciprocal chemotactic attraction (see p. 382), pair and fuse, so forming diploid zygotes. The cells or groups of cells which generate the gametes, or within which they are formed, are termed gametangia.

In the pennate diatoms (Fig. 402) and Conjugales (Fig. 425) copulation is between naked gametes, one of which lies in its parent cell (which has become a gametangium), while the other moves in an amoeboid manner towards it.

Not infrequently the gametes of different sex are of unequal size (anisogamy, Fig. 225B). Anisogamy is already encountered in the unicellular protophytes (e.g. *Chlamydomonas braunii*), and is widely distributed in the more highly organized algae (e.g. *Entero-*

morpha intestinalis, Fig. 418) and lower fungi (e.g. *Allomyces*, Fig. 466). The larger gametes in this instance are regarded as female, and the smaller male.

If the larger gamete remains completely non-motile, then by analogy with the situation in the animals, the female gamete is called the egg, and the smaller the spermatozoid (Fig. 225C). Spermatozoids are again attracted chemotactically to the egg cell. The female gametangium is now called the oogonium, and the male gametangium the antheridium (oogamy; e.g. *Chlorogonium oogamum*, Fig. 409; *Monoblepharis*, Fig. 467).

In the Protophyta, as well as in many lower algae and fungi, the oogonia and antheridia are single cells of the gametophyte in which the eggs and the spermatozoids are produced singly (e.g. *Oedogonium*, Fig. 421E) or severally (e.g. *Fucus*, Fig. 442). In aquatic forms the naked egg is often expelled from the oogonium (e.g. *Chlorogonium*, Fig. 409; *Laminaria*, Fig. 439). With higher development fertilization is usually, however, in the oogonium (e.g. *Chara*, Fig. 428). The spermatozoids, which are flagellate and freely motile, are attracted to the egg by various organic substances it secretes of chemotactic effect. As an adaptation to terrestrial life the gametangia of mosses and ferns are surrounded by a special sterile sheath. The flask-shaped female organ, which contains a single egg, is called the archegonium (e.g. *Marchantia*, Fig. 530J; Filices, Fig. 604B).

3. The zygote

The product of the fusion of the two gametes of different sex is termed the zygote. In the primitive algae and fungi the zygotes may be naked and flagellate as the gametes and zoospores (e.g. *Ulothrix*, Fig. 416G; *Allomyces*, Fig. 466E). Usually, however, the zygotes are non-motile. Indeed, they frequently become enclosed in a thick wall. This enables them to survive without damage such unfavourable conditions as drought and frost, and serve as a means of distribution (zygospores; e.g *Chlamydomonas*, Fig. 408F).

In the diatoms, as a consequence of their inextensible, siliceous, cell wall, asexual reproduction leads to a progressive reduction in cell size. The wall of the zygote, however, is extensible, and this cell grows to several times its initial volume before dividing, so providing new asexually-dividing individuals of restored size. Sexual reproduction thus has an important regulatory function in the life cycle (auxozygote or auxospore, (Fig. 401).

In many groups of fungi sexual reproduction involves the copulation of whole gametangia (gametangiogamy, e.g. *Saprolegnia*, Fig. 468D; *Peronospora*, Fig. 471), leading to the production of multinucleate zygotes (coenozygotes, e.g. *Mucor*, Fig. 476H). In the Eumycetes copulation is simply between vegetative cells (somatogamy, Fig. 513.3). In this instance the fusions of the plasmas and of the nuclei is separated by a shorter or longer dikaryon phase. Fusion of the nuclei is followed immediately by meiosis. The zygote becomes the meiosporangium. The primordia of the asci and basidia (Fig. 499B) are accordingly the zygotes of the Eumycetes.

D. Alternation of generations

An alternation of generations is characteristic of the life cycle of most of the more highly developed plants. Typically this is an alternation between two generations differing in their modes of reproduction, and often consisting of quite differently constructed and independent individuals. As a rule, an alternation of generations is accompanied by a simultaneous change in the phase of the nucleus (see p. 42).

The life cycle of a fern may be taken as a classical example of such an alternation of generations with accompanying nuclear change. The diploid, leafy 'fern', the sporophyte, produces haploid spores (meiospores or gonospores) by a process of reduction division. When these germinate they give rise not to a new leafy 'fern' but to a small haploid thallose structure, known as a prothallus (Fig. 88), upon which the gametes arise. Consequently, the prothallus is the gametophyte. Only when the fertilized egg cell develops is there a new diploid leafy sporophyte produced (Fig. 531). Thus a haploid gametophyte, giving rise to gametes, alternates with a diploid agamic generation which produces spores (antithetic or heterophasic alternation; see p. 42).

In many of the sexually reproducing lower plants there is an alternation in nuclear phase, but the haploid and diploid generations are externally quite similar. Consequently, an alternation in nuclear phase need not always be accompanied by a morphologically identifiable alternation of generations. On the other hand, there are several remarkable cases known in which from the vegetative cells of one generation the other can develop directly, and therefore with the same chromosome number In a form of the fern *Athyrium filix-femina* the prothallus is diploid and the sporophyte develops asexually from its vegetative cells. Meiosis and the production of spores do not occur, but instead diploid prothalli are produced, again asexually, from the diploid cells of the leaf margin (apospory). Also a sporophyte may arise directly and asexually from a single or from a group of haploid cells of a prothallus, as, for example, in *Dryopteris molle* (apogamy). Furthermore, a diploid gametophyte can be obtained experimentally by regeneration from fragments of the sporogonium of a bryophyte, which being sporophytic tissue is diploid. Such a gametophyte produces sexual cells capable of fertilization, and on fertilization they yield tetraploid capsules. Even tetraploid gametophytes can be obtained by repeated regeneration (see Fig. 26C). Related diploid, triploid, and tetraploid gametophytes and sporophytes are known in *Osmunda regalis* and *Woodwardia virginica*. In the algae and fungi there are numerous deviations from the normal cycle. There is therefore no evidence of a direct connection between the chromosome number and the morphology of the generation containing it, nor with the way in which it reproduces although generally there is a well-established relationship.

In bryophytes and seed plants the two generations are not, as in most ferns, independent individuals, but one generation remains permanently attached to the other and draws its nutrition from it, as parasite from its host.

While in the bryophytes the sporophyte (sporogonium) grows at the expense of the gametophyte, in the seed plants the reverse is the case. Here the female gametophyte, known as the embryo sac, remains closed in the tissue of the sporophyte throughout its life. Sexual reproduction in these plants is completed in the flowers, where there is first meiosis leading to the production of the spores and subsequently fertilization of the egg.

Part two: **Physiology**

While morphology is concerned with the form and structure of plants, the task of physiology is to investigate their particular life processes and the functioning of their organs. Physiology attempts to reveal the mechanisms of such typical vital phenomena as growth, development and reproduction, and those of irritability and movement. It also seeks to discover the nature of the forces of growth, and to detect causal relationships in the functioning of the plant. The ultimate aim of physiology therefore is to arrive at a detailed understanding of that phenomenon commonly known as 'life'. The physiologist's approach involves to an increasing extent the exact experimental methods of physics and chemistry, since there is no doubt that physical and chemical processes and laws play a fundamental role in the living substance, the protoplasm. Whether it will ever be possible to explain life completely in terms of a complex physico-chemical system need not be considered further here.

Plant physiology can conveniently be subdivided into three major fields, namely metabolism, development and movement.

The physiology of metabolism studies those physical and chemical changes in the cells and organs of plants which are inseparable from the phenomenon of life. Indeed, new materials must be constantly incorporated into an organism so that they can be used for its growth and reproduction. The production of the appropriate substances from unspecific food materials, and their incorporation into the structure of the plant, require profound chemical changes and syntheses. Some of these can take place only if energy is supplied to the system. Many other living processes also require energy, and usually the organism can obtain this only by the chemical breakdown of energy-rich foodstuffs (catabolism). The study of the complex chemical reactions occurring in the protoplasm throws a penetrating light on some of the fundamental phenomena of life, and it is therefore the principal concern of the physiology of metabolism.

The physiology of development occupies itself with the phenomena of growth, development and reproduction. Its goal is to try to understand the causal mechanisms underlying those problems of form and structure which morphology treats in a purely descriptive and comparative way. Particularly important contributions to this field are being made by contemporary genetics, a discipline which today forms a science in its own right and which can be touched on only briefly in this book.

The physiology of movement investigates the changes of place or position of whole plants or single organs. While in the lower plants, as in animals, there is often free movement from place to place, the higher plants are firmly rooted to one spot. They are nevertheless capable of many different kinds of movement, some of which bring about a definite orientation of their organs in space, and others serve different purposes Particularly striking in this connection is the phenomenon of irritability, and its physiology is often considered in isolation. Nevertheless, irritability is a quite general property of living things, and it also plays a role in development and metabolism. It must be made clear in fact that the subdivision of physiology proposed earlier is only for the sake of convenience of description. In reality all three aspects interact in a reciprocal manner. The phenomena of growth and movement, for example, cannot occur without simultaneous physico-chemical changes in the organism in question. For this reason our discussion of the physiology of plants must begin with the physiology of metabolism.

Section one: **Physiology of metabolism**[2]

I. The general composition of the plant body[3]

Knowledge of the chemical composition of the plant body is of fundamental importance for anything other than a superficial acquaintance with the physiology of metabolism. Although a mere knowledge of the chemical compounds present in a plant gives no information about the composition of the individual components of the cell, such as the nucleus and chromosomes, and the plastids, nevertheless it does reveal at once which substances are generally present in the body of the plant and the elements from which they are constructed. It also reveals which basic materials must be available to the living organism in its nutrition if it is to increase its mass by growth, or even merely to maintain itself in the same condition.

Water content. In all plants by far the greatest part of the fresh weight of the living material consists of water.

Thus the naked masses of protoplasm of certain slime fungi contain 70–94 per cent water, herbaceous parts of higher plants, e.g. leaves, 80–90 per cent and certain sappy fruits, such as the cucumber, even up to 95 per cent. Many algae contain 98 per cent water. Opposed to these, the water content of wood is about 50 per cent and that of dry seeds very small (13–14 per cent). Seeds are in fact an excellent example of how saturation with water is an absolute requirement for the regular cycle of metabolic processes and for that reason is general among living things. The small amount of water in seeds causes them to be in a resting condition, and before they can enter active life and germinate they must absorb large quantities of water.

Dry matter. The dry matter of the plant remaining after careful drying at something over 100°C contains a multitude of organic compounds. Some of these are to be regarded as providing the structure and framework of the plant, while others are probably food reserves, or intermediate or end-products of the metabolism. We cannot enumerate here all the substances so far isolated from plants; information about these can be obtained from textbooks of organic chemistry. Carbohydrates, organic acids proteins, fats, lipids, tannins, glycosides, alkaloids and many other classes of compounds occur as constituents of plant cells. Their significance, so far as this is known at present, will be considered to a certain extent later. Chemical analysis shows that all these compounds are built up from only a few elements, in fact from the six basic elements: carbon, oxygen, hydrogen, nitrogen, sulphur, phosphorus. As any piece of charred wood shows, of these elements carbon is present in the greatest quantity (exceeding 50 per cent).

Ash. Besides these basic elements present in the organic compounds, still other elements, amounting to a not inconsiderable percentage, are found in the dry matter. Their oxides form a substantial proportion of the ash that remains when the dry matter is burnt. Thus when the dry matter is heated with free access of air at a high temperature some of the basic elements escape in a gaseous form (e.g. carbon dioxide, water, ammonia, sulphur dioxide), while the oxides or carbonates of a large number of other elements are left behind in the ash. In leaves, e.g. those of tobacco (as in cigars and cigarettes), 10–20 per cent of the dry matter may consist of purely inorganic compounds. In seeds and fruits the ash content usually amounts to only 1–5 per cent of the dry weight. The following table gives what is admittedly only a general survey of the ash contents, and the proportions of the different elements in a number of plants:

Material	Ash as % dry wt	Amounts per 100 parts of ash								
		K_2O	Na_2O	CaO	MgO	Fe_2O_3	P_2O_5	SO_3	SiO_2	Cl_2
Tubercle bacillus	9.56	8.2	11.5	8.6	9.8	?	47.0	10.8	?	1.2
Boletus, fruit body	6.39	57.8	0.9	5.9	2.4	1.0	26.1	8.4	–	3.5
Rye grains	2.09	32.1	1.5	2.9	11.2	1.2	47.7	1.3	1.4	0.5
Apple fruits	1.44	35.7	26.2	4.1	8.7	1.4	13.7	6.1	4.3	–
Carrots	5.47	36.9	21.2	11.3	4.4	1.0	12.8	6.4	2.4	4.6
Potato tubers	3.79	60.1	2.9	2.6	4.9	1.1	16.9	6.5	2.0	3.5
Tobacco stems	7.89	43.6	10.3	19.1	0.8	1.9	14.2	3.5	2.4	3.6
Tobacco leaves	17.16	29.1	3.2	36.0	7.4	1.9	4.7	3.1	5.8	6.7
White cabbage, outer leaves	20.82	23.1	8.9	28.5	4.1	1.2	3.7	17.4	1.9	12.6

As can be seen the greatest percentages in the ash are provided by potassium, calcium and phosphorus. At the same time there also occur sodium, magnesium, iron, silicon and chlorine, and frequently aluminium, manganese, boron, copper and zinc, in addition to a whole number of even rarer elements, not mentioned here, in generally minute amounts. There is in fact hardly a chemical element that has not been found somewhere in a plant.

The analyses vary considerably with the species and its habitat. Thus it is not possible to decide from such analyses whether the elements encountered within a plant at any one time, and especially the quantities in which they occur, are essential for its life, or whether they should be regarded rather as substances absorbed into the plant accidentally. As will be considered in a later section (see p. 231), only careful nutritional experiments can give information on this point. However, to the expert the occurrence of certain plants provides a trustworthy indication of the presence of certain minerals in the soil, since such indicator plants will frequently grow successfully only in soils in which they occur. The violet *Viola calaminaria*, for example, grows principally on soils containing zinc. These facts can be of practical use in estimating the fertilizer requirements of agricultural soils, in the management of woodland and so on, and even in geological prospecting.

II. Water

A. The uptake of water throughout the plant

A plentiful supply of water throughout the plant body is apparently an essential prerequisite for the normal functioning of the vital processes, especially so for the continuous chemical transformations in the cell. Air-dried organs, such as seeds, accordingly show hardly any measurable metabolic phenomena. How the plant acquires the water essential for its well-being, and how it is able to maintain the optimal hydration of its tissues are important and fundamental problems in the physiology of metabolism.

1. Imbibition[4]

If air-dry seeds, such as peas, are covered with water in a glass cylinder it will be seen after some time that their volume has considerably increased; the seeds take up water and swell. This swelling or imbibition is an essential preliminary to the reawakening of the measurable metabolic activity in the hitherto resting seeds. It is, however, a purely physico-chemical process, shown to a similar degree by any dead colloidal substance, such as a piece of gelatin or gum arabic. Imbibition of this kind is merely a reversible increase in volume brought about by the incorporation of water into the material.

Bodies capable of swelling contain elongated macromolecules, such as the thread-like protein molecules of gelatin, and the carbohydrate chains of cellulose, starch, etc. In the air-dry condition they lie so close together that hardly any large spaces, and certainly no layers of air as are present in porous substances, come between them. As soon as they come into contact with water, these macromolecular units attract the water molecules with considerable force, so that the adhesive forces between the molecules or molecular aggregates of such materials are overcome, and they begin to move apart from each other. Films of water form within the material, the size of these 'hydration sheaths' being dependent upon the electrical properties of certain groups of molecules or molecular aggregates. Outstandingly hydrophilic, i.e. attractive to water, are the $-OH$, $=CO$, $-COOH$ and $-NH_2$ groups. On the other hand, some groups, e.g. $-CH_3$ and $-C_6H_5$, having no affinity at all for the dipolar water molecule, are termed hydrophobic. At the beginning of the swelling process, during the binding of the first water molecules (which is facilitated by warmth), the adhesive forces may be extraordinarily large, and a pressure of several hundred atmospheres develops. Substances which swell by imbibing water are used, for example, in forcing apart the skull bones in zoological preparations.

Some substances have a limited expansion, others unlimited. In the former, of which cellulose and starch are examples, an equilibrium condition is reached after a certain amount of water has been taken up. Although the substance has undergone in swelling a molecular loosening, and its molecules or molecular aggregates (also called micelles) have become separated from each other by layers of water, they are still held together here and there by valency-like bonds. Consequently, they form a framework or network, the meshes of which, the intermicellar spaces, are filled with water. Many such substances, e.g. agar-agar, are capable of forming stiff jellies or gels with water at low temperatures, although less than 1 per cent of the substance may be present. (Concerning anisotropic swelling, see p. 378.)

In substances with unlimited expansion, e.g. gum arabic and dehydrated egg albumin, the incorporation of water proceeds further and the substance becomes completely dispersed in the water, forming a solution.

Usually such solutions have a colloidal character, i.e. the individual particles dispersed in the solvent, the molecules or micelles, are so large that they remain suspended only because of their water sheaths. Such a colloidal solution is termed a sol. A change in the electric charge on the sheaths of solvent around the suspended particles in a sol may cause them to flocculate. Occasionally it is possible to transform a sol into a gel, and inversely a gel into a sol. Agar and gelatin gels have peculiar properties. On warming they behave as solutions, and the individual particles are able to move freely, while cooling apparently again causes loose aggregation and bonding, together with the incorporation of water, leading to the formation of a gel, and 'setting'.

The degree of hydration, and the extent of the swelling, is very much influenced by the presence of salts in solution, since their ions carry strong electric charges. They are not only capable of themselves binding water and thus, especially in strong concentration, removing it from gels and colloids (as, for example, in the 'salting out' of solution of egg albumin) but also of directly modifying the water-binding capacity of these substances by affecting their electric charge, either by neutralization or adsorption (see p. 234).

The strength and polarity of the charge, and the radius of the ion itself and of its hydrated shell all play an important part in determining the influence of ions on swelling. Since in biological material the most important of the constituents of cells capable of imbibing water (e.g. the protein of the cytoplasm) usually bear negative charges, it is understandable that cations, by neutralizing the charge, usually reduce the hydration, i.e. the swelling is usually weaker in the presence of cations than in pure water. Anions, on the other hand, by becoming adsorbed on to the particles of the gel, increase its capacity for swelling still further. Divalent ions have a greater effect than univalent.

If, as is usual in a physiological situation, there are several salts in solution at the same time the individual ions can act in opposite directions. Thus, for example, calcium reduces the hydration of gels to a greater extent than potassium, so when both are present an increase in calcium will reduce the hydration, whereas an increase in potassium will increase it. In a mixture of the two ions, therefore, the extent of the swelling of the gel will depend upon their relative concentrations (ionic antagonism). Such antagonisms occur between other ions, and in other phenomena (see p. 232). Naturally the behaviour of different gels towards salts varies in detail, depending upon the chemical composition of the gel and the nature of its charge.

The phenomenon of imbibition, here described only in outline, certainly plays an important role in the water economy of living organisms, since a great number of the most important components of the cell, prominent among which is the protoplasm itself, show this phenomenon. Moreover, the uptake of water and the increase in volume of air-dry seeds when they come into contact with water is purely a process of this kind. The dehydrated protoplasm and the cell walls absorb the water into themselves until the molecular loosening and degree of saturation essential for the normal functioning of the vital processes is reached. While cellulose and starch are certainly to be considered as substances capable of limited swelling, in protoplasm the degree of dissociation is somewhat greater. It is well known that protoplasm, especially in the phenomenon of 'streaming', often gives the impression of a fluid, although this 'fluidity' must not be taken to indicate that internal structure is lacking. Transitions from the gel to the sol condition, and vice versa, may well occur frequently, so that a completely free and random movement of all the particles, as in a true solution, is never reached. Here again, the extent of the hydration is certainly extensively determined and regulated by the capacity for imbibition. It is probable that any change in the degree of hydration of the protoplasm influences the extent and rate of many metabolic reactions. The facility of diffusion within the protoplasm, for example, is determined by the extent of its molecular loosening. The magnitude of the forces concerned with the imbibition of water ensures that the essential hydration can be maintained in the face of strong pressures. As will be explained in more detail in the next section, such pressures, deriving from osmotic phenomena, are always present, and are the principal reason why the actual suction-pressure developed by swelling biological material is usually less than that expected on the properties of the imbibing material alone.

2. Diffusion and osmosis[5]

Imbibition is not the only phenomenon which plays a part in the uptake of water into the plant cell. This is especially true of the differentiated plant cell, where usually only a thin peripheral layer of protoplasm, already fully hydrated, behind a similarly hydrated cell wall, surrounds a very large vacuole with watery contents. Here other processes must clearly determine the correct water level for the actively living cell. These are in fact predominantly diffusion phenomena. Such phenomena are fundamental to many physiological processes, and are also of great importance in water relations. Even the imbibition of water by a compact body is only possible, for example, if the water molecules can diffuse into the interior of the swelling material.

It is well known that all molecules possess a certain thermal or kinetic energy. In gases, where the cohesive forces between the molecules may become very reduced, the energy reveals itself in the tendency to expand and in the pressure which the gas exerts on the enclosing walls. A freely moving gas molecule may attain a velocity of several hundred metres a second. In fluids, where the cohesion of the molecules is much closer, their ability to move is naturally more severely limited. A molecule moving in a straight line always comes into collision with other molecules and is deflected by them. Visible evidence of molecular movement in aqueous media is given by the quivering movement, easily perceptible under the microscope, of small but still clearly visible particles, such as the smallest starch grains and minute crystals. This movement, known as Brownian movement, is brought about by the invisible water molecules colliding with the particles and displacing them. The kinetic energy of molecules is, moreover, the prime cause of the all-important phenomenon of diffusion.

Diffusion. Exactly as gas molecules, by virtue of their kinetic energy, uniformly fill a space provided for them, so also do molecules dissolved in a fluid such as water become uniformly distributed in the solution.

If a crystal of copper sulphate, or a granule of a readily soluble dye, be placed in a cylinder filled with water, the distribution of the dissolving coloured substance can be followed directly by eye. Diffusion takes place over small distances with great rapidity, but over large very slowly. This is because the distance travelled by a diffusing molecule is proportional to the square root of the time allowed for diffusion. Thus if a molecule diffuses a distance s in one day, and a distance s_x in x days, then $s_x = s\sqrt{x}$.

The quantity of material diffusing is proportional to the concentration gradient, the time, the cross-sectional area of the diffusion path and a constant which involves the specific molecular velocity of diffusion at a given temperature. Diffusion always takes place in a direction of falling concentration.

As an example of the order of size of the diffusion pathway, the following table gives the distance diffused by the dyestuff fluorescein (M.W. 332) in water at a concentration of 1 per cent, at 20°C, after increasing periods of time:

Time	1 sec	10 sec	30 sec	1 min	10 min	15 min	30 min	1 hr	24 hr	30 days	360 days
Distance diffused in mm	0.087	0.275	0.477	0.675	2.13	2.62	3.71	5.23	25.6	140.0	486.0

It is seen that already after one second 87 μm are traversed, after one minute 675 μm, after ten minutes 2 mm and after one hour 5 mm. However, in thirty days the distance is only 14 cm, and in a year hardly half a metre. At the dimensions of plant cells diffusion is of great rapidity, so that in a few seconds, for example, the whole vacuole can be traversed, a fact of fundamental importance in metabolism. The velocity can still be considerable in short chains of cells, provided there is no resistance in passing from one cell to the next. However, it is evident that over long, continuous stretches molecular kinetic energies alone will permit only slow movement.

The process of diffusion always tends to bring about uniformity of concentration, i.e. a uniform distribution of molecules in space. Of course, when this is accomplished the molecules do not cease to move, but in any interval as many molecules diffuse out of a given region as into it, so statistically the uniformity of concentration remains undisturbed. Naturally, in fluid solutions the molecules of both the solute and the solvent diffuse according to their concentration gradients. If two aqueous solutions of the same material, e.g. cane sugar, but of different concentrations, are carefully placed one on top of the other, then sugar molecules pass from the stronger solution to the weaker, and inversely water molecules from the weaker solution into the stronger, where the water molecules are fewer in number. The same will naturally occur if a sugar solution is brought into contact with pure water. Also, if a finely porous membrane, which is equally permeable to cane sugar and water, be placed between two solutions of different concentrations the equalizing of the concentrations still takes place without hindrance. Water molecules continue to pass through the membrane into the stronger sugar solution, and sugar molecules into the weaker until the concentration is the same on the two sides of the membrane. Such diffusion through a membrane is referred to as osmosis.

Fundamental laws of osmosis. Osmosis becomes of great physiological significance when the membrane offers resistance to the passage of molecules through it, the extent of the resistance depending upon the type of molecule.

In many membranes the diffusing molecules pass through the water-filled intermicellar spaces, which in the cellulose walls of most plants are so wide that most non-colloidal materials encounter no significant resistance. Such membranes do not, however, allow the passage of colloidal particles, the diameters of which may lie between 0.001 and 0.2 μm. In other membranes the pores may be so narrow that even of substances in true solution only the smaller molecules can get through, the larger molecules being held back as if by a sieve. Not only the diameter of the pores but also the physico-chemical properties of their walls (e.g. their electric charge) are of importance in this connection, since these can influence the free movement of charged particles. Sometimes with aqueous solutions only the small water molecules (with a diameter of about $2\frac{1}{2}$ Å (= 0.25 nm)) can pass through the membrane freely, the passage of the larger molecules of the solute, such as those of cane sugar (with a diameter of about 5 Å), being prevented.

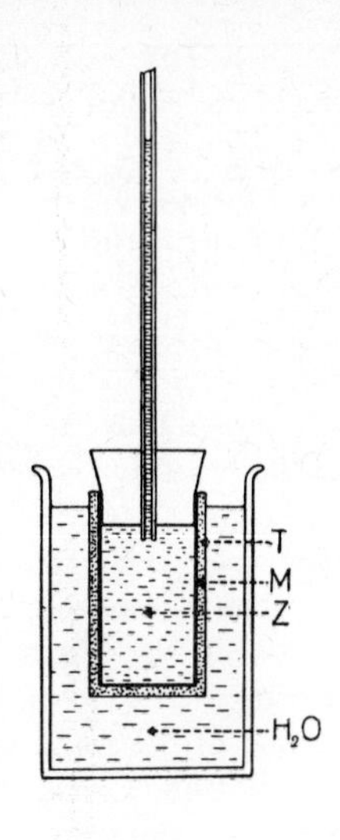

Fig. 226. Diagram of an osmometer. A clay vessel (T) is furnished with a stopper and an expansion tube, and placed in water: M, semi-permeable membrane deposited on the pot; Z, cane-sugar solution.

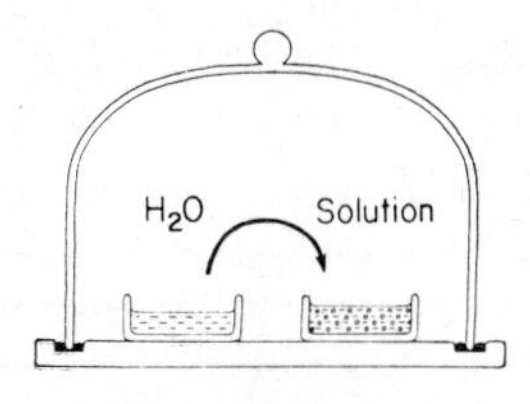

Fig. 227. Diagram illustrating isothermal distillation. Explanation in text.

Membranes which will not permit the passage of the solute to the same extent as the solvent are said to be semi-permeable.

A similar filtration effect may be given by a membrane where the intermicellar spaces are not acting as valves, but where the permeability is determined by the degree to which a substance is soluble in the material of the membrane. Thus if a solution of benzol in water comes into contact with rubber sheet or tube the rubber is impermeable to the water, but permeable to the benzol. This is because the latter is soluble in the rubber, and so is capable of diffusing through it.

Semi-permeable membranes of the first kind, where the permeability is determined by pore size, can be made artificially from many different materials, such as collodion, tannic acid–gelatin, precipitates of calcium phosphate, Prussian blue and so on. Membranes of copper ferrocyanide ($Cu_2[Fe(CN)_6]$), which are formed at the interface when solutions of copper sulphate and potassium ferrocyanide come into contact, have acquired special significance. Such precipitation membranes are, of course, very delicate, but if they are embedded in the wall of a porous pot, such as is used in electric cells, they are then quite satisfactory for quantitative experiments. By this means the plant physiologist Pfeffer was able in 1877 for the first time to study osmotic phenomena with considerable precision. These phenomena are of great interest to physical chemists as well as biologists.

If a cylindrical porous pot, the walls of which have been impregnated with a semi-permeable copper ferrocyanide membrane (a so-called *Pfeffer cell*; see Fig. 226), is filled with a concentrated sugar solution and immersed in water the water molecules can pass freely in both directions through the wall, while the sugar molecules remain confined within the pot. If the pot is closed and provided with a vertical tube it can then be seen that the volume of the solution in the interior of the cell increases and it begins to rise in the tube. Evidently there is an active movement of water from the outside into the interior of the cell. Thus the cell, often in this form referred to as an osmometer, behaves as if it were sucking up water. As the ascent of the diluted solution shows, the cell is capable of doing work against the force of gravity.

The physico-chemical explanation of this phenomenon is far from simple but can perhaps be made comprehensible in the following manner:

At the boundary between a water surface and the air the kinetic energy of the water molecules, a temperature-dependent property of fundamental importance in all diffusion phenomena, manifests itself in the constant tendency for some of the water molecules to leave the water and enter the air. In other words, the water evaporates, and at a given temperature the water has a definite vapour pressure. Thus in a closed space an equilibrium is eventually reached between the water molecules leaving the liquid phase and those returning to it. If there are substances dissolved in the water which are not themselves volatile the vapour pressure of the solution is depressed by their presence (and for the same reason the boiling point is raised). This is because not only water molecules are present at the bounding surface.

If two small dishes, one containing water and the other an aqueous solution of, for example, cane sugar, are placed side by side under a bell jar, then more water evaporates from the surface of the pure water than from the solution. As a result there are soon more water molecules present in the vapour phase than would correspond to the vapour pressure of the solution. Consequently, more water molecules return to the surface of the solution than leave it, i.e. the solution takes up water from the vapour phase and becomes more dilute. Water therefore distils from the dish with pure water into the solution (isothermal distillation; Fig. 227).

Since the vapour phase above the dishes is only passable by the solvent and not by the molecules of the solute, we can regard it as equivalent to a semi-permeable membrane, corresponding to the semi-permeable copper ferrocyanide membrane between the sugar solution and the water in the osmometer experiment. The events in the osmometer are in fact directly comparable with isothermal distillation. The diffusion pressure of the water molecules is, like the vapour pressure. greater in pure water than in a solution. Consequently, more water molecules arrive per unit time and per unit surface area on the water side of the semi-permeable wall and penetrate through it than on the side of the sugar solution. Here not only water molecules impinge on the wall but also sugar molecules, which are held back by it. But since diffusion will always tend to bring about a uniformity of concentration, water molecules pass from the region of higher molecular concentration (i.e. the pure water) to the lower (i.e. the solution), so bringing about a dilution of the solution. The poorer in water the solution, i.e. the more concentrated it is, the greater is the fall in diffusion pressure, exactly as the depression of vapour pressure depends upon the concentration of the solute.

The entry of water into the osmometer does not cease until the hydrostatic pressure generated by the dilution and increase in volume of the solution prevents any further entry. This pressure, which is registered by the height of the solution in the vertical tube and which prevents any further increase in volume, is referred to as the osmotic pressure. When this pressure is applied there is equilibrium between the number of water molecules passing through the membrane in each direction. Pressures above the osmotic pressure accordingly cause water to be expressed from the solution. Thus dilution of a solution by the osmotic absorption of water is capable of doing work, whereas the concentration of a solution is a process requiring work.

The extent of the capacity of an aqueous solution, exposed to water behind a semi-permeable membrane, to take up water and to develop a hydrostatic pressure is referred to as the osmotic pressure of the solution. At a given temperature the magnitude of the osmotic pressure of a solution depends principally upon the concentration, i.e. the number of dissolved molecules per unit volume. The chemical nature of the solute is only of relevance to the extent that electrolytes, because of their dissociation, behave differently from non-electrolytes (see below). The pressure generated is also inversely proportional to the molecular weight, so that macromolecular substances produce only a small osmotic pressure.

If a non-electrolyte is in true solution, and the concentration is low and not greater than about 0.1 g mol (more concentrated solutions have higher pressures), the osmotic pressure is numerically the same as the pressure the dissolved substance would exert in the same volume if it were present as a gas in the absence of the solvent. Thus 0.1 g mol of a gaseous substance enclosed within a volume of 1 litre exerts pressure of 2.24 atmospheres. If 0.1 g mol of a non-electrolyte is dissolved in a litre of water and placed behind a semi-permeable membrane the solution is capable of developing a hydrostatic pressure of 2.24 atmospheres. Even very dilute solutions can develop quite appreciable pressures. A 1 per cent solution of cane sugar, for example is capable of generating at room temperature a pressure of $\frac{2}{3}$ atmosphere, and a 1 per cent solution of potassium nitrate a pressure of over 4 atmospheres.

Since the osmotic pressure depends only on the number of the molecules in solution, equimolecular solutions of different substances develop the same pressure. They are said to be isosmotic or isotonic. Only salts, which dissociate in aqueous solution, provide exceptions to this rule. In these solutions the pressure developed is now proportional to the number of particles (consisting of both molecules and ions). For this reason a solution of potassium nitrate generates an osmotic pressure 1.69 times as great as that generated by a solution of non-ionizable cane sugar of the same molar concentration. Potassium nitrate is accordingly said to have an isosmotic coefficient of 1.69. That this falls short of 2.0 is attributable to mutal interaction between the ions, which diminishes their mobility.

The physico-chemical phenomena associated with partly impeded diffusion through semi-permeable membranes, and the development of osmotic pressures, have great significance for plant cells. Pfeffer constructed his porous pot with the embedded copper ferrocyanide membrane with the object of trying to find an explanation for the high pressures always to be observed in plant cells. If the easily permeable plant cell wall is regarded as being equivalent to the porous pot of the osmometer, and the layer of cytoplasm adjacent to the wall to the copper ferrocyanide membrane, then a cell coming into contact with water must behave exactly as the osmometer. Water will stream into the vacuole until the pressure generated within it prevents any further entry of water molecules.

Considerable pressures prevail generally within plant cells, causing the cytoplasmic layer adjacent to the wall to be pressed against it like a bladder against the case of a football. This condition is called the turgor of the cell, and it leads to the turgescence of cells and tissues.

Turgor plays an important role in maintaining the tautness and rigidity of the plant body. The wilting of the sappy parts of plants, which always occurs when the loss of water is too great, is caused by the loss of turgor and the consequent relaxing of the cells of the organ in question. If a plant tissue which is wilted and relaxed, but still living, is placed in water, then the cells absorb water exactly as an osmometer until they are again tautly filled. This, of course, holds only so long as the plasma is living. Death results in a rapid loss of semi-permeability, and accordingly of turgescence.

Plasmolysis. Actual evidence that the plant cell may legitimately be compared with an osmometer, and that turgor is therefore caused osmotically, and, moreover, that the role of the cytoplasmic layer adjacent to the wall in this connection is only that of a semi-permeable membrane, is provided by the phenomenon of plasmolysis. This phenomenon can be produced experimentally by placing a turgid cell in a concentrated solution of some substance such as cane sugar which is not capable of diffusing through the semi-permeable layer of cytoplasm. When the concentration of the cane sugar is greater than the total concentration of dissolved substances in the cell sap, then the elastic cell wall, which is under tension and extended, usually immediately shortens a little, indicating that the turgor of the cell diminishes. Following this, the cytoplasmic layer is seen here and there to detach itself from the wall, and to become thrown into numerous concave folds

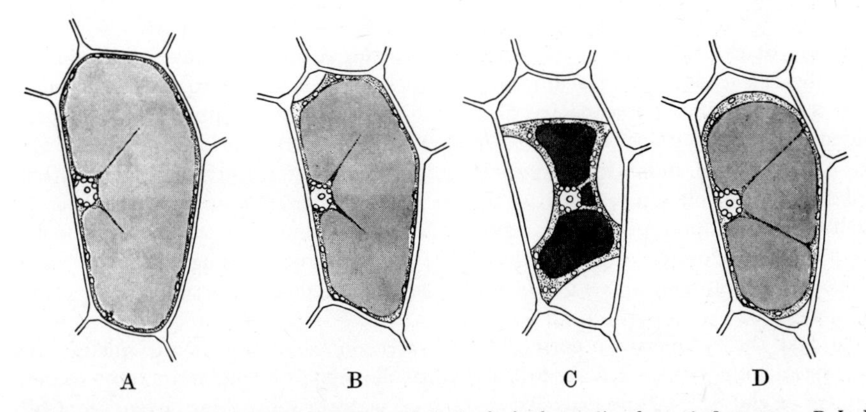

Fig. 228. Cells from the lower epidermis of the leaf of *Rhoeo discolor*. **A.** In water. **B.** In M/2 KNO_3, showing a decrease in volume facilitated by a contraction of the cell wall and the beginning of plasmolysis (upper left). **C.** Complete plasmolysis after prolonged action of M/2 KNO_3, the cell sap strongly concentrated. **D.** Deplasmolysis well advanced after transferring back to water. Diagrammatic.

around the shrinking vacuole. This is the process of plasmolysis (Fig. 228A–C). If the plasmolyte, i.e. the cane sugar, is removed and pure water substituted, then if the cell has not suffered too severely the reverse process, deplasmolysis, can often be observed (Fig. 228D). Here the vacuole takes up water, enlarges and becomes rounder, and finally presses the cytoplasmic layer against the wall until this is once again under tension and the cell is turgid.

The explanation of this phenomenon on osmotic grounds is very simple. The cane-sugar solution contains per unit volume more dissolved particles and less water than the cell sap. The cytoplasm provides a semi-permeable membrane through which only the water molecules are capable of diffusing freely in both directions. In consequence, the water molecules inevitably pass actively from the cell into the sugar solution, and they continue to do so until the concentration of osmotically effective particles in the vacuole has become as large as in the sugar solution. As the water leaves the vacuole it must become smaller, the cell wall loses its tension and finally even the protoplast is dragged away from the wall and the sugar solution is drawn in through it. The volume of the sugar solution within the cell (between the wall and the protoplast) continually increases as water is extracted from the protoplasm and the protoplast follows the shrinking vacuole. Conversely, if the sugar solution is replaced by water the now very concentrated vacuolar sap exerts its full osmotic pressure. Water molecules are drawn in, and the vacuole expands and drives the cytoplasm to the wall, so that any fluid between the cytoplasm and the wall is squeezed out of the cell. Finally, the wall is put under elastic tension, and the expansion of the cell continues until the tension in the wall balances the osmotic pressure and no further increase in volume can take place. The cell is then again fully turgid.

In view of the fact that, although the layer of cytoplasm adjacent to the wall is tenuous, it is nevertheless always of measurable thickness, it is difficult to say how much the semi-permeability of this layer is due to peculiar properties of its inner (tonoplast) and outer (plasmalemma) membranes alone. That the semi-fluid protoplasm is not broken up or crushed by the pressures it has to withstand (to be discussed later) suggests that it finds in the firm cellulose wall a support analogous to that which the copper ferrocyanide membrane receives from the porous pot of the Pfeffer cell. It also suggests that the cytoplasm has an imbibing power of its own, which will be closely related, but inversely, with the osmotic pressure of the cell sap. Also to be taken into account is the fact that substances will be dissolved in the watery phase of the protoplasm which will themselves have an osmotic effect. Indeed, it would probably not be wrong to ascribe, both in cells saturated with water, in which the cytoplasm is already fully turgid, and also in embryonic cells where there is no vacuole, considerable importance to the osmotic properties of the cytoplasm itself in maintaining the structure of the protoplasm and the water relations of the cell.

Suction pressure of the cell. The explanation of the turgor of the plant cell in terms of osmotic phenomena leads to several important conclusions in regard to the water relations of the cell in general. The capacity of a plant cell and tissue to provide themselves with an adequate supply of water, and to maintain their water relations in spite of all losses due to evaporation, must depend very largely on there being a sufficient concentration of solutes in the cell sap, as well as on the semi-permeability of the cytoplasmic layer. Death of the cytoplasm leads to a rapid decay of the semi-permeability and consequently to a complete loss of turgor by the cell. Even less violent changes in the permeability of the cytoplasmic layer can result in marked changes in the degree of turgor (see, for example, p. 370).

Assuming that the cytoplasm is permeable to water alone, but not to the substances dissolved in the vacuole (but see p. 233 seq.), then the following principles must apply to the water relations of the cell and its capacity to take up water from without. The extent of the tendency to take up water, the suction pressure of the cell, is dependent upon the molar or ionic concentration of the substances dissolved in the cell sap, this naturally increasing with the concentration of the sap. It is also dependent upon the extent of the already existing turgor, i.e. the extent to which the cell wall is under tension. Just as no

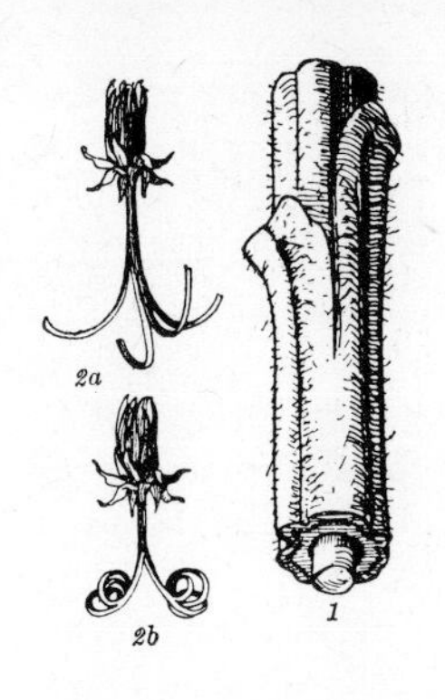

Fig. 229. Tissue tension. **1.** Portion of the stem of *Helianthus annuus*. The pith has been separated from the periphery by a cork borer and is extending when placed in water. **2.** Peduncle of *Taraxacum* cut into quarters longitudinally: a, immediately after splitting; b, after placing in water. After Jost, reduced.

more air can be pumped into a football or bicycle tube when the leather case or outer cover has reached its maximum extension, so the expansion of a plant cell is also not unlimited. Nevertheless, a cellulose wall is capable of considerable extension, amounting to 10–20 per cent and even in a few cases to 100 per cent, without bursting. The forces required for exceeding the limits of elasticity usually, however, exceed the suction pressure of the cell. The cell, therefore, can never take up water until there is a uniform concentration both in the cell and outside it, but only until the inward pressure of the elastically stretched wall, the wall pressure, prevents any further increase in volume. In Pfeffer's osmometer the entry of water into the pot is prevented by the hydrostatic pressure generated by the solution rising in the tube, and this hydrostatic pressure plays a role analogous to the wall pressure in the living cell. If the cell is not free, but lies in a tract of tissue, then the surrounding tissue will provide additional resistance to the expansion of the cell, even though the maximum expansion which the cell is capable is nowhere near reached.

This fact is shown particularly well by the phenomenon of tissue tension. This tension is made evident by the fact that when the tissues of a plant organ are separated from each other they extend to markedly different extents. If, for example, the peduncle of the dandelion is divided towards the summit by two cuts at right angles to each other (Fig. 229.2), and placed in water, it is seen that the segments become strongly recurved and roll up. This is because the inner layers of cells of the previously undivided peduncle were prevented from attaining their greatest possible extension under turgor because of the constraints imposed by the outer layers. Similarly, the pith of many shoots is subjected to pressure by the outer parts, contributing greatly to the rigidity of the whole organ, and explaining why, in the above experiment, the loosened cylinder of pith comes to project from the cut end of the stem (Fig. 229.1).

These opposing tendencies which govern the entry of water into the cell can be summarized in the following section pressure equation:

$$S_z = S_i - W$$

or

$$S_z = S_i - (W + A)$$

where

S_z = the suction pressure of the cell;

S_i = the osmotic pressure of the cell sap (expressed in atmospheres and determined by the concentration);

W = the inward pressure of the wall;

A = the pressure on the cell arising from any surrounding tissue.

Should the cell be in a completely collapsed and wilted condition, then W and A, if normally present, are zero, and the suction pressure of the cell is now equal to the osmotic pressure of the cell contents alone. As water enters the cell and it expands, so W and possibly A continually increase, until in the completely turgid cell they have reached the value of S_i. At this point S_z becomes zero and the cell is unable to absorb any more water.

Osmotic pressures in plants. It is essential for the understanding of the water relationships of plants to consider in more detail the actual osmotic pressure obtaining in the cell sap.

Osmotic pressures can be measured by two methods. It is possible, by subjecting with great care certain parts of plants to considerable pressures, to express almost pure cell sap. Its molar concentration can then be estimated by determining the depression of freezing point (the cryoscopic method). Plasmolytic methods can also be used (De Vries, 1883). Plasmolysis can be produced only by solutions which are more concentrated than the cell sap (hypertonic solutions). Weaker (hypotonic) solutions can never induce plasmolysis. It can then be discovered experimentally which concentration of a series of solutions causes the first beginnings of plasmolysis in, say, 50 per cent of the cells. This limiting concentration can then be taken to be isotonic with the cell sap, since the cell wall is no longer under tension. This osmotic pressure obtained by limiting plasmolysis (O_g) must naturally be somewhat higher than the osmotic pressure of the sap in the fully turgid cell (O_n). The plasmolytic method does, however, allow measurements to be made on individual cells, while only an average value for a large number of cells can be obtained by the cryoscopic method.

It has been found that considerable differences in the osmotic pressure of the cell sap exist not only between individual plants but also between organs and tissues of the same plant. This is quite apart from the fact that the osmotic pressure is variable within an individual cell and, what is very significant, can be varied according to its needs. The principal osmotically active substances appear to be sugars, organic acids and their salts. Sometimes, e.g. in halophytes, inorganic salts are also concerned. Osmotic values of between 5 and 15 atmospheres (but see p. 223) are usually found in the parenchyma cells of the cortex of the root. In the shoot the values usually increase with distance from the root, and in the cells of the leaf tissue may reach a maximum of 30–40 atmospheres.

Thus it was found that in the leaves of the beech tree the osmotic pressure of the sap in the cells of the lower epidermis was of the order of 13.7 atmospheres, while in the spongy parenchyma and palisade values of 21.1 and 37.6 were measured. Plants which live on very dry soils, as in deserts, or which grow in very saline situations, such as sea coasts, usually have cell saps with very high osmotic pressures. Values of over 100 atmospheres have been measured in these plants, and it is naturally tempting to assume that these high values are connected with the difficulties of obtaining water from the soils of such localities generally. Some moulds can grow on highly concentrated sugar solutions and even adapt themselves to increasing concentrations of over 200 atmospheres.

3. The uptake of water through the roots[6]

The foregoing discussion has clarified the processes, namely imbibition by the cytoplasm and osmosis, by which the plant cell obtains the water essential for its normal life, so long as it has direct contact with it. Such a direct contact, however, normally occurs only with the surfaces of submerged water plants, and with certain epiphytes which take up rainwater through absorption hairs (see p. 106). In addition, the parenchyma cells adjacent to the water-conducting elements should be mentioned. With the absorptive organs proper of the higher land plants, namely the roots, somewhat more complex relationships are found. This is related to the fact that the water in the soil is only partly available in a free form. The soil water is in fact already a dilute solution of salts, possessing a certain osmotic pressure. Moreover, a part of the water present is always more or less firmly bound by the soil particles and can be extracted from them only against considerable resistance. As shown in Fig. 230, the absorbing root hairs cling closely to the particles of soil. Each of these soil particles is furnished, as a result of either imbibition or absorption (see p. 233 seq.), with an envelope of water. The spaces between these envelopes contain partly air and partly water. As will be discussed later, this air is absolutely essential for the respiration of the root cells, and its absence in boggy soils usually has a detrimental effect on plant growth. The water is held by capillary attraction, a phenomenon of special importance for the requirements of the plant, since it provides a means by which water not immediately in contact with a plant can reach it over even comparatively long distances in the soil, provided the capillaries are wide enough (as in soils with a water

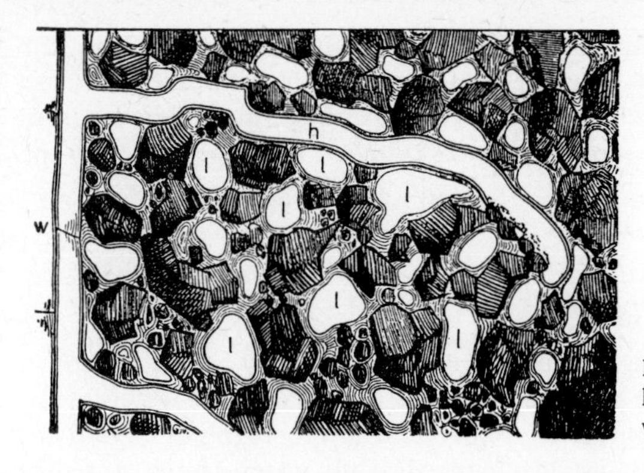

Fig. 230. Root hair in the soil: w, root; h, root hair; l, air-filled spaces in the soil surrounded by water. Diagrammatic. After Sachs.

content greater than 25 per cent). The physico-chemical composition, and especially the structure of a soil, are therefore of decisive importance in determining both its water content and the availability of the water to the plant (see p. 233 seq.). After rain, for example, a sandy soil retains a much smaller quantity of water than a soil rich in clay or humus. Soils of the latter kind are capable of imbibing water, but at the same time a greater proportion of the water they absorb is no longer held in a form in which it is available to the plant.

Generally speaking, the actual water-binding power of normal agricultural soils, arising from their ability to absorb and imbibe water and from the osmotic value of the soil solution, is fairly low and less than 5 atmospheres. In saline and desert soils it can naturally be very much higher; in the alkaline steppes of Hungary, for example, it reaches 30 atmospheres, and in the Algerian desert exceeds 100 atmospheres. Thus the osmotic suction of the root-hair cells always has to work against a certain amount of resistance in extracting water from the soil, so it is not surprising that quite different osmotic values have been found in root cells, depending upon the species and the habitat (e.g. *Phaseolus* c. 2–3.5 atmospheres; *Pelargonium* c. 5 atmospheres; halophytes >20 atmospheres; desert plants >100 atmospheres).

If now a root-hair cell, satisfying the conditions discussed above, has taken up water from the outside, then the parenchymatous cells of the cortex of the root can extract water from it, provided their own suction pressure is greater than that of the hair cell. The cell with the higher suction pressure will first take up water from the saturated cell wall. The cell wall in turn, by its capacity to imbibe water, will tend to make good its loss from the water in the adjoining cell with the lower suction pressure. Thus in practice the cell with the higher suction pressure will draw water from that with the lower until the pressures are equalized.

A gradient of suction pressures has in fact been found in the cortical cells of the root, extending up to the endodermis. The inner cortical cells, as the upper part of the accompanying table shows, possess a higher suction pressure than the outer. Thus it can be seen that not only individual parenchyma cells but also large tracts of tissue can obtain the

Suction pressures in a lateral root of ***Vicia faba*** (zone of absorption) (After Ursprung)

Layer of cells	Suction pressure (atm.)
Epidermis	0.7
First cortical layer	1.4
Third cortical layer	1.5
Fourth cortical layer	2.1
Fifth cortical layer	2.8
Sixth cortical layer	3.0
Endodermis	1.7
Pericycle	0.8
Vascular parenchyma	0.9

water essential for physiological activity by handing it on from cell to cell under the influence of purely osmotic forces. This gradient does not appear to be broken until the endodermis is reached, so that forces other than the purely osmotic must play a part in bringing about the entry of the water into the central cylinder, and especially into the vessels. This point will be discussed in more detail in the following sections (see p. 228 seq.)

B. The loss of water by transpiration[7, 8]

The uptake of water as described in the foregoing provides not merely a once and for all saturation of the cells and tissues of the plant but also a means whereby the water lost by evaporation from the plant body is continually replaced. The thoroughly hydrated cell walls of the aerial parts of the plant must inevitably give off water vapour until the vapour pressure is reduced to that of the atmosphere. The evaporated water has, of course, to be replaced, and it is clear that not only the extent of the water uptake but also that of this loss by transpiration is of great importance in the water economy of the living plant.

With the great surface area of leafy plants, the loss of water from the more or less completely saturated outer walls, and consequently the flow of water thereto from the roots, must be very considerable. It has been estimated that about 60 per cent of the total annual precipitation falling upon a large beech wood is returned to the atmosphere within the same period by transpiration. On a sunny day a sunflower is easily capable of evaporating 1 litre of water and a birch tree with 2×10^5 leaves some 60–70 litres, rising on particularly hot and dry days to as much as 400 litres. These quantities of water must be continuously replaced by the roots if serious wilting is to be prevented. These figures bring out clearly the intense activity of the roots, and the great importance which attaches to a large root area and the extremely fine ramification of the absorptive rootlets.

The quantity of water lost by transpiration can easily be demonstrated by repeated weighing of the whole plant. A still more impressive demonstration is provided by the potometer illustrated in Fig. 231. An air bubble is included in the thread of water in the capillary tube attached to the side of the water reservoir, and as the transpiring shoot takes up water, so the bubble moves along the tube. The rate of transpiration can thus be read off directly from the rate of movement of the bubble along the tube. For more exact measurements other methods must be used.

As already stated, the more or less fully hydrated outer walls of the plant, such as those of the epidermal cells of the leaves, will for purely physical reasons give off water molecules into the unsaturated atmosphere so long as any saturation deficit exists between the air and the walls. Here again we are concerned with a diffusion process

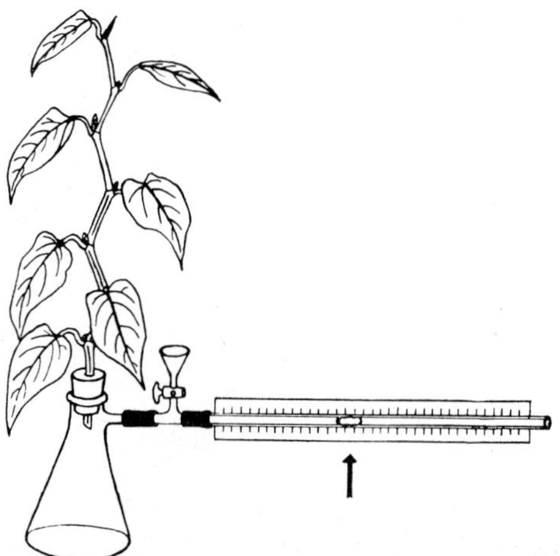

Fig. 231. Diagram of a simple potometer. Arrow points to the air bubble.

leading to the evening out of the vapour-pressure difference between the solution present in the cell walls and that of the moisture in the air. Thus it is clear that all external factors which change the relative humidity of the air (such as the heating of the air by the sun's rays and the moving in of masses of warmer air) must correspondingly influence the intensity of the transpiration. Evaporation causes the cell walls to lose some of their imbibed water, but because of their affinity for water they replace that lost by drawing upon the adjacent protoplasm. The protoplasm deprived of water in this way extracts in turn water from the vacuoles, thus raising the concentration of the vacuolar sap. All these processes, together with the very important relaxation of tension in the cell wall, lead to an increase in the suction pressure of the cells so that they begin to take up water from adjacent cells towards the inside of the tissue. In this way the tension created by the evaporation of water from the outer walls can be transmitted from cell to cell right up to the conducting elements, quite apart from water already travelling in the cell walls from the vessels by capillary means (see p. 230).

The properties of the outer walls are naturally of great importance in determining the amount of the water lost through them. Many of the features peculiar to the epidermis can only be understood from this point of view. It is certainly true that the fat-like cuticle is not completely impermeable to water, but there can be no question that it impedes the exit of the water molecules. Thus a conspicuous feature of the plants of dry places, the xerophytes, is often a massively developed cuticle, and cutinized layers of the wall which provide a further barrier to evaporation. In addition, the deposition of wax within or on the outer walls of the epidermal cells, as well as the tomentum of dead hairs found on many leaves, all tend to restrict transpiration.

The evaporation of water directly through the cuticle is referred to as cuticular transpiration. In plants with thin cuticles, such as those of damp habitats, the hygrophytes, this cuticular transpiration can amount to a considerable percentage of the total loss of water. Usually, however, it is quantitatively trifling, and in the leaves of birch, for example, amounts to only 6–8 per cent of the water actually lost by evaporation.

How effective wax, cutin and especially cork are in limiting evaporation can be seen especially well in fruits, such as apples and pears, and also in potato tubers. When unpeeled they can lie in the air many weeks without any great loss of water.

By far the greater proportion of water vapour which the plant loses to the atmosphere is not, however, given off through the cuticle of the epidermis, but escapes through the stomatal apertures of the leaves. This is referred to as stomatal transpiration.

The delicate mesophyll cells in the interior of the leaf, whose walls are covered by only a very thin cuticle, give off water molecules continuously into the extensively developed intercellular system. This in turn is connected with the external air through the stomatal apertures, so it can remain saturated with water vapour only when these apertures are closed.

It can be easily shown by a simple experiment that in leaves where the stomata are confined to the lower surface it is only here that any considerable quantity of water vapour escapes. A freshly picked leaf, such as that of a rose, is placed rapidly between filter-papers which have been soaked in a 5 per cent solution of cobalt chloride and thoroughly dried, and the system is enclosed between two glass plates. After a short time it will be seen that the paper adjacent to the lower surface, at first blue, has become pink owing to the uptake of water, whereas the paper adjacent to the upper side remains blue for some time.

The stomatal apertures are minute, the average diameter of an open pore being of the order of only a few μm, and although their frequency may rise to several hundred per square millimetre, the total area of the pores amounts to only about 1–2 per cent of the leaf surface. The view that they are nevertheless mainly responsible for the escape of water in transpiration has lost much of its implausibility since the process of evaporation has been studied in greater detail in purely physical models. It has been found that small pores allow a much more active evaporation than large, and that several small cardboard discs, for example, saturated with water dry much more rapidly than a complete piece of the same material of equal area. The explanation lies in a certain marginal effect. Water molecules leaving the middle of any given area can only diffuse vertically upwards because of interference from neighbouring molecules, while the molecules escaping at the margins are also able to diffuse freely in a lateral direction. Consequently, evaporation must take place more actively from the margin than from the centre. Now small pores will have a greater proportion of margin relative to their area than large pores, and it follows that a finely perforated membrane will be able to support very active evaporation. A leaf with fully open stomata can thus lose by transpiration some 50–70 per cent of that amount of water which would be given off by a free water surface of the same area. This is a good illustration of how remarkably well the leaf is organized from the point of view of physical principles.

In contrast to cuticular transpiration, the extent of the stomatal transpiration can be regulated by way of change in the stomatal aperture, and even completely stopped by a general closing of the stomata. This is of the greatest importance for the water relations of plants. The movement of the guard cells, the mechanics of which are discussed in greater detail on p. 371, can be brought about by various external and internal factors. Of great significance is the fact that a loss of water from the leaves greater than that which can be met by the upflow of water from the roots (some 3–10 per cent depending upon the circumstances) leads to a closing movement which vigorously suppresses transpiration. The plant is thus able actively to regulate its water relations and, at least in certain circumstances, to control according to the particular requirements of its physiological activities the purely physical process of transpiration which is the inevitable result of its construction.

If the environmental conditions are such that, despite the closing of the stomata, water continues to be lost to the atmosphere at a greater rate than can be supplied from the vessels, then the ensuing shrinkage of the cell walls and of the protoplasm of the epidermal cells provides a further barrier to the loss of water before complete wilting takes place.

Numerous peculiarities in anatomy and morphology indicate that in many plants there are specific adaptations related to the vitally important maintenance of their water economy. These adaptations make it possible for them to live in relatively dry climates where other less well-adapted plants are unable to exist. The detailed consideration of these specialized structures, already dealt with in the morphological section (see p. 180 seq.), forms a fascinating aspect of plant physiology. It is also of very great interest to plant geographers.

In normal atmospheric conditions the transpiration of the more highly organized plants displays a diurnal rhythm; during the morning it tends to rise, reaching a maximum as the intensity of the light and the temperature increase towards noon, and falling again towards evening. This is provided so that variations in the relative humidity following upon the daily march of temperature do not cause a precocious closing of the stomata. Not infrequently, for example, transpiration is reduced round about noon. If as a result of excessive transpiration during the day the plant enters a condition of water deficit, then this can be made good again during the cool and relatively more humid night, and if dew is formed the water equilibrium of the plant will be effectively restored.

Importance of transpiration. Transpiration should not be regarded merely as an encumbrance to the plant following inevitably upon its particular kind of organization. Unquestionably the plant is sometimes put to considerable difficulties in obtaining and replacing the water required, on account of which the availability of water plays a decisive role in determining the environmental conditions of vegetation. An adequate transpiration stream is of particular importance for the plant because of its cooling effect, which prevents a dangerous overheating of its organs by the sun's rays. Quantitatively, the rate of transpiration is such that the heat taken up by the evaporating water is approximately equal to that provided by the insolation. The temperature of wilting leaves, whose stomata are closed, may rise to some 6°C above the external temperature, whereas that of normal leaves is somewhat lower than the temperature outside. Furthermore, the transpiration stream provides the plant, albeit in one direction only, with a transport mechanism for all kinds of important substances. As will be considered in more detail later, it is the main means of transport of the essential minerals, and also perhaps occasionally of those organic substances formed principally in the roots (e.g. many alkaloids). The same holds good as well for the processes of guttation and bleeding, considered in the following section.

C. Guttation and bleeding

While the loss of water by transpiration involves no active participation by the plant, many plants are capable of actively excreting water when transpiration is at a complete standstill. After warm and humid nights large drops of water can often be seen hanging from the margins and tips of the leaves of certain plants. These give the impression of dew, but in fact they have emerged from water stomata or glands. This phenomenon is shown very beautifully by *Alchemilla* and *Tropaeolum* (Fig. 232), and it can also be seen

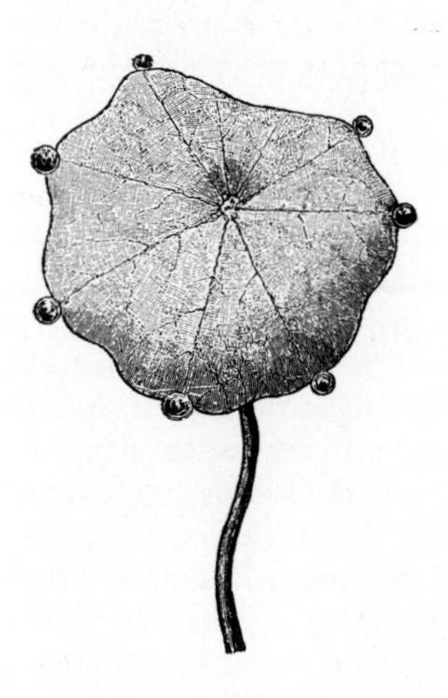

Fig. 232. Guttation droplets on a leaf of *Tropaeolum majus*. c. $\times\frac{1}{5}$. After Noll.

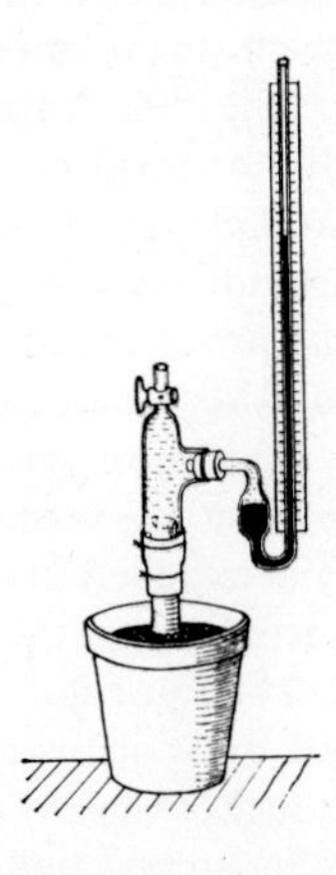

Fig. 233. Demonstration of root pressure. A mercury manometer is connected to the stump of a pot plant. The sap expelled from the plant forces the mercury up the tube.

at the tips of the leaves of certain grasses in the early morning. The excretion of water by certain Araceae of the humid primordial forest of tropical regions is particularly striking; one of the large leaves of *Colocasia nymphaeifolia* yields in a single night up to 100 ml of fluid. The lower plants also, especially the fungi, show a similar excretion of water droplets from their mycelia, e.g. the dry-rot fungus, *Merulius lachrymans*. In the unicellular fungus *Pilobolus* (see p. 228 and Fig. 321.1) it is immediately clear that a single cell possesses the capacity to excrete water. This phenomenon is known as guttation. It is evidently closely connected in the higher plants with the phenonemon of bleeding, i.e. the excretion of watery fluid after wounding. If the tips of the branches of birch, various species of maple and of vines are removed before the expansion of the leaves, or if certain pot plants are completely decapitated, fluid emerges from the remaining stumps. Apparently this is expelled from the vascular tracts under a pressure generated principally in the roots. If a manometer is attached to the stump, as shown in Fig. 233, this bleeding or root pressure can be measured. It is generally found to remain less than 1 atmosphere, but in extreme cases in birch it may amount to $2\frac{1}{3}$ atmospheres, and occasionally in cultivated tomatoes even to 6 atmospheres. The quantity of fluid which may flow out of a wound can be considerable. In twenty-four hours a vine may yield 1 litre, a birch 5 litres and a palm 10–15 litres.

The quantity of fluid excreted often shows a definite diurnal periodicity. Many plants bleed only in the spring and at a certain stage of development, in others bleeding can occur at almost any time. In bleeding occurring during the spring the sap often contains considerable quantities of organic substances, such as sugar, albuminous substances, enzymes and so forth. This is probably related to the mobilization of reserve substances and the manner in which they are pumped up in the vessels to the expanding buds. Such flushes of sap also occasionally find commercial uses, such as the obtaining of sugar from maple, 'palm wine' from various species of palm and other fermented drinks. Sap produced during summer bleeding, however, is in many plants usually free from organic substances and contains only a number of mineral salts in solution.

Guttation and bleeding probably both result from the same forces that are concerned with the normal movement of water in the plant. It has already been indicated earlier that there are difficulties in interpreting the entry of water into the roots on simple osmotic grounds, since the suction pressure of the cells does not increase steadily from the root hairs up to the vascular tracts. Instead, there is a fall in suction pressure from the endodermis to the vessels. Nevertheless, the water is passed into the vessels with a definite force, and it is possible that the cells adjacent to the vessels actively excrete the water as if they were glands. In certain conditions even portions of shoots lacking roots may also bleed, and in some instances the epithem lying beneath hydathodes also actively excretes water (see p. 121). Moreover, water plants living entirely submerged are said to maintain a slight passage of water through the plant body by guttation, caused in this instance principally by root pressure. The mechanism of this active excretion of water, like the mechanism of all glands, is still not finally settled.

Osmotic forces can lead to the expulsion of fluid from a cell if the permeability of the protoplasm on the excreting side of the cell is greater than the permeability on the absorbing side, that is as if the protoplasmic layer on one side is 'leaky'. Also local differences in the concentration of the osmotically effective materials within the cell might cause the cell to lose water. The sporangiophore of the fungus *Pilobolus* (Fig. 321.1) provides a beautiful example of such processes in a single cell. The lower part of the cell absorbs water, while above it is expelled in droplets. Recent investigations suggest that in roots mineral salts are transported across the cortex and secreted into the vessels independently of the intake of water. The salt solution thus formed within the vessels would then act as in an osmometer, and draw in water from the cortex. Certain electro-osmotic phenomena have also been thought relevant to the problem of the entry of water into the vessels. Recently, principally in animal cells, a number of 'pumping' mechanisms have been discovered, able, with the expenditure of energy, to transport water across membranes. These may also occur in plant cells. In general, however, guttation and bleeding, as any other process of active excretion, confront us with a still unsolved, but fundamentally important, problem of the physiology of metabolism (see p. 290).

D. The conduction of water[9]

Guttation and transpiration can only be carried on for long periods without serious disturbance to the water economy if the water lost is steadily replaced. This means that there must be a continuous movement and conduction of water from the place where the water is absorbed to that where it is being given off. This poses a special problem for multicellular plants.

With the more considerable lengths of axis, as in trees, the distances covered and the work performed in conduction are remarkable. In many trees the distances amount to 30–50 m, in species of *Sequoia* and *Eucalyptus*, as well as in many lianes, more than 100 m must be covered, not including the distance in the root. Thus water must be raised distances of this order against the force of gravity, and we must inquire into the forces and mechanisms by which this is accomplished. The quantities of water conducted are also considerable. To the facts already mentioned on p. 224 we may add that plants were observed on dry slopes in southern Germany which in one day were able to give off in transpiration an amount of water equal to twelve times the total amount they contained. Similarly, *Smirnovia turkestana* (Papilionaceae), growing in the sandy desert of Kara-Kum in western Asia, was found to lose in only one hour an amount of water equal to seven times its normal content. Such activity must imply a very vigorous movement of water up through the plant body.

It can be shown in various ways that this water flows principally in the vessels of the vascular bundles or of the xylem. If a branch or stem is ringed, i.e. a ring of bark some centimetres broad is removed, and the remaining wood is prevented from drying by some protective covering, then the leaves above the ring show no signs of wilting. On the other hand, if the wood is broken but the bark left undisturbed their wilting occurs immediately. If a branch bearing white flowers is cut off and placed in a solution of dye, then after some time it can be observed that the fine vascular bundles in the white petals have become coloured. If the stem is cut across it will be found that the dye has been taken up principally by the outer groups of vessels, and because of this we can conclude that it is mainly these that are utilized for the conduction of water. Especially beautiful demonstrations can be made with those plants (e.g. species of *Impatiens*) in which the vascular bundles can be observed with the microscope beneath the superficial parenchyma, either directly or after scraping away a little of the outside of the stem. Moreover, it is now possible with certain optical methods to make solutions of dyes ascending in undamaged vessels directly visible. By this means it has also been possible to demonstrate what is equally of fundamental importance, namely that intact vessels and columns of vessels contain no air bubbles, but are completely filled with water.

In view of the large quantities lost from the leaves, the water must move at considerable velocities within the plant. Speeds of 20–45 m per hour have been observed in wide-porous woods, such as the oak, whereas in certain lianes speeds of over 100 m per hour have been found. Even in small herbaceous plants values of up to 60 m per hour may be reached. In narrow-porous woods, such as beech, the water certainly moves more slowly, reaching speeds of the order of 1–4 m per hour, while in conifers the movement may amount to little more than 1 m per hour. Thus it becomes clear that despite the high frictional resistances which are present, especially in the tracheids, columns of water nevertheless move through the body of the plant with remarkable facility.

The question we must now consider is the nature of the forces which raise up this transpiration stream. Since the root pressure, as was considered in detail earlier, usually remains less than 1 atmosphere, it would be capable of raising a column of water at the most 10 m, and in trees at least there is no question of its being the principal force responsible. Nevertheless, it may play some part, especially in early spring before the expansion of the leaves. At this time it is possibly responsible for filling the fine vascular capillaries up to the tips of the highest branches. After the expansion of the leaves, however, transpiration soon reaches values which can no longer be met by the pumping action of the roots alone. Indeed, it can be shown that the transpiring surface now exerts a sucking action which may even lead to tensions and negative pressures in the vessels. This is especially so if, when transpiration is at its height, more water is evaporated than can be absorbed, and the plant goes into a temporary water deficit, amounting perhaps to 30 per cent of its normal content.

The sucking action of a damp surface on the interior of a body saturated with water can be clearly demonstrated by a model. A block of plaster of Paris is cast containing the mouth of a funnel. The block is saturated with water and a glass tube filled with boiled water, is attached to the stem of the funnel. As evaporation takes place from the damp surface of the plaster, so water is drawn up the tube. If the lower end of the tube is dipped into mercury, then this adheres to the water column and it may be raised up to over 760 mm (Fig. 234, a). A leafy branch attached to the glass tube in place of the plaster block will show exactly the same sucking action (Fig 234, b).

The tensions which arise during vigorous transpiration in the columns of water filling the vessels can be demonstrated in the following way. If a vigorously transpiring shoot is cut beneath the surface of mercury, then within a moment of the cut the mercury is pulled a considerable distance up into the vascular tracts. Also the fact that many plants wilt after cutting, even when rapidly placed in water, can be explained by air bubbles having been drawn into the vessels during the cutting, thus interrupting the continuity of the fine columns of water which is essential for the conducting mechanism. Since the water of the columns adheres to the lateral walls of the vessels, also saturated with water, it does not come away from the walls when the column is under tension, but the walls are drawn elastically inwards, principally between the bars of thickening. Indeed, the functional significance of the characteristic thickening of the walls of the vessels is only comprehensible in relation to strains of this kind. When the flow is halted, as, for example, by the cutting described above, these lateral indentations smooth themselves out, and in so doing draw in the mercury or air. This reduction in the diameter of the vessels during

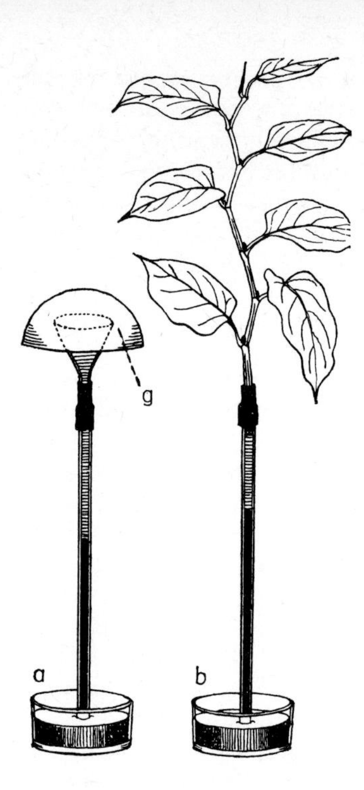

Fig. 234. **a.** Suction exerted by a gypsum block (g) saturated with water. **b.** Similar effect produced by a transpiring leafy branch.

vigorous transpiration has in fact been directly observed under the microscope. It has even been found that whole tree trunks show a slight reduction in their diameter during the midday hours.

If then it is accepted that the transpiring leaves draw up water through the vessels, the question immediately arises why the fine columns of water, running without interruption in the vascular tracts up from the roots into the nerves of the leaf, do not break when they are under tension. The answer is that the columns of water in the vessels are in effect within closed capillaries and thus cannot break because the cohesive forces between the water molecules are extremely high (of the order of 250 atmospheres; see also p. 379). The walls of the vessels are completely saturated with water, and this water is firmly attached by adhesion to the material of the wall. The vessels are also always enclosed within a mantle of parenchymatous cells lacking intercellular spaces, thereby preventing any lateral access of air. Thus, for the columns of water to break apart or disintegrate the cohesive forces between the molecules of water, forces which rarely have to be considered, must be overcome. The size of these forces is such that even in the highest trees the columns do not break.

This explanation of the mechanism by which water ascends the plant is accordingly known as the cohesion theory. It must, however, be made clear that the actual forces propelling the water are to be sought not in the cohesion of the water but in the difference in saturation which exists between the damp earth and the hydrated body of the plant, and the atmosphere, and for which ultimately the heat energy of the sun is responsible The plant with its capillary structures, which allow the cohesion of the water to become effective, is as it were merely intercalated into this saturation gradient. Consequently, even when large quantities of water are raised up no further expenditure of energy on the part of the plant is necessary.

Since the intermicellar spaces of the cellulose walls are saturated with water, and since these walls form unbroken strands of considerable length, perforated only by the numerous plasmodesmata, there is the theoretical possibility that the cohesive force of the water can be propagated through the walls over short distances. In leaf tissue especially, the cell walls appear to conduct water particularly rapidly from the vessels to the outer walls of the epidermis. When there is a plentiful supply of water from the roots it may therefore reach the epidermis by this means as well as by way of the protoplasts and vacuoles of the cells. It is, however, improbable that the tension of the cohering filaments of water is transmitted through the walls of the cortex of the root into the root hairs, and from there into the soil water. The Casparian strip in the endodermis (see p. 103) may in fact be a special barrier to such a conduction of water in the cell walls. Many observations, particularly those of root pressure, nevertheless show that an active participation of the protoplast is involved in at least the entry of water into the root. How the entering water flows through the cells of the cortex can not yet be stated with certainty.

III. The mineral nutrients[10,11,12]

The uptake, conduction and giving off of water are fundamental processes in the life of a plant, since a controlled water balance, by creating an internal environment in which the various metabolic processes can go to completion, is essential for normal physiological activity. The next most important processes are those concerned with nutrition. As an organism grows and develops, its increasing mass requires the supply and incorporation of the necessary structural materials, and these in turn must be taken up and synthesized from the food supply. The metabolic processes by which the various components of the cells are built up from inorganic and organic food materials, some of them quite simple, are said to be anabolic. At the same time a fully developed organism can remain alive only by continuously utilizing energy, so that the mere maintenance of life requires metabolic processes, just as an engine can remain going only if it is supplied with fuel. The fuel of the living organism is again drawn from its nutrition. Thus the nutrition of plants, and the origin and formation of all those substances used in both its construction and maintenance, are important problems in the physiology of metabolism. In the following we shall first consider how plants obtain their mineral nutrients, and the fate of the inorganic materials absorbed.

As was considered in more detail earlier, the plant body contains a certain percentage of minerals. In the higher plants, apart from the aquatics, these can be taken up from the soil only through the roots. It is clear that as mentioned on p. 213 an ash analysis by itself can give no information about the extent to which the elements found are essential for life. Therefore, if we wish to study the mechanism of uptake, significance and fate of the mineral nutrients in greater detail the first essential is to discover which elements of the ash are absolutely essential for the normal growth of the plant.

Nutrient solutions.[11] Since soil is a very complex substance (see p. 233), difficult to standardize chemically, physiologists have tried with success to grow plants in a simple inert substrate, such as fine quartz sand previously calcined and extracted with boiling water, irrigated with a solution of various chemically pure mineral salts. Indeed, well-aerated, simple aqueous solutions of this kind support vigorous root-growth, and the plants undergo completely normal development (Fig. 235). This technique known as 'hydroponics', is to some extent used in horticulture.

In this way it has been discovered that some of the elements frequently encountered in plant ash (e.g. sodium, so important for animals (see p. 237), chlorine aluminium, silicon) can in fact be dispensed with by many plants without damage. Apparently their presence in these plants represents nothing more than an inert component of the plant body.

Among the cations, usually only potassium, calcium, magnesium and iron are absolutely essential. Nitrogen, sulphur and phosphorus, generally taken up into the higher plants in the form of the anions NO_3', SO_4'' and PO_4''', are also necessary. In recent years it has, of course, been discovered that many plants require a number of other elements in addition, if only in minute quantities, for growth and development to be normal These discoveries have been so recent because the quantities of which these so-called trace elements are required are far beyond the limits of chemical analysis. They frequently get into nutrient solutions undetected simply as impurities in the other salts, or in other ways.

Fig. 235. *Fagopyrum esculentum* in water culture. **A.** Nutrient solution without potassium. **B.** Potassium present. After Pfeffer.

The amounts of minerals which, for example, go into solution from the walls of even good glass vessels may have undoubted physiological effects, so that essential prerequisites for the investigation of problems of mineral nutrition are the most careful cleaning of all apparatus and the use of the purest chemicals. Even quartz will hold on tenaciously to certain trace elements; any paraffin used to cover the test vessels must also be absolutely pure.

Among the trace elements, boron is especially important. Deficiency of this element in the soil may lead to serious damage in cultivated plants (see p. 238), and similarly with manganese, copper, zinc and molybdenum. Very different amounts of these trace elements are required by different plants.

Those salts which are quite indispensable need to be supplied to the plant only in relatively weak concentration. A large number of nutrient solutions capable of supporting the normal growth of green plants have been discovered by trial and error The following will serve as an example of the composition and concentration of such solutions:

Knop's solution: Distilled water 1000 ml; $Ca(NO_3)_2$ 1.0 g; $MgSO_4 . 7H_2O$ 0.25 g; KH_2PO_4 0.25 g; $FeSO_4$ trace. Von der Crone's solution: Distilled water 1000 ml; KNO_3 1.0 g; $MgSO_4 . 7H_2O$ 0.5 g; $CaSO_4 . 2H_2O$ 0.5 g; $Ca_3(PO_4)_2$ 0.25 g; $Fe_3(PO_4)_2$ 0.25 g. Nowadays a drop of the so-called A–Z solution of Hoagland is added to these solutions to provide trace elements. The composition of Hoagland's solution is: Distilled water 1000 ml; $Al_2(SO_4)_3$ 0.55 g; KI 0.028 g; KBr 0.028 g; TiO_2 0.055 g; $SnCl_2 . 2H_2O$ 0.028 g; LiCl 0.028 g; $MnCl_2 . 4H_2O$ 0.389 g; H_3BO_3 0.614 g; $ZnSO_4$ 0.055 g; $CuSO_4 . 5H_2O$ 0.055 g; $NiSO_4 . 7H_2O$ 0.059 g; $Co(NO_3)_2 . 6H_2O$ 0.055 g.

As can be seen, the total concentration of salts lies between 0.16 and 0.25 per cent. It is very interesting to note that these salts will only support growth when they are present in the balancing proportions given above. A solution of a single one of these salts by itself may even be poisonous. When present in the proportions given, an ionic antagonism is set up, similar to that briefly discussed in connection with hydration phenomena (see pp. 215 and 237).

Uptake of nutrient salts.[11, 12] The mineral component of the soil provides the normal source of the nutrient salts required by the plant. As a result of the weathering of rocks (brought about by atmospheric factors, hydrolysis, oxidation, solution by carbon dioxide, or by other acidic excretion products of the soil micro-organisms, plant roots or decaying plant remains), a certain proportion of the salts go into solution and then become absorbed on to the soil particles (see p. 235). From here they are taken up by a comparable process by the plant organs.

A soil capable of supporting plant growth displays a highly complex mixture of the most diverse substances, derived by the weathering of the minerals contained in the rocks. A particularly important component, besides sand (quartz) and chalk, is clay, produced by the breaking down of various rocks, especially the minerals of the group containing montmorillonite and mica. Clay particles are very small (less than 2 μm in diameter) and have undoubted colloidal properties. They are capable of imbibing water, and bind the rest of the soil particles into a plastic mass. The negatively charged colloidal particles are flocculated particularly by calcium ions, producing the crumb structure so desirable in cultivated soils. In this way the soil acquires the necessary empty spaces (see Fig. 230), and in good conditions about half this free space is filled with water and the remainder with air, essential for the respiration of the roots. The clay is also mainly responsible for the absorptive binding of the ions, but here the humus of the soil also plays a considerable part. This is a dark organic material of complex composition, produced principally by the decomposition of plant debris. A certain amount of humus must accompany the other components mentioned above to produce a fertile cultivated soil.

The binding of the soluble mineral salts by the soil colloids is an important feature because the danger of their being washed out (leached) by rain is thereby lessened. This is certainly true of the almost immovable phosphate, while calcium and nitrate may be comparatively easily leached away.

Although submerged water plants are able to absorb salts from the surrounding water not only by their roots but also to some extent over their whole surface, in the higher land plants, with the exception of certain rootless epiphytes, the root is the normal means of entry of the mineral nutrition. The root hairs, short-lived but constantly being formed, come into intimate contact with the soil particles (see Fig. 110D), and are especially well adapted for the uptake of salts.

The problem now before us is how the salts reach the interior of these cells, and from there the interior of the plant body. First of all it is clear that salts can only be taken up when in solution, since the finely porous cellulose walls of the cells cannot be penetrated by coarse and rigid particles. So far as the adjacent protoplasm is concerned, the discussion of osmosis and turgor in the preceding section may have given the impression that the cytoplasmic layer is completely impermeable to all osmotically active substances, and therefore to inorganic salts. This, however, is far from true, and the assumption can be legitimately made in considering the water balance of the cell only because water is always able to enter and leave the vacuole faster than any other substance. Here it should not be forgotten that the water traverses both the plasmalemma and the tonoplast, so that with respect to osmotic phenomena the whole cytoplasmic layer can be considered as a single semi-permeable membrane. In respect of the uptake of materials, however, we are concerned not so much with passage into the vacuole but rather entry into the plasma ('intrability' as opposed to permeability), since it is in the plasma that the essential metabolism occurs. Consequently we must consider passage through the plasmalemma itself. Detailed studies have long shown certain connections between the solubility of a substance in lipid and its ability to penetrate the plasmalemma. For many organic substances the relation is simply that the more lipophilic a substance is the more readily it enters the cell, but this clearly does not apply to inorganic salts, sugar and amino acids. Since the plasmalemma undoubtedly contains lipid (see p. 22), the rapid penetration of lipophilic substances can be readily accounted for (lipid theory of permeability). The control of the entry of water, sugar, and salts demands another explanation. This is afforded by the ultrafilter theory of permeability. This theory, based principally on studies made on the procaryote *Beggiatoa* (see p. 448), envisages a connection between the size and form of the molecule, and its ease of penetration. Small molecules penetrate more readily than large. Hence the plasmalemma must possess pores, which, as in a filter, will not allow those molecules through which are either too large, or possibly also electrically charged. The two ways of penetrating the membrane, namely passage by solubility in lipid, and the traversing of pores, are combined in the so-called mosaic theory. This envisages the plasmalemma as a mosaic of lipoidal regions, separated by areas of hydrophilic protein, the molecules of which surround minute pores.

Hitherto the electron microscope has provided firm evidence neither for the existence of pores, nor for the mosaic structure of the plasmalemma. Transverse sections usually reveal a uniform membrane with layer of lipid at the centre, and protein on both outer surfaces (see p. 22).

The thermal movement of the molecules making up the membrane might well lead to temporary local gaps or pores, which could provide an opportunity for movement through the membrane, but this is still purely hypothetical.

Nevertheless, the taking up of materials into the cell cannot be controlled solely by the permeability of the plasmalemma. It is well known that in many plants certain salts or ions (e.g. K^+, I') reach concentrations in the cell sap much higher than those in the external solutions from which they were taken up. The plant therefore possesses a well-marked storage capacity. Moreover, individual salts or ions are by no means taken up from a nutrient solution in equivalent quantities, but certain ions are clearly absorbed preferentially. There are also changes with time; often, for example, a plant will take up many more minerals in its juvenile stages than later. The plant is therefore able to vary the extent to which it takes up minerals as well as to accumulate them. Both these phenomena require special consideration.

The ability of a plant to absorb one ion more than another may lead to a marked change in the pH of a nutrient solution. If, for example, only the H^+ and NO_3' ions are absorbed from a solution of potassium nitrate, which, of course, contains the ions K^+, NO_3', OH' and H^+, then the solution becomes alkaline because of the relative fall in H^+ and rise in OH'. Potassium nitrate would thus be physiologically alkaline. Similarly, if NH_4^+ and OH' are taken up preferentially from a solution of ammonium chloride the solution remaining becomes more acid, and ammonium chloride would then be a physiologically acid salt. Such pH changes can be of great importance in the chemical dissolution of the soil.

While selectivity of absorption could possibly be explained by the different capacities of the substances concerned to dissolve in the lipid, or penetrate the pores of the bounding membrane, differences in the absolute amounts taken up could result from the entering substances being removed from solution by chemical bonding with components of the plasma or cell sap, or even by being precipitated as insoluble compounds. Penetration of the substances into the cell would continue so long as they were removed from the concentration and diffusion gradients in this way.

Such a process is seen, for example, when a cell whose vacuole contains tannin is placed in a solution of methylene blue. The dyestuff forms an insoluble compound with the tannin, and is consequently taken into the vacuole in large quantities even from dilute solution.

A substance entering the cell can also be removed from the concentration gradient by adsorption on to some component of the cytoplasm, or even on to the cell wall.

By adsorption is generally meant the loose and reversible uptake and binding of a substance on to some bounding surface. Thus, for example, certain colloidal silicates of the soil humus and so on, are capable of binding to their surfaces many gases and even substances from solution. Electrical forces are probably involved in the adhesion. Similar adsorption phenomena occur at the interface between two immiscible fluids and also on the bounding surface between suspended colloidal particles, e.g. of albumen, and the medium in which they are dispersed. In colloids where the total surface of the dispersed particles is enormous adsorption phenomena may be very highly developed. Many biochemical processes occurring in the cytoplasm can in fact be understood only if adsorptive bindings and concentrations of this kind are envisaged. Consequently, they must be taken into account in considering the uptake of materials.

Compared with many organic materials, the inorganic salts belong to the less easily adsorbed substances. With these salts, however, the strong electrical charges involved are of great importance. Since the separation of paired ions requires strong electrical forces, the adsorption of the ions is additive, i.e. the adsorbed ion draws its ionic partner with it. The tenacity of the binding, however, is diminished by the strength of the electric charge on the partner. With a common cation the anions are adsorbed in the following descending series of readiness: $OH' > CNS' > I' > NO_3' > Br' > Cl' > HPO_4'' > SO_4''$; and with a common anion the corresponding series of cations is: H^+, Ag^+, $HG^{++} > Al^{+++} > Zn^{++} > Mg^{++}$, $Ca^{++} > NH_4^+$, $Cs^{++} > Rb^{++} > K^+ > Na^+ > Li^+$. A more strongly adsorbed ion may dislodge one more weakly adsorbed and take its place, and in this the partner ion plays no part. This provides the basis for exchange adsorption, i.e. the process in which only one ion is taken up from the solution of a salt, and another weaker ion is simultaneously displaced from the adsorbent to become the partner of the ion of the opposite charge remaining in the solution. This gives the appearance of the adsorption of free individual ions. Adsorbents which bear strong electric charges, such as ferric hydroxide, clays, silica kaolin and also the protein of the cytoplasm, usually adsorb only by ionic exchange.

That ion-exchange plays a role in the uptake of salts by the root, and that this process is also important for the release of minerals in the soil, can be easily demonstrated experimentally. If roots are allowed to grow out on to a highly polished marble plate, after a time a fine etching of the marble can be observed at the points of contact of the root and the marble surface. The root has excreted carbon dioxide, and the otherwise insoluble calcium carbonate has been taken into solution. In certain cases yet other acids may be produced, and this process would be important in freeing salts in the soil. Occasionally K^+ and Ca^{++} ions are excreted, so conditions may be such that the exchange adsorption of both anions and cations can occur.

Since, as described earlier, the salts in soils are extensively bound by adsorption to the soil particles, they can always become free for diffusion as soon as the plant provides in exchange ions which are more strongly adsorbed. Among the cations the very strongly adsorbed H^+ ion is particularly important. Hydrogen ions are always available in the plant as a consequence of its respiratory metabolism (see p. 275), and are capable of displacing all other adsorbed cations. It is now widely accepted that respiratory processes are bound up with salt uptake, but the extent of this interaction cannot be considered further here.

In many instances penetration of the plasmalemma and uptake into the cytoplasm cannot be explained solely in terms of diffusion or adsorption. There appear to be undoubted so-called 'transport mechanisms', the characteristic of which is that substances can be taken into the cell against a concentration gradient. These 'pumps' naturally require a source of energy, normally provided by ATP (see p. 266). The fact that such mechanisms in aerobic organisms (both plant and animal) also depend upon a supply of oxygen is further evidence of their involving work done by the cell. Although their nature is not understood in detail, enzymes in the bounding membrane may play a decisive role. The so-called carrier hypothesis, for example, envisages the presence of a specific carrier molecule at the surface of the plasmalemma which binds with the substance concerned, the complex then passing through the membrane, possibly by diffusion. Once within, the complex is dissolved and the carrier molecule returns to its position on the cell surface, leaving its partner free in the plasma. The details of how this process may occur are, however, still unknown.

A further mechanism possibly involved in the uptake of materials into the cell is pinocytosis. Sac-like invaginations of the plasmalemma can often be seen in electron micrographs. These possibly detach themselves from the plasmalemma and pass into the interior of the cell as small vesicles, reminiscent of the feeding process of *Amoeba*. The extent to which pinocytosis occurs, and its significance (which is possibly confined to the uptake of fluids) are not known in any detail.

Sufficient has been said to show that the apparently simple process of the uptake of mineral salts is in fact extremely complex. It cannot be explained satisfactorily in terms of a simple diffusion process, as was found possible for the uptake of water. Purely physico-chemical principles, such as those concerned with permeability, and adsorptive binding and storage, enter very largely into salt uptake, but it is also true that it is intimately connected with the physiological activity of the cells. As with all living phenomena, these physiological processes are very difficult to sort out, and it is not possible to separate those concerned with salt uptake and consider them in isolation from the many other processes going on, by which they are influenced. This inter-relationship explains why the rate of salt uptake may change, and why the illumination or shading of the aerial parts of the plant can influence the uptake of salts by the root.

Transport of mineral salts. Once salts are taken up by the absorbing root hairs into the cytoplasm, there arises the further problem of how they are distributed from the place of absorption to the various regions of a multicellular organism. Thus if the young growing point of the aerial shoot is to be supplied with the essential minerals there must be transport of salts from the sites of absorption in the root hairs through the entire length of the plant body. At present very little is known about the mechanism of this transport and the exact places at which it takes place from cell to cell (see p. 288). That the entry of salts into a cell has nothing to do with the simultaneous entry of water is shown by the fact that the uptake of salts can take place quite independently of that of water. Nevertheless, there can be no question that the transpiration stream ascending in the vessels is utilized for the subsequent transport of most of the absorbed salts. Occasionally, however, it appears that transport of certain minerals also occurs in the sieve tubes. This is predominantly in a downward direction and is of importance, for example, in the movement of materials out of yellowing leaves (see p. 288).

After they have been transported, by a mechanism still unknown (either through the intermicellar spaces of the cell walls (the so-called free space), or through the plasma), across the cortex of the root, the minerals taken up at the surface are passed into the vessels, at least in part, by a kind of secretory process, and from here they ascend in the transpiration stream. Exactly as the parenchyma cells adjacent to the vessels extract water from them according to their needs, so now can the salts be gradually taken up from the water of the vessels, and possibly also from solutions penetrating the walls (see p. 230). Although the most vigorous uptake of salts generally occurs in the early stages of a plant's development, it may happen, particularly if the capacity for selective uptake is not well developed, that an excess of minerals, especially of calcium, arises in the transpiring tissues during vegetative growth. Thus, for example, the amount of calcium in leaves may progressively increase so that as transpiration goes on the surplus calcium salts are partly left behind in the cell walls, especially in the outer walls of the epidermal cells, and partly precipitated as insoluble salts of organic acids (e.g. as calcium oxalate). Heavy showers of rain are capable of washing out soluble salts, such as those of potassium, left behind in cell walls. The chalk glands seen on the leaf margins of many species of *Saxifraga*, so far as the salts are not actively excreted, owe their origin to the evaporation of droplets of guttated water containing calcium salts.

The importance of minerals for metabolism.[11] Salt uptake and salt transport are the prerequisites for the incorporation of the minerals into the metabolism proper. In many cases we are very poorly acquainted with the part played by individual salts. Generally speaking, two possibilities have to be considered. First, certain salts or elements may be essential for life because they are direct components of one or another important organic molecule, or may be essential for their activity (see p. 259 seq.). Often, as for example in the molecules of certain enzymes vitamins and hormones, only minute traces of these minerals are necessary. Secondly, salts may, since their ions are bearers of strong electric charges, exert a more general influence on the colloidal condition of important components of the cytoplasm, without undergoing a direct chemical combination with them.

It has already been mentioned in the discussion of hydration phenomena that salts or ions are capable of influencing to a very great degree the hydration sheaths of certain colloidal substances, such as the proteins. This they do by increasing or diminishing the charge of the colloid by adsorption or neutralization. Since not only the degree of hydration but also the capacity of the colloid to adsorb and enter into reactions is simultaneously affected, it is clear that individual salts can in this indirect way have a pronounced effect upon the course of metabolism. Thus the activity of enzymes can often be influenced by single ions (see p. 259). This profound effect of salts on colloids is the reason for the poisonous effect of ions when administered singly (e.g. the effect of magnesium salts), and also for the beneficial effect which mixtures of salts of definite composition have upon the organism. Such mixtures of salts are present in nutrient solutions (as described earlier), and also in cytoplasm and the cell sap. The too strong adsorption of an individual ion is here prevented by the tendency of the others to displace it.

In this connection it should be noted that the uptake of any one ion in preference to others may bring about indirectly changes in pH, i.e. increase in the relative concentrations of hydrogen or hydroxyl ions. Of all the factors of importance in colloid chemistry, changes in pH probably have the most drastic effect on the condition of the protoplasm. This is related to the fact that, as mentioned earlier, the hydrogen and hydroxyl ions are the most strongly adsorbed of all the ions, and accordingly are the most effective in changing the charge of the colloidal particles. Consequently, the physiologist must devote considerable attention to the concentration of the hydrogen and hydroxyl ions not only in nutrient solutions and soils but also within the living matter itself (see p. 280).

Our knowledge of the role of individual minerals, either as ions or elements, depends essentially upon observations of the deficiency symptoms which appear if the elements in question are omitted in culture experiments, or are replaced by others, as well as upon direct chemical analyses of certain vital components of cells. It is very striking how in general the various elements essential for life either cannot be replaced or can be replaced only with difficulty, even by elements closely related chemically. A brief account of each element individually follows:

Potassium. The amount of potassium in plant ash is particularly high (see table, p. 213). It seems that it occurs in plants not firmly bound in organic compounds, but entirely in ionic form. It is therefore probable that its importance lies principally in its effectiveness in promoting the hydration of the plasma colloids. Its activation of the enzymes concerned with protein synthesis may be related to this effect.

Young growing cells are particularly rich in potassium, and the greatest uptake of this element takes place in the young stages of the plant. Potassium influences specifically the water content of cells, and in turn the intensity of the assimilation of carbon dioxide. It cannot be replaced by lithium and sodium, although in certain instances rubidium and caesium may be effective, a difference probably due to greater radius of the hydrated ion of the first two elements. The most important mineral yielding potassium in the soil is felspar.

Sodium is usually present, although most plants seem able to dispense with this element. Where it does promote growth the manner of its action is quite obscure. Sodium is, however, an essential element for the blue-green alga *Anabaena*, as it is for all animals.

Calcium. This element, which is always to a certain extent in an antagonistic relationship with potassium, is usually found in soils as the carbonate, sulphate or phosphate. While potassium tends to cause the negatively charged plasma colloids to swell calcium, as mentioned earlier (p. 214), has the opposite effect, so that the degree of hydration of the protoplasm is strongly influenced by the relative concentrations of these two ions. Calcium does, however, appear to have some specific importance as a structural material over and above its general effect upon the colloids.

A calcium compound of pectin (see p. 65) enters into the structure of the middle lamellae of the cell walls. Accordingly, in culture experiments absence of calcium rapidly causes death of the growing point. Calcium salts of organic acids, especially calcium oxalate, occur in quantity in many plants, although it would not be legitimate to say that calcium was used entirely for the purpose of neutralizing and combining with organic acids.

While the higher plants require calcium in large quantities (and the chemically related strontium is no substitute for it), some algae and fungi appear able to grow satisfactorily with only traces of this element.

Magnesium. When alone in solution this element, which in soils is usually present as carbonate, acts as a strong poison. This poisonous effect is eliminated when the magnesium is accompanied by calcium and the relative concentrations of the two elements are maintained within a narrow range (ionic antagonism). This is probably explained by an effect on the cell colloids. At the same time it is undoubtedly true that magnesium forms complex organic compounds and is an important component of the cell. It cannot be replaced by any other of the soil bases.

Thus magnesium enters into the structure of the chlorophyll molecule, and is accordingly absolutely essential for the synthesis and activity of this important pigment. Protopectin contains magnesium in addition to calcium. Magnesium also acts as an activator of the enzymes involved in energy-transfer in the catabolism of carbohydrates (carboxylase, enolase, transphosphatases; see p. 266). It is also involved in maintaining the stability of ribosomes (see p. 284). Thus it is understandable that magnesium is an essential element even for those plants lacking chlorophyll.

Iron. The importance of this element for the plant also arises from the fact that it enters into complex molecules essential for the cell, even though only minute amounts of the metal are involved.

Iron in a divalent or trivalent form occurs principally in certain enzymes which play a part in respiration (see p. 272). A typical symptom of iron deficiency, which may arise, for example, through precipitation in soils rich in chalk, is the appearance in the plant of *chlorosis*, i.e. the imperfect formation of the chlorophyll. This is a secondary effect, since the chlorophyll itself contains no iron. The colourless ground substance (stroma) of the chloroplasts is, however, rich in iron and it is generally assumed that it is involved in photosynthesis as an electron-carrier (ferredoxin; see p. 249). Iron-containing cytochromes are also present in chloroplasts.

Manganese. This element also appears to be essential in small quantities for many plants. It probably plays a part similar to that of iron in metabolism, especially in relation to the dehydrogenase enzymes. Like iron, it seems to occur in the flavoprotein enzymes (see p. 271). Thus it is concerned in nitrate reduction, the decarboxylation of oxalosuccinic acid in the Krebs cycle (see p. 274), and in the evolution of oxygen in photosynthesis (see p. 249).

Deficiency of manganese can also cause chlorosis. The 'grey-speck' disease of oats and other plants, occurring principally on moorland soils, is a consequence of either a lack of manganese in the soil or the locking up of the manganese in a form no longer available to the plant (manganese can be taken up only in the divalent form). Citrus plantations often suffer manganese deficiency. Manganese is probably also concerned with nitrate reduction. Fungi, e.g. *Aspergillus niger*, like the higher plants, also require manganese unconditionally.

At present very little is known for certain of the action of the remaining trace elements.[12] They often appear to be concerned with certain definite enzymatic processes.

Lack of **copper** on acid heather moors causes cereals to give a very small yield of grain, and also gives rise to a deficiency disease in cattle. It is not improbable that copper (incorporated in a flavoprotein enzyme (see p. 271)) serves as an oxidation catalyst. An enzyme, polyphenol oxidase, found in potatoes, pumpkin seeds and certain fungi, consists of a copper proteid. The same is true of ascorbic acid oxidase (see p. 271). A copper-containing redox compound also plays a role in photosynthesis (see p. 249). Lack of **boron**, as mentioned earlier, causes the so-called 'heart rot' in sugar beet and mangolds, arising from an effect upon the meristematic tissue. Many other economic plants, such as barley, rape, kohlrabi, cauliflower, tomatoes, carrots, etc., require boron. The pollen of tomatoes and waterlilies will germinate only in the presence of minute traces of this element in the stigmatic secretion. **Zinc** appears to be essential for the growth of many higher (e.g. cereals, citrus, vines, etc.) and also for lower plants (e.g. fungi such as *Aspergillus niger* and algae). Those enzymes concerned with the transfer of hydrogen which work with the coenzyme nicotinamide adenine dinucleotide (NAD, see p. 266), as well as the lyase aldolase contain zinc, and it may also be a component of a further enzyme of photosynthesis. Some plants, e.g. *Viola calaminaria*, thrive better on soils containing zinc (see p. 213). Certain bacteria require **molybdenum** besides iron for the fixing of atmospheric nitrogen. Some higher plants, such as tomatoes, lettuce, cauliflower and many Leguminosae, demand molybdenum, perhaps in connection with the assimilation of nitrate. Molybdenum is also probably present in flavoprotein enzymes. **Chlorine** appears to be involved in some way in oxygen evolution in photosynthesis.

Our knowledge of the significance of those anions essential for growth (NO_3', SO_4'' and PO_4''') is somewhat more complete.

Nitrate. In autotrophic plants the nitrate anion is usually the starting-point for the whole of the nitrogen and protein metabolism (see p. 282).

The proteins which form an integral part of the protoplasm, owe their content of nitrogen principally to the $-NH_2$ groups of the amino acids. These $-NH_2$ groups are formed in the course of protein synthesis from the NO_3' ions taken up by the plant from the soil. Usually they can be replaced without any difficulty by NH_4^+ ions.

Sulphate. The sulphate ion is similarly present in the soil, and is essential for the synthesis of important amino acids and proteins containing sulphur (e.g. cystine, cysteine; see p. 279). Among other substances containing sulphur are vitamin B_1, biotin, and certain enzymes (see pp. 267, 273 and 286). In all cases the sulphate ion is apparently first reduced to the level of $=S$ or $-SH$.

Phosphate. This ion is also taken up from the soil (where it is mostly derived from the mineral apatite), but unlike the nitrate and sulphate ions, it is not reduced. Instead, in the course of the various metabolic processes, it is merely esterified.

Such phosphate compounds are found, for example, in the nucleotides and nucleic acids (see p. 280 seq.) of the nucleoproteins, in the prosthetic groups of many enzymes (see p. 271), in phosphatides, which are important components of the protoplasm (e.g. lecithin) and in the phytin of seeds, where the phosphoric acid is combined with inositol (which can act as a growth stimulator; see p. 300). Finally, phosphoric acid combines with sugars and their derivatives, and the phosphate bonds in these compounds have the important function of acting as energy carriers in respiratory metabolism (see p. 266 seq.).

Use of fertilizers.[6, 10, 14] These brief observations on the specific role of the various minerals in metabolism must suffice here. Although we know so little about them individually, it is nevertheless clear that these inorganic materials taken up from the soil by the plant through its roots are, despite their being relatively small in amount, essential parts of living matter, and without them it could not exist. The supplying of minerals to plants becomes therefore of great economic importance, especially since our intensive methods of cultivation drastically disturb the compensating balance that is always found in nature. Under natural conditions all the minerals present in plant ash are returned to the soil by the decay of fallen material, but with every harvest very considerable quantities of minerals are taken away from the soil by man. It is therefore necessary to make good this loss by the addition of minerals, and to dress the soil with appropriate artificial fertilizers. This is especially important also for the maintenance of a rich microflora (see p. 241), essential for a fertile soil. Nitrogen, phosphorus and potassium especially must be continually added to the soil if the high agricultural yields to which we are accustomed are to be maintained. Liming is also usually necessary to check acidity and to preserve other properties of the soil, e.g. its crumb structure, which is important for aeration and adsorption.

Although the use of inorganic fertilizers gives little cause for concern in relation to human health, the fact that roots can take up certain organic substances may lead to the use of animal dungs, now that animal feeds frequently contain drugs, such as penicillin, which are subsequently excreted, having quite serious consequences. Similar considerations hold with regard to the use of complex poisons and carcinogens in sprays which penetrate the cuticles of leaves.

It may be of importance in plant ecology and sociology that the roots of many plants excrete organic compounds into the soil (see p. 290).

IV. Carbohydrates

Water and salts are an essential part of every living cell. No less important for the building up of an organism are those substances which are justifiably referred to as 'organic', since they were originally known only as products of living forms. These are the intrinsic constituents and products of the protoplasm, such as the carbohydrates, fats, proteins and so on. Their formation and subsequent fate in the plant cell pose major problems in the physiology of metabolism, and indeed in physiology generally. The carbohydrates especially occupy a central place in metabolism. In plants, as in animals, they are the most widely distributed respiratory material, the oxidation of which provides the basal metabolism yielding the energy for the maintenance of the living system. In addition, carbohydrates are the source from which most of the other carbon compounds found in cells are derived. To investigate the origin and fate of the carbohydrates in the plant is thus to gain an insight into the fundamental processes of plant life.

To begin with, a fundamental distinction must be drawn between two large groups of plants. Although it has previously been stated that it is possible for plants to complete their development in a purely inorganic nutrient solution, this holds only for the greater part of the green plants. It is clear that these plants must produce all their carbohydrates within themselves; they are said to be completely autotrophic. There are, however, particularly among the so-called lower plants, e.g. in the bacteria and the fungi, very many organisms, usually colourless, which are not able to grow in such inorganic media. To support such organisms the medium has to be supplemented with already synthesized carbohydrates, and often with organic nitrogen as well. To distinguish them from the autotrophs, these plants are referred to as heterotrophs.

The heterotrophic plants and their special mode of nutrition are considered in more detail later (see p. 290). At this stage the autotrophic plants interest us particularly, since they present the problem of the primary origin and synthesis of the organic substances, espec ̇lly the carbohydrates, in a clear-cut fashion. When it is realized that the whole of the Animal Kingdom, including Man, as well as the heterotrophic plants and parts of plants, require for their nutrition already synthesized carbohydrates, and that the only source of these is the autotrophic plants, it is clear that we are dealing with a problem that concerns the very existence of life on this planet.

A. The assimilation of carbon dioxide

1. Photosynthesis[13]

When an autotrophic higher plant, with its roots in a purely inorganic nutrient solution, grows in a few weeks to a considerable size it is at once clear that the carbon of the organic mass generated before our eyes cannot have come from the nutrient solution. No other possibility remains but that the carbon, found not only in the carbohydrates proper, such as sugar, starch, cellulose, etc., but also in the numerous other organic compounds, such as in the amino acids of the proteins, comes from the air. It is in fact derived from the carbon dioxide of the atmosphere.

In order to make this concept, at first surprising, of the importance of the atmospheric carbon dioxide more evident, it is important to have a clear picture of the quantities of carbon continually present as carbon dioxide in our atmosphere, and of the amount locked up in plants. Numerous analyses have shown that the amount of carbon dioxide in normal air is remarkably constant. It amounts on the average to 0.03 per cent by volume, a slightly higher concentration being present directly above the soil surface, and, for obvious reasons, close to large towns and industrial areas.

A litre of air thus contains 0.5–0.6 mg carbon dioxide; a cubic metre 0.5–0.6 g carbon dioxide, or 0.13–0.16 g carbon. It has been estimated that some 300×10^{12} kg of carbon are bound up in the Plant Kingdom as a whole, while the total carbon dioxide of the atmosphere may amount to about 2100×10^{12} kg, equivalent to some 570×10^{12} kg carbon. In a year's growth $13-22 \times 10^{12}$ kg carbon might well be incorporated into plants, so that, even if these estimates are only approximately correct, the atmospheric store would be rapidly exhausted were there no replenishment. Uninterrupted growth of plant life at the expense of the atmospheric carbon dioxide can therefore only be possible if carbon is simultaneously returned to the atmosphere.

It is a remarkable fact that vital activity is itself responsible for this returning of carbon dioxide to the earth's atmosphere in large quantities. Living things characteristically undergo respiration, and in most cases, by a process to be described later, thereby give out carbon dioxide. A grown man, for example, breathes out about 1 kg carbon dioxide in twenty-four hours. The respiration of animals and of higher plants generally is of a similar high intensity. Even when supplemented by volcanic gases and industrial activity, the respiration of men, animals and higher plants would not be able to maintain the constant level of carbon dioxide observed in the atmosphere. The remainder is supplied by a host of so-called lower organisms, especially the soil bacteria, whose production of carbon dioxide must be quite enormous. In 1 ml of a good agricultural soil at least several thousand million bacteria are to be found, and it has been estimated that the hourly production of carbon dioxide per hectare is 2–5 kg, while in woodland soils it is substantially more. In a year this amounts to such a quantity of carbon dioxide that it is quite clear that these micro-organisms play a decisive role in maintaining the content of our atmosphere. It is the respiratory activity of these organisms which leads to the air at the soil surface having the highest content of carbon dioxide.

It is therefore actually possible for the carbon assimilated in a year's growth of the Plant Kingdom to be replaced by the respiratory activity of all the forms of life on our earth, and thus to explain the constancy of the amount of carbon dioxide in the atmosphere.

The process of carbon dioxide assimilation. That green plants, and especially leaves, actually extract carbon dioxide from the air (or in the case of submerged water plants, from the water) can be directly demonstrated. A stream of air or water containing an exactly known amount of carbon dioxide is passed over a plant, and after the passage re-analysed (see Fig. 237). It can then be shown quantitatively that the reduction in the amount of carbon dioxide corresponds exactly with the increase in the amount of carbon in the plant. This up-take of carbon dioxide by the plant is referred to as assimilation, because the carbon of the carbon dioxide apparently becomes incorporated into that of the plant. This process takes place, however, only if the plant is simultaneously illuminated. Light is therefore absolutely essential for this synthetic process, and it is accordingly also called photosynthesis. The green pigment chlorophyll must also be present, since non-green parts of plants, such as petals and roots, can be shown to be incapable of assimilating carbon dioxide in the light. Finally, yet another substance besides carbon dioxide must enter into the reaction, since the increase in dry weight as a consequence of assimilation is always greater than the weight of the carbon dioxide absorbed less the weight of the oxygen evolved. This substance is water, which, as will also be discussed in more detail later, becomes chemically involved.

Not only is material taken in, however, in carbon assimilation, but oxygen is also given out.

This excretion of gaseous oxygen during carbon assimilation can be made very evident with certain water plants, as shown in Fig. 236. Provided the water is saturated with air, a fine stream of gas bubbles ascends from the intercellular spaces of the cut stem as soon as the plant is illuminated. This rapidly ceases when the light is screened. A chemical analysis of the gas evolved shows that it is largely oxygen.

The ratio of the volume of oxygen evolved to that of the carbon dioxide absorbed, the so-called *assimilatory quotient*, has been shown by extensive investigation always to be unity.

If the experiment is so conducted that the oxygen liberated during assimilation is not evolved in a gaseous form, but remains in solution, then the amount of oxygen in the water can be accurately estimated titrimetrically. Knowing the assimilatory quotient to be unity, a quantitative estimate can now be made of the intensity of assimilation. With land plants a stream of air is usually passed at an appropriate velocity over a leaf enclosed in a small cuvette, and the changes in the carbon dioxide in the air determined. These changes can be quantitatively determined in a number of ways. An apparatus which can

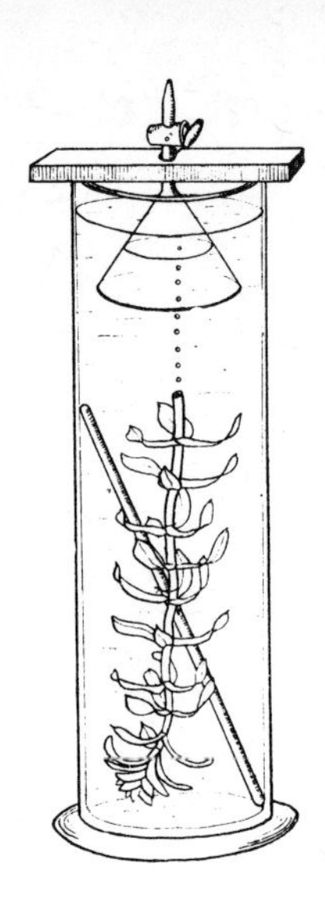

Fig. 236. Excretion of oxygen from a detached shoot of the water plant *Elodea* during assimilation. The ascending gas bubbles are collected in an inverted funnel.

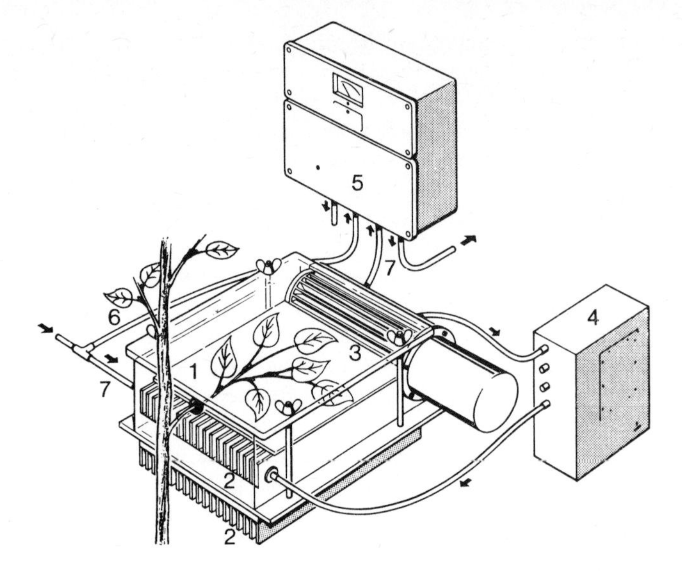

Fig. 237. Apparatus for measuring the carbon dioxide flux of a normally growing plant. A leafy shoot is enclosed in a transparent cuvette (1), in which the temperature and humidity are continuously equilibrated electronically with the external conditions. This is done by means of Peltier units (2), a blower (3), and an humidifier (4). The infra-red gas analyser (5) records the difference between the concentration of carbon dioxide in the external air (6) and that in the air leaving the chamber (7).

also be used for plants in the field is shown in Fig. 237. The carbon dioxide content of the air in the cuvette is here measured continuously by means of infra-red absorption. For more exact work volumetric methods are used; for example, the manometric determination of oxygen levels in the Warburg apparatus. Unicellular green algae (e.g. *Chlorella*) are usually used for these investigations because unavoidable errors, such as those introduced by changes in stomatal aperture, are not then encountered.

The important discovery that plants, by means of photosynthesis, can build organic substances from the carbon dioxide of the air, goes back to Priestley, who observed in 1771 that plants 'improve' the air (for animals). In 1779 Ingen-Housz, a Dutch doctor at the Viennese Court, showed that this was true only of green plants, and only when they were illuminated. It was also he who, according to Senebier, in 1796 first drew attention to the newly discovered carbon dioxide as a possible plant foodstuff. However, not until 1804 was the Swiss, de Saussure, able to prove that a green plant could live with the carbon dioxide of the air as its sole source of carbon, that oxygen was released during the process of carbon assimilation, and that water was involved in the reaction. Goethe knew nothing of these new discoveries, and they only slowly displaced the general belief that plants drew all their nourishment from the humus substances of the soil. The energy relationships of photosynthesis were first considered by Robert Mayer in 1845.

The oxygen is liberated from the chloroplasts embedded in the cytoplasm, so there must be the site of the photosynthetic reaction. Chloroplasts are very numerous: in a leaf of *Ricinus communis*, for example, it is estimated that 50×10^6 chloroplasts are present in 1 cm^2 of the leaf surface, of which perhaps 90 per cent are to be found in the palisade tissue.

The site of oxygen evolution can be demonstrated in an elegant manner by the use of chemotactic (see p. 382) bacteria sensitive to oxygen. If, for example, sharply delimited spots of light are allowed to fall on a *Spirogyra* cell immersed in water containing these bacteria they can soon be seen collecting at those parts of the wall immediately beneath which there is an illuminated part of the chloroplast. Clearly, as a consequence of assimilation, oxygen is being liberated at these places. Where a spot of light falls on an intermediate region of cytoplasm, no aggregation of bacteria occurs, since no oxygen is evolved (Fig. 238). With the same method it is possible to show that blue and red light are more effective than other wave lengths in promoting photosynthesis. If a spectrum of light is allowed to fall along an algal filament, the most marked accumulations of bacteria are to be found in these regions of the spectrum.

Even after only a few minutes' illumination, small starch grains can often be seen forming in the chloroplasts (Figs. 38G; 41D, E), referred to as assimilatory starch.

It can be shown that the appearance of this assimilatory starch is directly connected with the assimilation of carbon dioxide. If a leaf, which has been made starch-free by several days' shading, is covered with a stencil from which letters have been cut, or with a suitable photographic negative, it can be shown by means of the iodine reaction that, after a certain period of illumination, only where the light has penetrated has starch been formed (Fig. 239). Since the intensity of photosynthesis, and to a certain extent also the formation of starch, depends upon the intensity of the illumination, it is even possible by subsequent staining of the assimilation starch with iodine to obtain a clearly recognizable print from a photographic negative (Fig. 240).

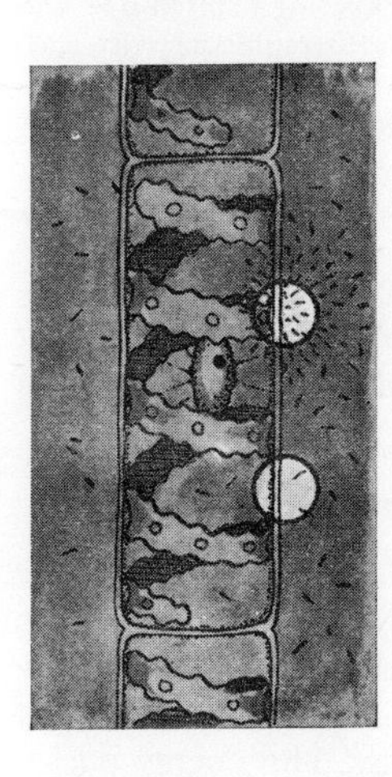

Fig. 238. Cell of *Spirogyra* showing oxyphilous bacteria aggregating where the band-like chloroplast is illuminated. Where the light falls between the bands of the chloroplast no bacteria collect. After Engelmann, modified.

Since starch is a substance insoluble in water which may be formed in the plastids at any time from the sugar conducted to them, even in darkness, we must assume that starch is formed secondarily in photosynthesis, the primary product being sugars. Also there are some leaves, e.g. in monocotyledons (*Tulipa, Allium*), in which even in intense assimilation starch can never be detected, but only sugars. Concerning the assimilatory products of the lower plants see p. 449.

As the formula of a hexose shows, the formation of a molecule of $C_6H_{12}O_6$ must at one time or another involve six molecules of CO_2. If we assume, as did de Saussure, that the hydrogen required in addition is taken from the water already present in the assimilating cell (as in fact can be demonstrated with the help of radioactive tracers), then the process of assimilation can be summarized in the following equation:

$$6CO_2 + 12H_2O \longrightarrow C_6H_{12}O_6 + 6O_2 + 6H_2O$$

According to this equation, for each molecule of carbon dioxide taken up a molecule of oxygen must be evolved thus accounting for the observed constant value of the assimilatory quotient. The equation also shows how, because of the utilization of part of the water, the increase of dry matter during assimilation will exceed in mass that of the carbon dioxide absorbed less the oxygen evolved. The manner of reformation of the water will become clear later (p. 245). This equation does not, however, reflect the actual reactions taking place, as will be explained in detail in the following.

As the heat of combustion of the sugar shows, a glucose molecule contains a considerable amount of energy. The complete combustion of 1 g mol of glucose (= 180 g) to carbon dioxide and water will liberate 675 kg cal. Such a reaction is said to be exothermic. Conversely, this amount of energy must be added to the system if sugar is to be synthesized from the simple compounds carbon dioxide and water. This is clearly the

Fig. 239. The iodine reaction done on a leaf which was covered by a stencil. Light was able to reach the leaf only through the stamped-out letters of the stencil, and only in these places was starch formed. After Sierp.

Fig. 240. The result of covering a starch-free leaf of *Tropaeolum* with a photographic negative and exposing it to the sun. After removal of the chlorophyll an aqueous iodine solution colours the assimilatory starch formed during the exposure and produces a positive print. After Molisch.

reason why light is essential for carbon assimilation by green plants. The light must supply the energy which is used in the plant to bring about the endothermic synthesis of sugar.

$$\underset{264\text{ g}}{6\text{ g mol }CO_2} + \underset{216\text{ g}}{12\text{ g mol }H_2O} + 675\text{ kg cal} \longrightarrow \underset{180\text{ g}}{1\text{ g mol }C_6H_{12}O_6} + \underset{192\text{ g}}{6\text{ g mol }O_2} + \underset{108\text{ g}}{6\text{ g mol }H_2O}$$

The peculiar nature of plant photosynthesis can now be seen. Not only does it bring about in plants in a unique way the formation of sugar from chemically simple precursors, a process without which life on our earth would be quite impossible, but also at the same time solar energy is trapped and stored as chemical energy. This energy is later utilized for the various activities of the cell.

This is the one process by which energy is secured in large measure by the living forms on the earth and continuously stored. It has been estimated that the quantity of solar energy acquired by plants in this way amounts to 2.5×10^{17} kg cal in a year. This energy, however, is made use of by all living things, animals as well as plants. The sugar in which the energy is stored is oxidized in respiration and the energy thereby released is utilized in other vital processes (see p. 262). Even one of the basic sources of industrial energy of today, coal, is derived from the products of assimilation of earlier epochs. It is therefore not surprising that this remarkable biological process increasingly attracts the attention not only of biologists but also of chemists and chemical engineers, and that attempts are repeatedly being made, so far in vain, to copy it artificially. Were these attempts successful, the consequences for human nutrition and the provision of industrial energy would be quite impossible to predict.

Chemistry of carbon assimilation. At present still very little is known about the physico-chemical peculiarities of carbon assimilation. Nevertheless, intensive research in the last few years has yielded much new information about this very complicated process. Only continued research, however, will show the extent to which the following account, in a number of respects tentative, has to be amended.

If the level of oxidation of the carbon in carbon dioxide is compared with that in carbohydrate

$$O{=}C{=}O \quad \text{and} \quad H{-}\overset{|}{\underset{|}{C}}{-}OH$$

it is at once clear that every molecule of carbon dioxide must chemically undergo reduction, i.e. hydrogenation, with the loss of one atom of oxygen. For this reason it was for long believed (and is still maintained by Warburg) that the oxygen evolved in photosynthesis is derived directly from the carbon dioxide Most workers today, however, take the view that the oxygen is generated by a so-called photolysis of water

$$H_2O \rightleftharpoons H^+ + OH'$$

The hydrogen ions then provide the agent reducing the carbon dioxide (the detached oxygen being reduced to water), and the hydroxyl ions, losing their electrons, yield water and oxygen (see Fig. 244)

$$2OH' \longrightarrow H_2O + \tfrac{1}{2}O_2$$

Although little is yet known in detail of the photolysis of water, the following observations support the validity of the concept.

1. There are certain pigmented bacteria (green and red sulphur bacteria) which live anaerobically in media rich in hydrogen sulphide. In photosynthesis these release sulphur, not oxygen. Apparently they utilize the hydrogen sulphide in place of water as the source of hydrogen for the reduction of the carbon dioxide:

$$CO_2 + 2H_2S \longrightarrow (H\,.\,C\,.\,OH) + 2S + H_2O$$

(The symbol H . C . OH here signifies merely a reduction product of carbon dioxide.) Photosynthesis in which there is no evolution of oxygen is also termed photoreduction.

By analogy with the process in these bacteria, normal photosynthesis can be denoted by

$$CO_2 + 2H_2O \longrightarrow (H\,.\,C\,.\,OH) + O_2 + H_2O$$

the water serving as the source of hydrogen.

Certain bacteria (e.g. the purple bacteria), even though grown aerobically, can utilize organic substances such as fatty acids and alcohols, and even molecular hydrogen itself, as sources of hydrogen ('hydrogen donors') for photosynthesis. Naturally no oxygen is evolved in these conditions.

2. The actual proof of a dissociation of water comes, however, from the observation that isolated chloroplasts, and even fragments of chloroplasts, are still able to yield oxygen in the light, even in the absence of carbon dioxide, provided that suitable hydrogen acceptors are present (e.g. potassium ferricyanide, quinone, or nicotinamide dinucleotide (NADP)). Since no carbon dioxide is assimilated, there is no formation of sugar. This important reaction is named after its discoverer, the Hill Reaction.

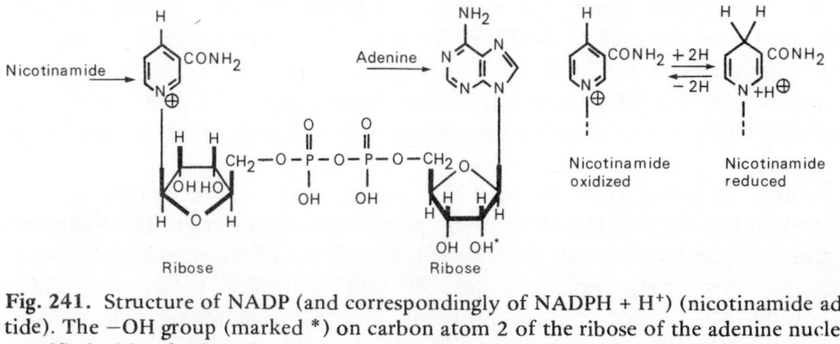

Fig. 241. Structure of NADP (and correspondingly of NADPH + H^+) (nicotinamide adenine dinucleotide). The –OH group (marked *) on carbon atom 2 of the ribose of the adenine nucleotide is esterified with a further phosphate. The relationship of the oxidized and reduced forms of the nicotinamide are shown on the right (cf. Fig. 255).

It was especially significant to discover that even NADP can act as a hydrogen acceptor, becoming thereby NADPH + H^+ (Fig. 241), since this substance had already been recognized as a hydrogen carrier in respiratory metabolism (see p. 270).

3. Labelling with the oxygen isotope $^{18}O_2$ has also shown that the oxygen evolved most probably comes from the water. If the isotope is present in the carbon dioxide, then the oxygen evolved lacks radioactivity. Conversely, if present in the water, then the oxygen is radioactive. When it is realized that, in the reduction of the carbon dioxide, water is

necessarily reformed, the success of such experiments depends upon the newly synthesized water not being immediately redissociated. This, and other technical difficulties, probably account for the results of such experiments not having been as clear cut as had been hoped.

A further important discovery was that isolated chloroplasts are capable, even without carbon dioxide and oxygen, of taking up inorganic phosphate and generating energy-rich phosphate bonds. Here we are concerned with ATP (adenosine triphosphate), formed from ADP (adenosine diphosphate) and phosphate

$$\mathrm{ADP} + \mathrm{P} \xrightarrow{h\nu} \mathrm{ATP}$$

($h\nu = \Delta E$). Evidently this is the compound in which the light energy is stored as chemical energy (Fig. 242). This substance also figures in respiration as an energy-bank (see p. 266). When a phosphate group is detached from the ATP, a certain quantity of energy is released (under standard conditions 7 kg cal per g mol) which is then available for the assimilation of the carbon.

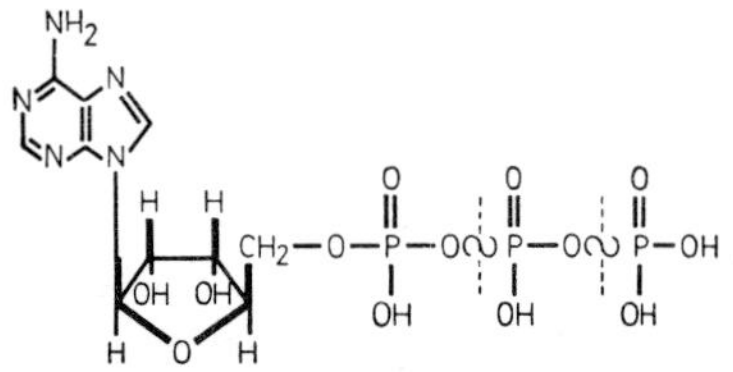

Fig. 242. Structural formula of adenosine triphosphate (ATP), showing the two energy-rich phosphate bonds (~).

We can thus recognize two fundamental processes of photosynthesis. First, the changing of a carrier molecule with hydrogen (NADPH + H^+), the hydrogen, generated by the photolysis of water, subsequently being used for the reduction of carbon dioxide, and, second, the transforming of light energy into chemical energy, and the temporary storage of this energy in ATP.

The pigments. The details of these two processes are not yet fully known. For light to be chemically effective it must first be absorbed by certain molecules in which, in a manner to be described later, the absorption brings about certain changes. The most important of these molecules, the photosynthetic pigments, is undoubtedly chlorophyll, in particular chlorophyll a. There are in fact many algae in which this is the only photosynthetic pigment present. Recently it has been shown that even within chlorophyll a itself there are several forms which, probably as a consequence of loose attachments to different proteins, differ in their absorption spectra and functions. Two groups of these variants are recognized, one, referred to as photosystem I, chlorophyll a_I, or sometimes as chlorophyll 700, shows absorption maxima at 682, 695 and 700 nm, and a second, referred to as photosystem II or chlorophyll a_{II}, shows maxima always a little lower than these values (e.g. 650, 674 nm, etc.). Both groups work together in an ordered manner, as will be described later.

Evidence of various kinds has led to the view that only a few of the very many chlorophyll molecules present in a chloroplast are actually photochemically active. The majority are believed to transmit the light energy they absorb by a form of resonance to the relatively few active molecules ('energy sinks'). Some 250–300 chlorophyll molecules, differing in kind according to the species, are probably grouped around a single photochemically active molecule of chlorophyll a. Every chloroplast naturally contains an enormous number (probably more than 10^6) of these 'phytosynthetic units'.

In many chloroplasts there is not only chlorophyll a in its various forms, but also yet other pigments. In higher plants about one-third of the chlorophyll is chlorophyll b, which differs in the higher state of oxidation of the pyrrole ring II and in its absorption spectrum (Fig. 243a). Other pigments include the carotinoids, e.g. the orange-red β-carotin (Fig. 243b, top) and the xanthophylls, similar to the carotins, but oxidized (Fig. 243b, bottom). Much autumn colouring of leaves is attributable to the carotinoids, since, although the chlorophyll of the deciduous leaves is degraded, the carotinoids remain.

The blue and red algae owe their colour to the presence of phycocyanin and phyco-

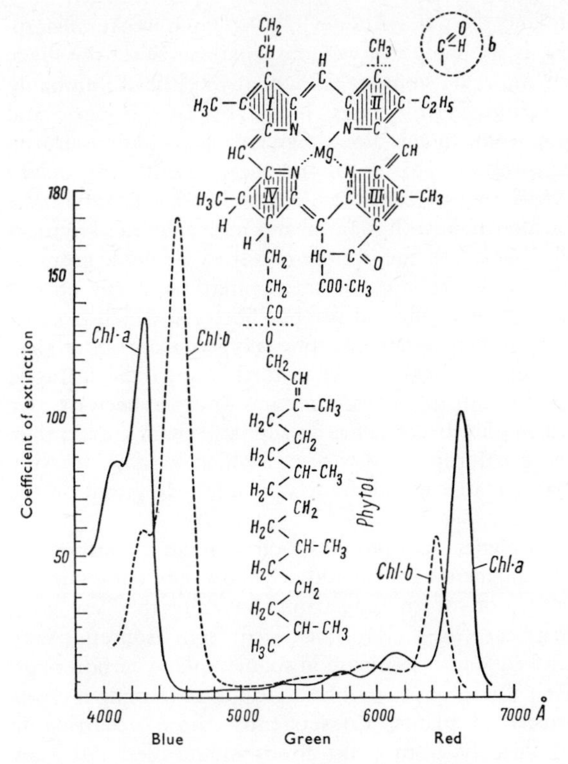

Fig. 243a. Structure and absorption spectra of chlorophylls *a* and *b* in solution in ether. After Zscheile and Comar.

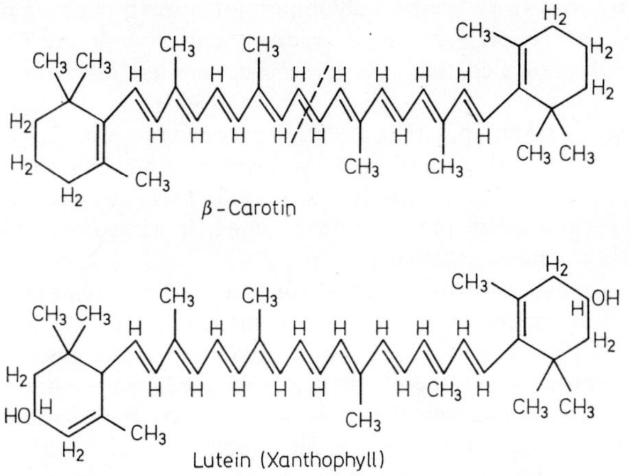

β-Carotin

Lutein (Xanthophyll)

Fig. 243b. Structural formulae of β-carotin and lutein (a xanthophyll).

erythrin (or their complexes with protein), which are related to the bile pigments (four pyrrole rings in an open chain, linked with protein). These accessory pigments mask the green of the chlorophyll. The accessory pigments as a whole (including chlorophyll b) probably pass on the light energy they absorb to the photochemically active chlorophyll a.

In the red algae, which live at depths at which most of the long-wave light has been absorbed by the water, the usefulness of an additional red pigment, absorbing the remaining shorter wavelengths, is easily understood. In the diatoms and brown algae the abundant carotinoids (e.g. the brown fucoxanthin, a carotinoid containing several oxygen atoms), which also often conceal the chlorophyll, may serve the same purpose. These pigments also absorb strongly in the blue-green region of the spectrum.

The structural formulae of chlorophylls a and b, together with the absorption spectra of these pigments in ethereal solution, are shown in Fig. 243a. The absorption spectra *in vivo*, which largely correspond to the intensity of carbon dioxide assimilation in different coloured lights (action spectra), are shifted a little to the right. Nevertheless, it is quite clear that the maxima lie in the blue and the red, with inevitably a much weaker absorption in the green. Even this, however, is probably not without significance for the shade flora of woodlands. The chlorophyll molecule consists of a plate-like 'head', probably rhombic in symmetry, consisting of four pyrrole rings, linked by HC bridges, and arranged around a central magnesium atom. In the blood pigment haemoglobin an iron atom is found in an exactly similar position. This porphyrin 'head' of the chlorophyll molecule is hydrophilic. To the pyrrole ring IV, however, is attached a long, aliphatic, diterpene chain, the so-called phytol, that is both lipophilic and hydrophobic. Phytol is also found in vitamin K. Chlorophyll b bears on the pyrrole ring II an aldehyde group in place of the $-CH_3$ group of chlorophyll a. Still further small variations occur on this ground plan. Bacteriochlorophyll, occurring in the red sulphur bacteria, which likewise carry out photosynthesis, has two hydrogen atoms and one oxygen atom more than chlorophyll a. The absorption spectrum reaches somewhat further into the infra-red region, even more so when several red carotinoids are present (purple bacteria; see p. 245). The chlorophyll of the green sulphur bacteria has an aliphatic chain shorter than that of phytol, while in the diatoms a chlorophyll c is found which wholly lacks the aliphatic chain. Chlorophyll d, occurring in the red algae, has an aldehyde group on the pyrrole ring I.

Many different carotinoids occur in plants. As examples β-carotin and a xanthophyll (lutein) are shown in Fig. 243b. These pigments are tile-red or yellow, and consequently absorb principally in the blue-green region of the spectrum. Basically they are long, unsaturated, hydrocarbon chains (e.g. carotin, $C_{40}H_{56}$), derived from isoprene, with conjugated double bonds. The chains frequently terminate in six-membered carbon rings, which in the xanthophylls may bear $-OH$ groups, or other groups containing oxygen. Carotinoids are synthesized exclusively in plants. Consequently those occurring in animals, including man, are derived initially from plant foods in the diet. For Man, β-carotin, which occurs abundantly in carrots, is especially important since, by halving of the molecule and the attachment of an $-OH$ group, it readily yields vitamin A. The carotinoids of tomatoes and rose hips, which lack a terminal ring, are less effective as precursors of the vitamin.

When a quantum of light energy is taken up, either directly or by conduction, into an active chlorophyll molecule, the effect is that an electron is raised to a higher state of energy, or, if the inflow of energy be sufficient, entirely dissociated. This electron then either returns immediately to the chlorophyll molecule from which it has come, or is trapped by a redox system which is at a more negative potential.

A redox system consists of two substances which differ from each other only in their state of oxidation, the transition from the one to the other occurring readily in the appropriate conditions (see, for example, Figs. 241 and 255). Thus, a reduced 'donor' gives up one or several electrons to an 'acceptor', which thereby becomes reduced, and the donor correspondingly oxidized. The oxidized donor can then take up electrons from another donor with a greater 'electron pressure' (i.e. at a more negative potential) than itself, so once again becoming reduced, while the original acceptor passes on its electrons to another system, and thereby becomes oxidized. A definite electrical potential exists between the partners of a redox system which can be measured, the potential of the hydrogen electrode at pH 7.0 usually being taken as the standard of reference (see p. 280). The more negative the potential, the greater the electron pressure, and thus the reducing power. Whole chains of redox systems with falling potentials are found in the cell organelles (plastids, mitochondria), along which electrons are transported (the movement of electrons often being accompanied by that of protons). In the movement of an electron from a strongly negative to a positive potential, part of the electron energy can be stored in ATP (adenosine triphosphate) by the addition of a third phosphate group by an energy-rich bond to ADP (adenosine diphosphate) (Fig. 244). It is inconceivable that the chemical reactions of photosynthesis could take place in anything other than an ordered system, and it is thus very probable the photosynthetic pigments, redox systems and enzymes are arranged in definite arrays within the thylakoid membranes.

The coupling of the two light reactions. Recent investigations have shown that photo-

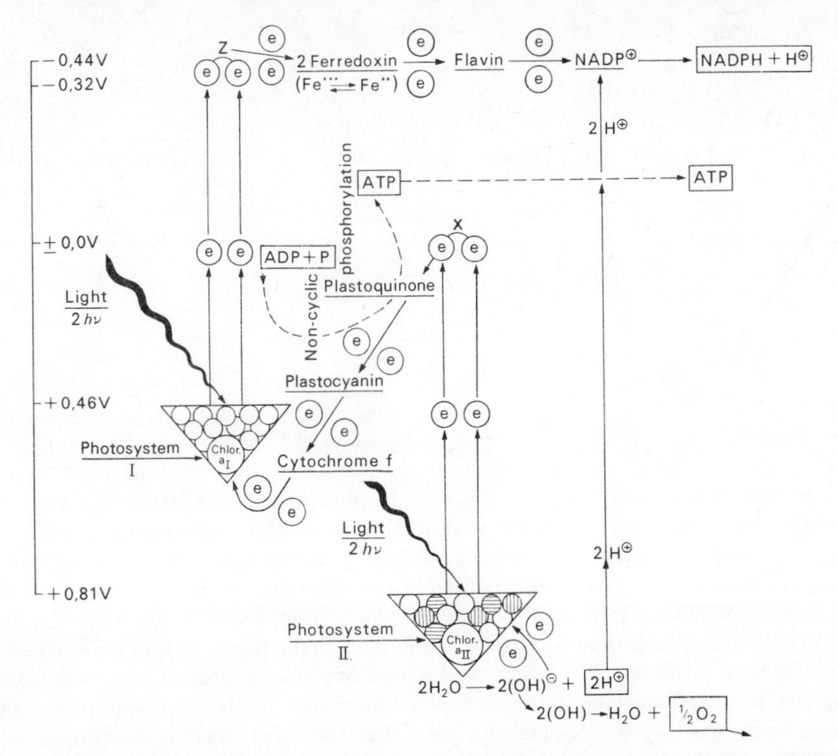

Fig. 244. Diagrammatic representation of the interrelationships of the two light reactions (photosystems I and II) in photosynthesis. The two triangles represent chlorophyll aggregates, each with an active chlorophyll a. The corresponding potential relationships (see p. 272) are shown on the left. ⓔ Represents an electron. The splitting of the water molecule and the liberation of oxygen are shown on the lower right, and the formation of NADPH + H^+ and ATP on the upper right.

synthesis involves two light reactions, occurring in photosystems I and II, but coupled with each other (Fig. 244). In photosystem I, illumination of chlorophyll a_I causes the displacement of two electrons which are then transported by way of an intermediate acceptor Z (not yet identified) to ferredoxin, a redox system containing iron and, like Z, with a strongly negative potential. Ferredoxin is a water-soluble protein, reddish-brown when oxidized, and its presence in leaves probably accounts for almost the whole of the iron they contain. The iron atom in the ferredoxin molecule can oscillate in valency between two and three. Electrons leave reduced ferredoxin by one of two pathways. In the first they move by way of a further redox system with a somewhat lower potential (flavin system) to $NADP^+$ (Fig. 244), where, together with two protons generated by the cleavage of water (Fig. 244, bottom), they form the hydrogen source $NADPH+H^+$, by means of which carbon dioxide is reduced. In the second the electrons return immediately to the chlorophyll, the energy being taken up into ATP. This form of phosphorylation (cyclic phosphorylation) is discussed further below.

When $NADPH+H^+$ is formed the oxidized chlorophyll a_I has to regain electrons from some other source in order to remain active. These electrons, together with the protons, come from the cleavage of water according to the following scheme: $2H_2O \rightarrow 2(OH)' + 2H^+$, yielding the protons, followed by $2(OH)' \rightarrow H_2O + \frac{1}{2}O_2 + 2e$ (as at the anode in the electrolysis of water), yielding both the electrons and the free oxygen of photosynthesis.

The mechanism of the splitting of the water molecule is obscure (it is known only that manganese is essential), but it seems to be dependent upon photosystem II (chlorophyll a_{II}). According to physical measurements on isolated chloroplasts, chlorophyll a_{II}, when illuminated, yields two electrons which are taken up by a redox system (X), with a negative potential of 0.8V. From here they pass over a whole series of redox systems with falling potentials (plastoquinone, plastocyanin (which contains copper), and cytochrome f), eventually reaching chlorophyll a_I (Fig. 244), which is thus regenerated. A part of the energy coming from these electrons is taken up in the formation of ATP (from ADP), this

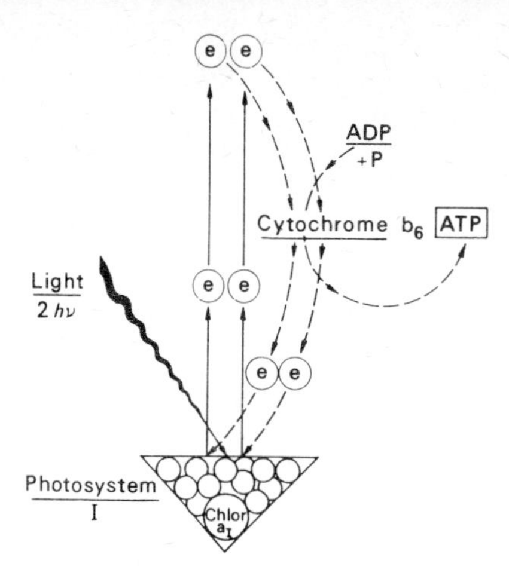

Fig. 245. Cyclic phosphorylation with the formation of ATP.

kind of phosphorylation being termed 'non-cyclic photo-phosphorylation' (see Fig. 244). As already mentioned, however, there is another way in which light energy can be transformed into chemical energy. This so-called 'cyclic phosphorylation' (Fig. 245) takes place when electrons, released by the excitation of chlorophyll a_I, escape being utilized in the formation of NADPH+H^+, and return to their source, probably by way of a redox chain (ferredoxin, cytochrome b_6), their energy also being taken up into ATP. Although it has so far been possible to demonstrate this kind of phosphorylation only with isolated chloroplasts in anaerobic conditions, it seems likely that it plays an important role in photosynthesis, since it is undoubtedly true that far more ATP is consumed in the synthesis of sugar than is generated by non-cyclic phosphorylation (twelve molecules ATP being required for the reduction of six molecules CO_2 to sugar, and six molecules ATP for the regeneration of the acceptors of the six molecules of CO_2 (see below)). It is this large requirement for ATP that has made Warburg's long-standing view that respiratory energy is used in photosynthesis appear very reasonable. The scheme shown in Fig. 244 summarizes the two light reactions and the subsequent electron transport, together with the intermediary and end products, NADPH+H^+, ATP and oxygen.

The assimilation and reduction of the carbon dioxide. After the formation of the reductant NADPH+H^+ (Fig. 244), made possible by the splitting of the water molecules and subsequent electron transport, and the accompanying transformation of the light energy into chemical energy in the form of energy-rich phosphate bonds in ATP, the conditions are set for the reduction of the carbon dioxide. Surprisingly, however, it has been found that the carbon dioxide molecule is not reduced directly, six reduction products then being assembled to a hexose. Instead, single carbon dioxide molecules first become bound to an acceptor which, at least in the alga *Chlorella*, is already a sugar, namely a phosphorylated pentose, ribulose diphosphate. A molecule of carbon dioxide becomes bound by a special enzyme, carboxydismutase (ribulose diphosphate carboxylase), to a molecule of the ribulose diphosphate, yielding a compound with six carbon atoms whose structure is still not exactly known. This molecule is unstable and immediately falls into two molecules of phosphoglyceric acid (Fig. 246). This phosphoglyceric acid is now reduced by the NADPH+H^+, the necessary energy being provided by the associated breakdown of ATP to ADP and phosphate, the exact converse of the reaction in respiration whereby glyceraldehyde is oxidized to glyceric acid with the formation of ATP (see p. 266). This, therefore, is the manner in which the carbon dioxide is reduced. In other plants other receptors of carbon dioxide may be present in addition to ribulose diphosphate, but this is still not clear.

The reduction of the phosphoglyceric acid in carbon assimilation is more complex than the oxidation of glyceraldehyde in respiration, and involves an enzyme with the sulphydryl (–SH) group. Noteworthy also is the formation of water, since the oxygen atom of the water comes from the carboxyl group. Two molecules of phosphoglyceric acid arise from the assimilation of one molecule of carbon dioxide. Both are reduced, but

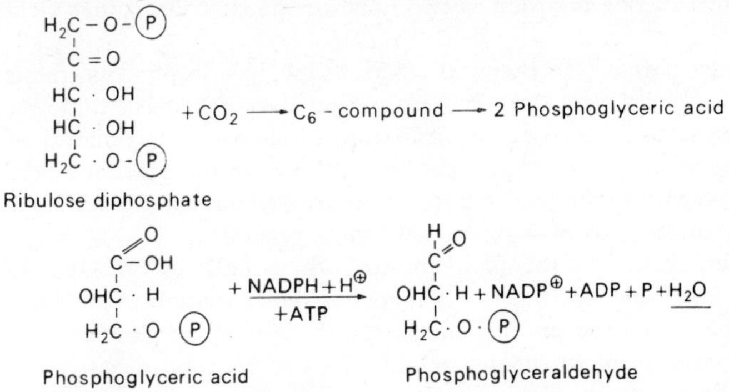

$$H_2C{-}O{-}(P),\ C{=}O,\ HC\cdot OH,\ HC\cdot OH,\ H_2C\cdot O{-}(P)\ (\text{Ribulose diphosphate}) + CO_2 \longrightarrow C_6\text{-compound} \longrightarrow 2\ \text{Phosphoglyceric acid}$$

$$\text{Phosphoglyceric acid} + NADPH + H^{\oplus} + ATP \longrightarrow \text{Phosphoglyceraldehyde} + NADP^{\oplus} + ADP + P + H_2O$$

Ribulose diphosphate

Phosphoglyceric acid

Phosphoglyceraldehyde

Fig. 246. The acceptance of carbon dioxide by ribulose diphosphate. Formation of phosphoglyceric acid, and reduction to phosphoglyceraldehyde.

in only one of the reductions will the oxygen of the water molecule come from the assimilated carbon dioxide. The two-fold requirement of NADPH+H$^+$ entails the splitting of four molecules of water (Fig. 244), with the formation of one molecule of oxygen. The assimilation of one molecule of carbon dioxide is thus accompanied by the release of one molecule of oxygen, agreeing exactly with the assimilatory quotient (see p. 241).

The reactions following the assimilation and reduction of the carbon dioxide are comparatively simple, and require no further addition of energy. The phosphoglyceraldehyde is partly transformed into dihydroxyacetone phosphate, and the two trioses combine to form fructose diphosphate (Fig. 247). One phosphate group is detached enzymatically, leaving fructose-6-phosphate, the sugar which can be regarded as the first product of photosynthesis. Fructose-6-phosphate can be readily transformed into glucose-6-phosphate (Fig. 247) and other metabolites.

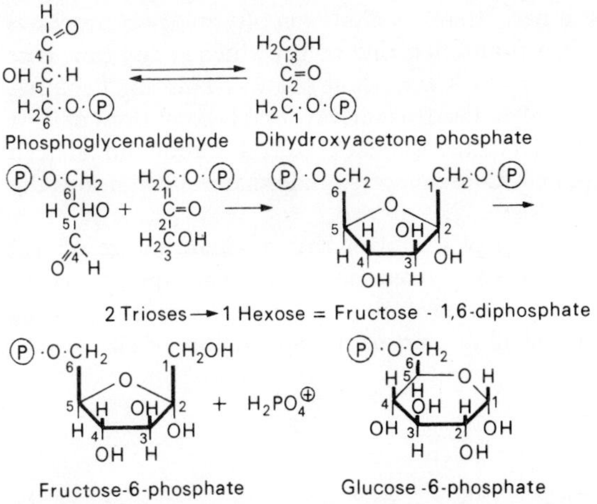

Phosphoglycenaldehyde ⇌ Dihydroxyacetone phosphate

2 Trioses → 1 Hexose = Fructose - 1,6-diphosphate

Fructose-6-phosphate + $H_2PO_4^{\oplus}$

Fructose-6-phosphate

Glucose -6-phosphate

Fig. 247. Equilibrium between phosphoglyceraldehyde and phosphodihydroxyacetone. Union of the two trioses to yield fructose diphosphate, leading, following the loss of a phosphate group, to the primary product of photosynthesis, fructose-6-phosphate. A further transformation results in glucose-6-phosphate. After Richter, modified.

The regeneration of the carbon dioxide acceptor. Naturally, if photosynthesis is to be continuous, the acceptor for the carbon dioxide (the ribulose diphosphate, or possibly some other substance) must be continually renewed. The greater part of the newly-formed sugar is, in fact, by way of a complex cyclic process (discovered in *Chlorella*, and named after its discoverer, the Calvin cycle) utilized for this purpose. Of every six molecules of hexose synthesized, only one can be regarded as a net gain, since five are required for the regeneration of six ribulose diphosphate molecules. The second phosphorylation of each of these molecules requires one molecule of ATP.

Indicating the sugars involved by the number of carbon atoms, the regeneration of the ribulose can be represented as follows. First, $C_6 + C_3 \rightarrow C_4 + C_5$. The erythrose C_4 reacts further, $C_4 + C_3 \rightarrow C_7$, and the seduheptulose then reacts with a triose, $C_7 + C_3 \rightarrow 2C_5$, a yield of three pentoses from one hexose and three trioses (whih are derived from the hexoses). Five hexoses (which can be regarded as two hexoses and six trioses) thus yield six pentoses.

As can be seen, the greater part of the chemical reactions involved in photosynthesis are independent of light (Blackman reactions), and are mediated by enzymes. Moreover, the transformation of fructose into glucose, the formation of sucrose, the removal of phosphate groups, and the synthesis of starch (which, formed between the lamellae of the chloroplasts, is often the first visible product of photosynthesis) do not belong to photosynthesis in the strict sense, but are part of sugar metabolism in general.

As was mentioned earlier, Warburg's theory of photosynthesis deviates considerably from that described in the foregoing. Central to his theory is the formation of a chlorophyll—carbon dioxide complex, and the photochemical splitting off of oxygen from the carbon dioxide. Warburg's concept of the energy relationships of photosynthesis is also quite different. He was the first to estimate the number of quanta of light energy required for the reduction of one molecule of carbon dioxide. In some instances his results showed up to three quanta, but in most other laboratories higher numbers (8—10) were found. Since 112.5 kg cal are necessary for the reduction of 1 g mol of carbon dioxide, and since the number of light quanta corresponding to 1 g mol (1 quantum per molecule in red light) corresponds to only about 42 kg cal, three quanta at least must be taken up and fully transformed into chemical energy. Warburg holds that photosynthesis, as other photochemical reactions, is a 'one quantum process' (i.e. one quantum of light energy activates one molecule). He envisages that part of the synthesized sugar is respired, the energy so released being transferred to ATP, so maintaining the energy balance. The experiments of the past few years do not, however, support this view.

The influence of individual factors on photosynthesis.[14] As is customary for a biological process, the intensity of photosynthesis is influenced by a number of quite different factors in a complex way. The general condition of the plant, the water supply, the extent to which the stomata are open, the intensity of the illumination, the temperature and the carbon dioxide supply all play a part. Here, as always in physiological processes influenced by a multiplicity of factors, it is found that that factor which at any time is at a minimum effectively limits the rate of the process as a whole (Law of Limiting Factors). Where, for example, the carbon dioxide supply is insufficient, favourable conditions of light and temperature cannot be fully utilized, since the assimilation of the carbon dioxide can only proceed to the extent that it is supplied. Consequently, under natural conditions the rate of assimilation is not uniform.

Under generally favourable conditions it may be taken that a square metre of leaf surface is capable of manufacturing 0.5—1.5 g glucose per hour. This approximately corresponds to the amount of carbon dioxide present in 3 m^3 air. A vigorously growing sunflower in the course of one day would fix all the carbon dioxide in 100 m^3 air.

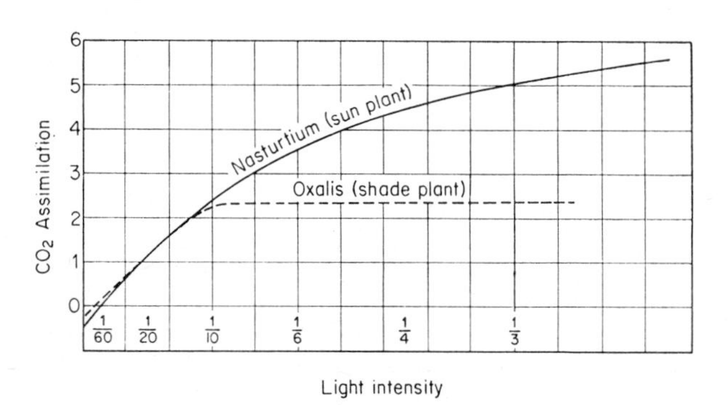

Fig. 248. Intensity of assimilation of a sun plant and a shade plant. Abscissa: fraction of full sunlight. Ordinate: uptake of carbon dioxide. The intersection of the curve with the zero ordinate represents the compensation point.

Light. In order to ascertain the effect of light intensity on the rate of photosynthesis, it is essential, in view of what has been said above, that all the other factors capable of influencing the rate, such as the temperature, carbon dioxide supply and so on, should be optimal. With these precautions it is found that photosynthesis increases in proportion with the light intensity. As Fig. 248 shows, however, at higher light intensities the increase in assimilation with increasing light becomes progressively less, until the light intensity becomes optimal, after which the rate of assimilation remains constant or may even fall with increasing light. Evidently at these higher light intensities other factors, such as lack of carbon dioxide, damaging effects of light and so on, become perceptibly limiting. (For discussion of the so-called optimum curve, see p. 334.)

In natural habitats individual plants are adapted to utilizing very different light intensities. Thus both shade plants and sun plants can be distinguished. Figure 248 shows the relationship between rate of assimilation and light intensity for a plant adapted to strong light (*Nasturtium palustre*) and a typical shade plant (*Oxalis acetosella*). It is seen that the maximum assimilation (as measured by the uptake of carbon dioxide) is considerably lower in the shade plant than in the sun plant, but that the maximum is reached at a substantially lower light intensity with the shade plant than with the sun plant. *Oxalis* has already reached its maximum assimilation rate at one-tenth of full daylight, while *Nasturtium* does not reach its maximum rate until considerably higher light intensities.

It is necessary, of course, to take into account their whole physiology to get a correct assessment of the sun and shade plants. Thus it is not only important to know how much carbohydrate a plant manufactures by assimilation in a given time, but also how much is used up again by the whole plant, including its non-assimilatory organs (e.g. roots), in the respiratory activity essential for the maintenance of life. Further, it should be noted that while photosynthesis is possible only in the relatively short periods of light, respiration continues also in the dark. Since at low light intensities photosynthesis is necessarily reduced, it is clear that there must be one intensity at which the amount of carbon dioxide assimilated is exactly equal to that generated in respiration. At this point, the so-called compensation point, there is no detectable change in the amounts of carbon dioxide and oxygen in the atmosphere surrounding the plant. At light intensities above the compensation point begins the so-called 'apparent' assimilation.

Generally speaking the respiration of shade plants is less vigorous than that of sun plants. Consequently the compensation point is exceeded at lower light intensities (Fig. 248), and apparent assimilation begins earlier. They are therefore capable of growing in situations in which sun plants would be incapable of making any net assimilatory gain.

These phenomena are naturally very important in relation to the establishment of plant cover, and the composition of the various plant communities. The more or less closed canopy of woodlands, for example, will have decisive effects on the ground flora. Even the individual leaves of a single plant can shade one another, causing such different light intensities that it may be possible to distinguish, not only functionally, but also often structurally, sun and shade leaves on the one plant (Fig. 290). In beech, for example, measurements have shown that the compensation point for fully insolated leaves on the south side may amount to 500 lux, while for shaded leaves on the north side the value may fall as low as 150 lux.

In dense beech, and also in *Abies* woods, only some 2–5 per cent of the incident light reaches the ground surface, but silver birch, larch and pine tend to allow more light (18–27 per cent) to penetrate. More important than sunlight which may reach the forest floor directly as 'flecks' is the more widely diffused 'sky' light, no matter whether the sky be clear or overcast (which may reduce the intensity by more than half). Summer clouds against a blue sky, as a result of reflection, may cause a higher intensity of diffused light than a cloudless sky. In temperate latitudes full sunlight at midday in summer reaches an intensity of $60-80 \times 10^3$ lux, and values of 10×10^3 may be obtained even when overcast. In polar regions the midnight sun yields an intensity of about 1000 lux, enough to exceed the compensation point. The light of the full moon, however, never exceeding 1 lux (usually about 0.25 lux), does not reach this value. The herbaceous flora of the forest floor cannot tolerate less than 1 per cent of full daylight, but in certain conditions bryophytes are said to thrive even at intensities as low as 0.05–0.01 per cent.

In water, light intensities, particularly of the longer wavelengths, diminish with increasing depth. For higher plants the compensation point is reached at depths of 5–8 m, naturally depending to some extent on the turbidity of the water. The adaptations peculiar to the red algae have already been mentioned (p. 247).

The following information will contribute to a wider understanding of these factors, so important in the fields of plant ecology and geography. Of the radiation incident upon the upper surface of a leaf, approximately 10 per cent of that within the spectral range 300–700 nm is immediately reflected. Longer wavelengths, for example, those in the infra-red range (700–800 nm), are reflected to an even greater extent (up to 45 per cent) by normal leaves, less so by the needle-like leaves of conifers. Consequently broad-leaved woodland shows up particularly strikingly in infra-red photography. About 10 per cent of the light passing right through the leaf. This transmitted light, the so-called 'green shade light' can be utilized by lower leaves or by the ground flora, since chlorophyll still absorbs, albeit weakly (see Fig. 243a), in the green region of the spectrum. A further amount of the incident light, probably about 10 per cent, is absorbed by the non-pigmented material in the leaf. The remaining 70 per cent is absorbed by the chlorophyll, and is thus available for photosynthesis. Nevertheless, only a very small proportion is in fact transformed into chemical energy, and stored in the form of sugar.

In one hour 1 m^2 of leaf surface can absorb about 200 kg cal from direct solar radiation. Since in this time on the average 1 g glucose will have been produced, utilizing 3.75 kg cal, the extent of the utilization of the radiant energy amounts to only 1.87 per cent. It is generally believed that in the spectral range 300–700 nm there is a gross gain of 2–3 per cent. Since in twenty-four hours the loss in respiration will amount to about 40 per cent, the net gain is much lower. In land plants the net gain probably represents the utilization of about 0.25 per cent of the radiant energy absorbed, in water plants (including the plankton) possibly as much as 0.5 per cent. The greater part is transformed into heat and dissipated, principally by transpiration.

Carbon dioxide. The concentration of carbon dioxide in the atmosphere is such that in many cases the photosynthetic apparatus is not working to full capacity. In full sunlight carbon dioxide is probably always a limiting factor, while with shade plants the carbon dioxide is fully utilized above $\frac{1}{10}$ full daylight. Consequently, it is possible with many plants to obtain an increased photosynthesis by raising the concentration of carbon dioxide, while maintaining all other conditions constant.

This fact can be put to practical use. In the culture of tomatoes, cucumbers, etc., in glasshouses the yield can in certain conditions be increased up to threefold by raising the concentration of carbon dioxide (up to a maximum of 0.1 per cent, higher concentrations cause damage in many plants). In populated places the concentration of carbon dioxide is already $2-2\frac{1}{2}$ times higher than in the open air, and is not so limiting to plant growth. The concentration of carbon dioxide immediately adjacent to the soil is also noticeably higher than in the open as a result of the intense respiration of the micro-organisms. This must be of considerable importance for the low-growing forms of plant life, since they obtain their carbon dioxide principally at this level, while the crowns of the higher trees, for example, in the absence of vertical air movements, are dependent almost entirely on supply from the atmosphere at large. Thus the stimulating effect of stable manure on plant growth depends not only on the direct action of the mineral salts it contains but also in part on the proliferation of the microflora and its respiratory activity.

The uptake of the carbon dioxide into the assimilating cells of the leaves—in land plants predominantly the palisade cells—is certainly a complicated process. Since at 15°C the percentage solubility of carbon dioxide in water is only about the same as its percentage presence in air, pure diffusion through the cell walls, covered by a cuticle and saturated with water, is too little to meet the continual and quite substantial demands of assimilation. The main path by which the carbon dioxide enters the interior of the leaf is consequently through the stomata.

As has already been mentioned in the discussion of stomatal transpiration, the capacity for diffusion through a finely porous surface is extraordinarily large. It has been determined by experiment that the amount of carbon dioxide diffusing through the open stomata of an epidermis reaches 66–90 per cent of the amount which would diffuse through the same area were the epidermis entirely absent. Naturally, closure of the stomata will impede this free passage. The regulation of the stomatal aperture, considered hitherto solely from the point of view of the water relations of the plant, effectively controls at the same time the intensity of the assimilation. If the plant is forced to close its stomata through lack of water the supply of carbon dioxide for assimilation is simultaneously removed. This is one of the many examples in physiology of how one metabolic process is closely interconnected with another. It appears, as already mentioned in relation to transpiration, that variation in stomatal aperture produces its greatest effect when the stomata are only partly open.

After traversing the stomata, the molecules of carbon dioxide are taken up by the cells of the spongy mesophyll or palisade tissue, where, before incorporation into the acceptor (see p. 250) and possibly transport to the reducing sites can take place, they are initially dissolved and bound as bicarbonate.

In many submerged water plants both dissolved carbon dioxide and calcium bicarbonate are absorbed over the whole leaf surface, while in others the absorption is confined to the lower surface of the leaf.

Where calcium bicarbonate is taken up, calcium hydroxide is excreted, after the extraction of the carbon dioxide, from the upper surface of the leaf. This becomes immediately converted again to calcium carbonate or bicarbonate. For this reason dirty, greyish-brown crusts of calcium carbonate are often seen on the upper sides of the leaves of *Helodea* or *Potamogeton*.

Temperature. Since the assimilation of carbon dioxide is a process bound up with the structure and vital activity of the protoplasm, and since it consists, except for the photochemical reaction, of a series of purely chemical events, it is inevitable that it will be affected by temperature (although the photochemical reaction itself is largely independent). In fact, with adequate illumination, the rate of assimilation at first increases with temperature. Eventually it reaches a maximum, beyond which increasing temperature either produces no further increase in rate or causes its decline because of damage to the protoplasm. Nevertheless, specific adaptations appear to exist in relation to the temperatures encountered in extreme habitats. Tropical plants, for example, show higher optimal temperatures than arctic.

For the majority of our native plants the temperature below which assimilation of carbon dioxide ceases is probably about 0°C or a little lower. In certain tropical plants, however, assimilation ceases at 5°C. On the other hand, many Arctic plants, lichens, Alpine shade plants and even native winter annuals (e.g. winter wheat) can still assimilate with a positive balance at −2 or −3°C, even under snow. With the evergreen conifers the minimum is likewise below zero (*Picea abies*, −3°C; *Abies alba, Picea maritima, Pinus silvestris* and *Pinus cembra* −6°C). Although the conifers of alpine regions appear to suspend photosynthesis completely during the winter months, short periods of freezing increase their resistance (frost-hardening, see p. 335). The yellowing of the needles is a visible indication of this change in the assimilatory apparatus, a slow return to normal not taking place until the spring. The nature of these changes is little known, as is also the precise manner in which freezing of the cell sap is prevented. Ice formation is more or less confined to the intercellular spaces. Alpine lichens (e.g. *Cladonia* spp.) are said to possess minima of the order of −22 or −24°C, at which temperatures their surfaces become dehydrated.

For many plants the optimum temperature for the assimilation of carbon dioxide lies between 20 and 30°C (but in potatoes is much lower, 18–20°C). Shade plants, however, and also many marine algae, have a very low optimum. The temperatures above which assimilation ceases generally lie between 35 and 50°C, but these values are little investigated and certainly vary in response to other factors. Maxima, and even optima, which exceed these values must clearly obtain in the blue-green algae which are confined to hot springs (50–80°C).

2. Chemosynthesis[14]

The essence of photosynthesis is that light energy is taken up and stored in carbohydrate molecules, the whole process being impossible without the supply of light energy. It has already been pointed out that the photochemical phase is only one small link in a long chain of dark reactions. Consequently, it is of great theoretical interest that there are some lower plants which assimilate carbon dioxide in complete darkness, and which are capable of living autotrophically in spite of the lack of pigments and of the photochemical phase facilitated by them. Here the energy necessary for the endothermal reduction of carbon dioxide must evidently come from the metabolism itself, or from certain special chemical reactions. Accordingly, this kind of assimilation of carbon dioxide is called chemosynthesis, in contradistinction to photosynthesis.

It has long been known that many colourless bacteria and other organisms have the capacity to incorporate carbon dioxide into their metabolisms. Thus *Escherichia coli*, for example, can form oxalacetic acid from pyruvic acid.

$$CH_3 . CO . COOH + CO_2 \longrightarrow COOH . CH_2 . CO . COOH$$

when energy is available in the form of ATP. Nevertheless, such processes, although widely distributed, do not yield carbohydrate. In the presence of ATP and NH_3 the energy-rich carbamyl phosphate may however be formed, from which, by a series of steps similar to those in animal metabolism, the storage product citrulline can be synthesized (see p. 287).

Chemosynthesis is found principally in certain colourless bacteria which are able to bring about the oxidation of a range of inorganic compounds, so obtaining energy (in the form of ATP). This quite different method of acquiring energy separates the chemosynthetic organisms sharply from the photosynthetic. The subsequent steps in sugar synthesis, however, appear fundamentally similar. The reducing agent is very probably $NADPH_2$, the CO_2-acceptor ribulose diphosphate, and the source of hydrogen appears, at least partly, to be the water. In many bacteria a polyhydroxy butyric acid is found as a food reserve. The synthetic pathways, however, are hardly known. The manner in which the energy liberated in the oxidation is utilized is described later (p. 276).

The very diverse group of sulphur bacteria, found on the floor of puddles rich in organic materials, and in great quantity in the filter beds of town sewage works, provide an example of the mechanisms involved. These colourless bacteria are capable of oxidizing by means of oxygen not only hydrogen sulphide, produced by the action of other bacteria on decaying protein or sulphates, but also other inorganic sulphur compounds to sulphates. The energy liberated in these oxidative reactions can be used in part for the chemosynthetic assimilation of carbon dioxide.

The bacterium *Thiobacillus thiooxydans*, found in agricultural soils, splits the water molecule in the same way in photosynthesis, but in darkness. The oxygen derived from the —OH radical serves for the oxidation of the sulphur. The energy becoming available as ATP is probably used partly in the reduction of $NADP^+$ to $NADPH_2$, and partly in the reduction of the carbon dioxide. The relatively large, colourless threads of *Beggiatoa* (see p. 448), although behaving heterotrophically in the presence of adequate nutrients, can also live wholly autotrophically by the oxidation of sulphur.

Other examples of chemosynthesis are provided by the nitrifying bacteria of soils. Some of these oxidize ammonia to nitrite, while others, often living associated with the first, oxidize the poisonous nitrite to nitrate (see p. 276). Only about 10 per cent of the energy liberated oxidatively is utilized in the assimilation of carbon dioxide. The iron bacteria, which oxidize ferrous to ferric salts (but which appear only facultatively autotrophic), the methane bacteria, which oxidize methane to carbon dioxide, and hydrogen bacteria which oxidize molecular hydrogen with atmospheric oxygen to yield water, and many other micro-organisms also possess the capacity for chemosynthesis (see also p. 276 seq.). The hydrogen bacteria can utilize the energy of oxidation very effectively (up to 30 per cent), but they can also live heterotrophically.

Especially interesting is the fact that the green alga *Scenedesmus*, which normally lives photosynthetically and uses water as a source of hydrogen, can be made experimentally to activate molecular hydrogen, which is then used to reduce $NADP^+$. This capacity depends upon the presence of the enzyme hydrogenase, as possessed, amongst other organisms, by the hydrogen bacteria. In the dark, *Scenedesmus* is even capable of oxidizing part of the hydrogen with atmospheric oxygen, and in this way to display truly chemosynthetic assimilation. This reveals very clearly the close connection between photosynthesis and chemosynthesis, the only distinguishing characteristic probably being the manner in which the energy is initially obtained.

Many investigators believe that the chemosynthetic organisms embody a kind of autotrophic life appearing very early in Earth's history, and that photosynthesis has developed by chance from chemosynthesis. This would have led to an accumulation of oxygen in our atmosphere, originally probably free from this gas.

B. The subsequent metabolism of the synthesized sugar[15]

In both photosynthesis and chemosynthesis the first compound to be synthesized from carbon dioxide and water, or other hydrogen compounds, is predominantly a hexose sugar, $C_6H_{12}O_6$. This sugar, however, occupies only an intermediate position in the sugar metabolism of the cell. This is true also of heterotrophic organisms, which do not manufacture hexoses, but which can absorb them into their metabolisms from without. It is the starting point for the synthesis of most of the organic carbon compounds going to form the structure of the plant, and also for those which are subsequently consumed in the respiratory metabolism. Thus part of the sun's energy which is stored in the plant is released again for the maintenance of its physiological activities. Only a few facts relating to the metabolism of the structural carbohydrates will be mentioned here in a preliminary way.

Under favourable conditions the production of sugar in the cell may be so great that its complete and immediate consumption or removal is not possible, and at least its temporary conversion into a physiologically less active form is absolutely essential, since the osmotic consequences alone of an accumulation of sugar in the cell would be injurious. The most frequent manner in which sugar is inactivated and stored is by its condensation into starch, which, being a macromolecular substance insoluble in water, is no longer immediately reactive. In fact, the first visible product of assimilation is often the starch accumulating in the chloroplasts. Before this the frequently occurring formation of disaccharides, e.g. cane sugar, from monosaccharides results in a halving of the osmotic activity. In many cases cane sugar also appears to be the form in which the carbohydrates arising from assimilation are transported away, often over long distances (see p. 288).

The individual macromolecules of starch laid down in a starch grain can be considered to have arisen by the linking together in a chain-like way of many molecules of α-glucose. The union of the individual glucose molecules must take place between the aldehydic carbon atom 1 of one molecule and the –OH group of carbon atom 4 of the adjacent molecule, together with loss of water (α-glycosidic linkage). Several hundred glucose molecules combined in this way and forming a simple, but probably spirally twisted, chain give rise to the amylose molecules of starch. These molecules, however, constitute only about 15–25 per cent of the starch grain. They are capable of dissolving in hot water, and yield with iodine the well-known blue-black coloration typical of starch. More usually, however, these amylose chains bear a number of side branches. These branches, of varying length, are similarly formed of glucose residues, and they are joined to the main chain by a bond between the carbon atom 1 of a terminal glucose molecule and a carbon atom 6 of one of the glucose molecules in the main chain (see Fig. 249). The so-called amylopectin molecules, consisting as it were of bundles of branched threads, are formed in this way. Such molecules may contain over 2000 glucose residues. They form the mass of the starch grain and they are responsible for the formation of mucilage when starch is treated with boiling water. Still more highly branched molecules occur in glycogen.

The formation of both the simple and the branched macromolecules of starch in the living cell appears to begin not from α-glucose itself but from the energy-rich glucose-1-phosphate, or possibly from the compounds derived from it, adenosine diphosphate–glucose or uridine diphosphate–glucose (in which the pyrimidine base uracil takes the place of adenine). Special enzymes (transglucosidases, see p. 260) transfer the glucose molecules, with release of energy, to shorter or longer chains ('starters') already in existence, so that these grow unit by unit. In Fig. 240 the process is shown beginning from glucose-1-phosphate, the phosphate being split off. The reverse process, phosphorylation

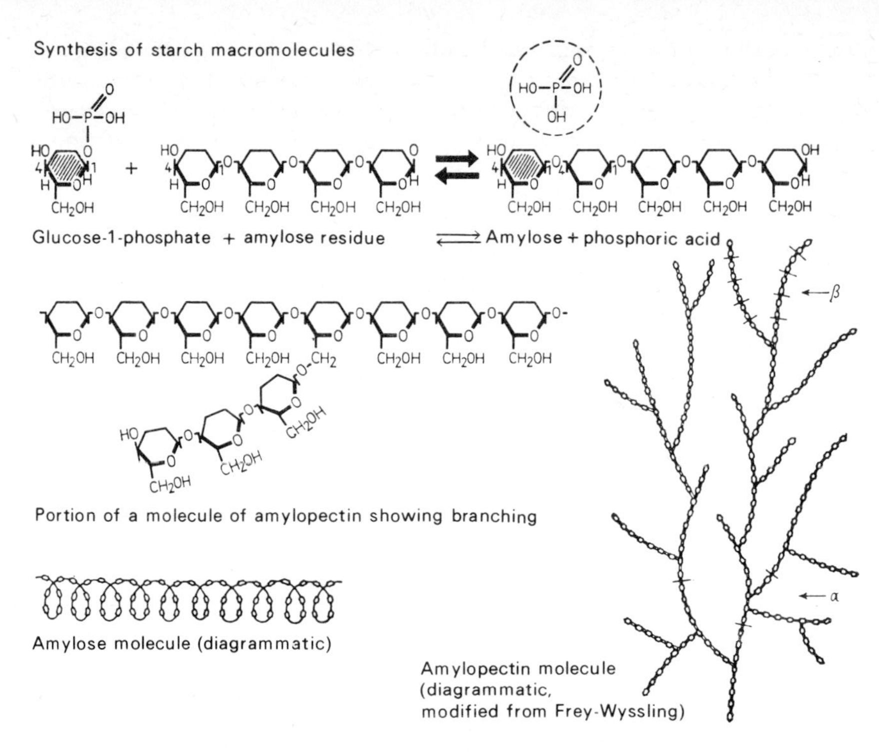

Fig. 249. Structure of the starch molecule. Explanation in text.

(see p. 260), probably plays a role in the several kinds of breakdown of starch. With the glucose-nucleodiphosphates mentioned above, probably the principal precursors of starch, it is naturally not phosphate which is split off as the glucose is added to the chain, but ADP or UDP. Branching of the chain might well require a special enzyme. In the synthesis of the disaccharide, sucrose, glucose is transferred in a similar way from the energy-rich uridine diphosphate-glucose to fructose.

During their formation in the interior of the plastids the macromolecules of starch always arrange themselves more or less radially around specific centres, the number and position (centric, eccentric, etc.) of which vary with species. The density of the deposition may periodically change, leading to a layering within the grain which often becomes visible under the microscope (see Fig. 53). The partly parallel arrangement of the branches of adjacent amylopectin molecules causes the birefringent behaviour of starch grains. An amazing number of glucose molecules can be physiologically inactivated in a single starch grain. The number in a starch grain of barley, for example, has been estimated as 12×10^{12}.

All the evidence in fact points to starch being merely the plant's physiologically inactivated reserve of sugar. In the germination of seeds, for example, and in the renewal of growth in bulbs and corms, there is an immediate and general catabolism of starch and reformation of sugar.

This dissolution of the starch grains can often be observed directly under the microscope. Canals and clefts penetrate from the margin of the grain to the centre, so that it appears corroded and eaten away (see Fig. 52).

The assimilatory starch in the chloroplasts of the photosynthetic parenchyma (concerning that in the guard cells see p. 371) seems to be especially labile. Apparently it is formed and persists only while the transport of sugar from the sites of photosynthesis is impeded. Consequently, in many leaves the starch disappears from the photosynthetic parenchyma during the night, so that the chloroplasts are again free and completely active in the morning.

This dissolution or mobilization of starch, like its synthesis, is brought about by means of specific cellular enzymes. These enzymes, as already mentioned in relation to photosynthesis, play an extremely important role generally in metabolism. Most biochemical reactions which will not take place outside the cell are to be explained by the occurrence of such specific catalysts. We must now consider their properties in general.

Enzymes.[16] In practical chemistry a catalyst is a substance present in only minute quantity which, without itself being consumed in the reaction or appearing in the end product, nevertheless by its mere presence accelerates the reaction, or even, by diminishing the energy of activation needed to start it, alone makes it possible. At normal temperatures the reaction then runs to the equilibrium determined by the law of mass action, this equilibrium not being in any way altered by the catalyst. In the presence of the catalyst the reaction can also run in the opposite direction.

The living cell possesses similar catalysts. They explain why, in a living organism, a multiplicity of biochemical reactions can take place which, outside the cell, perhaps in a test-tube, require quite extreme conditions (as, for example, very high temperatures). These biocatalysts are termed enzymes (ferments). While the catalysts of chemistry are often finely divided forms of heavy metals, such as iron, cobalt, nickel, platinum, all fairly unspecific in their action, the enzymes are complex proteins or proteids (see p. 280), formed in close relationship to the nuclear genes (see p. 284), and concerned with the catalysis of quite definite metabolic reactions. In certain conditions they distinguish between stereoisomers, so that the laevo-form of a compound enters into a reaction, while the dextro-form remains inert, or vice versa.

Every cell forms a wide range of very different enzymes. It has been estimated that a bacterium requires, for the building up of the cell, and for growth leading to the next division, about 1000 different enzymes. It has been possible to extract many enzymes from cells, and even to obtain them in crystalline condition. In just a few enzymes the sequence of amino acids in the protein component is already wholly or partly known (see p. 284). Still, however, the manner of enzyme catalysis is not fully understood. It is assumed that the enzyme forms a temporary complex with the substrate. This may come about by definite sites (active groups) of the enzyme, perhaps as a consequence of certain foldings of the polypeptide chain, displaying a particular pattern of charges or other reactive groups. The substrate or the reaction partner, or both together, are then so bound that they are caused to react. The reaction completed, the product dissociates from the enzyme, which is itself unchanged and capable of renewed catalysis. This can take place with such rapidity that even trifling quantities of enzyme can bring about substantial effects in a very short time.

In many enzymes, which are proteid (see p. 280) in nature, a fairly firmly bound active moiety, termed the prosthetic group, can be recognized, and may well be responsible for the actual catalysis. The specificity of the enzyme, which probably depends upon compatibility with the substrate, is determined, however, by the protein (see, for example, the flavoprotein enzymes, p. 271). In some enzymes the prosthetic group can be readily detached from the protein. Formerly it was generally believed that every enzyme could be separated into a prosthetic group (co-enzyme) and protein (apoenzyme, colloidal carrier), individually inactive, but able to effect catalysis as soon as complexed (holo-enzyme). Nowadays these readily separable groups are better considered as co-substrates (although partly still termed co-enzymes). Catalysis then requires a system of two enzymes and a co-substrate. For example (as described in more detail on p. 266), one enzyme transfers hydrogen from some substance to the co-substrate, a second enzyme then transfers the hydrogen further to some other receptor, a continual flow of hydrogen thus being possible. The chemical structure of some such co-substrates, and of some prosthetic groups, is today known exactly; prominent amongst them, for example, are derivatives of nicotinic acid, iron-containing porphyrin rings, and isoalloxazin compounds. Many vitamins are nothing more than such components of enzymes or co-enzymes. In addition there are yet other substances (the so-called activators and inhibitors) which increase or diminish the activity of an enzyme. Many minerals and trace elements are important in this respect, possibly facilitating the temporary association of enzyme and substrate. Such substances are frequently termed co-factors. Also, as with all proteins, hydrogen ion concentration plays an important role, and the activity of every enzyme reaches a maximum at a definite pH value (optimum pH; see p. 280). In many instances several enzymes appear to be aggregated in a large complex (multi-enzyme complex), making more understandable the catalysis of whole chains of reactions.

The name of an enzyme is usually constructed by adding the suffix -ase to the name of the substrate on which the enzyme acts, well demonstrated by the hydrolytic enzymes. Lipases, for example, hydrolyse fats, proteases proteins, cellulases cellulose, and amylases starch (amylum). The addition of -ase to the nature of a reaction indicates a group of enzymes with a particular catalytic effect. Dehydrogenases, for example, detach hydrogen, while transferases are concerned with the removal of whole radicals or other groups. More exact description is effected by beginning with the name of the substrate. For example, succinic dehydrogenase is the enzyme catalysing the removal of hydrogen from succinate. Several large groups of enzymes have been distinguished. Examples are:

1. *Oxidoreductases.* These transfer hydrogen or electrons and play a special role in respiratory metabolism (see p. 273 seq., dehydrogenases).
2. *Transferases.* These transfer whole groups from one substrate to another, e.g. $-NH_2$ groups, $-CH_3$ groups, glucose (as in starch synthesis), phosphate radicals (kinases), etc.
3. *Hydrolases.* These bring about hydrolysis, i.e. the attachment of the hydrogen and hydroxyl ions, coming from the dissociation of water, to the two parts of the simultaneously dissociated substrate. Examples are the enzymes hydrolysing esters, glycosides, peptides, etc.
4. *Lyases* (desmolases). These break the bond between two carbon atoms (e.g. aldolase, p. 266; decarboxylase, p. 266), and also in some instances between carbon and oxygen, and between carbon and nitrogen.
5. The enzymes catalysing these reactions in reverse, leading to synthesis, are termed ligases.
6. *Isomerases.* These bring about intramolecular rearrangement. Examples are the enzymes responsible for the transforming aldoses into ketoses, and α-glucose into β-glucose.

Although transferases (transglucosidases, see p. 257) seem to predominate in the synthesis of starch, the breakdown of starch appears to depend upon hydrolases, in many cases upon the so-called amylases. These are specifically concerned with the hydrolytic splitting of the aldehydic linkage of the glucose molecules in the α-position (see Fig. 249), along with the incorporation of the elements of water.

The so-called β-amylases sporadically detach a disaccharide, maltose, consisting of two glucose residues, from the non-aldehydic ends (carbon atom 4) of the chains of the macromolecule, provided these terminal residues do not bear branches as in the molecules of amylopectin. Maltose was consequently thought of for a long time as a particular structural unit of starch. The so-called α-amylases, on the other hand, split the amylopectin molecules indiscriminately between the side branches, giving rise to large fragments, known as dextrins. Another hydrolase, maltase, splits the maltose molecules produced by the action of the β-amylases into the basic glucose molecules. A whole group of different enzymes is thus concerned in the mobilization of starch.

At present it is still not at all clear how far in the higher plants starch can be broken down by the reversal of the process by which it is synthesized, with the re-formation of glucose-1-phosphate ('phosphorylation'), or possibly the nucleosides mentioned earlier (p. 257). Phosphorylation, rather than hydrolysis, would in some respects seem more reasonable since it would lead to the energy-rich glucose-1-phosphate instead of to simple glucose. This may occur in the guard cells of the stomata. The sites of branching of the amylopectins seem to be disjoined only by special phosphorylases. Bacteria and yeasts contain as a rule no amylases, so that here there can be no doubt that other methods of starch catabolism are present. The plant can probably activate or inactivate its enzymes according to its current needs, and so direct its metabolism towards anabolism or catabolism, but we have as yet very little information about these processes (see p. 330). It is very interesting to note that in both animals and man the saliva and pancreatic juice contain amylases for the breakdown of starch similar to those of plants (β-amylases, however, occur only in plants).

Starch is not the only substance produced by the transformation of the assimilatory sugar. The majority of the other carbohydrates of frequent occurrence in plant cells, the mono-, di- and polysaccharides, are ordinarily derived from the initial hexose molecules. The formation of cane sugar has already been mentioned. The reserve carbohydrate in the underground organs of many Compositae is inulin, a polymer of fructose. Mannans are polysaccharides derived principally from the hexose mannose; they are frequently stored temporarily in cell walls ('reserve cellulose'), as in the endosperm of the date. Pentoses and heptoses appear partly to be intermediate products in photosynthesis, but may also be derived from hexoses (see p. 276). As a final example of a product derived from assimilatory sugar we may mention cellulose. This polymer is the most important structural material of the cell wall, so far as it is not, as in the cell walls of many seeds, merely another form of carbohydrate reserve.

Cellulose. Like starch, cellulose is apparently formed by the linking together of numerous glucose molecules (100–3000, in cotton up to 14,000) to form thread-like macromolecules of variable length. Here the glucose molecules are not of the α-form as in starch (Fig. 249), but of the β-form (β-glycosidic linkage), and adjacent residues are always rotated through 180° with respect to each other (Fig. 250). This gives the chain a

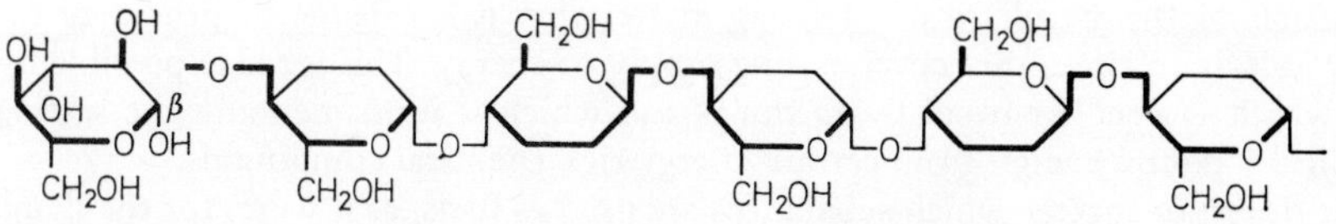

Fig. 250. β-Glucose, and portion of the macromolecule of cellulose.

certain rigidity and it does not, as in starch, take up the form of a helix. The linking together of the glucose molecules, which can be readily brought by isomerases into the appropriate configuration, is naturally dependent upon special enzymes. Glucose probably approaches the growing chain, not as β-glucose, but in the form of an energy-rich nucleotide (guanidine diphosphate-glucose or uridine diphosphate-glucose). Polymerization of these nucleotides (as with those leading to the formation of starch) has been effected *in vitro* in the presence of the appropriate enzymes. Although the width of the cellulose macromolecule is below the limits of optical resolution, its length often reaches visible dimensions. These macromolecules aggregate themselves into bundles, each of about 100 molecules, probably in part held together by hydrogen bonding. These bundles are termed micelles (see p. 66). The cellulose microfibril, visible in the electron microscope, consists of an aggregate of a few micelles (Figs. 60 and 61).

The formation of fats, which are found especially frequently as reserve materials in, for example, many seeds (rape, linseed, groundnut, cotton, among others), is closely bound up with the carbohydrate metabolism. The neutral fats are probably nothing more than a very effective form of carbohydrate storage, since it is known that they are easily transformed within the plant, following hydrolysis by lipase, into sugar, and conversely that they can be formed from sugar or from the products of the breakdown of sugar.

The fats are, as is well known, glycerol esters of the fatty acids. Glycerol can be very easily formed from the 3-carbon compounds produced in the normal catabolism of sugar (see p. 266), and in fact arises in a slight variant of alcoholic fermentation in sufficient quantities for this method of production to be used commercially. The fatty acids are probably derived from acetaldehyde or from the acetic acid linked to coenzyme A (see p. 274).

Fats, cellulose and starch are mentioned here merely as examples of the utilization of the assimilatory sugar in the metabolism of plant cells. In all probability the diversity of the exploitation of this sugar in metabolism is very much greater, and impossible to survey in detail. Coupled and esterified forms of sugars, giving rise to the so-called secondary plant products, e.g. the glycosides, tannins, etc., as well as the numberless other substances found in plants, especially cyclic aromatic compounds containing, for example, the benzene ring, are frequently found, and must owe their origin at some time to the products of assimilation of carbon dioxide. This cannot be gone into in more detail here.

Of great importance is the more extensive change in the sugar molecule occurring in respiration. This is so fundamental a phenomenon in living matter, and is in addition bound up biochemically with so many other metabolic reactions, that the fate of the sugar in respiration requires treatment in a separate section.

V. Respiration[17,18]

One of the most important facts about life is that it can only be maintained in the presence of a continuous supply of energy. Not only do directly visible acts of work, such as the contraction of an animal muscle or a rapid plant movement, consume energy but so also do many of the syntheses involved in building up the plant body. Indeed, the maintenance of the apparently labile submicroscopic structure of the protoplasm also probably entails the uptake of a certain amount of energy. Thus if the protoplasm is to retain its integrity, and syntheses are to take place, energy must be continually available. Pfeffer's image of a crew afloat on the ocean in a small leaking ship which can be kept above the water only by continuous work at the pumps well illustrates the precariousness of life. It is one of the peculiarities of living matter that it is capable of providing for itself, for an indefinite time, this never-ending supply of energy. This is made possible by respiration, which is peculiar to all living things and which is fundamentally the setting free of chemically bound energy from certain energy-rich chemical compounds.

Initially it does not matter which substances are used as fuels, as it were, for the living engine. It is only essential that in the course of certain chemical reactions, which usually bring about a breakdown of the substance concerned, energy is set free in a form which can subsequently satisfy the multifarious needs of the living machinery. At the end of this section types of respiration will be discussed in which this energy is obtained not from sugars, but from other respiratory materials, sometimes entirely inorganic. In the majority of plants and animals the central place in the respiratory mechanism is occupied by sugar. The green plant is indeed capable of producing it from inorganic materials, and a considerable amount of solar energy lies enclosed within it as a consequence of the assimilatory process. The freeing of this potential energy bound up in the sugar molecule is the function of the whole complex system of respiration. Eventually in both plants and animals, by processes which are either the same or very similar, the sugar is broken down again to its basic components, carbon dioxide and water.

As explained on p. 243, 675 kg cal are liberated in the complete combustion of 1 g mol of glucose to carbon dioxide and water, equal to that amount of energy taken up from the radiant energy of the sun in photosynthesis: 1 g mol $C_6H_{12}O_6$ + 6 g mol $O_2 \longrightarrow$ 6 g mol CO_2 + 6 g mol H_2O + 675 kg cal.

If this reverse process takes exactly the course indicated in the plant, then it should be possible to observe directly in respiration the consumption of oxygen and the evolution of carbon dioxide in quantities which are in the proportion of 1 : 1, i.e. the respiratory quotient should be unity.

It is in fact quite easy to show by gas analysis that green plants in the dark give out carbon dioxide and take up oxygen. It is simple to demonstrate qualitatively the evolution of carbon dioxide from germinating but not yet assimilating seeds by means of the apparatus shown in Fig. 251, where the carbon dioxide formed in respiration is precipitated as barium carbonate. For the more exact quantitative investigation of respiration apparatus is required in which the changes in the amounts of carbon dioxide and oxygen in the air passing over the specimen can be measured by gas analysis or in some other way. For this to be done on green plants they must naturally be darkened to prevent the precisely reverse process of photosynthesis.

Differentiated parts of higher plants often show a respiratory quotient which approximates to unity. This shows, just as the assimilatory quotient shows the reverse in photosynthesis, that carbohydrates, especially sugar, are in fact being 'burned' by means of the oxygen consumed. Substances poor in oxygen, such as fats and proteins, must take up

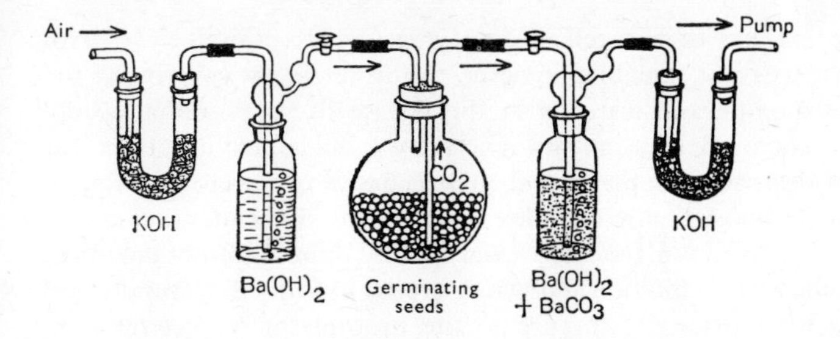

Fig. 251. Diagram of an apparatus to demonstrate the evolution of carbon dioxide by germinating seeds.

considerably more oxygen for complete decomposition into carbon dioxide and water, while substances rich in oxygen, such as organic acids, require less. Consequently, in the respiration of these substances the quotient must be respectively smaller or greater than unity. A respiratory quotient deviating from unity does not, however, exclude the possibility of the substrate being first transformed into carbohydrate and then respired as such. The quotient will also necessarily deviate from unity if, besides or instead of carbon dioxide, other respiratory products are formed, or if intermediary products are used for functions other than respiration.

It is more difficult to demonstrate the release of energy taking place in respiration, since the water evaporating in transpiration uses up most of that respiratory energy which appears as heat, while the useful part is stored in the form of ATP (see p. 266).

However, it is possible by careful thermal isolation (e.g. in a Thermos flask), and by selecting material having little transpiration, such as floral organs or germinating seeds, to demonstrate an increase in temperature corresponding to production of heat in animals. Only in special cases, e.g. in the spadix of the Araceae, can the production of heat by respiration be easily and directly detected. The spadix of *Arum italicum* reaches 17°C, the flower of *Victoria regia* 10°C and that of *Cucurbita* 5°C above the external temperature. Also in the interior of densely packed masses of plant material (e.g. haystacks, etc.) quite considerable increases in temperature can occur as a result of the respiratory activity of special thermophilous fungi and bacteria (see p. 335). The production of water in respiration is also very difficult to demonstrate amidst the great turnover of water generally in the plant. One way in which it can be inferred is the loss in dry weight during respiration is greater than that which corresponds to the weight of the carbon lost.

The intensity of respiration is naturally subject to great variations, according to the plant and to the prevailing conditions and needs of the organism, but it is of approximately the same order of magnitude in animals and plants. The respiration of green leaves, or example, at 25–30°C reaches about the same level (expressed on a basis of unit weight) as that of a man at rest. Lower plants, bacteria and fungi, however, for reasons which will become evident later, mostly have a much more vigorous respiratory metabolism.

As a representative figure it may be cited that a higher green plant gives off into the atmosphere in twenty-four hours about five to ten times its own volume of carbon dioxide. Altogether this is some $\frac{1}{5}$–$\frac{1}{3}$ of the quantity of carbon dioxide which the plant would fix during the day by photosynthesis. Respiration, therefore, uses only a fraction of the material produced by assimilation for the generation of energy (probably of the order of 40 per cent), the remainder being available for the building up of the plant body, for storage, and for the nutrition of the non-assimilating parts.

There can, of course, be no question of real combustion in the respiration of sugar. The internal structure of the protoplasm is such that any marked development of heat would lead to immediate death as a result of the coagulation of the protein. Also, in so finely adjusted a system a sudden massive liberation of energy could only have damaging effects. Thus there must exist a mechanism for the decomposition of the sugar and for the setting free of its energy which, as it were, makes possible in the cold reactions which the chemist can usually bring about only by the use of high temperatures, high pressures and so on.

Once again it is the biocatalysts of the cell, the enzymes, which effectively control these important conversions. It is not possible, however, to formulate the splitting of the sugar molecule into carbon dioxide and water with the release of its potential energy simply in terms of a single chemical reaction influenced by a single enzyme. In recent years the intensive study of respiration in plants and animals has shown ever more clearly that we are concerned with the most highly complex chains of interdependent reactions. Thus many partial reactions must take place, each catalysed by quite different enzymes, until finally the end products, carbon dioxide and water, are again formed. Naturally this presupposes a highly differentiated inner structure to the protoplasm so that the individual reactions may take place at different sites, but nevertheless, as in a large building operation, they must interlock in an ordered way (see p. 273). The reason for this protracted sequence of reactions is perhaps to be seen not merely in that each different partial reaction provides a recurrent opportunity for the organism to exert direction and control but also in that the energy is released only in a step-wise manner.

The biological breakdown of sugar. It would be too much to describe here in detail all the partial reactions known to occur in the biological breakdown of sugar; only the essential features of the process will be indicated. The fact that during respiration both plants and animals give off carbon dioxide has for a long time given the false impression of direct union of the respiratory material with the oxygen of the air. In fact not only can the higher plants respire more or less adequately for a certain time without oxygen but we also know that a whole range of lower plants (i.e. the anaerobes, e.g. bacteria in the muds of swamps, and those of the gut) are capable of existing the whole time without it; indeed, for these plants oxygen is actually a poison. Consequently, it can be concluded that no primary role can be attributed to oxygen, despite the fact of its being so essential in many cases for the normal course of respiration.

It is considered in chemistry that a substance is oxidized not only when it combines with oxygen but also if it is deprived of hydrogen. If, for example, hydrogen sulphide is oxidized by oxygen:

$$2H_2S + O_2 \longrightarrow 2S + 2H_2O$$

then the sulphur is said to have been oxidized although it has merely given up hydrogen to the oxygen. The oxidation of succinic acid to fumaric acid consists solely in a loss of hydrogen:

$$\begin{array}{l} COOH \\ | \\ CH_2 \\ | \\ CH_2 \\ | \\ COOH \end{array} \longrightarrow \begin{array}{l} COOH \\ | \\ CH \\ \| \\ CH \\ | \\ COOH \end{array} + 2H$$

The fate of the hydrogen split off is of no interest in relation to the oxidation. Finally, an oxidation may also take the form of the giving up of an electron and a consequent increase in positive charge, as occurs, for example, in the very important valency changes in the heavy metals, a notable instance of which is provided by iron (see p. 272):

$$Fe^{++} \xrightarrow{-e} Fe^{+++}$$

Trivalent iron is thus at a higher level of oxidation than divalent.

According to Wieland's theory of respiration, in all processes of biological respiration the primary process consists in the removal of hydrogen from the respiratory substrate (usually sugar) by means of special enzymes classified as dehydrases or dehydrogenases (oxido-reduction, see p. 260). In normal aerobic respiration this hydrogen is then combined with oxygen in a manner still to be discussed (see p. 273), and this 'burning' to water is the main source of energy in the respiration of sugar. The hydrogen can also be transferred to other substances, which act as hydrogen acceptors, and combine with these. In these cases we have types of respiration which proceed entirely without oxygen, and which are generally referred to as fermentations. They do not give as high yields of energy as aerobic respiration, but they satisfy nevertheless the energy requirements of the organism causing the fermentation, the quantitative deficiency being compensated for by a greater consumption of the respiratory material. Frequently, moreover, not all the hydrogen of the molecule is removed in fermentation, but only a part, so that the end-products of the respiration or fermentation are naturally not carbon dioxide and water, but, for example, as in alcoholic fermentation, carbon dioxide and ethyl alcohol.

These types of anaerobic respiration which do not completely use up the respiratory substrate are found principally in the heterotrophic plants. They are of immediate and great significance for understanding the normal respiration of higher plants, since such forms of respiration at once suggest themselves as component parts of the very much more extensive process of aerobic respiration. Apparently in aerobic respiration the first phase again consists of a fairly elaborate process of splitting up of molecules without any incorporation of oxygen. This is indicated by the fact, already mentioned above, that higher plants continue to give off carbon dioxide in the absence of oxygen. In these conditions, therefore, they are capable of a kind of fermentation. Consequently, our view of normal aerobic respiration is based very much on the study of the fermentation phenomena of lower heterotrophic plants, and for this reason these phenomena must be considered before the aerobic respiration of higher plants.

Alcoholic fermentation.[17] It is well known that sugary solutions become fermented as a result of the respiratory activity of microscopic forms of life, especially by the various yeasts. Ethyl alcohol and carbon dioxide are formed from the sugar taken up from without, a process which can be summarized in the equation:

$$1 \text{ g mol } C_6H_{12}O_6 \longrightarrow 2 \text{ g mol } CO_2 + 2 \text{ g mol } CH_3 . CH_2OH + 21 \text{ kg cal}$$

Oxygen does not enter into the reaction at all. Yeast, in fact, requires no oxygen for the maintenance of life. It can live vegetatively completely anaerobically, and the presence of oxygen has no significance for its metabolic fermentation. In well-aerated conditions, however, it is also capable of normal aerobic respiration, apparently essential for growth and for the formation and germination of the spores, but this cannot be considered further here. Aeration probably induces the formation of mitochondria (see p. 51), which contain all the enzymes of the Krebs cycle (see p. 273). Yeast is therefore not an obligate, but only a facultative anaerobe.

The equation shows furthermore that the yield of energy (equivalent to two molecules ATP) from such a decomposition of sugar into alcohol and carbon dioxide is minute compared with the total amount of energy bound up in the sugar molecule. In fact, the excreted respiratory product, the alcohol, is, as is seen when it is burned, still a very rich source of energy. Moreover, the yeast cell always ferments a relatively large amount of sugar in order to satisfy its vital requirements. Only about 3 per cent of the energy contained within the sugar is utilized, and only about 1 per cent of the total metabolized sugar is taken up into the synthesis of structural materials.

The chemical reactions involved in alcoholic fermentation, so far as they are at present known, are briefly as follows.

Yeast is capable of fermenting other sugars, e.g. disaccharides such as sucrose, besides glucose, but only after it has first transformed them into sugars similar to glucose. Sucrose, for example, is split by the enzyme saccharase (invertase) into one molecule of glucose and one of fructose.

In addition, the glucose has first to be as it were unlocked before it can be worked upon by the particular system of respiration of the plant concerned. In aqueous solution a number of different forms of glucose already exist side by side, among them a small amount of a labile and reactive form in which the oxygen bridge completes a ring with only four instead of five carbon atoms (furanose forms, h- or γ-glucose). It is probable that the first steps in fermentation, brought about by means of several enzymes, lead to the formation of such a labile and readily decomposable sugar. Esterification with phosphoric acid (controlled by the enzyme hexokinase) also occurs at this time, yielding after several intermediate stages a hexose diphosphate, probably 1,6-fructose diphosphate, which is now subject to further catabolism.

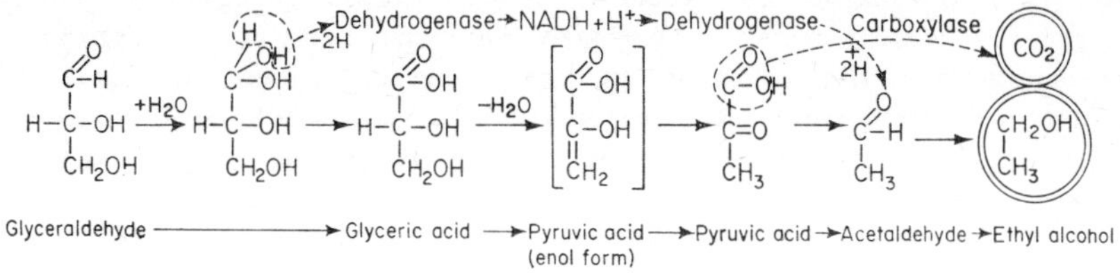

Glyceraldehyde → Glyceric acid → Pyruvic acid (enol form) → Pyruvic acid → Acetaldehyde → Ethyl alcohol

Fig. 252. The course of alcoholic fermentation. Note the transformation of one of the two glyceraldehyde molecules derived by the breakdown of glucose ($C_6H_{12}O_6 \rightarrow 2C_3H_6O_3$) into the end-products of the fermentation ($C_3H_6O_3 \rightarrow CH_3 \cdot CH_2OH + CO_2$).

In the following, for the sake of clarity, the main features of the chemistry of fermentation will be given without considering this formation of phosphates (see Fig. 252). The sugar also will be considered as an open chain and not in its cyclic form. Under the influence of the enzyme zymohexase (aldolase) the activated hexose breaks up into two molecules, each with three carbon atoms, dihydroxyacetone and glyceraldehyde. The dihydroxyacetone can be easily converted enzymatically into glyceraldehyde (Fig. 247). After forming a hydrate, glyceraldehyde (using the simplest formula; see, however, Fig. 254) gives up two hydrogen atoms to a dehydrogenase, from which they are transferred to the co-substrate NAD, the glyceraldehyde so becoming oxidized to glyceric acid. The enzyme enolase now cause the glyceric acid to lose water, so that it becomes, by way of an enol-form, pyruvic acid. Carbon dioxide can now be readily removed from this α-ketoacid by means of the enzyme decarboxylase (occurring in a form found only in plants; the co-substrate is the magnesium—requiring thiamine pyrophosphate, identical with vitamin B_1). Thus one of the end-products of anaerobic fermentation, carbon dioxide, is obtained. The acetaldehyde remaining now acts as a hydrogen acceptor and receives from the reduced form of the co-substrate, NADH + H^+, by way of another dehydrogenase, the hydrogen initially split off from the glyceraldehyde hydrate, and is thereby hydrogenated to yield the second end-product, ethyl alcohol. In this way a cycle is set up: the acetaldehyde arising from the decarboxylation of the pyruvic acid always reacts again oxido-reductively with the glyceraldehyde, so that in one and the same process one partner is oxidized and the other reduced. The total energy arising from the whole process of the breakdown of one sugar molecule into two molecules ethyl alcohol and two molecules carbon dioxide is, as the overall equation shows (p. 265), relatively small (about $\frac{1}{30}$ of the energy contained within the sugar molecule), since most of the energy remains behind in the alcohol.

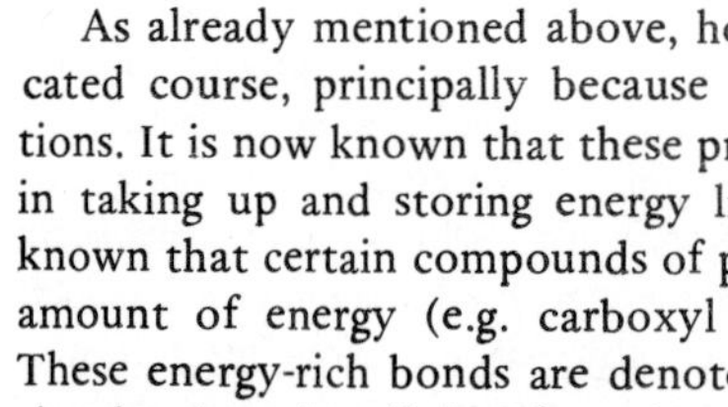

Fig. 253. Structure of adenosine triphosphate (ATP). The dotted vertical lines indicate where the phosphate bonds are broken hydrolytically, so releasing their energy. ATP breaks down to ADP + P with the release of 7 kg cal energy.

As already mentioned above, however, the actual reactions take an even more complicated course, principally because of the repeated phosphorylations and dephosphorylations. It is now known that these processes are essential, since they play an important role in taking up and storing energy liberated from the decomposing sugar molecule. It is known that certain compounds of phosphoric acid with organic substances contain a large amount of energy (e.g. carboxyl and carbonyl compounds, and the pyrophosphates). These energy-rich bonds are denoted by the sign ~. Especially important in this connection is adenosine diphosphate (ADP), a mononucleotide of the cell incorporating a pyrophosphate. If the necessary energy is at hand this substance readily takes on a third molecule of phosphoric acid, yielding the energy-rich adenosine triphosphate (ATP) (Fig.

253). If a molecule of phosphoric acid is split off from the latter with the re-formation of ADP, the energy of the bond is liberated (more than 7 kg cal/g mol under standard conditions). The reaction is thus exergonic, and ATP is consequently well suited to serve as a store and distributor of energy within the cell. The phosphorylation leading to the formation of hexose diphosphate is at the expense of the ATP, so that this is converted into ADP. Again, after the glucose has been split into two molecules of phosphorylated glyceraldehyde, hydrogen is extracted in reality not from a hydrate, as shown in the simplified scheme of Fig. 252, but from a combination of the aldehyde group with the –SH group of an enzyme. As shown in Fig. 254, with the transference of the hydrogen by a dehydrogenase to NAD, an energy-rich acyl-S bond is formed (corresponding to an acyl-CoA bond; see p. 273). Were there now a simple hydrolysis, the energy would be liberated as heat. Instead, by means of phosphorylation, the SH-enzyme is regenerated, and the energy is transferred to a phosphate bond. This kind of phosphorylation and its consequences, as opposed to that characteristic of the respiratory chain (see p. 273), is termed 'substrate phosphorylation'. This is because the phosphate radical can subsequently pass by way of a transferase (kinase) to ADP, the energy then being stored in ATP. Its simultaneous replacement by a hydroxyl group leads to the formation of phosphoglyceric acid (Fig. 254). Quite similar events occur when the enol-form of the pyruvic acid is formed

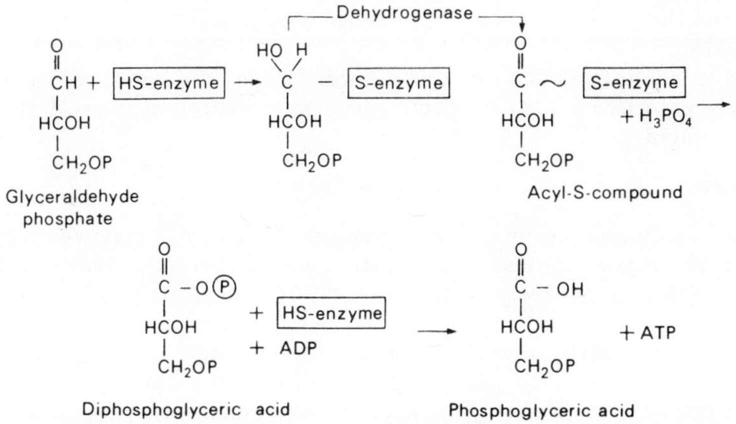

Dehydrogenase

$O{=}CH{-}HCOH{-}CH_2OP$ + [HS-enzyme] → $(HO)(H)C{-}$[S-enzyme] $-HCOH-CH_2OP$ → $O{=}C\sim$[S-enzyme] $-HCOH-CH_2OP$ + H_3PO_4 →

Glyceraldehyde phosphate — Acyl-S-compound

$O{=}C{-}O$Ⓟ $-HCOH-CH_2OP$ + [HS-enzyme] + ADP → $O{=}C{-}OH$ $-HCOH-CH_2OP$ + ATP

Diphosphoglyceric acid — Phosphoglyceric acid

Fig. 254. Transformation of glyceraldehyde diphosphate into phosphoglyceric acid. An enzyme with a free –SH group is involved. ATP is generated. Ⓟ Indicates a high-energy phosphate bond. See also Figs. 252 and 253.

with the loss of water from glyceric acid. Here again the migrating phosphate radical, after changing its position from carbon atom *3* to carbon atom *2*, is transferred with its energy to ADP, with the formation of ATP. In fact, therefore, these phosphorylations and the ultimate formation of energy-rich ATP (two molecules of ATP being formed for each molecule of pyruvic acid, i.e. four molecules of ATP per molecule of glucose fermented, of which two molecules are used for the phosphorylation of a new sugar molecule) are essential factors in the energy transformations involved in fermentation.

Lactic acid fermentation.[14] Not only does alcoholic fermentation provide a basis for understanding normal aerobic respiration but also for a whole series of other kinds of fermentation which occupy a very important place in nature. As the first example of these other fermentations we may mention lactic acid fermentation, which among other things accounts for the souring of milk. It also has great practical importance in the conservation of green fodder in silos and in the preparation of sauerkraut, since the formation of lactic acid prevents the development of foreign bacteria. The organisms responsible for lactic acid fermentation are various thermophilous, facultatively anaerobic bacteria, often living epiphytically on the leaves of higher plants (possibly utilizing substances secreted by the epidermis) or on the udders of cows. In their anaerobic respiration they break down the sugar contained in the green fodder or milk, and thereby obtain their vital energy, according to the following equation:

1 g mol $C_6H_{12}O_6 \longrightarrow$ 2 g mol $CH_3 . CHOH . COOH$ + 22 kg cal

Actually only a few simple bacteria, utilized in dairying and cheesemaking (e.g. *Bacterium delbrückii*), convert sugar entirely in this way. Lactic acid fermentation is usually said to be 'impure', and besides lactic acid other respiratory products, such as carbon dioxide, hydrogen, methane, acetic acid, propionic acid, etc., also rise. The 'coli' bacteria of the human gut and many staphylococci behave in this way, and the details of their fermentations cannot be considered further here.

We have reason to believe that, as in alcoholic fermentation, the first steps in lactic acid fermentation lead to the breakdown of the sugar into glyceraldehyde. These organisms, however, lack the enzyme carboxylase, with the result that the pyruvic acid is not decarboxylated and no acetaldehyde can be formed to serve as a hydrogen acceptor. The hydrogen split off from the hydrated glyceraldehyde probably becomes in consequence transferred to a molecule of pyruvic acid itself, converting this to lactic acid $CH_3 . CO . COOH + 2H \longrightarrow CH_3 . CHOH . COOH$. The formation of lactic acid in animal muscles depends upon a similar process.

Butyric acid fermentation.[14] A large number of obligately anaerobic bacteria (e.g. various forms of *Clostridium pasteurianum* (= *Bacterium amylobacter*), the pathogen causing malign oedema, etc.) produce, among other things, butyric acid as a result of their respiratory metabolism.

The butyric acid probably arises as a result of the acetaldehyde and 'activated' acetic acid bound to coenzyme A (see p. 274) reacting together to form aceto-acetic acid, this being subsequently reduced:

$$CH_3 . CO . CH_2 . COOH + 2H_2 \longrightarrow CH_3 . CH_2 . CH_2 . COOH$$

As with the 'impure' lactic acid organisms, other products are formed, among which is molecular hydrogen. This is given off freely into the atmosphere and is one of the sources utilized by the hydrogen bacteria (see p. 256). The yield of energy, some 17 kg cal, is minute.

Butyric acid fermentation plays an important role in nature. Most of the nitrogen-fixing bacteria of the soil form, among other things, butyric acid, although the substrate in most cases is not precisely known (see p. 285). The pectin of middle lamellae can also be fermented by a range of micro-organisms with the formation of butyric acid (pectin fermentation). This is of economic importance in the retting of flax and hemp, since, as a result of fermentation of this kind, the bundles of fibres can be separated from the parenchymatous tissue of the stem.

Cellulose fermentation.[14] After the fall of leaves in the autumn cellulose is decomposed in great quantities by bacterial action, both on and in the soil. It is still difficult at present to say how far this process should be included here. A part might be broken down by aerobic micro-organisms (certain fungi also destroy cellulose; species of *Ganoderma*, for example, the cause of the so-called 'white rot' of wood, decompose both cellulose and lignin), but a part at least probably undergoes anaerobic fermentation in which, besides acetic acid, some butyric acid is also formed. Naturally the cellulose must always be first converted by the enzyme cellulase into sugar.

Furthermore, the bacteria living symbiotically (see p. 292) and in great numbers in the stomach and gut of many animals, as in the fermentation sacs of wood-destroying insects, are capable of utilizing cellulose anaerobically. Certain of the products (e.g. butyric acid; in sheep, propionic acid) can be resorbed by the animal. In the ruminants, for example, cellulose fermentation goes on in the fore-stomach and reticulum, with the result that up to 80 per cent of the cellulose contained in hay and straw can be made available for nutrition. In man bacterial fermentation of cellulose occurs principally in the large intestine. Its significance lies less in the rendering of the cellulose available for nutrition (delicate cell walls, as in salads, can be utilized up to a maximum of 25 per cent), than in exposing the cell contents to attack by the body's own digestive juices. The gaseous subsidiary products of the fermentation (carbon dioxide, hydrogen and methane from acetic acid) cannot be utilized, and are excreted. Concerning the fermentation of nitrogenous substances in the gut, see p. 296.

Acetic acid fermentation.[14] In the anaerobic fermentations considered hitherto the atmospheric oxygen, by definition, does not enter into the system, and the energy obtained by respiration after the breakdown of the sugar into the 3-carbon compounds derives essentially from the intramolecular shifting of hydrogen, and the hydrogenation or dehydrogenation of intermediary products. Nevertheless, there are types of bacterial respiration known into which atmospheric oxygen does enter, just as in the normal respiration of higher plants and animals. An example of such a process is the so-called acetic acid fermentation of certain bacteria (such as *Acetobacter* and others) in which respiration involves the oxidation of ethyl alcohol to acetic acid:

$$1 \text{ g mol } CH_3 . CH_2OH + O_2 \longrightarrow 1 \text{ g mol } CH_3 . COOH + 1 \text{ g mol } H_2O + 118 \text{ kg cal}$$

Here not sugar, but the much simpler ethyl alcohol is serving as the substrate. Strictly speaking, this system should no longer be called a fermentation, since it utilizes as a normal procedure atmospheric oxygen. Even an oxidation process apparently so simple has nevertheless turned out to be a typical dehydrogenation, and has actually become a model for our whole interpretation of the mechanism of biological oxidations.

In this particular instance a specific dehydrogenase (with NAD as co-substrate) first removes hydrogen from the ethyl alcohol converting it to acetaldehyde:

$$CH_3 . CH_2OH \longrightarrow CH_3 . CHO + 2H$$

The acetaldehyde then forms a hydrate with water from which hydrogen is once more removed:

$$CH_3 . CHO + H_2O \longrightarrow CH_3 . C\begin{matrix} \diagup H \\ -OH \\ \diagdown OH \end{matrix} \longrightarrow CH_3 . COOH + 2H$$

The separated hydrogen, after being taken up by NAD^+ is transferred by yet another system of enzymes (probably the cytochrome system; see p. 272) to atmospheric oxygen which acts as an acceptor, forming hydrogen peroxide. This is immediately decomposed by the enzyme catalase into water and oxygen, since hydrogen peroxide is a dangerous poison:

$$4H + 2O_2 \longrightarrow 2H_2O_2 + O_2$$

The prosthetic group of catalase, like that of the cytochromes, is related to the blood pigment haemoglobin, and similarly contains iron. The net result is the same as if hydrogen is burnt to form water. Since this reaction, which is in fact used in technology, releases large quantities of energy, the very high yield of energy in this aerobic respiration compared with that obtained in anaerobic is understandable. The energy is again derived from a shifting of hydrogen within the system. As will emerge in the following discussion, this complicated manner of transferring hydrogen to oxygen found in the acetic acid bacteria is also present in the normal aerobic respiration of higher plants. The atmospheric oxygen thus merely serves as an acceptor for the dissociated hydrogen. In the acetic acid bacteria the oxygen can be quite easily replaced by another organic acceptor, and a genuine acetic acid fermentation may occur without access to any aerial oxygen. Oxygen of course cannot normally enter into respiration until it has first been 'activated' (see p. 272).

Fumaric acid fermentation, by which certain mildews form fumaric, succinic and malic acids (see the citric acid cycle, p. 273) from cane sugar, is a similar oxidative fermentation.

The respiration of the higher plants.[17, 18] We can conclude from the value of the respiratory quotient that sugar is the substrate in the normal aerobic respiration of higher plants. It is very probable that the initial biochemical transformations of the sugar molecule are in a large measure identical with those already described in alcoholic and lactic acid fermentation.

This is demonstrated by the fact that, as already mentioned, higher plants completely deprived of oxygen still continue to give off carbon dioxide for a while. During this time they produce, as well as carbon dioxide, ethyl alcohol or lactic acid (as in potatoes), i.e. precisely those substances which we have recognized as the end products of the fermentations described above. Also it has been possible to demonstrate the occurrence of most of the enzymes involved in alcoholic fermentation in higher plants. This kind of anaerobic respiration in higher plants was referred to formerly as 'intramolecular respiration', since the end-products can arise merely through the displacement of atoms and partitioning within the sugar molecule. In fact, according to modern views, very much the same occurs during fermentation in yeast.

It must therefore be assumed that in the normal aerobic respiration of higher plants the primary activation of the sugar, its phosphorylation, and the subsequent splitting of the glucose molecule into two 3-carbon compounds normally take place exactly as in anaerobic alcoholic fermentation. Furthermore, that glyceraldehyde, pyruvic acid and, after decarboxylation, acetaldehyde (or 'activated' acetic acid; see p. 273) and carbon dioxide are formed; and finally that instead of the dislodged hydrogen passing to the acetaldehyde (the fate of which will be considered later), it is merely oxidized by the oxygen of the air to form water, just as in the acetic acid bacteria.

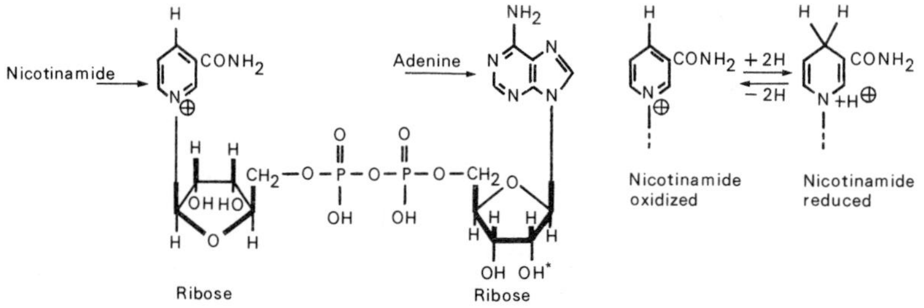

Fig. 255. *Left.* Structure of the co-enzyme NAD^+. *Right.* The active group (nicotinamide) in oxidized and reduced condition. Further explanation in text.

The transference of this hydrogen to molecular oxygen is of course a highly complicated process, because there is no direct affinity between oxygen and the dehydrogenase of the reduced form of the co-substrate. The structural formula of the co-substrate of the glyceraldehyde dehydrogenase is reproduced in Fig. 255 (although earlier regarded as a readily-dissociating prosthetic group of the enzyme, it is now better looked upon as a co-substrate). It consists of two nucleotides (see p. 280) joined together, one containing the pyridine base, nicotinamide, and the other the purine base, adenine, and consequently called nicotinamide-adenine-dinucleotide, abbreviated to NAD. Formerly this same substance was called, following its discoverer, diphosphopyridine nucleotide (DPN) or co-dehydrogenase (or co-enzyme) I. The important component is the nicotinamide which, as shown in Fig. 255 (right), can take up two hydrogen atoms (e.g. from glyceraldehyde, see p. 267), whereby the nitrogen loses its positive charge and the double bonds change. The accompanying shift in the absorption spectrum can be observed directly in the spectroscope. NADP, mentioned in relation to photosynthesis (p. 245), and also involved in other syntheses, is the same substance, but with a further phosphate radical in place of the –OH group of the ribose marked with an asterisk (Fig. 255).

This co-substrate can thus, in association with a dehydrogenase, become charged with hydrogen

$$NAD^+ + 2H \longrightarrow NADH + H^+$$

and with a further dehydrogenase the hydrogen can be once again removed and passed on to another acceptor. The hydrogen never passes from $NADH + H^+$ (sometimes written $NADH_2$) direct to atmospheric oxygen, but always by way of a whole series of interacting enzymes or co-substrates, all of which function as redox substances, readily oscillating between the oxidized and reduced conditions.

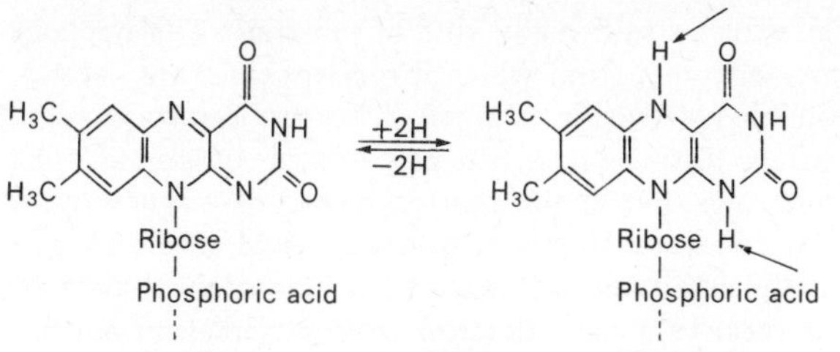

Fig. 256. Active group of a flavoprotein (isoalloxazin) in the oxidized and reduced condition.

Prominent amongst these substances are the yellow, or flavin, enzymes (flavoproteins), so called because in the oxidized condition they are yellow, but colourless when reduced. Common to them all as prosthetic group, firmly bound to the protein (see p. 259), is riboflavin (lactoflavin), a vitamin of the B_2 complex. As Fig. 256 shows, this is an isoalloxazin compound which can take up hydrogen reversibly at the sites indicated. It is in the nature of a nucleotide, the base being united with a five-carbon alcohol (ribitol), similar to ribose, and phosphoric acid. Besides riboflavin (not quite correctly termed flavin mononucleotide (FMN)), other prosthetic groups are found amongst the yellow enzymes in which an adenine-ribophosphate is linked to the phosphate of the riboflavin (thus yielding flavin-adenine-dinucleotide (FAD)). Many yellow enzymes are additionally loosely combined with certain heavy metals (iron, copper, molybdenum, manganese), which may partly influence the link with the protein, and partly the function. This indicates a role for the trace elements, even if it is not fully understood.

The first enzyme which interests us, the one which attaches itself to $NADH_2$ and which is alone able to transfer the hydrogen, appears to contain FMN and iron, and is not auto-oxidizable. Other yellow enzymes, however, can readily yield their hydrogen to atmospheric oxygen, with the formation of hydrogen peroxide. Others can take up hydrogen without the mediation of $NADH_2$, for example, succinic dehydrogenase of the citric acid cycle (see p. 274).

A further enzyme (perhaps better regarded as a co-substrate) which behaves as a redox substance, and which is possibly a member of the respiratory chain, is ubiquinone (co-enzyme Q). Quinone can readily take up hydrogen (thereby becoming hydroquinone), and lose it again. Structurally it is similar to vitamin K and plastoquinone (which possibly plays a similar role in photosynthesis (see p. 249)), consisting of a naphthoquinone and a side chain, made up of isoprene residues and reminiscent of the phytol of chlorophyll.

Ascorbic acid and phenols may also function as redox substances. When reduced they are colourless, but in the presence of air they are partly auto-oxidizable and become transformed into coloured oxidation products. This accounts for the well-known darkening of many freshly-cut plant tissues (e.g. apples, bananas, mushrooms, etc.). The enzymes involved mostly contain copper, and are discussed further later. It is, however, doubtful whether these systems are really involved in respiration; they may nevertheless be involved as hydrogen acceptors or donors in other metabolic processes. It is possible to measure the amounts of free hydrogen arising from dehydrogenation or hydrogen transport, and thus to obtain an estimate of the extent of the hydrogen flux or oxidation in the respiratory metabolism. The redox-potential, for which the symbol rH is used (rH being a negative logarithm), gives a measure of the pressure of hydrogen in atmospheres. Values of 20.5 correspond to the naturally very small amount of hydrogen normally present in water. Higher values (up to 41, corresponding to pure oxygen) indicate an effective removal of hydrogen, while lower values indicate an accumulation of unmetabolized hydrogen. In biochemistry it is customary to express the redox-potential in volts or millivolts, since the yield of electrons, which is equivalent to that of the freed protons, can be electrochemically measured. The standard of reference is a hydrogen ion concentration of unity (pH = 0). Against such a standard the hydrogen electrode at pH 7.0 has a potential difference of −0.42 V. The redox potential of the system $NADH + H^+ / NAD^+$, for example, amounts to −0.32 V, and that of riboflavin–H_2/riboflavin to −0.185 V.

None of the systems so far named, and known to stand in the respiratory chain, has a redox-potential which would allow the direct transference of hydrogen to atmospheric oxygen. According to Warburg's[12] theory of respiration, molecular atmospheric oxygen must first be 'activated' before it can attach itself to any kind of substrate. This happens by reaction with a reduced respiratory enzyme, from which it receives electrons. Activation thus takes the form of becoming electrically charged. The respiratory enzyme involved is cytochrome oxidase, which, as Warburg recognized, is a haem compound. Like chlorophyll, this contains four pyrrole rings, but at the centre, instead of magnesium, is an atom of iron. When in the divalent condition this iron atom can yield an electron to oxygen, whereby the oxygen becomes charged and the iron enters a trivalent (oxidized) condition. Immediately, however, it receives a new electron from an adjacent source, discussed below, and returns to the divalent (reduced) condition. These processes take place in rapid succession, so that the iron is continually oscillating between an oxidized and reduced condition.

$$4Fe^{++} - 4e + O_2 \longrightarrow 4Fe^{+++} + 2O^{--}$$
$$4Fe^{+++} + 4e \longrightarrow 4Fe^{++}$$

Recent evidence shows that cytochrome oxidase is polymerous, and surprisingly contains copper. In some plants (potato tubers, pumpkin seeds, fungi) other copper proteids have been found, but lacking the haem component. They can likewise activate oxygen, and bring about the oxidation certain hydrogen acceptors (polyphenoloxidase, ascorbic acid oxidase, tyrosinase). It is still not clear to what extent these particular copper-containing enzymes are involved in the main respiratory pathway.

The electron sources for the reduction of the trivalent oxidized iron of the cytochrome oxidase are certain iron-containing compounds closely associated with the enzyme. They are in fact of similar structure, and the iron is of oscillating valency, but perhaps as a consequence of a stronger attachment to the protein, they cannot react with oxygen directly. Since at high concentrations they appear green or red, they are termed cytochromes. A whole series of cytochromes can now be distinguished, a, b, c, . . . f (see also under photosynthesis, p. 249 seq.). Cytochrome oxidase, oxidized by the loss of an electron, takes up an electron from adjacent cytochrome a, the iron atom of cytochrome oxidase thus becoming divalent, and that of cytochrome a trivalent. Cytochrome a similarly receives an electron from cytochrome c, which is in turn supplied by cytochrome b. In this way an electron wanders over a whole chain of cytochromes to cytochrome oxidase, and from there to atmospheric oxygen. The sequence of these cytochromes in the chain, which is contained within the mitochondria, is still not completely known. Cytochrome c is also partly co-ordinated directly with the cytochrome oxidase. The chain receives its electrons from a yellow enzyme (flavoprotein), but perhaps ubiquinone acts as an intermediary between this and the first cytochrome. At this first flavoprotein the hydrogen, taken over from NADH + H^+, is split into protons and electrons, the hydrogen, since the electrons are transmitted along the chain to oxygen, thereby becoming charged. Ultimately the hydrogen and oxygen can combine to form water

$$2H^+ + O^{--} \longrightarrow H_2O$$

The entire path of the hydrogen, from the substrate (as, for example, glyceraldehyde) to its 'combustion' in the presence of oxygen is shown diagrammatically in Fig. 257.

There is however no instance in which all the steps in the respiratory chain, and their sequence, are fully known. It is not improbable that in different plants, even perhaps in successive developmental stages in one and the same plant, different enzyme systems are

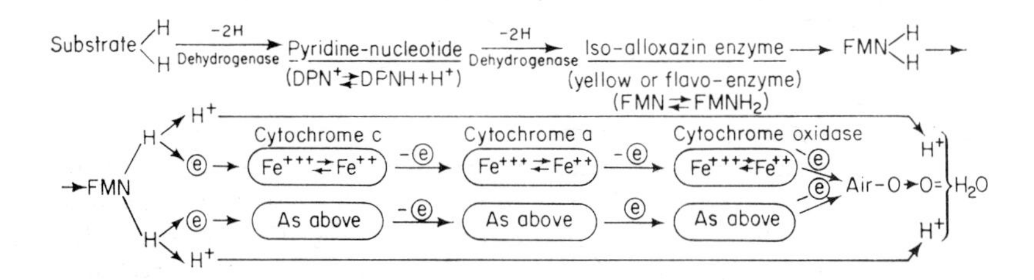

Fig. 257. Diagram showing the terminal oxidation of hydrogen. Explanation in text.

to be found. All, however, will function according to the same general principles. (See also gene activation and repression (p. 330).)

Phosphorylation in the respiratory chain. We have already described how in alcoholic fermentation (see p. 266) the energy released in the fermentation process is taken up into ATP by means of complicated phosphorylation and dephosphorylation processes. It is therefore hardly surprising that the further transmission of the hydrogen, and the handing on of the electrons, until the eventual union of the hydrogen and oxygen in aerobic respiration is also bound up with the formation of ATP. The mechanism, however, is still not completely known. In general it is considered that in the formation of one molecule of water, whereby half a molecule of oxygen is used, three molecules of ATP arise, one during the transference of the hydrogen from NAD to FMN, one during the splitting of the hydrogen into protons and electrons, and one during the activation of the oxygen by the cytochrome chain. This kind of phosphorylation, specifically associated with the respiratory chain, is to be distinguished from substrate phosphorylation (see p. 267). It can be 'uncoupled' by certain poisons, such as 2 : 4 dinitrophenol, so that the respiratory oxidation of hydrogen continues without the formation of ATP. Both the enzymes of the respiratory chain, and the formation of ATP, are sited within mitochondria, and this is perhaps correlated with the development of their internal lamellar systems, allowing complicated sequences of biochemical reactions to run as if on a conveyor belt. There is thus justification for referring to the mitochondria as the 'power house of the cell'. The aerobic break-down of sugar to the stage of pyruvic acid may, however, take place outside in the cytoplasm.

Terminal respiration. If the concept of a close relationship between aerobic and anaerobic respiration is correct, then not only the fate of the hydrogen, but also that of the pyruvic acid (or acetaldehyde) must be followed further, since this does not, as in anaerobic respiration, act as a hydrogen acceptor, but continues to be metabolized. This so-called terminal oxidation, in which two-carbon compounds like acetaldehyde are converted entirely into carbon dioxide and water, is for its part a very complicated cyclical process, in which even temporary resyntheses occur. In line with its best known intermediate product, this cycle is termed the 'citric acid cycle' or, after its discoverers, the Krebs or Krebs–Martius cycle.

The initial event is that acetaldehyde loses hydrogen, giving rise to an 'activated' acetic acid, i.e. an energy-rich coupling of acetic acid and co-enzyme A (acetyl CoA). The decarboxylation of the pyruvic acid and the removal of hydrogen from the acetaldehyde possibly, as in animal cells (where decarboxylase (see alcoholic fermentation, p. 266) is lacking), take place simultaneously in coupled reactions (co-enzymes, thiamin pyrophosphate and lipoic acid).

The active group of CoA is again similar to a nucleotide. Attached to adenine-ribose-pyrophosphate is pantothenic acid (well known as a member of the vitamin B_2 group), and thereto the amino acid cysteine, whose free –SH groups make possible the energy-rich combination with acetic acid. The co-enzyme and its complex with acetic acid are often denoted by HS–CoA and $CH_3 . CO \sim SCoA$ respectively. The latter represents the activated, i.e. the energy-rich, acetic acid. This is a very important intermediary product of metabolism, which, quite apart from the citric acid cycle, enters into many syntheses, such as of the fatty acids, ethereal oils and other isoprene derivatives. Here, however, we shall concern ourselves with the respiratory metabolism.

The activated acid with its two carbon atoms becomes linked with an acceptor with four carbon atoms (oxalacetic acid) to form citric acid, with six carbon atoms. This is then broken down in a cyclic process involving a four-fold dehydrogenation (with subsequent oxidation of the removed hydrogen), and the emission of two molecules of carbon dioxide (containing the carbon atoms originally in the acetic acid). The oxalacetic acid is simultaneously regenerated. As shown in Fig. 258, from one molecule of a triose (glyceraldehyde) are formed (with the addition of three molecules of water) 6 x 2 hydrogen atoms and 3 molecules of carbon dioxide, the yield from a hexose (glucose) correspondingly being 24 hydrogen atoms (which eventually becomes united with

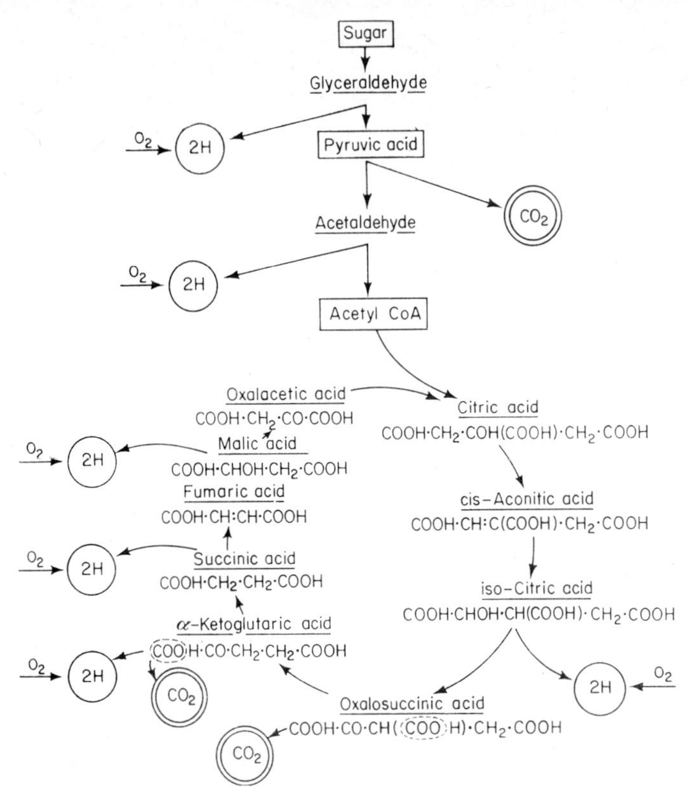

Fig. 258. Diagram showing the terminal oxidation of pyruvic acid (Krebs–Martius cycle). Explanation in text.

atmospheric oxygen, by way of the respiratory chain, to form water) and 6 molecules of carbon dioxide (see the equation of respiration, p. 262). Only in the oxidation of the pyruvic acid is the hydrogen accepted directly by a flavoprotein, instead of by way of $NADH_2$, resulting in the formation of only 2 instead of 3 molecules of ATP. As can be seen, most of the energy derived from the oxidation of sugar is not released until the citric acid cycle is reached. It is estimated that in the complete breakdown of a hexose sugar to carbon dioxide and water (during which theoretically 38 energy-rich ATP molecules should arise) some 35–40 per cent of the 675 kg cal of energy in the molecule, derived initially from photosynthesis, is taken up into ATP. This provides the only store of immediately available energy in the cell.

The fundamental principle of all these respiratory systems seems to be that hydrogen is repeatedly split off from the activated sugar by enzymes (about fifty different enzymes may be concerned in the respiration of glucose) with the formation of various intermediary products. The hydrogen is eventially oxidized by the aerial oxygen, while the carbon dioxide is returned to the atmosphere. Furthermore, the reaction which actually yields the energy, the oxidation of the hydrogen, is distributed over a whole chain of enzymes, thus offering the plant many opportunities for its direction and control.

This energy provides, amongst other things, for all the many syntheses which are involved in the metabolism of growth. Nevertheless, only a small fraction of the energy becomes actually bound up in the new substances. The greater part, after bringing about these processes in the organism, and after having undergone various transformations, eventually leaves the plant again in the form of heat in the transpiration stream. Nevertheless a living form is still superior to most man-made machines in the economy of its energy conversion.

Naturally it should not be overlooked that certain intermediary products are continually leaving this impressively controlled catabolism of sugar and, ceasing to be sources of energy, enter into the metabolism of growth. Thus not all the sugar consumed reappears as carbon dioxide and water, and biochemically there is often a close connection between the basal metabolism required for the maintenance of life and that concerned with growth. Intermediates of the Krebs cycle which drop out of the respiratory metabolism in this way must then be replaced from some other source (for the generation of oxalacetic acid from pyruvic acid and carbon dioxide, see p. 256). Energetically, these replacement processes, some of which are actually endergonic, have nothing to do with respiration, an activity which, by definition, is concerned solely with the winning of energy for the maintenance of life. Examples of metabolism of the carbon compounds which fall outside the concept of respiration are the utilization of keto-acids as the carbon skeleton of amino acids, and of activated acetic acid as the starting point for the synthesis of fatty acids. In fact many organic acids which are considered as normal components of the plant cell can be regarded as offshoots from the main pathway of sugar catabolism.

Acid metabolism of the Crassulaceae. In many succulent plants, e.g. of the Crassulaceae, a marked accumulation of organic acids, especially malic acid, occurs regularly during the night, to diminish again during the day. It appears that in these plants the nightly cessation of carbon assimilation, and the associated carbon dioxide accumulation in the tissue, causes a periodic inhibition of decarboxylation and therefore of the normal breakdown of the carbohydrate. It may be that the carbon dioxide becomes reincorporated (see p. 256) leading to a nocturnal accumulation or stabilization of the otherwise metabolized intermediary organic acids, which after the inhibition is lifted during the day can re-enter the metabolic system.

Chemoluminescence.[20] The occasionally observed production of light (chemoluminescence) by certain bacteria and fungi is only very indirectly connected with respiration. It does not involve any winning of energy. Luminescence of this kind is sometimes seen on dead fish and pieces of meat, places which favour the growth of the generally innocuous luminescent bacteria (*Bacterium phosphoreum, Pseudomonas lucifera* and others). Many members of the Peridiniales (Dinoflagellatae), and the mycelium of the tree-destroying 'honey fungus' (*Armillaria mellea*) also possess the capacity of luminescence. The luminescence of many animals appears to be emitted by bacteria which live symbiotically in special organs of the animal.

Nevertheless, there are certain animals (glow-worms, fresh-water snails, hermit crabs) which are themselves capable of synthesizing luciferin, and thus of emitting light. Chemoluminescence arises from the production, by metabolic pathways still unknown, of various species-specific substances (luciferins) which, in the presence of oxygen (acting as acceptor) undergo enzymic dehydrogenation, up to 80 per cent of the energy thus liberated taking the form of visible light. The presence of ATP is essential for chemoluminescence. An animal luciferin, the constitution of which is known, is now used as a very sensitive test for either ATP or oxygen.

The extent to which the complex processes of respiration are coupled with all the other vital activities of the organism is shown by the rapidity with which the rate of respiration responds to changes in its energy requirements. Respiration is in fact the most fundamental of all the activities of the cell. The onset of germination, salt uptake (see p. 235), growth, wounding and irritability phenomena all cause the intensity of respiration to increase markedly. The respiration of dormant parts of plants, on the other hand, especially dry seeds, is barely detectable. The mechanisms by which the rate of respiration is controlled remain, however, very little known.

Well-known poisons, such as cyanide, appear to act by paralysing respiration, bringing this about by binding with the heavy metals of the respiratory enzymes. Nevertheless, not all the enzymes, nor the cytochromes, are attacked in this way. Many plants show a surprising resistance to hydrogen cyanide. Whether the heavy metal enzymes are replaced in these plants by other systems is not known.

That **increase in temperature** increases the intensity of respiration, almost up to the point at which damage occurs, is readily understandable from the predominance of purely chemical processes in the respiratory system. This explains, for example, why flowers fade much more quickly at high temperatures than lower. The concentration of oxygen in the atmosphere, however, usually influences respiration far less than is the case with human respiration. The respiratory quotient does not begin to rise until the oxygen falls below 7 per cent, and in many plants not until below 3 per cent.

Respiration of other substances.[21] At the beginning of this section it was stressed that the significance of respiration is to be sought entirely in the freeing of bound energy, and that accordingly the chemical nature of the initial and final products is of secondary importance. Our account would be in no way altered if instead of starch the fat reserves, for example, were mobilized. The fat would first be split enzymatically (lipase) into glycerol and fatty acids, and now these components, after further corresponding transformations, would be introduced, possibly intermediately, into the normal course of sugar respiration. Similarly, a respiration of proteins, which occurs, however, in the higher plants only in conditions of severe starvation, could be joined on at some place, after the removal of the $-NH_2$ groups, to the normal respiration of carbohydrates. At the same time acetic acid fermentation provides an example of how simpler organic substances can also take the place of sugar without difficulty.

A direct oxidation of glucose to gluconic acid, followed by oxidative decarboxylation to yield a pentose (ribulose-5-phosphate) has been frequently observed in both animals and plants. It is not clear, however, to what extent this is related to respiration. Pentoses can be transformed by a complicated route once again into hexoses (the reverse of the way in which the carbon dioxide acceptor is regenerated in photosynthesis (see p. 251), the so-called pentose phosphate cycle). This manner of break-down of glucose is probably of importance in regard to the provision of pentoses (ribose, deoxyribose, etc.) which are an essential component of the nucleotides (see p. 280) of nucleic acids and certain enzymes, and which must be drawn continuously from the sugar metabolism. The compound $NADPH_2$ is also generated in this way.

Clearly very different is the method of winning energy of those micro-organisms which oxidize **inorganic substrates**, although the energy as usual is stored in ATP. We are concerned here with the chemosynthetic forms mentioned earlier (p. 255); they obtain at least the energy required for carbon assimilation from the oxidation of inorganic materials, and can consequently live autotrophically in the dark.

Thus the nitrifying bacteria, widely distributed in soils, oxidize ammonia to nitrite, and nitrite to nitrate, with the liberation of a considerable amount of energy, which becomes bound, but only to the extent of about 10 per cent, in the chemosynthetically generated carbohydrates:

$$2 \text{ g mol } NH_3 + 3 \text{ g mol } O_2 \longrightarrow 2 \text{ g mol } HNO_2 + 2 \text{ g mol } H_2O + 158 \text{ kg cal}$$
$$2 \text{ g mol } HNO_2 + 1 \text{ g mol } O_2 \longrightarrow 2 \text{ g mol } HNO_3 + 36 \text{ kg cal}$$

Nitrite-formers (*Nitrosomonas* spp.) and nitrate-formers (*Nitrobacter* spp.) live closely associated with one another; the nitrate-formers remove by their metabolism the poisonous products of the nitrite-formers. The initial material, ammonia, comes from the rotting of the nitrogenous components of organic matter by other bacteria. The oxidation, which has been shown to involve the cytochromes, probably leads to the production of hydroxylamine as an intermediate. The nitrate-formers provide one of the main sources of nitrate in normal soils, a mineral of great importance for the vigorous growth of higher plants.

The colourless sulphur bacteria (e.g. *Beggiatoa*; see p. 448), also mentioned earlier, oxidize hydrogen sulphide or sulphur, poisons for most other organisms, to sulphur or sulphuric acid respectively. So much energy is liberated by these reactions that here also the chemosynthesis of carbohydrates from the carbon dioxide of the air and water is made possible:

$$2 \text{ g mol } H_2S + 1 \text{ g mol } O_2 \longrightarrow 2 \text{ g mol } H_2O + 2 \text{ g mol } S + 118 \text{ kg cal}$$
$$2 \text{ g mol } S + 3 \text{ g mol } O_2 + 2 \text{ g mol } H_2O \longrightarrow 2 \text{ g mol } H_2SO_4 + 286 \text{ kg cal}$$

The initial material, hydrogen sulphide, is derived like ammonia from the bacterial decomposition and rotting of organic substances, principally the proteins of dead plants and animals (see p. 291). Some of it may also be produced by the reduction of sulphates (see below) by other bacteria.

In addition, other compounds containing sulphur, such as sulphides, sulphites, thiosulphates, etc., can be metabolized by many sulphur bacteria, a very diverse group of organisms. They all form an important factor in the so-called self-purification of water in nature. As a result of their activities, organically bound sulphur in particular becomes converted again into an inorganic form. The actual reactions involved may be a little different from those summarized in the equations above, but there is still little known about their biochemistry. In many sulphur bacteria (e.g. *Beggiatoa*) refractive droplets are seen to arise as intermediary products in the oxidation of hydrogen sulphide. These probably consist of elementary sulphur.

In the iron bacteria divalent iron in the form of ferrous carbonate becomes oxidized to ferric carbonate, which as a result of hydrolysis is precipitated as ferric hydroxide. These brown precipitations are not infrequently seen in ditches and puddles; in certain conditions they can prove a nuisance to water undertakings by blocking pipes. Here again it appears that part of the energy obtained can be used for chemosynthesis. In the presence of sugar, however, many iron bacteria appear to be capable of normal sugar respiration.

Many bacteria can also utilize manganese compounds oxidatively in a wholly similar manner. There are also methane bacteria which oxidize marsh gas to carbon dioxide and water. Indeed, even hydrogen itself, which as mentioned earlier is frequently generated together with methane in butyric acid fermentation in nature, can be oxidized by many soil bacteria to water. This reaction perhaps represents the most primitive manner of obtaining energy, but at the same time it serves as a model for the complex oxidation of hydrogen in normal sugar respiration. These hydrogen bacteria undertake chemosynthesis the necessary energy being obtained by the activation of hydrogen (by the enzyme hydrogenase) and its combination with atmospheric oxygen, leading to the release of considerable quantities of energy:

$$2H_2 + O_2 \longrightarrow 2H_2O + 114 \text{ kg cal}$$

It appears, however, as already mentioned, that they are also capable of living heterotrophically and undertaking normal sugar respiration. Their assimilation product is a complex polyhydroxybutyric acid. Polyphosphates often accumulate as storage products, but they appear to have no great significance in relation to energy.

Denitrification, desulphurication.[14] In anaerobic conditions certain bacteria are able to utilize activated molecular hydrogen to some extent to reduce nitrates and sulphates. Here it is not the oxygen of the air that is being used as the hydrogen acceptor, but the bound oxygen of the nitrate or sulphate radicle. Organic substances may also serve as a source of the activated hydrogen. The reduction of nitrate, which proceeds as far as the production of nitrous oxide or even of molecular nitrogen, can be harmful in very damp insufficiently cultivated soils, since it causes the destruction of the nitrate absolutely essential for the rest of the plant life. Desulphurication is, besides the decay of proteins, one of the main sources of origin of hydrogen sulphide. The frequently observed blackening of mud is due to the precipitation of iron following upon the formation of hydrogen sulphide. The Black Sea derives its name from this phenomenon.

A further example of the oxidation of hydrogen by bound oxygen is provided by those bacteria which are able to reduce carbon dioxide to methane (marsh gas). This is one of the sources of methane in nature.

In line with the views developed in this section, we interpret all reactions, even inorganic exergonic reactions, which lead to the formation of ATP, as respiratory in the wide sense. Since, however, chemosynthesis leads to sugar, and this sugar is inevitably further metabolized to yield the starting points of many different syntheses, it follows that a sugar metabolism exists and the enzymes of normal sugar catabolism are present. As indeed has been pointed out several times, many chemosynthetic organisms are also capable of living heterotrophically, showing thereby a normal sugar respiration. It is still not known to what extent such respiration contributes to the energy requirements of these organisms. If the energy from inorganic sources is used exclusively for the synthesis of sugar, then the energy coming from the breakdown of sugar would be considered as respiratory in the narrow sense. In all instances, however, the inorganic oxidation reactions predominate quantitatively, and give these micro-organisms their characteristic stamp.

VI. Protein metabolism[22]

The synthesis and breakdown of carbohydrate represent two of the most important metabolic processes in the life of the higher plant. While photosynthesis builds up material and fixes energy, which is subsequently released by respiration and utilized in various ways, there are other important processes going on in the cell at the same time, only some of which can be considered here in detail. Of these other metabolic processes, one of the greatest interest is undoubtedly protein metabolism, since protein forms one of the most important structural elements and components of the protoplasm. If a plant is to grow and produce new cells the proteins of the cytoplasm and cell nuclei must be continually reproduced. The fact that it is possible to grow a green plant in a nutrient solution containing only inorganic salts shows that protein can be formed from purely inorganic materials. Thus, with respect to the protein metabolism, there must be capacities for synthesis present in plants which far exceed those in animals. Unlike plants, no animal is capable of building up all the organic components of its cells from inorganic mineral salts. Not only sugar but also protein or its derivatives must first have been synthesized up to a certain level of organic complexity in the metabolism of plants, before animal life was possible on our earth. This demonstrates the great importance of plant life on the earth, and the intimate relationship all other living organisms, including mankind, have with the Plant Kingdom.

Chemistry of proteins. A knowledge of the chemistry of proteins is required before we can consider their metabolism. Unfortunately at present this is still rather limited, but one fact that is certain is that the essential units of the proteins are the amino acids. These are organic acids in which one of the hydrogen atoms of the carbon adjacent to the carboxyl group (the α-position) is replaced by an amino group; thus they have a general formula $R \,.\, CHNH_2 \,.\, COOH$, where R indicates any kind of radical, even in the simplest case hydrogen, as in glycine ($CH_2NH_2 \,.\, COOH$). Only twenty different amino acids have been isolated. Where R is a CH_3 group we have alanine ($CH_3 \,.\, CHNH_2 \,.\, COOH$), derived from propionic acid; where a methanol group, serine ($CH_2OH \,.\, CHNH_2 \,.\, COOH$); and where a $HS{-}CH_2$ group, cysteine ($HS \,.\, CH_2 \,.\, CHNH_2 \,.\, COOH$). Longer, simple or branched, aliphatic chains can also occur. Tyrosine (the presence of which in proteins accounts for Millon's reaction (see p. 18)) resembles alanine, but a hydrogen of the methyl group is replaced by phenol ($HO{-}C_6H_4{-}CH_2 \,.\, CHNH_2 \,.\, COOH$). Heterocyclic, nitrogen-containing rings are found in some amino acids, such as tryptophane and histidine, and in others further basic groups, as in lysine and arginine. Dibasic acids (especially in plant proteins) can also form amino acids, such as the important aspartic acid ($COOH \,.\, CH_2 \,.\, CHNH_2 \,.\, COOH$), which is an amino succinic acid, and glutamic acid, derivable from α-ketoglutaric acid. Textbooks of chemistry must be referred to for further details of the amino acids.

A common test for the presence of amino acids is currently the reagent ninhydrin, with which they give a blue colour.

The amino acids have the remarkable property of being able to form long chains (macromolecules) with other amino acids. This comes about by the carboxyl group of one molecule reacting with the amino group of the next molecule, with the loss of water. This condensation involves the uptake of a certain amount of energy since the natural equilibrium is towards dissociation (Fig. 259A), and this is provided by ATP, by which the amino acids become 'activated'. Two amino acids yield a dipetide, several or many (usually with a maximum of about 250–300), a polypeptide. In a natural protein the individual amino acids are arranged in a definite sequence, which determines the specificity of the protein. Since from twenty amino acids it is already possible to make not less

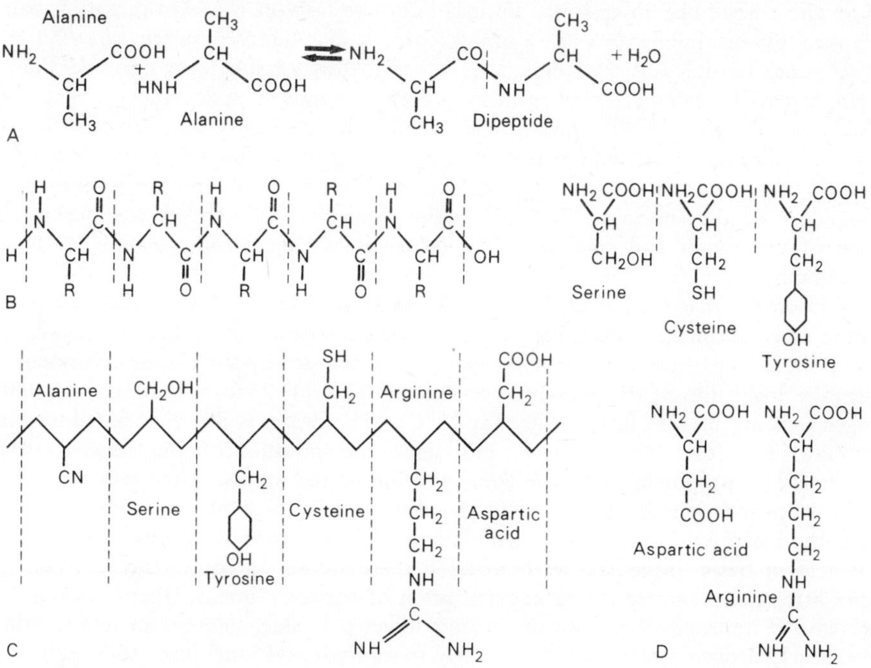

Fig. 259. Diagram showing the composition of a polypeptide chain. Explanation in text.

than 10^{1300} different polypeptides each with 1000 amino acids, it is understandable why many specific and subspecific differences between organisms should be reflected in and perhaps caused by their proteins.

A segment of a polypeptide chain with several amino acids is shown diagrammatically in Fig. 259B and C. This shows how the residues of the amino acids (R in Fig. 259B) project from the polypeptide as so-called side chains. The acidic or basic nature of these side chains is more responsible than the terminal groups of the main chain for the amphotytic character of the proteins, i.e. their ability to act sometimes as acids and sometimes as bases. The behaviour of the protein in water, and the presence or absence of a hydration sheath around the macromolecule and hence its ability to form a colloidal solution, are also determined by the nature of the side chains. This explains the great effect which salts can exert on solutions of protein (such as 'salting out' by concentrated solutions of ammonium sulphate), and also that of hydrogen ions. Heating usually causes an irreversible coagulation (denaturation) of protein, and its precipitation from solution (see below).

Polypeptides, even in their simplest form, already possess a chevron-like primary structure (Fig. 259B), but over and above this, in actual proteins, the chains take up quite definite conformations, the details of which are already becoming known in some cases. Fibrous (or 'linear') proteins play a less important part in plants than in animals, but are probably always present where fibrils are to be observed (e.g. nuclear spindles, flagella). They probably arise by the polypeptide chains taking up a parallel arrangement, the individual chains being linked by hydrogen bonds. These bonds (which possess only about a tenth of the energy of a covalent bond) form between a nitrogen or an oxygen atom and the hydrogen of a neighbouring –NH or –OH group (see Fig. 262). The bond distance is of the order of 2.8Å. In the sphaeroproteins, which are represented in many enzymes and storage proteins, the secondary structure takes the form of a right-handed helix (α-helix), often arising spontaneously, in which the individual gyres are held together by intramolecular hydrogen bonding, giving the whole macromolecule great stability. The side chains project perpendicularly from the helix. A tertiary structure then arises by the α-helix becoming rolled up in a knot-like fashion, the individual coils being cross-linked by disulphide (—S—S—) bridges or salt-like bonds between carboxyl and amino groups. In this way the molecule comes to form a sphere or ellipsoid (hence 'sphaeroproteins'). Many proteins seem to possess a quaternary structure in that they consist of several polypeptide chains, held together by various bonds between the side chains. Examples of such complex proteins are provided by haemoglobin and myoglobin, where not only the number of polypeptide chains is known (four), and the number (574) and sequence of the amino acids, but also the secondary and tertiary structures. The location of the twists and foldings in the α-helices has been made possible by X-ray crystallography.

In addition to what has already been said in relation to imbibition and absorption, the following must be added concerning the significance of the hydrogen ions. In physiology generally the concentration of the hydrogen ions is denoted by the pH, this representing the negative logarithm of the actual concentration. Thus pH 7 means that $10^{-7} \times 1.008$ g hydrogen ions are present in each litre (at 22°C). Since water is also dissociated to small extent into the ions H^+ and OH' and since the product of the concentrations $[H^+] \times [OH']$ is proportional to the concentration of the undissociated remainder (the constant of proportionality being called the dissociation constant), it follows that the value of the hydrogen-ion concentration alone is sufficient to show whether a given solution is acid or basic. In neutral water at 22°C there are 10^{-7} g equivalents of hydrogen ions per litre, and a corresponding concentration of hydroxyl ions, so that the pH is 7.0. An increase in hydrogen ions, and therefore a lower pH value, signifies an acid solution, while fewer hydrogen ions (and consequently more hydroxyl ions) leads to a higher pH value, which therefore signifies an alkaline solution.

The dissociation of an acid or base is suppressed by the addition of further hydrogen or hydroxyl ions respectively. It is therefore evident that the pH value of the solution must also be of influence on the ability of the acid and basic groups of the protein molecule to dissociate. Moreover, since these groups, by their electrical attraction of water molecules, are responsible not only for the solubility of the colloid but also for its important capacity for adsorption, it is clear that the behaviour and stability of a protein solution will be strongly influenced by its pH. Each protein has a so-called isoelectric point at which the acidic and basic groups of the protein are equally dissociated. At this point the stability of a protein solution is least, and many proteins are precipitated. It will now be clear why the activity of an enzyme is so strongly dependent on pH; this pH dependence betrays in fact the protein nature of enzymes.

It is customary to distinguish between proteins, which consist solely of polypeptides and amino acids, and proteids which contain, besides amino acids, other quite different molecular groups (prosthetic groups). Further, amongst the proteins, it is possible to differentiate on the basis of solubility, readiness of precipitation by salts, and other properties, between the highly basic histones (which are found particularly in chromosomes), albumins, globulins, and glutelins and prolamines (both found as storage proteins in seeds). These classes of protein are not, however, strictly defined in chemical terms. It seems almost certain that the storage proteins (as in aleurone grains) possess a simpler constitution than the structural proteins of the cell.

Prominent amongst the proteids are the lipoproteins, somewhat indefinite compounds of protein and lipoids (substances such as lecithin and cholesterol, similar to fats). The lipoproteins are essential components of the many membranes in the cell (plasmalemma, endoplasmic reticulum, envelopes of the nucleus, mitochondria and plastids, and the internal lamellae of the latter). Also important are the nucleoproteins, occurring particularly in the nucleus, but also found in the cytoplasm. They consist of protein and nucleic acid, and possess quite special significance in relation to both protein synthesis and inheritance. They are consequently treated in rather more detail in the following.

The nucleic acids are very large macromolecules, visible in the electron microscope, each built up from many thousands of mononucleotides. Mononucleotides are also present singly or in pairs in certain enzymes (e.g. NAD (p. 270), yellow enzymes (p. 271)). Each consists of a nitrogen-containing base, a pentose, and phosphoric acid. The combination of the base and the pentose without the phosphoric acid is termed a nucleoside.

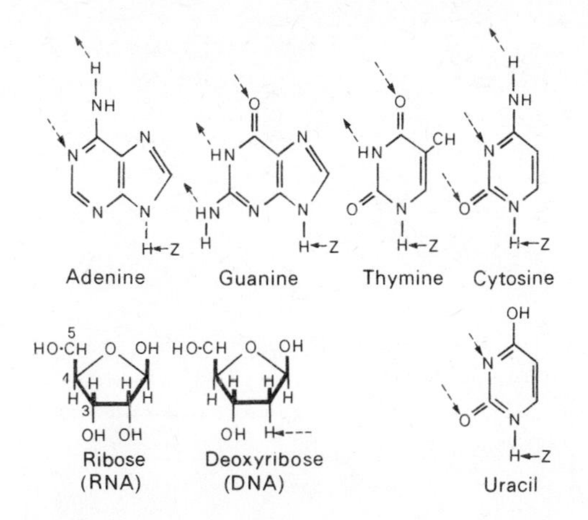

Fig. 260. The bases and sugars occurring in nucleic acids. The broken arrows by the bases show the sites at which hydrogen bonds form. Z and an entire arrow shows where the base combines with its pentose.

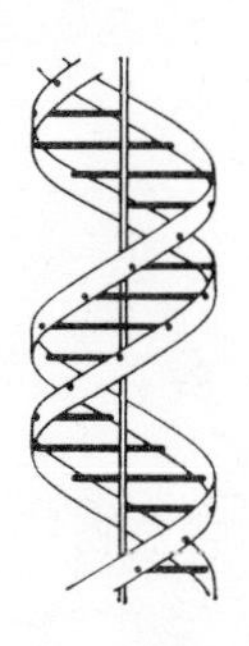

Fig. 261. Double helix of deoxyribonucleic acid (DNA), as envisaged by Watson and Crick. Two strands of DNA are held together by hydrogen bonds (black lines). After Steward, diagrammatic.

Although the thread-like nucleic acid molecule can reach a length of several μm, each nucleic acid contains only four different bases. Those are arranged in a definite, but irregular, sequence in the chain, and the combinations of the bases have, as will be seen, a quite special significance. The bases concerned are two purines, adenine and guanine (adenine is also present in ATP (see p. 266)), and two pyrimidines cytosine and thymine, or (in the case of ribonucleic acid) uracil (see Fig. 260). As shown in Fig. 262, the linking together of the nucleotides is always between the phosphoric acid and the third carbon atom of the pentose. The pentose is either ribose (yielding ribonucleic acid, present in the cytoplasm and nucleoli), or ribose lacking an oxygen atom (deoxyribose, yielding deoxyribonucleic acid, present in quantity in chromosomes and in many viruses (see p. 299)). A further difference between ribonucleic and deoxyribonucleic acids is that in the former the pyrimidine bases are cytosine and uracil, whereas in the latter they are cytosine and thymine.

The nucleic acids have the important property of being able to replicate themselves in the protoplasm. It has long been known that before mitosis the chromosomes split longitudinally, and that the chromatids before every division contain a complete set of the hereditary information characteristic of the species. Knowledge of the nucleic acids today gives us definite ideas about how this could happen. According to Watson and Crick native deoxyribonucleic acid, as in chromosomes, always has the configuration of a double helix (Fig. 261). The two sugar-phosphate chains are wound as round a cylinder, and the bases project towards the centre where the opposite pairs are linked together by hydrogen bridges. For spatial reasons thymine (or, in ribonucleic acid, uracil) is always paired with adenine, and cytosine with guanine (see Fig. 262) the bond distances being of the order of 2.8–3.0Å. Thus, with regard to the bases, the two strands of the nucleic acid are always complementary to each other. Further, analyses have shown that the ratio of the purine to pyrimidine bases is always 1 : 1. In the identical reproduction it is thought that the two strands become separated from each other at one end of the helix, and that new complementary bases with deoxyribose and phosphoric acid then become progressively incorporated as the separation proceeds along the double helix. Eventually two new double helices stand in place of the original one (semi-conservative replication; see Fig. 263). Many details of this process are still unknown. On the one hand it requires the continual synthesis of the bases and deoxyribose in the cell, and on the other quite special mechanisms and specific enzymes to bring about the opening of the parent helix, and the ordered incorporation of the new complementary nucleotides. Diagrams illustrating this basic biological process should not be allowed to conceal the formidable problems it poses. Nevertheless, it has been possible to bring about the synthesis of

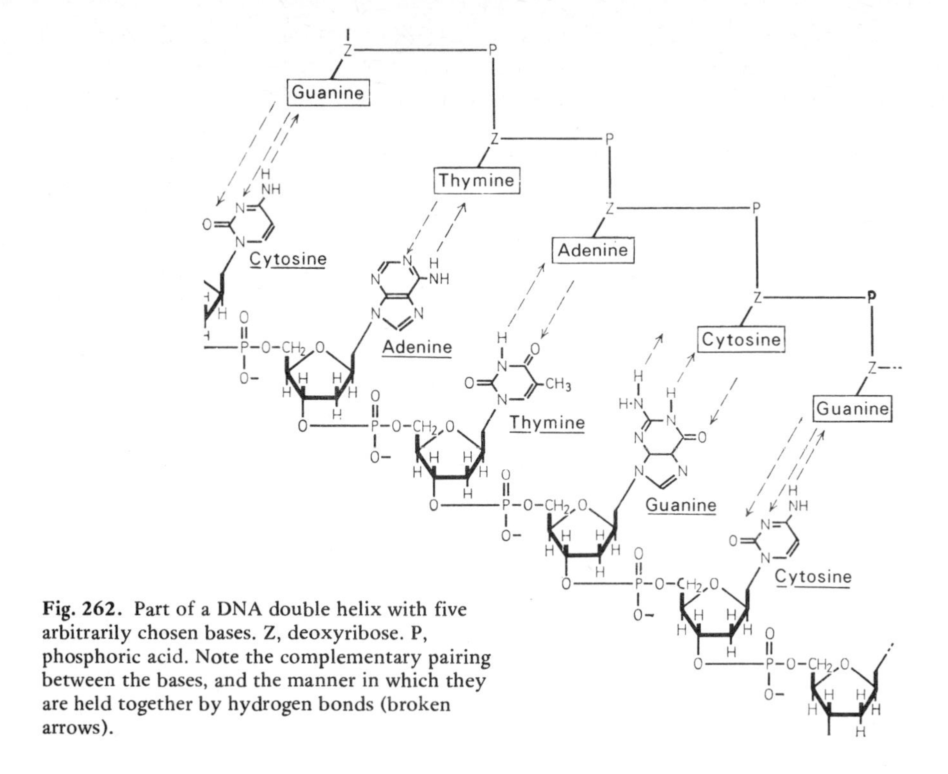

Fig. 262. Part of a DNA double helix with five arbitrarily chosen bases. Z, deoxyribose. P, phosphoric acid. Note the complementary pairing between the bases, and the manner in which they are held together by hydrogen bonds (broken arrows).

Fig. 263. Diagrammatic representation of the identical reproduction of a double-stranded DNA molecule. **A.** A portion of the molecule. The zig-zag line represents the chain of deoxyribose phosphates (see Fig. 262), and the circles the bases (A, adenine; C, cytosine; G, guanine; T, thymine). The dotted lines indicate the hydrogen bonds. **B.** Opening of the double strand at one end immediately preceding replication. **C.** The incorporation of nucleotides as the split moves along the parent molecule. Two new complementary strands are formed, resulting in two new double-stranded DNA molecules, each containing one strand from the parent (semi-conservative replication). After Weidel.

deoxyribonucleic acid in a cell-free system by means of isolated enzymes. A further problem, which will concern us in relation to protein synthesis, is that deoxyribonucleic acid is not only capable of replication, but also of generating adjacent to itself, large or small molecules of ribonucleic acid. These also show complementarity, in respect of the bases, with the parent deoxyribonucleic acid.

Protein synthesis. Besides stem apices, and other centres of growth and storage, green leaves appear to be sites of active protein metabolism. Protein is continuously synthesized in them, and also to some extent broken down again so that it may be distributed to other places where it is required, such as active stem apices, ripening fruits and so on. Active redistribution of proteins also occurs in the germination of seeds, sprouting of tubers, etc. If a green plant whose roots are immersed in a nutrient solution manufactures new proteins in its leaves or elsewhere after the pattern of those already present, then the necessary nitrogen, so far as it does not come from other sites of synthesis in the plant body, must be derived solely from the nitrate of the nutrient solution. The atmospheric nitrogen, the only other possible source, cannot be utilized by higher plants. (For the protein metabolism of heterotrophic plants see p. 290 seq.)

The amount of nitrate in the usual nutrient solutions and in natural soils is fairly small. In 100 g soil the nitrate amounts to some few milligrams, but this is continually being replaced by the activity of the nitrifying bacteria. A few milligrams are also usually present in a litre of rainwater, since traces of nitric oxide are always being formed by electrical discharges in the atmosphere. Owing to the large amount of water in plant tissue, the protein amounts to only about 3 per cent of the fresh weight, and the nitrogen (forming about one-sixth of the protein) to about 0.5 per cent of the fresh weight, so that the nitrate available suffices to meet that required for the essential proteins. As a rule, however, nitrogen is a limiting factor in respect of plant growth, and the effect of a thorough dressing with nitrogenous fertilizer is always very striking. Many plants, especially the ruderals, which grow on rubbish-dumps, are capable of storing considerable amounts of unused nitrate in their leaves.

The manner of synthesis of the amino acids, the basic units of protein structure, is not in every instance understood. It is certain that initially the nitrate must be reduced, and the necessary hydrogen is supplied from the respiratory metabolism. Nitrate reduction is always accompanied by the excretion of a certain amount of carbon dioxide (cf. denitrification, p. 270). This reveals at once the generally close connection between protein and carbohydrate metabolism.

This reduction can take place independently of light in both roots and leaves. There are apparently several intermediary products (nitrites, hydroxylamine (NH_2OH)) before ammonia is reached, and both flavoproteins and the heavy metals molybdenum and manganese are involved in the process.

Also still obscure is the manner in which the reduced nitrogen is joined on the carbon chains present in the amino acids. It appears that it is principally the α-ketoacids, which are continually arising in respiratory metabolism (especially in the Krebs–Martius cycle, p. 273), and which can apparently be easily withdrawn from it, that become combined with ammonia. This is brought about by simultaneous hydrogenation under the influence of a dehydrogenase, so forming amino acids. Thus alanine would be formed from pyruvic acid, aspartic acid from oxalacetic acid and glutamic acid from α-ketoglutaric acid.

$$\underset{(\alpha\text{-ketoglutaric acid})}{COOH.CH_2.CH_2.CO.COOH} + NH_3 + 2H \rightleftarrows \underset{(\text{glutamic acid})}{COOH.CH_2.CH_2.CHNH_2.COOH} + H_2O$$

In experiments with ammonia labelled with the radioisotope ^{15}N, and even after the assimilation of carbon dioxide labelled with ^{14}C, it has been possible to detect the label after only a few minutes in aspartic acid, glutamic acid and alanine. It is possible that aspartic acid might be formed from fumaric or malic acids, also intermediary products in respiratory metabolism, as well as from oxalacetic acid:

$$COOH.CH{:}CH.COOH + NH_3 \longrightarrow COOH.CH_2.CHNH_2.COOH$$

The view is frequently taken that in plants the first-formed amino acid is principally glutamic acid. This amino acid is believed to be capable, under the influence of special enzymes (transaminases, the active group of which is pyridoxal phosphate (vitamin B_6)), to transfer its $-NH_2$ group to other α-ketoacids or oxyacids. In this way most of the aliphatic amino acids can be built up by transamination. Heterocyclic amino acids and other nitrogen-containing compounds (such as pyrrole compounds, pyrimidines, purines, etc.) appear to be built up in a manner more complicated, but essentially the same, from simple amino acids (e.g. glycine), activated fumaric acid (co-enzyme folic acid, see p. 300), activated carbon dioxide (see p. 300, biotin), and $-NH_2$ groups from glutamine or asparagine. In purine synthesis, for example, the nitrogen atoms 3 and 9 are derived from the amide nitrogen of these last two compounds. The pyrimidine ring is synthesized from aspartic acid and carbamyl phosphate.

At all events all the amino acids can be synthesized in the plant cell from organic carbon chains, which can easily be provided by the intermediary products of the carbohydrate metabolism, and ammonia, derived by the reduction of nitrate. In this respect the plant is far superior to the animal. The favourable effect which a copious supply of carbohydrate exerts on protein formation reflects this close interconnection between the carbohydrate and protein metabolisms.

Besides nitrates, ammonium salts and even urea can be utilized by plants as a source of nitrogen. Nitrate in general produces a more favourable response, owing to the fact that ammonium salts, by the unilateral increase of NH_4^+ ions, may cause undesirable shifts of pH throughout the plant. The sensitivity of plants in this respect differs with the species and the stage of development.

After the synthesis of the amino acids, already a highly specialized aspect of the metabolism, they must be linked together to form the polypeptide chain. The different amino acids (all of which may occur more than once) are arranged in a definite order, the so-called sequence, which is specific in every instance. The kind, number, and sequence of the amino acids is completely known for only a few proteins, mostly enzymes and hormones (insulin, 51; ribonuclease (which depolymerizes ribonucleic acid) 124; tobacco mosaic virus protein (see p. 299) 158; papain (see p. 286) 180 amino acids). We have, however, some very clear ideas about how these sequences are reproduced identically. Although these ideas stem largely from the study of bacteria and animal cells, they are probably valid for all forms of life, and are of particular significance in relation to inheritance (see p. 329 seq.).

Protein synthesis takes place predominantly at the ribosomes in the cytoplasm (p. 24). Here are located the matrices which determine the sequences in which the different amino acids are assembled in the polypeptide chains. These patterns must, of course, be inherited. As has long been known, the Mendelian genes (see p. 329) are located on the chromosomes of the nucleus, so it follows that ultimate control of protein synthesis must be there. As was discussed earlier, the deoxyribonucleic acid of the chromosomes is capable of identical reproduction. It can also form short lengths of complementary ribonucleic acid (transcription). These ribonucleic acids (known as messenger ribonucleic acids), with definite (but complementary) base sequences, detach themselves from the chromosomes and enter the cytoplasm by way of the nuclear pores. Although the method of transport is still quite unknown, they reach, and become associated with, the ribosomes. With the longer lengths of messenger ribonucleic acid (m–RNA) the ribosomes become strung along the thread like beads, giving rise to circular or helical conformations (polysomes or polyribosomes). Generally messenger ribonucleic acids appear to be short-lived. The site of attachment of the ribosomes, which consist of two unequal parts held together by magnesium ions, to the ribonucleic acid is not absolutely clear. Somewhat hypothetical also is the manner in which the ribosomes bring about protein synthesis. Nevertheless, this certainly involves still smaller ribonucleic acids, consisting of seventy to eighty nucleotides, called soluble or transfer ribonucleic acids (s–RNA, or now more usually t–RNA). These also originate in the nucleus, and in some way bring the appropriate amino acids to the site of polymerization. Each of the twenty amino acids in proteins probably has its own specific transfer ribonucleic acid. The amino acid is attached by way of an energy-rich acyl bond (at the expense of ATP) to the ribose at one end of the strand of ribonucleic acid. The base sequences of some of the transfer ribonucleic acids are now known, and occasionally an unusual base is present. The strand of the transfer acid probably takes the form of a ring lobed like a clover leaf, the ends of the strand, at which amino acid is attached, winding together as a spiral, forming as it were the 'petiole' of the leaf. Three bases (known as the anticodon), protruding from one of the 'leaflets', are believed to 'feel out' the appropriate codon in the messenger ribonucleic acid, anticodon and codon then becoming united by hydrogen bridges, and the amino acid thereby held in the correct position in the sequence (Fig. 264). It is generally believed that the code ('codon') for every amino acid takes the form of three adjacent bases. Since, however, there are four different bases, sixty-four (4^3) different triplets are available to code for only twenty amino acids. Either, therefore, some amino acids are coded for by more than one triplet, or some sequences of bases are meaningless (which would imply that there was degeneracy in the code). The triplets of bases corresponding

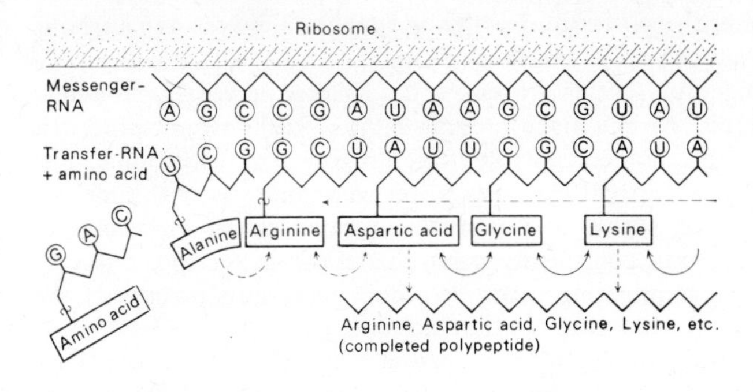

Fig. 264. Diagrammatic representation of protein synthesis on a ribosome (above dotted) attached to which is a length of messenger-RNA. Short lengths of transfer-RNA (here shown as short unfolded chains) lead individual amino acids (connected with the chain by an energy-rich bond) to the growing polypeptide. Each amino acid has a specific transfer-RNA, whose bases pair with those of the messenger-RNA. The sequence of bases in the messenger thus determines the sequence of amino acids in the polypeptide. Once in position peptide bonds form between the amino acids, and the completed polypeptide chain detaches itself (broken arrows, below).

to every amino acid have now been identified. Thus, although much remains to be discovered, it is now known how the individual amino acids are assembled, by means of transfer ribonucleic acids, according to the sequence encoded in the messenger ribonucleic acid (translation). With the assistance of enzymes, peptide bonds are now formed between the amino acids at the expense of the energy-rich bonds, by which they were initially linked to their transfer nucleic acids. Eventually the completed polypeptide detaches itself from the messenger ribonucleic acid. The base sequences of the messenger are of course a copy of those in part of the deoxyribonucleic acid of the chromosomes. Information in the genes is thus transcribed into the form of messenger ribonucleic acid, and then ultimately translated into proteins with definite amino acid sequences in the cytoplasm. Since these proteins are often enzymes, it is clear how the nucleus is able to control the metabolism and syntheses in the cell.

Only further research will show whether the scheme described here is of general validity. So far it is based largely on work with bacteria.

Fixation of atmospheric nitrogen.[14] As already mentioned, the immense reservoir of nitrogen in the atmosphere is not generally accessible to higher plants. Some of the bacteria, the Actinomycetes and Cyanophyceae, and probably also a few fungi, all fairly widely distributed in soils, are, however, capable of reducing and organically incorporating atmospheric nitrogen. The hydrogen they utilize for this purpose is either derived from the respiratory metabolism or, as in the purple bacteria, produced photolytically from water.

Among the heterotrophic bacteria, some anaerobic butyric acid fermenters (e.g. *Clostridium pasteurianum*), some aerobic bacteria of cultivated soils (e.g. *Azotobacter chroococcum*), and some symbiotic forms living principally in the roots of Leguminosae (see p. 293) are capable of carrying out these important syntheses. The chemical details are still obscure. Many organisms able to activate molecular hydrogen with hydrogenase (see p. 277) appear also able to bind atmospheric nitrogen. Often iron (ferredoxin), molybdenum and cobalt (vitamin B_{12}) appear to play a part. The yield of fixed nitrogen is quite considerable. It is estimated that 7–18 mg of nitrogen are bound for every gram of carbon compound used as a hydrogen donor. For a hectare of a natural soil this would amount to 20–40 kg per year, the yield being still greater if the symbiotic forms are present as well. These nitrogen-fixing micro-organisms are therefore of enormous importance for the nitrogen content of the soil.

The catabolism of protein. The synthesis of protein is only a part of the protein metabolism of the cell. As in all living systems, protein metabolism is never at a standstill, but only reaches an equilibrium, the position of which is variable, between the anabolic and the catabolic processes. An analysis of the protoplasm at any given time will show a certain percentage of the nitrogen present as protein, and a smaller percentage as amino acids, amides or even as ammonia. At any given time, in fact, a breakdown of protein is going on alongside synthesis. This is understandable if we imagine that, as in a machine, so also in the protoplasm the fine strutural elements suffer wear and tear with time and have to be renewed. Furthermore, a part of the protein formed at the main sites of protein synthesis, the leaves, has to be continually removed and conducted to the centres of growth. This requires, as in germinating seeds, a preliminary mobilization of the protein, i.e. its partial breakdown to transportable fission products.

The first steps in the breakdown of protein are apparently hydrolytic in nature, since the enzymes concerned, the proteases, are also hydrolases. The splitting of the peptide linkages apparently continues down to the structural units, the amino acids. The enzymes taking part in this process in the plant form the mixture called papain, similar to the kathepsin of animals.

Papain can be activated by various substances (hydrogen sulphide, hydrogen cyanide, cystein, ascorbic acid). The activity of the enzyme seems to depend on the formation of free –SH groups and at the same time to be determined by the redox potential of cells. Serious oxygen deficiency and interference with respiration can stimulate protein hydrolysis. Other factors are certainly also of importance.

The proteins undoubtedly become increasingly mobile and transportable as they are broken down to amino acids. During growth, however, a much more extensive catabolism of the nitrogenous substances is necessary. As already mentioned, the reserve protein of the aleurone grains usually has a composition quite different in respect of amino acids from that of the protein of the young growing cells of the seedling. Consequently, following the breakdown of the protein, there is frequently a further disruption of the amino acids, namely deamination, in which the $-NH_2$ group is removed from the carbon skeleton.

This deamination is in many cases oxidative. Thus under the influence of dehydrases (deaminases in association with NAD) hydrogen is first split off, and following separation of the amino group, α-ketoacids again arise:

$$R.CHNH_2.COOH - 2H \longrightarrow R.C{=}NH.COOH \xrightarrow{+H_2O} R.COH.NH_2.COOH \longrightarrow R.CO.COOH + NH_3$$

The ketoacids can then once again be acted upon by carboxylase, and the resulting aldehydes take up hydrogen to form alcohols containing one carbon atom less than the original acids. This, for example, is the way in which fermenting yeast, by breaking down the protein of bran mash, forms fusel oils. In aerobic respiration, however, in which the hydrogen is oxidized completely, the further oxidation usually leads to the production of fatty acids. This provides an illustration of how in plants organic acids can arise by the conversion of protein. Thus a large part of the oxalic acid of rhubarb, for example, may be derived from the protein, and not from the carbohydrate metabolism. With citric acid, besides its formation in respiratory metabolism, there is also a probable relationship with glutamic acid.

These oxidative deaminations, although they involve the uptake of oxygen and the giving off of carbon dioxide, should not be confused with respiration. We have no evidence that, when the supply of carbohydrate is adequate, protein is used as a respiratory substrate in order to obtain energy. Only in times of need can protein be drawn into the metabolism and its carbon skeleton utilized in respiration (see p. 291). Even though oxidative exothermic processes may occur in the normal breakdown of protein, it is not legitimate to speak of protein respiration, since these processes are not directed exclusively towards the obtaining of free energy, but towards the production of certain metabolites.

Besides oxidative deamination, there are other ways in which amino acids can be metabolized. Aspartic acid, for example, can be transformed reversibly by the enzyme aspartase into fumaric acid and ammonia. Many bacteria and fungi (see also p. 291) form amines by decarboxylation, the co-enzyme then being not thiamin, but pyridoxal phosphosphate, which with another enzyme system can bring about transamination (see p. 283). An example of an amine formed by decarboxylation is methylamine from glycine

$$CH_2NH_2 . COOH \longrightarrow CO_2 + CH_3 . NH_2$$

The bacteria of the gut form cadaverine and putrescine, the odoriferous substances of faeces, from the amino acids lysine and ornithine. The pharmacologically active amines of ergot, as well as many amines in the scents of flowers, may also be mentioned here as examples of this kind of metabolism. The histamine of the stinging hair of the nettle (see p. 101 is derived from histidine.

The detoxication of ammonia. Plants are distinguished from animals not only by their more extensive synthetic abilities but also by the striking fact that they lack any means of excreting nitrogenous substances. In the higher plants at least the whole of the valuable nitrogen is retained in an internal cycle. The ammonia arising from deamination is immediately re-intercepted and temporarily stored until the occasion when synthesis is required. Animals also detoxicate the ammonia arising in their protein metabolisms, but in the form of urea or uric acid (birds, terrestrial reptiles). These substances are subsequently excreted in the urine, since no further use can be made of them.

Analogous to the production of uric acid by animals is the formation by a number of plants (*Symphytum, Acer, Aesculus hippo-castanum* amongst others) of allantoin and allantoic acid. These are not excreted, but are stored in roots or stems. They ascend again with the sap in the spring and are used in the synthesis of protein in the new shoots. In the Betulaceae and Juglandaceae (but not in the Fagaceae) citrulline is the most important compound in which the ammonia is held, whereas in the Rosaceae (e.g. in the apple) and Saxifragaceae it is arginine. Both substances indeed occur as intermediates in the formation of urea in animals. In plants, however, with the exception of the fruiting bodies of certain fungi, there is no question of the accumulation and excretion of urea as in animals.

Other plants which store large quantities of organic acids in their cell sap, such as rhubarb, can use it to neutralize any ammonia arising in metabolism (acidic types). In most plants the already mentioned glutamic or aspartic acids act as detoxicators of ammonia by forming the corresponding amides glutamine or asparagine. These often accumulate in great quantities in the plant, and can apparently also be transported to sites of need where the nitrogen of the amide group can easily be detached and turned over to the renewed synthesis of amino acids, as has already been described (p. 286). In this way the cycle is completed. Nitrate is reduced and the nitrogen by way of ammonia becomes incorporated in glutamic acid, and from here passes to other amino acids. Finally, protein is synthesized. This, when broken down, gives rise first to amino acids, then to ammonia. This is bound in the form of amides, which enter again into the metabolism as required.

Other nitrogen compounds.[22] The discussion so far has considered only the main lines of the nitrogen metabolism in plants. In detail there is still much that is puzzling. Without doubt, too, there are complex sidepaths by which the synthesis and breakdown of a multitude of other nitrogenous compounds takes place. The pyrrol, pyridine and pyrimidine rings have already been mentioned. The important growth-substance indole-3-acetic acid (see p. 303) arises from the amino acid tryptophane. The alkaloids, a group of substances distinguished by their special pharmacological activity, appear in many plants to be formed particularly in the roots, and to pass from there into the leaves, although in some plants they are synthesized there as well. In a few plants alkaloids appear to be a means of eliminating certain amino acids. Very little is yet known of the significance of these substances for the plant itself.

VII. The transport of materials in the plant[24]

(Translocation)

The metabolic processes which have been described in the foregoing sections are necessarily bound up with continuous conduction to and away from the sites of metabolic activity. The initial materials for syntheses must flow into the cell, and the products be led away, either to make room for others or to supply those cells and tissues which have not the power to carry out the syntheses themselves. In the multicellular body of the higher plant there is in fact an extensive division of labour. The colourless cells throughout the whole subterranean root system, the pith cells of the interior of the shoot axis and many others are not, for example, capable of photosynthesis. To a certain extent they are, of course, capable of reducing nitrate and synthesizing protein, but with respect to carbohydrate they are absolutely dependent upon a supply from the main sites of synthesis, the leaves. Consequently, large quantities of sugar must flow continuously from these regions to the places where it is used, but there is probably as well as a brisk local exchange of various substances between the cells and tissues of the plant. The transport of protein and mineral nutrients (see pp. 236, 282) has already been discussed. In the autumn there is in addition a vigorous movement of nitrogen, phosphorus and potassium compounds and other substances, out of the yellowing leaves back into the perennial parts of the plant, where they remain stored during the dormant period.

The substances transported in the plant are not entirely concerned with synthesis and maintenance. Mention will be repeatedly made in later sections of hormone-like substances which play a prominent part in various developmental processes and irritability phenomena. These substances have not only to be formed in metabolism but also transported over relatively long distances (see p. 312).

If the material taken up from without or mobilized within the plant is to be transportable it must be in soluble form, and the size of its molecules must not be too large. Consequently, the movement of these substances in the plant is usually associated with intensive metabolic processes. At the points of supply there will be mobilization processes by which, for example, insoluble starch is transformed into soluble mobile sugars, and large molecules of protein are broken down into the small molecules of amino acids. At the same time the converse processes of condensation and conversion, leading to insoluble substances, will be taking place at the points of deposition. In many, if not all, instances movement seems to take place along a falling concentration gradient.

Still little is known in detail about the form in which substances are transported. Glucose is certainly transportable, yet sucrose, or even oligosaccharides such as stachyose, often seem to be preferred for transport over long distances. In protein metabolism amino acids and amides appear to be particularly mobile. The view is now growing that many substances, particularly in relation to their passage through membranes, are transported only when they are associated with specific carrier molecules.

The paths of transport and the forces responsible for the movement are also little known. In the local movement between the parenchyma cells, which quantitatively could be accounted for by diffusion alone, the energy coming from respiration within the cytoplasm seems to play an essential role. The fact that many substances, e.g. indole-3-acetic acid (β-indolyl acetic acid, IAA) (see p. 302), can be transported through a chain of cells only in one direction (i.e. in a polar manner) indicates that living protoplasm participates in the movement. It is still not certain whether the plasmodesmata between the cells are preferred as the pathways of movement. In polar transport, secretion through the whole plasmalemma at the basal end of the cell, diffusion through the cell wall and resorption through the plasmalemma of the adjacent cell all have to be considered. Current observations with the electron microscope upon the fine structure of protoplasm will probably be of importance in this connection.

With regard to transport over long distances, one fact that is certain is that speeds of many decimetres per hour can be achieved in living sieve tubes in both directions. It has already been mentioned (see p. 107) that the lumen of the sieve tube is filled with a highly hydrated plasma, lacking a nucleus, but still containing mitochondria, endoplasmic reticulum and plastids. The respiratory activity of the conducting cells, and of the surrounding cells, seems to bear some relationship to transport. Many investigators believe that the companion cells, which are very rich in mitochondria, play an important role in the movement of materials into and out of the sieve tubes, as well as in the actual mechanism of transport. Nevertheless, it must not be forgotten that the phloem of the conifers wholly lacks companion cells, although its efficiency is hardly less than that of many dicotyledons. There is in fact still no clear explanation of long distance transport in the phloem. The formation of callose on the sieve plants is controlled by enzymes in the plasma. The callose seems to restrict conduction, or even, as is customary in old sieve tubes, shut it off completely. Callose formation takes place very rapidly after wounding.

It should not be forgotten that the transpiration or guttation stream ascending continuously from the roots in the vessels may also be used for unidirectional transport, and also that transference of materials from phloem to xylem or conversely is apparently very easy. During the summer months, of course, predominantly inorganic salts are carried in this way, but in spring, especially in trees, large quantities of organic materials also ascend by this route (cf. 'bleeding', p. 227). It is not known whether this mode of transport is used occasionally at other times. That it is responsible for the movement of alkaloids from the roots to the leaves, and also of some amino acids, has already been demonstrated. The situation concerning ripening fruits is still doubtful.

VIII. Excretion and secretion in plants[25]

Characteristic of a continuing metabolism is not only that materials are taken up into the living cell, transformed and incorporated into it but also that certain substances are expelled from it. Some of these substances are waste products for which the organism can find no further use, but others have quite definite functions to fulfil in the organism. In the first case we are concerned with excretion, but the formation in and expulsion from the cell of functionally important substances is referred to as secretion. Excretion, which is so marked a feature of animal metabolism, hardly takes place at all in many spheres of plant metabolism, e.g. protein metabolism, but in others it is just as conspicuous a feature as in animals. The difficulty, with our very defective knowledge of the interrelationships in the physiology of metabolism, is to decide whether a substance passing out of a cell is truly a metabolic waste product or a substance which still has functions to fulfil. It is also conceivable that many of the substances laid down as waste products or excreted can occasionally acquire an ecological significance for the plant (e.g. protection against grazing by animals; the closing of wounds by latex, resins, etc.).

Undoubted excretions are the many end-products given off in respiratory metabolism, especially the numerous fermentation products of the lower plants. In pathogenic bacteria and fungi, metabolic excretions (toxins), the nature of which is still for the most part unknown, are often the direct causes of certain illnesses (see also p. 292). The deposition of calcium hydroxide on the upper sides of leaves of submerged water plants during the assimilation of carbon dioxide is again certainly excretory in nature (see p. 255). Elsewhere in metabolism, at least in the higher plants, the excretions are frequently not expelled externally but transported to definite places in the interior of the plant, and there laid down and encapsulated. This is true, for example, of the well-known oxalic acid crystals (oxalic acid being an end-product of the respiratory and protein metabolisms), of tannins, ethereal oils, resins, latex (the excretory character of which is not completely certain) and so on. In hardly any instance have we detailed knowledge of how these substances are formed. Many, e.g. the hydrocarbons (terpene derivatives) of the ethereal oils, must stem in some way from the carbohydrate metabolism (all isoprene derivatives, for example, are synthesized by way of mevalonic acid and acetyl-CoA). The excretion of these substances often takes place through special glands whose manner of functioning is again still unknown, although it is clear that (as in secretion) the golgi bodies are actively involved. The fall of leaves in the autumn from deciduous trees also provides a means of disposing of a whole range of unwanted substances.

Examples of secretions are, among others, many forms of guttation (see p. 227 seq.), and the sugary solutions produced by the nectaries present in many flowers and by the glands of insectivorous plants. These latter secretions serve partly to trap insects and partly to transform those caught into an absorbable form by means of the enzymes and other substances they contain. The saprophytic (see below) bacteria and fungi, as well as the parasites, also produce digestive secretions. This is probably the closest approach in plants to the kind of secretion found in animals. In addition, the roots of higher plants appear to give off a number of substances, not all inorganic in nature (see p. 235), such as amino acids. This is certainly of significance in maintaining the colonies of bacteria and fungi (rhizosphere) around the root (see also p. 294).

The formation and secretion of the antibiotics by bacteria and fungi (e.g. penicillin by the cheese-mould *Penicillium notatum*), by which these organisms in their natural habitats protect themselves from their competitors, have attracted general interest in recent years because of their medicinal uses. Even an 'inner secretion' in the sense of animal physiology might be said to occur in plants. The formation of important hormone-like substances (e.g. the growth-regulating substances) is limited to definite regions, from which they are transported throughout the plant body (see p. 312).

Thus both excretion and secretion form an important part of the plant's metabolism, and they must not be neglected despite the little that is currently known about them.

IX. Special features of heterotrophic nutrition

As has already been mentioned on p. 240, not all plants are capable of manufacturing by their own metabolic processes, from simple inorganic precursors, the organic substances, especially the all-important carbohydrates, essential for life. There are many examples in the lower organisms of forms in which photosynthesis, or the associated metabolism, is not adequate so that certain organic substances must be supplied from without. These are termed mixotrophs, or where there is the requirement of a specific vitamin or growth factor, auxotrophs (see p. 300). Even in autotrophic plants, the extensive division of labour between the various organs means that large tracts of tissue have frequently to be nourished by conduction from the sites of food production. This is true in respect of carbohydrates, for example, of all the cells of a higher plant lacking chlorophyll (epidermis, pith, root, etc.) and is probably true also in respect of many other substances, such as hormones, vitamins and the like (see p. 300). This represents no departure from the fundamental principles of metabolism set forth earlier, since from the point of view of the cell it does not matter, for example, whether it has itself photosynthesized the sugar it respires or whether it is conducted to it from elsewhere. Where, as in many bacteria and in the fungi, the whole organism has lost the capacity for autotrophic nutrition a special situation arises with particular features of its own. As in animals, the organic substances which are lacking must be obtained or taken up from without.

Many heterotrophic plants can thrive in an inorganic nutrient solution, such as that described previously for autotrophic organisms, if a source of organic carbon be added to the medium. Many need a source of organic nitrogen in addition.

Heterotrophic nutrition occurs in a number of quite different grades and forms. Saprophytes are usually distinguished from parasites. Saprophytes are those heterotrophs which extract their organic nutrition from dead substrates. They bring the carbon- or nitrogen-containing materials of the dead plant or animal into solution by means of secreted enzymes, and so resorb them. The continuous decomposition and decay of dead organic matter in nature is due to the feeding activities of such saprophytes. In the parasites the demands of their nutrition have become so specialized, extending to quite definite organic substances (even including growth substances), that they can be met only by direct union with and exploitation of another living organism. Thus a parasite needs a living host plant or animal.

Saprophytes.[26] The principal saprophytes are numerous bacteria and fungi, naturally excluding those capable of photo- or chemosynthesis. Their nutritive requirements differ remarkably in detail.

Besides the mineral salts, which are always essential, the carbon source in many cases may take the form not only of carbohydrate proper (including cellulose and starch which are brought into solution by fermentation) but also of many other carbon compounds, such as alcohols, fats and fatty acids. Indeed, certain specialized organisms can utilize even hydrocarbons, such as paraffin and naphthalene. It is almost possible to say that there is hardly a single natural organic compound which is not capable of serving as a source of nutrition to some form of life. The greatest, but not absolute, resistance to decay is shown by the cutinous substances. Accordingly cuticles, and pollen and spore walls are retained even in ancient rocks, and form important source material for palaeontologists and for the study of the history of floras. During this decay of biological materials any carbon dioxide liberated may be reincorporated into simple carbon compounds (see p. 256), and re-enter the metabolism of the organisms concerned.

Even proteins serve as a carbon source, the amino group first having been removed as ammonia and the carbon chain laid bare (see p. 286).

The processes of decay and putrefaction, which play so large a role in the circulation of materials in Nature (see p. 296), are brought about solely by the vital activities of those bacteria and fungi which draw their nutrition from the organic nitrogen compounds of dead plants and animals.

Many plants are capable of making very fine chemical distinctions. Certain moulds (*Penicillium glaucum* among others), for example, take up only the dextro-rotatory form of tartaric acid from a solution of the two isomers, leaving the laevo-rotatory form untouched. It can in fact be chemically isolated in this manner. Certain bacteria conversely prefer the laevo-rotatory form of tartaric acid.

After all these various organic nutrients have been brought into solution and absorbed, they must, of course, be transformed into a carbohydrate resembling glucose, or into substances which occur as intermediates in its metabolism. We have no reason to believe that the carbon metabolism of the saprophytes takes a course fundamentally different from that described earlier.

In Nature it is customary to find whole associations of different organisms working together. One species will utilize for its nutrition the intermediary metabolites or waste products of another, while the products of its own metabolism will serve as the nutrients of yet a third species. In such a complex metabolic flux it is, of course, usually difficult to decide whether a metabolic process occurs merely for the generation of a particular substance, or whether its function is to get at chemically bound energy, i.e. whether it is a fermentation or respiratory process of the kind already described for some heterotrophic organisms (p. 265 seq.). Often both kinds of reaction may be coupled together. The end-products of the putrefaction of protein are ammonia and hydrogen sulphide. This shows that the main aim of the process is not primarily to arrive at nitrogen or sulphur, but rather the carbon chains or the energy incorporated in them. In these extensive processes of decay other intermediary products are often also formed, among which may be mentioned the nitrogen-containing indole ring, derived from tryptophane, and the evil-smelling substance skatol, also responsible for the odour of faeces. Furthermore, the end-products ammonia and hydrogen sulphide can, as mentioned earlier, be used by certain bacteria as respiratory material.

The demands of many saprophytes in relation to their nitrogen metabolism are even more diverse. There are, of course, many bacteria and fungi which are heterotrophic only in respect of their carbohydrate metabolism, being with respect to nitrogen like the higher plants, i.e. completely autotrophic and satisfied with inorganically bound nitrogen.

Thus yeast, for example, can meet the demands of its protein metabolism from ammonium salts alone, and can therefore build up its amino acids and protein from the simplest units. Many saprophytes again can utilize nitrate, if somewhat less readily than the higher plants. The mould *Aspergillus niger*, for example, can be cultured quite well on a nutrient solution containing nitrate as used for higher plants, provided merely that some sugar is added. It has already been mentioned (p. 285) that many micro-organisms can even use the molecular nitrogen of the air.

There are also undoubtedly fungi and bacteria which for successful growth require organically bound nitrogen, such as amino acids, peptones or even complete proteins, probably because they lack the capacity to synthesize certain amino acids or other nitrogenous compounds.

Many believe that the existence of different grades of saprophytism indicate various degrees of loss of an originally much higher capacity for synthesis. Thus in the heterotrophic Ascomycete *Neurospora* it has been possible to create by artificial means mutations which have lost the capacity to synthesize certain metabolites, so that they have become heterotrophic with respect to these substances. The study of these mutants has thrown much new light on the normal course of various metabolic reactions.

Parasites.[27] There is a smooth transition from strongly heterotrophic saprophytes to parasites which for normal growth are completely dependent upon union with a living host.

There are both facultative and obligate parasites. Thus the bacteria giving rise to tetanus, anthrax, cholera and typhoid fever can live saprophytically in soil or water, and become parasites only when a favourable opportunity offers. The organism giving rise to diphtheria, on the other hand, is an obligate parasite. Parasitic bacteria and fungi are often the cause of diseases in their host plants or animals, as in man. The severe damage caused to cultivated plants by blights (e.g. *Phytophthora*), rust and smut fungi is well known. The parasite in these cases does not usually limit itself to the uptake of nutrients and larger metabolites from the body of its host (which may be partly destroyed in the process), but damages its host to an even greater extent by the excretion of poisonous metabolic products, the toxins (as, for example, in wilt disease), as in many bacterial diseases of Man.

Parasitic nutrition is not limited to the lower plants, such as bacteria and fungi, but occurs also in the higher plants, where again it is distinguished only by degree from saprophytism.

Mistletoe and also many Scrophulariaceae (*Rhinanthus, Melampyrum, Pedicularis* and others) possess green leaves, and are undoubtedly autotrophic in respect of their carbon metabolism. They are referred to as hemiparasites, and apparently they merely fasten on to the vascular tracts of the host and extract from them principally water and mineral salts. Nevertheless, in *Viscum* there are races which will attack only certain genera of plants, so they may in some way respond to certain metabolic products of their host plants.

Distinct from the foregoing are the complete parasites, represented particularly in our flora by *Orobanche, Cuscuta* spp. and *Lathraea*. Externally they immediately reveal their heterotrophic character by their lack of chlorophyll and the reduction of their leaf area. *Orobanche* and *Cuscuta* always develop connections with the sieve tubes of their host plants. They produce special absorptive cells which clamp on to the sieve tubes from outside and extract from them, in a manner still not understood, all the nutrients passing through them. The penetration of the parasite down to the sieve tubes of the host is facilitated by the secretion of enzymes which dissolve the cell walls, or at least their middle lamellae. *Lathraea* appears to nourish itself from sap drawn from the xylem of the roots of the host plant.

Symbiosis.[28] A phenomenon closely related to that of parasitism is symbiosis. This is the condition in which two different species live in intimate association, whereby each derives a certain advantage for at least part of the time from the metabolic partnership. In the majority of cases this commensalism can be recognized as originating from an alternating parasitism in which a kind of equilibrium has been reached in the aggressive and defensive behaviour of the two partners. As a result, each yields substances of use to the other, so that they may experience distinct assistance from the association. This reciprocal exchange is not limited to particular nutrients, and may even extend to the exchange of growth substances (see p. 300).

Such a symbiosis probably occurs, for example, in the lichens. Here a fungus is united with algal cells, giving rise to what appears externally as a new organism functioning as a unit. The fungus wraps itself around the algal cells in various ways (see p. 537 seq.), and partly even penetrates them by means of haustoria. The algae usually remain alive, but are no longer able to propagate themselves by means of their own reproductive cells. Apparently the principal substance extracted by the fungus from the algae is the carbohydrate generated by its photosynthesis. Indeed, in the barren habitats of many lichens the alga is the only possible source of carbohydrate. The fungus probably supplies the alga with water and mineral salts, but our knowledge of the reciprocal relationships of these organisms is still very limited. Of great interest also are the symbioses between unicellular Cyanophyceae and species of amoebae and colourless flagellates, in which the algal cell functions almost as the chromatophore of its host. A parallel situation is the presence of *Chlorella* and blue-green algae in certain polyps.

In the symbiosis between bacteria or fungi, on the one hand, and the higher plants, on the other, it is also possible to detect parasitic features. Thus in the development of the nitrogen-fixing nodules on the roots of the Leguminosae the first process is a normal infection of the rootlets by various races of the aerobic *Rhizobium leguminosarum* (*Bacterium radicicola*) living saprophytically in the soil. The bacteria usually penetrate through the root-hairs into the cortical tissue by way of a remarkable 'infection thread' bordered by cellulose laid down by the host plant. Within the cortical cells, the bacteria multiply at the expense of the nutrients and enzymes there present. At the same time the host cells divide and enlarge, apparently as a result of the stimulus of the infection (probably involving the secretion of indole-3-acetic acid; see p. 302), and their nuclei become tetraploid (so far as cells already with tetraploid nuclei are not preferentially attacked by the bacteria). A meristem is now formed, and vigorous cell division leads to the formation of the root nodules. These are visible to the naked eye, and the inner cells are usually densely filled with the bacteria which have moved into the nodule from the infection thread (Fig. 265C). Up to this point it is probably almost entirely simple parasitism on the part of the bacteria, but from now on it appears that the plant sets up defensive reactions. The infection does not spread beyond the confines of the nodule, and the bacteria soon change their outer form in a striking manner, giving rise to bacteroids. An accompanying change in the cytoplasm of the host cells is the appearance of 'digestion figures', indicating that the host is attacking and resorbing, or otherwise causing the release of the nitrogen compounds of the bacterial cells. In this way the host plant comes into possession of the valuable nitrogen which the bacteria took up from the air after they had colonized the nodule. Apparently contact with the cells of the host is essential for nitrogen fixation to occur. The cobalt-containing vitamin B_{12} appears to play a role in this connection. Within the nodule the dissolution of both the bacteria and the host cells proceeds rapidly, and when it reaches its climax, the whole nodule dies and the undigested bacteria are returned again to the soil. The ultimate advantage of the symbiosis lies undoubtedly with the host plant, since it directly parasitizes the bacteria, and so can provide itself with a copious supply of nitrogen compounds. On the other hand, always more bacteria are returned to the soil than took part in the original infection, so that the process has been referred to as a 'domestication' in which the higher plant 'tames' and exploits the bacteria without completely destroying them. In perennial Leguminosae individual nodules last for several years. The extent to which nitrogen is fixed by bacteria, and to which when fixed it is taken over by higher plants, is shown by the fact that a hectare of lupins is capable of fixing over 200 kg nitrogen in the course of one growing period. It is therefore possible to cultivate species of the Leguminosae on soils poor in nitrogen, and in certain circumstances it may be an advantage to attempt to promote the formation of the symbiosis by artificial inoculation of the soil or seed with pure cultures of bacteria.

Nitrogen-fixing bacteria have also been found in the leaves of tropical species of Dioscoreaceae, Myrsinaceae and Rubiaceae (*Psychotria, Pavetta*), and here they are even carried over with the seed. In the root nodules of alder, *Hippophae, Elaeagnus,* and also probably of *Myrica* and *Casuarina*, the organisms concerned are symbiotic actinomycetes, also capable of fixing atmospheric nitrogen.

The surfaces of many leaves, and especially those of the roots and their immediate vicinity (the rhizosphere, see p. 290), are normally colonized by micro-organisms which

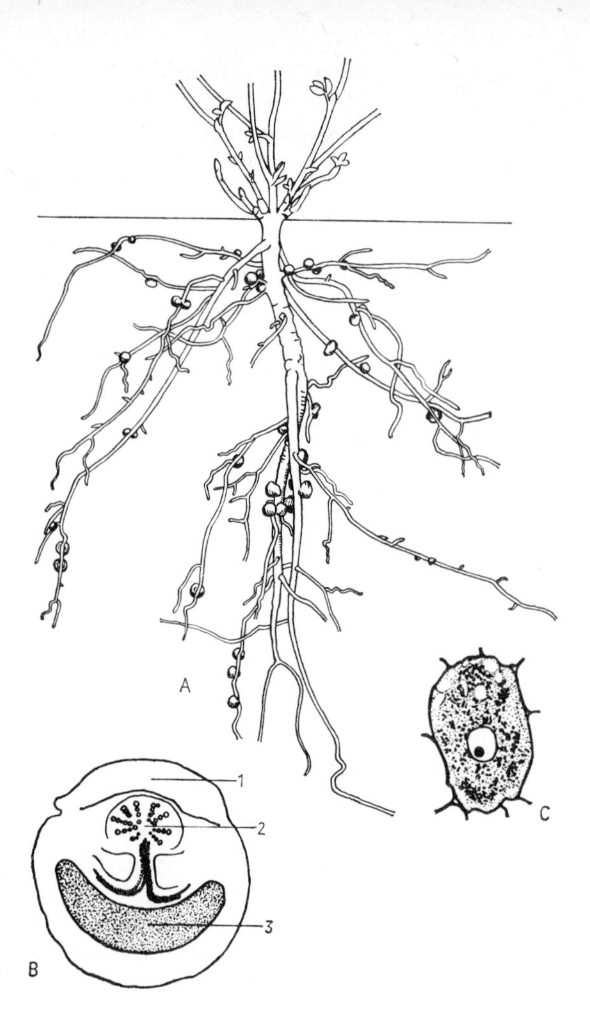

Fig. 265. A. Bacterial nodules on the roots of *Tetragonolobus siliquosus* (Leguminosae). *c.* $\times\frac{2}{3}$. **B.** Diagrammatic transverse section through a root nodule of *Lupinus luteus*: 1, cortex; 2, central cylinder of root; 3, tissue containing bacteria. **C.** Single cell from a root nodule of *Neptunia oleracea* (Leguminosae) showing numerous bacteria in the cytoplasm. **B.** *c.* ×10. After Tschirch, modified. **C.** *c.* ×450. After Schaede, modified.

have a loose and still little known relationship with the host plant. Also of wide distribution are symbiotic relationships between lower plants and animals. In man both the skin and the cavity of the mouth are continually occupied by bacteria. The well-developed bacterial flora of the gut considerably assists digestion by promoting the disintegration of food materials, and at the same time appears to be of great significance as a source of important vitamins. Similar processes occur in the stomach and gut of ruminants (see p. 268).

Mycorrhiza.[29] The mycorrhizal state can also be regarded as an alternating parasitism. The roots of very many plants (notable exceptions are provided by the Cyperaceae, Cruciferae and many Centrospermae) when growing freely in the wild are associated with fungi (principally species of Basidiomycetes). Thus, in most trees of temperate woodlands the absorptive rootlets remain short and thick and are surrounded by a dense network of fungal hyphae. This mycelium takes the place of the root hairs. Individual hyphae grow out from the sheath surrounding the root back into the soil and form connections with the mycelium present there (Fig. 266). The fungi penetrate into the cortex of the infected root (but not beyond the endodermis) in a number of different ways. They may grow more or less between the cells (ectotrophic mycorrhiza (Fig. 267), as in spruce and oak), or they may penetrate into the interior of the cortical cells (endotrophic mycorrhiza (Fig. 268), as in orchids). Transitions exist between the two types.

In many orchids, the minute seeds of which usually germinate and continue to develop only in the presence of the mycorrhizal fungus, it appears that the hyphae of the fungus, coiled up in the outer layers of the cortex of the root, parasitize the host plant. In the

Fig. 266. Portion of root of *Picea abies* with ectotrophic mycorrhiza. ×5. After Björkman.

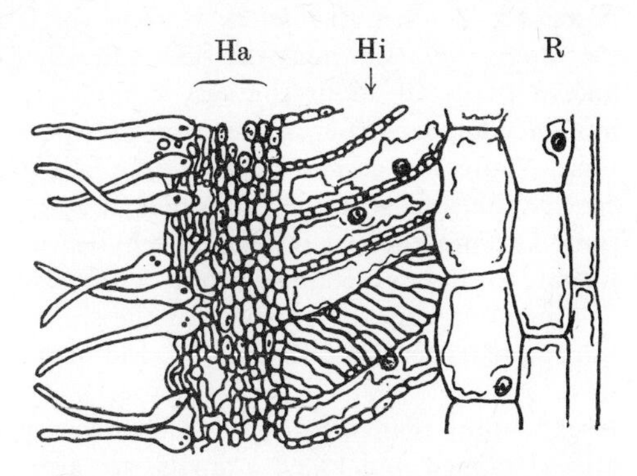

Fig. 267. Portion of a longitudinal section through a root of oak, showing the ectotrophic mycorrhiza. Ha, external mantle of hyphae; Hi, intercellular hyphae in the first layer of the cortex; R, inner layers of the cortex free from hyphae. After Burgeff, enlarged.

Fig. 268. Portion of a longitudinal section through a root of *Platanthera chlorantha* (Orchidaceae), showing the endotrophic mycorrhiza: e, epidermis; en, endodermis. The cells between the epi- and endodermises are filled with fungal hyphae. The black clumps are hyphae attacked and digested by the host plant. After Burgeff enlarged, partly diagrammatic.

deeper layers of tissue, however, the plant becomes master of the penetrating fungus and either digests it or brings it to a standstill (see Fig. 268). In those orchids, such as our *Neottia nidus-avis, Corallorhiza* and *Epipogium*, which either lack chlorophyll completely or possess only a trace inactive photosynthetically, it is clear that the higher plant must obtain all its essential nutrients, including carbohydrate, from the fungus just as a parasite. This is also true of *Monotropa* (Pyrolaceae), the fungus of which probably has simultaneous connections with the roots of trees. Whether the fungus in the case of these species enjoys any advantages from the metabolic association is still not clear. Mycorrhizas are also widespread in the autotrophic Pyrolaceae and Ericaceae. In those orchids which in the adult condition are autotrophic with respect to carbon, the fungus has probably to supply the young seedling not only with the usual nutrients but also with certain vitamins (see p. 300). When it has developed to a certain extent it can manufacture them for itself.

Many of our woodland trees have partly obligate mycorrhizas, mostly of the ectotrophic type (*Pinus, Picea, Larix, Quercus* (Figs. 266, 267)). Transitions to the endotrophic type can, however, also be recognized (*Fagus, Populus tremula*). It can be assumed that the fungus provides the tree with water and salts, especially with nitrates and phosphates, while the fungus for its part draws particularly carbohydrate from the tree. At all events many of our well-known agarics of the genera *Boletus, Amanita, Lactarius* and others will produce fructifications only if they are in connection with the roots of trees. As is well known to the fungus collector, many fungi prefer particular hosts, e.g. *Boletus grevillei* exclusively the larch, *Boletus scaber* the birch, *Boletus rufus* the aspen, *Boletus luteus* the pine and other conifers and so on. Hardly anything is yet known in detail about the very complex relations between the two symbionts. Some influence appears to emanate from the roots, directly promoting the germination of the fungal spores. In certain cases (e.g. orchid tubers) inhibitory effects are known, reminiscent of the formation of antibodies in animals, which serve as a protection against infection. The presence of other saprophytes may also affect the growth of the mycorrhizal fungus, so that in certain soils only a limited formation of mycorrhiza takes place.

Concerning the mycorrhiza of the shoot of the Psilotales and of the prothalli of the Lycopodiales see pp. 565 and 567, and of the eusporangiate ferns p. 582.

Insectivorous plants.[29] The insectivorous plants (see p. 199), to the extent that they utilize trapped and killed animals, are also to be considered heterotrophic. Since they ordinarily possess chlorophyll, are capable of photosynthesis and, in addition, with a good supply of minerals can be easily cultivated without any feeding with animal material, it appears that their heterotrophy is particularly directed towards the obtaining of adequate supplies of substances containing nitrogen or phosphorus. These elements, in the form of minerals, are often present in insufficient amounts in the natural habitats of these plants. Adaptations to particular animals do not appear to exist. The digestible parts of the trapped animals are brought into solution by enzymes, particularly proteolytic enzymes, secreted by certain glands. The dissolved materials are then resorbed into the plant by means of the same glands. In the Sarraceniaceae the digestion of the entrapped animals is brought about by symbiotic bacteria.

Cycling of metabolites.[30] The great range of heterotrophic plants, especially the saprophytic bacteria and fungi, play, as has already been mentioned several times, a major role in maintaining the circulation of materials in the biological world. It has already been remarked in relation to carbon assimilation that the astonishing constancy of the amount of carbon dioxide in the atmosphere is made possible only by the respiratory activity of all the lower forms of life on our earth. Since the autotrophic plants excrete, as a result of their respiratory metabolism, only at the most about one-third of the carbon dioxide they have fixed photosynthetically, it follows that the remaining carbon does not return to its original condition until after complex transformations and migration through other forms of life. Part of the consolidated material is taken up by the herbivorous animals, and part of this is respired to carbon dioxide, while the remainder is incorporated into the animal body. When the animal or plant dies the carbonaceous material serves to nourish the saprophytes, which, often acting in whole colonies, bring about progressive breakdown. As already mentioned earlier (see p. 267 seq.), the lactic acid and butyric acid bacteria, as well as many other soil bacteria and fungi, are responsible for the continued catabolism of carbonaceous material. The net result of these apparently wasteful respiratory and catabolic processes is that an essentially higher percentage of the carbon is excreted in the form of carbon dioxide or other carbon-containing metabolic breakdown products than is incorporated into the material of the saprophytes themselves. One species of the colony will often take up the metabolic products of another species and continue to decompose them, until finally the whole of the organically bound carbon has again reached the level of carbon dioxide, and is once again available for the photosynthesis of the autotrophic plants. In this way is completed an elaborate cycle which, as observation shows, is actually extensively coordinated, and in which the heterotrophic plants form an important component.

Most of the other vitally important elements show similar cycles. With nitrogen, for example, the following occurs. Nitrogen from the nitrates of the soil becomes incorporated by the autotrophic plants into their proteins. The utilizing of plants for food leads to part of the organically bound nitrogen being transformed into the bodily materials of men and animals. Part is subsequently excreted, and this, as are the dead plants and animals, is attacked by the organisms responsible for decay and putrefaction. As explained in greater detail earlier, these organisms use the organic nitrogen compounds in part merely as a source of carbon, and eventually excrete the nitrogen in the form of ammonia. Here again there are many stages in the breakdown, for which different members of a whole colony of heterotrophs are responsible. The cycle is completed when nitrifying bacteria re-oxidize the ammonia to nitrate. The extent to which denitrifying bacteria convert ammonia into molecular nitrogen is balanced by the activity of nitrogen-fixing bacteria, and by electrical discharges in the atmosphere, so that an equilibrium is maintained in the general economy of this element.

Section two: **Physiology of growth**[31, 32]

In the foregoing section we have dealt with those fundamental physico-chemical processes which, so far as they are yet known, are indissolubly bound up with the phenomenon of life. In this second section our main task is to subject the phenomena of growth and form in plants to a causal analysis. Growth and the development of form are as much fundamental phenomena of life as are those of metabolism, with which, of course, they are intimately connected. Every living thing shows a continuous modification of its form. It develops from a single cell to a multicellular or complex organism; it passes through a juvenile stage to maturity; in the formation of reproductive organs it again becomes the initiator of new cycles of development; and ultimately in old age it dies. The causes of these changes of form provide the main theme of the physiology of growth and development.

I. The growth of the cell[32, 34]

The fundamental phenomenon of any development, even in multicellular organisms, is the growth of the single cell. By *growth* is understood an irreversible modification of form, bound up with physiological activity of the protoplasm, and usually involving an irreversible increase in the volume or mass of the cell, principally of the protoplast or of the cell wall.

Increases in or alterations of volume are not always associated with an increase in dry matter, since often only water is taken up or displaced. If, for example, potato tubers in a cellar give rise to long shoots it can easily be shown by weighing that the total weight of the tuber and the shoots, as a result of respiration and transpiration, actually diminishes during this growth process. The increase in volume of one part at the cost of another is, however, irreversible, and consequently must be recognized as growth. On the other hand, an increase in volume brought about by imbibition, which can be reversed at any time by desiccation, has nothing to do with true growth. It should be noted that a true growth process need not always be accompanied by any change in volume particularly conspicuous externally. If, for example, the embryonic cells of the growing point undergo numerous divisions the volume of the daughter cells certainly increases before the next division ensues, but the increase in volume of such a portion of embryonic tissue as a whole is relatively minute. Consequently, it is hardly possible to detect any changes externally at the growing point with the naked eye. What in fact increases markedly at this stage is the living material itself, the protoplasm. Even differentiation can often take place without any further change in volume.

With reference to the normal development of the plant cell described earlier (p. 16), we must now distinguish several phases of growth. First, embryonic or plasmatic growth, in which it is principally the protoplasm which grows, the spatial enlargement of the cell remaining fairly limited. Secondly, the so-called extension growth of the cell, during which changes in the form and volume of the cell appear in a striking manner for the first time. During this process the mass of the protoplasm need not be further increased. The increase in volume may come about principally by the uptake of water, and the accompanying formation and enlargement of vacuoles. Nevertheless, an increase in the cytoplasm is not impossible even during this phase. Immediately succeeding any increase in volume is usually yet a third form of growth, namely differentiation, in which the particular characteristics of the cell are developed. Since in these three phases of growth we are apparently concerned with quite different processes, it will be appropriate to consider them separately in the following account.

1. **Plasmatic growth** provides one of the most puzzling phenomena of biology. So far as we know, living material is always generated from living material, and we have no evidence of its creation *de novo*. The formation of certain amino acids *in vitro* from ammonia, carbon dioxide, and hydrogen, under the influence of ultra-violet radiation, is still a long way from the creation of protoplasm. This latter involves the transformation of unspecific nutrients, such as sugar and inorganic salts, into specific organic substances after the pattern of those already present; that is, they become truly assimilated, and then incorporated in an ordered fashion into the architecture of the cell. Without doubt the protoplast invariably possesses an internal structure. Besides the nucleus and plastids there are mitochondria, internal membrane systems (such as the endoplasmic reticulum and dictyosomes), ribosomes, and probably even structures in the molecular region, still not visible in the electron microscope, which must all be replicated when a cell divides and the daughters grow to the size of the parent. All this is included in the concept of plasmatic growth. Thus it is not only a question of numerous, species-specific, syntheses of multifarious precursors, such as carbohydrates, amino acids and polypeptides, nitrogen-containing bases and nucleotides, and lipids, but also of their assembly in a species-specific manner, and incorporation into the structural elements of the protoplasm.

Multiplication of viruses. At the moment there is definite information only about the initial stages of this phenomenon of growth. This knowledge has arisen from a study of the observation, long puzzling, that in both the Animal and Plant Kingdoms there are certain infectious diseases in which the causative agent appears to have no connection with any of the hitherto known pathogenic micro-organisms. In infections of this kind the sap of an infected plant injected into a healthy plant will cause the symptoms of the disease to appear. In the case of the so-called tobacco mosaic disease, in which the leaves display a green mottled appearance, it was shown for the first time in 1935 that the expressed sap, such as would be carried by sucking insects, contained certain very large protein molecules. This proteinaceous material consists in fact of a nucleoproteid with a molecular weight of *c*. 40×10^6, and the molecules are visible in the electron microscope as rodlets. The actual infective agent is not, however, the protein, which forms the envelope of the particle, but a central helix of ribonucleic acid (consisting of about 6000 mononucleotides). An infective agent of this kind, which cannot itself be called living, is termed a virus. In many viruses, particularly those which affect animals, the infective agent is not a helix of ribonucleic acid, but a double helix of deoxyribonucleic acid, and the composite particle with its protein may take the form of a sphere or some other characteristic shape (concerning the still more complicated structure of the bacteriophages, see p. 445). Plant viruses are often transferred from plant to plant by insects or lice. Having entered the host cell by way of a minute incision or puncture they are then able to reproduce themselves, behaving almost like a chromosome. The study of viruses, apart from its practical importance, thus has great theoretical interest. They are often regarded as being equivalent to genes which have become free and degenerate.[22]

The principles governing the species-specific generation of new protoplasm will become evident from what has been said elsewhere about protein metabolism (p. 278) and gene mutation (p. 330). The fundamental phenomenon is clearly the ability of the nucleic acids to reproduce themselves identically, and of deoxyribonucleic acid in particular to generate ribonucleic acids whose base sequences determine the enzyme proteins synthesized. In the instance of tobacco mosaic virus the central thread of ribonucleic acid evidently codes for the protein of the sheath. The mode of replication of deoxyribonucleic acid has already been extensively discussed (p. 281). It should, however, be once again stressed that the formation, assembly, and correct pairing of the bases of the polynucleotides going to form a molecule of deoxyribonucleic acid still present many problems. It is clear, nevertheless, that plasmatic growth takes its origin in this basic phenomenon of the identical reproduction of the nucleic acids.

Still completely obscure is the second step in growth, namely the manner in which the newly-formed substances are assembled to yield species-specific structures. The bridge from molecular biology to morphology has yet to be built. The secondary, tertiary and quaternary structures of the proteins, apparently able to generate themselves spontaneously, insight into the origin and structure of starch and cellulose, and the connections between genes and enzymes, and the metabolic reactions they catalyse, are merely the first tentative steps to penetrate the field of organic diversity. Form is known to be genetically determined but how a given protoplast generates the characteristics of the cell it occupies remains largely unknown.

Of considerable importance is the discovery that, for normal plasmatic growth, besides the basic metabolites, such as carbohydrates, proteins, etc., traces of certain organic substances, known as vitamins or growth substances,[33] are also necessary.

Naturally this fact was first demonstrated with various kinds of heterotrophic plants, since they, like animals, no longer have the capacity to synthesize all the necessary metabolites in their own metabolisms from inorganic precursors. Now that it is possible to culture organs of autotrophic plants (such as isolated root tips, stem apices, etc.), and even single cells in artificial media (tissue cultures), it has been discovered that they too need growth factors. Evidently in the intact plant they are supplied from other regions or organs.

It was shown some time ago that with cultivated yeast traces of certain substances, known as 'bios' substances, were required for growth and cell division. The essential component appears to be biotin, the chemical constitution of which is now well known. This substance is effective at a dilution of 1 in 4×10^{11}. It is identical with vitamin H, known from animal metabolism, and forms the prosthetic group of an enzyme which catalyses the incorporation of carbon dioxide into certain carboxy acids (see p. 256). Other factors involved in the bios effect are pantothenic acid (a component of co-enzyme A (see p. 273)) and meso-inositol. Bios substances have also been discovered in higher plants in buds, cambia and in the aleurone layers of cereals. Many fungi (e.g. *Phycomyces, Tilletia tritici, Ustilago violacea*) are unable to grow without vitamin B_1 (thiamin) or at least one of the chemical components of this vitamin (pyrimidine, thiazole). Lactic acid and propionic acid bacteria, however, must be provided with riboflavin, a vitamin of the B_2 group. Another member of the B_2 group is *p*-aminobenzoic acid, regarded of especial significance in relation to bacterial growth. It is a component of so-called folic acid, concerned with the incorporation of 'activated' formic acid, a mechanism of methylation. The therapeutic effect of the sulphonamide drugs in bacterial infections appears to rest on their structural resemblance to *p*-aminobenzoic acid. They take the place of this acid in the folic acid, and so render the enzyme ineffective. The manner in which other antibiotics work is, however, less clear. Actinomycin appears to interfere quite centrally with the synthesis of messenger ribonucleic acid (transcription), whereas penicillin inhibits the formation of the bacterial cell wall.

The colourless organs and tissues of higher plants, not themselves capable of photosynthesis, are certainly, in relation to many growth factors, dependent upon supply from autotrophic regions, notably probably the leaves. Isolated roots, for example, usually need for their successful growth thiamin, nicotinic acid, pantothenic acid, pyridoxal phosphate (vitamin B_6, see p. 283), and occasionally yet other supplements. Vitamin B_{12} (the lack of which causes pernicious anaemia in Man) has also recently been detected in plants, and may play a distinct role in metabolism. It is particularly interesting that metabolic dependence of this kind may play a part in symbioses. In the symbiotic relationship of orchid seedlings and fungi already mentioned the latter provide the seedling, which has available none of the usual food reserves in cotyledon and endosperm, not only with food materials but also with growth substances belonging in certain instances to nicotinic acid derivatives. Not until a certain stage of development has been reached do the young orchid plants become autotrophic in respect of these growth substances. In other symbioses too the exchange or acquiring of growth substances which cannot be produced by one or other of the partners often appears to be of great importance.

As will be seen, all the growth factors so far named belong to the so-called **vitamins**.

This term derives from animal physiology, and was used for those substances, essential for growth and maintenance, but required in only very small quantity, which the animal was unable to make for itself. These substances must be obtained from plants, the metabolism of which is in many respects superior to that of animals. Clearly distinguished in animal physiology from the vitamins are the **hormones**. These are produced by the body itself, but, often after having been carried in the blood stream, produce their effects at a distance. In autotrophic plants both classes of substances are produced, but they cannot be clearly distinguished. Consequently it is probably better to refer to them collectively as growth factors or growth substances. Nevertheless, recently evidence has been accumulating that it may be possible to distinguish between vitamins and hormones according to the manner in which they act. In most instances so far investigated vitamins are co-substrates or prosthetic groups of enzymes. Hormones, however, appear to bring about their effects by the activation or inactivation of genes (see p. 330). For a discussion of plant hormones see pp. 302 and 314 seq.

The active syntheses associated with the processes of plasmatic growth set up an increased demand for energy. Consequently, it is not surprising, in view of the earlier discussion of the significance of respiration, that there is a close connection between the intensity of growth and that of respiration. Of course, only a very small part of the energy consumed in growth is bound up chemically in the synthesized molecules; the greater part serves, speaking figuratively, to drive the machinery and eventually flows out of the system as heat.

2. Extension growth.[34] Our knowledge of extension growth is somewhat less restricted than that of embryonic plasmatic growth. During extension growth water is taken into the cell and vacuoles develop, resulting in striking increases in volume. This is often the only kind of growth recognized by the layman.

When buds grow out in spring, and, as is the case in many trees, the flowers appear fully formed within a few days, we are concerned solely with extension growth. The protoplasts in the embryonic primordia of these flowers were formed long ago in the preceding summer, and need only to be as it were inflated by extension growth. In particular cases it has been accurately demonstrated that in these rapid growth processes, often detectable with the naked eye, the protein of the protoplasm itself undergoes no further increase. What is increased is the water content of the cell, especially in the vacuoles, and the amount of material in the wall forming the boundary of the cell. The increase in size of the cell is often so considerable that if the delicate wall surrounding the embryonic cell were merely subjected to extensive stretching it would very soon be ruptured. In a few instances, it is true, an increase in the protein of the cell has been detected during extension growth, but this is not necessarily connected directly with the growth itself, since it is possible that it represents only an increase in the enzymes involved.

The mechanics of this remarkable process of extension growth, with which there is nothing comparable in the Animal Kingdom, are still not wholly understood. A prerequisite of the sudden marked uptake of water is a similarly sudden increase in the suction pressure. It has been believed for a long time that a change in the elastic properties of the cell wall is to be regarded as the primary process in extension growth. If in a turgid cell the elasticity of the wall suddenly decreases, and the wall becomes as it were softened and plastic, it will be seen from the suction pressure equation (p. 221) that a diminution in wall pressure leads to an increase in suction pressure. During extension growth an increase in the plasticity of the region concerned has in fact been observed, and certain changes in the properties of the wall undoubtedly play an important part in the mechanism of extension growth. But the sudden enormous uptake of water at the beginning of the extension appears not to be completely explicable in so simple a manner.

A closer understanding of the processes involved naturally requires a detailed knowledge of the structure of the primary wall of the young embryonic cell. We shall here confine ourselves to the structure and growth of the walls of higher plants, neglecting those of bacteria and fungi which differ in a number of respects (bacterial cell walls contain mucopolysaccharides, incorporating amino-sugars and uronic acids (see p. 440), and fungal cell walls chitin). With reference to what has already been said on p. 60, we may again stress that the first delicate wall of an embryonic cell consists predominantly (other than water) of hemicellulose and pectin, these substances forming an amorphous (i.e. non-fibrillar) matrix. Not until later do the microfibrils of cellulose become laid down in this matrix, recalling the reinforcement of concrete with steel rod. The primary walls of young coleoptiles, before extension growth, consist, for example, of over 80 per cent matrix material (including some 30 per cent bound water), and only 12 per cent cellulose.

The matrix material, after its formation in the cytoplasm, is apparently secreted by way of Golgi vesicles through the plasmalemma into the young wall. The manner of synthesis and incorporation of the cellulose microfibrils (the glucose units of which are probably provided in the form of guanidine diphosphate-glucose) are still unknown. After removal of the matrix material the primary wall shows a loose mesh of interwoven microfibrils (Figs. 60 and 62). In proportion to the amount the cell stretches and its wall increases in surface area new matrix material and new cellulose microfibrils now become secreted into the wall. The microfibrils are first taken into the already-existing meshwork (intussusception), but subsequently, as they become increasingly frequent, they become applied to the inner surface of the wall in layers (apposition). The microfibrils in these layers tend to lie increasingly in parallel orientation, the direction often changing abruptly in successive layers so that the microfibrils cross, a clear indication of the decisive influence of the cytoplasm of the growth of the wall. The microfibrils first laid down in the primary wall are oriented at random, but as extension growth proceeds they slowly change their position and tend to become parallel to the longitudinal axis of the cell. They are thus evidently capable of gliding movements in the amorphous matrix. So long as extension growth is present, the orientation of the microfibrils, particularly of the more recent layers, continues to be affected. As growth ceases so does the extent to which the orientation of the microfibrils changes after deposition. Increase in thickness of the wall is then by simple apposition of layers of microfibrils oriented in parallel. This manner of growth of the wall, with the diminishing displacement of the later microfibrils is termed multi-net growth. The plasticity of the wall is accounted for by the ductility of the matrix, but the manner of deposition of the microfibrils is under the control of the protoplast.

The complexity of these processes is shown by the fact that the wall rarely grows uniformly over its whole surface. Growth is usually restricted to certain areas, and the adjacent cytoplasm is often denser than elsewhere. In many cells growth is predominantly, at the ends, referred to as 'tip growth'. In others particularly intense growth may be localized elsewhere. By these means the cells in a multicellular organism acquire, in relation to their function, their many and diverse shapes.

The extent and diversity of the participation of the protoplasm in wall growth are perhaps best illustrated by the fact that even extension growth can be decisively influenced by growth-regulating substances (phytohormones). Indeed, growth throughout the whole plant appears to be extensively controlled by means of these substances (see also p. 312). Sometimes they are conducted over long distances and, like the animal hormones, produce their effects distant from where they are manufactured.

Those substances which are especially important in relation to extension growth (although they may influence other processes in addition (see p. 314)) are frequently referred to as auxins. The most important auxin at present known is indole-3-acetic acid (β-indolyl acetic acid), often abbreviated as IAA and formerly known as heteroauxin. This auxin is structurally related to the amino acid tryptophane (Fig. 269). It has now been identified as a metabolite in many higher, as well as lower, plants. Human saliva and urine also always contain some indole-3-acetic acid.

The sensitivity of embryonic cells towards this auxin is very great and exceeds many of the hormone effects known from animal physiology.

Thus 1 g of indole-3-acetic acid, easily synthesized artificially, would produce a clearly visible growth effect if applied to a stand of some 13×10^{12} plants of *Cephalaria* (Dipsacaceae, resembling *Stachys*) growing close together on an area of 30 sq. km. A maize root reacts to a mere 10^{-12} g. The amounts of active growth-regulating substances

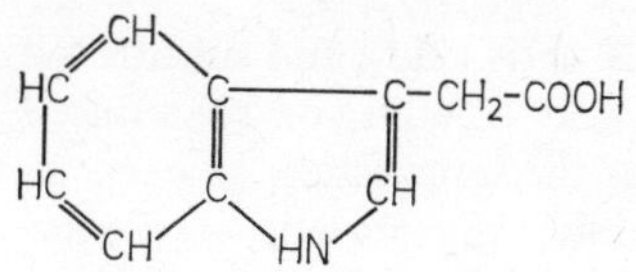

Fig. 269. Indole-3-acetic acid.

occurring in plants themselves are also very small. Only about 10^{-3} mg indole-3-acetic acid has been isolated from 10,000 tips of *Avena* (oat) coleoptiles.

A very important observation is that raising the concentration of the growth-regulating substance above a certain threshold (perhaps about 10^{-4} g) can in certain circumstances destroy its effect, indeed it may even become inhibitory. Moreover, the optimal concentrations are different for the different organs of the plant, so that, for example, the concentration optimal for the shoot is already inhibitory for the root (see p. 360). The same growth-regulating substance seems to be effective with plants of widely different genera, although the optimal concentrations for extension growth may vary.

It is not improbable that in certain cases the whole growth of a plant is controlled merely by variations in the concentrations of growth-regulating substances. Such substances can, for example, be destroyed by enzymes within the cell, deflected to one side of the axis (see p. 353), or be only reversibly inactivated by combination with various substances. In the coleoptiles of the Gramineae, for example, bound (and consequently inactive) indole-3-acetic acid appears to ascend from the endosperm to the apex, where activation ensues. Not until the auxin is liberated, and it is streaming downwards from the tip in a polar fashion, does it stimulate the growth of the coleoptile. In addition to the fact that growth is reduced when there is either too little or too much auxin, quite distinct substances have been discovered which specifically restrict growth. These substances are able, in a manner still not understood, to block the mechanism of growth, and they may well be concerned along with the auxins in the auto-regulation of plant growth. One of these substances is abscisin, concerned principally with the abscission of leaves. The effect of this growth substance, which proves to be a sesquiterpene, is opposed by IAA coming from the lamina of the leaf, an example of auxin antagonism. Ethylene, which can arise in metabolism, acts in a manner similar to abscisin (see p. 340). Abscisin is identical with dormin, an auxin which brings about the dormancy of buds, inhibits the germination of seeds, and is responsible for yet other physiological effects.

The actual point at which indole-3-acetic acid exerts its effect is not exactly known. It seems to occupy some central position in the metabolism of the cell, and is perhaps able to affect the formation of messenger ribonucleic acids. By the control of the synthesis of enzymic proteins it would then be able to influence widely different processes, among others the anabolism and catabolism of cellulose, respiration (which must be the source of the energy required), and the uptake of water (perhaps by means of changes in permeability). It is not known to what extent other components of the cell wall are affected.

In addition to indole-3-acetic acid and its derivatives (e.g. indole-3-acetonitrile), other growth substances present in the cell are probably concerned in extension growth, but little is known about their combined operation, which is probably partly antagonistic. The so-called gibberellins, for example, which were first isolated from a fungus parasitizing rice (see p. 513), and whose chemical constitution is now known, have now been found in higher plants and have the property, amongst others, of promoting internodal extension. Over twenty different gibberellins (designated GA_1, GA_2, etc.), all very similar to each other chemically, have already been reported. They have some practical uses (for example, the promotion of germination of barley in the preparation of malt, brought about by the activation of α-amylase in the aleurone layer). As with indole-3-acetic acid they seem to influence a variety of processes (for example, the promotion of flowering, and vernalization (see pp. 336 and 346)).

Moreover, a whole series of very diverse chemical compounds have now been produced synthetically which are able to produce effects on plants in every way similar to that of the natural indole-3-acetic acid. Besides relatives of indole-3-acetic acid (e.g. β-indolylbutyric acid), the principal ones that should be mentioned are naphthalene compounds (α-naphthylacetic acid, β-naphthoxyacetic acid, acenaphtheneacetic acid), and substituted phenolic compounds, such as 2,4-dichlor-phenoxyacetic acid (2,4-D), or substituted benzoic acids. These today find many practical uses, such as the suppression of weeds in corn fields (high concentrations of growth substances inhibit the growth of dicotyledonous plants more strongly than that of monocotyledons, or induce in them monstrous and destructive growth), and the prevention of the precocious shedding of ripening fruit, etc. (see p. 313). Also here we are quite ignorant of the manner in which these growth substances intervene in the growing system, and whether it is the same as that of the natural growth regulators. Nevertheless some interaction with the ribonucleic acid metabolism is again suspected.

3. **Differentiation.** In embryonic growth the cytoplasm increases in amount, while in extension growth the principal event is the increase in volume of the cell and its change of form. The final elaboration of the cell is not achieved until the onset of a third kind of growth, known as differentiation. Already in extension growth the plastic wall has ceased to increase uniformly in area over its whole surface. Were it to do so an isolated cell would inevitably assume a globular form. Usually growth is instead quite local. Thus a cell will perhaps expand to a cubical or cylindrical form, or show apical growth (filamentous algae, fungal hyphae, latex ducts, conducting elements, sieve tubes). It becomes a fibre or a tube, possibly branches, and ultimately reaches one of those specific forms which have been discussed in the morphological section. Now begins the more extensive differentiation, particularly of the cell wall. As already mentioned, the wall becomes thickened, usually by a process of apposition. In the conducting elements, for example, annular, spiral or reticulate thickenings are formed, and lignification sets in; in short, numerous growth processes bring about the final form of the cell (Fig. 65). The fact that these depositions on the wall are not uniform, but localized, reflects the inner structure of the protoplasm. This inner structure is unfortunately still largely unknown, but that it is effective in determining form is shown, for example, by the fact that even a naked cell (e.g. *Euglena*, p. 11) possesses a specific shape.

These formative forces in the protoplasm of the individual cell are in fact quite generally operative not only during the growth of an isolated cell but also in that of a whole group of cells. They must therefore be present in the growth of a multicellular organ or organism, although accompanied by the effects which adjacent cells exert upon one another. Thus, in the development of the specific forms of organs, e.g. of leaves, flowers and shoots, factors additional to the simple formative effects of the protoplasm are at work. These must now be considered in greater detail.

II. The growth of organs

A. Cell division[34]

The development of multicellular tissues and organs, general in the higher plants, is bound up with cell division, a process intimately connected with growth. A cell is not usually capable of growing to an unlimited size, but the size is confined within certain limits according to the species, and the maximum is soon reached. There are, of course, some cells, such as the phloem fibres or latex ducts, which can become very long. A form externally similar to that of a higher plant can in fact be acquired by a single cell, as in the large and highly differentiated unicellular (but multinucleate) alga *Caulerpa* (Fig. 9), and is not absolutely dependent on the formation of a cellular tissue. Nevertheless, it is a well-known fact that a more highly organized plant body is always composed of numerous, relatively small, individual cells fused harmoniously together and yielding the visible form. If a multicellular organism is to arise from a fertilized egg cell continuous cell division must accompany the growth processes. It is clear that the direction and intensity of this cell division can have a profound effect upon the form generated (see p. 73 seq.).

Little is known about the relationship between cell growth and cell division. The two processes are not necessarily connected. Normally there is a definite and constant proportion between the amounts of nuclear and cytoplasmic material in a cell. If the nucleus is increased in size by experimental means, such as occurs if the number of chromosomes is doubled by colchicine treatment (see p. 332), this usually results in a subsequent increase in the amount of cytoplasm and in the size of the cell. In general, a certain amount of plasmatic growth has to occur, although it may not be the immediate cause, before the division of the nucleus and cell will take place. This division can, on the other hand, act as a renewed stimulus to plasmatic growth. In fact, during active growth of a tissue we see a rhythmical alternation between growth and division of the cells.

It has not yet been possible to obtain any detailed information about the causes and forces which govern cell division. In recent years if has often been thought that the initiation of cell division may also depend upon growth substances of a hormonal nature.

It is in fact often possible to induce cell division by artificial stimulation. A number of substances are active in this way. Thus those previously mentioned in relation to extension growth, when applied in higher concentrations, also initiate cell divisions. Such substances may possibly play a role in this nature in the re-awakening of cambial activity in the spring, and in the formation of callus and galls (see p. 315), as well as in the initiation of organs. It seems very likely that throughout plant growth a whole series of hormone-like growth substances (phytohormones) continually interact, both synergistically and antagonistically. The so-called cytokinins form a special group of these substances. Following the discovery that kinetin (6-(furfuryl-amino) purine, isolated from samples of deoxyribonucleic acid from herring sperm) together with indole-3-acetic acid stimulated cell division (and had other effects such as inhibition of the breakdown of chlorophyll in yellowing leaves, and the attraction of a stream of amino acids to the treated site), substances with a similar nucleoside-like constitution and activity (termed cytokinins) have been isolated from both lower (bacteria) and higher plants (e.g. zeatin from *Zea mais*). They appear to be components of transfer ribonucleic acids (tRNAs), and are consequently involved in the ribonucleic acid metabolism and synthesis of enzymes. A substance has been obtained from wounded bean pods (traumatin), of unknown structure, very effective in inducing cell division. Nevertheless, despite all these observations, it is still not clear how cell division is hormonally controlled under normal conditions. It is probably a consequence of a complex of factors of the kind described in the foregoing. The earlier assumption that damaged cells produce substances directly capable of promoting cell division (wound hormones) has not been confirmed by experiments on potato tissue.

Generally speaking, nuclear and cell divisions occur particularly frequently in embryonic tissue, before the cells have undergone any kind of differentiation in the course of extension growth. As will be related subsequently, these divisions are of great importance in relation to morphogenesis. Divisions can still occur in fully developed cells, as is seen, for example, in the formation of secondary meristems (e.g. the cork cambium (see p. 93)) and as a result of wounding (see p. 311).

Cell divisions often take place with a definite frequency, the period sometimes being diurnal, but also occasionally such that several divisions occur within twenty-four hours. Light can apparently exert a controlling effect. This is also true of meiosis. In many algae divisions occur principally at night; *Spirogyra*, for example, usually divides itself around midnight. In multinucleate cells the nuclear divisions are often simultaneous, or proceed as a wave from one end of the cell to the other (e.g. in the embryo sac; Fig. 36).

B. Polarity[35]

It is evident that the position of the new transverse walls will be significant for the later morphology of the developing organs. In many cases the arrangement of the walls must be taken as the visible expression of a preceding determination and differentiation of the cells. Such a determination, i.e. the strict confinement of the development to one of the several courses originally possible, can be seen, for example, in the germination of the spores or zygotes of certain thallose plants. One of the daughter cells of the zygote forms the first rhizoid, while the other daughter cell gives rise to the remainder of the thallus. Even in the Cormophyta the germinating egg cell shows a similar polar differentiation (see Fig. 90). This polarity is visibly established after cell division by the position of the first wall, and it must lastingly affect all subsequent development. Thus it is clear that polarization is one of the basic features of cell organization, being either already present or developing before division occurs. It is hardly conceivable that ordered development could occur without it.

Even unicellular organisms may possess an undoubted polar structure, with clearly recognizable anterior and posterior ends. In certain germinating cells external factors, such as light and gravity, are capable of exerting a directing influence on the apparently still labile polarity of the cell. In the germinating spore of a moss (e.g. *Funaria*) the transverse wall of the first division will always form perpendicular to the direction of the incident light. The same is true of *Equisetum* spores, and also of the fertilized egg of *Fucus*, where the cell away from the light always gives rise to the rhizoid. Chemical influences, e.g. gradients of growth-regulating substances, can also determine polarity. In other cases, especially in that of the seed plants, where the egg cell is concealed deep within the tissue of the sporophyte, external factors probably have little effect. Nevertheless, perhaps influences emanating from the cells of the ovular tissue, or even other stimuli arising in the cell itself determine the position of the division spindle and thereby the polarity of the cells subsequently developing and dividing. Sometimes one of the daughter cells is very different in size from the other (unequal division), the inequality arising of course from an unequal division of the cytoplasm and not of the nuclear material. This can be interpreted as an expression of an already established polarity.

We still have little insight into the real nature of such a polarization of an individual cell, but it must in some way be concerned with a polarized arrangement of invisible structural elements, and the non-uniform distribution of certain substances in the cell. Surprisingly layering of the cytoplasm, or of its organelles (as visible in the light microscope), brought about, for example, by centrifugation, is usually without effect. Where polarity is established by light or gravity, the decisive events appear to take place in the outer layers of the cytoplasm. Photoreceptors (see p. 383), for example, seem to be located here. We can imagine that excitation of these molecules results in metabolic gradients which lead to more or less unequal division. As the cases mentioned above show, the polarity of germ cells appears to be still labile and displaceable before division, and its origin is not dependent upon the occurrence of a cell division. On the other hand, once a polarity has been established in the cells of a mature plant body, it is usually tenaciously retained. This can be seen particularly impressively in some hair cells in which the cytoplasm will conduct certain dyes only from the base to the tip of the cell, although there is all the time a vigorous circulation of the protoplasm. This polarized conduction, i.e. the peculiarity of a polarized cell that it will transport and accumulate certain substances only in one direction through the cell (cf. *Acetabularia*, p. 27), may well be of fundamental importance for the progressive elaboration and differentiation of multicellular organs. The transport of growth regulating substances, for example, is often strictly polar (see p. 303).

It is still not at all clear how far the polarity of whole organs is connected with this polarized structure of the individual cells. Just as the individual fragments of a magnet always display N and S poles, so with a fully formed polarized organ, polarity is often still shown by the individual parts, and even by the individual cells.

This is shown very beautifully, for example, by willow wands cut up into segments and suspended in a damp room. After some time, if the portion of stem is hanging in a normal orientation, the buds adjacent to the apical end shoot out, while at the basal end roots develop from existing primordia (Fig. 270), although bud primordia are present here as well. If the portion of stem is suspended in an inverted position roots again develop at the originally basal end, and at the apical end shoots, even though the basal part is now uppermost and the apical part below. This shows that the stem firmly retains its old polarity even though it is cut up into fragments.

The polarity within the cells and tissues also becomes evident in grafting: only correctly orientated parts grow together without interruption.

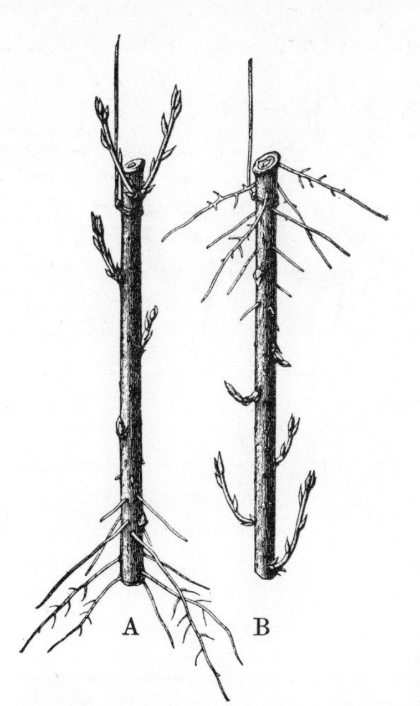

Fig. 270. Portion of willow producing shoots and roots in a damp atmosphere. **A.** In normal position. **B.** Inverted. After Pfeffer, reduced.

C. The growth zones of organs

When the fertilized egg cell of a seed plant has undergone one division, then, as related above, the fate of the two daughter cells is already determined by the position of the first dividing wall. The suspensor develops from the cell proximal to the micropyle, while the distal cell undergoes repeated divisions and gives rise to the embryo proper. A further polarization and differentiation soon occurs in these still completely embryonic cells, but henceforth it concerns whole complexes of cells. Repeated divisions lead to the formation of the root pole on the side adjacent to the suspensor, while at the opposite end arises the shoot pole, at which the primordia of the first leaves (cotyledons) soon become recognizable (see Fig. 131). The forces responsible for these early configurations of almost purely embryonic cells are completely unknown.

During further development of an embryo and young seedling, extension and differentiation begin to appear. Only some of the newly formed embryonic cells undergo this advanced development, another smaller group remaining continuously embryonic and showing only progressive plasma growth with accompanying cell divisions. Here is seen a fundamental difference between the organizations of animals and plants. Plants possess an 'open' form, i.e. they retain certain areas of embryonic tissue indefinitely and only the remainder differentiates completely. Development is never, as in the animal, completed, but the plant is always capable in certain conditions of growing out afresh and forming new parts (see p. 138 seq.).

This separation into persistently embryonic, extending and differentiating regions has as a consequence that it is always possible to find certain zones in which extension growth alone is occurring, even when the plant is old. Figure 271 shows quite schematically this separation of the zones of embryonic and extension growth in a higher plant.

The position of the extension zone can easily be demonstrated by making several small marks with Indian ink, about 1 mm apart, on an appropriate part of a plant (e.g. Fig. 272). The separation of these marks is then followed, possibly with the help of a horizontal microscope.

In this way it is found that in roots the zone of extension growth is only a few millimetres long and lies directly behind the tip (Fig. 272). In the region where the root hairs arise most of the cells have already reached their maximum size and have begun to differentiate. Longer zones of extension growth are present only in aerial roots. In shoots, however, the zone of extension growth is typically longer and can amount to over 50 cm (as in *Asparagus officinalis*). Where the axis is clearly divided into nodes and internodes,

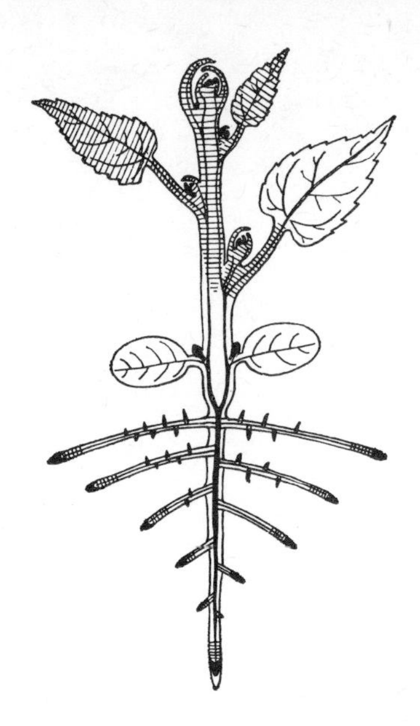

Fig. 271. Diagrammatic representation of the distribution of the different phases of growth in a dicotyledonous plant. The zones of embryonic growth at the growing points are represented as black, those of extension growth cross-hatched, and the fully grown regions white. After Sachs.

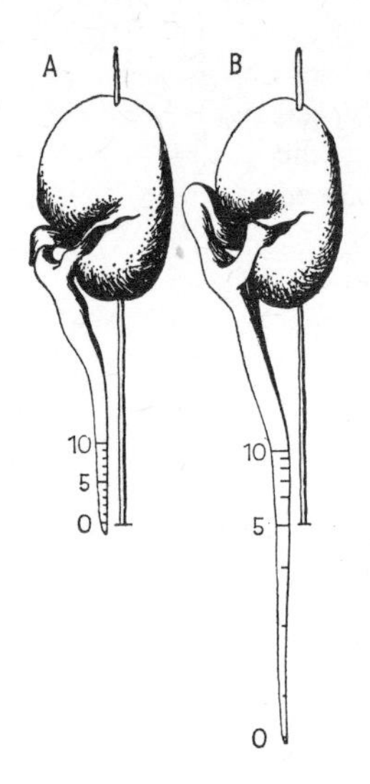

Fig. 272. Distribution of growth at the root tip of *Vicia faba*. **A.** The root tip marked with Indian ink, the marks 1 mm apart. **B.** The same root after 22 hr. The ink marks are now different distances apart depending upon amounts of growth in the individual segments. After Sachs.

active growth in the stem is usually interrupted by the former, at which hardly any elongation occurs. Each internode then possesses its own growing zone, usually located in the lower part of the internode. In the Gramineae this intercalary growth persists for a long time, and close above each node is not only a zone of extension growth, but preceding this also an embryonic zone of cell division and plasmatic growth. Such basal intercalary growing zones are found also in leaves, particularly clearly, for example, in conifers and monocotyledons, but also in some dicotyledons. The petiole, for example, is always interpolated in an intercalary manner between the lamina and the leaf base.

If the course of the growth of a segment of root about 1 mm long, taken from the zone of extension growth, be followed at fixed intervals of time, it will be found that the growth in length of the segment, and similarly of every cell in it, is at first slow. The intensity of growth rapidly increases, however, and reaches a maximum, after which it declines until all the cells of the segment are fully grown. This acceleration and deceleration of the growth rate is referred to as the 'grand period of growth'.

Thus the following figures were obtained for the growth of a segment originally 1 mm long, in a root of *Vicia faba*:

Day	1	2	3	4	5	6	7	8	Total growth in
Daily growth in mm	1.8	3.7	17.5	16.5	17.0	14.5	7.0	0	8 days 7.8 cm

In those roots and shoots where meristematic activity continues indefinitely at the growing point, the growth of the whole organ shows no such sigmoid curve. On the contrary, if a root tip is observed with a horizontal microscope it will be found that in a constant environment growth is quite steady. This depends upon the fact that the meristematic activity of the growing point leads to new cells being cut off continuously behind the apex. The extension growth of these cells begins while the older cells are passing the phase of maximum extension and completing their growth. The arrival of the new cells, and the beginning of their extension growth, is clearly so well coordinated with the decay of growth in the older parts that the organ as a whole appears to have a steady and continuous growth. Nevertheless, each cell passes through a grand period of growth and eventually is fully grown and reaches its maximum size. Occasionally, it is true, remarkable discontinuities in growth have been observed in shoots, so that growth proceeds by a series of jerks, but the cause of this phenomenon is unknown. Organs with limited growth, such as leaves, leaf sheaths (and coleoptiles of the Gramineae), filaments, etc., can individually display a grand period of growth, provided the embryonic growth is clearly separated from the extension growth and the organ is fully grown after the extension.

The rate of growth is very different in different plant organs. It may be so large that the movement of the advancing apex of the organ is observable with the naked eye. In almost all cases it can be seen directly with the horizontal microscope.

The filaments of the anthers of many grasses elongate at the opening of the flowers (anthesis) at a rate of about 0.5–1.0 mm/minute. Bamboo shoots can grow at about 0.3 mm/minute and vigorous shoots of cucumber at about 0.1 mm/minute. The pollen tubes of *Tradescantia* elongate at about 20 μm/minute and fungal hyphae at a maximum rate of about 52 μm/minute. Most of the higher plants, of course, remain well below this level and grow at a rate of some few μm/minute. This is understandable, since even extension growth, perhaps in connection with the incorporation of new components into the cell wall, involves active synthesis in the cytoplasm. This undoubtedly requires time, and from this point of view the figures given above acquire a special interest.

D. Correlation

As a result of the division and growth of the embryonic cells at the growing point, and the subsequent extension and differentiation of certain zones, specific organs begin to be developed. This is only possible because the continued development of the innumerable individual cells conforms to a general overriding plan. As already stated, the origin and functioning of the formative and directing forces, the existence of which is already evident in the embryonic growing point, is completely unknown. It is very significant that the embryonic cells initially possess the capacity to give rise to all the parts and tissues of the differentiated plant body. They are therefore, like the zygote, totipotent. After a certain stage of development, however, individual cells or groups of cells become determined. In consequence, only a few of the potentialities originally present in the cell are deployed and developed, and the division of labour between cells begins. We again have very little insight into the causes of this determination, but current genetic research into the suppression and activation of genes (see p. 330) is clearly relevant and informative. It is certain, however, that very close interrelationships, or correlations, must exist between the individual cells which go to make up an organized and differentiated structure. Without these correlations, it is difficult to see how a multicellular organism could exist. They must also, of course, play a controlling role during development. The harmonious and continued association of different tissue zones throughout growth, which has already been described, and which to some extent stands in opposition to the 'grand period' of growth of the individual cell, is itself an indication of such correlative influences.

It is clear that a number of very different and complex factors are involved in this question. Since almost all the cells of a higher plant are in close plasmatic contact with one another by virtue of the plasmodesmata, it is conceivable that correlation is effected either by the conduction of some kind of excitation or by the direct exchange of metabolites between cells.

So far as the metabolic relationships are concerned, the state of nutrition of individual cells or regions of cells, in regard, for example, to the supply of water, nutrients and especially certain hormonal growth substances, may play a very important part. When it is realized that not only the embryos of seed plants but also all growing points, since they are not autotrophic, are dependent upon a supply of nutrients, it is easy to see that conditions could arise in which different cells received different amounts of nourishment. This would inevitably have its effect upon the differentiation of these cells. Sometimes the peripheral cells are the better supplied and sometimes the central, depending upon the point of entry of the nutrients into the embryonic tissue. It must be remembered that, before the differentiation of specialized conducting tracts, the nutrients have to be passed from cell to cell. Consequently, cells lying distant from the entry of the nutrients can be supplied only when the needs of the cells lying between are satisfied, and the nutrients are allowed to pass through. The polarization of the cells, mentioned earlier, must be of considerable importance in relation to this transport of food materials.

Mechanical tensions and pressures can naturally have an influence on developing organs. If certain parts, possibly as a result of differences in nutrition, grow more extensively than others, tensions may arise in a tissue. Nevertheless, as a result of the controlling influence of correlation, the significance of such effects is probably only slight. In mosses, it is of course known that the pressure of the calyptra has a decisive influence on the form of the capsule. (For a consideration of the part played by the external environment in determination, see p. 334.)

1. Restitution phenomena[35, 36]

In many cases the differentiation of originally totipotent cells appears to consist, on the one hand, of a partial inhibition of this potency, resulting from the correlative effects of thc adjoining cells, and on the other, of a particular promotion of certain of their remaining capacities. This can be readily inferred from the study of the restitution phenomena which are displayed following wounding, and which are very informative in relation to many other problems of developmental physiology. Such phenomena are also known in the Animal Kingdom, and include those well-known examples where the organism will replace a lost or otherwise eliminated organ or part of an organ. The most important of the restitution phenomena is regeneration, in which replacement takes place either by the outgrowth of a resting primordium already present, or by a completely adventitious and new formation. The latter is of especial interest in the study of morphogenesis, since it often involves fully developed and differentiated cells becoming once again totipotent. It is relatively rare for the replacement to take place simply by the renewed meristematic activity of the damaged cells at the surface of the wound. This phenomenon, which does occur in the restitution of the damaged growing point of a root, and also in *Acetabularia* (see p. 27), is referred to as reparation.

Regeneration can easily be seen with leaves of *Begonia*. If a leaf together with its petiole is detached and laid on damp sand not only do roots develop at the end of the petiole but also complete vegetative buds are generated *de novo* on the lamina, especially if some of the veins have been transected. Entire new plants develop from these buds (Fig. 273). Microscopic examination shows that a normal epidermal cell of the leaf, a little above the site of the incision, has suddenly again become embryonic (Fig. 274). It regains its capacity to divide, and as a result of repeated divisions forms an adventitious bud. The adventitious roots usually originate from dividing cells adjacent to the phloem of the vascular bundle.

In this case it is clear that the cell initiating the regeneration must have carried within itself the capacity to give rise to an entire *Begonia* plant, although in the tissue of the leaf it was differentiated as an undoubted epidermal cell and therefore had to fulfil limited and specialized functions. So long as the leaf was connected with the parent plant there must have proceeded from it to the cells of the epidermis some influence which prevented their manifesting all the potentialities which lay within them, and only the capacity for developing into epidermal cells could become realized. As soon as this influence was interrupted by the cutting of the veins and petiole, the suppressed embryonic condition re-emerged in some of the cells, and they began to divide and give rise to completely new plants. The precise form of this inhibitory influence is still completely unknown, but it is clearly in the nature of a correlation. Another example is provided by kohlrabi, where the pith cells readily regenerate whole plants.

Not in all plants do the differentiated cells have the capacity to regenerate in this way. Markedly differentiated cells are frequently no longer capable of division. After wounding, the less differentiated parenchymatous cells adjacent to the wound frequently become meristematic again and form a mass of cells at first completely undifferentiated, the so-called callus. In woody plants the callus usually arises from the cambium. It has been observed in cultures of isolated tissues (e.g. pith of tobacco stem, explants from root of *Cichorium* (chicory)) that the particular balance of growth-regulating substances present in the medium determines the nature of the regeneration. A high concentration of

Fig. 273. Leaf-cutting of *Begonia* showing regeneration. After Stoppel, reduced.

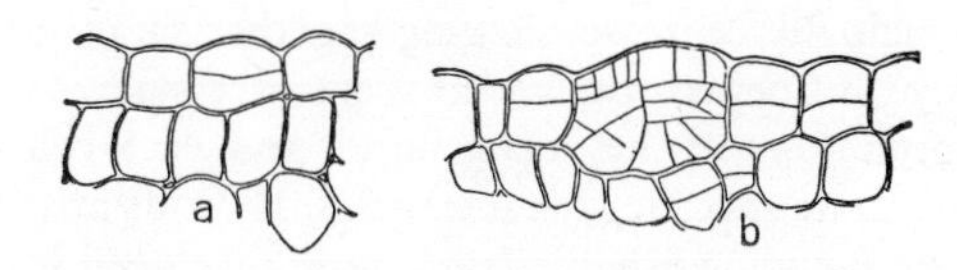

Fig. 274. Transverse section through thc epidermis of a leaf of *Begonia*. Formation of an adventitious bud from an epidermal cell. a. The epidermal cell after the first periclinal division. b. The epidermal cell has now given rise to a multicellular meristem from which the adventitious bud arises. X150. After Hansen.

kinetin in relation to that of IAA (indole-3-acetic acid) favours the production of shoot buds, the converse situation roots. At certain intermediate concentrations of kinetin and IAA the callus remains undifferentiated. With cultures of *Cichorium* the relationship between the concentrations of cane sugar (sucrose) and IAA was found to determine whether phloem or xylem was differentiated.

The extent to which so-called wound hormones, liberated from the damaged cells, are able to stimulate the resumption of meristematic activity and plasmatic growth of differentiated cells is quite uncertain. In the formation of cork cambium by injured potato tubers the changed water content of the exposed cells seems to play the decisive role.

Not until the callus has been formed does differentiation occur in some of its cells, leading to regeneration. A new root or shoot apex may be formed, as the case may be, or, if a vascular bundle has merely been cut through, the cells between the margins of the wound will differentiate to form vessels and sieve tubes, so restoring the broken continuity. This is in accordance with the general law of restitution, namely that which is formed is usually of the same kind as that which was lost.

There are, of course, differences of degree. In higher plants new shoot apices, and especially new root apices, are usually formed relatively easily. Lost laminae of leaves are only very rarely replaced; usually the whole leaf or petiole is shed. Also by no means all leaves are capable, like the *Begonia* leaf, of giving rise to new growing points.

Grafting.[37] Restitutions of a special kind are also involved in grafting. As is well known, grafting consists of causing two parts, known respectively as the scion, which bears buds, and the stock, bearing the roots, to grow together. The union of the two parts is brought about by the callus developing at the wound. New vessels and sieve tubes soon arise in this callus, connecting together the vascular bundles of the stock and scion. For a successful graft to occur there must usually be some systematic relationship between the two partners.

Grafting has very great importance in horticulture and agriculture. Varieties of fruit trees and roses, for example, which do not reproduce by seed, can be propagated and thus maintained in cultivation by grafting on to vigorously growing stocks. Again, correlative influences of an unknown kind must pass between the partners, since regeneration is suppressed, and instead of the scion giving rise to roots, and the stock to shoots, the two parts grow together and form a unified whole. Of course, following the graft union the hereditary properties of the stock and scion remain distinct, although a certain amount of metabolic exchange is possible, and consequent upon this one partner may influence the phenotype of the other.

This is particularly conspicuous in those grafts in which adventitious shoots arise from the callus at the site of the graft, consisting of tissues derived from both partners. Thus 'chimaeras' can arise, and even 'sectorial chimaeras' in which, for example, one-half of the shoot or leaf may be derived from the stock and the other half from the scion. Particularly remarkable are the 'periclinal chimaeras', in which the epidermis and possibly some other outer layers have been formed from one partner and the inner tissues from the other (examples are provided by grafts between *Cytisus* spp., between *Crataegus* and *Mespilus*, etc.). Such plants, referred to as graft hybrids, may have a superficial resemblance to true sexually produced hybrids, but in fact they are quite different. Even though the cells are growing together in an intimate manner, and the existence of correlative influences between layers of tissue of different origins is clearly revealed in the morphology of the organs developed, each cell and cell layer seems nevertheless to remain true to its specific characteristics. Hybrids arising from fused protoplasts, however, appear to be indistinguishable from those produced sexually.

Mechanism of regeneration. The fact that regeneration usually leads to the ordered repair of the damaged organism can be explained only by assuming that the organism exerts a correlative influence on the regenerating tissue. If, for example, parenchyma cells in a callus are transformed into vessels, forming a bridge between the separated ends of vascular bundles, as regularly occurs in grafting, then there must be some agent which stimulates these parenchyma cells and causes them to differentiate towards the damaged bundles. It is usually thought today that this correlation is due to the action of specific growth substances which must be formed by the wound, and in fact by the already differentiated parts of the damaged organism. Although such substances have not so far been isolated, their existence, as will appear again in the following section, is certainly possible (see p. 314). They are consequently of great interest, since relationships similar to those seen in regeneration must also be present in any normal development in a multicellular organism. If, for example, a shoot or root apex grows continuously, then the cells undergoing extension growth and subsequently beginning to differentiate must join on to the fully differentiated cells already present in an harmonious manner. Thus, when vascular bundles follow along behind the advancing growing point, the newly differentiating elements have to remain in direct contact with the older differentiated bundle succeeding them.

2. Hormonal action at a distance

a. The transport of growth substances.[34] The growth substances specific to any particular kind of growth must necessarily cooperate in all the morphogenetic phenomena which are consequent upon it. It has, however, now been shown that those substances stimulating particularly extension growth are not formed uniformly in all cells. Instead, in a growing multicellular organism a number of sites of synthesis, or activation centres, soon arise. Subsequently, a stream of growth substances (possibly also of inhibitory or inactivated growth substances) is given off from these regions into the neighbouring cells and tissues, and is capable of producing effects at a considerable distance. In this case the hormonal nature of the correlation is clearly evident.

In young seedlings the cotyledons seem to be the principal site of production of the growth substances, but active transport of growth substances (the formation of which is dependent upon light) also takes place from differentiated leaves into the axis of the plant. Amongst other things these may delay the formation of abscission tissue in the petiole and hence the fall of the leaf (see p. 303). In the seedlings of grasses the tip of the coleoptile is an activation centre, and growth substances ascending from the grain in an inactive form are, as already mentioned, here liberated. A stream of growth substances then flows down from the tip, and regulates the entire extension growth of the coleoptile. If, for example, the upper few millimetres are removed from the tip of the oat coleoptile (an object frequently used for research in plant physiology) the growth of the stump is completely suppressed for several hours. If a detached tip is replaced on the stump with the help of a drop of agar or gelatin, growth is soon resumed in the lower region. It is in fact sufficient to place on the stump a disc of agar on which a tip, with its cut surface downwards, has rested for several hours. The agar alone will then cause growth to be resumed, indicating clearly that a substance is given off from the tip which has growth-regulating properties (see p. 304).

In older plants streams of growth substances are also given off from the growing points of the shoots and roots. These not only control extension growth but also have other effects. Thus in the higher plants a stream of growth substances from the terminal bud seems to be concerned in the production of a specific inhibitor in the axillary buds, preventing them from growing out. This means, of course, that the terminal bud exerts a correlative influence on the production of lateral axes, and thereby plays a large part in determining the branching and symmetry of the whole plant. Usually, as soon as the leading shoot is removed, the lateral buds begin to develop. If, however, the stump is smeared with a growth-regulating substance worked up into a cream with lanolin the side buds remain inactive just as in an undamaged shoot. The explanation probably lies not in a direct inhibitory effect of the indole-3-acetic acid, but in a complex interplay between the growth substance and other factors, e.g. with an inhibitor arising under its influence.

Although the descending stream of growth substances here causes inhibition, in other instances it can, of course, have a stimulating effect. In early spring, for example, the expanding buds of trees send streams of growth substances into the cambium of the stem, thereby causing it to resume its meristematic activity. The growth substances, therefore, do not affect extension growth only, but have a very much more general stimulatory effect. This has been observed in other cases as well, and apparently extends to the promotion of cell division, the differentiation of xylem elements and so on.

Growth substances are also involved in the growth and development occurring after flowering, particularly in the formation of the fruit. It has already been mentioned (p. 303) how the formation of abscission tissue in the petiole can be initiated by the failure of the supply of growth-regulating substances from the leaf (as a consequence of its ageing or removal) or by the generation of a specific hormone (abscisin). Apparently the ovule and the developing embryo are a site of very intensive synthesis of auxins. It has been possible, for example, to induce the formation of a fruit without fertilization by the artificial addition of a growth substance to the wall of the ovary. In nature, therefore, the many cases of parthenocarpy, i.e. the formation of seedless fruits without fertilization, occurring principally in cultivated plants, such as bananas, seedless oranges and sultanas, may be traced to the generation of auxin or its activation. This may occur either spontaneously or following the deposition of pollen on the stigma. It is known that in some plants pollination alone, without any sexual fusion, can stimulate the growth of the ovary. In certain orchids this stimulus, which also causes wilting of the petals, comes from substances having the character of auxins adhering to the outside of the pollen grains. The curvatures often seen in the stalks of flowers and fruits frequently arise as a result of the unilateral displacement of growth substances.

There can therefore be no doubt that the formation of special centres of auxin production, and the streams of growth-regulating substances issuing from them, lead to important correlative relationships between the parts of the plant, and influence its whole development. It must be clearly understood that the streams of growth-regulating substances alone cannot bring about the building of specific organs. These questions of developmental physiology proper lie much deeper.

An extremely important phenomenon is that activated indole-3-acetic acid (in contrast to the inactivated) is transported in a polar fashion, i.e. only in one direction, through the cell. This means, as already stressed, that the polarization of the cells must be an event in development preceding the streaming of the growth-regulating substances. Moreover, the different sensitivities of different cells and organ primordia to growth substances might be the result of a certain amount of differentiation having already occurred within them. Thus it is clear that the correlation maintained by auxins cannot have any true determinative effect in morphogenesis, but serves at most to inhibit or stimulate the creative properties peculiar to the protoplasm. The quantity of active growth substance or inhibiting substance will merely determine the amount of extension growth, and perhaps to a limited extent that of cell division. The problem at the heart of developmental physiology is what causes the generation of specific forms. What is the nature of the fundamental factors which determine, at a growing point for example, that certain embryonic cells shall give rise to quite definite organs, such as leaves, flowers, roots, etc.?

b. Organogenic substances. It has long been thought, especially under the impact of the results of regeneration experiments, that specific organogenic substances are involved in the development of plants. These are envisaged as coming, like hormones, from certain centres of synthesis, and inducing the protoplasm of embryonic cells to generate specific organs.

First it should be said that auxins do not appear to be concerned solely with extension growth; they behave as if they can stimulate the production of organs as well. Thus in many plants, if an auxin is smeared on the stem in the form of a cream, or dusted on as a powder, or if the stem is immersed in a solution of the auxin (e.g. 0.05–0.10 per cent solution of indole-3-acetic acid (not, however of gibberellin)) roots readily emerge in great numbers. This may be either because embryonic root primordia, present in many plants, especially at the nodes, and invisible externally, are stimulated into development, or because new primordia are formed endogenously in the pericycle or cambium (Fig. 275). The readiness with which portions of stems produce roots when dipped in a solution of a growth-promoting substance, or smeared with a paste containing it, is utilized in horticulture for the striking of cuttings, especially for those which are otherwise very difficult to root.[34] This procedure is used extensively, for example, in the cultivation of cocoa.

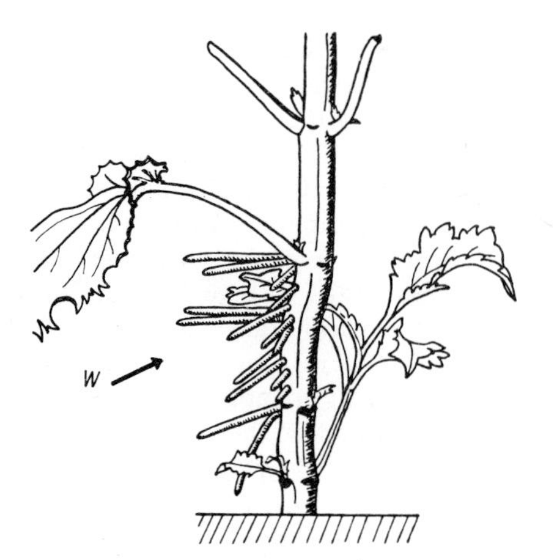

Fig. 275. *Coleus.* Basal internodes of the stem giving rise to adventitious roots after applying cream containing a growth substance to the left side. W, adventitious roots. *c.* $\times\frac{1}{2}$.

It is questionable whether these are really examples of growth-promoting substances having a specific organogenic effect. They may act merely by stimulating the protoplasm of the cell in an unspecific manner, so that cell divisions take place, the appropriate organs then arising by specific processes of differentiation, as in regeneration phenomena generally. That growth substances will be required for their development is self-evident. Though the 'organogenic' effects of the growth-promoting substances are striking and of great practical importance, they contribute little to the elucidation of the particular problems of growth physiology involved.

In *Dictyostelium*, belonging to the slime fungi (Myxomycetales), the aggregation of the freely motile amoeboid stage is brought about by a substance they themselves liberate called acrasin. Aggregation initiates the formation of the fruit body, which is raised above the substratum (see Fig. 69). The clearest evidence so far, however, for the existence of special organogenic and morphogenic substances comes from *Acetabularia*, an umbrella-shaped, unicellular, calcareous alga reaching several centimetres in height (Figs. 16, 423).[38] Grafting experiments between different species show that materials are given off from the nucleus into the cytoplasm and that these accumulate in definite places. According to their concentration, they stimulate the cytoplasm to generate the umbrella-like reproductive organ, the form of which is determined by the genetic constitution of the nucleus. The nucleus of one species transferred to another species can cause the cytoplasm of the second species to generate a reproductive structure of a form appropriate to the implanted nucleus, apparently by means of its genetically determined growth substances. Here we have for the first time some insight into the nature of the interrelationships of nucleus and cytoplasm.

Investigations of photoperiodism in the higher plants (see p. 344 seq.) have shown that here too there are probably mobile substances which influence the form of the growing organs, e.g. the length of the internodes or the form and degree of succulence of the leaves. Of even greater interest is the fact that numerous observations indicate the existence of flower-promoting substances (florigens), the production, migration and accumulation of which in certain places causes the development of flowers. It is supposed that they in some manner influence the nucleic acid metabolism.

Simple grafting experiments between annual and biennial races already suggest the existence of a diffusible principle inducing flowering. *Hyoscyamus niger*, for example, is entirely vegetative in its first year and does not produce flowers until the second. It can, however, be converted into an annual and made to flower in the first year if a shoot from a plant in flower (in certain circumstances even of another species, e.g. of tobacco) is grafted into it close to the growing point. Apparently substances of some kind pass from the flowering plant into the meristematic tissue of the plant hitherto growing only vegetatively, and though in minute quantity cause its apex to produce flowers in place of leaves. These substances appear to be common to a number of plants, and it has consequently been thought that they may be nothing other than growth-promoting and growth-inhibiting substances, acting, as in other kinds of morphogenesis, without regard to species. The presence of a large amount of growth substance does in fact seem to suppress the production of flowers in many species. Nevertheless, this does not exclude the possible existence of special flowering hormones. Gibberellins appear particularly to promote the elongation of the embryonic peduncle (see p. 303), although effects on cell division and also apparently a direct stimulation of flowering have also been reported. (Concerning other factors influencing flowering, see p. 344; concerning termones, see p. 327.)

Gall formation.[39] Another phenomenon often observed in nature, namely the formation of galls remarkably diverse in form, points to the existence of substances capable of exerting a profound morphogenetic effect upon young and little differentiated cells. Some galls arise from the irritation caused by plant parasites, particularly the parasitic fungi, while others are caused by the punctures or deposition of eggs of insects and mites.

Besides the nodular growths which arise when the roots of leguminous plants are attacked by nitrogen-fixing bacteria, and the club-root disease of cabbage caused by *Plasmodiophora brassicae* (see p. 496), the so-called witches' brooms are other examples of galls caused by plant parasites. These excrescences, consisting of numerous side branches bunched together in compact masses, are caused in hornbeam and cherry by species of *Taphrina*, while in *Abies* a similar effect is caused by rust fungi. When *Euphorbia cyparissias* is attacked by the rust *Uromyces pisi* its whole habit is so changed that it is hardly recognizable; it produces only short, thick leaves, and no flowers or side branches (Fig. 276). Finally, in *Melandrium album* infected with the smut fungus *Ustilago violacea*, anthers are formed in flowers of those plants normally only female. These are not, however, filled with pollen, but with the violet brand spores of the fungus.

Especially interesting are the galls which arise under the influence of insects, since here we see the stimulation of growths which are usually of no use to the plant itself, but which are often adapted in particular ways to the needs of animals. A striking problem in morphogenesis is posed by the fact that different animals on the same plant give rise to quite differently shaped galls (Fig. 278), making it quite clear that the deposition of the eggs does not merely have a general stimulatory effect. Specific stimuli, most probably of a metabolic nature, must pass from the eggs to the tissue of the plant. In the case of simple growths, such as bacterial nodules, indole-3-acetic acid, formed by the bacteria in their protein metabolism and subsequently excreted, may play some part. With differentiated galls it is considered likely that the gall insect causes its domicile to develop in an orderly manner by repeated and localized excretion of growth substances into the plastic tissue of the plant. It has not yet been possible to isolate and identify these substances.

It is customary to distinguish between organoid galls and histoid galls. Organoid galls

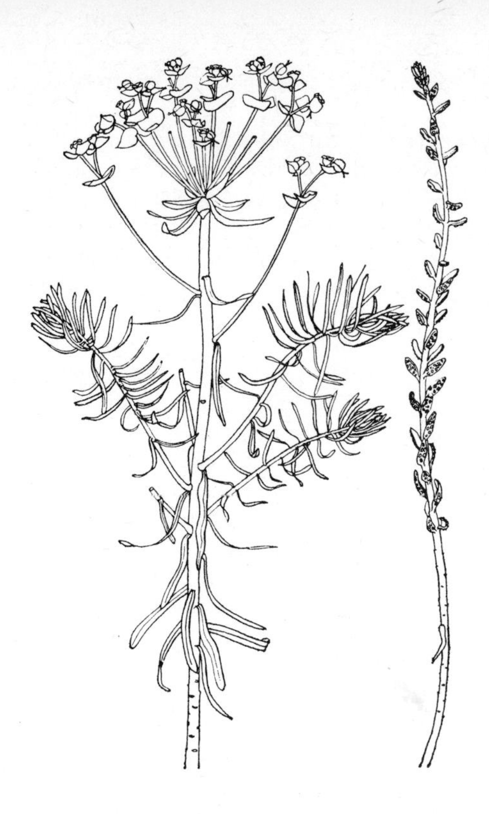

Fig. 276. *Euphorbia cyparissias*: left, normal; right, abnormal growth following infection with *Uromyces pisi*. *c.* $\times\frac{1}{2}$.

Fig. 277. 'Rose apple' on *Rosa canina* arising as the result of the rose gall wasp *Rhodites rosae* puncturing the young leaf primordium.
a. General view. The left leaf shows only small proliferations, that in the middle almost complete transformation of the primordium.
b. Section through the tip of the shoot with several larval chambers. After Ross and Hedicke.

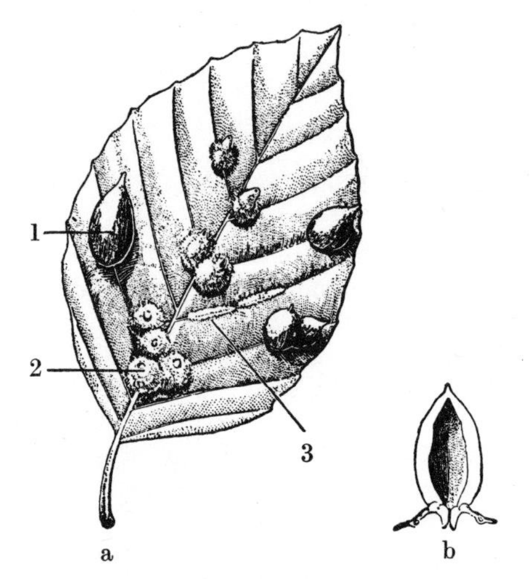

Fig. 278. **a.** Different kinds of sac galls on a leaf of *Fagus sylvatica*: 1, caused by the beech gall midge, *Mikiola fagi*; 2, hairy sac gall caused by the gall midge *Hartigiola annulipes*; 3, felt gall above the veins caused by the mite *Eriophyes nervisequens*. **b.** Section through a sac gall of type 1. After Ross and Hedicke.

consist of the basic organs of the host plant, strongly modified, but still clearly recognizable, such as abnormal branchings, aggregation of leaves at the tip of the shoot, 'witches' brooms', and the doubling and other leaf-like abnormalities of flowers. No organs can be recognized in the histoid galls; they arise as unique growths from parts of the stems, leaves or roots. An example of an organoid gall is the bushy 'rose apple' often seen on the wild rose. These arise following puncturing of the young leaf primordia and stems by the rose gall wasp, and they consist of a tightly packed mass of malformed leaves developed at the sites of puncture (Fig. 277). The cyst-like growths seen on the leaves of beech are an example of a histoid gall (Fig. 278b). They arise as a result of that portion of the lamina coming into contact with the beech gall midge or its excreta undergoing pronounced local surface growth. Ultimately these areas develop pouch-like concavities so that the source of the stimulation becomes completely enclosed in a cyst with an opening at its base. In many galls secondary thickening and differentiation of sclerenchymatous elements eventually occurs, so that a resistant chamber is formed for the developing animal. On the interior of the gall, numerous hairs and cells with delicate walls rich in nutrients serve for the nutrition of the parasite. The parasite therefore causes the plant to produce from its tissues a structure which no longer has any connection with the normal form and function of the plant.

All this goes to show the extreme plasticity of plant cells and tissues, and how extensively organogenesis and morphogenesis can be controlled by certain substances. Herein lies the significance of gall formation. Galls can be regarded as 'natural experiments', and they suggest very strongly that similar material influences play a part in normal development.

The hypothesis that specific organogenic substances are produced in metabolism, and are responsible for the normal morphology generated by the yet undifferentiated embryonic cells does not, of course, in any way solve the riddle of specialized development. Even though root- or flower-inducing substances may arise in the green leaves, they can only be effective when they are able to accumulate in adequate concentration in the active cytoplasm of the embryonic cells. This requires at least some process of transport to the sites of organogenesis, the mechanism of which must be the result of differentiation that has already occurred.

3. Ageing and death[40]

It follows from the preceding that the influences of cells and tissues upon each other, the phenomenon known as correlation, must be a decisive factor in any development and morphogenesis occurring in a multicellular organism. Only in a very few cases, however, are we yet able to state the nature of these correlations. Of great importance in the complex relationships which must necessarily exist between the parts of an organism are evidently polarization, and the interchange of metabolites and auxins. Although such correlations often cause the stimulation of development, they sometimes lead to inhibition. Inhibitions of this kind show in fact how the interrelationships between cells and tissues can in certain circumstances have a definite influence on the life span of the plant, since the phenomena of ageing and death can from certain points of view be interpreted as correlation effects. It appears that any living cell must inevitably change with time, and after a definite life span, different in different individuals, to die.

The causes of death may be many and various. Even inorganic colloidal solutions show an ageing effect, i.e. with time there is an increasing tendency for the particles to aggregate and precipitate. Also, perhaps, the necessary continuous removal of metabolic end-products from the cell is not always so complete as to prevent a slow accumulation of metabolic 'refuse'. This, even if it has no special deleterious effect, will eventually interfere with the fine structure of the protoplasm.

It is often seen that even the quite normal development and differentiation of a cell lead to its premature death, particularly when it is only in the dead condition that the cell fulfils the function proper to it in the organism. This is the case with those cells that differentiate to form vascular elements, sclerenchyma fibres, stone cells, cork cells and so on. Only with cork cells, where suberin lamellae impermeable to water are laid down, is there a clearly evident cause of the precocious death of the cells; in other cases the explanation is still unknown. It seems, especially with regard to reproduction and the formation of reproductive organs, that correlations can arise which limit the life span of certain cells, or even bring about their death. These correlations are at least partly metabolic in nature.

It is, for example, well known that cultures of lower organisms, such as bacteria, algae and fungi, can be maintained through many generations only if the nutrient medium be constantly renewed. This is not because any of the individual nutrients approach exhaustion, but because the growing organisms excrete metabolic end products which interfere with growth, and above a certain concentration prevent it completely. In multicellular organisms similar influences seem to arise between the cells, and to emanate especially from the reproductive organs, causing in certain conditions injurious effects in the vegetative regions.

In the lower unicellular organisms reproduction often merely involves a halving of the cell, following which each half again grows to the original size. There is no residue of any kind, and consequently in general no death. Unicellular organisms are thus potentially immortal, and succumb only to the kind of self-poisoning discussed above, and to other external catastrophes. As organisms become multicellular, so it is usual to find that only part of the mass of cells is devoted to reproduction, and differentiated to form the reproductive organs, the remaining so-called somatic cells undertaking only vegetative functions. After a certain time this vegetative part of the organism normally dies, the reproductive cells alone surviving and providing the starting-point for the succeeding generation. The death of higher organisms, which always affects only the somatic cells, can thus be legitimately regarded as a consequence of the division of labour within the organism, and there is now much evidence in favour of the view that its onset is brought about, or at least accelerated by, the reproductive organs. Thus in the higher plants there are many cases known where the differentiation of flowers and fruits shortens the life of the rest of the organism.

Certain annual plants, which normally die after flowering, can continue to live for two or more years if the formation of flowers is inhibited. Species of *Agave* grow purely vegetatively for several years, but rapidly die after they have eventually come into flower. The flowers of many orchids remain in perfect condition for weeks or even months, provided they are not pollinated. As soon as a pollinium is placed upon the stigma they rapidly wilt. There is no question here of an interruption of the nutrient supply. On the contrary, these degenerative phenomena affecting the perianth are caused in some curious way by growth substances adhering to the outside of the pollen grains, as already mentioned. Here is a clear example of how the death of certain organs is brought about correlatively by metabolic influences from other organs. Thus it is not impossible that the death of a multicellular organism is to be regarded as a similar correlation phenomenon.

On the other hand, the fact that life can in certain circumstances be continued indefinitely all over a multicellular organism, and not only in a few isolated cells, shows that the process of ageing, and the degenerative and inhibitory effects mentioned above, can sometimes be avoided.

In unicellular organisms cell division and the subsequent new growth appear to be an expedient by which each daughter cell ensures that at least half of its cytoplasm is new. It may even be that it is the maintenance or stimulation of the capacity to divide and grow, besides the suppression of any particular differentiation, that really ensures the continuation of life in the reproductive cells of all other plants. Sexual cells which fail to copulate usually rapidly die, but regeneration experiments have shown that even fully grown and differentiated cells can extend their life span if they are in some way stimulated to renewed meristematic activity and growth.

In agreement with this is the fact that the embryonic cells of the stem apex, capable of continuous division and growth, are, like unicellular organisms, potentially immortal. The death that eventually overtakes even the longest-lived trees may well not be due to the ageing of the growing point, but to the gradually increasing difficulty of supplying it with water, salts and other nutrients and growth substances. Thus it is often possible to maintain these apices in more or less perpetual growth by the continued taking of cuttings (as in the Lombardy poplar). In normal conditions death is again here seen to be a correlative effect.

Apart from the instances given above, the average life span of the somatic parts of plants varies with the species between extraordinarily wide limits.

In addition to the plants which go through a complete life cycle and die in a few weeks (ephemerals), and the annual and biennial flowering plants, are the perennials. These, as trees, may have a life span of up to several thousand years. There are authenticated counts of annual rings which show that specimens of yew may reach 900–3000 years, lime 800–1000 years, oak 500–1000 years, poplar and elm 300–600 years. At the extreme is probably the Big Tree of California (*Sequoia gigantea*), which is claimed to reach a maximum of 4000–5000 years, although *Pinus aristata* has been said to reach or exceed a similar age. Many other of our native trees attain an age of several hundred years, and even so lowly a plant as *Vaccinium myrtillus* may remain alive for up to twenty-eight years. It should be noted, of course, that there is a continual renewal of cells in the bodies of the long-lived plants, just as in animals. In trees, for example, this takes place through the activity of the cambium. The life span of an individual plant cell, such as a cell in the parenchymatous ray of a tree or in the interior of a succulent, would rarely extend for more than 100 years, provided there is no stimulus to renewed activity through cell division and growth. Even in that condition of dormancy which is induced in seeds and spores by extreme desiccation and suppression of metabolism, there seems to be a slow irresistible ageing, since observations suggest that viability rarely exceeds 100–200 years. Very long-lived seeds are found principally in the Leguminosae, Malvaceae, *Nelumbo nucifera* (lotus flower) (the viability of the seeds of which has recently been claimed to reach 1000 years) and others. Even the seeds of many lowly herbs (e.g. *Spergula arvensis*, *Chenopodium album*, amongst others) are said to remain viable for hundreds of years in anaerobic conditions. Persistent reports of the viability of the so-called 'mummy wheat' are, however, erroneous, since wheat is capable of germination for no more than ten years at the most. The viability of those seeds of tropical plants which are not adapted to tide over an unfavourable season may not even extend to one year.

III. Heredity (Genetics)[41, 42]

The reciprocal relationships between the cells of a tissue, referred to above as correlations, undoubtedly constitute important internal factors controlling growth and morphogenesis. They ensure that of the latent capabilities of a cell only a fraction in fact become manifest, the remainder being effectively suppressed, so that the individual cells become integrated in an harmonious way into the general plan of the organism. It does not, of course, follow from the fact that cells may occasionally become once again totipotent, as in restitution phenomena, that an embryonic cell from a bean, for example, removed from its normal correlations could become a cell in the tissue of a rose. In the developmental capacities of embryonic cells there are undoubtedly limits determining species and race, and in reproduction these are transferred from generation to generation, impressing upon the offspring not only the external form but also all the other properties of the parents. The detailed study of these factors influencing development, and their hereditary behaviour, is clearly of the greatest interest from the point of view of developmental physiology. The fact, for example, that the organism arising from a fertilized egg cell resembles its parents, in whole or even in part, indicates that something, which persists throughout morphogenesis, must be handed on unchanged from generation to generation, apparently accounting for the constancy of the general plan of construction of the developing organism. Since the paternal characteristics are fully transmitted in fertilization, despite the fact that the male nucleus is often accompanied by only a very little cytoplasm, it can be concluded that those stable factors which direct the pattern of development must be located principally in the cell nuclei. Although at present we still have very little knowledge of the mechanisms which, determined by the specific structure of the nuclei, govern the entire physiology of the cell, including growth and morphogenesis (see also p. 329 seq.), we must at least consider the general principles of heredity so far as they are known. A detailed discussion of heredity, already a specialized field of science, is, of course, beyond the scope of this book.

A. The Mendelian laws[43]

The study of inheritance in sexual reproduction has confirmed that the specific form and behaviour of a plant are effectively determined by definite structural elements in the nucleus which are passed from one generation to the next. The regularities observed in breeding are most satisfactorily explained by assuming that certain minute regions in the chromosomes, about which more will be said later (see p. 321 seq.), are the bearers of the individual genetic factors. Moreover, hereditary phenomena in general are dependent upon and the special properties of nuclear division which ensure that the genetic material of the chromosomes is partitioned equally between the daughter nuclei and cells. The recognition of this fact was one of the most important contributions of classical genetics.

In vegetative reproduction, involving no reduction division, the hereditary qualities located in the chromosomes are transferred unchanged. That this should be so is immediately evident from the mechanism of chromosomal cleavage in equational division (see p. 31 seq.). The genetically identical offspring of a single plant arising as a result of vegetative propagation are referred to as a clone. In sexual reproduction the situation is quite different, since here reduction division occurs and it is not chromatids, formed by the longitudinal splitting of the chromosomes, that are partitioned between the daughter nuclei, but whole chromosomes. After the formation of bivalents the homologous chromosomes, derived initially from the two parents, separate more or less in their entirety and pass into the daughter nuclei (see p. 37 seq.).

In the higher plants it is usually the case that the organism developing from a fertilized egg cell possesses diploid cells, i.e. the nuclei contain the doubled number of chromosomes, one set of chromosomes coming from each parent. There is no reduction of the chromosome number to the haploid condition until the meiosis which occurs in the formation of the pollen grains and embryo sac. At this time the bivalents arrange themselves at random at the equator, and it is a matter of chance which chromosomes go to which pole, so that each daughter nucleus contains some chromosomes derived from the female side and others from the male. Thus, where the two parents differ genetically, the genetic factors present in the sexual cells of the offspring are not necessarily in combinations identical with those of the parents. Consequently, in sexual reproduction there are quite definite, statistically detectable proportionalities in the genetic characteristics of successive generations.

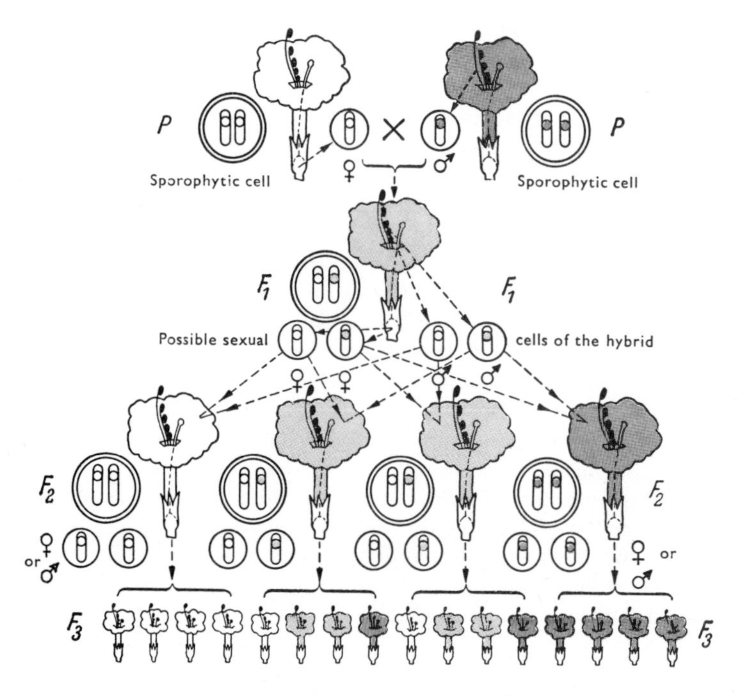

Fig. 279. The hybrid resulting from the crossing of a white-flowered and a red-flowered race (here indicated by shading) of *Mirabilis jalapa*, and its descendants through two generations. Diagram after Correns, modified.

These proportionalities, dependent upon the process of meiosis, were first clearly described by Gregor Mendel (1865), although the underlying cytological phenomena were not then known. Let us assume, as the simplest case, that the parental plants differ from each other in a single character. When these plants are crossed (i.e. the gametes, or sexual cells, which are caused to fuse are genetically different) the character concerned is inherited in the following manner.

As an example, two races (or pure lines, see p. 342) of *Mirabilis jalapa* may be considered, differing merely in the colour of their flowers, one being red and the other white. In practice, the pollen of one race is transferred to the stigmas of the other, ensuring subsequent fertilization. The two homologous chromosomes in the nuclei of the cells of the red-flowered plant each bear in corresponding positions factors (or genes) causing redness of the flowers, while in the other plant the equivalent factors cause whiteness (Fig. 279). Genes which occupy exactly the same place in homologous chromosomes, irrespective of whether they are identical or not, are termed alleles. When the red-flowered plant forms sexual cells subsequent to meiosis each gamete contains only one chromosome bearing the gene for redness, while the corresponding chromosome in the gametes of the white-flowered plant carries the allelic gene for whiteness. When copulation occurs, and the egg of one race fuses with the male nucleus of the other, the nucleus of the zygote contains, besides the other chromosomes which can be left out of consideration, the homologous chromosomes containing respectively the genes for redness and for whiteness. Similarly, in the nuclei of all the cells of the diploid organism which subsequently develops by meristematic activity from the zygote is one chromosome carrying the gene for redness and the homologous chromosome with the gene for whiteness (see Fig. 279). Thus such a crossing of two races gives rise to offspring, or hybrids, which must contain both genes. It is customary to designate the parental generation by the symbol P, the offspring or filial generation by F_1, and the next generation by F_2. We can then say that all the plants of the F_1 of a cross of the kind described above will be uniform in appearance (Mendel's first law of the uniformity of hybrids). This law is usually true even when the cross is made reciprocally, i.e. when both races are made to serve in turn as the female parent, the offspring being the same whichever way the cross is made (but see p. 333).

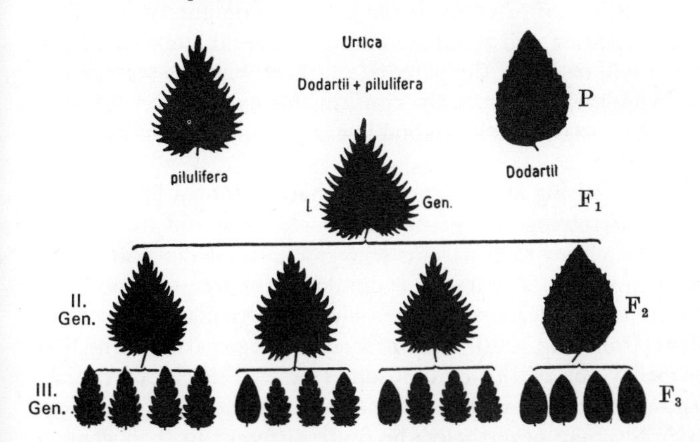

Fig. 280. Leaves of the hybrids of *Urtica pilulifera* and *U. dodartii* through three generations (the parental generation (P), followed by F_1, F_2 and F_3). Diagram after Correns.

In our particular example the flowers of the F_1 generation are pink, so that in respect of this character the hybrids are midway in appearance (or phenotype) between the two parents. Inheritance in this case is said to be intermediate, since apparently the two genes oppose or cooperate with each other with equal force. It is also possible for one gene to suppress the functioning of the other, i.e. one is dominant while its allele is recessive. With dominance (Fig. 280) the phenotype of the F_1 generation is determined by the dominant gene, although the recessive allele is in no way destroyed. It remains undetectable externally, but, as can be seen, immediately manifests itself again in the next generation. Thus, where one character is dominant, all the hybrids have the same phenotype as the parent containing the dominant gene. Where redness was dominant over white, for example, all the F_1 would have red flowers.

It can also happen that the F_1 generation repeats or resembles the phenotype of neither parent, but displays a wholly new appearance. Thus certain white-flowered races of pea crossed together give rise to red-flowered hybrids (see pp. 324 and 325). Such offspring are said to be hybrid novelties. Many hybrids are much more vigorous than their parents, a phenomenon known as hybrid vigour, or heterosis. Sterility is a frequent consequence of hybridity (see p. 325).

Often therefore the true character of hybrids cannot be known until the F_2 generation has also been considered. In our first example (Fig. 279) all the nuclei of the F_1 generation contain a pair of homologous chromosomes which are not identical (or homozygous), but differ in the genes for flower colour (and are thus said to be heterozygous in respect of this character). The sexual cells of the hybrid are thus not of all one kind, but of two, and since the probability of each kind being formed is the same, there will be as many containing the gene for redness as for whiteness. If now the flowers of the F_1 generation are pollinated among themselves (or simply, as is sometimes possible, e.g. in peas, beans, the flowers selfed) the following combinations, whose frequencies are predictable purely from the laws of chance, are possible. First, pollen with the gene for whiteness has an equal chance of causing the fertilization of an egg cell with the gene for whiteness as one with that for redness. In the first case the zygote gives rise to a plant which is homozygous, i.e. both chromosomes of the homologous pair contain the gene for whiteness, and this character alone is transmitted to its descendants. In the second the plant generated is heterozygous in respect of the colour character in exactly the same way as is the F_1 generation. It has identical hybrid character, and since the inheritance is intermediate, it possesses pink flowers. Secondly, pollen with the gene for redness has a similar equal chance of causing fertilization of the two kinds of egg cell, and of giving rise to plants which are either homozygous in respect of the colour character, possessing red flowers and breeding true, or heterozygous and identical in appearance and behaviour with the F_1 generation.

Thus in the F_2 generation there is a separation of the two genes, this separation manifesting itself phenotypically. Three kinds of plants are found, one group resembling one grandparent of the P generation, another group the other grandparent and finally a third group which are hybrids and identical with their parents in the F_1 generation. With a sufficiently large sample of gametes and of offspring it can be shown on purely statistical grounds that 25 per cent of the offspring will resemble one grandparent, another 25 per cent the other, while 50 per cent will resemble the parents, i.e. there will be a segregation in the ratio 1:2:1. Breeding experiments repeatedly confirm this expectation (cf. Fig. 279). This phenomenon is referred to as Mendel's second law of the integrity of gametes, or as Mendel's law of segregation.

It can easily be demonstrated by raising an F_3 generation that, as shown in Figs. 279 and 341, the plants of the F_2 resembling the grandparents are now and henceforth homozygous, while the hybrids resembling their parents segregate in the F_3 in exactly the same way as their parents did in the F_2. If a character is dominant the frequencies of the phenotypes are, of course, different, and segregation then always takes place in the ratio 1:3, as is immediately evident from Fig. 280. The process of segregation, which is entirely dependent upon the behaviour of the chromosomes at meiosis, is, of course, fundamentally the same. It shows very clearly that plants which are identical or closely resemble each other phenotypically may nevertheless be quite different in their genetic constitution, i.e. genotypically. Thus although the majority of F_2 generation of the *Urtica* experiment shown in Fig. 280 have a closely similar leaf form, only 25 per cent are homozygous, while 50 per cent are heterozygous and consequently will segregate in subsequent sexual reproduction. In such cases it is impossible to tell whether similarity of phenotype indicates identity of genotype without breeding experiments. This clearly demonstrates that what is inherited are not characteristics themselves, but merely genetic potentialities, i.e. the capacity to function in certain circumstances in a definite manner.

Back-crossing. In hermaphrodite plants the hybrids in the F_2 generation can be recognized after selfing by their segregation in the F_3. When the sexes are separated, however, a useful technique is to back-cross the suspected hybrids with that parent which is homozygous for the recessive gene. Hybrid plants of both sexes will have gametes of two kinds. In the case of the stinging nettle, for example, one will have the dominant gene for toothed leaves and the other the recessive gene for entire leaves. One parent, however, produces gametes containing only the chromosome with the recessive gene for entire leaves. Consequently, when the hybrid is back-crossed with this parent the resulting zygotes must contain either both the dominant and the recessive genes, i.e. they are heterozygous with respect to the leaf character, or only the recessive genes, i.e. they are homozygous. The offspring of such a back-cross will accordingly contain some 50 per cent of individuals again hybrid in character in which the dominance of the toothed condition is repeated, and some 50 per cent completely homozygous with entire-margined leaves. We thus obtain a segregation in the ratio 1 : 1. Back-crossing will be referred to again later on in consideration of the inheritance of sex (see p. 327).

Dihybrid crossing. The examples quoted hitherto have considered hybridization between partners differing only in a single character. Such hybrids are called monohybrids. The heredity becomes more complex when the parents differ simultaneously in two or more characters. It will first be assumed that the genes responsible for these characters are located in different non-homologous chromosomes. Hybrids heterozygous for two characters are known as *dihybrids*, and those heterozygous for several characters, polyhybrids. The genetic consequences of crossing both dihybrids and polyhybrids are directly dependent upon the manner in which the chromosomes are distributed at the reductional division.

Let us assume, for example, that two races of *Antirrhinum majus*, one with white and zygomorphic flowers and the other with red and radially symmetrical, are crossed. The genes for redness and for normal zygomorphic flowers are known to be dominant. It is customary in genetics to denote genes by letters, the dominant form of a gene being represented by a capital letter, and the recessive by the same letter in lower case. Thus the gene for redness would be represented by *R* and that for whiteness by *r*, and the gene for zygomorphic flowers by *Z* and that for radially symmetrical flowers by *z*. It is assumed, as mentioned above, that the genes for flower colour are not located in the same pair of homologous chromosomes as the genes for flower form. The nuclei of the *Antirrhinum* with red radially symmetrical flowers are therefore denoted by the symbol *RRzz*, and those of the race with white zygomorphic flowers by *rrZZ*, leaving the remaining chromosomes out of consideration. Of course, it is also possible, as is done in the combinational diagram of Fig. 281, to indicate the chromosomes and their genes pictorially. After the reduction division the gametes of the red race contain the alleles *Rz*, but those of the white race *rZ*. The nuclei of the hybrid plants of the F_1 resulting from the fusion of these two gametes consequently have the genetic constitution *RrZz*, and as a result of the dominance possess flowers which are both red and zygomorphic. When meiosis once again takes place in these hybrids and sexual cells are formed in their flowers, the following four combinations of the two pairs of alleles in question are present in both the male and female cells: *Rz, rZ, RZ* and *rz*. When such a plant is selfed the four different kinds of male and female cells are capable of forming sixteen different combinations. These are best represented in a 4 x 4 table (Fig. 281).

As will be seen, the F_2 generation contains nine plants with red zygomorphic flowers (squares 2, 4, 5, 6, 7, 8, 10, 13, 14), resembling in their phenotype the hybrids of the F_1 generation. In addition are two groups, each containing three plants with flowers resembling those of the grandparents, namely those with red radially symmetrical flowers (squares 1, 3, 9) and those with white zygomorphic flowers (squares 12, 15, 16). Finally, a completely new combination with white radially symmetrical flowers is now present (square 11). The frequencies of the different phenotypes segregating as a result of selfing the F_1 fall into the ratio 9 : 3 : 3 : 1. However, of each group of three with the same phenotype as a grandparent there is, as can be seen from the genetic formula and as can be readily verified by breeding experiments, only one member which is homozygous and actually genetically identical with the grandparent (squares 1, 16), the others being genotypically different (squares 3, 9, 12, 15). Of the group of nine with red zygomorphic flowers resembling the F_1 generation, one alone is homozygous (square 6) and therefore

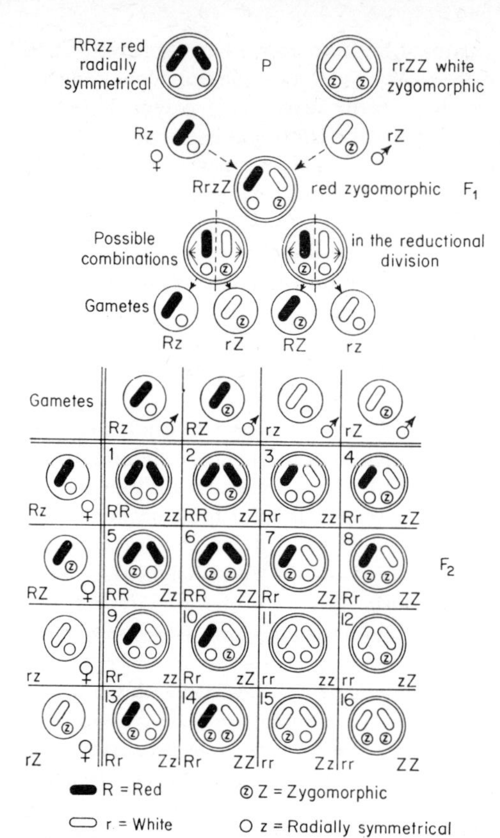

Fig. 281. Diagram showing the result of selfing a hybrid heterozygous for two independent characters. Explanation in text.

constitutes a new genetic combination. Likewise the new combination *rrzz* (square 11) is homozygous and true breeding. New combinations of this kind are of great importance, particularly in plant breeding, for the synthesis of new races. They provide a means of obtaining, by planned crossing experiments, new true-breeding combinations of desired genetic factors. If, for example, a desirable character occurs in one race among other characters of no value it is frequently possibly by hybridization to remove the desirable character and combine it with better ones. Contemporary plant breeding makes much use of this method.

Linkage. As will have been seen, the particular segregating properties of dihybrids also follow directly from the distribution of the chromosomes at the reductional division. This is also true of polyhybrids. So long as the pairs of genes concerned lie in different non-homologous chromosomes no other result is possible, and this leads to the third Mendelian law, that of the autonomy of genetic characters, or expressed in another way of the independent assortment of the genes. This law, also put forward by Mendel in ignorance of its cytological foundation, depends for its validity on the genes being located in different chromosomes. If two genes are located in the same chromosome, then they remain together in the reductional division; they are said to be linked. Numerous breeding experiments have shown that there are only as many groups of independently segregating characters as the haploid number of chromosomes.

The closer analysis of linkage and its complications has made it possible to speak with accuracy about the location of the genes in the chromosomes. Crossing experiments between races of stock, for example, differing in not less than eight characters, have shown that the pattern of segregation in the F_2 resembles that of a simple monohybrid. This is because the genes responsible for the eight characters are linked and lie together in a single chromosome.

Recombination. There are exceptions to this law of linkage. In both plants and animals it has been observed not infrequently that disturbances occur in the linkages of whole groups of genes. A gene or group of genes, for example, which hitherto has always been passed on as a member of a quite definite assemblage of genetic factors, will suddenly appear with another group. Such cases of genetic exchange between homologous chromosomes are accounted for by the phenomenon of crossing over in the bivalents at meiosis. The conjugated homologous chromatids break at the reductional division, and the fragments reunite in a new way (cf. Fig. 31). Thus in a heterozygote a group of genes located in one chromosome can be suddenly coupled with another group in the homologous chromosome, giving rise to a new combination which is handed on as such in subsequent heredity. The further apart two genes lie in a chromosome, the greater is the probability that they will be separated by crossing over, and appear in different chromosomes. The frequency of crossing over between two genes thus gives a measure of the distance between them. Another way by which pieces of chromosomes can be detached is by X-irradiation, and these fragments frequently attach themselves to other non-homologous chromosomes (see p. 332). Detailed investigations along these lines have made it possible to determine the sequence of certain genes along the chromosome, and have demonstrated that the genes, like the chromomeres, must be arranged in a linear series like a string of beads (Fig. 29). For several plants (e.g. *Zea mays*) it has even been possible to make chromosome maps indicating exactly the relative positions of many individual genes. The exact nature of a gene is not yet clear.

Pleiotropy, polygeny, polymery and lethal factors. One and the same gene can often influence several characters in the organism (pleiotropy, or polypheny of the gene). This is easily understandable, since a gene never influences a character directly, but always through a complex chain of reactions which may produce diverse external manifestations.

On the other hand, one single character of a plant may be dependent upon and influenced by several different genes (polygenic characters). If several non-allelic genes cooperate, so that it is merely the intensity of the character that is increased, the character is said to be polymeric. Thus the red colour of certain varieties of wheat can be polymeric in nature, since three different pairs of genes (*Rr, Ss, Tt*) cause red coloration, and any of the dominant genes alone causes a certain amount of redness. Plants with the constitution *rrsstt* have white grains, those with *Rrsstt* pale-red grains, the colour becoming intensified with *RrSstt*. The grains of plants with the constitution *RrSsTt* or *RRSSTT* are deep red.

A very remarkable example of genic interaction is provided by the hybridization of white-flowered and red-flowered races of *Linaria maroccana*. The F_1 generation possesses violet flowers, and is thus a hybrid novelty (see p. 322), while in the F_2 segregation takes place resembling that of a dihybrid. There are in fact two different pairs of genes involved, located in different chromosomes. One pair of alleles (*F, f*) decides whether anthocyanin will be present or not, while the other pair of alleles (*A, a*) determines whether the cell sap will be neutral (*A*) or acid (*a*). The white-flowered grandparent must be of the constitution *AAff*, i.e. the neutral cell sap lacks pigment, while the red-flowered grandparent has the constitution *aaFF*, i.e. anthocyanin is present in an acid cell sap and consequently has a red colour. In the F_1 generation, however, which will have the genetic constitution *AaFf*, anthocyanin will be present in a neutral cell sap, since *A* is dominant, and as a result the flower is violet.

In many diploid plants there are recessive alleles which, if homozygous, would deleteriously affect certain vital activities and might even bring about death. These are referred to as lethal genes, or since the situation is often complex as lethal factors. They may interfere, for example, with chlorophyll formation. So long as the diploid plant contains the normal dominant allele, the dangerous effect does not become apparent, but if the lethal gene is present twice, i.e. if the plant is homozygous for it, it is no longer viable. Recessive genes which when homozygous reduce, but do not destroy, viability account for the well-known dangers of in-breeding.

Sex-determination and inheritance.[44] Particularly interesting is the fact that in many plants the sex is determined by genes and shows genetic inheritance, obeying Mendelian laws.

a. Phenotypic sex determination. In the Plant Kingdom sex determination is governed by the following general principles. All the cells of animals and plants seem to possess at all times the capacity for both maleness and femaleness, i.e. they are bisexual. Under the appropriate conditions, therefore, they can produce both male and female sexual organs or cells. In many living forms the actual sex expressed can be influenced by external factors, or by certain processes within the organism. These may take the form of differential nutrition between cells or regions, or perhaps of metabolic processes arising during growth which promote the development of one sex and suppress that of the other. In instances such as these the sex determination is said to be phenotypic. It is genetic to the extent that the potentiality for the development of sex expression, and the time at which the expression begins to become manifest, are certainly inherited.

Examples of such phenotypic sex determination are provided by hermaphrodite plants, and by those which are monoecious, but have unisexual flowers (diclinous plants). In the typical hermaphrodite angiospermous flower the male and female organs develop at the apex in immediate proximity to each other, while in the diclinous plants the sex organs of the male and female flowers are separated by a considerable distance. The reductional division preceding the formation of the embryo sac and pollen grains has nothing whatever to do with determining the sex of these structures. Indeed, it is still possible, for example, in the pollen grains of *Hyacinthus orientalis*, by subjecting the grains to higher temperatures, to suppress the male tendency and to bring about the formation of an embryo sac with the egg apparatus.

In these cases the sex determination takes place in the diploid phase, but phenotypic determination can also occur in haploid organisms. Examples are provided by many algae (e.g. *Spirogyra, Vaucheria* and many other Chlorophyceae), certain members of the Mucoraceae and Saprolegniaceae, in the leafy gametophytes of hermaphrodite mosses (e.g. *Funaria hygrometrica*), and in the gametophytes of the homosporous ferns, in which antheridia and archegonia arise in adjacent regions of the prothallus as a phenotypic response.

b. Genotypic sex determination. Besides the plants with phenotypic sex determination there are others, often closely related species, in which the sex is determined by hereditary factors. The determination is then said to be genotypic.

Haplogenotypic sex determination. The simplest system of genotypic sex determination is found in haploid plants, such as many dioecious algae, fungi (e.g. many Mucoraceae (*Phycomyces*) and agarics), and bryophytes.

In certain dioecious races of the unicellular organism *Chlamydomonas eugametos* (Volvocales) the diploid zygote, formed by the fusion of two isogamous gametes, undergoes meiosis and gives rise to four haploid swarm cells. These are immediately capable of copulating again, but it becomes clear that the tetrad consists of two pairs of cells of different sex, indicated in this case by the symbols + and −. It follows that a segregation of the sex-determining factors must take place at the reductional division. In *Chlamydomonas* it is possible to infer this genetic division of sex only by the behaviour at copulation, but in some dioecious liverworts it is possible to correlate it cytologically with the separation of two dissimilar but partly homologous chromosomes. As shown in Fig. 282, the spore mother cells of the liverwort *Sphaerocarpus terrestris* contain, besides the usual pairs of homologous chromosomes (the autosomes), another pair distinguished by a remarkable difference in size. Consequently, at the reductional division one cell comes to contain the large chromosome, while the other receives the smaller Y chromosome. Only female plants develop from the two spores containing the X chromosome, while the two with the Y chromosome give rise to male plants. Thus the sex-determining factors appear to be located in the sex chromosomes or heterochromosomes. If a segment be broken off the X chromosome by X-rays (see p. 331) the female becomes transformed into a male. It is possible, therefore, that the Y chromosome has in fact no sex-determining function, the sex expressed depending merely upon the presence or absence of the X chromosome. Only in the diploid sporophyte, where both the X and Y chromosomes are present, is there any potential hermaphroditism, and in certain mosses this has been directly demon-

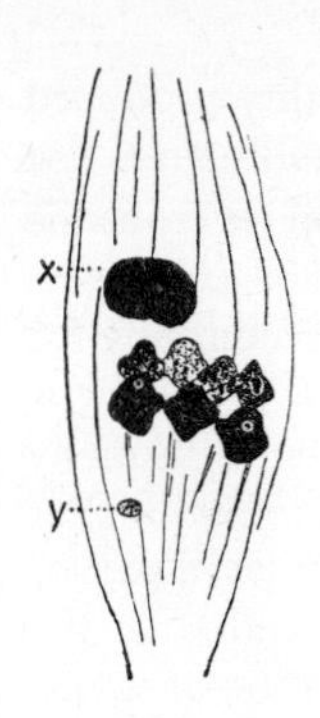

Fig. 282. Reduction division in the spore mother cells of *Sphaerocarpus terrestris*: x, y, the sex chromosomes. ×2300. After Lorbeer.

strated by regeneration experiments. In many other cases, however, the sex-determining chromosomes cannot be distinguished morphologically from the remaining chromosomes.

It must not be supposed that the sex chromosomes themselves contain the factors responsible for the elaboration of the male and female sex organs. They are more in the nature of determiners, which so influence the ambivalent sexuality persisting in all the cells of the haplophase and apparently located in the autosomes, that the expression of one sex is effectively suppressed. Very little is yet known about how these determiners work. The excretion of sexually specific metabolites, and their interaction with the growing system, seems to be one link in the chain of events which leads to the morphological differentiation distinguishing the sexes, and the activation of the sexual regions themselves. Substances which evoke the expression of sexual characters, and thus the development of sex organs, are called termones, while substances which chemotactically promote copulation and fusion of gametes are referred to as gamones. Sexual ambivalence still persists in the gametes, and perhaps as a consequence sexual differentiation is not always as great as it might be. In gametes of the same sex, for example, the strength of the sexuality, measured in terms of the amount of gamone excreted, may vary, so that it is possible to speak of both strong and weak male and female gametes. Since apparently only a certain amount of quantitative difference is necessary for sexual fusion, it is possible in such cases for gametes of the same sex, but of different strength, to copulate with one another. This phenomenon is referred to as relative sexuality.

Sexually specific termones and gamones have so far been demonstrated in a number of algae and fungi. In *Achlya bisexualis* (Saprolegniaceae), for example, the female apparently secretes a steroid which acts as a termone and stimulates the production of antheridia by the male. The female gametes of the brown alga *Ectocarpus siliculosus* liberate a volatile gas-like gamone, and a sesquiterpene (sirenin) has been identified as a gamone associated with the female gametes of fungus *Allomyces* (Phycomycetes). In isogamous *Chlamydomonas eugametos* the gamones, like those of many Metazoa, appear to be glycoproteids of high molecular weight, which act as agglutinins and promote, among other things, the interlacing and interlocking of the flagella of the sexual partners. (See also chemotoxis, p. 382.)

Diplogenotypic sex determination. The situation is different when genotypic sex determination takes place in a diploid organism, as occurs, for example, in the dioecious higher plants. The fundamental mechanism, which holds also for the higher animals including Man, was first elucidated by Correns, who crossed the dioecious *Bryonia dioica* with the monoecious *B. alba*. With haploid organisms sex determination takes place at the reductional division in sporogenesis, as a result of the segregation of the genetic determiners. In diplogenotypic sex determination the sex of the diploid plant is not decided until after the fusion of the two gametes. At subsequent meiosis only in the plant of one sex, usually the male plant, is there a production of two different kinds of gametes, the male-determining and the female-determining. In the other, usually female plant, there is no segregation of genetic tendency and the gametes are of uniform female sexuality. The male plant is consequently said to be heterogametic, and the female homogametic. Only in certain species of *Fragaria* (e.g. *F. elatior*) is it the female and not the male plant which is heterogametic, a situation similar to that in butterflies and birds. In the more usual situation in plants the fertilization of the gametes of uniform female sexuality produced by the female by gametes of female sexuality produced by the male inevitably gives rise to female zygotes, while the fertilization of the female gametes by male gametes from the male yields male zygotes. It is clear that if the male produces its two kinds of gametes in equal numbers the chances of getting male and female zygotes are equal, accounting for the fact that in nature the sexes occur in more or less equal numbers.

That one of the sexes is heterogametic can be confirmed in some plants, as in many animals, with the aid of the microscope. In the white campion, *Melandrium album*, for example, the diploid nuclei of the female plant contain, besides the autosomes, a pair of homologous and similar chromosomes of conspicuously large size (Fig. 283B). In the nuclei of the male plant, on the other hand, is a pair of dissimilar chromosomes, of which only one resembles the X chromosomes of the female, while the other Y chromosome is larger (Fig. 283A). In other organisms the Y chromosome is usually smaller or even, as in certain animals, wholly lacking (e.g. *Rumex acetosa*). These are again referred to as sex chromosomes or heterochromosomes. In the female plant all the spores resulting from the reductional division and the subsequent gametes (egg cells) must alike contain one X chromosome, while in the male plant one-half of the pollen grains contains an X chromosome and the other half a Y chromosome. If we assume that a gene is located in the X chromosome which when present twice (XX) evokes femaleness, but when present only once (XY) maleness, we immediately have a genetic mechanism effectively determining sex (Fig. 284). If an egg cell of the female plant receives an X chromosome from a sperm nucleus, the diploid plant generated has two X chromosomes and is consequently female, whereas an egg cell receiving a Y chromosome from the sperm nucleus gives rise to a diploid plant containing an X and a Y chromosome and is consequently male.

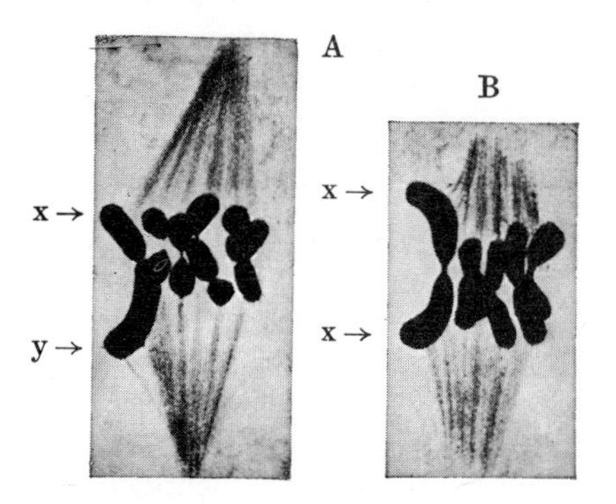

Fig. 283. Reduction division in the ♂ and ♀ plants of *Melandrium album*. **A.** Pollen mother cell with x/y chromosomes. **B.** Embryo sac mother cell with two x chromosomes. ×1800. After Belar.

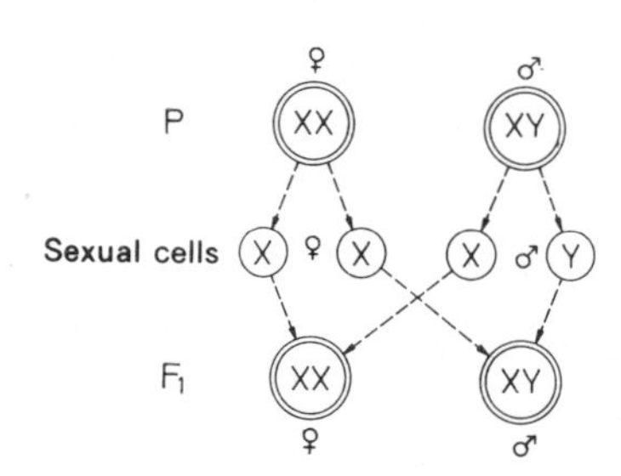

Fig. 284. Diagram showing the mechanism of sex determination in diploid monoecious plants with X/Y sex chromosomes.

In the numerous cases in which the sex chromosomes cannot be distinguished by difference in size or anything similar, there are other ways in which it can be shown that one sex produces two different kinds of gametes. Besides the crossing of dioecious with monoecious species, it can be easily demonstrated that the two kinds of pollen produced by a dioecious species differ in rapidity of germination and growth. If, for example, the stigma of *Melandrium album* is well coated with pollen, so that the number of grains far exceeds the number of ovules in the ovary, it will be found that only a fraction of the pollen tubes produced reaches the ovules, and that the seeds produced yield up to 69 per cent female plants, the remainder being male. This suggests that the pollen tubes from the grains containing the X chromosome grow more rapidly, and consequently reach the ovules earlier, than those from the grains with the Y chromosome. Also the pollen grains of heterogametic plants show a differential resistance to alcoholic poisoning. Thus if the pollen of *Melandrium album* is exposed to alcoholic vapour before it is put on the stigma the progeny contain a surplus of male plants of the order of 10–21 per cent. If pollen that has aged for 80–100 days is used almost only male plants are produced. This gives an experimental method of disturbing in certain circumstances the relative frequencies of the two sexes.

There are other genes in the sex chromosome besides the sex determiner, and the inheritance of these inevitably follows that of sex. These are referred to as sex-linked genes.

Such genes have been studied principally in animals. They are also important in human heredity; certain defects, e.g. colour-blindness and haemophilia, show sex-linked inheritance. The occurrence of such sex-linked characters provides further evidence for the correctness of the explanation of diplogenotypic sex determination advanced in the foregoing. Possible ways in which the sex determiners may function cannot be gone into here.

B. Genes and mutations[41, 45]

From the foregoing it will be seen that the genes, located in the chromosomes and effectively governing the whole of the morphological and physiological properties of the organism, are remarkably stable. Moreover, they remain unchanged throughout cell division and provide, together with a similar high immutability in chromosome form and number, the foundation for the immutability of species for many generations. Nevertheless, this immutability is not absolute, and this phenomenon must now be discussed in some detail.

It will first be necessary to consider the nature of the gene and its action, since in this field it has been possible to gain many new insights in recent years. Even though these have come largely from work on bacteriophages (see p. 445), bacteria, and fungi (since these offer many more promising lines of experimental inquiry than higher organisms), all our experience hitherto indicates that they are equally valid for all forms of life. There can be no doubt that the deoxyribonucleic acid in the chromosomes is the actual bearer of the hereditary information. It has been possible to bring about the incorporation of deoxyribonucleic acid from foreign cells into bacteria, and by this means to cause hereditary changes (transformation). This can also be done by the use of bacteriophages (transduction). In the discussion of protein synthesis (p. 282 seq.), we have already mentioned the ability of deoxyribonucleic acid identically to reproduce itself, and the phenomenon of transcription involved in the production of messenger and transfer ribonucleic acids. Genes appear to be nothing other than segments of the giant deoxyribonucleic acid molecule of the chromosome. Despite the shortness of these segments in relation to the whole chromosome, they nevertheless contain some thousands of nucleotides in a quite definite base sequence. Even for 1000 nucleotides, assuming the four bases to be equally available, there would already be 4^{1000} different possible sequences in the chain, and thus an almost infinite number of ways in which one gene could differ from another. The way in which such a segment of the deoxyribonucleic acid chain, referred to as a gene (or cistron), can influence a definite aspect of the cell, can be envisaged as by way of a corresponding messenger ribonucleic acid, subsequently translated into a particular protein, in all probability with enzymic properties, in the cytoplasm. The enzyme complement of a cell is thus determined directly by the genes. Enzymes, of course, bring about quite definite metabolic reactions, and are hence responsible for the appearance of the corresponding metabolic products. The further reactions which lead from there to the visible morphological characters are still largely a matter of speculation. Nevertheless, it is a decided advance to have obtained insight into the relationship between the structure of a gene and that of the protein in which it is ultimately expressed (transcription, translation; see p. 284 seq.). Using the language of electronics technology, it is customary today to speak of the transference of 'information' from gene to protein.

Now it must be clearly understood that, although every embryonic and meristematic cell contains all the information laid down in the base sequences of the deoxyribonucleic acid, no ordered development would be possible were it all realized at the same time. All development is in fact a sequence of individual steps, in the last analysis probably of individual chemical reactions. There are many indications that, besides the structural genes, whose function has already been described, there exist also regulator genes which are able to control if and when a structural gene is active. The regulator gene is believed to form a so-called 'repressor' (probably a histone protein) which somehow reacts with an 'operator', and opens or closes this like a lock. The operator appears to be a portion of the chromosome immediately adjacent to the structural gene. The operator and structural gene together are referred to as the 'operon'. The repressor, which, together with the operator, suppresses the functioning of the structural gene, can for its part be suppressed in a number of different ways, thus allowing the structural gene to become active. Many hormones appear to work in this way, and in certain conditions even external influences may be effective. Amongst microorganisms, for example, the phenomenon of adaptive enzyme formation, in which an organism can be stimulated to produce enzymes specifically related to unusual substrates offered in the medium, has been known for some time. This can be explained by supposing that the substrate itself reacts with the appropriate repressor, and in this way a previously existing blocking of the operator and its structural gene is set aside. The gene so activated forms a new enzyme, previously not present in the organism. The possibility of its production must, of course, have been present in the deoxyribonucleic acid chain, but up to this time been effectively masked. The long-known fact of developmental physiology that the developmental potentialities of an organism are in general much greater than those which in normal conditions become expressed can be explained in a similar manner. Apparently in some instances the repressor can be activated only by another substance (the 'effector'). Clearly we have only a glimpse into what is certainly a very complicated and finely coordinated regulation system, reaching right down into the molecular structure of the nucleotide sequences we recognize as genes. This branch of genetics is consequently often referred to as 'molecular' genetics. In relation to partial activation of the chromosome it is noteworthy that in the giant salivary gland chromosomes of insects, and in similar chromosomes in some regions of plants (Fig. 19), localized activation can be directly observed in the form of swellings or 'puffs' (see p. 31).

In the introduction it was stressed that a gene possesses a very high stability, and that for many generations identical reproduction is the rule. Nevertheless, it is evident that the hereditary information is not absolutely constant, and any such constancy would in fact be contrary to the concept of evolution. Further, careful observations on completely pure lines have shown that in Nature small changes appear sporadically and unpredictably in a few individuals. These changes are distinct from the variations which arise in response to environmental influences, since once they have appeared they remain hereditary. They must therefore be dependent upon some sort of change in the genetic constitution, and it is known that in certain conditions a change of a single base in the deoxyribonucleic acid chain can have consequences of this kind. Such spontaneous changes in heredity are known as mutations. Figure 285 shows a mutant of *Chelidonium majus* with laciniate leaves. This suddenly appeared in a Heidelberg garden in 1590, and the laciniate leaf has remained a constant feature of this form until the present day. The frequency of such mutations naturally depends upon the species. The percentage of haploid cells carrying a mutation at a given time and temperature is referred to as the mutation rate. In *Antirrhinum*, for example, the mutation rate can amount to several per cent, but is frequently much lower. Mutations may be caused by a sporadic change in a gene in one chromosome, the allele in the homologous chromosome remaining unchanged (gene mutation). When this occurs it is usual for only a very short length of the nucleotide sequence of the gene to be affected (in extreme cases, a single base, the so-called 'point mutation'), although occasionally the modification is more extensive ('block mutation'). Mutations can also be caused by a change in the structure of the chromosome (chromosome mutation), or to change in the number of chromosomes (genome mutation).

A mutated gene is usually recessive, thus making it difficult to detect. Some genes appear to be more labile than others, and thus to mutate more readily. The average mutation rate of a single gene is very low, and probably lies between 0.00005 and 0.005 per cent. It is possible for a gene to occur in several mutated forms, so that races may arise which are distinguished merely by their containing different mutations of one and the same gene (multiple alleles). It is also possible, though rare, for a mutant to change back to its original state. Certain external agents, especially short-wave irradiation such as

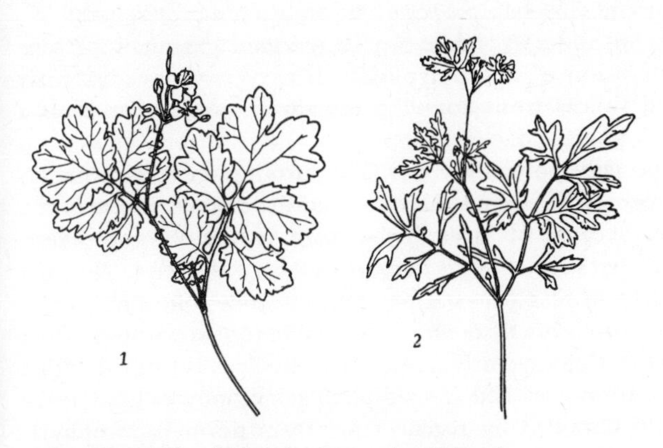

Fig. 285. Habit of *Chelidonium majus* (1.) and of the mutant *laciniatum* (2.). After Lehmann, reduced.

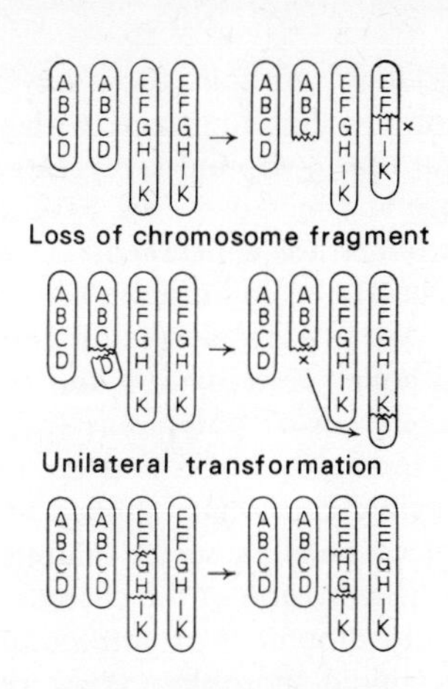

Fig. 286. Diagrams of various kinds of chromosome mutation. The letters indicate different genes in the chromosomes.

ultra-violet light, X-rays and emissions from radium, and also certain chemicals, considerably increase the mutation rate, and a relatively high percentage of gene mutations can be artificially created in this way. Such mutations, which are identical in nature with those occurring spontaneously in natural conditions, play an important part in plant breeding, and also in contemporary physiological genetics. They show at the same time the dangers to which all life is exposed when subjected to mutagenic radiations, such as those arising from atomic explosions and the careless use of X-rays, and to certain chemical substances.

The same holds for chromosome mutations in which whole chromosomes may be fragmented. If a fragment of a chromosome disappears entirely (usually lethal in the haploid condition) the effect is termed deletion. If a fragment attaches itself to another chromosome or portion of a chromosome, so that the total number of the genes in the nucleus remains unchanged, then this is termed translocation (Fig. 286). Sometimes the rejoining may take place such that certain genes are represented twice in the reconstituted chromosome (duplication). A translocation can also occur between two non-homologous chromosomes, either unilaterally or reciprocally, and must not be confused with chiasma formation (see p. 40). When a piece broken out of one chromosome subsequently unites with another, so that its chromomeres and genes lie in a sequence the reverse of that originally present, an inversion is said to be present. Changes of this kind in individual chromosomes can be observed directly with the microscope in suitable material. In some cases inversions and translocations affect not only the pairing behaviour of the chromosomes but also the physiognomy of the plant. This is despite the fact that only the position and spatial interrelationships of the genes have been changed, and not their condition in the cell. This so-called 'position effect' indicates the existence of interactions between neighbouring genes, and its explanation is perhaps to be found in what has already been said concerning regulator and structural genes (p. 330).

Genome mutation.[41] Finally, it is also possible for irregularities to occur at nuclear division in the distribution of the chromosomes to the daughter nuclei, so that cells arise which deviate from the norm in chromosome number (heteroploidy). Since the genes located in a haploid set of chromosomes are referred to collectively as the genome, such irregularities are spoken of as genome mutations. If individual chromosomes are lacking, or are represented more frequently than normally, the nucleus is said to be aneuploid. If structural modifications within the chromosomes lead to step-like changes in chromosome number, the condition is said to be one of dysploidy. If the whole genome is represented with a frequency higher than normal, the condition is termed polyploid. The causes of polyploidy lie in disturbances of mitosis or meiosis, partly also in endomitosis (see p. 36). Tetraploid individuals can arise from the copulation of diploid gametes produced without the intervention of meiosis, and triploid individuals from the copulation of a diploid gamete with a haploid. The haploid set of chromosomes is designated 'n', and thus a triploid is represented '3n' and a tetraploid '4n'. If the same genome is present several times over, we have a condition of autopolyploidy. However, where dissimilar genomes are concerned (as would follow from doubling the chromosome number of a hybrid), the condition is termed allopolyploid.

Aneuploidy, dysploidy and especially polyploidy can all be produced experimentally. Although X-irradiation and treatment with radium have been of little practical use in the generation of polyploids (perhaps because of inevitable damage to the chromosomes), temperature shocks and the use of certain chemicals have proved very effective. Thus the alkaloid colchicine, applied in the form of a 0.1–0.2 per cent aqueous solution to stem apices, buds, or seeds, inhibits the formation of the mitotic spindle (but is without effect on the division of the chromosomes). Consequently a restitution nucleus is formed with a doubled number of chromosomes. In mosses and ferns diploid gametophytes (and hence tetraploid sporophytes) are readily obtained by regeneration from normal sporophytic tissue.

From the close interrelations of nucleus and cytoplasm (particularly in regard to protein synthesis) it is not surprising that the increase in nuclear volume resulting from polyploidy has a decided effect on cell metabolism. Artificially created polyploids, for example, are usually distinguished by larger organs and cells. On the other hand, they often also show delayed development, reduced vigour, and impaired fertility. In natural polyploids these disadvantages have often been overcome by reduction in the sizes of the chromosomes, nuclei and cells.

Contrary to the situation in the Animal Kingdom, polyploidy, particularly alloploidy, plays a considerable role amongst plants in Nature. Alloploidy allows the combination of the hereditary traits of two different individuals resulting in an organism of increased adaptability, of considerable advantage in the colonizing of virgin areas (see p. 415 seq.). Allopolyploidy is thus not surprisingly very frequent in northern latitudes and amongst weeds (e.g. *Capsella bursa-pastoris, Scleranthus annuus*, etc.). Many of our cultivated plants are also polyploid. Rape, for example, is an allotetraploid derived from *Brassica oleracea* and *B. rapa*. Similarly the cultivated plum (*Prunus domestica*) is a hexaploid derived from a hybrid between *Prunus cerasifera* (2n) and *P. spinosa* (4n). Hybrids between species of the same genus or even of different genera (specific or generic hybrids) are frequently sterile, but following a doubling of their chromosome numbers become fertile. New species may even be generated in this way. Higher yielding varieties of wheat, oats, potatoes, tobacco, strawberries, etc., have been obtained by hybridization followed by induced polyploidy. Many large-flowered varieties of hyacinth, tulip, narcissus and so on are similarly polyploid.

Most of the mutations well known today, be they gene or chromosome mutations, cause only minor changes in the plant containing them. The extent of the phenotypic divergence they cause corresponds only to that of a variety. Moreover, the effect of mutation is usually deleterious or even lethal, especially if the mutated gene is present in the homozygous condition. Nevertheless, they possess very great theoretical importance for the study of evolution (cf. p. 397 seq.).

C. Extra-chromosomal inheritance[46]

Although there can be no doubt that the most important hereditary factors are localized in the chromosomes, and that heredity depends principally upon the way in which the chromosomes are redistributed at every cell division, it should not be assumed that that part of the cell lying outside the nucleus plays no part in inheritance. There must in fact be unique interrelationships, still little known, between the intimate structure of the gene and that of the cytoplasm. Only the initial interrelationships, namely those between the base sequences of the deoxyribonucleic acid and protein synthesis, are in any way understood, as has already been described. Of the subsequent events, leading up to morphogenesis, we are still largely ignorant, but it is quite evident that chromosomes can only be effective in the presence of the plasma. There is now, in both the Animal and Plant

Kingdoms, accumulating evidence that the plasma itself contains certain hereditary traits which, like the genes, are capable of self-reproduction. Although in most cases it is still unknown what kind of element is involved, in the Plant Kingdom the plastids, especially chloroplasts, undoubtedly form an independent, extra-nuclear, hereditary system. As is well known, the chloroplasts possess a certain autonomy, they reproduce themselves, possibly exclusively, by division, and always contain some deoxyribonucleic acid, which could well be capable of auto-replication. Ribonucleic acid and ribosomes are also found in plastids. Another site of hereditary traits might be mitochondria, since these also contain deoxyribonucleic acid, but little firm evidence on this point is yet available.

Plastid inheritance. The hereditary system located in plastids is termed the plastome or plastidome. It is unquestionably extra-nuclear and independent.

At cell division the plastids are not distributed between the two daughter cells with the exactness of the chromosomes, but nevertheless each receives more or less the same number. If little or no paternal cytoplasm enters the egg cell at fertilization, then all the plastids in the zygote and the plant developing from it must be derived from the mother. Cases are known, however, where besides the sperm nucleus, a few plastids are also brought in. Chloroplasts are also known in which, apparently by a kind of plastid mutation, the capacity to develop pigment is either wholly suppressed or can be realized only in the presence of a certain maternal or paternal genome. An egg cell may thus contain two sorts of plastids, one capable of developing chlorophyll and the other not, obtaining them either directly in the formation of the egg or subsequently from the paternal cytoplasm in fertilization. The distribution of the two kinds of plastids in the ensuing cell divisions is random, so that a mixture of cells may result. Cells arise which contain either wholly or predominantly pigmented plastids, while in others only the unpigmented are present. Veined or sectorially variegated forms arise in this way (although some variegated forms result from viral infections (see p. 299)). If the plastids are derived solely from the mother the ensuing purely maternal inheritance provides an explanation of certain instances in which reciprocal hybrids show plastid differences.

Cytoplasmic inheritance. Certain observations indicate that yet other independent and self-reproducing hereditary units exist in the apparently undifferentiated cytoplasm, perhaps partly to be identified with the mitochondria, but also possibly located in extra-chromosomal ribonucleic acid. The first Mendelian law (the uniformity of hybrids) does not hold without qualification for reciprocal hybridization (see p. 321). Cases are in fact known where hybrids not only partly resemble the mother but also differ reciprocally in quite other ways as well. This may be explained by the presence in the nucleus of genes which are activated or inhibited according to the specific nature of the cytoplasm in which the nucleus lies.

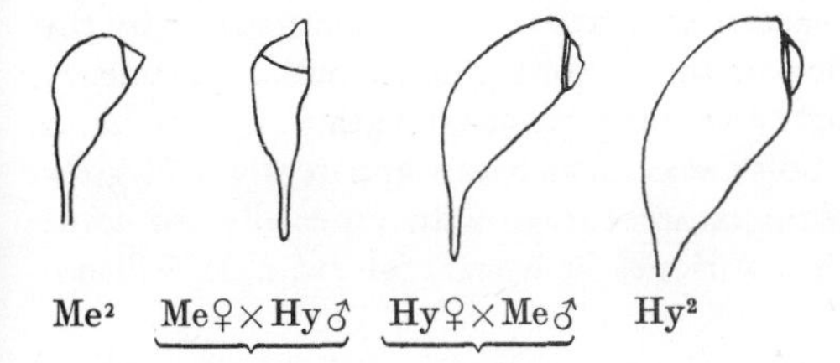

Fig. 287. Left, sporogonia of the pure species *Funaria mediterranea* (Me^2). Right, the same of *F. hygrometrica* (Hy^2). In the middle the sporogonia of the reciprocal hybrids. After F. von Wettstein.

Thus in mosses reciprocal hybrids have been produced, not only between species but also between genera, which differ quite clearly in respect of the resemblances they show to their maternal parents (Fig. 287). So far as observations have been made, these differences persist through many generations if the hybrid is propagated by back-crossing to the male parent. Reciprocal hybrids differing in a similar way are also known in both fungi and higher plants (e.g. *Epilobium, Streptocarpus*, etc.). The cytoplasm of the zygote derives principally, of course, from that of the egg cell. Since reciprocal hybrids are usually identical, it follows that specific genetic factors are rarely to be found there.

When a difference does appear in reciprocal hybrids it points, except where there is transference of plastids with the male cell, to a specific difference of the parental cytoplasms.

These phenomena are referred to as cytoplasmic (or plasmatic) inheritance, and the genetic factors located in the cytoplasm are referred to collectively as the plasmone. Only recently, e.g. with yeast cells, have investigations been undertaken to investigate plasmatic inheritance in greater detail.

IV. The influence of the environment[8]

Although in the preceding sections certain internal factors have been considered decisive in controlling the growth and development of living organisms, there is no doubt that growth can also be effectively influenced (though to a variable extent) by certain environmental factors. The extent to which this occurs in nature is in fact little known, and is being intensively investigated. We are here concerned principally with factors which are able to modify developmental processes or potentialities already present, not with those that cause fundamental genetic change. Variation of the former kind is called modification (see p. 340).

An uninterrupted development of all the potentialities within an organism naturally depends upon the environmental factors being optimal. An adequate supply of water, minerals, carbon dioxide and light are clearly the prerequisites for normal growth. Moreover the temperature, influencing the rates of biochemical reactions, will exert a profound effect upon growth.

Besides these general effects of the environment, individual environmental factors can in certain circumstances have specific morphogenetic effects upon plants. These formative influences often act almost imperceptibly, along with the other external and internal factors controlling growth. Some will be briefly considered in the following sections.

A. Temperature[8, 47]

Within a certain range, increase of temperature frequently causes an increase in the rate of growth. If the rate is plotted against temperature the curve obtained, as in many physiological processes, is not linear, but falls after reaching a maximum value (Fig. 288). Here, as with other physiological rate curves, three cardinal points can be distinguished. First is the minimum temperature, below which there is no growth, although this temperature is by no means the same as that below which life itself ceases. Second is the optimum or optimal temperature, at which the growth rate is at a maximum, and lastly the maximum temperature, above which growth generally ceases. The origin of such curves can be attributed to an initial acceleration of the reaction in a more or less linear fashion by the factor investigated, followed by a diminution of the factor's effectiveness as secondary inhibitory processes come into play with progressively increasing strength.

If plankton and certain other algae of polar seas (which can apparently still grow below 0°C) are excluded the minimum temperature for growth, especially for some cultivated plants, is surprisingly high. Summer varieties of wheat, for example, will not

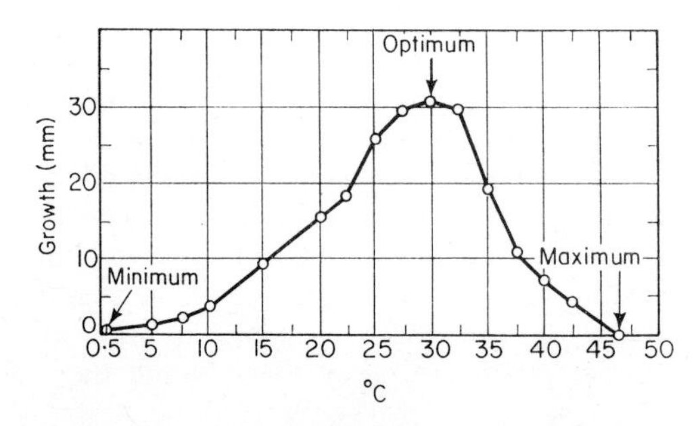

Fig. 288. Growth of a root of *Lupinus luteus* in 24 hr at different temperatures. Abseissa: temperature in °C; ordinate: growth in mm. After Vogt from Jost.

grow below 5°C, maize 8°C, cucumbers 12°C and tobacco not below 13°C. The tubercle bacillus is dormant below 30°C. Other bacteria and fungi, such as those causing heating in hay stacks (see p. 263), may not begin to grow until temperatures are reached which would be lethal for Man. The optimal temperature for growth is also extraordinarily variable. The same is true of maximum temperature; although for many plants it lies in the range 45–55°C, in thermophilous bacteria and blue-green algae (such as those in hot springs) it may be of the order of 70–85°C. These organisms must contain special proteins, with tertiary and quaternary structures (see p. 279 seq.) which are not denatured, as are most, by these high temperatures. Meristematic activity and extension growth likewise have an optimum temperature, above which the rate declines. The optimal temperatures for many plants show a diurnal variation, i.e. the plants are adapted to a change in temperature between day and night (thermoperiodism), and show optimal development only when such a fluctuation occurs.

We still have very little insight into the processes which are involved in the effect of temperature on growth and other physiological processes. Besides the acceleratory effect upon reactions, familiar from physical chemistry, changes in the degree of aggregation of the protoplasm and its components should probably also be taken into account. The fact, for example, that in many plants the minimum temperature for the formation of chlorophyll is actually higher than that for growth, so that they are yellowish if grown at a low temperature, evidently indicates that individual metabolic processes are influenced in quite different ways. The optimum temperature for growth and development can perhaps be regarded as that point at which the rates of the various processes going on in the cells are such that coordination is at its most effective. The serious damage, particularly in tropical plants, occurring at temperatures round about 0°C, probably results from their failure to tolerate changes in the rates of the individual, but interacting, physiological processes.

The particular damage caused by frost probably results from the freezing reducing the hydration of the plasma colloids, leading either to a denaturation of sensitive proteins, or to a change in the membranes of the chloroplasts, mitochondria, and other membrane-bound organelles. Visible ice crystals usually form in the intercellular spaces, not in the protoplast itself. In many cases, although water is taken up again on thawing, normal hydration is never regained, especially when the thawing is rapid. The very different frost resistance of different species (a property often showing variation throughout the year) seems to depend upon the extent to which complexes of protein and carbohydrate (particularly sucrose) are formed in the cytoplasm. This process can apparently be promoted artificially by slow acclimatization to low temperatures, the practice referred to as 'hardening'. The frost damage to unhardened plants seems to result principally from an inhibition of the formation of ATP during photosynthesis and respiration. Many plants can endure astonishingly low temperatures without showing any damage on thawing. Observations show that many forest trees are able to withstand winter temperatures of −60°C without harm, and *Cochlearia fenestrata*, a herb native to northern Siberia, can tolerate −46°C. In conifers night frosts often of course cause a prolonged suspension of carbon assimilation, even though the temperature may rise above 0°C during the day. It has been shown experimentally that certain diatoms can survive after cooling to −200°C, and similar results have been obtained with many fungal mycelia, and especially with bacterial spores, where the protoplast already contains only very little water. In this condition seeds and spores are also able to tolerate quite high temperatures, even over 100°C, whereas an active protoplast saturated with water would rapidly die at a much lower temperature through denaturation of the protein. The so-called pasteurization of milk or other heat-sensitive substances rests upon the fact that heating to about 65°C kills all the growing bacterial forms, but without doing damage to the spores. If the process is repeated after a few days the spores will in the meantime have germinated, and it is possible to secure an effective sterilization at quite low temperatures (fractional sterilization).

In many plants the effects of brief treatments with extreme temperatures are not immediately apparent, but marked changes in the rate and nature of development set in at a later stage.

The so-called biennials, for example, normally produce in the first year only a rosette of leaves and a subterranean storage organ (often only in the correct conditions of day length, e.g. the 'long day' plants (photoperiodism; see p. 344)), the flowers appearing in the following year. The development of the inflorescence does not take place unless the growing point has been subjected to low temperatures, and in natural conditions this occurs in the winter season between the two growing periods. In the tropics such plants can remain in the vegetative condition for several years, and may even entirely fail to flower. Examples are provided by species of *Beta* and *Brassica*, many Umbelliferae, the biennial race of *Hyoscyamus niger*, etc. In many plants this effect of cold can be reproduced by treatment with gibberellins, but the relationship between these growth substances and flowering is still rather obscure (see p. 315).

Winter varieties of cereals (e.g. winter rye) have to be sown in the autumn if they are to flower in time in the succeeding summer. Here the winter cold produces merely an acceleration of the later stages of development, so that the plants come into flower some two to three weeks earlier than they otherwise would, a factor which can be decisive for a successful harvest. Even the embryos in the soaked grains can be stimulated by cold, and despite subsequent higher temperatures, the stimulus is firmly retained for a considerable period. If grains of winter varieties of cereals, partially saturated with water, are kept for some weeks in well-aerated conditions at a temperature of 2—3°C, and subsequently sown in the spring, the plants developing from them come into flower as rapidly as spring-sown varieties. If the cold treatment is followed immediately by warming its effects are reversed, but if the warming is delayed the effects of the cold on the embryo become stabilized and persistent. This phenomenon is known as vernalization (or jarowization), and it finds some practical applications. On the other hand, to accelerate the onset of flowering and fruiting of cotton, soya bean and millet it is necessary to subject the seeds to a preliminary heat treatment.

Sometimes in the normal growth of a plant there are short phases when it is specially sensitive to temperature, and in general the different stages of development have different sensitivities. Thus in many plants, e.g. *Petunia*, the pigmentation of the fully expanded flowers can be influenced by the temperature of the bud during a brief preceding phase. Light may also play a part here. The gregarious flowering of certain orchids and other plants (such as coffee and many bamboos), often observed in the tropics, appears to follow a brief cold shock, such as cooling caused by a heavy thunder-shower. This stimulates large numbers of bud primordia which had become dormant to sudden simultaneous development. Many seeds (e.g. those of Alpine plants) will not germinate unless they are first frosted (see also p. 338).

We have no detailed knowledge of the mechanisms by which temperature causes any of these effects, either the immediate or the delayed.

B. Light[8]

Light, which is already of fundamental importance to autotrophic plants because of photosynthesis, also has manifold effects upon growth. While there is a minimum temperature below which growth will not occur, light is by no means essential for the growth of all plants. There are numerous organisms (e.g. bacteria and fungi) and organs (e.g. roots) which are capable of developing in the complete absence of light. Of course, where light is a normal component of the environment, very remarkable and profound effects follow its removal. These effects are not only concerned with the amount of growth but also with consequential changes in the whole morphology of the plant.

If for example two higher plants are reared in similar environments, but one in normal daylight and the other in complete darkness, certain characteristic changes appear in the darkened plant, referred to collectively as etiolation. In dicotyledonous plants the internodes, and frequently also the petioles, become very long, while the laminae of the leaves remain quite small and rudimentary (Fig. 289). Moreover, the whole plant is pale yellowish in colour, since chlorophyll formation is usually inhibited in the dark.

Differentiation within the plant, especially that of the sclerenchyma, is also extensively suppressed, so that etiolated shoots have very little mechanical strength. An endo-

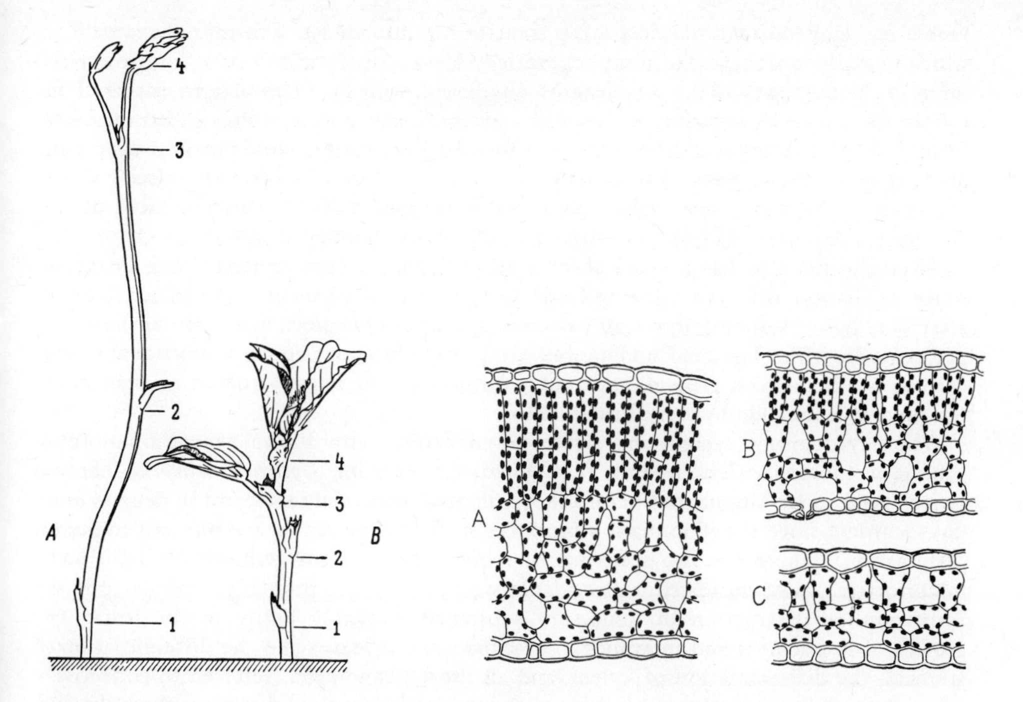

Fig. 289. Seedlings of *Vicia faba* at three weeks. **A.** Grown in dark. **B.** Grown in light. The numbers indicate the corresponding nodes.

Fig. 290. Transverse section through a leaf of *Fagus silvatica.* **A.** Sun leaf. **B.** Leaf enjoying moderate illumination. **C.** Shade leaf. ×340 approx. After Kienitz-Gerloff.

dermis is often formed. In monocotyledons it is the leaves rather than the shoots which become elongated. The typical etiolation phenomena of higher plants can be prevented by exposing them to light for only a few minutes each day, so that these phenomena cannot be caused by a shortage of carbohydrates following the deficient photosynthesis. Many (but not all) agarics, which are not usually dependent upon light for their nutrition, also show typical etiolation effects when deprived of light. The stalk of the fruit body elongates and the cap is reduced. Lack of blue light is particularly effective in this respect, while in higher plants a deficiency of red light also causes etiolation, especially the failure of the leaf laminae to expand. In weak light fern spores germinate to give long, undivided filaments; increasing the light intensity causes first transverse divisions and then longitudinal divisions, leading to a flattened plate of cells and ultimately to the cordate prothallus.

The fact that marked extension growth follows upon darkening, frequently enabling a plant in natural conditions to reach more adequate illumination, suggests that normal illumination actively inhibits such growth.

This seems to be the reason why plants of high altitudes, where the short-waved radiation is more intense, usually have a lowly habit. Typical Alpine plants if transplanted to the lowlands frequently undergo so much extension growth that their appearance is markedly changed (see also Fig. 292). That many plants grow more during the night than during the day may also be attributable in part to this effect, since plasmatic growth also appears to be inhibited in the light.

It has been possible in recent years to obtain deeper insights into the morphogenetic effects of light. The surprising fact has emerged that besides chlorophyll, with its special responsibility for photosynthesis, many other pigment systems, capable of absorbing light, exist in plants. Since these pigments are present in very low concentrations, they are not directly observable, even in the light microscope, but their ability to absorb certain wavelengths preferentially leads to their having quite specific effects on development. The most important of these pigments is probably phytochrome which, when irradiated with visible red light (660 nm wavelength), is transformed into an active form (designated P_{660}) which actually produces the morphogenetic effects. This form has an absorption maximum in the far-red (730 nm wavelength). Irradiation with such far-red light causes all but a trace of P_{660} to be transformed into the inactive form, designated P_{730}. When the active form is kept in darkness, there is spontaneous deactivation, and possibly also some destruction of the pigment. This is the primary reaction leading to etiolation (see p. 336), and it is perhaps also involved in photoperiodism (see p. 344). This reversible phytochrome (or red–far-red) system is often denoted by the symbols $P_{660} \rightleftharpoons P_{730}$.

Phytochrome also has a small absorption peak in the blue region of the spectrum. After irradiation with red light up to 80 per cent of the pigment is in the active form, after blue light (400 nm) about 50 per cent, and after blue-green light (440 nm) only 30 per cent. In white light red and far-red are present in about equal proportions. Phytochrome is a blue-green proteid, apparently furnished with four pyrrole in an open chain (cf. phycocyanin and phycoerythrin; p. 246).

The phytochrome system appears to be concerned with a whole range of morphogenetic phenomena. Certain seeds (e.g. *Nicotiana tabacum, Lythrum salicaria, Lactuca sativa*) need to be illuminated before they will germinate, and this seems to depend upon phytochrome since the illumination produces sufficient of the far-red pigment to permit germination. (There are also seeds which require darkness for germination, light being inhibitory (e.g. *Veronica persica, Phacelia tanacetifolia*).) This probably depends upon the formation or destruction of inhibitory substances, possibly partly in the testa. The growth of hypocotyls and internodes, the expansion of leaves and the differentiation of stomata, the differentiation of xylem, and all those phenomena (referred to collectively as etiolation) brought about by darkening are also evidently subject to phytochrome. Further, the synthesis of anthocyanin, protein and ribonucleic acid, and photoperiodic sensitivity (see p. 344) are all apparently dependent upon the presence of phytochrome in its red activated form. This indicates that it occupies a central position in the metabolism, perhaps even being concerned with the control of gene activation (see p. 330).

It is quite evident that these investigations into the physiological activity of light, in which chlorophyll is quite certainly not involved, have considerable practical significance. An example is provided by the artificial illumination of glasshouse crops, the success of which is dependent upon the light source emitting the correct spectral composition. For many economic plants a mixture of blue and red light, the intensities of the two wave bands being in the proportion 1 : 7, proves optimal. This indicates that blue wavelengths also have a significant effect.

It has already been mentioned (p. 306) that the **polarity** and **dorsiventrality** of a plant or plant organ can be determined by the direction of illumination. Here again light of short wavelength appears to be particularly effective.

Thus it is possible to fix the polarity of germinating *Fucus* zygotes by illuminating them unilaterally. In the gemmae of the liverwort *Marchantia* the side which is to become the morphologically upper and that the lower is again determined in the first place by light. In many fern prothalli the rhizoids and sex organs will develop only on the side turned from the light. The roots by which ivy climbs are produced only on the shaded side of the stem. In many trees the whole pattern of the branching is determined by the fact that the buds extend only on the illuminated side. The dorsiventrality of the side branches of many conifers (e.g. *Thuja, Thujopsis*, etc.) is induced by unilateral illumination, while in other instances (*Taxus, Picea*) gravity is the determining factor (see p. 339).

A striking relationship between anatomy and the extent of illumination can even be seen in the dorsiventral leaves of many trees. The outer leaves on the southern sunny side of a tree commonly possess a deeper palisade, sometimes even subdivided into several layers, than the 'shade leaves' of the northern side and the interior of the crown (Fig. 290). A difference is often already apparent in the primordia, and the buds are commonly fatter on the sunny side. The form of leaves can also be influenced by light. *Campanula rotundifolia*, for example, produces more or less circular leaves in weak light, but narrow leaves in strong.

The plasmodia of many Myxomycetes respond to light of quite different wavelengths with the formation of sporangia. The phenomenon of photomorphogenesis is thus widespread.

Particularly remarkable however is the discovery that not only the quality and quantity of the light, but also the time and periodicity with which it strikes the plant may also produce striking changes in its subsequent growth and development. Those so-called photoperiodic phenomena are discussed in greater detail later (p. 344).

With the exception of the phytochrome system described earlier, we still have hardly any understanding of the ways in which light is able to direct morphogenesis. What is basic to all of them must be that there is initially absorption, and subsequently a photochemical process, in some kind of photoreceptor. The nature of these pigments, and the subsequent members of the reaction chains are for the most part unknown.

C. Gravity

Besides temperature and light, gravity should also be briefly considered, since not only does it cause the plant to orientate itself in space (see p. 355) but it is also capable of exerting profound formative effects. The polarity and dorsiventrality of many organs, for example, can also be influenced by gravity, although this effect is usually obscured by that of light.

The orientation of the first division of the *Fucus* zygote can be determined by gravity in the absence of light. Dorsiventrality can be induced in the gemmae of *Marchantia* by allowing them to stay at rest for six hours in the gravitational field. The dorsiventral arrangement of the leaves of yew and *Abies* is similarly induced by gravity. Many dorsiventral flowers, such as those of *Epilobium, Gladiolus* and *Hemerocallis*, become radially symmetrical if the buds are removed from the unilateral action of gravity by rotation on a klinostat (Fig. 291). The torsion of orchid ovaries can be prevented by the same treatment. The wood on the lower side of a plagiotropic branch of a conifer is frequently found to be greater in quantity, and different in structure, from that on the upper side, another effect for which gravity appears to be responsible. In dicotyledons, it is often the upper sides of horizontal branches that are affected.

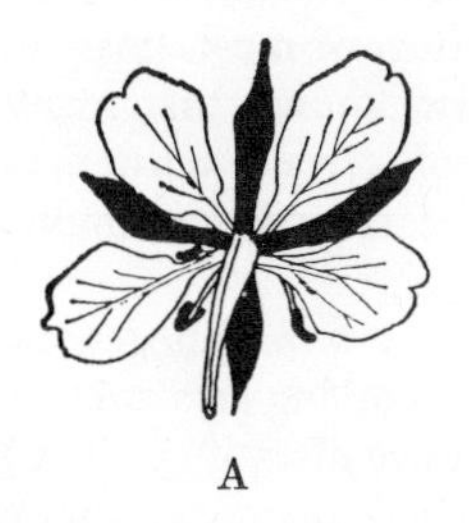

Fig. 291. View of the flowers of *Epilobium angustifolium* from behind. **A.** As formed in their natural position. **B.** Radial symmetry brought about by removal of the unilateral action of gravity. After Vöchting, modified.

D. Other environmental influences[8, 48]

The form of plant may be influenced by yet other environmental influences operating at a particular habitat. Shortage of water, for example, such as occurs in dry regions, can markedly affect the structure of a plant. It is also clearly evident from the mechanism of extension growth that a reduction of turgor will tend to limit development.

For this reason dwarfing (nanism) is frequently observed on dry soils. In a damp atmosphere, however, internodes and petioles become elongated, the laminae of the leaves large, thin and almost entire-margined, and any hairiness very reduced. Drying the atmosphere can lead to a thickening of the cuticle, a reduction in the size of the epidermal cells and an accompanying increase in the number of stomata per unit area. A more conspicuous development of conducting and strengthening tissue may also occur.

Not all these features, which are especially characteristic of xerophytes, need, of course, arise exclusively from a shortage of water. A shortage of minerals, particularly of nitrogen, may also be present in such habitats, and can give rise to similar morphological effects.

The availability of mineral salts, the pH of the soil and the extent of the photosynthesis all play a quite general part in the metabolism of the plant. Consequently, it is readily understandable that the nutrition available to the plant should influence its growth both quantitatively and qualitatively. It is often very difficult to work out the details of these complex effects with any clarity.

The effects of nutrition on development can be most readily studied in heterotrophic organisms. Thus the fungus *Basidiobolus ranarum* grown in a nutrient solution containing sugar and peptone forms branched hyphae with transverse walls, but in the presence of sugar and ammonium salts gives rise to rounded thick-walled cells which divide irregularly in all directions. In many lower plants the formation of reproductive organs, or the continuation of purely vegetative growth, as the case may be, can be brought about at will by varying the nutrition (see p. 344).

With higher plants in dense communities light, water and mineral nutrition naturally all play a part in controlling growth.

It is not known to what extent, if at all, other factors are concerned, such as root and leaf excretions, leaching of inorganic and organic materials from attached or fallen leaves and their return to the soil, formation of antibiotic substances (see p. 290), glandular secretions and other reciprocal effects between neighbouring plants. Much evidence points to the existence of such allelopathic relationships between neighbouring plants (see also p. 767). An interesting effect is that of ethylene, at certain times excreted from various fruits (such as ripening apples) and other parts of plants. This gas can not only accelerate the ripening of other fruits (and must consequently be taken into consideration in the storage of cultivated fruits) but also stimulate or inhibit extension growth, and with dorsiventral organs such as leaves cause exaggerated growth of the lower side (hyponasty) or the upper (epinasty). The deleterious effects (e.g. shedding of the leaves, etc.) which tobacco smoke and coal gas have on many plants are also probably due to the traces of ethylene they contain (see p. 303).

The strong reciprocal morphogenetic effects which symbionts exert upon each have already been mentioned. In addition, the spores of many parasitic fungi, and also the seeds of many parasitic angiosperms, such as *Orobanche* and *Lathraea*, germinate only if they are in the vicinity of the appropriate host. Some material influence must emanate from the host plant, stimulating the parasite to develop. It is also known that many pollen grains will not develop unless they are in the presence of certain substances normally secreted by the stigma. The pollen grains of *Nymphaea* spp., for example, require, among other things, boric acid.

Sometimes even mere contact with some external object will have a morphogenetic effect (thigmomorphosis). Thus many algae will develop rhizoids on coming into contact with the substratum, the tendrils of *Parthenocissus tricuspidata* adhesive discs (Fig. 213C) and the shoots of *Cuscuta* haustoria. Tendrils which have attached themselves to a support thicken on the side making contact with it. Many agarics will develop the normal cap in the dark only if the fruit body has made contact with some solid object for a short time. In all these instances we can exclude any chemical influence from the object or surface touched.

E. The inheritability of modifications[41, 42, 45, 46]

The capacity of the environment to influence growth and development, discussed in the foregoing, reveals a very considerable morphological lability in the plant. The science of genetics makes clear that within the species there must be limits to this variability. No matter how strong the environmental influences, a bean seed will never yield anything other than a recognizable bean plant, though it may be modified by these influences in numerous particulars. The only thing inherited is the reaction-norm, i.e. the capacity of an organism to react in a specific manner to a given assemblage of those internal and external factors which govern development. This capacity of the organism, and of the individual cells, is usually much more extensive than revealed in the customary appear-

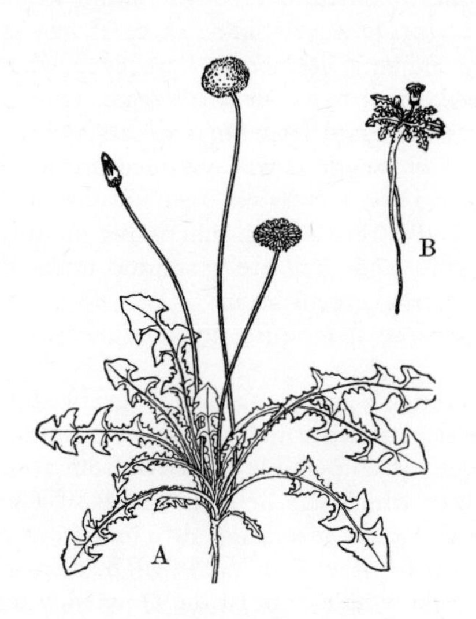

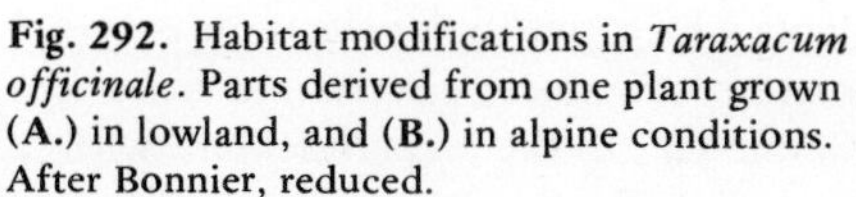
Fig. 292. Habitat modifications in *Taraxacum officinale*. Parts derived from one plant grown (**A.**) in lowland, and (**B.**) in alpine conditions. After Bonnier, reduced.

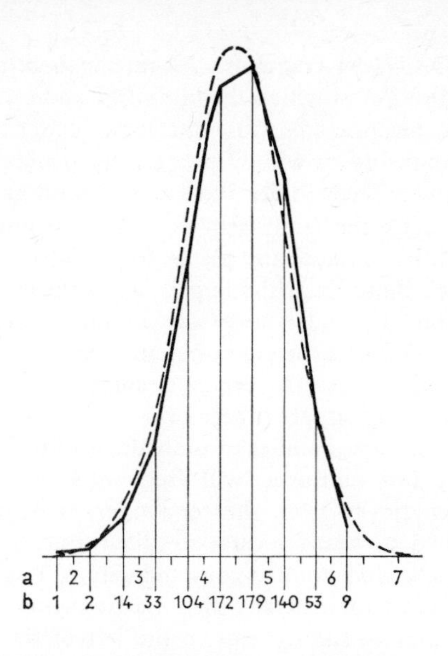

Fig. 293. Variation in weight of 712 beans of a pure line. Abscissa: weight in 10^{-1} g; ordinate: number of beans. Dotted line shows the theoretical Gaussian curve. After Johannsen, modified.

ance of the plants concerned. It is therefore not surprising that the 'normal' form of a plant grown under unusual environmental conditions can undergo marked changes, yielding a form now termed 'abnormal'.

So far as genetics is concerned, the important problem is the extent to which these changes induced by the environment are capable of influencing the genes, and of being propagated either sexually or vegetatively.

Figure 292 shows two specimens of *Taraxacum* originating from one plant whose rhizome was halved. One plant was grown in the lowlands, the other at a high altitude. The plants are drawn to the same scale, and the gross differences are striking. Even so, the modifications induced in the alpine plant, probably due principally to the increased amount of ultra-violet light it receives, are not inherited. Offspring of the alpine plant cultivated in the lowlands at once reassume the lowland form. The morphogenetic factor in the environment has not apparently modified the fine structure proper of the cell nucleus or cytoplasm, but has only directed the reaction of the plant to its environment. The reactivity of the plant is dependent upon both nucleus and cytoplasm, and under the influence of the morphogenetic factor some developmental potentialities have presumably been suppressed and others promoted. A further example is provided by the so-called amphibious plants (p. 169) in which the form of the leaf is quite different in the two habitats.

Since the morphogenetic factors in the environment can never, where a large number of plants or plant organs are concerned, act with exactly the same intensity throughout, it is not surprising that even plants which are uniform genetically show a certain amount of variability in an apparently constant environment. The size of beans, for example, varies considerably because the nutrition of the individual bean depends, among other things, upon the number of fertilized ovules in the pod, the distance between the pod and the assimilating leaves and the rate of assimilation itself. This variability is a general characteristic of living things, not only in morphological features but frequently in others as well.

If a large number of beans, gathered from a crop of genetically identical and uniformly treated bean plants, are individually weighed or measured, and the results expressed as a histogram, the frequency distribution of the dimension chosen is obtained (Fig. 293). This distribution is sometimes slightly asymmetrical, but in general closely approaches the normal curve of error (binomial curve). This is because the environmental factors tending to increase or diminish size are combined at random according to well-known statistical laws. Most frequently occurring is a definite mean dimension, corresponding to the situation in which the inhibiting and promoting factors have been in balance. Of very rare occurrence are those instances where the environmental influences have been wholly inhibitory or wholly promoting. If plants are now raised from the smallest and from the largest seeds under the same conditions the beans harvested from these plants will show exactly the same frequency distribution of weight or length as was obtained previously. This is because the plants from the smaller seeds do not themselves yield smaller beans, nor those from the larger, larger beans. Size of seed is thus not an inheritable modification. The differences are in fact only phenotypic, and they are generated under the influence of the environment from one constant internal organization.

Of course in certain circumstances it may appear as if modifications induced by the environment are inherited.

A large number of individuals of one species, even in a group appearing superficially to be very uniform, will frequently contain several varieties which are slightly different genetically from the remainder. It is, for example, often possible to select from a large field of beans several so-called 'pure lines', each of which will be descendants of a self-pollinated homozygous individual. Each of these lines is genetically different from the other to the extent that the frequency distribution for size of its seeds will be displaced either to the right or to the left of the curve for the whole population. Only such pure lines will yield, following self-pollination, descendants which give under the same external conditions identical frequency distributions, no matter how many generations be raised. If a breeding experiment has been begun, unbeknown to the experimenter, with a mixture of several pure lines (such as usually occurs in natural populations), misleading results will follow. In such conditions, for example, the larger beans may yield plants which in turn produce beans of a mean size larger than that of the mean of the population. This appears to indicate the inheritance of an environmentally induced feature, but in fact it is because what was originally selected was not a modification, but a pure line in which the tendency to produce large seeds was hereditary. It is thus essential that all experiments concerned with heredity be carried out on plants with a uniform genetic structure.

As a matter of fact, it is actually possible for modifications induced by certain external factors to be retained for several generations, apparently providing genuine examples of the inheritance of acquired characters. Careful investigations have shown, however, that although the modification may at first persist from one generation to the next, it nevertheless gradually disappears. Such modifications are termed dauer-modifications. What evidently happens is that the external factor brings about some change in a self-reproducing component of the cytoplasm (e.g. an enzyme). This would be hereditary to a certain extent, without the nucleus having been in any way affected. In sexual reproduction the modification usually vanishes abruptly, the descendants failing to show it, unless the cytoplasm of the egg cell has been involved. In this case the modification may persist for several generations, and with vegetative reproduction even longer.

Cuttings of ivy, for example, taken from a flowering shoot will continue to produce the form of leaf characteristic of this region (see p. 168), even though innumerable cell divisions may occur and large plants be generated. The leaf form is retained if these plants are propagated by cuttings, and the normal juvenile form of leaf does not reappear until seed is obtained and germinated. The sexual process apparently eliminates the modification.

Another example is provided by radishes which, as a result of poor nutrition (e.g. lack of water or an infertile soil), have elongated and remained unthickened in the region of the hypocotyl. If this malnutrition persists during flowering and the setting of the seed, then the embryos in the seeds are deficient in food reserves. As a result, the seedlings are starved in their early life, and because of these conditions peculiar to them, and not this time applied from without, will again fail to form 'bulbs'. The 'hunger' modification induced in the first generation thus reappears in the second. If the plants of the second generation are well nourished they will produce seeds which in good conditions of germination and growth yield plants in which the hypocotyl thickens normally. The modification here has clearly had nothing to do with the hereditary structure of the nucleus or cytoplasm of the embryo, but has been concerned with the environment in which it begins to develop.

All genetical investigations must take into account the possibility of modifications and dauer-modifications of the kind discussed in the foregoing.

F. The developmental process and its dependence upon external and internal factors[49]

The development of a plant is usually not uniform, but can be seen to consist of a number of distinct phases separated from each other qualitatively as well as in time, and usually following upon one another in regular sequence as the plant matures. Resting stages alternate with times of vigorous activity, leaves show sequences from cotyledons to foliage leaves and ultimately floral bracts, and the production of vegetative members by the growing points change over to that of flowers, fruits and seeds, either once and for all, as in hapaxanthous plants (e.g. *Agave*), or periodically, as in pollakanthous plants.

This raises the question of how far this change in form, and transition from one stage of development to the next, arises inevitably from some inner organization of the plant, and how far it is evoked, or at least influenced, by the external environment. In other words, to what extent is the course of growth autonomic (or endogenous), and to what extent aitionomic (see p. 349)?

Even the growth of a single cell under constant conditions is not uniform, but shows in the 'grand period' of growth (see p. 308) an acceleration of the growth rate to a maximum, and then its decline until the cell has reached its maximum size. Examples of a periodic alternation between activity and rest in multicellular organisms are provided in our latitudes by the broad-leaved trees. Their bursting into leaf, and in some cases into flower, in spring is familiar to all. By early summer, after the laying down of the buds for the next year's growth, they begin to stop growing, and after the fall of the leaves in the autumn spend the whole winter in a resting condition. The explanation that immediately suggests itself, namely that the periodicity of growth is imposed by the periodicity of the climate, and that under constant environmental conditions growth would be predominantly uniform and continuous, is not supported by observations in the constantly humid forests of the Tropics. Here many of the trees show a periodic shedding and renewal of leaves, but this is not strictly annual, and sometimes individual branches of a tree may shed their leaves at different times. This and other observations lead inevitably to the conclusion that the tendency towards a non-uniform process of development is a fundamental phenomenon of life. The markedly developed periodicity of the vegetation at our latitudes comes about as a result of the cosmically determined annual fluctuations of the climate regulating the onset of autonomic fluctuations within the plants, so that the latter become synchronized with the annual cosmic cycle.

That this remarkable alternation between growth and dormancy cannot be ascribed solely to external influences is shown also by seed-formation in higher plants. As soon as the embryo in the ripening seed has reached a certain stage of development (different in different species), its growth stops abruptly while the seed or the fruit, as the case may be, is shed from the parent plant. Again, although many seeds are capable of germination immediately they are fully formed (as in many cultivated plants), there are many others in which germination and resumption of growth by the embryo will not occur, even in favourable conditions, until after the seed has passed through a definite period of dormancy.

In these resting stages the cytoplasm is in a peculiar state of suspended activity. This is often characterized by considerable dehydration (see p. 21), which naturally causes widespread slowing down and suppression of the metabolism (e.g. respiration). Dry seeds and spores are consequently extraordinarily resistant to unfavourable environmental conditions, e.g. heat and cold. The ability to spend part of its life in a latent condition is undoubtedly of great value to an organism in the struggle for existence.

Embryonic tissue elsewhere, e.g. in resting buds, appears to undergo no such marked dehydration. Here other factors, possibly the formation and displacement of inhibitors (see p. 303), may account for the developmental behaviour. Whether, and at which point in time, cell division and differentiation within the growing point are suspended and subsequently re-initiated is probably different in different species. In many of our fruit trees the growing points appear to remain active even in the winter months.

The formation of the reproductive organs naturally occupies a special place in the life cycle of a plant, since they are essential for the maintenance of life and for the rhythmical continuation of the developmental cycle. The ability of external factors to influence this cycle has consequently been intensively studied, especially from the point of view of the inception of the reproductive phase. What has to occur, for example, in the growing point of a plant which has undergone prolonged vegetative growth to cause it to produce stamens and carpels? Is it inevitable that every organism must sooner or later reach a stage in which it produces reproductive organs, either vegetative or sexual? Or do external influences play a releasing or modifying role?

In many lower plants external influences are undoubtedly of considerable importance.

If the fungus *Saprolegnia*, for example, is grown in continuously renewed media it remains vegetative indefinitely, but if it is transferred to pure water it rapidly forms asexual swarm-spores. If, however, the fungus is fed with certain amino acids (e.g. leucine) it grows vegetatively for some time and then begins to produce sex organs. With the alga *Vaucheria*, a sudden reduction in the concentration of its medium is sufficient to bring about the formation of swarm spores. Similarly in many other fungi and algae various environmental factors affect the developmental cycle, so in these cases its sequence cannot be rigidly predetermined. Nevertheless, other evidence indicates that not all the behaviour of these organisms is determined by external influences.

Amongst higher plants the production of sexual reproductive organs, at least with all species, does not seem to be essential for the continuance of life. Many species, for example, are known which reproduce themselves vegetatively by rhizomes, aerial bulbils and so forth for an indefinite period. Such plants either produce no flowers at all or only sterile flowers.

For example, many cultivated bananas, species of *Dioscorea*, *Acorus calamus*, and varieties of cultivated vines, oranges and strawberries yield no progeny by sexual reproduction, yet they give no sign of undergoing any kind of degeneration. Both sexual and vegetative reproduction may perhaps have a rejuvenating and stimulating effect causally related to the phenomena discussed on p. 314, but the special significance of sexual reproduction is probably to be sought in the repeated opportunity it gives for new combinations of genetic factors to arise. The essential conditions for purely vegetative reproduction are still little known; they may be partly determined genetically.

In recent years many investigations have been made of the factors causing the onset of sexual reproduction in the higher plants. Here again it has been shown that the transformation of the vegetative into the reproductive apex, and the subsequent production of flowers, does not take place in all plants at one definite, internally determined, stage of development. In some plants, in fact, the transition to the reproductive phase can be considerably influenced by external factors.

Photoperiodism. Besides the effect of temperature mentioned earlier (see p. 334), there is the very surprising observation that in many plants the relative length of the light and dark periods exerts a decisive influence on the behaviour of the plant. The length of day a plant receives may determine if and when it passes from the vegetative to the reproductive conditions. The so-called long-day plants (Fig. 294) are those which either fail to flower, or flower only after prolonged delay, if the daily photoperiod falls below a certain level (critical day-length), which is different in different species. These plants flower the more readily the longer the daily illumination, the effect depending not so much upon the intensity as upon the duration of the illumination. This shows that the developmental consequences observed do not merely depend upon an enhanced nutrition following upon increased photosynthesis.

Many of our native plants are long-day plants, perhaps related to the fact that in our latitudes long days and short nights predominate during the main growing season. Many familiar cultivated plants also belong here, such as most of the cereals (rye, wheat, barley, oats), peas, mustard, spinach and many more. Spinach, for example, fails to flower with a

Fig. 294. *Nicotiana sylvestris* raised in long days (left) and short days (right). After Melchers and Lang.

Fig. 295. *Panicum* sp. raised in 18-hr days (left) and 12-hr days (right). After Maximow.

day-length of less than thirteen hours, and it flowers most rapidly in continuous light. *Hyoscyamus niger* flowers if the day-length reaches or exceeds $9\frac{1}{2}$—$10\frac{1}{2}$ hours (Fig. 296).

The long-day plants merely require a certain minimum daily photoperiod, and not a rhythmical alternation between light and dark in order to flower. In the so-called short-day plants a rhythmical alternation of light and darkness is normally essential for the production of flowers, and the dark period must equal or exceed a certain critical length. These species flower all the earlier the closer the duration of this dark period is to an optimum, which is different for different species (Fig. 295). In certain circumstances an interruption of this dark period, even though of short duration, is sufficient to prevent flowering.

Amongst the short-day plants are many familiar cultivated forms originally native to tropical regions, such as rice, millet, hemp and soya, as well as certain varieties of tobacco, beans and potatoes. Chrysanthemums are also short-day plants, and in our latitudes do not flower until the photoperiodic conditions of autumn. *Xanthium* (Compositae) does not flower unless the daily dark period extends for at least eight hours, while *Kalanchoe blossfeldiana* (Crassulaceae) requires at least twelve hours darkness (optimum fifteen to sixteen hours; Fig. 296). In the short days of the early part of the year *Viola odorata* forms normal flowers, whereas in the long days of later spring only cleistogamous flowers are produced. Particularly remarkable is the fact that different species of the same genus, or even races of the same species, may sometimes belong to different photoperiodic groups (e.g. *Nicotiana silvestris* is a long-day plant, *N. tabacum* var. Maryland Mammoth a short-day plant).

There are, however, in addition to the pronounced long- and short-day plants, a

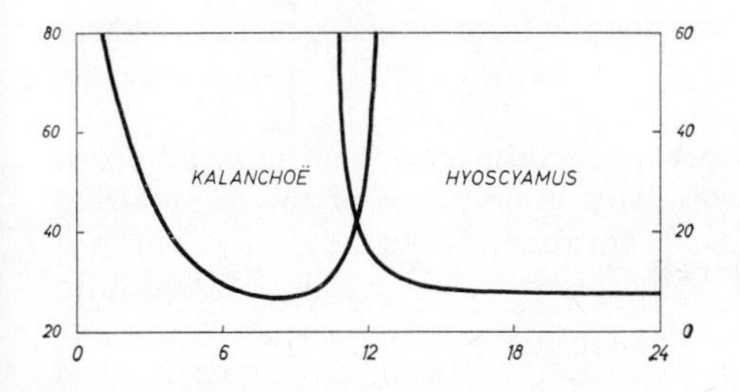

Fig. 296. Diagram showing the relationship between the development of a short-day plant (*Kalanchoe*) and of a long-day plant (*Hyoscyamus*) and the duration of the daily photoperiod. Abscissa: daily photoperiod in hours. Ordinate: left, days required for the primordia of the inflorescence to become visible in *Kalanchoe*; right, days required for the beginning of the growth of the flowering stem in *Hyoscyamus niger*. After Bünsow.

number of others which require for their normal development at first short days and then long days (short/long day plants) (corresponding to the first half of the year), and yet others which require the converse (long/short day plants) (corresponding to the second half of the year).

Often only the flower primordia are influenced by the photoperiod. Nevertheless, in certain cases the continued development of the whole flower may be dependent upon the illumination, either in the same way as the primordia or inversely.

Finally, it must be stressed that by no means all plants show this dependence of flower formation on the daily photoperiod. There are many, among them the so-called cosmopolitans (e.g. *Poa annua, Senecio vulgaris, Stellaria media*, etc.), which show little if any response to the duration of the illumination. These plants are termed 'day neutral'.

The ability of many plants to vary their manner of development in response to the length of the time for which they are illuminated each day is termed *photoperiodism*. Such effects are not confined only to the initiation or prevention of flowering, but influence many other details of vegetative and generative morphology. The form and succulence of the leaves, the length of internodes, the development of pigments, frost resistance, the formation of subterranean storage organs, the onset and termination of the dormancy of buds and so on can also all be influenced by day-length, showing how this environmental factor has an extraordinarily profound effect upon the plant, although the manner of its action remains obscure.

In spite of many investigations, particularly of the onset of flowering, the causation of photoperiodic effects is still little understood. The photoperiodic stimuli appear to be taken up principally by the still not completely developed foliage leaves, probably by way of the phytochrome system (see p. 337), and from these conducted to the growing point, itself photoperiodically insensitive. The stimuli are capable of traversing a graft union. Many have therefore considered the formation and conduction of 'flower hormones' (but see p. 315) to be responsible for the initiation of flowering, just as other morphogenetic substances are responsible for other effects. Long-day plants can in fact produce flowers in short days if they are treated with gibberellins. Nevertheless, little is known of the detailed interrelationships of these substances and the phenomenon of flowering.

A single leaf exposed to the appropriate conditions may be sufficient to influence the remainder of the plant, even if this receives quite a different photoperiod. Often the plant has to be exposed for only a very short time, at the extreme for only a single day, to the appropriate photoperiod to bring about a clear-cut effect (*photoperiodic induction*). In *Hyoscyamus niger* three days of continuous light, or eight to ten days with a twelve-hour photoperiod, will induce flowering. The flowering of long-day plants can be accelerated by supplementing short days with an additional period of illumination at a very low intensity.

It is well known that in both animals and plants numerous phenomena show a diurnal endogenous rhythm (see p. 371 seq.). This is in phase with, and probably controlled by, the twenty-four-hour period of the earth's rotation, but it is not actually caused by it. Nevertheless, interrelationships between the plant's autonomic rhythm and that of the normal alternation of light and dark seem not improbable. Thus in many plants, within each twenty-four-hour period, a so-called photophil phase, during which various diverse activities within the plant may be promoted, or at least not interfered with, by light, may be succeeded by a skotophil phase, during which light has a more profound effect. The duration of these two phases need not coincide with lengths of the light and dark periods to which the plant is subjected. It is clear that the greater the deviations from the 'normal' course of development of the plant which can be induced by the externally supplied photoperiods, the more this external artificial periodicity diverges from the endogenous rhythm of the plant and disturbs its natural course. Nothing is yet known in detail of these complex processes.

Naturally it is not believed that the photoperiodic sensitivity present in many plants alone determines their development. This sensitivity is only one of the many which enable a plant to react to, and be modified by, its environment. As the existence of day-neutral plants shows, in some plants developmental processes less readily influenced by external factors tend to predominate.

Flower formation is also undoubtedly influenced by other factors, such as the general level of the carbohydrate and mineral nutrition, etc. As is well known from horticultural practice, a lessening of the nitrogen and an increase of the supply of phosphate strongly promotes flowering. *Mimulus tilingii, Veronica chamaedrys, Hedera helix*, among others, form their flowers only in direct light and never in the shade, probably related to the greater photosynthesis in the former situations. Even these relationships, clearly of some practical importance, are certainly very complex in detail.

Section three: **The physiology of movement**[50]

The phenomena of movement, to be described in the following, are a directly observable and much more conspicuous aspect of the physiological activity of plants than metabolism, and the slow changes of morphogenesis. There is always, of course, some movement as organs are formed and expand, but in addition to this there are specialized processes of movement in plants of quite a different character. Some of the lower plants, e.g. some algae and bacteria, are capable of locomotion like animals, and by swimming or creeping move actively from place to place seeking a more favourable environment, or escaping from an unfavourable one. The higher plants, anchored firmly by their roots, are incapable of such locomotion, and in this respect are quite different from the freely moving animals. Nevertheless, individual organs of higher plants are often able to carry out movements. These sometimes serve to orientate the organ in space, but they may have other functions as well. Compared with free locomotion, these plant movements are usually very slow. They consist principally of curvatures or twistings, but sometimes individual organs will move as valves about a hinge. In most cases we are concerned with a true manifestation of the physiological activity of the plant, and of great importance for its survival. These movements often show particularly beautifully, and in a directly visible way, the capacity of the plant to react to certain environmental influences. This property of the plant is referred to as irritability, and it is often studied as a special field of physiology. It must be remembered, of course, that the environment influences to no less an extent the physiology of metabolism and growth, and indeed certain metabolic processes and physico-chemical changes are often directly responsible for the movements observed. Irritability in fact is a quite general and characteristic property of living things.

An irritability phenomenon always has the character of a release mechanism, i.e. some factor acts as a stimulus which releases a complicated series of reactions in the living material, the energy for which comes from the organism itself and not from the initial stimulus. Usually, therefore, there is no proportionality between the intensity of the stimulus and the degree of the reaction (see below and p. 367).

Thus, if a plant before a window turns its leaves to the light, or if a grass stem bent over by bad weather bends up until its axis is perpendicular again, the light or gravity is considered to act as a stimulus releasing the appropriate movement. Light is not, however, spoken of as a stimulus when it falls upon a green leaf and causes photosynthesis to take place. The difference lies in the fact that photosynthesis involves a definite chemical reaction continually utilizing the light energy absorbed into the leaf, whereas the movement of leaves towards light is brought about by a process drawing its energy from within the plant itself and not from the light absorbed. The amount of light energy able to cause a plant movement is also trifling compared with that utilized in photosynthesis. In an irritability response, therefore, the light releases a mechanism residing in the organism, just as pressure on an electric switch closes the contacts and causes the current to flow. Again, when a loaded gun is fired there is no proportionality between the pressure of the finger on the trigger and the force with which the bullet is discharged. The pressure of the finger merely releases the energy residing in the charge.

The concept of irritability is not therefore in any way vitalistic, but merely signifies that certain stimuli are able to set off complex reactions in the interior of the organism. These, of course, are open to causal analysis, but in no case has it yet been possible to elucidate them completely. Effective stimuli need not always be external in origin. Many irritability phenomena are brought about by stimuli, as yet little understood, arising within the organism. Phenomena dependent upon stimuli of this kind are said to be endogenous or autonomic, whereas those following from external stimuli are referred to as induced or aitionomic.

Irritability phenomena often, but by no means always, obey the so-called 'all or nothing' law, i.e. they either take place to their full extent or not at all (see p. 367), although with multicellular organs a limited proportionality is often observed between the extent of the stimulus and that of the reaction. One situation among others leading to this effect would be where only some of the cells received the stimulus directly, others receiving it by conduction. The speed of the conduction, as well as the number of cells the stimulus reached, would then determine the extent of the response. In the investigation of an irritability phenomenon it is usual to distinguish a number of consecutive phases. The first is the receiving of the stimulus, referred to as susception, then the excitation, referred to as induction or perception. This either leads directly to the reaction or there is an interpolated phase of conduction in which the stimulus is passed to other cells at various distances. These are then excited and produce the reaction.

Many of the movements seen in plants undoubtedly have all the characteristics of true irritability phenomena. There are some, such as the catapult action of many fruits and the hygroscopic movements of dead cells and membranes, which are either entirely different, or can only be classified as irritability phenomena with certain qualifications.

I. Movements of the organs of attached plants

The movements which can be induced in the various organs of plants, such as the stems, roots and leaves, are classified as follows according to their nature.

Movements which take place in a direction related to that of the stimulus are referred to as tropisms, and distinguished according to the nature of the stimulus, e.g. phototropism, geotropism, haptotropism, etc. Such tropisms facilitate in the first place the spatial orientation of the plant.

If, however, the movement is quite independent of the direction of the stimulus, but is hinge-like and fashioned solely by the structure of the organ, the movements are designated nasties. We can then speak of thermonasty, photonasty, etc.

In both kinds of movement the mechanical cause may be either a difference in growth rates on the two sides of an organ (nutational movements) or differential changes in the turgor of the cells (variational movements).

A. Tropisms[51]

1. Phototropism (heliotropism)[52]

The nature of the reaction. In unilateral light many parts of plants undergo curvature, referred to as phototropism. If the organ bends towards the source of light the reaction is termed positive phototropism.

This reaction is shown, for example, by the orthotropous aerial axes of many plants, which always tend to grow with their longitudinal axes in the direction of the light, and by many petioles. In normal illumination the coleoptile of grasses is strongly positively phototropic, and hence is much utilized in research in this field.

Instances of organs which turn away from the light, i.e. which are negatively phototropic, are less common.

This type of behaviour is shown principally by aerial roots and adhesive rootlets, and the primary roots of *Sinapis* (Fig. 297) and *Helianthus*. The unpigmented subterranean roots of most other plants are more or less insensitive to light, although their sensitivity to gravity is considerable (see p. 355). Other examples of organs which are negatively phototropic are the tendrils of the wild grape, which end in adhesive discs, the hypocotyl of germinating mistletoe and the rhizoids of fern prothalli and of liverworts.

In laminate leaves the upper side is usually turned to the light and the lower side from

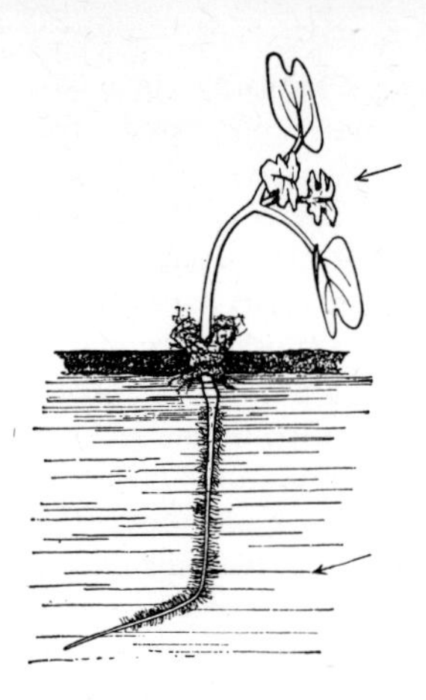

Fig. 297. Seedling of *Sinapis* in water culture illuminated unilaterally from the right. The stem is positively and the root negatively phototropic. The surfaces of the leaves arrange themselves perpendicular to the incident light (transversely phototropic). After Noll.

it, the whole organ being placed more or less transversely to the incident light, or at a certain angle to it. This orientation is termed transverse- or dia-phototropic, and it occurs principally with dorsiventral organs. The thallose liverworts also usually grow so that they are at right angles to the incident light.

Some plants rotate their leaves so that at midday, when the radiation is most intense, the edge of the leaf is presented to the light, the upper surface being held perpendicular only to the weaker light of morning and evening (compass plants, e.g. *Lactuca serriola, Silphium laciniatum*, etc. (see p. 181)). Factors other than light, such as the infra-red radiation, are also concerned in this phenomenon.

Phototorsion is the term used for the ability of some plants to orientate the laminae of their leaves more or less horizontally by torsion of the petiole. This phenomenon is little investigated, but seems to be distinguished from phototropism by its sensitivity to red light.

Phototropic behaviour is not confined to the higher plants; it occurs also in fungi and algae. The fruiting bodies of many *Coprinus* spp., for example, are strongly positively phototropic. Of particular interest are the positively phototropic unicellular sporangiophores of many Mucoraceae, the mycelia of which show hardly any phototropic sensitivity. Examples are those of *Phycomyces*, and of the small *Pilobolus crystallinus* (Fig. 298), a fungus of horse dung. If a culture of *Pilobolus* is placed in a dark box provided with only a small window in one side (Fig. 298), the sporangiophores bend in the direc-

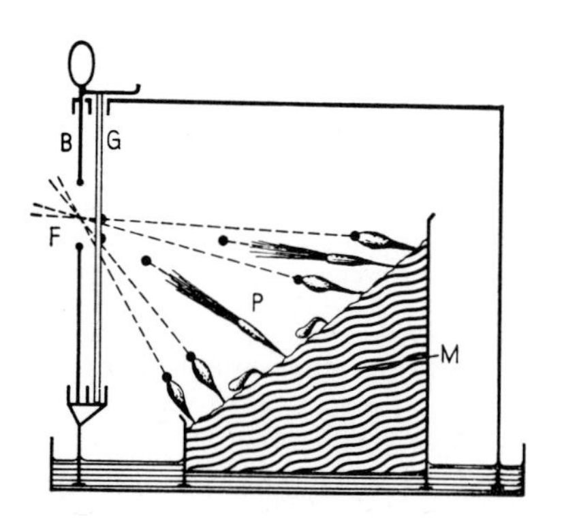

Fig. 298. *Pilobolus crystallinus*. Grown in a culture box filled with horse dung (M), the fungus (P) shoots its sporangia towards a window set in the side of the dark cabinet. G, glass plate; B, metal disc with window, F. After Noll, diagrammatic.

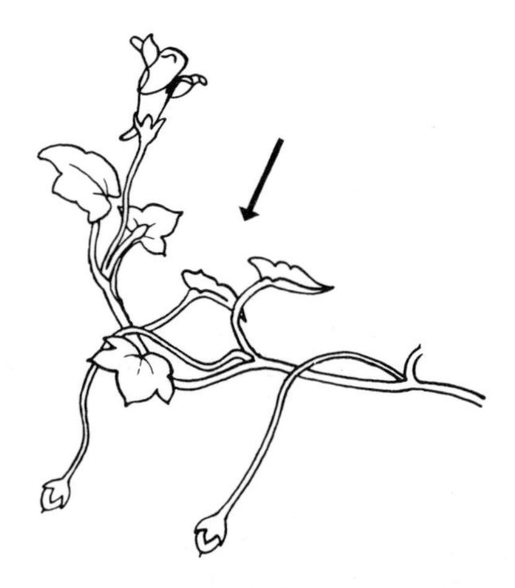

Fig. 299. *Linaria cymbalaria*: the flowers are positively, the fruits negatively phototropic. Arrow indicates the direction of the incident light. *c.* $\times 1\frac{1}{2}$.

tion of the entering light, so that when mature the black sporangia discharge themselves directly towards the opening (see p. 376).

Of great importance is the fact that the phototropic response of an organ can change. This may occur, for example, during development.

Thus the pedicels of *Linaria cymbalaria*, a plant common on walls (Fig. 299), are at first positively phototropic. After fertilization, as the fruit matures, they become negatively phototropic, at the same time elongating considerably so that the fruit is buried in dark crevices of the wall or other places suitable for germination of the seeds. The flowers and fruits of *Helianthemum vulgare* show a similar reversed behaviour to light. In some cases the amount of light (i.e. the product of the intensity of the light and the time of illumination) received by the plant determines its reaction. The familiar positive phototropism of the oat coleoptile, for example, takes place only with a weak illumination of less than 340 lux-seconds (metre-candle-seconds), and with illuminations in the range of 40,000–500,000 lux-seconds. Illumination with between 340 and 40,000 lux-seconds and with more than 500,000 lux-seconds causes a negative phototropism, so that with increasing quantities of light the phototropic response oscillates between positive and negative. Similar behaviour has been observed in *Phycomyces*. To relate these effects to natural conditions it may be mentioned that on a clear day in mid-summer the intensity of the sunlight at midday in our latitudes varies between 50,000 and 100,000 lux, an additional 10,000 lux being provided by the diffuse light. Intensities of at least 25–50 lux are required for reading and writing.

Another factor which influences the behaviour of the plant is the treatment it has received before illumination. A plant kept in the dark before it is illuminated unilaterally behaves differently from one that has been kept in the light. Just as the human eye adapted to darkness is much more sensitive to light, so also is a plant that has been previously darkened. This suggests that uniform illumination reduces the phototropic sensitivity, and such a desensitizing effect must be taken into account in all experiments with prolonged unilateral illumination. The degree of sensitivity to the stimulus is referred to as the 'tone' of the organ. Other factors, e.g. the temperature, may also affect the phototropic response. Above and below certain temperatures, which are not lethal, phototropic responses fail to occur. Narcotics, such as ether and chloroform, also cause the temporary loss of phototropic activity, and of other tropic responses.

The course of the reaction. Everything points to the fact that the phototropic response is a very complicated process. In Fig. 300 is reproduced diagrammatically the time-course of the reaction of an oat coleoptile illuminated from one side for only four seconds. Not until after almost an hour does the tip of the coleoptile begin to bend towards the source from which it was earlier illuminated. Subsequently the site of the curvature begins to travel down the coleoptile, until finally after twenty-four hours the curvature is limited to the base, the whole of the upper part of the coleoptile now lying in the direction of the illumination.

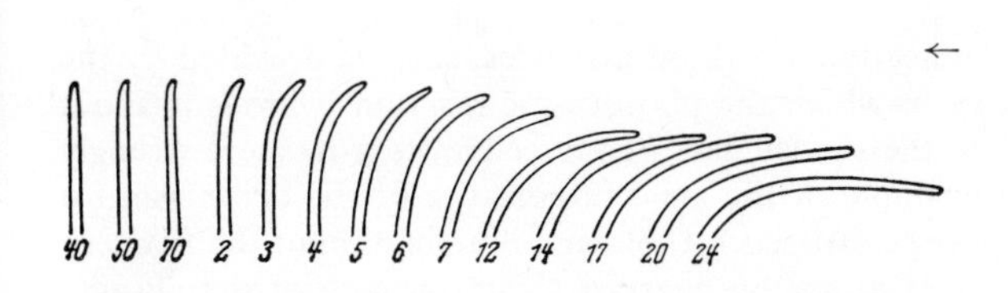

Fig. 300. Positive phototropic curvature of an oat coleoptile illuminated from the right for four seconds with light at an intensity of 30 lux. The figures indicate the time after the excitation, the first three in minutes, the others in hours. After Arisz.

From this several things follow. Evidently a single and quite brief stimulus is sufficient to evoke, after a latent period in which nothing appears to happen, the beginning of the phototropic movement. The time between the beginning of the stimulus and the first appearance of the reaction is called the reaction time. It has already been shown that a strong flash of light lasting only $\frac{1}{2000}$ second will produce a reaction, so that the sensitivity of a plant to unilateral illumination may be extraordinarily high. In experiments to determine this sensitivity the intensity of the illumination must, of course, be taken into account as well as its duration. The shortest time for which light of a given intensity must be supplied for a perceptible curvature to appear after the latent period is called the presentation time.

Within certain limits a perceptible curvature of this kind is produced by illumination with either an intense light for a short time or a weak light for a longer time. The product of the intensity and the duration of the illumination required to produce a perceptible curvature appears therefore to be constant. This product can be regarded as a measure of the amount of stimulation required, and its constancy, which appears principally to concern the phase of susception, is approximately valid only for the production of a just perceptible response. In *Avena* the smallest quantity of light which will cause a response, i.e. the threshold value, is of the order of from 3 to 25 lux-seconds. Up to about 100 lux-seconds there is a measure of proportionality between the amount of stimulation given and the response, both in respect of its size and the rapidity with which it is generated. With large quantities of light, however, the extent of the response becomes independent of the illumination.

The least intensity of light which, given over a long period, will produce a phototropic reaction is remarkably small. Young sporangiophores of *Pilobolus longipes* will respond to as little as 0.6×10^{-6} lux. Even the luminescence of cultures of certain bacteria (see p. 275) has been known to be effective. Nevertheless, such determinations still fail to give any complete picture of the actual sensitivity of the plant. It has been shown, for example, that the plant is quite able to detect individual stimulations too weak to produce any perceptible response. If stimulation below the threshold is repeated at short intervals, the intervals, of course, not exceeding a certain value, a perceptible response will eventually occur. Apparently in these conditions the individual subliminal stimuli summate until the threshold value is reached. A summation effect of this kind is possible only if the individual stimuli are detected in some way, and each by itself is incapable of producing a response. The actual ability of the plant to detect light must therefore be quite extraordinarily high.

Intermittent stimulation may produce an effect even larger than that produced by the same amount of light administered in one. This is probably explained by the fact that, as in all irritability phenomena, the stimulus begins to decline in the intervals of darkness, and at the same time the sensitivity to return. It has already been mentioned above that prolonged illumination has a blunting effect. For the same reason the constancy of amount of light required to produce a just perceptible response can only be approximate.

The mechanics of the movement. If an *Avena* coleoptile, raised in the dark, is brought into inactive red light (see below) in a dark room and marks are made upon it with Indian ink at intervals of about 1 mm it can easily be shown that the response to unilateral white light is in fact a growth movement. In a positive response the side adjacent to the light source grows more slowly, and that away from it more rapidly, than in an unilluminated coleoptile. Thus, although the total amount of growth remains about the same, a bending towards the source must inevitably take place. Phototropism in *Avena* therefore is a true growth movement. In many other plants as well phototropic behaviour is shown only by those organs still capable of growth.

Especially remarkable in this connection are those instances, such as provided by the sporangiophores of the Mucoraceae, in which the phototropic reaction is brought about by the growth of a single cell. Here the shaded side of the cell must grow more strongly than the illuminated, a beautiful example of the great importance of the cytoplasm for the mechanics of cell wall growth (see p. 301 seq.). Unilateral illumination of the filament of cells from a germinating fern spore causes by contrast greater growth of the illuminated side, an effect so far unexplained.

The phototropic reactions of fully developed leaves which possess pulvini (e.g. those of *Robinia pseudoacacia* and certain Malvaceae) are not, however, dependent upon growth, but upon turgor changes induced by the light. They will not be considered in greater detail here.

Closer investigation of the growth reaction in phototropism has provided some information about the way in which the stimulus is conducted. It is clear that the light must first of all produce some kind of change in the plant. This primary process has already been referred to as susception (see p. 349). Since light in general can only produce an effect if it is taken up by some substance, this first phase must consist of absorption, probably leading to a photochemical reaction.

The investigation of the sensitivity of the plant in different regions of the spectrum allows certain inferences to be drawn about the nature of the absorbing substance. In particular, it has been shown that light of long wavelength has hardly any effect, so that phototropic responses can be followed under a red safe-light in a dark room without their being disturbed. Blue-green and blue-violet light is especially active, but ultra-violet light, otherwise so active in biological processes, is here less so. This suggests that the absorbing material is a yellow-red pigment, and the carotinoids immediately come to mind. The presence of these pigments has even been demonstrated in the apparently colourless but phototropically sensitive fungi (*Pilobolus*, *Phycomyces*). The carotinoids, however, so far as is known, take part in no photochemical reactions. Either, therefore, the absorbed energy is passed on to some other acceptor by resonance (see p. 246), or it is conceivable that the carotinoid acts as a kind of yellow filter (as in photography). With very delicate organs, such as the sporangiophores of *Phycomyces*, this might produce the necessary difference in illumination between the lighted and the shaded side. Susception is actually brought about by the absorption of the light in a further yellow substance, possibly riboflavin (lactoflavin, see p. 263), which possesses a similar absorption, spectrum, and which is capable of the photo-oxidative destruction of indole-3-acetic acid *in vitro*. Recently an oxidation product of IAA (methoxyindol) has been detected which, after its formation, inhibits the further transport of IAA, resulting in its lateral displacement (see below).

Frequently the susception of the phototropic stimulus can take place only in certain definite places in a plant organ. In the leaves of *Tropaeolum* only the petioles are sensitive, the lamina acting merely as a source of growth-regulating substances. In the *Avena* coleoptile the very tip, only 0.25 mm in length, is especially sensitive. Only 2 mm behind the tip the sensitivity has already fallen to a mere $\frac{1}{36,000}$ of the value at the summit. It can be shown, by surrounding all but the tip of the coleoptile with a cardboard cylinder, so protecting it from the light, that unilateral illumination of the tip alone suffices to produce curvature towards the light in the lower part of the coleoptile. This shows that the stimulus must have been transmitted from the site of susception, i.e. the tip of the coleoptile, to the shaded basal portion. Nevertheless, even the base is not entirely insensitive, particularly to ultra-violet light.

This transmission of the stimulus apparently has a material basis. It is possible to obtain normal phototropic curvature in the lower part of the coleoptile if, after the tip alone has been illuminated unilaterally, it is immediately removed and then cemented back into position with a drop of agar or gelatin. The same result is obtained even if a tip is taken from one species and cemented on to the stump of another. Some substance must therefore diffuse down through the layer of gelatin or agar, and be the means by which the phototropic stimulus is transmitted. This substance has been found to be identical with indole-3-acetic acid which, as discussed elsewhere (p. 302), is an effective promoter of extension growth.

As already mentioned (p. 312), a stream of growth-regulating substances flows down uniformly from the tip of the unstimulated coleoptile, governing the growth of the whole organ. Since in phototropic curvature the shaded side grows faster than the illuminated, then the light must have caused some change in the conditions on the two sides of the coleoptile.

Theoretically there are several possible explanations. It could be that the growth-promoting substance is destroyed on the illuminated side, or inactivated or even transformed into an inhibitor, or merely that the sensitivity of the protoplasm and its ability to react to the growth substance diminishes on this side. It is also possible that the uniform distribution of the growth-promoting substance coming from the apex is disturbed, and that the unequal lighting causes the stream to be deflected towards the shaded side. In fact, it was first demonstrated by the experiment shown in Fig. 301 that when the tip of an *Avena* coleoptile is illuminated unilaterally more growth-promoting

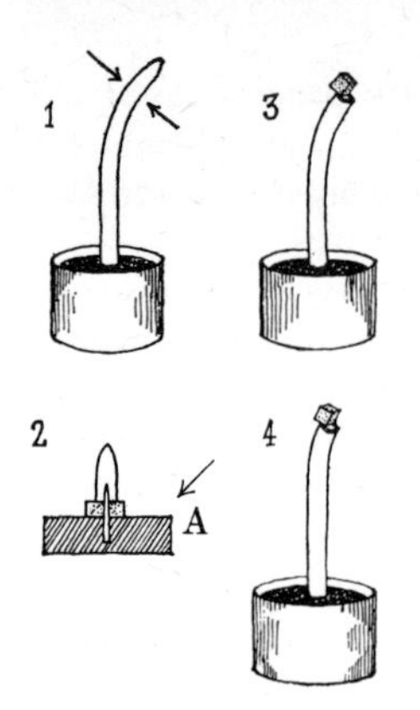

Fig. 301. 1. Normal phototropic reaction of an *Avena* coleoptile raised in a small pot and illuminated from the right. **2.** A tip detached from a coleoptile illuminated from one side for a short time and placed on a small cube of agar (A). A small disc of mica fixed vertically divides the agar and the lower part of the tip into two halves so that the growth substances coming from the illuminated and shaded sides can be trapped separately. **3.** The portion of agar from beneath the illuminated side set on the left of a stump of an unstimulated coleoptile kept in the dark causes only a weak curvature. **4.** The agar from beneath the shaded side similarly placed causes a strong curvature. Modified from Went.

substance is present on the shaded than on the illuminated side. (The amount of growth substance transmitted to the agar can be estimated by the extent of the curvature produced in the stump of an unilluminated coleoptile when the block is placed over one-half of the cut surface at its summit.) Since the total amount of growth substance collected separately from the illuminated and shaded halves is not significantly less than the amount which is transmitted to agar beneath the unbisected tip of an unstimulated coleoptile, it would appear that the stream of growth substance is indeed displaced towards the shaded side, thus enhancing its growth. The mechanics of this displacement, however, like those of the transport itself, are still completely obscure.

Recently it has been possible to demonstrate, by using ^{14}C-indole-3-acetic acid, the actual lateral displacement of downwardly diffusing growth substance towards the shaded side. There appears to be no measurable photo-oxidative destruction of the growth substance on the illuminated side, and the generation by photo-oxidation of the blocking product methoxyindol (see p. 353) is insignificant. Still uncertain is whether the decrease in sensitivity to the growth substance occasionally observed on the illuminated side (and a corresponding increase on the shaded) plays any substantial part in the process. Yet other physical and metabolic differences have been found between the two sides; for example, the convex side tends to be positively charged, while on the illuminated side the concentration of sugar, the catalase activity and acidity are all diminished, but the extent to which these changes are causal is quite unknown. They do reveal, however, the extensive changes which set in when an organ is illuminated unilaterally. It seems certain, nevertheless, that the first and most significant event is an accumulation of growth substance, probably fairly uniformly distributed in the transverse plane before unilateral illumination, on the shaded side of the organ. Where there is conduction of the stimulus, this most probably follows from the transmission, because of polarized movement, of the laterally displaced growth substance to the base.

Hitherto, however, too few objects have been studied for a generally valid account of the steps in the phototropic reaction to be put forward.

In summary therefore, susception and the process of induction just described are followed, possibly with the interpolation of transmission facilitated by a stream of growth-promoting substance, by the termination of the process, the visible reaction. This is such that the growth of the shaded side is promoted until the resulting curvature eliminates the unequal lighting, and consequently the stimulus. The whole process is of great value to the plant, and it enables its organs to occupy in a very short time, as is essential for the plant, the situation of optimal illumination. It is a true irritability phenomenon, since the light merely acts as a releasing factor. A minute amount of light energy is consumed in the photochemical process of susception, but the reaction proper comes about through the release of potential growth energy residing within the organism itself.

Multilateral illumination. A plant which is illuminated not from one side but from several sides simultaneously with light of equal or different intensities shows remarkable behaviour. If an orthotropous plant, illuminated from one side, is slowly rotated by clockwork (see Fig. 304, p. 357) about its vertical axis, so that all its parts in turn receive the light, there is generally no curvature. If a plant is illuminated from two directions simultaneously, provided the angle between them is not 180°, the plant bends in neither of the directions of illumination, but in that of the resultant. This will be given by the diagonal of a parallelogram in which two adjacent sides represent the direction and amount of the two stimuli. When the plant is illuminated from exactly opposed sources the curvature is towards the more intense source, provided the strengths of the two stimuli are in a certain proportion. In *Phycomyces*, for example, one of the intensities must be at least about a fifth greater than the other. Thus if one source is providing 100 lux the other must provide at least 120 lux, if one 200 lux, the other at least 240 lux, and so on, for the curvature to occur. This phenomenon, known as Weber's law, is also well known in animal and human neurophysiology. It seems as if, in spite of great differences in detail, the fundamental principles of excitation may be in a large measure identical in the two kingdoms.

2. Geotropism[53]

Many plants, besides being able to react to unilateral lighting, are also sensitive and able to react to the gravitational pull of the earth. This is referred to as geotropism, and is the cause of many characteristic orientation movements. If, for example, a steep slope is colonized by *Picea abies* (spruce) the stems do not grow perpendicular to the surface of the ground, but vertically as a plumb line. Other plants behave similarly. If they are artificially displaced from this position, then the stem curves until the growing region is once again vertical. This can be seen very beautifully with large inflorescences such as that of *Hyacinthus*. Grass haulms which have been lodged by bad weather are able to right themselves again by curvature at the nodes.

The nature of the reaction. Those movements which cause the organ concerned to move away from the centre of the earth are termed negatively geotropic. In contrast, main roots, for example, react in exactly the opposite way (Fig. 302). They grow towards the centre of the earth, and since they return to this direction of growth if they are displaced, they are said to be positively geotropic. Lateral roots of the first order grow either horizontally or at a certain angle obliquely downwards (Fig. 101A). This behaviour is referred to as transverse or plagiogeotropism. The lateral roots of the second order are usually insensitive to gravity.

Many side branches and leaves also show transverse geotropism. Particularly beautiful examples are provided by rhizomes which grow either horizontally or obliquely at a certain depth in the soil and always return to this position if they are artificially diverted (Fig. 205A). If dorsiventral organs such as leaves, and also many flowers, are displaced from their normal positions in space they also are able to assume them again by twisting of the petioles or pedicels (geotorsion). The spiralling of the ovaries of many orchids (see p. 339) also comes about under the influence of gravity.

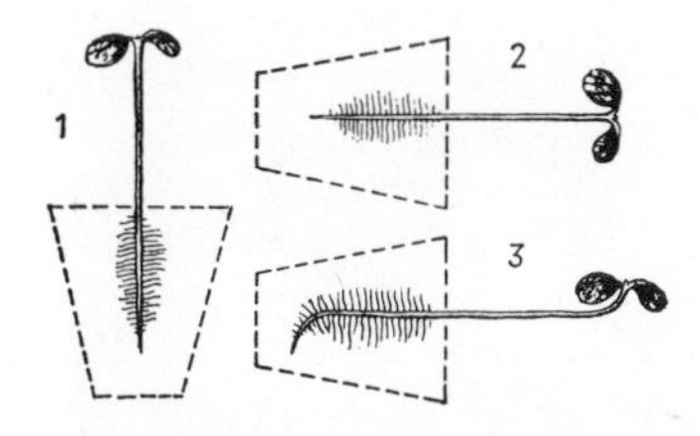

Fig. 302. Diagram showing the geotropic behaviour of a seedling. **1.** Normal position. **2.** Plant placed horizontally. **3.** Geotropic reaction. After Sierp.

Geotropic reactions are also widely distributed in lower plants. In contrast to the insensitive mycelia, the sporangiophores of *Mucor* spp. and the stalks of the fruit bodies of many agarics are negatively geotropic. The rhizoids of liverworts and fern prothalli react positively. Exactly as in phototropism, however, the geotropic reaction in many plants is variable, and internal or external factors may cause its reversal.

Thus the upper part of the flowering stem of a poppy is positively geotropic, and the flower bud nods, but it becomes negatively geotropic as soon as the flower opens. In many plants (e.g. *Holosteum umbellatum, Calandrinia*, etc.) a reversal of the geotropism of the pedicel can be seen after fertilization. In *Picea abies* and *Abies alba* the lateral branches are plagiogeotropic, i.e. they grow horizontally. If, however, the top of the negatively geotropic leading shoot is removed, then after some delay the upper side branches usually become negatively geotropic and turn upwards. One of them subsequently takes the place of the lost main axis, while the remainder return to their original positions. (An example of apical dominance, the frequently observed influence of the apical bud on the disposition of the laterals.) Environmental factors can also produce changes in geotropic response. The low temperatures of winter cause the shoots of many of our common weeds (e.g. *Senecio vulgaris, Sinapis arvensis, Lamium purpureum*, etc.), which in summer are negatively geotropic, to grow horizontally. The plagiogeotropic rhizomes of *Adoxa* and *Circaea* become positively geotropic when illuminated, and thus re-enter the soil. Those of *Aegopodium podagraria* behave similarly, even in red light, while an atmosphere of carbon dioxide causes a negative reaction. In darkness the plagiogeotropic shoots of *Vinca, Lysimachia nummularia*, etc., curve upwards. Traces of coal gas are often able to suppress geotropic responses completely.

Proof of the action of gravity. It can be demonstrated by a simple experiment, devised as long ago as 1806 by Knight, that in all the directional movements described above the earth's gravitational field is the effective agent. Knight fixed germinating beans to the periphery of a vertical wheel which he then rapidly rotated. He observed that all the roots grew outwards, while all the shoots grew towards the centre of the wheel. Evidently the growth of the plants is directed in these circumstances by the centrifugal force. If the wheel is placed horizontally and the speed of rotation is adjusted so that the centrifugal force is more or less equal to the gravitational force, then the roots grow obliquely outwards and downwards at an angle of approximately 45°, while the shoots grow at the same angle obliquely upwards and inwards. This is exactly in the direction of the resultant as calculated by the parallelogram of forces (Fig. 303), in which two adjacent sides represent the gravitational and centrifugal forces respectively. Gravitation and the centrifugal force are therefore detected in the same way by the plant. This is probably because the plant is extremely sensitive to any tendency for the components of its cells to be displaced by any kind of acceleration, no matter whether it arises naturally or artificially. A plant will therefore inevitably respond to the earth's gravitational field.

Besides the fact that gravity is responsible for initiating the orientation movements described above, it can also be shown that if gravity acts uniformly all around an organ no curvature develops. If, for example, a plant with a simple upright stem is placed in a horizontal position and rotated slowly about its longitudinal axis, one revolution lasting 3–30 min so as to avoid any significant centrifugal effect, then gravity acts on each side

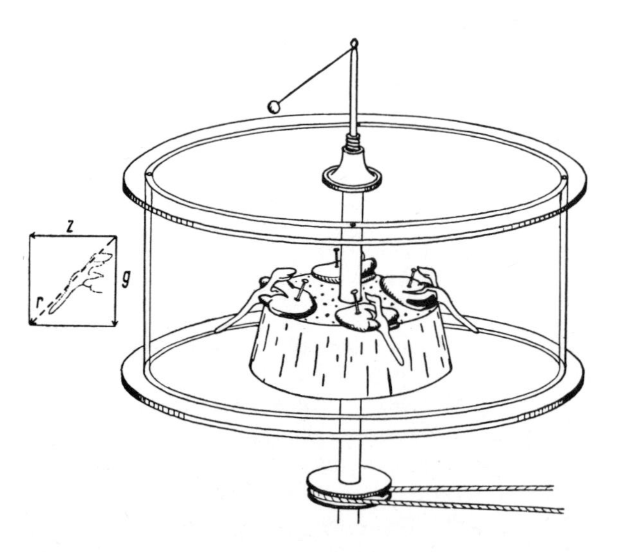

Fig. 303. Centrifugal apparatus rotating seedlings of *Vicia faba*. The roots grow obliquely downwards and outwards, and the shoots upwards and inwards, in the direction of the resultant (r) of the parallelogram of forces formed by the centrifugal force (z) and gravity (g).

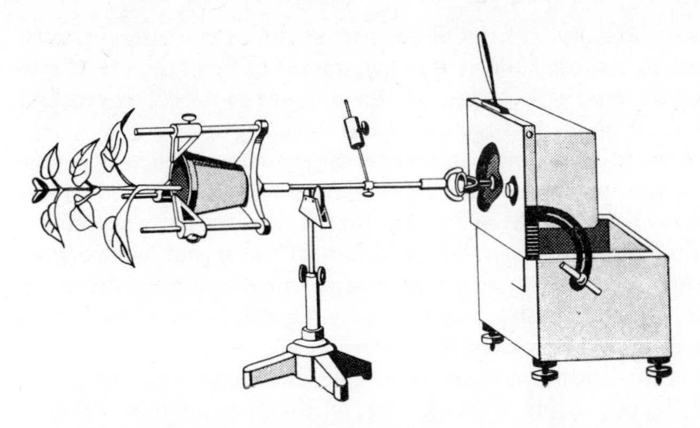

Fig. 304. Pfeffer's klinostat. The clockwork built into the lid of the box on the right can be fixed in any position. It is here seen turning a plant about a horizontal axis.

of the plant in turn. In these circumstances there is neither negative curvature of the stem nor positive of the root. The apparatus used in this experiment is termed a klinostat (Fig. 304). It must be stressed that this procedure does not suspend the action of gravity, but merely provides a means of eliminating its unidirectional action, and is useful in experiments on various kinds of irritability.

The mechanics of the movement. In geotropism, as in phototropism, the curvature is brought about by differences in growth on the two sides of an organ. In general, again only those regions still capable of growth show any reaction. Thus, if an orthotropous plant be placed horizontal, the curvature always develops in the main growing zone of the shoot or root directly below the tip, while the intermediate region remains unaffected (Fig. 302).

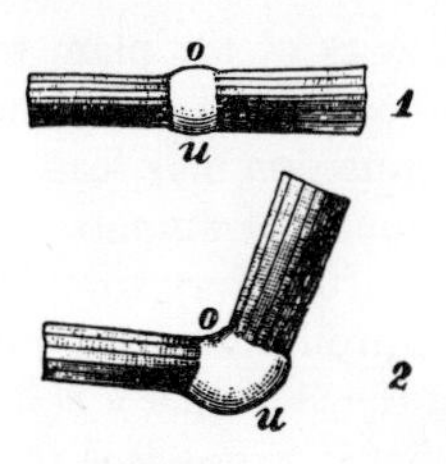

Fig. 305. A grass stem placed in a horizontal position (**1.**) shows an upwards geotropic curvature at the node. In (**2.**) it is seen how the lower side of the node elongates while the upper shortens, so that the stem becomes elevated at an angle of about 75°. After Noll.

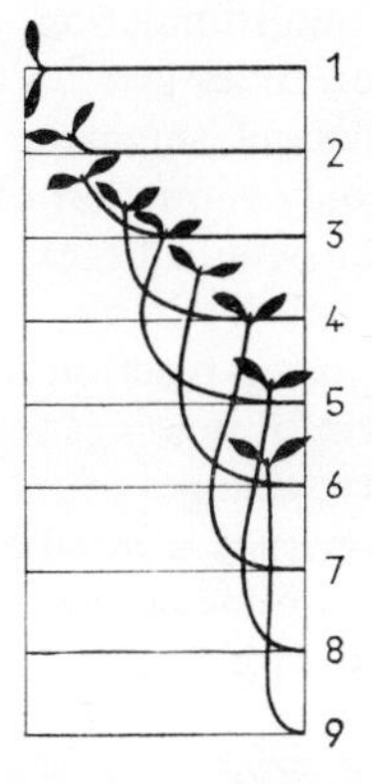

Fig. 306. Diagram showing the course of negative geotropic movement in a seedling. Explanation in text. After Noll.

Only in certain cases is gravity capable of stimulating differentiated or dormant cells to renewed growth. One example is shown by the haulms of Gramineae. If these are displaced from the vertical renewed growth occurs, but only at the nodes. The lower side of each node is stimulated into active growth, so that the haulm rights itself (Fig. 305). If a haulm is rotated horizontally in a klinostat growth occurs uniformly all around the node, showing that in these conditions gravity is still providing a stimulus. The trunks and branches of trees, in spite of their woodiness, can also occasionally show geotropic reactions, admittedly very slow, in the form of increased growth in length and thickness brought about by the cambium.

In those leaves of the Leguminosae provided with pulvini (e.g. of *Phaseolus*) geotropic curvature results not from growth, but from turgor changes on the two sides of the joint.

The successive phases of the bending of a stem of a seedling placed horizontally are shown diagrammatically in Fig. 306.

The curvature first appears directly behind the cotyledons (Fig. 306, 2), but then moves steadily towards the base. Finally, (9) the upper part of the stem is again vertical, while the bending is confined to a short zone at its base. Closer examination of phases 5–8, however, shows that the bending at first exceeds the amount required, whereupon the stem begins to curve back until finally it is again in the vertical.

It is tempting to assume that this bending back is merely the consequence of a new geotropic stimulus opposed to the first, since such must inevitably occur when the stem overshoots the vertical. It has, however, been shown that if the organ concerned is completely removed after stimulation from unilateral action of the stimulus concerned and rotated on a klinostat, then a geotropic curvature, just as one that is phototropic or caused by any other stimulus, always overshoots and returns. Therefore, in all the organs of a plant which for one reason or another develop curvature, a tendency exists for the organ to adjust itself to its normal position once the simple irritability phenomenon is exhausted. This tendency, indicating a sort of 'sensibility' in the plant, is referred to as autotropism. Together with the renewed geotropic stimulation, it is responsible for the peculiar, almost pendulum-like way in which the plant finally comes into the vertical after initially overshooting this position.

In some species, in certain conditions, the reaction time in a geotropic response can be much shorter than in a phototropic.

In favourable conditions, for example, the roots of *Lepidium* can show the beginning of a geotropic response in less than 20 min, and *Avena* coleoptiles in 14–49 min. The shoots and roots of *Vicia faba* do not react until after 85 min, and the nodes of grasses may require several hours.

The presentation time may also be very short.

The pedicels of *Capsella bursa-pastoris* will react if they are subjected to the influence of gravity for only 2 min, and then removed from its unilateral action by rotation on a klinostat. Likewise the pedicels of *Sisymbrium* and *Plantago*, and the hypocotyl of *Helianthus*, require only 3 min stimulation.

Again, these presentation times give no true picture of the sensitivity of the plant to gravity. Subliminal stimuli will summate until a visible reaction occurs, even if the individual stimuli last for only a fraction of a second. Here again summation may lead to a more marked effect than would be expected from the total amount of stimulation given. The interval between each stimulus must not be too long, since after every stimulation of every kind an opposing reaction immediately begins in the protoplasm and the initial effect dies away. Careful investigations of summation lead to the view that a plant can detect even the slightest changes of its position in space, but that it responds with a visible curvature only if gravity is allowed to stimulate it for a longer time. This failure to respond to shorter stimuli is of advantage to the plant, since otherwise the innumerable movements of the organs of a plant in wind, for example, would lead to continual growth curvatures.

As before, for the production of a certain curvature it is immaterial, within certain limits, whether a powerful stimulus acts for a short time or a weak one for a longer. The decisive factor is the amount of stimulation (intensity x time). Below a relatively small amount, there is again proportionality between the amount of stimulation and the size of

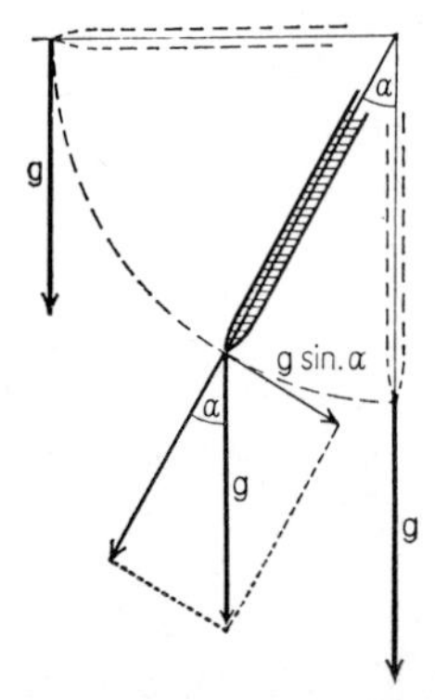

Fig. 307. The action of gravity on a root placed at an angle α to the vertical. The component $g \sin \alpha$ is geotropically active. Diagram after Sierp.

the reaction, but above this amount proportionality ceases. These quantitative relationships can be investigated by using a centrifugal force, which can be easily regulated, instead of gravity, or by allowing gravity to act at an angle of less than 90°. In the latter case it has been shown that the effectiveness of the gravitational force is proportional to the sine of the angle which the root makes with the vertical (Fig. 307). In a parallelogram of forces this would correspond to the component ($g \times \sin \alpha$) at right angles to the longitudinal axis of the obliquely placed organ. The component falling in line with the longitudinal axis does not remain completely ineffective, even though it initiates no geotropic curvature. Just as in phototropism, where light parallel to the axis has the effect of reducing the sensitivity of the organ, gravity acting in a longitudinal direction from the shoot apex downwards reduces the geotropic sensitivity. In the reverse direction this longitudinal component increases the sensitivity. In both phototropism and geotropism this moderating (or tonic) effect appears not to be identical with the irritability leading to the growth curvature, but to be an irritability phenomenon in its own right. In experiments continued for longer periods, in which the longitudinal component can become fully effective, the sine law is no longer strictly valid. In these circumstances gravity has its maximum effect when the root is placed not at an angle of 90° to the vertical, but at between 90° and 135°.

Susception. In the detailed analysis of geotropism we must again distinguish between the susception of the stimulus, induction, the possible transmission of the stimulus and the reaction itself. While it was possible with phototropism to make some assertions, at least about susception, in geotropism even this first step in the chain of events is beset with many uncertainties.

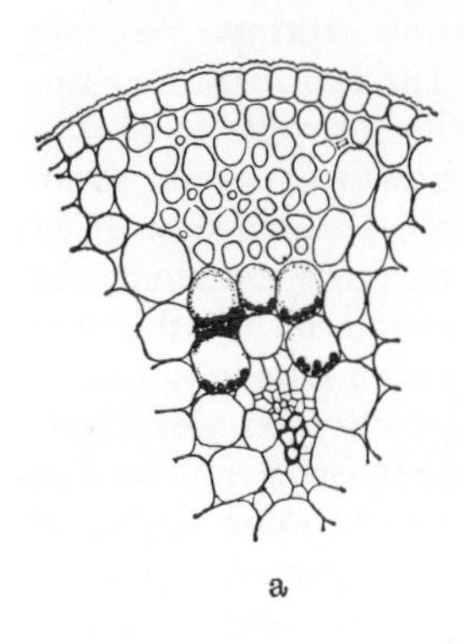

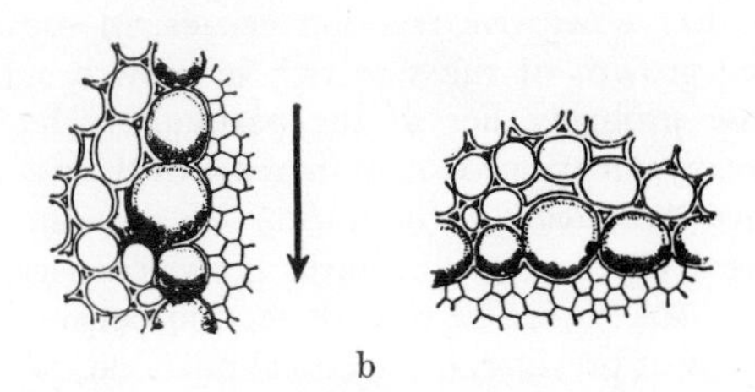

Fig. 308. Statolith starch. **a.** In the sheath of a vascular bundle in the peduncle of *Arum ternatum* placed and sectioned in a horizontal position. **b.** In a transverse section of a shoot of *Vinca minor* lying horizontally. On the right is the same shoot after rotating it 90° about its longitudinal axis. After Haberlandt.

At present two possibilities warrant detailed discussion. First, gravity must have in some way an acceleratory and compressing effect upon the sensitive protoplasm. By analogy with the statoliths of animals, which, e.g. in crabs, govern the equilibrium of the animal by the pressure they exert, specific heavy particles have been supposed to act similarly in plants. Movement of these particles in the cell under the influence of gravity would cause pressure, and so stimulate the protoplasm. In the cells of the root cap and the starch sheath of the stem (Fig. 94B), for example, there are always large starch granules which are remarkably easily movable in the attenuated cytoplasm of these cells. These granules in particular have been regarded as statoliths, and responsible for the susception of the gravitational stimulus. If a stem or root is placed horizontal these granules quickly accumulate on the lower sides of the cells (Fig. 308). Those fungi which react to gravity never possess starch, so here the initiation of the geotropic process would have to depend upon the displacement and pressure effect of other, possibly submicroscopic, particles. Only with the unicellular rhizoids of the stonewort *Chara* (see p. 474), where highly refractive bodies of unknown composition take on the role of statoliths, has it been possible to gain insight into the ultrastructural aspects of the geotropic response. When these bodies are displaced to one side, the approach of the Golgi bodies and vesicles on this side to the apical growing region is obstructed. The wall material is thus delivered principally to the other side of the tip, and curvature results. With multicellular organs, such as the root cap, so elegant an explanation of geotropism is not yet forthcoming.

Quite apart from the statolith theory, attempts have also been made to explain geotropism in terms of an electrical response, since it has been shown that the speed of diffusion of ions is differentially affected in the earth's gravitational field. Even if a dead membrane, such as a disc of filter paper, is saturated with salt solution and placed horizontally, the upper side is temporarily positive with respect to the lower, the potential difference amounting to a few mV (geoelectric effect). This is caused by cations migrating to the lower surface faster than the anions. Thus it is conceivable that the plant first perceives the geotropic stimulus as a displacement of ions within the cell which is being acted upon by gravity. It is still not possible to decide between these possibilities, but the evidence in favour of electrical theories is becoming increasingly doubted. It is nevertheless remarkable that when plant organs are laid horizontally, the lower side does become positively charged, and that curvatures can be produced by means of an electrostatic field. Roots and shoots behave inversely as in geotropism, roots bending towards the side which becomes positively charged, and shoots away from it (electrotropism).

Induction and conduction. Following susception is the induction of the stimulus. Here again it is well established, principally through the study of conduction, that growth-regulating substances are involved. In both shoots and roots a larger quantity of these substances collects on the lower side than on the upper, so that geotropism, like phototropism, is characterized by an unequal distribution of growth-regulating substances. In shoots it is immediately evident that the lower side simultaneously begins to grow vigorously, so that a negative reaction ensues. In roots the positive geotropic response implies reduced growth of the side rich in growth substances. The explanation of this inverse response probably lies in the particularly high sensitivity of roots to growth substances, already mentioned. Any increase of their content of growth substances above normal at once produces an inhibition of growth, so that roots supplied with an additional quantity of these substances grow less vigorously than those containing the normal amount. The nature of the connection between the susception, be it in the form of a displacement of pressure, or electrical polarization, and the flow of growth substance to the lower side of the organ is not yet known. Perhaps permease enzymes play a role. Were these locally activated by pressure they might allow growth substance to pass through from one cell to the next. As in phototropism this would allow transport between cells in the absence of plasmodesmata. Certainly, again as in phototropism, the concept of a laterally displaced stream of growth substance is adequate to explain all the phenomena observed.

In shoots any internode is capable of reacting to gravity, but in roots sensitivity is more or less confined to the extreme tip for a length of at most 2 mm, although the curvature takes place principally in the zone of extension growth, some 3–4 mm behind the tip. Consequently, there must be some conduction of the stimulus, albeit for only a few millimetres. Roots from which the extreme tip has been removed usually lose all their geotropic sensitivity, although it is regained if the tip is cemented back into position with gelatin. In the oat coleoptile the tip is also particularly sensitive for a length of about 3 mm, and it is here apparently that the lateral displacement of the stream of growth substance occurs. Nevertheless, there is, as in the root, a limited sensitivity elsewhere, but this seems to consist of a changed capacity to react to the growth substance. Especially clear cases of conduction of the stimulus are found in seedlings of *Panicum* and related genera. Here only the very short coleoptile is capable of susception, whereas the response occurs in the well-developed hypocotyl.

It is clear that geotropism, like phototropism, is a complex phenomenon, involving more than the mere displacement of growth substance to the lower side of the organ. There are other metabolic differences between the two sides. For example, the concentration of glucose increases on the lower side and the rate of respiration rises, and the side becoming convex shows a lower pH value in the cells and a smaller amount of catalase.

All these things indicate that geotropic stimulation causes marked physiological disturbances, the causal significance of which in the geotropic reaction is still unknown.

Other geotropic reactions. The plagiogeotropic organs present difficulties, since their behaviour cannot be explained simply by an accumulation of growth substances on their lower sides. A lateral root, for example, which normally grows obliquely with respect to the positively geotropic main root, shows positive geotropism if placed horizontally, but negative if placed vertically with the tip downwards. It therefore possesses both potentialities, and attempts have been made to explain its maintenance of the oblique position as a consequence of the simultaneous action of both kinds of geotropism. In dorsiventral organs such as leaves there is a tendency, arising from some unknown inner cause, for the upper side of the organ to grow more strongly than the lower, a so-called epinasty. If, for example, a specimen of *Coleus* is rotated in a horizontal position on a klinostat the removal of the unidirectional action of gravity causes all the leaves to bend down towards the shoot axis and cluster themselves around it. Apparently in the normal position of the leaf this epinasty works in opposition to a negative geotropic reaction, resulting in the precise plagiotropic position. This greater sensitivity of the upper side of the organ to growth substances may also account for the tendency for lateral branches to become erect when the apical bud is removed (p. 356). There are remarkable and unexplained instances of twining plants where gravity apparently stimulates the growth not of the upper or lower side but of one of the margins of the stem (lateral geotropism; see p. 373).

Although some steps of the process are understood, the whole phenomenon of geotropism is still in many respects puzzling. The plant's ability to detect gravitational force, and in consequence to orientate itself in space, is naturally of great importance for its growth. Because of the specific mechanisms the plant possesses for detecting them, light and gravity are by far the most important of the environmental factors determining the spatial arrangement of its organs.

In natural conditions a phototropic curvature in response to lateral light causes simultaneously a geotropic stimulation of the organ displaced from the vertical, and vice versa. In general, however, the phototropic stimulus appears to be stronger than the geotropic, so the former plays the predominant role. Even a very weak phototropic stimulus can counterbalance an antagonistic geotropic stimulus. Thus in *Avena* an intensity of only 0.04 lux is required to prevent the upward curvature of a coleoptile placed horizontal.

3. Haptotropism[54]

Besides light and gravity yet other environmental factors can cause orientation movements of individual plant organs. Thus many plants, as animals, are sensitive to touch. Many seedlings, for example, particularly if they are etiolated, react to being touched on one side (such as by rubbing with a matchstick) with a clear curvature towards the stimulated side. Many petioles (e.g. of several species of *Tropaeolum* and *Clematis*, *Fumaria officinalis*, etc.), leaf tips (*Gloriosa*), aerial roots (*Vanilla*), stems and inflorescences behave similarly after touching. This phenomenon is termed haptotropism (thigmotropism).

Tendrils. Particularly striking is the sensitivity to touch of the tendril climbers (see p. 192 seq.). This sensitivity causes them to move so that their climbing organs become attached to available supports, and the plant firmly established.

The thread-like tendrils of *Bryonia dioica* may be taken as an example. As Fig. 309 shows, these in their young state are coiled up towards their morphologically upper sides like clock springs. Later they unwind themselves, leaving only a slight downwards curvature. At the same time they begin a remarkable 'searching' movement, the tip describing a circle or an ellipse, and the whole tendril a conical surface. The axis of this cone is at first directed obliquely upwards, but later it may sink and fall below the horizontal. This nutational movement, which in many climbing plants may be intensified by a simultaneous nutation of the whole of the upper part of the shoot axis, comes about as a result of autonomic growth processes which rotate around the organ concerned (see p. 371 seq.). It has no direct connection with haptotropism itself.

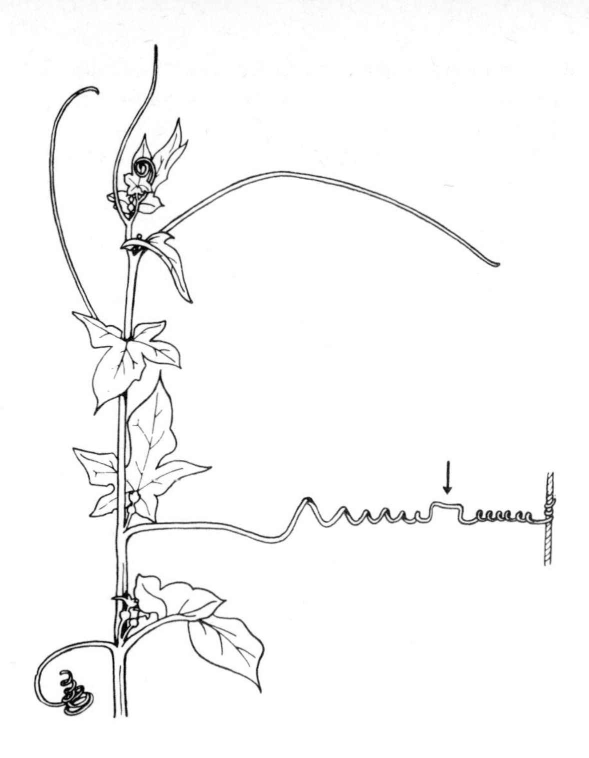

Fig. 309. Portion of a shoot of *Bryonia dioica* with tendrils in different stages of development. The uppermost tendril is still coiled up like a watch spring, in the middle is a tendril which has twined round a support and shows the 'inversion point' (arrow), and below on the left is a tendril unconnected with a support showing the coiling up with age. *c.* $\times\frac{1}{3}$.

As soon, however, as the tendril, as a result of this circular movement, strikes a support of not too large a diameter and in a suitable position, a typical haptotropic reaction ensues. The tendril begins to become concave at the site of contact and embrace the support, adapting itself to its shape, until its upper free end has wrapped itself several times around it. At the same time the proximal part of the tendril begins to coil up and form a helix, so that the plant becomes attached to the support as if by a spring, and is thereby anchored. On mechanical grounds this coiling up of the tendril requires one or more 'inversion points' (Fig. 309, arrow), so that the torsion is equalized.

Extension growth soon ceases in that part of the tendril in contact with the support. Growth in thickness, however, especially the laying down of mechanical tissue, frequently takes place, so that eventually it is very difficult to free the tendril again from the support. If a tendril encounters no support it eventually coils itself up again, if somewhat irregularly (Fig. 309, below), and no further differentiation occurs within it. Usually a tendril remaining functionless in this way has a distinctly shorter life than one which becomes attached.

Mechanism of haptotropic movement. If a tendril is touched for only an instant with, for example, a matchstick, a concavity develops, easily visible with the naked eye, at the site of contact. The reaction time, at optimal temperatures and with sensitive objects (*Cyclanthera, Sicyos, Passiflora*), may amount to less than thirty seconds. The concavity results from a sudden rapid growth of the side of the tendril opposite to that touched. This growth is of such intensity that even in old tendrils an amount of growth takes place within a few minutes that would hardly be achieved in twenty-four hours in the normal extension growth of the tendril. The place touched, however, shows no growth; it even contracts to a certain extent during the curvature. If the stimulation was very brief, the first reaction is usually followed by a reverse movement, revealing the phenomenon of autotropism already discussed in a preceding section. The tendril thus becomes straight again. When a support is being seized it usually happens that the incurving portion of the tendril makes further contacts, so that it receives continuous stimulation. This leads to new reactions and thus to the embracing of the support. Weak individual stimuli, rapidly repeated, will summate as with the other tropisms considered.

Even with a continued complex reaction, opposed autotropic reactions may still occur, so that after a tendril has wound round a support a temporary loosening of the coils is often observed. Soon, as a result of either further autonomic movement or wind agitating the coils, renewed contacts are made with the support, and the coils contract again. Finally, with extension growth setting in here and there, they become firmly attached to the support. The coiling up of the proximal part of the tendril also comes about as a result of strong unilateral extension growth.

The distal third of a tendril is usually the most sensitive region. Tendrils differ very much in different plants in respect of their haptotropic behaviour. In some plants (e.g. *Cobaea scandens*) they are equally sensitive all round and will bend towards whichever side is touched. Others are sensitive only in the morphologically lower side, and show no response if touched on the upper. The tendrils of most climbing plants are probably of this kind. Finally, some plants are also known in which the tendrils, if touched on the upper side, will bend only towards the lower side (*Sicyos, Momordica*). Such tendrils are not, strictly speaking, displaying a tropic, but a nastic movement. A very surprising discovery is that those tendrils which are capable of bending only towards one side, and which will react only when touched on the lower surface, are nevertheless sensitive on the upper surface. This sensitivity is in fact no less well developed than that of the lower surface. If such a tendril is lightly touched on both the upper and lower surfaces simultaneously there is usually no reaction. Such a situation can only be explained by the existence of haptotropic sensitivity on the upper surface. This observation shows that sensitivity to external stimuli may in some plants be developed in regions where stimulation gives no visible response.

The capacity to react haptotropically is bound up with a definite process of susception. Tendrils are never stimulated to curvature by rainfall, no matter how heavy. A fine jet of pure water, oil or even of mercury can be shot against a tendril, and even though the tendril may thereby suffer mechanical damage, no curvature results; but if very fine particles of clay are suspended in the water, or fine fatty acid crystals in the oil, the plant reacts immediately. Touching or striking with a glass rod that has been coated with damp gelatin is also without effect; but curvature begins immediately if a thread of wool, weighing no more than 0.00025 mg, and incapable of being detected by human touch, is drawn along a sensitive tendril such as that of *Sicyos*. These facts indicate that what is detected is not a static pressure, but a kind of touch or tickling stimulus. Touching the rough surface of a more or less firm organ always generates rapidly changing differences in pressure or tension between adjacent sites, and it is apparently this that is an effective stimulus. A blow probably fails to produce a haptotropic response because it depresses the whole surface to approximately the same extent.

Many attempts have been made to find an anatomical basis for sensitivity to contact, such as is present in the case of the human sense of touch. Possibly the plasmodesma-like fingers of cytoplasm in the outer walls of the epidermal cells (ectodesma, see Fig. 58), demonstrated in many tendrils, play a part in the susception of the stimulus similar to that of the touch points in human skin. Particularly conspicuous and isolated pits in the outer walls of the epidermal cells (called sensory pits) are present in only some tendrils, and cannot therefore provide a general basis for haptotropic sensitivity.

Following the reception of the stimulus at certain sites in the protoplasm of the epidermal cells is the excitation process proper, the induction. The nature of this is still not clear. What is certain is that there is a rapid conduction of the stimulus across the whole tendril, since the actual reaction is on the side opposite to that touched. In very sensitive tendrils the speed of this conduction is estimated as at least 4 mm/min, and thus it exceeds phototropic and geotropic conduction more than tenfold. It results in the promotion of growth of the opposed side, as already described.

A certain amount of conduction also occurs longitudinally in the tendril, but this is confined to a few millimetres on each side of the site of contact. It plays some part in the seizing and embracing of supports, in addition to the continued stimulation of new parts of the tendril already described. It is probable that the subsequent spring-like coiling of the proximal parts of the tendril is caused by more extensive conduction which arises as the tendril makes itself fast around the support.

Since the reaction of the tendril is a growth movement, a redistribution of growth substances is probably involved. No further insight has yet been gained into the individual steps of the reaction.

4. Chemotropism[54]

Among the external factors which can cause a directed growth movement of individual plant organs are, in addition to light, gravity and contact, certain gases and water-soluble chemical substances. If these substances are not distributed uniformly in a medium, curvature may occur in the direction of the concentration gradient. If the curvature is towards the higher concentration it is said to be a positive chemotropic reaction, and if towards the lower, negative. In the first instance the organ turns towards the source of the substance, in the second away from it.

The hyphae of fungi, particularly at germination, react in a positively chemotropic manner to concentration gradients of sucrose, glucose, peptone, asparagine and meat extract, and often also to those of phosphates and ammonium salts. Organic and inorganic acids and certain of their own metabolic products produce a negative response. Chemotropic reactions, brought about by specific gamones, also appear to play a part in sexual fusion (*Mucor*, see p. 505). The optimal effect often occurs at a certain concentration, positive responses occurring with lower concentrations and negative with higher. The significance of such reactions, particularly in the nutrition of heterotrophic organisms, is immediately clear. Chemotropism probably also enters into the penetration of the absorptive organs of parasitic fungi and higher parasites into the tissues of the hosts, and for the detection within them of specific sources of nourishment, e.g. the detection of the sieve tubes of the host by species of *Cuscuta*. Hyphae of *Saprolegnia* are very sensitive to mixtures of amino acids.

In autotrophic higher plants chemotropic reactions are well known in the germination and growth of the pollen tube, which in this way is facilitated in its passage from the stigma through the style to the ovule.

If pollen grains, e.g. of *Scilla patula*, are germinated on a plate of nutrient agar into which a piece of stigmatic tissue has been inserted the pollen tubes will always grow towards it, apparently because of certain sugars, and in other cases of proteins and other substances, diffusing from it. Pollen tubes usually react negatively towards the oxygen of the air. Inhibition of the growth of the pollen tube in the tissue of the style is probably responsible for certain forms of self-sterility in plants. In the Cruciferae and Compositae inhibition may even take place on the papillae of the stigma, a lack of the enzyme cutinase preventing penetration of the cuticle. Whole series of multiple allelomorphic self-sterility genes are known in several plants. If the same allele is present in both the diploid tissue of the style and in the haploid pollen tube the growth of the tube is inhibited, while when the alleles of the partners are dissimilar the style is traversed without hindrance (see p. 402). Specific substances, the nature of which is determined genetically, also appear to be excreted by the ovules, and to direct the approach and penetration of the pollen tubes chemotropically. Amongst the species of *Oenothera* various mixtures of sugars and amino acids play this role. Hereditary differences in the behaviour of the tubes may also be present at this stage. Such phenomena are clearly of considerable importance in the reproduction of plants.

In certain circumstances roots also appear to behave chemotropically. Thus differences in humidity will often cause roots to curve towards the damper side, a phenomenon known as hydrotropism. This evidently makes it possible for roots, and also for the rhizoids of liverworts and fern prothalli, to seek out by active growth the moister regions of the substrate. In roots it is only the extreme tip which is sensitive to, and able to transmit, this stimulus.

Specific excretions possibly cause the roots of different genera and species to be chemotropically attracted to each other, or conversely repelled (see p. 340), a factor of considerable importance in plant sociology. How far the known capacity of root hairs to react to dissolved minerals affects the growth of the root in the soil is little known. Gases can also be detected by roots. Oxygen and even carbon dioxide usually produce a positive chemotropic response.

Mechanically all these chemotropic curvatures appear again to be brought about by differential growth. The individual steps in the reaction are in no case known.

Mention can only be made here of yet other stimuli, such as those provided by electricity, wounding and heat, which bring about tropic movements. These tropisms, referred to as galvanotropism, traumatotropism and thermotropism respectively, are possibly only special forms of chemotropism.

B. Nastic movements[55]

As has already been pointed out (p. 349), beside the tropic movements there exist others in which the direction of the releasing stimulus is of no account. The stimulus in these cases often consists of nothing more than a general change in some external factor, such as temperature or light intensity. These nastic movements always take certain courses, determined by the particular dorsiventral structure of the organ concerned. They are not responsible therefore for the spatial orientation of the plant. With regard to the mechanics of these movements, we must distinguish between those nasties which are brought about by growth, resembling in this respect the tropisms already considered, and those which are caused by turgor changes, usually reversible. Some tropic movements are in fact caused by turgor changes, but they are comparatively rare.

1. Thermonasty

Nastic movements, depending on growth differences on the two sides of the organ concerned, are seen particularly clearly in the perianth segments of the many flowers which open or close according to the temperature. Such a phenomenon is referred to as thermonasty.

If a tulip or crocus flower is brought from the open air into a room some 10°C warmer the flower opens in the matter of a few minutes, and the reverse movement takes place when the temperature falls. Closer inspection shows that the dorsiventrality of the perianth segment expresses itself in a different optimum temperature for growth on the two sides of the segment. The optimum for the upper side is higher than that for the lower, so that a rise in temperature promotes the growth of the former and a fall in temperature that of the latter. The perianth segments are capable of reacting repeatedly, and in the course of a single thermonastic curvature may, as for example in the tulip, elongate by as much as 7 per cent of their length. In the whole flowering period, therefore, repeated thermonastic curvatures may cause an increase in length of over 100 per cent. The sensitivity of many flowers is extremely high. Crocus flowers can detect changes in temperature of as little as 0.2°C, and tulip flowers of 1°C.

Pedicels and peduncles, e.g. of *Anemone, Oxalis, Geranium*, etc., often show thermonastic responses. Similarly, tendrils may respond nastically to variations in temperature by undergoing coiling. Leaves in general show hardly any response to changes in temperature, but where the petioles are furnished with pulvini (as in *Oxalis acetosella, Desmodium, Mimosa*) thermonastic turgor changes may occur.

2. Photonasty

Changes in light intensity can also give rise to nastic growth movements. Many flowers show such photonastic opening and closing movements, so that the appearance of a meadow full of flowers, for example, in full sunshine is often very different from that on a dull day. Of course, it is not possible to say without further investigation whether light or temperature is the principal factor involved; often both are acting simultaneously.

The perianth members of many waterlilies, cacti and species of Oxalidaceae show photonastic behaviour, as also do the capitula of many Compositae (Fig. 310), in which the ligulate ray florets simulate single petals. Even the shadows of passing clouds can cause a response in sensitive plants (e.g. *Gentiana* spp.). Illumination usually causes an opening of flowers, and shading a closing, but night-flowering species (e.g. *Silene nutans, Melandrium album*, etc.) behave inversely. The developing leaves of various plants, provided they are still capable, in some parts at least, of growth, may also show photonastic responses. Thus the young leaves of *Impatiens* spp. sink when they are darkened as

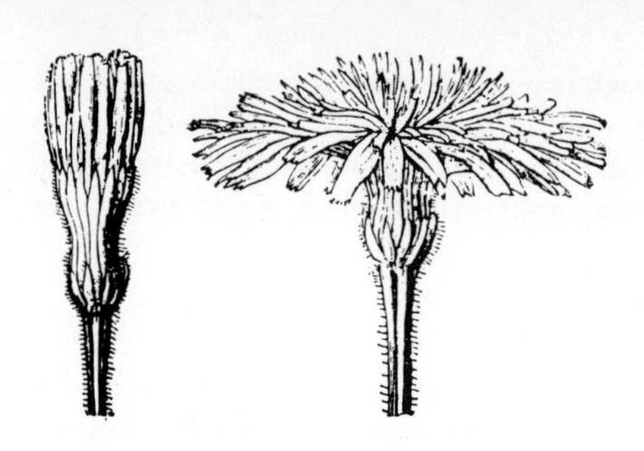

Fig. 310. Capitula of *Leontodon hastilis* (Compositae): left, closed in the dark; right, opened in the light. After Detmer.

a result of the growth of the upper side accelerating. This is later compensated for by growth of the lower side, even in continued darkness. Fully developed leaves only show nastic movements when a pulvinus is present at the junction of the petiole and axis, as in the Leguminosae. Here we are concerned with movements brought about not by growth, but by turgor changes (see p. 367).

Very little is yet known of the mechanism of the photonastic response. The changing of the light intensity seems to have some delayed effect. Thus in *Cereus nycticalus* ('Queen of the Night') the flowers always open twelve hours after a preceding change from darkness to light. Thus they normally open in the evening, and have withered by morning.

In nature the diurnal cycle of day and night causes all thermonastically and photonastically sensitive flowers rhythmically to open and close, a cycle generally termed nyctinastic. The analysis of these periodic movements is made difficult by the fact that autonomic movements often coincide with and contribute to those induced by external factors. The closer examination of nyctinasty is accordingly deferred (see p. 374).

3. Hapto- and chemonasty

In certain insectivorous plants, especially with the tentacles of *Drosera* spp., nastic growth movements occur in response to touch and chemical stimuli. Haptonasty and chemonasty are closely related to haptotropism and chemotropism, in fact the nastic movements are sometimes almost tropic, and vice versa.

The marginal tentacles of the leaf of *Drosera* (Fig. 221) react like tendrils to a light touch, but contact with damp gelatin and so forth produces no similar haptonastic response. The only sensitive region is the glandular head producing the sticky secretion. When this is touched the stimulus is rapidly transmitted to the base of the tentacle. The growth of the tentacle is then immediately stimulated, particularly on the abaxial side, so that the tentacle curves in towards the middle of the adaxial surface of the leaf.

The response, especially after chemical stimulation, may appear as little as ten seconds after contact with the glandular head, indicating a rate of conduction of some 8 mm/min. The sensitivity is also high. Contact with a hair weighing only 0.0008 mg can cause curvature of a tentacle. If the shorter centrally placed tentacles be touched instead of those at the margin the stimulus is transmitted to the latter. Now we no longer see a purely nastic response, since the marginal tentacles quite clearly bend towards the touched place, indicating a tropic curvature. An insect which is at first held by the gluey secretion of only a few tentacles is thus, as a result of these responses, seized by an increasing number of glandular heads. Eventually it dies in the secretion and the corpse is digested away.

The tentacles respond not only to touching but even more strongly to chemical stimulation, and stimulation of this kind might well arise from the insects caught. Even pure protein, fragments of cheese and meat, droplets of water containing traces of ammonium salts or phosphate and so on can cause the stalks of the tentacles to undergo very marked chemonastic growth reactions. In certain conditions these may cause curvatures of up to 180° within no more than one minute.

Only the glandular head of the tentacle is sensitive to the chemonastic stimulus, and there is again a smooth transition from nastic to tropic responses. If the chemical or haptic stimulus was of very short duration a reverse autonastic curvature, corresponding to an autotropic movement, may be observed after its cessation. Usually the capacity of the tentacles to react to stimuli and to show growth movements is somewhat limited, and after the excitation has been repeated two or three times they become exhausted.

In some of the foreign narrow-leaved species of *Drosera* the whole leaf blade may roll up in a nastic manner over the imprisoned animal. In *Pinguicula*, on the other hand, only the leaf margins show a weak nastic incurving.

4. Seismonasty[32]

The tropic and nastic phenomena considered hitherto have been brought about largely by growth processes. There are, however, a series of very striking movements displayed by certain fully developed parts of plants which are caused not by growth, but by more or less reversible turgor changes in cells located in definite regions of the tissue. These variation movements are usually very much more rapid than those caused by growth, and consequently are often particularly striking.

A whole series of plants displays movements of this kind after touching, a phenomenon termed seismonasty. Whereas with haptonastic and haptotropic movements only a light touch or tickling stimulus is effective, a stream of fluid, for example, always failing to produce any movement, seismonastic movements can be caused by any kind of pressure. Thus a blow from a raindrop, provided the stimulus exceeds the threshold value, will cause an immediate response, usually reaching full intensity. In fully developed plants at a sufficiently high temperature there is ordinarily no proportionality between the size of the stimulus and that of the response ('all-or-nothing' law).

Many petals, stamens and stigmata show seismonastic reactions. They also occur in leaves of *Mimosa pudica* (the 'sensitive plant'), and in those of some other species and genera, including many insectivorous plants.

Movements of stamens. In the unstimulated condition the stamens of *Berberis* (Fig. 311) diverge widely within the petals. As soon as the filament is touched on its inner side at the base the whole stamen moves rapidly upwards and inwards about its point of attachment, so that it becomes adpressed to the ovary. In favourable environmental conditions there is an interval of only 0.04 sec between the touching and the reaction, and the reaction itself requires only 0.15 sec.

In the flowers of *Opuntia* the stamens move in towards the stigma in similar manner,

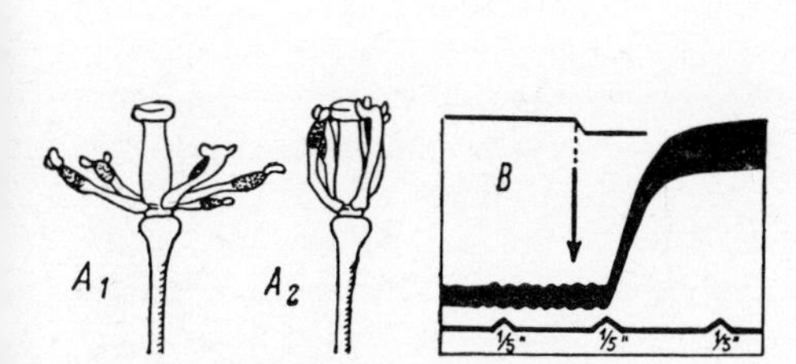

Fig. 311. Flowers of *Berberis vulgaris*, the perianth removed. **A$_1$.** Stamens in the unstimulated condition. **A$_2$.** After touching. **B.** Photographic record of the movement of a stamen (black band). The time marks below indicate intervals of $\frac{1}{5}$ sec. The arrow indicates the moment of touching. **A.** *c.* ×2. **B.** After Colla, modified.

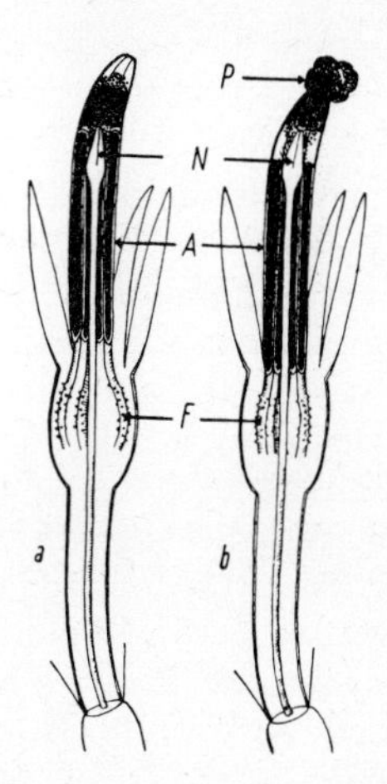

Fig. 312. Disc florets of *Centaurea jacea* cut open. **a.** Filaments (F) turgescent and bent outwards. **b.** Following stimulation the filaments unbent and contracted. The anther tube (A) is thereby drawn down over the stigma (N), so that the pollen (P) is pushed out above. *c.* ×3.

but in *Sparmannia africana* and species of *Helianthemum*, where the filaments are also excitable, the stamens move not inwards after touching, but outwards.

In some species of *Centaurea* (e.g. *C. jacea, montana*, etc.) the filaments of the stamens are turgid and bent outwards (Fig. 312). When they are touched they show a rapid elastic contraction, thus drawing the closed anther tube downwards. The stigmatic head of the style, which lies within the anther tube like the piston of a pump, pushes the pollen out of the top of the tube, from where it is able to adhere to the bodies of insects visiting the flowers. The contraction of the filaments often amounts to 20–30 per cent of their initial length and it takes place in a few seconds. In *Centaurea*, as in *Berberis*, recovery sets in immediately after the reaction. Often the turgor is restored in no more than a minute, and the organ returns to its original position. It is now capable of renewed stimulation and reaction.

Movements of stigmas. Quite similar phenomena are shown by the excitable stigmas found in species of *Mimulus, Incarvillea, Catalpa, Torenia*, etc. (Fig. 313). In the flowers of these species the lobes of the unstimulated stigmas gape widely apart. As soon as the inner sides are touched, however, the lobes snap together like the jaws of an animal, and enclose any pollen deposited between them. Here again rapid recovery occurs, and the stigmatic lobes reopen.

All these striking and rapid movements are mechanically similar. They can all be traced back to a sudden fall in turgor in certain groups of cells within the organ concerned. Conspicuous in this connection is the extraordinary elasticity of the cell walls of the motor tissue. As soon as the turgor is reduced (cf. the plasmolysis of the cells), the cell walls undergo marked contraction. Supporting evidence comes from the fact that in, for example, the filaments of *Centaurea* and *Sparmannia* fluid leaves the motor cells at the moment of contraction. This fluid consists not simply of water, but is apparently the content of the vacuoles.

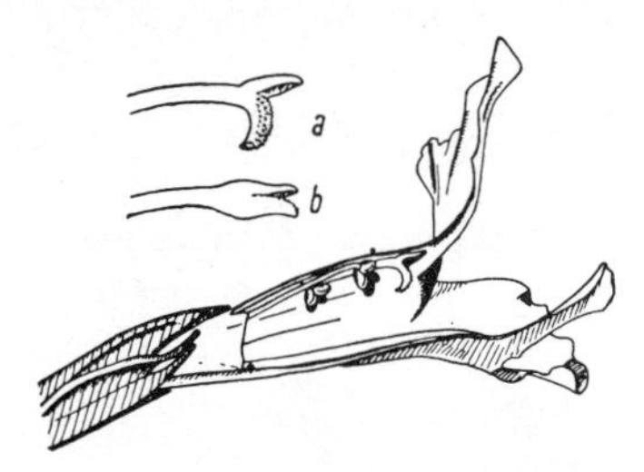

Fig. 313. *Mimulus luteus*. Flower cut open so that the positions of the stamens and of the unstimulated stigma are visible.
a. Unstimulated stigma. **b.** Stimulated stigma. **a.** and **b.** *c.* ×2.

As will be seen from the discussion on p. 320 seq., a turgor change may come about in various ways. On the other hand, there are no precedents for a rapid change in the elastic properties of the cell wall, and consequently of the wall pressure. The suddenness with which the fluid leaves the cells rules out even a rapid change in the osmotic concentration of the cell sap as the causal factor. The only possible explanation remaining is a sudden change in the permeability of the protoplasm. Even in plants not capable of showing seismonastic reactions, mechanical shock may increase permeability and transpiration, probably indicating changes in the bounding layer of the cytoplasm. It seems likely that in a seismonastic reaction, following the susception of the stimulus, there ensues, from causes as yet unknown, an almost explosive breakdown of the semi-permeability of the protoplasmic sac, so that the elastic force in the expanded cell wall can immediately manifest itself. As a result of the protoplasm becoming suddenly permeable, the cell wall contracts and the cell sap is expressed. It is still not clear whether contractions in the protoplasm play any part in this process. Immediately after this event, however, recovery sets in; the protoplasm becomes dense again and its semi-permeability is regained. Simultaneously there may well be a renewed manufacture and secretion into the vacuole of osmotically active substances, so that the suction pressure of the cells is regenerated. The water that was earlier expressed is generally, as for example in *Sparmannia*, soaked up by the strongly absorbent cell walls. This is taken up again by the recovering cells, and their former turgor and reactivity is fully restored. It is thus understandable that immediately after a reaction further stimulation is without effect (refractory stage), but that after a short time renewed stimulation will cause a response, although possibly much weaker.

Since the maintenance and restoration of the semi-permeability of the protoplasm are typical vital phenomena, it is not surprising that oxygen deficiency, narcosis with ether or chloroform, and temperatures outside the physiological range either inhibit seismonastic reactions or prevent subsequent recovery.

Mimosa. Conspicuous seismonastic movements are shown by various tropical *Mimosa* spp. Here the leaves and petioles, all of which are furnished at their bases with joints, rapidly close together after touching. Wounding, burning, and stimulation by electric current from an induction coil will also cause the movement to take place (traumatonasty, thermonasty, electronasty).

In *Mimosa pudica* (Fig. 314) there is a joint at the insertion of the primary rachis on the axis. The four secondary rachides, themselves similarly jointed at the base, bear pinnules in opposite pairs, and each pinnule is jointed at its insertion. These joints, usually appearing externally as a slight basal swelling, consist of isodiametric parenchymatous cells, usually somewhat larger on the lower side, and all with delicate walls. The ring of vascular bundles running into the petiole coalesces in this region to a central strand which shows very little resistance to bending (see Fig. 315).

When such a leaf is knocked, the pinnules fold obliquely upwards in pairs until their surfaces touch. At the same time the secondary rachides approach each other, and finally the primary rachis itself collapses downwards (Fig. 314). At a sufficiently high temperature and humidity the whole process can take place very rapidly, and the latent time up to the beginning of the reaction may take as little as 0.08 sec. If after stimulation the plant is allowed to rest, complete recovery takes place in 15–20 min, and the leaves return to their original position.

Particularly striking in *Mimosa* is the extensive conduction of the stimulus which occurs after wounding. This can be demonstrated by slightly scorching a leaflet with a

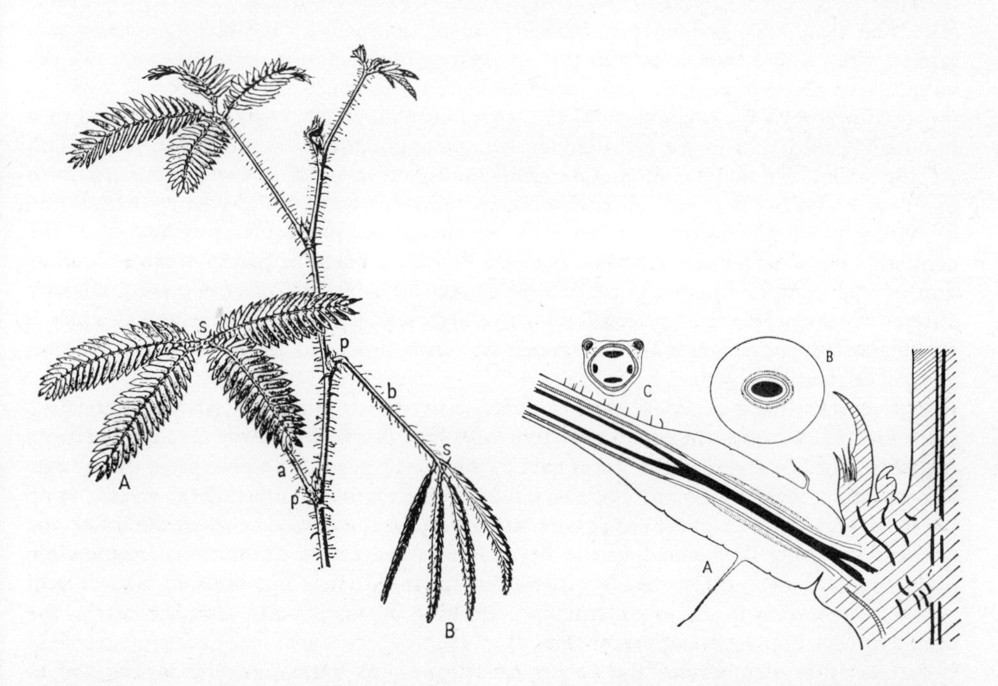

Fig. 314. Shoot of *Mimulus pudica*. Leaf **A.** in unstimulated condition, leaf **B.** showing a seismonastic reaction: p', p, joints of the petioles a and b respectively; s, joints of the pinnae. After Pfeffer, reduced.

Fig. 315. Joint in the petiole of *Mimosa pudica*. **A.** Longitudinal section showing the course of the vascular bundles (black). **B.** Transverse section through a similar joint. **C.** Transverse section through the petiole above the joint. *c.* ×6. Diagrammatic.

burning glass, taking care not to disturb the plant by shaking. It is then seen how, starting from the place of wounding, successive pairs of leaflets fold together. The stimulus passes down into the joint of the pinna, and from there both into the neighbouring pinnae, and also down into the primary rachis, until the main joint itself responds. From here the stimulus may even travel both up and down in the stem, in certain circumstances covering 50 cm or more, and leaves above and below that originally stimulated begin to collapse. In these leaves the movement starts first in the main joint, and then ascends to the joints at the bases of the pinnae and pinnules. The rate of conduction of the stimulus can be easily noted from the successive reactions of the individual joints, and in normal circumstances reaches from 4 to 30 mm/sec. After bad wounding a maximum rate of conduction of some 10 cm/sec has been observed, a value approaching that observed in the nerves of animals (in the freshwater mussel only 1 cm/sec).

Nastic reactions can be produced in detached shoots by treating them with dilute solutions of certain organic substances (excitatory substances). If a shoot of *Mimosa* is detached and reconnected with the stump by means of a glass tube filled with water it is then possible for a stimulus to be conducted from a leaf on the stump into the detached part of the shoot. Following stimulation certain substances evidently diffuse from the upper surface of the stump into the water of the glass tube. They are then drawn by the transpiration stream into the leaves of the detached portion of the shoot and there provoke nastic reactions. Detailed analyses have shown that a number of quite dissimilar chemical substances (e.g. anthraquinones, amino acids, etc.) are able, even in very great dilution, to release such chemonastic reactions. Indeed, a keto-acid has been isolated from *Mimosa*, which is still effective at a dilution of $1/10^8$. It appears that these excitatory substances can be conducted not only in the transpiration stream of the vessels but also in the phloem and even in parenchyma.

Whether and to what extent these excitatory substances play a part in the natural conduction of the stimulus and subsequent response is still not well understood. It is conceivable that mechanical shock, wounding or electrical stimulation disturb some labile chemical system, perhaps of the same nature as a thixotropic gel, in the sensitive protoplast, and that in consequence, either directly or indirectly, a stimulatory substance is formed. This would then release in the cell, and by conduction in neighbouring cells, the seismonastic reaction proper. Many workers believe that electrical currents play a part in the distribution of the stimulus, and that these currents might cause the formation of a stimulatory substance in the cells they excite. It is undoubtedly true that any stimulation of protoplasm, not only seismonastic, results in the setting up of weak electric currents in the plant body, but it is still obscure whether these currents play any active part in the irritability phenomenon, or whether they are merely an inevitable consequence of the numerous physicochemical changes involved. It is also possible that in *Mimosa* conduction of the stimulus may take place simultaneously in several different ways, and by different routes. The rates of conduction are higher than would be expected if conduction depended on the transport of organic substances alone, and they indicate some other method of spread.

The reaction itself depends, as in the previous instances discussed, on a sudden fall in the turgor of the cells of the joint, accompanied by a loss of fluid from their vacuoles. At the instant of the reaction the lower part of the joint is seen to become dark, this change in appearance being caused by the cell sap infiltrating into the intercellular spaces. Again the mechanics of the reaction appear to depend upon the relaxing cells of the lower side of the joint losing their semi-permeability, followed by a slow restoration of the suction pressure as the semi-permeable layers are reconstituted. It is not clear to what extent contractile proteins in the cytoplasm are concerned in this process, although many have assumed them to play an important role.

A few other members of the Leguminosae, and also some Oxalidaceae, respond to mechanical shock and wounding in the same way as *Mimosa*. In most, however, such as *Robinia pseudoacacia* and *Oxalis acetosella*, the irritability is much less markedly developed. The stimulus must be both much stronger and preferably repetitive before a slow reaction sets in. Here there is also some proportionality between the size of the stimulus and that of the response, usually lacking in the highly sensitive *Mimosa*.

Insectivorous plants. It is very difficult to see any ecological significance in the seismonastic movements of *Mimosa*. With the North American *Dionaea muscipula* and the rare European *Aldrovanda vesiculosa* (both Droseraceae), however, the seismonastic properties of the leaves are related to their trapping of insects.

In *Dionaea* (Fig. 221C, D) in favourable environmental conditions a light blow causes the two halves of the leaf to fold together about the midrib in 0.01–0.02 sec, thus firmly enclosing the animal initiating the stimulus. Weeks pass before the trap opens again, the corpse of the incarcerated animal being meanwhile digested away in the fluid secreted from the glands on the inside of the leaf. Chemonastic stimuli presumably keep the leaf closed while this is going on. The response is particularly rapid if a few of the hairs on the upper side of the leaf are bent down. The movement of the two halves of the leaf appears to be brought about by a rapid increase in turgor in the cells on the lower side of the midrib. Later, however, growth processes are also involved, so the mechanics of the whole reaction are rather complex. In the aquatic *Aldrovanda*, although similarly constructed, the movement is dependent upon a fall in turgor in the cells of the upper epidermis.

5. The movement of stomata[58]

The movement of the stomatal guard cells is also a nastic process depending upon reversible turgor changes. As has already been mentioned on p. 226 seq., the stomata open when the turgor in the guard cells increases, and close when it falls. Although there are instances where changes in stomatal aperture are brought about passively by changes in the turgor of surrounding tissue, stomatal movement is principally a question of turgor changes in the guard cells themselves. Naturally there are reciprocal relationships with the surrounding tissue, particularly with the subsidiary cells.

It is clear that stomatal movement can occur in response to several quite different stimuli. The guard cells with their chloroplasts can, for example, react photonastically, causing the stomata to open in moderately bright light and close in darkness. They also show particularly important hydronastic reactions, so that the stomata close when the leaf is becoming deficient in water (the deficiency amounting possibly to only a few per cent) and open when it is saturated, thus regulating the transpiration. It seems reasonable to suppose that in these movements, which serve to regulate transpiration and the water-content of the leaf, stimuli reach the guard cells from the remainder of the leaf tissue, but the extent to which the guard cells themselves serve as receptors requires further research. Thermonastic and chemonastic movements of guard cells are also known.

The mechanism of the turgor changes involved in the photonastic reaction has often been investigated, but is still not completely explained. The old view, arising from observations on fluctuations in the amount of starch in the guard cells, that conversion of starch into sugar, leading to an increase of osmotic pressure, and in consequence of turgor, caused opening, and the converse process closing, of the stomata has proved far too simple. These reactions are much too slow to account for the small rapid movements which take place between complete closure and partial opening. Nevertheless, turgor changes do have a disproportionately great effect, but it is still quite obscure whether they involve active changes in the permeability of the cytoplasm of the guard cells. A substantial increase in the carbon dioxide concentration in the intercellular spaces usually leads to an opening of the stomata.

C. Autonomic movements[56]

All the movements discussed so far are responses to stimuli originating in the external environment. Besides these tropic and nastic movements there is yet another group of movements which, so far as is known at present, are independent of any external stimuli. They seem in fact to arise from certain processes going on within the organism, and they are therefore termed endogenous or autonomic. These autonomic movements are also dependent mechanically upon differential changes in rate of growth or turgor.

Autonomic nutations. Characteristic of many growing plants, of seedlings, and of young shoots and inflorescences are fluctuating curvatures, caused by transient irregularities in growth around the axis concerned. These movements are called nutations. These are of a number of different forms, and they can readily be demonstrated cinematographically.

Young shoots, for example, often show circular or pendulum-like movements. In the cultivated onion the swaying movement of the peduncle is so marked that the tip occasionally touches the soil. The coleoptiles of the Gramineae show a pendulum-like motion in the plane of seedling and seed. This may interfere with the results when these plants are used, as they often are, for physiological experiments of various kinds.

Circumnutation is a widespread phenomenon. As has already been briefly described on p. 361, young tendrils especially display peculiar 'searching' movements in which their tips describe more or less a circle, and the whole axis a conical surface. At normal temperatures one revolution takes from forty minutes to several hours, according to the species. The revolving movement is again brought about by one side of the tendril growing faster than the other. The site of promoted growth, however, continually changes its position, affecting in turn all sides of the axis, moving round in the same sense as that in which the tendril circumnutates. During the nutation the tendril must twist in contrary sense about its longitudinal axis if torsion is to be avoided.

This twisting movement can be demonstrated by firmly fixing a piece of rubber tubing at one end and describing a circle with the other in the same manner as a tendril, observing meanwhile a mark at one point on the periphery of the tube. We are quite ignorant of what causes this promotion of growth to rotate about the axis of the organ. It is autonomic, and apparently as characteristic of the organization of some species as the local growth that leads to the differentiation of their specific cell forms. In many tendril-bearing plants the shoots also circumnutate in an irregular manner.

Twining plants.[53] Growth promotion of a similar cyclical nature occurs in twining plants. The actual twining, of course, is a very complex process, involving not only autonomic nutation of the shoot but also irritability movements, gravity being an important factor (see p. 361).

Fig. 316. Twining plants. **I.** Left-handed twiner. **II.** Right-handed twiner. After Noll.

Fig. 317. Diagram showing the beginning of nutational movements in a seedling of *Pharbitis hispida*. After Rawitscher, modified.

Twining plants are found in widely different families. They climb up more or less vertical supports, provided they are not too thick, by twining their slender stems around them in helical fashion. In most plants the direction of twining is constant, and is such that the helix ascends with a left-handed rotation (Fig. 316I). Only a few plants, such as the hop and the honeysuckle, are right-handed twiners (Fig. 316II). In *Polygonum convolvulus* and a few other plants the direction of twining may vary, and in *Loasa* and *Bowiea* the direction may change along one stem.

The young seedlings are at first negatively geotropic and consequently orthotropous. At a certain stage of development the tip of the shoot becomes plagiogeotropic, develops a lateral growth curvature and begins to describe a circle like the tip of a tendril. The movement is actually helical, since the shoot continues to grow and the older parts become negatively geotropic and upright, but exactly as in the tendril the movement is brought about by a local promotion of growth that rotates about the stem (Fig. 317). In the left-handed twiners the growth promotion rotates in a clockwise direction if looked at from above. In the hop the circle described by the tip of the shoot can attain a diameter of over 50 cm, and in *Hoya carnosa* of over 150 cm. The time taken to complete a circle differs in different plants, and averages 3–9 hours, but in the scarlet runner bean and *Convolvulus sepium* less than two hours suffice. These nutations are purely autonomic. If the 'seeking' tip comes into contact with a more or less vertical support it begins to twine around it, the manner in which the helix is formed apparently depending upon a different set of reactions in different species. Contrary to the behaviour of tendrils, the shoots, because of their geotropic sensitivity, will not twine around supports that are horizontal or nearly so.

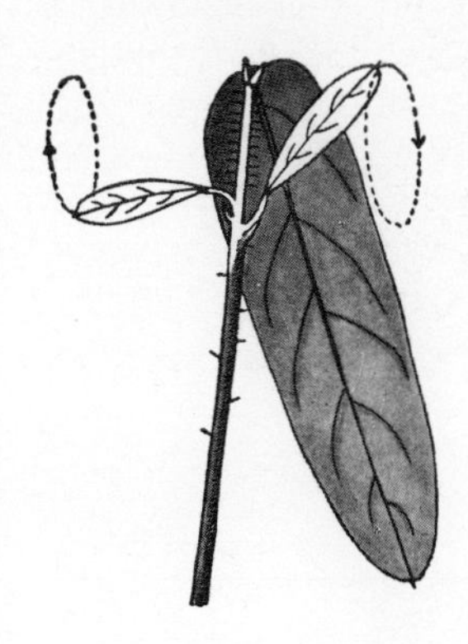

Fig. 318. Leaf of *Desmodium gyrans*. The arrows indicate the direction of the movement of the small lateral pinnae. Approx. natural size.

There is no direct sensitivity to touch as in a tendril. Very thin supports, such as wires and strings, can be climbed only if there is a continuous circumnutation. On thicker supports many plants display in addition a special form of geotropism. If, for example, young shoots of *Pharbitis* are placed horizontally they not only become upright as a result of negative geotropism, but at the same time they show a curvature towards one side, coinciding with the direction of the circumnutation. This indicates that the gravitational stimulus causes the promotion of the growth not only of the lower side but also of one of the flanks (lateral geotropism, see p. 361). If a nutating shoot meets a support the circumnutation is impeded, but the lateral geotropism now becomes effective. This leads to the promotion of the growth of the outer side of the shoot causing it to apply itself more closely to the support. In those plants which lack lateral geotropism autonomic torsions within the stem assist in the formation of the helix.

The older parts of the stem, which are once again negatively geotropic, often show torsions of this kind. They cause the gyres of the helix to straighten slightly, and so to tighten themselves firmly upon the support.

Twining, therefore, although it at first appears simple, is in all probability a very complex process, involving large numbers of growth movements, both autonomic and induced. Of these movements only the circumnutation is known at present to be certainly autonomic, and this may well be the basic phenomenon of the whole process.

Flowers. Rhythmically fluctuating and unilateral promotions of growth frequently occur in perianth segments.

Thus in *Calendula officinalis* (Compositae) the capitula open and close with a twelve-hour rhythm. The flowers of this species are also photonastically sensitive, and consequently in nature their opening and closing corresponds with the diurnal cycle of light and dark. The existence of the autonomic twelve-hour rhythm is shown by keeping the flowers in continuous darkness. As in the photonastic and thermonastic responses of these flowers, the movement is dependent upon a difference in the growth rates on the lower and upper sides of the marginal ray florets.

Autonomic movements dependent upon turgor change. Autonomic movements of individual plant organs can also result from changes in turgor, arising, like the differences in growth rate considered in the foregoing, from causes internal to the organism. They are to be found particularly in those leaves furnished with pulvini or joints, and they constitute some of the most rapid and striking movements to be seen in the whole of the Plant Kingdom.

In the Malaysian *Desmodium gyrans* (Papilionaceae), for example, the two small lateral leaflets move so rapidly at a sufficiently high temperature (35°C) that each tip completes an ellipse in about half a minute (Fig. 318). Although the movement is occasionally somewhat jerky, it goes on constantly in light. In *Oxalis hedysaroides* the leaf tips in certain conditions travel a distance of 0.5–1.5 cm in a single second. These movements are caused by the joint, and result from variations in turgor which rotate rhythmically about the petiole. They are thus almost in the nature of pulsations, such as are well known in animal physiology. The leaves of *Trifolium pratense*, by contrast, move more slowly. In darkness, but not in light, they swing up and down with a two to four-hour rhythm. Although these movements are so conspicuous, their value to the plant is unknown.

Nyctinastic movements.[57] Many jointed leaves also show movements which have an approximately twelve-hour period corresponding to the diurnal cycle. These are consequently referred to as 'sleep' or nyctinastic movements.

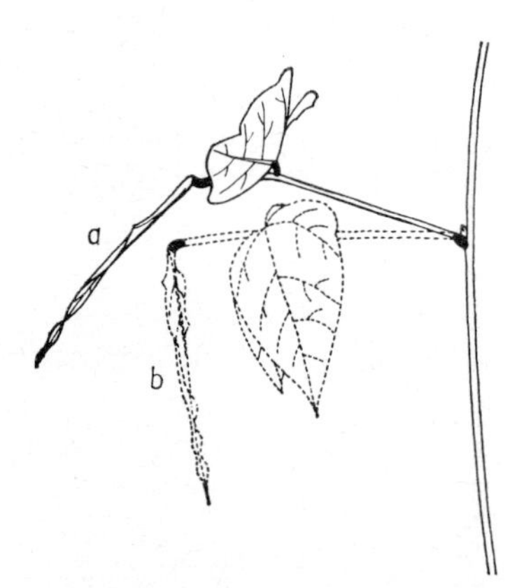

Fig. 319. Leaf of *Rhynchosia phaseoloides* (Leguminosae). **a.** Day position. **b.** Night position. The joints are represented as black. *c.* $\times\frac{1}{3}$.

The pinnae of the runner bean, for example, sink downwards at the joint in the evening and rise up again in the morning. The laminae or petioles of the leaves of many other plants (e.g. *Robinia pseudoacacia, Oxalis acetosella*) collapse in a similar way in the evening (see Fig. 319), although in *Trifolium pratense* the terminal leaflet remains stiffly erect. The seismonastically sensitive *Mimosa* also shows sleep movements, and at nightfall appears almost as if stimulated by mechanical shock. It is still, nevertheless, capable of showing a seismonastic response. Though popularly and aptly referred to as 'sleep', this diurnal rearrangement of the leaves has nothing whatever to do with the fatigue and sleep of animals.

As already mentioned on p. 367, nyctinastic movements may be caused by both growth and turgor changes. The fact that the rhythmical movements of the leaves coincide with the diurnal cycle immediately suggests that nyctinastic movements are induced from outside, and are therefore photonastic or possibly thermonastic in nature (see p. 365). The plants concerned are in fact usually photo- and thermonastically sensitive. It is therefore easily possible to bring about the 'sleep' position of the leaves prematurely during the day, especially in the early afternoon, by shading, and correspondingly the daytime position of the leaves at night by illumination. The artificially induced periods need not alternate with a twelve-hour rhythm, but may be shorter or longer.

In the tropical *Albizia lophantha* (Leguminosae), for example, the leaves will go on rising and sinking with light and dark periods of three hours or even one hour, and at the other extreme the rhythm is no less perfect with twenty-four-hour periods. Other plants, generally speaking, will not tolerate such large deviations from their customary rhythm. If the light and dark periods depart too far from the natural duration of twelve hours they return by themselves to the inherited norm despite the continuation of the unnatural rhythm of illumination.

The movements cease in constant light or darkness. They do not stop suddenly in darkness, but die away gradually. These continuing movements, which persist for some time, cannot be directly induced by the cessation of illumination, and they must be regarded either as after-affects, or as autonomic movements. If *Phaseolus* is raised in an etiolated condition in continuous darkness and at a constant temperature movements of the leaf joints can be initiated at any time by temperature changes of the order of only 1°C, and they continue rhythmically independent of the hour of day. This indicates an autonomic component in their causation. The natural period of these movements is about twelve hours, being shorter or longer according to the temperature. Under these laboratory conditions the periodicity of the movements is quite independent of the diurnal cycle.

These observations, so far made in detail on only a very few plants, make it very probable that, contrary to what was formerly believed, nyctinastic movements are fundamentally autonomic. The normal rhythmical external stimuli, particularly the diurnal cycle of light and dark and the accompanying rise and fall in temperature, would then merely have a steering influence, so that the movements, though autonomic, come in general to coincide with the daily rhythm. In many plants it seems that yet other processes, not connected with nyctinastic movements show an autonomic daily rhythm. Examples are seen in many metabolic reactions and in the sensitivity of plants towards light. Nyctinastic movements, therefore, are probably only a part of a very much more extensive endogenous daily rhythm within the plant.

Mechanism of nyctinastic movements. It is still not known in detail which processes in the interior of a plant take place according to an autonomic rhythm. A number of different metabolic processes have been discovered relating to turgor mechanisms which might be causally involved in autonomic movements. It is known that the joint of *Phaseolus*, on which the most intensive investigations have been done so far, possesses a physiological dorsiventrality. The cells of the lower side possess by both day and night a higher osmotic pressure than the cells of the upper side. Contrary to the situation in other plants, in *Phaseolus* this dorsiventrality appears to be induced by gravity, since inverting the plant reverses both the dorsiventrality and the direction of movement of the leaves. It is also known that the concentration of the cell sap remains constant on both the upper and lower sides of the joint during movement. This is despite the fact that when the leaves sink, for example, the volume of the cells on the lower side diminishes considerably, while that of the cells on the upper side increases. This is only possible if there is a simultaneous regulation of the absolute amount of osmotically active substances, possibly organic acids, in solution. For a long time these variations in the amounts of substances dissolved in the cell sap were considered to play a predominant role in turgor mechanisms. Recently this has been questioned, and there is the possibility that they are more a consequence than a primary cause of the movement. The view is favoured today that here, just as in the seismonastic and photonastic turgor reactions, the primary responsibility lies in permeability changes in the protoplasm. These might in turn be connected with rhythmical metabolic processes, such as the carbon dioxide content of the protoplasm, which fluctuates with the daily march of carbon assimilation, and perhaps also the flow of growth-regulating substances from the leaves, which likewise shows a diurnal variation. The indole-3-acetic acid moving out of the leaf might possibly, as in extension growth, cause changes in the osmotic pressure of the cells of the joint.

D. Discharge mechanisms depending upon turgor[59]

The turgor movements considered in the last section depend upon changes in the turgor of one side of an organ, leading to its undergoing reversible curvatures. In certain other cases, however, concerned principally with the distribution of propagules, differences in the turgor of particular layers of tissue are utilized in the production of a movement which can only very rarely be regarded as a response to a stimulus. Usually it is the result of a natural process of development and maturation, and is not reversible.

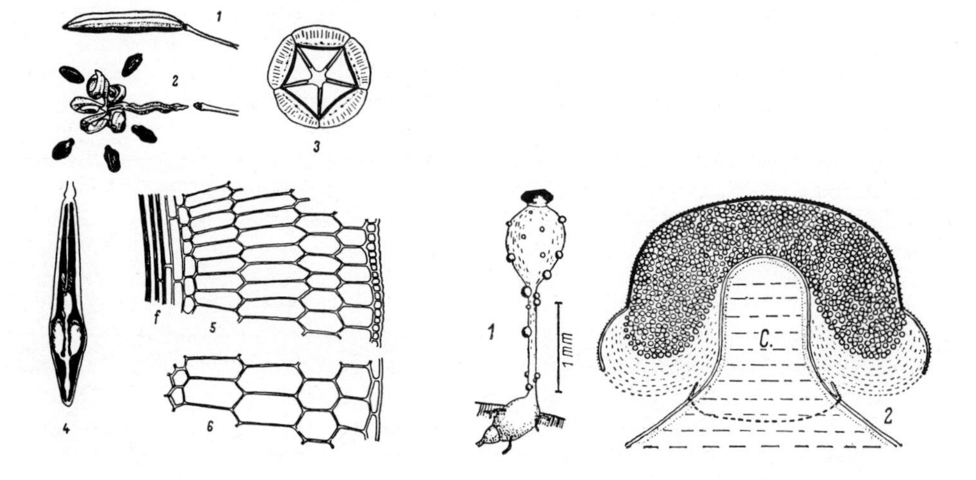

Fig. 320. *Impatiens* sp. **1.** Ripe fruit. **2.** Exploding fruit. **3.** Fruit in transverse section. **4.** In longitudinal section. **5.** Part of a longitudinal section through the fruit wall more strongly magnified (f, fibres). **6.** Part of a transverse section without fibres. **1.** and **2.** After Troll, modified. *c.* $\times\frac{1}{2}$. **3.** to **6.** After Overbeck.

Fig. 321. *Pilobolus* sp. **1.** Sporangiophore shortly before the discharge of the sporangium (indicated as black). **2.** Section through the sporangium and the upper part of the sporangiophore with the columella (C) and the site of separation (dotted). The wall of the sporangium is already ruptured by swelling mucilage. After Ingold, enlarged and modified.

It has already been mentioned (p. 221) that tensions within tissues serve to maintain the rigidity of organs. If now, as a result of some process or other, tissues in which such tensions exist cease to be connected together the tendency of the tissues concerned to expand or contract can now be realized. As a result, sudden jerky curvatures are produced which may, for example, cause seeds to be ejected. A well-known example of such a catapult mechanism is provided by the fruits of *Impatiens* spp. Thus if a ripe fruit of *Impatiens parviflora* is given only a slight touch it breaks open explosively. The individual valves of the fruit roll in upon themselves like watch springs, and the seeds are simultaneously ejected a considerable distance.

This process depends upon the fact that the outer part of the wall of the fruit (Fig. 320.5 and 6) consists of thin-walled parenchyma which, because of the form of its cells and the extent of their turgor pressure (of the order of 20 atm), tends to expand longitudinally. The two innermost layers of the wall consist of elongated fibre cells (f), which resist the tendency of the outer expansive tissue to elongate. Since the five carpels forming the ovary coalesce to form a cylindrical chamber, the tendency for them to roll up is resisted, and the chamber is consequently in a state of tension. In maturation the middle lamellae along the sutures of the carpels break down. The mechanism of this breakdown is unknown, but the process is similar to that occurring in the fall of leaves in the autumn (see p. 303), and of petals. In the ovary of *Impatiens* it results in the individual valves now being free to roll up and equalize the pressure set up by the outer expansive tissue. The valves separate from the pedicel at the base, each rolls in upon itself and coils up like a watch spring, and the still adhering seeds are flung far and wide. It has been observed that as a result of this process the outer part of the valve increases its length by about 32 per cent, while the fibrous layer actually contracts by about 10 per cent.

Similar mechanisms occur in other plants. In many the dissolution of the middle lamellae, which enables the tissue tensions to express themselves, arises naturally as a result of ageing and maturation, but in others it is the result of a typical irritability mechanism. Such, for example, is the shedding of the petals that occurs in many flowers after pollination, mechanical shock, excessive heat or exposure to tobacco smoke or traces of coal gas (see p. 340), and so on. Contrary to nastic turgor movements, however, these movements are non-recurrent, since they lead to the irreversible destruction of an organ or tissue.

Ejection movements frequently come about as the result of the bursting of individual turgid cells.

Thus in the Ascomycetes the spores are shot out of the ripe ascus as a result of the elastic wall of the ascus, which is under tension as a result of turgor, suddenly rupturing at a predetermined site at the tip. The energy in the cell wall is then released by contraction, and the whole contents of the ascus are ejected. The firing off of the sporangia of *Pilobolus* (see p. 350) depends upon the same mechanism. The unicellular sporangiophore develops a turgor pressure of about 5.5 atm, causing the upper end to swell up like a club (Fig. 321.1) and the elastic cell wall to increase its length by as much as 100 per cent. The wall of the uppermost portion of sporangiophore which arches up into the interior of the sporangium as a columella (Fig. 321.2, C) differs from the remainder in being inelastic. It is here that the sporangiophore breaks down at maturity, and the rapid expulsion of the cell contents shoots the sporangium up to 1 m distant. In these instances, therefore, the irreversible destruction of certain cells fulfils an important function with respect to the whole organism.

E. Hygroscopic movements[60]

Finally is a group of movements termed hygroscopic in which living cells no longer actively participate. Here the purely physical process of imbibition, or conversely loss of water, and the drying out of dead membranes or of whole cells, lead to definite movements, often capable of being repeated many times. These movements may nevertheless be of considerable importance for the organism as a whole.

Imbibition mechanisms. The opening and closing of seed vessels and sporangia often depend upon the loss or gain of water.

Many seed vessels, for example, open themselves as soon as the protoplasts of the wall cells die and the membranes begin to dry out. Moisture usually causes the vessels to close, and this may happen repeatedly. In some cases precisely the reverse occurs, and hydration and swelling of the membranes by rain or dew causes the fruit to open (*Mesembryanthemum, Veronica, Sedum*, etc.). In fruiting plants of *Anastatica hierochuntina* (N. Africa, Cruciferae) ('Rose of Jericho') in the dry condition the whole system of fruiting branches bends in over the fruits. In *Asteriscus pygmaeus* (Compositae) of the same region the dead bracts enclose the fruiting capitulum, and they do not expand and release the fruits until they are moistened.

The outer peristome teeth of the sporangia of mosses, which consist only of parts of the walls of two adjacent layers of cells (see p. 556, Fig. 547, and Fig. 324), move hygroscopically on drying either inwards or outwards according to their fine structure. These movements, which follow changes in the atmospheric humidity, facilitate the distribution of the spores.

All these movements follow from the fact that the walls of the cells composing the organ, or even different walls of the same cell, show, as a result of their particular

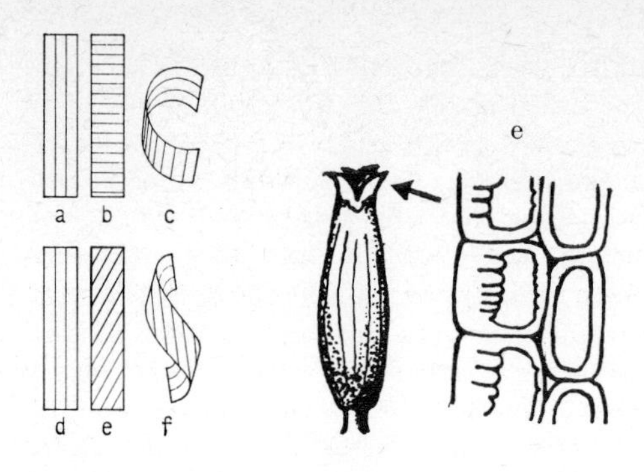

(Left)
Fig. 322. Curvatures of the paper strips **c.** and **f.**, each composed of two different strips (**a.** and **b.**, and **d.** and **e.** respectively) glued together. After Jost.

Fig. 323. Fruit capsule of *Saponaria officinalis*. Right: radial longitudinal section through the outer layer of cells of a tooth of the capsule (arrow). **e.** Epidermis. Right after Steinbrinck.

submicroscopic structure, unequal amounts of swelling or contraction in a given direction when the degree of the hydration changes (anisotropic swelling). The direction of the greatest expansion is in fact determined by the orientation of the cellulose microfibrils (micelles) in the wall, since the greatest change in size on imbibition is always perpendicular to the longitudinal axis of the fibrils (Fig. 324). If the directions of maximum expansion or contraction of the various layers of a cell wall, of the walls themselves or of whole cells do not coincide, the members concerned remaining nevertheless firmly attached to one another, then changes in hydration must lead to curvatures or spiral coiling.

A model producing hygroscopic movements of this kind can be made by sticking together two pieces of writing paper, since this usually expands and contracts to different extents in the transverse and longitudinal directions. Curvatures then appear on drying (Fig. 322) similar to those obtained in nature with other materials.

In the opening of the fruit of *Saponaria* (Fig. 323) the thick outer walls of the epidermal cells of the teeth of the capsule shrink in the longitudinal direction much more than the other membranes, thus causing the teeth to bend outwards. In the peristome tooth of a moss, depicted in Fig. 324, the movement on drying is caused by the fact that the fibrils in the outer lamella (a) are orientated transversely to the axis of the tooth, so

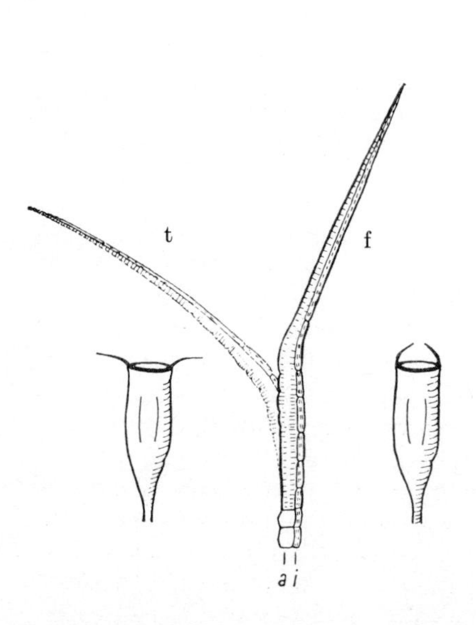

Fig. 324. Outer peristome tooth of the capsule of the moss *Orthotrichum diaphanum* in the dry (t) and moist (f) condition: a, outer; and i, inner lamellae of the tooth with a diagrammatic representation of the direction of the component fibrils. Left and right, capsules with open and closed peristomes respectively, diagrammatic. After Steinbrinck, modified.

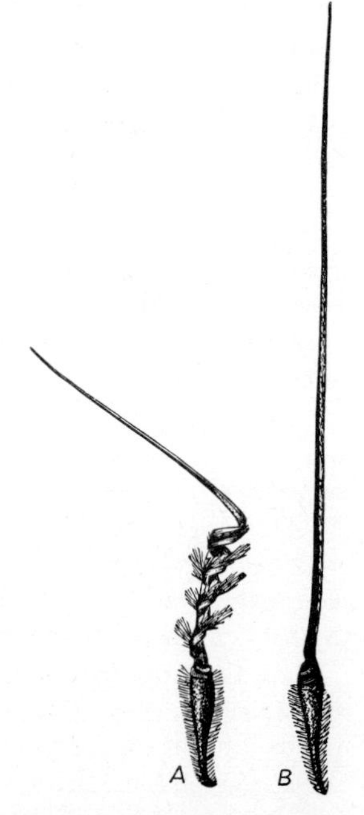

Fig. 325. Partial fruits of *Erodium gruinium*. **A.** Dry. **B.** Damp. After Noll.

that this layer contracts principally in a longitudinal direction. The inner lamella (i), because of its fibrils lying parallel to the axis of the tooth, merely diminishes somewhat in thickness with no decrease in length. Since it is firmly attached to the outer lamella, it is impossible for the tooth as a whole to contract, and an outward curvature inevitably develops. The structure of the peristome teeth differs very much in different genera, and the directions of movement are accordingly very variable. The elaters of *Equisetum* spores (see Fig. 575) and the capillitia of many Myxomycetes, both of which also consist only of wall material, show similar hygroscopic movements.

In the partial fruits of *Erodium* spp. (Fig. 325) the fibres in the awn are in part so arranged that the main directions of contraction of the individual layers lie at an acute angle to each other. Consequently, on drying the awn must roll up to form a helix, corresponding to the system shown in Fig. 322 f. On moistening, the awn tends to unwind again, and if the free end has some point of purchase the fruit is drilled into the soil. The spiralling of the valves of the pod of the common broom and other legumes depends upon similar principles of construction.

F. Cohesion mechanisms[59, 60]

Desiccation can also have mechanical consequences in another way. The cohesive force of water, already considered in relation to the ascent of sap, can cause, provided the cells are filled with water, a curvature of dead, and more rarely of living, tissues. These are called cohesion mechanisms.

The annuli of the sporangia of certain ferns provide an example of such an apparatus (see p. 589). The inner and dividing walls of the cells of the annulus, arranged in an arc over the sporangium (Fig. 326), are thickened, whereas the outer walls remain thin. At maturity these cells gradually lose water to the atmosphere. Since the water adheres so firmly to the hydrated walls, and because of the high cohesive force of the water itself, no space appears in the lumina of the cells (over 250 atm would be necessary to rupture the droplet). Instead, as the volume of water in the cell diminishes, the outer ends of the

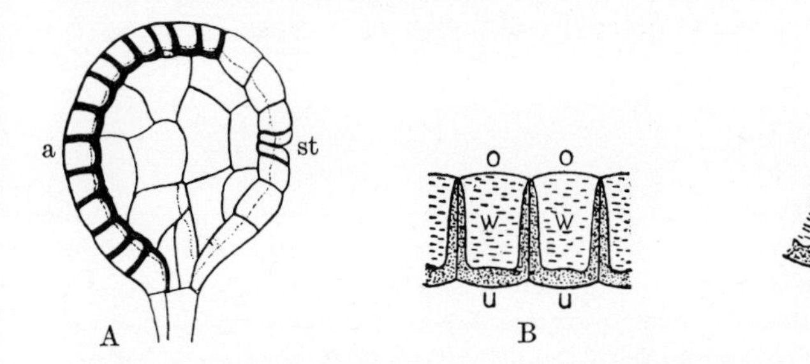

Fig. 326. A. Sporangium of *Polypodium vulgare*: a, annulus; st, stomium. **B.** Cells of annulus filled with water (w), showing the outer (o) thin, and inner (u) thickened walls. **C.** Following partial evaporation of the water the outer thin walls become indented. **A.** ×1000. After Haider. **B.** and **C.** After Noll, diagrammatic.

anticlinal walls approach each other, the outer walls becoming simultaneously indented (see Fig. 326C). In this way a tangential force is generated on the surface of the sporangium. A similar process causes at the same time two specialized cells (stomium: Fig. 326A st) to pull apart from each other, and the sporangium wall then immediately begins to split slowly open from this point. When the deformation of the annular cells has gone a certain way there finally comes a point when the water droplets within break. The elastic annulus, which is now bent back, then rapidly returns to its original position, together with upper part of the sporangium wall, so causing the adhering spores to be widely scattered.

A precisely similar mechanism is responsible for the opening of anthers. The fibrous cells of the endothecium, lying in the anther wall, possess U-shaped or annular thickenings, and behave exactly like the cells of the fern annulus. Similar cohesion mechanisms

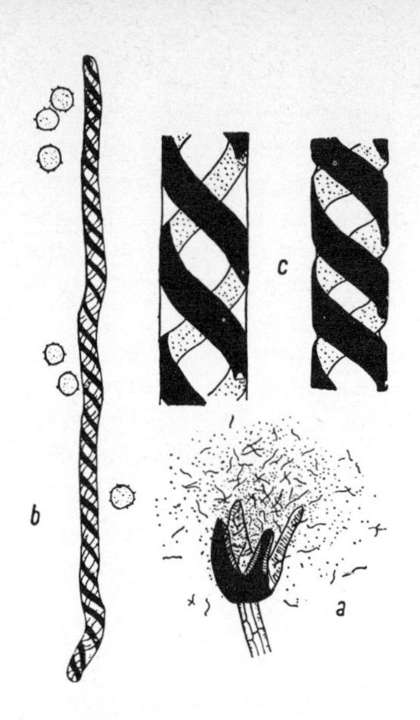

Fig. 327. Elaters of the liverwort *Cephalozia bicuspidata*. a. Capsule burst open. b. Isolated elater and spores. c. Portion of an elater, left filled with water, right after partial evaporation of the contained water. After Ingold. a. ×6. b. ×100. c. ×425.

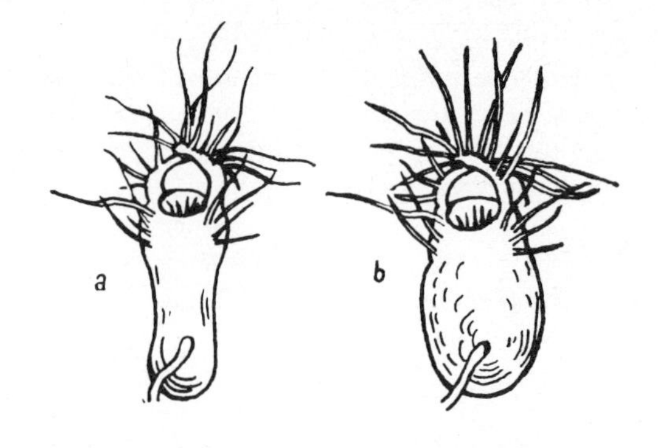

Fig. 328. *Utricularia exoleta*. View of the bladders from below. a. Before the gulping movement. b. After the movement. Both c. ×10. After Bünning.

are developed in the walls of the capsules and in the elaters of many liverworts (Fig. 327).

In the bladders of *Utricularia* (Fig. 328; see also p. 199) part of the water is resorbed through the inner surface of the bladder and expelled to the exterior. The remaining water adheres to the inner surface, and this adhesion is responsible for the indentation and tension in the lateral walls which precedes the sucking movement. This is an instance of a cohesion mechanism depending upon the activity of living cells.

II. Free locomotion[61]

All the movements discussed so far affect only the organs of a generally immobile and firmly rooted plant. There is also a whole series of plants which, like animals, are capable of free locomotion. They can therefore actively escape unfavourable habitats and environments, and search for the more favourable. With the exception of rhizomes which move slowly forward through the earth and die away behind, free locomotion is found principally in the lower plants, e.g. bacteria, Cyanophyceae, Flagellatae, Diatomales, Volvocales and Myxomycetes. In many plants locomotion may be confined to the germ cells, e.g. the swarm spores of many algae and fungi, and also the sexual cells, the male of which is still motile even in the Pteridophyta and some gymnosperms. These motile cells swim actively through the medium, often with the help of flagella, or, e.g. the amoeboid and plasmodium stages of the Myxomycetes, creep over and through the substrate. There are many instances where locomotion is brought about in other ways, such as the unilateral excretion of mucilage (Desmidiaceae), protoplasmic streaming (Diatomales), etc. In these respects the plant cells differ in no way from animal cells of a comparable level of organization.

However, in higher plants quite incapable of locomotion the protoplasm at least of the individual cells is able to move within its firm container. A large part of it, especially the nucleus and plastids, varies its position (see p. 19). This phenomenon can reasonably be considered along with free locomotion.

Little is known for certain of the mechanics of these movements. The flagella are extraordinarily fine, and are hyaline emergences of the cytoplasm, connected with the

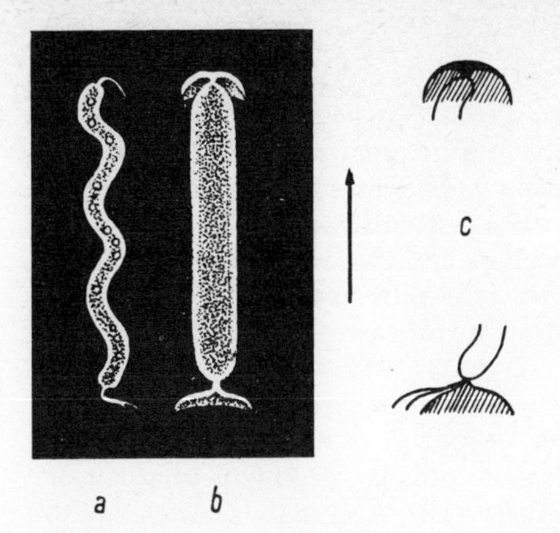

Fig. 329. *Spirillum volutans.* **a.** Seen stroboscopically in apparent rest. **b.** View in the dark field during the movement. **c.** Action of the anterior and posterior tufts of cilia during movement in the direction of the arrow. After Metzner, diagrammatic.

inner cytoplasm, in those forms possessing firm cell walls, through small pores in the wall. Many flagella revolve like propellers, or ships' screws, performing some thirty to sixty revolutions a minute. The cell bearing them is also set in motion and is driven freely through the water, often with a peculiar rotating motion, caused by the screw-like form of the cell (see Fig. 329). Oar-like movements in one plane also occur. When tufts of flagella are present the movements of the individual flagella seem to be coordinated, and by contracting one after the other the whole tuft is set in spiral motion. The rapidity with which flagellated organisms move can be very considerable. That of certain dinoflagellates reaches 200 μm/sec (see p. 451), and that of the swarm cells of the slime fungus *Fuligo varians* as much as 1 mm/sec, i.e. in one second they cover many times their own length.

Although the structure of bacterial flagella is still not known in detail, in numerous algae and various spermatozoids the ultrastructure of the flagella has proved surprisingly uniform. In section, nine pairs of fibrils (microtubules) are seen at the periphery, and two single fibrils at the centre (Fig. 385). Minute hairs (flimmer) are often attached to the outside of the flagellum, possibly in register with the peripheral fibrils (see Figs. 456 and 457). The movements of flagella are fundamentally rhythmical contractions, partly limited to the basal zone, and partly advancing as contraction waves over the whole flagellum (as seen, for example, in the undulating transverse flagellum of the dinoflagellates, Fig. 388A). Such movements require a special ultrastructure of the minute cytoplasmic thread, or of the component fibrils. The mechanism may be similar to that occurring in the contraction of animal muscle. In any event the energy is derived from ATP.

The mechanics of the amoeboid creeping movement of naked masses of protoplasm, and of the phenomenon of protoplasmic streaming in the interior of cells, are also as yet unknown.

Suggested mechanisms have depended upon local changes in the surface tension of the bounding layer of the protoplasm, reversible sol–gel transformations and contractions of invisible protein fibrils or molecules, as may occur in flagellar movements.

All these movements may well be partly autonomic, but they also appear to require for their release certain external stimuli. Generally unfavourable environmental conditions may in certain conditions result in loss of the ability to move, or even in rigidity.

A. Tactic movements[61]

Locomotory movements are often by no means random, but are capable of being influenced by certain external factors, the particular factors and the nature of the influence varying widely with the organism. Such directed movements are referred to as taxis. They are typical irritability phenomena, following the same laws as the tropisms and nastic movements considered earlier, and they are classified according to the nature of the releasing stimulus. We can therefore speak of chemotaxis, phototaxis, etc.

1. Chemotaxis

Chemotaxis is an important phenomenon in nature. It enables the saprophytic and parasitic (pathogenic) bacteria to find suitable substrates, and the gametes of algae and fungi, and the spermatozoids of higher plants, to locate and fuse with the appropriate sexual partner. The direction of these movements is determined by the concentration gradients of the active substances involved. In some instances the higher concentrations attract (positive chemotaxis), in others repel (negative chemotaxis) the moving organism.

According to the organism, a wide range of quite different substances can act as stimulants. The bacteria involved in rotting and decay, for example, can be attracted not only by organic but in certain circumstances by inorganic substances as well, such as all the neutral salts of the alkalis and alkaline earths. Among organic compounds, positive chemotaxis is also produced by glucose, lactose, mannitol and dextrin, furthermore by urea, asparagine, creatine, peptone and meat extract, and also by sodium salicylate and even morphine and ether. Other bacteria react specifically to other substances, and this may well be the reason why pathogens attack only certain organs and not others. Certain flagellates are attracted by carbon dioxide, phosphate and nitrate. Swarm cells of Myxomycetes react positively to low concentrations of hydrogen ions, negatively to high. In the formation of the fruit body of the Acrasiales (Myxomycetes) a substance secreted by the myxamoebae (acrasin) causes the amoeboid cells to aggregate closely together, without any cell fusion, to form the dense sporangium. This in turn shows a negative chemotactic response to some substance (possibly carbon dioxide) produced by the periplasmodium and consequently becomes elevated above it. Oxygen also can evoke, according to the requirements of the organism, a positive or a negative response (aerotaxis, see p. 240). Among sexual cells there are often specific substances attracting the appropriate gametes together (gamones, see p. 327), facilitating their mating. The spermatozoids of ferns, *Equisetum, Isoetes, Selaginella* and *Salvinia* react to traces of malic acid, which may well be excreted from archegonia. In *Lycopodium* citric acid appears to be active, whereas the chemotactic agent in mosses is cane sugar, and in *Marchantia* a protein.

The susception of the stimulus probably involves definite chemical reactions between the penetrating stimulant and certain components of the protoplasm or flagella, as the case may be. As yet we have little information about the individual steps in the tactic reaction.

Only very small quantities of active substances may be required to produce chemotaxis. Fern spermatozoids can still detect a concentration of malic acid of only 0.001 per cent (in one recorded case an absolute amount of 0.028×10^{-6} mg). With gamones even weaker concentrations can probably be detected. In other cases the weakest concentration that will produce a response is of a higher order. The directing factor is always a concentration difference. For a concentration gradient to be detected its steepness must, of course, equal or exceed a certain definite threshold. In a solution of malic acid of 0.001 per cent concentration a concentration of the same acid of 0.03 per cent can be perceived, but in a solution of 0.01 per cent there is no response until the higher concentration reaches 0.3 per cent. The ratio of the two concentrations needed to produce a chemotactical response thus remains at a constant factor of 30. This is nothing other than Weber's law, already discussed in relation to phototropism (see p. 349). After a time exposure to a uniform solution depresses the sensitivity.

These phenomena make it possible to determine whether an organism can distinguish one chemotactically active substance from another. If the threshold concentration of a substance is determined in pure aqueous solution, and then determined again in a solution containing a second active substance, the threshold value is raised or lowered only if the two substances are not detected independently. It has been found in this way that organisms are in fact quite capable of distinguishing between different substances. Thus the NH_2 groups of amino acids can be clearly distinguished from the NH_4 groups of ammonium salts, the ammonium salts themselves all producing an identical response. Stereoisomers can even be distinguished from one another.

The reaction eventually released by the stimulus, and in consequence its orientation, may take one of two quite different forms. In very many bacteria a phobic reaction is

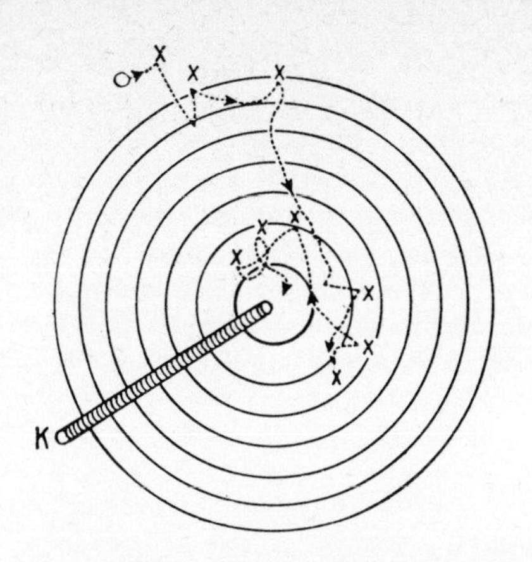

Fig. 330. Course of the positively chemotactic movement of a swarmer of a myxomycete. K, capillary containing M/2 malic acid. The concentration of the acid diminishes uniformly in a radial manner, indicated by the concentric circles. Phobic reactions occur at the sites marked X. After Kusano, modified.

developed, i.e. the organism draws back with random swimming movements every time it approaches a zone of diminishing concentration. The reaction always in fact influences the motile apparatus in the same manner. The final orientation in the direction of the stimulus is thus merely the result of numerous random attempts, of which all those contrary to the concentration gradient are abruptly terminated. The direction of the movement by the stimulus is consequently only indirect.

If a solution of the chemotactic substance is enclosed in a fine capillary, and the tip introduced into a drop of water so that diffusion can take place from it, a result is obtained similar to that shown in Fig. 330. As soon as the organisms freely moving in the water enter a region of falling concentration, either the direction of movement of the flagella, and of the whole organism, reverses or the movement comes jerkily to a short stop, so that the organism turns from its initial direction into a more favourable one. When the normal movement of the flagella recommences, the organism swims away on its changed course, to be checked again immediately if it encounters a falling concentration on its new path. In this indirect manner, and in spite of much trial and error, the cell finally finds the source of diffusion, the whole process taking an astonishingly short time. By this means a whole population of small organisms may eventually be drawn from the drop and trapped in the capillary.

Many organisms can apparently orientate themselves directly in a concentration gradient and swim along the shortest path to the source of diffusion. A reaction of this nature is termed 'topic'.

With fern spermatozoids, for example, the axis of the spermatozoid is apparently polarized, and it is seen to align itself with the concentration gradient and swim directly to the capillary. It seems that in this instance the two sides of the spermatozoid detect a concentration difference. The flagella on the side of the weaker concentration beat somewhat more rapidly than those on the side of the stronger. Thus the organism in moving forward constantly checks its alignment, so that the concentration, and consequently the rate of movement of the flagella, is the same on each side of its axis.

2. Phototaxis

Both phobic and topic reactions occur in response not only to chemical substances but also to light falling in a particular direction, a phenomenon known as phototaxis. Such phototactic movements are found almost entirely in those forms which utilize light for the assimilation of carbon dioxide. Apparently the ability to be stimulated by light makes possible the seeking out of the location most favourable for photosynthesis. If a vessel filled with water containing, for example, green flagellates, species of the Volvocales, or the swarm spores of algae is placed in a window so that it is illuminated from only one side, then it is possible to see with the naked eye after only a short time that the initially uniform green coloration of the water has shifted to the illuminated side. This is because all the organisms have topotactically assembled there. With stronger illumination the positive phototaxis may become negative, and the organisms then move away from the source of light.

Among those organisms which, although not photosynthetic, nevertheless clearly show phototactic responses, are some non-pigmented flagellates and the plasmodia of the Myxomycetes. These at first react negatively, but the phototaxis reverses to positive as soon as a plasmodium enters the fruiting stage.

The only thing known about the mechanism of phototaxis is that the susception of the stimulus in flagellated forms is frequently located in the bases of the flagella. Where a carotin-containing eye-spot is present this may play some part in susception. In this respect it is less likely that the light absorbed by the carotin causes a photochemical reaction than that the pigmented spot shades sensitive regions of the cytoplasm. Evidence in this direction is provided by the fact that sensitivity of an organism in light of different wavelengths does not always coincide with the absorption spectrum of β-carotin, and the carotin may therefore be acting as it does in phototropism. In *Euglena, Oscillatoria*, etc., the reaction is phobic.

In the photosynthesizing purple bacteria, especially sensitive to red and infra-red radiation, the absorption of light by the bacteriochlorophyll probably facilitates phototaxis. The reaction of these organisms is purely phobic. As soon as the light intensity falls only a minute amount they recoil from the darker zone. If they are suspended in a drop of water on a slide they can be held in a spot of light as if in a trap.

3. Other tactic movements

Besides the capacity to react to chemical and photic stimuli, many freely motile organisms respond similarly to differences in humidity (hydrotaxis, see p. 495), in temperature (e.g. the thermotaxis of plankton), to tactile stimuli (thigmotaxis), etc.

B. Movement of protoplasm in cells[61]

As has already been mentioned, the cytoplasm, nucleus and plastids, enclosed within the rigid walls of a cell, often show movement. In many respects this is not far removed from the free locomotion of the individual cells of lower organisms. The streaming of the protoplasm in integumented cells is probably closely related to those processes which facilitate the creeping of a naked amoeba or of a plasmodium. The stimulatory processes which govern the movements of the cytoplasm, nucleus and chloroplasts likewise follow those laws of which we have already spoken in connection with tactic movements.

The circulatory and rotational streaming of protoplasm may well be partly autonomic in nature, but it is also clear that they are often not initiated until there is some external stimulus. Thus the protoplasmic streaming of *Vallisneria*, for example, stops on darkening, but is induced again on illumination, particularly with red light. Chemical stimulation, which may arise from a number of quite different substances, is especially effective. Thus certain amino acids, particularly *l*-histidine, will cause vigorous protoplasmic streaming in leaf cells, even at a dilution of $1/80 \times 10^6$. It is probable that the intense streaming which is often seen to follow wounding depends upon the setting free of traces of such amino acids.

Here again the protoplasm is very sensitive to minute differences in molecular structure in the excitatory substances, and stereoisomers can be clearly differentiated. In individual plants the rapidity of the streaming can become very considerable. On the average it is of the order of 0.2–0.6 mm/min, but in the large internodal cells of *Nitella* at a high temperature the velocity may reach 6 mm/min.

The significance of plasmatic streaming is still unknown. It has not so far been possible to demonstrate that it plays any decisive part in the metabolism of the cell, or in transport between cells. Streaming may often be nothing more than a visible expression of a certain excitation, or of certain metabolic processes, in the protoplasm. Since the internal structure and the polarity of the cell remain unchanged despite the streaming, it must be assumed that only certain parts of the protoplasm are in motion, the outer layers remaining at rest. The mechanism of protoplasmic streaming is also still obscure. Proteins similar to actomyosin have been found in the plasmodia of Myxomycetes, so a contraction process similar to that occurring in animal muscle is not inconceivable. Fine fibrils or tubules (microtubules) are in fact frequently seen in electron microscope preparations. Naturally energy must be consumed in the movement (ATP from respiration; see p. 274).

The cell nuclei, which generally occupy a definite position in cells, can also occasionally move in a characteristic manner.

With local wounding it can be seen that the nuclei of the cells adjacent to the wound move towards the wall which is nearest to the wound. Also certain chemical substances, such as salts, organic acids and carbohydrates, applied externally to the cell are able to cause a chemical 'attraction' of the nucleus. After a short time the nucleus returns to its old position.

Still more striking are some of the migrations made by chloroplasts. They are able, for example, to take up positions in the cell in which they can enjoy optimal illumination for the assimilation of carbon dioxide. As with cell nuclei, marked changes in form may also sometimes occur.

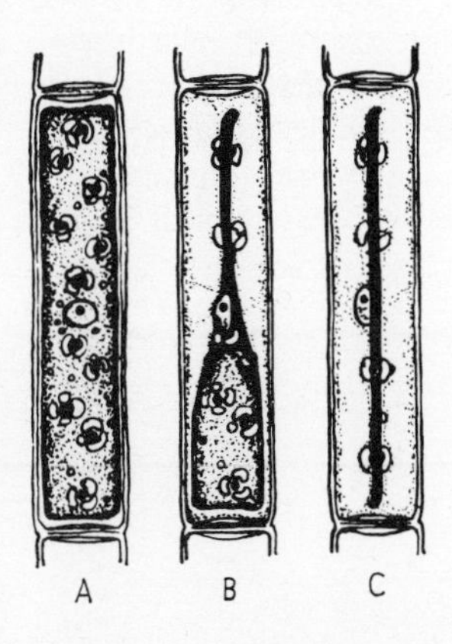

Fig. 331. Cell of *Mougeotia scalaris* showing the position of the plate-like chromatophore. **A.** In weak light. **B.** In a transitional phase. **C.** In strong light. *c.* ×400. After Palla, from Oltmanns.

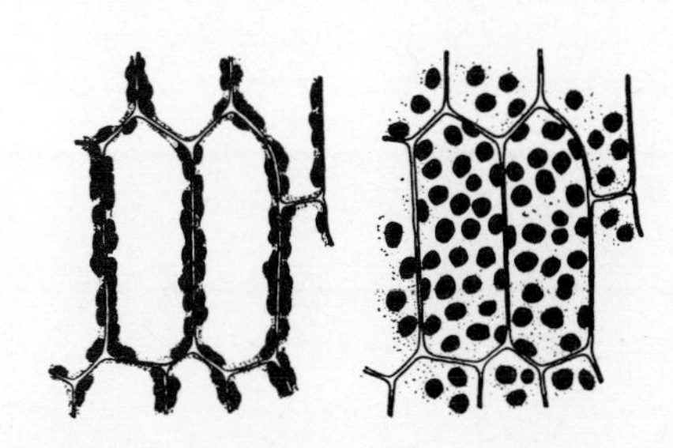

Fig. 332. Position of the chloroplasts in the cells of a moss leaf; left, in strong, right, in diffuse light. Diagrammatic.

In the alga *Mougeotia* (Conjugatae), for example, the plate-like chromatophores in weak light place themselves so that their flat surfaces are presented to the incident rays, but in strong light they turn so that only the edges are illuminated (Fig. 332). Especially effective in this respect is the visible red region of the spectrum (see p. 338). The absorption, however, is not in chromatophores, but in the peripheral layers of the cytoplasm, where the receptor molecules (in all probability phytochrome) must be present in a definite spatial orientation. Polarized red light is more effective when the plane of polarization is transverse to the longitudinal axis of the cell than when parallel, pointing to a transverse or nearly transverse arrangement of the receptor molecules. In the leaves of mosses and in fern gametophytes the lens-shaped chloroplasts often collect on the upper or lower walls in weak light, thus trapping all the available illumination, but in strong insolation they lie along the side walls (Fig. 332). Again in weak light the chloroplasts are often flattened and disc-like, contracting to spherical form in stronger illumination. A change in position of the chromatophores can in certain circumstances cause visible changes in the depth of greenness of a leaf. The pigment concerned in this response is not chlorophyll, but in all probability riboflavin. In leaves rich in intercellular spaces the plastids often lie along the side walls at night, probably a result of the gases excreted into the intercellular spaces setting up chemotactic effects in adjacent cells. Carbon dioxide, sugar, organic acids and many other substances can in fact cause chemotactic movements of chloroplasts.

The mechanism of these movements of nuclei and chromatophores is not completely clear. In any cell of *Vallisneria* it can easily be seen that the nucleus and chloroplasts may be pushed along passively by plasmatic streaming, but in other cases it appears that a capacity for autonomic movement should not be excluded.

III. The significance of irritability phenomena in plants

A recurrent problem arising from the irritability phenomena and movements described in the foregoing may be briefly alluded to here. Irritability is common to plants and animals, and is characteristic of all active life. Many of the responses of plants which have just been described, especially the tactic movements, undoubtedly give the impression of a conscious and purposeful organization. The question repeatedly arises of whether and how far activities analogous to nervous mechanisms should be ascribed to plants. It cannot be questioned that a plant objectively detects very diverse stimuli (optic, chemical, gravitational, etc.), i.e. it responds to them with a physicochemical sequence of reactions. We have, however, no evidence at all that such stimulation is accompanied in the plant by a subjective element, as it is in man, or that anything resembling a psychical experience, even of the simplest kind, exists. Even the lower animals provide no real suggestion of such experience. Our contemporary methods of scientific investigation enable us to identify in both the lower animals and in plants a chain of reactions, beginning with the susception of the stimulus and ending with a visible response. These responses are, of course, often, but by no means always, useful to the organism. To conclude that the plant possesses some kind of 'soul', or even consciousness, is not justified by the scientific evidence.

Part three: **Systematics and evolution**

Section one: **General principles**

The diversity of living organisms is almost overwhelming: there are perhaps half a million plant species and over a million kinds of animal. Nevertheless, one can generally recognize circumscribed groups which can themselves be further grouped or subdivided: this great range of organic forms is both discontinuous and hierarchical. The concept of hierarchy is pre-scientific: thus Scots pine, Austrian pine and mountain pine form the group 'pines'; pine, larch, fir and spruce are all conifers. The following questions arise in consequence:

1(*a*). Has this multiplicity of living things arisen by the differentiation of related organisms, and if so (*b*) what processes and laws have governed this historical development?

2. How can (*a*) the degrees of similarity and (*b*) the history of descent of living things be explained?

3. How can (*a*) plant and animal groups be arranged, and (*b*) how can they be unambiguously named?

Attempts to answer these questions are made by 1(*a*) the theory of evolution and 1(*b*) the study of evolutionary processes; by 2(*a*) systematics and 2(*b*) phylogenetics; and by 3(*a*) taxonomy and 3(*b*) nomenclature (Fig. 333).

1. Relationship or affinity

The concrete basis of relationship (consanguineity, genetic relationship, affinity, kinship) is afforded by the corporeal, cellular germ-lines which unite all the individuals of a reproductive community (population) in space and time. 'Germ-line' in the broad sense means

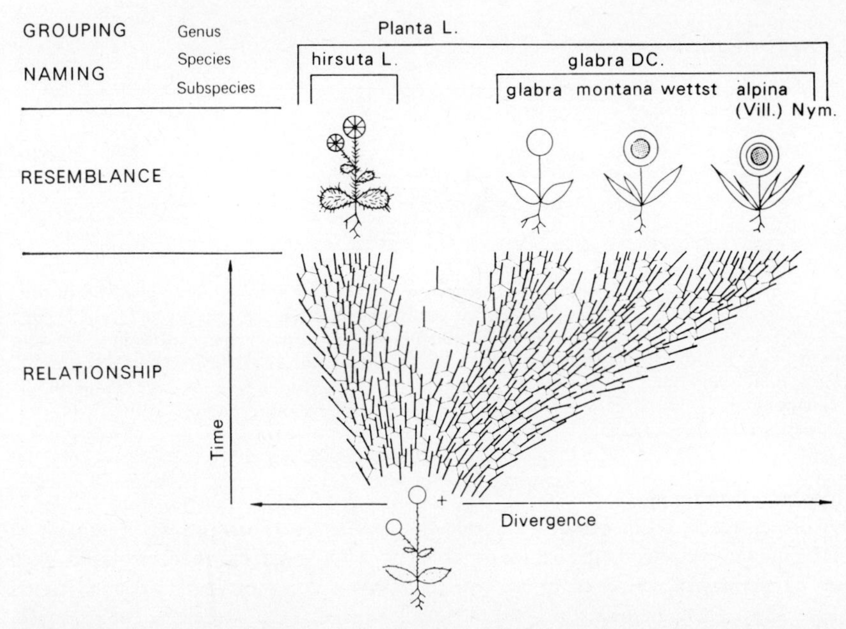

Fig. 333. A hypothetical group of related plants posing the basic questions of systematics. Relationship or affinity: the genealogical tree with germ-lines plotted against time and divergence, with an extinct ancestral form (+) and living descendants. Resemblance or similarity; hierarchical classification and nomenclature; taxa of different rank (with imaginary names).

that sequence of cells which connects successive germ-cells (cf. p. 202; Fig. 333). The ontogeny of each individual begins with the fusion of the parental gametes. In the course of time the reproductive community undergoes differentiation, diversification and isolation (Fig. 333: incipient in 'subsp. *glabra, montana* and *alpina*'; advanced in '*Planta hirsuta*' and '*P. glabra*', which are linked only by occasional sterile hybrids). In the course of phylogeny (history of descent) the divergent evolutionary branches (groups) are increasingly separated by insuperable barriers to cross-fertilization, and intermediate forms disappear, giving rise to discontinuity. Early separation of the connections between germ-lines results in distant relationships between diverging evolutionary groups; late separation retains closer relationships; so hierarchy arises. Ontogeny and phylogeny together make up hologeny (Fig. 334; cf. p. 9). In this picture of the 'evolutionary tree', the 'cross-section' of the ultimate 'branches' shows the present-day groups, more or less clearly defined; these are united to ancestral groups in the recent or distant past, forming a hierarchy of groups with a real existence in space and time. Although these principles of the **theory of evolution** are now generally accepted, the causes and forces which bring about this evolutionary change are still insufficiently understood, but the **study of evolutionary processes** is fundamental to all biology.

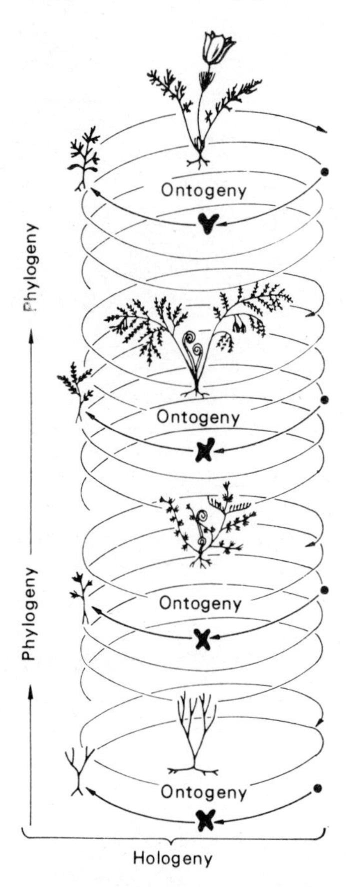

Fig. 334. Ontogeny + phylogeny = hologeny. A diagram of the development of cormophytes from the psilophyte level (Upper Silurian to Middle Devonian), through the pteridophyte (Middle to Upper Devonian) and seed fern (Carboniferous to Permian) levels, to the angiosperm level (to the present day). Time-span about 400 million years. After Zimmermann.

2. Resemblance or similarity

We have no direct access to information about the genetic relationships of organisms in the past. Of course, genetic relationship or affinity is to some extent correlated with resemblance or similarity since each individual shares a common inheritance with its relations (Fig. 333). Nevertheless, this correlation is not perfect, as shown, for example, by the occurrence of human 'doubles', or by the superficial similarity of sharks and dolphins, or by the resemblances between various stem succulents (Fig. 203; analogous convergence; cf. pp. 9, 184). In general one can draw few conclusions about relationships from a single characteristic (for example, hair or eye colour in man, or leaf length in '*Planta*', Fig. 333: in this respect '*P. hirsuta*' and '*P. glabra* subsp. *glabra*' are identical!). Firm conclusions about the degree of relationship are only reached by considering and comparing as many characters as possible, ranging from macroscopic external form (e.g. growth form, branching, hairiness, shape and size of leaves, structure and size of flowers in '*Planta*') and the molecular level of chemical composition to physiological and ecological behaviour. All this is the field of **systematics** which deals with the description, delimitation and comparison of living groups and the elucidation of their common structure. Furthermore, study of historical remains (fossils), the arrangement of series of characters (progressions), the spatial distribution of groups (as an expression of temporal existence) and the use of cytogenetical methods (for example in the experimental reconstruction of a group) all lead to an understanding of evolutionary history or **phylogenetics**, that is to say, the historical and causal interpretation of groups and their basic morphology. Systematics and phylogeny together afford a continual synthesis of all our knowledge, however incomplete, of the organic world.

3. Arrangement and naming

Systematics is concerned with the delimitation and the natural relationships of 'groups'.* Its findings form the basis for the arrangement of these 'groups' by means of taxonomy using an abstract system of hierarchical categories (e.g. genus, species, subspecies, Fig. 333) related to the concrete 'groups' and indicating so far as possible their genetic relationships in a hierarchical natural system. A 'group' taxonomically graded, whatever its rank, is called a taxon (plural taxa; 'group' + category = taxon) and given a scientific name (in latin form) according to definite rules. Since 1753 the naming of species has followed the binomial **nomenclature** with a generic name (e.g. '*Planta*') and a specific name (e.g. '*Planta hirsuta*'). Supraspecific taxa (e.g. families) are placed above the rank of species; infraspecific taxa (e.g. subspecies; subsp. '*montana*') are placed below. Where necessary the specific name is followed by abbreviations for the original author of the taxon (e.g. L. = C. Linnaeus; DC. = A. P. de Candolle).

The hierarchical taxonomical system and its associated nomenclature are the expression of our present knowledge of the degrees of similarity between plants. They give us both a workable synopsis of the diversity of the Plant Kingdom and the means (scarcely obtainable otherwise) for generalization, for the repetition of and prediction from experimental observations, and for the analysis of characters of a group or similar groups. In such a system, fresh information brings about further changes in arrangement and nomenclature, but with this exception, taxonomy must try to attain the greatest possible stability of scientific plant names since these provide the most important guide to botanical literature on the one hand and international understanding on the other. Thus studies of evolution, phylogenetics, systematics and taxonomy aspire not only to produce a synthesis of our entire knowledge of plants, but also to provide the basis for continuing research into and utilization of the vegetable world.

I. The theory and processes of evolution

The earlier assumption that plant and animal groups arose independently and individually by creation or spontaneous generation was replaced under the influence of Darwin by the theory of evolution (theory of descent), according to which all forms of life are genetically related: simpler prehistoric ancestors have undergone changes in form and behaviour to give rise to the present plants and animals. The most important causes of the evolutionary change of organisms and their adaptations to environment are now recognized in the modern synthesis of evolutionary theories as:

* The German word 'Sippe' has no English equivalent which in the present context is both accurate and unambiguous. It is translated here as 'group' to imply a more or less uniform assemblage of genetically related individuals. In some, but not all, passages 'Sippe' approximates to 'taxon' or 'biotype'; in certain compound words such as 'Ausgangssippe' it might be translated as 'stock'.

1. Mutation and recombination modifying the genotype (the hereditary complement) of the individual and conditioning the phenotypic variation of populations.
2. Selection and isolation acting on the phenotypic variations, producing adaptation, differentiation and finally divergence of populations.

Until the eighteenth century, species were thought to have arisen by spontaneous generation, sudden transformation (e.g. worms from corpses) or divine creation. This was often associated with the idea of the constancy of species (e.g. in the earlier writings of C. Linnaeus, 1707–78). From the middle of the eighteenth century the ideas gradually developed that the degrees of similarity between organisms arose from their degree of relationship (as indeed held by Linnaeus in his later writings), that the 'ladder' from simple to complex forms was due to evolution, and that variability within species was the first step in speciation. J. B. Lamarck (1744–1829) was an important pioneer of evolutionary ideas, and Charles Darwin (1802–82) brought about their final acceptance with the publication in 1859 of *On the Origin of Species*. Even in ancient times an active 'self-adaptation' to environment had been postulated as an agent of evolution, as well as the origin of new organs as a response to 'need'; eventually it was thought that such 'acquired characteristics' were inherited (Lamarckism). In contrast, Darwin considered that undirected variation with the selection of the individuals best fitted to survive in the 'struggle for existence' was the mainspring of evolution (Darwinism). While the postulates of Lamarckism generally remain unproven, the selection theory continues to be regarded as valid. Of course, in the meantime other evolutionary factors have been recognized, even if their significance was over-emphasized at first, for example geographical isolation (championed by M. Wagner and A. Kerner von Marilaun), mutation of particulate heredit-able factors (H. de Vries), hybridization (J. P. Lotsy, E. Anderson), recombination within the gene-pool of populations (S. Wright), genetic isolation as a result of barriers to crossing (T. Dobzhansky), etc. These ideas are now incorporated in the generally accepted synthetic theory of evolution of which the main features are considered in sections B–D.

A. Evidence for the theory of evolution

Since evolution is an historical phenomenon its main features cannot be studied either directly or experimentally. Nevertheless, much evidence supports the validity of the theory of evolution; this will be illustrated by some botanical examples.

1. **Palaeobotany** demonstrates the appearance and extinction of various groups in the course of the Earth's history. Furthermore, the time sequence shows many progressive modifications, increasing differentiation and complexity of organization, and the gradual approach towards present-day forms. Many groups which are at present completely isolated from each other are linked by fossil forms ('missing links'; cf. Figs. 553, 628).

The Dasycladaceae, a family of green algae, are particularly well preserved as fossils (Fig. 335; cf. also p. 470). The main periods of their evolutionary development were in the Silurian, Permian, Jurassic and Cretaceous; numerous lines are now extinct. The Dasycladaceae live in the surface zone of the sea and are reef-formers. At a very early stage they developed calcareous incrustations and dense aggregations of lateral branches (cf. *Primocorallina* and *Vermiporella*). Further important evolutionary modifications included the transition from irregular to whorled lateral branches (cf. *Vermiporella–Diplopora*); the development of cysts in the main axis (cf. *Primocorallina–Vermiporella*) and subsequently in laterals of the first order (cf. *Triploporella*) and later of the second order (cf. *Neomeris*); the division of labour between vegetative and fertile laterals (cf. *Halicoryne*); and finally the fusion of the fertile lateral branches to form a cap (*Acetabularia*, Fig. 423D, E).

Among the cormophytes we may refer to the psilophytes (Upper Silurian to Middle Devonian), with undifferentiated axial organs, as the original forms which gave rise to the later pteridophytes and seed plants with stems, leaves and roots (cf. pp. 83 et seq., 562 et seq.); or to the historical sequence from the sporogenous progymnosperms to gymno-sperms and finally to the angiosperms which are now dominant (cf. pp. 612 et seq.); indeed, psilophytes and progymnosperms are perfect examples of 'missing links'.

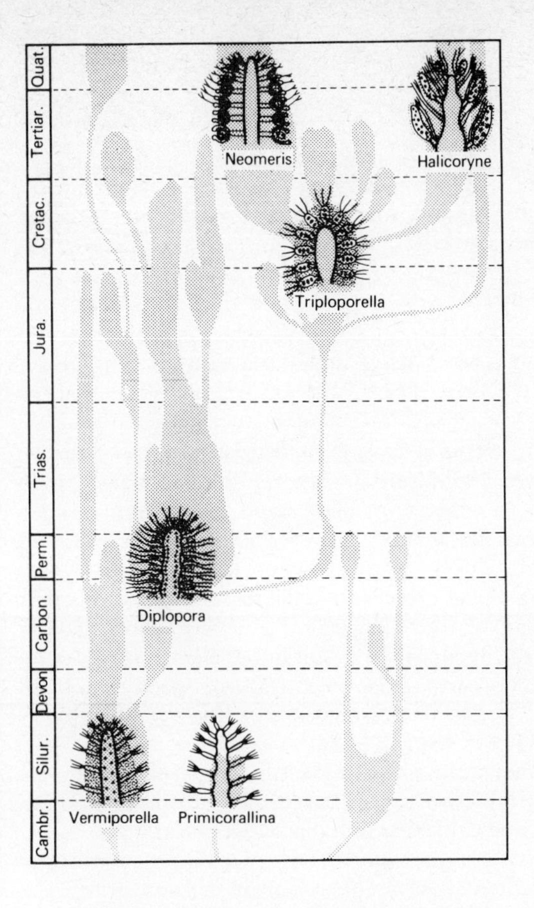

Fig. 335. Evolutionary tree of the green algal family Dasycladaceae indicating diagrammatically some typical examples. Note: main axis, lateral branches, calcareous incrustation (dotted) and cysts (black). Partly after Pia, Kamptner and Zimmermann.

2. **Molecular biology and physiology**; these show that the complex life-processes of all living things have much in common, for example the role of nucleic acids as carriers of the genetic code for protein synthesis and the complex enzyme systems involved in fermentation, respiration and photosynthesis (see Part Two, Physiology). It is extremely improbable that these complex structures and processes arose more than once.

3. **Comparative morphology, anatomy and cytology** reveal the basic patterns common to 'groups' of varying size. The common structure of very differently adapted 'groups' may be revealed by series of homologous characters, by the occasional reversion to primitive characters (atavism), or by the occurrence of functionless organs (rudiments). All this can only be explained by common descent and common genetic factors.

For example, the cormophytes, despite their great morphological diversity, ranging from mosses and ferns to the seed plants, show a similar alternation of generations and similar gametangia (antheridia and archegonia). Some ancient groups of seed plants (cycads and *Ginkgo*) still have motile spermatozoids (pp. 607, 616).

Adaptation may result in great differences between homologous organs (e.g. leaves may become tendrils, thorns, storage organs for water and nutrients, traps for animals, attractive devices, etc.; (cf. Figs. 199, 208, 212, 213A, B, 655). Juvenile stages may recall ancestral forms (e.g. the juvenile needle-leaves of *Thuja* contrasted with the adult scale leaves, or the pinnate juvenile foliage of *Acacia* contrasted with the adult phyllodes, Fig. 180); this is the basis of E. Haeckel's 'biogenetic law'. On the other hand, adaptation to particular environmental conditions or functions may give rise to analogous structures (e.g. stem succulents in unrelated families, Fig. 203; leaf-like shoots or platyclades, Fig. 197A; or subterranean storage organs, Figs. 207–10).

The Scrophulariaceae afford an example of modifications of the flower (Figs. 336, 727). Divergent specialization in floral biology has led to the elongation, flattening, or palate-like closing of the corolla-tube, the formation of a spur, dorsiventrality, or the progressive reduction of the stamens: A5 → A2. Thus *Verbascum* has five stamens, one of

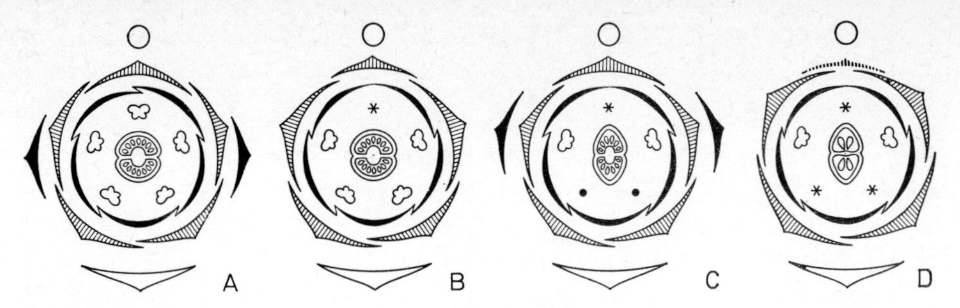

Fig. 336. Floral diagrams of the Scrophulariaceae. **A.** *Verbascum.* **B.** *Digitalis purpurea.* **C.** *Gratiola officinalis.* **D.** *Veronica officinalis.* Stamens infertile, dots; absent, asterisks. Partly after Eichler, drawn by Hartl.

which is rudimentary in *Antirrhinum* and *Scrophularia*; *Digitalis* has only four stamens; *Gratiola* two fertile and two reduced ones; *Veronica* only two fertile ones, while certain species of *Calceolaria* have merely two half-stamens. In snapdragon (*Antirrhinum*), however, reversion may again give rise to radially symmetrical (actinomorphic), unspurred flowers with five stamens (Fig. 340II, C).

4. **Systematics** demonstrates the hierarchical degrees of similarity between organisms which can only be interpreted as the expression of hierarchical genetic relationships (cf. pp. 425 et seq.). Positive indications of the dynamics of evolutionary change are given by the frequent occurrence of transitions between infra-specific taxa, of 'good' and 'bad' species and higher taxa (cf. p. 433), and by the decreasing capacity for hybridization as one passes from species to genus (cf. pp. 411 et seq.).

5. **Floristics** and **plant geography** demonstrate close connections between the age, the geographical isolation and the peculiar features of a flora (particularly the proportion of progressive endemics). This, too, can only be explained in terms of evolution.

For example, the flora of the Hawaiian Islands (which have never been connected to any continent) has some 20 per cent of the native vascular plants in endemic genera and 90 per cent in endemic species. In contrast the circumpolar Arctic flora has far fewer special 'groups' since it is relatively modern and was not isolated during the Pleistocene glaciations (cf. Figs. 751, 752).

6. **Cytogenetics** shows that both spontaneous and induced changes in hereditable characteristics in wild and cultivated plants often resemble the differences between races, species or even genera (cf. p. 399). Polyploid populations have often been produced artificially by hybridization and chromosome doubling, thus reproducing a natural process of species-formation.

The experimental synthesis of polyploid wheats is particularly significant (Figs. 337, 739C–E). Taken in conjunction with archaeological evidence this shows that cultivated wheats with a firm axis to the spike or ear were selected as early as the seventh millenium B.C. from wild plants with a brittle axis. At the diploid level (*Triticum monococcum*) the cultivated small spelt ('monococcum') was selected from the wild taxon 'boeoticum'; at the tetraploid level (*T. turgidum*) the cultivated emmer ('dicoccon') was selected from the wild 'dicoccoides'. Eventually at the turn of the third century B.C. the hexaploid bread wheats (*T. aestivum*) arose by allopolyploidy from the tetraploid cultivated emmer and a

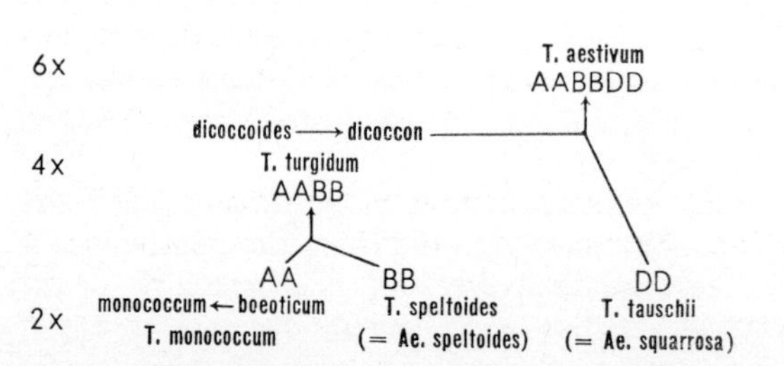

Fig. 337. The evolution of the more important wild and cultivated diploid, tetraploid and hexaploid wheats (*Triticum monococcum* 'monococcum', small spelt; *T. turgidum* 'dicoccon', emmer; *T. aestivum*, bread wheat); Ae. = *Aegilops*; large capitals indicate the genomes; basic chromosome number, $x = 7$. (The taxonomic rank of the various groups is somewhat debatable.)

diploid weed taxon (*T. tauschii*). Because of their high yields, the bread wheats gradually ousted the diploid and tetraploid cultivated wheats, and are now the only ones of general economic importance.

Completely new populations can be produced experimentally which are distinct from the parental populations and separated by barriers to hybridization (cf. p. 416). The theory of evolution can be tested directly by such methods.

B. The causes of variation

We are very familiar with the diversity of characters by which the individuals of a reproductive community can be distinguished, and by the ways in which these characters are transmitted by heredity. This is the case both in human populations (for example, hair and eye colour) and in plants. This variation may be continuous, for example, in the various crown-forms in the spruce, the leaf shapes of our native oaks, or in the size of seeds from a field of beans (Fig. 293), or it may be discontinuous variation, as in the number of petals in the lesser celandine (*Ranunculus ficaria*), or the flower colour in the milkwort (blue, red or white in *Polygala vulgaris*). Furthermore, characters may vary within an individual (intra-individual), or within a population (intra-population), or between populations (inter-population) (Fig. 338). We shall now consider the causes of these variations and the extent to which they form the basis for the origin of 'groups' and for the evolution of plants.

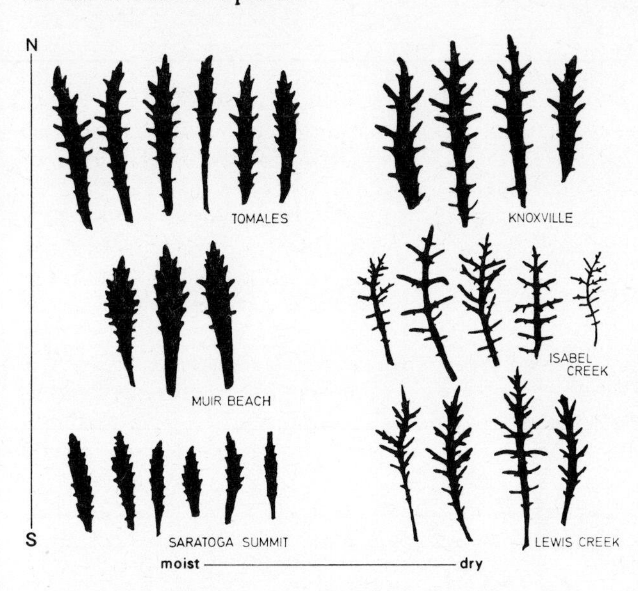

Fig. 338. Variation in shape of the basal leaves within and between six populations of a Californian composite, *Laya gaillardioides*. Left: populations from the outer coastal mountain range; right: from the inner. Plants cultivated under uniform conditions; each leaf is from a different individual. After J. Clausen.

1. Ontogeny, phenotype and genotype

Part of this variation in plant populations is determined ontogenetically.

Thus in assessing the crown-forms of the spruce we must take into account the different ages of the individuals. The enormous differences determined by the stages of development are exemplified by the slime-moulds (Fig. 69), the rust fungi with their alternations both of generations and host-plants (pp. 520 et seq.), the gametophytes and sporophytes of various brown algae which have been described as belonging to different genera (Figs. 437, 439), or the very different leaf shapes at different stages in the development of seed plants (Figs. 173, 180, 184). In order to exclude the ontogenetic component of variation between different individuals, comparable organs and similar stages of development must be studied.

A further component of variation, whether intra-individual, intra-population or inter-population, is that determined by different environmental conditions, i.e. modificatory (cf. p. 334 et seq.).

Examples are provided by the sun and shade leaves of the oak (cf. Fig. 290), the seed weight of a pure line of beans (Fig. 293) and to some extent by the variation in petal number in *Ranunculus ficaria*. Striking modifications can be produced in *Euglena* (e.g. *E. gracilis*; cf. pp. 397, 450) by growing it in the light to produce an autotrophic photosynthetic clone with numerous chloroplasts, or alternatively in the dark (but with a sufficient supply of organic nutrients) to produce a clone with heterotrophic nutrition and extremely reduced chloroplasts. The modifications induced by growing clones from the same individual in the lowlands and in the mountains (Figs. 292, 339), or in long and short days (Figs. 294, 295) may be mentioned again here.

The hereditable, fixed or genetic components of variation and the modificatory ones can be distinguished by excluding the effects of variation in the environment. This can be done by comparing individuals found growing in the same habitat, by cultivation under conditions as uniform as possible in experimental gardens (Figs. 338, 339, 351), or even better in controlled environment cabinets—if possible over several generations. Genetically controlled differences are then obvious. Genetically similar individuals constitute a biotype. Populations of most organisms with sexual reproduction and cross-fertilization show much genetic variation and include numerous biotypes (Figs. 338, 339, 351).

We have already considered (p. 322) the complex interplay between the totality of all the genetic material (genotype) and the environment during the ontogeny of an individual, and hence the progressive expression of its actual appearance (phenotype). It follows that it is not characters or differences between characters which are inherited, but reaction-norms (or standard patterns of behaviour).

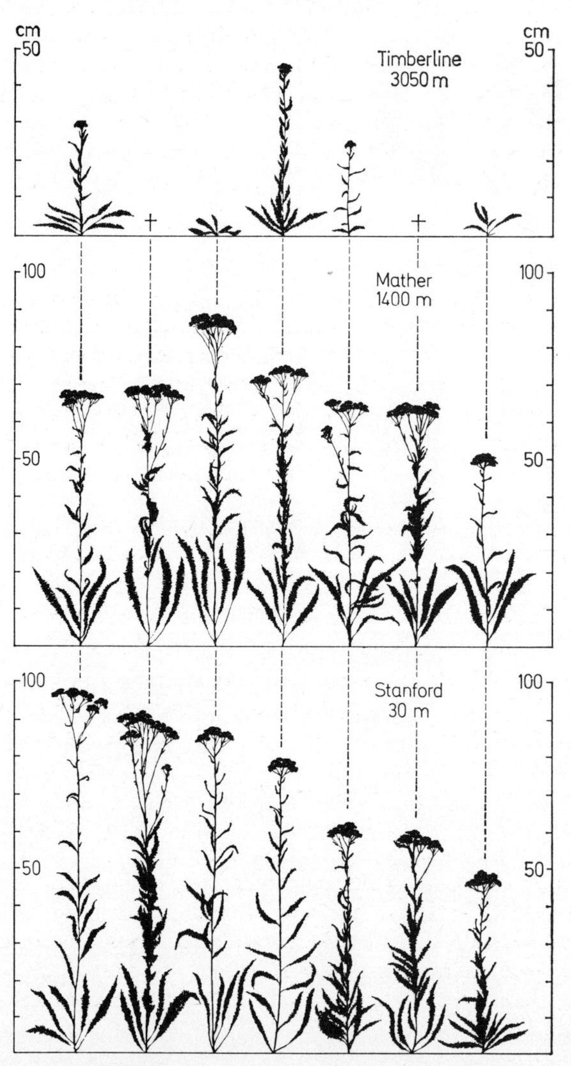

Fig. 339. Experimentally produced modifications of a Californian yarrow (*Achillea lanulosa*, tetraploid). Vegetatively propagated pieces (clones) of seven individuals from one population from the montane zone of the Sierra Nevada (Mather) grown in three experimental gardens, at Stanford (30 m altitude), at Mather (1400 m) and at Timberline (3050 m), showing hereditable differences between individuals and different reaction-norms of the same individual at different altitudes. After Clausen, Keck and Hiesey.

In water crowfoot (*Ranunculus peltatus* and other species, Fig. 182A) the floating leaves only develop at the interface between water and air, and there only under long days; more divided land-leaves are also formed only in long days when a fall in water level exposes the developing leaves to air; while the underwater leaves form only in short days. Some biotypes of *Primula sinensis* develop red flowers only at temperatures below 20°C, remaining white at 30°C, while other biotypes are always white or always red. It follows that under certain temperature regimes these biotypes are not phenotypically distinguishable. Such phenocopies are seen when the high alpine races (Fig. 351) of the Californian *Achillea lanulosa* are compared with montane races grown under alpine conditions. Similarly, apparently uniform dwarf populations of coastal herbs reveal both genetically fixed and modificatory variations when grown side by side. The reaction-norms of a genotype (i.e. its patterns of behaviour) only manifest themselves in a range of environmental conditions (Fig. 339).

The different organs of a plant may show different degrees of modificatory plasticity. Leaf-size, branching and numbers of flowers and fruits of annual herbs of dry habitats are 'euryplastic', varying with the environmental conditions, while seed-size, for example, is very uniform (stenoplastic). Dry conditions produce dwarf plants, moist conditions luxuriant ones; seed number is the maximum possible in the circumstances, while the individual seed as the unit of propagation is uninfluenced by this plasticity. Modificatory plasticity as a form of adaptation appears to be advantageous where fluctuations in the environment are irregular and occur within the life-span of a generation. In other cases genotypic changes in the population are more important as the basis of adaptation. The related question as to whether persistent modifications (dauermodifications) and phenocopies can result in hereditable changes in the genotype has generally been answered in the negative. Nevertheless one must remember that modifying influences may more readily fix similarly directed mutations ('genetic assimilation', e.g. the complete mutational loss or inactivation of chloroplasts in darkened cultures of *Euglena gracilis*). The possibility of ultimately inheritable alterations of the cytoplasm (cf. pp. 332, 401) is still debated.

2. Mutation

In general all hereditable changes are called mutations; they are the basis of evolution. All cell organelles containing DNA may be involved in mutational change, and in particular, of course, the chromosomes (the genome; changes by gene-, chromosome- and genome-mutations), the plastids (the plastome) and certain cytoplasmic structures (the plasmone) (cf. pp. 332 et seq.). All these are components of the genotype. Analysis of these changes depends primarily on experiments on hybridization, on comparative cytological studies of the chromosomes and their behaviour in parental and hybrid plants, and on the experimental induction of mutations.

Mutational changes can be demonstrated even in viruses (some with RNA) as well as in all other groups of plants ranging from bacteria, algae and fungi to bryophytes, pteridophytes and seed plants.

Spontaneous mutation is a very rare event. Very large differences in the rate of mutation are found between different genes and different chromosomal regions within an individual, between the various genotypes within a group and between various groups. 'Mutator genes' can greatly increase either the rate of mutation of other genes or the rate of breakage of chromosomes. The frequency of mutation of a single gene is rarely more than 0.05 per cent per generation, and frequently much less. Nevertheless, bearing in mind the large number of genes in a higher organism (at least 10,000!), up to 10 per cent of the individuals of a progeny may carry some new mutations. The rate of change of a population through mutation is therefore highly dependent on generation-time.

Mutations, and gene-mutations in particular, may affect any structure and process of an organism, but the difference between the original form and the mutant (the degree of mutational change) varies from scarcely detectable micro-mutations to drastic macro-mutations which can alter basic organizational characters (cf. Fig. 340). Of course, micro-mutations have significantly better chances of survival and are therefore more frequently encountered than macro-mutations.

Fig. 340. Gene mutants in snapdragons (*Antirrhinum majus*). **I.** Development of the whole plant: A, normal; B, dwarf; C, early flowering. **II.** Shapes of the flower: A, normal, zygomorphic; B, radial (actinomorphic); C, spurred. After Stubbe.

Most mutations, whether spontaneous or experimentally induced, act deleteriously on the vitality or fertility of their carriers. This is understandable when one thinks that every organism is a highly complex and balanced system which has become adapted to its environment over countless generations; hence the probability is very small that a given mutation will confer an additional benefit. Nevertheless, many mutations have been observed which are superior to the originals in normal situations—and especially in novel ones: for example the phage- and streptomycin-resistant bacteria, the increasingly aggressive mutants of parasitic rust fungi (against which increasingly resistant varieties of cereals have to be bred), the more vigorous, earlier or dwarfer mutants of vascular plants (Fig. 340), etc.

Mutations are non-directional with regard to the particular requirements and the environment of an organism; that is to say, mutations are not themselves oriented towards better adaptation than before. Of course, one cannot properly speak of 'chance' mutations since the structure of the genotype influences and limits its own possibilities for mutation: hence the frequent occurrence of homologous mutations in related groups (e.g. the laciniate or red-leaved, anthocyanin-containing, mutants in a wide range of angiosperms). Furthermore, numerous mutations are suppressed since their expression is incompatible with the normal processes of development. This leads to an endogenous orientation of mutational change. Some environmental or exogenous orientation is possible through 'genetic assimilation' and perhaps through the effects of the environment on the plasmone.

Similar mutational changes occur within populations, in different races and in even more widely separated groups. In addition there seems to be no significant different between spontaneous and induced mutations with respect to their origin, sphere of influence or method of action.

(*a*) **Gene mutations** originate as submicroscopic molecular changes in the chromosomal DNA (cf. pp. 329 et seq.). They affect all biological structures and processes regardless of whether they are spontaneous or induced (Figs. 340, 341). They are clearly responsible for the greater part of evolutionary change.

For example, in bacteria, flagellates and fungi gene mutations have been shown to control metabolism (e.g. respiration and amino-acid synthesis), the structure of flagella, the form of the thallus, (para-) sexuality (cf. p. 402), etc. Among bryophytes, *Marchantia* shows a spectrum of mutations which reflects the characters of related genera and even includes the development of radial, shoot-like forms. Gene mutation also proves that enzyme systems (e.g. esterase, catalase, peroxidase) are under genetic control, as well as chlorophyll synthesis—if this fails as a result of mutation only heterotrophic existence is possible. In *Datura* mutation switches the production of alkaloid from hyoscyamine to scopolamine, the latter being characteristic of other solanaceous genera.

Vegetative features of angiosperms drastically modified by macro-mutations include the change from two cotyledons to one, various transformations of stipules, leaflets and tendrils (analogous in the garden pea to the differences between other Leguminosae), laciniation of the leaves (Fig. 285), and the switch from the annual to the perennial habit (e.g. in maize). Simple gene mutations in the snapdragon determine dwarfness or early flowering (related to a reduction in internode length) (Fig. 340I). Extensive studies of hybridization between ecological races of the Californian *Achillea lanulosa* (Fig. 351) and *Potentilla glandulosa* have shown that small mutational steps due to a few or many genes have led to large differences in stem height, flowering time, winter dormancy, leaf shape and hairiness as well as in floral and fruit characters. The two species *Nicotiana alata* and *N. langsdorffii* (Fig. 341) differ mainly in leaf-width and the relative lengths of calyx, corolla tube, corolla limb and style, all of which depend on the degree of cell elongation and ultimately on the pleiotropic effects of genic differences on the activity of growth substances (cf. p. 300). Among floral characters modified by gene mutation are sexuality (e.g. dioecious or monoecious flowers), symmetry, presence or absence of perianth, number of floral parts, free or united petals, presence or absence of a spur, and flower colour. Figure 340II shows an example in *Antirrhinum*: the macro-mutant with a radial corolla corresponds in floral diagram to the scrophulariaceous genus *Verbascum*, that with a spur to *Linaria* (Figs. 336, 727), and mutants with only two stamens to *Ixianthes*. Crossing and recombination experiments enable one to determine the location of the mutant genes on the chromosomes; the resultant genetic maps have been made for certain bacteriophages, bacteria and higher plants (e.g. maize, peas and snapdragons).

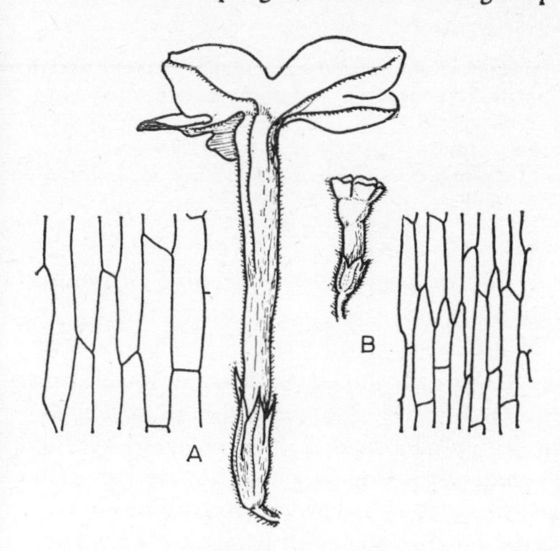

Fig. 341. Specific differences between *Nicotiana alata* (**A.**) and *N. langsdorffii* (**B.**) determined by gene mutations: different shapes and sizes of the flowers (calyx, corolla) and of the cells of the throat of the corolla tube. After Anderson and Ownbey.

(*b*) **Chromosome and genome mutation** involve alteration of the structure or number of the chromosomes and are therefore often visible under the microscope (cf. pp. 331 et seq.). Chromosome mutations are caused by the breakage or rejoining of the chromosomes in various ways, giving rise to deletions, duplications, inversions or translocations (Figs. 286, 342A, B, 358). Such chromosomal rearrangements may bring about a stepwise change in chromosome number (dysploidy; Fig. 342C, D). Irregular division at mitosis or meiosis may result in the addition or loss of complete chromosomes (aneuploidy). Finally, irregularities at mitosis or meiosis may result in the failure of the daughter chromatids or chromosomes to pass to separate nuclei resulting in the multiplication of sets of chromosomes (polyploidy; Figs. 26, 343; cf. also pp. 36–7, 331, 418 et seq.). Polyploid individuals may arise either from polyploid vegetative cells or tissues or by the fusion of unreduced gametes.

Chromosome and genome mutations occur spontaneously in all groups of organisms but are particularly frequent in hybrids with irregular mitosis or meiosis. Various experimental treatments greatly enhance their frequency.

Since chromosome and genome mutations do not in general directly modify the genes but only their position or number, they mostly produce only small or moderate morphological or physiological effects. Yet they are very important in evolution because they may lead to the suppression of recombination and set up barriers to hybridization

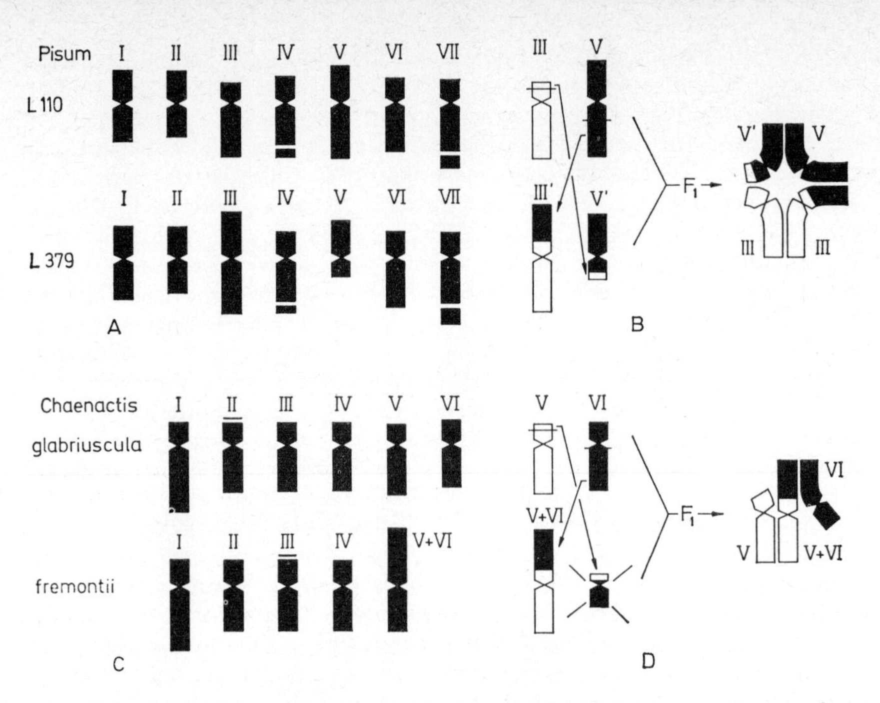

Fig. 342. Chromosome mutations and dysploidy. **A.** Haploid chromosome diagram of two cultivated kinds of garden pea (*Pisum sativum*). **B.** Diagrams of the differentiating reciprocal translocation and of meiotic chromosome pairing in the structurally heterozygotic F_1. **C.** Haploid chromosome diagram of two closely related species of *Chaenactis* (Asteraceae) with 2n = 12 and 2n = 10. **D.** Diagram of the differentiating reciprocal translocation with loss of chromosome segments and of meiotic chromosome pairing in the F_1. **A.** After Lamprecht. **B.** After Kyhos.

pp. 402, 413 et seq.). Polyploidy may also facilitate both the hybridization of groups otherwise isolated genetically (pp. 418 et seq.) and the increase of the genic material.

The genus *Pisum* affords a good example of differentiation by means of chromosome mutation. Thus the cultivated races L110 and L379 differ by a reciprocal translocation between chromosomes III and V. This can be seen in the chromosome diagram (caryogram; Fig. 342A), in the chromosome pairing during meiosis in heterozygous F_1 hybrids (Fig. 342B), in the modified segregation of characters in the F_2 generation and in their sterility (up to about 50 per cent), all as a result of defective chromosome combinations (III + V′ and III′ + V instead of III + V and III′ + V′). Geographical races of *Pisum* are similarly differentiated; thus the *abyssinicum* group differs *inter alia* from the normal by several translocations (in chromosomes I, II and apparently also VI); hybrids with the normal group show sterility up to 75 per cent. Fairly frequently groups are also separated by inversions (cf. Fig. 358). Sex chromosomes (cf. Figs. 282, 283) carrying sex-determining genes also show structural modifications.

Aneuploidy usually leads to defective development either because of meiotic irregularities arising from the superfluity of chromosomes or from the loss of chromosomes and therefore of genes. Consequently aneuploidy is generally only a transitory phenomenon. However, if the excess chromosomes become structurally modified, or if all the essential segments of two chromosomes are combined into one by translocations with the loss of residual segments (cf. Fig. 342C, D), the result may be ascending or descending dysploidy. Dysploid changes in chromosome number have often been brought about experimentally (e.g. in rye, 2n = 14 → 16) but they often occur naturally in populations (e.g. in *Nigella*, Ranunculaceae, 2n = 12 ⇌ 14) and occasionally in geographic races (e.g. *Myosotis sylvatica*, Boraginaceae, 2n = 22 → 20 → 18), in closely related species (e.g. *Chaenactis*, Asteraceae, from western N. America; Fig. 342C, D; 2n = 12 → 10, and in addition there is a further translocation between chromosomes II and III and an inversion within chromosome V), or within genera or families (e.g. Dipsacaceae, Fig. 365). The lowest known plant chromosome number has arisen in the N. American composite

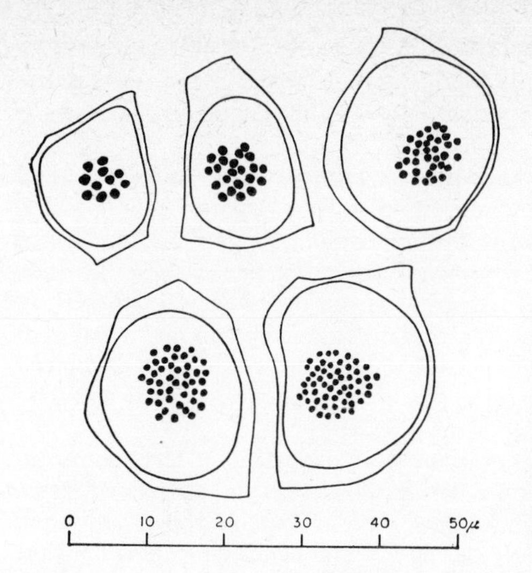

Fig. 343. Infraspecific polyploid series in a bedstraw (*Galium anisophyllum*). Meiosis (metaphase I) in pollen mother cells in di-, tetra-, hexa-, octo- and decaploid races ($2x$, $4x$, $6x$, $8x$ and $10x$ with the basic chromosome number $x = 11$); the size of the chromosomes decreases with the increasing grade of ploidy.

Haplopappus gracilis by dysploidy: $2n = 8 \rightarrow 6 \rightarrow 4$. In groups with polycentric chromosomes (spindle fibres attached to many points instead of to a single one on each chromosome; cf. p. 33) changes in chromosome number readily arise from the unhindered transmission of chromosome fragments (agmatoploidy, e.g. in the Juncales).

The multiplication of chromosome sets in non-hybrids (autopolyploidy, cf. pp. 36, 331) can be artificially induced (cf. Fig. 26) but despite its occasional spontaneous occurrence in almost all plant groups it is of relatively little evolutionary importance. Changes resulting from autopolyploidy are generally of a quantitative kind (and often disadvantageous) while fertility is often reduced by meiotic disturbances (especially when more than two homologous chromosomes pair). In contrast, allopolyploidy, the multiplication of non-homologous chromosome sets in hybrids, is exceedingly important in plant evolution (pp. 418 et seq.).

(*c*) **Plastome and plasmone mutations** have been insufficiently studied (cf. pp. 332–4) but their importance for evolution should not be underestimated. The resultant hereditable changes are shown by reciprocal differences in crossing experiments and have been demonstrated, and to some extent produced artificially, in plastids, mitochondria, centrioles, etc., among bacteria, flagellates, fungi, bryophytes (Fig. 287) and angiosperms (Fig. 360).

Their expression is manifest, for example, in the shape of plastids (e.g. in *Chlamydomonas*), in mating systems (e.g. in *Podospora*, Ascomycetes), in leaf characters (e.g. in *Epilobium*, Onagraceae) and in sexuality or pollen sterility (e.g. gynodioecism and in many Lamiaceae). Crosses between different races and species (e.g. in Onagraceae, *Achillea*, etc.) have shown that divergent changes in genome, plastome and plasmone can upset their balanced action together; defective gene action, sterility and diminished vigour are the result and act as barriers to hybridization (cf. pp. 413 et seq.). Conversely, the co-action of particularly harmonious plasmones or genomes contributes to the phenomenon of heterosis in hybrids between various biotypes or other 'groups' (cf. pp. 322, 419).

3. Recombination and reproduction

Every population maintains a balance between genetic plasticity and stability: adaptive characters must be genetically stabilized and yet be capable of variation under changing environmental conditions. Important as mutation is as a basic ingredient of evolution, it is not in general sufficient by itself to ensure the essential genetic variation and capacity for adaptation of populations. Take the wild strawberry (*Fragaria vesca*) as an example: it reproduces vegetatively by vigorous runners (cf. p. 203) and each clone can only develop the desirable genetic plasticity by means of stepwise somatic mutations (bud sports). The recombination of advantageous mutations arising in different clones is made possible only by sexual reproduction, with meiosis, which significantly increases the genetic plasticity and capacity for adaptation (cf. p. 205). Such advantageous combinations may then again be extensively reproduced vegetatively (as in cultivated perennials). Hence the balance between vegetative and sexual reproduction influences both the genetic equilibrium and the rate of recombination.

The greater part of the hereditable and phenotypically expressed variation in natural populations depends on Mendelian genes and alleles which have arisen by mutation (cf. pp. 319–25). The gene-pool or gene-reservoir of a population of higher plants consists, however, both of those genes expressed in the phenotype and of the recessive or otherwise masked (and sometimes disadvantageous) genes; as the occasion arises these may separate and facilitate adaptation to new environmental conditions. This maintenance of a reserve of recessive alleles is of course only possible in the diplophase or dicaryophase and not in the haplophase (cf. p. 42). Furthermore, the balance between genetic plasticity and stability is influenced by the organization of the genotype.

The processes of pollination and fertilization are also important in recombination: *Fragaria vesca* has hermaphrodite flowers and is self-fertile so that pollination and fertilization of an individual either with its own pollen (autogamy) or foreign pollen (allogamy) are both possible (cf. pp. 657 et seq.). Clearly, autogamy and hence inbreeding limit the rate of recombination and the extent of genetic variation; allogamy has the opposite effect. Finally, the effectiveness of dispersal of pollen and fruit determines the extent to which gene-flow is possible in space while the generation time of the population determines gene-flow in time. These factors working together form the recombination system of a population and control the integration and mobilization of its genetic variation, whether existing or arising *de novo* by mutation.

(*a*) **Organization and transmission of inheritance** have undergone profound changes in the course of evolution. Bacteria and blue-green algae are procaryotic and have no true chromosomes, nucleus or mitosis, while only a few forms exhibit parasexuality (transformation, transduction and conjugation in bacteria; cf. pp. 329 et seq., 441); similarly there is no meiosis and vegetative reproduction predominates. Only the eucaryotic organisms have precise control of genetic inheritance by means of the chromosomes which divide exactly in mitosis and facilitate recombination with true sexuality (mating with plasmogamy and caryogamy, meiosis with crossing-over; cf. p. 325). A diplophase (or dicaryophase) has been interpolated and progressively promoted at the expense of the haplophase in the most diverse lines of evolution (e.g. in the Chlorophyceae, Fig. 429; Phaeophyceae, Fig. 443; Mycophyta, Pteridophyta and Spermatophyta; cf. also p. 42). A diplo- or dicaryophase permits both dominance and recessiveness on the one hand and better buffering (homeostasis) against the environment (by the cooperation of two more or less different genomes) and against deleterious recessive mutations on the other (reserve function of the second normal allele or normal chromosome). Finally the amount of recombination depends on the number of chromosome or linkage groups (p. 324) among which the genes are distributed (more chromosomes, more recombination), on the frequency of crossing-over and on the extent to which crossing-over is suppressed in certain chromosome-segments or 'super-genes' (for example as a result of structural heterozygosity between chromosomes).

(*b*) The various kinds of **sexual reproduction** and their **break-down** are most important in the recombination systems of eucaryotic plants. Fertilization between gametes from different parents (allogamy, cross-fertilization) is made obligatory by the genotypic sex determination and dioecism of the gametophytes of many algae, fungi, bryophytes and ferns and of the sporophytes of many seed plants (cf. pp. 326 et seq.). In monoecious or hermaphrodite plants the same result is produced by genetic incompatibility since gametes from any one individual cannot unite (e.g. in many fungi, ferns and seed plants). Self-sterility genes (S) with many multiple alleles (S_1, S_2, S_3 . . .) are often found in seed plants: given a plant with the constitution S_1S_2, pollen grains with S_1 or S_2 are inhibited and only pollen with S_3 or S_4, etc., can germinate (cf. pp. 364, 657). Even where compatibility exists, temporal or spatial separation of the reproductive organs commonly promotes allogamy, as in many angiosperms (cf. pp. 657 et seq., Fig. 670). A combination of the last two phenomena is seen in heterostyly (p. 657) in those angiosperms in which incompatibility is combined with different positions of the anthers and stigmas and

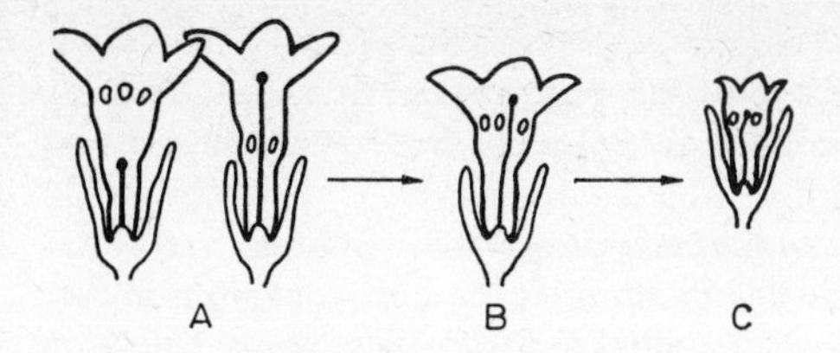

Fig. 344. The series from cross-pollination (allogamy) to self-pollination (autogamy) in the Californian *Amsinckia spectabilis* (Boraginaceae). **A.** Heterostylous allogamous form. **B.** Long homostyle, large-flowered form. **C.** Short homostyle, small-flowered autogamous form. After Ray and Chisaki.

often with different forms of pollen and stigmatic papillae (e.g. in *Primula*, Fig. 669). Cross-fertilization in all these cases enhances the rate of recombination and hence the variability of the population.

Many hermaphrodite plants develop facultative (occasional) or obligatory (invariable) self-fertilization or pollination, i.e. they become autogamous. This happens, for example, in some diatoms (p. 458), certain fungi, bryophytes and pteridophytes, and especially in various angiosperms (p. 658); the consequent inbreeding leads to less recombination and reduction in the variability of the population. It can be seen in *Amsinckia* (Fig. 344), for example, that with the loss of heterostyly from the original form (A) the relatively large-flowered intermediate form (B) is still insect-pollinated and because of the spatial separation of stamens and stigma may still undergo some cross-pollination, while the inconspicuous small-flowered form (C) is mostly self-pollinated since the stamens and stigma are in contact. Autogamy is particularly frequent among annual weeds, e.g. shepherd's purse (*Capsella bursa-pastoris*), heart's-ease (*Viola arvensis*), goose-grass (*Galium aparine*), groundsel (*Senecio vulgaris*), etc. For these plants of intermittently available habitats it is advantageous for a large population to be built up quickly from a few individuals without dependence on external pollinating agents; their relative genetic uniformity may be compensated by phenotypic plasticity.

Finally, apomixis is the total loss of sexuality. Thus in many ferns (e.g. our native *Dryopteris borreri*) new sporophytes develop asexually from gametophytes at the same level of ploidy. In many angiosperms embryos develop from diploid unfertilized egg-cells (parthenogenesis and agamospermy; p. 666), but sometimes this development requires the stimulation of pollination (pseudogamy). A diploid embryo sac results either from defective meiosis in the embryo sac mother cell (diplospory) or from the suppression of the original haploid embryo sac and its replacement by an adventitious diploid one (apospory; Fig. 345). This kind of vegetative reproduction precludes normal genetic recombination but a haploid cell is occasionally fertilized and develops normally by way of compensation. Agamospermy allows the perpetuation of favourable genetic combinations especially in groups of hybrid and polyploid origin which have faulty sexual reproduction (e.g. many Rosaceae, Cichoriaceae and Poaceae; cf. p. 421). Of course sexual

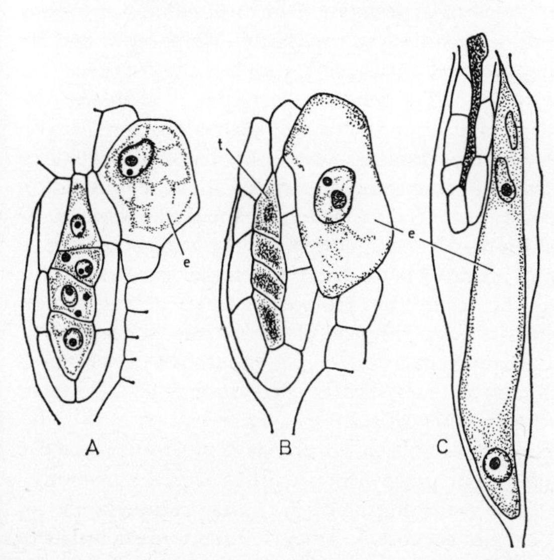

Fig. 345. Apomixis and agamospermy in angiosperms: *Hieracium flagellare* (Cichoriaceae). The nucellus of the ovule points downwards towards the micropyle. The megaspore tetrad (t) of which the lowermost cell would normally form a haploid embryo sac aborts. In its place an enlarged cell of the integument (e) develops into a diploid aposporous embryo sac. After Rosenberg.

reproduction may be replaced by any kind of vegetative reproduction (e.g. by conidia in many fungi or by bulbils in the practically sterile *Dentaria bulbifera*; Figs. 223A, 224B). In a N. American representative of the otherwise tropical fern genus *Vittaria* the suppression of the sporophyte is particularly noteworthy: the gametophyte is a liverwort-like prothallus which propagates itself exclusively by means of vegetative gemmae.

(*c*) **Gene-flow** in space within reproductive communities or between neighbouring ones is dependent on the exchange of individuals, conidia, spores, pollen, seeds, fruits and so on. In the case of pollen transfer, gene-flow is more extensive in wind-pollinated plants with copious pollen production than in self-pollinated plants with little pollen. Gene-flow and recombination are also enhanced in large continuous populations without internal differences in chromosome number or structure, especially where hybridization takes place between related groups, but both are limited where populations are small and disjunct or where there are chromosomal differences and barriers to hybridity. The close relationship between gene-flow and rate of recombination holds good also in a temporal context: short and rapidly repeated life cycles increase rates of recombination (e.g. in bacteria and annuals) while long ones decrease them (e.g. in perennial woody plants which are late in reaching sexual maturity).

(*d*) The **recombination system** is thus of crucial importance for evolution. Long-lived, complex multicellular organisms could only develop with the combination of mitosis, sexuality + meiosis, and the dominance of the diplophase. Both this evolutionary development and the present-day control of the essential genetic plasticity and stability of localized populations is only possible because all components of the recombination systems themselves are under genetic control (e.g. the frequency of crossing-over and both sexual and vegetative reproduction). Mutational changes and the coordination of the components of the recombination system, both among themselves and with the environment, depend on cybernetic feed-back and selection.

C. Adaptation, differentiation and divergence

Only a small fraction of the variation in a group of organisms which is theoretically possible by mutation and recombination can survive the limitations and selection imposed by the physical and biological environment. This is evident if we compare the diversity of cultivated forms selected by man with their less variable wild progenitors (Fig. 350). The course of evolution is canalized by selection and isolation; the following sections consider how these factors interact with mutation and recombination and lead to adaptation, the formation of races and the divergence of 'groups'.

1. Selection, drift and population structure

The genic structure and frequency of alleles in a population of outbreeding organisms cannot be changed by recombination alone. According to the Hardy–Weinberg law, if we have two allelles A and a with frequency p and q (where $p + q = 1$), the frequency of genotypes in the population is given by $p^2(AA) + 2pq(Aa) + q^2(aa) = 1$. Whatever the original values of p and q, the frequencies of the genotypes will never alter from generation to generation. This assumes that no new mutations occur, that inbreeding is excluded (otherwise the homozygotes AA and aa would come to predominate), that the population is large, and that no genotype has the advantage over another. In small populations, or when large ones are reduced catastrophically, or in the course of migration, the chance elimination or perpetuation of alleles may occur (genetic drift). Much more important for the modification of gene frequency is, however, the promotion of or discrimination against the reproduction and increase of particular genotypes by selection: the carriers of favourable genes increase, of unfavourable ones decrease in the population. Selection is thus a particular case of competition or interference. Natural selection may be contrasted with artificial selection by Man (e.g. in the breeding of cultivated varieties).

An example is provided by the competition between normal dark green plants of the small stinging nettle (*Urtica urens*) and a light green mutant with reduced chlorophyll content (Fig. 346). Under the same conditions the growth of the mutant is much inferior to that of the normal form in a mixed culture because of its greatly reduced assimilatory

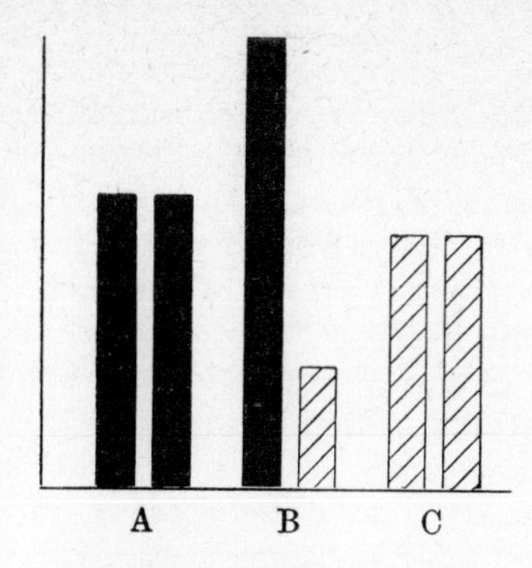

Fig. 346. Competition between normal plants of the small stinging nettle (*Urtica urens*), rich in chlorophyll (black columns), and a mutant poor in chlorophyll (cross-hatched). Fresh weights when: **A.** The normal plant, is competing with itself. **B.** When the normal plant is competing with the mutant. **C.** The mutant competing with itself. After Correns.

capacity (B), while the difference between the two forms in pure cultures is much less (A, C). Obviously the difference will affect the reproduction of the two forms and so bring about a selective change in the corresponding gene frequencies in the population. The recessive allele for such a chlorophyll deficiency (*a*) is not completely eliminated because the heterozygote (*Aa*) is often not inferior to the homozygous normal form (*AA*). Indeed, in certain habitats it may even be superior; this has been demonstrated for various grasses.

This superiority of heterozygotes (with respect to gene and chromosome mutations) compared with homozygotes is related to the phenomenon of heterosis (cf. pp. 322, 401) and may be largely responsible for 'balanced' genetic polymorphism and the frequency of disadvantageous homozygous alleles in populations. Many angiosperms show polymorphism with respect to hairiness or flower colour (e.g. red and white flowers in *Corydalis cava*). Polymorphism is also promoted by fluctuations in the environment which favour first one, than another genotype (for example, the early unfolding of the leaves of the beech permits extra photosynthesis in mild springs but is disadvantageous if there are late spring frosts). Various biotypes in a population may also be favoured by the juxtaposition of different micro-habitats. Finally, mixed stands of different biotypes may show greater productivity as a result of positive cooperation. All these factors contribute to polymorphism and increase the size of the gene-pool of a population.

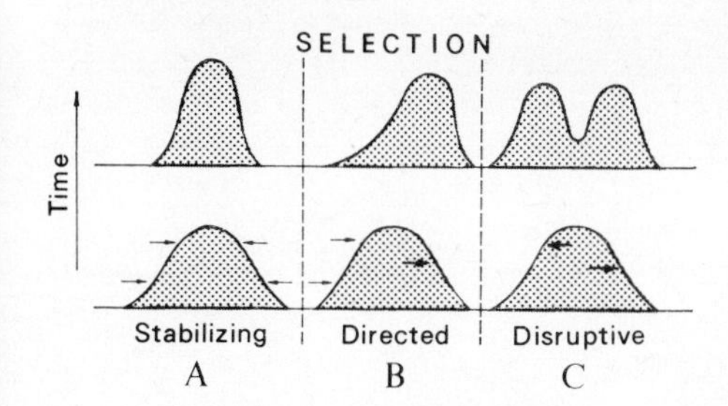

Fig. 347. Stabilizing, directed and disruptive selection. The range of variation of the initial population (below) is determined by the frequency of genetically distinct individuals and is either narrowed, displaced or broken up by the various kinds of selection (arrows). After Mather.

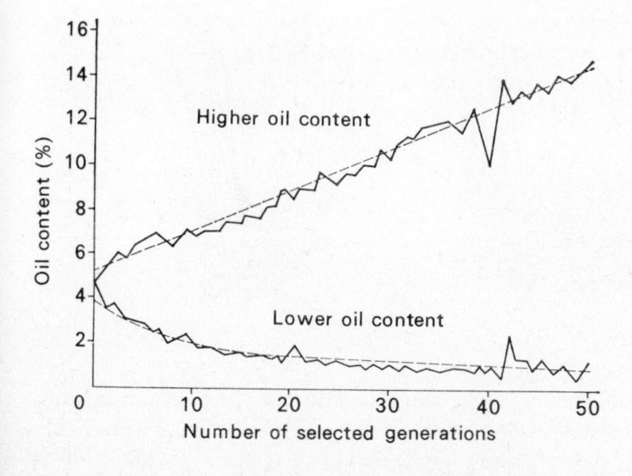

Fig. 348. The effects of directed (artificial) selection in increasing or decreasing oil content in maize over fifty generations. After Woodworth *et al.*

The phenotypes of a population and hence its spectrum of genotypes (or gene pool) are continuously subject to control by selection. Figure 347 shows the various resultant changes which are possible. Stabilizing selection (A) eliminates only the extreme forms under uniformly intense environmental conditions and hence decreases the variability. Thus the beech in our example coming into leaf too early or too late would be gradually eliminated. Directed selection (B) displaces the range of variation as a result of unilateral environmental pressure. The selection of maize varieties with a higher or lower oil content (Fig. 348) provides an example. The biotype spectrum of a grass population can be drastically modified within a few months by the natural grazing of large herbivores (Fig. 349). The frequent parallel selection of 'improved' globular stem succulents among cacti (Fig. 202) must also be regarded as the result of long term selection in the direction of the reduction of transpiration in extremely dry conditions. Finally, disruptive selection (C) produces a curve of variation with several peaks as a result of the simultaneous action of different environmental factors. The selection by Man of edible stem, leaf or inflorescence tissues in the various common brassicas provides an example (Fig. 350). Under natural conditions this kind of divergent selection occurs, for example, at the environmental limits of a population, or when floral visitors preferentially pollinate particular colour forms in polymorphic populations (cf. pp. 412, 659).

In addition selection may promote neutral (indifferent) or even disadvantageous (negative) characters when these are associated with the characters directly influenced by selection by pleiotropic gene action, by genetic linkage or by developmental correlations.

Fig. 349. The effects of directed selection due to grazing or mowing on the biotype spectrum in meadow grass (*Poa pratensis*). **A.** Low growing biotype from a pasture. **B.** Tall growing biotype from a meadow. After Kemp.

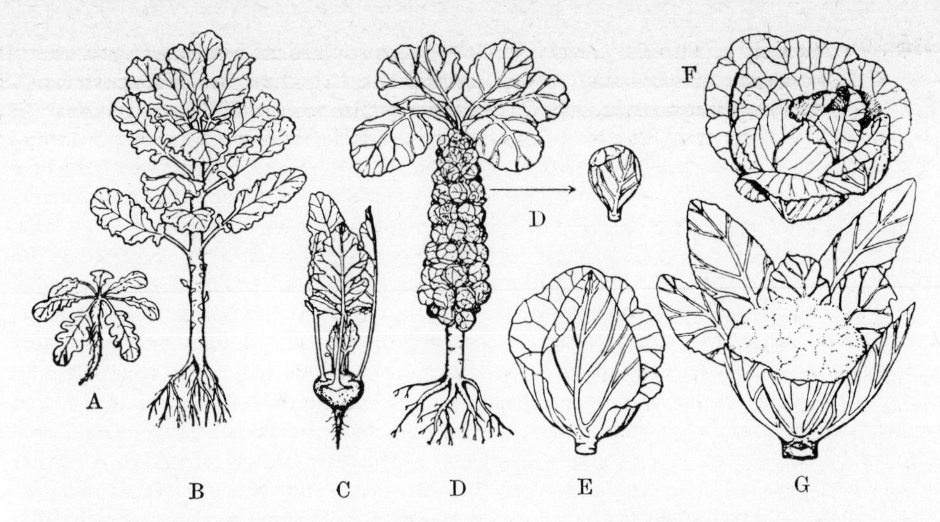

Fig. 350. The effects of disruptive (artificial) selection in: **A.** Wild cabbage (*Brassica oleracea* var. *oleracea* and related taxa). **B.** Kale (var. *viridis*). **C.** Kohlrabi (var. *gongylodes*). **D.** Brussels sprout (var. *gemmifera*). **E.** White, red or summer cabbage (var. *capitata*). **F.** Savoy cabbage (var. *sabauda*). **G.** Broccoli or cauliflower (var. *botrytis*). After Transeau, Sampson and Tiffany.

In the case of *Nicotiana* (p. 399, Fig. 341) selection for the length of the corolla tube brings in its wake changes in the length of the petals, calyx and fruit.

Whether a particular gene or gene-combination in a polymorphic population is advantageous or disadvantageous depends very much on the conditions. This may be expressed as a selection coefficient which may be combined with factors for mutation pressure, gene-flow, population size, generation time, etc., and built into the recombination formula to give complex mathematical population models. Studies of the range of variation and genetic structure of natural plant populations promise a deeper understanding in the future of the actual significance of these factors under natural conditions.

It appears in retrospect that the adaptation of populations to their environment is an extremely complex phenomenon. Indeed, adaptation is only possible through (1) modification, (2) change in the percentage occurrence of various biotypes, (3) new recombinations of the genes in the gene-pool, and (4) the formation of new mutants.

2. Spatial isolation and the evolution of races

Many widespread 'groups' include genetically distinct races which are differentiated geographically and are adapted to particular environments. This is the case not only for Man and animals but also for plants, from unicells (e.g. diatoms) to seed plants. Thus in the Alps the mountain pine (*Pinus mugo*) occurs mainly as the erect form in the west and as the shrubby form in the east. A similar differentiation into races was described for our model taxon '*Planta glabra*' (Fig. 333: lowland, montane and alpine races).

The evolution of ecological and geographical races can be illustrated by experimental models. Thus extremely rare resistant mutants of bacteria (e.g. *Staphylococcus*) can be selected and promoted on culture media containing antibiotics. In recent years numerous resistant races of bacteria have arisen in this way from forms originally sensitive to antibiotics. If a particular mixture of various biotypes of a cereal (e.g. of various autogamous barleys) is grown in different places, after a few years of competition only one or a few biotypes survive in a given place, the others being eliminated. Similarly, in populations of Californian yarrows from different altitudes only those biotypes survive which are sufficiently adapted to the prevailing environmental conditions by stem height, hairiness, periodicity, rate of respiration, etc. (Fig. 351).

These examples suggest that the evolution of races arises from the 'canalization' of the gene-pool of the local population, which is fed by mutation and recombination, by means of selection, genetic drift and spatial isolation. Directed selection of particular genes is particularly important in ecological adaptation. In contrast, geographical differentiation is dependent more on the chance occurrence and fixing of selectivity neutral genes as a result of genetic drift (p. 404). Geographical isolation reduces the gene flow between the original races and hence contributes to their stabilization and further development. All these phenomena are illustrated by the races of '*Planta glabra*' (Fig. 333).

The ecological or geographical differentiation of characters along a gradually varying environmental gradient may be more or less continuous. In such a case we speak of a cline, either an ecocline or a topocline. Often, however, the differences in characters between sharply separated localities or environments are more marked and it is then possible to distinguish ecological races (ecotypes) or geographical races. A glance at the population pattern shown by the Californian *Laya gaillardioides* (Fig. 338) shows how closely ecological and geographical variation are interlocked: the populations from the outer coastal ranges may be regarded as a geographical race, but with their wider leaf segments they are better adapted to moist climatic conditions and can be regarded as an ecotype; similarly for the drier inner coastal range populations with their narrower leaf segments. Furthermore, each population clearly forms a morphologically recognizable sub-unit of the 'group' (or a portion of a cline). This interlocking structure is further complicated because other characters (e.g. flower colour) vary to some extent independently of leaf characters since they may be subject to different selective control. Such relationships are characteristic of the differentiation of 'groups'.

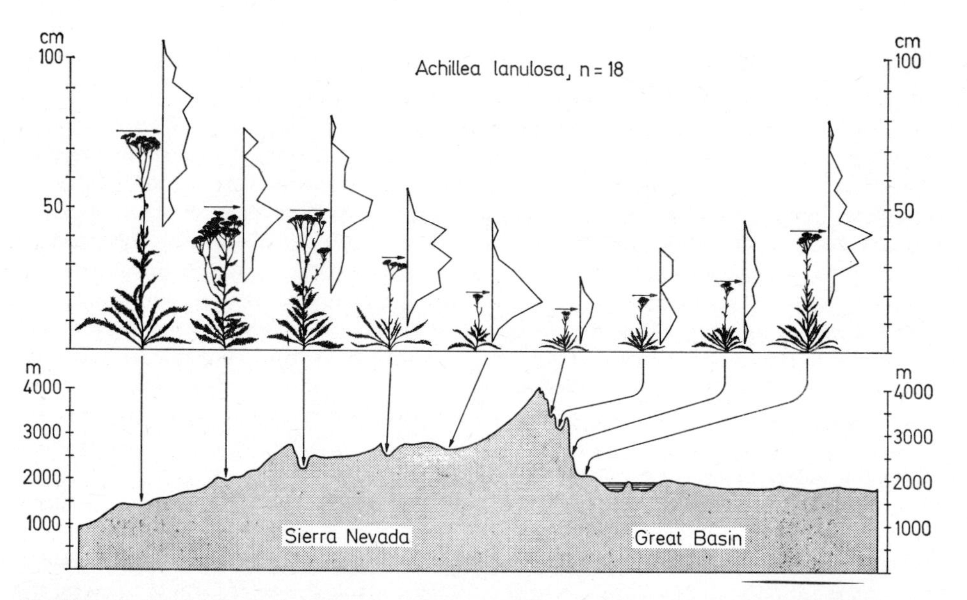

Fig. 351. Ecological races of a Californian yarrow (*Achillea lanulosa*, tetraploid) from various altitudes (1400–3350–2100 m) along a transect of some 60 km across the Sierra Nevada and the neighbouring Great Basin at about 38° N. Some sixty individuals were grown at Stanford (30 m) from seed from each population. The diagrams show the genetic variability in stem height, the mean values (arrows), and a typical individual from each population. After Clausen, Keck and Hiesey.

A relatively simple case of clinal differentiation in cell chemistry has been worked out for wild white clover (*Trifolium repens*) in which a dominant allele *A* controls the synthesis of a cyanogenetic glucoside; when the plant is injured this liberates hydrocyanic acid (prussic acid) by enzymatic hydrolysis. The genotypes *AA* and *Aa* produce cyanide; *aa* does not. Figure 352 shows the varying frequencies of the two alleles of which *A* decreases in populations from south to north and from low to high altitudes. This must either be due to temperature-dependent genes for growth activities or to the better protection of *A* genotypes in areas with many herbivores. Many groups of plants show genetic differentiation with respect to frost resistance; Figure 353B shows such a cline for Scots pine (*Pinus sylvestris*). The rate of respiration too, is generally dependent genetically on the mean temperature of the environment; thus at the same temperature the respiration rates of alpine ecotypes of yarrow are higher than those of montane ecotypes

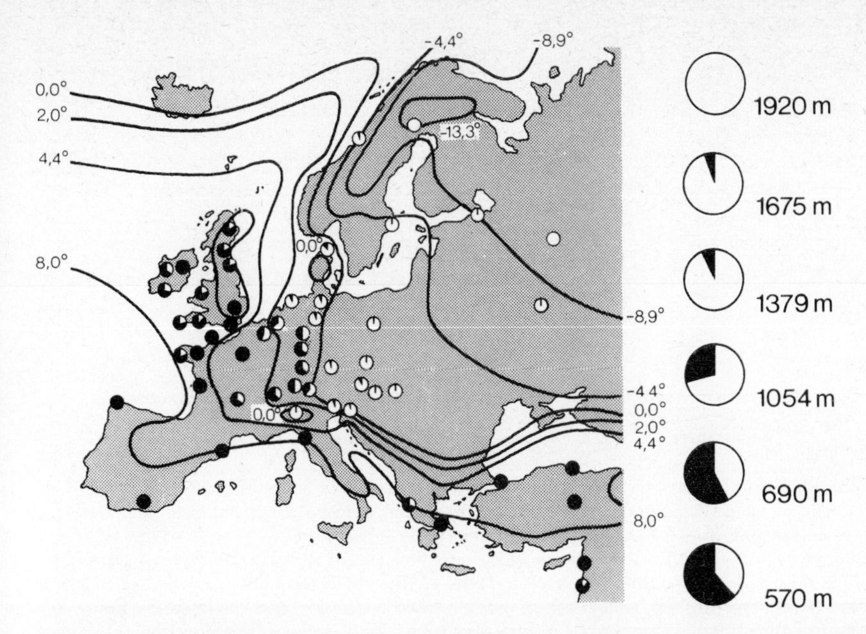

Fig. 352. Clinal variation in white clover (*Trifolium repens*). The population frequency of the alleles A and a (black sectors in the circles *A*, white *a*) controlling the synthesis of glycoside is correlated with the January isotherms from the Mediterranean region to northern Europe (map, left); in the Alps, with altitude (circles, right). After Daday, partly modified.

(Fig. 351). Alpine races grown in the lowlands where the temperature is higher respire their reserves faster and consequently die. Similar adaptations occur in photosynthesis: thus alpine ecotypes of mountain sorrel (*Oxyria digyna*) are better adapted at high temperatures, arctic ones at lower (Fig. 353A).

(*a*) **Ecological differentiation** in ecoclines or ecotypes at the level of the species may involve any aspect of morphology or physiology: mode of nutrition (e.g. autotrophy, myxotrophy or heterotrophy in flagellates); genetic adaptation of parasites to particular hosts (e.g. in rust fungi or in mistletoe); edaphic specialization (e.g. adaptation to various saline, calcareous, serpentine or heavy metal soils; edaphic ecotypes often occur in close juxtaposition); utilization of light (e.g. shade forms with lower enzymatic activity, sun forms with higher enzymatic activity in photosynthetic CO_2-fixation); drought resistance (different rates of transpiration, leaf areas (Fig. 338), hairiness, waxy coatings, etc.); rhythms (short- and long-day forms, temporal variation in leafing, flowering (Fig. 340), leaf-fall, dormancy, germination, etc.); life- and growth-forms (e.g. variation in development of flotation processes in flagellates, regulating the rate of sinking; Fig. 338); annual and perennial habit in angiosperms; height growth and stem form (Figs. 340, 349, 351); and the form of flowers, fruits and seeds (adaptations to various methods of pollination or dispersal; Fig. 344).

It is significant that such ecological differentiation can also be observed between species or even higher taxa: thus, soil indicators (*Gentiana clusii* on calcareous soils, *G. kochiana* on silicate); evergreen and deciduous habits (e.g. larch and cedar); woody and herbaceous (e.g. Magnoliales and Ranunculales); freshwater and marine (e.g. Oedogoniales and Siphonales).

(*b*) The fundamentals of **geographical differentiation** can be illustrated by wild forms of wallflowers (*Erysimum* section Cheiranthus). These generally exist in small populations in rocky habitats in the Aegean archipelago. During the last 5 million years a system of progressively diverging and spatially vicarious (replacing) species, subspecies and local races has developed as a result of spatial isolation and genetic drift (Fig. 354). Furthermore, crossing experiments show that reproductive barriers have gradually developed by genic and chromosomal changes: thus, for example, *E. naxense* will scarcely cross with neighbouring taxa. Further examples of the widespread phenomenon of the formation of geographical races as a result of spatial isolation with little or no genetic isolation are

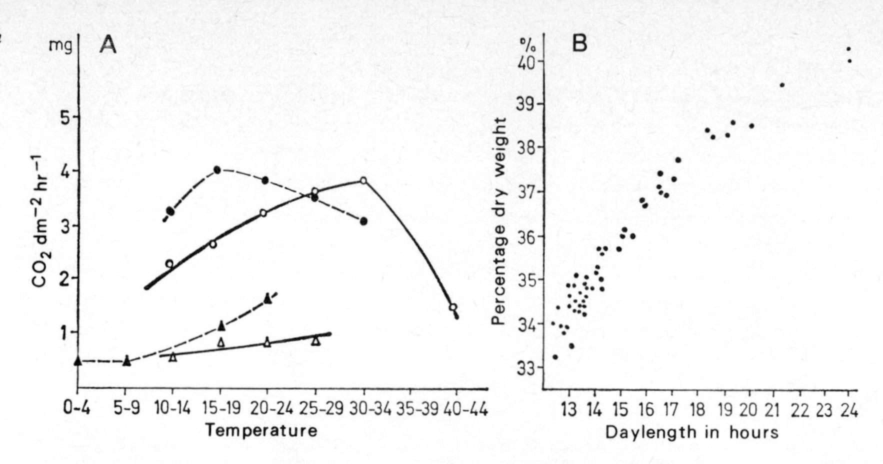

Fig. 353. Ecological differentiation in seed plants. **A.** Ecotypes of the mountain sorrel (*Oxyria digyna*, Polygonaceae) and their different patterns of physiological response: mean rates of photosynthesis (○ ●) and respiration (△ ▲) in relation to temperature in a southern alpine (○ △) and a northern arctic (● ▲) race. **B.** Clinal variation in the Scots pine (*Pinus sylvestris*); fifty-two plants of European origin under identical cultural conditions show a close correlation between the dry weight of the needles as a percentage of fresh weight (as an index of cold resistance) and the length in hours of the first spring day (mean temperature 6°C) in their natural habitats (as an index of latitude, continentality and the length of the growing season). **A.** After Mooney and Billings. **B.** After Langlet.

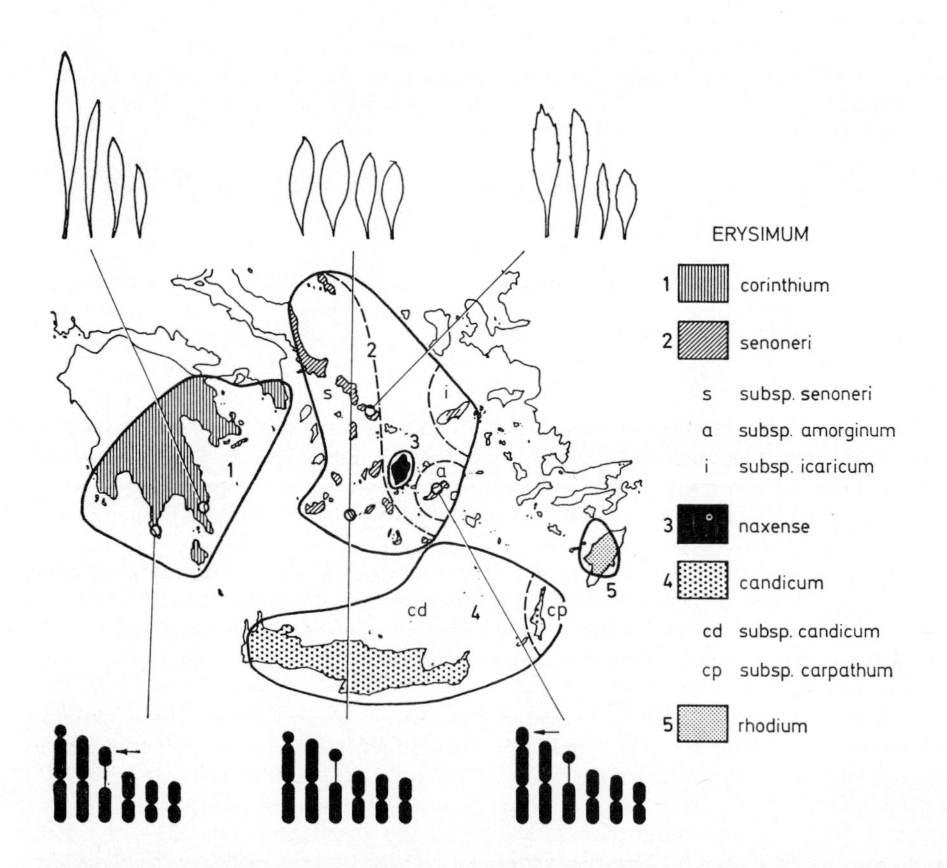

Fig. 354. The development of geographical races in wallflowers (*Erysimum* section Cheiranthus) in the Aegean region: vicarious distribution of species and subspecies, and examples of local races with their morphological (leaves, above) and caryological variation (shown by arrows in the caryograms, below). After Snogerup.

provided by the circle of affinity of the Mediterranean-montane black pine (Fig. 355) and the genus *Hepatica* which has a very disjunct distribution in the deciduous forest regions of the Northern Hemisphere (Fig. 356): the European *H. nobilis* is represented in E. Asia by two races within the same species and is closely related to the two N. American species, the areas of which have become secondarily superimposed: despite occasional hybridization these retain their identities since they have different habitat requirements. The divergence is much greater between the Carpathian *H. transsilvanica* and *H. nobilis* which occur in the same area.

In addition, the principles of geographical replacement and of the differentiation of races can often be recognized at higher taxonomic levels: for example, in the Fagaceae, *Fagus* and related genera occur in the Northern Hemisphere, *Nothofagus* in the Southern; or in the Caryophyllidae in which the Cactaceae have their centre of distribution in the arid regions of the New World, the Aizoaceae in the Old.

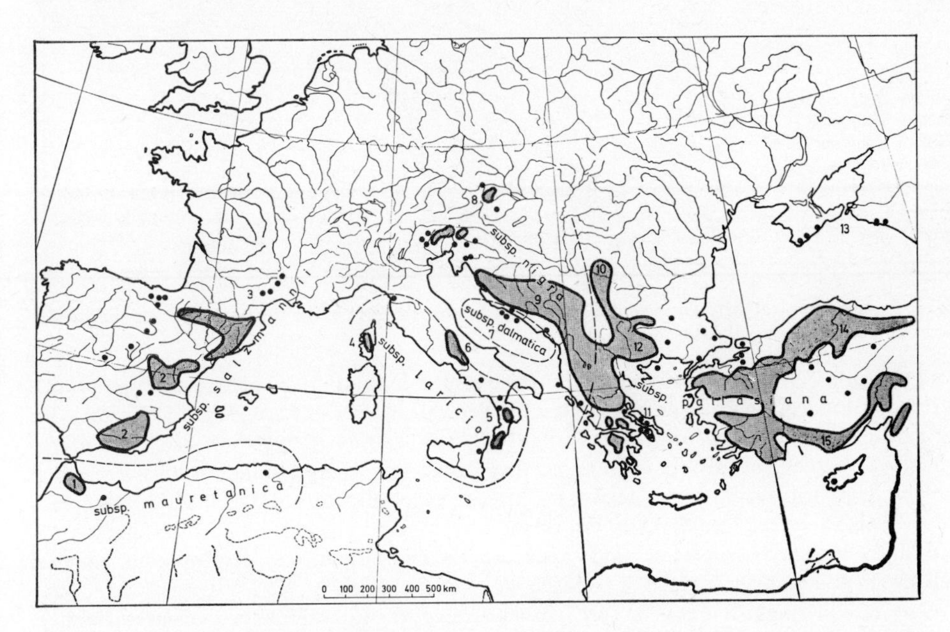

Fig. 355. Geographical differentiation in the circle of affinity of the Mediterranean-montane black pine (*Pinus nigra*). The names of subspecies are shown; subordinate local races are indicated by numbers. After Critchfield and Little, Meusel, Jäger and Weinert, and Niklfeld.

Thus ecological and geographical differentiation commonly form an essential first stage in the evolutionary process. Spatial isolation facilitates divergent genetic adaptation and the formation of races under the influence of selection and genetic drift, preventing the blending by crossing and recombination of the incipient races and excluding direct competition between them. Thus in various neighbouring areas (allopatric), closely-related races are separately represented (vicarious). Taxa of common descent can only survive in the same area (sympatric) after the development of reproductive isolation mechanisms (e.g. barriers to hybridization; see below).

These principles of the formation of allopatric taxa apply especially to sexually reproducing plants with high rates of recombination. More or less sympatric distribution arises when there are particularly sharp habitat boundaries, or when recombination is restricted by autogamy, apomixis, or the abrupt formation of barriers to hybridization (cf. Fig. 364).

3. Reproductive isolation and the formation of species

Further phylogenetic divergence and sympatric co-existence are only possible following the erection of barriers to hybridization between reproductive populations: in Fig. 333 '*Planta hirsuta*' and '*P. glabra*' are connected only by sterile hybrids (but compare

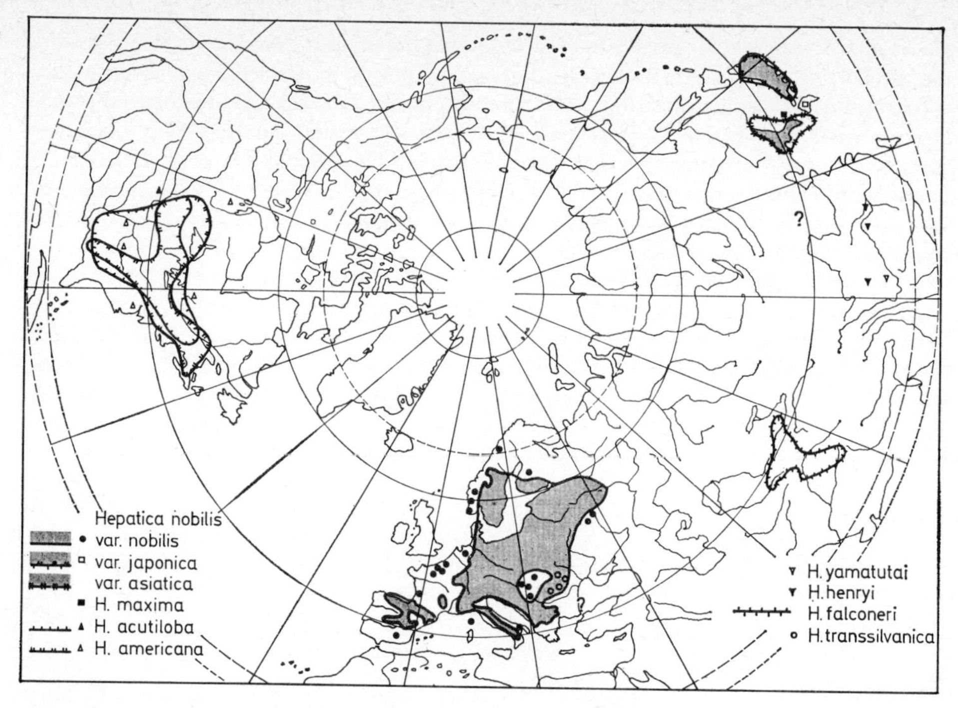

Fig. 356. Geographical differentiation in *Hepatica* (Ranunculaceae) in the deciduous forest regions of the Northern Hemisphere. After Meusel, Jäger and Weinert.

Platanus, p. 414). Reproductive isolation frequently follows after ecological-geographical differentiation (cf., e.g., *Erysimum* section Cheiranthus or *Hepatica*, Figs. 354, 356). In our various native yellow primroses (*Primula veris, P. elatior, P. vulgaris*), which are often sympatric, ecological differentiation of the species is greatly enhanced by complex barriers to hybridization: cross-fertilization leads especially to poor seed production and the hybrids show reduced fertility; all this is determined by genes, by chromosome structure and by the cytoplasm, and operates through disturbed development of the endosperm and embryo, defective meiosis, etc.

One can distinguish between pre- and post-zygotic isolation mechanisms (effective before and after fertilization), and between environmentally (exogenous) and internally determined (endogenous) ones. The spatial isolation already considered is pre-zygotic and exogenous; temporal, floral-biological and gametic isolation or incompatibility are pre-zygotic and endogenous; while reduced viability and sterility of hybrids are post-zygotic. Barriers to hybridization are usually determined by the operation together of a number of these mechanisms.

(*a*) The significance of **pre-zygotic isolation mechanisms** may be illustrated firstly by the timing of reproduction in related taxa. While the common snowdrop (*Galanthus nivalis* subsp. *nivalis*) blooms early in the spring, subspecies *cilicicus* from S.W. Asia flowers in the depths of winter and subspecies *reginae-olgae*, also from S.W. Asia, flowers in the autumn. Similarly, of two eastern Mediterranean *Cyclamen* species, *C. coum* flowers in the spring, *C. cilicicum* in the autumn. Sympatric races or species of grasses often liberate their pollen at different times of the day.

Furthermore, different specializations in floral biology function as isolating factors since pollination takes place only (or preferentially) between individuals of the same pollination type. This is true, for example, of related allogamous and autogamous races (e.g. of *Amsinckia*, Fig. 344). The relative fidelity of visiting insects (especially honey bees) to particular flowers may contribute to divergent differentiation: thus in populations with mixed flower colours, flowers of the same colour are preferentially visited and pollinated (e.g. in *Phlox*). In such a way the differentiation of otherwise completely interfertile taxa may be maintained, for example in snapdragon biotypes with different colours and forms of flowers. Floral-biological isolation is especially effective when taxa are adapted to different floral visitors. Thus closely-related montane and alpine species of Californian monkey musk (*Mimulus cardinalis* and *M. lewisii*) are pollinated respectively by humming birds and bumble bees, and hence isolated. The same is true of *Aquilegia* (cf. p. 431, Fig. 369) and many orchids, e.g. those Mediterranean *Ophrys* species with deceptive flowers imitating female Hymenoptera of which the males effect pollination.

Gametic isolation or hybrid incompatibility occurs when chemical attraction fails between the gametes of different biotypes (e.g. in flagellates and algae), or when pollen tube development is inhibited (e.g. on the stigmas of angiosperms; often associated with incompatible *S*-gene systems; cf. pp. 364, 402).

(*b*) The causes and operation of **post-zygotic isolation mechanisms** are investigated mainly by experimental hybridization. Experiments on the annual Californian composite genus *Laya* may be considered (Fig. 357) as one example out of many. (1) Within both the *L. platyglossa* group and the *L. glandulosa* group more or less fertile F_1 hybrids are possible but the F_2 generation shows various depressive tendencies; (2) F_1 hybrids between the two groups are sterile; (3) *L. heterotricha* crossed with other species produces only (sub-)lethal F_1 hybrids (non-flowering seedlings or rosettes); (4) *L. septentrionalis* will not cross with any other species. These progressively greater barriers to hybridization are in this, as in many other groups, determined by the divergent differentiation of genes, chromosome structure, genomes (dysploidy: n = 7,8; polyploidy: n = 16) and the resulting lack of genetic or physiological balance in the hybrids (e.g. defective chromosome pairing or defective cooperation of embryo and endosperm).

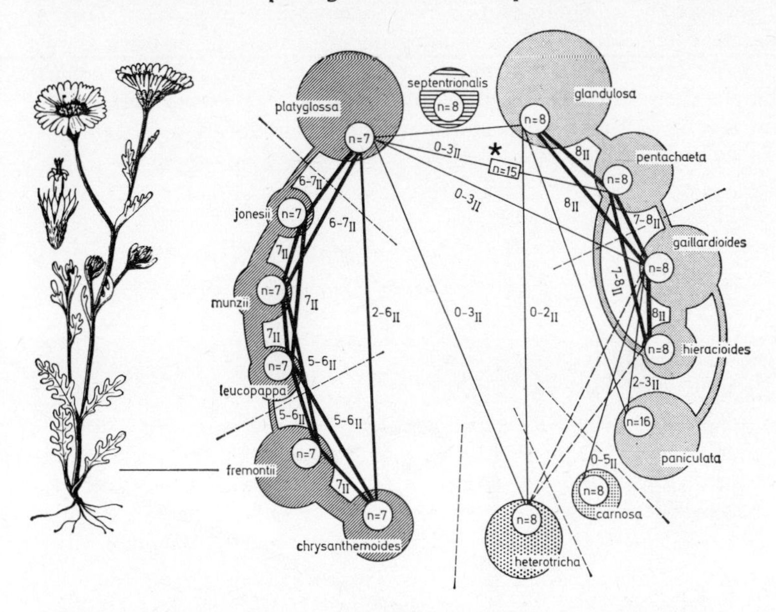

Fig. 357. Hybridization polygon for the Californian genus *Laya* (Asteraceae). The following are indicated: specific names; haploid chromosome numbers; fertility of the experimental F_1 hybrids (broken lines linking the species = (sub-)lethal; thin lines = sterile; thick lines ± fertile) and chromosome pairing in their pollen mother cells (mean number of bivalents = II); a synthetic allopolyploid (asterisk); extent of natural gene-flow (mechanically stippled connections); the most important morphological discontinuities (broken transverse lines between species). Left: whole plant and disc floret of *L. fremontii*. After Clausen, Keck and Hiesey, and Abrams and Ferris.

Genetic barriers are usually built up gradually, but polyploidy sets them up abruptly, often with little or no accompanying morphological changes ('cryptic barriers'). For example, some populations within the species *Laya glandulosa* (Fig. 357) already show sterility when crossed. In the garden pea (*Pisum*, Fig. 342A), in rye and in many other species isolating differentiation in chromosome structure is brought into play at the population level and may be followed at the level of the race or species (Fig. 364). The same also holds for polyploidy (p. 418).

The post-zygotic operation of sterility factors ranges from disturbances in mitosis and the defective development of F_1 embryos to the breakdown of the endosperm in the F_1, and the consequent failure of nutrition of the embryo, and to the defective development of the sexual organs. When meiosis is impaired in F_1 hybrids the faulty interaction of the different genomes or plasmones may be expressed in numerous ways: increased rate of breakage of the chromosomes, failure of pairing of non-homologous chromosome sets, of individual chromosomes or of chromosome segments, faulty division of paired but structurally heterozygous chromosomes (e.g. after translocations, Fig. 342, or inversions, Fig. 358), defects in the spindle apparatus and in anaphase (often associated with the incorporation of all the daughter chromosomes into one nucleus: restitution nucleus formation), etc. While in the F_1 each of the two genomes acting together is individually balanced, in the hybridogenous F_2 generation there is breakdown (when, for example, chromosome segments are lacking or duplicated), and hence increased genetic disharmony: the vitality and fertility are generally thus further reduced and recombinants deviating markedly from the parental or F_1 types are often completely suppressed (in scatter diagrams this is expressed by a 'recombination spindle', Fig. 360A).

Between genetic changes with post-zygotic isolating action and those with morphological effects there is only a partial correlation (for example as a result of pleiotropy or close linkage). The morphological discontinuities in *Laya* do not always correspond to the effectiveness of the barriers to hybridization (*L. septentrionalis* for example is morphologically close to *L. glandulosa* and *L. pentachaeta*, Fig. 357). It follows that hybridization patterns do not exactly mirror evolutionary relationships (cf. p. 426). That in addition the development of genetic barriers is independent of the time factor is shown by the N. American and Mediterranean planes (*Platanus occidentalis* and *P. orientalis*) which have been phylogenetically separate since at least the early Tertiary without the development of any genetic barriers to hybridization (numerous hybrids in cultivation!).

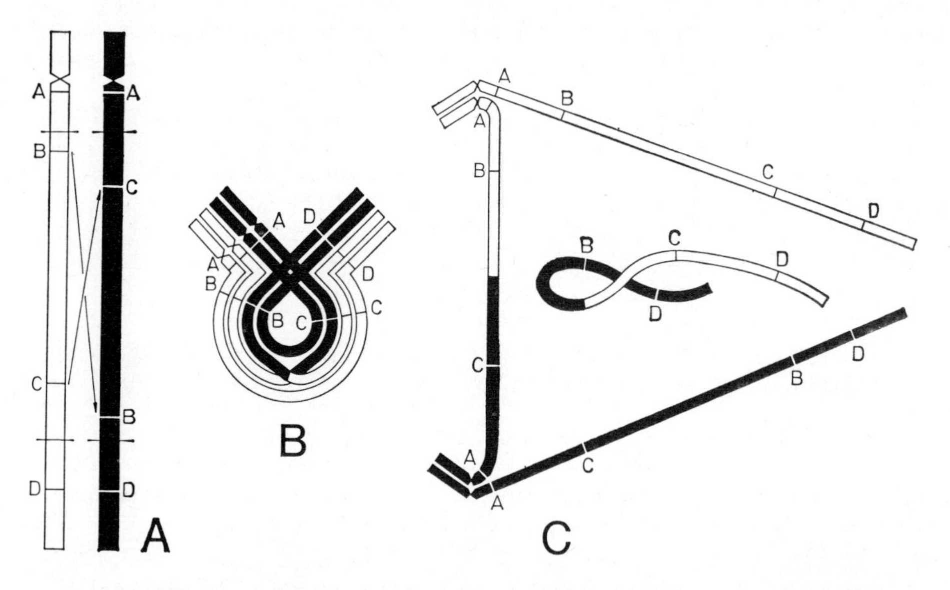

Fig. 358. A chromosome mutation (inversion) as a barrier to hybridization. **A.** Diagram of a chromosome pair with one normal (white) chromosome and one with an inverted segment (black); marker genes A, B, C, D; points of breakage; and inversion of the affected chromosome segment. **B.** Meiosis in the F_1: pairing of the structurally heterozygous chromosomes and crossing-over within the inverted segment. **C.** The resultant bridge with two centromeres at anaphase I and a fragment without a centromere (acentric): both are eliminated and only gametes with either the unaltered chromosomes of the normal form or of the inverted form are viable. Modified after Stebbins.

The formation of reproductive isolating mechanisms is not, however, a chance by-product of divergent evolution. It has been shown several times recently that the formation of barriers is under selective control: for example, in races of extreme habitats the formation of barriers is advantageous in preventing 'dilution' with the normal race, so that adaptive characters are 'shielded'. In sympatric biotypes a strengthening of prezygotic barriers (e.g. through floral-biological isolation) limits the 'squandering' of gametes on less viable hybrids. Very commonly the action of barriers is largely independent of the environment: when strong competition operates against hybrids (for example in stable habitats where they cannot gain a foothold among the parental forms) even a weak isolating mechanism will be functional; if, however, hybrids are promoted by selection (for example in unstable, open, and still 'unoccupied' habitats where the parental forms are in recession) they will increase in spite of strong barriers. Thus hybrids between pedunculate and sessile oaks (*Quercus robur* and *Q. petraea*) occur only occasionally in C. Europe but predominate in Scotland.

The development of various, often complex reproductive isolating mechanisms is thus essential, particularly for the internal cohesion, the mutual delimitation, and ultimately the phylogenetic divergence of biotypes and their gene-pools. These processes are very important in the formation of 'species' (p. 434); of course it must be remembered that the formation of barriers, morphological-physiological differentiation and evolutionary relationships are only loosely correlated. Reproductive isolation among related biotypes is often only relative and is under the control of selection. Reproductive barriers promote genetic stability; their breakdown by hybridization leads to greater genetic plasticity (see below). Thus there is obviously a close connection between isolation and recombination.

4. Hybridization and allopolyploidy

Processes of crossing between incompletely isolated 'groups' with different genetic compositions and adaptations are called hybridization (bastardization). If these are coupled with chromosome doubling we speak of allopolyploidy. These are processes of genetic recombination which reach beyond the limits of the normal reproductive community (the population). The significance of these processes (which were often underestimated in the past) lies in the fact that the origins of almost all cultivated plants, of at least a third of all cormophytes, and of a great many thallophytes, are associated with hybridization and allopolyploidy.

Compared with crosses between 'groups' with few gene differences (pp. 319–25), hybridization between those with many liberates an enormous amount of variability in F_2 and later generations as a result of recombination. Thus in biotypes with differences in two gene loci only $3^2 = 9$ recombinant genotypes (compared with the F_1) occur in the F_2 (cf. Fig. 281), but with differences in ten genes, $3^{10} = 59{,}049$ recombinants are possible! Variability due to hybridization is thus very rapidly mobilized and facilitates rapid adaptation after sifting by selection, even in drastically modified environments.

These natural processes have been imitated in model experiments, e.g. by crossing ecologically very distinct races of *Achillea* with other 'groups', propagating vegetatively the very diverse F_2 individuals, and cultivating each clone in experimental gardens at low, montane and alpine altitudes. After a very short time numerous inadequately adapted recombinants dropped out, while selection favoured other novel biotypes superior to the parental plants. Similar examples are provided by cultivated plants: thus our multicoloured diploid and tetraploid petunias (*Petunia hybrida*) arose by selection from crosses between wild diploid white and violet flowered species (*P. axillaris* and *P. violacea*). Similarly the diverse cultivated brassicas (Fig. 350) arose at the diploid level from various wild Mediterranean forms; and the numerous wheats arose at polyploid levels ($4x$, $6x$; Fig. 337).

Successful establishment of hybridogenous populations is of course only possible under certain conditions: (1) in general new areas must be available for colonization so that competition with the parental plants is avoided; in recent geological times such areas have been provided by Man in particular, but previously they arose during ice ages or in periods of volcanic activity; (2) to counteract sterility phenomena (resulting from postzygotic barriers to hybridity in the parental 'groups') and too much genetic lability (resulting from recombination) a certain degree of reproductive capacity and genetic stability must be attained; this is reached in homogamous, heterogamous, allopolyploid and apomictic hybrid populations by means of various cytogenetic mechanisms.

(*a*) In **homogamous hybrids** the chromosomes agree in number (and often also in structure) with those of the parental biotypes; such hybrids may either be sterile or sexual reproduction is more or less normal.

The following series demonstrates various stages in the breakdown of barriers to hybridization. Only solitary hybrid individuals are produced, for example, between various genera of orchids (e.g. *Nigritella nigra* x *Gymnadenia odoratissima*), or at the species level between the black and Scots pine (*Pinus nigra* x *P. sylvestris*) or between bilberry and cowberry (*Vaccinium myrtillus* x *V. vitis-idaea*). Larger hybrid populations with greater variation (occasional back-crosses, F_2 individuals, etc.) arise from the aspen and white poplar (*Populus tremula* x *P. alba*), from the crosses between *Rhododendron hirsutum* and *R. ferrugineum*, or from the sweet and hairy violets (*Viola odorata* x *V. hirta*), etc. Polymorphic and more or less fertile hybrid swarms sometimes join our native oaks (*Quercus robur* and *Q. petraea*; and on the Continent, *Q. pubescens*); occasionally such swarms occur in the absence of the parents. Other hybrid swarms occur in willows (*Salix alba* x *S. fragilis*), in red and white campions (*Silene dioica* x *S. alba*), and in the medicks (*Medicago falcata* x *M. sativa*). In the Danube basin in Bavaria and Austria the pasque flower appears at first sight to afford an example of a cline (Fig. 359), but crossing experiments and quantitative analysis of characters demonstrate the hybridogenous origin of the continuous transitional series between the tetraploid *Pulsatilla vulgaris* and *P. grandis* which in the Post-glacial migrated from the west and east respectively.

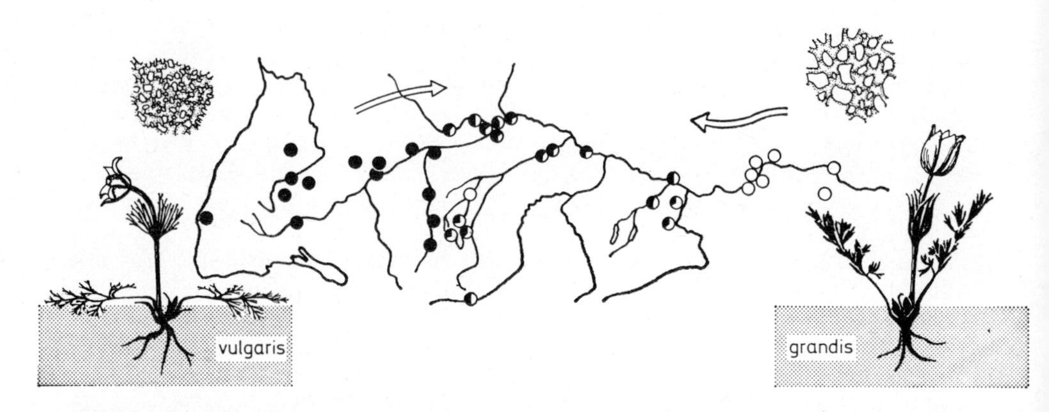

Fig. 359. Hybridogenous intermingling of the western *Pulsatilla vulgaris* (black circles) and the eastern *P. grandis* (white circles) in the Danube basin in Bavaria and Austria. General habit; spongy parenchyma; direction of migration; localities of the populations studied. In the intermediate populations the black and white sectors indicate the proportions of parental characters. After Voelter-Hedke and Zimmermann.

Hybrid offspring generally exhibit a certain 'cohesion' of the characters of the parental 'groups'. The scatter diagram in Fig. 360A shows this for experimentally produced hybrids between two diploid species of yarrow (*Achillea*) which are strongly isolated reproductively. Similarly, a corresponding correlation of characters in a natural population suggests a hybridogenous origin (Fig. 360B). This kind of analysis suggests that *A. roseo-alba*, a woodland and meadow plant of the geologically recent northern Italian plains and neighbouring areas, has arisen from the geologically older *A. setacea*, a pontic-pannonian steppe plant, and *A. asplenifolia*, a pannonian plant of lowland meadows, by hybridization and back-crossing with unidirectional gene flow towards *A. asplenifolia*; this can be confirmed by experimental hybridization and other means. Such limited infiltration of genes and characters (introgression) in hybrids between strongly isolated species is probably widespread.

Even strong and complex reproductive barriers can be broken down by hybridization and selection pressure. Crossing experiments (on species of *Nicotiana*, for example) show that fertile recombinants can be selected from generally extremely sterile hybrid progeny;

in some cases the segregation of sterility factors or chromosomal modifications isolates these new types from the parental ones ('formation of species'). Since most characters are under polygenic control, the genetic stabilization of diploid hybrid populations is also possible (for example, *AABBCCDD* x *aabbccdd* → *AABBccdd*). Furthermore, certain combinations of genome and plasmone may result in entirely new characters being expressed, or in enhanced rates of mutation. It is not surprising, therefore, that hybridization frequently leads to novel differentiation (cf. Fig. 364).

For example, experimental crossing of blue and salmon-red flowered 'groups' of *Streptocarpus* yields F_2 descendants with entirely new anthocyanins and hence new flower colours. During the Quaternary migrations of the flora the fir *Abies borisii-regis* arose by hybrid-contact between the C. European *A. alba* and the Greek *A. cephalonica*; while in S. Poland the local *Betula oycoviensis* arose from a subarctic dwarf birch (section Nanae) and the widespread arborescent silver birch (*B. pendula*). Man's methods of cultivation have led to the hybridization of wild Mediterranean forms of the carrot (*Daucus*) and radish (*Raphanus*) and the parallel development of new horticultural, agricultural and ruderal 'groups'.

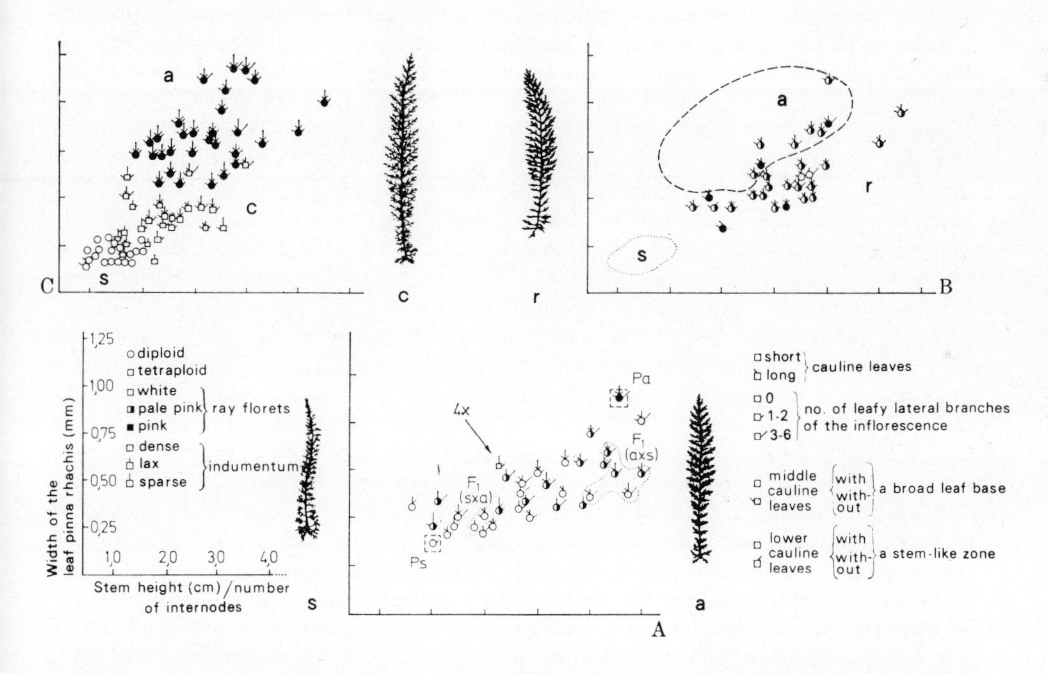

Fig. 360. Experimental analysis of the phylogeny and evolution of some microspecies from the hybrid and polyploid complex in yarrow, *Achillea millefolium* agg. ($x = 9$): *A. setacea* (s, 2 *x*), *A. asplenifolia* (a, 2 *x*), *A. roseo-alba* (r, mostly 2 *x*), *A collina* (c, 4 *x*). Leaf silhouettes and scatter diagrams (differentiation of characters represented by ordinates and abscissae and by symbols for single individuals). **A.** Experimental hybridization of the diploid *A. setacea* (P_S) and *A. asplenifolia* (P_a); reciprocal crosses producing differing F_1 individuals (s x a and a x s; plasmatic differentiation!); limited recombination in the sub-viable F_2 (recombination spindle!); and a spontaneous allotetraploid (→ 4 *x* = synthetic *A. collina*). **B.** Thirty individuals of a very variable population of the hybridogenous *A. roseo-alba* approximating closely to *A. aspleniifolia* (as a result of back-crossing and introgression!). **C.** Individuals from various populations of the diploid *A. setacea* and *A. aspleniifolia*, and the allotetraploid *A. collina* which has arisen from them (intermediate position!).

(*b*) **Heterogamous hybrids** transmit the two structurally different haploid chromosome sets derived from the parents to their offspring in an unaltered form, although they reproduce sexually (permanent structural heterozygosity or complex heterozygosity). The two chromosome sets differ by a series of translocations in the arms of the chromosomes, for example 1,2; 3,4; 5,6; 7,8 in one set, 2,3; 4,5; 6,7; 8,1 in the other. In metaphase I of meiosis, pairing of homologous terminal portions of the chromosomes causes them to be arranged in a zigzag chain:

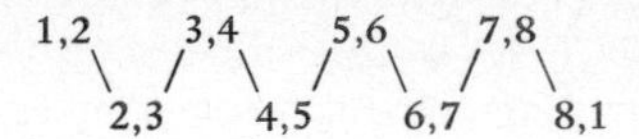

and at anaphase I each of the two original chromosome sets passes *en bloc* to a gamete. Homozygous combinations of these chromosome sets are rendered inviable by the incorporation of lethal factors (p. 325). This rather rare type of structural hybridity is found, for example, in the evening primrose (*Oenothera*).

A special case transitional to the next group is found in the hybridogenous dog-roses of the *Rosa canina* group. These usually have five sets of chromosomes of which, however, only two are homologous, pairing as bivalents in meiosis, while the others remain as univalents and are eliminated from the ♂ but are retained in the ♀, thus maintaining the pentaploid state (pollen, n = 7; egg cell, n = 28; zygote, 2n = 35).

(*c*) **Allopolyploid hybrids** have two chromosome sets from each parental 'group', the sets differing in genes, in structure or in number (pp. 37, 331, 401). Here we shall consider those very numerous and important allopolyploids which reproduce sexually.

Allopolyploidy is very widespread, particularly among pteridophytes and angiosperms but it also occurs among algae, fungi, bryophytes and gymnosperms. The evolutionary significance of hybridogenous polyploidy for the origin of related groups of the most diverse extent may be briefly illustrated by the following: polyploid races are found for example in charophytes (*Chara zeylanica*: 2 *x*, 3 *x*, 4 *x*, 5 *x*), crucifers (*Biscutella laevigata*: 2 *x*, 4 *x*; Fig. 363), annual mercury (*Mercurialis annua*: 2 *x*, 4 *x*, 6 *x*, 8 *x*, 10 *x*, 12 *x*, 14 *x*), bedstraws (e.g. *Galium anisophyllum*: 2 *x*, 4 *x*, 6 *x*, 8 *x*, 10 *x*; Fig. 343E), and within many other species. Well-known examples of related diploid and polyploid species occur in the moss genus *Mnium* (2 *x*, 4 *x*), in the fern *Polypodium* (basic chromosome number *x* = 37: 2 *x*, 4 *x*, 6 *x*), in docks (*Rumex*: 2 *x*, 4 *x*, 6 *x*, 8 *x*, 10 *x*, 12 *x*, 14 *x*, 20 *x*), in oats (*Avena*: 2 *x*, 4 *x*, 6 *x*), in wheats (2 *x*, 4 *x*, 6 *x*; Fig. 337), etc. Genera, sub-families and families with polyploid basic numbers include, for example, *Sequoia* (2n = 66, 6 *x*), *Platanus* (2n = 42, 6 *x*), *Soldanella* (2n = 40, 4 *x*), Rosaceae-Pomoideae (2n = 68, ? 4 *x* from 8 + 9), Equisetaceae (2n = 216, ?? 14 *x*) and Salicaceae (2n = 38, 6 *x*). Finally, the Psilotales, all eusporangiate ferns and most leptosporangiate ferns have such high chromosome numbers that they must be regarded as polyploid.

Almost all successful polyploids in nature or among cultivated plants prove on analysis to be of hybrid origin. From the diploid yarrows *Achillea setacea* and *A. aspleniifolia* already mentioned (p. 416) the allotetraploid *A. collina*, for example, has originated (Fig. 360C). In moderately dry (and ruderal) habitats this latter species successfully penetrates between the parental species and indeed has spread beyond them into continental C. and S.E. Europe. Between 1870 and 1890 on the coast of southern England near Southampton the native saltmarsh grass *Spartina maritima* (2n = 60) and a species escaped from N. America, *S. alterniflora* (2n = 62), gave rise first to an almost sterile hybrid, *S.* x *townsendii* (2n = 61), and from this originated the allopolyploid *S. anglica* (2n = 122) which is now widespread in southern England and on the Continent. Similarly in N. America the genesis of allotetraploid species of goat's-beard (*Tragopogon*) has been traced from introduced European species. In many cases allopolyploids have arisen spontaneously from artificially produced F_1 hybrids (e.g. in *Laya*, Fig. 357; and *Achillea*, Fig. 360A). This is true also of intergeneric hybrids of cultivated plants such as *Raphanus* x *Brassica* (x *Raphanobrassica*) and *Triticum* x *Secale* (x *Triticale*). This preferentially hybridogenous origin of polyploids arises from the greater susceptibility of diploid hybrids to mitotic and meiotic disturbances and hence to the formation of polyploid restitution nuclei and unreduced gametes.

How does the superiority of allopolyploids to autopolyploids and frequently also to diploid homogamous hybrids arise? If we represent the chromosome sets or genomes of allopolyploids by the formula AABB, of autopolyploids by AAAA, and of diploid hybrids by AB, it is obvious why meiosis and fertility in the two last are often upset: in autopolyploids particularly by the formation of multivalents or univalents (pairing of more than two homologous chromosomes, or unpaired single chromosomes: A–A–A–A or A–A–A/A); in diploid hybrids, on the other hand, by the failure of pairing or defective pairing (A/B, as a result of structural differences); in both frequently because of defective genetical or physiological balance. These defects in meiosis, fertility and viability are circumvented in allopolyploids by homogenetic chromosome pairing (i.e. A–A/B–B) and by superior gene balance. While the variability of homogamous hybrids is quickly expressed as a result of heterogenetic chromosome pairing (i.e. A/B) and recombination (segregation!), homogenetic pairing in allopolyploids effects a stabilization of phenotypes similar to the F_1 (and heterosis also operates and is fixed; p. 322). If the two genome-pairs of an allopolyploid are structurally very dissimilar then practically only homogenetic pairing will occur, but if they are alike then heterogenetic pairing (i.e. A–B/B–A) will become more frequent. Thus between the recombination characteristics of allopolyploids on the one hand and autopolyploids on the other there is every gradation and every possibility of genetic control. The essential point is that allopolyploids can both store their potential for variation by recombination, and also mobilize it. They have the further advantage that their potential for recombination is greater with several genomes than it is with only two. Divergent genetic differentiation of the duplicated genetic material can lead ultimately to the 'diploidization' of polyploids.

To elucidate the genesis of polyploids the quantitative analysis of morphological characters (for example as scatter diagrams, Fig. 360C) and of chemical content (Fig. 361) are both used. The homologies or differences between genomes can be established by means of caryograms (Fig. 362), by the pairing of meiotic chromosomes in experimental hybrids, and by genetical analysis. For example, in the triploid *Asplenium* hybrid RMM (Fig. 361) the two homologous chromosome sets M–M form bivalents while R remains unpaired (univalents); RMP has only univalents, while the tetraploid RMPM

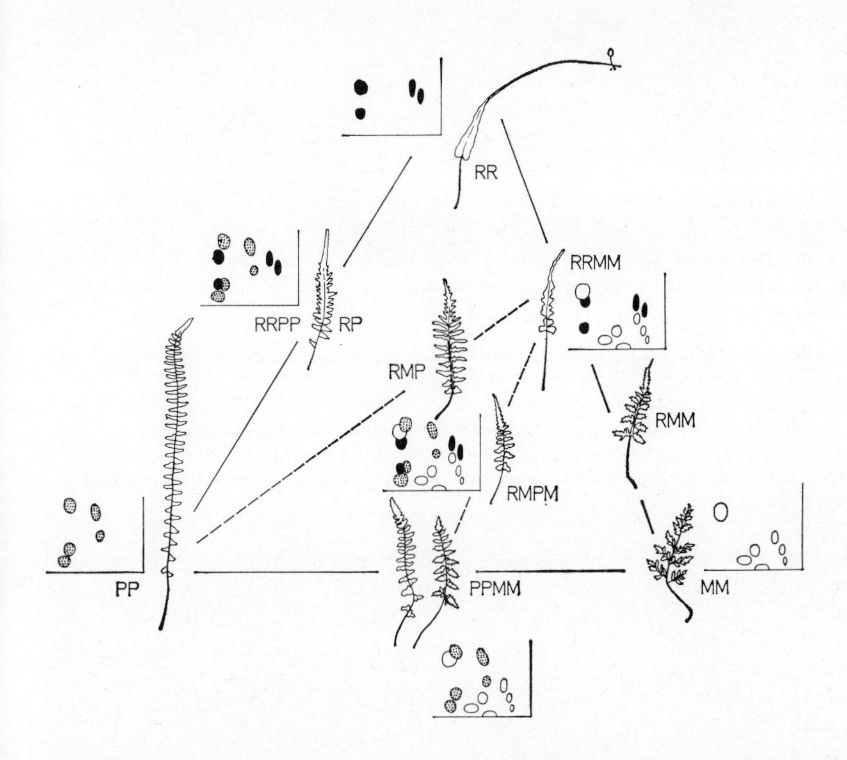

Fig. 361. Analysis of the origins of a N. American fern complex (*Asplenium* species). Genome-formulae established by chromosome numbers and chromosome pairing in hybrids: diploid basic species *A. platyneuron* (PP), *A. rhizophyllum* (RR), *A. montanum* (MM); di, tri- and tetraploid hybrids (RP, RMM, RMP and RMPM); and the allotetraploid daughter species *A. ebenoides* (RRPP), *A. pinnatifidum* (RRMM) and *A. bradleyi* (PPMM). Confirmation of these origins from morphology (e.g. frond shape) and comparative phytochemistry (phenolic contents as shown in two-dimensional chromatograms). Modified after Wagner, Smith and Levin.

produces both univalents (R, P) and bivalents (M–M). Of course, a degree of genic control of chromosome pairing sets limitations on this method of genome identification. The most convincing evidence is the experimental synthesis of allopolyploids from hybrids between their hypothetical diploid ancestral forms by polyploidization, either spontaneously or induced by colchicine (p. 332); this has been achieved, for example, in *Triticum* (p. 394, Fig. 337), *Achillea collina* (Fig. 360A), and in many other groups.

In the development of polyploid complexes from diploids lower polyploids are always formed first, followed by higher ones, thus $2x + 2x = 4x$, $2x + 4x = 6x$, $4x + 4x = 8x$, etc. This sequence provides one of the most reliable criteria for the spatial–temporal development (phylogeny) of a particular group of related forms.

Thus the evolutionary development of a polyploid complex can be assessed: it begins with more or less isolated diploids and scattered hybridogenous 'neopolyploids' (e.g. *Laya*, Fig. 357), and progresses to ever higher levels of ploidy with the gradual extinction of the ancestral 'groups'. Meanwhile the barriers to hybridization between the polyploids are weakened and hybrid contact takes place between 'groups' at the same level of ploidy—or even at different levels. Aneuploid variations in chromosome number also take place, and the diversity of forms reaches an optimum (e.g. *Galium anisophyllum*: $2x$–$10x$; *Achillea*: $2x$–$8x$; *Triticum*: $2x$–$6x$). Finally, however, impoverishment of forms sets in, until only isolated 'palaeoploids' tell the tale of the gradual decline of a polyploid complex. In the course of their history going back to the Palaeozoic, the nowadays relict eusporangiate fern group Ophioglossales has lost all 'groups' with less than 2n = 90, and indeed *Ophioglossum reticulatum* with 2n = *c*. 1260 has the highest chromosome number known in any organism. Among the angiosperms the ancient Magnoliaceae, for example, are palaeopolyploid (derived basic number = 19; probably $6x$; cf. also Fig. 745).

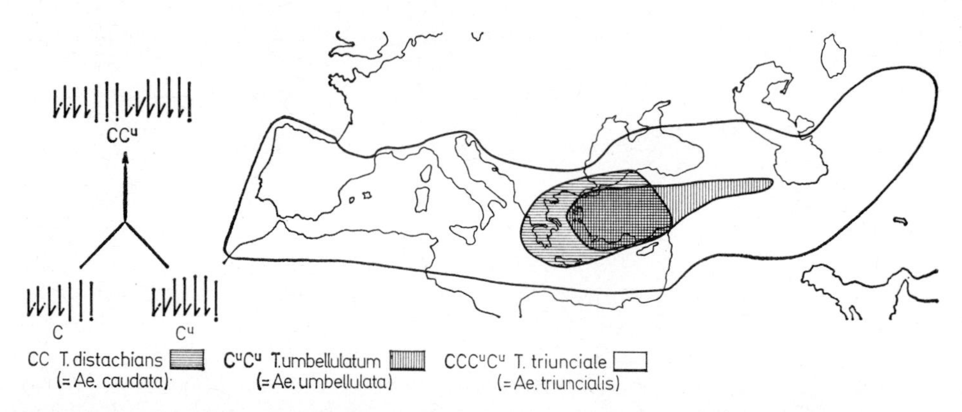

Fig. 362. Origin and distribution of a polyploid complex in a group of annual grasses (*Triticum* = *Aegilops*). Left: haploid chromosome sets (shown as caryograms) and the experimental synthesis of *T. triunciale*; right: the more extensive distribution of the allotetraploid compared with the diploid taxa. Somewhat modified after Kihara and Stebbins.

Like homogamous hybrids, allopolyploids succeed best in supplanting their ancestral 'groups' in newly available and rapidly changing habitats because of their greater capacity for adaptation. To this the many polyploid cultivated plants (cf. p. 332) and weeds bear eloquent witness (e.g. *Stellaria media, Urtica dioica, Polygonum aviculare, Capsella bursa-pastoris, Solanum nigrum, Agropyron repens* and *Triticum* (*Aegilops*) *triunciale*, Fig. 362). Polyploids were also very important in the recolonization of C. Europe and the Alps after the last glaciation: their diploid progenitors frequently survived in refugia beyond the ice sheets or in the south. At the same time alpine and (sub)Mediterranean diploids apparently participated in the formation of widespread tetraploid 'groups' of bird's-foot trefoil (*Lotus corniculatus*) or sweet vernal grass (*Anthoxanthum odoratum*). The *Biscutella laevigata* group of forms is almost exclusively represented in the heavily glaciated Alps by $4x$ races while the $2x$ progenitors have survived principally in the unglaciated areas of C. and W. Europe and the Carpathians (Fig. 363). In montane

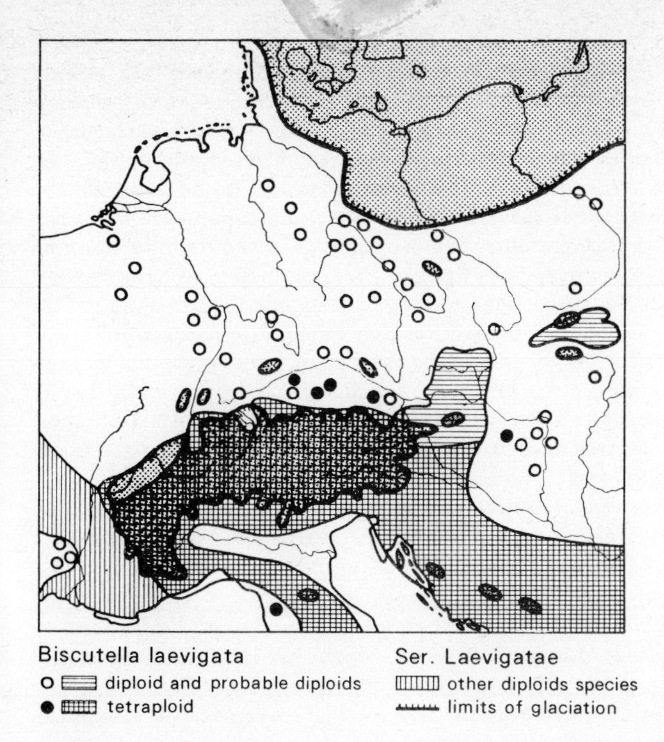

Fig. 363. Distribution of diploid and tetraploid biotypes of *Biscutella* series Laevigatae in C. Europe: the diploids mainly grow in those areas unglaciated during the Wurm Glacial, the tetraploids the formerly heavily glaciated Alpine areas. After Manton *et al.*, diagrammatically modified and supplemented.

populations of *Galium anisophyllum* (Fig. 343) the areas of the diploids and polyploids reflect the repeated retreats and advances of the ice during, between and after the Glacial Period.

(*d*) **Apomictic hybrids** are characterized predominantly or exclusively by asexual reproduction. Examples are found especially among ferns and angiosperms. A great variety of forms of apomixis (p. 403) is associated generally with structural heterozygosity, polyploidy (including uneven-numbered degrees of ploidy, e.g. $3x$, $5x$) and aneuploidy. Apomictic groups rarely extend beyond the limits of genera: this pattern of variation has no evolutionary significance beyond this level.

The wide distribution of the diploid enchanter's nightshade hybrid *Circaea* x *intermedia* (from *C. lutetiana* x *C. alpina*) arises from its vegetative reproduction. Many hybridogenous and polyploid races of the mints (e.g. the peppermint, *Mentha* x *piperita*) and the usually triploid *Acorus* also spread vegetatively; cultivated races of many bananas, the sugar cane and potatoes are also propagated in this way. Structurally heterozygous diploid, polyploid and aneuploid hybrids of *Allium* (garlic) and *Ornithogalum* (star of Bethlehem) spread by bulbils. The high polyploid *Dentaria bulbifera* ($12x$; Fig. 223A), widespread in C. Europe, is probably hybridogenous from $6x$ progenitors of Glacial forest refugia.

Many hybridogenous and polyploid groups of ferns have unreduced spores and 2n-gametophytes; parthenogenesis of the egg-cells leads to the development once again of 2n-sporophytes. The asexual development of seeds (agamospermy) in conjunction with hybridization, polyploidy and aneuploidy in various angiosperm groups (cf. p. 666 and Fig. 345) leads to the production of swarms of extremely polymorphic forms, for example the group of alpine meadow grasses (*Poa alpina*) in which more or less apomictic polyploids and aneuploids ($2n = 21-61$) with agamospermy or leafy bulbils (Fig. 223C) have arisen from sexual diploids (and tetraploids) in heavily glaciated Arctic–Alpine areas. The apomictic-polyploid swarms of the spring cinquefoil (*Potentilla verna* agg., $4x-12x$) have also arisen from $2x$ and $4x$ biotypes from the Mediterranean, the Alps and the eastern steppes. Further genera in the Poaceae (e.g. *Calamagrostis*), Rosaceae (e.g. *Rubus, Alchemilla, Sorbus*), Cichoriaceae (e.g. *Taraxacum, Hieracium*), etc., are similar.

The change from sexuality to apomixis, especially agamospermy, is a complex process controlled by numerous genes. However, the necessary mutations occur occasionally in non-hybrids at the diploid level, and their combination is greatly facilitated by hybridization and polyploidization. Selection operates very powerfully: for many vegetatively competitive but sterile combination types (for example, structural hybrids, uneven numbered polyploids—3 *x*, 5 *x*—aneuploids, etc.) apomixis affords the only chance of survival. Apomixis is thus an important factor in genetic stabilization and permits the maintenance and increase of numerous hybrid derivatives which would otherwise be eliminated. This is not the only cause of the great variety of forms in apomictic groups: in addition there is often the partial retention of the capacity for sexual reproduction, the occasional fertile pollen grain or embryo sac permitting the continuation of hybridization and the appearance of new forms. When this capacity fails apomictic—polyploid groups also enter a terminal phase of impoverishment (as for example in various groups of lady's-mantle (*Alchemilla*)). But prior to this—in the early phase with sexual diploids and only a few polyploids and apomicts (as in the apple, *Malus*), and particularly in the optimal phase with very numerous apogamous forms (as in the European blackberries, *Rubus* section Moriferi)—this pattern of variation permits rapid adaptation and successful distribution, especially (as before) in new and labile habitats.

The various forms of hybridization thus permit a rapid mobilization of genetic variability in the intermediate area between the formation of relative and absolute barriers to crossing. This is of enormous importance, particularly for adaptation to new and labile environments. Various cytogenetic mechanisms (particularly polyploidy and apomixis) permit the surmounting of barriers to hybridization and ensure the necessary genetical stabilization. Polyploidy furthermore brings about an increase in DNA and progressive 'division of labour' and differentiation between originally identical genes. Finally, the processes of hybridization lead not only to the convergence of 'groups'; in various ways they also catalyse phylogenetic divergence.

D. Micro- and macro-evolution

The processes of differentiation and divergence of populations up to the level of the species are denoted as micro-evolution, while the formation of larger, more comprehensive groups of related forms (at and above the rank of genus, for example) is called macro-evolution.

Some fundamentals of micro-evolution are summarized in Fig. 364. (A) Variation based on mutation and recombination, allopatric geographical-ecological initial differentiation (west/east cline). (B) Spatial isolation of a marginal eastern biotype, adaptation and unification under the influence of selection, formation of 'cryptic barriers' (for example as a result of chromosome mutations) in the north-west. (C) Hybridization between the eastern and parental biotypes at the diploid level, spatial separation of the north-western biotype. (D) Allotetraploid, abruptly isolated biotype from the hybrid zone in the east, hybridogenous introgression in the south, north-western and parental biotypes again sympatric in part, but isolated through the gradual raising of barriers to crossing.

In the previous sections we have indicated more than once the fundamental similarity of the action of mutation, recombination, selection and isolation within and beyond the level of the species. That is the basis for the nowadays widely accepted hypothesis that these factors also operate at the level of macro-evolution. This can be exemplified within the angiosperm order Dipsacales with the teasel family Dipsacaceae (Fig. 365).

Within the Dipsacales the predominantly herbaceous Dipsacaceae have apparently arisen from the generally woody Caprifoliaceae. The latter are well represented in the deciduous forests of the Northern Hemisphere while the Dipsacaceae occur mainly in open habitats in the Mediterranean and Near East (but also in C. and N. Europe). The Dipsacaceae have a characteristic combination of features which are approached in the Caprifoliaceae: the flowers are borne in dense thyrsoid capitula surrounded by bracts; the marginal florets are often radially elongated; and despite the small size of the individual florets the optical attraction of visiting pollinators is thus guaranteed. In addition a so-called epicalyx or outer calyx is formed of four united bracteoles close to the inferior ovary.

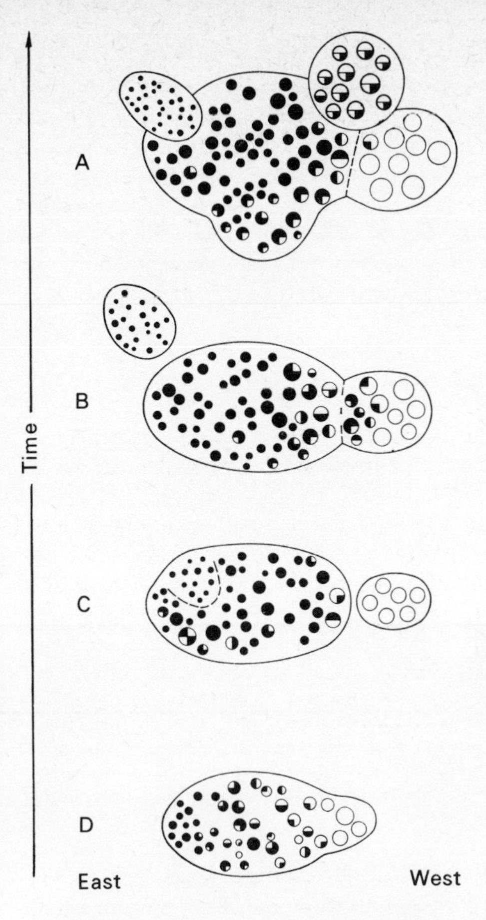

Fig. 364. Diagram of the evolution and phylogeny of a group of related forms at four moments in time (**A.** to **D.**). The circles represent individuals, their size and marking their genetical and physiological constitutions, the distance between them their geographical-ecological positions. Broken or continuous boundary lines indicate partial or complete barriers to hybridization. Further explanation in the text.

The development of the Dipsacaceae is closely linked to the biological differentiation of the fruit, with manifold modifications of the floral bracteoles and the inner calyx and epicalyx. It appears that originally there were herbaceous bracteoles, a short four-lobed epicalyx and a calyx of five bristles; similar undifferentiated achenes occur for example in *Succisa*. In *Dipsacus* the stiffening of the bracteoles, and their elongation as hooks in some cases, together with the thistle-like, springy habit, facilitate the catapulting of the achenes when the plant is brushed against by animals. In the anemochorous genera with winged fruits the bracteoles are in contrast reduced. In *Pterocephalus* the inner calyx bristles are increased in number and length and are clothed with hairs, forming a pappus, while *Scabiosa* develops an enlarged membranous epicalyx as a parachute. In *Knautia* dispersal by ants (myrmecochory) is promoted by a nutritious elaiosome (p. 669) at the base of the fruit. Finally, some of the annual species of *Cephalaria* and *Scabiosa* adhere to animals by means of enlarged and barbed outer or inner calyx teeth (epizoochory). The various genera and other taxa of the Dipsacaceae have thus specialized in various forms of fruit dispersal, corresponding to their diverse habitats. The formation and progressive modification of the necessary mechanisms under the influence of selection can be traced throughout.

In the evolution of the Dipsacaceae increasing dysploidy ($n = 9 \rightarrow 10$) took place at an early stage. In the differentiation of the genera allopolyploidy (and aneuploidy) and allogamy played a great part in the perennial taxa, while in the annuals chromosomal rearrangements, decreasing dysploidy and autogamy were important. Close parallels in different genera indicate selective control of this 'evolutionary strategy'. The genus *Morina* (Morinaceae) has few species and forms an ancient link between the Dipsacaceae and Caprifoliaceae; it only survives as a palaeopolyploid ($n = 17$, probably $8 + 9$).

Even at the level of macro-evolution we can therefore recognize interlocking phases: (1) Anagenesis describes the origin of novel types of construction and substantial phylogenetic progression. (2) Cladogenesis characterizes the modification and differentiation of

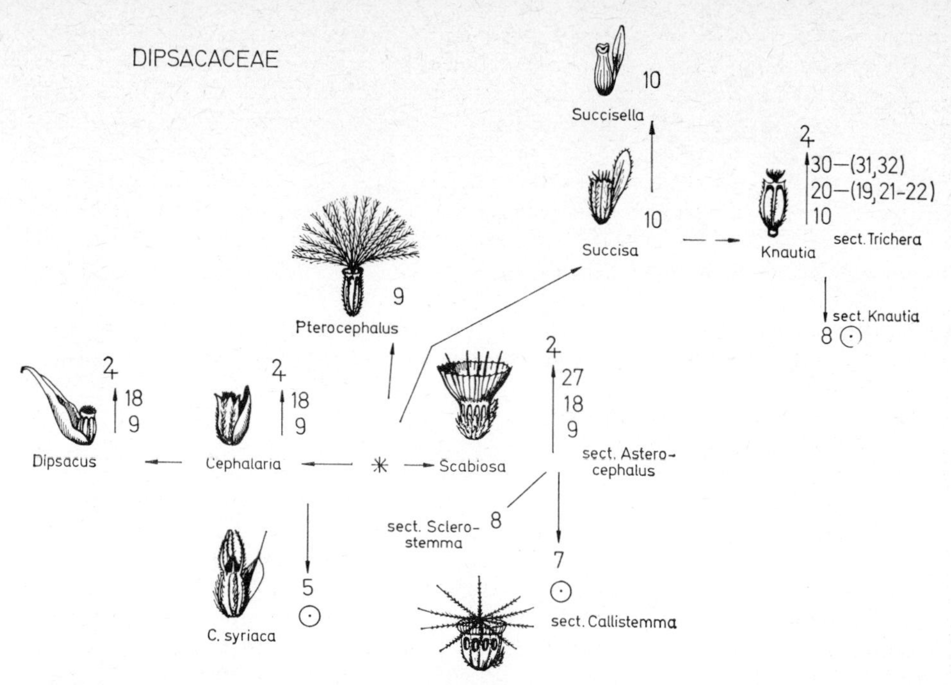

Fig. 365. Macro- and micro-evolution in the Dipsacaceae. Diagram of the hypothetical phylogenetic connections between the principal taxa (arrows); hypothetical ancestor (asterisk); differentiation of the subtending bracts (when present) and fruits (with epicalyx and calyx); life-forms (♃, perennial; ⊙, annual); haploid chromosome numbers (beside or at the ends of the arrows; dysploidy; polyploidy: $2x$, $4x$, $6x$; aneuploidy). Further explanation in the text.

certain basic types and produces a great diversity of forms by means of specialization. (3) Stasigenesis indicates phylogenetic consolidation, stabilization and conservation. Comparison is frequently made with the development of types of mechanical construction: the combination of existing structural elements to form a new type (e.g. wheel, carriage, internal combustion (piston) engine = motor car); modification and differentiation (heavy goods vehicle, private vehicle, different makes, etc.); the gradual replacement by new types (Wankel engine, turbine, hovercraft, etc.).

Progressive evolution and **anagenesis** frequently depend on the combination of processes, cell-types or organs to give novel or improved complex processes, tissues or systems of organs. Thus the processes of respiration apparently developed in stages by the combination of the anaerobic breakdown of sugars (fermentation), the aerobic liberation of CO_2 from pyruvic acid, the citric acid cycle, and endoxidation (the enzymatic combination of hydrogen and oxygen; cf. pp. 262 et seq.). The same is true for photosynthesis, especially the all-important difference between bacteria, in which H_2S is the hydrogen donor, and the blue-green algae, in which it is H_2O (cf. pp. 240 et seq.). Fundamental progressions in the Plant Kingdom include the step from Procaryota to Eucaryota, for example, at the level of cell organization (and hence true sexuality); and at the levels of tissue and organ differentiation, the genesis of the Cormobionta with their numerous adaptations to terrestrial existence (epidermis with cuticle, conducting and mechanical tissues with lignin, archegonia and antheridia etc.: cf. pp. 83 et seq.), and the development of the seed and hence the independence of the processes of fertilization of atmospheric water. These progressions, which are documented by the fossil record, have undoubtedly increased the capacity of plants for self-regulation and for utilization of the environment. In these developments the decisive control of selection cannot be doubted.

In the phase of **cladogenesis** new types of construction branch out in numerous directions as a result of differentiation, specialization and adaptation (adaptive radiation). In **stasigenesis** the great diversity of forms developed in this way undergoes considerable reduction, and ultimately entire groups may die out. A sequence of cladogenesis followed by stasigenesis can be seen, for example, in the Devonian Psilophytatae which had a great multiplicity of shoot forms compared with their few surviving descendents (Psilotales); or in the Mesozoic Ginkgoatae with large numbers compared with their only surviving representative (*Ginkgo biloba*); or by the contrast between the serried ranks of the crucifers and their fragmented, more primitive relatives, the members of the caper family.

Orthogenesis, the frequent parallel or unidirectional course of evolutionary development, has often led to the assumption of special evolutionary factors. Examples include the parallel lines of development from flagellates to filamentous and then complex thalli in the various algal groups, and the progressive increase in size of vascular plants. Indeed, terminal members, such as giant forms like *Sequoia*, may run into difficulties of adaptation. However, these lines of development merely indicate a certain 'canalization' of mutational changes (p. 398), especially by the orientation in the same direction of selective forces in similar environmental conditions. Thus, an increase in size of attached plants frequently confers an advantage in the 'struggle for light' and is favoured by selection. Of course numerous correlated factors prevent 'trees from growing into the skies'. Although specialization generally confers benefits in the utilization of the environment, they are bought at the expense of diminished adaptability.

A striking feature of *complex* evolutionary processes is that they show **irreversibility**: a return to a life-form once abandoned can only be made by a different route. Thus the loss of chromatophores with the change from autotrophy to heterotrophy in various lower animals and fungi can only be compensated for by intracellular symbiosis with algae or even by the direct incorporation of chloroplasts. The monocotyledons have evolved a new kind of thickening to replace the typical secondary thickening of seed plants which they have lost (pp. 145–7). The flowers of many wind-pollinated Euphorbiaceae are greatly reduced and unisexual: zoophily is regained in *Euphorbia* by monoecious inflorescences surrounded by coloured bracts and glandular nectaries; these inflorescences function like hermaphrodite flowers (cyathia). On the other hand, *simple* phylogenetic changes can be reversed, for example, the reversion from zygomorphic to actinomorphic flowers (Fig. 340II, B). As the number of mutations required to develop a complex structure or process increases, the less likely is an exact repetition of its evolutionary development once it has been lost.

Thus micro- and macro-evolution basically appear to follow the same natural laws: autonomous mutations, not primarily induced by the environment, but equally not entirely a matter of chance (p. 398) form the raw material; recombination and hybridization are responsible for the organization and mobilization of variation. Selection leads to the promotion of biotypes with greater competitive ability and higher reproductive capacity, and hence to adaptation and progression (but without immediately sacrificing individuals or biotypes with neutral or even moderately disadvantageous characters); isolation prevents hybridogenous 'fusion' from retarding differentiation and so ultimately determines evolutionary divergence. However, our understanding of the interplay and the environmental relationships of these evolutionary factors is in many respects still inadequate.

II. Systematics and phylogenetics

The problems of systematics and phylogenetics are centred on the description, delimitation (discontinuity), comparison, ground plan, hierarchy and descent of natural groups. We have already seen (Fig. 333) that the historical relationships of the germ-line, including its transmitted genes, provide the concrete basis for such studies of the relationships of past and present individuals. Systematics and phylogenetics are principally concerned with the *effects* of such relationships, while evolutionary studies primarily investigate their general *causes* (cf. pp. 389 et seq.). The findings of systematics and phylogenetics form the basis for the groupings of taxonomy (the 'natural system') and will be dealt with in the section on the classification of the Plant Kingdom. While only a very simplified account can be given there, often without supporting evidence, we will here present some indications of the basic principles with examples of the techniques of the systematic–phylogenetic discipline.

The historical development of systematics as 'the study of similarities' has closely followed other discoveries in botany and biology. Since ancient times the most important basis for comparison has been provided by the general habit; from the sixteenth and seventeenth centuries up to the time of Linnaeus and well into the nineteenth century the macroscopic characters of the flower and fruit formed the most important basis for comparison. The study of thallophytes and their reproductive organs started in the nineteenth century when the microscope was first generally used (E. M. Fries, H. A. de Bary, A. Pascher and others); the microscopical study of the anatomy of cormophytes was also begun (e.g. H. Solereder, C. R. Metcalfe). Since the acceptance of the theory of evolution, palaeobotany has assumed ever-increasing importance for the study of historical relationships (phylogeny; H. Graf zu Solms-Laubach, R. Kidston, W. Zimmermann and others). In the first half of the twentieth century cytology and genetics became the basis of experimental systematics (e.g. G. Turesson, A. Müntzing, G. L. Stebbins); systematic embryology (K. Schnarf and others) and palynology (especially G. Erdtmann) were also added. In recent decades comparative serology and phytochemistry (C. Mez, R. E. Alston, R. Hegnauer and others), and the quantitative and statistical treatment of characters ('numerical taxonomy', often with the aid of computers; R. R. Sokal, P. H. A. Sneath) have made considerable progress. Finally the electron microscope is of increasing importance, e.g. in the study of the characters of Protobionta (I. Manton and others). Thus systematics and phylogenetics generally rest on a broad foundation, but in most groups much remains to be discovered.

1. Characters, similarity and relationship

One of the most important problems of any systematic–phylogenetic analysis is to infer the underlying degrees of relationship between 'groups'* from their degrees of similarity with respect to structure, development and other characteristics.

(*a*) The **degree of relationship** ('consanguineity', 'affinity') is determined by the closer or more distant genealogical or phylogenetic connections, i.e. by the later or earlier separation (divergence) of the germ-lines (or by the development of discontinuities between the branches of the evolutionary tree).

'Groups' 3 and 4 in Fig. 366 are thus more closely related to each other than are 5 and 6; 2 and 1 are even less close. Brackets join 'groups' which are 'horizontally' related at the present time, but these relationships are complicated by reticulations (e.g. by convergent evolution and the abrupt appearance of allopolyploids such as 9). The 'vertical' relationships' link ancestors and descendants in time (for example 0 and 1). While in the horizontal direction discontinuous 'groups' (biotypes, taxa) are seen, the vertical dimension of time shows a continuum of forms. 'Groups' belonging to one hereditary community (e.g. 3–4 and 5–6) are monophyletic, i.e. they arise from a single ancestral 'group', whereas a 'group' based on 6 and 7 would be polyphyletic. 'Derivation' in the phylogenetic sense means that a daughter 'group' has descended from an ancestral one: thus 1 is 'derived' from 0. Understanding of the relationships of the evolutionary tree leads to the reconstruction of the phylogeny of 'groups'. Since the links of the historical germ-line are not preserved one often runs into considerable difficulties; even the arrangement of fossil forms in phylogenetic lines of development is only possible on the basis of the comparison of characters. It follows that representatives of side branches may easily be mistaken for ancestral forms (cf. + and 12). Even the geographical proximity of ancestral and daughter forms (cf. 3–4, 8–9–10) only gives indirect information (p. 431), while hybridization phenomena (i.e. the absence, presence of degree of effectiveness of genetical isolating mechanisms) give no infallible criteria to assess degrees of relationship. The formation of genetical barriers is often independent of other differentiation (cf. p. 414 and Fig. 357); this contradiction will be obvious if barriers exist between groups 9 and 10 (as a result of polyploidy) or between 11 and 12 (as a result of dysploidy), while 10 and 11 remain crossable.

* See footnote, p. 391.

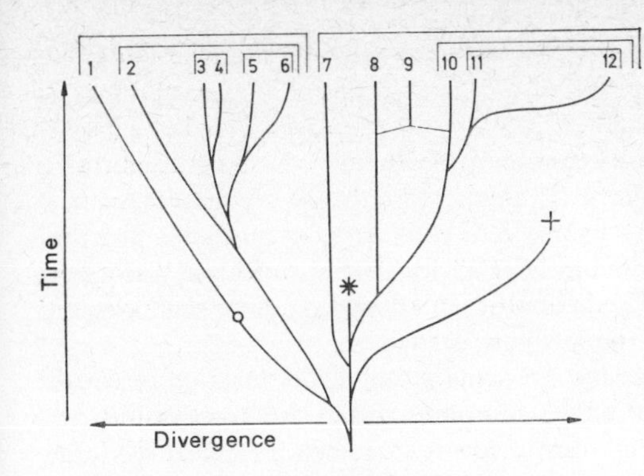

Fig. 366. A model evolutionary tree of a group of related organisms. Axes, time against divergence; the network of germ-line relationships (Fig. 333) is simplified as lines of development; O, an ancient biotype which has survived to the present; *, +, ones which are extinct; 1–12, recently evolved biotypes; 'natural' relationships (phylogenetic relationships) indicated by brackets; similarity (phenetic relationships) shown by relative distances horizontally.

(*b*) The **degree of similarity** (resemblance, i.e. the phenetic affinity based on the comparison of phenotypes) depends on the extent of agreement between characters. In this context, systematic characters are logically indivisible concepts relating to particular features of organs, structures or patterns of behaviour.

Systematic characters in this sense are exemplified by the red or black colours of the fruits of cowberries or bilberries; asci or basidia as forms of sporangia in Ascomycetes or Basidiomycetes; the various combinations of chlorophylls a, b, c, d and e, fucoxanthin, phycoerythrin and other photosynthetic pigments in the various algal groups (p. 449); or the aerobic or anaerobic modes of existence in various bacteria. But in addition, tendencies are also important characteristics for populations or higher taxa, e.g. the polymorphism of colour in *Corydalis cava*, the frequent occurrence of succulence in the Caryophyllidae, or the transition from spiral to whorled insertion of the floral organs in the more primitive angiosperms. Obviously, systematic comparisons only make sense between features with similar origins, i.e. homologous organs, structures or patterns of behaviour. In general one tries to augment qualitative by quantitative analysis.

Any phenetic comparison must eventually lead to the attempt to infer genotypic from phenotypic similarity, and from the extent of common heredity to the degree of relationship: hence the interest in the genetical basis of diagnostic characters. Modifications based on simple mutational changes (e.g. the absence of anthocyanin in albino forms; laciniate leaves, Fig. 285; the presence or absence of receptacular scales in composites) have, therefore, a lesser indicator value in assessing relationships than differences with a complex genetical basis (e.g. the dichotomous or pinnate venation of leaves, Figs. 174, 177; thyrsoid capitula in the teasel family compared with racemose capitula in the composites).

Study of the ontogeny and function of characters is essential both in understanding the causal basis of differentiation (Figs. 341, 365) and in assessing the value of characters as indicators of relationship: thus adaptive characters under intense control by selection (e.g. growth form or stem height, Fig. 351) have in general less indicator value than organizational characters less dependent on selection and external environment (e.g. the various 'internal' structures of pollen grains, Fig. 619; embryo sacs, Fig. 668, etc.).

As with a proof of paternity, the validity of determining relationship from simil ity depends on the number of characters considered: single differences or resemblances may have arisen by chance or by convergence; when many such are taken into account this is less likely. The close relationship of the taxon '*montana*' to '*glabra*' and '*alpina*' (Fig. 333) is based on its short, unbranched stem, but is made more plausible by the possession of longer, acute leaves, larger flowers and the lack of hairs (contrasted with '*hirsuta*'). In the past the Rubiaceae were placed near the Caprifoliaceae, mainly because of their inferior ovaries; it is now clear that with many similarities in vegetative, floral and chemical characters they are closer to the Loganiaceae and Apocynaceae.

Difficulties in reconstructing phylogenetic relationships will occur especially if the evolutionary tree has branched more than once within a short space of time (Fig. 366, groups 7, 8, 10), particularly during periods of cladogenesis (for example when the dicotyledons were evolving). Furthermore, there are numerous examples of imperfect correlation between similarity and relationship, for example when the differentiation of characters accelerates (Fig. 366, groups 11, 12), slows down (groups 7, 8), runs parallel (1, 2) or even becomes convergent (groups 6 and 7, or 8, 9 and 10). Parallel lines of development from motile flagellates through capsalean and coccalean colonial types to trichalean and parenchymatous types of construction occur in diverse algal groups. These striking external resemblances (analogies) have led in the past to the systematic recognition of these levels of organization. A good example of convergent evolution is provided by the lichens which have arisen by the symbiosis of various groups of algae and fungi; both on practical grounds and because of the difficulties in classifying them phylogenetically, they are usually treated together taxonomically as 'Lichenes'.

Finally, it frequently happens that in one and the same group differentiation of certain characters is accelerated while that of others is retarded; this is 'heterobathmy'. For example, the Psilotales (p. 564 et seq.) are in most respects extremely primitive, yet their subterranean gametophytes have mycorrhiza; the ancient *Ginkgo biloba* (p. 616) has nevertheless a specialized long and short shoot system; and the Magnoliidae, which are pre-eminently primitive amongst the angiosperms in floral structure, possess specialized alkaloids.

All these difficulties mean that the reconstruction of concrete evolutionary lines (the phylogeny of 'groups') is quite impossible in most circles of affinity, or can only be sketched in broad outline. The situation is more promising in the case of the so-called phylogeny of characters. This means the general historical tendency of development of a character, organ, or organ-complex, as, for example, in the basic tendencies of development of the shoot system, of the inflorescence or of leaf form in the angiosperms, or of the evolution of the cactus habit (as shown in Figs. 86, 146, 175 and 202). The following principles of the phylogeny of characters are equally important for morphology and systematics: (1) Fossils provide information about the early appearance of characters and their historical changes (e.g. the gradual loss of dichotomous branching in cormophytes, pp. 129 et seq.). (2) Characters widespread in major groups are more primitive than those which occur in minor sub-groups (e.g. the loss of flagellate gametes and zoospores in the Conjugales, pp. 471 et seq.). (3) Juvenile stages often exhibit primitive characters (e.g. the sequence from pinnate leaves to phyllodes in *Acacia*, Fig. 180A). (4) Unspecialized features are generally more primitive than specialized (e.g. homospory compared with heterospory, p. 562). Thus characters also can be said to be 'derived' in the phylogenetic sense.

2. Methods and materials for the study of similarity and relationship

Any systematic–phylogenetic study requires the following at the outset: observations on the variation, habitat and distribution of populations and 'groups' in the field; the gathering of living material for cultivation in botanic gardens, in thallophyte or bryophyte culture collections, in controlled environmental conditions, or in growth cabinets; collection of dried or otherwise fixed and preserved material in the herbarium, or for specialized study; and fossil material. With these materials on hand (and they are all too often inadequate and unrepresentative), systematic studies proper can begin.

Modern systematics depends on the results of study in numerous disciplines; these of course to some extent overlap: very different fields consider for example some aspects of developmental history (ontogeny), of function, of heredity, or of the historical past (palaeobotany). The following remarks and examples present only a broad outline.

(*a*) **Morphology** is undoubtedly the most important 'handmaid' of systematics. Differences in ground plan and progressions largely form the bases for classification and ordination, whether on a large scale (e.g. Protobionta–Cormobionta; Bryophyta–Pteridophyta–Spermatophyta; the inflorescences of angiosperms, Fig. 146) or a small one (e.g. the orders of the Chlorophyceae; the genera of the Rosaceae, Figs. 693–4, or of the Betulaceae, Fig. 688). Of course the relevant information is often very superficial and precise data on the functional aspects are almost entirely lacking. The importance of morphological–ontogenetical studies is exemplified by the angiosperm androecium: the distinctions between primary and secondary polyandry and between centrifugal and centripetal development of the initials has revolutionized angiosperm systematics in recent times (pp. 647 et seq.). The significance of abbreviated ontogeny with reproduction early in development (neoteny) has barely been studied in relation to the differentiation of taxa (cf. for example the progression from Araceae and Lemnaceae).

(*b*) **Anatomy** (or **histology**) has also made contributions of great systematic importance. Thus the classification of the Ascomycetes has been extensively modified on the basis of the structure of the wall and apex of the ascus (Figs. 484, 495). The systematics of lichens is largely based on the anatomy of the thallus and the form of the spores, while the arrangements and progressions shown by vascular bundles (Fig. 152) are of great importance for the phylogeny of vascular plants. The same is true of the phylogenetic changes from tracheids to vessels in the secondary xylem of angiosperms (Fig. 120 h → i). At the level of the family the Solanaceae are distinguished from their nearest relatives by their bicollateral vascular bundles, amongst other things, while the Elaeagnaceae have characteristic peltate hairs (Fig. 106E and F). Amongst *Eucalyptus* species the fine structure of the cuticular wax layer revealed by the electron microscope is diagnostic.

(*c*) The findings of **cytology**, especially those made with the electron microscope, have revealed the great divide in cell structure between the Procaryota and Eucaryota (nucleoid versus nucleus, thylacoids versus plastids, etc.). Differences in the fine structure of the flagella, plastids, eye-spot, etc., are of growing importance in the classification of flagellate and algal groups, and for the further question of their relationships with various groups of fungi. That part of cytology called *caryology* (which deals with the nucleus and chromosomes) has already shown us many examples, particularly the significance of polyploidy as an indicator of phylogenetic progression (cf. pp. 419 et seq.). These can be supplemented by mentioning increasing dysploidy ($n = 8, 9, 11, 13$) as a guide-line of chromosomal differentiation in the Cycadales; the inclusion of the genera *Agave* and *Yucca* (formerly placed in the Amaryllidaceae and Liliaceae respectively) in the Agavaceae on the basis of their very characteristic caryogram (five large and twenty-five minute chromosome pairs); and the significance of the fine structure of the interphase nucleus for the systematics of the Onagraceae.

(*d*) **Embryology**, by comparative studies of the development of sporangia, gametophytes, endosperm and embryo, points, for example, to the fundamental differences between the pteridophyte groups; to the necessity for transferring the Callitrichaceae (which have no perianth) to the sympetalous Lamiales; or to an improved classification of grasses (Poaceae). Tendencies for the reduction of the integuments and nucellus, and progressions among embryo sac types (Fig. 668) are important pointers to the phylogeny of characters in the angiosperms.

(*e*) **Palynology** reveals details of the coarse and fine structure of pollen grains (and spores) which are of growing importance in the systematics of seed plants (and to some extent of ferns). Figure 367 shows an example for the Cactaceae, where the progressions are characteristic of other angiosperm groups as well (3-colpate → pantocolpate → pantoporate; coarsening of the sexine structure; see pp. 649 et seq. for the technical terms). Ultra-thin sections and studies with the electron microscope have given important clues for the classification of the Asteraceae, for example.

(*f*) **Comparative phytochemistry** and **serology** have made particularly far-reaching contributions to the study of relationships. A familiar example is the systematic value of the occurrence of cellulose or chitin in the cell walls of lower fungi (pp. 493 et seq.). Paper-, thin-film- and gas-chromatography nowadays permit the routine chemosystematic analysis of flavonoids (e.g. in ferns, Fig. 361), terpenoids (e.g. in the classification of *Pinus, Citrus*, etc.) or alkaloids (e.g. in the Papaverales, now separated from the Capparales and placed in the Magnoliidae on account of the very similar isoquinoline alkaloids), etc. Very remarkable is the mutually exclusive occurrence of anthocyanins and the structurally quite different betacyanins in the Caryophyllidae (Fig. 368); there is, however, at present no other valid differentiating character to split up this probably heterogeneous subclass. Furthermore, chemical constituents may arise by convergence, although this may be explained by alternative biosynthetic pathways: thus the caffeine in tea (Theaceae) and in coffee (Rubiaceae) is synthesized by different routes. Finally, current advances in understanding the distribution and biosynthesis of porphyrins (including the haem group and chlorophylls) and phycobilins are of great significance in studying the relationships of the bacteria, blue-green algae and higher algae. The differentiation of

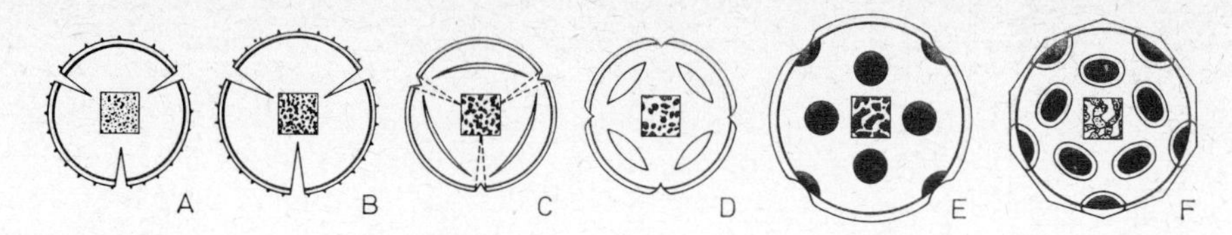

Fig. 367. Pollen grain types and their progressions in the Cactaceae. Optical section and apertures; colpi or pores black. *c.* ×400. Fine structure of the sexine (in the square insets, *c.* ×800). **A.** and **B.** tricolpate. **C.** 6-pantocolpate. **D.** 12-pantocolpate. **E.** 12-pantoporate. **F.** 15-pantoporate. After Tsukada.

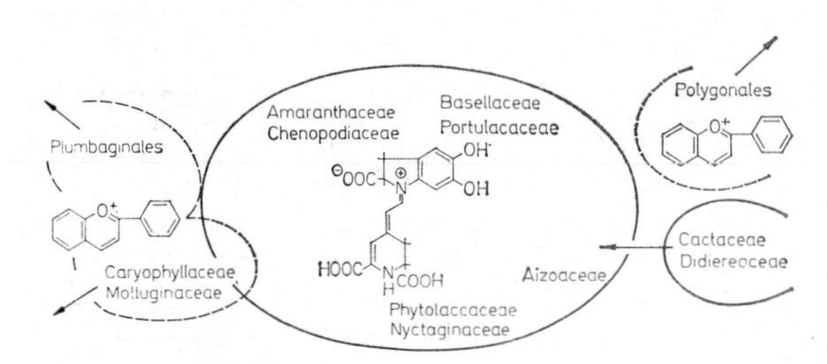

Fig. 368. The occurrence of betacyanins (centre, solid line) and anthocyanins (broken lines) in the Caryophyllidae. Arrows indicate suggested modifications to earlier taxonomic arrangements. Modified after Merxmüller.

these pigments is related to the evolution and stepwise 'improvement' of photosynthesis and respiration in these pioneer groups of the Plant Kingdom.

The methods of serology have been used for a long time to elucidate relationships at the macromolecular level. They depend on the fact that a foreign protein (A) injected into an experimental animal causes the formation of antibodies. The antiserum obtained from the blood produces with another protein (B, as antigen) a precipitation reaction which increases in intensity the more alike A and B are. Such methods of assessing the degree of similarity of proteins have been improved in recent years and used with success in the Ranunculaceae, for example (Fig. 371). If antiserum and antigen are subjected to gel-diffusion or electrophoresis the precipitates are produced in bands (this method has been used, for example, in the Solanaceae). Proteins can also be analysed directly by electrophoresis. Interesting conclusions relevant to questions of relationship have been made by elucidation of the amino-acid sequences in homologous enzymes (e.g. cytochromes) and in determining the nucleotide homologies in DNA and RNA.

(*g*) **Phytopathology** can also contribute to systematics since various bacteria, fungi and phytophagous insects are specific to definite groups of plants. Thus the larvae of butterflies of the group Pierinae feed on Capparaceae and Brassicaceae, but not on the Papaveraceae with which they were formerly grouped. Rust fungi (Uredinales) differentiate between the various sub-families of the Rosaceae: thus the Phragmideae occur only on the Rosoideae, *Gymnosporangium* only on Pomoideae, and *Thecospora* only on Prunoideae.

(*h*) The innumerable connections between **physiology**, **ecology** and systematics are still too little studied although evolution is intimately related to functional differentiation. Thus the evolution and systematics of the Schizomycetes depend to a large extent on the differentiation of the physiology of nutrition (e.g. substrate dependence, fermentation, chemo- and photosynthesis, etc.). The Protobionta are broadly divided by the distinction between autotrophy (Phycophyta) and heterotrophy (Mycophyta, Protozoa). Examples of the multiplicity of ecological adaptations as a basis for the differentiation of groups and for the progressive invasion of new habitats have been considered in the chapter on evolution (pp. 407 et seq.). At the level of orders and families, physiological–ecological 'centres of gravity' can be recognized with respect, for example, to water-relations (Nymphaeales: hydrophytes; Cactaceae: xerophytes), nutrient uptake (Chenopodiaceae: requiring copious minerals; Ericaceae, raw humus plants), carnivory (Sarraceniales), or parasitism (Loranthaceae, Orobanchaceae), etc.

(*i*) **Reproductive biology** is also of great significance for the analysis and understanding of the differentiation of groups. The formation of spores in certain bacteria

(Bacillaceae), the progression from isogamy to anisogamy and oogamy, and further to gametangiogamy and somatogamy in most groups of algae and fungi, or the differentiation of the fruit-bodies and hymenophores in relation to spore dispersal in the Holobasidiomycetes, are all of systematic as well as functional importance. A similarly fundamental guiding principle is afforded in cormophytes by the development of heterospory from homospory and further to the development of the seed. Angiosperm evolution could not be understood without considering the adaptations of the primitive flower to insect pollination, or the development of catkin-bearing plants (Amentiferae) without taking secondary anemophily into account (pp. 657 et seq.). The classification of the Ranunculaceae is largely based on specialization in floral biology (e.g. honey-leaves, Fig. 684I–M; dorsiventrality in *Aconitum*, Fig. 684D–H; secondary anemophily in *Thalictrum*). This is expressed in the genus *Aquilegia* by the development of honey-leaf spurs of various lengths and colours which are adaptations to various hymenoptera, moths or hummingbirds with mouthparts of various lengths (Fig. 369). The differentiation of *Ficus* is linked with its peculiar mode of pollination by gall wasps, that of *Salix* with its recent reversion to entomophily from anemophily (as still seen in *Populus*). Other important guidelines are provided by progressions to heterostyly, dioecism, autogamy and apomixis (cf. pp. 657 et seq. and p. 667). The biology of seeds and fruits is no less important for the phylogeny and systematics of Spermatophyta. The progression from dehiscent to indehiscent fruits is reflected in the distinction between the Scrophulariales and Lamiales, and in the classification of the Rosaceae (Fig. 694), while the provision of various appendages to the indehiscent fruits affords basic characters in the Dipsacaceae (Fig. 365).

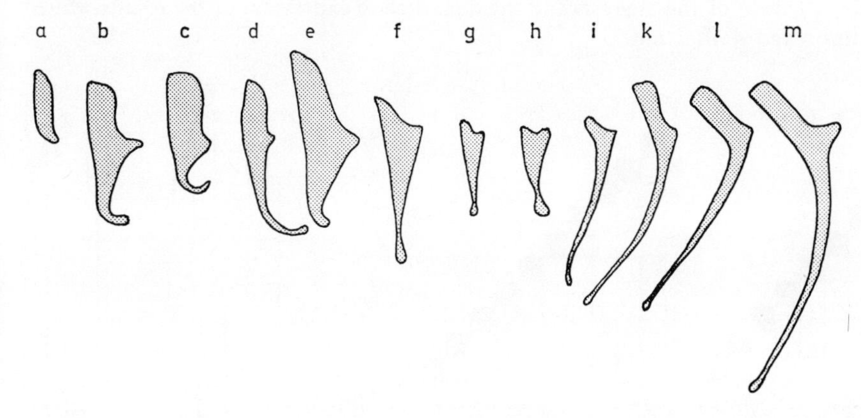

Fig. 369. Honey-leaves in species of the genus *Aquilegia* (columbine, Ranunculaceae). **a.** Without spur. **e.** to **k.** Broad to narrow spur. **c.** to **m.** Short to very long. The diversity reflects adaptations to various floral visitors. After Praźmo.

(*j*) **Genetics** and **cytogenetics** form an integral part of modern systematics as indicated repeatedly in the section on heredity and evolution (pp. 391 et seq.). Data of especial importance concern natural variation, the transmission of differentiating characters, chromosome affinity (meiosis), reproductive isolation and the formation of other barriers; the experimental reconstruction of the course of evolution is also very desirable. Apart from a few groups subjected to relatively intense study (e.g. *Escherichia, Chlamydomonas, Neurospora, Arabidopsis, Potentilla glandulosa, Clarkia, Oenothera, Nicotiana, Zea, Triticum*), genetic analysis still leaves many gaps.

(*k*) **Plant geography** (**areography**) assists in understanding the formation of taxa in time and space in ways which have already been considered (pp. 409 et seq.). Geographical–morphological analyses mainly depend on the fact that closely-related taxa are generally allopatric initially but may become sympatric later with the development of barriers (cf. p. 411 and Figs. 355, 356, 364). Unfortunately the areas of distribution of even conspicuous taxa in well studied regions are frequently known only imperfectly.

(*l*) **Palaeobotany** gives the only direct evidence of ancestors or ancestral forms, and it has provided our knowledge of such important but extinct key groups as the Psilophytales, the progenitors of the pteridophytes (pp. 562 et seq.), the Progymnospermae as the ancestors of the seed plants, the Cordaitidae as the initiators of the conifers, and the Pteridospermae as the precursors of the cycads, bennettites and angiosperms (pp. 612 et seq., 629, 640). Fossil plants give evidence for many sequences of characters and provide documentation for anagenesis, cladogenesis and stasigenesis as characteristic phases of evolution (cf. Figs. 335, 628), and for the historical expansions and contractions of areas of distribution (Fig. 745). Further important contributions to the phylogeny of plants are made by studies of fossil spores and pollen.

3. The synthesis of phylogenetics and systematics

The wealth of data provided by studies of similarity and relationship requires arrangement and synthesis. Numerical and statistical methods are of growing importance in constructing taxonomic groups. The synthetic approach attempts to improve understanding of how the various kinds of differentiation are related within a group, how they are causally related to the environment, and how they have arisen in the course of time.

(*a*) In the quantitative statistical determination of similarity and grouping (numerical taxonomy), as many characters as possible (at least fifty) are studied in a given taxonomic unit (individuals, populations, species, etc.) and quantitatively classified according to some uniform scheme. A comparison of all units, taking all characters into account (generally by means of a computer), produces coefficients of similarity (generally expressed in percentages). The taxonomic units can then be grouped in various ways and represented in the form of a phenetic 'dendrogram' (Fig. 370). The advantage of these methods over the conventional, more or less intuitive methods of grouping lies in the predetermined course of the analysis and the quantitative expression of the results which allow repetition and verification.

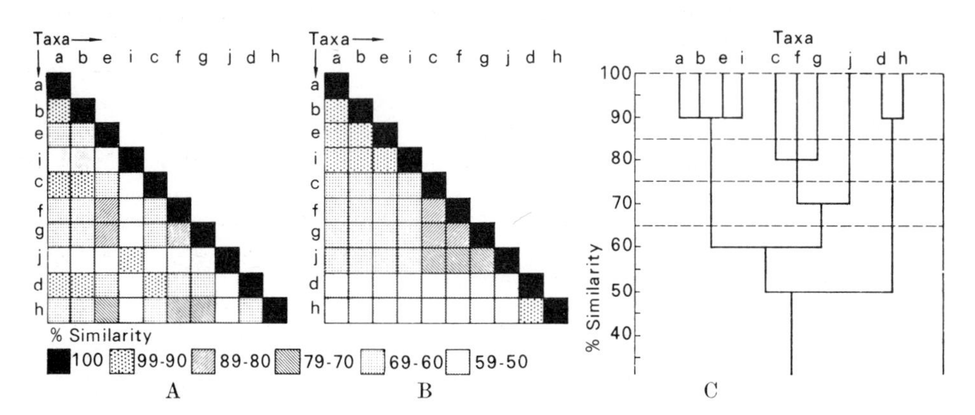

Fig. 370. Quantitative estimation of similarity and grouping in a model group with the taxa a–h. Coefficients of similarity (%) between all taxa, unordered in **A.**, ordered in **B. C.** The dendrogram based on these showing phenetic similarity; groups with similar values are linked by horizontal lines. After Sneath.

(*b*) The synthesis of morphological, anatomical, embryological, palynological, caryological, phytochemical and particularly serological data now gives a good idea of the relationships within the Ranunculaceae (Fig. 371). The phylogeny of the family is connected with the development of habitats in the Northern Hemisphere during the Tertiary and Quaternary and is characterized by adaptive radiation and expansion from deciduous forests into open xeric, hygric and Arctic–Alpine habitats, with parallel changes in life- and growth-form, leaf-shape, and also in the kind of fruit. The extremely diverse differentiation of the flowers is related to the utilization of progressively differing modes of pollination (cf. Fig. 369). Relict localities harbour species-poor groups in stasigenesis, new habitats facilitate the cladogenesis of swarms of new forms, while the cytogenetic relationships accurately reflect these different evolutionary phases (cf. also the *Anemone*-group: *Hepatica*, Fig. 356; *Pulsatilla*, Fig. 359).

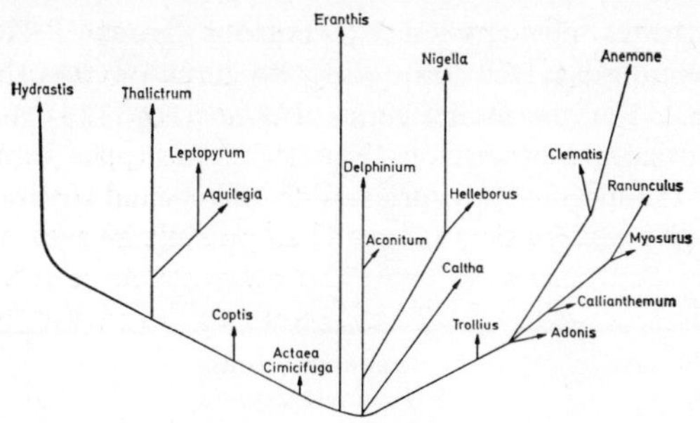

Fig. 371. Synthetic scheme of the relationships between the most important genera of Ranunculaceae (with particular referencc serological resemblances). After Jensen.

III. Taxonomy and nomenclature

A pack of playing cards can be arranged either by suits or by numbers and court-cards. The diversity or organisms can also be arranged on various principles. The frequently overlooked distinction between inanimate objects and organisms is that because of their evolutionary relationships, a hierarchical principle of classification, independent of the observer, already exists for organisms. In consequence, starting from the degrees of similarity between organisms, it is possible to construct progressively better, 'natural' hierarchical-taxonomic systems. Because they contain the maximum content of information they are to be preferred to any other possible system.

Changing views of the principles of classification are reflected in successive systems of the Plant Kingdom (cf. p. 437). Artificial systems used arbitrarily selected characters for classification (e.g. growth form, numbers of parts of the floral organs). Consideration of a larger number of characters led to subsequent improvements but many groups were based on levels of organization (and not on reproductive communities), giving purely formal systems. After the acceptance of the theory of evolution all resemblances were rather rashly interpreted as expressions of kinship, and numerous phylogenetic systems were put forward. Nowadays one is conscious of numerous difficulties: the dearth of fossil remains; the varying rates of evolution; parallel, convergent and reticulate lines of development (pp. 426–8). One therefore tries to consider natural groups on the broadest possible basis, and to reconstruct their histories, although one is not afraid of treating certain convergent groups as taxonomic units (for example, the fungi = Mycophyta which has arisen from diverse flagellate or algal groups). Thus the phase of synthetic systems has been reached.

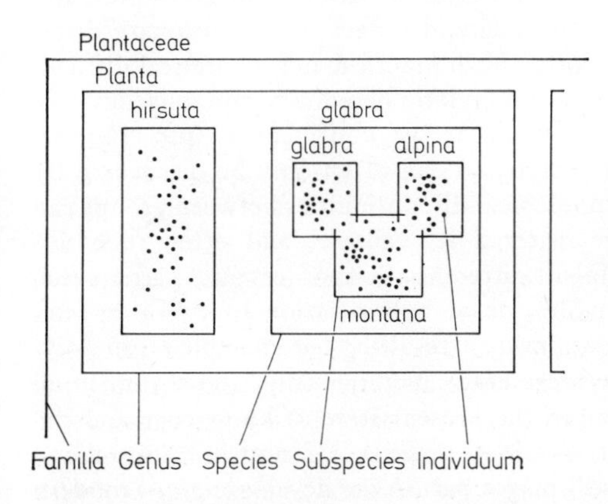

Fig. 372. Model of the relationship group '*Planta*' (Fig. 333) in diagrammatic horizontal projection. On the concrete individuals and the equally concrete hereditary communities, abstract taxonomic categories are based (subspecies, species, genus, familia); the resultant taxa of various rank are shown with imaginary names (e.g. subspecies *montana, Planta glabra, Planta*, Plantaceae).

1. **Taxonomic categories and units.** Obligatory taxonomic categories are used in classifying plants; initially these categories are 'vacant' abstractions which are arranged within the framework of a hierarchy of definite ranks. Thus the taxonomic category 'species' is subordinate to that of 'genus'; between them lies the 'series', while below the species lie 'subspecies' and 'variety'. Taxonomic units or 'taxa' (singular 'taxon')

are formed when these abstract categories are applied to concrete 'groups'. The hierarchy of taxa, that is to say, the taxonomic system, expresses to a certain extent the delimitation and relationships of 'groups'. For the model genus '*Planta*' (Fig. 333) this can be represented in the form of a horizontal projection (Fig. 372), while the table below sets out the more important taxonomic categories with their standardized terminations, and uses the yarrow (*Achillea millefolium* L.) as an example of the system.

Taxonomic categories (English, Latin, abbreviations)	**Endings**	**Taxonomic units** (Examples, synonyms)
Kingdom (regnum)	-ota	Eucaryota
Subkingdom (subregnum)	-bionta	Cormobionta
Division (phylum)	-phyta	Spermatophyta
Subdivision (subphylum)	-phytina	Magnoliophytina (= Angiospermae)
Class (classis)	-phyceae, -mycetes or -atae (also -opsida)	Magnoliatae (= Dicotyledones)
Subclass (subclassis)	-idae	Asteridae
Superorder (superordo)	-anae	Asteranae (= Synandrae)
Order (ordo)	-ales	Asterales
Family (familia)	-aceae	Asteraceae (= Compositae p.p.)
Subfamily (subfamilia)	-oideae	–
Tribe (tribus)	-eae	Anthemideae
Genus (genus)		*Achillea*
Section (sectio, sect.)		sect. Achillea
Series (series, ser.)		–
(Aggregate, agg.)		*Achillea millefolium* agg.
Species (species, spec. or sp.)		*Achillea millefolium* L.
Subspecies (subspecies, subsp. or ssp.)		subsp. *sudetica* (Opiz) Weiss (= *A. sudetica* Opiz)
Variety (varietas, var.)		–
Form (forma, f.)		f. *rosea*

Taxonomic units should so far as possible coincide with hereditary communities and be recognizable by genetically stable characters. For the delimitation of taxa, isolation and morphological discontinuity are more important than the amount of variation or any one character in particular. The hierarchy of taxa should reflect the evolutionary divergence of the groups which it contains. Very often both practical and theoretical difficulties conflict with these demands: apparently monophyletic hereditary communities may have arisen as a result of superficial similarities or by convergence (pp. 426–8); whether differentiating characters are genetically based can often only be discovered by experiment; the presence or absence of morphological discontinuities between groups can only be ascertained if fully representative material is available, and often such discontinuities are indistinct as a result of incipient differentiation, of 'cryptic' barriers (i.e. ones which are not morphologically detectable), or of hybridization (pp. 411 et seq., pp. 415 et seq.); reticulate evolutionary relationships (resulting for example from polyploidy, pp. 418 et seq.) cannot actually be represented hierarchically, and so on. In all these cases a compromise is necessary between the present state of knowledge and the desirability of producing a synopsis which is usable in practice. Of course, tradition and attempts to maintain a degree of stability still play a part in the development of modern systems.

In all attempts to make taxonomic categories standard and comparable the species has always held a key position. In this age of the synthetic system one draws as much as possible on phenetic, genetic and genealogical criteria (cf. pp. 425 et seq.). The category species is applied to the lowest unit 'group' (hereditary community) which is distinguished from all other unit 'groups' by constant hereditable characters and by reproductive isolation.

The species thus marks a decisive stage in the course of evolution in which the germ-line relationships between the reproductive and hereditary communities separate. Of course, this statement is only true for groups with sexual reproduction (which is often lacking in both lower and higher plants). Furthermore, the rank of species embraces very different kinds of 'group'. In many 'groups' the unequivocal use of this category is made difficult by deficient knowledge and by poor correlation between or the obscure expression of phenetic, genetic and genealogical criteria. All this allows scope for many different views of the species concept, whether broad or narrow. Nevertheless, the 'species' is still the most important unit of biological taxonomy and on it are based all other taxonomic units, whether ranking at supra- or infra-specific levels.

In very variable species it is useful to distinguish infra-specific taxa. 'Groups' within a species which are not sharply delimited from each other are generally treated as subspecies. Such 'groups' are often geographical or ecological allopatric races linked by intermediate populations, but they may also be autogamous, polyploid or otherwise more or less isolated in reproduction, and are then often sympatric. The category of variety is nowadays used rather rarely, for example when it is desirable to subdivide subspecies, or when an infra-specific 'group' is inadequately understood. The corresponding unit among cultivated plants is the 'cultivar' (abbreviated 'cv.'). Finally, noteworthy biotypes or mutants can be designated taxonomically as 'forms' ('forma', 'f.').

At the supra-specific level the genus is the most important taxonomic unit and as such is incorporated in binomial nomenclature. A genus consists of a group of species which shows distinct discontinuities between itself and all other groups of species. If necessary sections, series, etc., can be distinguished within genera. Very closely related species which are only distinguishable with difficulty (so-called 'micro-species') may be united in 'aggregates' ('agg.') which are not taxonomically binding.

Between the genus and the kingdom (regnum, the highest unit in the system of organisms) further categories are inserted, of which the most important in ascending series are family (familia), order (ordo), class (classis) and division (phylum). Despite many difficulties, one aims at a certain comparability with respect to differentiation and extent even in these higher taxa. However, on practical grounds particularly diverse groups are often given a relatively high and independent taxonomic rank (for example, the Cormobionta, which are really only a developmental branch of the Chlorophyceae, or the Apiaceae which have attained much less differentiation than the Araliaceae from which they have arisen).

2. **Nomenclature**. The *International Code of Botanical Nomenclature* lays down binding rules for the description and naming of plant taxa. For new taxa a Latin diagnosis and 'effective' publication are obligatory. All scientific plant names must be in latinized form. The names of genera and higher taxa are nouns (with an initial capital) while specific and infra-specific names are generally adjectival (with a small initial letter). Names of hybrids have x prefixed (e.g. x *Raphanobrassica, Mentha* x *piperita*). The interpretation of a name (from the rank of family downwards) is based on a specified 'nomenclatural type' (generally a herbarium specimen, or a leading taxon). Taxa subordinate to a genus or species and including the 'type' repeat the generic or specific name, as appropriate (e.g. *Achillea* sect. Achillea; *A. millefolium* subsp. *millefolium*). At a given rank, only the oldest 'legitimate' name (i.e. one conforming to the Code) is valid for a taxon (the priority rule); the starting point for this is the first edition of *Species Plantarum* by Linnaeus (1753). Synonyms are different names for the same taxon, homonyms are identical names for different taxa. Names of families and genera in common use, although illegitimate, may exceptionally be 'conserved'; unfortunately this is never the case for illegitimate specific names. For ease of reference the name of the first describer ('author') or a taxon is appended, generally in an abbreviated form (in specialist works, but not in textbooks). If the rank of a taxon is changed the name of the author of the 'basionym' appears in parentheses, followed by that of the author of the new combination, e.g. *Achillea sudetica* Opiz → *Achillea millefolium* subsp. *sudetica* (Opiz) Weiss; the same is true if a taxon is transferred to another species or genus. Since the correct name of a

taxon depends on the Code of Nomenclature, on the correct interpretation of the 'type' and on taxonomic grouping and ranking, changes in nomenclature can unfortunately never be totally avoided.

3. **Documentation**. The wealth of systematic and taxonomic information about the Plant Kingdom reposes in monographs and revisions of various groups, in the Floras of particular regions, in countless individual publications, in herbaria, and so on. Classification and scientific names make this information accessible and allow both the identification of new plant material and continuing familiarization with fresh knowledge. Modern data processing machines and computers are beginning to facilitate the collection and evaluation of all kinds of old and new information in all fields of systematics and taxonomy, and will no doubt revolutionize them in the future.

Section two: **Survey of the Plant Kingdom**

The concepts 'plant' and 'animal' were once regarded as fundamental systematic–taxonomic units. Nowadays we recognize that these two groupings are based primarily on modes of nutrition rather than on evolutionary relationships. The most fundamental division of living things is into Prokaryota (Akaryobionta) on the one hand and Eukaryota (Karyobionta) on the other. Prokaryota comprise the division Schizophyta with two classes, Schizomycetes (bacteria) and Cyanophyceae (blue-green algae). Within the Eukaryota the Protobionta (protista) represent a common basis for both higher plants (Cormobionta) and multicellular animals (Zoobionta). The Protobionta can be divided, more on practical and physiological grounds than from a phylogenetic viewpoint, into Phycophyta (algae), Mycophyta (fungi), and Protozoa; Lichenes (lichens) are symbionts of algae and fungi. The Cormobionta (Archegoniatae, Embryophyta) comprise the Bryophyta (mosses and liverworts), Pteridophyta (ferns) and Spermatophyta (seed plants). Within these divisions we further distinguish a large number of often very independent classes; the mutual relationships and probable affinities of these will be discussed at the end of each relevant account, and finally reviewed retrospectively (Fig. 743 and the accompanying text). A synopsis of the divisions and classes is provided by the table of contents.

There have been many changes over the centuries in the points of view from which the Plant Kingdom has been classified (cf. p. 433). The most familiar of the artificial systems of classification is the sexual system of Linnaeus (1735). He recognized twenty-three classes of flowering plants and a twenty-fourth class, the Cryptogamia, in which he included not only the then little-known ferns, mosses and liverworts, algae and fungi but also some higher plants with not very obvious flowers (*Ficus, Lemna*) and even corals and sponges. He divided the flowering plants (Phanerogamia) principally on the distribution of the sexes in their flowers and on the number, mode of insertion, relative length and degree of fusion of the stamens. We now recognize the cryptogams as spore plants, since they generally develop from unicellular spores (Divisions 1–6 below), whereas the phanerogams are seed plants (Division 7).

Linnaeus had already attempted to set up a natural system in 1738, but A. L. de Jussieu (1789), A. P. de Candolle (1819), S. Endlicher (1836) and others can be regarded as the founders of the most important formal systems. After the acceptance of the theory of evolution, the systems of A. Braun (1864), G. Bentham and J. D. Hooker (1862–83), A. Eichler (1883), and particularly the still widely-used arrangement of A. Engler, were dominated by the taxonomic use of levels of organization and grades of development. The first truly phylogenetic scheme was proposed by R. von Wettstein (1901–8). The systems in common use today represent various stages on the route from formal arrangements, via phylogenetic to synthetic ones. Important contributions in recent decades are those of C. van Niel and R. Y. Stanier for bacteria; L. Geitler for Cyanophyceae; A. Pascher, F. E. Fritsch, B. Fott and T. Christensen for algae; E. Gäumann, M. Chadefaud and R. Singer for fungi; A. W. Evans and M. Fleischer for bryophytes; D. H. Scott, F. O. Bower and E. B. Copeland for ferns; C. J. Chamberlain and R. Florin for gymnosperms; and H. Hallier, C. E. Bessey, A. A. Pulle, A. Takhtajan, amongst others, for angiosperms.

Even now, comparison of modern systems shows many, often fundamental disagreements, an expression of the great state of flux of modern systematics and taxonomy, and of the continuing need for basic research before a generally acceptable arrangement can be achieved. The arrangement which follows is merely an attempt to show the broad relationships fairly clearly. Bearing in mind the purposes of a textbook, certain simplifications have been made intentionally.

So far some 370,000 species of living plants have been described. Of these about two-thirds are seed plants (about 800 gymnosperms and 226,000 angiosperms, of which 172,000 are dicotyledons, 54,000 monocotyledons), some 12,000 are ferns, and 26,000 mosses and liverworts. Amongst the Protobionta, it is estimated that there are over 33,000 algae, about 50,000 fungi and some 20,000 lichens. Finally there are some 1600 bacteria and 2000 Cyanophyceae. The large number of species (especially of fungi and angiosperms) described each year leads one to suppose that more than half a million members of the Plant Kingdom will eventually be described.

Division 1: **Schizophyta**

The Schizophyta display the lowest level of organization found in plants. The cells are solitary or united in colonies (coenobia). They reproduce by simple division of the cells (fission). There is no true nucleus surrounded by a nuclear membrane; they are thus called Prokaryota in contrast to the Eukaryota which have true nuclei. Furthermore, those parts of the cytoplasm which contain photosynthetic pigments also lack surrounding membranes. Finally, the cell wall structure is essentially different from that of the Eukaryota. The two classes of the Schizophyta differ principally in their modes of nutrition. The bacteria include both heterotrophic and autotrophic types, but those which can photosynthesize never do so with the production of oxygen. The blue-green algae are predominantly autotrophic; as in other green plants, molecular oxygen is liberated during photosynthesis.

Class I: Schizomycetes, bacteria

The bacteria are primitive organisms, lacking typical plastids and true nuclei; chlorophyll is normally absent. They are extraordinarily small (the smallest less than 1 μm in diameter; cf. pp. 11–13). The majority are unicellular and show very little morphological differentiation (Fig. 373). The cells may be spherical (cocci), rod-shaped (bacilli), or spirally twisted (vibrios, spirilla). The cells of some remain united after division forming colonies, packets (Fig. 373B), filaments (Fig. 377) or nets. Some bacteria are distinguished by club-shaped cells (corynebacteria) or branched cells (mycobacteria), or may even form a mycelium (Fig. 378).

The 'nucleus' (nuclear equivalent, nucleoid) of the bacterial cell is not enclosed in a membrane. The genetic material (DNA) lies free in the cell, embedded in the central nucleoplasm (Fig. 374).

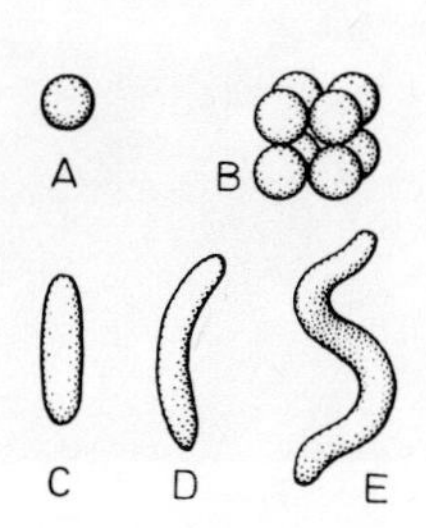

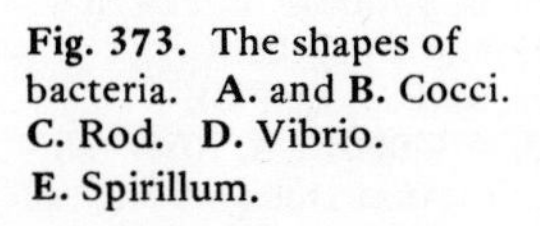

Fig. 373. The shapes of bacteria. **A.** and **B.** Cocci. **C.** Rod. **D.** Vibrio. **E.** Spirillum.

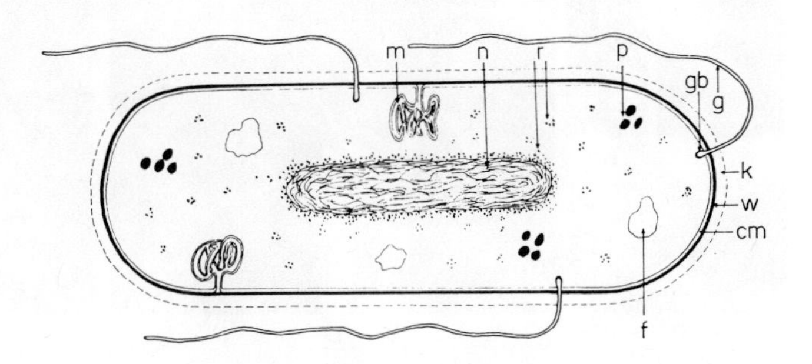

Fig. 374. Diagrammatic section of a bacterial cell. k, capsule; w, cell wall; cm, cytoplasmic membrane; g, flagellum; gb, base of the flagellum; m, mitochondrion-like structure; p, polyphosphate body; f, fat granule; r, ribosomes; n, nucleoid. ×25,000 approx. Based on Drews and Schlegel.

DNA can be recognized under the light microscope by the Feulgen reaction (staining after acid hydrolysis with fuchsin decolorized with sulphurous acid). Under the electron microscope ultra-thin sections of suitably fixed cells show the nuclear material as a fine filamentous network. Radioactive labelling with tritiated thymidine followed by radiography of the extended DNA shows that in *Escherichia coli* the 'chromosome' consists of a closed, ring-shaped double strand with a total length of about 1 mm. Cell-division is preceded by duplication of the DNA; with rapid division new DNA is produced at a rate of 35 μm chain-length per minute. DNA is the carrier of the genetic information of the cell. Circular genetic maps have been drawn for *Escherichia coli, Salmonella typhimurium* and others.

Some bacteria (green Chlorobacteriaceae and purple bacteria; p. 245) possess photosynthetic pigments, e.g. green bacteriochlorophyll (p. 248) and several red carotenoids which in red species mask the green pigment. The pigments are borne on membranes in thylacoids (Fig. 376), but there are no enclosing membranes, so no true plastids.

Various other inclusions deposited intracellularly can be regarded as reserve materials. Many bacteria store starch-like or glycogen-like polysaccharides. Staining with Sudan III reveals fat globules consisting of poly-β-hydroxybutyric acid. Mycobacteria and actinomycetes, on the other hand, store neutral fats and waxes. Phosphoric acid is stored in the form of polyphosphate granules (volutin; Fig. 374). Gas vacuoles occur in some bacteria.

The surface of the protoplast consists of a delicate cytoplasmic membrane of lipoprotein. It is the site of certain enzymes and responsible for semi-permeability. The inner surface has a mitochondrion-like appearance, but true mitochondria are absent from bacteria. Cell turgor presses the membrane against the surrounding cell wall.

The bacterial cell wall is about 20 nm thick and unlike the cellulose cell wall of higher plants shows no fibrillar structure. The 'supporting-membrane' which gives the cell its shape consists of muropeptides built up from units of amino sugars and certain amino acids which are joined by peptide and glucoside linkages to form a macromolecular net (murein sacculus). The two big groups of bacteria, the so-called Gram-positive and Gram-negative bacteria, differ in the amino-acid composition of their cell walls. Aniline dyes taken up by Gram-positive cells cannot be washed out again by alcohol, whereas they can

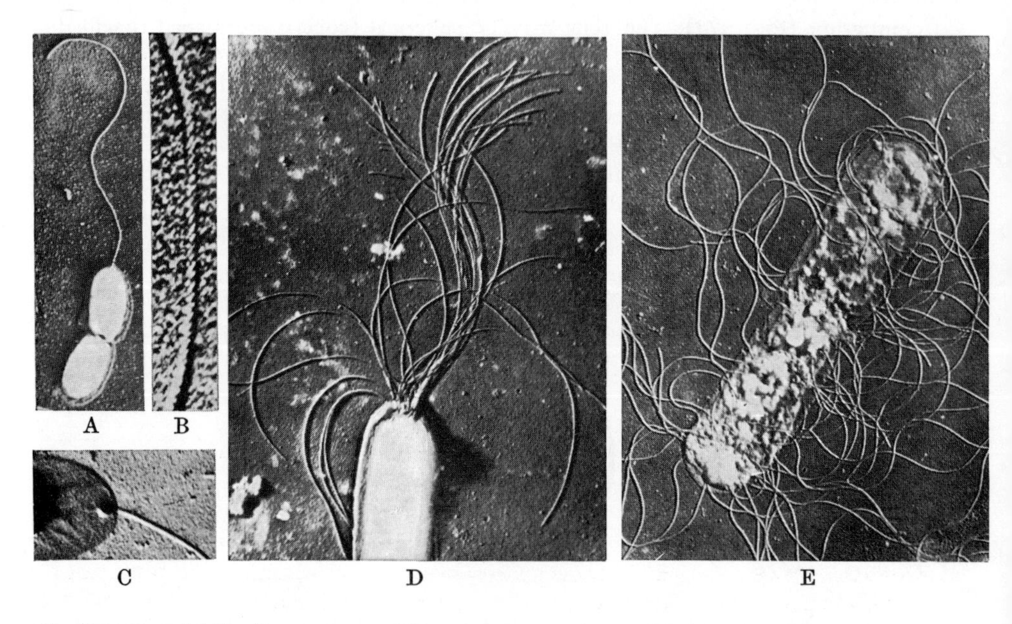

Fig. 375. Bacterial flagella. **A.** Monotrichous flagellum (*Vibrio metchnikovii*; ×7000). **B.** Part of a flagellum (*Brucella bronchisepta*; ×60,000). **C.** Basal granule at the insertion of a flagellum (*Rhizobium radicicola*; ×20,000). **D.** Lophotrichous flagellation (*Spirillum undula*; ×8000). **E.** Peritrichous flagellation (*Proteus vulgaris*; cell contents partly autolysed; ×10,000). **A.** After van Iterson. **B.** After Labaw and Mosley. **C.** After Ziegler. **D.** After Scanga. **E.** After Houwink and van Iterson.

be in Gram-negative ones. Some bacteria develop greatly-swollen masses of slime (zoogloea-stage; p. 69) or capsules of varying composition (usually polysaccharides, or polypeptides). The cells of *Acetobacter xylinum* are held together in a film of cellulose (mother of vinegar); in *Sarcina ventriculi* the cells are held together in chains by cellulose. Certain bacteria may lose their walls, either by mutation or by chemical treatment, and in certain circumstances can continue to grow and divide as naked protoplasts (L-forms).

At certain stages of development many bacteria possess extremely delicate flagella (Fig. 375) which permit active swimming which is reversible in direction (pp. 52, 381). Flagella are either terminal and solitary (monotrichous, Fig. 375A), or in bunches (lophotrichous, as in *Spirillum*, Fig. 375D), or are distributed over the whole surface (peritrichous, Fig. 375E). A monotrichous flagellum is either polar (Fig. 375C) or is inserted laterally a little below the end of the cell (subpolar). In many species the number of flagella is not constant; thus *Proteus vulgaris* has only two subpolar flagella when starved, but many peritrichous flagella when well-fed. If lost, flagella can be replaced by new ones. Each flagellum arises—so far as is known—from a basal granule (Fig. 375C) or basal body within the cell (cf. p. 52). Electron microscopy reveals a spirally patterned surface (Fig. 375B); each flagellum consists of a few extremely fine fibrils of contractile protein (similar to the myosin of muscle) twisted together; there is no 9 + 2 fibrillar structure as in all other organisms with flagella (cf. Fig. 385). Besides flagella, some bacteria have numerous even finer threads (fimbriae) the function of which is unknown.

The rate of movement by means of flagella reaches 200 μm sec^{-1} in *Bacillus megatherium*—fifty times its own length. *Spirillum* rotates on its axis thirteen times per second while the flagella perform forty rotations—approximately the same speed of rotation as an electric motor. Movement is generally effected by propulsion, as by a ship's propellor; but it can at any time be suddenly changed to a traction movement, as by an aircraft screw.

As a rule, reproduction is vegetative by cell division, which in elongate forms is always at right-angles to the long axis (Fig. 375A).

In division a cross wall first grows inwards from the margin towards the centre of the cell, which later splits along the new wall so that the cells separate; the cells may also remain united in loose chains. The division of cells without true nuclei is called fission; because of this bacteria have been called 'fission fungi' or Schizomycetes.

Even in bacteria transfer of genetic material is possible (parasexuality). Characters, generally recognized by a defect in biochemical activity, may be transferred in three ways. In conjugation a piece of DNA from a donor cell is transferred directly to a recipient cell in which it undergoes recombination. In transduction the DNA is transferred from donor to recipient by means of a bacteriophage (p. 445). In transformation the transfer to the recipient is effected by means of free DNA extracted from the donor; this process was by far the earliest to be observed as a means of transfer of characters in bacteria, and it provided the knowledge that DNA is the genetic material of organisms. Meiosis has never been observed in bacteria.

When conditions are unfavourable many species of *Bacillus* and *Clostridium* form resting cells (spores). In these the protoplast lays down reserves in the middle of the cell, or towards one end, together with a nucleoid. It then passes into an inactive stage, undergoing dehydration and contraction, and surrounding itself with a many-layered, sometimes sculptured, tough wall. When the spore is mature, the wall of the parental cell usually swells and disintegrates. On germination a new bacterium grows out of the spore.

Distribution and ecology. Some 1600 species of bacteria occur in enormous numbers over the whole world, in water, in the soil, even in the dust in the air and lying on everything. Their wide distribution is mainly a result of the following factors: their small size and consequent very large surface in relation to mass, so that very high levels of metabolic and physiological activity are possible (e.g. very rapid reproduction); the resistance of their vegetative cells, and in particular of their spores (p. 335), to unfavourable environmental factors; and the diversity of their modes of nutrition. Under optimal conditions some species (e.g. *Vibrio comma*) can divide several times in an hour, so that within twenty-four hours a single cell can produce many million descendants (cf. p. 203).

The spores of bacteria are, by virtue of the special physical and chemical properties of their protoplasm, very resistant to desiccation and extremes of temperature (p. 335); some can survive several hours (up to thirty at the maximum) in boiling water. Even the vegetative cells of many species are very resistant to desiccation. Some can even live at high temperatures, e.g. in hot springs, while some actively produce considerable heat by respiration (spontaneous heating of damp hay, dung, tobacco or cotton-wool; cf. pp. 263, 335).

The bacteria have a larger number of types of metabolism than the Eukaryota. Most bacteria are heterotrophic, either saprophytic or parasitic. Obligate parasitism is rare: most pathogenic species can be grown in the absence of animal or human bodies. Culture is suitable nutrient solutions (e.g. meat extract with peptone) generally presents no difficulty. On solid nutrient media bacteria often form slimy colonies (coenobia, p. 72) of various shapes; these are usually colourless, but sometimes are coloured by the secretion of pigments; pigments occur inside the cells only in the photosynthetic green and purple bacteria. Aerobic bacteria need atmospheric oxygen for growth; anaerobic ones can grow without it. The enzymes secreted by bacteria produce extensive decomposition of the substrate (decay by aerobes; putrefaction by anaerobes). Pathogenic bacteria generally produce poisonous toxins.

The special modes of nutrition of bacteria include: autotrophy, either by photosynthesis (red and green sulphur bacteria, pp. 245, 248) or by chemosynthesis (pp. 255, 256); heterotrophy in saprophytes, parasites or in symbiosis (pp. 291–4); aerobic and anaerobic metabolism (many fermentations, pp. 267–9 et seq.); denitrification and desulphurification (pp. 277–8); and fixation of molecular nitrogen (p. 285).

Luminescent bacteria produce a substance (luciferin) which on oxidation gives off light (p. 275). Some marine animals (e.g. pyrosomes, cephalopods, abyssal fishes) have luminescent bacteria in special organs.

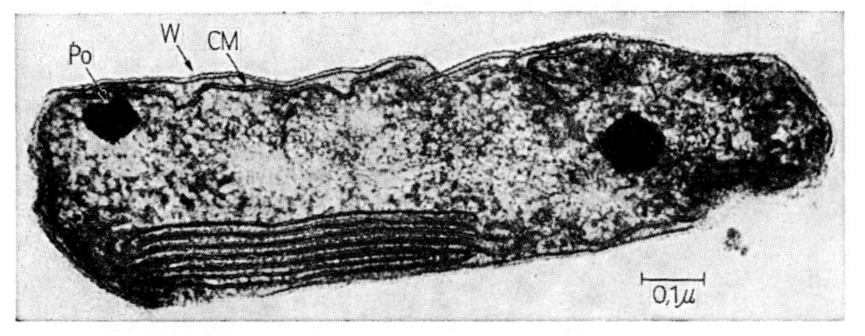

Fig. 376. *Rhodopseudomonas* with stack of thylacoids; CM, cytoplasmic membrane; Po, polyphosphate body; W, cell wall. After Drews and Giesbrecht.

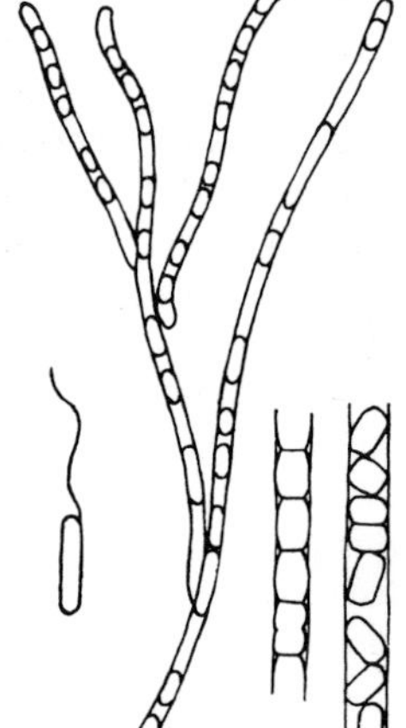

Fig. 377. *Sphaerotilus natans*. Motile stage, ×700; *Sphaerotilus* stage, ×330; start of cell division, ×800. After Pringsheim.

Bacteria are responsible for many processes of fermentation: lactic acid and butyric acid fermentations; cellulose, pectin and protein fermentations; and aerobic acetic acid fermentation (p. 269). Practically all natural products can be broken down by bacteria, even mineral oil, paraffin and asphalt. Only some synthetic resins and plastics are highly resistant to bacterial destruction.

Numerous bacteria produce diseases of plants as well as of animals and man. The plant pathogens either enter through stomata, hydathodes, etc. (particularly species of *Pseudomonas* and *Xanthomonas*), or they infect wounds (caused by frost, insect damage, etc., e.g. *Erwinia carotovora*). The presence or absence of flagella is not correlated with pathogenicity. It is remarkable that only rod-shaped and sporeless forms are plant pathogens; these generally live between the cells of the host (intercellular), where they dissolve the middle lamella (cf. p. 61) and the host tissue is converted into a pulpy rotten mass (soft rot); the death of the cells separated in this way is occasionally accelerated by toxins. Only relatively few bacteria enter plant cells (e.g. *Pseudomonas tabaci*). Sometimes they cause the wilting and death of the host by blocking the vessels; generally wilt toxins (p. 292) are also produced (*Corynebacterium michiganense*). More than 200 bacterial diseases of plants are known.

Systematics. The taxonomy of the bacteria is made extremely difficult because of their minute size and the paucity of morphological characteristics. In consequence physiological attributes are used, especially in the delimitation of species. We must content ourselves by briefly indicating the orders and naming a few representatives since at present there is little agreement on the delimitation and arrangement of the taxa.

Order 1: **Pseudomonadales**. Rod-shaped or spiral, rarely spherical bacteria, Gram-negative, monotrichous or lophotrichous, only exceptionally united in chains, and never spore-forming. They include the nitrifying (*Nitrobacter, Nitrosomonas*) and denitrifying bacteria (*Pseudomonas denitrificans*, amongst others), the vinegar bacteria (*Acetobacter*) and the luminescent bacteria (*Photobacterium*). Some species of *Pseudomonas* are saprophytic (*P. fluorescens* liquefies gelatine and forms a diffusible pigment, *P. syncyanea* colours milk blue) or parasitic on plants (producing rots, wilts and tumours). The genera *Vibrio* (comma-shaped, e.g. the cholera pathogen, *V. comma*, the sulphate-reducing species *Desulphovibrio desulphuricans*) and *Spirillum* (spiral, frequent in polluted waters) are generally included in this order.

Order 2: **Eubacteriales**. Rod-shaped or spherical, peritrichous bacteria. *Azotobacter* and *Rhizobium* fix nitrogen (p. 293). *Agrobacterium tumefaciens* produces crown galls on flowering plants, *Escherichia coli* lives in the gut of warm-blooded animals, *Chromobacterium prodigiosum* (*Serratia marcescens*) grows in colonies resembling drops of blood on flour, bread, etc., producing an insoluble red pigment. *Salmonella paratyphi* produces paratyphoid in man. The Coccaceae have spherical cells forming chains (*Streptococcus*), plates (*Micrococcus*), or packets (*Sarcina*, Fig. 373B). The rod-shaped, spore-forming species of *Bacillus* are Gram-positive; they include *B. anthracis*, the anthrax bacillus, which was the first bacterium shown to be the agent of a disease by Robert Koch in 1876; and the protein destroying hay bacillus, *B. subtilis*. In *Clostridium* the cells enlarge in spore-formation (*C. botulinum* on decaying meat, *C. tetani* produces tetanus).

Order 3: **Chlamydobacteriales**. The attached filaments live in water and are enclosed in a delicate sheath within which the cells may separate and emerge either singly or in groups, either in the non-motile state or as swarmers with flagella (Fig. 377). Subsequently, these free cells become attached and grow into new filaments. Very common is the falsely branched *Sphaerotilus natans* which forms soft white furry growths in polluted flowing water; a habitat-modification was previously known as *Cladothrix dichotoma*. The iron bacterium *Leptothrix ochracea* deposits ferric hydroxide in its sheaths and produces bog iron ore; *Crenothrix polyspora*, another iron bacterium (p. 277), liberates numerous small non-motile globular cells during reproduction; its vigorous growths can block water pipes.

Order 4: **Actinomycetales**. The rod-shaped non-motile cells show true branching under certain conditions of culture: this is the rule in *Actinomyces* (ray fungi), but occurs only in young cultures of *Mycobacterium. M. tuberculosis*, the tubercle bacillus, usually grows as unbranched, slender, non-motile rods. The actinomycetes, which are nearly as abundant in the soil as the true bacteria, are extraordinarily rich in species. In artificial media they generally grow as a mycelium several centimetres across, often consisting of a single, richly branched cell (diameter 0.5–1 μm) with numerous nucleoids but lacking cross walls, cellulose and chitin (Fig. 402). In some circumstances the filaments fragment into rods remarkably similar to true bacteria; in addition, they can produce, especially on hyphae projecting into the air, chains of spores of various kinds. *Actinomyces bovis* causes festering swellings in animals and man (actinomycosis); *Streptomyces scabies* stimulates the formation of visible wound cork in potatoes and beets (common scab disease); the nitrogen-fixing symbiont (pp. 285, 293) in the root nodules of the alder probably belongs to the actinomycetes. Many actinomycetes produce important antibiotics used in medicine against pathogenic bacteria.

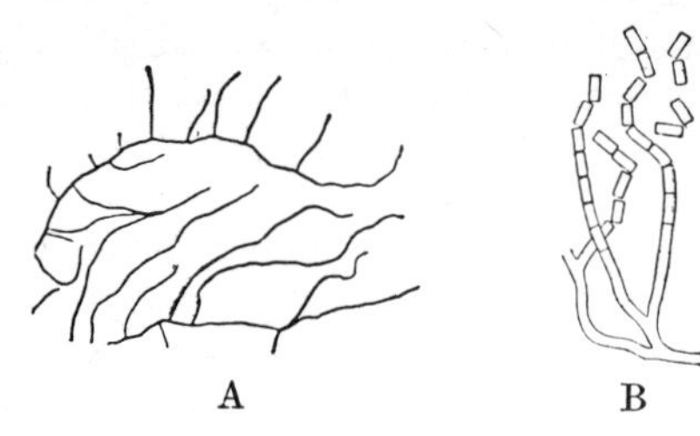

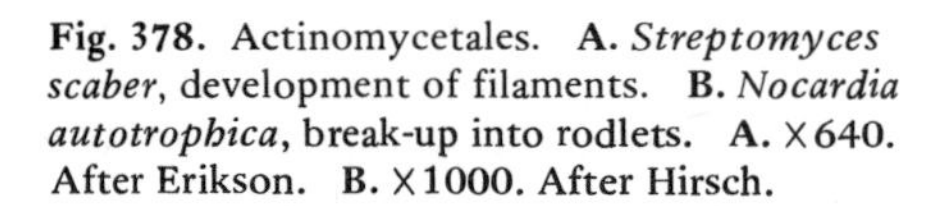
Fig. 378. Actinomycetales. **A.** *Streptomyces scaber*, development of filaments. **B.** *Nocardia autotrophica*, break-up into rodlets. **A.** ×640. After Erikson. **B.** ×1000. After Hirsch.

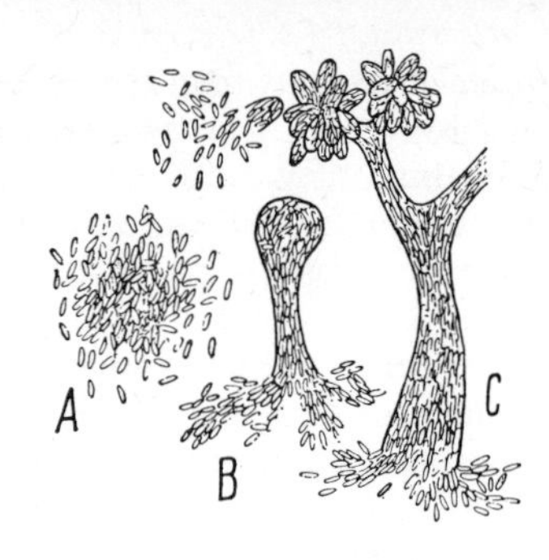

Fig. 379. *Chondromyces* (Myxobacterium); diagrammatic. **A.** Swarm of rodlets; ×200. **B.** Young. **C.** Branched fruit body; ×30. After Kühlwein.

Order 5: **Myxobacteriales.** The red or otherwise coloured cell aggregates live in the soil or on manure; they consist of a swarm (pseudoplasmodium) of small, flexible rods, without cell walls or flagella, which can creep about, apparently by the active contraction of the cells. In many of them the rods creep together to a spot where they form characteristic accumulations, cemented together by mucilage, of various colours and shapes (fructifications or cystophores; Fig. 379) within which rods are converted into resting cells (cysts); from these new swarmers are eventually formed (e.g. in the widespread *Myxococcus* spp.). In artificial cultures some myxobacteria can feed on other living organisms (e.g. on bacteria).

Order 6: **Spirochaetales.** The slender, often very long body differs from other bacteria in its fine structure; it lacks a rigid wall and is flexible, showing active, snake-like movements. *Spirochaete* (*Treponema*) *pallida* causes syphilis.

By way of an appendix we may mention the **rickettsias**; very small rod-, coccus-like or irregular intracellular parasites in man and animals.

Phylogeny. Some have supposed that the cocci are the most primitive, from which the flagellate forms and (perhaps on becoming terrestrial) the spore-forming bacteria may have been derived. Others suppose that the cocci are derived from actinomycetes by reduction. Actinomycetes and myxobacteria must be markedly advanced. Other relationships are extremely obscure. There may be connections with the most primitive Cyanophyceae, but affinities with other classes of plants are not discernible.

Bacteria and blue-green algae differ from all other plants in being Prokaryota. The lack of chloroplasts, mitochondria and a nuclear membrane, the peculiar structure of the flagellum, and the basic cell wall materials (murein) set the bacteria apart from the Eukaryota.

The chemical apparatus of the cell differs little from that of higher plants. The Prokaryota have anabolic and catabolic enzymes, electron transport and phosphorylation systems in common with the Eukaryota. Nevertheless we find amongst the Prokaryota a diversity of metabolic types of which in the course of evolution all but the heterotrophic and phototrophic aerobionts have been lost. The Prokaryota are relicts of the dawn of organic evolution, and many of them have also remained at an early stage of biochemical evolution. They probably held sway in the primeval swamps when there was no oxygen in the air; today they only populate extreme environments.

Fossil bacteria (in iron ores, oil shales, limestones and in fossil plants) occur as far back as the Palaeozoic and indeed even in the Precambrian (principally filamentous bacteria).

Uses. Bacteria are employed to a considerable extent in industry in the preparation of foodstuffs, solvents, raw materials and pharmaceuticals (lactic acid, acetic acid, glutamic acid, enzymes, vitamins, antibiotics).

Appendix: Bacteriophages. Despite their minute size bacteria can contract lethal diseases caused by bacteriophages (p. 299), very highly organized, relatively large viruses (length 0.02–0.1 μm) consisting essentially of a 'head' containing DNA and a 'tail', both enclosed in a common protein membrane (Fig. 380A). As a result of specific chemical attraction, the tip of the tail attaches itself to the surface of the bacterial cell (Fig. 380B) whereupon the contents of the head are injected into the bacterium; the delicate, empty protein membrane remains outside. After a short time (much less than an hour) the first indications of new phage particles become visible, and after a similar interval a batch of a few to several dozens of phage particles are liberated, and the bacterium has completely disintegrated. The phage particles do not increase by division: instead they develop *de novo* from the bacterial protoplast; the phage diverts the normal metabolic processes of the host so that instead of the usual cell constituents of the bacterium the specific components of the phage are synthesized, just as happens with the viruses. Phages can also alter the biochemical properties of their host by mutation; they can cross and

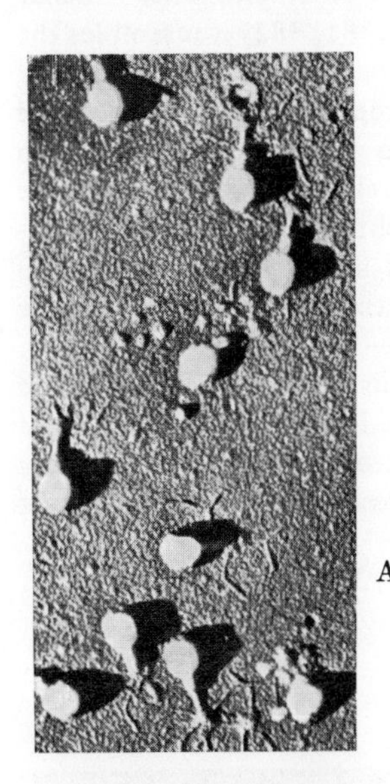

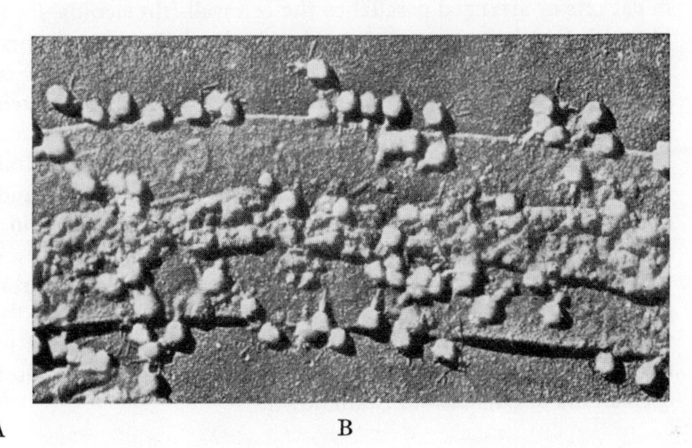

A B

Fig. 380. Bacteriophage. **A.** Separate T_2-phages; ×40,000. **B.** Bacterium attacked by T_2-phages; ×26,000. After Hellenberger and Arber.

recombine. This had lead them to be regarded as precursors of life, but they exhibit no metabolism or energy exchange (e.g. they do not respire), so that they are rather to be regarded as independent constituents of the genome (p. 299) which can propagate themselves in foreign cytoplasm, and which have the capacity to persist outside the cell in a completely inactive state until they are again incorporated into the metabolism of a host. This hypothesis is supported *inter alia* by the observation that not all bacteriophages are lethal to bacteria: the tempered or modified phages can exist harmlessly for a long time in the bacterial cell. Their genetic material must therefore by very similar to that of the bacteria.

Class II: Cyanophyceae, blue-green algae

The Cyanophyceae (also known as Schizophyceae or Myxophyceae) are predominantly blue-green, often very small, unicellular or filamentous algae (Fig. 384) with primitive organization. They have no true nucleus (only a nuclear-equivalent) and no chromatophores demarcated from the cytoplasm by cell-membranes (cf. p. 46).

The Cyanophyceae are autotrophic, but can also take up organic nutrients.

In the central, honeycomb-like, colourless or nearly colourless part of the cell (centroplasm) are granular, reticulate, filamentous or rod-shaped bodies containing DNA; together they make up the chromatin-apparatus which functions as a nucleus. A delimited nucleus with membrane, nucleolus, chromosomes and spindle is absent. The whole complex divides transversely by constriction (Fig. 68B); it has been claimed that the chromatin strands arranged lengthwise in the cell divide longitudinally and move apart. The centroplasm passes gradually into the peripheral pigmented chromatoplasm which forms a hollow sphere (or other shape corresponding to the shape of the cell). The chromatoplasm has no true vacuoles (although gas-vacuoles occur in some species), and there is no cytoplasmic streaming; it contains diffusely distributed RNA and the photosynthetic pigments chlorophyll *a*, carotenoids (particularly β-carotene) and two water-soluble chromoproteins (phycobilins) in which the prosthetic groups are related to bile pigments: the blue phycocyanin and (only in some species) the red phycoerythrin. To some extent the proportions of these pigments are labile, so that the Cyanophyceae sometimes look more blue, sometimes more red; this is particularly marked in those (rare) species which exhibit chromatic adaptation, forming mainly red pigments in green light and mainly blue and green pigments in red light, so that the same species develops a colour complementary to the light it receives. In so far as the chromatoplasm has lamellae in packets or arranged parallel to the cell wall (thylacoids, Figs. 381, 382) it resembles the plastids of higher plants, but unlike the latter has no membrane separating the groups of lamellae from the rest of the cytoplasm, and hence no true plastids. Glycogen-like cyanophyte starch is deposited as small granules (cyanophycin granules) and is the main photosynthetic product; it is probably similar to floridean starch. The significance of the volutin granules (polyphosphate) in the centroplasm is still doubtful.

The multistratose cell wall, of pectin, hemicellulose and sometimes cellulose, frequently swells to become gelatinous or mucilaginous; as in the bacterial cell wall the innermost layer forms a supporting membrane.

Most Cyanophyceae are non-motile, but some filamentous ones show creeping or gliding movements on a moist substratum (up to 4 μm sec^{-1}). Flagella never occur, and the movement is due to the swelling of mucilage extruded into the water through very fine pores in the cell wall (10 nm diameter, Fig. 383). The mechanism of the regular oscillations of *Oscillatoria* is unknown.

Fig. 381. *Oscillatoria*, stacked thylacoids. ×11,000. After Jost.

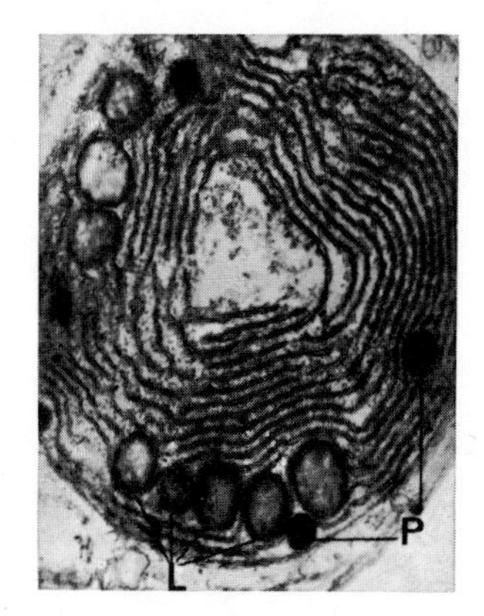

Fig. 382. *Cyanota*, concentric thylacoids. L, lipid bodies; P, phosphate bodies. ×25,000. After Hall and Claus.

Fig. 383. *Cylindrospermum*, band of pores. ×26,000. After Drawert.

Reproduction is exclusively vegetative by rapid cell division; new cross walls grow inwards like a closing iris diaphragm (Fig. 384M). Sexual reproduction and flagellate stages are unknown.

The rows of cells of some filamentous species separate into few-celled pieces (hormogonia) which creep forward and produce new plants (Fig. 384K). In a few unicellular forms the cell contents divide successively, after the enlargement of the parent cell, to form a large number of globular endospores; these are naked at first, but after leaving the parent cell they grow into new individuals complete with cell walls; in some species with elongated cells the basal part remains sterile while the apex repeatedly regenerates to produce endospores (Fig. 384D). Exospores also occur. Akinetes (cysts), which permit survival through unfavourable periods, are formed (especially in the Hormogonales) by laying down reserve foods, by enlargement and by marked thickening of the wall of single cells (Fig. 384H); they germinate to produce hormogonia. Further, a complete short side branch can become a resting organ (hormocyst) by the development of a thick enclosing sheath. There are thus many methods of asexual reproduction and of the formation of resting organs.

Distribution and ecology. Some 2000 species of Cyanophyceae are found throughout the world as gelatinous masses or films of slender filaments in water, especially fresh water (even at 75°C in hot springs), on or in both moist and arid soils, on tree-trunks and bare rocks, even in the Arctic and Antarctic. The Cyanophyceae of 'ink-streaks' on vertical calcareous rock-faces are exposed to great variations of temperature and water-supply; they grow partly on the surface (epilithic), partly in capillary crevices (endolithic). Some dissolve limestone; others secrete calcium carbonate in their mucilaginous sheaths (Fig. 384G), giving rise in fresh water to lake chalk and calcareous tufa or in the tidal region of the warmer seas to the deposition of stratified calcareous crusts (stromatoliths). Some species inhabit the surface of ponds, others the open seas where they cause water-blooms; some of these species, e.g. *Microcystis aeruginosa* and *Anabaena flos-aquae* are poisonous to fish. Many genera participate in the formation of lichens, which consist of algae and fungi (p. 537). Some live endophytically in cavities inside other plants, e.g. *Anabaena* in the leaves of *Azolla* (Fig. 610D), *Nostoc* in the thallus of some liverworts (*Blasia*, *Anthoceros*, Fig. 534B), in the roots of cycads and the rhizome of *Gunnera*. Some with certain abnormalities of structure occur endosymbiotically as so-called cyanellae within the protoplast of certain colourless flagellates and Chlorococcales where they function as 'plastids'. Several genera (*Nostoc*, *Anabaena*, etc.) have species which fix atmospheric nitrogen: as much as 50 kg N ha^{-1} $year^{-1}$ in rice fields.

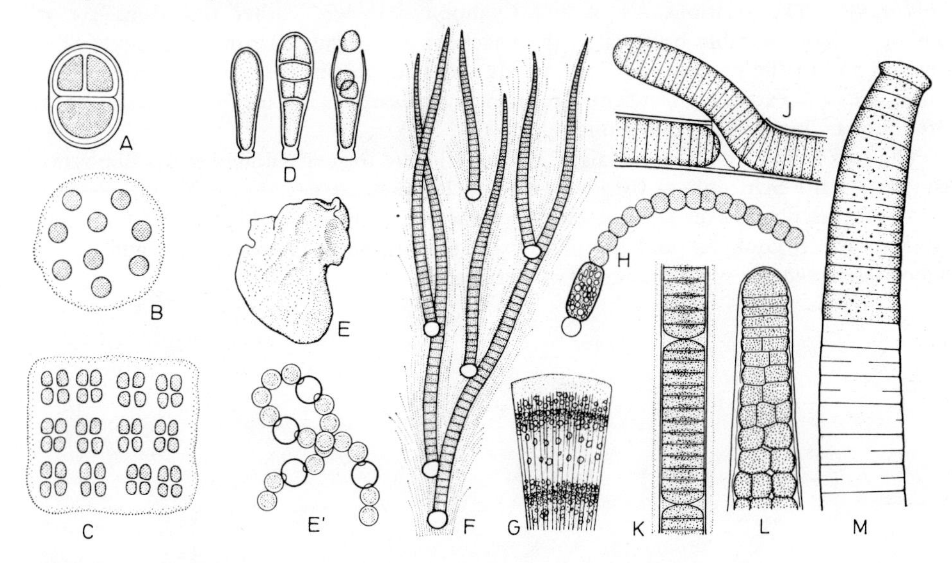

Fig. 384. Cyanophyceae. **A.** *Chroococcus turgidus.* ×400. **B.** *Aphanocapsa pulchra.* ×500. **C.** *Merismopedia punctata.* ×600. **D.** *Dermocarpa clavata*, formation of endospores. ×450. **E.** *Nostoc commune*, colony. ×1. **E'.** The same, chain of cells with four heterocysts. ×400. **F.** *Rivularia polyotis*, part of a colony. ×200. **G.** *Rivularia haematites*, part of a colony in section with deposition of lime in annual layers. ×15. **H.** *Cylindrospermum stagnale* with a cylindrical akinete and a spherical, terminal heterocyst. ×500. **J.** *Plectonema wollei* showing false branching. ×200. **K.** *Lyngbya aestuarii*, formation of hormogonia. ×500. **L.** *Stigonema mamillosum*, apex of filament. ×250. **M.** *Oscillatoria princeps*, apex of filament. ×300. **A.**, **D.**, **H.** and **L.** After Geitler. **C.** After Smith. **E'.** and **F.** After Thuret. **G.** After Brehm. **J.** and **K.** After Kirchner. **M.** After Gomont.

Systematics. The most primitive Cyanophyceae (**Chroococcales**) have roundish blue-green cells, e.g. *Chroococcus* and *Gloeocapsa* which occur chiefly as gelatinous films on rocks and walls; the cells cohere after division as multicellular colonies (coenobia, p. 72) in stratified mucilaginous envelopes (Fig. 384A). The tabulate colonies of *Merismopedia* (Fig. 384C) float in fresh water.

In the **Hormogonales** the cells are in filaments, often enclosed in mucilage-sheaths; growth is intercalary. The generally unbranched filaments lie close to their substrate; they rarely show true branching, more frequently false branching (Fig. 384C). In the Oscillatoriaceae, which are abundant everywhere in water and on mud, the filament consists of similar disc-shaped cells (Fig. 384M). In *Nostoc*, on the other hand, the necklace-like threads embedded in mucilaginous cellulose (Fig. 384E) form spherical or irregular gelatinous masses (Fig. 384E) in water or on damp soil; at regular intervals there are, as in many other Hormogonales, heterocysts, yellowish, empty-looking single cells with thick cellulose walls and devoid of photosynthetic pigments; these appear to be the seat of particularly intense metabolic activity (nitrogen-fixation). In *Rivularia* (Fig. 384F, G) the base and apex of the multicellular filament differ markedly in structure; at the base is a heterocyst, while the filament tapers above into a colourless hair so that the filament is constructed on a definite ground plan. *Stigonema* (Fig. 384L) also shows differentiation into base and apex: an apical cell cuts off towards the base segments which further divide length- and cross-wise; the multiseriate filaments also develop true branches.

Some genera have lost their pigments; amongst these should probably be included *Beggiatoa*, which was formerly classified with the sulphur bacteria, to which many botanists still relate it. The delicate *Oscillatoria*-like filaments contain sulphur and form whitish growths in water containing H_2S; they live autotrophically by chemosynthesis (cf. p. 256). The colourless *Leucothrix* found on moribund marine algae probably also belongs to the Cyanophyceae.

Since both the colour and form of the Cyanophyceae are greatly modified by the environment, the delimitation of species and even genera is often problematic.

Phylogeny. The relationships of the Cyanophyceae with other organisms are still obscure. They resemble bacteria in their lack of a true nucleus, in their mode of cell division and in the possession of thylacoids and a supporting membrane in the cell wall. They are an extremely ancient group with calcareous forms going back to the Precambrian, more than 1000 million years ago.

The two following divisions (algae and fungi) were formerly united as **Thallophyta.** In contrast to the Schizophyta they have true nuclei: they are eucaryotic. Nevertheless they lack both vascular tissues and the characteristic ♀ sex organs (archegonia, cf. Figs. 530J, 604) of the bryophytes and pteridophytes: the organs producing spores and gametes almost always lack a special covering of sterile cells.

Division 2: **Phycophyta, algae**

The Phycophyta or algae are uni- or multicellular, variously pigmented, autotrophic water plants. Their gamete- and spore-bearing organs are generally unicellular and lack a jacket of sterile cells. The zygote never develops into a multicellular embryo inside the female sex organ. In the lower algal groups the reproductive cells (gametes, spores) are flagellate, but in the higher groups generally only the male gamete has flagella. A few groups (certain diatoms, the Conjugales and the Rhodophyceae) have no flagellate stages at all.

The flagella of all plants (except bacteria) and animals (including the mammalian spermatozoid) have two central, thinner fibrils and nine outer, thicker ones (Fig. 385). The fibrils extend without twisting right up to the apex; they are enclosed in a sheath. The flagella may point forwards (anterior flagella), rarely backwards (posterior). They commonly occur in pairs, either two equal ones, or a longer and a shorter. They may be smooth, and then often tapering like a whip-lash, or they may have flimmer hairs which arise from the sheath ('tinsel-type' flagella) (Figs. 456, 457; cf. p. 52).

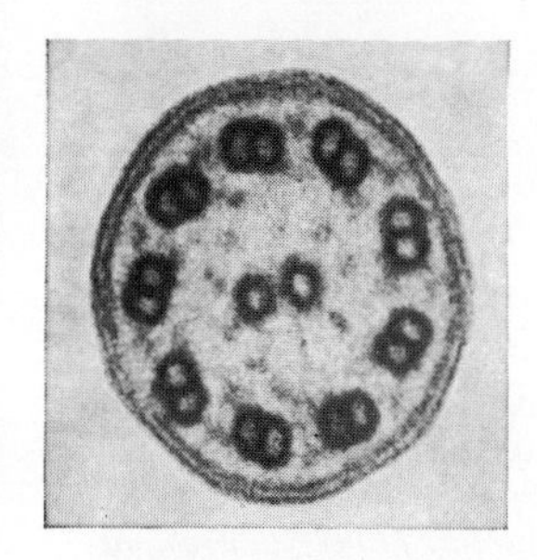

Fig. 385. Transverse section of a flagellum of *Pseudotrichonympha.* ×65,000. After Gibbons and Grimstone.

The algae are divided into several classes on the basis of their plastid pigments (see the following Table); the classes are divided into orders according to the level of organization. While the lower orders ('Flagellatae' or 'Monadales') are always motile and flagellate,

Some chemical characteristics of the algal classes (Compiled by H. Metzner)

	Chlorophylls					Phycobilins	Carotenes			Xanthophylls			Principal reserves (other than aliphatic oil)
	a	*b*	*c*	*d*	*e*		α	β	Others	Lutein	Fucoxanthin	Others	
Cyanophyceae	+	−	−	−	−	+	−	+	+	(+)	−	+	Cyanophycean starch
Euglenophyceae	+	+	−	−	−	−	(+)	+	−	+	−	+	Paramylon
Pyrrophyceae	+	−	+	−	−	+[1]	+[1]	+	−	−	−	+	Starch, polyglucans
Chrysophyceae	+	−	+	?	?	−	?	+	?	?	+	+	Chrysolaminarin (leucosin)
Xanthophyceae	+	−	+	−	+	−	−	+	−	−	−	+	Chrysolaminarin (leucosin)
Chlorophyceae	+	+	−	−	−	−	+	+	+	+	−	+	Starch
Phaeophyceae	+	−	+	−	−	−	−	+	−	+	+	+	Laminarin, mannitol
Rhodophyceae	+	−	−	+	−	+	+	+	−	+	−	+	Floridean starch

[1] In the Cryptophycales

others exist as non-motile unicells or colonies ('Coccales', Fig. 412). In the 'Trichales' the cells are joined in simple or branching filaments (Fig. 416). The most complex forms have true tissues (Fig. 436). Some groups have tubular, multinucleate cells ('Siphonales', Fig. 407).

The various textbooks and monographs of systematics differ greatly in their arrangements of the algal classes. The sequence here is determined by the highest levels of organization reached in each class: thus Euglenophyceae includes only unicells and colonies, while the Pyrrophyceae and Chrysophyceae include filaments, and the Chlorophyceae has both complex filamentous and siphonaceous forms. Finally, both the Phaeophyceae and Rhodophyceae include advanced types of tissue.

Class I: Euglenophyceae

The most important genus, *Euglena*, is unicellular, motile and flagellate, and either naked or with a characteristic periplast (pellicle). The green chromatophores contain pigments similar to those in the green algae (chlorophyll *a* and *b*, β-carotene, traces of α-carotene), but include a xanthophyll not known elsewhere in the Plant Kingdom. Reserves include aliphatic oil and a starch-like carbohydrate, paramylon, which does not stain blue with iodine; it occurs as grains or discs in the cytoplasm. *Euglena* reproduces by longitudinal division. When conditions are unfavourable *Euglena* loses its flagella and lays down a thick mucilaginous sheath; these cysts can survive prolonged desiccation.

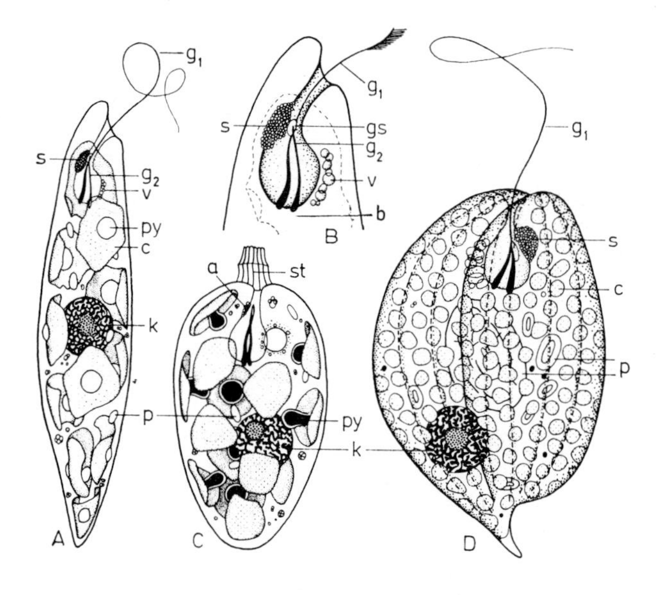

Fig. 386. Euglenophyceae. **A.** *Euglena gracilis.* ×600. **B.** Anterior end of the same. ×10,000. **C.** *Colacium mucronatum.* ×500. **D.** *Phacus triqueter.* ×600. a, eye-spot; b, basal body; c, chloroplast; g_1, motile flagellum; g_2, second flagellum; gs, flagellar swelling; k, nucleus; p, free paramylon; py, pyrenoid with paramylon-sheath; s (= a), eye-spot; st, mucilage-stalk; v, contractile vacuole. After Leedale.

The periplast of *Euglena* allows it to change its shape (metaboly) but that of *Phacus* is rigid; in both it shows spiral sculpturing. Near the front of the cell (Fig. 386A, B) is the flagellar sac (reservoir) which opens outwards through the gullet (canal). Each of the two flagella arises from a basal body at the base of the sac. One flagellum is long; the shorter one terminates in the gullet where it is fused to the longer one, at which point there is a light-sensitive organ, the photoreceptor. At the side of the gullet is an 'eye-spot', coloured red with carotene (Fig. 386B, cf. p. 384). The longer anterior flagellum has flimmer-hairs and describes during its motion the surface of a cone. As the cell revolves around its long axis it moves forward by two to three times the length of its body per second.

Species of *Euglena* (Fig. 386A) occur mainly in stagnant, nutrient-rich water (making village ponds and other polluted waters green). In contrast, *Phacus* (Fig. 386B) prefers nutrient-poor waters. Some forms are colourless and saprophytic. *Colacium* (Fig. 386C) is attached by a mucilage-stalk to free-swimming micro-organisms; it only becomes flagellate and motile during reproduction.

Class II: Pyrrophyceae (Dinoflagellatae)

The Pyrrophyceae are almost always unicellular, have two long flagella, and bear yellow-brown or reddish chromatophores which contain chlorophyll *a* (and a little *c*), β-carotene and various xanthophylls. Oil, starch and starch-like polyglucans are the products of assimilation. The chromosomes are often visible even in the interphase nucleus.

In the Dinophycales (Peridiniales), the most important order, the cell wall generally consists of polygonal, porose plates of cellulose. Fine cytoplasmic threads project through the pores and may engulf and consume micro-organisms (Fig. 3). Forward propulsion is effected by a posterior flagellum of the whip-lash type, rotation about the long axis by a flagellum with flimmer-hairs in an equatorial furrow. A cell of *Peridinium* moves forward spirally about four times its length per second, revolving on its axis meanwhile.

Vegetative reproduction takes place by longitudinal division while still motile. In armoured forms the cell wall breaks into halves along a line oblique to the equatorial furrow, and two new half-walls are laid down (Fig. 388D). After several such divisions two nearly naked flagellate cells develop inside the armoured cell wall, from which they are liberated and develop new armour. Under unfavourable conditions a thick wall is laid down inside the armour to form a cyst (Fig. 388E). In *Glenodinium* sexual reproduction takes place between isogametes; in *Ceratium* anisogamy is reported. Meiosis takes place on germination of the zygote.

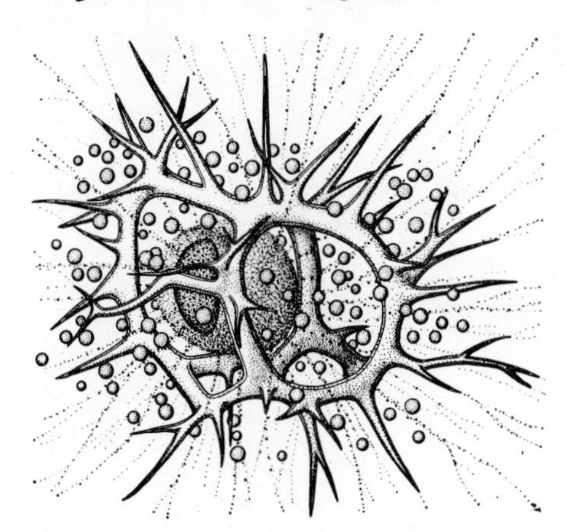

Fig. 387. Radiolarian (*Eucornis challengeri*) with zooxanthellae. ×260. After Haeckel.

Distribution and ecology. Only a few Pyrrophyceae live in fresh water; most are marine, and along with the diatoms and coccolithophores form the bulk of marine phytoplankton. Many have remarkable flotation-processes (Fig. 388C, F–J). *Noctiluca miliaris* and various species of *Ceratium, Gonyaulax* and *Peridinium* produce luminescent sea water. Some Peridiniales secrete toxins and when numerous poison fish. Several are parasites of marine animals.

Systematics. The **Cryptophycales**, which doubtfully belong here, have a very thin pellicle, contain phycobilins and bear two flagella. They include the 'zooxanthellae' (Fig. 387) which live symbiotically in marine animals (radiolaria, sponges, molluscs). The **Desmocontales** have a longitudinal suture in the cell wall; the longitudinal and transverse furrows are often bordered with widely projecting wings (Fig. 388C). In the **Dinophycales** (Peridiniales) the cell wall is either a thin cellulose membrane (Gymnodiniaceae, Fig. 388B) or is armoured (Peridiniaceae, Fig. 388A); the two flagella arise at the intersection of the longitudinal and equatorial furrows. In a few genera several cells cohere to form a filamentous colony. Numerous fossil forms are known from the Permian onwards.

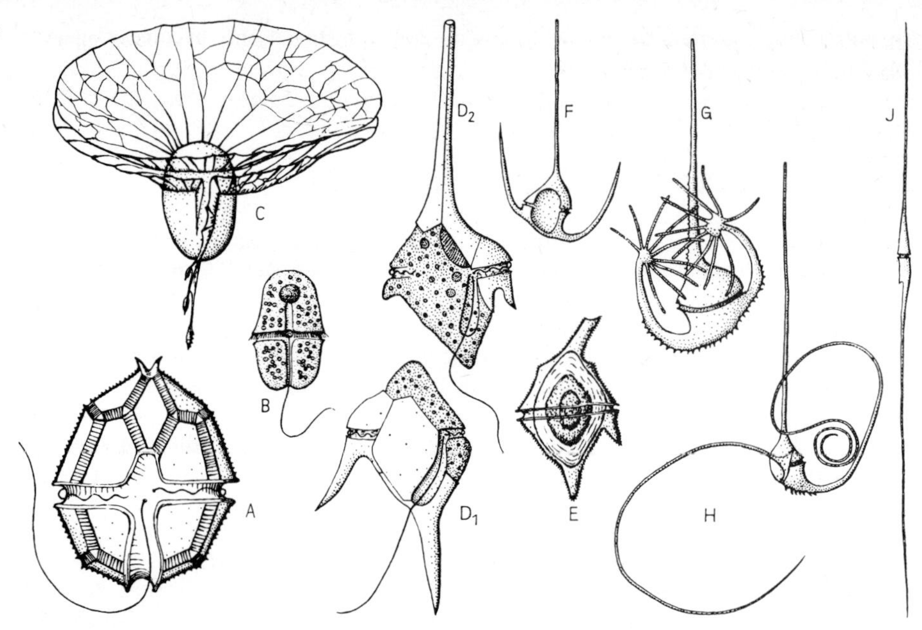

Fig. 388. Dinophycales (Peridiniales). **A.** *Peridinium tabulatum.* ×600. **B.** *Gymnodinium aeruginosum.* ×300. **C.** *Ornithocercus splendidus.* ×125. $\mathbf{D_1}$. and $\mathbf{D_2}$. *Ceratium hirundella* after division. ×350. **E.** *Ceratium cornutum*, cyst. ×150. **F.** *Ceratium tripos.* ×125. **G.** *C. palmatum.* ×125. **H.** *C. reticulatum.* ×65. **J.** *C. fuscus.* ×50. **A.** and **E.** After Schilling. **B.** After Stein. **C.** and **J.** After Schütt. **D.** After Lauterborn. **F., G.** and **H.** After Karsten.

Class III: Chrysophyceae

The Chrysophyceae are predominantly unicellular and golden-brown to brown, with chlorophylls *a* and *c* (no *b*), β-carotene and various xanthophylls (e.g. lutein, fucoxanthin). Reserves are chrysolaminarin (in the vacuole) and oil (in the cytoplasm, occasionally in a vacuole).

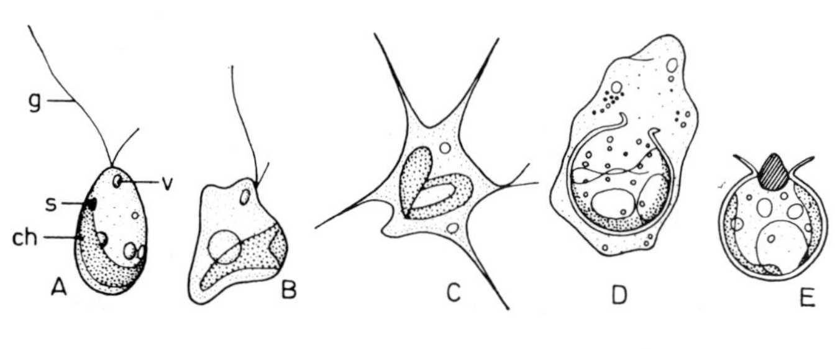

Fig. 389. *Ochromonas.* **A.** to **C.** Transition from the normal form with two flagella to the amoeboid state with pseudopodia. **D.** Cyst formation in an amoeboid protoplast. **E.** Cyst with plug (shaded) and pores, eye-spot; ch, yellow-brown chromatophore; v, vacuole; g, flagellum. ×1000. After Pascher.

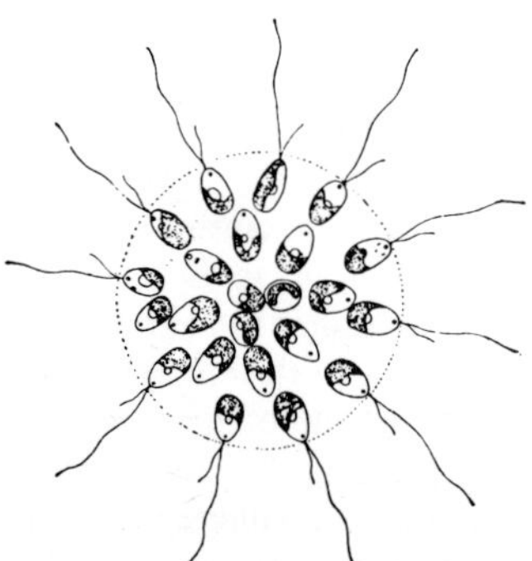

Fig. 390. *Uroglena americana.* ×400. After Pascher.

Order 1: **Chrysomonadales.** These golden-brown unicells usually have two unequal flagella, the longer with flimmer-hairs. Some form cysts, usually with silicified walls and a plugged aperture (Fig. 389E). Many chrysomonads have at the point of attachment of the two flagella a third thread or haptonema; although resembling a flagellum it serves for attachment and not for locomotion. The haptonema may be longer than the flagella or reduced to a short stump (Fig. 393).

The simplest forms like *Chromulina* are naked, amoeboid and can ingest solid food (cf. Fig. 389). *Uroglena* (Fig. 390) and *Synura*, both frequent in fresh water plankton, have numerous cells radially arranged in a spherical colony; in *Synura* the cells are covered with delicate silicified scales (Fig. 391). In *Dinobryon*, common in both fresh and sea water, each elongated cell produces by rotatory movements a characteristic 'cornet' shaped envelope (lorica). After division the daughter cells attach themselves to the margin of the parental lorica, and each then forms a new one so that a branching colony develops. Copulation occurs when two cells, each with its lorica, swim together, fuse, and produce a resting zygote. *Prymnesium* becomes attached by its haptonema to the gills of fish, to which it is lethally toxic when abundant.

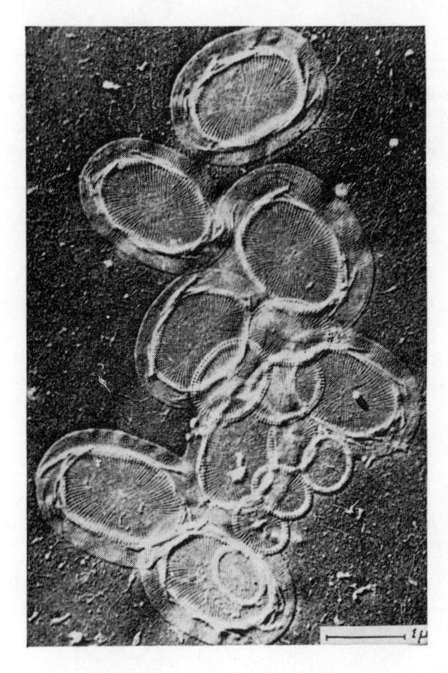

(Left)
Fig. 391. Scales of a chrysomonad. ×10,000. After Parke, Manton and Clarke.

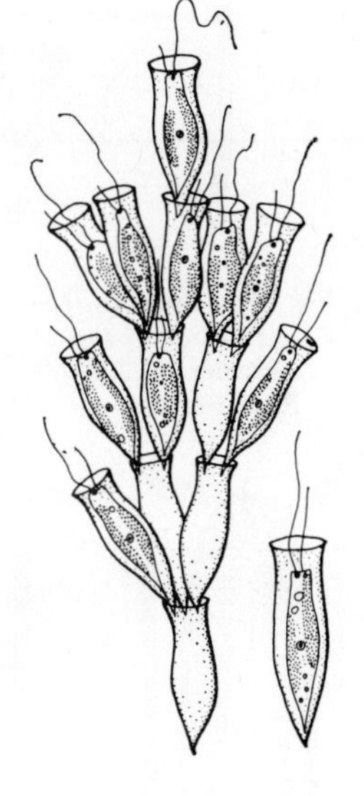

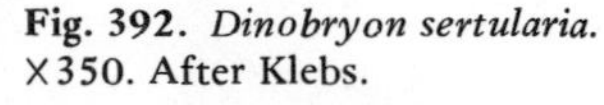

Fig. 392. *Dinobryon sertularia.* ×350. After Klebs.

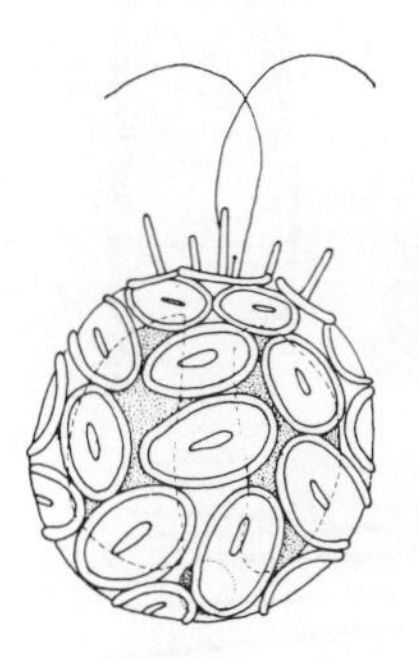

Fig. 393. *Syracosphaera pulchra*. Reduced haptonema between the flagella. ×1500. After Lohmann and von Stosch.

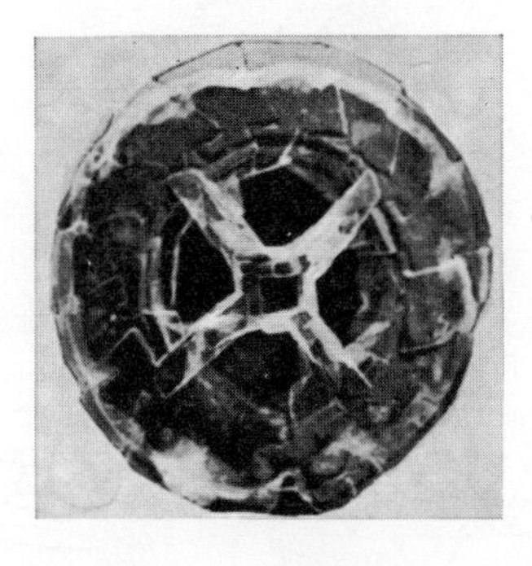

Fig. 394. Fossil coccolith consisting of rhombohedral calcite (*Deflandrius* sp.). Lower Cretaceous. ×7000. After Black.

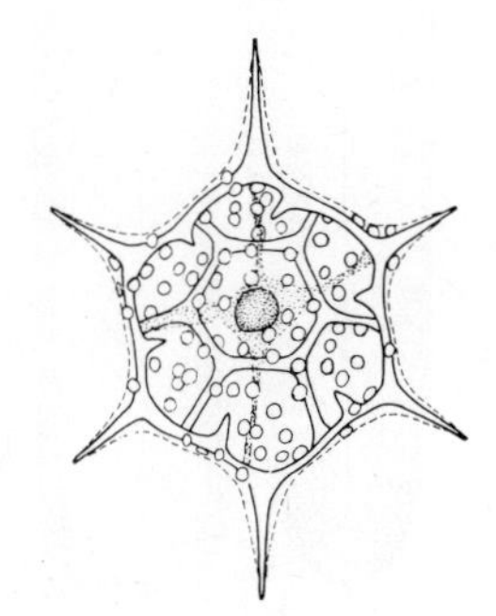

Fig. 395. *Distephanus speculum*. Chromatophores chiefly in the ectoplasm outside the siliceous skeleton. Flagella omitted. ×1000. After Gemeinhardt.

The naked Silicoflagellineae are exclusively marine; the protoplast contains an elegant siliceous skeleton (Fig. 395). Also marine are the minute Coccolithineae (coccolithophores, Fig. 393) which produce internally very diverse scales, platelets or rodlets (coccoliths, Fig. 394); these are moved to the surface where they form an armour. Fossil coccoliths are known from the Jurassic onwards, and play an important part in the formation of calcareous sediments, e.g. the Chalk (up to 8×10^8 cm^{-3} of rock).

Order 2: **Chrysocapsales.** Here the non-motile vegetative cells form gelatinous colonies, e.g. *Hydrurus* which occurs as mossy growths on stones in mountain streams (Fig. 396).

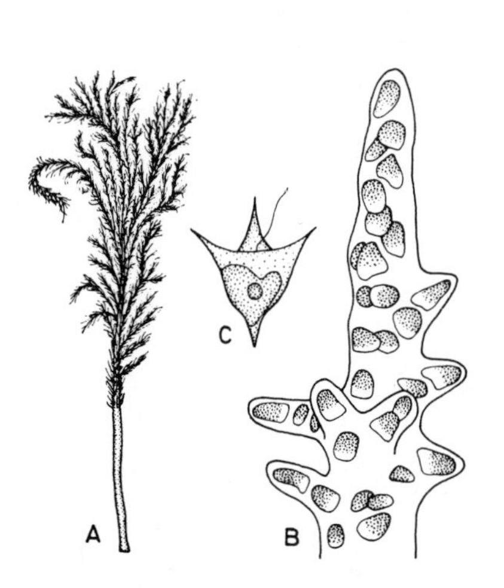

Fig. 396. *Hydrurus foetidus.* **A.** Young colony. ×1. **B.** Apex of a gelatinous branch. ×450. **C.** Swarmer. ×1200. **A.** After Rostafinsky. **B.** After Berthold. **C.** After Klebs.

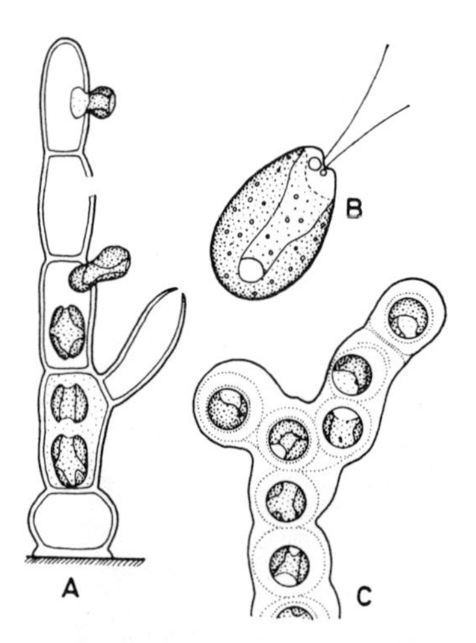

Fig. 397. *Phaeothamnion borzianum.* **A.** Thallus with developing swarmers. ×400. **B.** Swarmer. ×750. **C.** Palmella stage. ×400. After Pascher.

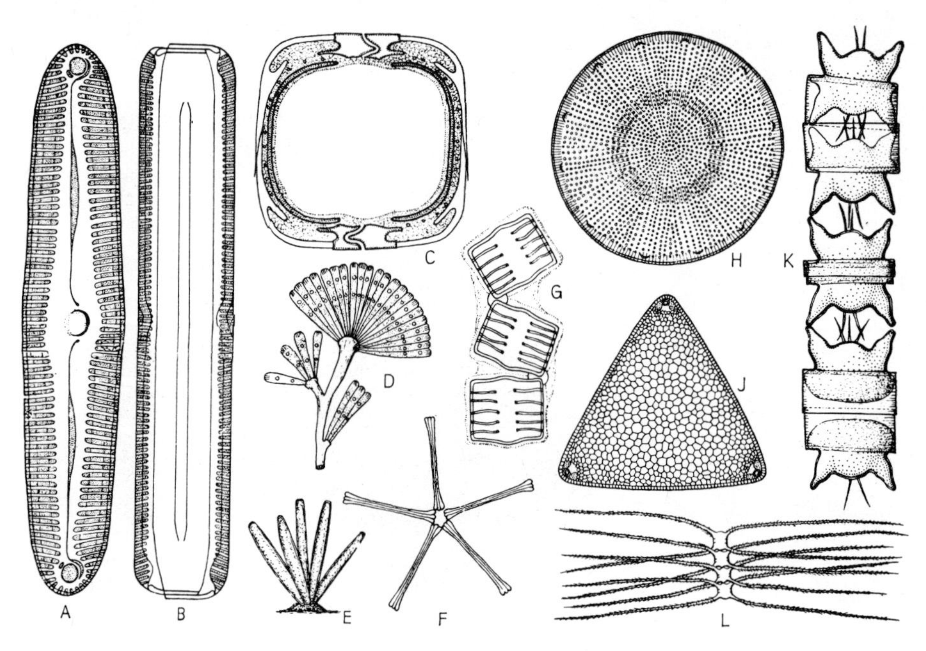

Fig. 398. Diatomales. **A.** to **C.** *Pinnularia viridus.* **A.** Valve-view, with raphe. ×600. **B.** Girdle-view. ×600. **C.** Transverse section. ×1200. **D.** *Licmophora flabellata.* ×200. **E.** *Synedra gracilis.* ×200. **F.** *Asterionella formosa.* ×200. **G.** *Tabellaria flocculosa.* ×400. **H.** *Coscinodiscus pantocseki.* ×200. **J.** *Triceratium distinctum.* ×200. **K.** *Biddulphia aurita.* ×400. **L.** *Chaetoceras castracanei.* ×250. **A.** and **B.** After Pfitzer. **C.** After Lauterborn. **D.**, **E.** and **K.** After Smith. **F.** After van Heurck. **G.** After Schröder. **H.** After Pantocsek. **L.** After Karsten.

Order 3: **Chrysotrichales.** Here the cells are joined in simple or branched filaments as in the fresh water *Phaeothamnion* (Fig. 397). Sometimes the swarmers lose their flagella, surround themselves with a thick wall, but continue to divide. Such a 'palmella' stage (Fig. 397C) probably enhances survival in unfavourable environments, and occurs in other algae (e.g. Chlamydomonaceae, Ulotrichaceae). *Stichochrys immobilis* has totally lost the capacity to produce flagellate swarmers.

Order 4: **Diatomales** (Bacillariales). The diatoms are a group of unicells of great diversity of shape; they also exist as linear or fan-shaped colonies. They have the characteristic pigments and reserves of the Chrysophyceae. They are notable for the possession of two silicified walls (frustules) inside the external cell membrane: one of these frustules (epitheca) overlaps the other (hypotheca) like the lid of a box or tin (Fig. 398B, C). The cell looks very different according to whether it is viewed from above, or below—the valve-view (Fig. 398A)—or from the sides—the girdle-view (Fig. 398B).

Between the valve and girdle intercalary bands sometimes occur (Fig. 398G); in some species these bear septa projecting into the cell. The siliceous wall, especially in valve-view, is seen to bear extremely fine and complicated structures, often arranged in rows. These may consist of minute chambers, either open or closed above or below; if closed they are perforated by minute pores (Fig. 399).

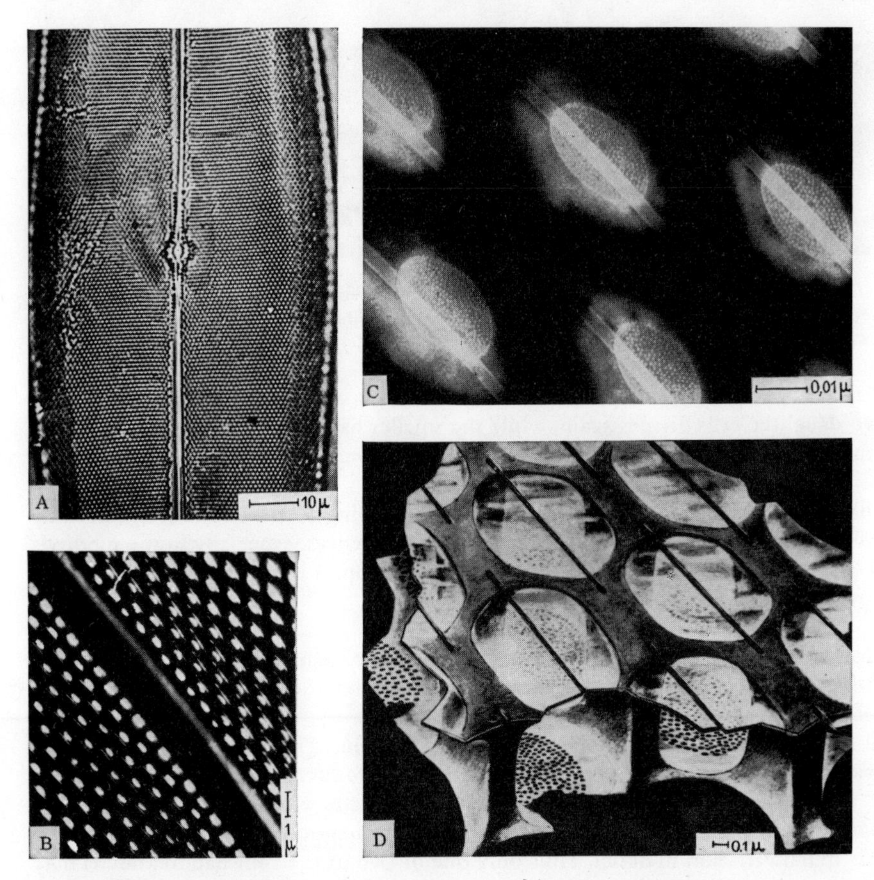

Fig. 399. ***Pleurosigma angulatum*****.** Frustule structure. **A.** General view of the central part of the valve with raphe. **B.** Raphe and pores. **C.** Pores. **D.** Reconstruction of the valve structure from electron microscope pictures. After Helmcke and Krieger.

The diatom cell contains one diploid nucleus and either one or two large brownish chromatophores or numerous small ones; the pyrenoids, however, form no starch. Products of assimilation are deposited outside the chromatophores: chrysolaminarin in the vacuolar sap, and the copious oil formed by the pyrenoids in special vacuoles. In addition the cell contains mitochondria and dictyosomes (Fig. 400).

Diatoms reproduce by fission. The two frustules are pushed apart at the girdle by the enlarging protoplasts, and each daughter cell forms a new hypotheca (Fig. 400). Hence one daughter cell is the same size as the parental cell but the other is smaller. This leads to a progressive diminution in size after each division until a minimal size is reached; thereafter sexual reproduction is associated with a marked increase in size of the zygote.

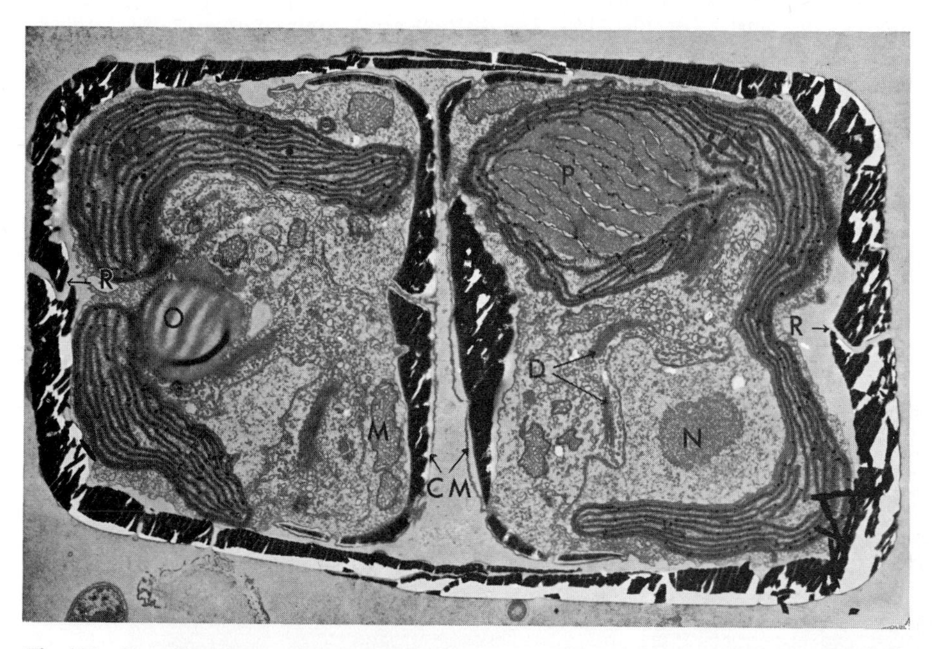

Fig. 400. *Gomphonema parvulum.* Section through a cell after division. ×10,000. CM, cytoplasm membrane; D, dictyosome; M, mitochondrium; N, nucleolus; O, oil-drop; P, pyrenoid in chromatophore; R, raphe. After Drum and Pankratz.

Some species counteract this progressive diminution in size in as far as only the larger of the two daughter cells divides again, while the smaller has some plasticity in the girdle-zone so that ultimately the size difference between epitheca and hypotheca remains very small.

The Diatomales are divided into two suborders according to their symmetry: in some the frustule is radial, in others bilateral. Since other differences, especially in the methods of sexual reproduction, are correlated with this distinction, it is desirable to consider the two groups separately.

1. Centricae. Their frustules have a circular or bluntly triangular outline (Fig. 398H–L) with a radial or concentric arrangement of sculpturing on the wall. This group has no method of vegetative locomotion.

Sexual reproduction is effected by meiogametes (p. 206). The male cells produce four spermatozoids after meiosis (Fig. 401D–F) and ova are produced in a generally larger cell modified as an oogonium. A flagellum with flimmer-hairs effects locomotion of the spermatozoid (cf. Fig. 457). Sex determination operates phenotypically (cf. p. 326).

There is much variation in detail. Thus only one or two of the four haploid nuclei may give rise to ova in the oogonium. The ova may either be discharged or remain in the oogonium in which case the spermatozoids enter through a cleft between the frustules.

In the male, a vegetative cell either directly becomes an antheridium (Fig. 401) or, as in *Biddulphia*, divides mitotically into smaller cells which can be considered as reduced antheridia; in extreme cases these may be naked. Each antheridium forms four spermatozoids.

After fertilization, whether inside or outside the oogonium, the zygote surrounds itself with a pectin membrane (perizonium) and germinates forthwith increasing many times in size as the pectin membrane stretches, forming an auxozygote. Meanwhile the two frustules of the oogonium are progressively pushed apart and a new pair is formed. Thus a new, diploid 'firstling cell' is formed, from which, as previously described, a new diploid generation of daughter cells arises with concomitant diminution in size. Diatoms are thus purely diplontic with a gametic change of nuclear phase.

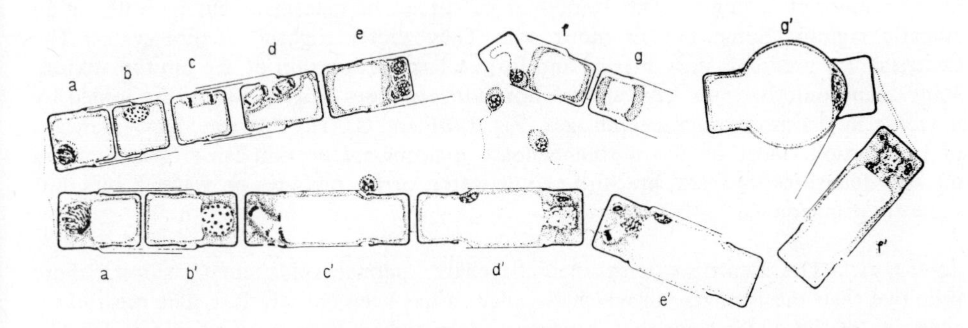

Fig. 401. *Melosira varians*. Sexual reproduction (schematic). **a.** to **g.** Male. **a'.** to **g'.** Female filaments. a. to e. and a'. to e'. Meiosis. **f.** Antheridium discharging. **g.** Antheridium empty. **d'.** Male nucleus after entering the fertilization cleft. **f'.** Fertilization. **g'.** Young auxozygote. After von Stosch.

2. Pennatae. The cells are rod- or boat-shaped, more rarely wedge-shaped (Fig. 398A–G); they have bilateral symmetry and their sculpturing makes a pinnate pattern.

Very frequently a groove or 'raphe' runs along the line of symmetry of the valve; the fine structure of this varies greatly in different genera (Figs. 398A, 400). It is supposed that the extrusion of cytoplasm through the raphe causes the characteristic gliding motion which is peculiar to the Pennatae.

Generally during sexual reproduction in the Pennatae two isogametes without flagella conjugate (Fig. 402).

Conjugation begins when two vegetative cells come together and secrete copious pectic mucilage. The nucleus of each undergoes meiosis, producing four haploid nuclei of which, however, two degenerate. Each cell thus produces two gametes as naked protoplasts. The frustules separate, the two pairs of gametes fuse, giving rise to two zygotes which, as in the Centricae, immediately enlarge as auxozygotes. Each of these ultimately lays down a pair of frustules, forming a new firstling cell several times larger than the parental cells. Firstling and parental cells either lie at right angles (Fig. 402) or parallel to each other.

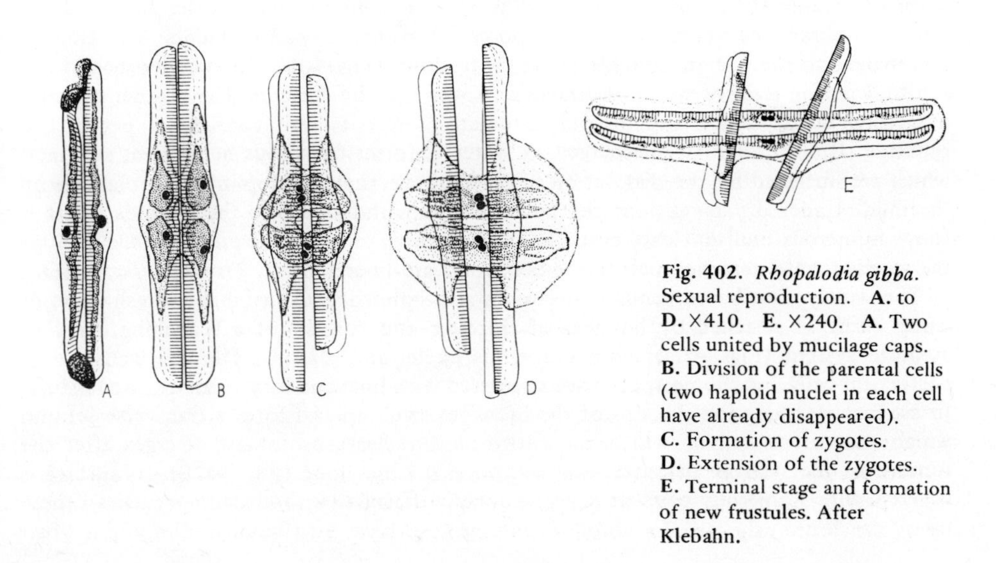

Fig. 402. *Rhopalodia gibba*. Sexual reproduction. **A.** to **D.** ×410. **E.** ×240. **A.** Two cells united by mucilage caps. **B.** Division of the parental cells (two haploid nuclei in each cell have already disappeared). **C.** Formation of zygotes. **D.** Extension of the zygotes. **E.** Terminal stage and formation of new frustules. After Klebahn.

There are many departures from this normal process, such as the formation of only one gamete in the parental cell, and thus only one zygote; the fusion of two sexual nuclei from the same cell (autogamy); or the complete omission of the sexual process (apomixis).

Distribution and ecology. The 10,000 or so species of diatoms occur in water in all climatic regions. Some live in moist soil. They prefer slightly alkaline water. The Centricae are predominantly marine and form a large proportion of the phytoplankton. Many planktonic diatoms have special flotation processes (Fig. 398L) or are joined by mucilage in chains or other assemblages (Fig. 398F and G). The Centricae have no means of locomotion. Most of the motile pennate diatoms are bottom-living (sometimes in massive quantities) in fresh, brackish or salt water, or as epiphytes on water plants, but some are planktonic.

Phylogeny. The Centricae with their flagellate spermatozoids are considered more primitive than the Pennatae in which flagellation has been entirely lost. The most likely ancestors of the diatoms are the Chrysomonadales (which have silicified cyst walls). The diatoms remain at the Coccales level, although the formation of cell-chains in many points towards the Trichales.

The oldest diatoms are indeed Centricae from the Jurassic. Very numerous diatom species occur from the Cretaceous onwards. In the Tertiary and Interglacials they formed massive deposits (diatomaceous earth, Kieselguhr).

Class IV: Xanthophyceae

Like the Chrysophyceae the Xanthophyceae have chlorophyll *a* (no *b*) and β-carotene as plastid pigments, but they differ in the lack of lutein and fucoxanthin amongst their xanthophylls; instead they have heteroxanthin, vaucheriaxanthin and diadinoxanthin. The chromatophores look green and turn bluish with hydrochloric acid. Reserves are oil and chrysolaminarin; there is no starch. The cell wall generally gives a pectin reaction and is often composed of two pieces; it becomes silicified in cysts. The motile stages bear two somewhat laterally inserted unequal flagella (hence the earlier name Heterocontae) of which the longer has flimmer-hairs.

The Xanthophyceae include all levels of organization up to the siphonaceous form. The simplest forms (Heterochloridales) are always flagellate (Fig. 403). The Heterococcales include planktonic and unattached forms with a rigid cell wall. The Heterotrichales include the genus *Tribonema* (Fig. 404), common in fresh water and on damp soil; its unbranched filaments are composed of H-shaped wall segments. The Heterosiphonales include the remarkable genus *Botrydium* (Fig. 406), the bladder-shaped cell (up to 2 mm in size) of which is attached to wet mud by rhizoids. The bladder contains numerous nuclei and many discoid chromatophores; its wall consists of pectin and cellulose. If *Botrydium* is submerged in water it forms numerous heterocont swarmers which are liberated by the dissolution of the cell wall; they develop into new bladders on the mud. Under dry conditions the protoplast is withdrawn into the rhizoids where it forms numerous multinucleate cysts. In the branching terrestrial *Capitulariella* (Fig. 405) the sporangia are detached before the zoospores are liberated (cf. *Trentepohlia*, p. 468).

The widely distributed genus *Vaucheria* is also siphonaceous. It inhabits fresh water or damp earth, is attached by bunches of rhizoids, and consists of a branching, tubular, aseptate system (Fig. 407D) with numerous nuclei and plastids. The cell walls are of pectin and cellulose. Some species are encrusted with lime and may form calcareous tufa. In asexual reproduction the tips of the branches swell up and form a transverse septum which cuts off a cell of which the entire multinucleate protoplast emerges after the rupture of the wall as an ovate swarmer some 0.1 mm long (Fig. 407B). Its surface is covered with numerous pairs of slightly unequal flagella (without flimmer hairs); these move synchronously. In the colourless outermost layer two basal bodies and a pear-

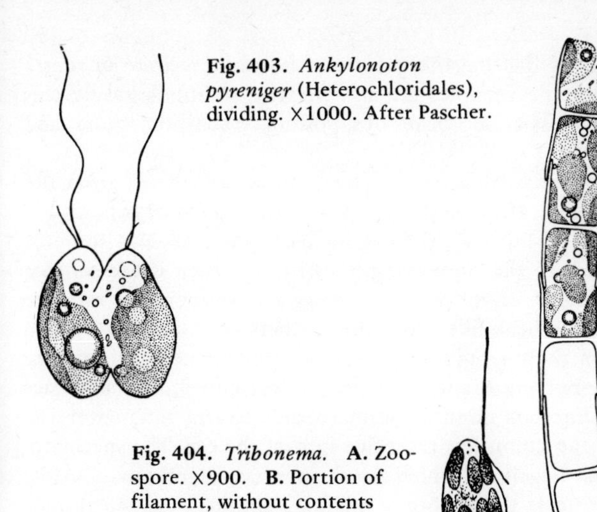

Fig. 403. *Ankylonoton pyreniger* (Heterochloridales), dividing. ×1000. After Pascher.

Fig. 404. *Tribonema.* **A.** Zoospore. ×900. **B.** Portion of filament, without contents below. ×600. **A.** After Luther. **B.** After Pascher.

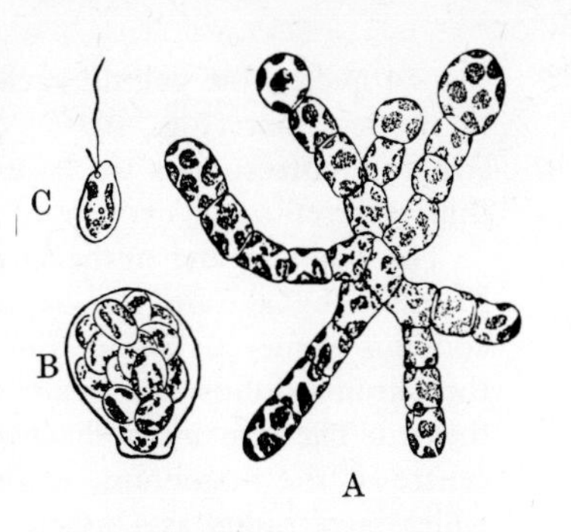

Fig. 405. *Capitulariella radians.* **A.** Thallus with terminal sporangia. **B.** Liberated sporangium. **C.** Zoospore. After Pascher.

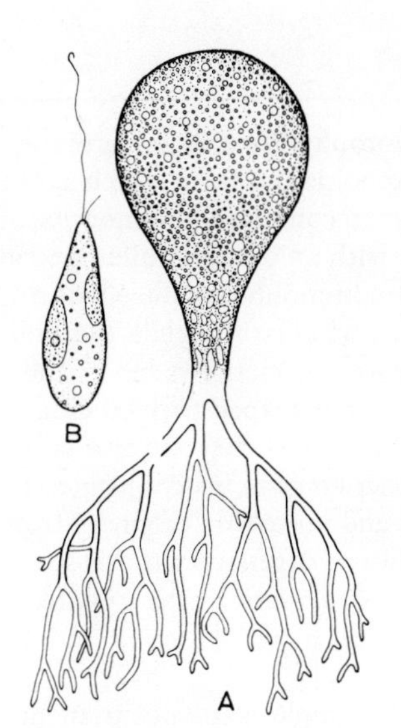

Fig. 406. *Botrydium granulatum.* **A.** Whole plant. ×30. **B.** Zoospore. ×1000. **A.** After Rostafinsky and Woronin. **B.** After Kolwitz.

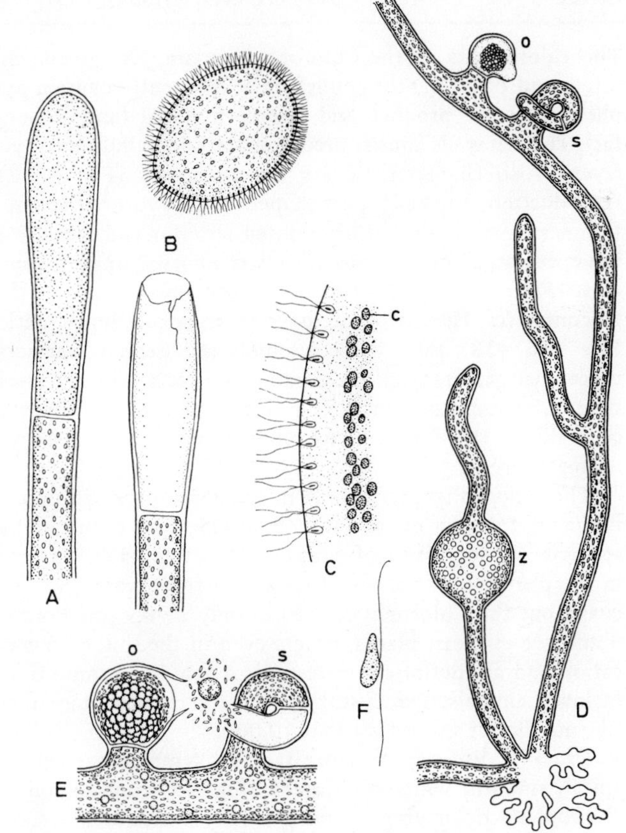

Fig. 407. *Vaucheria* (**A.** to **C.** *V. repens.* **D.** to **E.** *V. sessilis.* **F.** *V. synandra*). **A.** Sporangium. ×150. **B.** Zoospore liberated from the sporangium. ×150. **C.** Margin of a zoospore. ×900. **D.** A young plant developed from a zoospore with rhizoid and gametangia. ×50. **E.** Part of a filament with gametangia. ×150. **F.** Spermatazoid. ×700. c, chromatophores; o, oogonium; s, antheridium; z, zoospore. **A.** and **B.** After Goetz. **C.** After Strasburger. **D.** After Sachs, modified. **E.** After Oltmanns. **F.** After Woronin.

shaped nucleus lie behind each pair of flagella; the chromatophores lie nearer the centre of the swarmer (Fig. 407C); contractile vacuoles are also present. Morphologically this structure corresponds to the entire mass of zoospores of an ordinary zoosporangium, and thus represents a 'synzoospore'.

The oogonia and antheridia of *Vaucheria* arise as lateral processes cut off from the filaments by a transverse septum (Fig. 407D and E). At first the young oogonium (o) contains numerous nuclei; however, all but one, the egg nucleus, together with some of the chromatophores, migrate back into the supporting branch; only then is the septum formed. The remaining chromatophores, oil-drops, and the egg nucleus contract into the centre of the oogonium, while in the beak-like projection gathers colourless cytoplasm which is extruded as a sphere when the oogonium opens. The apex of the multinucleate antheridium (s), which like its supporting branch is more or less coiled, also becomes mucilaginous at maturity. The numerous minute spermatozoids swarm out, enter the oogonial aperture and gather near the colourless receptive spot of the egg. The spermatozoids (Fig. 407F) have two unequal flagella of which the shorter has flimmer hairs. After fertilization by one of the male gametes the zygote, containing abundant oil, surrounds itself with a many-layered wall and becomes dormant. Eventually on germination it undergoes meiosis and develops directly into a new haploid filament.

Class V: Chlorophyceae, green algae

The chloroplasts of the Chlorophyceae are pure green (chlorophylls *a* and *b*, carotenes, lutein and other xanthophylls); they frequently contain pyrenoids and form starch as the photosynthetic product and reserve material (and sometimes considerable amounts of fat). The cell walls consist predominantly of cellulose (often with an outer, swollen pectin layer). Also characteristic are the zoospores (Fig. 420B) commonly produced during reproduction; typically pear-shaped, with two or four equal, whip-lash flagella (without flimmer hairs) at the rather pointed anterior end, usually two contractile vacuoles and a red eye-spot, and in the posterior part a curved or even cup- or pot-shaped parietal chloroplast (Figs. 408A, 411). Because of the equal length of their flagella they are also called Isocontae (cf. Heterocontae, p. 458). In sexual reproduction two gametes conjugate (cf. Figs. 416, 418); these often resemble the asexual swarmers and invariably originate from unicellular gametangia. The ♂ gametes at least are nearly always flagellate; the ♀ is sometimes a non-motile ovum (e.g. Fig. 421). The product of conjugation, the zygote, is generally a thick-walled, roundish resting cell (cystozygote) often coloured red by haematochrome.

The Chlorophyceae include microscopically small unicells, simple filaments, branched filaments forming dense tufts (Fig. 419) and even complex bodies, e.g. leaf-like thalli, superficially suggestive of cormophytes (Fig. 417). The majority (about 90 per cent) live in the plankton or benthos (p. 489) of fresh water. Some larger species occur on the seacoast, but the Chlorophyceae form only a very small proportion of the marine plankton. Some live in moist places, others even in the soil; certain species tolerate complete desiccation and are definitely terrestrial. A few live symbiotically in lichens or inside the cells of lower animals. The cystozygote is the chief resting organ during unfavourable periods. The number of species is about 10,000.

Order 1: **Volvocales.** Their radially symmetrical cells have two, four or eight equal, apical, smooth, whip-lash flagella (Fig. 457) and contain a single beaker- or pot-shaped chloroplast, open above, and generally containing a starch-producing pyrenoid (p. 49, Figs. 408A, 410B) at the base and a red eye-spot (stigma, Fig. 408A) near the apex. When a wall is present it is predominantly of cellulose (but according to the species and stage of development variously mixed with hemicelluloses, pentosans and pectins).

The Volvocales occur widely in freshwater plankton and may be so abundant as to turn the water quite green; they are absent from the sea. Organic substances promote the growth of many species, which therefore tend to occur in organically polluted waters. A few, e.g. *Polytoma uvella*, lack chlorophyll and are entirely saprophytic. Under phototactic stimulus *Chlamydomonas* can move some ten times the length of its body per second.

The small family **Polyblepharidaceae** is probably the most primitive, with 2–8 flagella; all members are naked. While *Polyblepharides*, so far as is known, only divides asexually into two by longitudinal division, higher forms exhibit sexual reproduction, which may be either phenotypically or genotypically (*Dunaliella salina*) determined (cf. p. 326). *Dunaliella salina* lives in the concentrated brine of salt works; its red colour is due to haematochrome.

The **Chlamydomonadaceae** differ from the Polyblepharidaceae in the possession of a cellulose cell wall. The common genus *Chlamydomonas* (Fig. 408A) has numerous species. Asexual reproduction is effected by zoospores, 2–16 of which are produced inside the mother cell (sporangium), generally by repeated longitudinal divisions of the contents (Fig. 423B); they are liberated by the rupture of the original wall. Sexual reproduction also occurs by the union of small, biflagellate gametes. In the most primitive forms (e.g. *Polytoma uvella*) the gametes are identical to vegetative cells and can either fuse in random pairs or reproduce vegetatively (facultative determination of function). We are close to the origins of sexuality here. In other cases the gametes look identical but are genotypically different (*Chlamydomonas*). They are formed mostly in large numbers (2–64) in a mother cell by repeated longitudinal division and unite in pairs by their anterior ends to form the zygote (Fig. 408C–E). Generally the two gametes of different strains come into contact at the tips of their flagella, which subsequently become spirally entwined (Fig. 408C). The zygote is motile at first with four flagella (planozygote), but later these are withdrawn and a thick-walled, non-motile resting-stage (cystozygote) develops (F). Gametes are invariably naked at first, but may develop a wall later, in which case the protoplast is liberated during sexual union. Meiosis occurs in the germinating zygote (G), producing four swarmers, two of either strain; the swarmers are therefore meiospores (p. 202).

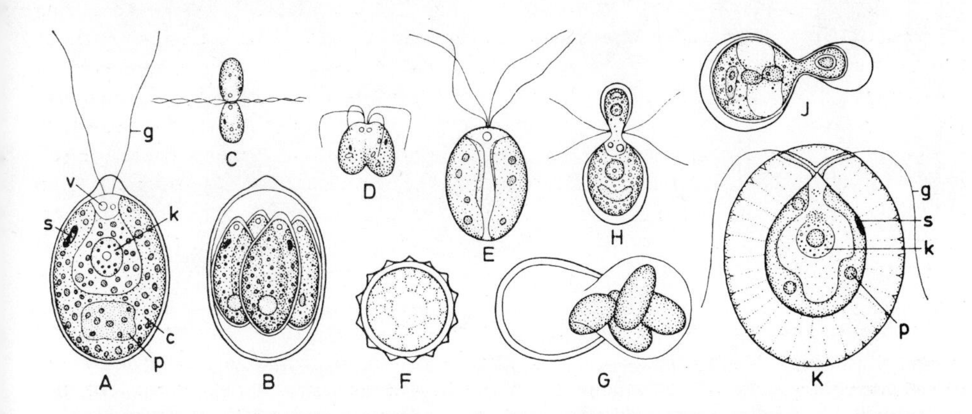

Fig. 408. Chlamydomonadaceae. **A.** *Chlamydomonas angulosa.* ×1100. **B.** The same, with four daughter cells in the mother cell. ×1100. **C.** and **D.** *Chlamydomonas botryoides*, fusion of two isogametes. ×250. **E.** *Chlamydomonas paradoxa*, zygote. ×500. **F.** *Chlamydomonas monoica*, resting zygote. ×500. **G.** *Stephanosphaera pluvialis*, germinating zygote. ×300. **H.** and **J.** *Chlamydomonas braunii*, fusion of anisogametes. ×400. **K.** *Haematococcus pluvialis*, cell enclosed in a thick gelatinous wall. ×330. c, chloroplast; g, flagellum; k, nucleus; p, pyrenoid; s, eye-spot; v, contractile vacuole. **A.** and **B.** After Dill. **C.** to **G.** After Strehlow. **H.** and **J.** After Goroschankin. **K.** After Reichenow.

Besides the species with isogamy (p. 206, Fig. 408C), others show anisogamy (Fig. 408H, J), in which a smaller ♂ gamete unites with a larger ♀ one. In *Chlamydomonas suboogama* the flagella of the ♀ gametes are functionless, while in *Chlorogonium oogamum* they are completely absent, so that the ♀ gamete is an amoeboid egg or ovum (Fig. 409), fertilized after liberation from the ♀ cell by a naked spermatozoid, 64 or 128 of which are produced by successive divisions in the ♂ cell as light-green, biflagellate needle-shaped objects (oogamy). Finally, in *Chlamydomonas coccifera* an entire ♀ cell, complete with wall, is fertilized, after losing its flagella and without liberating the gamete, by a ♂ spermatozoid which also has a cell wall (oogoniogamy). Thus an ascending series of sexual processes can be traced even in these unicellular plants.

Similarly, the plastids show a series from the most primitive *Chlamydomonas* species with a central chloroplast, through the majority with a parietal chloroplast, to higher forms in which it is reticulate, while yet others have numerous discoid chloroplasts.

Haematochrome (p. 460) reddens many species, e.g. *Haematococcus pluvialis* (Fig. 408K), which reddens puddles, and *Chlamydomonas nivalis*, which produces 'red snow' of Alpine and Arctic regions. Some chlamydomonads (and other flagellates) even colonize melting snow and ice during our winters.

The family **Volvocaceae**, with colonial forms in which the cells are united (at least by mucilage), is closely related to the unicellular Chlamydomonadaceae. In *Oltmannsiella* four cells form a row; in *Gonium* 4–16 cells form a flat plate in which all the flagella are directed to one side; each individual cell is similar to *Chlamydomonas* in construction. In *Pandorina* (Fig. 409G, H) sixteen such cells form a sphere, while *Eudorina* and *Pleodorina* have 32 and 128 cells respectively united in a hollow sphere. The colonies of *Stephanosphaera* (Fig. 409F), which lives in puddles, consist of a crown of 4–16 cells with processes looking like pseudopodia, but quite rigid; the chloroplasts generally contain two pyrenoids.

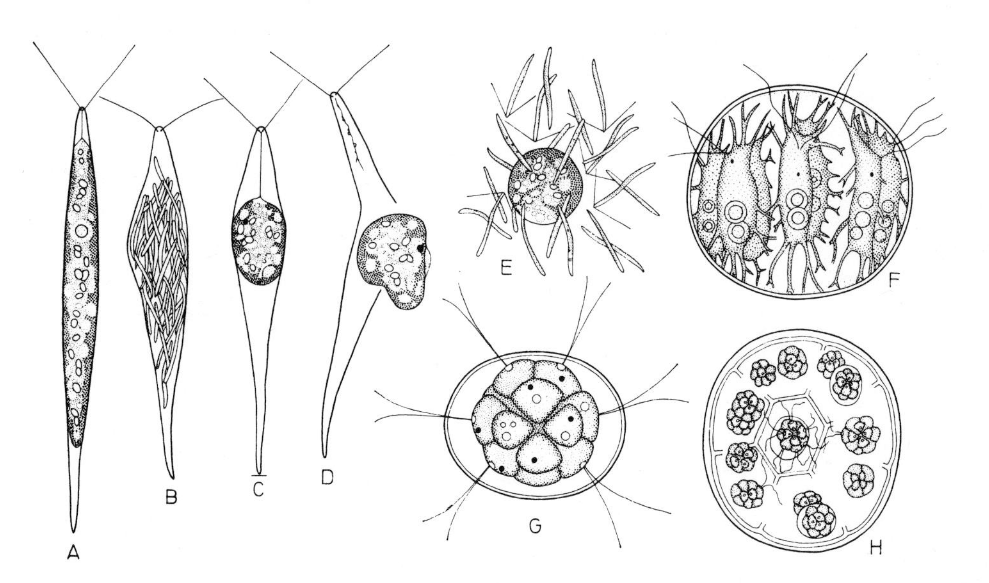

Fig. 409. Volvocales. **A.** to **E.** *Chlorogonium oogamum.* ×240. **A.** Vegetative cell. **B.** Male cell with spermatozoids. **C.** Female cell with ovum. **D.** Discharge of the ovum. **E.** Ovum surrounded by spermatozoids. **F.** *Stephanosphaera pluvialis.* ×250. **G.** *Pandorina morum.* ×160. **H.** The same, formation of daughter colonies (the mother cell wall already partially broken down). ×150. **A.** to **E.** After Pascher. **F.** After Hieronymus. **G.** After Stein. **H.** After N. Pringsheim.

In vegetative reproduction *Oltmansiella* divides to produce four-celled band-shaped colonies. In *Gonium* 2–16 zoospores arise by repeated division in each cell; these lie in the mother-cell in the form of a bowl, and on liberation remain united in the form of a flat plate. Exactly the same processes of division and cell-orientation take place in the spherical forms, but the bowl turns itself inside out and backwards, so that the flagella come to lie on the outside of the sphere. The marginal cells of the original plate thus become the rear and the central cells the front of the new colony as it swims along. The synchronous beating of the flagella in all these coenobia is facilitated by extremely fine protoplasmic connections between the cells (plasmodesmata, p. 62). Otherwise each cell is independent of its neighbours, in most cases even in reproduction.

In *Volvox* (Fig. 410) a large number of cells (up to 20,000 in *V. globator*), each with two flagella, an eye-spot and a chloroplast, forms a hollow sphere filled with mucilage and visible to the naked eye; the cells are interconnected by broad protoplasmic processes (Fig. 410B, C). Only a small proportion of the cells, scattered over the hinder part of the sphere, is reproductive. The majority are restricted to photosynthesis and locomotion, but even these show polar differentiation, especially in the gradual decrease in the size of the eye-spot from the front to the rear. The sphere of *Volvox* is therefore not so much a colony as a multicellular individual.

The protoplasmic connections originate during division of the embryonic cells by the failure of their pseudopodia-like processes to separate completely. In asexual reproduction daughter colonies arise by the successive division of single cells; in this case, too, a flat plate of cells develops at first, later becoming a concave bowl; this next (as in *Pandorina*) turns inside out and backwards, and falls into the central cavity of the sphere, where further growth takes place (Fig. 410D–J); the daughter colonies are only liberated when the parental individual disintegrates. Sexual reproduction in *Volvox* (and *Eudorina*) is oogamous. The 6–8 green ova arise singly by the enlargement of a vegetative cell (Fig. 410K); the numerous small, yellowish spermatozoids are produced by the repeated division of other vegetative cells; before they are liberated they, too, form a flat plate (Fig. 410K, M). After fertilization the ovum becomes a thick-walled resting zygote; meiosis occurs on germination. All the cells not involved in reproduction die, so that *Volvox* provides the first example of the regular formation of a corpse. It should be noted that in the Volvocales all the cells of a colony invariably arise from a single cell.

Even in the few-celled spherical colonies, polarity of movement, of cell size and of reproductive capacity adumbrate the development of a multicellular individual from a colony.

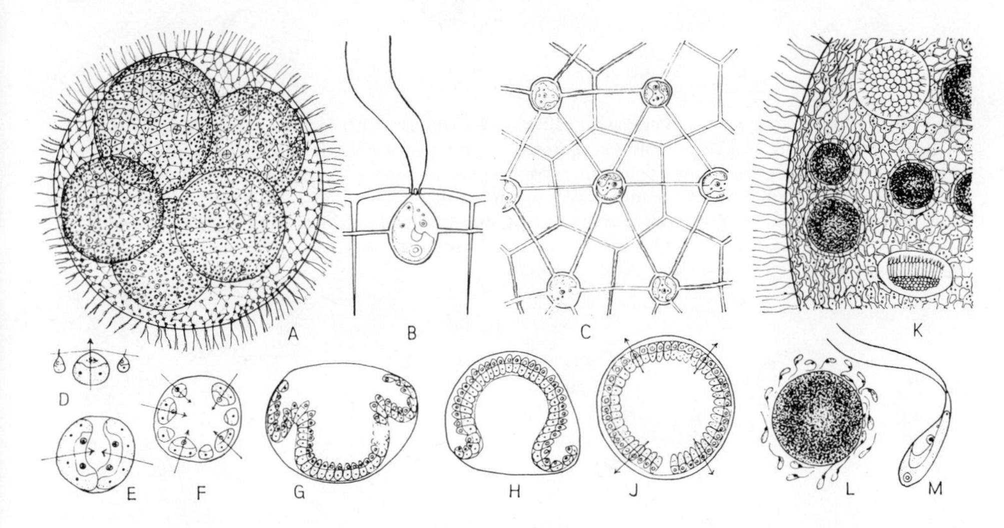

Fig. 410. *Volvox.* **A.** Individual with six daughters. ×50. **B.** Single cell with the plasmodesmata running laterally to adjacent cells. ×1000. **C.** View of the cell assemblage. ×500. **D.** to **J.** Development and inversion of a daughter colony. **D.** ×250. **E.** to **F.** ×350. **G.** ×250. **K.** Part of an individual with five ova and two plates of spermatozoids. ×200. **L.** Ovum surrounded by spermatozoids. ×265. **M.** Spermatozoid. ×1000. **A.** to **J.** *Volvox aureus.* **K.** and **L.** *V. globator.* **A.** After Klein. **B.**, **C.** and **M.** After Janet. **D.** to **J.** After Zimmermann. **K.** and **L.** After Cohn.

Order 2: **Chlorococcales** (Protococcales). The vegetative cells have no flagella and are therefore non-motile; each has generally one nucleus and one chloroplast. Apart from the unicellular forms, they produce characteristically shaped aggregates (p. 73).

Biflagellate zoospores (Fig. 411) are formed only in reproduction, when they are naked; after swarming they surround themselves with a wall (encyst). Vegetative cells never divide into two; instead the cytoplasm invariably cleaves to form numerous zoospores (Fig. 411, 2–4, 8). The same cell is therefore successively zoospore, vegetative cell and zoosporangium. Frequently the spores surround themselves with a wall while still inside the mother cell and are liberated without flagella as aplanospores (Fig. 412). Sexual reproduction, in the few cases where it is known, is effected by flagellate isogametes (e.g. in *Pediastrum* and *Hydrodictyon*); oogamy does not occur.

The Chlorococcales occur mainly as freshwater plankton, but some live in moist soil; some (along with other algae) are regular constituents of the green coating on walls and the bark of trees. Others live as symbionts in lichens or even in the protoplasts of lower animals (*Chlorella vulgaris* in infusoria, *Hydra*, etc.).

As in the Volvocales, there is an ascending series from unicellular forms to disc-shaped and ultimately globular aggregates (p. 73). *Chlorococcum* and *Chlorella* (Figs. 411, 412) are spherical unicells; *Scenedesmus*, a common freshwater alga, forms simple aggregates of four (or at most eight) cells united transversely in a row (Fig. 413). More complex is the equally frequent *Pediastrum* (Fig. 414), which forms an elegant, free-floating, flat plate, like a *Gonium* without flagella (p. 462). In the water-net, *Hydrodictyon utriculatum* (Fig. 415), a free-floating freshwater alga, the cylindrical cells meet at their ends in groups of three or four and produce a sack-shaped multicellular aggregate, up to 50 cm long in some cases, in the form of an elongated hollow net of numerous meshes. Sexual reproduction is effected by isogametes which are smaller than the zoospores.

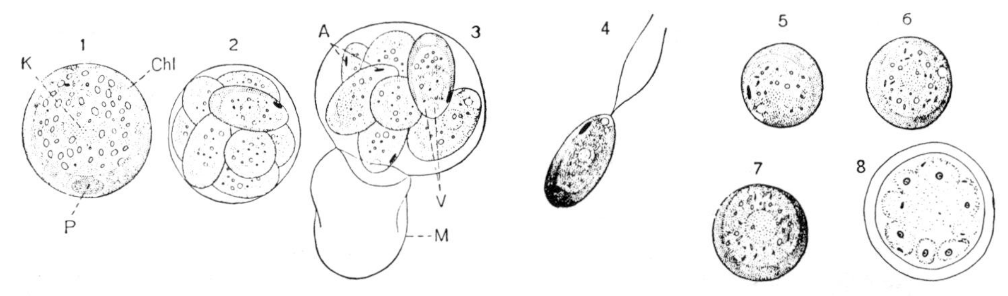

Fig. 411. *Chlorococcum* sp. **1.** Vegetative cell; the chloroplast (Chl) forms a spherical shell with a small anterior opening and a pyrenoid (P) opposite; K, nucleus visible through the chloroplast. **2.** Division into eight cells. **3.** Discharge of zoospores (A, eye-spot; V, contractile vacuoles) into an evanescent vesicle formed from the inner layer of the mother-cell wall (M). **4.** Free zoospore with equal anterior flagella. **5.** Zoospore coming to rest, eye-spot and vacuoles still present. **6.** and **7.** Development of stage **1.** by the loss of eye-spot and vacuoles. **8.** Cleavage of the protoplast (intermediate between **1.** and **2.**). ×1200. After Pascher.

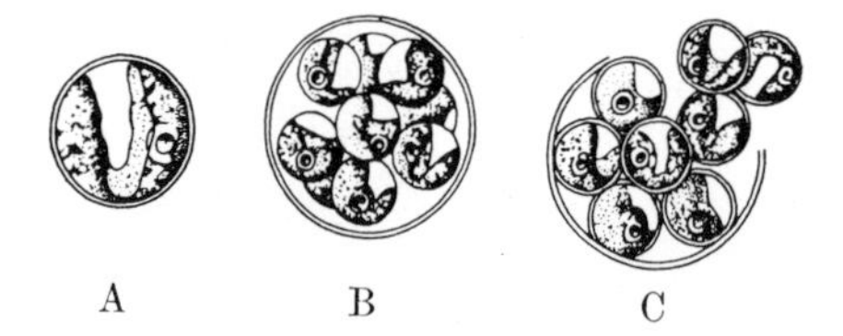

Fig. 412. *Chlorella vulgaris.* **A.** Vegetative cell. **B.** and **C.** Division into eight aplanospores. ×500. After Grintzesco.

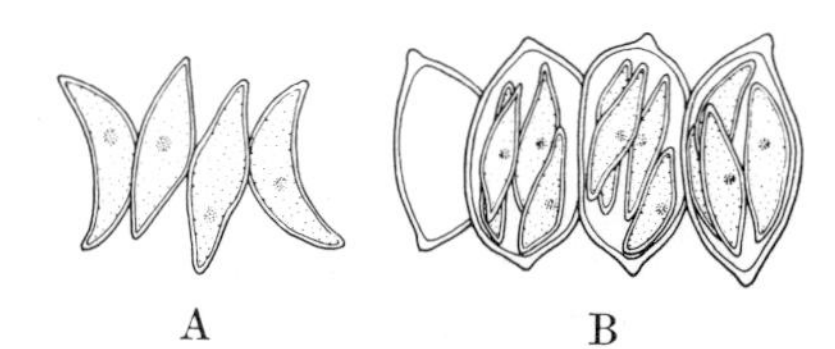

Fig. 413. *Scenedesmus acutus.* **A.** Four-celled aggregate. **B.** Division. ×1000. After Senn.

In asexual reproduction in these genera the zoospores or non-motile aplanospores are not liberated separately, but come together with the fusion of their walls at an early stage to form an aggregate with the cell number and form characteristic of the species (Figs. 413–15). This union may take place soon after liberation from the mother cell in a mucilaginous vesicle (Fig. 414A), or even in the mother cell itself (Figs. 413, 415C, D, E, A), which therefore liberates an initially small plant with the complete number of cells; no further cell division then takes place in the aggregate until reproduction occurs.

The similarity to the corresponding series in the Volvocales is obviously only superficial: the development is quite different. In the Volvocales the colony arises by the repeated longitudinal division of the cells, so that the position of every cell in the colony is fixed from the outset, whereas in the Chlorococcales the whole brood of young cells

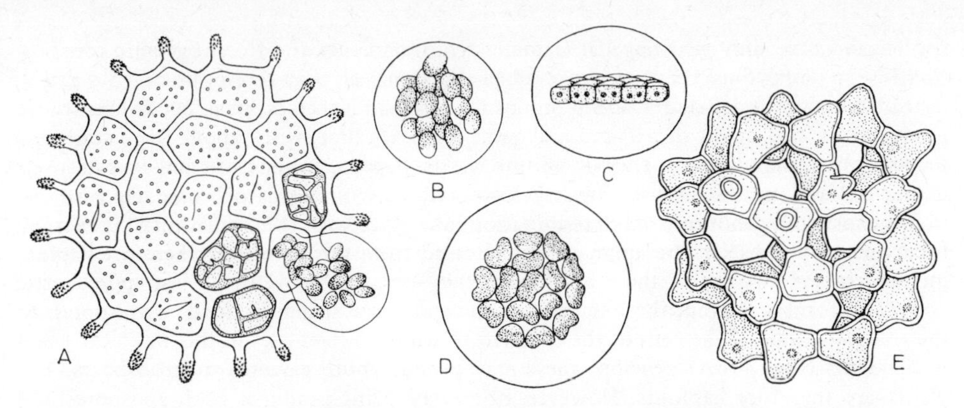

Fig. 414. **A.** to **D.** *Pediastrum granulatum.* **A.** Disc-shaped aggregate, empty except for three dividing cells; one cell is liberating a vesicle with sixteen zoospores (flagella omitted). **B.** Zoospores after liberation of the vesicle. **C.** 4–5 hr later. **D.** In side view. ×300. After Braun. **E.** *Coelastrum proboscideum.* ×550. After Senn.

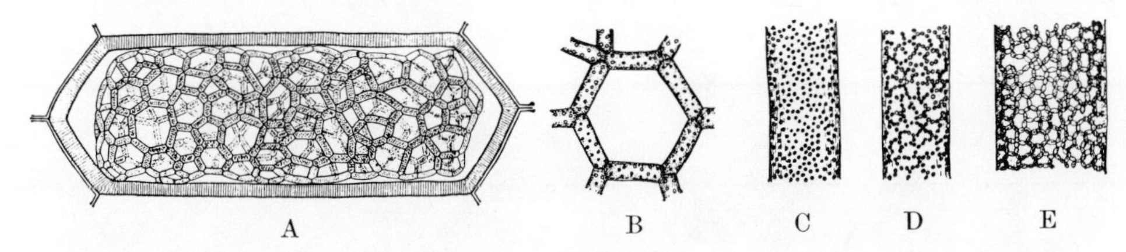

Fig. 415. *Hydrodictyon utriculatum.* **A.** Young net in a cell of the mother-net. ×50. **B.** Part of the mesh, highly magnified; the dots are pyrenoids. **C.** Part of an old cell with zoospores. **D.** and **E.** Zoospores arranging themselves to form a new net. **A.** and **B.** After Klebs. **C.** to **E.** After Harper.

formed by the cleavage of one protoplast can freely intermingle (Fig. 414B) and the cells only subsequently come together randomly (aggregation colony, p. 73).

Chlorella, Scenedesmus, Ankistrodesmus and *Hydrodictyon* are frequently grown in pure culture for physiological researches.

Order 3: **Ulotrichales.** The simplest members of this group are filamentous (unbranched, Fig. 416, or branched); they elongate by transverse division of the cells. In some genera the cells of the older filaments divide transversely, mainly in one plane, giving a flat expanse of tissue, while in the sea lettuce, *Ulva lactuca*, the thallus consists of a large, leaf-like expanse of parenchymatous tissue two cells thick (Figs. 89, 417). *Enteromorpha*, which also is coastal, forms hollow cylinders or flattened bands. The thalli are—at least at the outset—generally attached to the substratum.

There is almost always only one nucleus in each cell and usually only a single chloroplast. Asexual reproduction is effected by zoospores, sexual reproduction by the fusion of flagellate gametes or by oogamy. Every cell, apart from the rhizoid, may take part in reproduction, and when all the cells do so the parent plant is left empty and terminates its individual existence.

The common freshwater *Ulothrix zonata* (Fig. 416A) has unbranched filaments with intercalary growth. The short cells contain a band-shaped chloroplast arranged like an open ring or collar just inside the cell wall (A). The filament is attached to stones, etc., by a slender, elongate, generally colourless rhizoid cell. The asexual swarmers (zoospores, C) are produced singly from a cell (zoosporangium), or more generally the zoosporangium becomes multinucleate and cleaves into a number of zoospores corresponding to the number of nuclei. Each zoospore has four flagella, one eye-spot and one chloroplast. The zoospores are liberated through a hole arising on the side wall (B); they swim about, and each settles down with the secretion of mucilage and the loss of flagella and eye-spot, grows a cell wall and then gives rise to a new filament. At the onset of conditions unfavourable for growth, isogametes are formed in a similar way, but in much greater numbers, from cells of the filament (gametangia, D, E); they resemble zoospores, but are smaller and have only two flagella. Gametes from filaments of different genetic constitution fuse in pairs. Since they are morphologically identical, they cannot be called ♂ and ♀; instead they are designated + and − strains (as in all similar cases; cf. p. 206). The zygote produced by the union of a + and a − gamete (F–H) first swims about for a time as a four-flagellate planozygote; then it withdraws its flagella, rounds off and invests itself with a cell wall, producing a resting-stage coloured red with haematochrome. On germination, meiosis and genotypic sex-determination take place with the formation of (generally four) aplanospores (K). The aplanospores released from the zygote are therefore haploid meiospores and two are of the + and two of the − strain (Fig. 429A). These spores settle down and form a rhizoid; the long axis of the spore as it begins to divide corresponds to the transverse axis of the cells of the filament to which it gives rise. In this way new + and − filaments (Fig. 416A) develop; these may produce both gametes and zoospores. The plants are therefore haplonts. However, not every plant produces both zoospores and gametes: a number of generations may reproduce solely by zoospores.

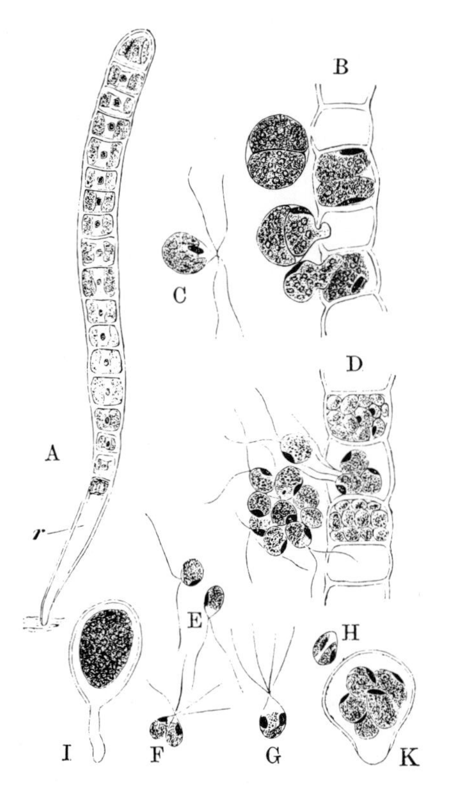

Fig. 417. *Ulva lactuca* on a stone. Marginal cells colourless after the liberation of zoospores. ×0.5. After Kuckuck.

Fig. 416. *Ulothrix zonata.* **A.** Young filament with rhizoid cell, r. **B.** Part of a filament with each cell liberating two swarmers. **C.** A swarmer. **D.** Formation and discharge of gametes. **E.** Gametes. **F.** Fusing. **G.** and **H.** Zygote. **J.** Zygote germinating after the resting period. **K.** Formation of aplanospores in the zygote. **A.** ×300. **B.** to **K.** ×480. After Dodel.

While in *Ulothrix* both vegetative and sexual reproduction may occur in the same individual, in *Ulva* (sea lettuce, Figs. 88, 417) and *Enteromorpha* (sea guts), both frequent round the coast, and also found in brackish water, there is isomorphic alternation of generations (p. 207). The diploid zygote develops into the sporophyte; this undergoes meiosis, with genotypic sex-determination, to produce zoospores. These give rise to the haploid gametophytes whih look identical to the sporophyte, but are either + or −, and give rise to isogametes (similar to the *Cladophora* scheme, Fig. 429B). In these, as in all other marine algae, the zygotes germinate immediately, conditions being constantly favourable for growth, in contrast to freshwater habitats.

Not all members of the Ulotrichales with alternation of generations have the two generations identical in form (isomorphic). In laboratory cultures of some forms with a

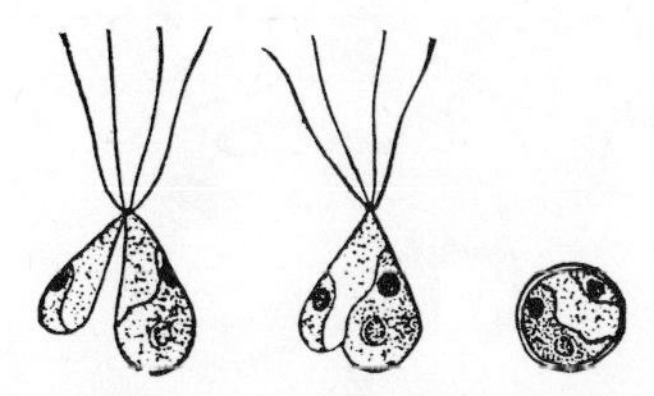

Fig. 418. *Enteromorpha intestinalis*. Fusion of anisogametes, and zygote. ×1800. After Kylin.

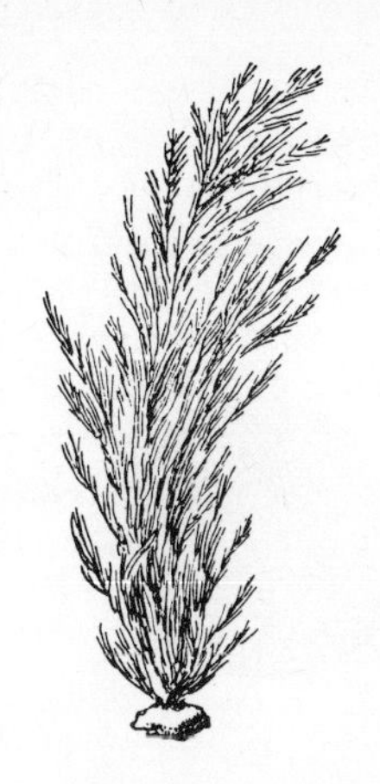

Fig. 419. *Cladophora*. Habit. ×0.3. After Oltmanns.

macroscopic, well-developed, foliaceous gametophyte (*Monostroma* and others) the zygotes can grow for months on end, increasing in diameter twenty-fold or more (but without forming cross walls), thus constituting a sporophytic generation (*Codiolum* stage); eventually—apparently after meiosis—numerous zoospores are formed. The same cell is thus successively zygote, sporophyte and meiosporangium. In another case, which has not been fully elucidated, the zygote develops into an independent, bladder-shaped plant, anchored to the substrate; the zoospores again give rise to gametophytes. All such cases in which the gametophyte and sporophyte differ in form are said to be dimorphic (heteromorphic) (p. 207). In the life-cycle of *Ulothrix* described above the zygote which germinates after meiosis also corresponds to a sporophyte (p. 476).

The unbranched filamentous *Sphaeroplea annulina*, a rather rare algae of flooded ground with multinucleate cells and of doubtful affinity, is oogamous.

Order 4: **Cladophorales.** The cells of the tassel-shaped thalli are multinucleate.

The Cladophorales can also be grouped with certain other algae with multinucleate cells in the Siphonocladales.

The numerous species of *Cladophora* (Fig. 419), with branching filaments forming tassels attached to solid substrata, are frequent in fresh water (especially flowing) and in the sea. They are attached by a rhizoidal basal cell and show apical growth. *C. glomerata* often grows to a foot in length in rivers. The chloroplasts are reticulate and parietal; the cell walls are characteristically lamellose. The formation of cross walls is, as in many other multinucleate plants, independent of nuclear division. As in the Ulotrichales, the reproductive swarmers (zoospores and isogametes) develop from unspecialized cells, although these tend to be restricted to the terminal cells of side branches.

Cladophora exhibits isomorphic alternation of generations (Fig. 429B). In contrast, in other filamentous forms believed to be allied (although there is no general agreement among algologists that they belong to this order) the sporophyte is small and unicellular, so that the alternation of generations is dimorphic. Indeed, in several cases the zygotes develop into plants which were once included under the generic name *Codiolum* in the Chlorococcales, so that each of a number of gametophytes (e.g. *Spongomorpha lanosa*, *Chlorochytrium inclusum* and *Urospora penicilliformis*) has a sporophyte described as a separate species of *Codiolum*. It would appear that *Codiolum* is not an independent genus, but consists merely of the sporophytic generation of various filamentous algae. In all these cases the incidence of meiosis is not known with certainty, but it can scarcely be doubted that it occurs in the formation of zoospores in the sporophyte.

Order 5: **Chaetophorales.** The cells are uninucleate, generally with a single chloroplast. The thallus, highly organized compared with most Chlorophyceae, is heterotrichous, i.e. it consists of a prostrate system (basal plate) of creeping, branched, often pseudoparenchymatous filaments, closely appressed to the substratum, and of a more or less richly branched erect system of filaments bearing the reproductive organs (Fig. 420A).

Both parts are well developed in *Stigeoclonium*, *Coleochaete* and *Trentepohlia*, but in

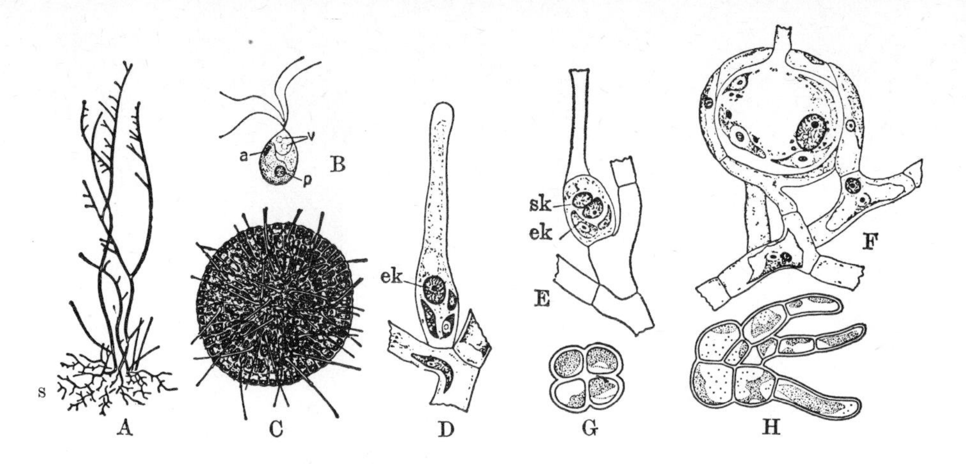

Fig. 420. Chaetophorales. **A.** *Stigeoclonium tenue*. Prostrate system s with erect filaments above. ×4. **B.** *Stigeoclonium subspinosum*, zoospore; a, eye-spot; p, pyrenoid; v, contractile vacuole. ×900. **C.** *Coleochaete scutata*, prostrate disc. ×80. **D.** to **F.** *Coleochaete pulvinata*. **D.** Oogonium shortly before opening. **E.** Fertilized; ek, egg nucleus; sk, sperm nucleus. **F.** Zygote enclosed as a 'fructification'. ×500. **G.** and **H.** *Pleurococcus naegelii*. ×600. **A.** After Huber. **B.** After Juller. **C.** After Jost. **D.** to **F.** After Oltmanns. **G.** and **H.** After Chodat.

many genera either the prostrate or the erect system is developed at the partial or complete expense of the other.

Stigeoclonium, a common small freshwater alga, has both four-flagellate zoospores (Fig. 420B) and two-flagellate isogametes, while *Coleochaete* (in which the disc-shaped prostrate system (Fig. 420C) generally grows epiphytically on other algae and on aquatic higher plants) has biflagellate zoospores but is oogamous, and shows the most highly developed sex organs amongst the Chlorophyceae. The flask-shaped oogonium has a colourless neck (trichogyne, Fig. 420D) which opens at the apex to receive the colourless spermatozoids. After fertilization the spherical zygote enlarges, while filaments arising from the underlying and neighbouring cells grow around it, investing it in a plectenchymatous envelope, one cell thick, thus forming a fructification or sporocarp (Fig. 420F). On the germination of this resting-organ meiozoospores are not formed directly; instead meiosis is followed by the production of a sixteen- to thirty-two-celled body, from each cell of which a haploid swarmer is liberated. *Gomontia* lives endolithically in mollusc shells.

Trentepohlia is completely adapted to a terrestrial existence (in Europe on rocks and tree-trunks, in the Tropics on leathery leaves); it is often coloured red with haematochrome. The zoosporangia are mostly liberated intact and distributed by wind. The biflagellate swarmers generally behave as zoospores, but may also conjugate in pairs—a primitive characteristic. *Pleurococcus* is greatly reduced and is probably a collective genus; it also is frequent on the bark of trees but does not produce sporangia (Fig. 420G, H). *P. vulgaris* is probably the commonest terrestrial alga throughout the world. In *Fritschiella* (India, Africa) erect, branching aerial filaments arise from creeping rows of cells buried in the soil. One can imagine higher land plants arising from green algae of this kind.

The Chaetophorales are pure haplonts without alternation of generations.

Order 5: **Oedogoniales.** The very numerous species of *Oedogonium* are common in fresh water. The unbranched oogamous filaments possess uninucleate cells with an elaborate hollow-cylindrical, reticulate chloroplast (Fig. 421A), and produce swarmers with a characteristic subapical crown of numerous flagella which are not arranged in pairs (Fig. 421C).

The entire contents of a cell become one relatively large zoospore in which the crown of flagella is near the colourless anterior end. On the same filament single cells swell to form barrel-shaped oogonia containing one large ovum (Fig. 421E) which remains in the oogonium. In other parts of the same or another filament spermatozoids are produced, generally in pairs, in relatively small cells (antheridia). The yellowish spermatozoids

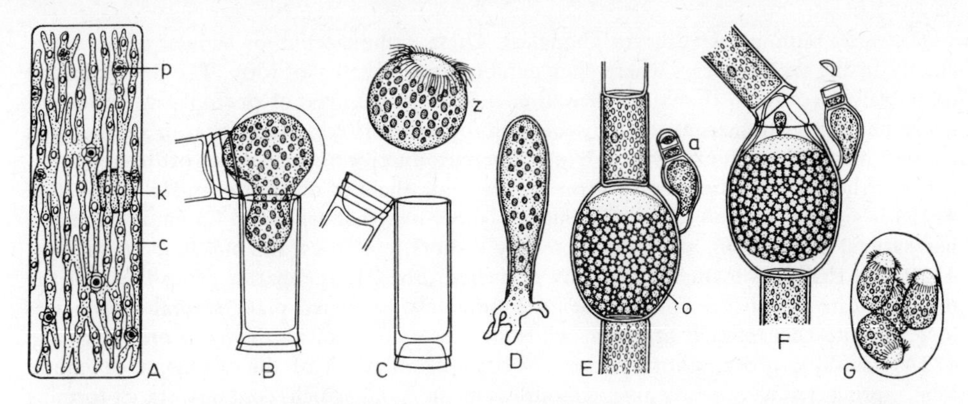

Fig. 421. *Oedogonium.* **A.** A single cell. ×600. **B.** to **D.** *Oedogonium concatenatum*, liberation and germination of a zoospore. ×300. **E.** and **F.** *Oedogonium ciliatum*, fertilization. ×350. **G.** The same, germination of z the zygote. ×350. a, dwarf male; c, chromatophore; k, nucleus; o, oogonium; p, pyrenoid; z, zoospore. **A.** After Schmitz. **B.** to **D.** After Hirn. **E.** and **F.** After N. Pringsheim. **G.** After Juranyi.

resemble zoospores but are smaller; they swim through a pore into the oogonium, where one fuses with the ovum, which develops *in situ* into a large, thick-walled red resting zygote. On germination this produces four large haploid meiozoospores which are liberated and develop into new filaments (Fig. 421D).

In some species the antheridia liberate swarmers (androspores) similar to spermatozoids, but which do not fertilize the ovum directly; instead they become attached to the female filament, where they develop into small, few-celled plants, the so-called dwarf males (Fig. 421E, F). The actual spermatozoids are produced from the upper cells of these.

Characteristic caps ocur at the upper ends of some of the cells (Fig. 421B, C); these arise during the peculiar processes of division and growth of the cells which have the walls in two layers.

It is manifest that the Oedogoniales is a very advanced order.

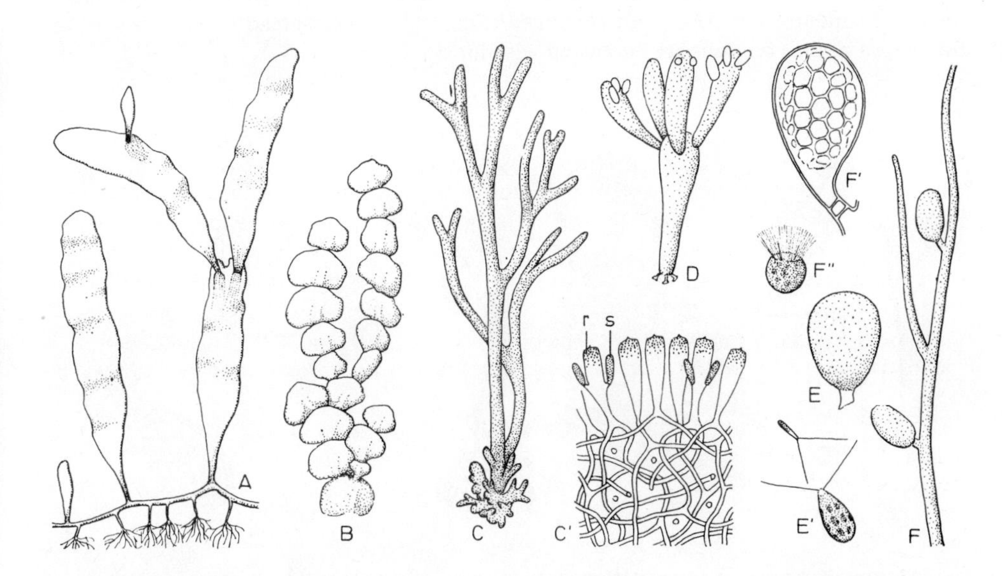

Fig. 422. Siphonales. **A.** *Caulerpa prolifera*, thallus. ×0.5. **B.** *Halimeda tuna*, thallus. ×0.5. **C.** *Codium tomentosum*, thallus. ×0.5. **C'.** The same, transverse section of thallus. ×15. **D.** *Valonia utricularis*, thallus. ×1.5. **E.** *Halicystis ovalis*, thallus. ×3. **E'.** The same, male and female gametes. ×500. **F.** *Derbesia marina*, part of a thallus. ×30. **F'.** The same, sporangium. ×120. **F''.** Zoospore. ×400. r, cortical vesicle; s, gametangium. **A.** After Schenck. **B.** After Oltmanns. **D.** After Schmitz. **E.**, **E'.** and **F'.** After Kuckuck. **F.** After Harder. **F''.** After Davis.

Order 7: **Siphonales** (Chlorosiphonales). These siphonaceous or tubular algae occur chiefly in the warmer seas, where they exhibit great diversity of form. The thallus has no cross walls (septa), so that the cell wall encloses a single mass of protoplasm containing many nuclei and numerous small discoid chloroplasts (differing somewhat in xanthophyll content from other Chlorophyceae); only the reproductive organs are cut off by septa.

Nevertheless, they frequently form macroscopic thalli of diverse form. *Halicystis* (Fig. 422E) is simply constructed: its globular, bladder-like thallus, about 1.5 cm in diameter, has parietal protoplasm, and is anchored by a short, colourless rhizoid. In *Derbesia* (Fig. 422F) the thallus is made up of richly branched tubes. In the marine genus *Bryopsis* the multinucleate coenocyte forms a delicate pinnately branched plant several centimetres long, despite the absence of septa, while in the Mediterranean *Caulerpa prolifera* (Fig. 422A) leaf-like, green, photosynthetic lobes (assimilators) of the thallus, a decimetre long, spring from the creeping, colourless main axis, which contains starch-forming leucoplasts and sends colourless rhizoids into the substratum.

In thc **Dasycladaccac** the thallus consists of a long vesicular central axis attached by rhizoid-like branches and bearing whorls of simple or branching laterals with terminal gametangia (Fig. 423B). The highest degree of differentiation is seen in the Mediterranean *Acetabularia* (Fig. 423D–H): its unbranched stalk bears an umbrella-like cap composed of radially arranged chambers and two whorls of sterile branches, one above and one below the cap. Initially the plant has only a single diploid nucleus (primary nucleus) situated in the rhizoid; this persists for a considerable time (Fig. 16C–E; p. 314) until the cap has developed. It then divides into numerous secondary nuclei which migrate into the chambers where thick-walled cysts develop (Fig. 423E). The cysts are freed when the cap disintegrates; they open by a lid and liberate the gametes formed by meiosis (Fig. 423F). Two gametes fuse to form a zygote (Fig. 423G, H) which settles down and forms a new thallus. The outer cell wall layers of the central axis of the Dasycladaceae are very heavily calcified (Fig. 423B) so that a perforated calcareous tube remains after the death of the plant (Fig. 423C), hence the importance of fossil Dasycladaceae as rock-formers, e.g. in the Trias in the Alps. From the Cambrian onwards eighty-six genera of Dasycladaceae are known fossil, yet only ten survive today. One can trace their evolution from simple forms with irregular branching to complex forms like *Acetabularia* (cf. Fig. 335).

The thallus of the **Codiaceae** consists of a felted mass of interwoven, branching, aseptate filaments (Fig. 422C). In *Halimeda* (Fig. 422B), widespread in warmer seas, the disc-shaped thallus sections are encrusted with lime.

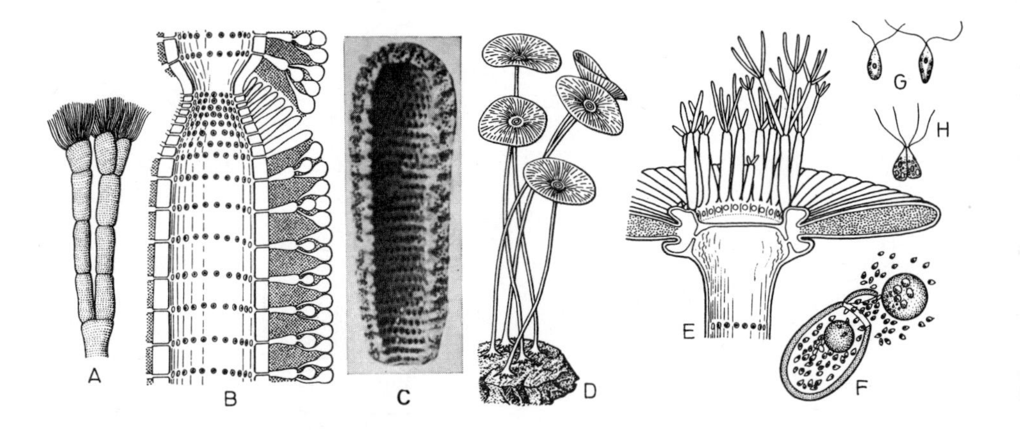

Fig. 423. Dasycladaceae. **A.** *Cymopolia barbata*, upper part of a plant. ×4. **B.** The same, longitudinal section of part of the thallus; shaded: calcareous encrustation. ×40. **C.** *Dactylopora cylindracea* (Tertiary), calcareous tubule. ×4. **D.** to **H.** *Acetabularia mediterranea.* **D.** Mature thalli. ×1. **E.** Longitudinal section of the cap, with crown of sterile branches above, and below the scars of the lower whorl of sterile branches. **F.** Cyst liberating gametes. ×100. **G.** Gametes. ×300. **H.** Fusion of gametes. ×300. **A.** and **B.** After Solms-Laubach. **C.** After Morellet. **D.** and **E.** After Oltmanns. **F.** to **H.** After de Bary and Strasburger.

In the **Valoniaceae** (Fig. 422D) a large club-shaped cell cuts off smaller cells much constricted at the base; these may grow to the size of the parental cell, giving a cushion-like plant. The cells of *Valonia*, with their one large vacuole, many nuclei and numerous parietal chloroplasts, make an excellent subject for permeability and cell wall studies (Fig. 60).

With few exceptions the Siphonales reproduce only sexually, meiosis preceding the formation of gametes (generally anisogamous; gametes biflagellate, Fig. 422E; parthenogenesis also occurs); the plants are therefore diplonts. In *Caulerpa* the entire thallus of both male and female plants is devoted to the formation of gametes, and releases whole clouds of green sex cells, whereupon the empty plant quickly dies. On the other hand, *Acetabularia* produces its sex cells from the secondary nuclei in a complicated manner in the radial chambers of the cap.

An exception to the rule is provided by *Derbesia* (*inter alia*), which is sporophytic. The ovate sporangia form short side branches delimited by a septum (Fig. 422F, F′). They liberate zoospores with a crown of numerous flagella (F″). The filamentous *D. marina* shows heteromorphic alternation of generations (Fig. 422E, F), its gametophyte being the bladder-like dioecious *Halicystis ovalis* (Fig. 422E). Another heteromorphic *Derbesia* species, *D. tenuissima*, is the sporophyte of *Halicystis parvula*, while *D. neglecta* is the sporophyte, not of another *Halicystis* species but of *Bryopsis halymenia*; the various *Derbesia* species belong therefore to quite different siphonaceous gametophytes. It is very probable that meiosis takes place during the formation of swarmers.

Order 7: **Conjugales** (Conjugatae). These have neither zoospores nor flagellate gametes; they are therefore sometimes known as Acontae. In sexual reproduction the zygote is formed from two similar non-flagellate gametes each derived from the entire protoplast of a vegetative cell (Fig. 425). After a long period of dormancy the resting zygote germinates and undergoes meiosis: the Conjugales are therefore haploid organisms. The cells are uninucleate, but the nucleus has an aberrant fine structure. Several thousand species of diverse form occur throughout the world, almost entirely in fresh water, as benthos or to a lesser extent as plankton. They are either unicellular (Desmidiaceae) or form unbranched, unattached filaments (coenobia) which readily break up into unicells (Zygnemaceae).

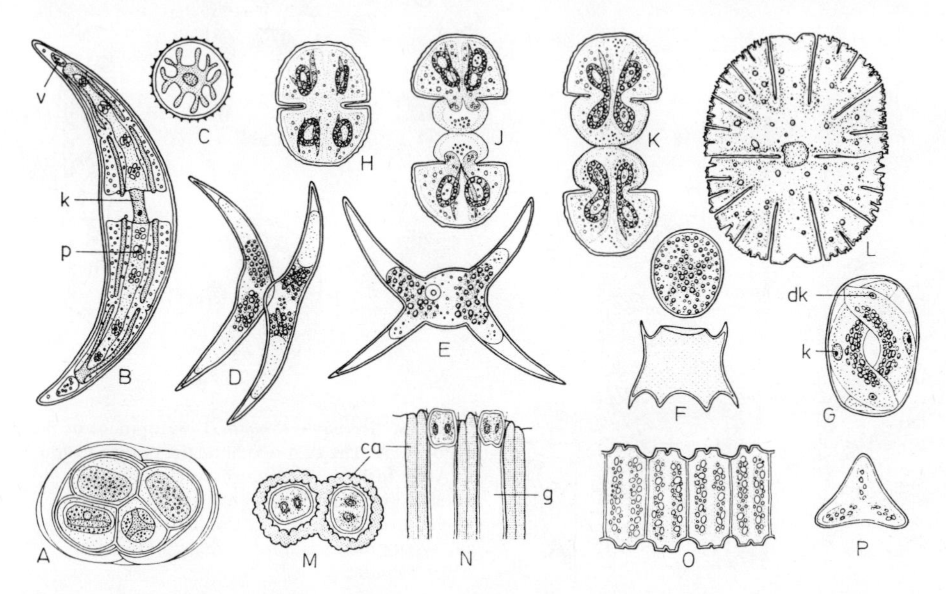

Fig. 424. Mesotaeniaceae and Desmidiaceae. **A.** *Mesotaenium braunii.* ×280. **B.** *Closterium moniliferum.* ×400. **C.** *Closterium regulare*, transverse section. ×400. **D.** and **E.** *Closterium parvulum*, conjugation. ×300. **F.** *Closterium rostratum*, liberation of the zygote from the spore-wall. ×200. **G.** *Closterium* sp., division of the zygote. ×200. **H.** *Cosmarium botrytis.* ×280. **J.** and **K.** The same dividing. ×280. **L.** *Micrasterias denticulata.* ×125. **M.** and **N.** *Oocardium stratum*, seen from above and in longitudinal section. ×320. **O.** *Desmidium swartzii*, part of a cell chain. **P.** The same, transverse section of one cell. ×350. ca, calcareous sheath; dk, degenerating nucleus; g, mucilage stalk; k, nucleus; p, pyrenoid; v, vacuole. **A.**, **D.** to **F.** and **H.** to **K.** After de Bary. **B.** After Palla. **C.** and **L.** After Carter. **G.** After Klebahn. **M.** and **N.** After Senn. **O.** and **P.** After Delponte.

The **Mesotaeniaceae** occur as unicells or in gelatinous colonies (Fig. 424A). The cell wall consists of a single piece and is not sculptured. *Mesotaenium berggrenii* and *Ankylonema nordenskioeldii* both have red cell-sap and contribute to 'red snow' on glaciers in the Alps, Arctic and Antarctic (cf. pp. 462, 491).

The **Desmidiaceae**, which occur particularly in the acid waters of peat-bogs, are among the most beautiful of the algae and show a great variety of forms. For example, the cells may be crescent-shaped (*Closterium*, Fig. 424B), or very often they are constricted in the middle and biscuit-shaped (*Cosmarium*, Fig. 424B) or stellate (*Micrasterias*, Fig. 424L). The cell wall is generally sculptured and consists of two equal portions. The cell contains in each of the two perfectly symmetrical halves a large, very complex, central (not parietal) chloroplast with one or more pyrenoids (Fig. 424B, C); the nucleus lies in the middle of the cell. Many desmids can move: mucilaginous threads extruded through pores in the cell wall slowly push them along. The cell wall is multistratose (cellulose, pectin, mucilage) and is sometimes impregnated with iron, or incrusted with a calcareous sheath (Fig. 424M, N).

Vegetative reproduction takes place by division: as in the diatoms each half cell wall has to be laid down afresh (Fig. 424J, K). The two half-walls are therefore of different ages. In some genera the daughter cells cohere in chains. In conjugation two genetically different cells come together and surround themselves with mucilage; the cell walls rupture at the constriction (Fig. 424D–F), the protoplasts emerge as amoeboid gametes through a projecting conjugation canal (which soon becomes mucilaginous) and unite to form the zygote (E); the four empty semi-cell walls of the parents remain beside the mature zygote, the wall of which often has prominent spines. On germination, in most desmids, two at least of the four haploid cells produced by meiosis degenerate so that no more than two daughter plants are produced.

The most familiar genus of the filamentous, invariably unbranched **Zygnemaceae** is *Spirogyra* (Figs. 34, 425). The numerous species frequently occur in the spring as green,

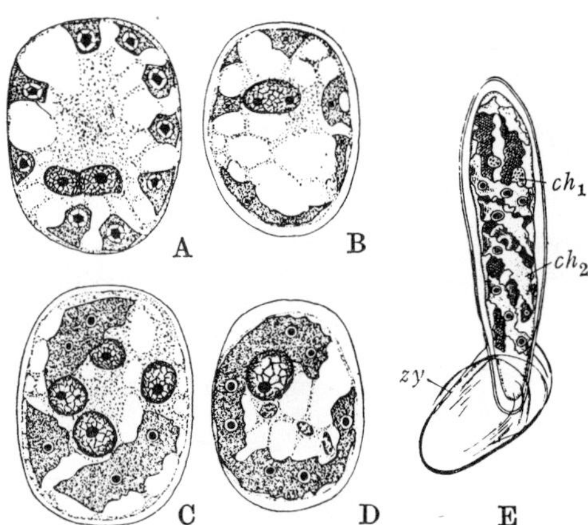

Fig. 426. *Spirogyra longata*. Development of the zygote. **A.** The two sexual nuclei before fusion. **B.** After fusion. **C.** Division of the zygotic nucleus into four haploid nuclei. **D.** Three nuclei disintegrating. **E.** Uninucleate germling: zy, zygote membrane; ch_1, ch_2, chloroplasts. After Tröndle.

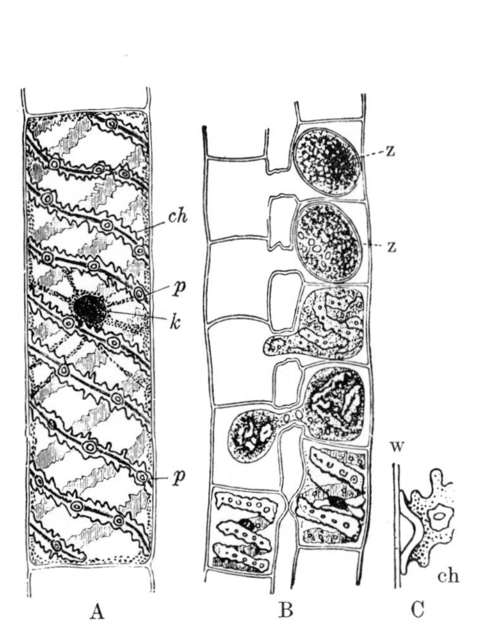

Fig. 425. Spirogyra. **A.** *S. jugalis*, cell. ×250. **B.** *S.quinina*, conjugation. ×240. **C.** *S. longata*, part of a chloroplast against the cell wall. ×750. ch, chloroplast; k, nucleus; p, pyrenoid; w, cell wall; z, zygote. **A.** and **B.** After Schenck. **C.** After Kolkwitz.

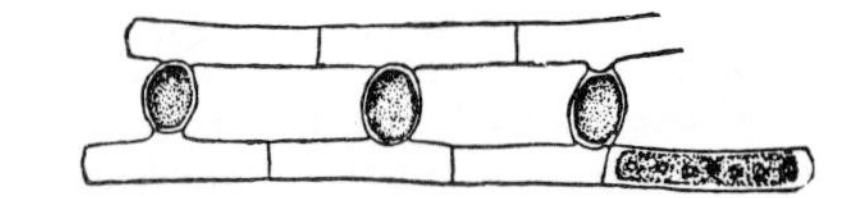

Fig. 427. *Mougeotia pulchella*. Zygotes in the middle between the gametangia. ×100. After Transeau.

free-floating tangled masses in still water. The filaments grow by the transverse division and elongation of all the cells; every cell is identical, and the filaments lack polarity. Their smooth cellulose walls have no pores and are covered with a delicate sheath or 'cuticle' containing pectin which swells to become slightly mucilaginous and which holds the cell colony together (coenobium). The cross walls develop by growth inwards like a closing iris-diaphragm—a mode of growth rarely seen in other uninucleate cells. Vegetative reproduction is effected by the dissociation of the filaments at the cross walls into short lengths or even into unicells (p. 203). Strictly speaking, *Spirogyra* is unicellular, but the cells are held together as filaments by the superficial membrane (p. 73).

Each cell contains a single nucleus and one or more chloroplasts (Fig. 425A, C, ch) in the form of a spiral, channelled band lying against the wall; pyrenoids (p) are embedded in each chloroplast. *Zygnema* has two stellate chloroplasts in each cell; *Mougeotia* (Figs. 331, 427) has one in the form of a flat axial plate which reacts phototactically.

At the onset of conjugation two sexually different (phenotypically determined) but morphologically identical filaments become parallel and closely stuck together; papillae swell up along the line of contact between the cells so that the two filaments are pushed apart again and assume a ladder-like (scalariform) appearance (Fig. 425B). By the dissolution of the walls at the point of contact two adjacent papillae form a conjugation canal between the gametangia; the protoplast of each ♂ cell then moves as an amoeboid gamete through the canal and fuses with the protoplast of the ♀ cell. After contraction and the loss of water a resting zygote (z) is formed with a multistratose, thick, brown wall and densely filled with starch and oil. The chloroplast of the ♂ gamete degenerates. On germination three of the four nuclei formed by meiosis also degenerate (Fig. 426D) so that a single haploid germling results; this gives rise to a new filament by elongation and cell division. Any cell of a filament may become a gametangium.

In some *Spirogyra* species two adjacent cells form a lateral conjugation canal through which one gamete passes into the neighbouring cell, while some species of *Mougeotia* and *Zygnema* form the zygote midway in the conjugation canal (Fig. 427), thus recalling the desmids.

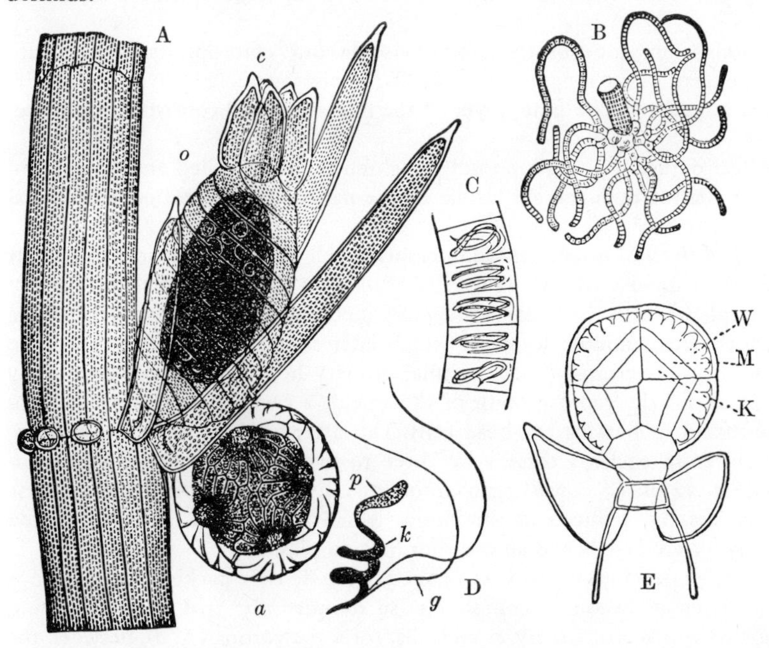

Fig. 428. Characeae. **A.** and **D.** *Chara fragilis.* **B.**, **C.** and **E.** *Nitella flexilis.* **A.** Side view with 'antheridium', a, and oogonium, o, with its enveloping filaments and corona. ×50. **B.** Manubrium with head-cell and spermatogenous filaments. **C.** Cells of the spermatogenous filaments, each with one spermatozoid. **D.** Spermatozoid; g, flagellum; k, spirally-twisted, elongate nucleus; p, cytoplasm with reserve starch. ×540. **E.** Longitudinal section through a young 'antheridium'; K, head-cell; M, manubrium; W, wall. **A.**, **B.**, **C.** and **E.** After Sachs. **D.** After Strasburger.

The cell structure and method of reproduction of the Conjugales sharply delimit them as a group which must have diverged very early from primitive green algae, with the total loss of all flagellate stages.

Many systematists treat the Conjugales as a separate class.

Order 9: **Charales** (stoneworts). The few genera of the very highly organized family **Characeae**, the only living family, with some 300 species, are benthonic and often form 'meadows' several decimetres high in ponds and streams. They are characterized by their candelabrum-like thalli (Figs. 73, 428) with whorls of branches separated by long internodes (p. 76) and by their gametangia: the erect oogonia (or nucules) are invested in spiral filaments, while the antheridia are arranged in filaments inside a hollow spherical structure (antheridiophore or globule; Fig. 428A). No spores are produced and there is no alternation of generations. The Charales are thus pure haplonts with a zygotic change of nuclear phase (cf. p. 42).

The short shoots arising at the nodes are in turn composed of nodes and internodes; they are either simple or bear at the nodes short second-order lateral branches. In each whorl a long shoot similar to the main axis arises in the axil of a single one of the short shoots (Fig. 73). The main axis is anchored to the substratum by colourless, branched filamentous rhizoids arising from the basal nodes. In contrast to the majority of benthonic algae, the Characeae colonize sand and mud rather than hard substrata (rocks, wood). Some produce on the lower parts of their axes bulbil- or tuber-like overwintering organs packed with starch.

Both main and lateral branches grow by an apical cell (Fig. 73B) which cuts off segments by the formation of transverse walls. Each segment again divides transversely into two cells, the lower of which elongates to form an internode (i), while the upper becomes a nodal cell which develops by longitudinal divisions as a nodal disc. The nodal cells remain short but continue to divide lengthwise in a regular sequence producing the lateral branches and, on the lower part of the axis, the rhizoids. While in *Nitella* each internode consists of a single naked cell, in *Chara* each internodal cell is invested in a cortical layer of longitudinal rows of cells which develop at the nodes from the basal cells of the lateral axes.

Every cell contains, in addition to numerous discoid chloroplasts, a normally developed nucleus, except in the long internodal cells, where the nucleus divides amitotically into numerous fragments. The inner layer of the bistratose wall consists of cellulose, the outer of callose.

Both the spherical orange antheridiophores (or globules; often called antheridia) and the ovate green oogonia (or nucules) are visible to the naked eye at the nodes of lateral branches.

The ♂ organ (Fig. 428A, E) originates from a mother cell which first divides into eight cells. Each octant then divides by two tangential walls into three cells (E). In this way arise the eight peripheral cells (shields) which are incompletely partitioned by septa extending inwards; the eight middle segments which later elongate radially and form the manubria; and the eight innermost cells (capitula, primary head cells) which eventually round off. As a result of the rapid growth of the shields a cavity develops inside, into which project the manubria bearing the head cells. The latter produce three to six secondary head cells, and from each of these arise three to five long unbranched spermatogenous filaments (Fig. 428B, C) consisting of numerous disc-shaped cells. Each of these cells constitutes a greatly reduced antheridium which liberates a spirally twisted spermatozoid (D) with two flagella and an eye-spot, but no plastids.

The oogonium (A, o) develops initially as a naked ovum densely packed with oil-drops and starch grains; later it becomes completely surrounded by five spirally wound filaments, the ends of which, cut off by cross walls, form the corona (A, c), between the cells of which the spermatozoids enter. After fertilization the zygote (oospore) surrounds itself with a thick, colourless membrane. The inner walls of the enveloping filaments also thicken and become brown; they commonly become encrusted with calcium carbonate, while the outer walls disintegrate soon after the shedding of the 'fruit' (resting organ). Meiosis takes place on germination of the zygote; three of the four haploid nuclei degenerate so that only one germling develops.

The encrusted Charophyceae are important contributors to calcareous tufas.

Fossil Characeae are known since the Devonian, particularly as zygotes; of the three families formerly existing only one survives today.

The peculiar structure of the thallus, above all of the antheridiophores with their characteristic shields (without parallel in the entire Plant Kingdom), and the spiral spermatozoids (not found in any other algae) indicate that the Charophyceae occupy an extremely isolated position without close relationships to the other green algae.

In retrospect the **Chlorophyceae** appear as a naturally related group of thallophytes ascending gradually from the primitive Flagellatae to highly developed forms.

In the algae which are provided with cell walls reproduction (formation of zoospores and gametes) does not proceed by repeated division into two as in the naked flagellates; instead the cytoplasm undergoes cleavage after the completion of nuclear division, just as in the germination of the cysts of flagellates (Fig. 408G). It is therefore likely that algal evolution has proceeded from the flagellate cysts—an idea which can be developed further for the filamentous habit.

Within each order development can be traced from simple to more highly differentiated forms. This is particularly clear in the Volvocales, where primitive naked unicells lead up to multicellular colonies with identical cells, each able to live and reproduce independently, and ultimately to multicellular individuals with definite division of labour and with plasmodesmata between the cells; but a similar ascent from simple to complex is generally evident in the other orders. The complexity of the heterotrichous thallus, differentiated into a prostrate base and erect filaments reaching up from the substratum, is only rarely exceeded in more advanced types of organization (e.g. the parenchymatous). The Charales have attained an advanced level in another manner. The chloroplast, cup-shaped in primitive forms, breaks up into nets and discs in advanced ones. The sexual processes also show progressive differentiation from isogamy through anisogamy to simple oogamy and ultimately to the most advanced level in which the ovum is no longer released, but is fertilized in the oogonium; even gametangiogamy occurs. This trend can be followed independently in most of the orders. Particularly noteworthy is *Coleochaete*, in which fertilization stimulates the filaments supporting the oogonium to the production of a primitive 'fruit'; the Charophyceae also exhibit a kind of 'fruit', although very different in development.

Those Chlorophyceae which are primitive develop as haploid plants and the zygote alone is diploid. Meiosis occurs on germination of the zygote and can be regarded as the ultimate phase of sexual reproduction, usually leading to the production of four haploid zoospores which are therefore meiozoospores, in contrast to the asexual swarmers liberated from the haploid plant. In these haploid species both asexual and sexual organs of reproduction are very commonly produced from the same plant (*Ulothrix* scheme, Fig. 429A), but each individual does not necessarily produce both kinds of reproductive organ—frequently an extended cycle of asexual reproduction precedes the eventual appearance of sexually reproducing individuals. Thus both haploid asexual and haploid sexual stages may occur. If meiosis in the zygote were deferred until the end of the asexual stage there would be alternation between a diploid sporophyte and a haploid gametophyte.

This may indeed have been the origin of alternation of generations, as can be observed now in *Stigeoclonium subspinosum* (Chaetophorales): the life cycle is normally of the *Ulothrix*-type (Fig. 429A), but in certain circumstances meiosis fails to take place in the zygote, which instead grows into a microscopic, few-celled filament, the cells of which form four to eight spores by meiosis, thus paving the way for the development of a diploid sporophyte.

Meiosis is generally coupled with the genetic determination of sex (Fig. 426). The two generations are sometimes identical or they may be different (heteromorphic, *Urospora-Codiolum, Halicystis-Derbesia,* etc., (Fig. 429C).

Haploid species without alternation of generations would appear to be the most primitive. Alternation of generations has arisen independently in the different orders.

There are many deviations from the normal types of alternation of generations, e.g. the postponement of meiosis, loss of a generation, etc.; the latter is the case in, for

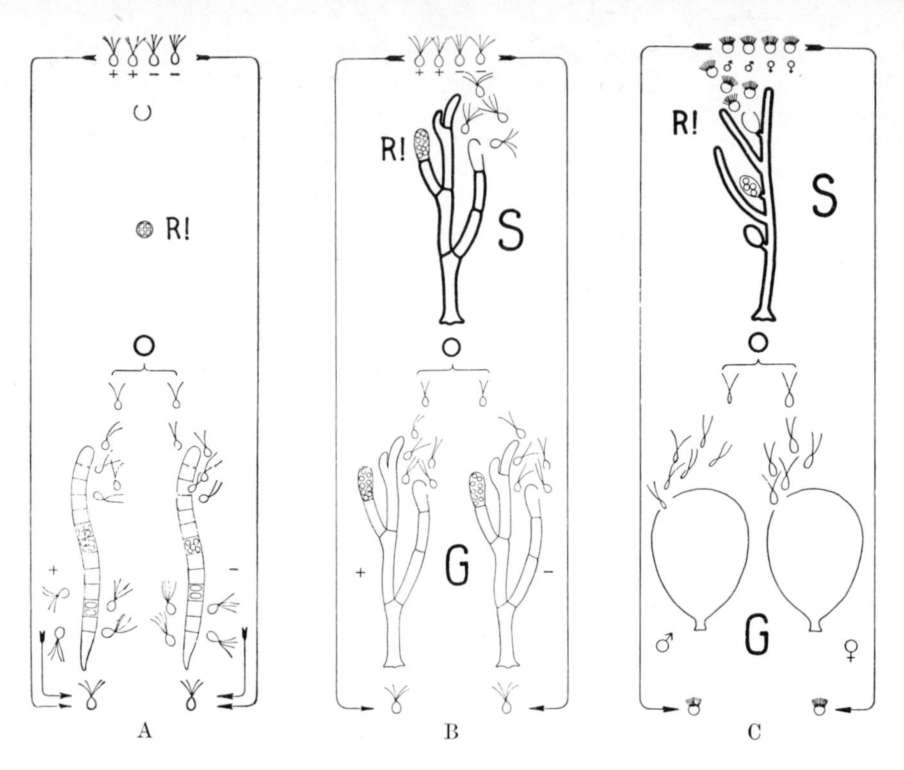

Fig. 429. Diagrammatic representation of the main types of nuclear cycle and alternation of generations in the green algae. **A.** *Ulothrix.* **B.** *Cladophora.* **C.** *Halicystis-Derbesia.* Diploid phase, thick lines; haploid phase, thin lines. G, gametophyte. S, sporophyte. O, zygote. R! meiosis. After Harder.

example, the Siphonales, which are generally entirely diplont, meiosis occurring during gametogenesis. The intercalation of a number of similar generations (accessory reproduction) is not uncommon. In this case diploid sporophytes produce, instead of zoospores as a result of meiosis, diploid swarmers which grow into new sporophytes, while haploid gametophytes produce swarmers which develop without sexual union into new gametophytes. Either generation can therefore propagate itself without alternation of generations—but sooner or later this occurs.

Even more extreme deviations from the normal expression of alternation of generations occur. Thus it has been shown that the following three algae, formerly regarded as independent, are alternative forms of each other: *Schizochlamys* (Volvocales, with pseudocilia), *Phacosphaera* (Chlorococcales, with a peculiar sheath of calcite) and a *Urococcus*-like form of doubtful systematic position with very remarkable laminated gelatinous masses secreted on one side only of the cell.

The green algae are undoubtedly a very ancient group of lower plants, but only the resistant calcified thalli of marine Siphonales (especially Dasycladaceae) have been recognized as far back as the Cambrian. Since the Dasycladaceae were already present in great variety in the Ordovician, they must have originated very much earlier. Of eighty-six genera which have occurred over the course of 500 million years only ten still survive, *Acetabularia* among them. The unmistakable zygotes of the Charales are found as far back as the Devonian.

Class VI: Phaeophyceae, brown algae

The almost exclusively marine Phaeophyceae are brown, often very large algae, known in their robuster forms as kelp, oarweed or wrack. The chromatophores (phaeoplasts, usually numerous in each cell) contain chlorophyll *a* (no *b*, a little *c*), *β*-carotene and a number of xanthophylls (especially the brown fucoxanthin which masks the other pigments).

Starch never occurs as a photosynthetic or reserve product; instead there are (*inter alia*) mannitol (an alcohol related to the sugars), oil and especially (up to half the dry weight) the dextrin-like polysaccharide laminarin; this is formed in minute pyrenoids just outside the chromatophores and diffuses in solution into the vacuoles. Cellulose has been detected in the inner cell wall, pectic substances in the outer, and in some families the gel-like polysaccharide algin (calcium salt of alginic acid; 20–60 per cent of the dry weight in *Laminaria*) and the carbohydrate fucoidin also occur. The cells are uninucleate.

Almost without exception the motile stages (zoospores and gametes) have two unequal flagella laterally inserted on the pear- or spindle-shaped body; the longer flagellum, beset with flimmer hairs, is directed forwards in motion, the other backwards (Fig. 430). Near the flagella is a red-brown eye-spot and at the broader, hinder end is one (rarely more) brown chromatophore.

The brown algae are remarkably diverse in form (cf. Figs. 72, 79, 433, 437, 440). Their habit ranges from minute, branching filaments only one cell thick and from heterotrichous thalli to massive plants many cells thick, sometimes tree-like in form and reaching many metres in length; the organs of these look superficially like leaves, stems and roots (for phylloid, cauloid and rhizoid see p. 80). There are, however, no unicellular forms.

The highly differentiated form is reflected histologically. Growth is intercalary or by means of a large apical cell (Figs. 70E, 72, 79), and the tissues are divided as a rule into an outer photosynthetic tissue and an inner region of colourless storage cells (Fig. 436). In some orders the cells divide longitudinally as well as transversely giving rise to true parenchyma (as in Cutleriales, Dictyotales, Laminariales and Fucales; pp. 79–80). The walls bear pits in Laminariales and Fucales, and in some there are primitive mechanical tissues. Tubes resembling the sieve-tubes of cormophytes, and like them responsible for translocation, occur in the Laminariales (Fig. 431).

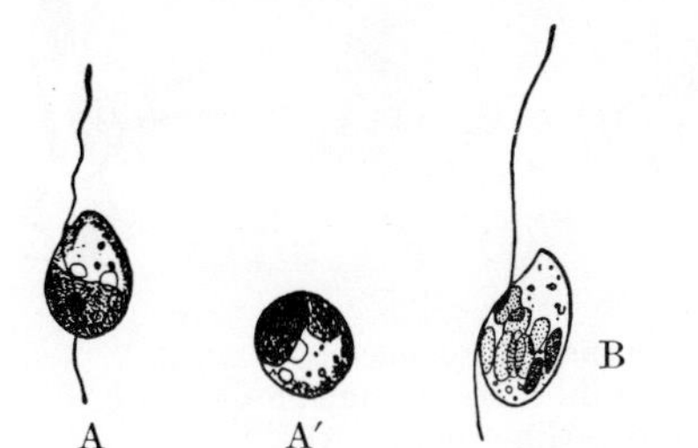

Fig. 430. Zoospores, one of which (**A′**.) is rounded off before germination. **A′**. *Chorda filum.* ×1200. After Reinke. **B.** *Ectocarpus globifer.* After Kuckuck.

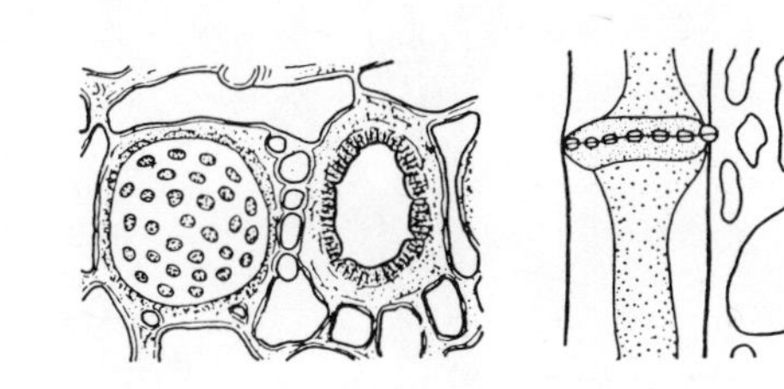

Fig. 431. Sieve-tube cells of *Macrocystis* in transverse and longitudinal section. ×250. After Will and Oliver.

Unicells (motile or non-motile) and unbranched filaments are completely lacking and there are no primitive haplonts undergoing meiosis on the germination of the zygote (as in the green algae *Volvox* and *Ulothrix*). Practically all brown algae undergo alternation of generations with meiosis during spore formation; they therefore have a haploid gametophyte and a diploid sporophyte.

Conditions in the sea are always relatively uniform, so that resting organs are unnecessary: the zygote germinates directly (with a germination tube and not swarmers).

Order 1: **Ectocarpales.** The majority of the Phaeophyceae belong here.

The brown, branching tufts of *Ectocarpus* recall in habit the green alga *Cladophora* (Fig. 419); the genus is very widely distributed (*E. siliculosus* is cosmopolitan) and common in shallow water round our coasts. The plant is attached to solid substrata (rocks, larger algae) by creeping rhizoids. There is no apical cell and growth is intercalary, but reproduction is limited to certain cells.

Zoospores are produced in large numbers in initially uninucleate, bladder-shaped sporangia (Fig. 432B) by repeated division of the nucleus and chromatophores (cf. Fig.

438); these so-called unilocular sporangia are only borne on diploid plants, and since meiosis invariably occurs in them, the zoospores produced are haploid. Gametes are typically formed in compound, plurilocular structures (Figs. 432A, 433, 435). It is no longer the case that every cell is capable of becoming reproductive. The internal walls break down when the gametes are released (Fig. 433).

The zygote resulting from the fusion of two isogametes (Fig. 434, 2–5) germinates directly, without meiosis or the formation of swarmers (contrast the Chlorophyceae), to produce a diploid plant (very similar to the haploid one) which only produces unilocular sporangia (sporophyte). After meiosis the zoospores develop into rather less robust gametophytes with plurilocular sporangia, and so on. The alternation of generations therefore corresponds to the *Cladophora* scheme (Fig. 429B).

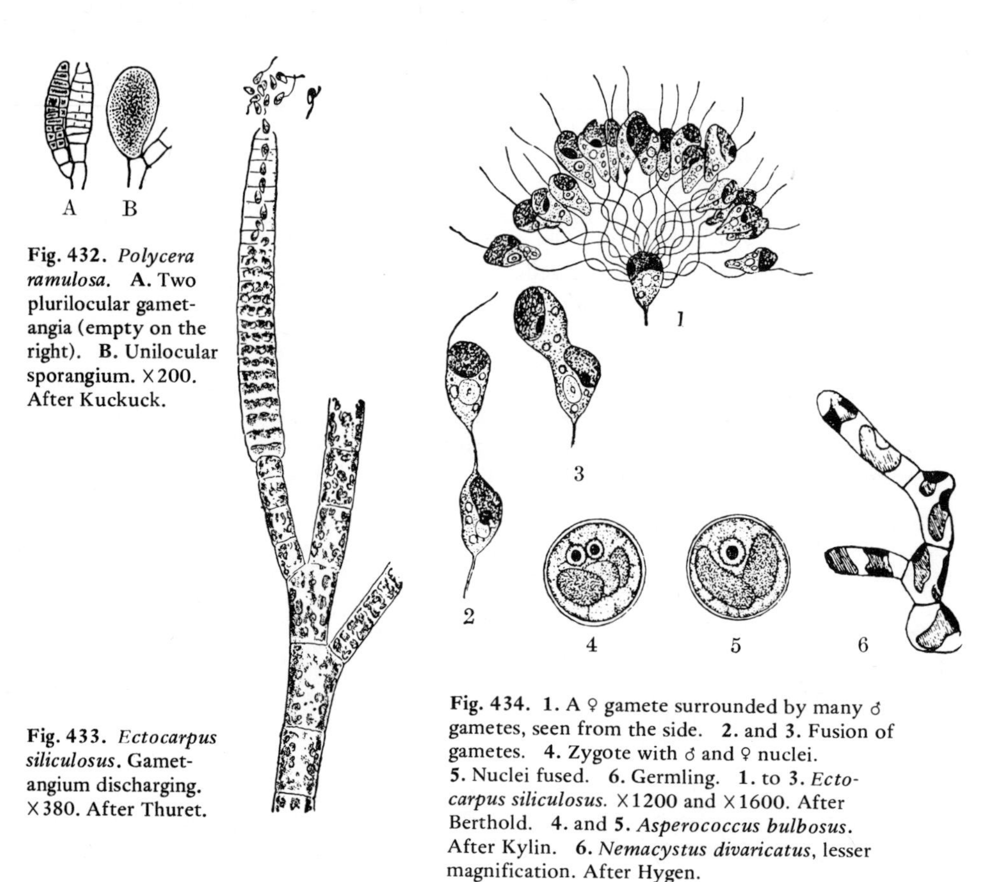

Fig. 432. *Polycera ramulosa.* **A.** Two plurilocular gametangia (empty on the right). **B.** Unilocular sporangium. ×200. After Kuckuck.

Fig. 433. *Ectocarpus siliculosus.* Gametangium discharging. ×380. After Thuret.

Fig. 434. **1.** A ♀ gamete surrounded by many ♂ gametes, seen from the side. **2.** and **3.** Fusion of gametes. **4.** Zygote with ♂ and ♀ nuclei. **5.** Nuclei fused. **6.** Germling. **1.** to **3.** *Ectocarpus siliculosus.* ×1200 and ×1600. After Berthold. **4.** and **5.** *Asperococcus bulbosus.* After Kylin. **6.** *Nemacystus divaricatus*, lesser magnification. After Hygen.

In addition to unilocular sporangia the sporophyte of *Ectocarpus* may also develop plurilocular sporangia with diploid swarmers (mitozoospores) which again give rise to sporophytes.

In some cases (e.g. *Ectocarpus confervoides*) the sporophyte is considerably more robust than the gametophyte. In some epiphytic species of *Ectocarpus* and *Pylaiella* the two generations live on quite different genera of host plants (e.g. the sporophyte of *P. litoralis* on *Fucus*, the gametophyte on *Ascophyllum*). *Elachista fucicola* is a very common epiphyte on *Fucus*.

The primitive Ectocarpales have gametes identical in form and behaviour (isogamous, Fig. 225A), but in species rather more advanced sexually, although the gametes are morphologically identical they differ in motility, the ♀ quickly settling down (Fig. 434). In even higher forms morphological anisogamy also occurs.

From *Order 2:* **Sphacelariales** (distinguished *inter alia* from the Ectocarpales by their apical cell, Fig. 70E) we pass directly to:

Order 3: **Cutleriales.** *Cutleria multifida*, a medium-sized alga of the warmer European seas, is anisogamous with small ♂ and large ♀ gametes produced in microgametangia on ♂ plants and megagametangia on ♀ ones (Fig. 435). Alternation of generations is heteromorphic: the gametophyte consists of erect, forking ribbons, fimbriate at the ends (Fig. 443), while the sporophyte is smaller, flat, lobed, prostrate and crustose (known as *Aglaozonia*, Fig. 443).

Order 4: **Dictyotales.** These are widespread marine algae. *Dictyota dichotoma* (Fig. 79A) is found on our coasts. The thallus is medium-sized, flat, and typically dichotomous as a result of the activity of a large apical cell (Fig. 79B, E); it is differentiated into peripheral assimilatory cells and central storage cells (Fig. 436). The two generations are identical in form (Fig. 443). Meiosis in the unilocular sporangium of the diploid sporophyte invariably results in only four non-flagellate meiospores (tetraspores; Fig. 436D).

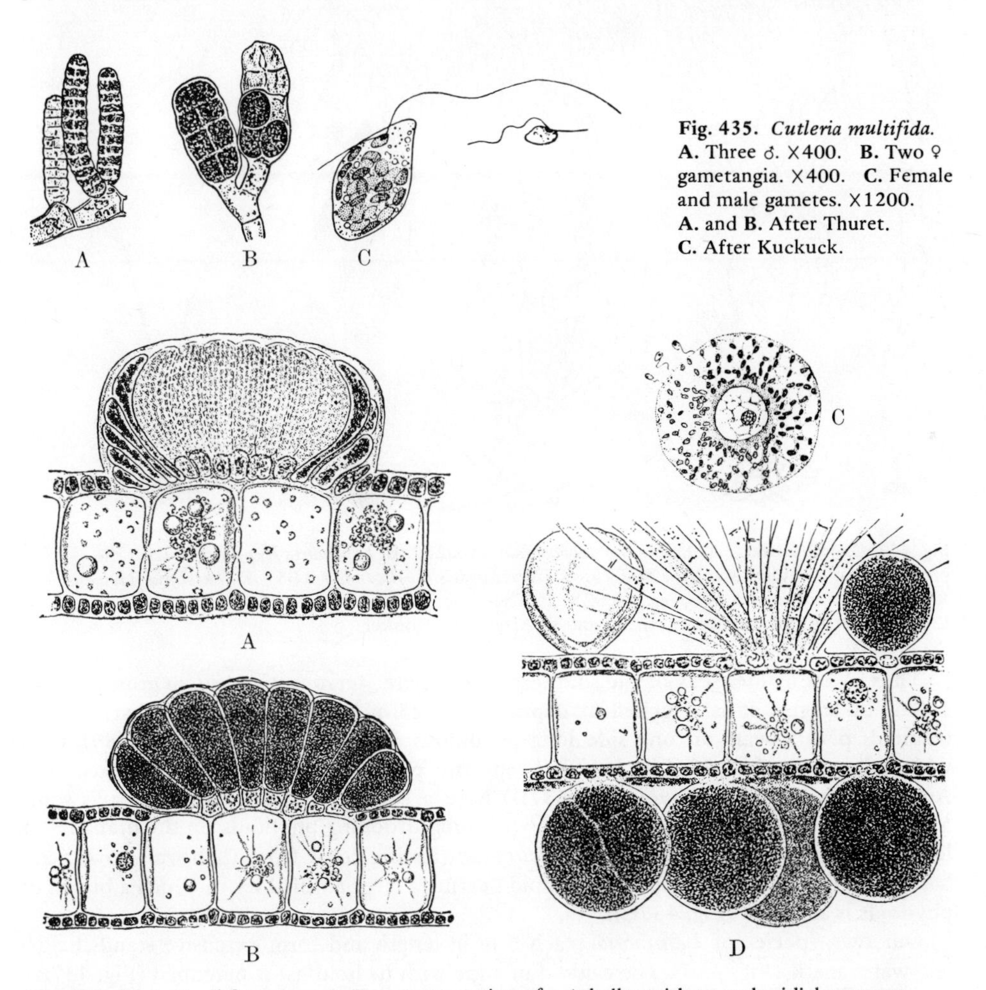

Fig. 435. *Cutleria multifida.* **A.** Three ♂. ×400. **B.** Two ♀ gametangia. ×400. **C.** Female and male gametes. ×1200. **A.** and **B.** After Thuret. **C.** After Kuckuck.

Fig. 436. *Dictyota dichotoma.* **A.** Transverse section of a ♂ thallus with an antheridial sorus surrounded by a cup-like wall of sterile cells. ×200. **B.** Transverse section of a ♀ thallus with oogonial sorus. ×200. **C.** Ovum with spermatozoids. ×400. **D.** Thallus section with tetrasporangia, one of them empty. ×200. **A.**, **B.** and **D.** After Thuret. **C.** After Williams.

Sexual reproduction has progressed to oogamy. The oogonia and the plurilocular antheridia always occur in groups (sori) on different plants (A, B). Each oogonium contains a single, large, brown, non-flagellate ovum, which, like the tetraspores, passively floats in the water. It is fertilized by a pear-shaped spermatozoid (C) with a much reduced chromatophore and only one lateral flagellum (with flimmer hairs). A second reduced flagellum with its own basal granule exists as a minute stub hidden in the cytoplasm. The gametangia are only produced during the summer months. The release of gametes depends both on the sun and moon, taking place in the first hour after dawn on only two days in the month.

The fan-shaped *Padina*, frequent in warmer seas, grows by means of a marginal meristem.

We pass over the Chordariales (with *Chordaria*), Sporochnales, Desmarestiales (with *Desmarestia*, sea sorrel, landlady's wig) as of lesser importance and turn to:

Order 5: **Laminariales.** The sporophytes show a high degree of morphological and anatomical differentiation (Fig. 437A–F) and often reach enormous size.

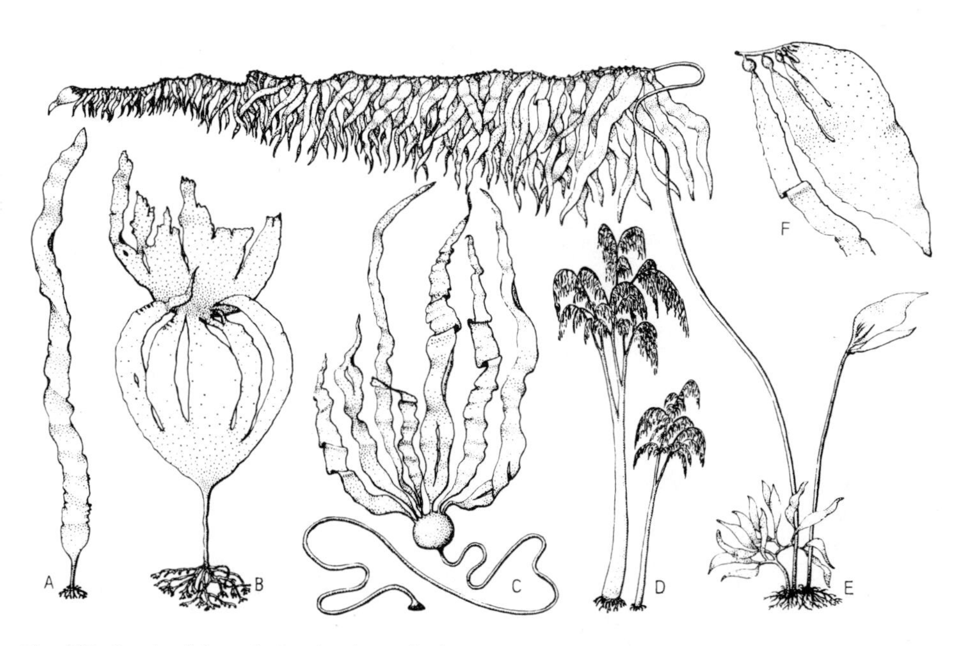

Fig. 437. Laminariales. **A.** *Laminaria saccharina.* ×0.025. **B.** *Laminaria hyperborea* with the previous year's phylloids above. ×0.025. **C.** *Nereocystis luetkeana.* ×0.05. **D.** *Lessonia flavicans.* ×0.025. **E.** *Macrocystis pyrifera.* ×0.025. **F.** The same, apex of thallus. ×0.05. **B.** After Schenck. **C.** After Postels and Ruprecht. **D.**, **E.** and **F.** After J. D. Hooker.

Thus in the colder seas of the southern hemisphere *Macrocystis pyrifera* grows to over 100 m in length; it is attached at depths of 2–25 m by a claw-like holdfast; its axis (cauloid, p. 80) bears on one side long pendulous thallus lobes (phylloids, p. 80), each with a large basal air-bladder which keeps the phylloid afloat near the surface. The Antarctic species of *Lessonia* (Fig. 437D) have a branching, trunk-like main axis some 5 m long and as thick as a man's thigh, with long drooping phylloids on the branches; it looks remarkably like a palm tree. *Nereocystis* (Pacific coast from California to Alaska) has a long (up to 100 m) rope-like cauloid bearing a large air-bladder to which a bunch of phylloids is attached (Fig. 437C).

Our own species of *Laminaria* reach 5 m in length and form extensive stands below low-water mark (Fig. 450). The cauloid or stipe with its holdfast is perennial (Fig. 437B). It bears a phylloid or lamina, many cells thick, which is renewed at the base each year (towards the end of the winter) by an intercalary meristem. The old lamina is pushed away and gradually dies (Fig. 437B). The lamina of *L. saccharina* (sea belt; Fig. 437A) is simple, that of *L. digitata* and others (oarweed, tangle; Fig. 437B) is palmately divided. *Chorda filum* (sea laces, dead men's ropes) has a stringy, undivided thallus several metres long.

The sporophytes regularly alternate with microscopically small ♂ and ♀ gametophytes which grow in felted masses (Fig. 439). The difference in structure of the ♂ and ♀ plants is characteristic of the Laminariales (secondary sexual characteristic).

The gametophytes arise from zoospores. The male plants are relatively richly branched, grow quickly and consist of many small cells (Fig. 439B); they bear apical antheridia, each with a biflagellate spermatozoid. The female gametophytes grow more slowly and have relatively few but larger cells (A, C)—they may even be reduced to a single tubular cell (D)—and bear oogonia, each with one ovum. The naked ovum emerges through a pore at the apex of the oogonium, where it generally remains (A, o) and develops after fertilization into a diploid sporophyte (A, k_1–k_3). The surface of the latter bears extensive areas of both tubular sterile cells (paraphyses) and club-shaped unilocular sporangia (Fig. 438) in which numerous biflagellate meiozoospores are formed after meiosis (with genotypic determination of sex).

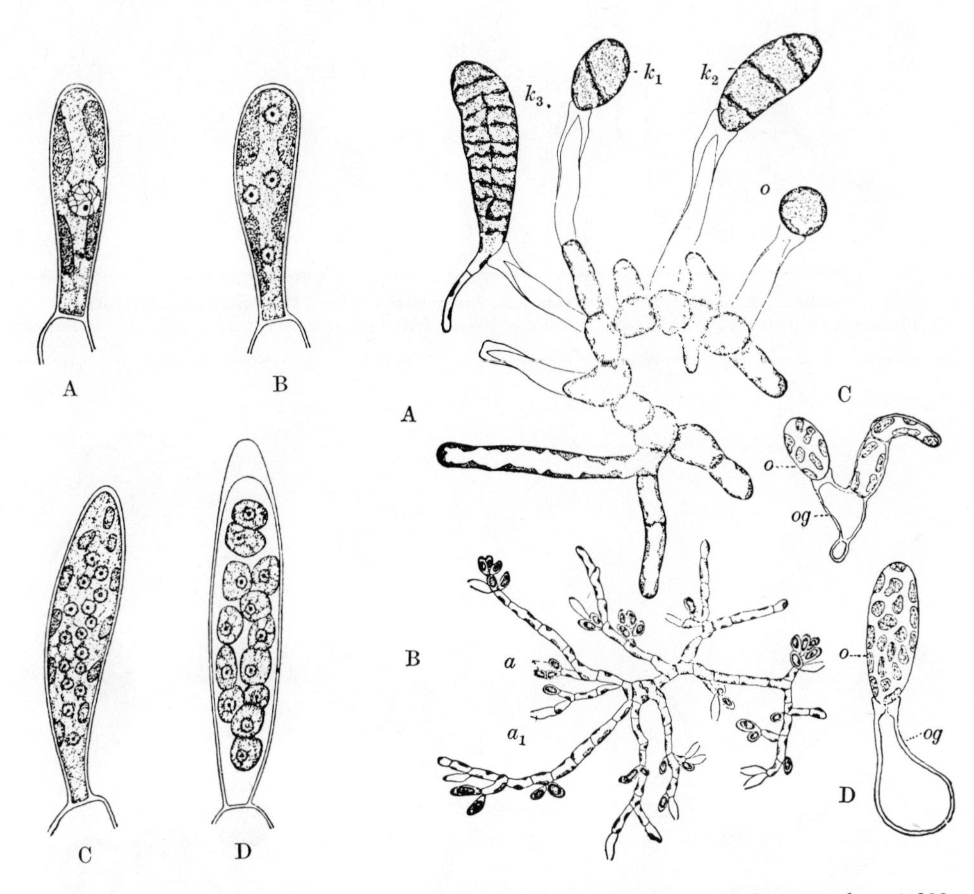

Fig. 438. *Chorda filum.* Development of sporangium. **A.** Uninucleate. **B.** 4-nucleate. **C.** 16-nucleate. **D.** Nearly mature zoospores. After Kylin.

Fig. 439. *Laminaria.* Gametophytes. **A.** ♀ gametophyte. ×300. **B.** ♂ gametophyte. ×300. **C.** and **D.** Reduced ♀ gametophytes. ×300 and ×600. a, antheridium (a_1, empty); og, oogonium; o, ovum; k_1–k_2, young sporophytes still attached to the empty oogonia. **A.** and **B.** After Schreiber. **C.** and **D.** After Kylin.

Order 6: **Fucales.** Together with the Laminariales these constitute the dominant intertidal algae of the colder seas. They include the abundant *Fucus* species, which reach to over a metre in length and are familiar on our coasts (Fig. 440D, E). Reproduction is by spermatozoids and ova produced in flask-shaped depressions at the ends of the thallus (Figs. 441, 442). They have no obvious alternation of generations (p. 483).

The long-lived *Fucus* species have band-shaped, leathery, dichotomous thalli reinforced with midribs (Figs. 440D, E) and attached to rocks by disc-shaped holdfasts (Fig. 440D). In the shallow waters of the seas of northern Europe they form extensive stands exposed at low water; but the thallus is prevented from drying out by the secretion of mucilage

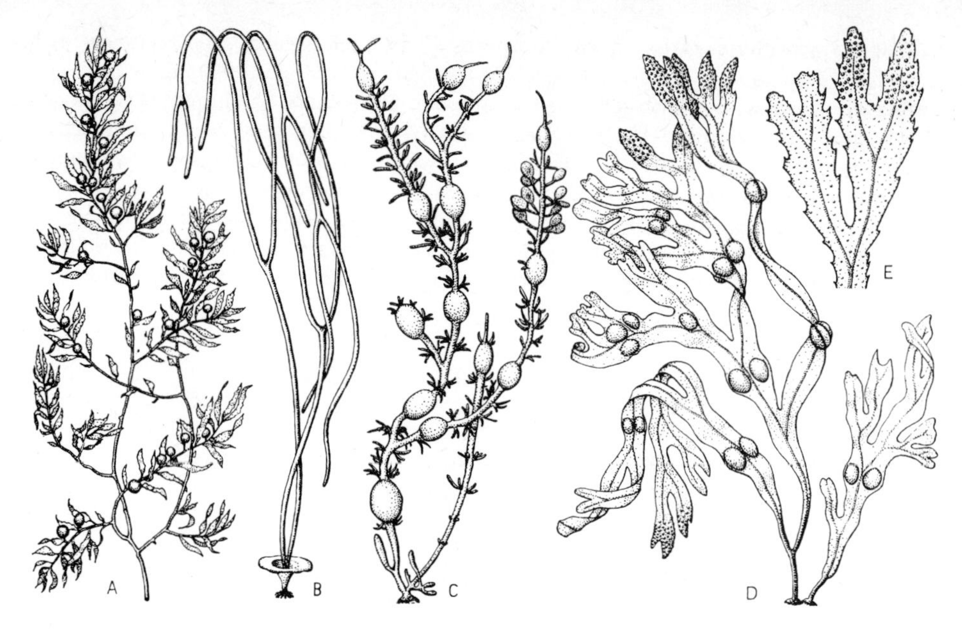

Fig. 440. Fucales. **A.** *Sargassum bacciferum.* **B.** *Himanthalia lorea.* **C.** *Ascophyllum nodosum.* **D.** *Fucus vesiculosus.* **E.** *Fucus serratus*, apex of thallus. All ×0.25.

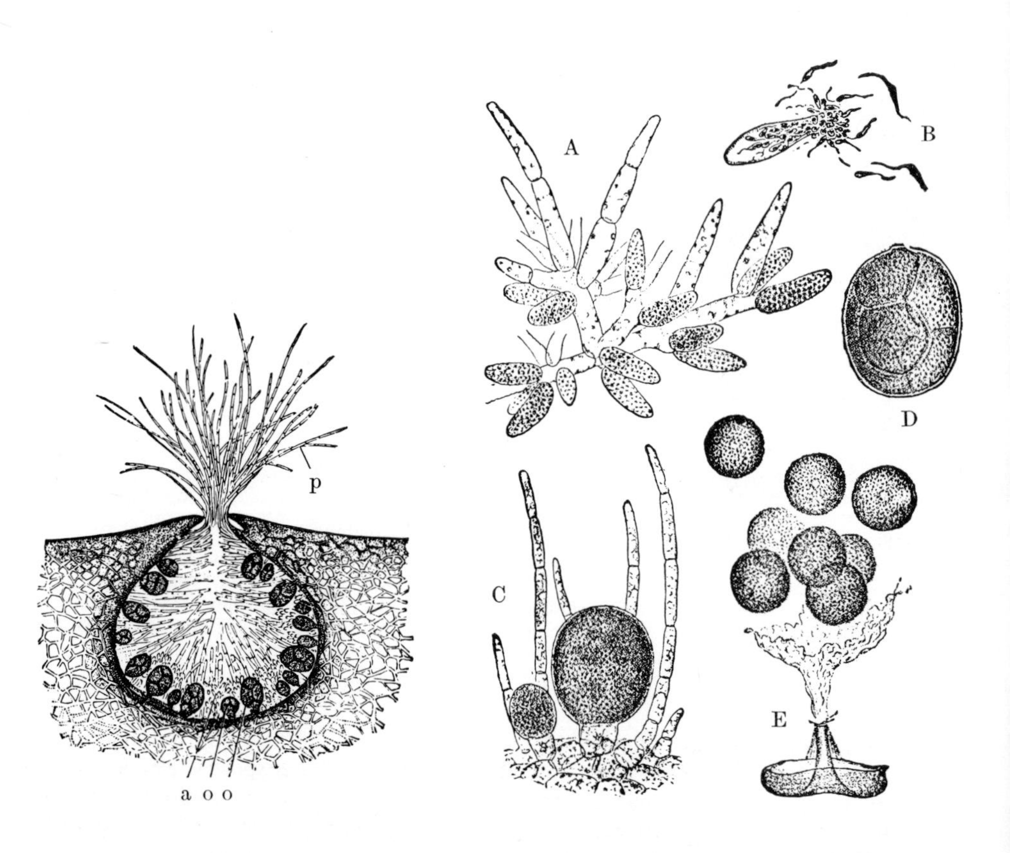

Fig. 441. *Fucus spiralis.* Conceptacle with oogonia of various ages (o) and tufts of antheridia (a). p, paraphyses. ×25. After Thuret.

Fig. 442. *Fucus vesiculosus.* **A.** Antheridial branch. ×200. **B.** Discharging antheridium. ×250. **C.** Young oogonia. **D.** After exit from the oogonium, divided into eight ova. **E.** Liberation of the ova. **C.** to **E.** ×120. After Thuret.

(fucoidin), and in consequence continues to photosynthesize. *F. serratus* (saw wrack) has a serrate margin; *F. vesiculosus* (bladder wrack) and the related *Ascophyllum nodosum* (knotted wrack, found in similar habitats, Fig. 440C) both have roundish air-bladders in the thallus. *Himanthalia* (Fig. 440C) has a button-shaped base to the thallus; this bears one or more forked thongs. The thallus is more elaborate in the tropical American and Australian genus *Sargassum*; some of the 250 species occur as free-floating purely vegetative branches in the mid-Atlantic Sargasso Sea, carried by their air-bladders with the Gulf Stream between the West Indies and the Azores; they reproduce entirely by fragmentation of the thallus (Fig. 440A).

In *Fucus* the ends of the branches, which are shed yearly (see p. 79 for the apical cell), are somewhat dilated and bear close-set, flask-shaped depressions or conceptacles (Fig. 441) containing oogonia and antheridia interspersed with sterile hairs or paraphyses. Some species are dioecious, but in others both sex organs are found in the same conceptacle (Fig. 441). The oval antheridial cells are clustered on short, much-branched filaments (Figs. 441 a, 442A). The contents of each antheridium divide after meiosis into sixty-four spermatozoids which are discharged in a mass still enclosed in the thin, inner wall of the antheridium. Eventually the pear-shaped spermatozoids are set free (Fig. 442B); these consist largely of nuclear material with an eye-spot (Fig. 409) attached to the rudimentary chromatophore; they have two lateral flagella, of which the forwardly directed tinsel-type flagellum is in contrast to other Phaeophyceae, the shorter. The oogonia (Figs. 441 o, 442C) are large, nearly globular, and borne on a short unicellular stalk; the single mother nucleus divides after meiosis into eight, followed by the cleavage of the protoplast to form eight ova. The oogonial wall ruptures to liberate the contents still surrounded by a thin but several-layered membrane (Fig. 442D); the outer layer of this membrane deliquesces at the apex and turns partly inside out, while the inner membrane ruptures and the brown ova, each surrounded by a very delicate pellicle of protein, are liberated into the water (Fig. 442E) where many float away and are lost. After fertilization the ovum surrounds itself with a membrane partly consisting of cellulose, settles down and gives rise by division to a new diploid plant. Meiosis occurs only in the formation of gametes.

Distribution and ecology. The brown algae are exclusively marine except for a very few and rare freshwater species (three genera). They are most abundant in the temperate and colder oceans. With about 2000 species they are mostly benthonic (p. 489), being attached to rocks, stones, piers, etc., but some lie unattached at low water. Others are epiphytic on other algae, or indeed to a certain extent endophytic. They form the bulk of the vegetation where the coast is not sandy, exhibiting marked zonation on rocky coasts (Fig. 450).

Survey of the Phaeophyceae. Like the Chlorophyceae, the Phaeophyceae show a progression from isogamy to oogamy (Fig. 225). In the lower forms (*Ectocarpus*) all the gametangia are alike and multicellular; in more advanced forms (*Cutleria*) the size of the ♀ gametes and the size of the gametangium increase together, while simultaneously the number of loculi is reduced; finally, in *Dictyota* and *Laminaria* only a single ovum is produced by each oogonium. Although the antheridia are still multicellular in *Dictyota*, in *Laminaria* they are unicellular with a single spermatozoid. This is not to say that the Laminariales can be derived from the Dictyotales. The Dictyotales and Laminariales are the most advanced orders of two developmental series, the so-called Isogeneratae and Heterogeneratae respectively.

Alternation of generations is isomorphic or heteromorphic (Fig. 445). The former can be regarded as the more primitive state from which heteromorphy has been derived by the greater development of the sporophyte, as already adumbrated in the Ectocarpales.

The Fucales occupy an isolated position in that there is no apparent alternation of generations. We have seen, however, that in the Laminariales the ♀ gametophyte occasionally consists of a single cell produced from the contents of a zoospore which has come to rest and developed directly into an ovum (Fig. 439D). We can therefore imagine that in *Fucus* reduction of the gametophyte has gone even further, so that the asexual spore has itself become an ovum or ♂ gamete. This hypothesis is supported by the unilocular condition (in contrast to other Phaeophyceae) of the antheridia and oogonia. The Fucales thus demonstrate the ultimate reduction of the gametophyte and the diploid *Fucus* plant can be regarded as a sporophyte on which the gametophyte reduced to gametes is dependent.

In *Dictyota* and *Laminaria* alternation of generations appears to be obligatory, but in many others, especially in the relatively primitive Ectocarpales, there are numerous deviations from simple alternation. This is particularly so in the easily cultured *Ectocarpus confervoides*, the heteromorphic generations of which are readily distinguished and in which very many aberrations have been observed, e.g. relative sexuality (p. 327), parthenogenetic development of the gametes, formation of zoospores without meiosis and even fusion of the zoospores. In general, it appears that: (i) both generations can reproduce themselves, and (ii) the location of the sexual processes is not completely

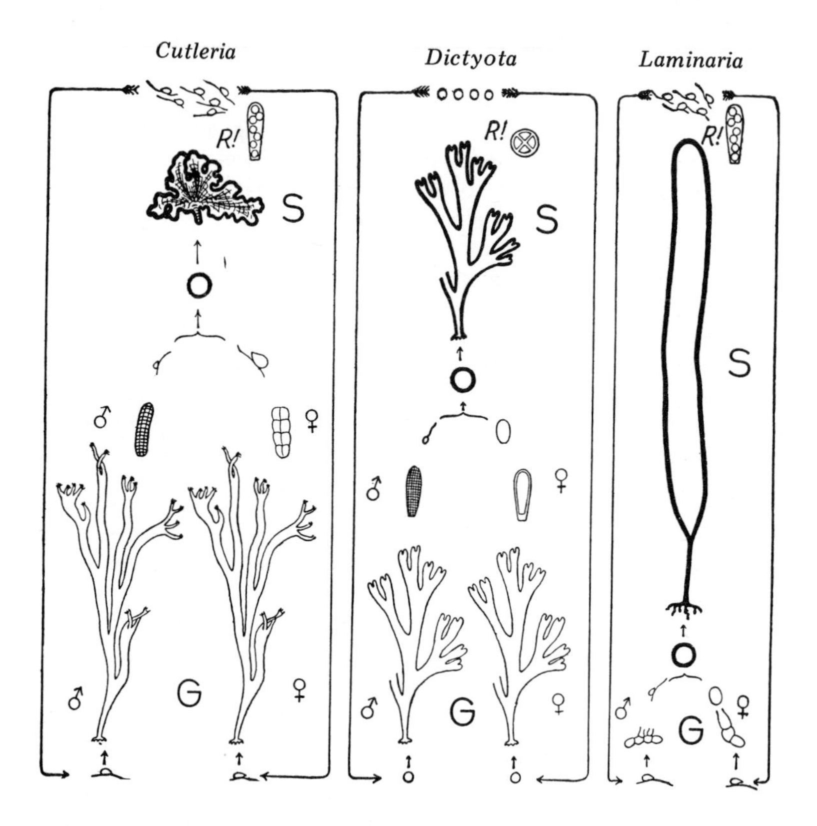

Fig. 443. Diagrammatic representation of alternation of generations and nuclear phases in some Phaeophyceae. G, gametophyte; S, sporophyte; O, zygote; R!, meiosis. Haploid phase: thin lines. Diploid phase: thick lines. After Harder.

fixed even in these relatively advanced algae. Nevertheless, in culture all swarmers are produced in plurilocular reproductive structures (i.e. zoosporangia), excepting only in unilocular structures (in which meiosis invariably occurs) on diploid sporophytes. If, as is very often the case in other Ectocarpales, the diploid sporophyte produces both haploid zoospores following meiosis as well as unreduced diploid zoospores, then unilocular and plurilocular structures occur together on the same plant probably as a function of temperature. In many species (e.g. *Ectocarpus siliculosus*) regular alternation of generations very rarely takes place.

Phylogeny of the Phaeophyceae. The flimmer hairs on one of the two unequal flagella, the photosynthetic pigments and the reserve materials indicate affinity with the Chrysomonadales. Although they could not have been derived directly from chrysophytes, the

Phaeophyceae must have arisen at an early stage from a common stock and then evolved independently. Some systematists include the Phaeophyceae and Chrysophyceae in the great brown group Chromatophyta, which thus forms a major branch comparable to the red and green algae. Primitive forms are lacking in the Phaeophyceae: even the simplest are highly differentiated branching plants with sporangia, gametangia and alternation of generations. Possibly the ancestral forms die out as the salt content of the primeval oceans increased.

Despite their often considerable dimensions, the Phaeophyceae are generally less well preserved as fossils than the calcareous Chlorophyceae, but it is very likely that they were already present in the Silurian and Devonian. Certain Lower Devonian and Silurian 'stems' as thick as a thigh (*Nematophycus* = *Prototaxites*), consisting of an aggregation of tubular filaments and ending in large tufts of *Laminaria*-like thallus-lobes, possibly belong here.

Economic uses. The ash of various Laminariceae and Fucaceae (kelp, varec) yields iodine, while in recent years salts of the mucilaginous alginic acid have been used in many ways (textile, food and cosmetic industries); soda has also been obtained from them. Many Laminariaceae yield abundant mannitol (e.g. *Laminaria saccharina*); they are used, especially in China and Japan, as a food (kobu).

Class VII: Rhodophyceae, red algae

The predominantly marine Rhodophyceae are usually red or violet, rarely dark purple or reddish-brown. Almost without exception the cells have a single nucleus and contain numerous simple chromatophores or rhodoplasts (disc-shaped, oval or lobed (Fig. 38D) but never cup-shaped) in which chlorophyll (*a*, no *b*, in some species a little *d*) and the associated carotenoids in the lamellae of the plastids are masked by a red, strongly fluorescent, water-soluble pigment, phycoerythrin (a phycobilin, cf. p. 246, with absorption-bands differing from those in the Cyanophyceae). In certain species the blue phycocyanin is also present, and there are many modifications of both pigments.

True starch is not formed as a product of assimilation; instead, another carbohydrate, floridean starch, is synthesized from glucose in the form of roundish, insoluble, often stratified grains which stain red with iodine; chemically it is intermediate between starch and glycogen. The grains are not formed inside, but on the surface of the chromatophores, and in the cytoplasm. Certain other substances known only in the red algae also occur (floridosides, compounds of galactose and glycerol), as well as drops of oil, Pyrenoids, probably functionless, are present only in a few lower forms. The red algae are generally autotrophic; relatively few are colourless parasites, and of these a dozen or two exist in very reduced forms exclusively on other red algae, to which they are apparently closely related (adelphoparasites, putatively derived from their hosts by the genetic loss of pigments). The cell wall usually has an inner cellulose layer and an outer one of often highly mucilaginous pectin.

Morphology. Unicells only occur in the rather isolated small order Bangiales, e.g. the common terrestrial *Porphyridium cruentum*; but this group also includes filamentous forms and extended sheets (*Porphyra*). The other orders, grouped together as Florideae, include forms with relatively simple but branched filaments (e.g. *Polysiphonia, Callithamnion,* Fig. 445), pinnate or dichotomous broad or narrow ribbons (e.g. *Chondrus crispus, Gigartina mammilosa*), extended sheets, crusts (*Hildenbrandtia*), etc. The Atlantic *Delesseria sanguinea* (Fig. 77) is particularly complex: the leaf-like thallus has a midrib and lateral veins and springs from a disc-shaped holdfast; in autumn the lamina disappears, leaving only the midrib as a persistent axis which produces new foliose thallus-branches the following spring. Even the simplest of the Florideae are already heterotrichous (i.e. composed of prostrate and erect systems of filaments), but in contrast to the Phaeophyceae even the most complex are never parenchymatous, and at the most plectenchymatous (p. 77), building the thallus solely by modifications of the uniaxial or multiaxial (fountain-type) filamentous construction (Figs. 75, 76). All Florideae have well-developed pits.

Life history. It is characteristic of the Rhodophyceae that the reproductive cells, both spores and gametes, are invariably non-motile. Sexual reproduction is predominantly oogamous, and the female gametangium (carpogonium) of the Florideae always bears a long, generally slender receptive organ, the trichogyne (Figs. 444, 449D, t). After fertilization by a non-motile spermatium, the zygote develops in a very characteristic manner, germinating immediately without leaving the carpogonium to form sporogenous filaments (or gonimoblasts) which produce naked mitospores. In the great majority of red algae these diploid carpospores give rise to diploid plants similar in appearance to the gametophyte. The diploid plant then produces haploid tetraspores by meiosis (Figs. 444, 445), and is therefore called the tetrasporophyte. The tetraspores germinate to give new sexual plants. Thus three generations develop in succession: 1, the haploid gametophyte; 2, the diploid carposporophyte, attached to the gametophyte; 3, the independent diploid tetrasporophyte. The development of the three generations therefore proceeds on two plant bodies (diplobiontic). Most of the red algae (except the Nemalionales) belong to this type.

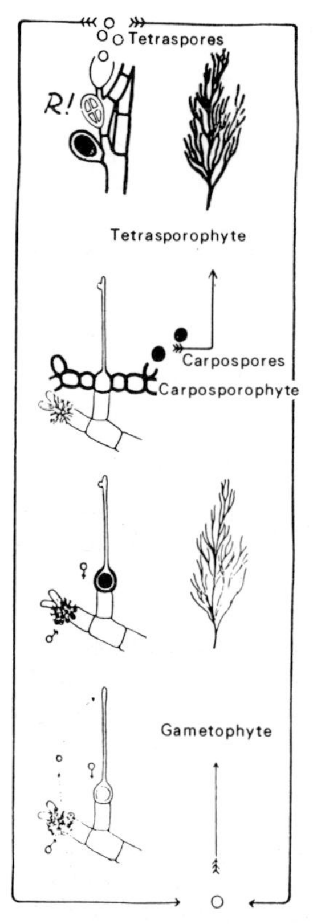

Fig. 444. Alternation of generations and nuclear phases in the Rhodophyceae. Haploid phase: thin lines. Diploid phase: thick lines. R!, meiosis. After Harder.

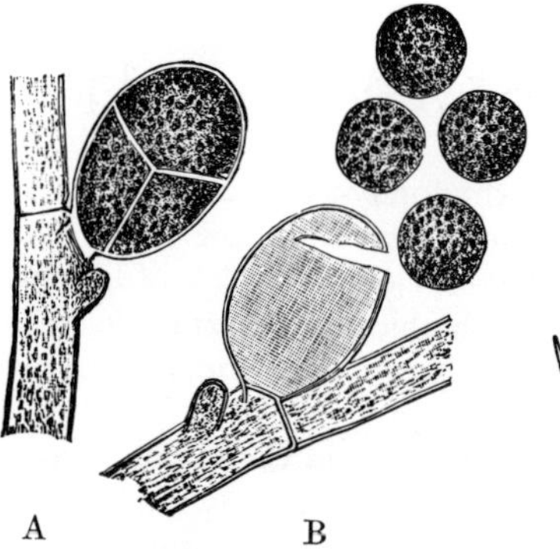

Fig. 445. *Callithamnion corymbosum.* Formation of tetraspores. **A.** Intact sporangium. **B.** After liberation of four tetraspores. ×300. After Thuret.

Fig. 446. Gametophyte and tetrasporophyte of *Bonnemaisonia hamifera.* **A.** Gametophyte with cystocarps. **B.** Sporophyte, known as *Trailliella intricata.* ×5. After Koch.

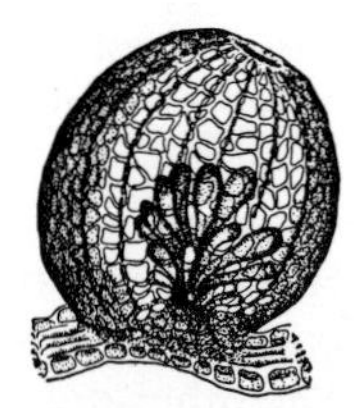

Fig. 447. *Platysiphonia miniata,* cystocarp with gonimoblasts visible inside. ×100. After Börgesen.

Gametophyte and tetrasporophyte are generally similar in appearance; rarely they are so dissimilar (Fig. 446) that they have been assigned in the past to different genera or even to distinct orders. In some cases the dependent carposporophyte appears so peculiar that it has been described as a truly parasitic, distinct genus, and given a separate name. Occasionally secondary sexual characteristics distinguish the structure of ♂ and ♀ plants (the sex distribution may be synoicous or heteroicous). Commonly the gonimoblasts are enveloped in special gametophytic filaments which form an enclosed 'fruit' or cystocarp (Fig. 497); but not in *Batrachospermum*.

Many other peculiarities of development could be mentioned. Thus the carposporophyte is often nourished by special auxiliary cells which are particularly evident in *Dudresnaya coccinea* (Fig. 448A, a_1-a_6) as gametophyte cells with dense contents. The

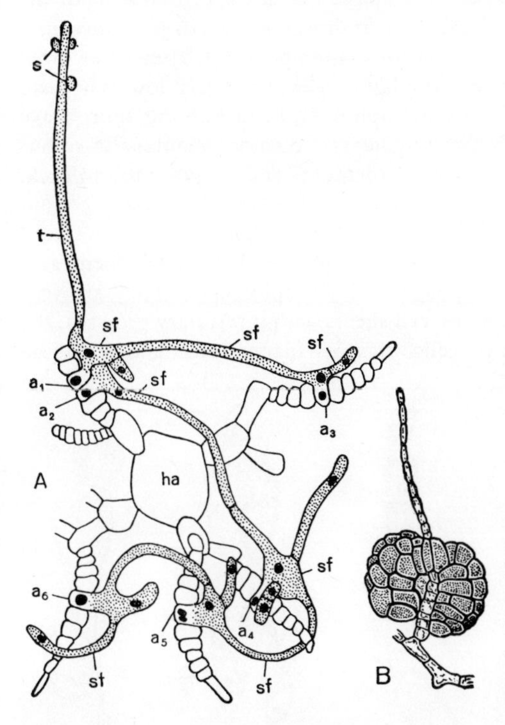

Fig. 448. *Dudresnaya.* **A.** The fertilized carpogonium, with spermatia (s) attached to the trichogyne (t), has grown out as branched sporogenous filaments (sf) which have united with six auxiliary cells (a_1-a_6); the cells a_1-a_6 are borne on branches arising from the axis ha. **B.** Cluster of mature carpospores. ×250. **A.** After Oltmanns. **B.** After Bornet.

very long, branching sporogenous filaments grow towards the auxiliary cells with which they then unite (without nuclear fusion); at the point of union the sporogenous filaments swell up and produce short outgrowths which divide further to form roundish masses of cells (Fig. 448B) from which the carpospores are ultimately liberated.

The life-cycle of the frog-spawn alga *Batrachospermum* (Nemalionales) differs from the usual type. It occurs, for example, on stones in freshwater streams as brownish or dark violet spawn-like masses embedded in mucilage (p. 69); it is constructed of filaments with whorls of branches (Fig. 449A). The antheridia (Fig. 449B) are borne in large numbers, generally in pairs, on the terminal cells of the lateral branches. Each consists of a single cell, the entire protoplast of which develops into one colourless spherical spermatium with a large nucleus and very delicate wall. The ♀ carpogonia are also apical, occurring among the antheridial branches; each consists (Fig. 449C, D) of an elongate cell with a basal flask-shaped portion and an upper, club-shaped prolongation, the trichogyne. The non-motile spermatium is carried passively by water currents to the trichogyne, to which it adheres, and liberates its entire contents through an opening into the carpogonium. After fusion of the ♂ and ♀ nuclei the basal part of the carpogonium containing the zygote is separated from the trichogyne by a plug of mucilage (Fig. 449E). Branching filaments then grow out from the side of the carpogonium as the carposporophyte (Fig. 449F). In the dilated apical cells of these filaments mitospores are produced (the carpospores), each containing a nucleus and chromatophore. The carpospores are liberated as naked, spherical, non-motile bodies, leaving behind the empty walls of the terminal cells (Fig. 449G, k_1, k_2). They develop as a still diploid juvenile stage (*Chantransia* stage), attached to the substratum and consisting of branching filaments. In some cells of the *Chantransia* filaments meiosis takes place, but without the formation of meiospores and hence for long overlooked. The resulting haploid cells then develop into the adult gametophyte with whorled branches. Thus *Batrachospermum* also has a tripatite, heteromorphic and heterophasic alternation of generations, but the three generations are united for life in one plant: 1, the diploid *Chantransia* stage (juvenile stage); 2, the whorled haploid gametophyte; 3, the diploid carposporophyte. Development takes place in one vegetative body (haplobiontic).

Distribution and ecology. Some 4000 species of red algae live in the littoral zone of the seas, especially in the warmer ones; only a few live in fresh water, e.g. *Batrachospermum* in streams. Many species are very sensitive to variations in temperature. They often grow at considerable depths (down to 200 m) where the light intensity is very low. They are indeed not merely shade plants, but their chromatophores can utilize the short wave radiation prevalent in deep water, to which their pigments are complementary in colour (p. 247). They are invariably benthonic, attached by filaments or discs generally to rock, but many are epiphytic on larger algae.

Systematics. The **Bangiales** have a number of peculiarities which separate them as a sub-class (Bangiophycidae) from the other red algae (Florideophycidae). They are very simply constructed (unicells, simple filaments or cell-sheets with intercalary growth); the carpogonia develop directly from vegetative cells. *Bangia* (filamentous) and *Porphyra*

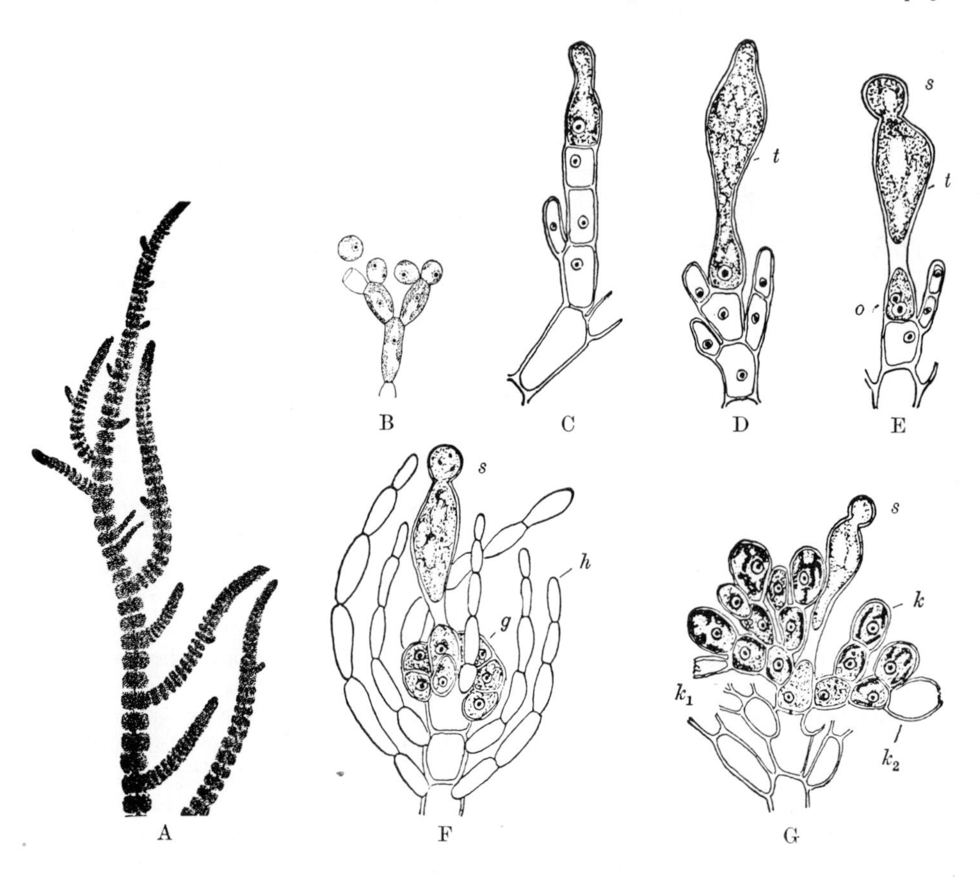

Fig. 449. *Batrachospermum moniliforme.* **A.** Habit. ×3. **B.** Part of a branch with four antheridia, on the left releasing a spermatium. ×540. **C.** A young carpogonium. **D.** Mature carpogonium with trichogyne (t). **E.** Carpogonium after fertilization by a spermatium (s). **F.** Developing sporogenous filaments (g) and enveloping filaments (h). **C.** to **F.** ×960. **G.** Mature sporogenous filaments with carposporangia (k); k_1 and k_2 empty. ×720. **A.** After Sirodot. **B.** After Strasburger. **C.** to **G.** After Kylin.

(very thin leaf-like expanses) live near the high water mark of spring tides (Fig. 450). Whether the filamentous *Conchocoelis* is an alternate stage of *Porphyra* is still not quite clear.

In the **Nemalionales** (*Batrachospermum* and *Lemanea* in fresh water, *Nemalion* in the sea) there is no tetrasporophyte; meiosis occurs in the *Chantransia* juvenile stage.

The **Cryptonemiales** develop their auxiliary cells in special groups of branches. The Corallinaceae are calcified (*Corallina, Lithothamnion, Lithophyllum*); fossil representatives are significant rock-formers.

The **Gigartinales** and **Rhodymeniales** are also distinguished by peculiarities of the auxiliary cells. The former includes *Plocamium*, with comb-like branches; *Furcellaria*, with a fountain-type of construction (Fig. 75C); and *Chondrus*, with a repeatedly dichotomous flat thallus. The latter includes the Atlantic *Rhodymenia* with a leaf-like thallus.

In the **Ceramiales** the thallus consists of richly-branched, corticated filaments. Representatives on our coasts include *Polysiphonia, Ceramium, Plumaria* and the leaf-like *Delesseria* (Fig. 77C).

Phylogeny. The evolutionary relationships of the red algae are obscure. The ancestors of this highly developed class have often been sought amongst the green algae. The immediately obvious derivation from *Coleochaete* (p. 467) is generally objected to on the grounds that the trichogyne of the red algae was apparently originally an independent cell above the carpogonium (cf. the nucleus in the trichogyne), whereas in *Coleochaete* it is merely an outgrowth of the oogonium. The lack of chlorophyll *b* in the red algae also militates against this derivation. Furthermore, the chromatophore is stellate in primitive red algae, cup-shaped in the green algae. On biochemical grounds (red and blue chromatophore pigments, no chlorophyll *b*) a connection with the Cyanophyceae is possible, by way of the primitive Bangiales, in which, as well as more complex forms, unicells and falsely branched filaments occur. The absence of a true nucleus and clearly differentiated chromatophores from the Cyanophyceae sharply separates them from the Rhodophyceae. The possibility has also been considered that the red algal plastids are endosymbiotic Cyanophyceae. Finally, derivation from red Cryptophycales has been considered.

The Rhodophyceae occur fossil (especially the corallines) in all formations from the Ordovician (*Solenopora*) onwards.

Uses. Various polysaccharides used in medicine and industry are extracted from the cell walls of a number of red algae: *Chondrus crispus* and *Gigartina mamillosa* from the coasts of N.W. Europe (dried as carrageen or 'Irish moss'); various Florideae of the Pacific Ocean (e.g. *Gelidium* and *Gracilaria* species) but more recently some European species yield agar-agar from their walls (used in culturing micro-organisms, and also in the food and pharmaceutical industries). *Porphyra* is eaten especially in Eastern Asia ('nori') where it is grown in plantations on the coast.

Distribution and ecology of the algae

Algae are found in almost all biotopes (habitats) although most are confined to water where they live either as benthos, attached to substrata, or as floating plankton. Aquatic environments fall into two very different groups according to whether the water is salt or fresh.

The benthonic plants of the sea, with the exception of *Zostera* and related angiosperms, consist entirely of algae amongst which the Phaeophyceae and Rhodophyceae predominate. Most are attached to a stable surface (e.g. rock) by means of attachment discs or holdfasts (Figs. 437, 440). Unstable substrata (mud, sand) are colonized by only a few genera, e.g. *Caulerpa* (Fig. 422A). Benthonic algae extend from the spray-zone of the coast down to the greatest depth (200 m) at which photosynthesis is possible.

The algal vegetation of tropical seas is less luxuriant than in temperate regions.

Phaeophyceae are less abundant than Rhodophyceae and those families of the Chlorophyceae which favour warmer waters: Caulerpaceae, Dasycladaceae, Codiaceae, Valoniaceae. The vegetation of tropical coral reefs is also luxuriant; indeed, some algae (*Halimeda*, Dasycladaceae, *Lithothamnion* species) may play a greater part in the deposition of lime than the corals themselves. A unique phenomenon is the Sargasso Sea between the West Indies and the Azores where the floating oceanic brown alga *Sargassum* (Fig. 440A) forms a considerable mass of vegetation (up to 15 kg plant mass per ha).

In warm temperate seas, e.g. the Mediterranean, the benthos consists chiefly of Rhodophyceae and the smaller Phaeophyceae. The tropical Siphonales are represented by a few species and *Lithothamnion* is still well developed. As a result of the seasonal change in light intensity in shallow water the main period of growth is in the spring, in deeper water in the summer and autumn.

In cold temperate seas, e.g. around our coasts, the Phaeophyceae dominate both in size and mass. Many species show marked seasonal growth. Thus in the autumn *Desmarestia* loses all but the overwintering ribs of its delicate thallus, and the massive *Laminaria* species renew their phylloids each year. Figure 450 shows an example of the marked vertical zonation of algae in relation to tidal levels on a rocky coast in the Channel. Species of the upper zones (e.g. *Bangia, Porphyra, Fucus*) can withstand temperatures as low as −20°C, while the inhabitants of the lower zones (which never dry out), e.g. *Laminaria* and *Delesseria*, are killed by even a few degrees of frost.

The coldest seas are poor in species but those brown algae that do occur reach their largest dimensions there, e.g. *Macrocystis* (Fig. 437E), *Lessonia* (Fig. 437D), *Nereocystis* (Fig. 437C) (all Laminariales), and *Durvillea* (Fucales); these are nearly as big as the largest land plants.

Marine phytoplankton (plant plankton) consists predominantly of Pyrrophyceae and Chrysophyceae, particularly diatoms, Peridiniales, and the minute Coccolithineae and silicoflagellates. Individuals of these last two groups pass through the mesh of a plankton-net and can only be obtained by centrifuging (nannoplankton). The greatest density of phytoplankton (up to 10^6 cells per litre) is found in the illuminated upper layers of water. One litre of surface water from near the Faroes contained the following numbers of cells: 32,000 Pyrrophyceae, 1600 diatoms and 54,000 Coccolithineae. The density drops markedly below 100 m. Nevertheless, Coccolithineae and 'olive-green cells', the systematic position of which remains obscure, have been found at great depths (4–5 km).

The regional distribution of phytoplankton is similar to that of the benthos (Fig. 451). The greatest density is found in the colder seas and in regions of cold oceanic currents; this is the result of the higher levels of nitrogen and phosphorus in the water. These are exhausted in the upper layers but increase at depth following the sinking of dead cells. In cold regions cooling of the surface of the sea at night and during the winter leads to a more thorough mixing of the layers of water than in the Tropics, and hence to a more luxuriant development of the plankton. The same happens when cold, deep water rich in nitrogen and phosphorus is brought to the surface by ocean currents.

The floating of plankton is in reality very slow sinking. An object sinks more quickly the more its specific weight exceeds that of the fluid medium; sinking will be slowed down by friction and the viscosity of the fluid. This explains many of the peculiarities of planktonic algae: the presence of oil as a reserve, and of gas vacuoles (reducing the specific weight); the development of processes or wings from the cell wall, and the cohesion of cells in chains (Figs. 388, 398); and the observation that in warmer waters flotation processes are longer than in colder ones. Gas vacuoles can indeed cause the cells to rise (water bloom, p. 447).

Mineral skeletons of planktonic algae accumulate in the marine sediments. Since calcium carbonate goes into solution at depths below 4–5 km, the deepest sediments contain only the siliceous remains of diatoms, silicoflagellates and radiolaria, whereas at moderate depths (1–5 km) calcareous sediments of coccoliths, *Globigerina*, etc., are deposited; but it takes a thousand years to accumulate a layer 1.5 cm thick.

Freshwater benthos consists mainly of flowering plants, but occasionally some algae predominate, e.g. diatoms and Characeae in brackish water.

The specific composition of freshwater phytoplankton is closely related to the

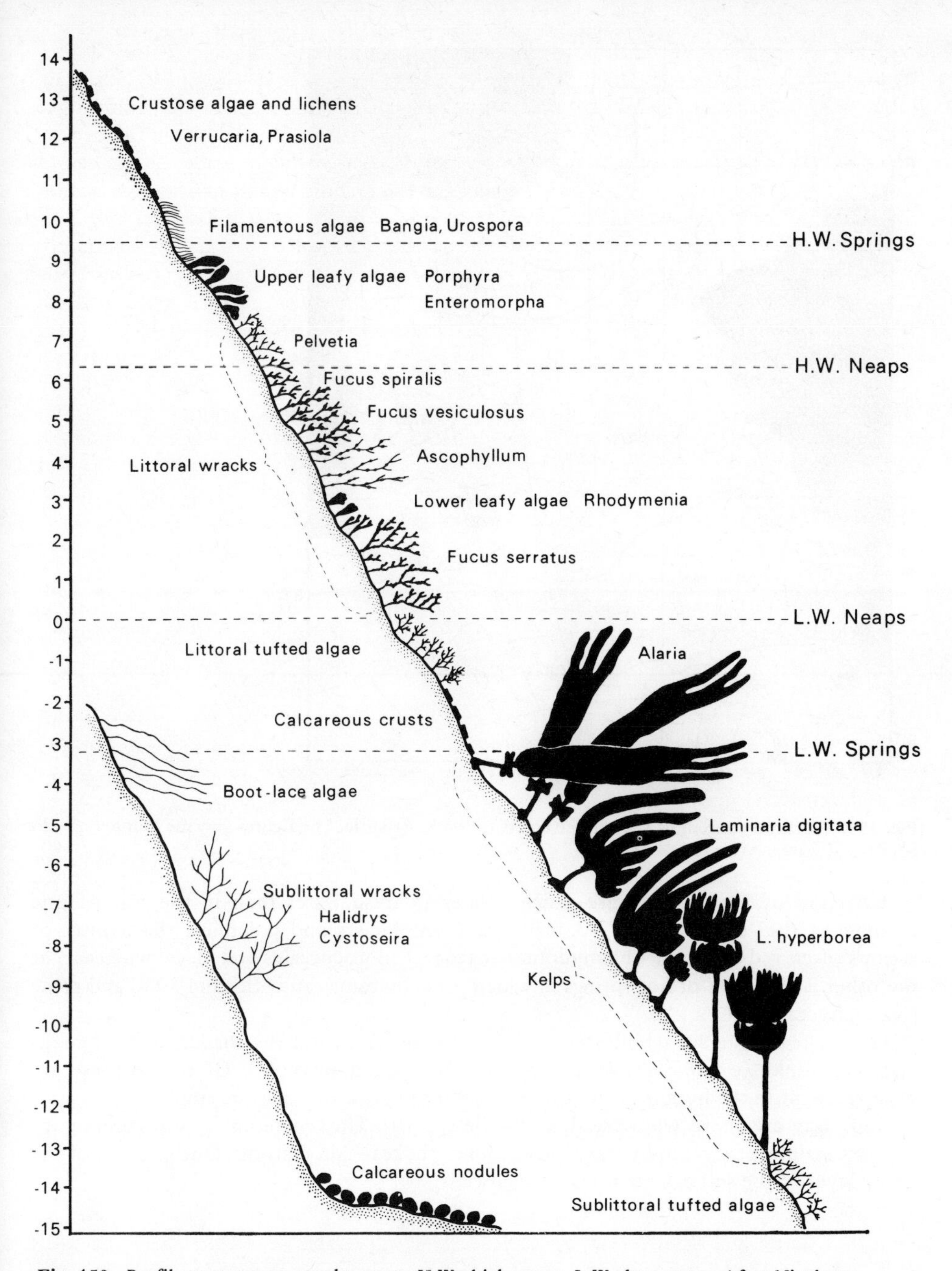

Fig. 450. Profile transect on a rocky coast. H.W., high water; L.W., low water. After Nienburg.

nutrient level of the water which, conversely, can be recognized by the occurrence of certain indicator species. In nutrient-rich (eutrophic) waters the phytoplankton may utilize organic substances (myxotrophy); such waters naturally favour bacterial growth. Indicators of eutrophic waters include *Euglena viridis* and *Polytoma uvella*. *Asterionella*, *Dinobryon* and *Peridinium* are characteristic of nutrient-poor (oligotrophic) waters, whereas *Phacus*, *Pediastrum* and *Melosira* occur in mesotrophic (intermediate) ones. Many lakes and rivers in developed countries are becoming progressively more eutrophic as a result of all kinds of pollution.

Neuston comprises those organisms living at the interface of water and air. It generally consists of unicellular algae, e.g. *Euglena* and *Chromulina*; 40,000 cells per mm^2 of the latter have been recorded, giving a golden sheen to the surface of the water.

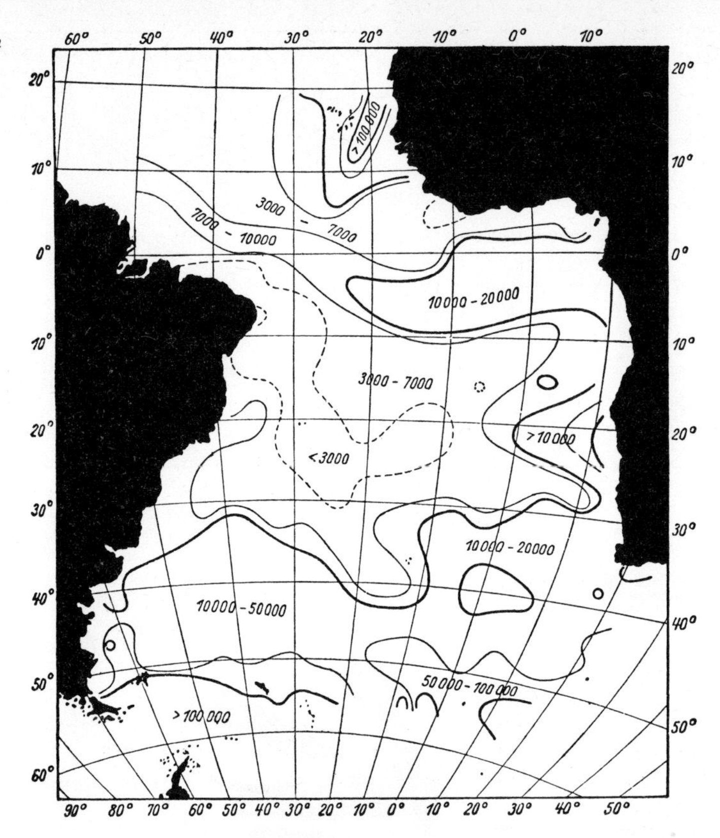

Fig. 451. Plankton content of the surface layers of the S. Atlantic. The figures give the number of cells per litre of water. After Hentschel.

Extremes of temperature are much greater in fresh water than in the sea. At one extreme is the 'cryoplankton' of meltwaters on glaciers and polar ice, the habitat of various often red-coloured Chlamydomonadaceae, Chlorococcales and Mesotaeniaceae; at the other is the flora of hot springs in which some diatoms can withstand 50°C and some Cyanophyceae 70°C.

Only a few algae are wholly terrestrial, but some occur on the shaded sides of rocks and tree-trunks (e.g. *Pleurococcus, Trentepohlia,* Cyanophyceae). Of course these are most frequent in the humid Tropics where they even grow on leaves as epiphylls.

Soil algae are more widespread, but little studied. The 'edaphon' or soil community consists mainly of Cyanophyceae, lower Chlorophyceae and diatoms. One gramme of the upper layer of the soil can contain 10^6 algal cells.

Division 3: **Mycophyta, fungi**

The fungi lack chromatophores and are commonly colourless; they are heterotrophic. However, various pigments (mostly nitrogen-free cyclic compounds) occur, especially in the fructifications of the higher fungi. In the simplest groups the thallus is naked and amoeboid; in all others the cells are surrounded by a wall which generally gives a chitin (rarely cellulose) reaction. Occasionally the vegetative body is bladder-shaped; typically it consists of slender filaments, sometimes simple but far more commonly branched; it may be unicellular or multicellular (septate). Each single fungal filament is called a hypha; collectively the hyphae form a mycelium.

Various kinds of spores effect reproduction: in aquatic fungi the spores are naked, flagellate, generally radially symmetrical swarmers (zoospores); in terrestrial fungi they have a wall and may arise endogenously (endospores) or exogenously (conidia, cf. p. 205). In sexual reproduction union takes place between gametes (isogamy, anisogamy or oogamy (cf. pp. 206, 207)), between entire gametangia (gametangiogamy, p. 505), or between two thallus cells not specifically differentiated as sexual cells (somatogamy, pp. 207, 527). However, asexual reproduction tends to predominate.

The fungi exhibit considerable diversity in their sexual relations. In dioecism an individual can only function as a nuclear-donor (male) or as a nuclear-acceptor (female). Dioecism may be expressed morphologically, but in some cases (e.g. many zygomycetes, some yeasts) it only operates physiologically. In monoecism each individual can function both as a nuclear-donor and a nuclear-acceptor; such fungi may have differentiated male and female sex-organs, or each mycelium may have the physiological potentiality of both sexes. Amongst monoecious fungi there are many in which fusion of nuclei from the same individual cannot take place; instead, two 'mating-types' are necessary. This genetically determined sexual barrier is known as incompatibility. In the simplest case the two mating-types are determined by a single pair of alleles designated + and −. Fertilization can only take place between a + and a − mycelium and genetically identical mycelia are incompatible (Fig. 452); consequently this system is called homogenetic incompatibility in contrast to the system of heterogenetic incompatibility found in some fungi (e.g. *Podospora*), in animals and in higher plants, which is based on the incompatibility of genetically distinct gametes or gametic nuclei. So far as the intermixing of genetic

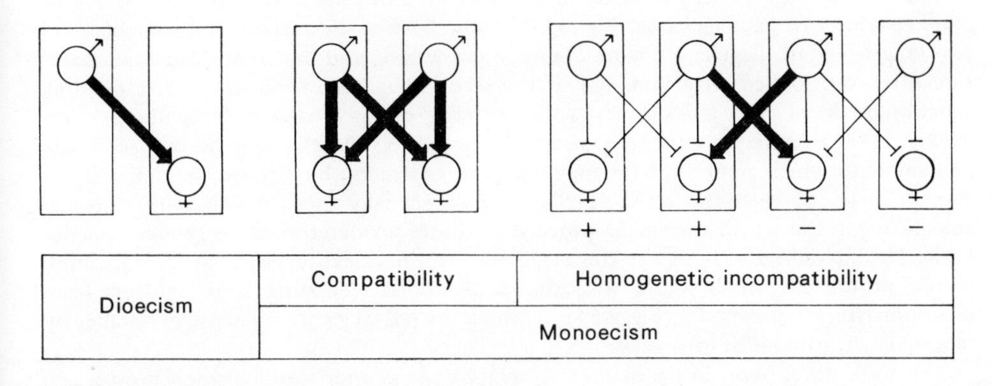

Fig. 452. Sexual relationships in the fungi. After Esser.

material is concerned, homogenetic incompatibility in monoecious plants produces the same effect as dioecism (Fig. 452), i.e. the suppression of inbreeding, while heterogenetic incompatibility has the reverse effect: it is an isolation mechanism.

The mycelia of many fungi can reproduce by breaking up into single cells (oidia); very frequently resting stages are formed as hard, tuberous masses of hyphae (sclerotia). Interwoven hyphae may also form cord-like rhizomorphs several metres long which bring about the spread of the fungus (e.g. in *Armillaria mellea*).

Fungi are saprophytic or parasitic, freshwater or terrestrial, but rarely marine. Most of the saprophytes can be cultured on artificial media; many require not only a supply of organic carbon and nitrogen but also of certain growth substances. Drops of glycogen and fat are commonly present as reserve materials, as well as mannitol and other substances; starch never occurs.

Class I: Myxomycetes, slime moulds

Order 1: **Myxomycetales.** With the myxomycetes we return to the lower levels of the evolutionary scale among primitive, phylogenetically ancient organisms. They have no chlorophyll, and in the vegetative state exist as naked, multinucleate, amoeboid masses of protoplasm (plasmodia, p. 17, Figs. 8, 454) which live saprophytically or even (like animals) phagotrophically (taking solid particles into the cytoplasm). Plasmodia are formed after sexual fusion and develop into sporangia in which each spore is surrounded by a wall. Glycogen is the reserve material.

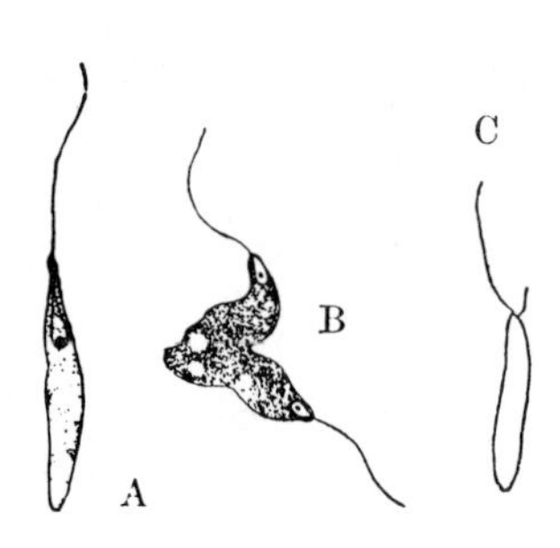

Fig. 453. Myxoflagellatae. In **A.** and **B.** the short flagellum is not shown. ×1500. **A.** After Gilbert. **B.** After von Stosch and von Wettstein. **C.** After Elliot.

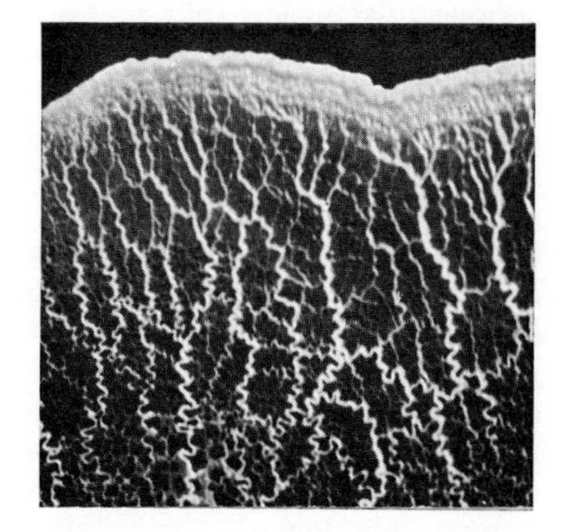

Fig. 454. Margin of the plasmodium of *Badhamia utricularis*. ×2. After Jahn.

The spores (Fig. 455E) germinate in water or on moist substrates to produce one or more swarmers (myxoflagellatae, Fig. 453). These have at the anterior end a nucleus and two long whip-lash flagella, or more commonly one long and one very short (and therefore often overlooked) flagellum (C); if the second flagellum is absent a corresponding functionless basal body still occurs; at the posterior end is a contractile vacuole (shown in Fig. 453B). After a time the swarmers lose their flagella and become creeping myxamoebae which reproduce (as may also the swarmers) by fission. The uninucleate myxamoebae (or even the myxoflagellatae) fuse sexually in pairs (Fig. 453) forming amoebozygotes in which the nuclei proceed to unite; sex-determination geno- or phenotypic. The diploid state is not a resting stage, but continues existence as naked, amoeboid, motile protoplasts which unite with similar protoplasts forming large, multinucleate plasmodia (fusion-plasmodia, Fig. 454) in which the nuclei greatly increase in number by repeated synchronous mitotic division.

Cell walls never occur in the plasmodia, which exist as undivided masses of protoplasm with a very high water content (p. 17). They can be grown on nutrient agar, and (like the haploid stage) ingest bacteria, fungal hyphae and other solid nutrients (fragments of agar) which they digest in the vacuoles or elsewhere; they can also break down substrates exoplasmatically by the secretion of digestive enzymes and thus take up nutrients in soluble form. Saprophytic species can be grown in absolutely pure (axenic) culture, but parasitic species can only exist indefinitely with the addition of living food (e.g. bacteria). The plasmodia occur especially on the ground in woods, on fallen leaves or on decaying timber, where they slowly move about, branching and changing their form. The advancing margin (Fig. 454) consists of denser cytoplasm; at the rear the plasmodium breaks up into a network of strands. They reach situations favourable for nutrition principally by chemotactic, hydrotactic and negatively phototactic movements. The plasmodia of certain species attain over 30 cm in diameter (e.g. *Fuligo varians*).

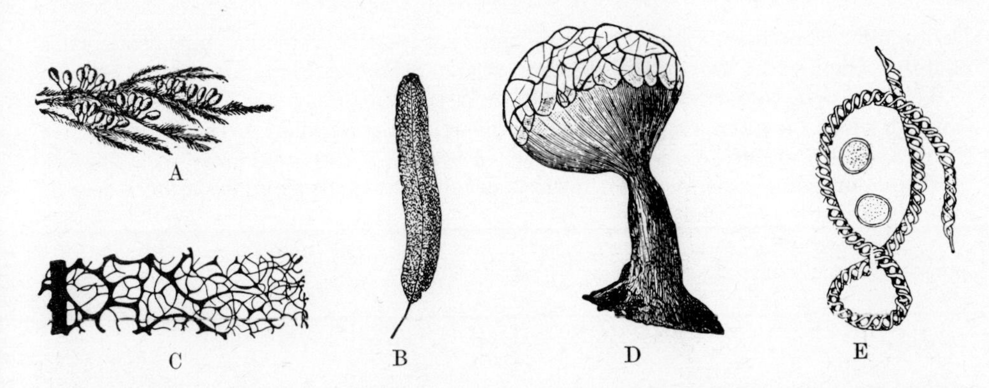

Fig. 455. Myxomycetales. **A.** *Leocarpus fragilis*. Several sporangia on moss. Nat. size. **B.** *Stemonites fusca*, sporangium. ×5. **C.** *Comatricha typhoides*, part of the capillitium. ×180. **D.** *Cribraria rufa*, sporangium. ×30. **E.** *Trichia varia*, capillitium fibres and spores. ×300. **A.**, **B.** and **D.** After Schenck. **C.** and **E.** After Lister.

The first stage of plasmodium development is associated with high humidities. The second stage follows in a drier environment: the plasmodium develops into numerous sporangia (fructifications; Fig. 455A), and in so doing changes its irritability, creeping away from moist substrates and towards the light; there is also a considerable reduction in water content. Each fructification has a firm, often calcified, peripheral investment (peridium) and contains numerous, small, uninucleate spores, each with a cell wall of a protein-like substance (keratin); sometimes cellulose is present, but chitin (widespread in the true fungi) is absent. Meiosis probably takes place during the formation of spores in the sporangium, but this has not been conclusively established. The myxoflagellatae and those myxamoebae which have not undergone fusion are thus haploid, the plasmodium and fructification diploid; thus the diploid phase is dominant in the life cycle.

In the fructifications of many genera the diploid protoplasm between the spores coagulates to form a capillitium consisting either of a continuous network of fine tubes or threads (Fig. 455C) or of unattached fibres which taper at the ends (Fig. 455E). At maturity the peridium of the sporangium ruptures and the spores are blown out of the capillitium by the wind, or in some species the capillitium assists in their liberation by means of hygroscopic movements just as happens with the elaters of liverworts (Fig. 327).

Systematics. The structure and characteristics of the sporangia (generally a few millimetres in height), and their bright colours, afford the means of diagnosis of the orders, families, etc., which comprise about 500 species. In *Fuligo varians* (= *Aethalium septicum*), the yellow 'flowers of tan' on tan-bark (also in woods), many sporangia are united in a common, brown, cake-like fructification several centimetres in diameter, but generally the sporangia are solitary and of great delicacy, e.g. the brown *Leocarpus fragilis* (Fig. 455A), widely distributed on leaves and conifer needles. These two have calcified peridia, unlike *Stemonites fusca* (Fig. 455B, brown-purple, widespread on wood) and *Comatricha typhoides* (Fig. 455C, lilac-brown). In *Chibraria rufa* (Fig. 455D, orange-red, on decaying pine-wood) and *Trichia varia* (ovate, unstalked sporangia on rotten wood) the capillitium is not reticulate, but fibrous (Fig. 455E) and moves hygroscopically. The spores of the cosmopolitan *Ceratiomyxa* are not formed within a fructification but are liberated from the surface.

Order 2: **Acrasiales**, cellular slime moulds. On germination the spores produce myxamoebae (not flagellates), which creep together to form a pseudoplasmodium without fusion (aggregation-plasmodium, p. 73, Fig. 69), in which a stream of haploid myxamoebae creep up in a highly organized manner to form a fructification with a cylindrical stalk and ovate head. In the head the peripheral cells form a cortex while the inner round off as haploid spores (cysts); cellulose walls are laid down in both stalk and head. The occurrence of sexual union and meiosis in the amoeboid stage is disputed. *Dictyostelium* is a familiar laboratory subject.

Phylogeny. The myxomycetes are phylogenetically very primitive organisms. It is doubtful whether the two orders are at all related. The Myxomycetales appear to have connections with colourless flagellates, the Acrasiales, on the other hand, to Amoebae (through which they, too, may have arisen ultimately from flagellates). The myxomycetes (formerly called mycetozoa) are a very isolated branch of the evolutionary tree because of their morphological and developmental peculiarities; indeed, some authors do not include them in the Mycophyta.

Class II: Phycomycetes, algal fungi, lower fungi

In the lowest Phycomycetes the thallus is microscopically small and uninucleate; in the higher representatives it is well developed, branched and multinucleate, and (with very few exceptions amongst the highest forms) it lacks cross walls (aseptate); it is therefore tubular (siphoneous). The diploid phase is restricted, with few exceptions, to the zygote.

These organisms were called algal fungi by earlier investigators because of the similarity of the forms then known to certain algae (cf. p. 507).

The motile reproductive cells of the Phycomycetes have three different types of flagellation (Figs. 456, 457): (1) opisthocont, i.e. with one whip-lash flagellum directed to the rear (in the Chytridiales, Blastocladiales, Monoblepharidales); (2) acrocont, i.e. with one tinsel-type flagellum (with flimmer hairs) directed forwards (Hyphochytriales); (3) biflagellate, i.e. with two anterior or lateral flagella, of which one is of the whip-lash type and the other of the tinsel-type (Oomycetales), or both are of the whip-lash type (Plasmodiophorales).

In some orders (Chytridiales, Blastocladiales) a 'nuclear cap' rich in RNA lies alongside the nucleus (Fig. 466); it disappears during nuclear division.

Order 1: **Plasmodiophorales.** In *Plasmodiophora brassicae*, which causes club root or finger-and-toe disease of brassicas, the spores germinate in the soil, liberating biflagellate swarmers which shed their flagella on entering the root-hairs of a brassica, where they develop into multinucleate, amoeboid protoplasts from which biflagellate gametes are differentiated. The ensuing zygotes also cast off their flagella and penetrate into the cells of the root where they form diploid, multinucleate protoplasts (plasmodia); at the same time they cause the root to produce swollen excrescences. Spores originate within the host cells by meiosis, and in the spring they are liberated by the rotting of the root tissues. Swarmers and gametes both have two anterior, very unequal, smooth (whip-lash) flagella (Fig. 456). The spore wall is of chitin. Some related genera parasitize various land and water plants in which they produce similar swellings of the organs.

Order 2: **Chytridiales.** These live as parasites or saprophytes, generally in or on aquatic lower plants and animals, but also in the soil. The simplest forms are intracellular parasites with naked protoplasts; the more advanced ones form a simple or branched mycelium. The wall material is chitin.

Quite extraordinary are those which capture and feed upon unicellular organisms by means of rhizopodial processes or very fine hyphae (apparently aided by the secretion of

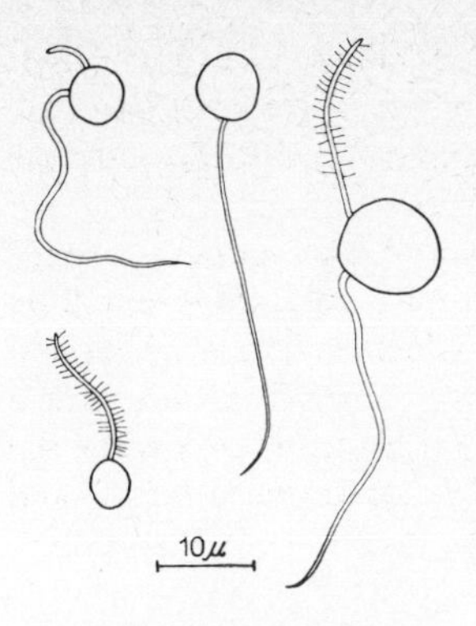

Fig. 456. Types of flagella in the Phycomycetes. Above: *Plasmodiophora, Cladochytrium, Achlya*. Below: *Rhizidiomyces*. After Kole and Gielink, and Couch.

Fig. 457. Tinsel-type flagellum (above) and whip-lash flagellum (below) of a zoospore of *Phytophthora infestans*. ×8000. After Kole and Horstra.

an adhesive). An individual of *Polyphagus euglenae* (Fig. 458A) can attack over fifty cells of *Euglena*. *Arnaudovia* lives at the surface of water as a constituent of the neuston (p. 491) and catches unicells with the aid of six long, finely pinnate hyphae (Fig. 458B). This is clearly comparable with carnivory in higher plants (pp. 199, 296).

Olpidium provides an example of a life-cycle amongst the lower Chytridiales. The naked protoplast is parasitic within the cells of the host plant, enlarges, surrounds itself with a wall of chitin (Fig. 459C, D), and forms numerous opisthocont swarmers by nuclear division and fission of the entire protoplast. These swarmers either infect new host cells as zoospores, or they can fuse in pairs as gametes to form naked, biflagellate zygotes (Fig. 459F) which penetrate the host cells and then develop into resistant resting cells. The two sexual nuclei only fuse during the following spring (Fig. 459H) and then form (presumably after meiosis) numerous swarmers which are liberated through a projecting papilla. The related Synchytriaceae also live as endoparasites in flowering plants but have a more complicated life-cycle. *Synchytrium endobioticum* causes wart disease of potatoes.

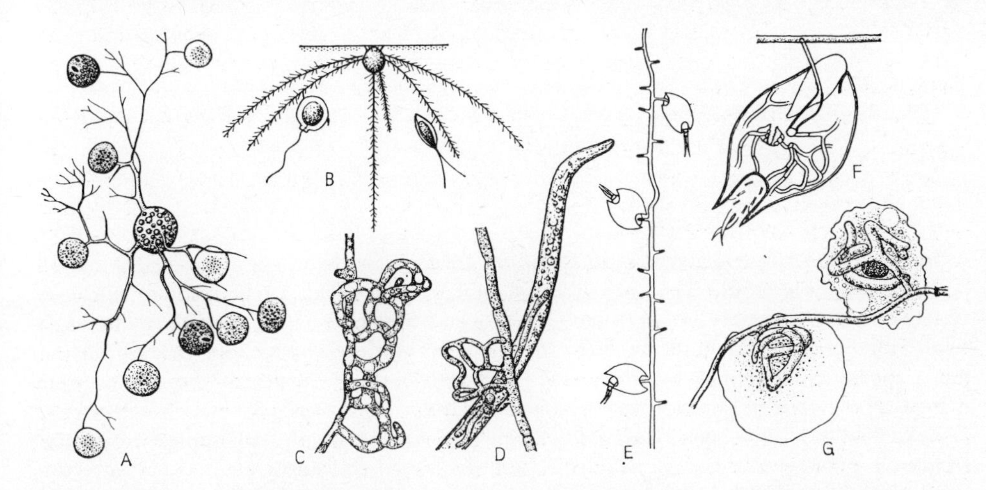

Fig. 458. Carnivorous fungi. **A.** *Polyphagus euglenae* with ten individuals of *Euglena* in various stages of decomposition. ×200. **B.** *Arnaudovia hyponeustica* with entrapped *Trachelomonas* and *Phacotus*. ×250. **C.** *Arthrobotrys oligospora* with hyphal nooses. **D.** With entrapped *Tylenchus*. ×150. **E.** *Zoophagus insidians* with three entrapped rotifers. ×90. **F.** A rotifer entrapped and invaded by *Zoophagus*. ×125. **G.** *Zoopage thamnospira* with two amoebae. ×500. **A.** After Nowakowsky. **B.** After Valkanow. **C.** and **D.** After Zopf. **E.** and **F.** After Sommerstorff. **G.** After Drechsler.

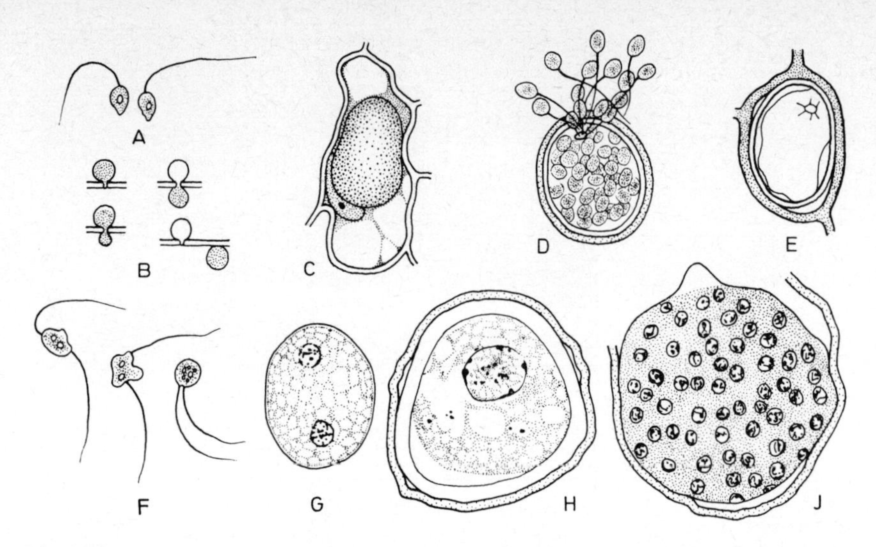

Fig. 459. *Olpidium viciae.* **A.** Zoospores. **B.** Penetration of the host cell. **C.** Naked protoplast of the fungus in the host cell. **D.** Zoosporangium or gametangium. **E.** The same, discharged. **F.** Fusion of two gametes. **G.** Young, still binucleate zygote. **H.** Encysted zygote. **J.** The same, germinating. **A.** to **F.** ×500. **G.** ×600. **H.** and **J.** ×120. After Kusano.

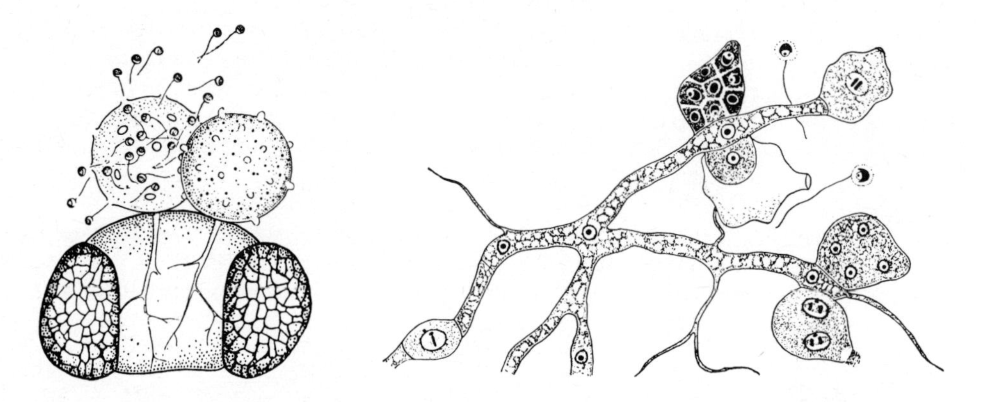

Fig. 460. *Rhizophydium halophilum.* Zoosporangia with exit papillae, one with zoospores issuing; on a pollen grain of *Pinus* with internal haustoria. ×400. After Uebelmesser.

Fig. 461. *Polychytrium aggregatum.* Small multinucleate siphoneous mycelium with sporangia at various stages. ×400. After Ajello.

In the higher Chytridiales, e.g. in *Rhizophydium*, parasitic on planktonic algae and on pollen grains (Fig. 460), there is a division of labour between a reproductive chitinous vesicle and a nutrient-absorbing rhizoid. The highest members of the order even develop a small, usually irregular, multinucleate, siphoneous, chitinous mycelium bearing one or more sporangia delimited by cross walls (Fig. 461). In *Physoderma* there is even an indication of heterophasic alternation of generations.

Sexuality has only been observed in relatively few forms. In the simplest cases (e.g. *Olpidium brassicae*) there is no distinction between zoospores and sexual gametes. Similarly in *Synchytrium*, a sexually mature swarmer behaves at first as a male gamete, but if it fails to find a gamete of the other sex it becomes non-motile and then behaves as a female. This sex-change, and indeed the fusion of gametes, may take place even inside the structure which produces the gametes. Besides these species with very labile sexuality there are genera (e.g. *Rozella*) with genotypic sex determination.

Besides the fusion of naked swarmers (Fig. 459F), the nucleus of a small ♂ plant may pass over to an adjacent ♀, in some cases (e.g. in *Polyphagus euglenae*, Fig. 462) through a connecting conjugation tube. In *Polyphagus* the ♂ protoplast does not enter the ♀ cell; instead the gametes unite in a swelling of the conjugation tube near the ♀ gametangium (D), forming a binucleate thick-walled resting spore (E). Nuclear fusion only takes place on germination when the ensuing sporangium (F) undergoes nuclear division—presumably after meiosis—with simultaneous cleavage of the cytoplasm, and liberates numerous zoospores (presumably zoomeiospores). In *Zygochytrium* two identical conjugation tubes come together and after fusion at their tips (Fig. 463) form a thick-walled zygote; nothing is known about the nuclear cycle in this case. Sometimes the conjugating gametangia (with or without tubes) are multinucleate, very much as in the Zygomycetales (p. 506).

Resting organs in the Chytridiales, when formed, do not develop inside a mother cell or receptable; instead the outer wall of the cell or zygote becomes thickened, as in *Polyphagus* and *Zygochytrium*.

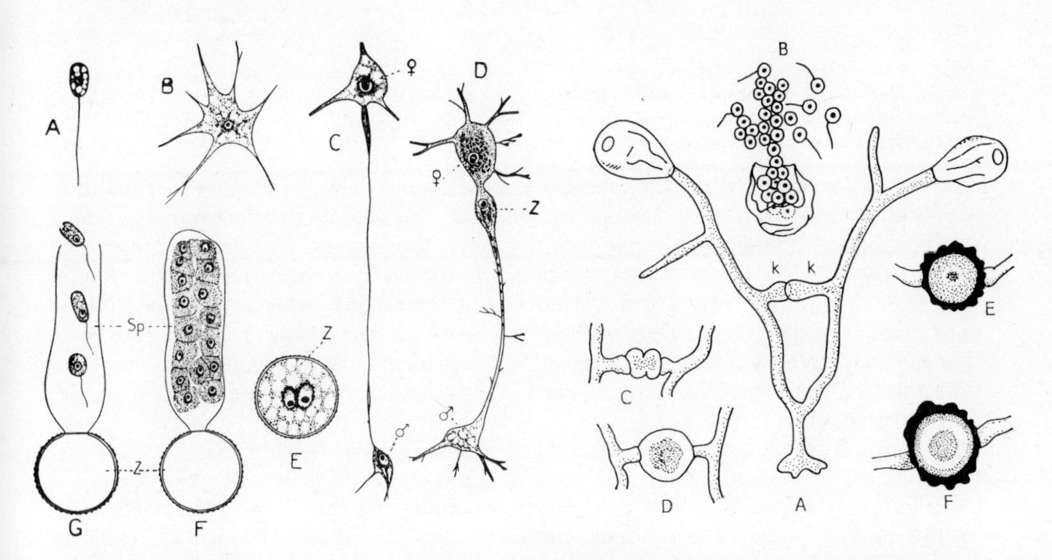

Fig. 462. *Polyphagus euglenae.* **A.** Zoospore. **B.** Plantlet emitting rhizoids. **C.** Sexual fusion between the smaller ♂ and larger ♀ individuals. **D.** ♂ nucleus in the future zygote (z). **E.** Zygote with still un-fused ♂ and ♀ nuclei. **F.** and **G.** Development and discharge of the zoosporangium (Sp). About ×450. After Wager.

Fig. 463. *Zygochytrium aurantiacum.* **A.** Plantlet with two terminal, empty zoosporangia and two conjugating gametangia (k). **B.** Discharging zoosporangium. **C.** to **F.** Conjugating gametangia forming a zygote. ×350. After Sorokin.

The Chytridiales are thus very primitive fungi in which the swarmers have one opisthocont flagellum, a nuclear cap and (a point not previously mentioned) a lipid-globule (reserve material). They show on the one hand a progression from a naked protoplast by way of a simple thallus enclosed in a chitinous wall, exhibiting division of labour, to the beginnings of a mycelium; and on the other hand a series from isogamy to gametangiogamy.

Order 3: **Blastocladiales**. The simplest members are superficially very similar to the Chytridiales, e.g. *Blastocladiella* (Figs. 464, 465), which lives in the soil. *Allomyces*, also a soil-dweller, has, however, richly branched multinucleate hyphae (a branch shown in Fig. 466A) which form mycelia a centimetre or more in diameter in artificial cultures. The wall is of chitin. The physiology of their nutrition is more specialized (more advanced) than that of the Chytridiales.

They reproduce sexually by means of uniflagellate, opisthocont isogametes (Fig. 465) or anisogametes (Fig. 466). After sexual fusion (Fig. 466D, E) the zygote grows into a sporophytic plantlet (Fig. 465) morphologically identical to the gametophyte. Meiosis takes place in the thick-walled, 'resting sporangia' (Fig. 464C), which germinate only after

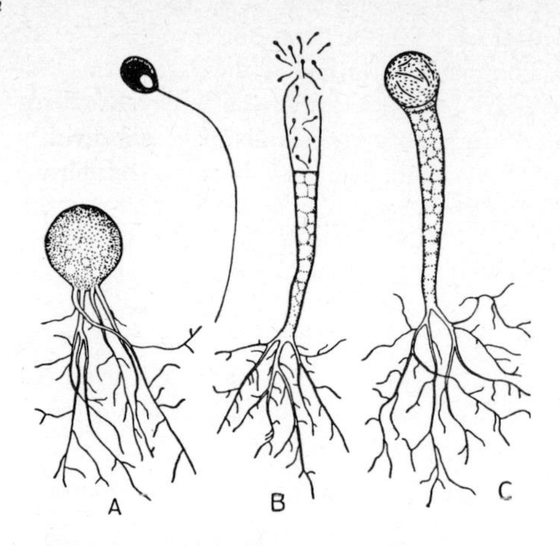

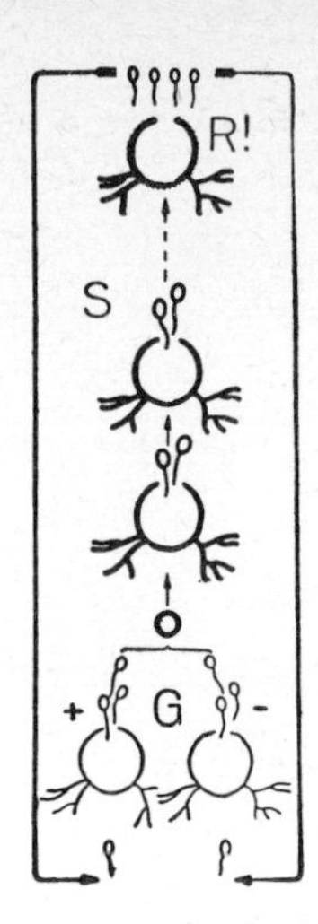

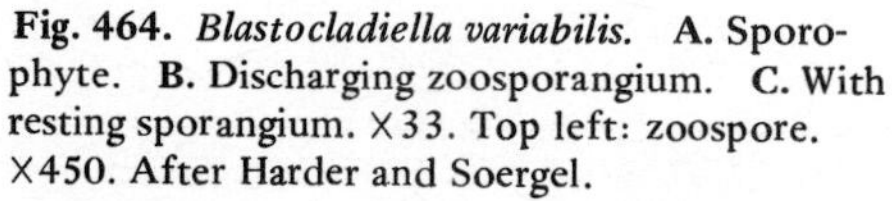

Fig. 464. *Blastocladiella variabilis.* **A.** Sporophyte. **B.** Discharging zoosporangium. **C.** With resting sporangium. ×33. Top left: zoospore. ×450. After Harder and Soergel.

Fig. 465. *Blastocladiella*: alternation of generations. For explanation see Fig. 429. After Harder.

a period of dormancy, when the meiospores develop into new gametophytes. Thus alternation of generations occurs. The sporophyte may also bear thin-walled sporangia which produce diploid zoospores without meiosis; these develop into further sporophytes, so that the regular alternation of generations is interrupted by a succession of sporophytes (Fig. 465). In the branched *Allomyces* thick- and thin-walled sporangia are found on the same plant (Fig. 466A), but in *Blastocladiella*, in which each thallus forms only a single sporangium (Fig. 464), they occur on different plants. There are numerous deviations from the regular alternation of generations in the Blastocladiales according to the species or to environmental factors.

Thus alone amongst the Phycomycetes the Blastocladiales exhibit sexual and asexual reproduction on different individuals which alternate from generation to generation. Otherwise, in their uniflagellate, opisthocont swarmers and the composition of the walls (chitin) of their more or less tubular mycelium they closely resemble the Chytridiales, from which they have apparently originated.

Order 4: **Monoblepharidales.** The species of *Monoblepharis* form small whitish pustules on plant remains under water. They consist of insignificant branched, non-septate, multinucleate hyphae with walls of glucans and cellulose. Vegetative reproduction is by uniflagellate opisthocont zoospores produced in large numbers in club-shaped sporangia (Fig.

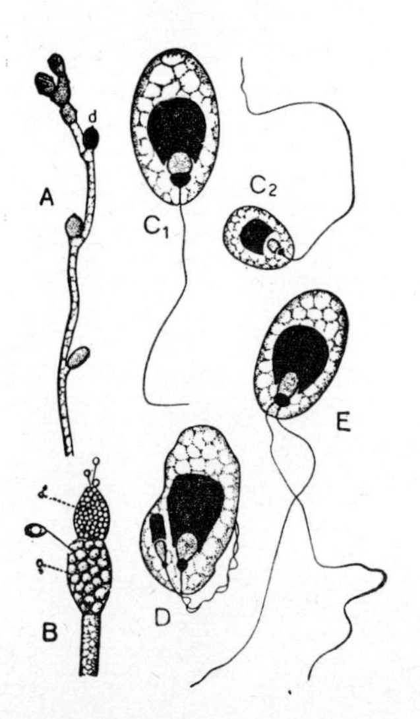

Fig. 466. *Allomyces javanicus.* **A.** Sporophyte with sporangia and resting sporangium (d). **B.** Part of the gametophyte with discharging ♀ and ♂ gametangia. **C_1.** ♀, **C_2.** ♂ gametes. **D.** Fusing. **E.** Planozygote. **C.** to **E.** ×100. After Kniep.

467a); sexual reproduction is advanced and oogamous. The swollen uninucleate oogonia are usually terminal (Fig. 467b) and separated from the rest of the thallus by a cross wall; their contents are reduced to a single uninucleate ovum. The antheridia, situated below the oogonia, are also delimited by a cross wall. They produce a number of uninucleate and uniflagellate spermatozoids (b). One of these enters through an opening in the oogonium and fertilizes the ovum (c, d), which either remains *in situ* or (in most species) squeezes through the opening, where it becomes a thick-walled spiny zygote (f); or the zygote swims away by means of the persistent ♂ flagellum. The amoeboid motility of the fertilized egg suggests a phylogenetic origin from a more actively motile megagamete. Meiosis takes place on germination, but, instead of gonozoospores, a germ tube (g) is produced—an advanced characteristic. Despite the higher level of sexuality compared with the Blastocladiales, alternation of generations is absent and meiosis occurs when the zygote germinates.

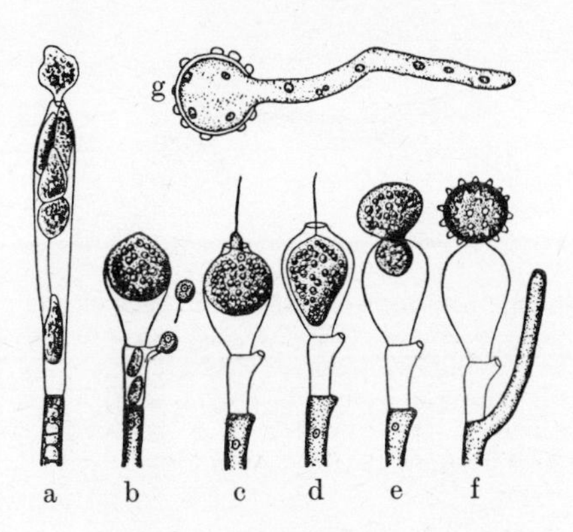

Fig. 467. *Monoblepharis.* **a.** Sporangium with emerging zoospores. **b.** Tip of a hypha with an oogonium and, just below it, an antheridium from which spermatozoids are escaping. **c.** A spermatozoid has entered the apical pore and is fusing with the ovum. **d.** Fusion completed. **e.** The fertilized egg sliding out of the oogonium. **f.** Zygote with thick spiny wall. **g.** Germination of the zygote. ×300. **a.** to **f.** After Woronin. **g.** After Laibach.

Order 5: **Hyphochytriales.** The members of this group are mainly aquatic parasites on algae or other fungi. The cell walls contain both chitin and cellulose. The motile stages have a single anterior tinsel-type flagellum (Fig. 456).

Order 6: **Oomycetales.**

Family 1. **Saprolegniaceae.** The tubular, non-septate, multinucleate mycelia of the Saprolegniaceae (Fig. 468C) mostly live as saprophytes in water on decaying plant remains and dead insects, more rarely as parasites on fishes. The cell wall consists of glucans and cellulose. Vegetative reproduction is effected by club-shaped sporangia (A) which produce uninucleate zoospores by cleavage; these have two unequal flagella of which one (tinsel-type) is furnished with two rows of flimmer-hairs (Fig. 456, right).

The pear-shaped zoospores of *Saprolegnia* possess two nearly axial, slightly unequal flagella (Fig. 468A), of which the shorter (tinsel-flagellum, with flimmer hairs) is directed forwards, the longer backwards during locomotion. After the swarming phase the flagella are withdrawn; the spore rounds off, lays down a wall and on suitable substrates extrudes a germ tube, forming a new plant. With unfavourable nutrition a further zoospore is released; this secondary zoospore (B) is bean-shaped and has two unequal lateral flagella, of which the shorter is still of the tinsel-type and directed forwards. Thus the zoospores of *Saprolegnia* are dimorphic. In other genera (*Achlya*) the zoospores of all swarming phases are similar (monomorphic) and of the secondary (B) form, while certain soil-dwelling forms lack flagella.

At first the spherical oogonia (Fig. 468D) contain numerous nuclei, but many of these degenerate, whereupon the cytoplasm aggregates and contracts round each of the remaining ones, forming a naked ovum or oosphere; hence there is a greater or lesser number of ova in each oogonium.

The multinucleate antheridia do not form separate ♂ cells; instead an entire antheridium applies itself chemotropically to the oogonium and sends simple or branched fertilization tubes (Fig. 468E, F) to the ova, discharging into each a ♂ nucleus which fuses with the ♀ nucleus. Each ovum thus becomes a cystozygote (oospore) with a thick wall resistant to attack by micro-organisms.

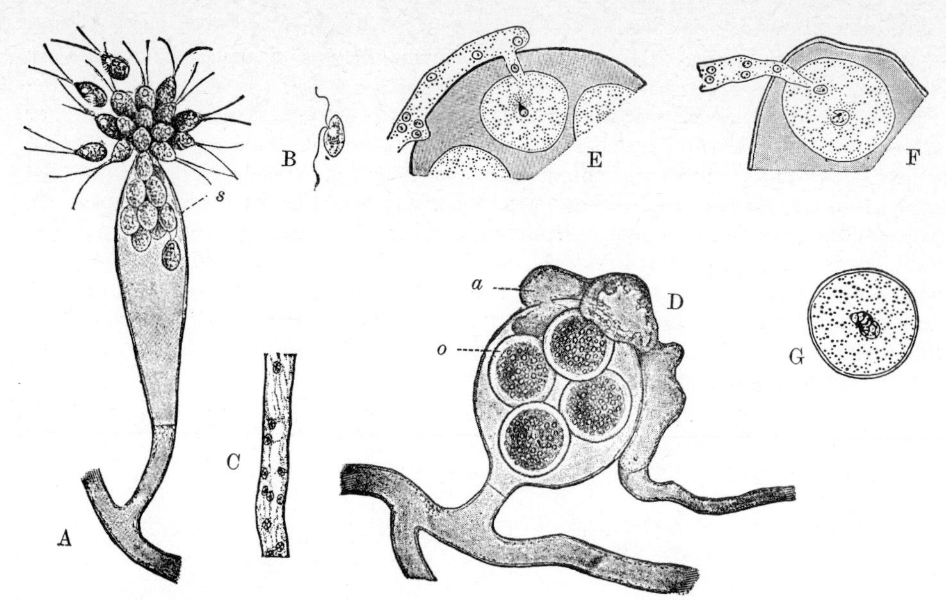

Fig. 468. Saprolegniaceae. **A.** Sporangium liberating biflagellate zoospores (s). ×200. **B.** Second type of zoospore. About ×350. **C.** Part of a hypha with numerous nuclei. ×500. **D.** Hyphae with sex-organs: a, antheridium which has sent a fertilization tube into the oogonium; o, fertilized eggs. ×600. **E.** Fertilization tube with ♂ nuclei. **F.** ♂ nucleus penetrating an ovum. **G.** Zygote with nuclei fusing. **E.** to **G.** ×600. **A., B.** and **D.** *Saprolegnia mixta.* **C.** *Thraustotheca.* **E.** to **G.** *Achlya flagellata.* **A.** and **D.** After Klebs. **B.** After Höhnk. **C.** After Schrader. **E.** to **G.** After Moreau.

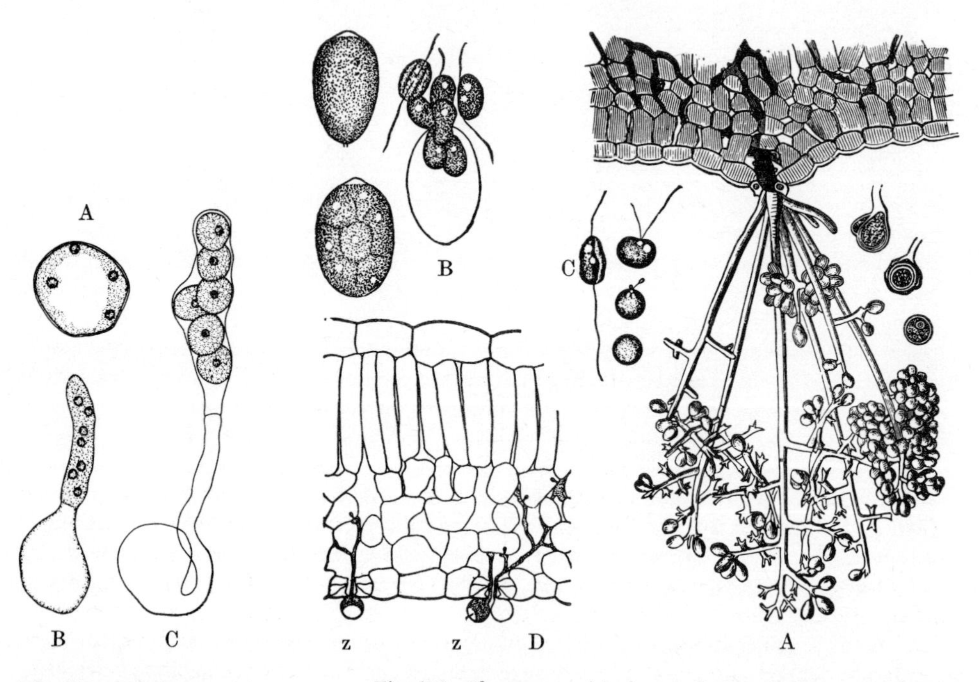

Fig. 469. Germination of zygotes in the Saprolegniaceae. **A.** With four haploid nuclei. **B.** Germ-tube. **C.** Germ-sporangium with zoospores. **A.** *Achlya recurvata.* **B.** *Isoachlya intermedia.* **C.** *Thraustotheca primoachlya.* ×1400. After A. W. Ziegler.

Fig. 470. *Plasmopara viticola.* **A.** Sporangiophore emerging from a stoma; to the right oogonia (with antheridium) and zygotes. **B.** Formation and liberation of zoospores. **C.** Withdrawal of flagella. **D.** Germinating zoospores (z) penetrating stomata and intercellular spaces. **A.** ×100. **B.** and **C.** ×600. **A.** After Millardet. **B.** to **D.** After Arens.

After a resting phase the zygotes germinate, undergoing meiosis, to form a germ tube (at first with four, later with many, nuclei) which usually soon develops a sporangium (Fig. 469). *Saprolegnia* is synoicous, but others (*Achlya*) are heteroicous.

Family 2. **Peronosporaceae**. These are predominantly parasites of higher plants. Thus *Plasmopara viticola* is an intercellular parasite in the tissues of the leaves and fruit of the grape vine, where it sends short processes (haustoria; Fig. 470D) into the living cells. Branches of the mycelium grow out through stomata into the air, forming mildew-like growths visible to the naked eye. These consist of branched sporangiophores (Fig. 470A) bearing numerous zoosporangia, which are shed intact and are carried by wind to the leaves of other plants. Meanwhile the contents have divided to form a number of kidney-shaped swarmers with two unequal lateral flagella (the shorter with flimmer hairs); these are released in drops of rain or dew (B).

The swarmers gather by chemotaxis around the stomata, withdraw their flagella (C) and produce hyphae which enter the leaf through the stomata (D) or directly through the epidermis.

Sex organs develop inside the host, the oogonia as swellings at the hyphal tips, the antheridia as tubular outgrowths (Fig. 471A, an). The gametes are not sharply delimited in either oogonium or antheridium.

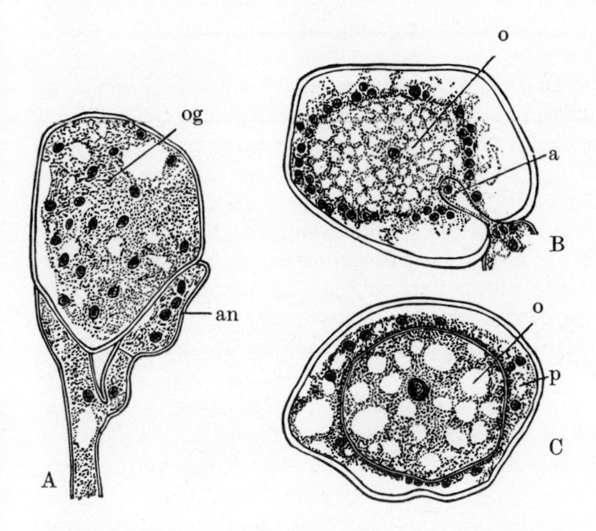

Fig. 471. Peronosporaceae. Fertilization. **A.** Young multinucleate oogonium (og) and antheridium (an). **B.** Oogonium with the central uninucleate part (o) and the antheridial fertilization tube (a) introducing the ♂ nucleus. **C.** Zygote in the oogonium (o) surrounded by the young zygote wall and the periplasm (p). **A.** *Peronospora parasitica.* **B.** and **C.** *Albugo candida.* ×600. After Wager.

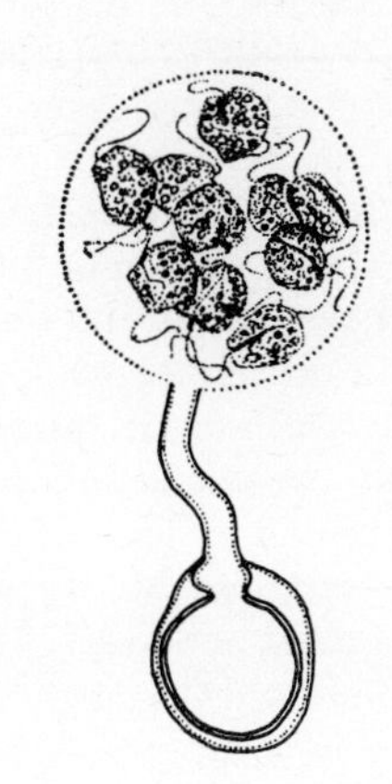

Fig. 472. *Pythium ultimum.* Zygote germinating with zoospores. ×800. After Drechsler.

Both organs are delimited by cross walls and contain many nuclei. In the oogonium the nuclei collect at the margin in the so-called periplasm, leaving only one in the middle to form the centre of the single ovum (o) which is not sharply demarcated from the surrounding cytoplasm. The antheridium sends into the oogonium a process (a) which opens at the tip and discharges a single ♂ nucleus into the central region which now becomes bounded by a wall. After a time the sexual nuclei fuse. The periplasm is used up in the formation of the outer wall of the oospore (C). The antheridia generally contain many potentially sexual nuclei, but in some Peronosporaceae (*Albugo tragopogonis, Pythium*) the number is reduced to one. In a few (e.g. *Basidiophora entospora*) functional antheridia are absent and two of the oogonial nuclei fuse together.

Reduction has not proceeded so far in *Albugo bliti*, in which there are numerous functional ♀ nuclei in the central part of the oogonium. Each of these is fertilized by a ♂ nucleus, and a common wall then surrounds all the fused nuclei and their cytoplasm, giving a zygote externally like a normal one but containing many diploid nuclei (coenozygote). In primitive forms the zygote germinates liberating zoospores, but more commonly it produces a germ tube which either forms a terminal sporangium (Fig. 472) containing zoospores, or directly penetrates the tissues of the host. This last is regarded as the most advanced condition, comparable to the similar transformation of a zoosporangium into a conidium (see below).

Vegetative reproduction shows adaptation to terrestrial life: as in the case of certain algae (pp. 458, 468) the zoosporangia are dispersed intact and only subsequently liberate zoospores (*Plasmopara*), or the formation of zoospores is suppressed and the entire wind-borne sporangium develops an infective germ tube (*Peronospora*); the sporangium has become a conidium (p. 205; these are produced in chains in *Albugo*); traces of blepharoplasts beside the nuclei indicate, however, the flagella of the ancestral forms. The sporangium of *Phytophthora infestans* germinates according to the humidity of the environment either with swarmers or with a germ tube. In the more primitive *Pythium* (common on water plants, also saprophytic in soil), on the other hand, zoospores are always liberated from sporangia still attached to the thallus. *Zoophagus* (Fig. 458E, F) lives in water among algae and duckweeds and captures rotifera, on which it feeds, by means of short, sticky hyphal branches.

Many widespread plant diseases are caused by the Peronosporaceae.

Phytophthora infestans causes blight, a serious potato disease which can extend to the tubers since during rain sporangia are washed down from the leaves and into the soil, where the tubers are infected via lenticels. In wet seasons more than 20 per cent of the potato crop may be lost in this way. In 1959 *Peronospora tabacina*, the blue mould of tobacco (so called from its bluish-white conidia), appeared for the first time in Europe (previously in America and Australia) and already in the wet summer of 1960 destroyed a large part of the tobacco crop throughout Germany. The downy or false mildew of the vine, caused by *Plasmopara viticola* (Fig. 470), is also important economically. In humid weather this is epidemic on the leaves, which fall prematurely, and the grapes shrivel. Some 20 per cent of the vintage is lost each year by this and other, less important, fungal diseases (a further 20 per cent by animal pests). Related fungi attack beet, hops, onions and other economic plants. *Pythium* also causes plant diseases. *P. debaryanum*, commonly present in soil, produces lethal damping-off and foot-rot of various seedlings. Diseases caused by Peronosporaceae can be successfully controlled by spraying the foliage with a suspension of copper sulphate and lime (Bordeaux mixture); this prevents the germination of sporangia. In spite of all preventive measures, the average annual loss of plant products in Germany by plant diseases and animal pests is some 15 per cent.

It follows from the preceding account that the Oomycetales comprise tubular, non-septate, branching, multinucleate fungi, partly saprophytic and partly parasitic, which have progressed from aquatic to terrestrial habitats. Their cell walls give a cellulose reaction. Vegetative dispersal in the mainly aquatic Saprolegniaceae is by zoospores with two unequal, generally lateral, flagella, while in the terrestrial Peronosporaceae dispersal is effected by detached intact sporangia or conidia. Sexual reproduction is always oogamous. Free ♂ gametes never occur: their function is taken over by ♂ nuclei, which are carried to the ova by a fertilization tube which grows from the antheridium into the oogonium. In extreme cases antheridia are lacking and two oogonial nuclei fuse together. Further, in the higher forms ova are not always fully differentiated, and in extreme cases the central region of the oogonium contains a large number of ♀ nuclei; after these have been fertilized by a corresponding number of ♂ nuclei the whole structure (diploid nuclei and the cytoplasm containing them) is surrounded by a wall to form a multinucleate coenozygote. While in the Saprolegniaceae all the ♀ cytoplasm takes part in the formation of gametes, in the Peronosporaceae a part remains sexually functionless.

Order 7: **Zygomycetales.**

The first family, the **Mucoraceae**, comprises terrestrial moulds which live mainly as saprophytes on plant or animal matter; they are rarely parasitic.

One of the most widespread species is *Mucor mucedo*, the pin mould, the richly branched mycelium of which forms a white cottony growth on dung, bread, etc. From the nutritive hyphae growing in the substrate arise erect hyphal tubes, each bearing at the apex a spherical sporangium (Fig. 473), containing instead of zoospores numerous roundish, multinucleate spores which are adapted to terrestrial life in being non-motile, enclosed in a wall and resistant to desiccation (Figs. 473, 474).

The sporangium is separated from its sporangiophore by a domed cross wall which projects into the sporangium as the columella (Figs. 473, 474). The multinucleate protoplast of the sporangium divides by cleavage into multinucleate spores (Fig. 474) which are liberated by the rupture of the sporangium wall and may germinate at once to form new mycelia, or which may remain dormant and viable for long periods. It seems that vegetative reproduction by spores may be repeated indefinitely.

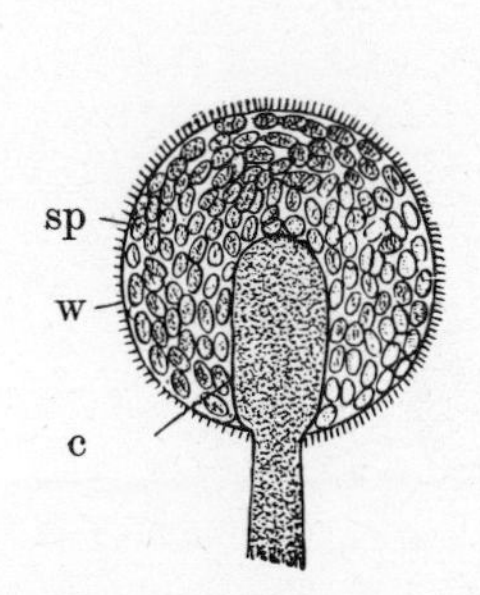

Fig. 473. *Mucor mucedo.* Sporangium in optical section. c, columella; w, wall; sp, spores. ×225. After Brefeld.

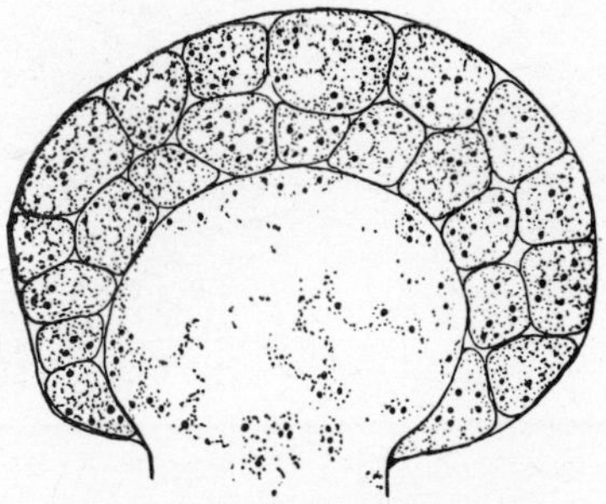

Fig. 474. *Sporodinia grandis.* Section of a mature sporangium; spores multinucleate. ×425. After Harper.

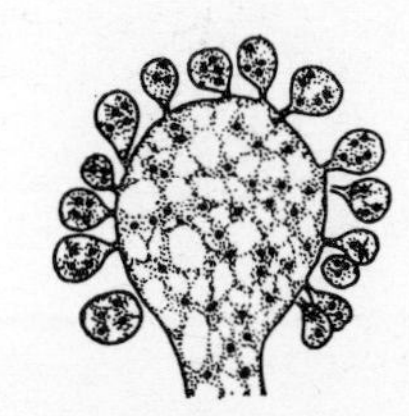

Fig. 475. *Cunninghamella echinulata.* Formation of conidia. ×370. After Moreau.

Mucor mucedo is physiologically dioecious, i.e. heterothallic: sexual reproduction only occurs when two mycelia of different mating-types (+ and −) meet. They then develop club-shaped branches, the ends of which come together in pairs chemotropically, whereupon each cuts off an identical multinucleate gametangium (Fig. 476A, B); antheridia and oogonia can no longer be distinguished (but some zygomycetes have anisogametangia). Gametes are not differentiated; instead the entire gametangia fuse together forming a warty zygote (gametangiogamy, p. 207, Fig. 476D, F). Thus the zygote is not formed inside a ♀ gametangium as in the oomycetes; instead the walls of the gametangia become the wall of the zygote, inside which the numerous nuclei derived from the two parents come together and unite in pairs (Fig. 476H).

Meiosis and the determination of the sex of the nuclei take place on germination, but only after a resting period. Zoospores are never produced; instead a germ tube grows out, and in most Mucoraceae immediately forms a sporangium at its tip (Fig. 476J). The thousands of meiospores in this germ-sporangium are generally uninucleate in contrast to the ordinary multinucleate spores; all the nuclei in the mycelium originating from such a spore must therefore be of the same sex; some of these spores give mycelia of the + strain, others of − strain.

The sporangia are not always of the *Mucor* type; some genera bear few-spored sporangia in addition to the normal ones; other sporangia contain a single spore which is shed together with the sporangium wall to which it is fused: the sporangium is now a conidium. Alternatively, conidia may be produced by the protrusion from the surface of the young sporangium of lobes containing cytoplasm and several nuclei; these later separate as exospores (Fig. 475). *Choanephora* forms either conidia or endospores, depending on the environment (nutrition, temperature). Some species produce only conidia. *Pilobolus* is well known for its method of spore discharge (Figs. 298, 321), and *Phycomyces* (similar in structure to *Mucor*) is much used in physiological studies (cf. p. 350). In the monoecious *Zygorrhynchus* the two gametangia are of different sizes (Fig. 476E).

Sexual reproduction is similar in all other Mucoraceae. In the soil-dwelling *Endogone* (fam. **Endogonaceae**) the gametangia conjugate to form an arched structure (Fig. 477A)

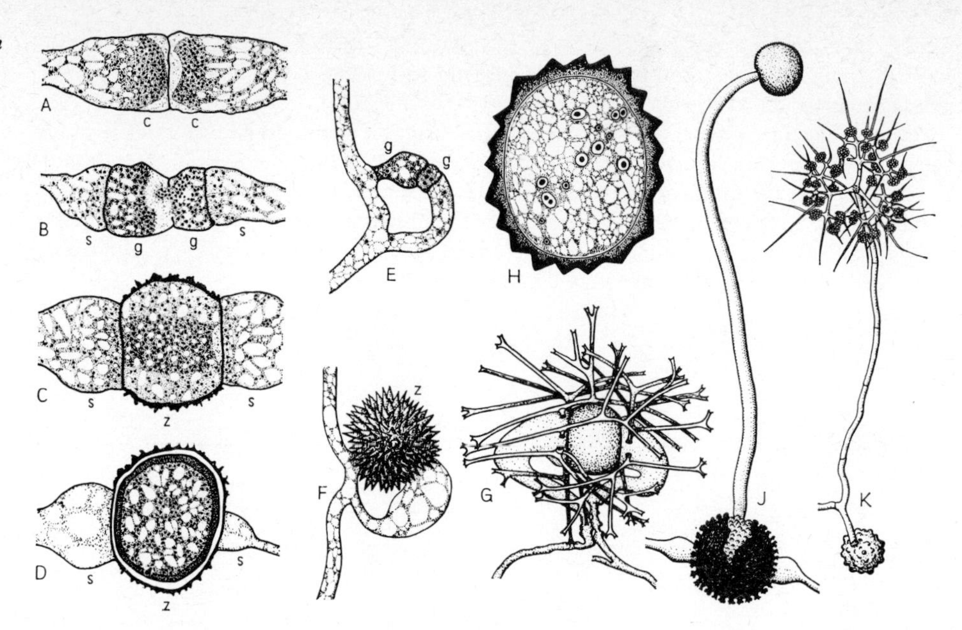

Fig. 476. Zygomycetales. **A.** to **D.** Organs of fertilization and formation of the coenozygote in *Sporodinia grandis.* ×50. c, conjugating branch; g, gametangium; s, suspensor; z, zygote. **E.** and **F.** The same in *Zygorrhynchus moelleri.* ×75. **G.** *Phycomyces blakesleeanus*, zygote with surrounding filaments. ×30. **H.** *Mucor hiemalis*, zygote with haploid nuclei, nuclear fusion and diploid nuclei. ×550. **J.** *Mucor mucedo*, germ-sporangium. ×60. **K.** *Chaetocladium jonesii*, germination of zygote with conidiophores. ×75. **A.** to **D.** After Keene. **E.** and **F.** After Green. **G.** After Gwynne-Vaughan. **H.** After Moreau. **J.** and **K.** After Brefeld.

from the apex of which the spherical zygote buds out (B). All nuclei except one move out of each young gametangium so that only one pair of nuclei fuses in the zygote. In certain species the nucleus of a smaller, ♂ gametangium passes into the larger, ♀ gametangium; then the two sexual nuclei move into the zygote which grows out of the ♀ gametangium (Fig. 477C, D). In species of *Endogone* numerous zygotes are embedded in a mass of investing hyphae forming fruit-bodies (sporocarps) which may be as big as a hazelnut. The mycelium is initially multinucleate and non-septate; cross walls form later, but their development is quite different from that in plants with uninucleate cells, in which wall-formation is linked with nuclear division.

The family **Entomophthoraceae** (parasitic on insects or on desmids) also has tranverse walls in the hyphae of older mycelia, and some are septate even when young. The nuclei vary from many to one in each cell (Fig. 479D); indeed, *Basidiobolus ranarum* has mostly uninucleate cells. Sexual reproduction proceeds by the conjugation of two gametangia, either by the formation of a conjugation bridge between two hyphae or laterally between adjacent cells of the same hypha (Fig. 478). Conidia homologous with sporangia effect asexual reproduction.

In the most familiar example, *Empusa muscae*, which causes an epidemic disease of flies, the conidiophore forcibly shoots off the multinucleate conidium (Fig. 479B, C), the germ tube from which penetrates the body of a fly, where it forms a lethal parasitic mycelium. Great numbers of conidiophores project from the dead fly and surround it (e.g. on a window-pane) with a white halo of conidia. In dried-out flies thick-walled cysts are formed inside the hyphae; these are perhaps to be regarded as parthenogenetic zygotes.

The **Zoopagaceae** are closely related to the Entomophthoraceae and parasitize amoebae and nematodes by means of haustorial hyphal branches (Fig. 458G).

Thus the **Zygomycetales** are mostly saprophytic fungi with a well-developed mycelium which is non-septate and multinucleate in the Mucoraceae, but septate in certain Entomophthoraceae; the cell walls consist of chitin. Reproduction is adapted to terrestrial life, but in a manner different from that in the Oomycetales, in which the entire sporangium is

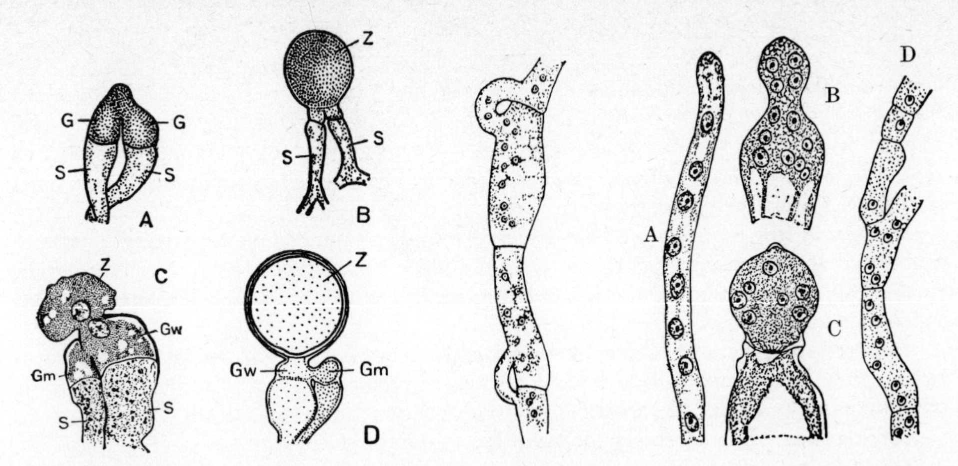

Fig. 477. Fertilization of *Endogone*. **A.** Conjugation. **B.** Mature zygote of *E. pisiformis*. **C.** Growth of the zygote after the transfer of the nucleus from the male (gm) to the ♀ (Gw) gametangium. **D.** Mature zygote of *E. lactiflua*. G, gametangium; S, suspensor. **A.** and **B.** After Thaxter. **C.** and **D.** After Buchholz.

Fig. 478. *Ancylistes closterii.* Fertilization between neighbouring cells. After Dangeard.

Fig. 479. A. to **C.** *Empusa muscae.* **A.** Hyphal tip in a fly. **B.** The developing conidiophore emerging from the fly. **C.** Formation of a conidium. **D.** Young hypha of *Entomophthora sciarea.* **A.** to **C.** ×450. **D.** ×180. After Olive.

detached and gives rise to zoospores where it germinates. The Zygomycetales, on the other hand, produce instead of zoospores resistant structures (p. 505) which can be transported by air. However, the reduction seen in the Oomycetales of the sporangium to a conidium producing a germ tube also occurs in the Zygomycetales, although in a somewhat different form. The Zygomycetales never produce gametes in sexual reproduction; conjugation invariably proceeds by the fusion of two entire, frequently identical, generally multinucleate gametangia (gametangiogamy), forming a zygote which is almost always multinucleate; this also is an arrangement well adapted to life on land.

Phylogeny. The Phycomycetes include a number of groups which are probably not related to each other. The Chytridiales, with opisthocont smooth flagella and chitinous walls, are probably close to the flagellates. There is a progressive developmental series from the Chytridiales to the Blastocladiales, to which the Monoblepharidales might be added if these did not have cellulose cell walls. There is, however, the briefly-mentioned Hyphochytriales in which the walls have both cellulose and chitin so that a connection between fungi with cellulose walls and those with chitin cannot be wholly excluded.

The Plasmodiophorales have the same flagellation as the Myxomycetes, with which in consequence many authorities unite them, but their spore walls consist of chitin which is absent from the Myxomycetes.

The cell wall chemistry of the Oomycetales (glucans and cellulose) and their oogamy would suggest a connection between the Saprolegniales–Peronosporales (which certainly form a coherent series) and the Monoblepharidales if the flagellation of their swarmers were not entirely different. The swarmers of the Oomycetales are bilaterally symmetrical with a shorter tinsel-type flagellum and a longer whip-lash flagellum, and are strongly reminiscent of the Heterosiphonales from which they might have been derived by loss of chlorophyll.

The non-flagellate Zygomycetales with chitinous walls and gametangiogamy may be related to those Chytridiales with conjugating gametangia, and may form a series parallel to the Blastocladiales–Monoblepharidales. Forms such as *Zygochytrium* amongst the Chytridiales indicate how the characteristic conjugation of gametangia could have arisen.

Class III: Ascomycetes

The Ascomycetes are predominantly terrestrial like the next class, the Basidiomycetes. Except in certain reduced forms, the mycelium consists of copiously branched septate hyphae with chitinous cell walls. The septa are perforated by a simple pore. Sexual reproduction is always gametangiogamous. Being terrestrial (saprophytes or parasites) they never form flagellate reproductive cells.

The spores are borne in asci: an ascus is a tubular sporangium containing a definite number of spores (usually eight) which are produced endogenously by free cell formation (p. 44). Spore-formation in the ascus is preceded by the fusion of two nuclei followed immediately by meiosis.

The asci are generally united in special plectenchymatous (Fig. 78, p. 77) fructifications. They are arranged side by side, usually interspersed with sterile filaments or paraphyses, as an extended palisade-like surface or hymenium (Fig. 494B, C).

Conidia occur as an accessory means of reproduction (p. 205).

The Ascomycetes number some 20,000 species. The form and mode of development of the fructification and the structure and mechanism of dehiscence of the ascus serve to distinguish the subclasses and orders. The systematics of the Ascomycetes is at present in a state of flux, being greatly complicated by many overlapping diagnostic characters. The orders are delimited here according to current practice. The grouping of related orders in subclasses should facilitate understanding of the great diversity within the class.

A. Protoascomycetidae

The Protoascomycetidae (or Hemiascomycetidae) comprise those ascomycetes which form no fructification and in which the zygote develops directly into an ascus.

Order 1: **Saccharomycetales** (Protoascales, Endomycetales). The yeasts (Saccharomycetaceae) provide the first examples of this group in which the mycelium is poorly developed; the members generally grow in or on sugary fluids. The spherical or ovate cells usually increase by budding (Fig. 37) and remain united in more or less branched chains of variable length (Fig. 480G). In *Schizosaccharomyces* the cells divide transversely (Fig. 480A). The cell wall consists chiefly of carbohydrates (glucans and mannans), but with some protein and a very little chitin. The reserve material is glycogen and numerous vitamins are present, especially of the B-group.

Sexual reproduction in the Saccharomycetaceae (Fig. 480) is effected by the conjugation of two cells (sometimes through a short conjugation bridge). The zygote develops directly (or with the interpolation of a budding phase) into an ascus in which four or eight ascospores are formed after meiosis. These are liberated after the breakdown of the ascus wall as new vegetative cells.

The yeasts show three different types of life-cyle. In the haplontic type (*Schizosaccharomyces*, Fig. 480A–F) vegetative reproduction takes place in the haploid phase: the zygote immediately undergoes meiosis to form the ascus. In the haplo-diplontic type (*Saccharomyces cerevisiae*, Fig. 480G–L) both the haploid cells and the zygote can increase by budding; asci can thus be formed by diploid bud-cells. Finally, in the diploid type (*Saccharomyces ludwigii*, Fig. 480M–S) the ascospores conjugate forthwith in the ascus and vegetative budding only takes place in the diplophase.

While the yeasts named above may be thought of as reduced in their vegetative structure and reproduction, other genera of the Saccharomycetales show more primitive characteristics. In *Dipodascus albidus*, which grows on the mucilaginous exudates of trees, the mycelium consists of multinucleate cells. Adjacent cells form hooked conjugation branches which fuse at their apices while their bases remain separated by a cross wall (Fig. 480W–Y). The nucleus of a male gametangium passes into the female one where fusion of the two nuclei takes place; the female gametangium then elongates to form the ascus. While the remaining nuclei disintegrate, the fusion nucleus undergoes meiosis and forms numerous haploid nuclei each of which becomes an ascospore by means of free cell formation (p. 44).

In *Eremascus albidus*, found in fruit juices, the hyphal cells are multinucleate at first

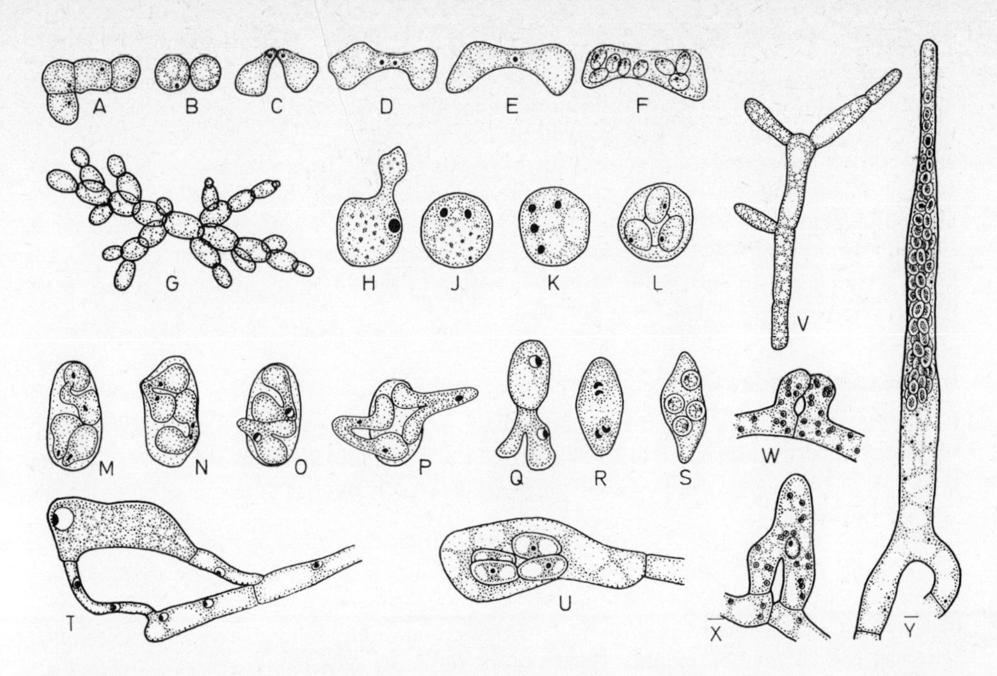

Fig. 480. Saccharomycetaceae. **A.** to **F.** *Schizosaccharomyces octosporus.* ×350. **A.** Cell colony. **B.** to **F.** Conjugation and formation of the ascus. **G.** to **L.** *Saccharomyces cerevisiae.* **G.** Chains of buds. ×200. **H.** to **L.** Formation of the ascus. ×550. **M.** to **S.** *Saccharomyces ludwigii.* ×375. **M.** to **P.** Conjugation of germinating spores in the ascus. **Q.** Budding of the diploid cell. **R.** and **S.** Formation of ascospores. **T.** and **U.** *Endomyces magnusii*, conjugation and formation of the ascus. ×375. **V.** *Candida reukaufii.* ×375. **W.** to **Y.** *Dipodascus albidus*, conjugation and formation of the ascus. ×275. **A.** and **V.** After Lodder and Kreger. **B.** to **I.**, **H.** to **U.** After Guillermond. **G.** After Lindau. **W.** to **Y.** After Juel.

but uninucleate later. The nuclei of two gametangia fuse while the conjugation bridge swells to form a spherical ascus; the diploid nucleus undergoes meiosis, forming eight haploid ascospores. In *Endomyces magnusii*, found in the slime flux of oak trees, the male and female conjugation branches show a significant difference in size; after nuclear fusion and meiosis the female develops a four-spored ascus (Fig. 480T, U). In all Saccharomycetales the ascus wall disintegrates or becomes mucilaginous at maturity; the spores are thus not discharged actively.

The Cryptococcaceae include the food yeast, *Candida utilis*; *Candida reukaufii* (Fig. 480V) which lives in floral nectar; and *Rhodotorula*, coloured salmon-red with carotinoids. In none of these are asci known; only the budding stage has been observed. Consequently the systematic position of the family is uncertain.

For the uses of yeasts see p. 519.

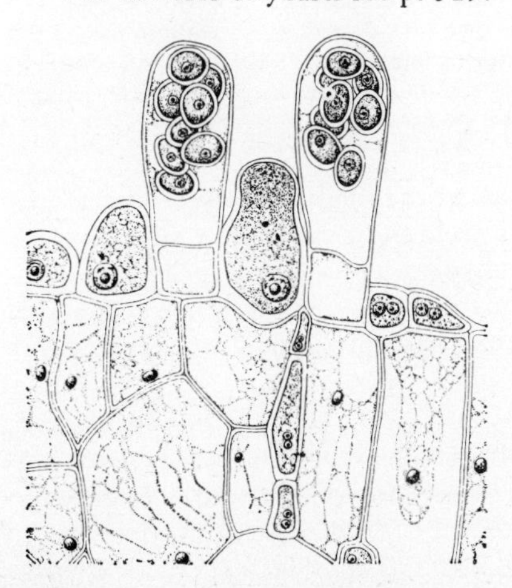

Fig. 481. *Taphrina deformans.* Caryogamy and mature asci. ×800. After Martin.

Order 2: **Taphrinales.** The genus *Taphrina* (*Exoascus*) is a plant parasite producing various deformities. Some species cause witches' brooms (p. 315) on cherry-trees, birches and hornbeam; *T. deformans* is responsible for peach leaf curl; *T. pruni* transforms the ovaries of plums into hollow galls without a stone (the so-called pocket plums). The asci erupt between the epidermal cells of the host, but form no fructifications (Fig. 481). The ascospores develop into a haploid, saprophytic mycelium which grows by budding. Only after conjugation does the mycelium with paired nuclei lead to infection. In some species the diploid nuclear stage is reached without the subsequent formation of cross walls, a process which can in fact result from the division of the nuclei in the ascus. In such cases the haploid phase is completely suppressed.

B. Plectomycetidae

In the Plectomycetidae the asci generally develop in closed fructifications (cleistothecia) so that the ascospores are only liberated when the cleistothecium disintegrates. After fertilization the female gametangium develops dicaryotic hyphae at the ends of which the asci are formed.

Order 1: **Plectascales.** *Aspergillus* and *Penicillium* are amongst the most frequent mould fungi ('mould' is not a systematic concept, but a collective term for surface-living fungal mycelia). Vegetative reproduction takes place extremely freely by conidia, mostly bluish-green and produced on closely crowded conidiophores. In *Aspergillus* the swollen spherical end of the conidiophore bears short, radiating cells or sterigmata which cut off successive conidia in coherent chains. In *Penicillium* the spores are similarly borne in necklace-like chains, but the conidophore is branched.

In the simplest members the sex organs consist of two club-shaped gametangia lying side by side. In *Penicillium* the gametangia become spirally entwined (Fig. 482D), while in *Aspergillus* the coiled female gametangium (ascogonium) is surrounded by the branching antheridium (Fig. 482C). In fertilization only plasmogamy occurs initially and

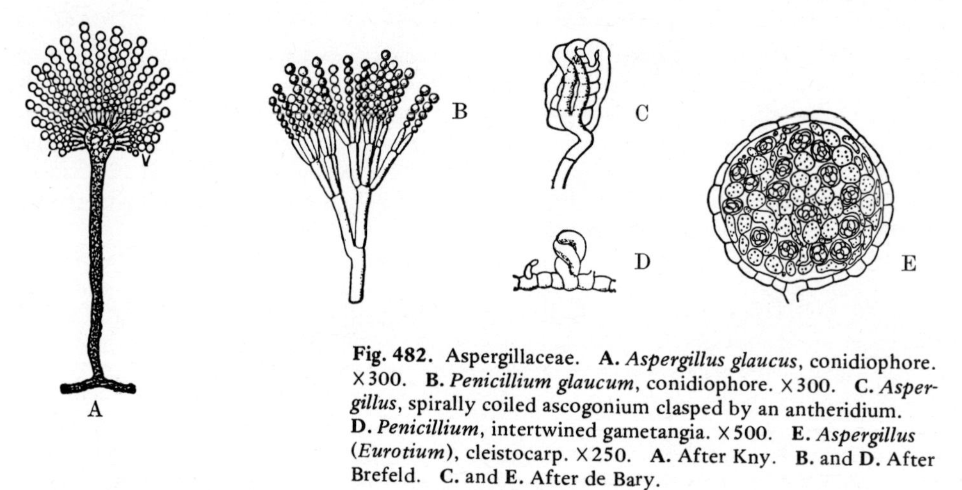

Fig. 482. Aspergillaceae. **A.** *Aspergillus glaucus*, conidiophore. ×300. **B.** *Penicillium glaucum*, conidiophore. ×300. **C.** *Aspergillus*, spirally coiled ascogonium clasped by an antheridium. **D.** *Penicillium*, intertwined gametangia. ×500. **E.** *Aspergillus* (*Eurotium*), cleistocarp. ×250. **A.** After Kny. **B.** and **D.** After Brefeld. **C.** and **E.** After de Bary.

dicaryotic hyphae spring from the ascogonium. The paired nuclei only fuse in the asci which develop at the tips of these hyphae. After meiosis in the ascus four or eight haploid spores are formed. The dicaryotic hyphae become invested by haploid hyphae forming a fructification with a completely closed plectenchymatous covering (peridium). The asci are arranged randomly in the fructification. Since this cleistothecium has no opening the ascospores are only liberated when it disintegrates.

Many of the numerous species of *Aspergillus* and *Penicillium* are of economic importance. *A. fumigatus* can be a human pathogen (lungs, the ears). The antibiotic penicillin (p. 290) is obtained from *P. notatum* and others; it is secreted into the nutrient solution. *P. roquefortii* is used in cheese manufacture. *Histoplasma capsulatum* produces a widespread lung-disease of man in warmer countries. *Elaphomyces cervinus* lives underground

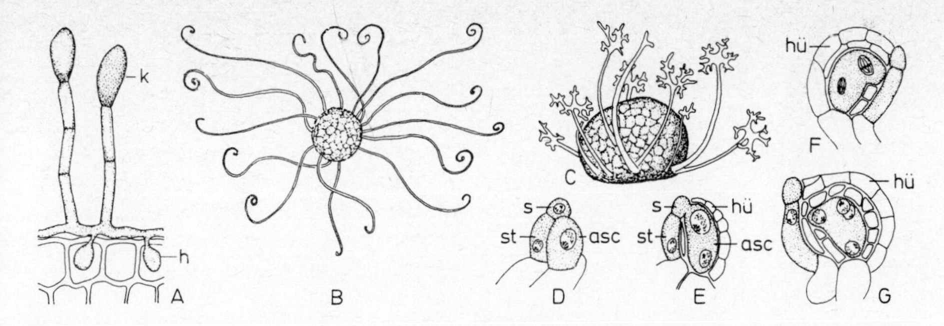

Fig. 483. Erysiphales. **A.** *Uncinula necator*, formation of conidia. ×100. **B.** The same, cleistothecium with appendages. ×30. **C.** *Microsphaera alphitoides*, cleistothecium with appendages. ×30. **D.** to **G.** Fertilization of *Sphaerotheca fuliginosa.* ×250. asc, ascogonium; h, haustorium; hü, peridial hyphae; k, conidia; s, antheridium; st, stalk-cell. **A.** and **B.** After Sorauer. **C.** After Blumer. **D.** to **G.** After Bergman.

like the truffles (p. 517) and forms mycorrhizal associations with *Pinus* and *Picea*.

Order 2: **Erysiphales.** The parasitic 'true' or powdery mildews grow on the leaf surfaces of higher plants as a web-like mycelium which sends haustoria into the epidermal cells of the host (Fig. 483A). During the summer conidia (Fig. 483A) are produced in such vast quantities that infected leaves appear to be dusted with flour. Sexual reproduction proceeds much as in the Plectascales. The male conjugation branch is divided into a stalk-cell and a uninucleate antheridium and lies alongside the ascogonium which is also uninucleate (Fig. 483G). The male nucleus passes into the ascogonium (F). After a conjugate division (G, H), one pair of nuclei fuse to form the zygotic nucleus which divides by meiosis into four or eight ascospore nuclei. The ascogonium is enveloped by investing hyphae which eventually form the light-coloured ground tissue and the dark peridium of the cleistothecium which is forced open at maturity by the pressure of the swelling asci. The spores are generally liberated by the rupture of the ascus. Very often the base of the cleistothecium is surrounded by a crown of simple, dichotomous or hooked hyphae which assist in dispersal (Fig. 483B, C).

Uncinula necator (Fig. 483A, B) attacks the leaves and berries of the vine (conidial stage: *Oidium tuckeri*). *Sphaerotheca mors-uvae* (with only one ascus in the cleistothecium) parasitizes the gooseberry. *Microsphaera alphitoides* lives on the leaves of oak. *Erysiphe* parasitizes numerous species of herbs. The downy mildews are controlled with sulphur preparations.

C. Loculomycetidae

While in the Plectomycetidae the sex organs are naked at first, and only later are the dicaryotic ascogenous hyphae enclosed in a covering, in the Loculomycetidae the fructification is formed first and the gametangia subsequently develop within its plectenchyma (pseudothecium, Fig. 485). Asci or groups of asci come to lie within cavities (loculi) which develop around the ascogenous hyphae. The asci have a two-layered wall (bitunicate). The outer wall is thin but rigid; at maturity it bursts open at the apex. The thick but extensible inner wall elongates through the ruptured outer wall by means of turgor; the spores are successively expelled through an elastic pore (Fig. 484).

Order 1: **Myriangiales.** The asci are not grouped and lie at random in the plectenchyma. Predominantly tropical and parasitic on plants and insects.

Order 2: **Dothiorales.** The asci are arranged in groups. The fructification has a substantial opening produced by the rupture or breakdown of the surface layer; frequently several fructifications are embedded in a large vegetative hyphal mass (stroma). Saprophytes or parasites on bark, stems, leaves and fruits.

Order 3: **Pseudosphaeriales.** The fructification opens only by a pore or canal formed by the localized dissolution of the plectenchyma (Fig. 485); very often the opening is surrounded by bristles. Many agents of plant disease belong here. *Venturia* (conidial stage: *Fusicladium*) is the scab of apples and pears, producing dark flecks covered with

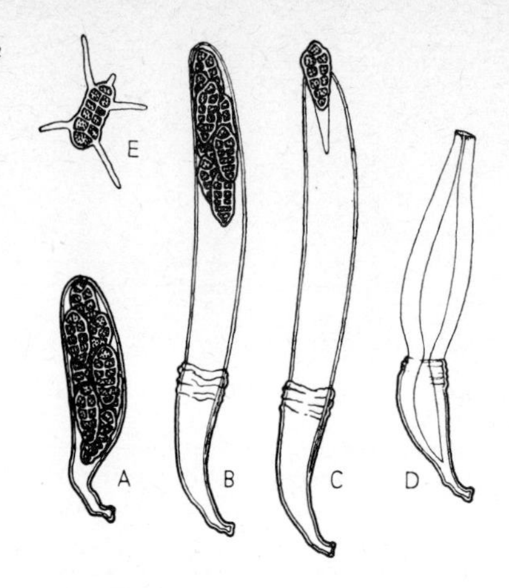

Fig. 484. *Pyrenophora scirpi*. Mode of dehiscence of the bitunicate ascus. A. Mature ascus with eight multinucleate spores. B. The outer wall of the ascus ruptured, the inner elongated. C. Shortly before the ejection of the last spore. D. The empty ascus. E. Germinating spore. ×175. After Pringsheim.

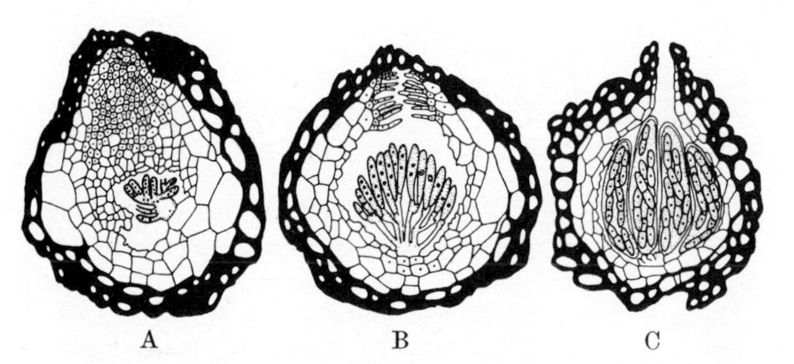

Fig. 485. *Mycosphaerella tulipifera*. Development of the pseudothecium. A. Young stage with a branched ascogonium. B. With asci of various ages. C. Mature pseudothecium. ×400. After Higgins.

Fig. 486. *Fusicladium*-scab on a pear. After Kirchner and Boltshauser.

inextensible cork, so that the growing fruits later crack open (Fig. 486). *Capnodium* is one of the 'sooty moulds' on leaves. In alpine regions *Herpotrichia* covers the twigs of conifers with a dark brown hyphal felt and brings about the death of the needles.

D. Pyrenomycetidae

The flask-shaped fructifications, often embedded in an extensive stroma, only develop after the formation of the sex organs; they have a pre-formed aperture and are called perithecia, but at maturity they are scarcely distinguishable from the pseudothecia mentioned above. The palisade-like fertile layer or hymenium covers the base of the pre-formed cavity; the asci are interspersed with numerous haploid hyphae (paraphyses; Fig. 487).

The ascus has a single wall, but has at the apex various structures involved in spore-discharge. At maturity an ascus stretches so far beyond its neighbours that its tip reaches the aperture of the perithecium, whereupon all eight spores are discharged simultaneously to a distance of 20 cm or more—some 500 times the length of the ascus. After discharge the empty ascus collapses leaving the mouth of the perithecium free for the next ascus.

Order 1: **Sphaeriales.** The members of this very large group (over 500 genera!) include both parasites on and in plants and saprophytes on rotting wood, dung, etc. The species of *Neurospora* are important subjects in biochemical genetics; their tiny perithecia (scarcely 0.5 mm) occur singly. In other genera the perithecia are aggregated or are embedded in a cushion- or club-shaped stroma. Thus in *Xylaria hypoxylon*, common on the stumps of deciduous trees, the black lower part of the antler-like stroma bears perithecia whilst the whitish upper part bears conidia. The cinnabar red stromata of *Nectria cinnabarina*, found on dead twigs, first produce conidia, later perithecia. *Nectria*

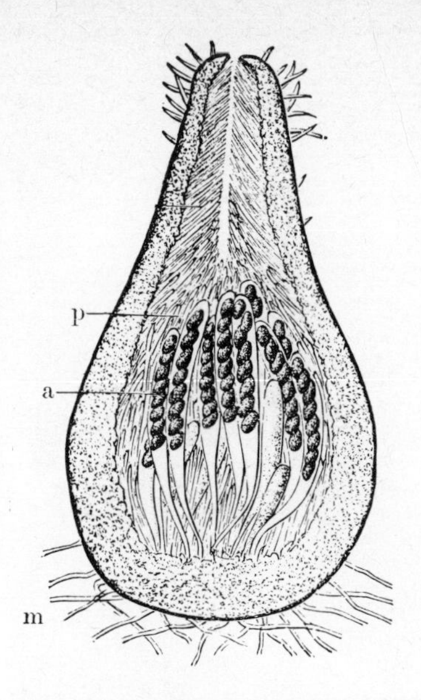

Fig. 487. *Podospora fimiseda.* Perithecium. a, asci; p, paraphyses; m, hyphae. ×90. After von Tavel.

Fig. 488. *Nectria*-canker on the branch of a fruit-tree. After Braun and Riehm.

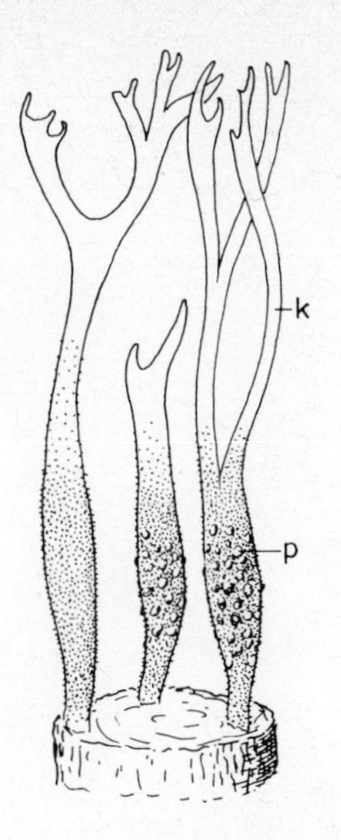

Fig. 489. *Xylaria hypoxylon.* Nat. size. k, conidial region; p, perithecial region.

galligena (with white conidia) parasitizes the bark of fruit-trees, producing canker; portions of the bark killed by the fungus are overgrown by wound callus; then the callus itself is killed so that new hypertrophied excrescences are constantly being formed (Fig. 488); finally the tree may be completely killed. *Gibberella*, parasitic on rice, in which it causes excessive extension growth, is a source of the growth-substance gibberellin (p. 303). *Ceratocystis* (*Ceratostomella, Graphium*) *ulmi* lives in the vessels of the youngest annual ring of elms in which it causes the formation of tyloses (Fig. 168) which block the vessels and hence kill the tree (Dutch elm disease). Related species cause a blue coloration of unseasoned pine timber.

Some Sphaeriales (e.g. *Neurospora, Sordaria* and others) form no antheridia; the trichogyne (or a functionally similar 'receptive hypha') receives a male nucleus from a conidium which has thus taken over the function of a spermatium.

Order 2: **Clavicipitales.** This is distinguished from the preceding order by its filamentous ascospores; initially unicellular, they later become septate. *Claviceps purpurea* parasitizes the young ovaries of the Gramineae, especially rye, where it forms conidia (Fig. 490A, B). At the same time it secretes a sugary fluid (honeydew); this attracts insects which carry the conidia to other florets. When the ovary tissue has been consumed the mycelium forms a sclerotium (p. 78) in which the closely-packed hyphae develop transverse walls, especially towards the periphery, forming a pseudoparenchyma (Fig. 78B). The sclerotia project from the ear of the rye as hard, dark-violet bodies known as ergot (Fig. 470C, D). Their alkaloid content renders them poisonous but they are used in medicine as 'Secale cornutum'. They fall to the ground, over-winter, and grow out at the flowering time of the host as stromata in the form of pink stalked heads in which numerous perithecia are embedded (E). The long asci contain eight filamentous spores which are carried by the wind to the stigmata of their host. *Cordyceps* lives either on insects, which after infection bury themselves in the soil, or on hypogeous fungi (e.g. *Elaphomyces*). The club-shaped stromata emerge from the soil and have on their upper part numerous perithecia. The filamentous spores in the ascus are multicellular with transverse septa; they break into separate fragments. *Epichloe typhina* parasitizes grasses; the yellow stroma encircles the stem, first producing conidia, later perithecia.

Order 3: **Laboulbeniales.** These minute, bristle-like fungi parasitize insects; they have virtually no hyphae and consist of little except sex organs and perithecia (Fig. 491).

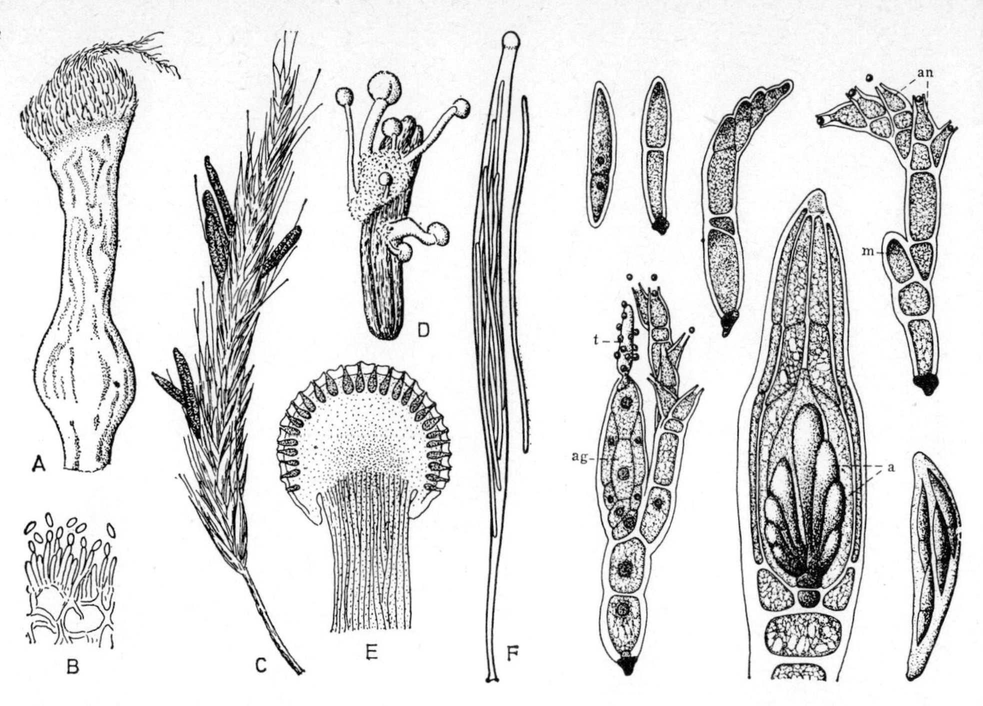

Fig. 490. *Claviceps purpurea*. A. Infected ovary of rye. ×15; *below*, the beginning of the formation of the sclerotium, *above* conidial hyphae, and at the top the remains of the stigmata. B. Formation of conidia. ×300. C. Ear of rye with mature sclerotia. ×$\frac{2}{3}$. D. Germinated sclerotium with stalked fructifications. ×2. E. L.S. through a fructification with numerous perithecia. ×25. F. Ascus and ascospore. ×400. A., B., D. to F. After Tulasne. C. After Schenck.

Fig. 491. *Stigmatomyces baerii*. Upper row: development up to the formation of antheridia (an) and the ascogonium mother-cell (m). Lower row: fertilization (ag, ascogonium; t, trichogyne), immature perithecium with young asci (a), and ascus with four two-celled spores. ×400. After Thaxter.

Nevertheless they are very numerous (1500 species). Instead of a male gametangium they form free male cells (spermatia). They generally penetrate the chitinous cuticle of their host by means of a short, dark-coloured 'foot'. From infection of a host to the maturity of the spores is only ten to twenty days; accordingly their development is remarkably rapid.

E. Discomycetidae

The typical fructification of the Discomycetidae is the cup- or disc-shaped apothecium bearing a superficial hymenium of asci and haploid paraphyses. In some genera the hymenium is exposed from the outset, but generally it develops inside the young fructification of which the apex is ruptured at maturity by vigorous lateral extension. The asci are unitunicate, i.e. although the wall consists of several layers they do not separate as in the bitunicate Loculomycetidae. The asci open in various ways in the different orders.

The development of a discomycete can be represented by the simply constructed and much-studied *Pyronema confluens*. Its dull red apothecia a few millimetres in diameter occur in woods on the sites of old fires. In the developing fructification some of the hyphal tips become ♀ organs; each consists of a dilated multinucleate ♀ gametangium, known as the ascogonium (Fig. 492A og), borne on a stalk cell, and bearing at its apex a curved, multinucleate papilla, the trichogyne (t). In the immediate neighbourhood of the ascogonium arises a club-shaped, multinucleate antheridium (a) with which the tip of the trichogyne comes into contact. The sex organs occur in groups. The trichogyne opens at the point of contact, whereupon its nuclei degenerate and the ♂ nuclei pass by way of the

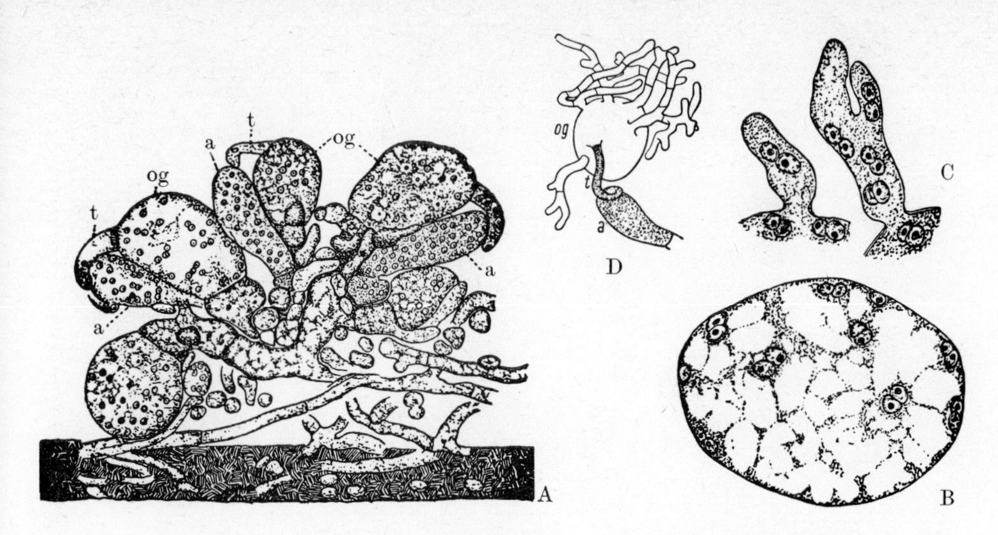

Fig. 492. *Pyronema confluens.* **A.** Immature fructification with ascogonia (og), trichogynes (t) and antheridia (a). **B.** T.S. of an ascogonium with conjugating ♂ and ♀ nuclei. **C.** Migration of the paired nuclei into the ascogenous hyphae developing from the ascogonium. **D.** Ascogonium with ascogenous hyphae. **A.** ×450. **B.** and **C.** ×1000. **D.** ×150. **A.** to **C.** After Claussen. **D.** After de Bary.

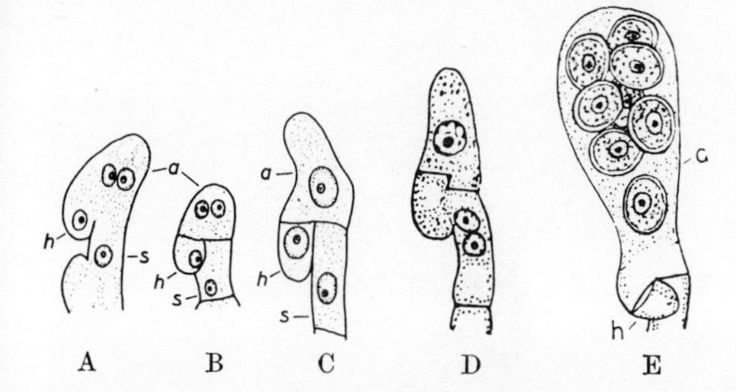

Fig. 493. A. to **D.** Development of the ascus of *Pyronema confluens*. After Harper. **E.** Young ascus (c) of *Boudiera* with eight spores. s, stalk-cell; h, hook; a, ascus initial. After Claussen. Explanation in the text. ×1000.

trichogyne into the ascogonium, the wall between the latter developing a transitory pore (plasmogamy). ♂ and ♀ nuclei associate together in pairs in the ascogonium (Fig. 492B). Numerous tubes now grow out from the ascogonium as ascogenous hyphae into which the pairs of nuclei pass (Fig. 492C). These hyphae, representing the sporophyte, undergo branching (Fig. 492D), while the pairs of nuclei increase in number by conjugate division (i.e. simultaneous division of paired ♂ and ♀ nuclei). Cross walls now cut off cells each with two nuclei of opposite sex. An ascus is formed from the terminal cell of each of the ascogenous hyphae in the following way (Fig. 493). The tip of the terminal cell bends backwards, the pair of nuclei divide conjugately so that of the four daughter cells two of opposite sex remain in the apical cell whilst the third goes into the hook (h) and the fourth goes into the stalk (s). The hook and stalk are now cut off from the apical cell by a cross wall (Fig. 493B). The binucleate apical cell becomes an ascus-initial in which the two nuclei eventually fuse (caryogamy), forming a club-shaped, initially uninucleate diploid sporangium, the ascus. The fusion nucleus now undergoes three divisions, of which the first two constitute meiosis, forming eight nuclei; the eight haploid meiospores (ascospores) are delimited by their walls by means of free cell formation (p. 44).

The cytoplasm not incorporated in the spores (the periplasm) is frequently active in the deposition of a further, variously sculptured layer on the spore wall.

The hook-cell and the stalk-cell usually fuse, the nucleus of the hook-cell passing into the stalk-cell (Fig. 492D); the new dicaryotic cell can give rise by hook-formation to a new ascus-initial. Repetition of this process leads to a whole cluster of asci at the end of an ascogenous branch.

The sex organs are already enclosed at the time of conjugation in a loose mass of

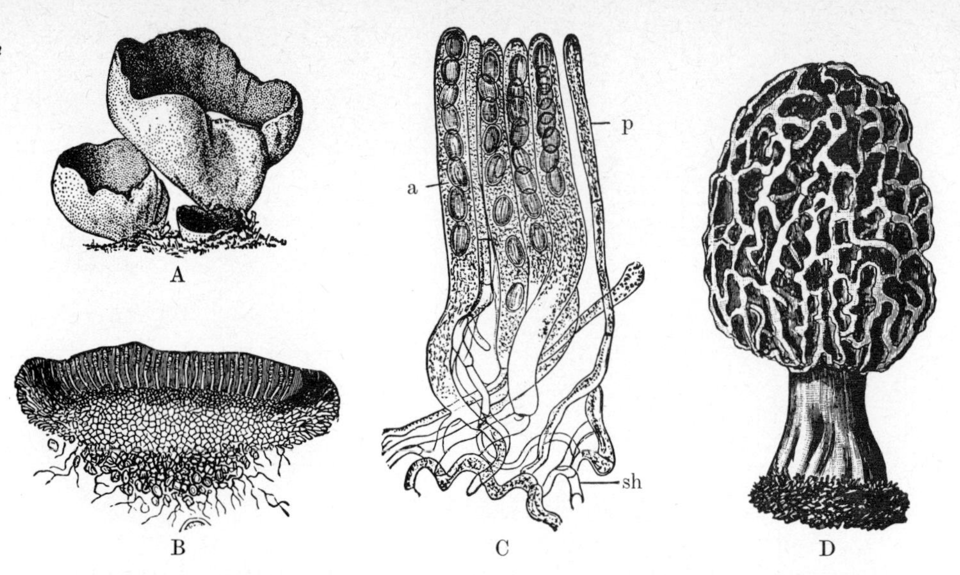

Fig. 494. Pezizales. **A.** *Peziza (Otidea) leporina.* $\times\frac{2}{3}$. **B.** *Pulvinula convexula,* T.S. of an apothecium. $\times$20. **C.** *Morchella esculenta,* part of the hymenium; a, asci; p, paraphyses; sh, subhymenial tissue. $\times$240. **D.** *Morchella esculenta,* fructification. $\times\frac{3}{4}$. **A.** After Michael. **B.** After Sachs. **C.** After Strasburger. **D.** After Schenck.

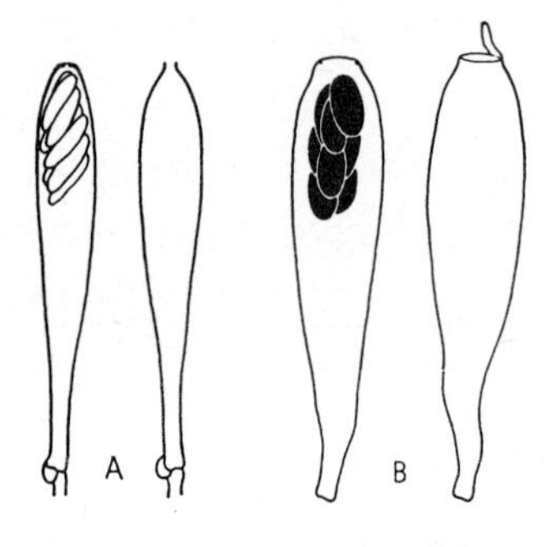

Fig. 495. Asci before and after the discharge of spores. **A.** Inoperculate. **B.** Operculate asci. After Oberwinkler.

Fig. 496. *Monilia*-rot on a pear originating from a wound. $\times\frac{1}{2}$. After Kotte.

haploid investing hyphae. They develop simultaneously with the ascogenous hyphae to form a fructification (Fig. 494B) and send out sterile filaments (paraphyses, Fig. 494C) which together with the asci form the fertile layer, the hymenium (Fig. 494B).

Order 1: **Pezizales.** The asci open by a pre-formed lid (operculate; Fig. 495B). They are almost exclusively saprophytes of the woodland floor, of decaying wood and the sites of bonfires. The fructification of *Peziza* (Fig. 494A) is cup- or disc-shaped; in *Morchella* (morel; Fig. 494D) and *Helvella* the fructification is differentiated into a stalk and a fertile cap.

Order 2: **Helotiales.** The asci open not by a lid but by a complex 'apical-apparatus' forming a pore (inoperculate, Fig. 495A). The Helotiales have cup- or saucer-shaped fructifications which are in general smaller (like their asci and spores) than in the Pezizales. Most of the many genera are saprophytic although some, e.g. *Trichoscyphella willkommii*, the agent of larch canker, are parasitic. *Sclerotinia fructigena* grows on apples and pears. It often develops at first in concentric circles (determined by the diurnal light-dark cycle) of conidial pustules (*Monilia*; Fig. 496); in the spring it develops long-stalked apothecia on mummified fruits. *Sclerotinia fuckeliana* or its conidial form *Botrytis cinerea* causes grapes to fall off in wet years, yet in dry seasons is responsible, as 'noble

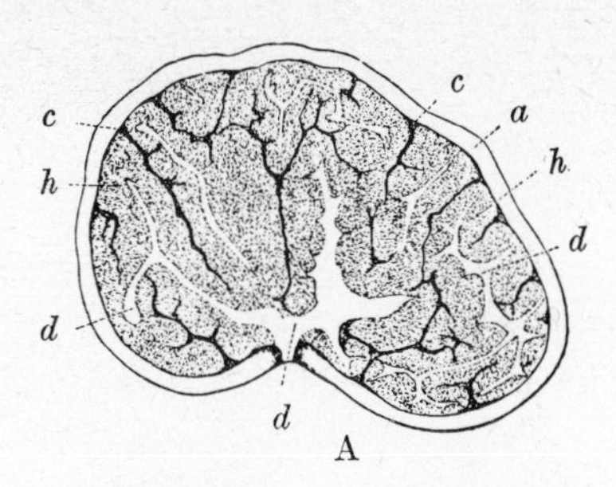

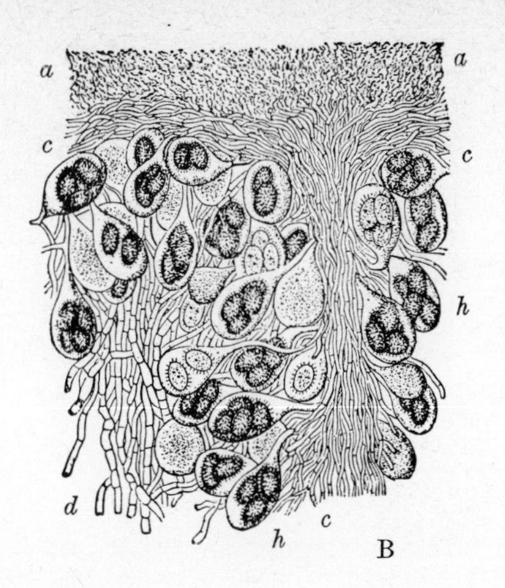

Fig. 497. *Tuber rufum.* **A.** A fructification in vertical section. **B.** A section of the hymenium. a, cortex; d, air-containing tissue; c, dark veins of compact hyphae; h, hymenium. **A.** ×3. **B.** ×300. After Tulasne.

decay', 'pourriture noble' or 'Edelfäule', for the particularly high sugar content of the ripe grapes (used in Beerenauslese hocks). Many lichen fungi (p. 537) are related to the Helotiales but are mostly included in the order Lecanorales.

Order 3: **Phacidiales.** This order was previously united with the preceding one but shows relationships with the Pseudosphaeriales in that the asci are developed in cavities and only subsequently form a hymenium; however, the asci are unitunicate. They are mainly parasites. Members include *Rhytisma acerinum* which produces black spots on sycamore leaves in the autumn (apothecia in the spring), and *Lophodermium pinastri* on the needles of young conifers (pine leaf-cast).

Order 4: **Tuberales.** Known as truffles, these generally live subterraneously in woods (as mycorrhizal fungi); in the simplest cases the fructifications are saucer-shaped with an apical aperture, but they are mostly tuberous and permeated by passages which at least in the early stages open to the exterior. These spaces are lined with the hymenium (Fig. 497). Many species of the genus *Tuber* (with four-spored asci) have been prized as esculent fungi since antiquity.

Survey of the Ascomycetes. Typical ascomycetes have asci with eight meiospores; only the most primitive forms have many.

In the subclass Protoascomycetidae the ascus almost always arises directly after the conjugation of the gametangia. In the higher Ascomycetes, on the other hand, the sexual act which typically takes place between a multinucleate ascogonium equipped with a trichogyne and a multinucleate antheridium (gametangiogamy) leads first to the formation of ascogenous hyphae, from which the asci arise later. The function of the antheridium is only occasionally taken over by a free unicell. ♂ and ♀ nuclei do not fuse in the ascogonium but pass in pairs into the ascogenous hyphae, increase in number by conjugate division and only fuse after hook-formation in the ascus initial. Plasmogamy and caryogamy are thus widely separated both in space and time by the dicaryotic stage with paired nuclei. The dicaryotic cell is functionally diploid even though the nuclei are still separate. As in the Protoascomycetidae, meiosis takes place during the first two divisions of the diploid nucleus in the ascus so that the ascospores are haploid.

The haploid mycelium bearing antheridia and ascogonia represents the gametophyte on which the sex organs may be synoicous (intermixed) or heteroeicous (separate). The binucleate (dicaryotic) sporophyte develops from the fertilized ascogonium in the form of ascogenous hyphae. This phase is terminated with the production of sporangia (the asci), in which haploid ascospores are produced after nuclear fusion and reduction division. The plectenchyma of the fructification consists of haploid gametophytic mycelium (Fig. 498). The two generations are thus united in a common structure in which the gametophyte is almost always the more strongly represented.

There is a remarkable number of deviations from this type, especially by the reduction, replacement or even complete suppression of sexual organs, by the conjugation of

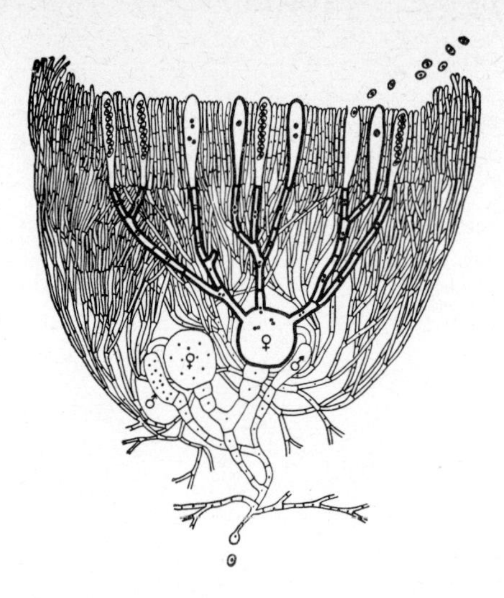

Fig. 498. Fructification of a monoecious discomycete (schematic). Thin lines: haploid phase; thick lines: diploid. Hooks not shown. After Harder.

two vegetative hyphal cells or indeed of two ascospores (Fig. 480M–P). Often only the antheridium is lacking; the ♂ nucleus then comes from a vegetative hypha or even from a spermatium which is either constricted exogenously or produced in a special structure (pycnidium). In many such cases the trichogyne is very long, spirally coiled, sometimes even multicellular or branching, growing towards the cells which function as ♂ gametes. Furthermore, sexual union may be entirely absent and the nucleus of the ascogonium (which is not always multinucleate) divides, giving rise to a dicaryotic state, which may indeed be achieved only when the nuclei at the ends of uninucleate ascogenous hyphae divide. Alternatively, the haploid ascospore divides on germinating into two nuclei the progeny of which remain in pairs in all cells of the mycelium up to the time of nuclear fusion in the ascus. These and other reductions in the sexual processes occur not only within families but even in one and the same individual.

Of special interest are those cases in which ♂ nuclei are liberated from generally exogenous conidia which take over the function of the missing antheridia; these may be called androconidia. However, in some species these have lost the capacity to germinate, while in others they can grow into vegetative mycelia; they are thus asexual spores which have assumed the functions of the ♂ organ, just as a vegetative hyphal cell can behave sexually. In a few other examples, e.g. *Bombardia lunata* and the so-called microconidia of *Neurospora*, the androconidia are not constricted exogenously, but are formed in succession as small uninucleate spores inside flask-shaped receptacles, from the apex of which they are liberated by the rupture of the wall. These endogenously produced fertilizing spores may possibly represent the last reduced vestige of an ancestral form of ascomycete sexuality with antheridia producing free spermatozoids; in response to terrestrial life these have become reduced to non-motile spores with walls, yet still capable of fertilization (spermiospores).

Not uncommonly the cells of the ascogenous hyphae are multinucleate at first, and pairing of the nuclei only takes place shortly before hook formation.

When the ascospores conjugate a gametophyte is inevitably absent, but the sporophyte can also be reduced when asci are formed directly from the ascogonium. The spores may even be formed in the ♀ gametangium, in which case both sporophyte and sporangium are absent. Further deviations in the sporophyte include the absence of hook formation (e.g. in *Peziza*) or, more rarely, hooks may be produced at every cross wall (e.g. in *Tuber*). These and other variations need not have arisen exclusively by reduction, but may sometimes be primitive. The remarkable diversity often makes it difficult to decide whether a character is primitive or advanced.

Phylogeny. Views of the origin and evolution of the Ascomycetes are extremely divergent. The preceding account is based on the assumption that the Protoascomycetidae with their conjugating gametangia are related to the Zygomycetales (or their relatives or ancestors) and hence to the conjugating Chytridiales. The transition from the gametangiogamous Phycomycetes to the Protoascomycetidae is so gradual that some phylogeneticists do not recognize the latter as belonging to the Ascomycetes, but regard them as Phycomycetes (and then, of course, give them some other name). But in a few of the Protoascomycetidae the ascus is not formed directly from the zygote; instead a small but still diploid mycelium is first produced on which the asci only develop later, thereby reaching the level of development which predominates in all higher Ascomycetes, i.e. the heterophasic alternation between a haploid gametophyte and a diploid sporophyte. It is true that in these groups union of the sexual nuclei takes place not in the zygote but in the sporangium (the ascus), so that the sporophyte is dicaryotic. This is indeed a very striking distinction, but it is one which could develop gradually. According to this view there is a continuous series from the Chytridiales–Zygomycetales to the higher Ascomycetes.

An entirely different idea has been proposed more recently that the Ascomycetes have originated not by way of the Protoascomycetidae but directly from extinct Phycomycetes. In this theory an important part is played by the formation of hooks. These are certainly very characteristic structures, but it is difficult to explain convincingly why they have arisen or how they function. The new explanation is that they are rudimentary sex organs: the ancestors of the Ascomycetes are supposed to have formed lateral conjugation bridges (rather as shown in Fig. 478) during sexual reproduction, a process lost during further development but replaced by the present form of gametangiogamy. The original process is thus retained in living Ascomycetes only in the hooks, which no longer function as they once did. If this be so, the Protoascomycetidae form a blind alley leading from the Phycomycetes and have nothing to do with the Ascomycetes, since they show no hook formation. On the contrary, the Ascomycetes would have branched off from some very primitive phycomycete stock and developed later parallel to present-day Phycomycetes.

One can also base phylogenetic speculation on the conidia which function as spermatia (p. 518), suggesting derivation from ancestors with free ♂ gametes and an origin not from the lower Phycomycetes but from Rhodophyceae (♀ gametangium with trichogyne, nonflagellate ♂ gametes and the outgrowth of sporogenous filaments) by the loss of photosynthetic pigments. But there is no convincing evidence for this: even the most recent attempt, relying on the parasitic dwarf forms amongst the red algae, has not confirmed this hypothesis.

Uses. The yeasts (*Saccharomyces*) have many applications as agents of alcoholic fermentation. Whilst the wine yeast (*S. ellipsoideus*) occurs naturally on grapes, the brewers' yeasts (*S. cerevisiae* and *S. carlsbergensis*) are only known in culture. Brewers' yeasts are also used in baking to raise the dough; the leaven used in breadmaking includes both yeasts and lactic acid bacteria. The fat- and protein-rich food yeast (*Candida utilis*) can even be cultivated in waste sulphite liquors. Antibiotics (penicillin) are obtained from *Penicillium notatum* and other species. Other species (*P. roquefortii, P. camembertii*) are necessary in the preparation of different kinds of cheese. Some of the larger Pezizales and Tuberales are esculent fungi (cup-fungi, morels, truffles). Many moulds produce toxins, e.g. aflatoxin, which damages the liver, from *Aspergillus flavus* and other species. The sclerotia of *Claviceps purpurea* contain alkaloids which, although poisonous, are used medicinally.

Class IV: Basidiomycetes

The characteristic organ of the Basidiomycetes, which comprise some 15,000 species, is the basidium, a structure which almost invariably abstricts four separate spores by a process of budding. It is club-shaped and unicellular in the Holobasidiomycetidae (Fig. 499C), septate in the Phragmobasidiomycetidae (Fig. 505D). As in the ascus of the Ascomycetes, the fusion of two nuclei in the basidium is followed by meiosis (Fig. 499).

The Basidiomycetes are also distinguished from the Ascomycetes in the structure of the pores in their cross walls. Whilst in the Ascomycetes the pore is a simple perforation

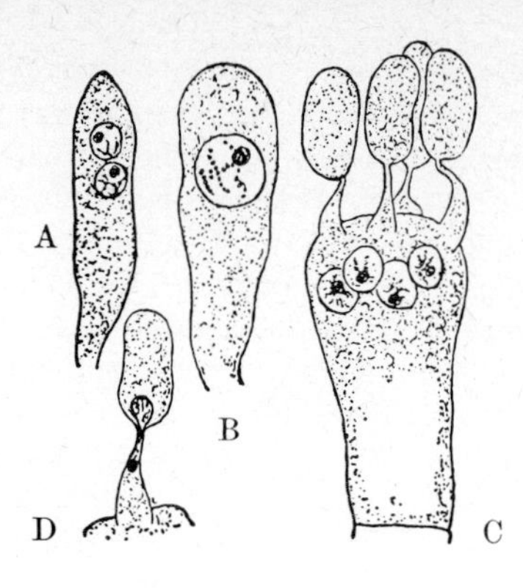

Fig. 499. A. Young basidium with two primary nuclei. B. After their fusion. C. Basidium with four haploid nuclei derived from the fusion nucleus before their migration into the young apical basidiospores. D. Migration of a nucleus through the sterigma into the basidiospore. A. and B. *Armillaria mellea.* C. and D. *Hypholoma appendiculatum*. After Ruhland.

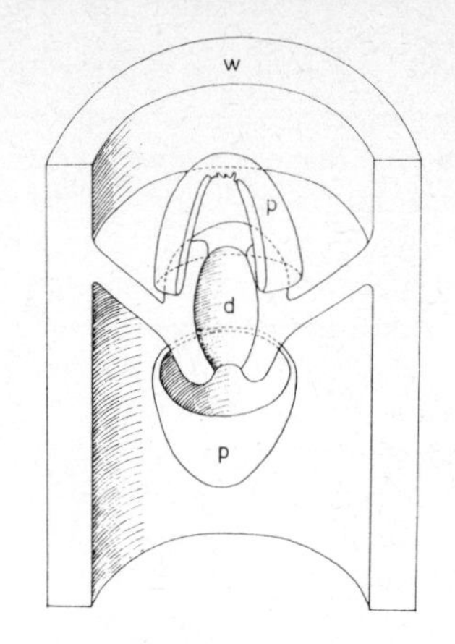

Fig. 500. Transverse septum of a basidiomycete hypha. d, dolipore; p, parenthosome; w, cell-wall. After Moore and McAlear.

of the wall, in the Basidiomycetes it is a barrel-shaped structure (dolipore) and on each side is covered by a 'parenthosome' formed from the endoplasmic reticulum (Fig. 500). Such dolipores have been demonstrated in the Auriculariales, Tremellales and Holobasidiomycetidae; the pore structure in the Uredinales and Ustilaginales is not yet understood.

A. Phragmobasidiomycetidae

The basidia are divided into four cells by transverse septa; rarely the septa are longitudinal (Tremellales) or absent (Tilletiaceae).

Order 1: **Uredinales**, rust fungi. The thousands of species of this order, which produce widespread rust diseases, have four-celled basidia with cross walls (Figs. 505D, 507A). They live parasitically in intercellular spaces, mainly in the leaves of higher plants, where they send haustoria into the cells (Fig. 501 a). Usually their growth is confined to the neighbourhood of the original site of infection. They form no fructifications, but are remarkable for the great variety of form shown by their spores (Fig. 508), which develop in a regular sequence. The typical course of development is as follows.

The sexually differentiated basidiospores germinate in the spring on the leaves of a host. The germ tube penetrates the epidermis and develops an intercellular mycelium, the haustoria from which penetrate the host cells, abstracting food but without immediately killing them. The cells of this mycelium are uninucleate. The hyphae group together in places under the epidermis to form pustulate, flask-shaped spermogonia or pycnidia (Figs. 501, 503), which eventually break through the epidermis; they contain both sterile hairs (paraphyses) and a dense central mass of short hyphae which cut off minute, uninucleate, elliptical conidia, the spermatia or pycnospores (Fig. 503B). These will grow in nutrient solutions, forming a short germ tube, but are generally incapable of infecting a healthy leaf.

Other hyphae grow in the spongy mesophyll and aggregate near the lower epidermis, where they form rounded, partly plectenchymatous masses of hyphae called protoaecidia, which develop further only if nuclei of opposite sex reach them.

The mycelium developing directly from a basidiospore is not immediately capable of conjugation, but must first undergo certain physiological changes (a kind of maturation), whereupon it forms terminal sexual cells, known as basal cells.

These basal cells are the true sex organs of the Uredinales. If a leaf has been infected by basidiospores of different sexes the basal cells of opposite sex conjugate laterally in pairs in the protoaecidium (Fig. 504). The conjugation bridge formed is dicaryotic; it cuts off a whole chain of binucleate, rust-coloured aecidiospores. Since numerous basal cells

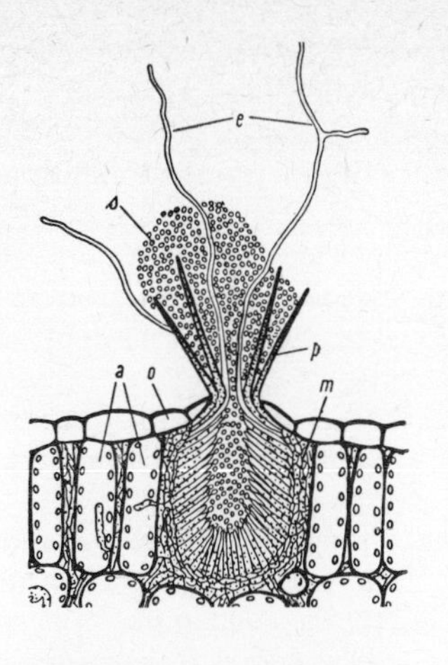

Fig. 501. *Puccinia graminis.* Spermogonium on *Berberis.* o, epidermis; a, palisade cells with haustoria; m, intercellular mycelium; p, paraphyses; e, receptive hyphae; s, spermatia. After Buller.

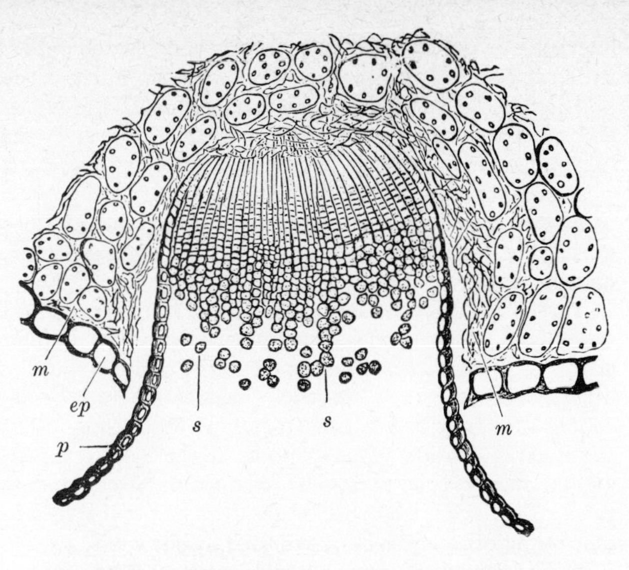

Fig. 502. *Puccinia graminis.* Aecidium on *Berberis vulgaris.* ep, lower epidermis of the leaf; m, intercellular mycelium; p, pseudoperidium; s, chains of aecidiospores. ×140. After Schenck.

are present in each protoaecidium, a mass of parallel chains of aecidiospores break out of the lower side of the leaf as cup-shaped aecidia (Fig. 502).

If all the infecting basidiospores are of one mating-type (p. 493), the other may be introduced later by means of spermatia.

In this case the haploid mycelium sends out hyphae to the exterior through epidermal cells, through stomata or even out of the pycnidia themselves (Fig. 501 e). These so-called receptive hyphae (Fig. 501 e) are generally very much longer than the paraphyses of the pycnidia, and they have no cross walls; each has one nucleus at the base. They are readily brought into contact with pycnospores of the opposite sex by insects; indeed, the pycnidia secrete nectar, and the insects attracted by it carry the pycnospores all over the leaves. If contact fails in this way the receptive hyphae can still send out processes towards nearby pycnospores. Fusion then takes place between a receptive hypha and a pycnospore which discharges its contents into the hypha, down which the pycnospore nucleus passes through performations in the cross walls to the protoaecidium on the

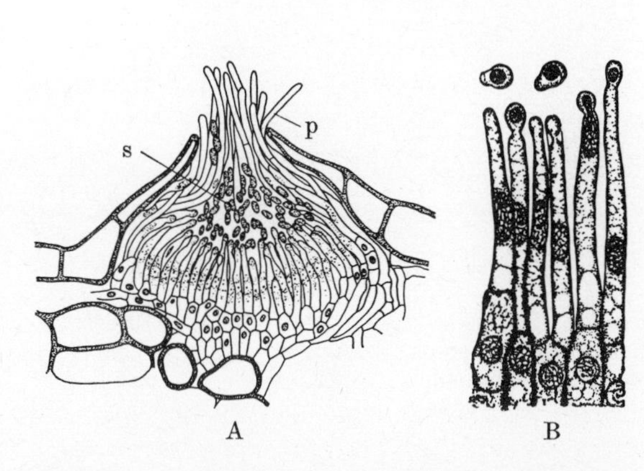

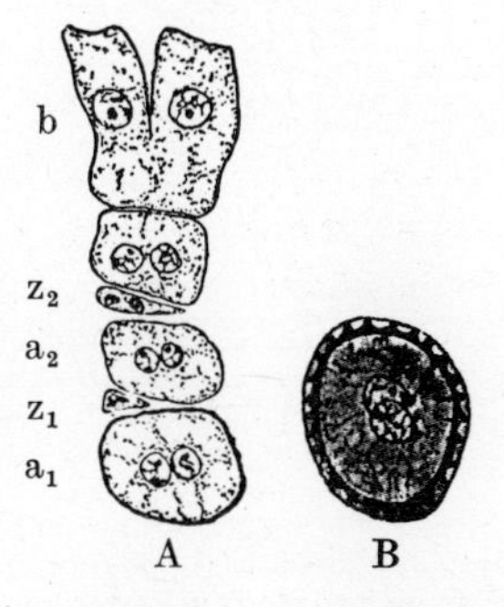

Fig. 503. A. *Gymnosporangium clavariaeforme.* Spermogonium on the leaf of *Crataegus*, bursting through the upper epidermis. s, spermatia; p, paraphyses. **B.** *Peridermium strobi.* Abstriction of the uninucleate spermatia. **A.** ×450. After Blackman. **B.** ×1200. After Colley.

Fig. 504. A. *Phragmidium speciosum.* b, basal cells with conjugation bridge; a_1 and a_2, binucleate aecidiospores; z_1 and z_2, intercalary cells. After Christman. **B.** Mature aecidiospore of *Phragmidium violaceum.* ×800. After Blackman.

underside of the leaf, where it divides and together with basal cells of opposite sex gives rise to aecidia and chains of dicaryotic aecidiospores as before.

The haploid mycelium is monoecious since it produces both spermatia (pycnospores) and protoaecidia with receptive hyphae. However, a genetically determined inhibition (incompatibility) prevents fertilization between individuals of the same mating type.

In some species the dicaryotic phase follows from infection by a single basidiospore; in these the spermogonia are often feebly developed or absent. Indeed, in certain species a nucleus of opposite mating-type can be introduced into a haploid mycelium by a dicaryotic uredospore (see below).

The chains of aecidiospores generally consist of alternate true spores and small intercalary cells which later become mucilaginous and disappear (Fig. 504A, z_1, z_2). In some genera (e.g. *Puccinia*) all the spores of the peripheral chains and the terminal spores of the other chains lose their spore-like character before breaking through the epidermis and cohere as a firm investment (pseudoperidium, Fig. 502 p). The pressure set up by the continual formation of new spores at the base of the chains (in *Puccinia graminis* over 10,000 spores in an aecidium) ruptures the pseudoperidium and the epidermis, pushing them to the sides, while the spores, which at first were compressed into polygonal shapes, now round off and can be distributed by the wind.

The aecidiospores germinate on a new host; the germ tube enters through a stoma and develops into an intercellular, binucleate mycelium of limited growth and without clamp connections. It proceeds without delay to form numerous conidia, here called summer spores or uredospores, each of which arises from the swollen end of a hypha and contains a pair of nuclei (Fig. 506). They are grouped in small, linear rust-coloured sori which break through the epidermis; they spread the fungus during the summer.

Every infected plant bears millions of uredospores, each one capable of infecting new host plants by means of a germinal hypha issuing from the germination pore (Fig. 505A, p), which enters a stoma, grows into a new mycelium, and within a week forms new uredospores; hence the disease spreads with great rapidity.

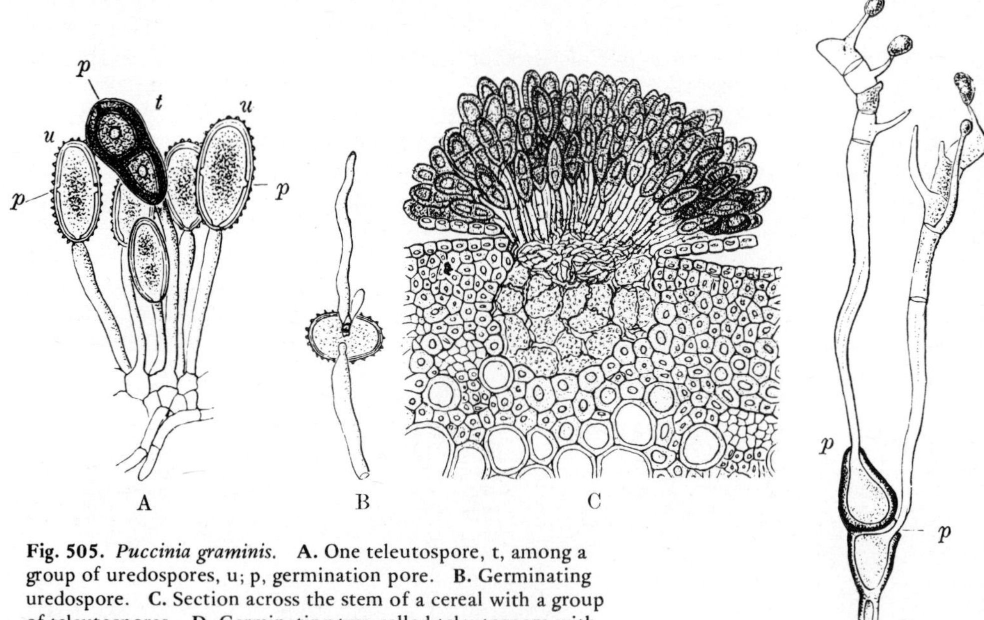

Fig. 505. *Puccinia graminis.* **A.** One teleutospore, t, among a group of uredospores, u; p, germination pore. **B.** Germinating uredospore. **C.** Section across the stem of a cereal with a group of teleutospores. **D.** Germinating two-celled teleutospore with two basidia. **A.**, **B.** and **D.** ×300. **C.** ×150. **A.** and **B.** After de Bary. **C.** After Tavel. **D.** After Tulasne.

As the growing season of the host draws to a close the fungus sets about forming basidia. This is a protracted business. First comes the development of the zygote, here called a probasidium, in which nuclear fusion takes place, as in certain Auriculariaceae, but whereas in the latter the probasidium immediately forms a basidium, this happens in

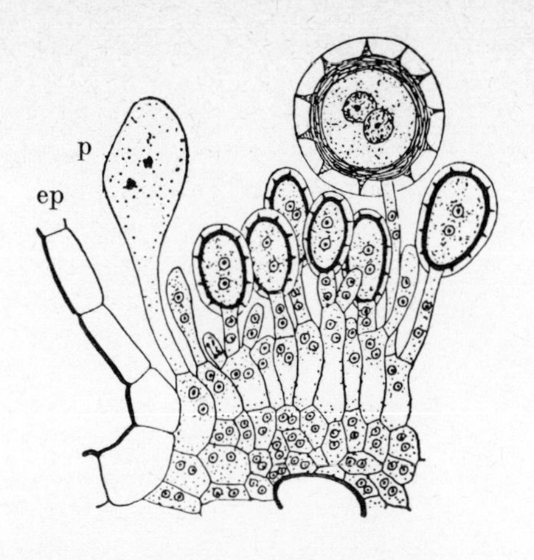

Fig. 506. *Phragmidium rubi.* Margin of a nearly mature sorus of binucleate uredospores in various stages after it has ruptured the epidermis (ep) of the host; p, paraphysis. ×565. After Sappin-Trouffy.

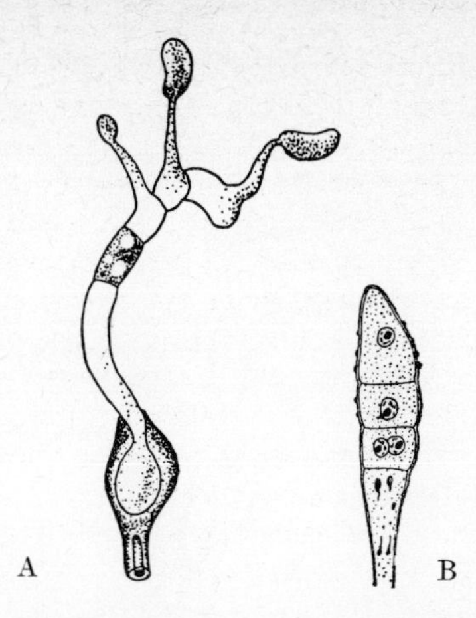

Fig. 507. Teleutospores. **A.** *Uromyces appendiculatus*, unicellular with basidium (nuclei omitted). **B.** *Phragmidium violaceum*: the lowest cell has two nuclei which have fused in the two upper cells. ×500. **A.** After Tulasne. **B.** After Blackman.

the Uredinales only in the following spring (Figs. 505D, 507A), except in a few species (e.g. *Cronartium ribicola*) in which the probasidium grows forthwith; usually it overwinters as a perennating organ called a winter spore or teleutospore. The teleutospores occur in the same sori as the uredospores (Fig. 505A, t) or in separate ones (C). They are thick-walled and generally dark, with one (*Uromyces*, Fig. 507A), two (*Puccinia*, Fig. 505A, D) or more (*Phragmidium*, Fig. 507B) cells. At first the cells are binucleate, but at maturity the pairs of nuclei fuse (Figs. 507B, 508). In the following spring each diploid cell puts out a tubular basidium (or promycelium; Figs. 505, 507A). The four haploid nuclei formed by meiosis are separated by cross walls, and each cell produces a haploid basidiospore borne on a sterigma (Fig. 505D). The basidiospores (meiospores) are forcibly shot off from the sterigmata, wafted by the wind to the leaves of a host, and the life-cycle is complete (Fig. 508).

The life-cycle is not so complicated in all Uredinales; in many cases one or more of the five forms of spores are regularly absent. For example, *Uromyces fabae* forms uredospores immediately after nuclear fusion, while *Puccinia malvacearum* lacks pycnidia, aecidia and uredospores, and produces teleutospores directly. Teleutospores are almost always present and are absent from only a few species, e.g. *Endophyllum sempervivi*; in this the basidia develop from aecidiospores in which nuclear fusion (normally occurring in the teleutospores) takes place. Further deviations occur. The forms with shorter life-cycles may be the more primitive; such abbreviated development is of ecological significance in regions with a short growing season, e.g. in the Arctic.

Many rust fungi go through their entire life-cycle on the same species of host. Others develop the haploid and the binucleate phase on different species. An example of such alternation of hosts is afforded by *Puccinia graminis*, one of the commonest rusts of cereals. The basidiospores germinate only on the leaves of *Berberis*, which bear the aecidia as small rusty dots, as well as the spermogonia; the aecidiospores, on the other hand, will germinate only on Gramineae which bear linear sori of rusty uredospores or blackish teleutospores.

Similarly in *Uromyces pisi*, the spermogonia and aecidia occur on *Euphorbia cyparissias* and *E. esula*, the uredo- and one-celled teleutospores on *Pisum sativum* and *Lathyrus* species. (The infected *Euphorbia* plants remain unbranched, have yellowish, abnormally thick and short leaves, and usually fail to flower; Fig. 276.) Such rusts are thus highly selective, but in some heteroecious species the haplonts or dicaryonts grow on a wide range of hosts.

Although the rust fungi do not grow extensively in the tissues of their hosts, they nevertheless cause many serious diseases. In particular, they severely lower the yield of

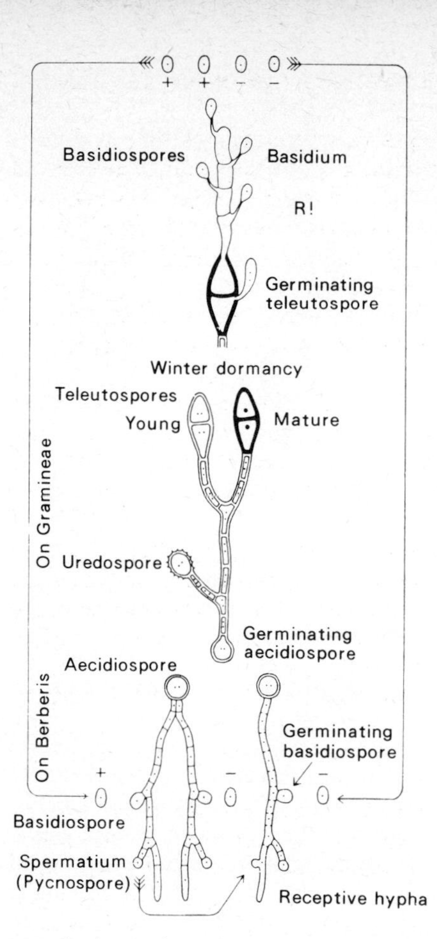

Fig. 508. Life-cycle of *Puccinia graminis* (schematic). Mycelium between the spores and number of spores greatly reduced; cf. Figs. 501–7. Haploid phase, thin lines; dicaryotic phase, double lines; diploid phase, thick lines. R!, meiosis. After Harder.

cereals (sometimes by 25 per cent, but generally not much more than 5 per cent). The cosmopolitan *Puccinia graminis*, called black rust from its teleutospore sori, attacks all kinds of cereals and numerous wild grasses. In Germany it is less injurious than in warmer countries, where this relatively warmth-demanding species undergoes more rapid development. With us the yellow rust, *P. glumarum*, with bright lemon yellow uredospore sori, is particularly injurious, especially on wheat, but also occurring epidemically on barley, rye and various wild grasses; its alternate host is unknown. *P. coronata*, the brown rust of oats, has *Rhamnus catharticus* as the alternate host, while *P. simplex*, the brown rust of barley, forms aecidia on species of *Ornithogalum*. This by no means exhausts the list of cereal rusts. Further species of *Puccinia* occur on asparagus, carrots, onions, gooseberries and other cultivated plants; species of *Uromyces* on peas, beans and beetroot; *Gymnosporangium* on the leaves of the pear. Belonging to other families of the Uredinales are the autoecious *Melampsora lini*, which destroys the fibres of flax; and *Melampsorella caryophyllacearum* (canker and witches' brooms on *Abies*, uredo- and teleutospores on Caryophyllaceae) and *Cronartium ribicola* (the aecidial stage kills Weymouth pine, the inflated aecidia bursting through the bark; alternate host *Ribes*), both injurious in forestry. Attempts to eliminate heteroecious rusts by eradicating the alternate host have only had a very limited success since in the great majority of species the uredospores either can over-winter or they may infect recently sown winter wheat or various cultivated grasses (*P. glumarum* also on *Agropyron*) in the autumn; furthermore, uredospores can be brought in from great distances by the wind. Since effective chemical means of control have not been found, attempts are made to breed rust-resistant varieties, but these run into the difficulty that each rust species has a large number of morphologically indistinguishable 'physiologic races' which are constantly evolving anew by mutation and genetic recombination, or which are brought in from other areas. The formation of physiologic races is a matter of great concern to the plant pathologist, and the search of the plant breeder for resistant varieties is never-ending.

Typical Uredinales thus show a clear alternation of nuclear phases and hence alternation of generations (Fig. 508).

The haploid basidiospore develops into the haploid gametophyte; this stage terminates with the conjugation of the basal cells which occurs prior to the formation of aecidia and represents a reduced sexual act (somatogamy). The resulting binucleate cells mark the beginning of the dicaryotic sporophyte, which reproduces copiously by means of conidia (the aecidiospores and uredospores); this phase is completed in the basidium when meiosis takes place. The teleutospores are to be regarded not so much as organs of dispersal as over-wintering zygotes. In many Uredinales the gametophyte and sporophyte have very different nutritional requirements and hence grow on different host plants.

Phylogeny. The Uredinales probably lived originally as parasites on tropical ferns in the Upper Palaeozoic, later attacking gymnosperms, especially conifers in the Mesozoic, and finally growing on angiosperms from the Upper Cretaceous onwards, when with the change to cooler climates and the transition from forest to grassland they developed alternation of hosts.

Order 2: **Ustilaginales**, smut fungi. Like the rusts, they form no fructifications. They are generally intercellular parasites of higher plants, producing smut diseases, but they can also be grown on artificial media. The dicaryotic mycelium breaks up into single cells, the brand spores (Fig. 509E–G), which behave as zygotes homologous with the teleutospores of the Uredinales; thus fusion of their paired nuclei (Fig. 509F–G) is followed by the outgrowth of an elongated basidium, generally with cross walls (Fig. 509B), in which meiosis takes place. The basidiospores (gonospores; Fig. 509B) are borne directly on the basidium (without sterigmata); on germination they conjugate, forming a dicaryotic mycelium once again.

The basidiospores conjugate immediately on germination or a little later (Fig. 509C) to form a dicaryotic, initially saprophytic mycelium, which in some species is provided with clamp connections (Fig. 509D, E). as in the hymenomycetes. Whilst the haploid phase generally shows little or no capacity for infection (except, for example, in the smut of maize), the dicaryotic mycelium penetrates to the growing point of the seedling host plant; host and parasite then grow up together. No external symptoms of disease are manifest at first, while the parasite is intercellular; later it becomes intracellular at definite points, e.g. in the ovaries of cereals, where the host tissues are completely destroyed and dense masses of mycelium are formed. The hyphal cells then swell up like strings of beads (Fig. 509E) and separate as dark, thick-walled spores which are shed like coal dust (hence the names brand spore and smut).

The two families Ustilaginaceae and Tilletiaceae and the genera within them show many differences in detail.

In the **Ustilaginaceae** the brand spore puts out (generally after over-wintering) a hypha into which the diploid nucleus passes and divides into four haploid nuclei (Fig. 509A), between which transverse septa are formed, giving a basidium (or promycelium) similar to those of the Auriculariales and Uredinales. Each cell of the basidium now abstricts a lateral, spindle-shaped basidiospore; during this process the basidial nuclei themselves remain *in situ* and divide; one daughter nucleus enters each spore (Fig. 509B). Under favourable conditions of nutrition further spores may be abstricted; these either separate and grow into new filamentous mycelia or much more commonly form yeast-like chains of budding cells. In certain circumstances the basidia produce hyphal filaments directly instead of spores. Mycelia (or isolated cells of the chains) of different mating-type then conjugate. The basidiospores may also conjugate directly (Fig. 509C), or basidia of opposite mating-type form lateral conjugation bridges. There are even species in which the basidium is absent and meiosis takes place in the brand spore.

In the **Tilletiaceae** there are no cross walls in the basidium (? derived from the basidium of the Ustilaginales by reduction), and four or eight elongate basidiospores are borne at its apex (Fig. 509J, K). Conjugation, generally between pairs of basidiospores, is in principle the same as in the Ustilaginaceae. Both haploid and dicaryotic phases can reproduce by means of conidia (K, c).

The smut diseases of cereals are of great economic importance. *Ustilago zeae* produces

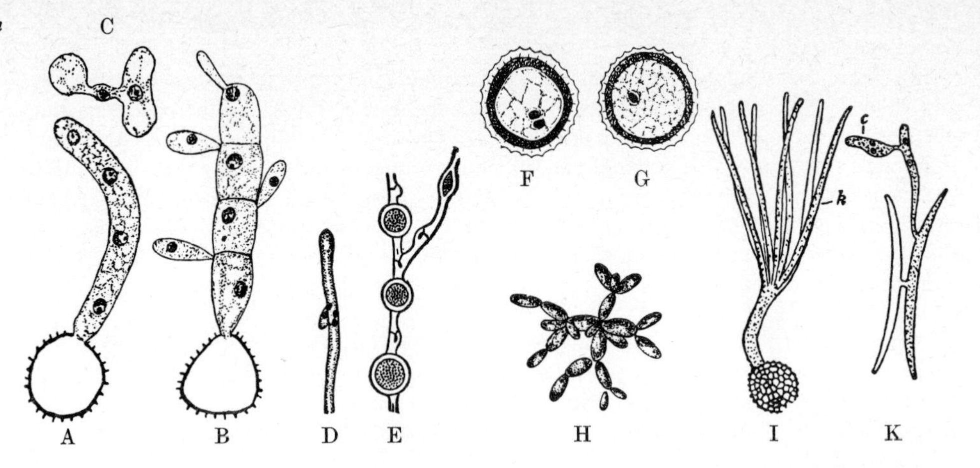

Fig. 509. Ustilaginales. **A.** and **B.** *Ustilago scabiosae*, germinated brand spore and spore-formation on the four-celled basidium. ×1100. **C.** *U. carbo*, conjugating basidiospores. ×1200. **D.** *Entyloma calendulae*, hypha with clamp connection and paired nuclei. **E.**, **F.** and **G.** *Ustilago vuijckii*, formation of brand spores, dicaryotic and diploid brand spores. **H.** *U. carbo*, budding brand spore in a nutrient solution. ×350. **J.** *Tilletia tritici*, basidium produced by a brand spore with four pairs of terminal spores, k. ×300. **K.** *T. tritici*, two conjugated basidiospores producing a dicaryotic mycelium from which a conidium (c) is developing. ×650. **A.** and **B.** After Harper. **C.** and **K.** After Rawitscher. **D.** After Stempell. **E.**, **F.** and **G.** After Seyfert. **H.** and **J.** After Brefeld.

ulcerous swellings filled with brand spores on the stems, leaves and inflorescences of maize; other species of *Ustilago* fill the ovaries and neighbouring parts of the spikelets of oats, barley and wheat with a dusty mass of brand spores (loose smut) at harvest time, while species of *Tilletia* form compact masses of hyphae in the wheat grain which smell of pickled herrings (stinking smut or bunt). Since a single grain of wheat infected with *T. tritici* contains several million spores, which on threshing are dusted over the seed, hence infecting young seedlings, the disease rapidly spreads at the next sowing. In former times 20–60 per cent of the corn harvest was lost in this way. Nowadays smuts are successfully controlled by immersing the seed for a short period in poisonous 'disinfectants' (usually solutions of organic mercury compounds), or dusting it with similar substances, thereby killing the adhering brand spores. This, however, is ineffective against the loose smuts of barley (*Ustilago hordei*) and wheat (*U. tritici*), in which the brand spores are already formed in the young ovaries before the florets open; they disperse and germinate, and the basidiospores infect the stigmas and carpel walls of previously uninfected florets; mycelium develops from the basidiospore in the maturing grain and overwinters in its embryo. Control consists of steeping the seed in hot water, whereby apparently alcohol and acetaldehyde are formed by anaerobic respiration and kill the mycelium. *Ustilago violacea* causes anther smut in Caryophyllaceae; the fungus induces the female flowers of *Melandrium rubrum* to develop anthers in which the brand spores then develop.

Order 3: **Auriculariales.** This order shows a series from primitive forms with loose hyphal wefts to forms with complex fructifications, e.g. *Auricularia auricula-judae* (Jew's-ear) which usually grows on the stems of old elder bushes; it has gelatinous, dark reddish-brown, ear-like fructifications, with the hymenium on the concave lower surface. The basidia in all Auriculariales are divided by cross walls into four superposed cells each of which bears a lateral sterigma with a single spore (Figs. 510B, 518A). In many Auriculariales the basidium has at its base a spherical swelling, the probasidium. For a time this forms the terminal cell of a dicaryotic hypha (Fig. 510A) in which nuclear fusion takes place; after meiosis its grows out to form the basidium.

Order 4: **Tremellales.** These mostly live on dead branches and tree stumps. The simplest members have no fructification. *Tremella* and *Exidia* form brain-like or leaf-like, gelatinous, yellow, brownish or black fructifications. The *Hydnum*-like (cf. Fig. 520A) fructifications of *Tremellodon* bear spines covered with the hymenium on the lower surface. The basidia generally have four longitudinal septa (Figs. 511, 518B). *Dacrymyces*

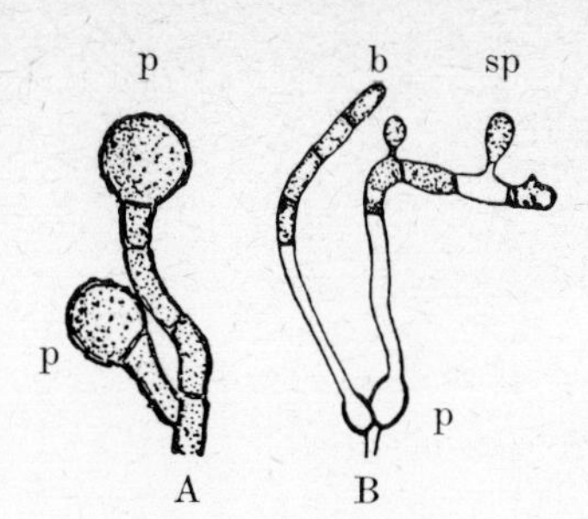

(Left)
Fig. 510. Auriculariales. Probasidia (p). **A.** of *Septobasidium pilosum.* **B.** Of *Cytobasidium lasioboli*, developed into basidia (b); to the right the beginning of spore formation (sp). **A.** After Boedijn and Steinmann. **B.** After Lagerheim.

(Right)
Fig. 511. *Tremella lutescens.* Basidium. ×300. After Brefeld.

has forked, two-spored basidia (Fig. 518D). In *Bourdotia* the longitudinal septa of the basidium do not extend to the base; such forms are transitional to the holobasidium of the next subclass.

B. Holobasidiomycetidae

The majority of those plants known to the layman as mushrooms or toadstools in woods and fields belong to this group (Figs. 520–2). The mycelium is almost always perennial, overwintering generally in the soil but also in wood; each year it can produce new fructifications. Many species develop, especially in the late summer, fast-growing, conspicuous fructifications which sometimes reach considerable dimensions. The tissues of the fructifications (Fig. 512) sometimes show a degree of differentiation into structural, thick-walled (skeletal) hyphae serving for support, other thick-walled, richly-branched, binding hyphae which clasp the others, and thin-walled hyphae which produce basidia. In the robust mycelial strands of some fungi there are water conducting hyphae with wide lumina, but tissues preventing water loss are absent. The mycelia of many Holobasidiomycetidae can be grown on artificial media but then rarely produce fructifications.

Although development differs between members of the group, in principle it follows the pattern to be described for *Corticium serum* (Fig. 513). The basidiospores germinate to form a mycelium with uninucleate cells and of practically unlimited growth. No sex organs are produced, which in this case is a considerably advanced condition and not primitive. If this mycelium meets another of the opposite mating-type fusion takes place where two vegetative cells are in contact (somatogamy, Fig. 513.3), and the nuclei of different mating-types pair off, but do not fuse. The fusion cell now grows out to form a new hypha, into which the pair of nuclei passes. The first and subsequent cross walls are formed in the following way. Just beside the pair of nuclei there develops a lateral, hook-shaped outgrowth of the cell, pointing backwards; this is the beginning of a clamp connection (Fig. 513.4), into which one of the nuclei passes and then divides (Fig. 513.5); one of the daughter nuclei remains in the hook, whilst the other moves out and towards the apex of the cell. Meanwhile the other nucleus has also divided, one of its daughter nuclei moving to the apex, the other to the base, of the cell (Figs. 513.5, 513.6); between these two daughter nuclei and immediately below the hook a cross wall now develops, whilst a second cross wall separates the hook from the terminal cell, which therefore now contains two daughter nuclei of different mating-type. At the same time the subterminal cell also becomes binucleate, since the nucleus passes out of the hook and into it via an opening at the tip (Figs. 513.6, 513.7), forming a clamp connection which is homologous in structure with the hook of the Ascomycetes. The terminal cell continues growth, and clamp formation is repeated at every cell division, resulting in a much-branched mycelium in which every cell is dicaryotic and which has a clamp connection at every cross wall (Fig. 513.8). The mycelium may continue to grow in this condition for years in the soil or in wood until under certain conditions—possibly involving the presence of certain growth substances—the hyphae interweave to form fructifications. These, in contrast to those of the Ascomycetes (Fig. 498), consist of plectenchyma of dicaryotic hyphae (Fig. 514). The hyphae which form basidia aggregate in palisade-like hymenia on or in the fructification (commonly on the lower side): the terminal cells of these hyphae enlarge, forming club-shaped basidia (Fig. 499), in which at last the two nuclei fuse (Fig. 499B); immediately meiosis (with determination of mating-type)

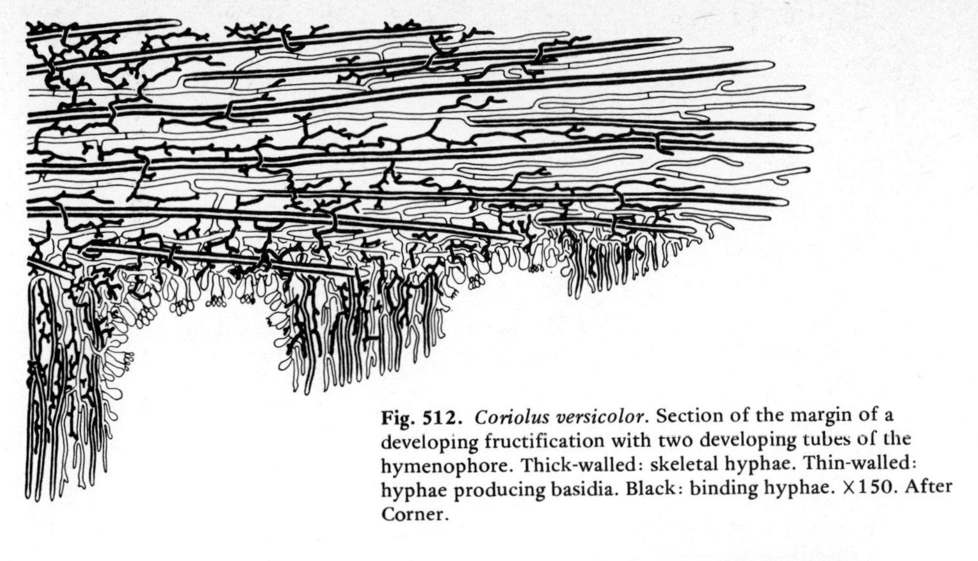

Fig. 512. *Coriolus versicolor.* Section of the margin of a developing fructification with two developing tubes of the hymenophore. Thick-walled: skeletal hyphae. Thin-walled: hyphae producing basidia. Black: binding hyphae. ×150. After Corner.

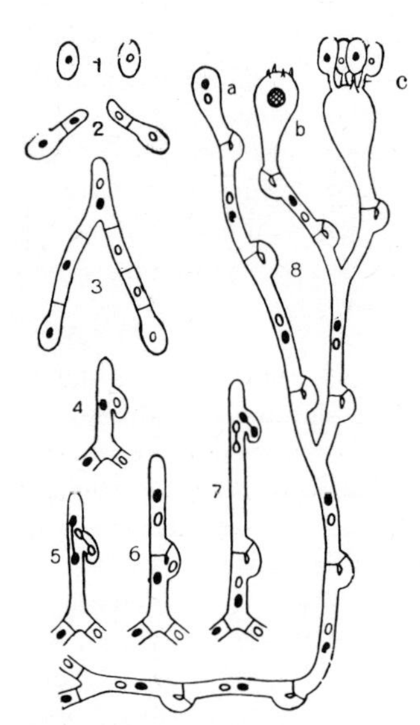

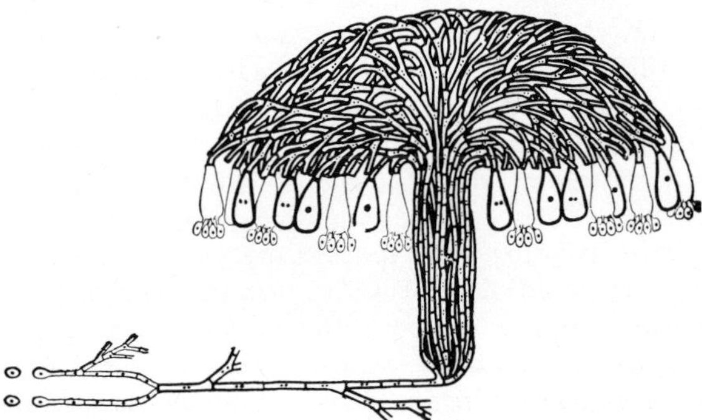

Fig. 514. Diagrammatic representation of the development of an agaric. Thin lines: haploid phase. Thick lines: dicaryotic or diploid phase. Clamp connections omitted, basidia shown very large in relation to the cap. After Harder.

Fig. 513. Diagrammatic scheme of the development of the dicaryotic mycelium of a hymenomycete. **1.** Spores of different mating-type. **2.** Germination to form mycelia without clamp connections. **3.** Conjugation. **4.** to **6.** Formation of the first clamp connection; and **7.** of the next one. **8.** Mycelium with clamp connections bearing (a) a young binucleate basidium, (b) a young basidium with fusion nucleus and (c) a mature basidium with two spores of each mating-type. After Harder.

follows, giving four haploid nuclei (Fig. 515.7). From the top of each basidium arise four short, slender processes, the sterigmata (Fig. 499D), at the end of each of which a spore sac swells up (Fig. 499C). Each of the four haploid nuclei passes along a sterigma (Fig. 499D) and into the spore sac, where it develops into a spore (Fig. 517). Almost invariably the spore wall fuses at an early stage with the wall of the sac, and the double nature of the wall is then not apparent, the sac wall forming the perispore. The spores thus develop endogenously but are abstricted at maturity. They are usually ellipsoidal and flattened on one side. Two spores are of one mating-type, two of the other.

The spores are thrown only a short distance when the highly turgid basidium suddenly expresses a liquid droplet from the tip of the sterigma, thus forcing off the spore (Fig. 516). The discharge distance in those fungi with tubular pores is about half the diameter of the tubes; in fungi with gills about half the distance between them. The trajectory of the spores is finally vertically downwards until they meet the air currents below the tubes

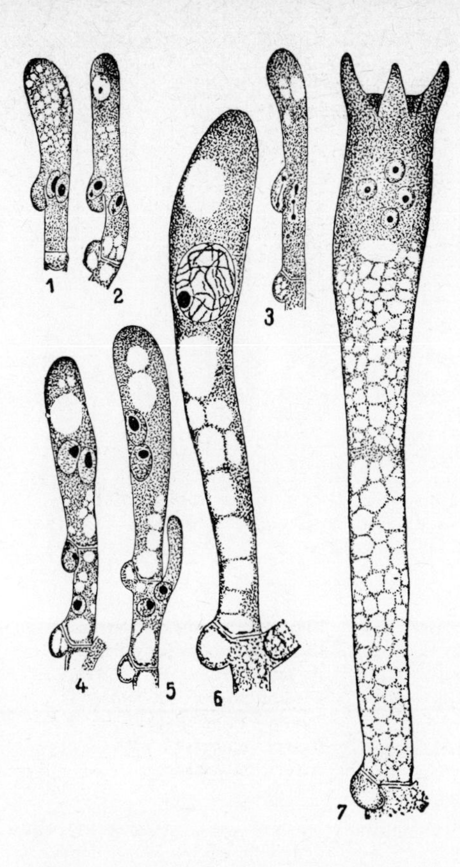

Fig. 515. *Oudemansiella mucida.* Clamp formation and development of the basidium. **1.** Beginning of clamp formation in a binucleate terminal cell. **2.** One nucleus passing into the clamp. **3.** Conjugate nuclear division. **4.** Wall formation in and beside the clamp; young basidium cut off from the stalk-cell. **6.** Basidium with fusion nucleus. **7.** Young basidium with four basidiospore nuclei and, above, four developing sterigmata (one hidden). ×620. After Kniep.

or gills; long-distance dispersal is thus by wind. If the cap of a gill-fungus is laid on a sheet of paper, after a few hours the descending spores produce a faithful representation of the arrangement of the gills. It has been estimated that a mature fructification some 10 cm in diameter of a field mushroom (*Psalliota campestris*) has a hymenial surface of 1200 cm^2, which produces some 16,000 million spores; some 40 million spores are liberated each hour.

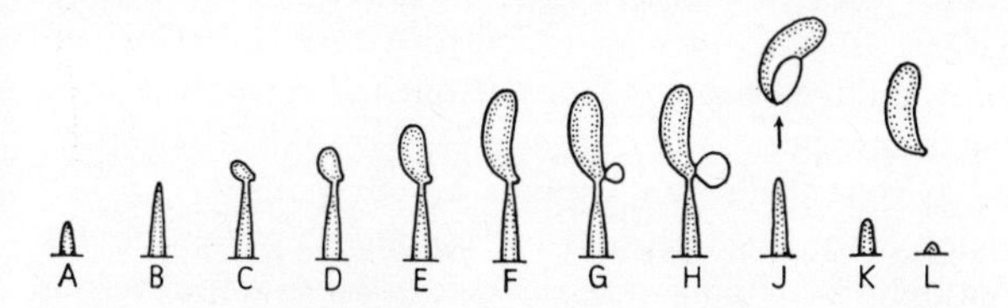

Fig. 516. *Calocera cornea.* Basidiospore discharge. **A.** and **B.** Extension of the sterigma. **C.** to **F.** Abstriction of the basidiophore (time: 40 min). **G.** and **H.** Formation of a droplet at the point of attachment of the spore (time: 10 sec). **J.** Discharge of the spore together with the droplet. **K.** and **L.** Collapse of the sterigma. ×900. After Buller.

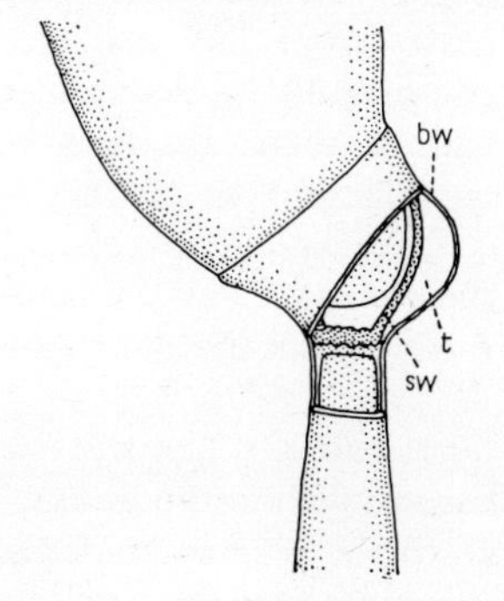

Fig. 517. *Schizophyllum commune.* Point of attachment of the basidiospore to the sterigma. bw, wall of basidium; sw, basidiospore wall; t, fluid drop. ×15,000. After Wells.

Holobasidia are very variable in size and structure (Fig. 518). The cylindrical form is the commonest (F), but there are also dilated flask-shaped ones (E), those borne laterally on a hypha (pleurobasidia, H), and repeatedly proliferous forms (repetobasidia, J).

Besides basidia the hymenium also bears sterile hyphae with degenerate pairs of nuclei, the pseudoparaphyses, and often much larger, equally sterile terminal hyphae of diverse form, the cystidia (Fig. 519). These latter act as organs of protection and secretion (e.g.

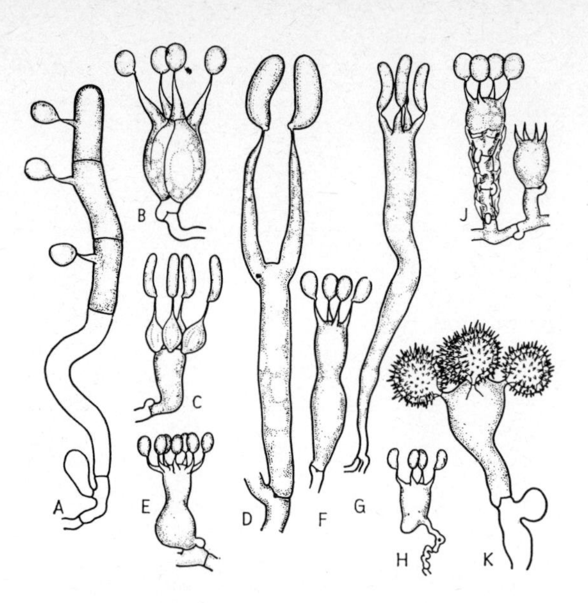

Fig. 518. Forms of basidia. A. *Platygloea* (Auriculariales). B. *Sebacina* (Tremellales). C. *Tulasnella* (Tremellales). D. *Dacrymyces* (Tremellales). E. *Sistotrema* (Poriales). F. *Hyphoderma* (Poriales). G. *Exobasidium* (Exobasidiales). H. *Xenasma* (Poriales). J. *Repetobasidium* (Poriales). K. *Scleroderma* (Gastromycetales). ×750. After Oberwinkler.

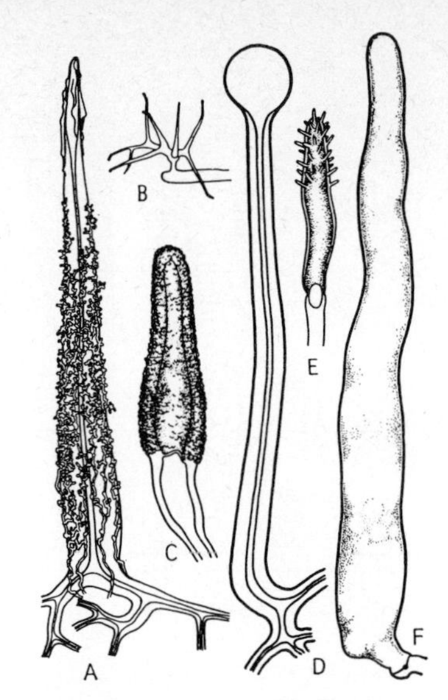

Fig. 519. Forms of cystidia in the Corticiaceae (Poriales). A. *Tubulicium*. B. *Vararia*. C. *Peniophora*. D. *Tubulicrinus*. E. *Stereum*. F. *Hyphoderma*. ×750. After Oberwinkler.

in the Poriales), or may possibly prevent the gills sticking together (e.g. in *Coprinus*); their form is important as a taxonomic character.

The origin of the dicaryotic phase is generally not quite as simple as in the preceding account; for example, a nucleus which has entered a mycelium often passes through a long series of cells, the cross walls of which are broken down whilst the nucleus repeatedly divides, before pairing and ultimately clamp formation take over. Other deviations also occur. Many species are of mixed mating-type and form dicaryotic mycelia without conjugating with another mycelium. Rarely the mycelium of one mating-type is larger than that of the other (a secondary sexual characteristic), in which case the nucleus passes from the smaller to the larger. In some genera (*Schizophyllum, Coprinus*) incompatibility is determined by two non-linked pairs of genes. If one indicates these by A and B, there are four possible mating-types: A_1B_1, A_2B_2, A_1B_2, A_2B_1 (tetrapolar incompatibility). Neither of the two pairs of alleles can combine homozygously.

The haploid mycelia which develop from basidiospores correspond to gametophytes; the dicaryotic mycelium arising from them by conjugation represents the sporophyte, and the basidium the sporangium; the basidiospores are meiospores.

Systematics. The structure and development of the fructification and hymenium are used to delimit the orders.

Order 1: **Exobasidiales**. These are plant parasites. *Exobasidium* produces brightly coloured galls on the leaves of Ericaceae. The intercellular mycelium produces basidia (Fig. 518G) which burst out to form a continuous hymenium on the leaf surface (convergence with the Taphrinales).

Order 2: **Poriales** (Aphyllophorales). These predominantly saprophytic fungi have very diverse but always open (gymnocarpous) fructifications of which the surface is wholly or partially clothed with a hymenium. It is characteristic of the whole order that the hymenium lies exposed on the surface of the fructification, and is indeed developed at an early stage, so that as the fructification enlarges it undergoes continuous extension.

In *Corticium* the fructification grows over the substrate (dead twigs, etc.) as thin, smooth crusts clothed with hymenia; the related *Coniophora* has a warty surface. *Serpula* (*Merulius*) *lachrymans*, the dry-rot fungus, rapidly develops fructifications up to a metre

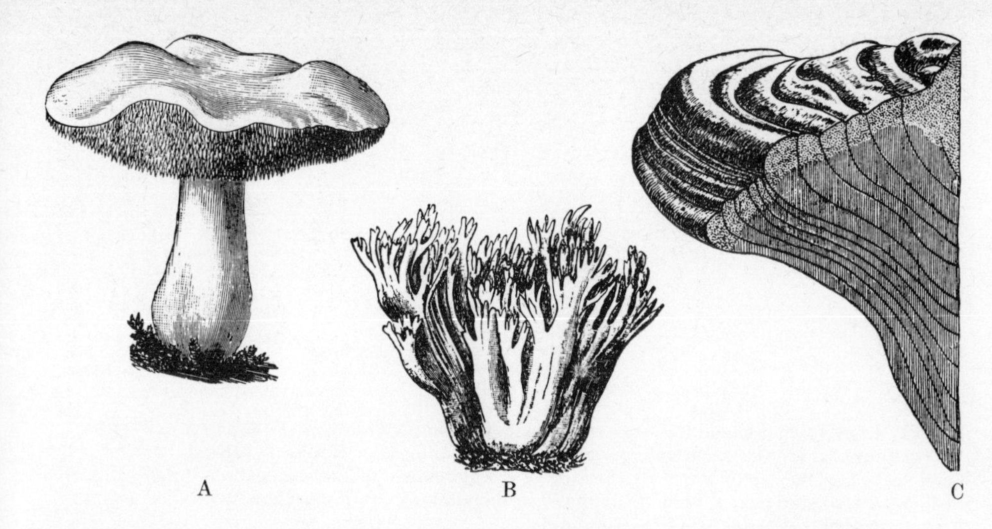

Fig. 520. Poriales. **A.** *Hydnum repandum.* $\times\frac{1}{2}$. **B.** *Clavaria botrytis.* $\times\frac{1}{2}$. **C.** *Phellinus igniarius*, section through an old fructification with annual zones of growth. $\times\frac{1}{2}$. **A.** and **B.** After Schenck. **C.** After Harder.

across; the surface is thrown into shallow irregular grooves and folds. It is extremely destructive to building timbers (p. 535); it forms robust mycelial strands in which vessel-like hyphae with wide lumina (but still containing cytoplasm) conduct water and nutrients so that the fungus can extend from a damp site of primary infection into the drier parts of a building. The many-layered fructification of *Stereum* is partly free from its substratum (rotting wood); in *Thelephora* it is lobed, and in *Craterellus* funnel-shaped; in these cases the hymenium is on the lower surface. *Cantharellus* (chantarelle) resembles the agarics in the forking ridges on the underside of its obconical fructification, but differs in being gymnocarpous. The coralloid branches of the species of *Clavaria* are covered all over with a smooth hymenium (Fig. 520B), whereas the fertile lower surface (hymenophore) of *Hydnum* consists of warts or spines which greatly increase the surface area (Fig. 520A).

The hymenophore of the great majority of the Poriales consists of closely spaced tubes lined with hymenium (the former collective genus *Polyporus*). The annual fructifications of *Poria* are flat and completely appressed to the substrate; *P. vaporaria* destroys structural timber in the same way as the dry-rot fungus. The species of *Coriolus* (*Polystictus*), frequent on tree-stumps, have leathery or corky fructifications, e.g. *C. versicolor*, striped in various colours. *Fomes*, parasitic on deciduous trees, forms woody, perennial bracket-like fructifications (*F. fomentarius*, German tinder, soft amadou or punk, on beech); *Phellinus igniarius*, hard amadou or touchwood, on apple trees, is similar; the hymenophore consisting of fine tubes occurs on the underside of the bracket which in section clearly shows the annual growth increments (Fig. 520C). The tubes exhibit a very precise positive geotropism which allows the dispersal of the spores as previously described (p. 528). The hymenophore of *Daedalea quercina* (on oaks) is formed of remarkable broad labyrinthine channels. The genus *Polyporus* as currently understood (e.g. *P. squamosus*, parasitic on deciduous trees) shows peculiarities in the development and anatomy of the annual, tough and fleshy fructifications which link it to the next order, the Agaricales.

Order 3: **Agaricales.** The hymenium is almost always formed inside the fructification in schizogenous cavities which later open to expose the hymenium (hemiangiocarpous). The whole hymenium develops simultaneously in the Agaricales, and not gradually as in the Poriales. The hymenophore usually takes the form of leaf-like, radially arranged, vertical gills (more rarely tubes) which at maturity cover the underside of the stalked cap or pileus. The gills consist of ground tissue (trama) which passes outwards into a delicate plectenchyma (subhymenium); this bears the cells which compose the hymenium: basidia, cystidia and pseudoparaphyses (Fig. 521C).

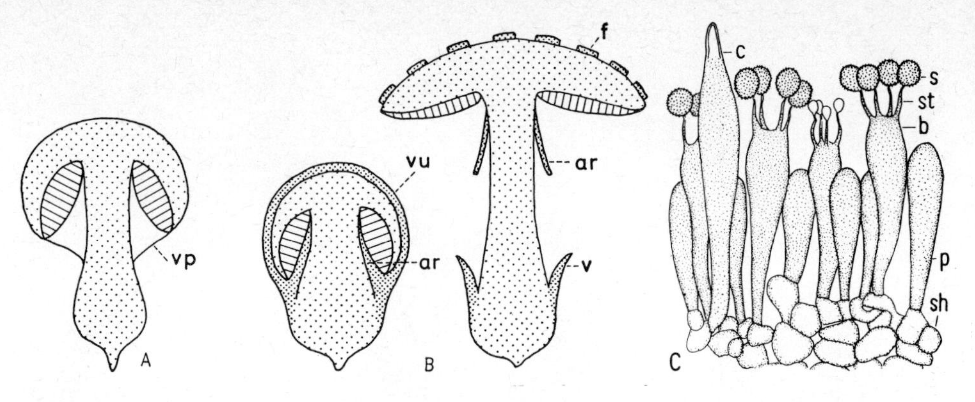

Fig. 521. Agaricales. **A.** and **B.** Diagrammatic L.S. of fructifications. **A.** With partial veil (vp). **B.** With universal veil (vu); left, young; right, mature; ar, annulus or armilla; v, volva; f, remains of the universal veil on the cap or pileus. **C.** Section of the hymenium of *Russula rubra*; b, basidium; s, basidiospore; st, sterigma; p, pseudoparaphysis; c, cystidium; sh, subhymenium. ×500. **A.** and **B.** After E. Fischer. **C.** After Strasburger, modified.

In most Agaricales the marginal tissue of the young cap is joined by the partial veil to the stipe or stalk (Fig. 521A). In the annular cavity between the cap and the stalk the gills or tubes develop beneath the cap. As the cap enlarges the partial veil is torn apart at the margin, remaining either as a ring or inferior annulus on the stalk, or it stays a short time forming a delicate web (cortina) between cap and stalk; or it completely disappears (e.g. in *Boletus*). In some gill-fungi (e.g. *Amanita muscaria*, the fly-agaric) the young cap and stalk are enclosed in a common membrane (the universal veil) which after the expansion of the fructification remains at the base of the stipe as a volva, while fragments frequently persist on the cap as white flecks (Fig. 521B). The connective tissue between the stalk and the gills separates from them on both sides and hangs down as an armilla or superior annulus from the junction of cap and stalk. Physiologically noteworthy is the luminescence (p. 275) of the mycelium of *Armillaria mellea* (the honey-fungus) (see also pp. 494, 536); even the fructification is luminescent in *Omphalotus olearius* which grows on old olive trees. Species of *Lactarius* produce white or orange latex in broad anastomosing hyphae; this contains saturated fatty acids, e.g. stearic acid.

In C. Europe there are some 2000 species in the Agaricales. Their identification often requires careful microscopical study, especially of the components of the hymenium. The recent inclusion of *Boletus* (e.g. *Boletus edulis*, cèpe) and related genera in the Agaricales (and not in the Polyporaceae) is based on their hemiangiocarpous structure and the mode of development of the hymenium.

Order 4: **Gastromycetales.** These have closed fructifications of which the outer part or peridium often dehisces or breaks down in a highly characteristic fashion when the spores are ripe: they are thus angiocarpous. The spore-forming inner tissue, the gleba, is variously chambered or divided up (areolate) in the different families (cf. Fig. 522D). The basidiospores are usually spherical (Fig. 518K). The nucleus divides at once in the basidiospore and the mycelium is dicaryotic from the outset.

The giant puff-ball (*Calvatia gigantea*) has a fructification which may reach a diameter of 0.5 m or more, and the gleba produces some 7.5×10^{12} spores. If each spore developed into a fructification then in the second generation a volume of fungus 800 times greater than that of the Earth would be produced.

The peridium is often differentiated into several layers. In *Bovista* and *Lycoperdon* (puff-balls) the outer layer disintegrates at maturity, or breaks up into granules, while the inner layer ruptures at the apex, liberating the brown spores as a cloud of dust. In *Geastrum* (earth-star, Fig. 522G) the exoperidium has three distinct layers of which the outer remains in the soil at maturity while the middle and inner ones split, turning inside out and lifting up the endoperidium. A similar turning inside out of the inner exoperidium occurs in *Sphaerobolus*, which is about the size of a mustard seed (Fig. 522F), but here it takes place explosively and discharges the endoperidium to the height of a metre. In *Cyathus* (bird's-nest fungus) the chambers of the gleba separate to form several

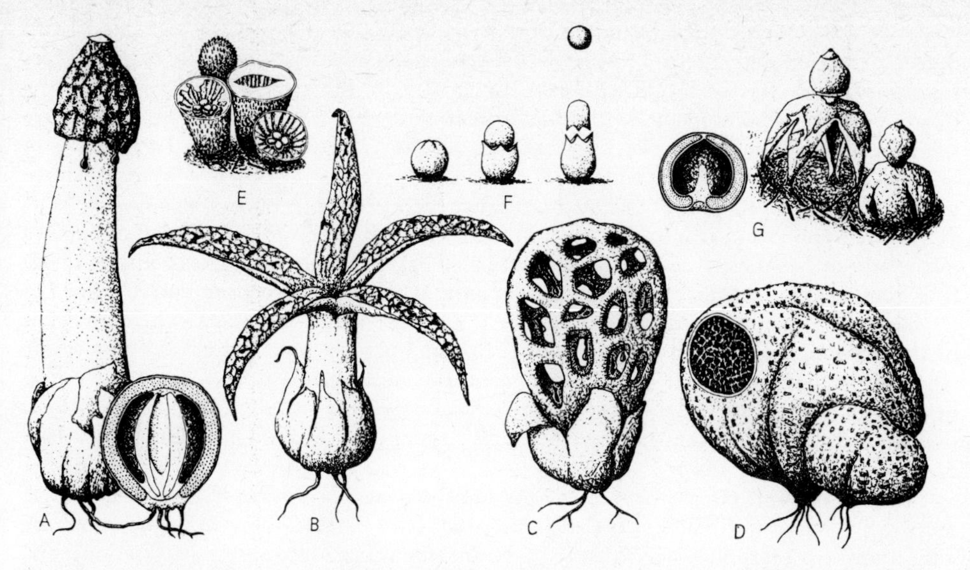

Fig. 522. Gastromycetales. **A.** *Phallus impudicus*; mature fructification with trickling droplets of the gleba and an immature fructification in L.S. ×½. **B.** *Anthurus archeri*. ×½. **C.** *Clathrus ruber*. ×½. **D.** *Scleroderma aurantium*, cut away to show the areolate gleba. ×½. **E.** *Cyathus striatus*. Nat. size. **F.** *Sphaerobolus stellatus*; to the right, discharge of the endoperidium. ×3. **G.** *Geastrum quadrifidum*. ×½. **A.** After Lange. **B., D.** and **G.** After Poelt, Jahn and Caspari. **C.** After Fayod. **E.** After Gramberg. **F.** After Michael and Hennig.

independent peridiola, which at maturity lie like minute eggs in the cup-shaped peridium (Fig. 522E). *Phallus impudicus*, the stink-horn (Fig. 522A) looks superficially similar to the morels (Ascomycetes), but its development is quite different. The young fructification (or devil's-egg) is surrounded by a peridium with a mucilaginous inner layer. Within the peridium the hyphal tissues differentiate into an axial stalk and a bell-shaped cap; the latter has a delicate membrane under which is the gleba, surrounding a tissue homologous with the partial veil of the Agaricales. At maturity the white stalk rapidly elongates (in a matter of hours), carrying the cap with it, to some 15 cm high, meanwhile rupturing the peridium which remains at the base as a volva. The outer membrane of the cap then breaks down exposing the green gleba which has a carrion-like stench. This then deliquesces and drops from the reticulately sculptured partial veil or receptacle. Blowflies and dung-flies disperse the spores endozooicly. In *Clathrus* (lattice fungus, Fig. 522C) the receptacle is reticulate, in *Anthurus* (cuttle fish fungus, Fig. 522B) it has a number of separate arms; in both it is reddish. Otherwise development is similar to *Phallus*. The families Secotiaceae and Podaxaceae which live mainly in the steppes are intermediate in structure between the Agaricales and Gastromycetales.

Compared with the Agaricales the Gastromycetales are fewer in number (some 190 species in C. Europe) but show a greater diversity in form.

Economic uses. Many species of Holobasidiomycetes are collected as esculent fungi; a few, e.g. the mushroom (*Agaricus bisporus*, with only two spores on the basidium) and *Kuehneromyces* (*Pholiota*) *mutabilis*, grown on logs, are cultivated. Some Agaricales, e.g. the death-cap *Amanita phalloides*, contain toxins (especially peptides) which act, often in a complex manner, on the stomach, nerves, liver and heart and are frequently lethal. In addition, old fructifications of esculent species develop toxins (as in putrid meat). In various species of *Amanita* and *Psilocybe* hallucinogenic indole derivatives occur which are used in religious rites, e.g. in Mexico (cf. p. 55). Antibiotics are derived from various genera. Tinder prepared from the tissues of *Fomes fomentarius* was used from Neolithic times until the middle of the nineteenth century in kindling fire.

Survey of the Basidiomycetes. The Basidiomycetes are related phylogenetically to the Ascomycetes.

As in the Ascomycetes fusion of the sexual cells (plasmogamy) is separated from fusion of the sexual nuclei (caryogamy) by a characteristic phase with paired nuclei (dicaryophase). The formation of clamp connections in the dicaryotic mycelium of the Basidiomycetes is homologous with hook formation in the Ascomycetes, but in the latter the process is restricted to the terminal cells of the ascogenous hyphae (very rarely in other cells of these hyphae), whereas in the Basidiomycetes it generally occurs at every cell division. In both groups fusion of the paired nuclei only occurs in apical cells of dicaryotic hyphae and is immediately followed by meiosis and the formation of haploid ascospores or basidiospores. The basidium (which exceptionally has eight spores) is therefore homologous with the ascus and differs only in that the meiospore nuclei formed in the basidium pass into the dilated spore sacs at the ends of the sterigmata, where the walls of the developing spore and of the spore sac fuse and become indistinguishable; the whole structure is then abstricted exotopically. The fructifications of the two groups are not, however, homologous, since those of the Basidiomycetes are composed of dicaryotic hyphae whilst those of the Ascomycetes consist mainly of haploid mycelium (Figs. 498, 514).

Fungi imperfecti (Deuteromycetes). As indicated above, the natural classification of the fungi is based on their life-cycles and reproductive organs. Many fungi, however, are only known to reproduce vegetatively by means of conidia (accessory reproduction), and it is not obvious whether the main organs of reproduction are yet to be discovered, or whether the capacity to develop them has been lost. All these fungi (some 20,000 'species') are included in the artificial group of Fungi imperfecti or Deuteromyces; they are classified by the form of their conidia. Three orders are so recognized:

1. **Sphaeropsidales**: conidia borne in structures resembling perithecia (pycnidia), or in other cavities.
2. **Melanconiales**: conidia borne in layers on stromata.
3. **Moniliales**: conidia not in layers but usually borne on richly branched hyphae which are either separate or united into bundles (coremia).

Most Fungi imperfecti appear to belong to the Ascomycetes, and only a few to the Basidiomycetes. Numerous species of *Penicillium* and *Aspergillus* are Fungi imperfecti, although in others we know that the conidial forms belong to various 'perfect' genera, e.g. of *Aspergillus* to *Eurotium*, *Emericella* and *Sartroya*, of *Penicillium* to *Talaromyces* and *Carpenteles*. Very many Fungi imperfecti are agents of disease: thus *Septoria apii* (Sphaeropsidales) produces leaf spot of celery; *Gloeosporium fructigenum* (Melanconiales) produces bitter rot of apples; *Trichophyton* (Moniliales) causes skin disease in man; while *Fusarium oxysporum* forma *lycopersici* causes tomato wilt.

Survey of the Mycophyta. Considering sexuality in fungi as a whole we find first of all no sharp distinction between zoospores and gametes in the most primitive of the Chytridiales. Specialized sex cells first occur in the related lower Phycomycetes, and indeed only in these, since amongst the Oomycetales (cf. p. 501) the ♂ gametes are reduced to nuclei in the Saprolegniaceae, as are both ♂ and ♀ gametes in the Peronosporaceae; the contents of the ♂ gametangium in the Saprolegniaceae are transferred to the ♀ through a fertilization tube. In the Zygomycetales (and even in certain Chytridiales) reduction has gone further: entire gametangia conjugate. In the Ascomycetes, too, free sex cells never occur, and even the sex organs, although still well developed, e.g. in *Pyronema*, are much reduced in many Ascomycetes (cf. p. 517): hyphal tips or even conidia (e.g. *Bombardia*) function as the ♂ organ, whilst the ♀ may be replaced by any vegetative hypha (e.g. in *Tuber*); or a single nucleus may even divide spontaneously to give two sexual nuclei. In the Holobasidiomycetes there are no sex organs whatsoever; only sexuality (somatogamy) remains.

Parallel with the reduction of the sex organs goes an ever-increasing separation between plasmogamy and caryogamy by the interpolation of a dicaryotic phase, an extraordinarily characteristic feature of the higher fungi. Plasmogamy does not of necessity give rise at once to a binucleate stage, since the more active of a pair of sexual nuclei can migrate through many cells of the mycelium of opposite sex before it finds its partner (e.g. in *Neurospora* [Ascomycetes], *Typhula* [Hymenomycetales] and certain Uredinales). In the Ascomycetes the dicaryotic phase is generally restricted to a few cells which are nourished by the haploid mycelium; in the Uredinales, in contrast, the dicaryon is distributed to various hosts by several kinds of spores, whilst in the Holobasidiomycetes it lives for years on end and forms the bulk of the vegetative body. Fructifications are wholly absent from the lower forms; in the Ascomycetes they are composed of gametophytic hyphae, whereas in the sexually reduced Holobasidiomycetes they arise entirely from sporophytic hyphae under the influence of external factors and quite independently of any sexual process.

The conversion of swarmers into resistant spores and the replacement of gametogamy by gametangiogamy have enabled the fungi to progress from aquatic to terrestrial existence.

Fungi are rarely preserved as fossils. The oldest finds going back to the Cambrian are Chytridiales in fragments of the carapace of marine animals. Non-septate hyphae have been found in the remains of Devonian land plants; Uredinales (and possibly Ascomycetes) and mycorrhizal tree roots occurred in the Carboniferous.

Distribution and ecology of fungi. There are some 55,000 species of fungi: about 1000 Phycomycetes, 20,000 Ascomycetes, 15,000 Basidiomycetes and 20,000 Deuteromycetes. They are exclusively heterotrophic, and, in contrast to the algae, are predominantly terrestrial. Aquatic fungi are mainly members of the Oomycetales amongst which numerous marine representatives have recently been discovered. The terrestrial fungi (Zygomycetales, Ascomycetes, Basidiomycetes) are parasites or saprophytes.

Out of the enormous subject of mycology we can only deal with three main topics of ecological significance: fungi as the agents of the decay of wood; fungi as symbiotic organisms; and parasitic fungi as the agents of disease.

1. *Fungi as agents of the decay of wood.* Under natural conditions the wood of dead trees and stumps is broken down by fungi, mainly by Polyporaceae and Agaricaceae. Some of these wood-destroyers attack living trees, e.g. *Trametes pini*, causing red ring-rot, and *Fomes annosus*, causing butt-rot or heart-rot, both on conifers; whilst *Fomes fomentarius* occurs on beech and *Phellinus igniarius* on various deciduous trees. However, many fungi only grow on dead wood, e.g. species of *Polystictus, Coriolus, Daedalea* and *Lenzites*. Many cause serious destruction of building timbers, such as the cellar-rot, *Coniophora*, and above all the dry rot fungus, *Serpula lachrymans*, which, starting from damp places, can cause enormous havoc. In general, one can distinguish between white-rots and brown-rots. Many white-rot fungi (e.g. *Phellinus igniarius*) attack lignin more rapidly than cellulose; the rotten wood is longitudinally fibrous and is white as the result of a bleaching process. The brown-rot fungi (e.g. *Serpula, Coniophora*) on the other hand preferentially destroy cellulose; the wood cracks transversely and becomes brown. A third type of rot has only recently been recognized: this occurs particularly under very moist conditions, giving rise to a soft-rot caused by species of *Chaetomium* (Ascomycetes); the hyphae attack the secondary walls of tracheids and fibres. In all three cases both the compressive and bending strength of the timber are reduced almost to nil. Recently timber partly decomposed by Basidiomycetes has been impregnated with synthetic resin and widely used as an easily worked 'Mycoholz'. Timber which has been used wrongly in very wet situations is liable to a surface rot caused by cellulose-destroying Ascomycetes and Deuteromycetes. The greying of timber exposed to the weather is caused by Basidiomycetes. The blueing of deal by species of *Ceratocystis* has no effect on its mechanical properties since the fungus only destroys the contents of the xylem parenchyma.

2. *Fungi as symbionts.* Most fungi of the woodland floor live symbiotically with the roots of deciduous and coniferous trees, and many also in association with herbaceous plants (mycorrhiza, p. 294). The close symbiotic relationship between fungi and algae in the form of lichens will be dealt with in the next section. Symbioses between fungi and animals are remarkably diverse. Nutritionally specialized animals, such as blood-sucking, timber-destroying and plant sap feeding insects have so-called mycetomes (bacteria and

yeasts) in definite parts of their intestinal tracts as vegetable symbionts; their transmission to following generations is ensured by various arrangements. These symbionts probably provide these nutritionally specialized animals with vitamins, especially of the B group. The tropical leaf-cutting ants cultivate the mycelium of Agaricaceae (*Rhozites*) in their underground nests where the dilated, nutrient-rich ends of the hyphae are used as food; the fungus is cultivated by the ants on a substrate of chewed leaf fragments and is transported each time to a fresh nest. Termites only feed their carefully cultivated fungal mycelium to the queen and to larvae. Less well studied are the so-called ambrosia fungi which live in the passages of native bark-boring beetles (Ipidae) which graze upon them.

3. *Parasitic fungi as agents of plant diseases.* Pathological modifications in the morphology and metabolism of plants can be produced by external factors (nutrient deficiencies or excesses, lack of light, extreme temperatures, etc.), by parasitic animals and by parasitic plants. Paramount amongst these last are the fungi, of which many groups (e.g. Peronosporaceae, Erysiphales, Uredinales, Ustilaginales) are almost exclusively parasitic on plants from which they derive all essential nutrients. Development of the parasite on the host begins with infection: this may originate through a wound, through a stoma, or directly through the epidermal cells. Temperature and even more so humidity are crucial for the development of the parasitic fungus on the host by means of spores or conidia; hence epidemics of fungal diseases in wet years. Often a single spore or conidium is sufficient to produce a heavy infection. Sources of infection include overwintering stromata which develop fructifications in the spring (e.g. *Venturia, Claviceps, Rhytisma*); overwintering zygotes which germinate in the spring (Uredinales, Peronosporaceae); or mycelia overwintering in the rhizomes, tubers or winter buds of the host. Infection can also start from a saprophytic stage as in the honey fungus (*Armillaria mellea*) which can live on dead tree-stumps but which can also attack living trees especially when these have been physiologically weakened, e.g. by a period of drought. The dispersal of pathogens is similar to that of the fruits and seeds of higher plants: mostly by wind, often by animals, and in recent times often by man, who has transported many plant diseases from one part of the world to another. Epidemic diseases of phytopathogens only occur under conditions of monoculture, when the total destruction of a single cultivated species may occur (e.g. of the vine in Tenerife by *Uncinula necator*).

Whereas therapeutic measures are important in the control of bacterial disease in man, they play a minor role in phytopathology. More important are various prophylactic practices such as attempting to prevent contact between host and parasite by means of appropriate methods of cultivation, rotation of crops, eradication of the alternate host (in rusts), or by killing the parasite before or during the infection stage (by seed disinfection, spraying or dusting with fungicides). The breeding of resistant varieties is also very important. Successful control depends above all on knowledge of the life-cycle and environmental requirements of the parasite.

Division 4: **Lichenes, lichens**

A lichen consists of a morphological and physiological unit formed by the association of fungal hyphae with a lower alga. The algae (formerly called gonidia) found in lichens are unicellular or filamentous Cyanophyceae (e.g. *Chroococcus, Gloeocapsa, Scytonema, Nostoc*) or Chlorophyceae (e.g. *Coccomyxa* [Volvocales], *Cystococcus* and *Chlorella* [Chlorococcales], *Trentepohlia* [Chaetophorales]). The associated fungi are almost always Ascomycetes (Helotiales, more rarely Sphaeriales) and only very rarely Basidiomycetes (Thelephoraceae).

Morphology. In some cases the form of a lichen is determined by the algal component, but generally by the fungus. The former is the case in the gelatinous lichens such as *Collema*, which live on soil or bark; in this a mucilaginous colony of *Nostoc* is permeated by fungal hyphae; in such filamentous lichens as *Ephebe* the fungus entwines filamentous Cyanophyceae. In the vast majority of genera, however, the fungus determines the form. In the very slow-growing crustose lichens which live on rocks, earth or bark the thallus is firmly attached to the substratum or even penetrates it to a certain extent and has no very definite shape (Fig. 523H). Foliose lichens have flat, generally lobed thalli (Fig.

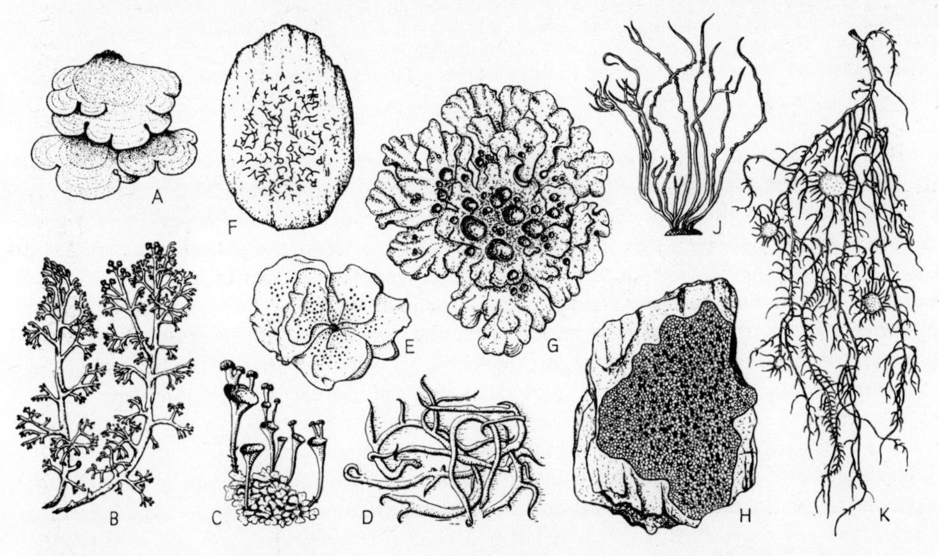

Fig. 523. Lichenes. **A.** *Cora pavonia.* **B.** *Cladonia rangiferina.* **C.** *Cladonia pyxidata* (with cup-shaped podetia). **D.** *Thamniola vermicularis.* **E.** *Dermatocarpon miniatum.* **F.** *Graphis stricta.* **G.** *Parmelia acetabulum.* **H.** *Rhizocarpon geographicum.* **J.** *Roccella boergesenii.* **K.** *Usnea florida.* All ×½.

523G) attached to the substratum by hyphal strands (rhizinae). Umbilicate lichens (Fig. 523E) have a disc-shaped thallus only attached at the mid-point. Lastly, the fruticose lichens are attached to the substratum only by a narrow base and show shrub-like branching (Fig. 523J). The arctic and high alpine *Thamniola* (Fig. 523D) lies loose on the ground or is at most attached by a few hyphal strands. In the genus *Cladonia* (Fig. 523B, C) the basal, foliose part of the thallus is generally poorly developed but from it arise erect cup-shaped or fruticose podetia which bear the apothecia.

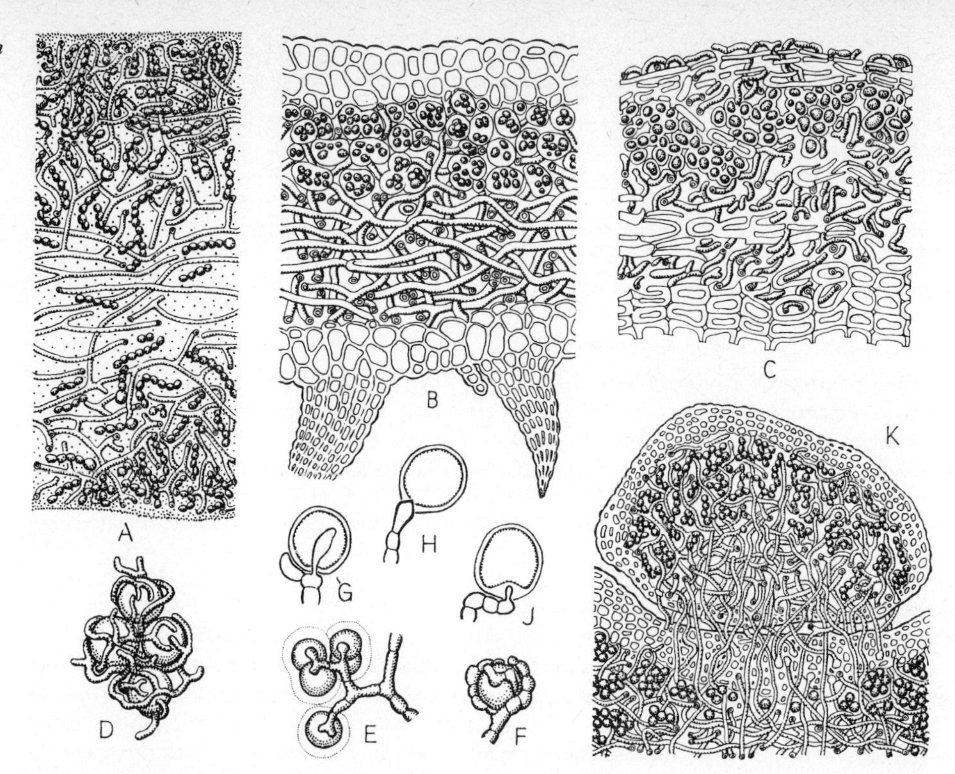

Fig. 524. Lichenes. **A.** *Collema pulposum*, section of a thallus. ×200. **B.** *Sticta fuliginosa*, section of a thallus. ×250. **C.** *Graphis dendritica*, section of a thallus. ×200. **D.** Soredium of *Parmelia sulcata.* ×450. **E.** to **J.** Haustoria (**E.** appressorium; **F.** clasping hyphae; **G.** intracellular haustorium; **H.** intra-membrane haustorium; **J.** intra-membrane haustorium excluded by cellulose deposition. **E.** to **F.** ×450. **G.** to **J.** ×600). **K.** Cephalodium of *Peltigera aphthosa.* ×200. **A.** After Des Abbayes. **B.** After Sachs. **C.** After Bioret. **E.** and **F.** After Bornet. **G.** and **H.** After Tschermak. **J.** After Plessel.

Histology. A section of a gelatinous lichen (Fig. 524A) shows a more or less uniform distribution of the alga through the thallus (homoiomerous structure). The fungal hyphae are often denser on the upper and lower surfaces, forming a cortical layer. In fruticose and foliose lichens (Fig. 524B), and in some crustose lichens, the algae lie in a definite layer parallel to the surface of the thallus (heteromerous structure); because of their situation in the thallus the algae are adapted to low light intensities. The fungal hyphae of the upper cortex often form a compact pseudoparenchyma. In crustose lichens the association of the two partners is less intimate, the algal layer is often discontinuous, and the cortex less well developed (Fig. 524C). In hypophloeadal lichens which live in the bark of trees and in endolithic lichens immersed in the surface of rocks the thallus penetrates the substratum so deeply that it no longer reaches the exterior.

The fungus and alga live symbiotically, the fungus growing around the alga (Fig. 524E, F), penetrating it in most lichens by means of haustoria; these either enter the lumen of the algal cell (in most crustose lichens; Fig. 524G) or merely into the cell wall (in the more highly organized lichens; Fig. 524H). Often the fungal hypha is 'switched off' in the autumn by the thickening of the algal cell wall (Fig. 524J). A characteristic feature of this symbiosis is the formation of very many 'lichen substances' which are only formed by the lichen but not by the partners in isolation. They usually occur as tiny crystals on the outer surface of the hyphae and they give a characteristic coloration to many lichens. They belong to diverse classes of organic chemicals: aliphatic acids, depsides, quinones, dibenzofuranes, etc.

Some lichens contain in addition to the predominant alga a second one quite unrelated to the first. This secondary alga either occurs in small tubercles on the surface of the thallus (cephalodia; Fig. 524K), but occasionally also in particular locations in the thallus itself. Furthermore, in addition to the true fungal symbiont a second associated fungus can also be present; no doubt this is parasitic; many examples of these 'lichen-parasites' are known. Finally, some lichens regularly live as parasites within the thallus of other species.

Reproduction. Lichen algae only reproduce vegetatively; their cells are often larger than in the free-living state since division is apparently retarded. The fungi, however, develop their characteristic fructifications, either perithecia or apothecia; the hymenium usually is destitute of algae. A new lichen thallus can therefore only arise when the fungal spore germinates by chance in the neighbourhood of the appropriate algal cell. Such resyntheses of lichens have been repeatedly produced experimentally. Only a few lichens (e.g. *Endocarpon*) also have in the fungal hymenium algal cells which are carried off when the spores are discharged so that the two symbionts are present when the spores germinate. The function of the pycnidia seen in many lichens is still obscure.

Reproduction of foliose and fruticose lichens is predominantly vegetative. This is generally effected by soredia (Fig. 524D), groups of algal cells entwined by fungal hyphae, which are commonly borne on definite areas of the thallus (soralia) and are dispersed by wind, giving rise to a new lichen on a favourable substratum. In other species the surface of the thallus gives rise to small cylindrical or coralloid outgrowths (isidia) which readily break off and also serve for vegetative propagation. Finally, any broken-off portion of a lichen thallus can freely regenerate.

Distribution and ecology. Lichens grow on a wide variety of substrates: on rocks, on earth, on the bark of deciduous and coniferous trees, on dead wood and so on. In the Tropics small lichens even live on leaves and are indeed partially parasitic. Saxicolous crustose lichens can dissolve limestone (but not quartz) and as pioneers pave the way for higher plants. A few lichens live in fresh water, others in the coastal spray-zone (Fig. 450). The growth of lichens is most luxuriant in humid montane forests of the temperate zone and of high tropical mountains, and also in tundra where they often cover most of the ground over large areas, giving rise to their own vegetation formations (p. 768), which fungi alone are not capable of doing. Lichens avoid the stony wastes of great cities where growth is inhibited by atmospheric pollution and low humidities. Growth rates of lichens are very low compared with other thallophytes. Even our large native foliose and fruticose lichens never put on more than 1–2 cm per year. Saxicolous crustose lichens of the alpine zone, e.g. *Rhizocarpon geographicum* (the 'geographical lichen'; Fig. 523H) grow radially only 0.5 mm per year; the age of post-glacial moraines can be determined from the diameters of such lichens. The life-span of lichens ranges from one year (in the tropical epiphylls) to many hundreds or perhaps thousands (in the Arctic–Alpine saxicolous crustose lichens).

Water (and water vapour) are taken up by the fungal hyphae. However, particularly in the larger foliose lichens, a proportion of the hyphae are often unwettable so that even when the thallus is thoroughly moistened aeration is still possible; sometimes the underside of the thallus bears regular air vents (cyphellae). Lichens living on sunny rocks can withstand not only the high temperatures (up to 70°C) of the habitat but also complete desiccation for months on end. On moistening, photosynthesis is resumed within a few minutes.

Lichens occur as the furthest outposts of life in the cold deserts of high mountains and polar regions; they can withstand freezing to −196°C without suffering harm and can even show net photosynthesis at −24°C.

Systematics. The orders and families of the Lichenes are characterized by the structure of the fungus fructification. Some 400 genera with more than 20,000 species are recognized. Identification of species involves chemical characters.

Lichens do not occur fossil before the Tertiary, where they are found in amber in highly developed forms scarcely different from various present-day species. Two classes are recognized.

While some authorities group the various orders of lichens with the most closely related fungi, the Lichenes are treated here as a separate division. One can justify this because the lichen fungi can only grow in nature when united with the appropriate algae, because the symbiosis of fungus and alga leads to the appearance of quite definite structural characteristics and metabolic products (lichen substances), and because lichen fructifications show structural peculiarities.

Class I: **Ascolichenes.** Only a few genera have flask-shaped perithecia (subclass **Pyrenocarpeae**), e.g. *Dermatocarpon* (Fig. 523E) and the many species of *Verrucaria* endolithic in calcareous rocks. The majority (subclass **Gymnocarpeae**) have apothecia, but in contrast to those of the fungi they are firm in texture and long-lived, and the asci have thicker walls. In the **Caliciales** the asci disintegrate so that the spores form a loose mass together with the paraphyses which continue to elongate; included here are *Calicium* living on tree-bark and *Sphaerophorus* on siliceous rocks (both with stalked fructifications). The apothecia of the crustose **Graphidales** (e.g. *Graphis scripta*; Fig. 523F; occurring on the bark of beech) are elongate, resembling runic inscriptions. All lichens containing blue-green algae are included in the **Cyanophilales**; these include the felted lichen *Ephebe*, the gelatinous lichen *Collema* (Fig. 524A) and the large foliaceous genera *Lobaria* and *Peltigera*. The **Lecideales** contain green algae, but the margin of the apothecium remains free of them: the enormous genus *Lecidea* belongs here, as do *Rhizocarpum geographicum* (Fig. 523H) and *Cladonia* (Fig. 523) which bears podetia as previously described. The **Lecanorales** have green algae even in the margin of the apothecium: examples include the crustose genus *Lecanora* (with over 1000 species), the foliose *Parmelia* (Fig. 523G), the 'Iceland moss', *Cetraria islandica*, and the bearded lichens, *Usnea*, which exceptionally reach 8 m in length (Fig. 523K). Finally, the **Caloplacales** have thick-walled, generally two-celled spores: the orange-yellow foliose *Xanthoria parietina*, frequent on wood, bark, rocks and roofs, and the firmly crustose, usually yellow *Caloplaca* belong here.

Class II: **Basidiolichenes.** For a long time the only known examples were tropical symbionts between Thelephoraceae and Cyanophyceae, e.g. the pantropical, terrestrial *Cora pavonia* (Fig. 523A). More recently Basidiolichenes have been found both in the Tropics and in the Temperate zones; in these Clavariaceae or Agaricaceae are associated with Chlorophyceae.

Economic uses. *Cetraria islandica* (Iceland moss) was used for a long time medicinally as a demulcent; it grows in dry woods and on heaths from the tundras of the north to the Alps. Recently antibiotics have been extracted from a number of lichens. The manna lichen, *Lecanora esculenta*, has small lobes or nodules which are edible; it occurs in the steppes of N. Africa and the Orient. Species of *Roccella* (Fig. 523J) from N. Africa and the Canary Islands yield litmus. *Cladonia alpestris*, mainly obtained from northern Europe, is used as an everlasting plant. A perfume (mousse de chêne) is obtained from *Evernia prunastri*. The reindeer moss, *Cladonia rangiferina* (Fig. 523B), together with other fruticose lichens forms the staple diet of the reindeer. *Letharia vulpina*, a yellow, epiphytic fruticose lichen with an Arctic–Alpine distribution is the only poisonous lichen in Europe; it was formerly used to kill wolves.

Division 5: **Bryophyta, liverworts and mosses**

All plants above the thallophyte level of organization are generally green land plants with chlorophylls *a* and *b* and with cellulose cell walls (p. 449); they always exhibit heterophasic alternation of generations and usually have a shoot differentiated into stem and leaves (hence the names Cormobionta or Cormophyta). In the Bryophyta and Pteridophyta considered next, both sporangia and gametangia are always multicellular and differ from those of all thallophytes by being invariably enclosed in a jacket of sterile cells. The basic structure of both the male and the female gametangia is essentially similar in all the classes belonging here. This is particularly true of the female gametangium, the archegonium, whereby the Bryophyta and Pteridophyta may be grouped as Archegoniatae. However, even in the Spermatophyta reduced archegonia can still be recognized.

The archegonium (Figs. 530J, 604) is a flask-shaped organ generally with a single layer of cells forming the venter and neck. The venter surrounds a large central cell which divides shortly before maturity to give an egg cell and a ventral canal cell; this latter is situated at the base of the neck just below the neck canal cells. Bryophyta have a whole row of these (Fig. 530J), Pteridophyta mostly only a single one (Fig. 604). The archegonium is further reduced in the Spermatophyta. The egg cell is fertilized in the archegonium (Fig. 525).

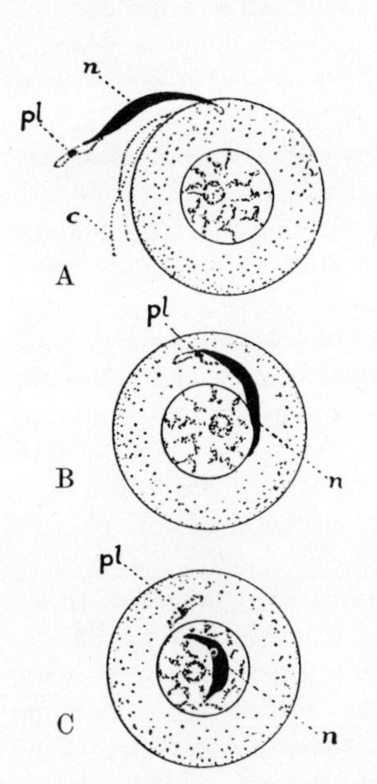

(Left)
Fig. 525. Fertilization in *Anthoceros laevis*. **A.** Spermatozoid arriving at the egg-cell. **B.** Penetration of the spermatozoid. **C.** Sperm nucleus in the egg nucleus, its cytoplasm remaining in the egg cytoplasm. pl, plastid; n, nucleus; c, flagella. ×900. After Yuasa.

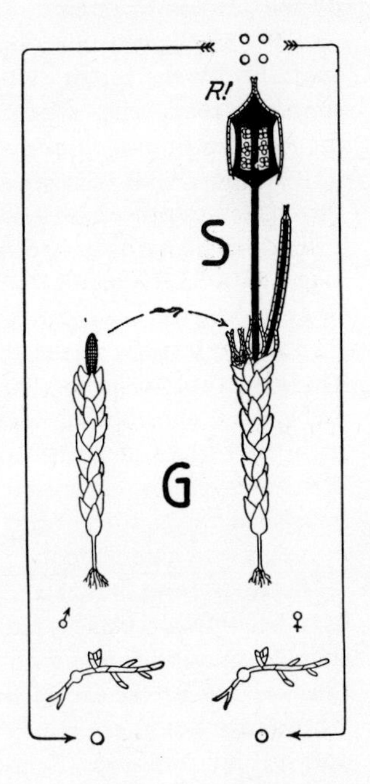

(Right)
Fig. 526. Life-cycle of a dioecious moss (spore; protonemata; gametophytes, G; fertilization; sporophyte, S; spores). Thin lines: haploid phase; thick lines: diploid phase; R!: meiosis. After Harder.

The antheridium (male gametangium, Fig. 530E) is a spherical or club-shaped structure; a jacket one cell thick encloses the spermatogenous tissue (Fig. 535). The small, more or less numerous spermatogenous cells each divide into two spermatocytes which separate, each forming one spermatozoid (or sperm). The spermatozoids are always short, twisted filaments consisting mainly of nucleus and bearing long, whip-lash flagella (Fig.

530F) near the anterior end; in some a minute plastid is present in the remnant of the cytoplasm (Fig. 525). Archegonia and antheridia are homologous, and structures transitional between them occasionally occur. Fertilization is effected, even in the land forms, only in the presence of water (rain, dew). The archegonium opens at the apex and the canal cells become mucilaginous and release certain substances which attract the spermatozoids chemotactically (p. 382). The embryo develops from the fertilized egg cell without a resting period (Fig. 530K). Further development of the embryo follows different courses in the two archegoniate groups: in the Bryophyta it becomes a relatively small spore-producing structure which is not independent; in the Pteridophyta it grows into a plant which also bears spores but which is large and independent. Spore-formation is always associated with meiosis. In both groups the sexual processes depend on the presence of water but the spores are dispersed in the air.

The life history of a bryophyte is briefly as follows (Fig. 526).

The unicellular haploid spore germinates to form a small green protonema which is highly developed in some groups but only slightly in others. The plants arise from the protonemata, sometimes from special buds, either as a ribbon-shaped or lobed thallus (e.g. Figs. 530A, 534A), or divided into stems and leaves (Figs. 533D–G); they bear gametangia but never possess roots, only rhizoids: long, simple, tubular cells resembling root hairs, or filaments of cells (Figs. 85, 546E). This haploid plant bears the gametangia.

After fertilization by an invariably biflagellate, corkscrew-like spermatozoid (Fig. 530F) the egg cell develops without a pause into the diploid embryo. The basal part of the embryo (the 'foot', Figs. 530A, 534A) usually penetrates the tissue below the archegonium, but the main growth is towards the top of the archegonium, where a spherical or ovate, stalked capsule (Figs. 533, 546E) containing numerous spores is formed; foot, stalk and capsule together form the sporogonium. Since the neck of the archegonium is too narrow for the growing embryo and the wall of the venter can only keep pace with the growth of the embryo for a limited time (Fig. 526, right), the archegonial wall is pierced by the developing embryo or the ventral wall is torn apart.

The spores arise in groups of four (tetrads) from the inner tissue of the capsule (archesporium) by the meiotic division (in some species associated with sex determination) of spore mother cells which have separated and rounded off. The spores give rise to monoecious or dioecious plants, depending on the species.

The spore wall consists of an inner delicate endospore and an outer resistant exospore; the latter is ruptured on germination.

Besides being dispersed by spores, bryophytes very commonly reproduce by gemmae (Figs. 529, 533G; p. 203) which are produced in various ways on the thallus, stems, leaves or protonemata; they separate and produce new plants.

The life history of the Bryophyta shows a clear alternation of generations (Fig. 526). The gametophyte is the haploid plant (including the protonema) bearing antheridia and archegonia; the diploid sporophyte develops from the fertilized egg cell and terminates with the formation of meiospores by meiosis. The sporophyte does not become independent but remains attached as the sporogonium to the gametophyte, by which it is nourished through the foot. Despite their chlorophyll content, sporogonia fail to develop normally when detached.

As their form indicates, the bryophytes are terrestrial plants except for a few which have become secondarily aquatic (p. 558). This is expressed in their morphology. Besides simple, creeping forms with a lobed thallus, many have erect stems provided with leaves, thus making better use of the available light. The leaves are one cell thick, or more along the midrib, but there is no mesophyll. Most bryophytes show geotropic and some show phototropic responses. Even the thallose forms are differentiated into photosynthetic and storage tissues, and some have so-called 'air pores' permitting gaseous exchange (Fig. 528; see below). The jackets of sterile cells surrounding both gametangia and sporangia are also terrestrial adaptations. Furthermore, a cuticle is present, but generally it is very delicate and the plants quickly and completely dry out when water is lacking. Many foliose kinds possess very simple conducting bundles (Fig. 546H) composed of elongated living and dead cells with thick walls (hydroids, p. 82) and of thin walled cells (leptoids), but true roots are lacking and only rhizoids are present, unbranched in the liverworts, branched in the mosses. For apical cells and the development of tissues see p. 80 et seq.

The Bryophyta comprise two sharply separated classes: the Hepaticae (liverworts) and the Musci (mosses), distinguished both by vegetative structure and the mode of development of gametangia and sporogonia. Both classes are evergreen (apart from annual species), autotrophic and produce starch.

Class I: Hepaticae, liverworts

The germlings (protonemata) of thallose liverworts are usually very short tubes; only in the leafy 'foliose' forms are they multicellular and diverse in form. In *Protocephalozia* the filamentous protonema becomes sexually mature, bearing antheridia and archegonia. Most liverworts contain oil-bodies of specific form in some of their cells (most characteristically as spherical droplets containing one or more ethereal oils; such bodies occur in no other plants; Fig. 528A ö). Unequal division of the spore mother cells in the capsules produces both spores and sterile cells called elaters. The class includes some 10,000 species and is divided into four orders; the structure of the gametophyte shows a great diversity.

Order 1: **Sphaerocarpales.** The simple thallus either consists of a small rosette growing on soil (*Sphaerocarpus*, Fig. 527A) or of an erect axis bearing an undulate wing in the submerged aquatic *Riella* (Fig. 527B). The archegonia and antheridia are enclosed in pear-shaped perianths which have an apical opening. The capsule wall consists of a single layer of cells which disintegrate at maturity. *Sphaerocarpus*, much used in genetical research, was the first plant in which sex-chromosomes were discovered (by C. Allen in 1917; see p. 326).

Order 2: **Marchantiales.** Many members of this order have a relatively complex thallus. *Marchantia polymorpha*, frequent in damp places, may be described as an example (fam. **Marchantiaceae**); it has flat, rather fleshy, ribbon-shaped thalli some 2 cm broad which branch dichotomously and have a weak midrib (Fig. 530A, G).

The underside bears ventral scales consisting of a single layer of cells, and negatively phototropic unicellular filamentous rhizoids (Fig. 530G), some of which have peg-like thickenings of the wall projecting into the lumen; these rhizoids attach the plant and conduct water to it, mainly by capillarity between the wick-like rhizoids, but partly by internal conduction.

The upper epidermis has an almost impermeable cuticle; beneath it lie large intercellular spaces, the air chambers (Fig. 528A, C), which are separated laterally by walls one or two cells thick and which are apparent on the surface of the thallus as rhomboidal or hexagonal areas. Numerous short branching filaments of roundish cells containing chlorophyll arise from the floor of each chamber (a) and function as photosynthetic tissue; occasionally these are joined to the epidermis. Each chamber communicates with the exterior through a barrel-shaped air-pore (Fig. 528A, B) consisting in *Marchantia polymorpha* of four ring-shaped tiers, each of four cells. In some cases these may show a certain degree of opening and closing, but this has probably little functional significance. These gametophytes are unique in their specialized development for photosynthesis and transpiration. Large parenchymatous storage cells below, with few chloroplasts, include oil-bodies (ö). Two rows of scales, one cell thick, arise from the median line of the lower epidermis; these enclose a longitudinally arranged mass of filamentous rhizoids.

The upper surface of the thallus usually bears along the midrib cup-shaped outgrowths with a toothed margin; these gemma-cups (Fig. 530A) contain a number of flat gemmae which arise, as shown in Fig. 529A–C, by the outgrowth and division of single epidermal cells, and are attached by a stalk-cell (st) from which they later absciss (at X in Fig. 529D). There is a growing point in each of the two notches, and some colourless superficial cells of the many layered structure form the initials of rhizoids (r). The gemmae which develop into new thalli (cf. p. 203) effect vegetative reproduction very copiously.

The gametangia are borne on special erect branches of the thallus (gametangiophores,

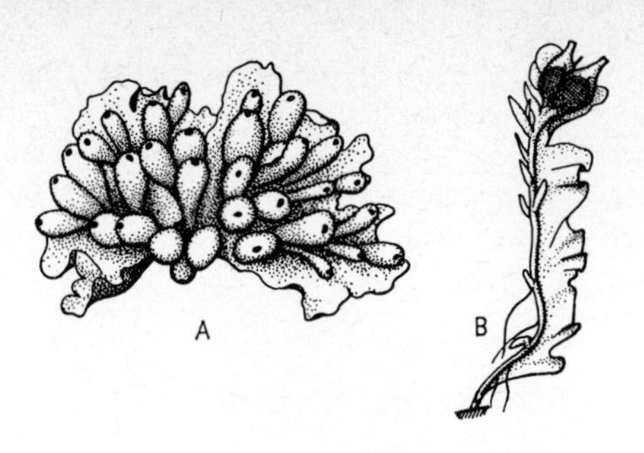

Fig. 527. Sphaerocarpales. **A.** *Sphaerocarpus michelii*, ♀ thallus with perianths. **B.** *Riella helicophylla*, ♀ thallus. **A.** ×5. **B.** ×2.5. After K. Müller.

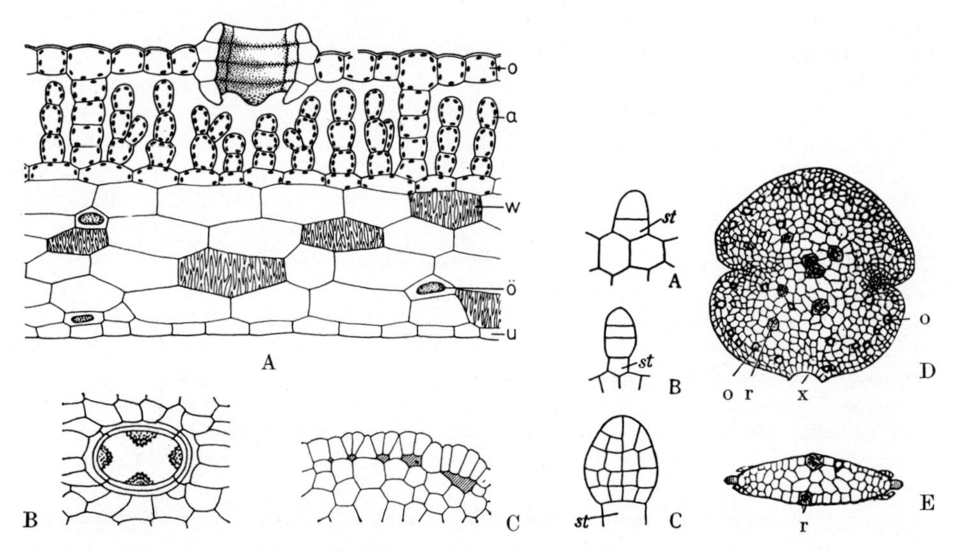

Fig. 528. *Marchantia polymorpha.* **A.** Transverse section of the thallus. ×200. **B.** Air pore seen from above. ×200. **C.** Development of the air chambers. ×270. a, assimilators; o, upper epidermis with air pores; ö, oil-bodies; u, lower epidermis; w, wall thickenings. **B.** After Kny. **C.** After Leitger.

Fig. 529. *Marchantia polymorpha.* **A.** to **C.** Development of a gemma; st, stalk cell. **D.** Gemma in side view. **E.** In T.S. x, point of detachment; o, oil-cells; r, rhizoid initials. **A.** to **C.** ×275. **D.** and **E.** ×65. After Kny.

Fig. 530A, G). The lower parts of these thallus lobes are revolute and resemble stems, whilst the upper parts repeatedly branch dichotomously forming a stellate 'umbrella' or disc. Antheridia and archegonia occur on different plants. As in many other bryophytes, sex is determined by sex chromosomes (p. 326).

The margins of the thallus forming the stalk of the gametangiophore are revolute and enclose a channel containing rhizoids (Fig. 530B, C) which conduct water upwards by capillarity, a mechanism which evidently works efficiently.

The antheridiophores terminate in a horizontal, eight-lobed disc (Fig. 530A) formed by three successive dichotomous divisions; each of the several antheridia is sunk in a flask-shaped cavity which opens to the upper surface by a narrow canal (Fig. 530C). These cavities are separated by tissue containing air chambers.

The liverwort antheridium (Fig. 530E) generally arises as follows (Fig. 530D): A superficial cell divides into transverse disc-shaped segments, each of which divides into four cells by intersecting vertical walls. The resulting quadrants of this cylindrical structure now divide by tangential walls forming peripheral jacket cells and inner cells which give rise to spermatogenous tissue. The antheridia open and discharge after rain, when the walls swell and become mucilaginous. The spermatozoids (Fig. 530F) collect in the water (rain or dew) which is retained on the upper side of the disc by its lobed margin. The spermatozoids may be splashed to a height of several centimetres.

The archegoniophores (Fig. 530G) closely resemble the antheridiophores in their early

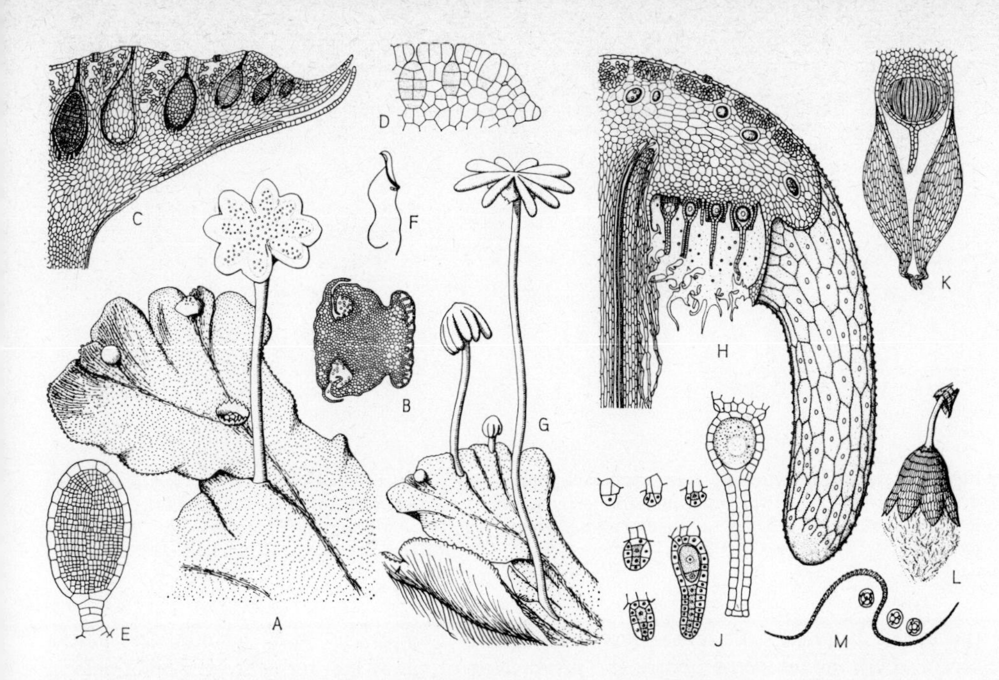

Fig. 530. Reproduction of *Marchantia polymorpha*. **A.** ♂ plant with gemma-cups and antheridiophore; the dots on the upper surface of the thallus are air-pores. ×1.5. **B.** T.S. of the stalk of an antheridiophore just below the 'umbrella' (disc). ×13. **C.** L.S. of an antheridiophore. ×18. **D.** Development of the antheridia. ×160. **E.** L.S. of a nearly mature antheridium. ×160. **F.** Spermatozoid. ×400. **G.** ♀ plant with archegoniophores. ×1.5. **H.** L.S. of an archegoniophore with the perichaetium behind the row of archegonia. ×25. **J.** Development of the archegonium. ×160. **K.** L.S. of a young sporogonium still enclosed in the archegonium and surrounded by a perigynium. ×35. **L.** Dehisced sporogonium liberating spores and elaters, with the remains of the archegonium at the base of the seta. ×10. **M.** Spores and elater. ×160. **A., C., D., E., G., H., K., L.** and **M.** After Kny. **F.** After Ikeno. **J.** After Durand.

development. The archegonia develop in eight radial rows of which the two lying nearest the upper side of the stalk are further apart than the others. The margin of the young disc grows downwards so that the rows of archegonia are carried underneath, and the originally acropetal sequence of archegonia becomes basipetal. Finally, the tissues lying between the rows of archegonia and those alongside the two outermost rows grow out to form the nine rays of the disc.

In the development of the archegonium a superficial cell divides into a basal cell, which gives rise to the stalk, and an outer cell which is divided by three longitudinal walls into three peripheral cells and one axial cell; this latter divides transversely into a cover cell and a central cell. The outer cells (excluding the cover cell) ultimately form the neck and venter, whilst the central cell forms the four to eight neck canal cells, the ventral canal cell and the egg cell.

Fertilization takes place during rain, the raindrops splashing the water containing the spermatozoids from the ♂ to the ♀ umbrellas. The epidermal cells of the latter are papillose and form a superficial capillary system through which the spermatozoids are conducted to the archegonia which attract them chemotactically.

A few days after fertilization the egg cell begins to develop into a multicellular embryo and subsequently into a very short-stalked, small, ovate, chlorophyllose sporogonium (Fig. 530K, L). As in *Anthoceros* (p. 548), the embryo is exoscopic, i.e. of the two cells formed at the first division of the zygote the upper (directed towards the neck) develops into the round capsule and the lower into the foot and, in this case, the stalk or seta of the capsule as well (Fig. 530L). The early stages of development differ somewhat in the various families and genera. Periclinal walls divide the capsule into outer and inner cells (Fig. 530K), the latter forming the archesporium, a multicellular sporogenous tissue. Each archesporial cell divides into one narrow and one broad daughter cell. In certain genera

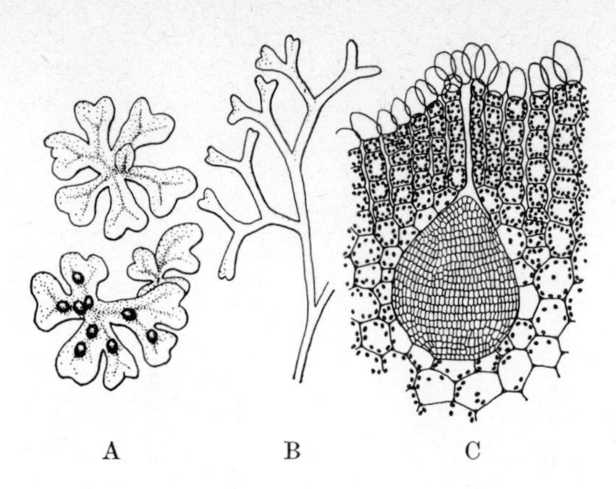

Fig. 531. *Riccia.* **A.** *Riccia glauca*, the lower plant with sporogonia. ×2. **B.** *R. fluitans*, submersed water form. ×2. **C.** *R. glauca*, T.S. of thallus with antheridium. ×125. **C.** After Kny.

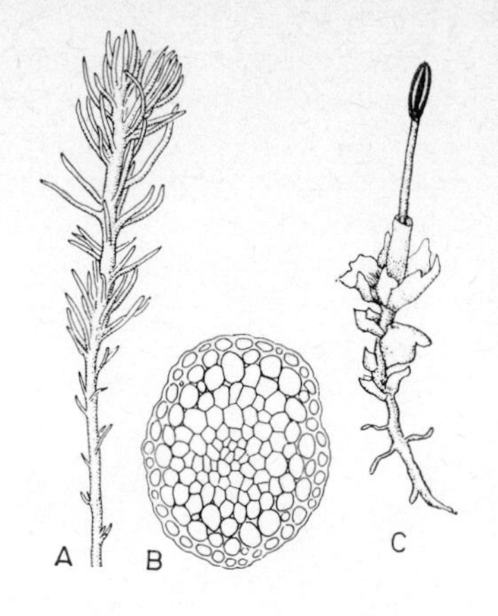

Fig. 532. Calobryales. **A.** *Takakia lycopodioides.* ×6. **B.** T.S. of its stem. ×100. **C.** *Haplomitrium hookeri.* ×6. **A.** and **B.** After Schuster. **C.** After K. Müller.

the broad daughter cells form spore mother cells directly, but in *Marchantia* and others several divisions intervene before the production of spore mother cells, in which meiosis and spore formation take place. The narrow cells develop as undivided, thin-walled, fibre-like tubes with spiral ribbon-like thickenings on the walls; these cells are the elaters (Fig. 530M), which after the dehiscence of the capsule undergo movements by a hygroscopic cohesion mechanism, loosening and scattering the spores (Fig. 327).

In *Marchantia* the capsule wall has a single layer of cells with annular thickenings except at the apex where the wall is bistratose and dehiscence of the capsule begins by the disintegration of the cap, the wall splitting into a number of recurved teeth. At first the capsule is covered by the expanded archegonial wall (Fig. 530K); later this is ruptured by the elongating seta and remains as a sheath at the base. Outside this is a four- or five-lobed, membranous perigynium which develops as a sac-like envelope prior to fertilization from the stalk of the archegonium (Fig. 530H, K). Finally, each radial row of archegonia is enclosed in a further outgrowth of the thallus, a finely toothed, slot-like perichaetium (Fig. 530H). The capsule liberates several hundred thousand spores (Fig. 530L, M) which germinate as very short photosynthetic filaments (protonemata); at first these have a wedge-shaped apical cell, but subsequent growth is more complex, ultimately resulting in mature thalli. (See pp. 80–1.)

Conocephalum conicum, also frequent on rocks and moist earth, resembles *Marchantia* in thallus structure, but has simpler air-pores and no gemma-cups.

The smallest thallose liverwort, *Monocarpus sphaerocarpus* (Australia), has a much reduced thallus bearing a single spherical sporogonium enclosed in a relatively well-developed sheath containing air chambers which indicate its relationship to the Marchantiaceae.

The family **Ricciaceae** shows a simpler structure (Fig. 531). The thallus generally forks at frequent intervals so that small rosettes are formed (Fig. 531A). In some the thallus is chambered with simple pores, but in the majority the upper surface divides into vertical rows of cells which terminate in a large, colourless cell (Fig. 531C). Both gametangia and sporophyte (which lacks both foot and seta) are embedded in the thallus (Fig. 531C). Most *Riccia* species live on soil (Fig. 531A); the dichotomous, ribbon-like *R. fluitans* is submersed in water (Fig. 531B) while *Ricciocarpus natans* floats like duckweed on the water surface.

Order 3: **Calobryales**. The erect stems of the Calobryales, which lack rhizoids, bear three rows of identical leaves. The capsule opens with four valves. Whilst the Asiatic *Takakia* (Fig. 532A) has cylindrical leaves, *Haplomitrium* (also found in Europe) (Fig. 532C) has flat leaves several cells thick at the base. The stems of *Takakia* possess a thin-walled conducting strand (Fig. 532B).

Order 4: **Jungermanniales.** These are predominantly tropical, mostly small liverworts, living on the ground or on tree-trunks, or even on leaves in tropical forests. The numerous families with 9000 or so species comprise some 90 per cent of the Hepaticae. The simplest forms, e.g. *Pellia epiphylla*, frequent on moist soil, have a thallus which is superficially like that of *Marchantia* (but without its internal differentiation) or is strap-shaped and dichotomous like that of *Riccia*, e.g. *Metzgeria furcata* (Fig. 533A, B), found on tree-trunks and rocks. Others have a broad, lobed thallus with a midrib and leaf-like lateral appendages, e.g. *Blasia pusilla* (on soil; Fig. 533C).

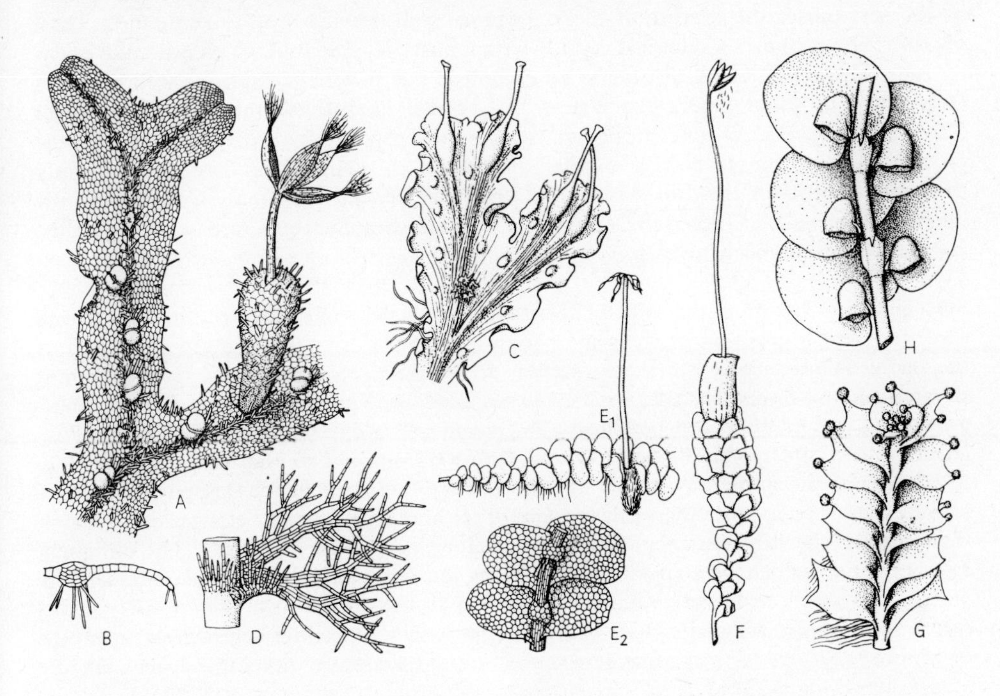

Fig. 533. Jungermanniales. **A.** *Metzgeria conjugata* (under side) with one ♀ and several ♂ branches; tufts of elaters at the tips of the four valves of the capsule; base of the seta enclosed in a perichaetium. ×15. **B.** *M. conjugata*, T.S. of the thallus. ×30. **C.** *Blasia pusilla* with flask-shaped gemmae receptacles and auricles on the upper side of the thallus containing *Nostoc*. **D.** *Trichocolea tomentella*, leaf and underleaf. ×7. **E.** *Calypogeia trichomanis*. **E_1**, Plant from above with marsupium and mature sporogonium. ×2; portion from below with four leaves and two underleaves. ×6. **F.** *Scapania undulata* with perianth and mature sporogonium. ×2. **G.** *Lophozia ventricosa* from above with gemmae on the tips of the leaves. ×10. **H.** *Frullania dilatata* from below with water-retaining postical leaf lobes. ×25. **A.** and **C.** After Schiffner. **B.** After Lindberg. **D., E.** and **F.** After W. J. Hooker. **G.** and **H.** After K. Müller.

The majority are clearly differentiated into a prostrate or ascending stem and leaves, the latter being one cell thick with no midrib; the laminae are obliquely inserted in two lateral ranks on the stem (Fig. 533D–H), which has no conducting tissues.

The lateral leaves are frequently divided into upper (antical) and lower (postical) lobes (Fig. 533F). In some species of habitats liable to periodic drought the postical lobe forms a sac which retains water by capillarity, e.g. *Frullania tamarisci* (Fig. 533H), a delicately branched, brown liverwort frequent on rocks and tree-trunks. In most genera there is a further row of smaller central leaves, the amphigastria or underleaves, differing in shape from the others (e.g. in *Frullania, Calypogeia*, Fig. 533E_2, H). The formation of three rows of leaves depends on the presence of a tetrahedral apical cell, one of the three dividing faces of which forms smaller leaves, or in species without underleaves none at all. Side branches originate beside the leaves.

Unlike other liverworts, the Jungermanniales have no stomata or air pores.

Development of the antheridia and the early development of the embryo deviate somewhat from the marchantiaceous type. The archegonia of the thallose forms are surrounded by a perichaetium (p. 546), of the foliose by variously modified leaves forming a 'perianth' (Fig. 533A, F) which consist of three united leaves.

The protonema differs from genus to genus, but usually has very few cells; in *Metzgeriopsis pusilla*, however, it is expanded and represents the main vegetative organ, producing minute, few-leaved plantlets which serve merely as the bearers of sex organs.

As usual, the lower of the two cells arising from the zygote becomes the foot of the sporogonium whilst the upper gives rise to the capsule and (in contrast to *Marchantia*) the long, fragile seta. The sporogonium is already fully developed before the rapid elongation of the seta causes the rupture of the archegonial wall, leaving it as a membranous basal sheath. The seta bears a spherical capsule with a many-layered wall which generally opens by four valves (Fig. 533A, F); it has no columella but produces elaters as well as spores (Fig. 327). In some genera the elaters are attached in tufts at the tips of the valves (*Metzgeria*, Fig. 333A) or centrally at the base of the capsule (*Pellia*). The cells of the capsule wall are provided with annular or rod-like thickenings, or they are uniformly thickened except for the thin outer wall. Dehiscence depends on cohesion of water in the lumina to the walls of these cells; as its volume diminishes, the thin, outer walls are pulled inwards (cohesion mechanism, p. 379).

Systematics. The great majority of the Jungermanniales can be divided into two groups according to the position of their sporogonia. (1) In the Anacrogynae the apical cell of the thallus is not involved in the formation of the archegonium, and the sporogonium is situated on the dorsal surface; its base is enclosed in many genera by a sheathing outgrowth of the thallus, the perichaetium (involucre). Representatives include thallose genera (*Pellia*, *Metzgeria*, Fig. 533A) and forms transitional to the foliose (*Blasia*, Fig. 533C). (2) In the numerous Acrogynae the archegonium (and hence the sporogonium) is formed from the apex of the stem (or one of its branches), further elongation of which then ceases. The dorsiventral foliose forms belong here (Fig. 533D–F). The obliquely inserted leaves which overlap like roofing tiles are very variable in shape: simple (Fig. 533E), two or more lobed (Fig. 533G), divided almost to the base (Fig. 533F), or dissected into slender filaments (Fig. 533D). Some species reproduce vegetatively by means of gemmae (Fig. 533G). In some genera the young sporophyte develops inside a sac-like 'marsupium' (Fig. $533E_1$).

The Jungermanniales give the impression of a series progressing from thallose to foliose forms, but it is equally possible to regard the leafy liverworts as primitive and to derive the thallose from them by the fusion of the overlapping leaves, by broadening of the axis and by displacement of the archegonium from the apex to the dorsal side. Both the anacrogynous Jungermanniales and the Marchantiales could have arisen in this way.

Order 5: **Anthocerotales.** These include only a few calcifuge species. The sporogonium has a more complex internal structure than that of other liverworts.

The gametophyte is a disc-shaped, lobed thallus a few centimetres in diameter, firmly attached to the soil by rhizoids; its structure is very simple (Fig. 534A). In contrast to all other Bryophyta, each cell contains only one large basin-shaped chloroplast with a pyrenoid; it resembles the chloroplast of the Chlorophyceae and tempts one to speculate that the Anthocerotales must be of great phylogenetic age. The lower surface of the thallus bears stomata with two bean-shaped guard cells (B); in fact, the intercellular cavities behind them are filled with mucilage and are generally colonized by *Nostoc* (Fig. 534B). Below the upper surface are numerous antheridia inside cavities which are closed at first and only open later; the archegonia are also sunk in the upper surface of the thallus. As in all the Hepaticae, the fertilized egg cell divides by a transverse wall into two cells, of which the upper (i.e. towards the neck of the archegonium) divides further to form the capsule, while the lower becomes the swollen foot (haustorium, Fig. 534D) which is attached to the thallus by rhizoid-like cells. The unstalked, horn-like sporogonium has a pod-shaped capsule (Fig. 534A) one or more centimetres long; this opens by two longitudinal valves, its central axis being an elongate columella (Fig. 534C, D, c) composed of a few rows of sterile cells. The columella is completely surrounded laterally and apically by a thin layer of sporogenous cells (the archesporium, a) which produce both meiospores and sterile cells which continue to divide (elaters); spore mother cells

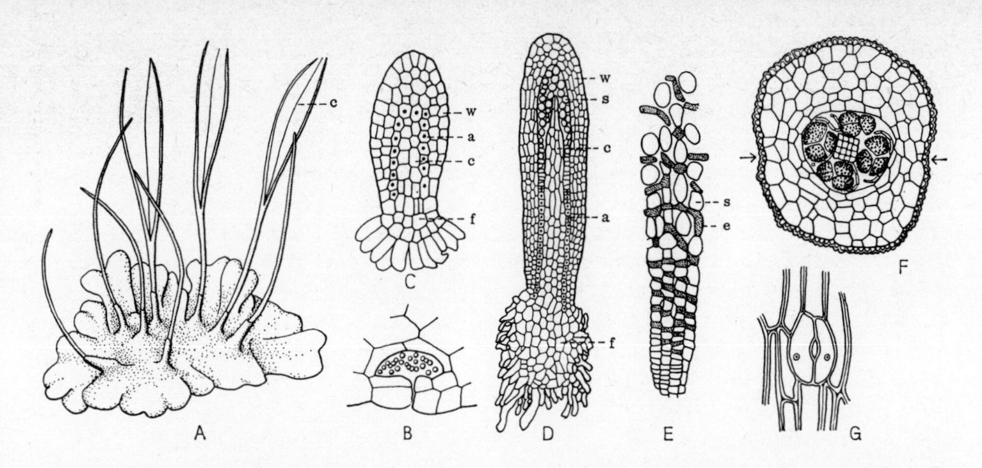

Fig. 534. Anthocerotales. **A.** *Anthoceros laevis*, thallus with immature and dehisced sporogonia; c, columella. ×2. **B.** *Anthoceros vicentianus*, stoma on the ventral surface of the thallus with the air-chamber colonized by *Nostoc.* ×270. **C.** *Anthoceros punctatus*, young sporogonium in L.S. ×130. **D.** *Dendrocerus crispus*, nearly mature sporogonium in L.S.; a, archesporium; c, columella; f, foot; s, spores; s, wall of the sporogonium. ×80. **E.** *Anthoceros punctatus*, unequal cell division in the archesporium; e, elaters; s, spore mother cells. ×100. **F.** *Anthoceros husnotii*, T.S. of the sporogonium with tetrads of spores and columella; → points of dehiscence of the wall of the sporogonium. ×100. **G.** *Anthoceros pearsonii*, stoma on the sporogonium. ×125. **B., C.** and **D.** After Leitgeb. **E.** After Goebel. **F.** After K. Müller. **G.** After Campbell.

and elater cells are always sister cells (Fig. 534E). In contrast to all other liverworts, the parts of the capsule do not mature simultaneously; instead it continues to elongate from a basal meristem. The wall of the sporogonium contains chloroplasts and, unlike other liverworts, bears stomata.

It has been suggested that the Anthocerotales should form an independent class comparable to the Hepaticae and Musci, but the development of the elaters indicates a close affinity with other Hepaticae.

Class II: Musci, mosses

Moss spores germinate as much-branched, positively phototropic, green filaments, the protonema (Fig. 546A), which when abundant form a green felt visible to the naked eye. At first the filaments have abundant chloroplasts and transverse walls perpendicular to the axis (chloronema). Later they develop as filaments with few chloroplasts, appressed to the substrate, and with oblique walls (caulonema). If the illumination is adequate, the caulonemata develop buds (Fig. 546A), generally on short lateral branches; the buds give rise to the moss plant. The caulonema also gives rise to numerous, generally erect, side branches similar to the chloronema (Fig. 546A). The bud first cuts off one or two stalk cells; then the enlarged terminal cell develops oblique walls which delimit a three-sided pyramidal apical cell (Fig. 546B, C); its three faces cut off segments to form the leafy moss plant. If many such buds are formed the moss grows as a sward. The plant is always differentiated into stem and leaves; filamentous branching rhizoids with oblique cross walls attach it to the soil (Fig. 546E). Lateral branches arise below the leaves (cf. Fig. 84A).

The perennial protonemata of the 'luminous' moss *Schistostega pennata* (Fig. 548E, F, H) grow in rock crevices and consist of spherical cells which focus and partly reflect the incident light; it is propagated by multicellular gemmae (Fig. 548G).

Mosses have a remarkable capacity for regeneration (p. 312). Detached fragments of stem or leaf can grow into new plants directly or via protonemata. Many species produce multicellular gemmae along the axis of the leaves or at their tips (Fig. 549J″).

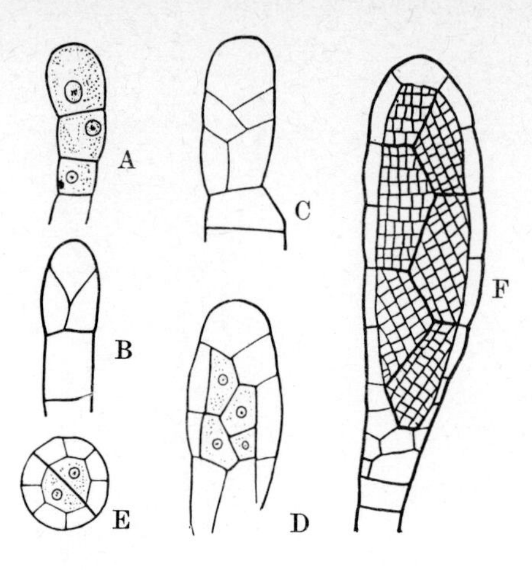

Fig. 535. Development of the antheridium of *Funaria hygrometrica*. **A.** Transverse division of the initial. **B.** Formation of the apical cell. **C.** Division of the apical cell. **D.** Division into wall and spermatogenous initials. **E.** The same in T.S. **F.** Nearly mature antheridium. **A.** to **E.** ×600. **F.** ×300. After Campbell.

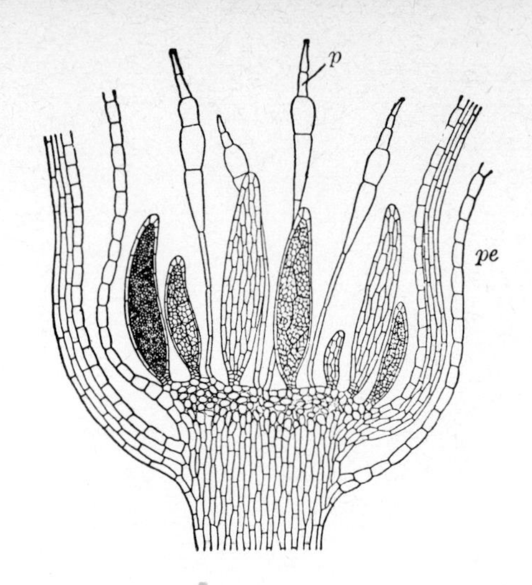

Fig. 536. L.S. of a group of antheridia of *Mnium hornum*, antheridia partly in side view, partly in L.S. p, paraphysis; pe, perianth. ×100. After Harder.

Mosses are distinguished from the dorsiventral, foliose, acrogynous Jungermanniales by the spiral arrangement of their leaves which are rarely borne in two rows (e.g. *Fissidens*, Fig. 549C). In those mosses with prostrate stems the leaves, although arising in spirals, may point to one side, or be parted in two rows, so that they have an upper and a lower side, but in a manner quite different from that in liverworts. The leaves of *Schistostega* are inserted obliquely and spirally at first, but as they develop they come to lie in a vertical plane perpendicular to the incident light (Fig. 548A, B).

The sex organs are borne in groups at the tip of the main axis or of small lateral branches, surrounded by the upper leaves, which are frequently modified to form a perianth (Fig. 536 pe).

The ♂ and ♀ sex organs are either borne on different plants (dioecious) or on the same plant (monoecious); in the latter case ♂ and ♀ organs may occur together in the same perianth or separately in different perianths. Numbers of multicellular hairs (paraphyses; Fig. 536 p), often with a spherical terminal cell, commonly occur between the gametangia. In many dioecious species the ♀ plants have a more compact growth and larger leaves than the ♂, and the groups of archegonia are tightly enclosed in bud-like perianths, whereas the ♂ perianth is expanded (e.g. in *Splachnum luteum*, in which even the protonemata show sexual dimorphism). In some the minute ♂ plants have only a few leaves below the antheridia; indeed, in *Buxbaumia aphylla* the ♂ only bears a single, concave, bladder-like, colourless leaf, whereas the ♀ has numerous leaves (Fig. 537), thus showing secondary sexual characters. Even the spores which produce the dwarf ♂ plants may be smaller than those producing the ♀ (e.g. in *Macromitrium*); thus even in the mosses there is a sort of heterospory, which in the ferns has arisen repeatedly and to a much greater extent.

The stalked antheridia and archegonia differ from those of other Archegoniatae in their complicated development from apical cells. In the antheridium the apical cell is wedge-shaped with two segmenting faces (Fig. 535B); each segment cut off (C) divides into peripheral jacket cells and an inner cell (D) which later forms spermatogenous tissue. The early stages of development of the archegonium (Fig. 538) are the same as in the antheridium (up to stage C in Fig. 535); subsequently, the apical cell with two segmenting faces becomes the true archegonial mother cell and forms periclinal walls separating an inner tetrahedral cell from three peripheral cells. After formation of a transverse wall in the tetrahedral cell there are now a cover cell (d), a central cell (dotted) and jacket cells

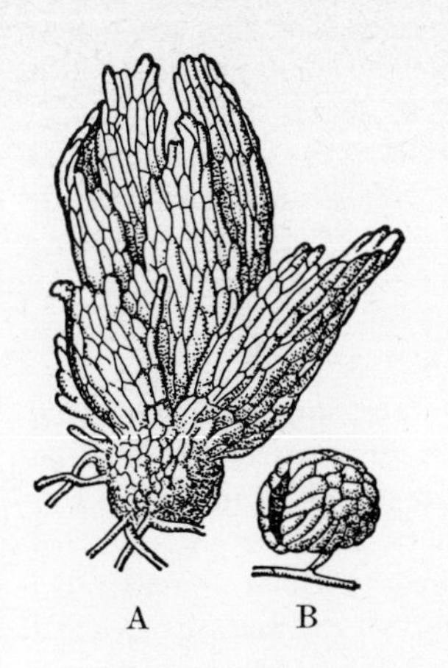

Fig. 537. *Buxbaumia aphylla.* **A.** ♀. **B.** ♂. ×35. **A.** After Dening. **B.** After Goebel.

Fig. 538. Development of the archegonium of *Mnium undulatum.* **A.** Stalk still without archegonial initial. **B.** Archegonium (a) initiated by the formation of the central cell (dotted), cover cell (d) and jacket cells. **C.** Central cell divided into egg cell and ventral canal cell; st, stalk. **D.** Numerous ventral canal cells cut off from the cover cell. ×250. After Goebel.

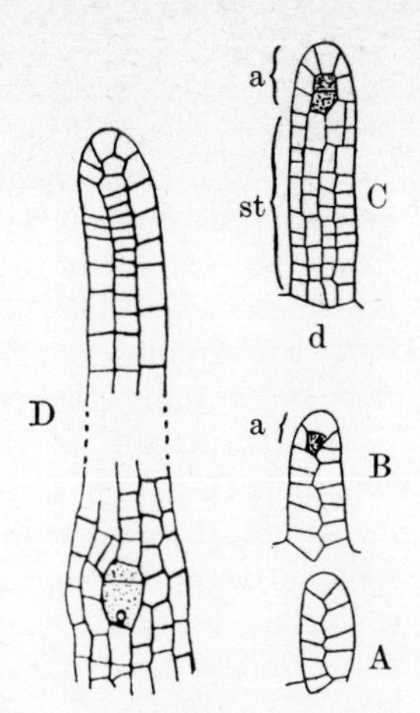

(Fig. 538B). The central cell produces the egg cell and the ventral canal cell (C); of the numerous (ten to thirty or more) neck canal cells the lower are derived from the central cell, whilst the upper ones originate by segmentation of the three-sided cover cell (D).

The antheridia open at the apex when the contents of the upper cells become mucilaginous and swell, rupturing the cuticle. The uppermost cells of the archegonial neck rupture in the same way and roll back in the form of a funnel, or often as four lobes. Spermatozoids and egg cells both contain a very few minute chloroplasts.

After fertilization by the chemotactically attracted spermatozoids (p. 382), the zygote first divides transversely and then forms an elongated embryo of which the upper cells typically have oblique walls, with a wedge-shaped apical cell with two dividing faces (Fig. 539A, B). This cuts off segments, which divide further, on both sides. In those segments which give rise to the capsule a radial wall is formed at right angles to the wall of each segment so that a transverse section of the embryo shows four quadrants; these divide by periclinal walls into outer cells (amphithecium) and inner cells (endothecium) (a and e in

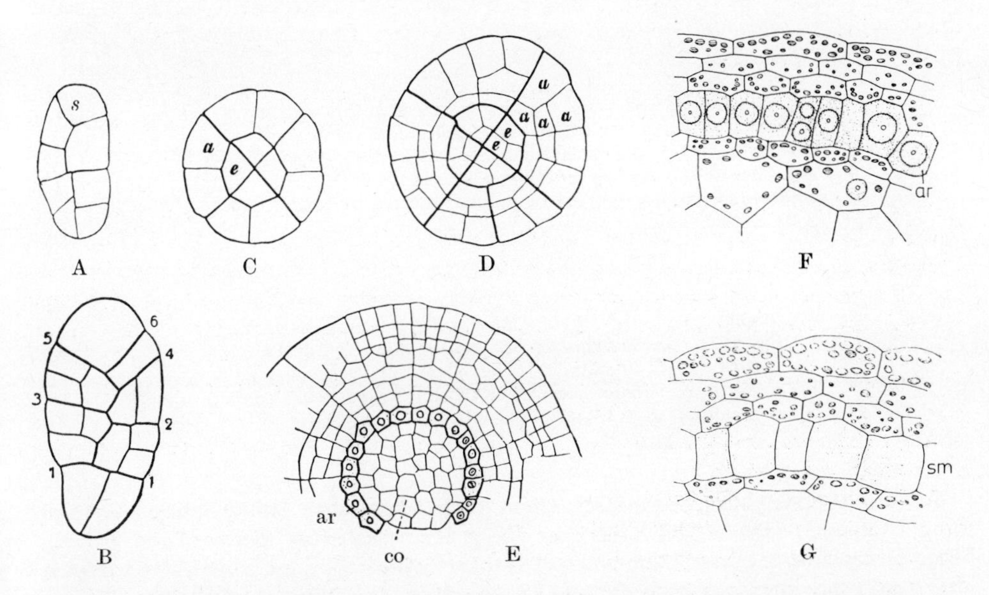

Fig. 539. Development of the sporogonium of *Funaria hygrometrica.* **A.** and **B.** In L.S.; early divisions of the zygote; s, apical cell. **C.** and **D.** In T.S. (**C.** division into endothecium e and amphithecium a. **D.** further divisions.) **E.** Older sporogonium; the outermost layer of the endothecium forming the archesporium ar, distinct from the columella co. **F.** and **G.** T.S. through the archesporium (ar) and the still united spore mother cells (sm). **A.** to **E.** ×400. After Cambell. **F.** and **G.** ×350. After Sachs.

Fig. 539C, D). The outermost layer of the endothecium generally becomes the archesporium (E, F ar) which forms only spore mother cells (G sm); there are no sterile cells producing elaters. Each spore mother cell forms four haploid spores by meiosis. In contrast to the Marchantiales and Jungermanniales, the inner cells of the endothecium do not participate in the formation of the archesporium, but generally form a core of sterile tissue, the columella (co in Figs. 539E, 541) which is surrounded by the spore-forming tissue (s in Fig. 541). The lowest part of the embryo forms an organ of absorption, the foot (haustorium), which penetrates the often greatly enlarged tissues of the archegonial stalk and in some cases the tissues of the stem.

In the capsule of the moss sporogonium the spore sac thus surrounds a central columella (Figs. 541 s, 542C) which stores water and conducts nutrients to the developing spores; the cells with dense cytoplasm adjacent to the spore sac also function in this way. In the young sporogonium there is an active photosynthetic tissue, bounded by an epidermis, outside the spore sac. In most mosses stomata (usually with two guard cells) occur on the lower part of the capsule (Fig. 102). Near the apex of the mature capsule a ring of several characteristic structures allows the capsule to open and the spores to be scattered. The stalk (seta) carries up the capsule, enabling the ready dispersal of the spores by wind.

The young sporophyte (embryo) is initially surrounded by a jacket (embryotheca) derived from the venter and stalk of the archegonium, or even from the stem tissues. As the sporophyte increases in length the jacket fails to keep pace with it, so that it breaks transversely, the upper part being carried up as the calyptra, whilst the lower remains behind as the vaginula (Fig. 540).

Although the sporophyte may partly depend on the chlorophyllose tissue of the apophysis (Fig. 541) for nutrition and growth, it is greatly dependent on the gametophyte. It is not permissible to speak of this as parasitism since the two participants belong to the same species (cf. p. 292). Nutrition of one generation by another, in this case of the sporophyte by the gametophyte, is frequently encountered amongst plants (e.g. amongst Rhodophyceae, heterosporous ferns, spermatophytes), and is best termed gonotrophy (parental nutrition).

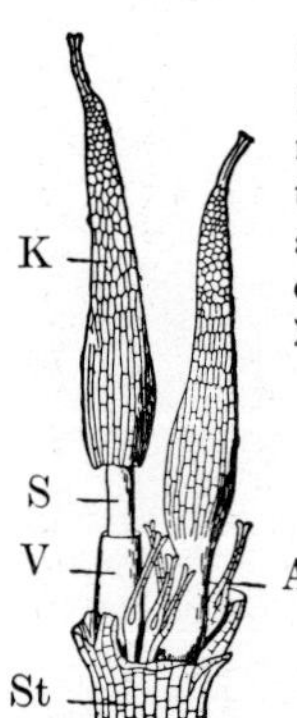

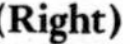

(Left)
Fig. 540. *Pottia lanceolata.* Upper part of a stem (St) with leaves removed. Two archegonia have been fertilized; the embryo of that to the left (S) has raised the upper part of the archegonium as a calyptra (K), leaving the lower part as a sheath (V). The embryotheca to the right is still intact. A, unfertilized archegonia. ×20. After Leunis.

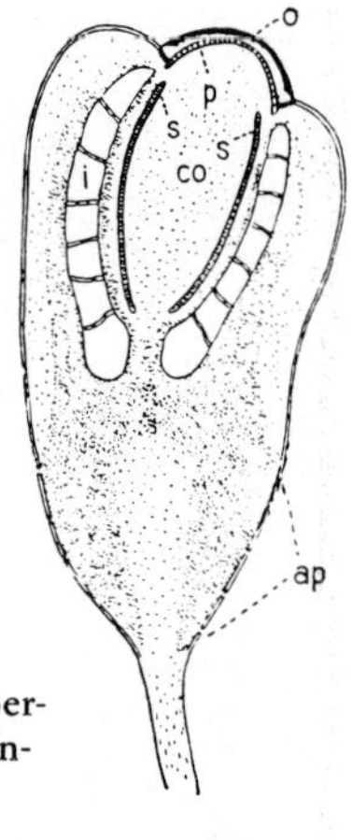

(Right)
Fig. 541. *Funaria hygrometrica.* Sporogonium in L.S. ap, apophysis with stomata; co, columella; i, annular lacuna; o, operculum; p, peristome; s, spore-sac. Closely dotted: tissue containing chloroplasts. ×15. After Haberlandt.

Subclass 1: **Sphagnidae**, bog mosses. These comprise the single family **Sphagnaceae** and the one genus *Sphagnum* which has over 300 generally calcifuge species. These grow in boggy places, forming large hummocks or blankets which continue growth upwards year after year, whilst the lower parts die off and are eventually converted into peat.

The tetrahedral spores germinate only in the presence of a certain mycorrhizal fungus to form a protonema which is not filamentous, but lobed, thallose and one cell thick; it produces a single leafy gametophyte with a tuft of rhizoids at its base (Fig. 542D).

The erect stems have no rhizoids; they always grow in dense hummocks and bear regular groups of lateral branches of which some spread out whilst others grown downwards, closely appressed to the stem (Fig. 542A). The branches form a dense rosette at the apex. Many species, especially those of raised bogs, are coloured brown or bright red by pigments in the cell walls. Each year a branch just below the apex grows as fast as the parental axis, leading to a false dichotomy. As the stems progressively die off from below upwards, adjacent daughter shoots become independent plants.

The cortex of the stem consists of one or more layers of dead, empty cells which readily take up water by capillarity since their longitudinal and transverse walls are frequently perforated by circular pores (Fig. 542E). Similar perforated cells with annular or spiral thickenings on the walls occur in the leaves, which are one cell thick, occupying the meshes between a network of elongate, living, chlorophyllose cells (Figs. 83, 542G, H). This peculiar structure ensures a supply of water and nutrients; plants can hold up to twenty times their dry weight of water in this way. The leaves have no midrib, the axes no central strand.

Certain branches of the apical rosette are distinct in form and colour; these produce the sex organs. In the axils of the leaves of the ♂ branches are long-stalked, spherical antheridia (their gametes, observed in 1822, were the earliest known plant spermatozoids); the ♀ branches bear terminal archegonia which do not develop from an apical cell (thus resembling those of the Hepaticae). The stalk of the sporogonium is very short with a swollen foot; the sporogonium remains enclosed for some time in the embryotheca (p. 552), eventually breaking through this at the apex so that it remains as a vaginula (Fig. 542B aw). In the globular capsule the hemispherical columella is covered by the dome-shaped sporogenous tissue (C, s), possibly indicating an affinity with the Anthocerotales. The archesporium arises not from the endothecium but from the innermost layer of the amphithecium. The expanded foot (haustorium) of the mature sporogonium is embedded in the dilated apex of a prolongation of the stem (pseudopodium, ps), which develops after fertilization. Excess air pressure in the capsule forces off the lid, discharging the spores more than a decimetre.

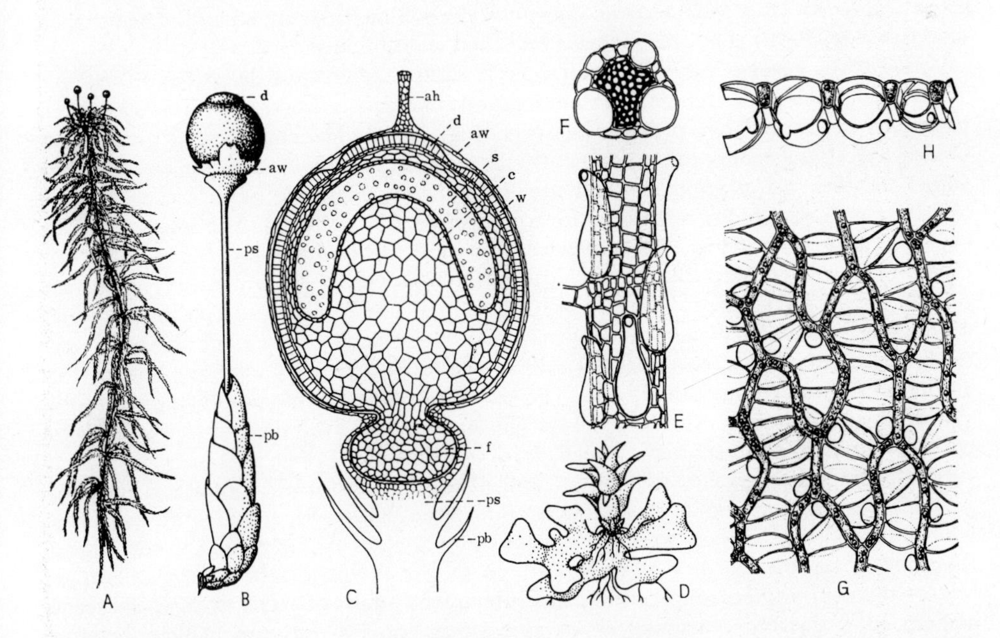

Fig. 542. *Sphagnum.* **A.** *S. acutifolium*, plant with sporogonia. $\times\frac{2}{3}$. **B.** *S. squarrosum*, mature sporogonium at the apex of a branch; pb, perichaetial leaves; ps, pseudopodium; aw, vaginula; d, operculum. ×10. **C.** *S. acutifolium*, L.S. of young sporogonium; f, foot, w, wall; c, columella; s, spores; ah, neck of the archegonium. ×17. **D.** *S. acutifolium*, protonema with a plantlet. ×100. **E.** *S. molluscum*, part of a defoliated stem with flask-shaped water storage cells. ×100. **F.** The same in T.S. ×100. **G.** *S. acutifolium*, the cell network of a leaf in surface view; large water storage cells with spiral thickenings and pores; between them narrow chlorophyllose cells. ×300. **H.** The same in T.S. ×300. **B.** and **D.** After W. P. Schimper. **C.** After Waldner.

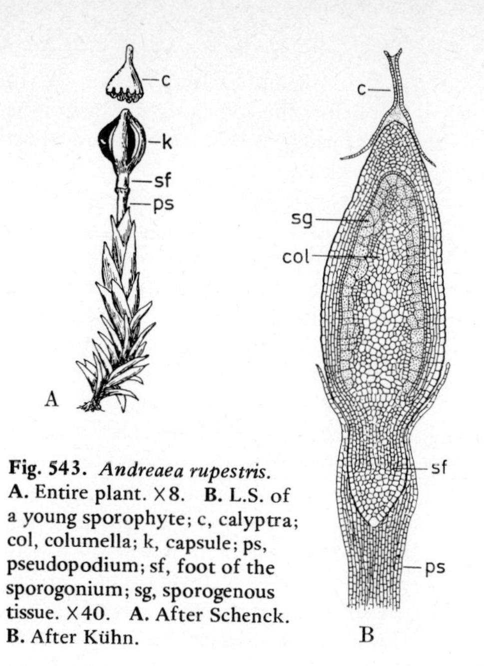

Fig. 543. *Andreaea rupestris.* **A.** Entire plant. ×8. **B.** L.S. of a young sporophyte; c, calyptra; col, columella; k, capsule; ps, pseudopodium; sf, foot of the sporogonium; sg, sporogenous tissue. ×40. **A.** After Schenck. **B.** After Kühn.

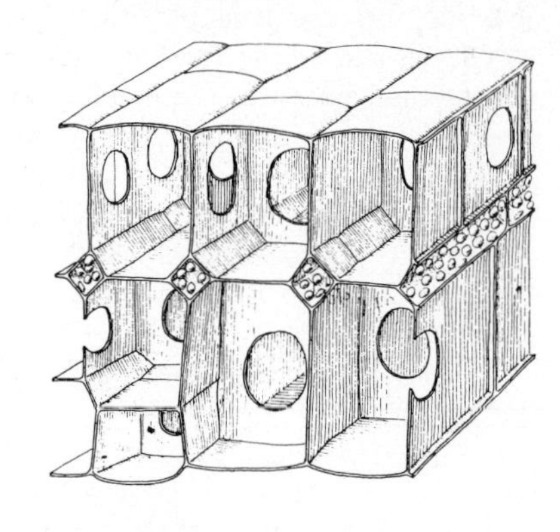

Fig. 544. *Leucobryum glaucum*, leaf structure. Two layers of empty cells with large pores connecting them; between them, small, elongated photosynthetic cells. ×350.

Subclass 2: **Andreaeidae.** The only family, **Andreaeaceae**, has three genera. *Andreaea* has 120 species forming small, dense, dark-brown cushions on non-calcareous rocks in the Alps, the Arctic and Antarctic. As in *Sphagnum*, the sporogonium is raised up on a pseudopodium formed from the archegonial stalk. The capsule, initially covered with a conical calyptra opens by four longitudinal splits into four valves which remain united above and below (Fig. 543A). As in *Sphagnum* the columella is capped by the bell-shaped spore-sac (Fig. 543B). The protonema is branched and ribbon-shaped.

Subclass 3: **Bryidae.** Here the gametophyte shows the greatest diversity of form and the highest levels of differentiation amongst the mosses; however, in a few cases it is practically restricted to a protonemal stage (Fig. 537B). The stems either grow erect, bearing apical archegonia and subsequently the stalked capsules (acrocarpous mosses, Fig. 546E), or they are plagiotropic and usually pinnately branched, with archegonia and capsules on short side branches (pleurocarpous mosses, Fig. 549M). The stems generally have a central strand (Fig. 546H) which in the most highly developed forms (e.g. *Polytrichum*) shows considerable histological differentiation (p. 83). In some genera the central strand gives rise to leaf traces (Fig. 546H) and hence to a continuous water conducting system. The leaves are one cell thick but generally have a midrib several cells thick (Fig. 546J, K, L). The marginal cells of the leaf often form a definite border (Fig. 546K), or the margin may bear teeth. The leaf of *Leucobryum* has both living green cells and dead water storage cells (Fig. 544). The upper surface of the leaf of *Polytrichum* bears longitudinal photosynthetic lamellae rich in chloroplasts (Fig. 545).

The capsule also reaches its highest level of organization in the Bryidae. The sporogonium consists of a flexible stalk or seta (Figs. 546E, 549B–P), of which the basal foot is embedded in the tissues of the female parent, and of a radial (Fig. 546E) or dorsiventral (Fig. 549O) capsule which is initially covered by the calyptra (formed from the upper part of the embryotheca; Fig. 540K); this subsequently falls off. The neck of the archegonium soon shrivels, remaining as an apical beak on the calyptra, which therefore consists not of diploid sporophytic tissue but is haploid and gametophytic (Fig. 526). The calyptra is, however, nourished by the sporophyte and develops several cell layers; in some mosses (e.g. *Polytrichum*) the calyptra is covered with hairs which are homologous with protonemata of limited growth. In certain mosses (e.g. *Funaria*) the young calyptra becomes inflated and forms a reservoir of water for the developing capsule. The uppermost part of the seta, just below the capsule, is the apophysis (Fig. 541) which in some species is strikingly coloured and shaped (e.g. *Splachnum*, Fig. 549H). The columella runs

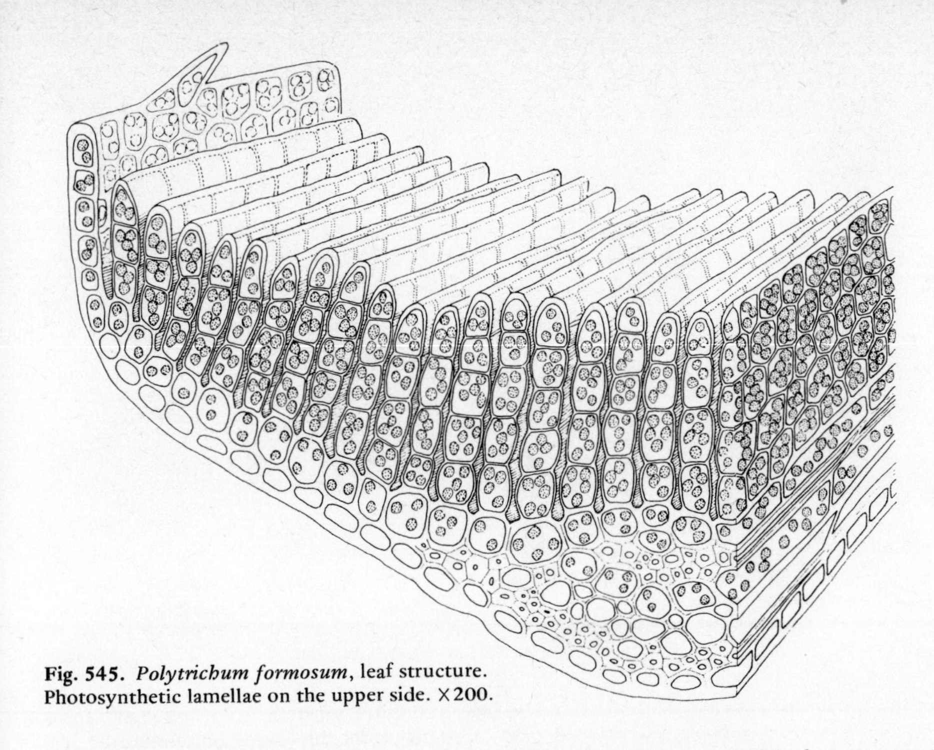

Fig. 545. *Polytrichum formosum*, leaf structure. Photosynthetic lamellae on the upper side. ×200.

the length of the capsule and is surrounded by the hollow-cylindrical spore sac (Fig. 541 s); outside again is an intercellular air space (Fig. 541 i) which is formed from the amphithecium and becomes especially evident at maturity. The meiospores are generally spherical and contain numerous chloroplasts (Fig. 546O).

The upper part of the capsule forms an operculum or lid (Figs. 541, 546M). Below the edge of the lid lies the annulus, a narrow, ring-shaped band of cells containing mucilage; at maturity these swell, facilitating the dehiscence of the lid. Meanwhile the calyptra has fallen. In the majority of Bryidae the lid is thrown off to reveal at the mouth of the urn-shaped capsule a ring of generally tooth-like appendages, the peristome (Figs. 541, 546N); in the minority this is lacking.

A few mosses (Polytrichales, Tetraphidales, Fig. 549J′) have peristome teeth consisting of rows of cells. In all other mosses the peristome develops beneath the lid from the three innermost layers of cells of the amphithecium; Fig. 547 shows a nearly mature peristome in T.S. The tangential walls between layers 1 and 2 are strongly and characteristically thickened whilst the walls between layers 2 and 3 are less strongly thickened. The thickened walls persist when the radial walls and the unthickened parts of the tangential walls have broken down, thus giving rise to a double peristome (Fig. 546N) consisting of the thickened tangential walls and not of whole cells. The outer peristome consists of sixteen transversely ridged teeth inserted on the inner margin of the capsule wall (Fig. 546N), whilst the inner peristome lies immediately inside it and consists of narrow lamellae (cilia) and filaments with transverse ridges on the inner side, and joined together at the base forming a continuous membrane. Two cilia of the inner peristome always stand between a pair of outer peristome teeth (in the group Diplolepideae as opposed to the Haplolepideae with only a single peristome).

The outer peristome teeth exhibit hygroscopic movements inwards and outwards (Fig. 324), closing or opening the capsule (Fig. 546N) according to the weather (generally opening when dry), thus effecting the gradual scattering of the spores. Inclined capsules and those with a wide mouth generally have a well-developed peristome, whilst in those genera with erect, narrow-mouthed capsules it is often reduced (Fig. 548D). Peristome structure varies considerably. The sporangium is greatly reduced in some minute, annual mosses (*Archidium, Phascum, Ephemerum*): lid, annulus and peristome are lacking and the capsule wall opens irregularly by rotting away (cleistocarpy, Fig. 549A).

Young sporophytes can be made to regenerate as protonemata; hence it is possible to

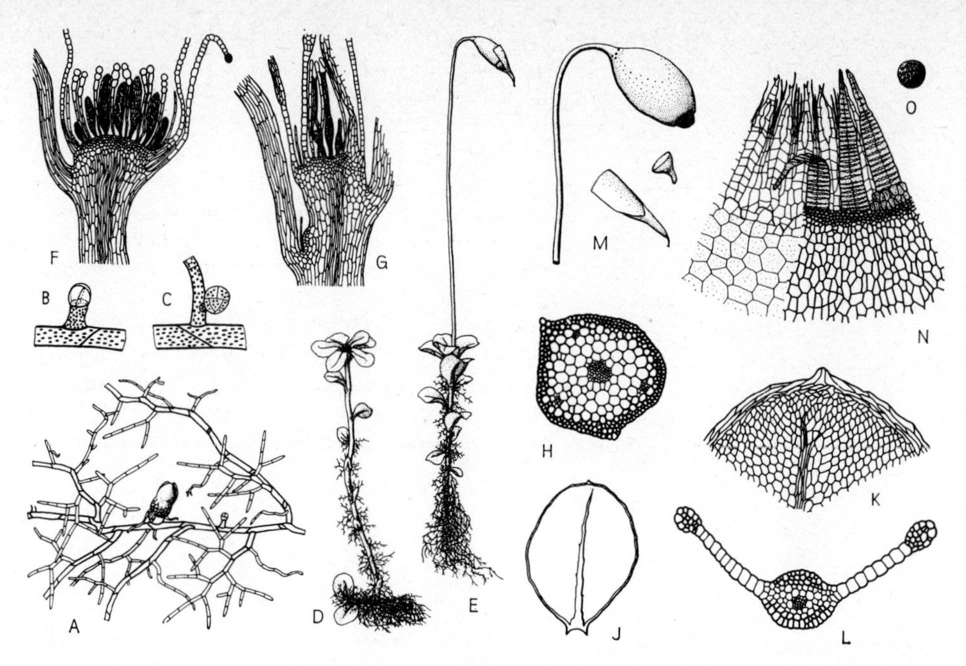

Fig. 546. *Mnium punctatum.* **A.** Protonema with bud. ×20. **B.** Formation of the bud on the protonema; chloroplasts of the upper cells omitted. ×80. **C.** Initial of the three-sided apical cell. ×85. **D.** ♂ plant. Nat. size. **E.** ♀ plant with sporophyte. Nat. size. **F.** ♂ 'inflorescence' in L.S. ×15. **G.** ♀ 'inflorescence' in L.S. ×15. **H.** T.S. of stem with central strand and three leaf traces. ×40. **J.** Leaf (perichaetial leaf). ×4. **K.** Leaf apex. ×25. **L.** T.S. of the lower part of a leaf. ×50. **M.** Mature capsule with operculum and calyptra. ×4. **N.** Peristome; to the left, outer peristome removed; one of the three outer peristome teeth recurved in the dry state. ×30. **O.** spore. ×100.

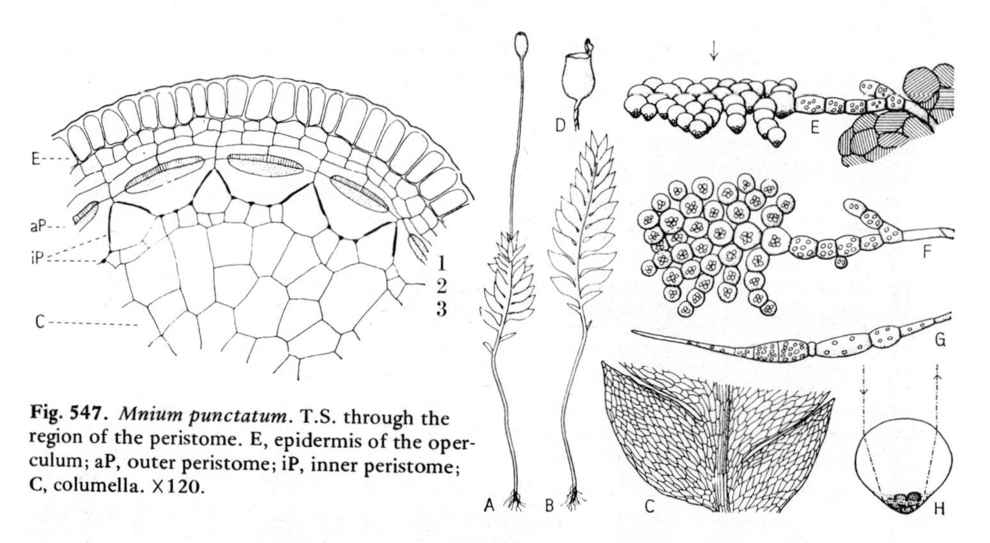

Fig. 547. *Mnium punctatum.* T.S. through the region of the peristome. E, epidermis of the operculum; aP, outer peristome; iP, inner peristome; C, columella. ×120.

Fig. 548. *Schistostega pennata.* **A.** Plant with capsule. ×10. **B.** Sterile plant. ×10. **C.** Part of the latter. ×50. **D.** Empty capsule. ×25. **E.** Protonema of the 'luminous moss' from the side; the arrow shows the direction of incident light. ×150. **F.** The same from above. ×150. **G.** Protonemal gemma. **H.** Path of a ray in a protonemal cell. **A.**, **B.** and **D.** After W. P. Schimper. **H.** After Noll.

produce diploid gametophytes which then produce tetraploid sporophytes. By repeating this process, 16-ploid gametophytes have been produced (p. 36, Fig. 26A–C).

Systematics. Some 15,000 species of Bryidae are classified by characters both of the gametophyte and sporophyte (especially the peristome). The **Archidiales** (with the widespread genus *Archidium*) have an unstalked capsule without columella, lid or peristome (Fig. 549A). The **Dicranales** are distinguished by having sixteen bifid peristome teeth (*Ceratodon purpureus*, cosmopolitan on sandy soils; *Dicranum* frequent in woods, Fig. 549B). The **Fissidentales** have two rows of leaves with a wing projecting from the back of the midrib. In the **Pottiales** the leaf cells and the outer side of the peristome teeth are generally papillose (*Tortula*, frequent on walls, with a long, twisted peristome, Fig. 549D). The leaves of the **Grimmiales** (predominantly in cushions on rocks, e.g. *Grimmia*, Fig. 549E) have a long colourless hair as a continuation of the midrib. The four preceding orders are acrocarpous and have a simple peristome. The **Funariales** (*Funaria hygrometrica*, cosmopolitan, Fig. 549F) and the **Eubryales** (*Bryum*, the largest genus with 800 species, Fig. 549H; *Mnium*, frequent in woods, Fig. 546) are also acrocarpous but with a double peristome. The **Schistostegales** (*Schistostega pennata*, 'luminous' moss, Fig. 548) and the **Tetraphidales** (*Tetraphis*, frequent on rotting wood, Fig. 549J) are of uncertain systematic position. Pleurocarpous growth and a double peristome are shown by the Isobryales, predominantly in woods on the ground or on tree-trunks (*Climacium*, Fig.

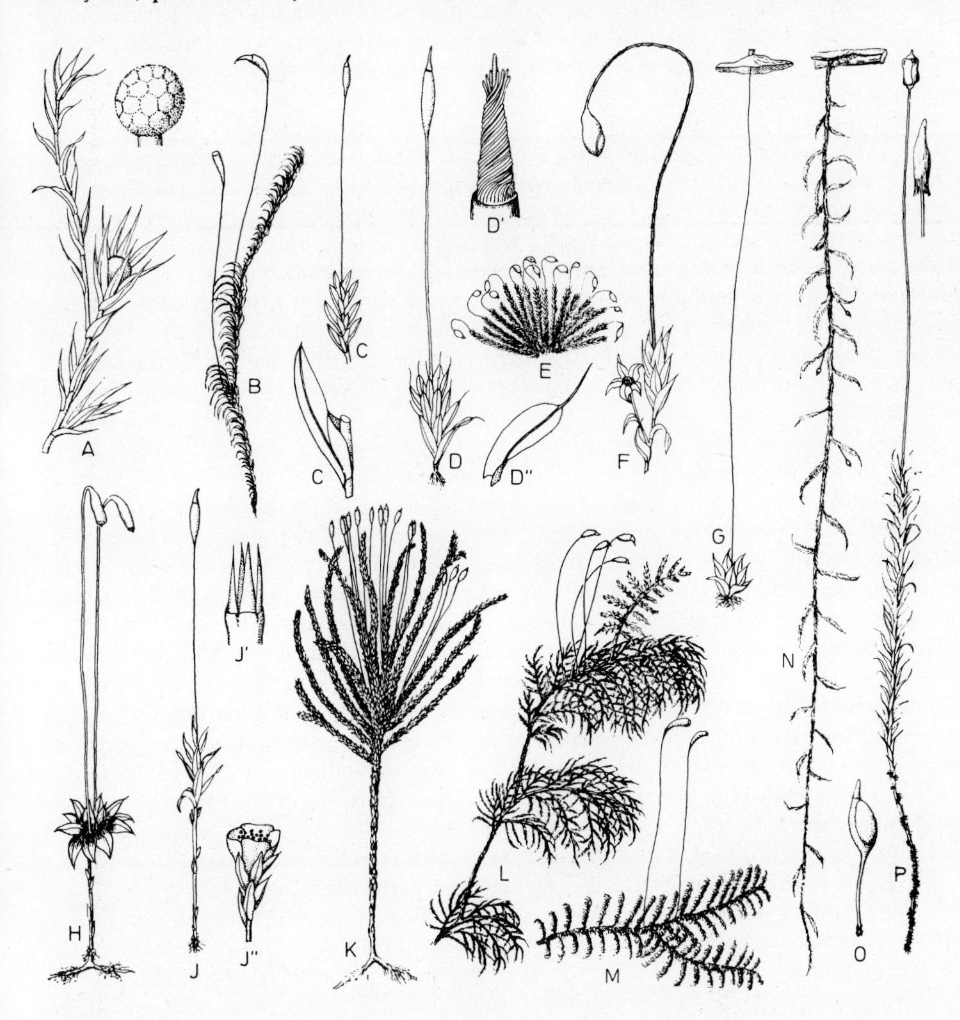

Fig. 549. Bryidae. **A.** *Archidium phascoides*, entire plant (×5) and capsule (×20). **B.** *Dicranum scoparium*, three-year-old plant. Nat. size. **C.** *Fissidens bryoides*. ×4. **C'.** Leaf. ×15. **D.** *Tortula muralis*. ×4. **D'.** Peristome. ×30. **D''.** Leaf with hyaline hair. ×10. **E.** *Grimmia pulvinata*. Nat. size. **F.** *Funaria hygrometrica*. ×2. **G.** *Splachnum luteum*. Nat. size. **H.** *Rhodobryum roseum*. Nat. size. **J.** *Tetraphis pellucida*. ×2. **J'.** Peristome. **J''.** Gemma cup. ×8. **K.** *Climacium dendroides*. Nat. size. **L.** *Hylocomium splendens*, four-year-old plant. ×½. **M.** *Cratoneuron commutatum*. ×½. **N.** *Papillaria deppei*. ×½. **O.** *Buxbaumia aphylla*. Nat. size. **P.** *Polytrichum commune*, with young sporogonium and calyptra. ×½.

549K; and *Thamnobryum*, 'tree mosses'), by the predominantly tropical **Hookeriales**, and by the numerous Hypnobryales (e.g. *Hypnum, Brachythecium, Hylocomium, Cratoneuron*, Fig. 549L, M). The **Buxbaumiales** (*Buxbaumia*, with greatly reduced gametophytes, Figs. 537 and 549O) have three to six rows of peristome teeth which, as in the preceding orders, consist of membranous plates. In contrast, the peristome teeth of the **Polytrichales** (*Polytrichum*, Fig. 549P) are made up of horseshoe-shaped cells. The last two orders are united by some systematists in a separate subclass on account of their peculiar peristome structure.

Distribution and ecology of the bryophytes. Apart from the sea and extreme deserts, there is scarcely a habitat not colonized by bryophytes, but they are most abundant where humidities are high: in woods and mires. In general the liverworts demand more moisture than the mosses. The bryophytes attain their greatest diversity of form in the Tropics. Plant formations consisting exclusively of bryophytes occur in the Arctic (tundra), or occasionally on raised bogs; otherwise they form associations, often in competition with lichens, subordinate to vascular plants. The geographical distribution of bryophytes is generally similar to that of flowering plants; the cosmopolitan distribution of some species (*Marchantia polymorpha, Bryum argenteum, Funaria hygrometrica*) is possibly due to Man.

Water uptake by foliose species takes place, with few exceptions, over the whole surface. The capillary spaces between stem and leaves permit considerable storage of water; in some liverworts this is enhanced by water sacs (Fig. 533H), in a few mosses by specialized water storage cells (Fig. 542G, 544). The capacity of bryophytes to store considerable quantities of water is largely responsible for the buffering effect of forests on the water balance of the landscape. The capillary system also serves as a path of water movement in those bryophytes which have no central strand in the stem, and no internal conduction. Most acrocarpous mosses can conduct water taken up by the rhizoids through the central strand (Fig. 546H). In the Marchantiales the thallus is supplied with water by strands of rhizoids running the length of the underside (rate of transport up to 1 mm per second).

Xerophytic bryophytes are greatly resistant to desiccation and high temperatures, enduring long periods in the dry state (up to fourteen years in *Tortula muralis*) without losing viability. On the other hand spores are much less resistant. Mosses on sunny rocks generally form dense cushions (Fig. 549E), often appearing silvery grey because of their long, dead leaf apices. These hyaline hairs (Fig. 549D′) possibly protect the plant against excess light. Aquatic bryophytes, e.g. *Fontinalis antipyretica*, have reduced external and internal conducting systems; and they are very sensitive to desiccation.

The mosses of temperate climates often show marked annual growth rhythms (Fig. 549B, L).

Many bryophytes are restricted to particular soils; thus the bog mosses prefer acid reactions (pH 3–4), whilst the calcareous tufa-forming mosses (*Eucladium, Cratoneuron*) prefer basic conditions (pH 7–8.5). Others are indifferent (e.g. *Bryum argenteum*, pH 5–8). A few mosses (e.g. species of *Pottia*) are halophytes both on the coast and on inland saline soils.

Bryophytes reach their greatest diversity as epiphytes in the mist and montane forests of the Tropics where they form huge carpets and metre-long festoons (Fig. 549N); a remarkable number of species also grows on leaf surfaces (epiphyllous bryophytes).

Many liverworts regularly have fungal hyphae in their rhizoids or in thallus or stem cells, but it is difficult to say whether this is parasitism or mycorrhizal symbiosis (p. 294). The latter is certainly the case with *Cryptothallus mirabilis*, a liverwort with no chlorophyll which lives underneath mosses.

Survey of the bryophytes. All bryophytes exhibit a characteristic alternation of generations; the haploid gametophyte bears antheridia and archegonia, and supports the diploid sporophyte (the sporogonium) which completely differs in form; the sporogonium is largely nourished by the gametophyte and terminates its development with meiosis and the production of meiospores (Fig. 526).

Fossil bryophytes occur sporadically as far back as the Upper Devonian; thus both thallose and foliose liverworts occur in the Carboniferous in England and highly developed mosses in the Permian of the Saar and S. Russia. The bryophytes are therefore very ancient and appear not to have developed further in the Mesozoic and Tertiary. Most fossil bryophytes are Tertiary and can be assigned to living genera; there are no known primitive forms linking the existing groups. This great dearth of fossils makes the phylogeny of the Bryophyta even more speculative than that of higher plants.

Phylogeny. In their biochemistry (photosynthetic pigments, reserve substances) the bryophytes, like the pteridophytes and spermatophytes, resemble only the Chlorophyceae amongst the thallophytes, and they must have arisen from this algal class, but intermediate forms are unknown, either fossil or living. The bryophytes thus form a phylogenetic 'awkward squad'. To derive them from lower Pteridophyta by reduction is a council of despair which takes no account of their range of characteristics and is not supported by fossil evidence.

Of the two classes of Bryophyta, the Hepaticae are more primitive than the Musci. The Hepaticae exhibit numerous thallose forms, and their elaters, found nowhere else in the Plant Kingdom, unite the various orders. Furthermore, the moss sporophyte, both in comparison with the gametophyte and with the liverwort sporophyte, is more complex, having stomata, conducting and photosynthetic tissues and a highly developed dehiscence mechanism. This generally higher organization of the mosses enables them to grow in more diverse habitats than the predominantly hygrophytic liverworts. The Marchantiales and Sphaerocarpales, the Jungermanniales and Calobryales, the Anthocerotales, and the three subclasses of mosses must have developed independently very early on.

Division 6: **Pteridophyta, ferns and fern-allies**

The Pteridophyta comprise the club-mosses, the horse-tails, the true ferns and a number of extinct groups.

Like the Bryophyta they exhibit heteromorphic and heterophasic alternation of generations (Fig. 550; for deviations see p. 208).

The gametophyte or prothallus (Fig. 550G) is haploid, as in the bryophytes. It generally only lives for a few weeks, reaching at most a few centimetres in diameter and superficially resembling a simple thallose liverwort. Although there are many exceptions, typically it consists of a simple green thallus attached to the soil by unicellular, filamentous rhizoids borne on the underside (Fig. 551); from it arise numerous antheridia and archegonia. As in the bryophytes, fertilization only proceeds in the presence of water, i.e. when the prothallus is wetted.

The fertilized egg cell develops into a diploid asexual generation, the sporophyte (Figs. 550.3, 550.4), which is structurally quite different from that in the bryophytes, and much more highly differentiated. Only the earliest stages of development resemble those of the bryophytes. In most pteridophytes the prothallus is short-lived (persisting for some years only if fertilization is prevented) whilst the young sporophyte becomes an independent, perennial fern plant with roots, stems and leaves (Figs. 550.4, 592, 595, 598).

In all the surviving classes of pteridophytes the fertilized egg cell generally develops, in addition to a foot (haustorium), a root apex, a stem apex and a leaf apex, which give rise respectively to the first root, the stem and the first leaf (cotyledon) of the sporeling (Figs. 550.3, 551, 605).

The possession of roots is characteristic of Pteridophyta and Spermatophyta. In the Spermatophyta the first root (radicle) develops from the 'root pole' of the embryo at the opposite point of the axis to the 'shoot pole', but in the Pteridophyta the first root arises as an endogenous, adventitious structure lateral to the axis (Fig. 605B). The fern sporeling is thus unipolar in contrast to the bipolar spermatophyte embyro. The first root (Fig. 551 w) soon dies and is subsequently replaced by numerous slender, adventitious roots (see p. 177 for the origin of lateral roots).

The root, stem and leaf of most pteridophytes grow by means of apical cells (cf. pp. 90 et seq.; Figs. 91A, 91C, 94). The stem branches (cf. p. 129) dichotomously or laterally (but never from the leaf axils) and bears numerous leaves. The roots have a root cap (Fig. 94); lateral roots arise not from the pericycle but from the innermost cortical layers (p. 175).

The epidermis of the aerial parts is generally furnished with a cuticle (an important prerequisite if living tissues are to be maintained at some elevation above the surface of the land) and stomata (p. 97); however, the epidermis generally still possesses chloroplasts (p. 96). The leaves agree in structure, at least in the most advanced pteridophytes, with those of the spermatophytes. The stems, roots and leaves are traversed by well-differentiated vascular bundles consisting of xylem and phloem, encountered here for the first time in typical form in the Plant Kingdom. The water-conducting elements are lignified tracheids; very rarely (in *Pteridium*) even vessels are present (Fig. 118, p. 111). Supporting fibres are not present. One or more concentric bundles (with internal xylem) are usually present, but other bundle types occur. The entire phylogenetic series of arrangements of vascular bundles shown in Fig. 152 can be traced in the pteridophytes. The lignified tracheids so enhance the conduction of water, and at the same time the mechanical strength of the shoots, that the pteridophytes, unlike the bryophytes, have been able to

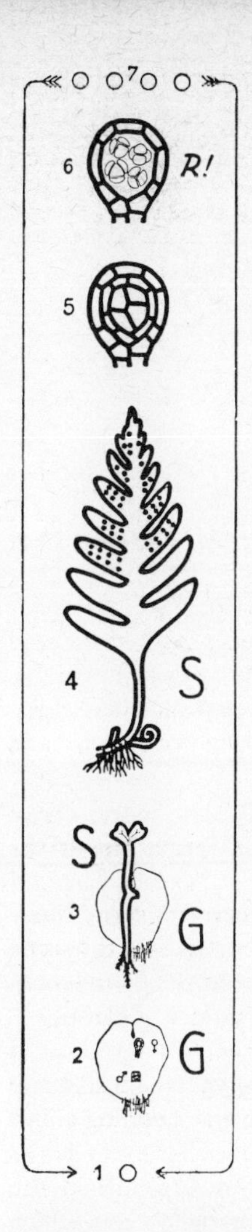

Fig. 550. Diagram of the life-cycle of a fern. G, gametophyte; S, sporophyte. Haploid phase, thin lines; diploid, thick. R! meiosis. **1.** Spore. **2.** Prothallus with ♀ and ♂ gametangia. **3.** Prothallus with young sporophyte. **4.** Sporophyte (reduced) with sporangial sori. **5.** Immature sporangium (greatly enlarged) from a sorus. **6.** Mature sporangium with spore-tetrads. **7.** Spores. After Harder.

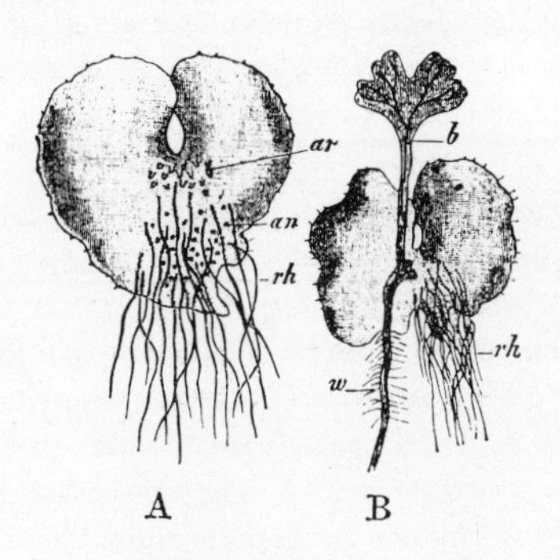

Fig. 551. *Dryopteris filix-mas.* **A.** Prothallus (under side) with archegonia ar, antheridia an and rhizoids rh. **B.** Prothallus with young fern plant; b, first leaf; w, root. ×8. After Schenck.

develop as highly organized and even tree-like land plants. The possession of roots also ensures an adequate supply of water and permits the development of large leaves which can supply the necessary assimilates. Assimilates are transported in sieve tubes (p. 107). Cambial activity and secondary thickening occur only sporadically and to a limited extent amongst living families, but were characteristic of certain extinct groups. The pteridophyte sporophyte, the 'fern', is thus a true cormus, and the Pteridophyta (along with the Spermatophyta) are true Cormophyta.

The spores are borne in sporangia (Figs. 550.6, 552E) on the leaves, or in some cases in the leaf axils, or, in certain very primitive, leafless classes, directly on the shoot axis. Sporangia differ greatly in structure. Leaves bearing them are called sporophylls; these frequently have simpler shapes than the assimilatory leaves (trophophylls) and may be united in special groups which can be termed 'cones' or 'flowers'; these are often raised above the ground, so assisting spore dispersal.

The sporangium contains an archesporium including sporogenous tissue (Figs. 550.5, 552C sp), the cells of which round off, separate and form spore mother cells (usually sixteen), each of which undergoes meiosis to produce four haploid meiospores which are often arranged in tetrahedral tetrads (Figs. 550.6, 552D, E).

Not all the cells of the archesporium are sporogenous; in many pteridophytes one or more layers of cells, rich in cytoplasm, surround the sporogenous tissue; these are the tapetal cells (Fig. 552C t) which supply nutrients to the spores, either by giving up their contents to the inner cells (secretory tapetum) or by their walls breaking down and the protoplasts fusing to form a periplasmodium (amoeboid or plasmodial tapetum) around the spore mother cells. As the young spores are released from their tetrads the

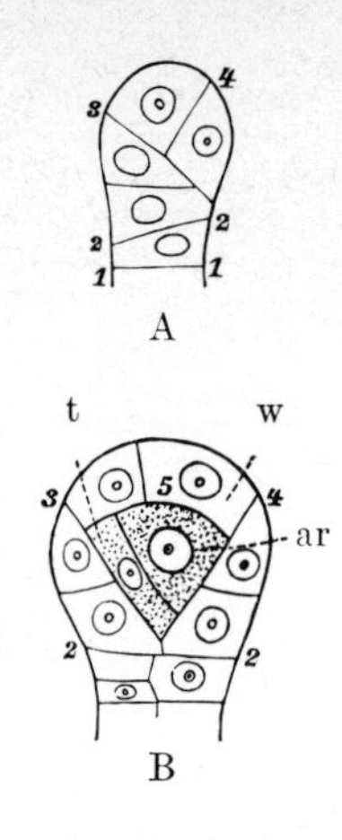

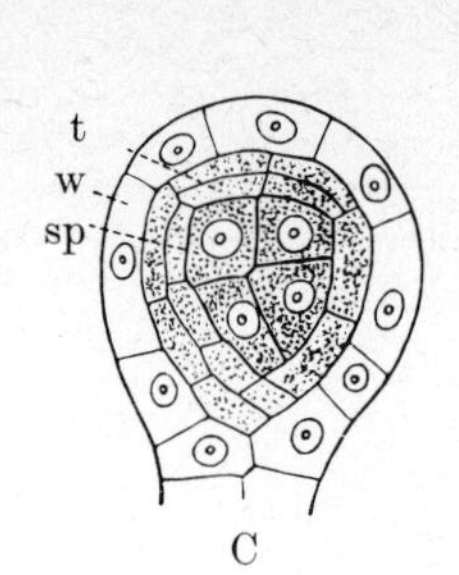

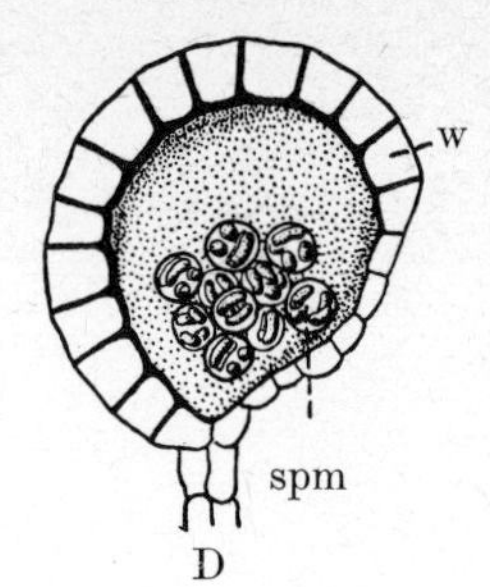

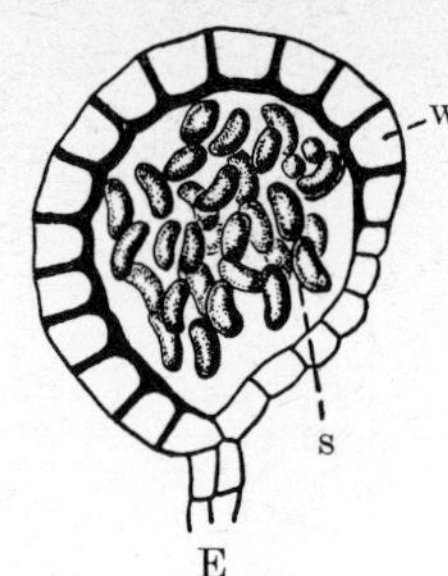

Fig. 552. Development of the fern sporangium. (**A.** to **C.** *Asplenium*; **D.** and **E.** *Polypodium*). **A.** The earliest divisions of a superficial initial cell. **B.** Division into peripheral wall cells (w) and a central (ar) cell (archesporium) which has already cut off a tapetal cell (t); 1 to 5, successive walls. **C.** Differentiation of the archesporium into tapetum (t) and sporogenous tissue (sp). **D.** Annulus cells (w) with thickened walls; tapetum broken down; spore mother cells (spm) forming tetrads of spores. **E.** Mature sporangium with spores (s); stomium cells not shown (cf. Fig. 326). **A.** to **C.** ×300. After Sadebeck. **D.** and **E.** ×200. After Harder.

periplasmodium diffuses between them, its nuclei dividing amitotically. The spores derive nourishment from the periplasmodium, which also takes part in the formation of the spore wall (perispore) and is thus used up (Fig. 552D, E).

The young spores first surround themselves with an extraordinarily resistant membrane, the exospore, within which a thin cellulose layer, the endospore, is laid down. In many cases (probably to be regarded as advanced) the periplasmodium deposits a perispore of various forms around the exospore. The spores almost always lack chlorophyll, but are often coloured brown or yellow by carotenoids.

In most pteridophytes all the spores are alike and on germination give rise to prothalli which bear both antheridia and archegonia; this is probably the primitive condition, but in certain cases the prothalli are dioecious. Such a separation of the sexes has extended in some groups to the formation of two kinds of spores: megaspores (macrospores) with copious reserve substances which arise in megasporangia (macrosporangia), and which germinate to produce relatively large ♀ prothalli; and microspores, borne in microsporangia, which form smaller ♂ prothalli (Figs. 565–7). One can therefore distinguish between homosporous (or isosporous) and heterosporous orders, but this distinction is of little value for the primary division of the Pteridophyta since heterospory has arisen repeatedly in unrelated classes.

Class I: Psilophytatae (Psilopsida)

Order 1: **Psilophytales.** These were the most ancient land plants with vascular bundles and stomata. They appeared at the Silurian/Devonian transition about 400 million years ago and became extinct at the beginning of the Upper Devonian.

The psilophytes are the lowest pteridophytes so far known. The most primitive members (**Rhyniaceae**) had dichotomous stems (Figs. 553A, 554) with a single vascular bundle but completely lacked leaves and roots (cf. p. 83). They bore terminal sporangia (sp) and were thus ancestral telomes (cf. p. 84).

Rhynia (Fig. 533A), the ancestral land plant, had two wholly leafless species up to 50 cm high. They consisted of subterranean, horizontal, rootless rhizomes with unicellular rhizoids; the rhizomes bore aerial, erect, cylindrical, dichotomous shoots with no leaves but with a cuticle and stomata (C) of relatively simple form; these axes were presumably the photosynthetic organs. *Rhynia* was thus a terrestrial plant, growing *en masse* like a *Juncus*. The vascular bundle consisted of tracheids with only very simple

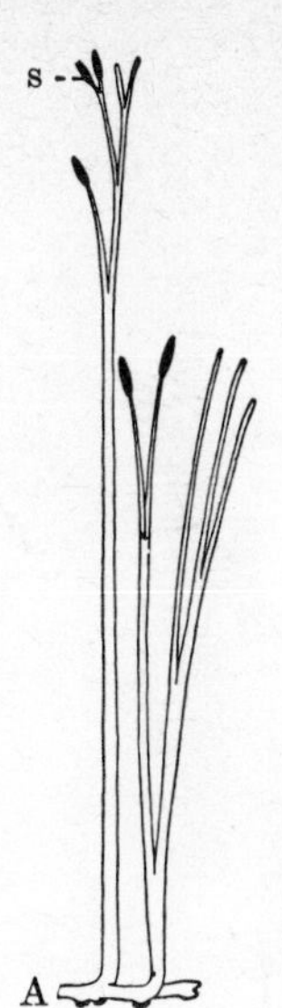

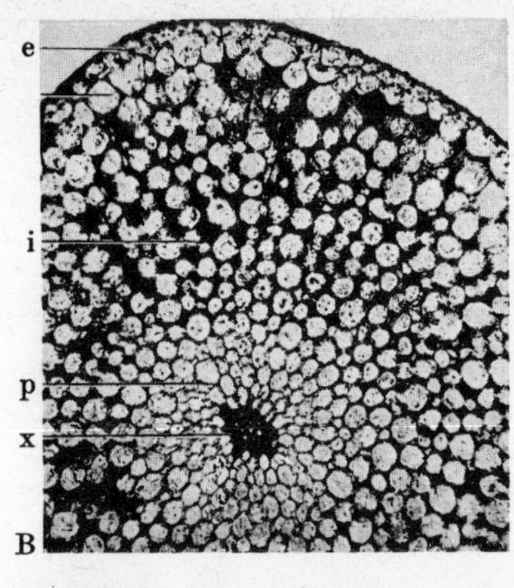

Fig. 553. *Rhynia.* **A.** Shoot with sporangia. ×174. **B.** T.S. of shoot. ×50. **C.** L.S. of sporangium. ×2. **D.** Spore tetrad. ×100. a, outer cortex; e, epidermis; i, inner cortex; p, phloem; s, sporangium; x, xylem. After Kidston and Lang.

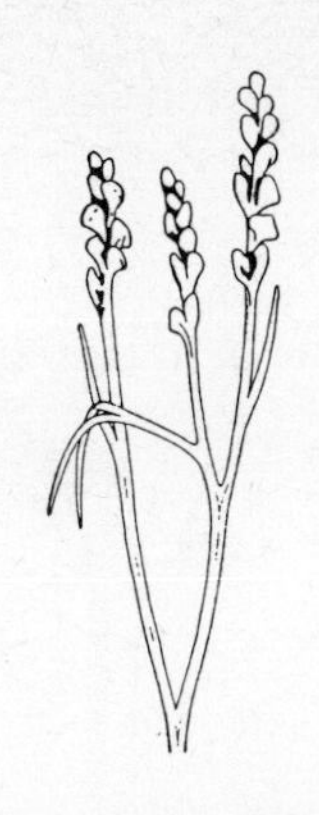

Fig. 554. *Zosterophyllum rhenanum.* Reconstruction. ×½. After Kräusel and Weyland.

thickening (annular and spiral) and was a protostele (Figs. 152A, 553B). Sometimes metaxylem was present, but secondary thickening and typical phloem with sieve tubes and sieve plates were absent from the stele. The growing point consisted of numerous similar cells; there was no apical cell. The relatively large, cylindrical or club-shaped sporangia were terminal on the axis and had a wall several cells thick, but lacked any dehiscence mechanism; they were filled with homosporous tetrads (Fig. 533D, E). In *Horneophyton* there was a central sterile pillar of tissue similar to the columella of some bryophyte sporogonia. Other completely leafless genera included *Taeniocrada, Zosterophyllum* and *Yarravia*; these probably lived in part in shallow water, but the higher forms had aerial shoots. In some the sporangia were fused laterally into synangia by the shortening of their stalks (e.g. *Zosterophyllum*, Fig. 554).

In another type (**Pseudosporochnaceae**) a number of equal branches, forking only a few times, sprang from the apex of the unbranched main axis; in consequence of the suppression of the axes between the forks, the branches formed a cluster (Fig. 555) which terminated in numerous slender, dichotomous axes. Some of these bore club-shaped sporangia, but the majority no doubt served for assimilation. In some cases—and this is a fundamental point—their ends were somewhat expanded (B); this represents the beginning of planation and fusion in the sense of the telome theory (p. 84). In these plants, which scarcely exceeded a metre in height, the lateral branches with their expanded photosynthetic surfaces can be considered the forerunners of large leaves, fronds or megaphylls.

More advanced forms like the Middle Devonian *Asteroxylon* (**Asteroxylaceae**, Fig. 556), with stems about 1 m high, bore widely or closely spaced small photosynthetic leaves (microphylls or needles, Fig. 556A, B) a few millimetres long with numerous stomata.

In many species (e.g. *Asteroxylon elberfeldense*) a lateral off-shoot of the vascular bundle extended to the base of the leaf as a leaf trace (Fig. 556C), or even entered it (e.g. in *Arthrostigma*). In transverse section the stele of the stem appeared stellate (actinostele, Figs. 556C, 152B), the rays of the star being the departing leaf traces; very young shoots still had a simple protostele. In some species the stele already possessed a medulla (siphonostele, Fig. 152C), and sometimes the tracheids even had bordered pits. The sporangia, which already had some indication of a dehiscence mechanism, were apparently borne at the tips of leafless lateral branches; but it is possible, since the fossil material consists only of fragments, that the sporangia were borne on the leaves (Fig. 556A), in which case *Asteroxylon* would belong not to the Psilophytatae, but to the Lycopodiatae.

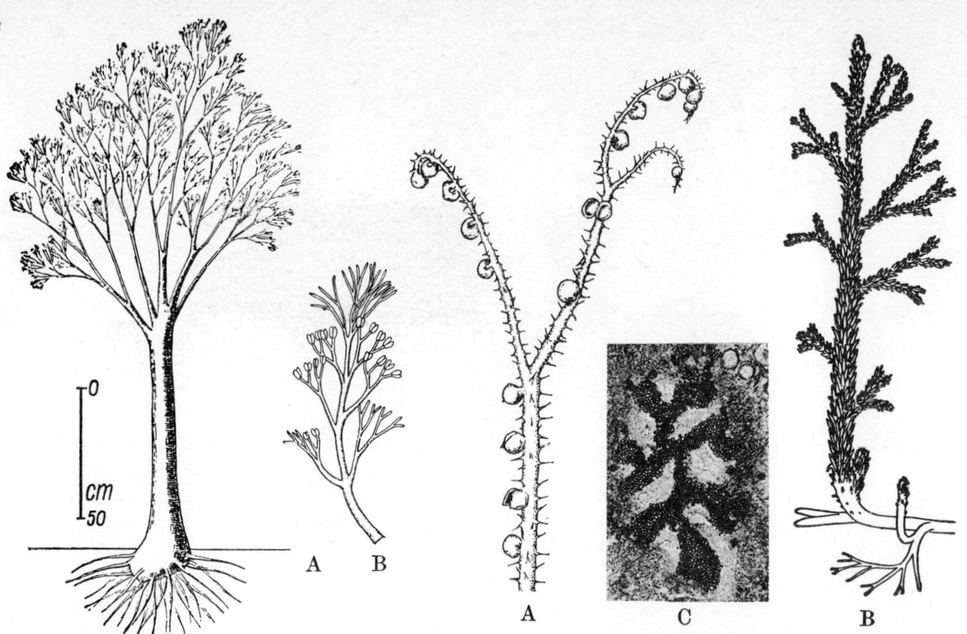

Fig. 555. *Pseudosporochnus* with megaphylls. **A.** Reconstruction. **B.** End of a branch. Nat. size. **A.** After Zimmermann. **B.** After Leclercq and Banks.

Fig. 556. **A.** *Psilophyton princeps*, shoot bearing sporangia. ×¾. **B.** *Asteroxylon mackiei*, reconstruction. ×⅓. **C.** T.S. of the stele of **B.**; dark, xylem; light, phloem. ×10. **A.** After Ananiew. **B.** and **C.** After Kidston and Lang.

The diversity of leaf form in the Psilophytales provides an extremely important basis for the derivation of the other Pteridophyta and perhaps even for certain Gymnospermae (p. 596).

The leafless Rhyniaceae have also been considered, not as primitive, but as reduced forms of the *Asteroxylon* type in which leaves originally present have been suppressed, as in many leafless parasites among present-day flowering plants.

Gametophytes of the Psilophytales have not yet been discovered, unless, as some authors have conjectured, the 'rhizomes' described under *Rhynia* represent the haploid generation (cf. the *Psilotum* prothallus described below).

The Psilophytales were apparently only thinly scattered over the earth's surface; they must have been restricted to the neighbourhood of water at a time when coastal waters were still thickly populated by dense algal 'forests' (e.g. *Prototaxites*, p. 485). Some twenty-five genera have been described. At first the naked forms predominated: in the Silurian nearly 90 per cent were leafless, in the Upper Devonian only 4 per cent. Recent claims that tetrahedral, pyramidal spores (like those in Figs. 582D, 558J, K) have been found in the Lower Cambrian, from which the presence of terrestrial plants at this very early epoch has been deduced, are regarded by most palaeontologists as erroneous.

Order 2: **Psilotales.** The living *Psilotum* species are lowly, perennial, dichotomous, widely spreading herbs (Fig. 557A) with multicellular growing points or an apical cell. They are actinostelic (Fig. 557B) and without roots, having only a leafless, protostelic rhizome with filamentous rhizoids and a mycorrhizal fungus. The leaves are minute, nerveless scales arranged in a lax spiral. Sporangia are thick walled and are united in threes as synangia (Fig. 557C, D). There is no tapetum, but the homospores receive nutrition from scattered sterile cells of the archesporium. The synangia are borne on very short stalks near the fork of the widely spaced bifid leaves (sporophylls, Fig. 557C).

The prothalli are a few centimetres long, cylindrical and branched (Fig. 557H); they are colourless and live underground with the aid of a mycorrhizal fungus (J my). They bear superficial, multilocular antheridia which liberate many multiflagellate spermato-

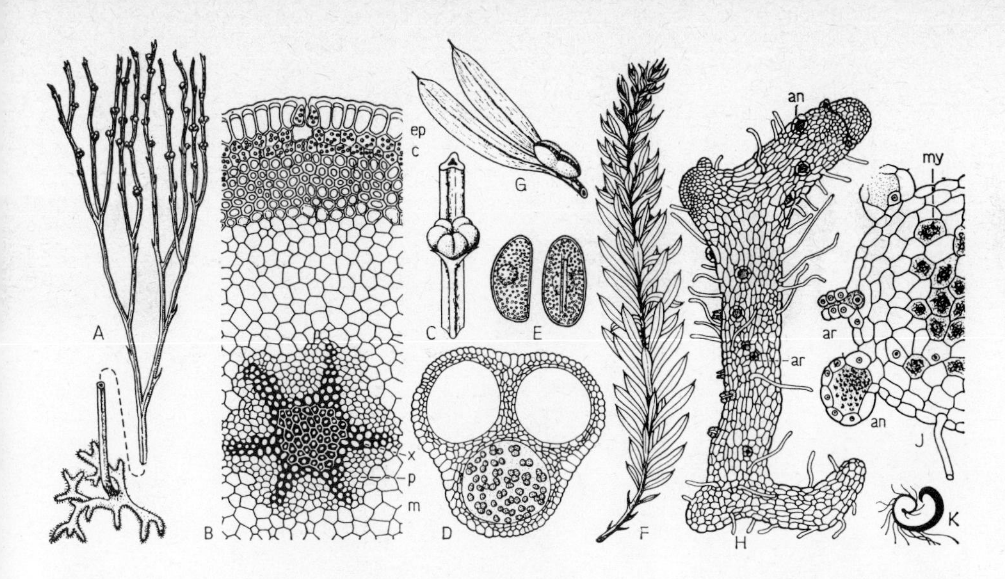

Fig. 557. Psilotaceae. **A.** *Psilotum triquetrum*, habit. ×½. **B.** *P. triquetrum*, stem in T.S. ×40. ep, epidermis; c, outer green cortical layer; x, xylem; p, phloem; m, inner cortex. **C.** *P. triquetrum*, part of a shoot with synangium. ×2.5. **D.** The same, T.S. of synangium. ×8. **E.** The same, spores. ×250. **F.** *Tmesipteris tannensis*, habit. ×½. **G.** The same, sporophyll. ×2.5. **H.** Prothallus of *Psilotum triquetrum*. ×15. **J.** The same in T.S. ×40. ar, archegonia; an, antheridia; my, mycorrhizal cells. **K.** *Psilotum triquetrum*, spermatozoid. ×900. **A.** After Wettstein and Pritzel. **B., C., E.** and **F.** After Pritzel. **D.** and **G.** After Wettstein. **H., J.** and **K.** After Lawson.

zoids (K). The small, somewhat sunken archegonia have only one, or rarely two, neck canal cells. The embryo has no suspensor (cf. p. 571) and is exoscopic (i.e. shoot apex towards the archegonial neck). Particularly large prothalli possess vascular bundles with an endodermis and lignified tracheids with annular thickening; thus the gametophyte might have attained the dimensions of the sporophyte but for the necessity of external water for fertilization by the motile ♂ gametes; this keeps it near the ground. The Psilotales comprise only *Psilotum* and *Tmesipteris*, each with only two tropical species. *Tmesipteris* (Fig. 557F, G) has larger leaves which run down the stem as a wing; unlike normal leaves, they are flattened in a vertical plane. No fossil Psilotales are known. In spite of this, they must be ancient relicts which may best be placed in the Psilophytatae. Nevertheless they show on the one hand distinct resemblances to the next class, on the other via the New Caledonian *Stomatopteris* to the Schizaeaceae.

The origins of the Psilophytatae, like the Pteridophyta in general, remain obscure.

Class II: Lycopodiatae (Lycopsida), clubmosses

Order 1: **Lycopodiales.** The genus *Lycopodium* (clubmoss) has some 400 species, but only six are native in Germany and five in Britain. They are evergreen herbs with no secondary thickening. Amongst the commonest European species is *Lycopodium clavatum* (stag's-horn moss, Fig. 558). One of the branches of the dichotomous stem shows apical dominance so that growth appears monopodial (p. 132). Instead of an apical cell there is a group of similar initials (Fig. 91 and p. 90). The vascular system is a complex plectostele (Fig. 152, p. 143) which is derived from an actinostele; it has sieve tubes in the phloem and sieve fields on the longitudinal walls, but no true sieve plates. The stems creep extensively over the ground and are thickly clothed with small, awl-shaped, spirally arranged leaves (Figs. 87, 558) which are homologous with the microphylls of the Asteroxylaceae; they have only a single unbranched midrib.

The mesophyll of *L. clavatum* is uniform; only a few species show a differentiation

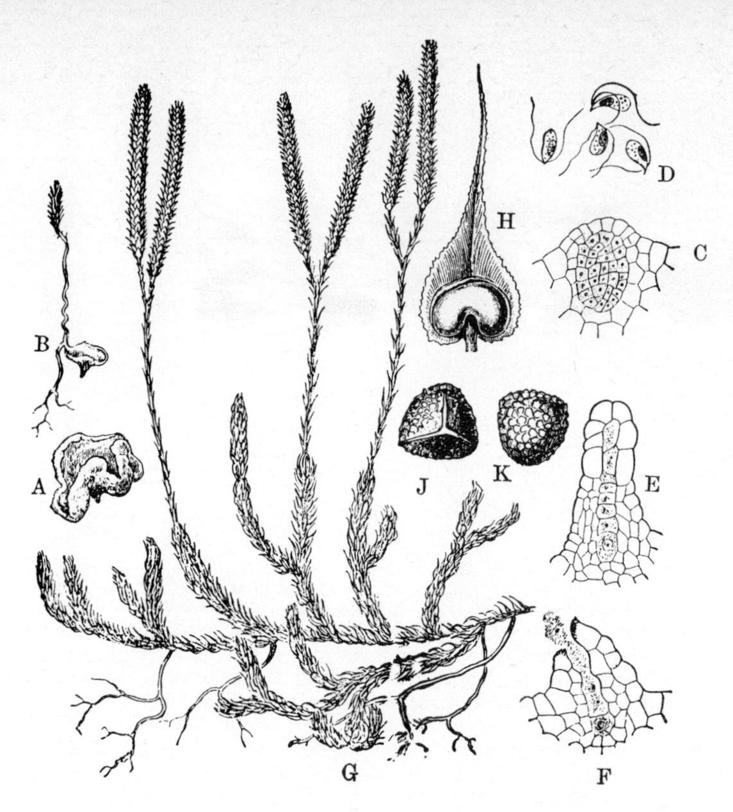

Fig. 558. *Lycopodium clavatum.* **A.** Old prothallus. **B.** Prothallus with young plant. **C.** Antheridium, not yet open. **D.** Spermatozoids. **E.** Young archegonium, neck closed. **F.** Open archegonium ready for fertilization. **G.** Plant bearing cones. **H.** Sporophyll with opened sporangium. **J.** and **K.** Spores from two aspects. **A.** ×2. **B.** ×$\frac{2}{5}$. **C., E.** and **F.** ×75. **D.** ×400. **A.** to **F.** After Bruchmann. **G.** to **K.** After Schenck.

into palisade and spongy mesophyll. As always with dichotomy, stem branching is independent of leaf insertion. For apical growth see p. 90.

Dichotomously branched roots arise from the lower side of the stems. Some of the branches are negatively geotropic and bear, frequently above a region with few leaves, dense, spike-like aggregations of sporophylls ('cones' or 'flowers', Fig. 558G); the apical growing point is used up in their formation so that the cone is apical. The sporophylls (H) are broadly scale-like, and each bears at the base of its upper surface a large, compressed, kidney-shaped sporangium containing numerous identical meiospores (homospores or isospores: J, K).

Whilst in *Lycopodium* the sporophylls are aggregated in cones on short, erect, lateral branches of the creeping main stem, in *Huperzia* (=*Urostachys*, Fig. 87C) trophophylls and sporophylls are borne in seasonal alternation on erect dichotomous axes. *Diphasium* has sporophylls in cones like *Lycopodium* but the scale-like microphylls are borne on flattened, dorsiventral shoots.

According to the telome theory (p. 83) the transition from terminal to foliar sporangia can be thought of as shown in Fig. 562.

The wall of the *Lycopodium* sporangium consists of several outer layers of cells; within this is a tapetum, the cells of which give up their contents to the developing spores without their walls actually breaking down; this is a secretory tapetum as opposed to the typical periplasmodial tapetum. The sporangium opens by two valves along an apical line already marked out by anatomically distinct cells. The spores remain in tetrads until they are ripe. The exospore consists of several layers and bears reticulate sculpturing (Fig. 558J, K). In nature the spores germinate only after six or seven years, initially forming at the expense of the reserve materials a germling of five cells (Fig. 559A) which, after a resting period, develops further only when the hyphae of a phycomycete fungus form a mycorrhizal association with the lower cells (Fig. 559B p).

The prothalli (Figs. 558A, B, 559, 560) live underground as whitish, saprophytic, lobed tubers which reach up to 2 cm in diameter and bear long, tubular, water-absorbing rhizoids; their shape varies from species to species. No doubt the mycorrhizal fungus housed in the peripheral cells plays an important part in their nutrition (Fig. 559B p).

The prothallus reaches sexual maturity only after twelve to fifteen years, and its total life span may be some twenty years. In artificial, bacteria-free pure culture, however, the whole development takes only a few weeks. In some species the upper part of the prothallus projects above the surface of the soil and becomes green.

The prothalli are monoecious and bear numerous sex organs, generally on the crown (Figs. 558C–F, 560 an, ar). The multicellular antheridia (an) are somewhat sunken. Each cell (excluding the wall cells) produces an oval spermatozoid with only two flagella just below its apex (Fig. 558D). The archegonia (Figs. 558E, F, 560 ar) are also sunken and have numerous neck canal cells (up to twenty, a primitive character; but sometimes reduced to one). The upper neck cells degenerate on opening.

The fertilized egg cell divides by a so-called basal wall into two cells, of which the lower divides first into quadrants and then into octants which give rise to the embryo, while the upper (directed towards the archegonial neck) forms the suspensor (Fig. 561 et). The apex of the embryo is thus directed away from the neck (endoscopic orientation). The suspensor forces the embryo into the prothallial tissue. The embryo grows out of the prothallus by curving upwards (Fig. 561A, B), its lower convex side swelling up as a haustorium (f) facing the main body of the prothallus from which it draws sustenance. The first, scale-like leaf (B, b) develops at the apex of the shoot; the first root arises

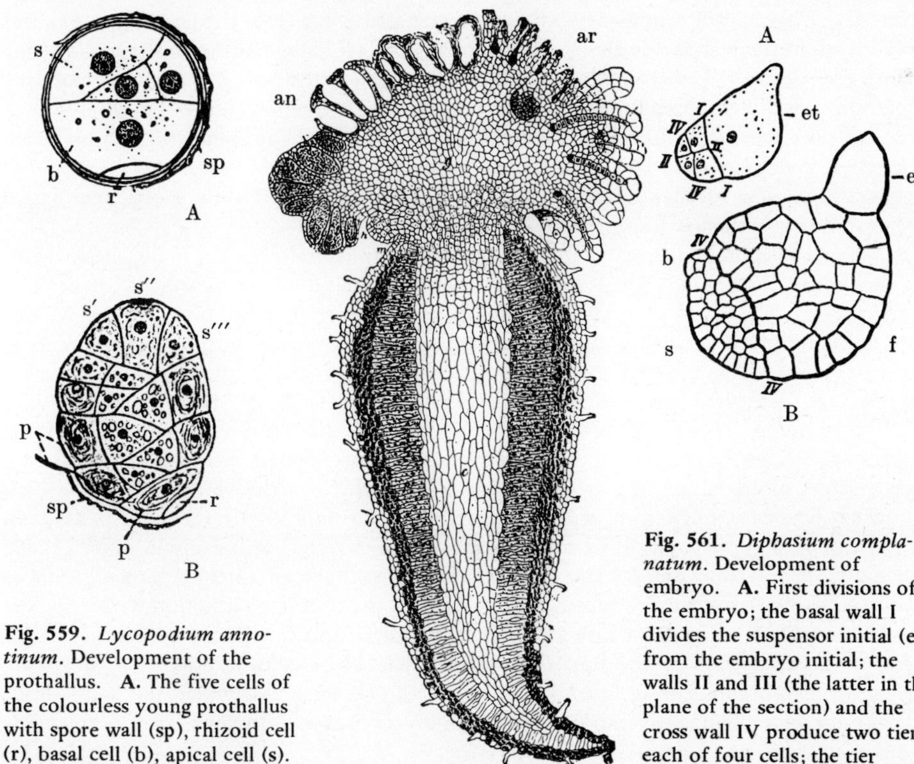

Fig. 559. *Lycopodium annotinum.* Development of the prothallus. **A.** The five cells of the colourless young prothallus with spore wall (sp), rhizoid cell (r), basal cell (b), apical cell (s). ×580. **B.** Young prothallus with an endophytic fungus (p) in its lower cells; the apical cell has divided into three meristematic cells, s', s'', s'''. ×470. After Bruchmann.

Fig. 560. *Diphasium complanatum.* Mature prothallus with antheridia, an; archegonia, ar; and mycorrhizal cells (dark). ×24. After Bruchmann.

Fig. 561. *Diphasium complanatum.* Development of embryo. **A.** First divisions of the embryo; the basal wall I divides the suspensor initial (et) from the embryo initial; the walls II and III (the latter in the plane of the section) and the cross wall IV produce two tiers each of four cells; the tier between I and IV gives rise to the haustorium, the other to the shoot. ×112. **B.** A later stage. s, stem apex; b, leaf primordium; f, haustorium. ×112. After Bruchmann.

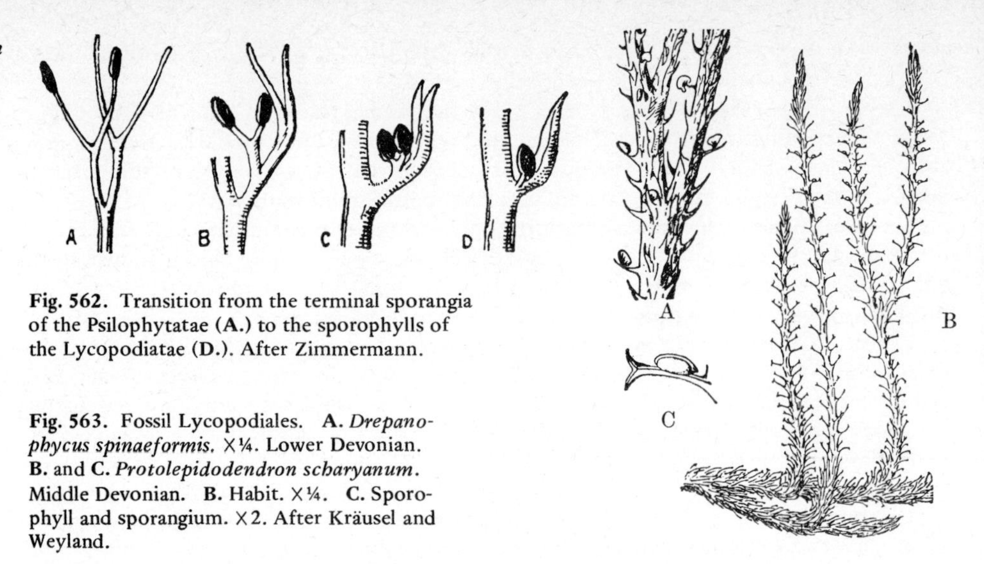

Fig. 562. Transition from the terminal sporangia of the Psilophytatae (**A.**) to the sporophylls of the Lycopodiatae (**D.**). After Zimmermann.

Fig. 563. Fossil Lycopodiales. **A.** *Drepanophycus spinaeformis.* ×¼. Lower Devonian. **B.** and **C.** *Protolepidodendron scharyanum.* Middle Devonian. **B.** Habit. ×¼. **C.** Sporophyll and sporangium. ×2. After Kräusel and Weyland.

laterally and adventitiously near the suspensor and opposite the foot. There is no definite apical cell; the growing point is multicellular (Fig. 91C).

The **Protolepidodendrales** are fossil plants similar to the lycopods; they occur in the Lower and Middle Devonian; in their microphylls (with numerous stomata) even plastids can sometimes be recognized. *Drepanophycus* (Fig. 563A), the oldest known C. European land plant, older even than *Asteroxylon* which it resembles, and *Protolepidodendron*, with leaves still bifurcate at the apex (Fig. 563B), bore sporangia on the upper side of the leaves (C), in which the vascular bundle was single as it arose from the common stem, but forked both in the leaf and in the sporangium; as in a few living clubmosses (e.g. *Huperzia selago*), the sporophylls were not restricted to cones. The species of *Lycopodites* from the Upper Devonian were already very similar to present-day lycopods; the typical clubmoss form has thus persisted through many geological formations for some 300 million years.

Order 2: **Selaginellales.** The habit of the herbaceous species of *Selaginella* (Fig. 564A) is generally similar to that of the lycopods. Although there are only a very few native species in Europe, there are some 700 in the Tropics.

Selaginella has prostrate or erect, much branched, unequally dichotomous stems without secondary thickening; some form mats, others scramble over bushes to a length of several metres. The stem bears small, scale-like leaves which may be spirally arranged, but more commonly are decussate and dorsiventral in four rows, e.g. in the Alpine species *S. helvetica* (Fig. 564A), in which the stems bear two rows of smaller, dorsal or upper leaves, and opposite them, two rows of larger, ventral or under leaves (Fig. 181C–E; anisophylly, p. 169). In many species there arises at a fork of the stem a colourless, leafless, exogenous, elongate, cylindrical, dichotomous structure which grows downwards and bears clusters of adventitious roots at the distal end; this is a rhizophore (Fig. 564A w). The leaves possess a single unbranched central vein and only very rarely have palisade in addition to spongy mesophyll. In some species the mesophyll cells contain only a single large bowl-shaped chloroplast. The growing point may have an apical cell or a group of dividing cells. Stelar structure ranges from a central protostele to a siphonostele; there is no secondary thickening. Very rarely vessels with scalariform thickening occur.

From the epidermis of the upper side of the leaf of *Selaginella* arises a small, basal, colourless, membranous scale, the ligule (Fig. 564C) by means of which the leafy shoot can take up water extremely rapidly from rain-drops (p. 105). In some species tracheids connect it to the vascular bundle (Fig. 564C).

Selaginella is characterized by heterospory and extremely reduced prothalli.

The terminal cones (flowers) (Fig. 564A, C) are simple or branched, four-angled, and radially symmetrical or dorsiventral. Each sporophyll bears a single axillary sporangium. Large megaspores are borne in megasporangia, small microspores in microsporangia (Fig. 565); both kinds of sporangium occur in the same cone (Fig. 564D).

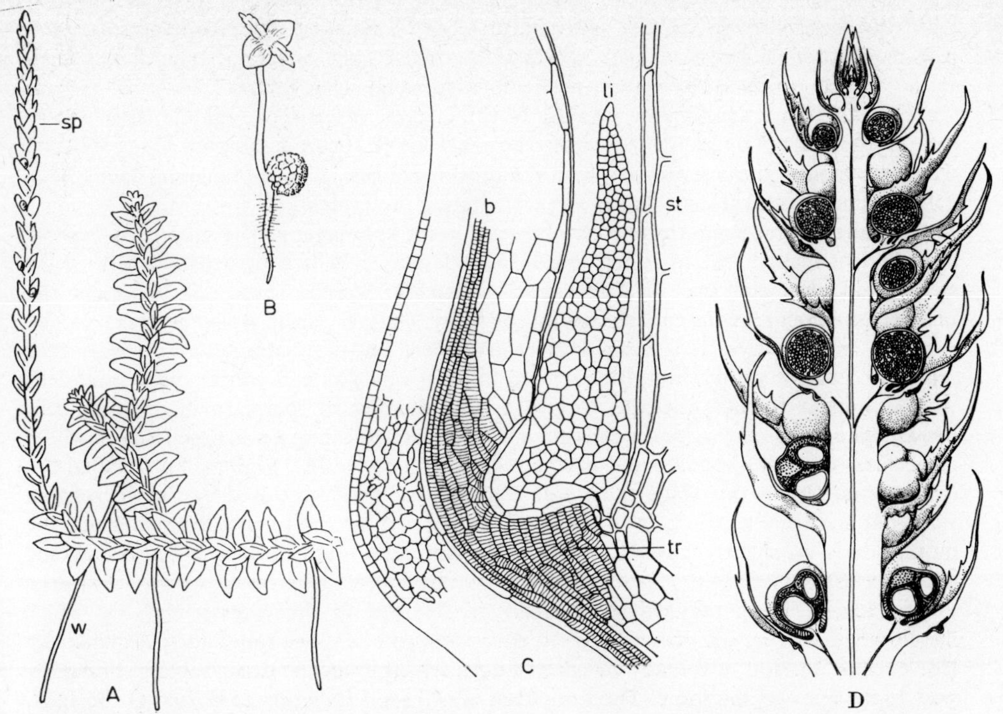

Fig. 564. *Selaginella.* **A.** *S. helvetica*, plant with cone. ×2. **B.** *S. kraussiana*, megaspore with sporeling. ×10. **C.** *S. lyallii*, L.S. through the base of a leaf. ×250. **D.** *S. selaginoides*, L.S. of cone with megasporangia below and microsporangia above; the ligule is visible above the point of insertion of those sporangia shown in the median plane. ×6. b, leaf base; li, ligule; sp, cone; st, epidermis of the stem; tr, tracheids; w, rhizophore. **A.** After Luerssen. **B.** After Bischoff. **C.** After Harvey-Gibson.

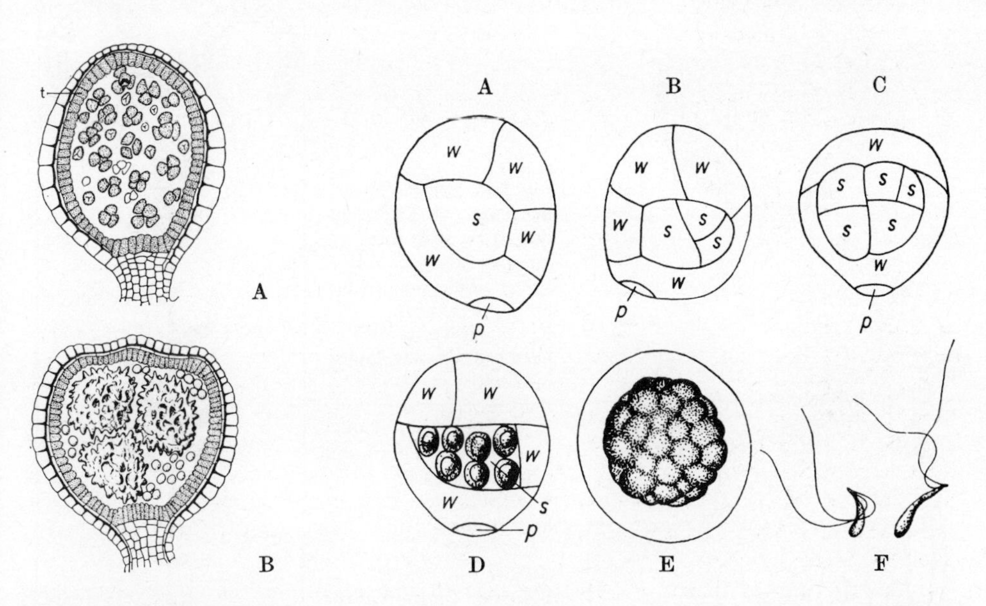

Fig. 565. *Selaginella inaequifolia.* **A.** Microsporangium with tetrads of microspores; t, tapetal cells. **B.** Megasporangium with one megaspore tetrad and several abortive spore mother cells. ×70. After Sachs.

Fig. 566. A. to **E.** *Selaginella stolonifera.* ×640. Successive stages in the development of microspores; p, prothallial cell, equivalent to a rhizoid; w, wall cells of antheridium; s, spermatogenous cells. **A., B.** and **D.** From the side. **C.** From the back. In **E.** the prothallial cell not visible, the wall cells broken down. **F.** *S. cuspidata*, spermatozoids. ×780. After Belajeff.

In the megasporangia all the spore mother cells degenerate except one which undergoes meiosis, producing four large spores with nodular walls (Fig. 565B); numerous small spores are also produced by meiosis in the microsporangia (Fig. 565A).

The wall of the sporangium consists of three layers of cells; the middle layer is very thin in the mature sporangium, whilst the inner layer forms a tapetum (Fig. 565A t) which nourishes the spores, but without breaking down. The sporangium opens by a cohesion mechanism along a definite line, scattering the spores.

The microspores begin to develop while still in the sporangium. The spore first cuts off a small lens-shaped cell (p in Fig. 566) and a large cell which then divides into eight sterile wall cells and two or four central cells (Fig. 566A). These cells represent the prothallus, which remains entirely within the spore. Only the small lens-shaped cell can be regarded as vegetative; it is termed the rhizoid-cell, but is functionless. The other cells represent a single antheridium, surrounded by wall cells (w) and containing central cells which undergo further divisions to give a large number of spermatozoid mother cells (B–D). The walls of the jacket cells then disintegrate, forming a mucilage layer in which the central mass of spermatozoid mother cells is embedded (E). The small prothallial cell (p) persists. Up to this stage the whole ♂ prothallus is still enclosed by the microspore wall, but eventually this breaks open, liberating the spermatozoids formed from the mother cells as slightly curved, club-shaped structures with two long flagella at the pointed end (F).

The less strongly reduced ♀ prothalli are formed in the megaspores (Fig. 567). Development differs somewhat between the various species. The spore nucleus divides by free nuclear division into many daughter nuclei which gather in the cytoplasm lining the wall at the apex of the spore. The formation of cell walls then proceeds, first at the apex, then towards the base until almost the whole spore is filled with large prothallial cells; at the same time further division of the large cells into smaller ones proceeds in the same direction. A few archegonia form near the apex of the prothallus.

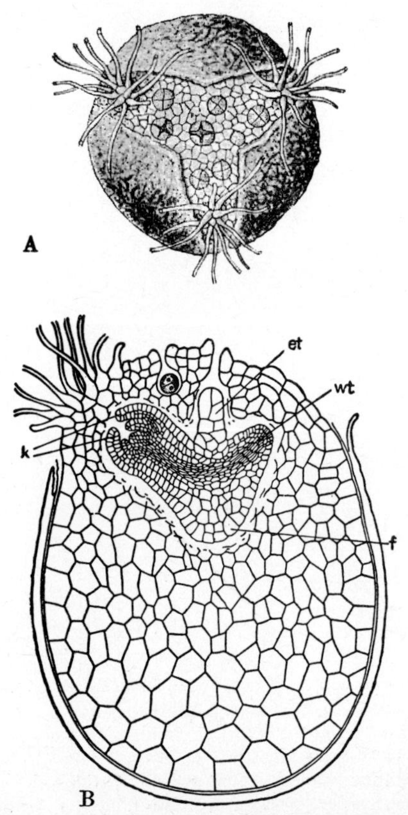

(Left)
Fig. 567. *Selaginella martensii.* **A.** Ruptured megaspore with three groups of rhizoids and several archegonia visible on the prothallus. **B.** L.S. with two fertilized egg cells developing into embryos; et, suspensor; f, foot (haustorium); wt, rhizophore; k, cotyledons with ligules. **A.** ×112. **B.** ×150. After Bruchmann.

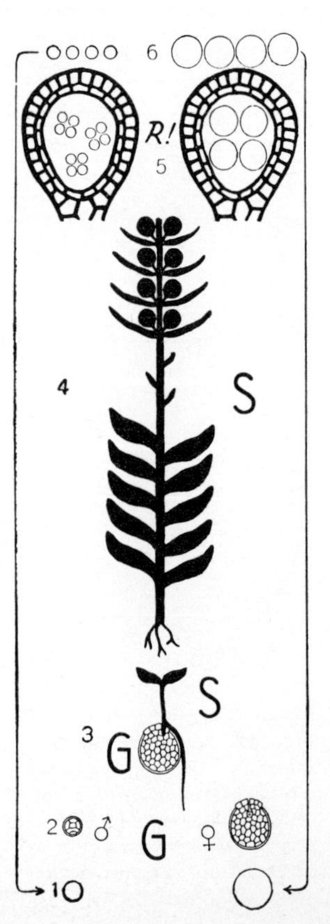

(Right)
Fig. 568. Diagram of the life-cycle of *Selaginella.* G, gametophyte; S, sporophyte. Haploid phase, thin lines; diploid, thick or entirely black. R!, meiosis. After Harder.

The spore wall opens along the three triradiate scars (Fig. 567A) and the colourless prothallus projects slightly, with water-absorbing rhizoids on three mounds of tissue. One or two archegonia are now fertilized. The first division of the zygote separates a suspensor (Fig. 567B), which is directed towards the archegonial neck, from the endoscopic embryo which has to curve outwards to get free from the prothallus (B). Most species grow by an apical cell. The herbaceous species of *Selaginellites* of the Carboniferous resembled *Selaginella* and were already heterosporous; plants looking very like present-day *Selaginella* thus existed some 260 million years ago.

Most *Selaginella* species carpet the ground in moist tropical forests. Only a few are adapted to dry habitats, e.g. the Mexican *S. lepidophylla* of which the shoots form an inrolled rosette when dry (false 'Rose of Jericho').

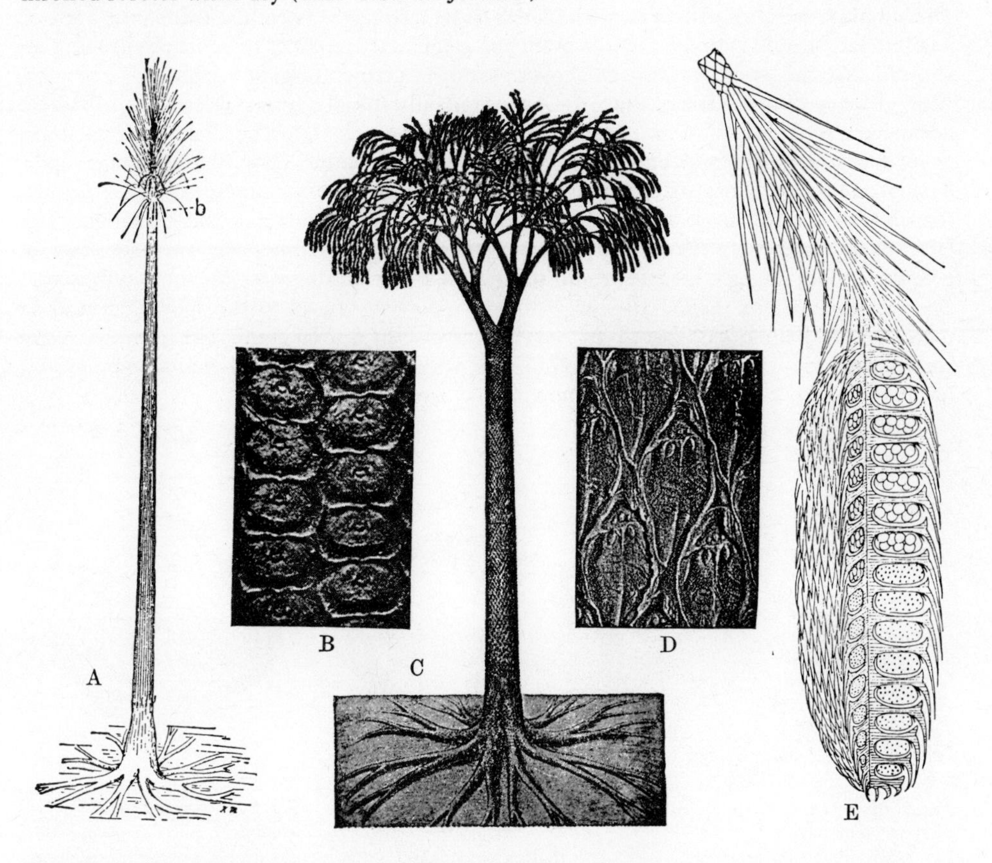

Fig. 569. Lepidodrendrales. **A.** and **B.** *Sigillaria.* **C.** to **E.** *Lepidodendron.* **A.** *Sigillaria*, reconstruction. $\times\frac{1}{80}$; b, cones. **B.** Leaf scars. $\times$ 2.5. **C.** *Lepidodendron*, reconstruction. $\times\frac{1}{200}$. **D.** Leaf scars. Nat. size. **E.** Cone. Nat. size. **C.** After Hirmer. **D.** After Stur.

Order 3: **Lepidodendrales.** These extinct 'clubmoss trees', some 40 m high and up to 5 m in diameter (Fig. 569), reached their maximum development in the Carboniferous and played a major part in the formation of coal. Their spirally arranged, needle-like leaves were microphyllous; only one simple (rarely forked) vascular bundle penetrated the mesophyll (no palisade). Stomata occurred in two longitudinal dorsal furrows. On abscission the leaves left characteristic scars and leaf cushions on the stem (Fig. 569B, D). The stems were siphonostelic; the phloem was thin-walled and poorly differentiated. A ring of cambium (Fig. 570) gave rise to a very limited amount of secondary thickening; the wood, including scalariform tracheids (with somewhat aberrant bars on the walls) of very uniform diameter and sometimes uniseriate vascular rays, recalls modern coniferous wood (but no bordered pits and, as in almost all Carboniferous plants of the Northern Hemisphere, no annual rings). The entire secondary xylem was apparently of little significance for support or water conduction. The stems had a secondary meristem corresponding to a cork cambium; this cut off cells very actively, especially on the inner side, forming a periderm of considerable thickness relative to the xylem (in *Lepidodendron* up to 99 per cent of the section was bark; Fig. 570). The periderm consisted principally of mechanical tissue, but was probably also active in supplying water, since the ligule persisted long after the fall of the leaf. The trees were anchored by shallow (cf. the waterlogged soils), rhizome-like, repeatedly dichotomous rhizomorphs (Fig. 569A, C); these underwent secondary thickening and produced numerous exogenous, rather slender roots of characteristic structure ('appendices'); these subsequently broke off, leaving numerous scars, from which the rhizomorphs are known as *Stigmaria*.

The **Sigillariaceae** (Fig. 569A) had longitudinal rows of more or less hexagonal leaf cushions (B) which enlarged by dilatation as the stem underwent secondary thickening. The simple leaves, up to 1 m long and 1 dm wide, were aggregated in a tuft at the apex of a columnar, simple or only slightly branched stem. Massive cones of sporophylls hung on very short lateral branches from the lower part of the crown (Fig. 569A b).

In the **Lepidodendraceae** (Fig. 569C) the generally spirally arranged leaves, up to a few decimetres long, were inserted on rhombic leaf cushions (D). The dichotomous stems were much branched and bore terminal sporophylls in cones up to 0.75 m long, superficially similar to those of conifers (Fig. 569C, E). The very numerous, broadly scale-shaped and spirally overlapping sporophylls covered and protected the sporangia. The latter were heterosporous almost without exception; sometimes the megasporangium contained only a single megaspore up to at least 6 mm in diameter; in some representatives (*Lepidostrobus major*) the megaspore was partially fused to the sporangial wall so that the prothallus was forced to develop inside the sporangium. The prothalli were similar to those of *Selaginella* (Fig. 571). The spermatozoids of all Lepidodendrales and hence the number of flagella are unknown.

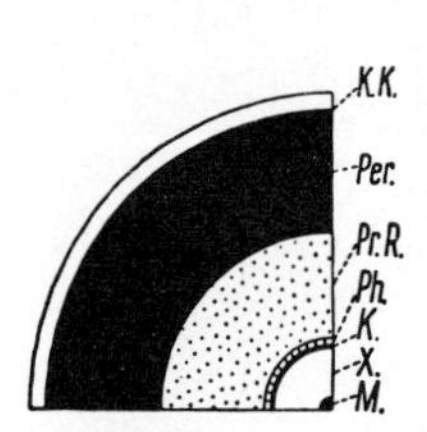

Fig. 570. *Lepidodendron*. T.S. of stem (schematic). K, cambium; KK, cork cambium; M, medulla; Per, periderm; Ph, phloem; PrR, primary cortex; X, xylem. $\times\frac{1}{160}$. After Hirmer.

Fig. 571. *Bothostrobus mundus*. Megaspore with prothallus. ×35. After McLean.

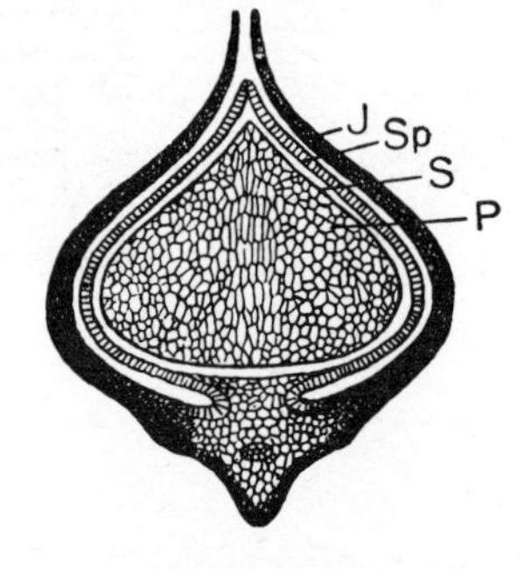

Fig. 572. *Lepidocarpon lomaxii*. L.S. of megasporangium. P, prothallus; S, spore wall; Sp, sporangium wall; J, integument. ×8. After Scott.

Of great interest are a few Carboniferous Lepidodendrales (*Miadesmia*, herbaceous, *Selaginella*-like; *Lepidocarpon*, tree-like) which had seed-like structures; although they are not closely related they can be grouped together as **Lepidospermae**. In these the megasporophyll surrounded the sporangium as an integument (Fig. 572 J) with an apical opening by which the microspores could enter, fertilizing the prothallus (P) of the single megaspore in some unknown manner. The whole organ remained attached to the parent plant and developed into a seed, the seed coats being formed both by the sporangial wall and the integument. The megasporophylls were arranged in cones very similar to those of present-day seed-bearing Gymnospermae.

Order 4: **Isoetales.** The species of *Isoetes*, the quillworts (Fig. 573), live as perennial herbs either submerged or on moist soil; they have a short, tuberous, rarely dichotomously branched axis, and can grow to a great age. There is a single vascular bundle which undergoes a very limited amount of somewhat anomalous secondary thickening. Rows of dichotomously branched roots arise on the axis from two or three longitudinal grooves; the long (up to 1 m in certain species) subulate leaves, with a single unbranched main vein and four air passages, are borne in a dense rosette. The leaves, which have a uniform mesophyll, are dilated at the base, where there is an elongated depression, the fovea, on the upper side. Only the innermost leaves of the rosette are sterile; all the rest have a sporangium sunken in the fovea. Otherwise there is no difference between sterile leaves and sporophylls. Above the fovea is the ligule, a triangular membrane, its base embedded in the leaf, which produces mucilage and serves as an organ of absorption of water and nutrients (Fig. 573B, C). For phyllotaxis, see p. 128.

The outer sporophylls bear megasporangia with numerous rounded tetrahedral megaspores; the inner, younger ones bear microsporangia with hundreds of thousands of ellipsoidal microspores which are somewhat flattened on one side. The sporangia may be up to 5 mm long and are crossed by strands of sterile tissue (trabeculae). The sporangium wall has many layers of cells and a secretory tapetum. The spores are liberated only when the sporangium wall disintegrates.

The dioecious prothalli are extremely reduced and develop inside the spores. The ♂ prothallus (Fig. 573) develops like that of *Selaginella* (and shows a remarkable similarity to the first divisions of the spores of *Lycopodium*, Fig. 559A), but produces only four spermatozoids. These are spirally coiled with a cluster of flagella at the anterior end. The ♀ prothallus (Fig. 574) also develops like that of *Selaginella* and entirely fills the megaspore. A few archegonia are formed on the part exposed by the rupture of the wall.

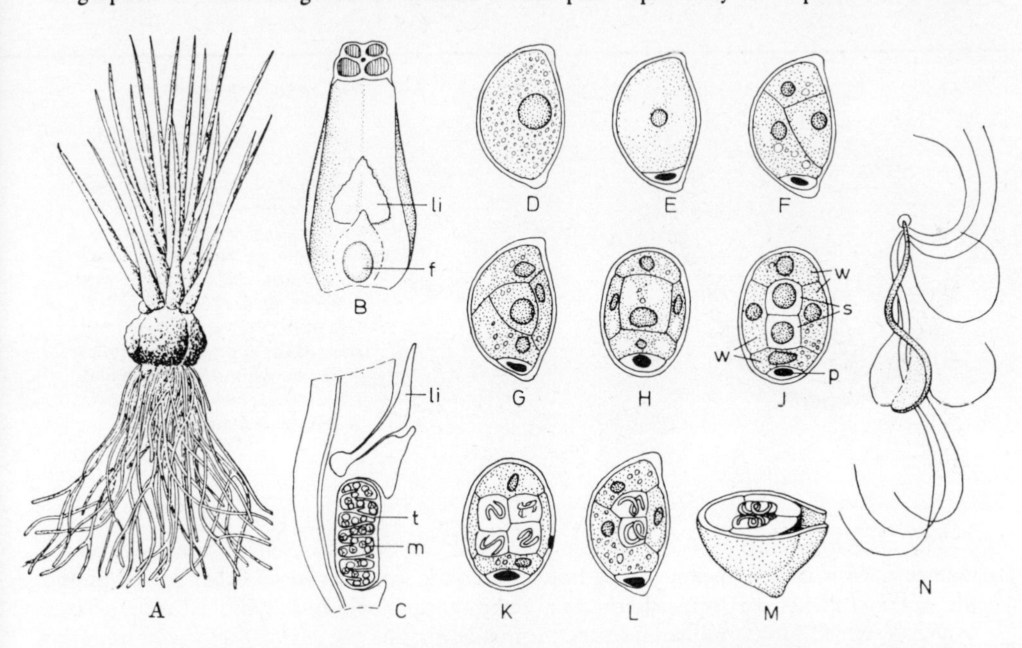

Fig. 573. *Isoetes.* **A.** to **C.** *I. lacustris.* **D.** to **M.** *I. setacea.* **N.** *I. malinverniana.* **A.** Entire plant. ×½. **B.** Basal part of a leaf with ligule and fovea. ×2. **C.** The same in L.S. ×4. **D.** to **M.** Development of the ♂ prothallus and spermatozoids. ×500. **N.** Spermatozoid. ×1100. f, fovea; li, ligule; m, microspores; p, prothallial cell; s, spermatogenous cells; t, trabeculae; w, wall cells. **A.** to **C.** After Wettstein. **D.** to **M.** After Liebig. **N.** After Belajeff.

Two cross walls perpendicular to each other divide the zygote into quadrants, of which two adjacent ones form the shoot apex and the first leaf with its ligule, whilst the other two form the first adventitious root and the haustorium. There is no suspensor, and the embryo is exoscopic; no apical cell is formed.

The only other living genus of the Isoetales, *Stylites* (two species in Peru), has only recently been discovered; it is probably phylogenetically older than *Isoetes*, having a longer stem (15 cm) covered with leaf scars, a single groove bearing roots, and a greater tendency towards dichotomous branching.

Despite the different number of flagella and development of the embryo, the Isoetales are undoubtedly reduced Lycopodiatae which have become secondarily aquatic. The living species are relicts of a more diverse group living in past geological periods (Cretaceous and earlier). These extinct forms were distinctly larger; this is true to some extent of the Lower Cretaceous *Nathorstiana* and certainly of the Lower Triassic *Pleuromeia* in which the unbranched stems reached 2 m in height and were as thick as an arm; this genus was heterosporous, with a terminal tuft of relatively long but simple leaves. The precursor may have been the very similar but much larger *Sigillaria*. The living *Isoetes* must therefore have evolved from the Lepidodendrales (even if not quite so straightforwardly as indicated here) by the progressive shortening of the stem: the series *Sigillaria–Pleuromeia–Nathorstiana–Stylites–Isoetes* shows this gradual reduction.

The preceding account shows that all **Lycopodiatae** are characterized by dichotomously branching stems and roots, and by the simple form of the numerous sessile leaves which can be regarded as advanced microphylls. The leaves, which in several orders possess ligules, are closely set in spirals on the stems. The sporophylls differ little in form from the assimilating leaves and are usually aggregated into spike-like cones which are terminal on the branches (exceptions: *Isoetes, Huperzia selago*). The solitary sporangium is borne on the upper side of the sporophyll at its base. The microphyll of the Lycopodiatae indicates their relationship to the Psilophytatae (cf. *Asteroxylon*, p. 563), from which very probably they have evolved.

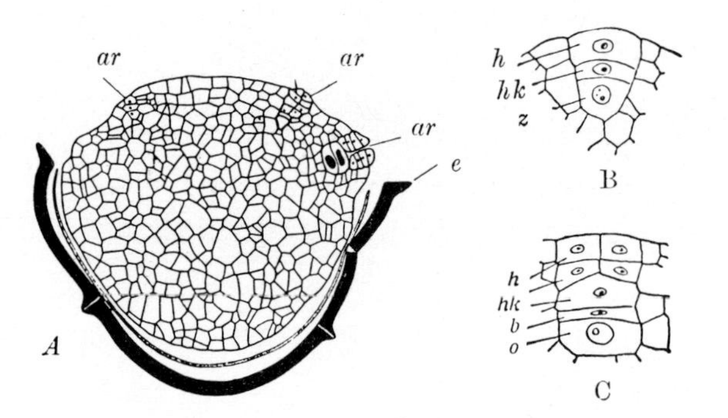

Fig. 574. Isoetales. ♀ prothallus. **A.** Prothallus in the ruptured megaspore wall with archegonia (ar); on the right with ventral canal cells and egg cell; e, exine with intine inside. **B.** and **C.** Development of the archegonium from a superficial cell; h, neck cells; hk, neck canal cell; z, central cell which gives rise to (b) the ventral canal cell and (o) the egg cell. **A.** *Stylites andicola.* ×60. After Rauh and Falk. **B.** *Isoetes echinospora.* ×250. After Campbell.

The innermost layer of the sporangium wall, the tapetum, never breaks down at maturity. With the exception of the Isoetales, and in contrast to all other Pteridophyta, the spermatozoids are biflagellate; the embryo—again excepting the Isoetales—has a suspensor which forces it down into the prothallial tissue.

Devonian forms such as *Protolepidodendron* and *Drepanocphycus* (Fig. 563A) may indicate the common stock from which the Lycopodiatae arose. They were much more highly developed in the Carboniferous, with numerous arborescent genera, than today (Fig. 614; they had even developed the seed habit (Lepidospermae). Their inefficiency in water uptake and conduction may have contributed to the decline of the arborescent forms when the climate became drier at the end of the Palaeozoic, or they may simply have been superseded in the struggle for existence by new forms with more adequate conducting systems (e.g. the Cordaitatae; cf. Fig. 614). The herbaceous lycopods and *Selaginella* have remained virtually unchanged for some 300 million years ('persistent types'). Of course they now play a very subordinate role in the world's vegetation, yet in the Carboniferous the arborescent lycopods, calamites (p. 578) and some tree ferns (p. 584) dominated the landscape. The Psilophytatae were virtually the only land plants in the Silurian/Devonian, but became extinct when the lycopods and calamites reached their peak.

Class III: Equisetatae (Articulatae, Sphenopsida), horse-tails

Living horse-tails are only represented by the thirty-two species of the genus *Equisetum*; these are much alike in structure and development. Their perennial, subterranean rhizomes often creep at considerable depths in the soil, and bear erect stems with an apical cell (Fig. 91). These stems are usually annual and either simple or with whorls of branches of the second or higher orders (Fig. 575).

The ridged axis consists of elongated internodes; the original protostele has divided to form a ring of collateral (p. 113) vascular bundles with very little xylem. The earliest formed xylem soon disappears and is replaced by a ring of intercellular passages, the carinal canals (Fig. 576). The medulla also breaks down to form a large air cavity (the central canal), whilst in the cortex is a further ring of vallecular canals (opposite the grooves on the stem). The vascular bundle of the sporophyll is concentric.

At each node, separated by long internodes, is a whorl (p. 124) of pointed scale leaves which are united at the base to form a sheath surrounding the stem (Fig. 575A). Each leaf has a single slender vascular bundle.

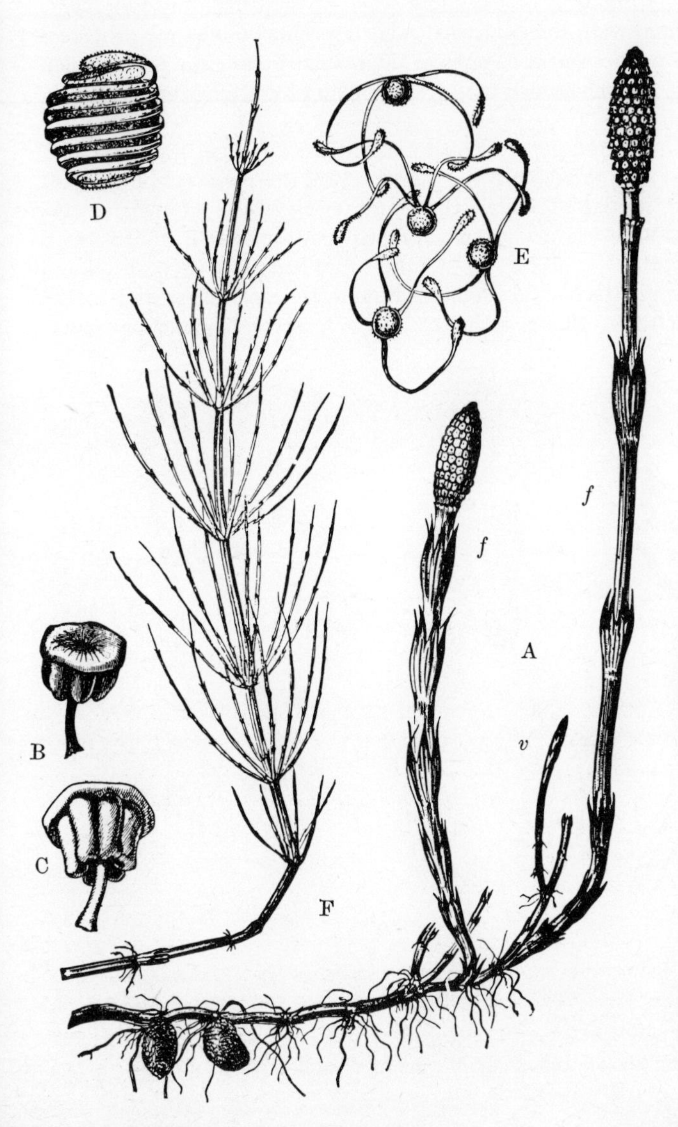

Fig. 575. *Equisetum arvense.* **A.** Fertile shoot (f) arising from the rhizome which bears tubers; vegetative shoot (v) still in bud. **B.** and **C.** Sporophylls with sporangia, dehisced in **C.** **D.** Spore with the two spiral bands (haptera) of the perispore. **E.** Dry spores with extended haptera. **F.** Sterile vegetative shoot. **A.** and **F.** ×½. **B.** and **C.** ×6. **D.** ×360. **E.** ×100. After Schenck.

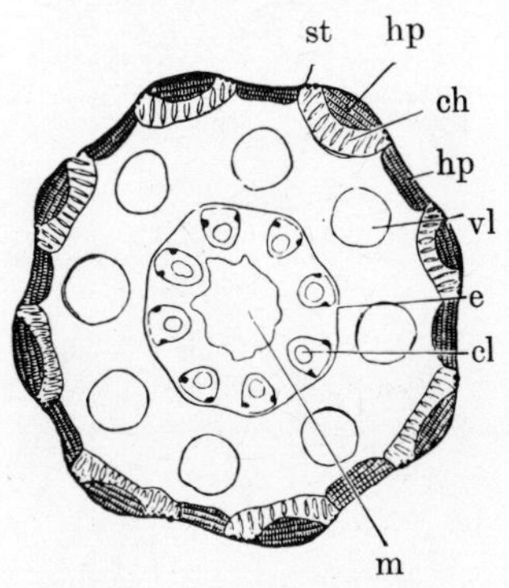

Fig. 576. *Equisetum arvense.* T.S. of stem. m, lysigenous central cavity; e, endodermis; cl, carinal canals in the collateral bundles; vl, vallecular canals; hp, strands of sclerenchyma in the grooves and ridges; ch, chlorophyllose tissue of the cortex; st, rows of stomata. ×11. After Strasburger.

The outer wall of the epidermis of the stem is more or less heavily impregnated with silica (hence the use of horse-tails as 'scouring rushes' for cleaning pewter, etc.).

The basal part of the internode, which exhibits intercalary growth, is enclosed by a leaf sheath. Lateral branches do not arise in the leaf axils (cf. p. 130) but between the leaves which thus alternate with the branches; they then break out through the sheaths. Photosynthesis is carried out by the chlorophyllose tissue of the stems, not by the reduced leaves. In *Equisetum arvense* (Fig. 575) and some other species which die down in the winter lateral branches of the rhizome are modified as roundish overwintering tubers rich in reserve materials. Evergreen species also occur (e.g. *E. hiemale*).

In certain horse-tails some of the stems remain sterile, branching freely, whilst other stems bear terminal cones and do not branch, or they produce only a few sterile side branches later (Fig. 575A, F).

The sporangia are borne on peculiar sporangiophores which are aggregated in many alternating whorls separated by short internodes, forming terminal cones (Fig. 575A). The sporophylls are peltate, i.e. they have the form of a one-legged table, with five to ten sac-like sporangia on the underside (B, C).

In the young sporangium the sporogenous tissue is surrounded by a wall of several layers of cells. The innermost form the tapetum; their walls break down and their protoplasts unite as a periplasmodium which penetrates between the spores as they round off, and which is used up in the formation of the spore wall. At maturity only two layers remain as a definite sporangium wall; the epidermal cells have spiral and annular thickenings, and as the water in them dries out it coheres to their walls, pulling the unthickened parts inwards and so causing the dehiscence of the sporangium by a longitudinal slit on the inner side (cf. p. 379).

The sporangium liberates numerous green meiospores with a wall of peculiar structure. Around the endospore and exospore which form the true wall the periplasmodium lays down a perispore of several layers, of which the outermost consists of two narrow, parallel bands which are spathulate at the ends; when moist they are spirally coiled round the spore. These are the haptera (Fig. 575D, E). As the spore dries out they unroll and become extended, but remain attached at their mid-point to each other and to the exospore; when remoistened they roll up again (cf. p. 379). These hygroscopic move-

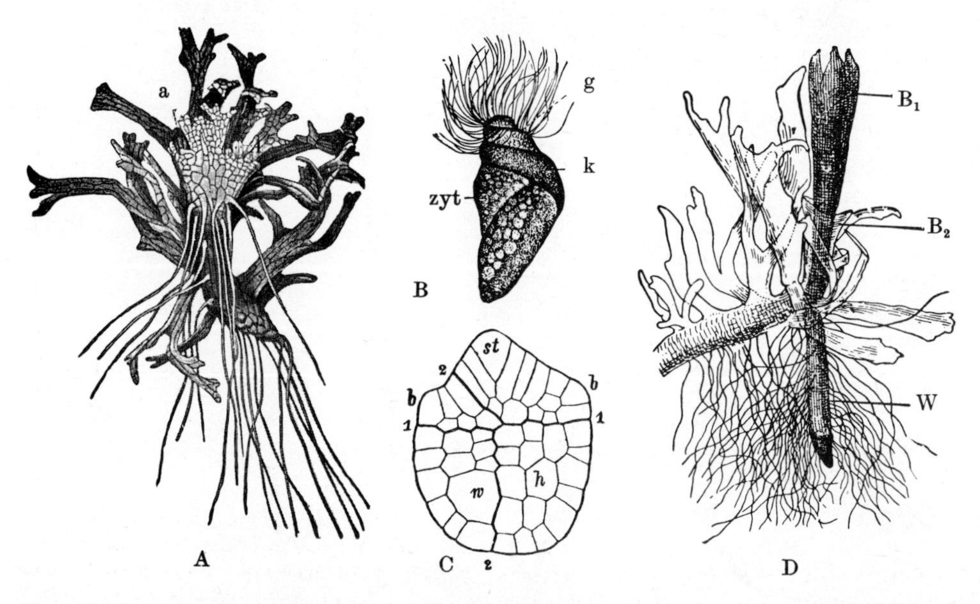

Fig. 577. A. *Equisetum pratense*, ♀ prothallus from below, with archegonia (a). ×17. **B.** *E. arvense*, spermatozoid; k, nucleus; g, flagella; zyt, cytoplasm. ×1250. **C.** *E. arvense*, embryo; 1, 2, walls forming quadrants; from the half above the basal wall (1) arise the stem (st) and the first whorl of leaves (b); from the lower half the root (w) and the haustorium (h). ×165. **D.** *E. maximum*, ♀ prothallus in lateral view with sporeling (shown darker); B_1, B_2, the first whorl of leaves; W, root. **A.** and **D.** After Goebel. **B.** After Sharp. **C.** After Sadebeck.

ments not only disseminate the spores but also link them together; this may be of significance for the fertilization of the sometimes dioecious prothalli produced by them. The spores remain viable only for a few days.

All the meiospores are identical and germinate to form much-lobed, green prothalli (Fig. 577A).

On germination the first rhizoid is cut off by a watch-glass-shaped wall on the side away from the light; the other cell develops into the green part of the prothallus, which forms further rhizoids. The monoecious or dioecious prothalli consist of considerably branched, dorsiventral, curling lobes. The antheridia are sunk in the prothallus whilst the archegonia project from its surface. The spiral spermatozoids have numerous flagella (Fig. 577B, Fig. 23). At the first division of the zygote the basal wall (1–1 in Fig. 577C) demarcates two halves, both of which—unlike those in *Lycopodium* (Fig. 561)—take part in the further development of the embryo; no suspensor is formed. The embryo is exoscopic, i.e. the shoot pole is directed towards the archegonial neck from the start; the first leaves arise in a whorl encircling the stem apex which grows by an apical cell with three dividing faces (Fig. 91A). The primordium of the first root is lateral to the main axis (Fig. 577C) and the root grows downwards through the prothallus (D).

The species of the genus *Equisetum* occur in generally moist habitats from the Tropics to the colder regions. The teeth of the sheathing leaves often exhibit guttation. Whilst our native species reach 2 m in height (*E. telmateia*), some tropical representatives (e.g. *E. giganteum* in S. America) are scrambling climbers some 12 m in length.

Systematics. The morphological uniformity of the living horse-tails contrasts strikingly with the diversity of the fossil Equisetatae. Only the most important groups are considered below.

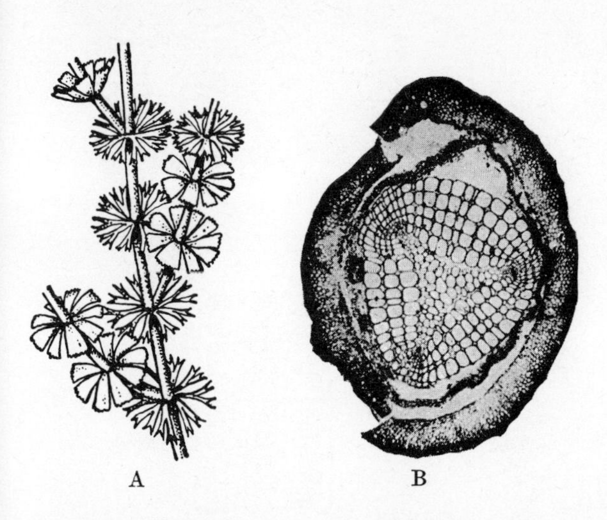

Fig. 578. *Sphenophyllum.* **A.** *S. cuneifolium*, part of shoot with dichotomous and entire leaves. $\times\frac{1}{3}$. **B.** *S. plurifoliatum*, T.S. of the shoot axis; at the centre the triangular primary xylem with three protoxylem groups, surrounded by secondary xylem. $\times 7$. **A.** After Hirmer.

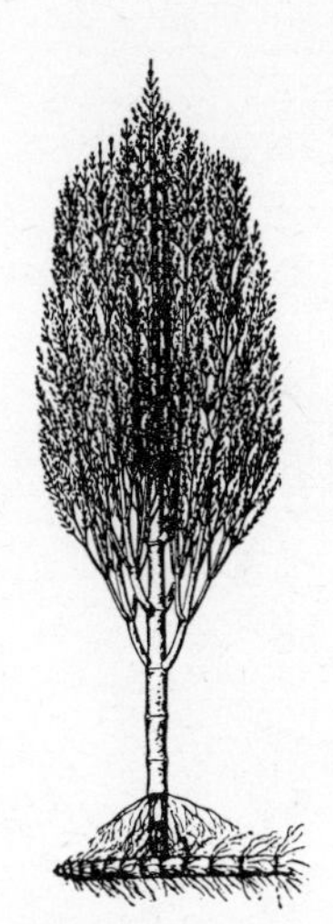

Fig. 579. *Calamites carinatus.* Reconstruction. $\times\frac{1}{200}$. After Hirmer.

Order 1: **Sphenophyllales.** These Palaeozoic fossils (from the Upper Devonian to the Permian) were characterized by whorls of (generally six) forking or wedge-shaped leaves with many dichotomous veins (Fig. 578A). The sphenophylls were herbaceous plants about a metre long, probably existing as scrambling climbers comparable in habit to the living *Galium* species. The slender, little-branched stems had long internodes; the vascular bundle was triarch with secondary thickening; the tracheids had reticulate and bordered pits (Fig. 578B). The rather complicated cones were either homosporous or heterosporous.

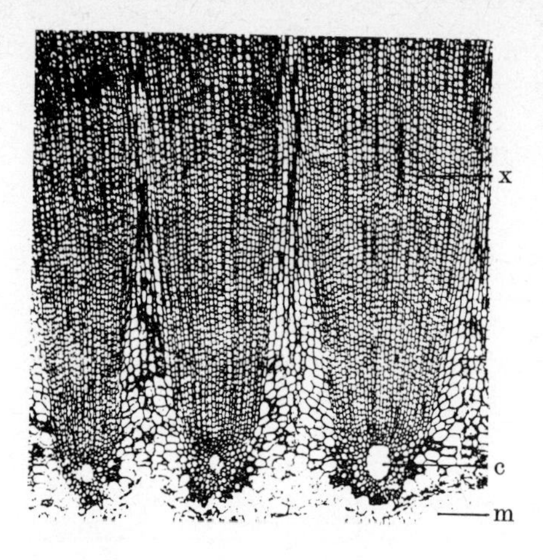

Fig. 580. *Arthropitys communis*. T.S. of part of the wood. ×10. c, carinal canal; m, medulla; x, secondary xylem. After Knoell.

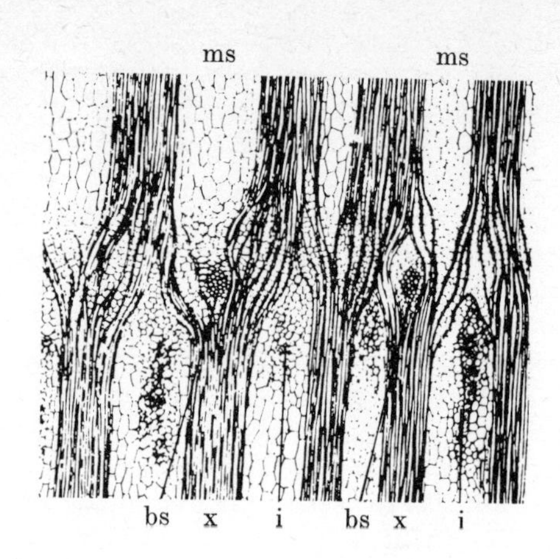

Fig. 581. *Arthropitys communis*. Tangential section of a young stem. ×10. bs, leaf trace; i, infranodal canal; ms, medullary ray; x, xylem. After Scott.

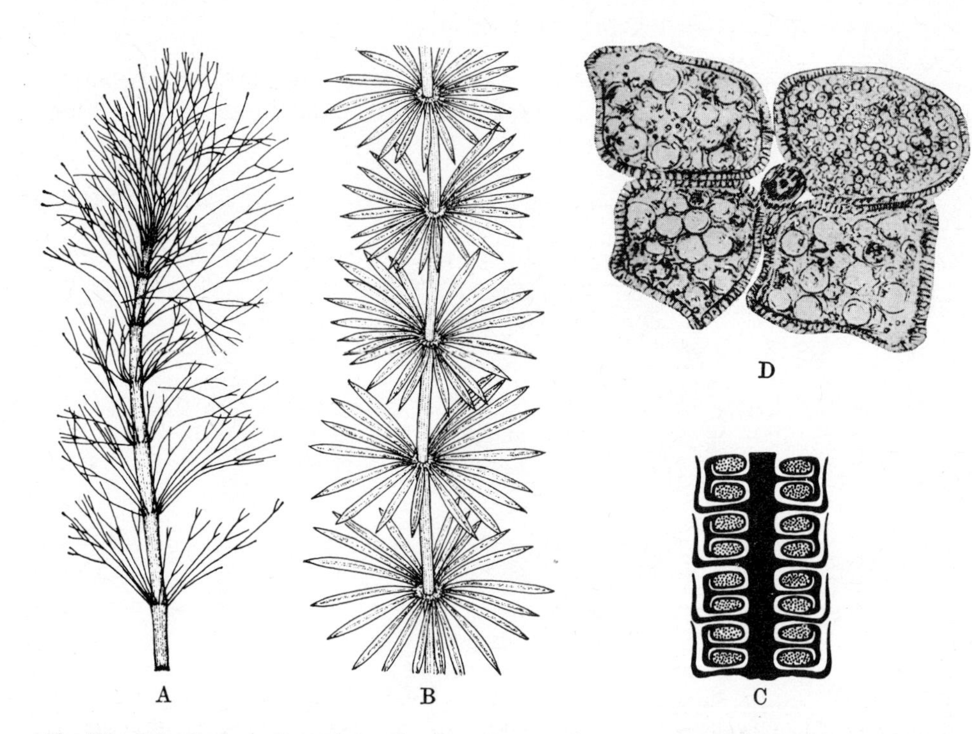

Fig. 582. Calamitaceae. **A.** *Archaeocalamites radiatus*. ×$\frac{1}{3}$. **B.** *Annularia stellata*. ×½. **C.** *Calamostachys binneyana*, cone in L.S. with sterile leaves. ×4. **D.** *Calamostachys casheana*, tangential section of sporangiophore bearing three megasporangia and one microsporangium. ×22. **A.** and **B.** After Stur. **C.** After Harder, from Scott. **D.** After Williamson and Scott.

Order 2: **Equisetales.** From the end of the Devonian to the present these were the principal group; the central medullary cavity was characteristic and was surrounded by a ring of collateral vascular bundles; in arborescent Palaeozoic representatives secondary thickening occurred.

Family 1: **Archaeocalamitaceae.** The genus *Archaeocalamites*, confined to the Lower Carboniferous, had dichotomous leaves (Fig. 582A) standing in superposed whorls corresponding to the vascular bundles which ran straight across the nodes.

Family 2: **Calamitaceae.** *Calamites*, widespread in the Upper Carboniferous and

Permian, formed a major component of the coal-forming forests along with the lepidodendroids and sigillarias. Some species attained a height of 30 m, and massive secondary thickening allowed diameters of up to 1 m (Figs. 579, 580). Nevertheless, as in *Equisetum*, there was a large central medullary cavity ('hollow trees'). The stems of most species bore whorls of branches, but a few were simple. The vascular bundles (as in *Equisetum*) forked at the top of each internode; two neighbouring forks, one from each of two neighbouring bundles, united at the node to form a bundle of the next internode, whilst a third bundle passed outside as a leaf-trace (Fig. 581). Infranodal canals, formed by the dissolution of thin walled cells, ran radially and probably served for aeration. The leaves (Fig. 582B) were simple and lanceolate with a single nerve and, like the leaf sheath teeth of modern horse-tails, had a hydathode at the tip. Corresponding to the alternation of the vascular bundles in successive internodes, the leaves stood in alternate whorls. In the cones, whorls of peltate sporophylls alternated with whorls of lanceolate bracts (Fig. 582C). Both homospory and heterospory occurred (Fig. 582D). The species had no haptera.

Family 3: **Equisetaceae.** First occurring in the Carboniferous and continuing to the present-day, this family reached its peak of development in the Mesozoic. It differs from the calamites in the absence of bracts in the cones and in the possession of haptera on the spores.

The class **Equisetatae** was most extensively developed in the Palaeozoic; it is extinct except for the genus *Equisetum* (cf. Fig. 614). Even this is only the relict of a once greater development since in the Mesozoic there were arborescent forms of *Equisetites* with secondary thickening. Although relict, our living horse-tails cannot be regarded as derived from heterosporous Palaeozoic representatives since heterospory has developed from homospory, but never the reverse. The recent horse-tails must be derived from extinct homosporous ancestors. Although many extinct forms (*Calamites, Sphenophyllum*) were heterosporous, the seed habit (as in the Lepidospermae) was never, as far as is known, developed in the Equisetatae.

The Equisetatae and Lycopodiatae thus represent two parallel branches of the evolutionary tree, characterized by the possession of simple microphylls (or derivative forms), and both very probably stemming from the Psilophytatae.

The common characteristics of the Equisetatae are: leaves which are small in comparison with the stem and which in contrast to all other Pteridophyta are arranged in whorls; a stem distinctly divided into nodes and internodes, generally bearing whorls of branches; sporophylls always different from the vegetative leaves and generally peltate, with a number of pendulous sporangia on the underside, and aggregated into terminal cones. In living species the prothalli are green and develop outside the spores.

Class IV: Filicatae (Filicopsida), ferns

All Filicatae have large, usually stalked, megaphylls (macrophylls) or fronds with abundant veins and numerous sporangia on the underside; the young frond is inrolled at the tip (circinate; exception Ophioglossales). The stem is usually monopodial (p. 132). The evolution of the lamina from telomes, and the migration of the sporangia to the dorsal surface by the greater development of the upper side of the leaf can be visualized as in Fig. 583. A fossil intermediate stage is shown in Fig. 584. Even the large pinnate frond could have arisen in this way (see the Primofilices). The living ferns embrace three subclasses: **Eusporangiatae, Leptosporangiatae** and **Hydropterides.**

The Eusporangiatae are so named because the mature sporangia have a thick wall composed of several layers of cells (Fig. 591C), whilst in the Leptosporangiatae the wall has a single layer (Fig. 552). In the Eusporangiatae the sporangium develops from a multicellular primordium whereas in the Leptosporangiatae it arises from a single epidermal cell (Fig. 552A). The sporangium wall also has a single layer in the Hydropterides (Fig. 608F).

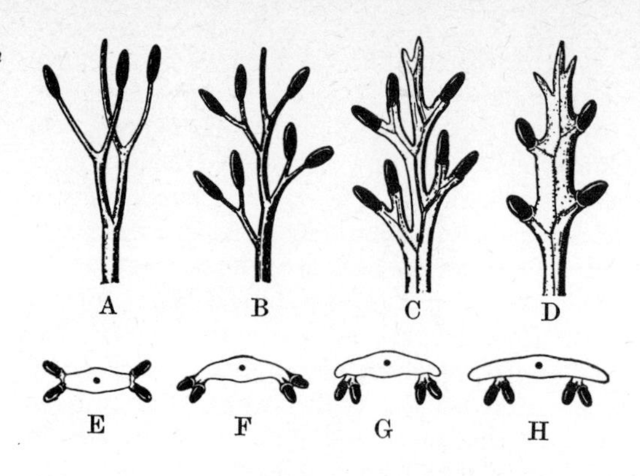

Fig. 583. Formation of the fern megaphyll from telomes (**A.** to **D.**) and the shift of the sporangia to the dorsal side of the leaf (**E.** to **H.**). After Zimmermann.

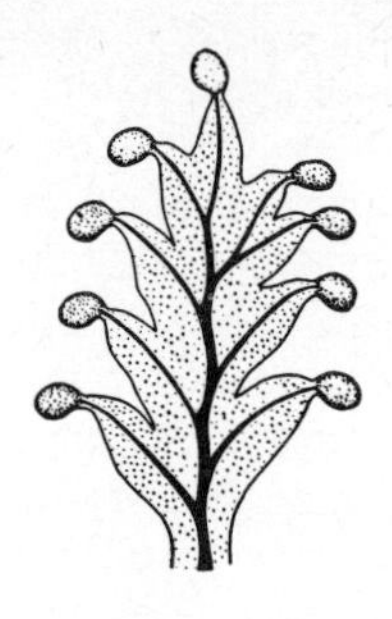

Fig. 584. Sporophyll of *Acrangiophyllum* from the Upper Carboniferous. ×7. After Mamay.

The Eusporangiatae form no periplasmodium; furthermore, this is absent from a number of families of the Leptosporangiatae.

A. Primofilices

The Psilophytatae are regarded as the ancestors of the Filicatae. The Primofilices formed the link between these two classes, having on the one hand certain features recalling the psilophytes, on the other some strikingly advanced characters. All Primofilices had terminal sporangia (Figs. 586, 588) and they usually had three-dimensional fronds, i.e. the pinnae did not lie in one plane. The transition from the Psilophytatae to the Primofilices was so gradual that one can be in doubt whether a form like *Protopteridium* (Fig. 585) belongs to one class or the other. The phylogenetic position of the Primofilices corresponds to their position in time: they appeared in the Middle Devonian and became extinct in the Lower Permian.

In *Cladoxylon* (Fig. 588) the terminal branches of the telomes were already developed as small, flat, wedge-shaped, irregularly dichotomous leaflets (B) or as flat fan-shaped groups of sporophylls with marginal sporangia (C). More extensive lateral fusion of the telomes resulted in larger flat leaves with dichotomous venation, as seen today in *Adiantum* (Fig. 177), but which already occurred in the Upper Devonian in some Primofilices (*Archaeopteris hibernica*, Fig. 589).

Venation gradually progressed from dichotomous to reticulate; in the Upper Devonian only flabellate venation with forking branching of the veins occurred, in the Lower Carboniferous was pinnate venation, whilst in the Upper Carboniferous appeared reticulate venation, which is the most efficient in supplying the leaf with water and nutrients (Fig. 587).

It is essential for the development of such flat leaf surfaces that the telomes should come to lie in one plane, but in the primitive forms the telomes were sometimes disposed at right angles to each other (as still occurs in the Ophioglossales!); furthermore the 'leaves' were often terete and unflattened. Both of these features existed in the Upper Carboniferous *Stauropteris* (Fig. 586); nevertheless such cylindrical leaf telomes already showed palisade parenchyma in some cases.

The Primofilices were generally homosporous, but heterospory occurred very early in a few (*Archaeopteris*, Upper Devonian, Fig. 589). All had thick-walled sporangia, i.e. were eusporangiate; dehiscence mechanisms like annuli occurred very seldom (Coenopteridales).

The primitive sporangia were solitary and terminal (Figs. 586, 588), but comparatively early they became aggregated on narrow sporophylls interspersed with sterile pinnae (Fig. 589), or the sporophylls themselves were grouped together, although scarcely into 'cones' or 'flowers'. A similar progression of diverse forms is manifested in stelar structure (from protostele to eustele).

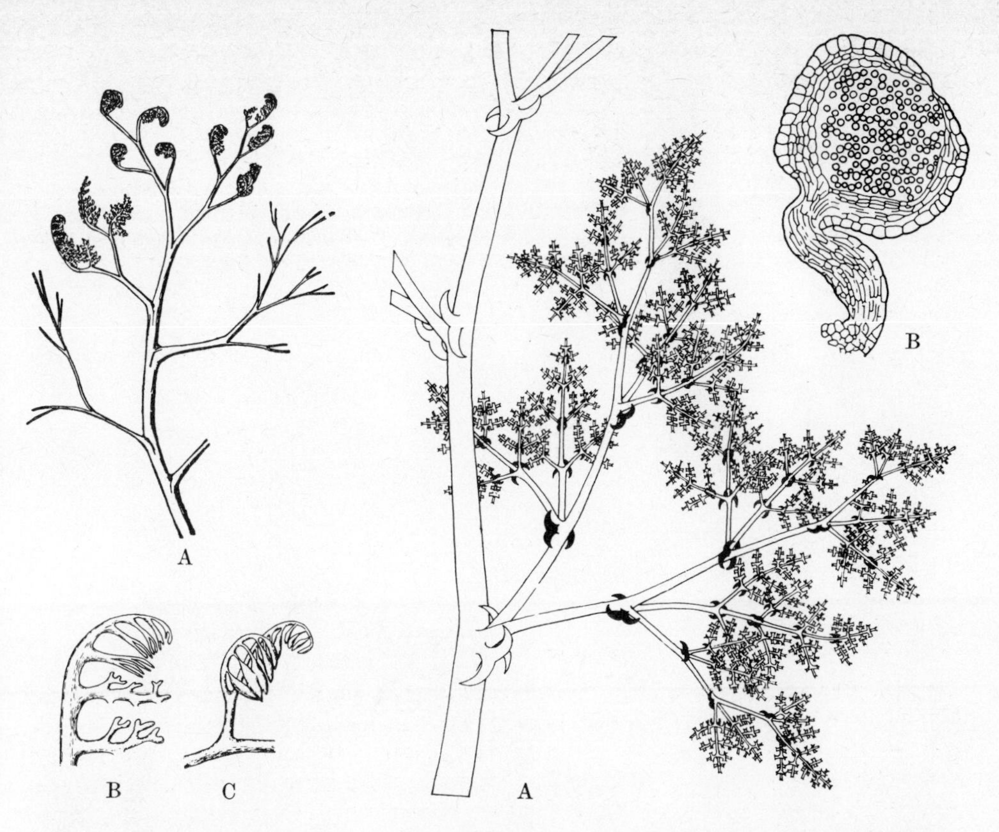

Fig. 585. *Protopteridium hostimense*. Devonian. **A.** Frond. ×¼. **B.** Sterile pinnae. **C.** Fertile pinnae. ×3. After Kräusel and Weyland.

Fig. 586. *Stauropteris oldhamia*. Carboniferous. **A.** Portion of sterile frond; reconstruction. Nat. size. **B.** Sporangium with point of dehiscence. ×35. **A.** After Chaphekar, modified. **B.** After Scott.

The origins of all the features characteristic of the living Filicatae can thus be seen amongst the amazing diversity of form shown by the Primofilices; the megaphyllous Filicatae can be derived from the Psilophytatae as a branch parallel to the microphyllous Lycopodiatae and Equisetatae.

Systematics. The Primofilices were an extremely heterogeneous group, and can be divided into four orders. The Gymnospermae must have arisen from amongst these (cf. p. 612).

The Lower to Middle Devonian **Protopteridales** (with *Protopteridium*, Fig. 585; *Aneurophytum*) more or less resembled the psilophytes in frond structure. The **Coenopteridales** (Upper Devonian to Lower Permian, with maximal development in the Lower Carboniferous) invariably had three-dimensional fronds (*Stauropteris*, Fig. 586; *Botryopteris*; the climbing fern *Ankyropteris*; and many others). The sporangia of some forms already had an annulus. The shrubby **Cladoxylales** (*Cladoxylon*, Fig. 588) lived from the Middle Devonian to the Lower Carboniferous; the structure of the stele, consisting of numerous V-shaped bundles in transverse section, differs from that of all other vascular plants. The **Archaeopteridales** (Upper Devonian to Lower Carboniferous) included the then world-wide genus *Archaeopteris* (Fig. 589), still primitive in the form of its pinnae and the position of the sporangia, but nevertheless highly advanced in its quite flat fronds, its heterospory and in its arborescent habit (with secondary thickening).

B. Eusporangiatae

Order 1: **Ophioglossales.** The single family **Ophioglossaceae** contains about eighty species,

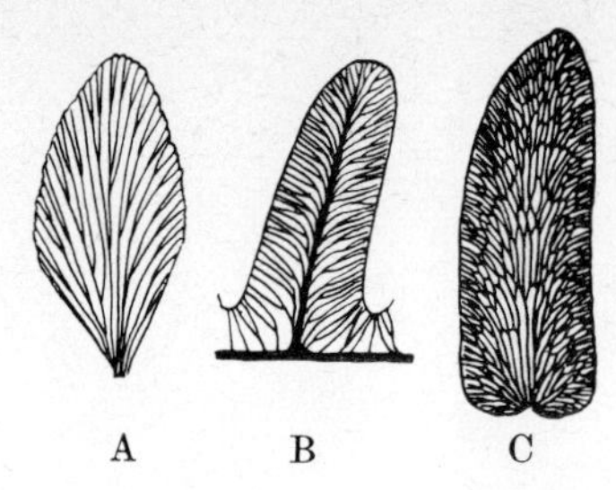

Fig. 587. Venation of ferns. **A.** Flabellate (*Archaeopteris hibernica*, Upper Devonian). **B.** Pinnate (*Alethopteris costei*, Upper Carboniferous). **C.** Reticulate (*Linopteris obliqua*, Upper Carboniferous). ×½. After Seward and Gothan.

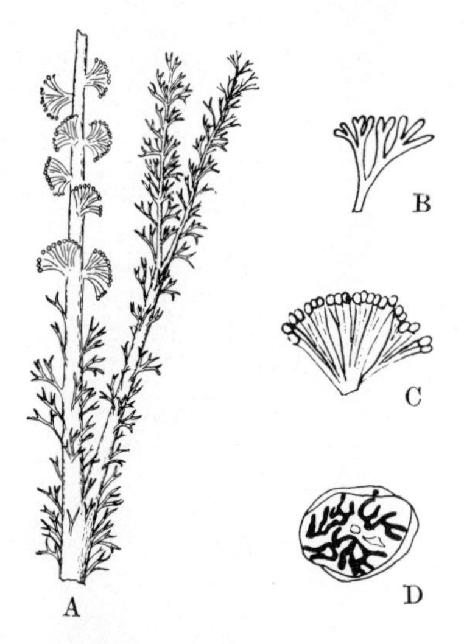

Fig. 588. *Cladoxylon scoparium*. Middle Devonian. **A.** Portion of branch. ×⅔. **B.** Leaflet. ×2. **C.** Group of sporangia. ×2. **D.** T.S. of the plectostele. ×4. After Kräusel and Weyland.

Fig. 589. *Archaeopteris*. Upper Devonian. Vegetative (v) and fertile (f) pinnae. ×½. Groups of micro- and megasporangia (mi, ma). ×10. After W. P. Schimper and Arnold.

e.g. *Ophioglossum vulgatum*, adder's-tongue (Fig. 590A), and *Botrychium lunaria*, moonwort (Fig. 591A), which are native in meadows and pastures. Both have short underground stems, the oldest parts of which are protostelic whilst the younger are siphonostelic with leaf gaps. *Botrychium* is unique amongst living ferns in undergoing a little secondary thickening. The growing point consists of a group of meristematic cells and not of a single large apical cell. Generally the stem produces annually a single long-stalked leaf which is provided with a membranous sheath. The leaf of *Ophioglossum* is tongue-shaped, of *Botrychium* pinnate; in neither is it circinate when young. The mycorrhizal fungus always present in the roots probably augments the supply of nutrients; this is suggested by *O. simplex*, in which the leaf generally has no photosynthetic tissue, merely bearing sporangia. Normally the leaf consists of an assimilatory portion (but without palisade) and a fertile portion which arises from the base of the lamina and perpendicular to it (Figs. 590A, 591A); the whole leaf is thus three-dimensional and shows a very ancient type of branching (cf. Primofilices, p. 580). The spore-bearing part has the lamina suppressed and in *Ophioglossum* is unbranched and cylindrical, with two marginal rows of sunken and laterally fused sporangia (Fig. 590B); in *Botrychium* it is pinnately branched with separate large, spherical, marginal sporangia (Fig. 591B, C), each containing 1500–2000 spores. The thick-walled sporangium opens by a transverse slit (B). The spores are not forcibly ejected (in contrast to the Leptosporangiatae).

The Ophioglossaceae are homosporous. The subterranean, monoecious prothalli are several millimetres long; they lack chlorophyll and live symbiotically with a mycorrhizal fungus, forming bulky, often perennial tubers, cylindrical and radial in *Ophioglossum*

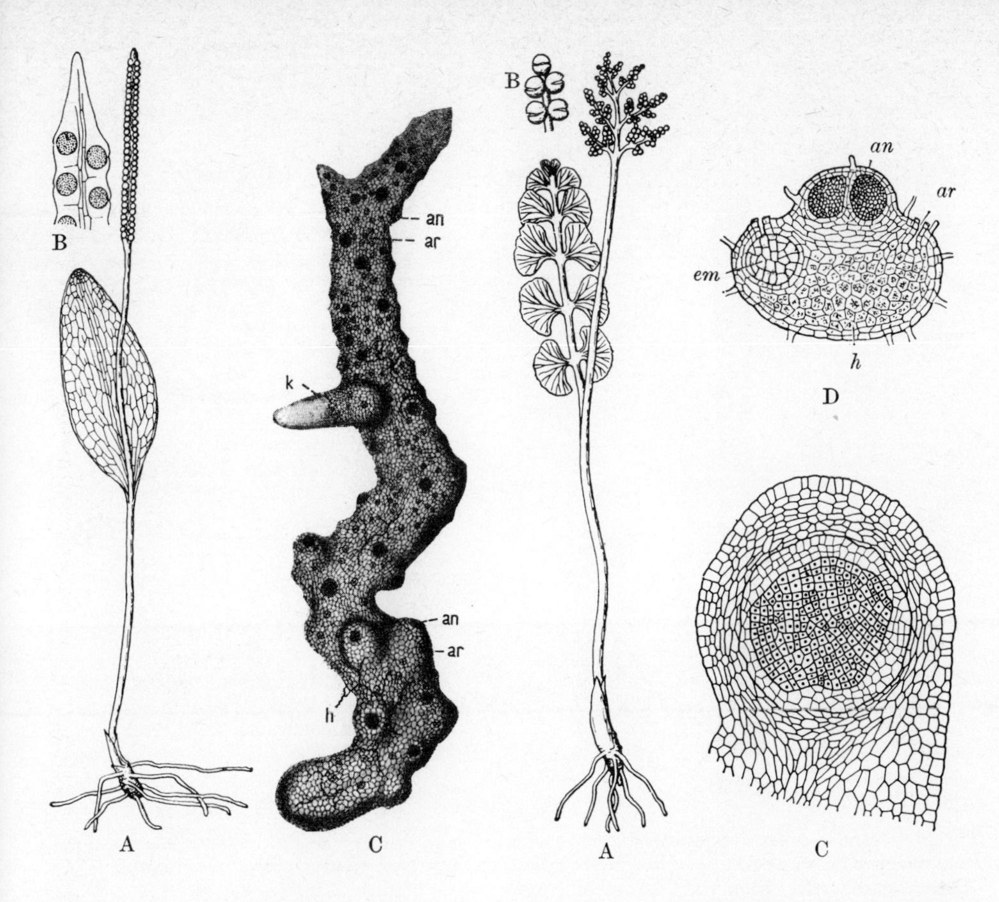

Fig. 590. *Ophioglossum vulgatum.* **A.** Sporophyte. ×½. **B.** L.S. of the tip of the fertile portion of the frond. ×2. **C.** Prothallus; an, antheridium; ar, archegonia; k, young sporophyte with first root; h, fungal hyphae. ×15. **C.** After Bruchmann.

Fig. 591. *Botrychium lunaria.* **A.** Sporophyte. ×$\frac{2}{3}$. **B.** Dorsal view of sporangia. **C.** L.S. of an immature sporangium with wall of many layers and, in the centre, spore mother cells surrounded by tapetal cells. ×15. **D.** Section of a prothallus with antheridia (an), archegonia (ar), embryo (em) and fungal hyphae (h). ×45. **C.** After Goebel. **D.** After Bruchmann.

(Fig. 590C), ovate or heart-shaped and bilateral in *Botrychium*. Archegonia and antheridia are embedded in the superficial tissues (Fig. 591D)—a primitive characteristic. In contrast to the Leptosporangiatae the antheridia contain a large mass of spermatogenous cells. The spirally coiled spermatozoids bear numerous flagella. In some species the exoscopic embryo which develops from the zygote leads a subterranean existence for several years. The root develops first, soon bursting out through the prothallial tissues (Fig. 590C k), whereas the first leaf and the growing point of the stem only develop much later. The first root is lateral to the axis of the embryo and not opposite the shoot pole. In *Ophioglossum* more copious and rapid reproduction than by spores is effected by adventitious endogenous buds on the roots (p. 178). *Botrychium* is not known fossil, but *Ophioglossum* occurs from the Tertiary to the present. However, the pinnate leaves and dichotomous venation of *Botrychium* suggest that it is phylogenetically older than *Ophioglossum* with its reduced entire leaf and reticulate venation. *Ophioglossum* has remarkably high chromosome numbers (*O. vulgatum*, n = 256; *O. reticulatum*, n = 630).

Order 2: **Marattiales.** This more primitive and geologically older order has a few tropical genera. The short, tuberous stem typically bears a bunch of pinnately compound fronds reaching several metres in length; they are circinately coiled when young. Venation is generally open. The thick-walled homosporous sporangia are borne in groups (sori) on the back of the frond; they contain over 1000 spores but have no annulus, or at most a very rudimentary one, and are generally fused laterally to form a capsule-like chambered

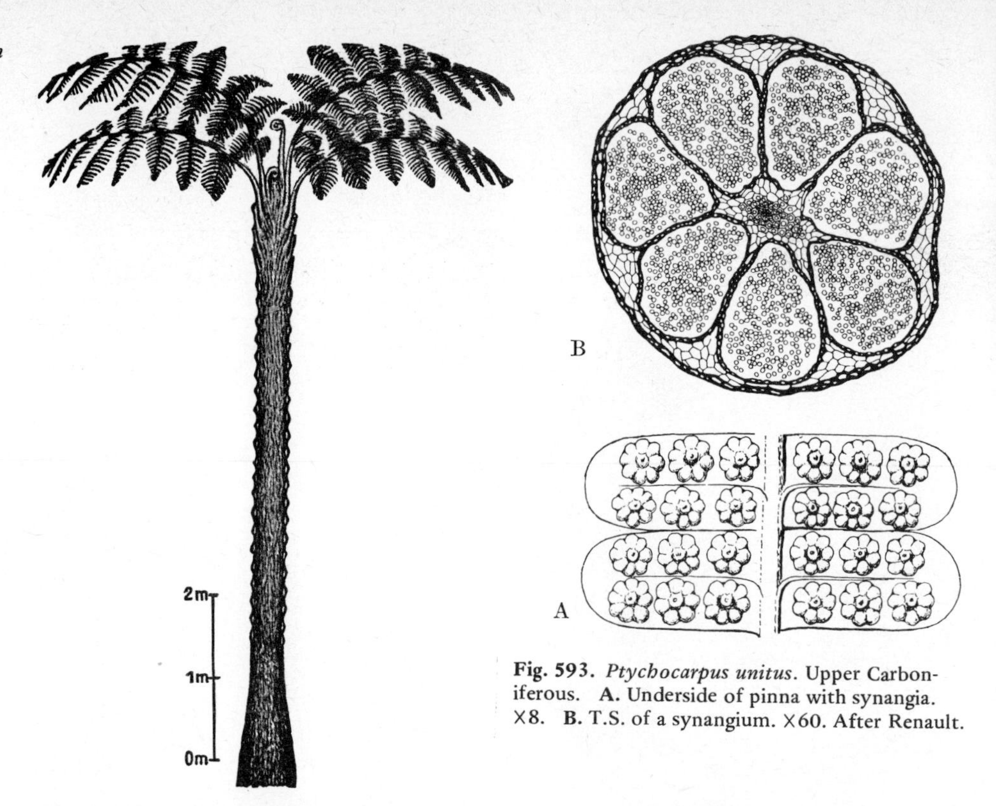

Fig. 593. *Ptychocarpus unitus*. Upper Carboniferous. **A.** Underside of pinna with synangia. ×8. **B.** T.S. of a synangium. ×60. After Renault.

Fig. 592. Marattiaceous tree-fern *Megaphyton* with two rows of leaf scars on the trunk. Base of the stem thickened with a covering of downward growing adventitious roots. Upper Carboniferous. Reconstruction. After Hirmer.

structure (synangium; Fig. 593) which later dehisces. The long-lived prothallus harbours an endophytic fungus but grows above ground as a green, autotrophic, massive structure resembling a thallose liverwort. The sex organs are embedded in the underside; the embryo is endoscopic. The Marattiales thus differ in a number of characters from the Ophioglossales and more closely resemble the Leptosporangiatae. A well-known genus is *Angiopteris*. In the Carboniferous and Lower Permian the Marattiaceae were distinctly more numerous and widespread than at the present-day, entirely predominating then over the Leptosporangiatae; some were trees of up to 10 m in height (the largest and commonest was *Asterotheca arborescens*). Particularly striking was *Megaphyton*, with distichous fronds (Fig. 592). Today the Marattiales have some 200 species in tropical forest regions, e.g. *Angiopteris* in Asia (fronds to 5 m!), *Danaea* in S. America and *Marattia* throughout the Tropics.

The first Marattiales occurred in the Carboniferous; they must have originated from homosporous Primofilices which almost invariably had thick-walled sporangia.

C. Leptosporangiatae (Filices)

The Leptosporangiatae are distributed throughout the world, especially in shady places; there is a remarkable number of species (some 9000 in about 90 per cent of all the genera of Filicatae) which reach their highest development in the Tropics, where they vary greatly in form from minute, reduced plants of a few mm (e.g. *Didymoglossum minutissimum*, fam. Hymenophyllaceae) to tree ferns some 15 m high (Fig. 595); xerophilous forms are uncommon. The unbranched woody trunks of the tree ferns (families Cyatheaceae and Dicksoniaceae, genera *Cyathea, Alsophila, Dicksonia, Cibotium*) bear a terminal, spiral rosette of pinnately compound fronds up to 3 m or more in length. Our native ferns are generally herbaceous and have a perennial, creeping or ascending, little branched rhizome (for lateral branching see p. 129). In contrast to the Eusporangiatae,

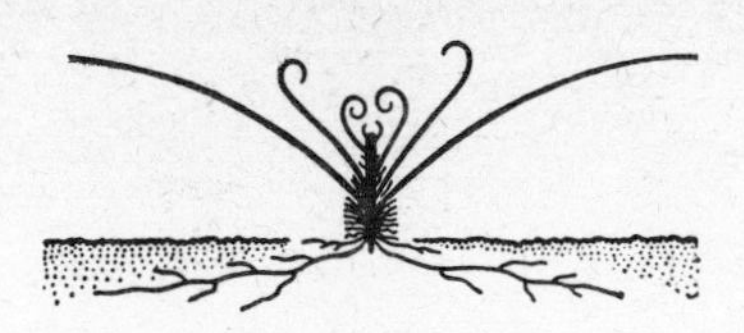

Fig. 594. *Asplenium nidus.* Habit, diagrammatic. After Troll.

Fig. 595. *Alsophila crinita.* Tree-fern from Ceylon. $\times\frac{1}{120}$. After Schenck.

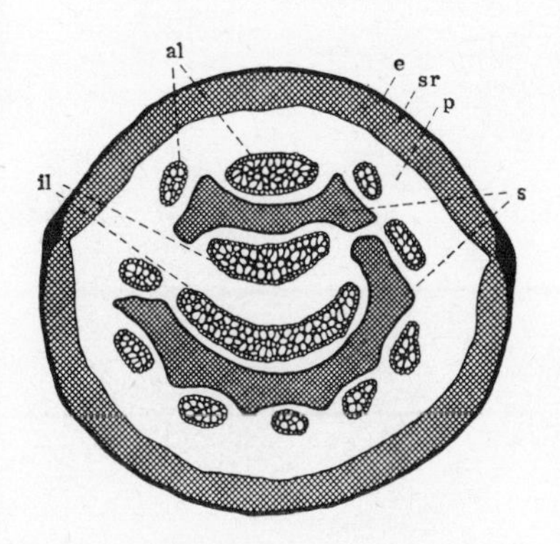

Fig. 596. *Pteridium aquilinum.* T.S. of rhizome. al, outer, il, inner vascular bundles; s, band of sclerenchyma; p, parenchyma; sr, ring of sclerenchyma; e, epidermis. ×7.

the Filices grow by a large apical cell (Fig. 94A), probably an advanced characteristic.

Some of the Filices have distinct long shoots and lateral, leaf-bearing short shoots. The stems—or in the herbaceous forms, the rhizomes—generally have a protostele at first; later this gives way to diverse types of siphonostele or polystele (Fig. 152C, D), generally with central xylem (scalariform tracheids; Fig. 118) and peripheral phloem (Fig. 596; cf. p. 113). Rarely vessels also occur (e.g. in *Pteridium aquilinum*, Fig. 118).

The stems are not supported, as they may be in the Lycopodiatae and Equisetatae, by secondary thickening, which never occurs in the Filices; instead, numerous leaf traces generally run for considerable distances in the cortex and, together with bands of sclerenchyma (Fig. 596), help to stabilize the axis (see also p. 117). In some species a remarkably thick sheath (to a few decimetres) of stiff adventitious roots enhances rigidity.

Numerous roots arise close to the shoot apex; only the underground ones become long, the aerial roots remaining short and generally thickly clothing the stem (Fig. 594). All roots except the first embryonic one are adventitious (homorrhizy, p. 177). Lateral roots arise endogenously from the inner cortex.

Whilst only a single central vein is present in the microphylls of the clubmosses and horse-tails, the megaphylls of the ferns consist of many telomes and the veins branch in various ways. Primitive, dichotomous leaves occur only rarely amongst the Filices, and in leaves with extended laminae dichotomous venation is generally confined to sporeling and juvenile leaves (Fig. 551B), whereas in adult leaves over-topping soon leads to other types of venation (but see Fig. 177), and the whole frond generally becomes pinnately compound (e.g. bipinnate, *Dryopteris filix-mas*, male fern, Fig. 598; once pinnate, *Polypodium vulgare*; but undivided in *Phyllitis scolopendrium*, hart's-tongue, Fig. 597).

The development of the leaf often extends over several years, e.g. in *Pteridium*

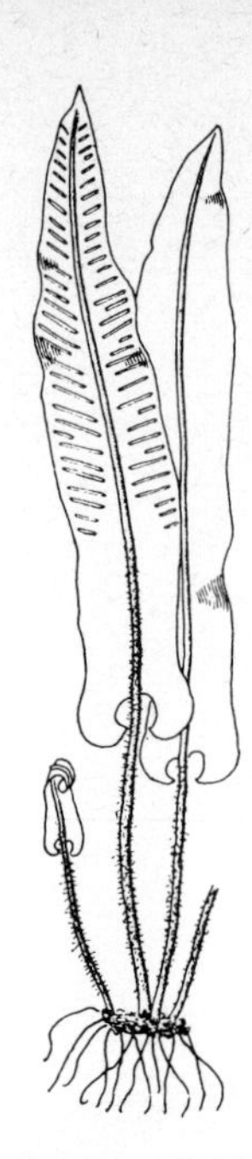

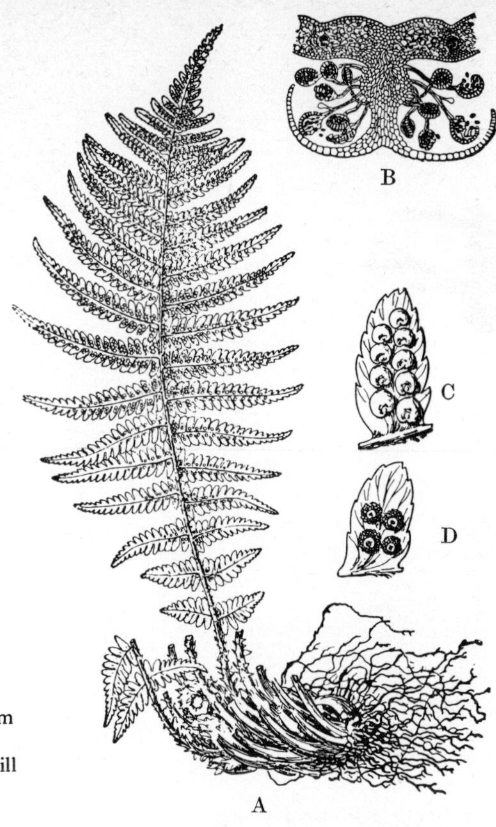

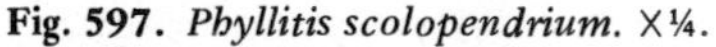
Fig. 597. *Phyllitis scolopendrium.* ×¼.

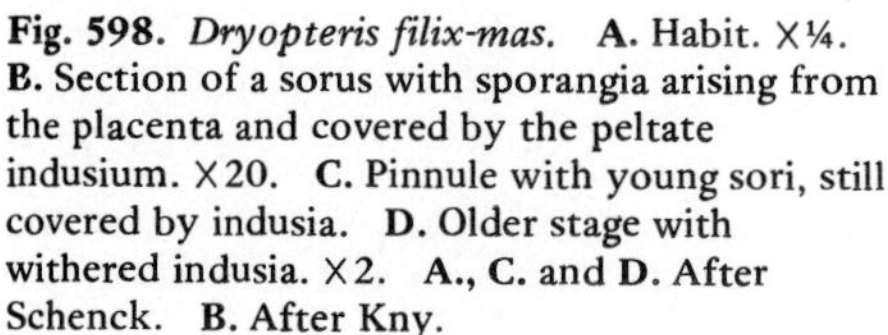
Fig. 598. *Dryopteris filix-mas.* **A.** Habit. ×¼. **B.** Section of a sorus with sporangia arising from the placenta and covered by the peltate indusium. ×20. **C.** Pinnule with young sori, still covered by indusia. **D.** Older stage with withered indusia. ×2. **A., C.** and **D.** After Schenck. **B.** After Kny.

aquilinum each underground short shoot produces annually a single aerial frond which has taken three years to reach this stage. After their death the fronds leave large, conspicuous petiolar scars, especially in the tree ferns, the fronds of which live for a number of years (Fig. 595).

As shown in Figs. 594, 595 and 597, the leaves are coiled inwards in bud (circinate) and develop acropetally, a peculiarity of all Filicatae (except the Ophioglossales): such inrolling is seen also in the shoot apices of Psilophytatae (Fig. 556A). This is one of the indications that the megaphyll has arisen from an entire shoot system.

This coiling is caused by the more rapid growth of the abaxial (under) side of the young leaf; only as the lamina expands does the growth of the adaxial side catch up. In contrast to most phanerogams, fern fronds exhibit long-continued apical growth; indeed, in a few it is unlimited, continuing for many years. The fern frond is further distinguished by the possession at first of an apical cell with two dividing faces; this is replaced by a group of dividing cells. However, the anatomy of the fern frond is very similar to that of the higher terrestrial plants, with palisade and spongy mesophyll, but the epidermis usually contains chloroplasts.

Numerous sporangia are borne on the underside of the fronds (Fig. 598B–D). As a rule, sporophylls and sterile leaves are superficially similar; only in a few genera are they essentially different (dimorphic), usually by the suppression of the lamina in the sporophyll.

Dimorphism is shown by the ostrich fern, *Matteucia struthiopteris* (native in C. and E. Europe), and the hard fern, *Blechnum spicant* (more western, including Britain), in which numerous dark-brown crowded sporophylls stand in the centre of a rosette of green fronds (trophophylls). In the royal fern (*Osmunda regalis*) only the fertile portions of the frond are modified; the rest of the lamina is normal.

The sporangia, which after the early breakdown of the tapetal cells have a wall only one cell thick, differ in detailed structure between the various families. In the Poly-

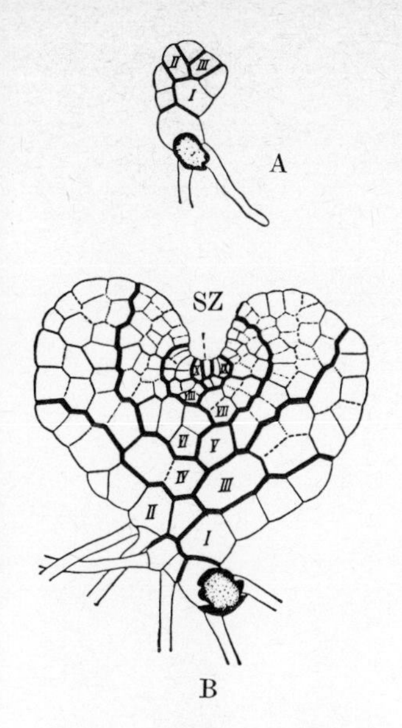

Fig. 599. Development of the prothallus of *Matteuccia struthiopteris* from the spore.
A. Eleven days old.
B. Twenty-one days. SZ, apical cell; I–X, segments cut off from it. ×70. After Dopp.

Fig. 600. *Trichomanes rigidum.* Filamentous prothallus with archegoniophores (a), one of which bears a sporeling. After Goebel.

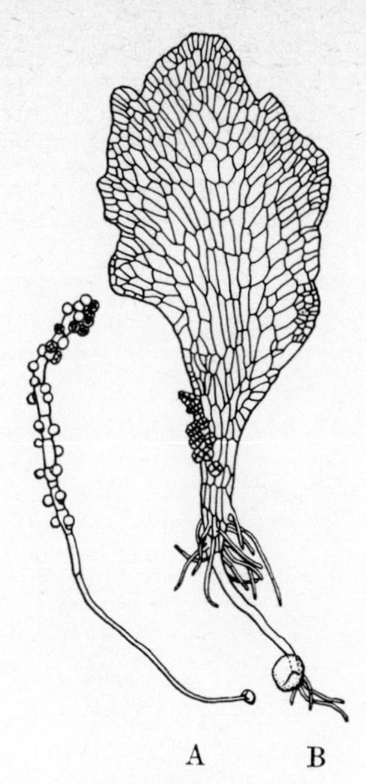

Fig. 601. *Platyzoma microphyllum.* **A.** ♂ prothallus. **B.** ♀. ×20. After Tryon.

podiaceae, to which the great majority of our native ferns belong, they are united in groups (sori) of various shapes. These arise from a projecting mound of leaf tissue, the placenta or receptacle (Fig. 598B), and in many species the developing sporangia are covered by a protective membrane, also arising from leaf tissue, the indusium (Fig. 598B, D). Each sporangium (Fig. 598B) is formed by the division of a single epidermal cell (Fig. 552), and at maturity consists of a small capsule with a wall one cell thick, borne on a multicellular stalk (Fig. 552E); it contains a large number of spores, which in some genera (*Asplenium, Dryopteris, Acrostichum* and related forms) have a perispore. Highly characteristic is the annulus which in the Polypodiaceae runs over the top and back of the sporangium to the middle of the lower side (Fig. 606D) as a series of projecting cells with strongly thickened radial and inner walls. Together with the cells of the stomium, which break apart, the annulus constitutes a cohesion mechanism which opens the sporangium and ejects the spores (cf. p. 379 and Fig. 326).

The Filices are almost entirely homosporous. The spores germinate to form a short-lived prothallus (Figs. 551, 599) never exceeding a centimetre in length and bearing both kinds of gametangia (antheridia and archegonia). Only the Australian *Platyzoma* (Polypodiaceae) is heterosporous with separate ♂ and ♀ prothalli (Fig. 601).

At first the prothallus develops as a filamentous protonema bearing rhizoids; in the rare cases in which this stage is well developed, e.g. in *Trichomanes* (Hymenophyllaceae) and *Schizaea* (Schizaeaceae) it bears antheridia directly, and archegonia on special multicellular side branches (Fig. 600). Much more commonly the filamentous stage is very brief and after forming a few cells develops a wedge-shaped apical cell, the segments from which divide further (Fig. 599A, B) forming usually a cordate, prostrate, membranous prothallus; finally, the apical cell is replaced by a group of meristematic cells (Fig. 88).

Antheridia and archegonia generally occur on the same prothallus, arising normally on the underside, which is turned away from the light and is in contact with moist soil; they are not or scarcely embedded in the tissues. The antheridia are usually formed first, the archegonia developing only when the prothallus has built up a considerable quantity of nutrients; under very poor conditions of nutrition the archegonia completely fail to develop. The sex organs differ in certain respects from those of the Eusporangiatae. The

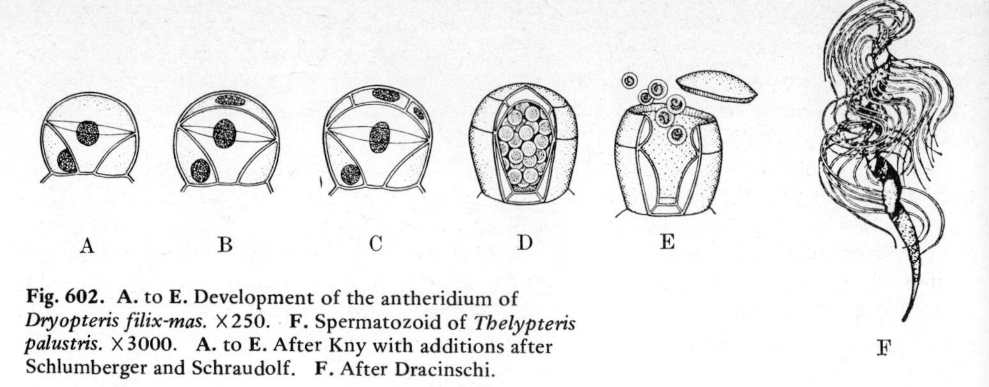

Fig. 602. **A.** to **E.** Development of the antheridium of *Dryopteris filix-mas.* ×250. **F.** Spermatozoid of *Thelypteris palustris.* ×3000. **A.** to **E.** After Kny with additions after Schlumberger and Schraudolf. **F.** After Dracinschi.

Fig. 603. Development of the archegonium from a superficial cell. (*Ophioglossum vulgatum*). ×150. After Bruchmann. Key to lettering in Fig. 604.

Fig. 604. Archegonium. **A.** Closed. **B.** Open (*Polypodium vulgare*). h, wall cells of the neck; b, basal cell; hk, neck canal cell; z, central cell; o, egg cell; bk, ventral canal cell. ×240. After Strasburger.

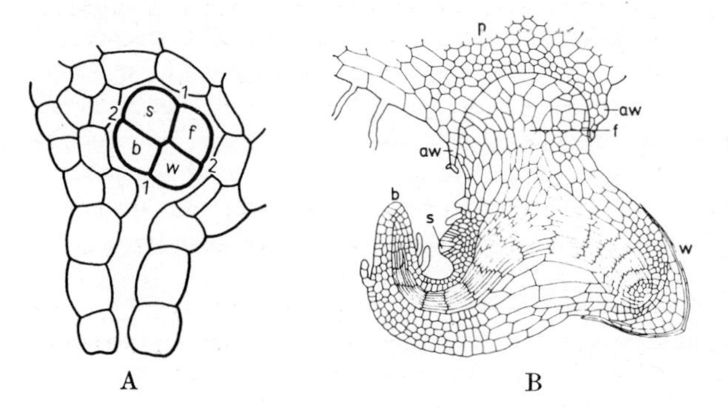

Fig. 605. Development of the embryo of *Pteridium aquilinum.* **A.** After the formation of the first walls, still in the archegonium. **B.** A later stage with the foot embedded in the expanded venter (aw) of the archegonium. f, foot; s, stem apex; b, first leaf; w, root; p, prothallus. **A.** After Zimmermann. **B.** After Hofmeister.

antheridium (Fig. 602) is an unstalked, spherical, projecting structure which sits directly on the epidermal cell from which it arose as a papillate projection cut off by a cross wall.

In the development of the antheridium, the papilla first forms a funnel shaped wall, thus dividing an outer, lower ring shaped cell from an inner, upper one (Fig. 602A); this latter cuts off by a periclinal wall (B) a helmet shaped cell from which eventually a cap cell and another ring shaped wall cell (C) are separated.

The mature polypodiaceous antheridium thus consists of two annular cells and a cap cell and contains a central cell from which the spermatogenous cells arise by division. In the ripe antheridium the wall cells are forced apart by the pressure of the cells inside (Fig. 602D, E), and the spherical sperm mother cells themselves swell up and are liberated into the water, each producing after a time a tightly wound, spiral, multiflagellate spermatozoid.

As in all the Archegoniatae, the spermatozoid (Fig. 602F) consists largely of nucleus. Initially it bears a posterior vesicle of cytoplasm containing small plastids and reserve starch, but this is cast off before it enters the archegonium. It has several dozen flagella, which arise from a delicate band of cytoplasm towards the anterior end; each terminates in a fine, whiplash-like process.

The archegonia (Fig. 604) arise by the division of a superficial cell from the thicker, multistratose part of the older prothalli.

A cross wall separates an upper and a lower cell (Fig. 603A, B). The upper cell divides by two perpendicular walls into four cells (h in C and D) which after further transverse divisions project as the archegonial neck consisting of superimposed tiers each of four cells (Fig. 604). The lower cell divides into an elongated neck canal cell and a central cell (z); the latter again divides, forming the egg cell and the ventral canal cell (bk). All the neck cells thus arise from the upper cells (otherwise than in the development of *Marchantia*, cf. p. 545). In primitive forms several neck canal cells may be present.

The lower part of the archegonium is embedded in the prothallus. The ventral and neck canal cells disintegrate by the swelling of the mucilage contained in them, filling the canal with a substance which greatly expands on the access of water. The archegonium opens at the apex (Fig. 604B) and the spermatozoids are attracted by chemotaxis to its neck.

After the formation of the first walls in the zygote (Fig. 605A) the stem apex (s) lies beside the future foot (f) of the endoscopic embryo, whilst the primordia of the first leaf (b) and of the root (w) face the neck of the archegonium. There is no suspensor, and the root arises laterally, not opposite the shoot apex. Since the archegonium is situated on the lower side of the prothallus, the shoot and first leaf of the embryo have to curve upwards geotropically after their exit from the archegonium (B). The sporophyte remains for some time united by its haustorium (f) to the prothallus until the latter dies (Fig. 605). The primary root is replaced later by adventitious roots. The orientation of the axis of the embryo can be altered neither by light nor gravity; the prothallus of the Leptosporangiatae must itself impose polarity on the cytoplasm of the egg cell. The prothalli of various families exhibit only minor differences (e.g. in the distribution of the sex organs); hence the gametophytes have little systematic importance.

Adventitious buds or bulbils (p. 204) are not uncommon on the leaves; they separate and propagate the plant vegetatively; shoots or even leaves may be modified as stolons with the same result. The normal course of alternation of generations is upset in some species by apogamy and apospory (p. 208); this occurs especially in polyploid forms, many of which in the Filicatae have high chromosome numbers.

Systematics. The leptosporangiate ferns, with some 9000 species in 250 genera, are by far the largest group of living pteridophytes. The most important families are the following:

Osmundaceae: The sporangia are not grouped in sori and have no annulus; a group of thickened cells effects apical dehiscence (Fig. 606A). Indusium and ramentum absent;

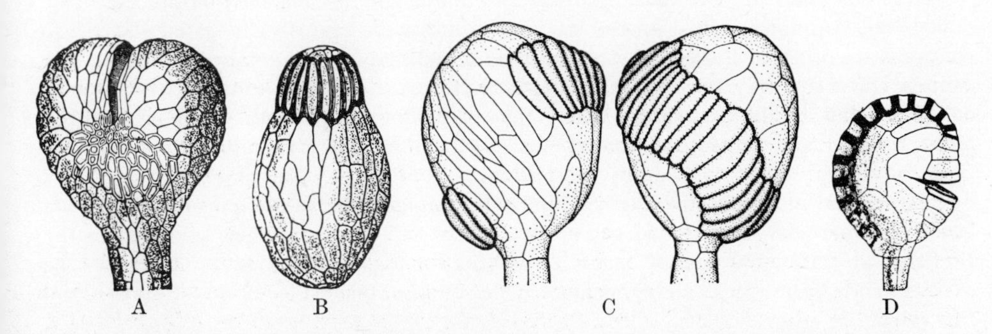

Fig. 606. Fern sporangia. **A.** *Osmunda regalis* (Osmundaceae). **B.** *Aneimia caudata* (Schizaeaceae). **C.** *Hymenophyllum dilatatum* (Hymenophyllaceae). **D.** *Dryopteris filix-mas* (Polypodiaceae). **A.** ×40. **B.** to **D.** ×70. **A.** and **B.** After Luerssen. **C.** After Bower. **D.** After Harder.

prothallus perennial and often long-lived. They first occurred in the Upper Carboniferous and exist today as a few genera, e.g. *Osmunda*, with a short rhizome, and *Todea* (S. Africa, Australia) with a thick 1 m high stem.

Schizaeaceae: The marginal, sessile sporangia dehisce by means of a longitudinal slit and an apical, transverse annulus (Fig. 606B). This family first appeared in the Upper Carboniferous and is now mainly tropical. The leaves are grass-like and dichotomous in *Schizaea*, attain lengths of 15 m (!) in the twining *Lygodium*, and in *Aneimia* are pinnate with the lowest pair of pinnae fertile.

Gleicheniaceae: Their sessile sporangia have a transverse annulus above the middle; the sori have few sporangia and lack an indusium. They are known from the Upper Carboniferous; at the present-day they are widespread in the Tropics. The frond is pseudo-dichotomous with abortive buds in the forks (Fig. 607).

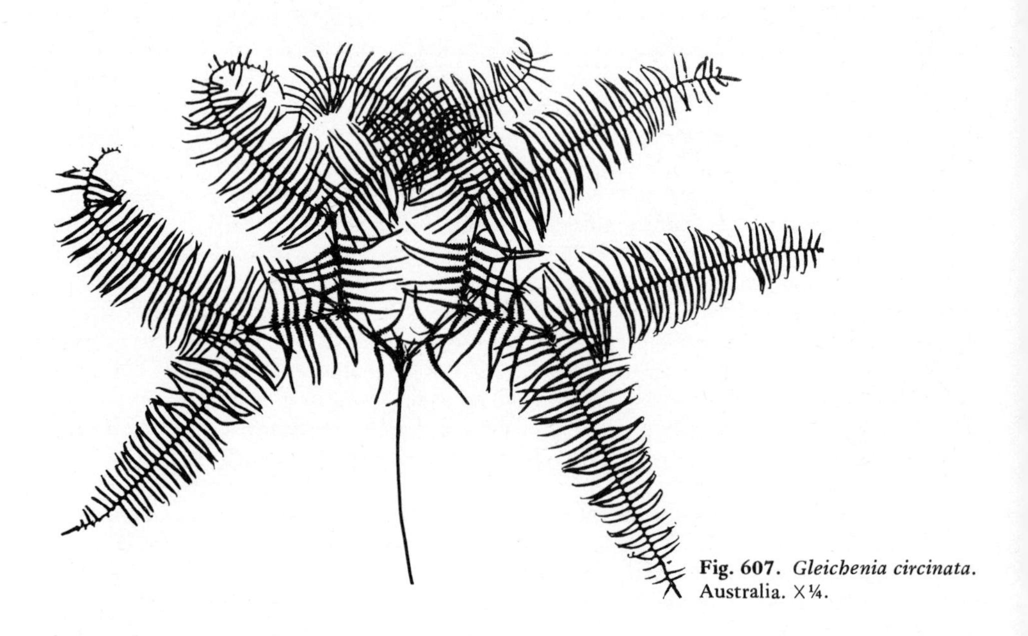

Fig. 607. *Gleichenia circinata.* Australia. ×¼.

Matoniaceae: Their sori contain a few sporangia with an oblique annulus and have a peltate indusium. They were widespread in the Mesozoic, but are now reduced to three species in Malaysia.

Dipteridaceae: Sporangia as in Matoniaceae but lacking an indusium. Fronds pedate (cf. Fig. 184E). Likewise widespread in the Mesozoic but now confined to a single genus (*Dipteris*) in S. and E. Asia.

Hymenophyllaceae: The sporangia have an oblique annulus and are almost sessile (Fig. 606C). The sori are marginal on the leaf, sometimes with a greatly elongated placenta (a continuation of a vein); they are protected by a cup-shaped or two-lobed indusium. The delicate leaves are only one cell thick and lack stomata. They are known with certainty only from the Tertiary onwards. Today there are some 650 species, predominantly in moist tropical and subtropical forests. Large genera include *Hymenophyllum* and *Trichomanes*. *Didymoglossum* is a tiny rootless epiphyte of tropical mist forests.

Cyatheaceae and **Dicksoniaceae:** The stalked sporangia have an oblique annulus and are borne in superficial or marginal sori protected by an indusium. They appeared in the Jurassic and exist today mainly as tree ferns (to 18 m high) in the montane forests of the Tropics and Subtropics. Large genera include *Alsophila* (Fig. 595), *Cyathea* and *Dicksonia*.

Polypodiaceae: The stalked sporangia open transversely by means of a vertical annulus (Fig. 606D); they occur in superficial sori, generally covered by an indusium. They are found fossil from the Lower Cretaceous onwards. Now there are some 7000 species in all

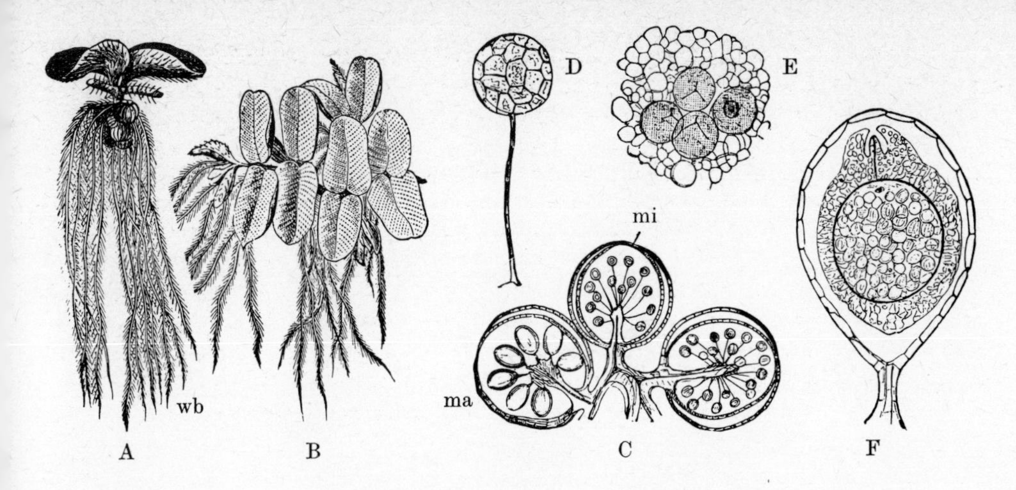

Fig. 608. *Salvinia natans.* **A.** Side view, with spherical sporocarps; wb, submerged leaves. ×¾. **B.** From above. ×¾. **C.** Megasporocarp (ma) and microsporocarps (mi) in L.S. ×8. **D.** Microsporangium. ×55. **E.** Microspores embedded in foam-like periplasmodium. ×250. **F.** Megasporangium with megaspore surrounded by perispore, in L.S. ×55. **A.** and **B.** After Bischoff. **C.** to **F.** After Strasburger.

climatic regions, but mostly in warmer areas. The majority of our native species belong here, e.g. *Dryopteris* (with a kidney-shaped indusium, Fig. 598C); *Polystichum* (indusium peltate); *Polypodium* (indusium lacking); *Phyllitis* (frond entire and tongue-shaped, Fig. 597); *Asplenium* (sorus linear); *Pteridium* (sori marginal on the pinnules, covered on one side by the indusium, on the other by the inrolled leaf margin).

Parkeriaceae: Sporangia occurring singly along the leaf veins, with a vertical annulus. *Ceratopteris* (*Parkeria*) *thalictroides* in shallow waters in the Tropics and Subtropics (p. 595).

In the first four families the sporangia all develop at the same time ('Simplices'); in the Hymenophyllaceae, Cyatheaceae and Dicksoniaceae they develop within the sorus in basipetal succession ('Gradatae'). In most Polypodiaceae the sporangia develop irregularly amongst each other ('Mixtae'). The Dipteridaceae are intermediate between the 'Gradatae' and 'Mixtae'.

D. Hydropterides, water ferns

The water ferns comprise only a few genera of herbs living in water or in marshy places; all of them are heterosporous. The mega- and microsporangia have thin walls and no annulus; they are enclosed in special sporocarps at the base of the leaves. The spores are surrounded by a peculiar perispore.

The water ferns comprise two families: the secondarily aquatic Salviniaceae (Figs. 608, 610) and the Marsileaceae (Fig. 611) which are still terrestrial.

The **Salviniaceae** are free-floating aquatics. The genus *Salvinia* is represented in C. Europe (and as an Interglacial fossil in Britain) by *S. natans*. At each node of the sparingly branched stems are three leaves of which the two upper (Fig. 608A) are ovate and floating, with abundant large intercellular spaces, whilst the lower one (wb) is divided into numerous filiform hairy segments which hang down in the water where they assume the functions of the absent roots (heterophylly, p. 169). At the base of the submerged leaf are several globose sporocarps (A); these enclose the sporangia which arise from a columnar placenta (Fig. 608C). This corresponds in origin to a modified segment of the submerged leaf, whilst the wall of the sporocarp is equivalent to an indusium two cells thick, arising below the placenta as a ring-shaped outgrowth which becomes cup-shaped and finally envelops the placenta and its sorus in a hollow sphere tightly closed at the apex.

The sporocarps each contain one sorus of either a large number of microsporangia or a

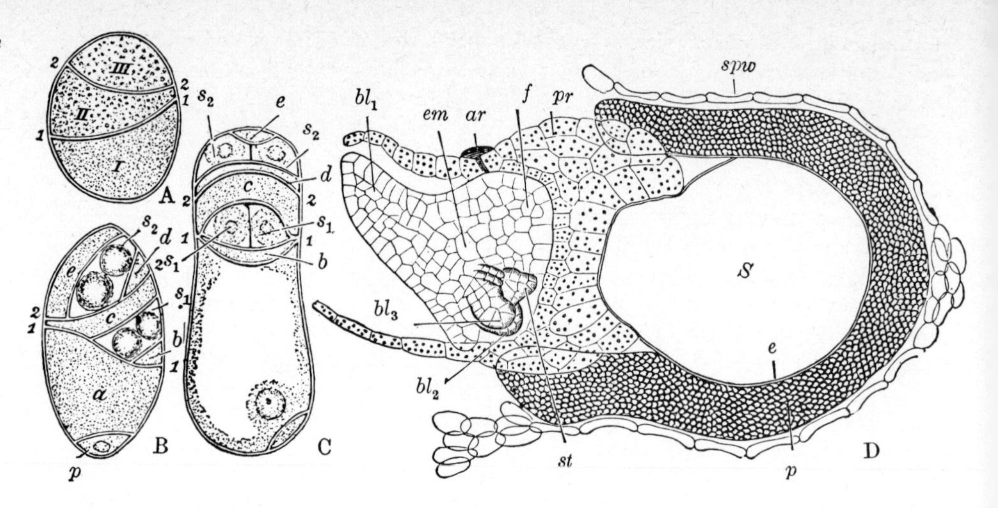

Fig. 609. *Salvinia natans.* **A.** to **C.** ♂ prothallus. **A.** Division of the microspore into three cells, I–III. ×860. **B.** Mature prothallus from the side. **C.** Seen ventrally. Cell I has divided into the prothallial cells a and p (p, functionless rhizoidal cell); cell II into sterile cells c, b and the two spermatogenous cells s_1, s_1, from each of which two spermatozoids are formed; cell III into the sterile cells e, d and two spermatogenous cells s_2, s_2. The cells s_1 s_1 and s_2 s_2 are two antheridia, the cells b, c, d, e, their wall cells. ×640. **D.** Embryo em in L.S.; prothallus pr with chloroplasts; S, spore cell; e, exospore; p, perispore. spw, spore wall; f, haustorium; bl_1, bl_2, bl_3, the first leaves; st, stem apex; ar, remains of archegonium. ×100. **A.** to **C.** After Belajeff. **D.** After N. Pringsheim.

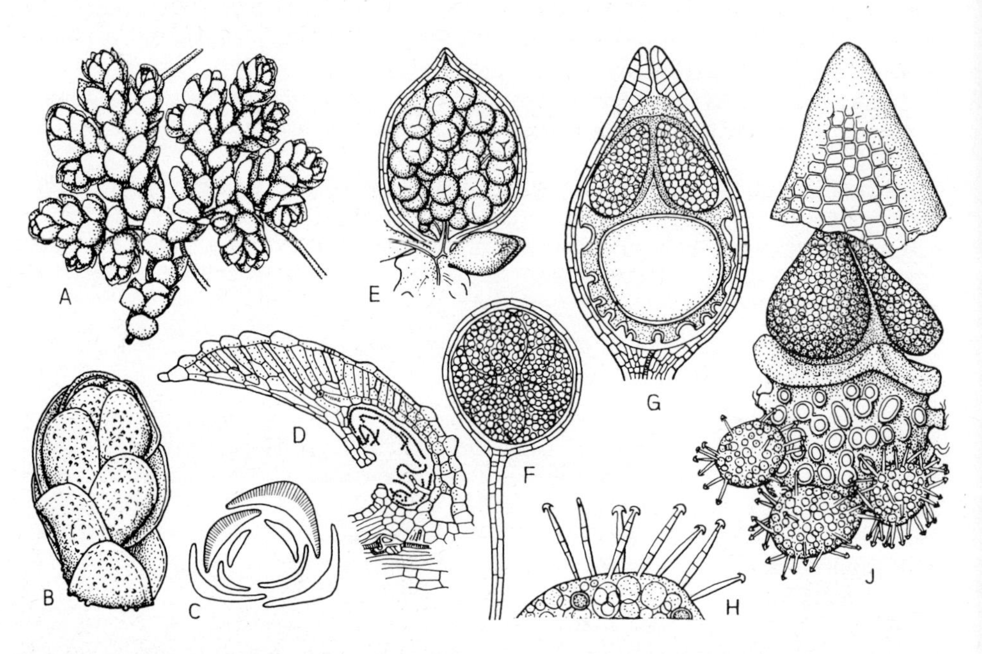

Fig. 610. *Azolla* (**A.** and **H.** *A. caroliniana*; the rest *A. filiculoides*). **A.** Plant from above. ×4. **B.** Shoot tip, from above. ×12. **C.** The same in T.S. ×12. **D.** L.S. of the upper lobe of a leaf; *Anabaena azollae* in the cavity. ×70. **E.** ♂ (above) and ♀ (below) sori. ×20. **F.** Microsporangium. ×65. **G.** Megasporangium enclosed in the indusium and containing a megaspore with 'swimming apparatus'. ×65. **H.** Part of a massula with glochidia. ×160. **J.** Megaspore removed from the upper half of the indusium to render the 'swimming apparatus' visible; three massula are anchored to the epispore by their glochidia. ×65. **A., D.** to **J.** After Strasburger. **B.** and **C.** After Goebel.

smaller number of megasporangia (Fig. 608C mi, ma); the determination of sex is phenotypic. Both kinds of sporangia are stalked, and when mature have a wall one cell thick (D, F) with no annulus; the spores develop in them after meiosis.

The microsporangia contain sixty-four microspores formed in tetrads. These lie embedded in a hardened, foam-like interstitial substance (Fig. 608E) formed from the periplasmodium. Each microspore develops as a short, tubular, few-celled prothallus which only produces two antheridia (Fig. 609B). Each antheridium has only two spermatogenous cells, each of which divides into two spermatozoid mother cells, so that one antheridium produces only four spermatozoids; these are liberated by the breakdown of the cell walls. Thus the ♂ prothallus is greatly reduced. This development takes place inside the sporangium, which does not dehisce; instead the ♂ prothalli, which are elongated almost like pollen tubes, pierce the sporangial wall in a number of places, so liberating the spermatozoids.

The megasporangia (which also have a wall one cell thick, Fig. 608F) are larger than the microsporangia, but only contain a single megaspore which has developed at the expense of the rest of the thirty-two spores originally formed. The megaspore is densely packed with protein granules, oil drops and starch grains; denser cytoplasm containing the nucleus lies at its apex. The brown spore wall (exospore) is surrounded by a dense frothy envelope (perispore), which like the interstitial substance of the microsporangium is formed from the plasmodial tapetum. The megaspore remains inside the sporangium which is liberated from the parent plant and floats horizontally on the surface of the water. On germination it forms an apical ♀ prothallus with small cells (Fig. 609D pr), whilst the large cell (S) behind it nourishes the prothallus from its store of reserve materials and does not divide further, although its nucleus undergoes free nuclear division, producing many peripheral daughter nuclei.

The spore wall opens by three valves, the sporangium wall also ruptures, and the prothallus projects as a small dorsiventral structure. Although it contains chloroplasts, it is still dependent on the reserve food of the large cell (S). A number of archegonia are formed, but only in one of these does the zygote develop into an embryo, the foot of which is embedded in the distended and eventually disrupted venter of the archegonium (Fig. 609D). If fertilization fails to occur new archegonia are formed.

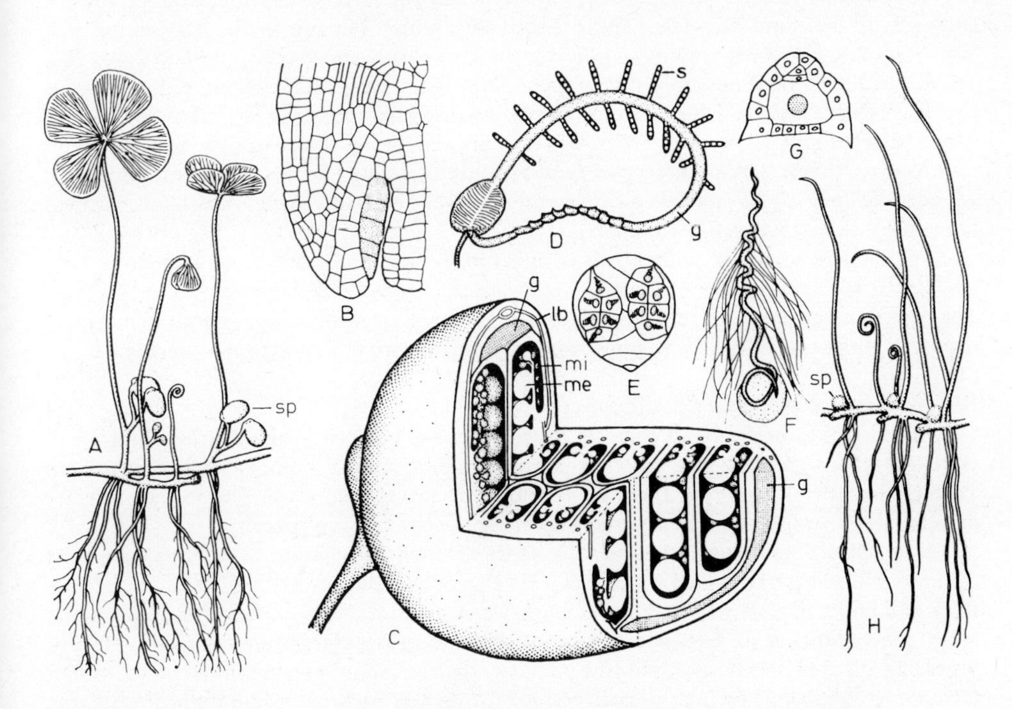

Fig. 611. Marsileaceae. **A.** *Marsilea quadrifolia*, habit. $\times\frac{2}{3}$. **B.** Section of a young sporocarp (dotted: sorus initial). $\times 200$. **C.** Mature sporocarp. $\times 8$. **D.** Dehisced sporocarp of *M. salvatrix*. Nat. size. **E.** Germinated microspore with two antheridia. $\times 150$. **F.** Spermatozoid. $\times 700$. **G.** Archegonium. $\times 150$. **H.** *Pilularia globulifera*, habit. $\times\frac{2}{3}$. g, ring of mucilage; lb, vascular bundle; me, megasporangium; mi, microsporangium; s, sporangial sacs; sp, sporocarp. **A.** and **H.** After Bischoff. **B.** After Johnson. **D.** After Hanstein. **E.** After Belajeff. **F.** After Sharp. **G.** After Campbell.

The other genus, *Azolla*, is predominantly tropical (but *A. filiculoides* is naturalized in Europe, where it is also found as an Interglacial fossil); the small, profusely branched floating plants bear two ranks of densely crowded leaves and long slender rootlets on the underside of the stem (Fig. 610A). Each leaf is divided into two lobes, of which the upper floats and photosynthesizes whilst the lower is submerged and assists in the uptake of water (Fig. 610B). On certain lateral branches the lower leaf lobes are modified as sporocarps which are covered with an outgrowth from one of the leaf lobes. The nitrogen-fixing blue-green alga *Anabaena azollae* (Fig. 610D) lives as a symbiont in a cavity of the upper leaf lobe; in consequence *Azolla* is used as a green manure in rice fields (cf. p. 447).

Azolla is of interest for the arrangements which ensure fertilization. After liberation from the microsporangium the sixty-four microspores are aggregated into five to eight roundish floating balls (massulae) formed of the frothy periplasmodium. Arising from the surface of each massula are a number of stalked, barbed hooks or glochidia (Fig. 610H, J) which are also formed by the periplasmodium. These hooks anchor the massulae to the megaspore, which is buoyed up by a special 'swimming apparatus' containing air and attached to the apex of the sporangium; this is formed from the strongly vacuolated periplasmodium (Fig. 610G, J). The macrospore then develops a prothallus as in *Salvinia*.

The **Marsileaceae**, which grow in marshy places, include the genus *Marsilea*, of which *M. quadrifolia* is one of the European species (Fig. 611A). The creeping, branched axis bears at intervals leaves with long petioles; the lamina consists of two pairs of pinnate leaflets arising in close proximity. *Marsilea* is the only pteridophyte in which the leaflets exhibit sleep movements (nyctinasty, p. 374). A pair of stalked, ovate sporocarps arises just above the base of the petiole; in some species they are more numerous. In contrast to the Salviniaceae, the sporocarp wall is homologous, not with an indusium, but with an assimilatory leaflet which by the enhanced growth of the under (dorsal) side encloses the developing sorus (Fig. 611B). The sori, consisting each of one megasporangium and numerous microsporangia, are borne in rows and enclosed in a cavity (C). A mucilaginous ring which runs through the sporocarp swells when the latter is mature, pulling out the contents of the soral sacs (D). The ♂ prothallus, which remains in the microspore (E), contains only two antheridia which develop a few corkscrew-shaped spermatozoids (F). The ♀ prothallus develops only one archegonium (G). The genus *Pilularia* with the native species *P. globulifera*, pillwort, differs from *Marsilea* in its simple, linear leaf at the base of which arises a single sporocarp (Fig. 611H); this also corresponds in development to an assimilatory leaflet. The sporocarp of *Pilularia* only contains four soral cavities. The leaves of both genera show acroplastic growth (p. 162), and like other ferns are inrolled (circinate) when young (Fig. 611A, H).

Fossil Marsileaceae are not known with certainty, but the Salviniaceae are represented by *Azolla* from the Lower Cretaceous, and *Salvinia* from the Palaeocene, to the present. The Marsileaceae may be related to the Schizaeaceae (leaf structure, marginal origin of the sori), while the Salviniaceae show some resemblance to the Hymenophyllaceae.

Distribution and ecology of pteridophytes

The Pteridophyta occur in all climatic zones but reach (especially the Filicatae and Lycopodiatae) their greatest size (tree-ferns!) and their greatest number of species in the Tropics; like the Bryophyta they are favoured by moist habitats; a few species tolerate dry climates, but not the true deserts. Boreal regions are poor in species.

Some species are cosmopolitan, e.g. the bracken (*Pteridium aquilinum*) and the common clubmoss (*Lycopodium clavatum*). The pteridophytes exhibit disjunct distributions and endemism as strikingly as the angiosperms. Since they avoid cultivated ground (with the exception of *Equisetum arvense*) their distribution—in contrast with that of flowering plants—is little modified by man's activities. Some pteridophytes occur under favourable conditions in such quantity as to form their own plant communities, e.g. the bracken (*Pteridium aquilinum*) on the edges of woodlands and the water horse-tail (*Equisetum fluviatile*) on shallow lake margins.

The fern sporophyte closely resembles the flowering plant in its water relations. The relatively few xerophytic ferns can be recognized by their waxy coverings or a clothing of scales (ramenta) or hairs of diverse forms. Those that live in moist habitats often guttate,

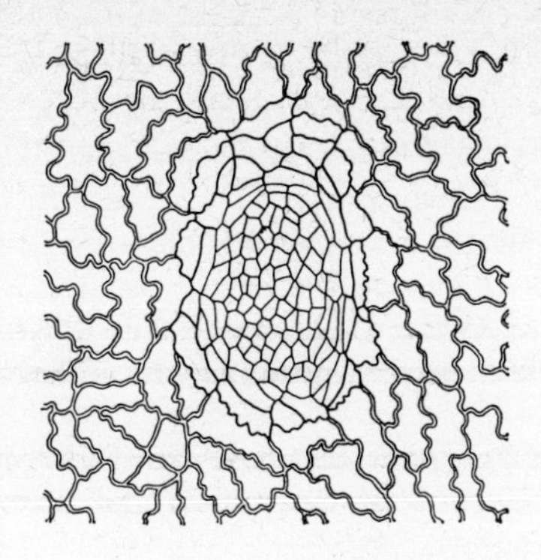

Fig. 612. *Polypodium vulgare*. Water-gland. ×80.

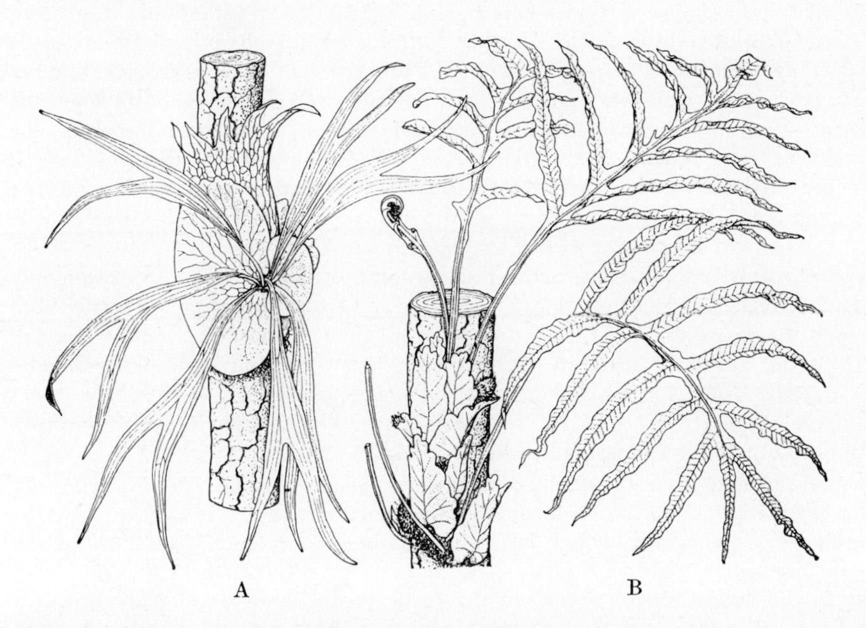

Fig. 613. Epiphytic ferns with dimorphic leaves (heterophylly; humus-retaining 'nest-leaves' and sporotrophophylls). **A.** *Platycerium alcicorne.* **B.** *Drynaria quercifolia.* $\times\frac{1}{6}$.

either through hydathodes as on the leaf-sheath teeth of *Equisetum* or from peculiar 'water-glands' in some ferns (Fig. 612). Completely aquatic ferns are few, e.g. the floating *Salvinia* and *Azolla*, and *Ceratopteris* which grows on mud or floats on water, or lives submersed. Some species of *Isoetes* grow on intermittently wet ground, others live at depths of 1–3 m in lakes where they can form extensive stands.

The Filicatae prefer shady places and correspondingly penetrate further into caves than flowering plants. Ferns can climb by various means: the tropical *Gleichenia* species and the bracken are scramblers (the latter to a length of 6 m in European forests); *Lygodium* and *Salpichlaena* climb to 15 m by means of their rhachises; tropical polypods ascend tree-trunks as root climbers. Epiphytes (p. 194) can collect humus in clasping or nest-like groups of leaves, e.g. *Platycerium* (Fig. 613A) and the polypodiaceous *Drynaria* (Fig. 613B).

Although most ferns are favoured by humus-rich habitats or more or less acid soils, some are markedly calcicolous (e.g. *Asplenium viride*). Still not understood is the existence of characteristic forms exclusive to the rock serpentine. Ferns are absent from all saline habitats with the exception of *Acrostichum aureum* which lives in mangrove swamps throughout the Tropics.

Survey of the Pteridophyta

The most primitive members of the class Psilophytatae are leafless (fam. Rhyniaceae), but the higher forms (fam. Asteroxylaceae) link up with the Lycopodiatae and Equisetae by possessing microphylls, and with the megaphyllous Filicatae by the gradual conversion of the branch system into fronds (fam. Pseudosporochnaceae).

The Lycopodiatae and Equisetatae had their greatest display both of form and number in the Palaeozoic. The Filicatae were strongly developed in the Mesozoic and have maintained themselves to a greater extent than the other two classes up to the present (Fig. 614); forms which predominated from the Carboniferous to the Trias survive only as a few relict species, whereas the families which arose in the Mesozoic are dominant at the present time.

Unlike all lower plants, the pteridophytes (except the Psilophytatae) have true roots. The first root primordium is always lateral to the axis of polarity of the embryo, and not opposite the shoot pole as in the seed plants (Figs. 567B, 605B).

The Pteridophyta exhibit a regular heterophasic alternation of generations (Fig. 550).

The considerable development and diverse form of the diploid sporophyte are made possible (as they are not in the bryophytes) by the possession of lignified vascular bundles which have become progressively more highly differentiated in the course of phylogenetic development, both for support and for the conduction of water and nutrients; the development of true roots assists this tendency. Since, moreover, the epidermis is cutinized, the shoot can grow up into the air and light, can form leaves and can assimilate carbon dioxide; it is no longer dependent on the gametophyte for the supply of organic materials, and the way is open for its further development.

The spores formed by meiosis arise in sporangia on sporophylls which very commonly differ more or less in form from the sterile leaves and which are often aggregated into special cones. In the primitive forms all the spores are alike (homosporous or isosporous), but in the more advanced they are differentiated as micro- and megaspores (heterosporous). Heterospory has arisen several times quite independently in the various classes (Lycopodiatae, Equisetatae—both in the calamites and sphenophylls—and Filicatae). Heterospory results in the formation of small ♂ and large ♀ prothalli.

The haploid gametophyte (prothallus) invariably remains thalloid; quite exceptionally it forms a few tracheids (*Psilotum*). Its development soon terminates with the production of antheridia and archegonia, which are often simpler in structure than those of the bryophytes; large gametangia with many cells are more primitive than smaller ones with few cells. In the heterosporous forms no free-living prothallus is developed; instead it grows as a few-celled structure more or less enclosed in the spore, and its entire development is gone through in a few hours. The reduction is particularly marked in the ♂ gametophytes: in extreme cases these remain enclosed throughout in the microspore wall (e.g. *Selaginella, Isoetes,* Hydropterides); they may even be reduced to a single vegetative cell bearing a single antheridium; only the spermatozoids escape from the spores. The ♀ gametophytes are never so greatly reduced since they must have at their disposal sufficient reserves to nourish the embryo; sometimes they are still green, but with extreme reduction they remain inside the megaspore (e.g. *Selaginella, Isoetes,* Hydropterides), and only after fertilization does the developing embryo emerge from the spore. This reduction of the gametophyte is emphasized in those species in which the megaspore is not liberated from the megasporangium, which in a few extreme cases (Lepidospermae, Pteridospermae) remains firmly enclosed by a sporophytic integument so that fertilization and even the formation of the embryo takes place on the parent plant, after which the embryo, sporangium and integument are shed as a unit ('seed'). The development of the seed has, like heterospory, arisen more than once. The first Pteridospermae (p. 630) occurred certainly in the Carboniferous and very possibly as early as the Devonian, i.e. they arose prior to the great mass of ferns and not as the endpoint of pteridophyte development. This development of the seed clearly indicates how other Spermatophyta evolved, probably from both groups of seed ferns (Lepidospermae and Pteridospermae; see pp. 572, 630); the Pteridophyta thus form the starting-point for all higher plants. The Spermatophyta have since conquered the earth: up to the Middle Devonian the only land plants were homosporous; today less than 3 per cent are homosporous, some 0.3 per cent heterosporous and 97 per cent are seed plants.

The phylogenetic connections of the Pteridophyta with more primitive plants are less

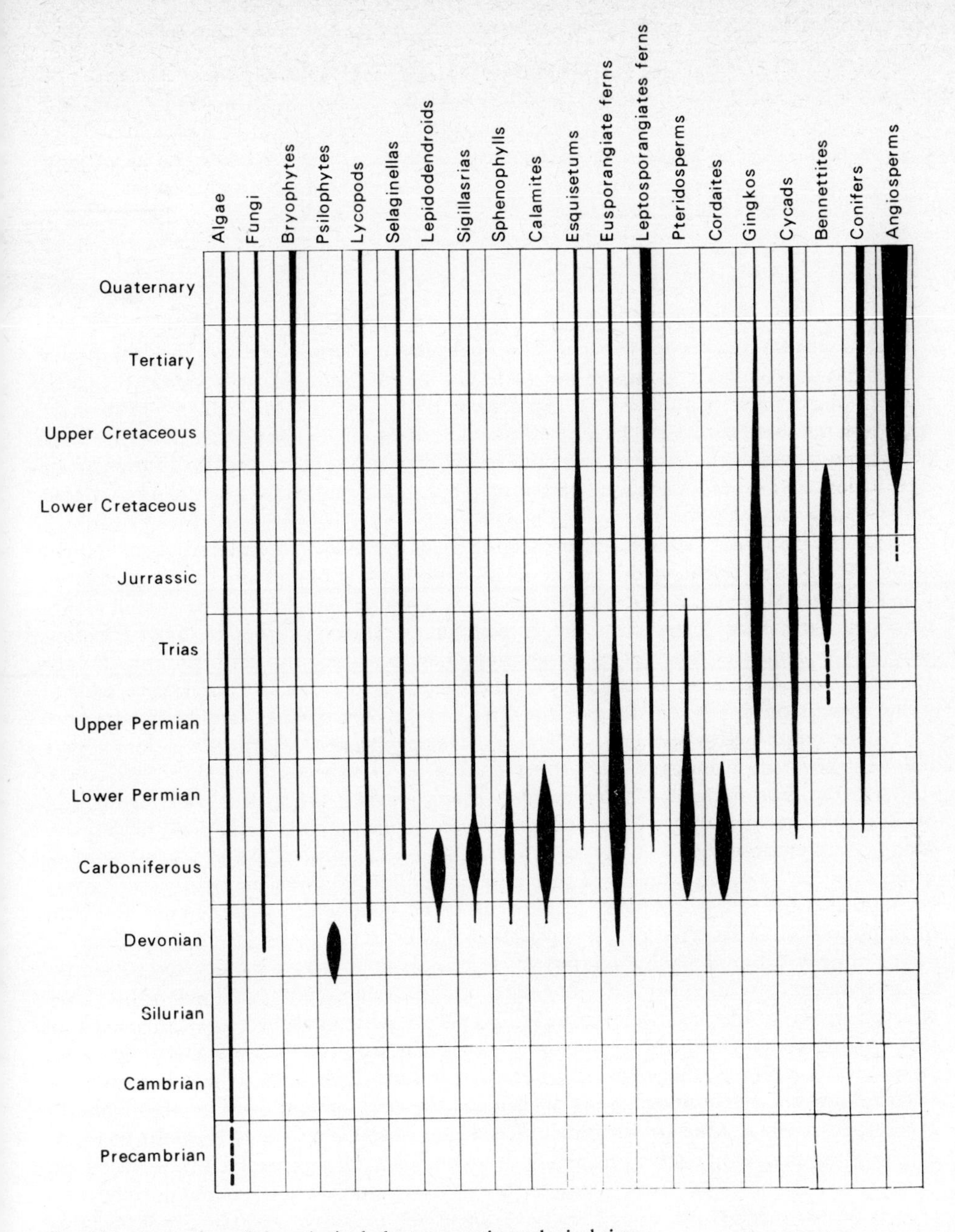

Fig. 614. Expansion of the principal plant groups in geological time.

clear. The simple, leafless form with terminal sporangia of the most primitive Pteridophyta (Rhyniaceae) is certainly reminiscent of the habit of the larger brown algae. The Pteridophyta may have arisen as a branch parallel to the Bryophyta from some yet unknown common algal ancestor. Only the Chlorophyceae can be considered in this respect, since they alone have in common with all Archegoniatae chlorophylls *a* and *b*, the same carotenoids, and flagella without flimmer hairs. While the Bryophyta form a dead end of evolution and indeed have scarcely developed further since they played themselves out in the Carboniferous some 250 million years ago, the Pteridophyta have made their biggest surge upwards since that time (Fig. 614).

Division 7: **Spermatophyta, seed plants**

The seed plants, like bryophytes and pteridophytes, undergo heteromorphic alternation of generations between gametophyte and sporophyte with a corresponding change in nuclear phase from haplophase to diplophase (Fig. 615 and pp. 207–8). The more primitive representatives still have recognizable archegonia (♀) and antheridia (♂) in the gametophyte. As in the modern fern, the sporophyte shows characteristic differentiation into shoot, with axes and leaves, and root. Seed plants are thus Cormobionta (cormophytes; pp. 83 et seq.).

Not until 1851 did Wilhelm Hofmeister discover the hidden alternation of generations in the seed plants, revealing their close relationship to bryophytes and pteridophytes. At that time special terms already existed for the various reproductive organs of seed plants. Although their homologies with the corresponding organs of pteridophytes are now generally established, two sets of terminology have remained in use to the present-day, and indeed are both used alternatively in the following account. A comparative table of terms appears on p. 754. In the past the Greek prefix 'macro-' has frequently been used for female reproductive cells or organs; nowadays the equivalent prefix 'mega-' is adopted following international usage.

As in the most advanced heterosporous pteridophytes, the spermatophytes undergo meiosis forming microspores (uninucleate pollen grains or pollen cells) and megaspores (uninucleate embryo sacs or embryo sac cells). The reduction of the ♂ or ♀ gametophytes or prothalli is, however, so marked that they are no longer visible externally and have to be nourished by the sporophyte. In particular, the mature megaspore never leaves the megasporangium (nucellus of the ovule) and thus remains within the sporophytic maternal parent. Similarly the microspores (uninucleate pollen grains) develop within the microsporangium (pollen sac); development of the ♂ gametophyte follows with at least one cell division. The multicellular pollen grains are next transported to the vicinity of the megasporangium with its ♀ gametophyte (pollination); there they grow a pollen tube, sometimes containing spermatozoids, but more frequently sperm cells lacking flagella. Then follows the fertilization of the egg cell and the development of an embryo from the zygote. At the same time an integument (seed coat or testa) is formed from the envelope of the megasporangium (the one or two integuments of the ovule), so enclosing the embryo and its nutritive tissue (endosperm). A new dispersal unit, the seed, thus arises instead of the megaspore. These modifications of the processes occurring in ferns make the reproduction of the seed plant independent of the presence of atmospheric water and give the young sporophyte a better start in life.

In the Spermatophyta the megasporangia with sterile envelopes (ovules consisting of nucellus and integument(s)) are homologous with groups of microsporangia (groups of pollen sacs; cf. p. 601). These fundamental units of reproduction are borne singly or in groups on simple or more or less elaborately branched supports which can be called mega- or microsporophylls (carpels and stamens).

The sporophylls of the Spermatophyta are almost invariably borne on short shoots of limited growth: these are spoken of as 'flowers' (or 'cones'). The seed plants can thus be called flowering plants ('Anthophyta'). A flower (or cone) is unisexual if it includes only microsporophylls or megasporophylls, hermaphrodite if both. A floral envelope (perianth) is present, especially in hermaphrodite flowers. The arrangement in the flower of the micro- or megasporophylls, which are separate from the vegetative organs, facilitates pollination as the flower opens (anthesis). Organs which enclose mature seeds and

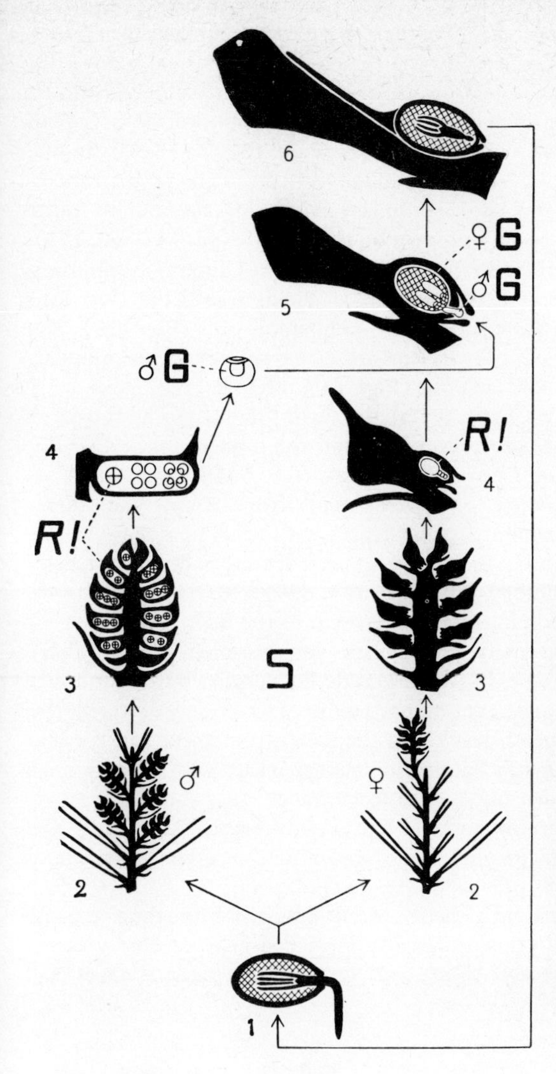

Fig. 615. Diagrammatic life-cycle of a gymnospermous seed plant (*Pinus*) with alternation of generations. Sporophyte (S) and gametophyte (G). Diplophase, solid black, meiosis (R), and haplophase (thin lines). **1.** Germinating seed with testa, primary endosperm (haploid, cross-hatched) and embryo. **2.** Shoot with axes, leaves, and ♂ and ♀ inflorescences. **3.** ♂ flower and ♀ inflorescence (young cones). **4.** *Left*: ♂ cone scale with pollen mother cell and one- and two-celled pollen grains (air-bladders not shown); *right*: cone scale of the ♀ flower below the fused 'carpels' (ovuliferous scales) with a naked ovule and its embryo sac cells (only one of four developed). **5.** ♀ flower and ovule at the time of fertilization with germinating pollen grain (♂) and ♀ gametophyte (cross-hatched; egg cells white). **6.** Ripe cone scale with winged seed. After Firbas.

sometimes ensure dispersal are called fruits.

The young seed plant can be recognized even in embryo by having opposite root and shoot poles (Fig. 131). From the root pole develops the primary root which is absent from modern ferns. Axillary branching of the shoot is also characteristic of seed plants, as are the possession of a eustele and the potential for secondary thickening. The seed plant is thus primarily a woody plant with an effective system for water uptake and transport.

Seed plants have dominated the terrestrial flora since the beginning of the Mesophytic (Upper Permian). Although traces of similar characteristics can be seen in various groups of advanced pteridophytes (e.g. reduction of the gametophytes, heterospory, flower-like aggregations of sporophylls, and even the formation of seeds; cf. p. 572), the origins of the spermatophytes must go back to psilophyte-like ancestors in the Middle Devonian and have followed a course parallel to that of the pteridophytes.

Vegetative organs. The bipolar seedling of the Spermatophyta has a shoot apex and a root apex (Fig. 131A–C). The primitive number of cotyledons is two, but they can be more (Fig. 626) or reduced to one. The apical meristems of shoot and root are multicellular and undergo progressive differentiation into layers (cf. tunica and corpus, pp. 90–1). Leaf insertion is spiral in all the more primitive seed plants, but often becomes distichous, decussate or whorled (pp. 123–9). Lateral branches are invariably borne in leaf axils (Fig. 131D). Monopodial shoots are more primitive than sympodial, undifferentiated systems more primitive than those with long and short shoots.

In the stem an arrangement of open, collateral vascular bundles in a eustele is primitive (Fig. 150). Fossil seed plants show how lobed proto- or actinosteles developed medulla and medullary rays, finally breaking up into eusteles; subsidiary trends lead to poly- and atactosteles with closed vascular bundles (cf. pp. 143–4, Figs. 152, 153B–C). In the root, medullated actinosteles with radial vascular bundles (cf. pp. 172–5) are characteristic. All the more primitive seed plants show secondary thickening: a cambium forms secondary xylem on the inside and secondary phloem on the outside (pp. 147–50, 175). Progressive differentiation also takes place in these tissues (e.g. conducting, supporting and parenchymatous cells in the xylem and phloem); tracheids finally develop into vessels, sieve-cells with fine pores into sieve-tubes with companion cells (Fig. 113). The phylogenetic series in the thickening of xylem elements evidently runs from annular, spiral and reticulate to scalariform and bordered pits (Figs. 117, 120).

The leaves of seed plants fall into two basic types: the dichotomous (forked) type in the Coniferophytina (Fig. 616A, e.g. *Ginkgo*, Fig. 177B) and the pinnate in the Cycadophytina (Fig. 616B, e.g. *Tetrastichia* or *Lyginopteris*, Figs. 641, 642B) and Magnoliophytina (Figs. 174, 183G). The sequence of development, from telome-like, three-dimensional and more or less radial segments to flat leaves with veins branching in one plane, with or without petiole, and finally to needle- or scale-leaves, is documented by fossil material and comparative morphology (cf. Figs. 627, 629, 630, 634A, 635, 641, 642B, C). This runs parallel to a reduction in the apical growth of the leaf.

The change from evergreen to seasonally green (especially summer green) foliage should also be noted, and also the progressive ontogenetic differentiation of the foliar members (prophylls and bracts, bud scales, etc., cf. pp. 169–70). In connection with the vascular supply to the leaf, holes (lacunae) develop in the stele; from these run the leaf traces (Fig. 152D). Apparently in the gymnosperms unilacunar nodes with a single trace are primitive, in the angiosperms trilacunar nodes with three traces.

All primitive Spermatophyta are shrubby or tree-like woody plants. Apparently the tree-form has given rise many times to parallel liane-, shrub- and all other growth- and life-forms of the seed plants (cf. pp. 134, 178–201; Fig. 211).

Whilst the Coniferophytina were evidently richly branched right from the start, the earliest Cycadophytina and Magnoliophytina apparently branched little or hardly at all; the stems were originally slender but became thick, bearing numerous pinnate leaves (e.g.

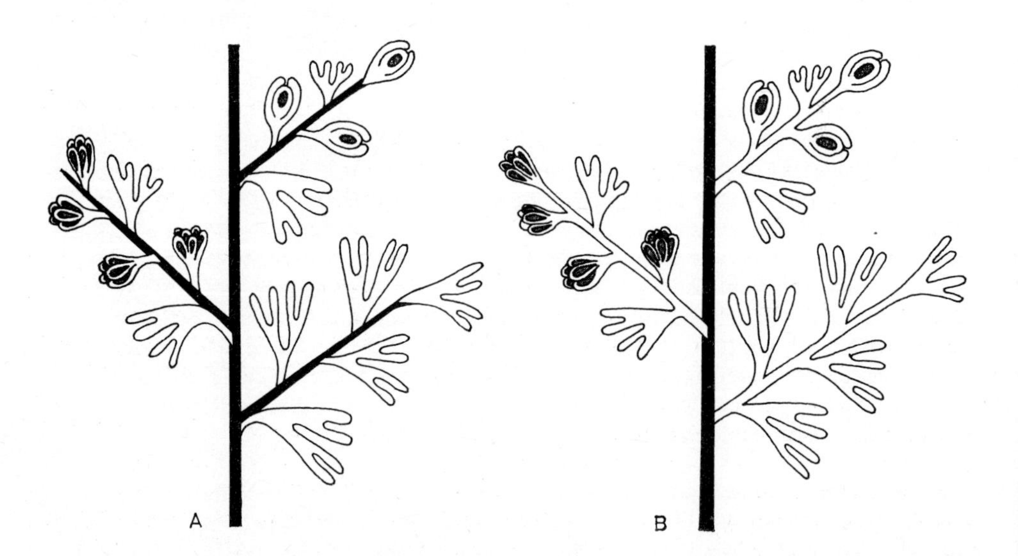

Fig. 616. Diagram of the shoot architecture of primitive seed plants of the subdivisions Coniferophytina (**A.**) and Cycadophytina (**B.**) with axes (solid black) and simple or complex vegetative and spore-bearing leaf organs (thin lines: sporogenous tissue grey): trophophylls, microsporophylls (with groups of pollen sacs) and megasporophylls (with ovules). Original.

Figs. 641, 645A). These thick-stemmed (pachycaulous) plants gave rise many times to slender-branched (leptocaulous) woody plants with repeated branching and ever thinner twigs, and correspondingly smaller and simpler leaves (e.g. Figs. 647D, 211C). Possibly this progression is related to accelerated growth (stem and leaf development quicker on smaller shoot apices than on larger), mitigation of hazards (many small shoot apices less liable to damage, e.g. by phytophagous (plant-eating) insects, than fewer larger ones) and better utilization of light (smaller leaves forming a mosaic better than larger ones).

Flowers. The flowers (or cones) of seed plants take part in sexual reproduction. The actual sexual generation (♂ and ♀ gametophytes) is progressively reduced and is transferred (along with fertilization) to the sporophyte which plays an increasing part in preparing for the sexual process, in the welfare of the developing embryo, and in its dispersal within the seed.

It is hardly surprising that the floral members were formerly regarded as the actual sex organs, and that seed plants were correspondingly known as 'Phanerogamae' (i.e. 'visibly wedded').

Flowers (or cones) are aggregations of sporophylls, i.e. short shoots of limited growth (determinate) bearing micro- or megasporophylls, or both. The development of the flower can be followed step by step in the Cycadophytina: first we see sporotrophophylls scattered on a main axis which continues to grow (as in the seed-fern *Tetrastichia*, Fig. 641); then sporophylls and trophophylls alternating on an axis with similarly unlimited growth (shown by the megasporophylls of *Cycas*, Fig. 645A–C); finally, numerous megasporophylls arranged spirally on lateral branches of limited growth (as in *Encephalartos*, Fig. 646A). There is thus a progressive division of labour between reproductive shoots (flowers or cones) and vegetative shoots as seen earlier in certain pteridophytes (e.g. in *Lycopodium, Selaginella* and *Equisetum*). In the course of evolutionary development of the flower of the seed plant, the originally numerous floral organs frequently became fewer and relatively fixed in number (oligomerization); spiral arrangement became whorled; at the same time the floral axis became greatly condensed; and a floral envelope (perianth) commonly developed, especially in hermaphrodite flowers.

The flowers of seed plants were originally unisexual (with only stamens, male = ♂, or with only carpels, female = ♀) and wind-pollinated; hermaphrodite flowers (with both stamens and carpels, bisexual = ⚥) arose later in conjunction with pollination by animals; (see pp. 658 et seq.); finally, secondary unisexuality frequently developed. Plants with ♂ and ♀ flowers on the same individual are called monoecious (♂♀, e.g. pine or hazel); those with ♂ and ♀ flowers on separate individuals are dioecious (♂/♀, e.g. yew or willows). Plants with both unisexual and hermaphrodite flowers variously distributed are termed polygamous (thus, andromonoecious = ♂/⚥, e.g. *Veratrum album*; gynodioecious = ♀/⚥, e.g. *Thymus serphyllum*; or trioecious = ♀/♂/⚥, e.g. *Fraxinus*).

In various groups of seed plants the relatively large flowers were originally solitary. With a general tendency towards reduction in the size of flowers went a compensatory development of inflorescences (cf. pp. 136 et seq.; Fig. 146).

Stamens and carpels. Comparative study of the sporangiophores of the oldest groups of seed plants (especially Ginkgoatae, Cordaitidae and Lyginopteridales) indicates that the fundamental morphological unit was the more or less radially organized group of pollen sacs and radial (or more or less compressed) ovules with a nucellus and a sterile envelope of one or two integuments (Fig. 616).

The situation in the forerunners of the seed plants ('Progymnospermae'; cf. pp. 612 et seq. and Figs. 585C, 589, 627) suggests that the groups of pollen sacs have arisen by the contraction of dichotomously or more or less pinnately branched groups of microsporangia. Corresponding tendencies amongst groups of megasporangia were apparently the differentiation of a nucellus from a central fertile sporangium whilst outer sterile branches (enveloping telomes) formed the (inner) integument; thus the ovule may have originated (Fig. 617).

In all the primitive Spermatophyta groups of pollen sacs and ovules are homologous: both are terminal on more or less radial telome-like axes, as in the psilophytes and other

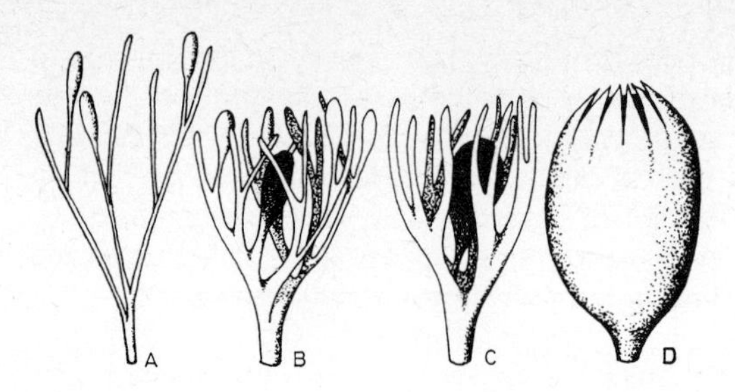

Fig. 617. Hypothetical phylogeny of the ovule: a *Rhynia*-like telome system with some sterile and some fertile axes (**A.**) becomes a fertile nucellus (black) surrounded by a sterile integument (white) (**D.**). After Walton, from Andrews.

primitive pteridophytes. In the oldest Coniferophytina the groups of pollen sacs and the ovules are always solitary and borne on short, unbranched axes directly arising from the floral axis (Fig. 616A), resembling in this the more or less dichotomous vegetative organs; but in the oldest Cycadophytina (and also the more ancient of the Pteridospermae) they appear as numerous components of complex, pinnately-branched foliar organs (Fig. 616B; p. 613). Despite the very great diversity of sporangiophores in the Coniferophytina and Cycadophytina, and despite their very slight resemblance to leaves (at least in the primitive representatives) it is customary to speak of micro- and megasporophylls, or stamens and carpels (pollen-leaves and fruit-leaves). The fossil record shows clearly that strongly flattened leaf-like 'sporophylls' in the various groups of seed plants only developed relatively late or, in consequence of various special modifications, not at all. The original homology between axes bearing groups of pollen sacs and those bearing ovules was completely obscured in many groups (e.g. in the bennettites, Fig. 647) by divergent development of the ♂ and ♀ organs.

Pollen sacs. These are homologous with the eusporangia, i.e. those microsporangia with many layers of cells found in the pteridophytes (cf. Figs. 557D, 565A, 591C, 618). A central archesporium forms the pollen mother cells; after meiosis each of these produces four uninucleate or one-celled pollen grains (pollen cell, haploid microspore); these quickly become multicellular. *En masse* the pollen grains are simply called pollen. Around the pollen mother cells lies a layer of cells, the tapetum (Fig. 618), which functions principally in the nutrition of the developing pollen grains, but also in the formation of the wall of the grain. The multi-layered wall of the pollen sac dehisces by a cohesion mechanism (p. 379) operated by a fibrous layer which in the majority of the gymnosperms is an epidermal exothecium (e.g. Fig. 646E), but is a subepidermal endothecium in the majority of angiosperms (Fig. 657E).

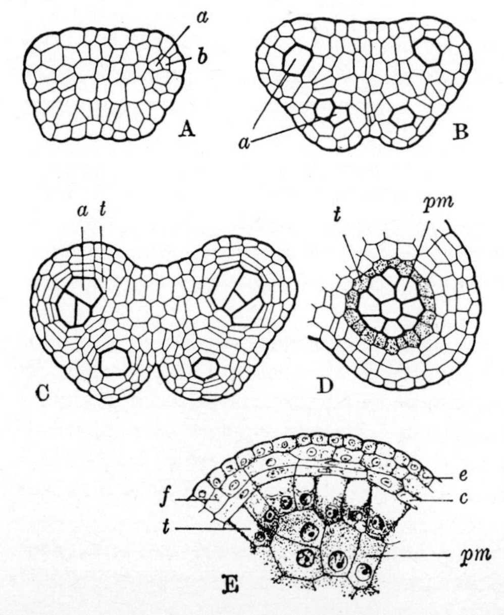

Fig. 618. Development of the pollen sacs in an angiosperm stamen. a, archesporium; b, parietal cell layer; e, epidermis; f, fibrous layer (endothecium); c, intermediate layer; t, tapetum; pm, pollen mother cells. Complete T.S. anthers of *Chrysanthemum* (**A.** to **C.**); partial T.S. of *Menyanthes* (**D.**) and *Hemerocallis* (**E.**). Cf. p. 648. **A.** to **D.** After Warming. **E.** After Strasburger.

Pollen. During transport from the pollen sacs through the air to the female floral organs the pollen grains are exposed to extreme conditions, often for long periods. The protection of their contents (the paternal genetic information!) is of course essential for reproduction; this is largely ensured by the wall of the pollen grain, the sporoderm. This consists of a complex of two layers, the exine (outer) and intine (inner); topographically they correspond to the exospore and endospore of pteridophyte spores. The exine in particular may be greatly differentiated in relation to various types of pollination (Fig. 619).

The intine surrounds the protoplast without any gaps; it is generally delicate and shows little resistance to chemical reagents. Two or three layers can usually be distinguished, the outermost often being rich in pectins so that the intine is readily separable from the extine by solution; cellulose fibrils are a significant component of the inner or middle layers. The pollen tube developed on germination consists exclusively of intine. The complex of outer layers, the exine, consists largely of the very resistant sporopollenins. These can be broken down only by oxidation; they consist of terpenes which recent studies suggest are produced by the oxidative polymerization of carotenoids and carotenoid esters. The exine is easily extracted chemically but its structure is diverse and complex. The basic units are granules some 6 nm in diameter. The exine of the gymnosperms generally has three layers: an innermost lamellar one, a middle granular and sculptured layer and an outer orbicular one; differences arise from the various configurations or even absence of one of the two outer layers. In the angiosperms (Fig. 619) one can distinguish, on a purely topographical basis, an inner, denser and more homogeneous nexine and an outer, usually more conspicuously structured and sculptured sexine; the latter in particular is often very complex in its make-up (pp. 649 et seq.).

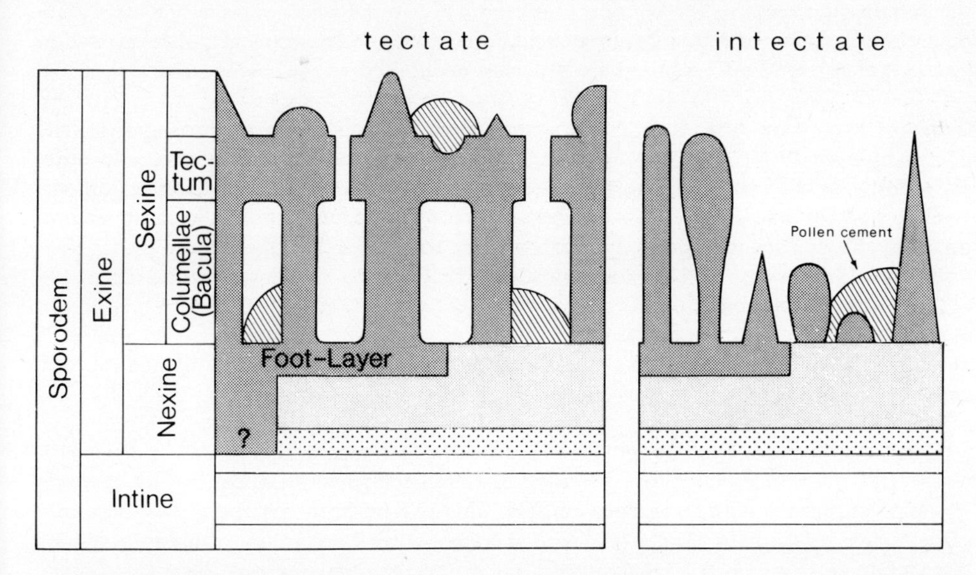

Fig. 619. Diagrams of the fine structure of the pollen grain wall of angiosperms with various forms of element. Ektexine dark grey, endexine dotted, intine white. For further explanation see pp. 603 et seq. and 649 et seq. Sketch by Teppner, after Erdtman, Faegri and others.

Germination areas or apertures are generally preformed in the solid exine; the intine often protrudes through these in the form of a papilla when the pollen grain is both young and moist, eventually growing out as a pollen tube after pollination (Figs. 660A, 623, 632, 666).

The pollen grains of primitive seed plants often had only indistinct areas for germination (leptomata); only later did definite apertures develop, i.e. holes through part or the whole of the exine. Apertures were originally elongate (colpi), but they eventually developed as round pores and other complex germination areas. Quite commonly apertures became reduced again, and such pollen grains are called atreme. Pollen grains

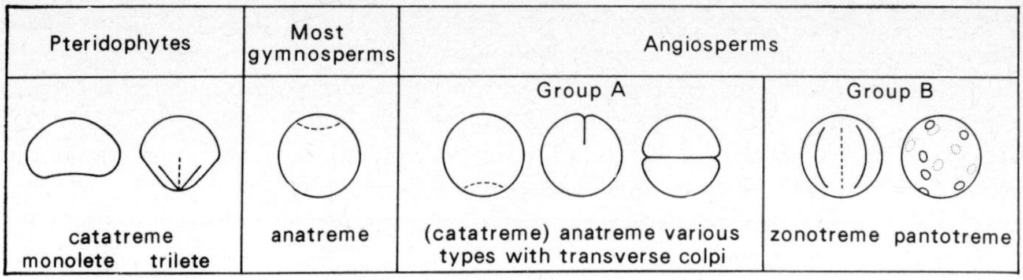

Pteridophytes	Most gymnosperms	Angiosperms	
		Group A	Group B
catatreme monolete trilete	anatreme	(catatreme) anatreme various types with transverse colpi	zonotreme pantotreme

Fig. 620. Arrangement of the germination areas of the spores or pollen grains of vascular plants; proximal pole below, distal pole above. For further explanation see pp. 603 et seq. and 649 et. seq. (Monolete, trilete: with simple or three-fold germination areas.) Sketch by Teppner, after Erdtman, Canright and others.

without apertures or only with leptomata are called inaperturate; with apertures, aperturate.

The arrangement and number of the apertures are important pollen characters (Fig. 620). The pole of a pollen grain pointing towards the centre of a tetrad is called proximal, the opposite pole is distal. The line through the two poles is the polar axis; perpendicular to it is the equatorial plane. When the germination area of a pollen grain or spore lies at the proximal pole, it is called catatreme; this type is evidently primitive, being found in many ferns, but only very rarely amongst seed plants (e.g. in pteridosperms and Annonaceae). In contrast anatreme pollen grains with a germination area at the distal pole are common in the gymnosperms and especially amongst primitive angiosperms. The shift of the germination areas to the equator and eventually to the whole surface, the increase in their number from one to more than a hundred, and the cohesion of pollen tetrads or even larger aggregations of pollen are progressions limited to the angiosperms (Fig. 659).

Ovules. These are the highly characteristic integumented megasporangia of the Spermatophyta (Figs. 621, 624A). Some 0.1–1 mm in length, they are generally ovate and consist of a stalk zone, the funicle; a solid core, the nucellus; a basal region, the chalaza; and the one or two coats or integuments. The integuments arise from the base of the ovule, leaving at the opposite pole an entrance to the nucellus, the micropyle.

The posture of the ovule on its funicle may be (1) erect (atropous or orthotropous), (2) inverted (anatropous), or (3) transversely bent (campylotropous) (Fig. 621E, F, G). Modifications generally appear to go in the direction 1 → 2, or 1 or 2 → 3. As the ovule develops, the nucellus is formed first, then the integuments which grow up and over the nucellus (Fig. 621A–D). If two integuments are present the inner generally develops before the outer.

Corresponding to their origin from investing telomes (pp. 602, 633; Figs. 617, 644A–D), the integuments of primitive seed plants sometimes consist of separate segments furnished with a vascular supply; this is true both for the phylogenetically earlier single integument (Fig. 644A) and the later outer (second) integument developed from a cupule (Fig. 644D). Furthermore, the telome-segments become totally fused, the vascular bundles are reduced, and sometimes the inner and outer integuments become united. Those seed plants with spermatozoids have a pollen chamber at the apex of the nucellus (Fig. 624, 644D); this chamber is reduced in advanced groups in which fertilization is by means of a pollen tube.

As in the pollen sac, so in the nucellus an archesporium gives rise to several embryo sac mother cells; but only one is functional. Meiosis leads first to the formation of four uninucleate embryo sac cells (haploid megaspores) of which three usually degenerate (Fig. 622). Although as a result of neoteny (p. 606) the remaining megaspore does not leave its sporangium (the nucellus) or the ovule, in all the more primitive seed plants it still shows a distinction between exospore and endospore, even though the cell wall is thin. It therefore corresponds exactly to the megaspore of heterosporous pteridophytes. The maternal parent has to provide for the megaprothallus and the embryo through this spore wall. One can thus actually see the physiological barrier which has to be overcome in the develop-

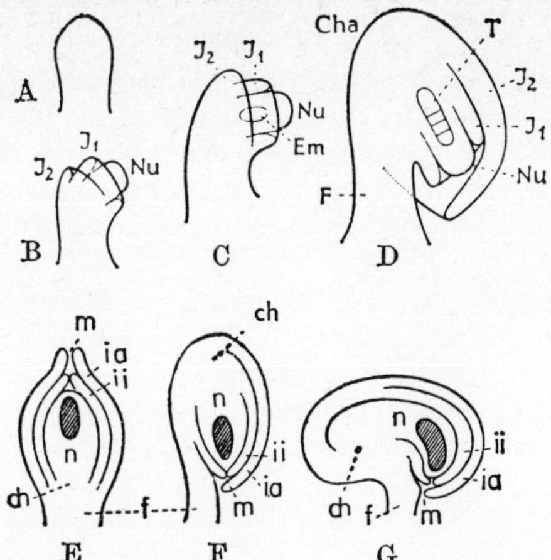

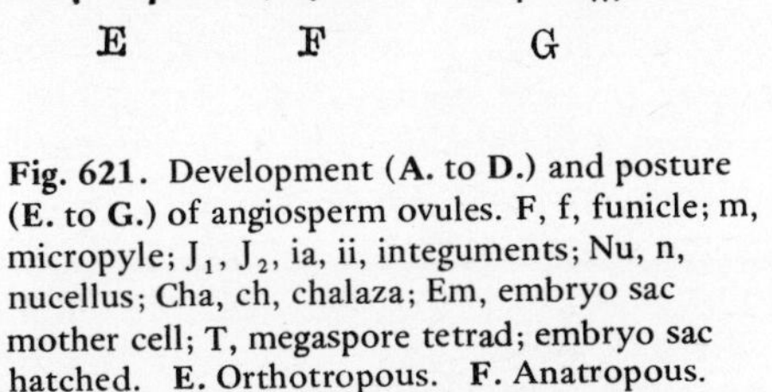

Fig. 621. Development (A. to D.) and posture (E. to G.) of angiosperm ovules. F, f, funicle; m, micropyle; J_1, J_2, ia, ii, integuments; Nu, n, nucellus; Cha, ch, chalaza; Em, embryo sac mother cell; T, megaspore tetrad; embryo sac hatched. E. Orthotropous. F. Anatropous. G. Campylotropous. A. to D. After W. Troll, diagrammatic. E. and F. After Karsten.

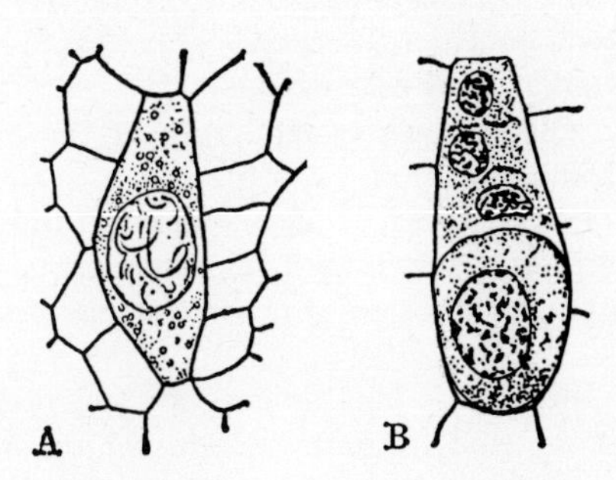

Fig. 622. Embryo sac mother cell (*Pinus nigra*) in prophase (**A.** ×400) and after meiosis (**B.** ×534). Only the lowest megaspore becomes an embryo sac. After Coulter and Chamberlain, somewhat modified.

ment of the seed and the transfer of the function of dispersal from the megaspore to the seed (cf. p. 610).

Gametophytes. The development of the ♂ gametophyte (microprothallus) begins whilst the uninucleate pollen grain (microspore) is still in the pollen sac (microsporangium) and ends after pollination of the ♀ organs. The ♂ gametophyte is even more simplified and reduced than in the heterosporous pteridophytes (Figs. 566, 573D–M, 609A–C). Development is as follows in the most primitive gymnospermous seed plants (Fig. 623). First the uninucleate pollen grain undergoes unequal cell division to form several (in *Araucaria* up to 40!) or generally only two (and eventually only one) lens-shaped prothallial cells which lie against the pollen grain wall. The residual antheridium mother cell divides into a large vegetative or pollen tube cell (with pollen tube nucleus), which fills most of the pollen grain, and a small generative or antheridial cell which lies against the prothallial cell or cells. Whilst the pollen tube cell ultimately forms the pollen tube (p. 608), the generative cell divides again into a basal stalk cell and a spermatogenous cell. The stalk cell is homologous with a sterile antheridial cell or an antheridial wall cell and functions as a 'dislocator': when it breaks down it frees the two sperm cells which arise from the spermatogenous cell and which eventually liberate two multiciliate spermatozoids. The very numerous flagella (perhaps 20,000!) have the typical 9 + 2 fibrillar structure (cf. Fig. 385) and are inserted on a spiral band (Figs. 623J, 629C). Exceptionally further divisions of the stalk cell may produce over twenty additional spermatozoids (e.g. in *Microcycas*, Fig. 623F). However, in most gymnosperms and all angiosperms the sperm cells function directly as non-flagellate ♂ gametes. Finally in *Taxus* and the advanced Gnetatae, but particularly in the angiosperms, the ♂ prothallus is reduced to three or four cells: one vegetative or pollen tube cell, one stalk cell (which is ultimately lacking) and two sperm cells (Fig. 666).

The ♀ gametophyte is not reduced so much. It develops from a uninucleate embryo sac (megaspore) in the ovule or nucellus (megasporangium). In the primitive gymnospermous seed plant the homologies with the ♀ gametophyte of the heterosporous pteridophyte (Figs. 567, 574) are clear: the megaprothallus (primary endosperm, p. 609) develops inside the large embryo sac cell by free nuclear division in the peripheral cytoplasm (some hundreds or thousands of nuclei!) and the formation of cell walls. At the micropylar pole, adjacent to the future archegonial chamber, several sunken archegonia develop, each consisting of a large egg cell, a number of neck wall cells (true neck canal cells are absent), and often also a transitory ventral canal cell (or at least a ventral canal nucleus). Already the advanced gymnosperms (Gnetaceae), but in particular the angiosperms (pp. 654 et seq.; Fig. 668), show an abbreviated development: the megaprothallus may arise from all four megaspores (tetrasporous embryo sac); sometimes the formation of cell walls and the development of archegonia are suppressed; and in extreme cases the embryo sac only contains four cells or nuclei.

The progressive reduction of the ♂ and ♀ gametophytes of the seed plants affords a striking example of neoteny, i.e. sexual maturity in progressively earlier and less differentiated stages of development. Along with the reduction of other floral organs this facilitates accelerated development, significantly quicker reproduction, and hence very often the invasion of otherwise inaccessible habitats with extreme climatic conditions.

Pollination. The pollen grain (microspore or microprothallus) of the seed plant must be transported from the pollen sac to the receptive spot of the ovule (megasporangium + integument, Figs. 615.5, 624A) or of the ovary wall (i.e. the stigma of the carpel, Figs. 651.2, 673A), where it germinates; this process is called pollination. Whilst the spores of pteridophytes are scattered in various habitats and germinate where they can, in the spermatophytes microspore transport has to be more precise. Many arrangements and structural modifications of the flowers of seed plants can be understood only as adaptations involved in the liberation, transfer and reception of pollen.

Floral biology or ecology deals with the enormous diversity of pollination mechanisms in seed plants. This field of study was founded by the penetrating observations of J. G. Koelreuter (1733–1806) and C. K. Sprengel (1750–1816). The latter wrote the famous work *The Mystery of Nature Discovered in the Structure and Fertilization of Flowers* (1793). Charles Darwin (1802–82) decisively influenced floral biology by making many original studies and by linking it to his theories of selection and evolution. After further fundamental contributions by H. and F. Müller, F. Delpino, P. Knuth and others, floral biology has advanced in recent decades especially in experimental studies and in observations in the Tropics.

Pollination may be effected between flowers on different individuals of a species (cross pollination, allogamy or xenogamy), or between flowers on the same individual (self pollination or autogamy; either within one flower, or between different flowers: geitonogamy, cf. p. 658). In monoecious or hermaphrodite plants devices are commonly present which reduce or prevent autogamy and hence inbreeding, for example, genetic incompatibility (pp. 364, 402), often occurring in gymnosperms, but particularly in angiosperms, or spatial or temporal separation (in terms of development) of ♂ and ♀ flowers or organs (e.g. in *Pinus*, Fig. 631, and many angiosperms, pp. 657 et seq.).

The biologically functional unit in pollination in seed plants is the anthium ('bloom'). This frequently corresponds to the flower, which is the morphological unit (euanthium, e.g. Figs. 647B, 652A, 671A–E), but occasionally the functional unit is only part of a flower (meranthium, partial flower, e.g. *Iris*, Fig. 734C, in which each flower consists of three functional units, each with an arching style and outer perianth segment). The functional unit may also be composed of more than one flower (plus other supplementary organs), and is then a pseudanthium (e.g. the female cone of the conifers, Figs. 631B, 639D; the cyathium of *Euphorbia*, Fig. 704H–K; or the capitula (heads) of the Compositae, Fig. 728F, G; cf. also pp. 661 et seq.).

Wind pollination (anemophily, anemogamy) is undoubtedly primitive amongst seed plants; this is indeed the basic type of spore dispersal amongst pteridophytes. The small probability of a pollen grain reaching an ovule is compensated by the massive production of pollen (cf. the 'rain of sulphur' when conifers are at anthesis); by enhanced buoyancy of the grains as a result of small size and increase in surface area, e.g. by air-bladders (Figs. 631K, 632A–D); by the secretion of a pollination droplet as a pollen trap at the micropyle of the ovule (Fig. 639D); and by the fully exposed positions of the ♂ and ♀ flowers at the tips of the branches (Fig. 631A). Indeed, anemophilous pollen can be scattered by the wind in significant quantities over hundreds of kilometres and to altitudes of

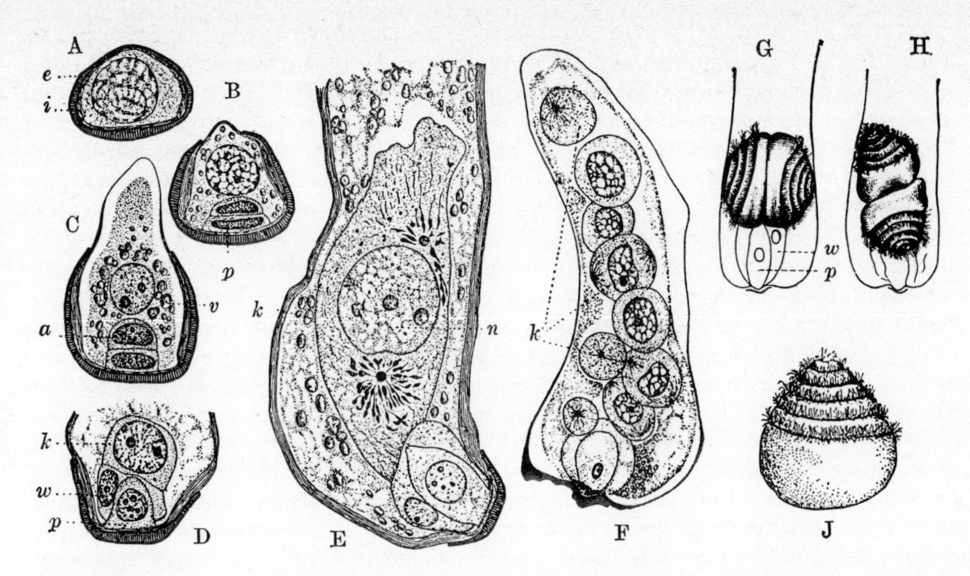

Fig. 623. Development of the ♂ gametophyte in primitive seed plants (Cycadales). **A.** to **E.** Germination of the pollen grain (wall with exine, e, and intine, i) in *Dioon edule*. **F.** Germinated pollen grain of *Microcycas calocoma* with nine spermatogenous cells. **G.** to **J.** Pollen tube and spermatozoids of *Zamia floridana*. p, prothallial cell; v, pollen tube cell; a, antheridial cell; w, stalk cell; k, spermatogenous cells (their nucleus, n; in mitosis two centrioles are formed; cf. E, F and p. 34). (See also pp. 635 et seq.). **A.** to **C.** ×840. **D.** ×667. **E.** ×420; after Chamberlain. **F.** About ×200; after Caldwell. **G.** to **H.** ×50. **J.** ×75; after H. J. Weber.

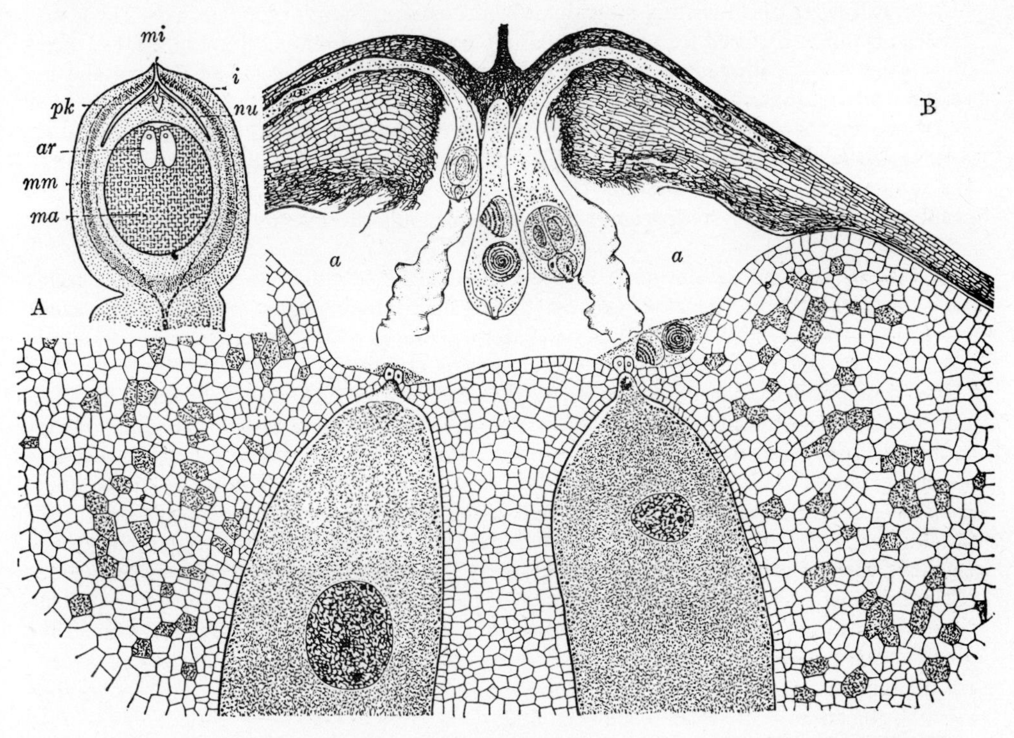

Fig. 624. Ovule and fertilization in primitive seed plants (Cycadales). **A.** L.S. of an ovule of *Ceratozamia* with micropyle (mi), integument (i), nucellus (nu) and pollen chamber (pk) with germinating pollen grains; germinated megaspore: megaprothallus (embryo sac, (ma)) with wall (mm) and two archegonia (ar, each with two neck wall cells and egg nucleus). ×2.5. **B.** Upper part of the nucellus of *Dioon edule* at the time of fertilization: pollen tubes anchored in the nucellar tissue and penetrating the archegonial chamber (a); some spermatozoids discharged; the left-hand archegonium already fertilized. About ×100. See also pp. 635 et seq. **A.** After Firbas. **B.** After Chamberlain.

1000–1500 m above the surface. Of course these devices only function when wind pollinated plants grow in exposed and windy places, are themselves tall in growth, and occur in large populations, with small distances between individuals, so enhancing the probability of successful pollination.

Even the flowers of primitive and normally wind pollinated gymnosperms are occasionally visited (more or less by chance) by animals, especially by insects with biting mouth-parts (e.g. beetles). These feed for example on the nutritious pollen or on pollination droplets containing mucilage or sugar, and perhaps use the female flowers as a place in which to lay their eggs (e.g. in the cycad *Encephalartos*). These initially very loose plant–animal relationships may be intensified and improved by selection if the plant thereby gains any advantage in pollination and seed production. Thus even in the cycads a strong pollen scent occasionally operates as an attractant. Furthermore, initially inconspicuous flowers may become optically conspicuous, whilst the dispersal of pollen may be enhanced by the stickiness of pollen (pollen cement, cf. p. 650) which then is carried more readily by the animal. Finally, the development of hermaphrodite flowers (or hermaphrodite pseudanthia) avoids the disadvantages of the insect feeding selectively or separately on ♂ and ♀ flowers and facilitates the simultaneous removal and delivery of the pollen. In certain gymnosperms (Bennettitatae, Gnetatae) and above all in the angiosperms (Magnoliophytina) this development has led to the regular and obligatory symbiotic relationship (cf. p. 292, pp. 658 et seq.) between floral visitors and flowering plants, that is, to animal pollination (zoophily, zoogamy). This goes parallel with the development of protective devices against the floral visitors, especially perianth segments around the floral buds, or interseminal scales (Bennettitatae, Fig. 647), or carpels (Magnoliophytina, Fig. 661) protecting the tender ovules. The great advantage of zoophily obviously lies in the much more effective transfer of pollen by animals which fly from flower to flower, often visiting individuals of the same species for long periods. The great masses of pollen required for wind pollination can then be significantly reduced. Furthermore growth is facilitated in windless habitats, e.g. the undergrowth of forests, and in very scattered populations.

Of course, the progression from (primary) anemophily to zoophily is not irreversible. Amongst the angiosperms many examples of secondary anemophily occur (cf. p. 663), and this may apply to the Gnetatae (cf. pp. 637 et seq.). Finally the progression to water pollination (hydrophily, hydrogamy) amongst some angiosperms should be mentioned.

Fertilization and the development of seed and fruit. Germination of the pollen grain and hence the further development of the ♂ gametophyte begins either in the pollen chamber at the apex of the nucellus or in the micropyle of the ovule (in gymnosperms, Fig. 624), or on the stigma of the carpel (in angiosperms, Fig. 673A). Frequently the exine of the pollen grain opens up by means of pre-formed, thin-walled areas (germination areas) and the pollen tube cell forms a pollen tube by the extensive stretching of the intine (p. 364; Figs. 632, 666). In the most primitive seed plants with fertilization by spermatozoids (zoidiogamy) the pollen tube serves only in the nutrition and rhizoid-like anchoring of the ♂ gametophyte in the wall of the pollen chamber. The emerging spermatozoids swim actively in a fluid secreted by the maternal parent to an archegonial chamber formed at the apex of the nucellus by the breakdown of tissue, and thence to the archegonia (Fig. 624B). In the majority of the more advanced seed plants with pollen tube fertilization (siphonogamy), the pollen tube now has the new task of conducting the more or less passive sperm cells to the ♀ gametophyte; this is achieved by the often very considerable elongation of the pollen tube and by the breakdown of and nutrition by the sporophytic nucellar or stylar tissue (Fig. 632).

Fertilization now follows. A spermatozoid or a sperm cell emerging from the ruptured pollen tube penetrates the egg cell; its membrane disintegrates and the protoplasts unite (plasmogamy); finally the nuclei fuse (caryogamy), forming the zygote. In the gymnosperms a period of months or even more than a year passes between pollination and fertilization, but in angiosperms generally only days or hours.

In the more primitive gymnosperms the first phase of development of the embryo is of free nuclear division leading to the formation of a few or numerous (around 1000!)

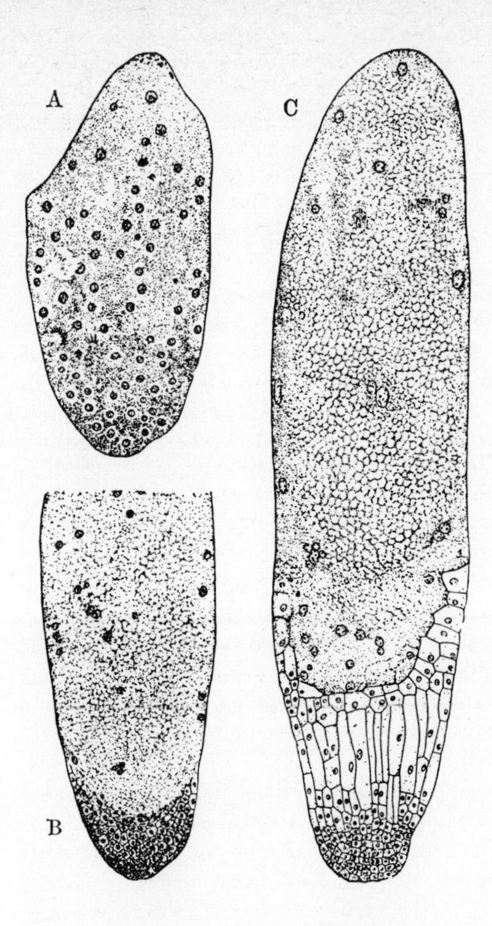

Fig. 625. Development of the embryo in the Cycadales (*Zamia floridana*). **A.** Free nuclear division in the zygote. ×12. **B.** Cell wall and tissue formation at the base. ×18. **C.** The start of differentiation into suspensor (elongated cells) and basal embryo. ×22. Cf. p. 635. After Coulter and Chamberlain.

nuclei; formation of cell walls follows later (Fig. 625). In contrast, almost all angiosperms (and only a very few gymnosperms) form cells right from the start (Figs. 90, 674). The proembryo thus developed forms a suspensor directed towards the micropyle and the true embryo directed towards the base of the embryo sac (i.e. the chalazal end). Its exogenous root pole with the primary root primordium (radicle) is directed towards the micropyle, its shoot pole with the primordia of the cotyledons towards the chalaza. The embryo of seed plants is thus endoscopic and bipolar from the outset (Fig. 131) in contrast to the embryo of modern ferns.

The growing embryo is generally surrounded by a nutritive tissue (endosperm). In the primitive gymnosperms the endosperm is formed from the ♀ prothallus prior to fertilization so that it is primary and haploid (Figs. 626, 640F). In the angiosperms this function is generally undertaken by a secondary (triploid) endosperm usually formed by the fusion of two embryo sac nuclei and one sperm cell (Fig. 674I, K). However, diploid nucellar tissue and even tissue of the embryo itself (e.g. its cotyledons) can also develop as nutritive and storage tissues.

The integuments of the maturing ovule develop into a usually many-layered seed-coat (testa). In primitive forms this testa often consists of outer fleshy layers (sarcotesta) and inner lignified layers (sclerotesta); otherwise it becomes hard, dry, and reduced when the seed remains enclosed in it. Commonly the area of micropyle remains thinner, facilitating the emergence of the radicle. The place where the seed breaks away from the funicle is called the hilum. Anatropous ovules develop into seeds in which the funicular strand (supplied with vascular bundles) is still recognizable as a raphe.

The mature ovule separated from the maternal parent is the seed. Surrounded by the testa, it generally contains the temporarily dormant embryo and endosperm. Originally the seed was itself the basic organ of dispersal of seed plants. Subsequently the seed became united with other maternal organs, forming fruits as compound units of dispersal. Fruits are thus portions of flowers or whole flowers, sometimes with additional organs, even inflorescences, in a state of maturity; they either liberate the seeds or break away with them.

Seed and fruit also show many devices and adaptations which can only be understood as adaptations for the dispersal and welfare of the young sporophyte; this is the subject matter of seed and fruit biology or ecology (carpology). The most important dispersal agents of seeds and fruits of the Spermatophyta are (as with pollen) animals (no doubt originally reptiles: saurochory), wind (anemochory) and water (hydrochory). See pp. 673 et seq.

The formation of seeds is confined to the Spermatophyta amongst living plants, but fossil representatives of the Lycopodiatae (e.g. *Lepidocarpon, Miadesmia*, p. 572) show that the seed has originated many times amongst land plants by parallel evolution. Amongst living Lycopodiatae a passable substitute for the seed is found in those few species of *Selaginella* in which the young sporophyte develops within the megaspore inside the megasporangium on the maternal plant, which does not, however, nourish its offspring. The shift of the processes of fertilization from the free-living gametophyte to the sporophytic maternal parent gives important advantages to the seed plant as compared with the fern: (1) Pollen grains (microspores or microprothalli) with solid walls and not spermatozoids are transported to the neighbourhood of the ♀ gametophyte; in as far as spermatozoids occur they only move inside the maternal plant, and then in an aqueous medium secreted by it. In pollen tube fertilization even this dependence on moisture is absent and the critical necessity for the presence of atmospheric water is avoided. (2) Instead of a megaspore, a seed (megasporangium + megaprothallus, generally also with zygote or embryo) is the unit of dispersal. Consequently the protection and care of the zygote and embryo sporophyte are greatly enhanced and the development of an independent and unprotected embryo is no longer critical.

This transfer of function from the spermatozoid to the pollen grain and from the megaspore to the seed certainly presented difficulties in the nutrition of the ♀ prothallus (p. 604) and in the novel severance of the 'mature' megasporangium. As they were overcome, so the course of development was speeded up at an ever-increasing pace: whilst in the most primitive gymnosperms (e.g. *Ginkgo*, cordaites, pteridosperms, cycads) fertilization only took place (if at all) in the fallen seeds, and the development of the embryo only proceeded on the ground, these processes take place in the more advanced seed plants on the maternal plant, although in other gymnosperms at a very leisurely rate (seed maturation commonly a year or more!); but in the angiosperms these processes are much quicker (sometimes within weeks). This advantageous acceleration of reproduction can be attributed to the progressive reduction and neoteny of the gametophytes, sporangia, sporangiophores and indeed the whole flower.

Germination of the seed. Dispersal results in at least a proportion of the seeds, either naked or enclosed in the fruit, reaching the surface layer of the ground. Under the conditions for germination described on p. 163, the seed takes up water, swells, and the inner tissues burst open the testa (and, if present, the fruit); at the same time the embryo begins to grow and the reserve foods are mobilized; in this the cotyledons in particular produce enzymes and remain at least for a time in the seed.

Since the embryo is always orientated in the seed so that the radicle points towards the micropyle, on germination the radicle and hypocotyl always emerge first from the seed, and indeed through the micropyle (Fig. 626). In the more primitive epigeal germination (Fig. 172), the cotyledons are drawn out of the testa and raised above the ground by the hypocotyl. The cycle of development is now complete.

Alternation of generations and nuclear phases. A review of the course of development of the seed plant, as described above, gives the following picture (cf. also pp. 207–8 and the Table on p. 754). The generations alternate as in all other Cormobionta between gametophyte and sporophyte. The gametophyte finally produces ♂ and ♀ sex cells: spermatozoids or sperm cells, and egg cells. The sporophyte starts with the zygote and finishes with the formation of meiospores: uninucleate pollen grains and embryo sacs which again develop into ♂ and ♀ gametophytes. In contrast to the Pteridophyta, the

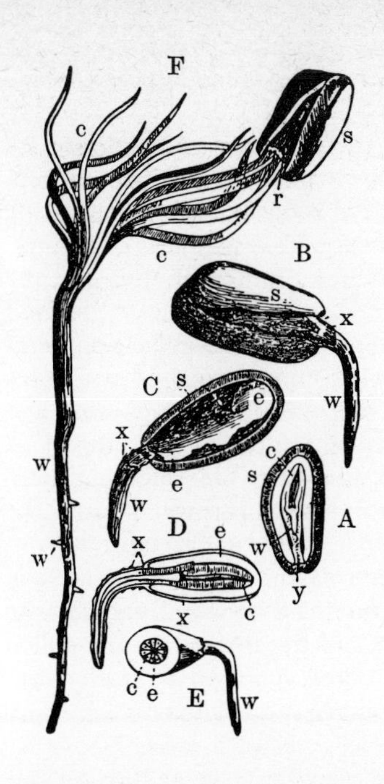

Fig. 626. Seed (in L.S. in **A.**) and germination (**B.** to **F.**) in the Pinales (*Pinus pinea*). s, testa; y, micropyle; e, primary endosperm. c, embryo or seedling with cotyledons; h, hypocotyl; w, w′, primary and lateral roots; x, protruding torn embryo sac. After Sachs.

sporophyte of the Spermatophyta shows a marked pause in development with the dormancy of the young sporophyte in the seed. The juvenile stage constitutes the embryo, or embryonic sporophyte. Alternation of nuclear phases normally (but not entirely) coincides with the alternation of generations: the haplophase extends from the uninucleate pollen grain or embryo sac to the sex cells, the diplophase from the zygote to the pollen or embryo sac mother cells.

Evolution and systematics. With some 227,000 known species, the Spermatophyta dominate the terrestrial communities of our planet today. Of these only about 600 species are Coniferophytina and some 200 Cycadophytina, so that the overwhelming majority are Magnoliophytina. This dominance of the seed plants, and especially of the angiosperms, compared with the spore-dispersing pteridophytes, only gradually developed in the course of geological time (Fig. 614).

The oldest unequivocal fossil spermatophytes (remains of seeds) have recently been recognized from the Upper Devonian (*Archaeosperma*). Since then the proportion of seed plants amongst terrestrial plants has continuously increased. The evolutionary tree (Fig. 628) shows that at least since the Lower Carboniferous the two gymnospermous lines of development, the Coniferophytina (Cordaitidae, ? Ginkgoatae) and the Cycadophytina (Lyginopteridatae = pteridosperms), occurred side by side, although still subordinate to the dominant pteridophyte groups, Lycopodiatae, Equisetatae and Filicatae (younger 'Fern Age', or 'Palaeophytic'). This only changed with climatic changes (drought, etc.) at the transition from the Lower to Upper Permian, when the dominance of these pteridophyte groups was broken and the older gymnospermous groups diminished or died off, but the younger groups of gymnosperms (particularly Ginkgoatae, Pinidae = conifers, Cycadatae and Bennettitatae) became so markedly prevalent that one can speak of a 'Gymnosperm Age' (Mesophytic). In the Middle Cretaceous the rapid spread of the Magnoliophytina (Angiospermae) again produced a major change in the flora: the 'Angiosperm Age' (Cainophytic) dawned and many gymnosperms became extinct whilst others have persisted as relict groups (e.g. Ginkgoatae, Cycadatae, Gnetatae) up to the present-day. Only a few lines of development in the Pinidae (conifers) to some extent held out against the angiosperms.

Previous classifications of the Spermatophyta have almost invariably contrasted two equal subdivisions 'Gymnospermae' and 'Angiospermae'. This division was based chiefly on the distinction between the 'naked' ovules of the gymnosperms and the carpels (with stigmas) enclosing the ovules of the angiosperms. The less differentiated flowers, the less reduced ♂ and ♀ gametophytes and the more primitive xylem and phloem structure of the gymnosperms were also used in separating the two groups.

Earlier ideas of spermatophyte phylogeny emphasized the similarity of eusporangiate ferns and pteridosperms, quite frequently postulating a direct phylogenetic relationship, but it remained a problem why the pteridosperms occurred both as early as and along with the eusporangiate ferns, and why the two groups had very different stelar structures. But the greatest difficulty all along was the derivation of the Coniferophytina and the morphological interpretation of the arrangement whereby their ovules were inserted essentially directly on the floral axis. It was assumed that there was reduction from pteridosperm-like Cycadophytina and drastic suppression of richly-branched structures bearing numerous ovules. It remained inexplicable why the oldest Cycadophytina and Coniferophytina occurred simultaneously, and why intermediate forms were absent. A polyphyletic origin of the Spermatophyta was frequently discussed: the 'microphyllous' Coniferophytina might have arisen from the Lycopodiatae in which, as is well known, the seed level of organization was reached (pp. 572, 610). Against this second hypothesis was, on the one hand, the great similarity between Coniferophytina and Cycadophytina (e.g. in root development and in axillary branching; the structure of xylem and phloem; the similar groups of pollen sacs and ovules; the pollen grains and the formation of a pollen tube, etc.), and, on the other, the radical differences from the Lycopodiatae (with rhizophores and different modes of branching; the invariably solitary sporangium on the upper side of the sporophyll, etc.).

Only in the last decade have important new points of view been opened up for our understanding of the relationships between and evolution of the Spermatophyta; these have been reached by the increased knowledge of the immediate precursors, both homosporous and heterosporous, of the seed plants from the (? Lower), Middle and Upper Devonian (or Lower Carboniferous); these are the so-called **'Progymnospermae'**. It is fundamental that the progymnosperms form an immediate link between yet older Psilophytatae, which consist of telomes, with no clear distinction between stem, leaf and root, and the younger Spermatophyta, both Coniferophytina and Cycadophytina, in which the basic organs are differentiated, including the seed.

It was a sensational discovery when in 1960 the completely unexpected connection was established between large, fern-like, heterosporous frond-systems (*Archaeopteris*, Figs. 587A, 589) and arborescent gymnosperm-like trunks (*Callixylon*; up to 1.5 m in diameter and probably up to 20 m high!). Since then the picture of the Progymnospermae as a group has become clearer: they are (? Lower) Middle Devonian to Lower Carboniferous woody plants with secondary thickening (with tracheids) and terminal thick-walled sporangia arranged in rows or tufts (Fig. 627C); they are either homosporous or heterosporous. The Progymnospermae were formerly generally put in the Primofilices (pp. 580–1), but these characters rather indicate relationships with the Psilophytatae (e.g. *Trimerophyton*, Fig. 627A) and early Spermatophyta (e.g. Ginkgoatae, Cordaitidae and Lyginopteridatae). The oldest progymnosperms (e.g. *Protopteridium*, Fig. 585; *Aneurophyton* or *Actinoxylon*, Fig. 627B) from the (Lower or) Middle Devonian were still branched in three dimensions, so the telomes were scarcely planate and showed little distinction between stem and leaf; the anatomical structure of protostele or actinostele was simplicity itself. In later representatives the distal branches extended in a plane and showed 'fusion' or 'webbing' whilst leaves and stem were distinguishable; in addition a medulla was present (transition to eustele). This development appears to lead from the Archaeopteridales (= Pityales) on the one hand to several forms resembling Coniferophytina with moderately complex lateral organs and compact secondary wood (e.g. *Barrandeina*, Fig. 627D, and Ginkgoatae, Fig. 629), whilst on the other the Aneurophytales (= Protopteridales), with complex pinnate lateral organs and xylem more strongly dissected by medullary rays, more closely resemble the Cycadophytina (e.g. *Tetraxylopteris*, Fig. 627C, and Lyginopteridatae, Figs. 641–4). The Middle Devonian progymnosperms were still homosporous, but in the Upper Devonian heterospory frequently occurred (e.g. *Archaeopteris*, Fig. 589). The hypothetical development of ovules with one or two integuments has already been referred to (p. 604; Figs. 617, 644); an almost complete record exists from extreme heterospory, with one tetrad of mega-

Fig. 627. Extinct precursors of the seed plants. **A.** Psilotatae: *Trimerophyton robustius* (Lower Devonian, telome system with groups of sporangia). **B.** to **D.** 'Progymnospermae'. **B.** *Actinoxylon banksii* (Middle Devonian, vegetative shoot system). **C.** *Tetraxylopteris schmidtii* (Upper Devonian, fertile portion of complex sporophylls with groups of sporangia). **D.** *Barrandeina dusliana* (Middle Devonian, shoot system with dichotomous leaves and simple groups of sporangia). Approx. × ¾. **A.** After Hopping. **B.** After Matten. **C.** After Bonamo and Banks. **D.** After Kräusel and Weyland; somewhat modified.

spores, to the seed in which the single tetrahedric megaspore in the nucellus shows only traces of the other three megaspores. Nevertheless it remains possible that the seed of the Coniferophytina (often flattened, with one integument of few (? 2) telomes) and that of the Cycadophytina (radial seed with one or two integuments, each of several telomes) have arisen in parallel.

At least in the oldest progymnosperms the undifferentiated, telome-like shoot organs were strongly reminiscent of the Psilophytatae. We have seen that in the differentiation of stem and leaf in the Lycopodiatae, Equisetatae and Filicatae there has clearly been a parallel development of foliar telome systems of greater or lesser extent and derivable from the Psilophytatae (Figs. 562, 583). It is an obvious hypothesis that by similar processes of differentiation both the simpler and the more complex vegetative and fertile organs of the Coniferophytina and Cycadophytina could have originated from undifferentiated shoot systems in the older Progymnospermae. Figure 616 indicates this diagrammatically.

This hypothesis, which seems most likely at the present time, avoids the difficulties of spermatophyte evolution and phylogeny mentioned on p. 612; it can be summarized as follows: (1) the seed plants arose in the Devonian by way of homosporous and heterosporous progymnosperms directly from psilophytes; they did not originate from eusporangiate pteridophytes or from Lycopods, but developed parallel with these and other higher pteridophytes; (2) the numerous similarities between the Coniferophytina and the

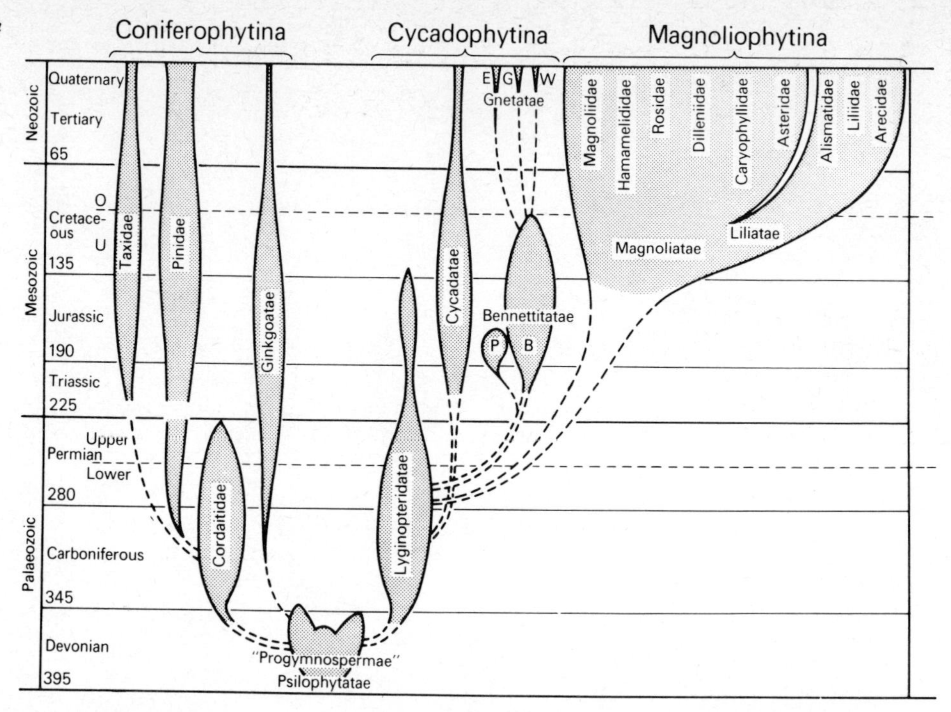

Fig. 628. Hypothetical phylogenetic connections between related groups of seed plants and their development in geological time (the figures at the beginning of the geological formations indicate millions of years). Dubious connections not documented by fossil evidence: pecked lines or left white. B, Bennettitidae; P, Pentoxylidae; E, Ephedridae; G, Gnetidae; W, Welwitschiidae. Original.

Cycadophytina are comprehensible if they had a common origin in the Progymnospermae; furthermore, this justifies the retention of the taxon 'Spermatophyta'; (3) both Coniferophytina and Cycadophytina have developed independently from psilophyte-like progymnosperms, and neither can be derived from the other. Similarly the sporophylls and trophophylls of different degrees of complexity in the two groups have developed independently from the telome-like shoots of the older progymnosperms; they are only to a limited extent homologous and cannot be derived from each other.

The recognition that the 'Gymnospermae' include two groups, the Coniferophytina and Cycadophytina, which were independent as far back as the Middle Devonian and probably arose separately from spore-bearing precursors, has recently been taken into account by giving them equal rank as subdivisions. The Magnoliophytina (= Angiospermae) make the third group of the Spermatophyta; as the most advanced and most recent subdivision of the seed plants they show phylogenetic relationships only with the Cycadophytina (pp. 676 et seq.). The 'Gymnospermae' exhibit a very ancient level of organization within the Spermatophyta, but do not constitute a natural group.

Subdivision 1: **Coniferophytina (Pinicae), fork- and needle-leaved gymnosperms**

The first group of gymnospermous seed plants is characterized principally by the possession of simply constructed vegetative and fertile lateral organs: the leaves (trophophylls) have basically a dichotomous ground-plan, the stamens (microsporophylls) bear single groups of pollen sacs and the 'carpels' ('megasporophylls') consist from the start merely of simple (very rarely forked) seed-bearing structures (Fig. 616A).

The Coniferophytina are richly branched woody plants with monopodial shoot systems; right from the beginning they were leptocaulous. The secondary xylem consists mainly of densely packed tracheids with bordered pits (Fig. 57) and narrow medullary rays (pycnoxylous, Figs. 163–5); vessels are lacking. The ground plan of the leaf is dichotomous and forked, but is frequently reduced to linear (Fig. 630B), needle-shaped or scale-like (Fig. 179; but still with a pair of vascular bundles; Figs. 180B, 635). Phyllotaxis is spiral, or whorled or decussate in advanced groups.

The flowers are always unisexual, monoecious or more rarely dioecious and very simply constructed: stamens or one or more 'carpels', and sometimes sterile scales, arise from the floral axis; there is no true perianth. The stamens occur in single, stalked, originally radial groups of pollen sacs (e.g. Figs. 629A–B, 630D, 640C); the uppermost may become sterile and vegetative, giving rise to a dorsiventral structure (Figs. 631H, 636D). The 'carpels' consist simply of a single stalked or sessile naked ovule (very rarely two as the result of dichotomy); several 'carpels' (or possibly a single terminal one) are borne laterally and directly on the floral axis (stachyosporous). The integument is always single; sometimes its origin from two telomes is discernible (Figs. 630F, 634C–D). The much reduced female flowers are often borne in open, cone-like inflorescences. The pollen grains have an obscure or distinct distal germination area (analeptic or anacolpate), or are atreme; sometimes the exine is inflated forming annular or bladder-shaped air sacs. The ♂ gametophyte may have many cells (Figs. 630E), but sometimes is reduced to four cells or four nuclei. The ♀ gametophyte consists of a multicellular prothallus with archegonia. Pollination is by wind. Fertilization is effected either by a pollen tube (siphonogamy) or by spermatozoids (zoidiogamy). The seeds are either fleshy (with sarcotesta and sclerotesta) or dry, in which case the scales or axes may be fleshy, or the seed has a membranous wing; they are often borne in cone-like, woody or fleshy infructescences.

The Coniferophytina go back to the Lower Carboniferous, perhaps even to the Upper Devonian, and clearly arose from the Devonian Progymnospermae; today, with some 600 species, they are forest trees of world-wide distribution and importance.

Class I: Ginkgoatae

The ♂ and ♀ flowers of the Ginkgoatae stand in the axils of bracts, the rather long floral axes bear lax lateral or terminal stalked groups of pollen sacs (stamens) or stalked or sessile ovules ('carpels'); there are no sterile organs (Fig. 629). This corresponds closely with the basic ground plan of the coniferophyte flower (Fig. 616A).

The class is certainly known as far back as the Lower Permian (*Trichopitys*), and may go back to the Upper Devonian. Some Progymnospermae show striking similarities (e.g. the Middle Devonian *Barrandeina*, Fig. 627D). The greatest diversity of forms existed

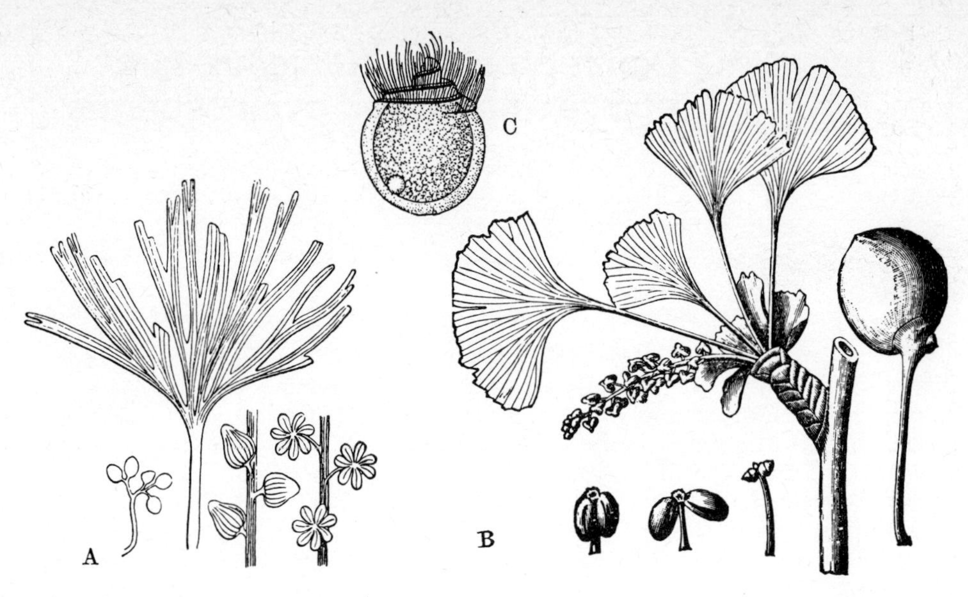

Fig. 629. Ginkgoatae. **A.** *Baiera muensterana* (Rhaetic–Liassic): leaf; ovules on the ♀ floral axis (somewhat reduced); radial pollen sac groups, undehisced and dehisced, on the ♂ floral axis (×2 approx.) **B.** *Ginkgo biloba* (Recent): short shoot with ♂ flower and young leaves (nat. size); dorsally reduced groups, each of two pollen sacs (stamens) (enlarged); ovules and seed (somewhat reduced). **C.** Spermatozoid. ×200. **A.** After Schenck. **B.** After Richard and Eichler. **C.** After Shimamura, modified and somewhat diagrammatic.

from the Triassic to the Cretaceous. In the Jurassic the genus *Ginkgo* is represented by forms closely similar to the species *G. biloba* (Fig. 629B) living today. Fossil remains indicate a cosmopolitan distribution in the Jurassic and Cretaceous and a progressive diminution of area up to the present: *G. biloba* was saved from extinction as an ornamental tree in China and Japan and is now grown in gardens throughout the world: a perfect example of a 'living fossil'.

From a seedling with two cotyledons *Ginkgo biloba* grows into a much-branched deciduous tree with long and short shoots. The leaves are fan-shaped with conspicuous dichotomous venation (Figs. 177B, 629B). Older representatives (Fig. 629A) had deeply divided leaves, but Mesozoic genera also demonstrate the development of narrow, linear leaves.

Ginkgo biloba is dioecious. The ♂ flowers bear numerous dorsiventral stamens, each with two pollen sacs; in the Mesozoic *Baiera* (Fig. 629A) the stamens were still more or less radially constructed. In *G. biloba* the ♀ flowers have generally only two ovules, but were more extensively branched in *Baiera*. The ♂ and ♀ gametophytes have a relatively large number of cells; fertilization is effected by means of large spermatozoids (zoidiogamy, Fig. 629C). Pollination and fertilization are separated by some months, and some of the fallen 'seeds' are still unfertilized (and hence virtually still ovules). The single integument forms an outer fleshy sarcotesta, smelling strongly of butyric acid, and an inner sclerotesta (endozoochorous distribution).

Class II: Pinatae

The ♂ and ♀ flowers consist of shortened axes bearing lateral (rarely terminal), closely set, stalked groups of pollen sacs (stamens) and stalked or sessile ovules ('carpels'); in addition there are almost always sterile foliar organs. The ♀ flowers in particular are often aggregated in catkin- or cone-like inflorescences.

There are three subclasses primarily distinguished by the structure of the ♂ and ♀ flowers.

A. Subclass: Cordaitidae

In their ♂ and ♀ flowers the extinct Cordaitidae had, besides lateral stamens and 'carpels' (ovules), numerous sterile scale leaves. Apart from the condensed floral axis, there is close agreement with the basic ground plan of the coniferophyte flower (Fig. 616A).

The cordaites formed forests in the Carboniferous but apparently became extinct in the Permian. They were trees up to 30 m high with a richly branched crown (Fig. 630A); they had secondary thickening, tracheids with 'araucarioid' bordered pits (p. 621), and a transversely septate medulla; their spirally arranged leaves were linear or lanceolate with dichotomously branching, parallel veins (Fig. 630B). The mesophyll was in some already differentiated into palisade and spongy tissue. The flowers were borne in the axils of bracts and aggregated in catkin-like inflorescences.

The ♂ flowers consisted of short axes bearing spirally a few perianth leaves below and then many stamens, each with several terminal pollen sacs (Fig. 630C–D). In the ♀ flowers (Fig. 630F) was also a spiral arrangement, first of several scale-like perianth leaves, then of a few carpels each with one (or, as a result of dichotomy, rarely two) terminal, erect ovule(s). The pollen grains, which have been observed several times in the pollen chamber of the ovule, were notable for containing, as in the pteridosperms, numerous cells corresponding to a microprothallus or antheridium (Fig. 630E). The presence of pollen chambers suggests that spermatozoids were formed.

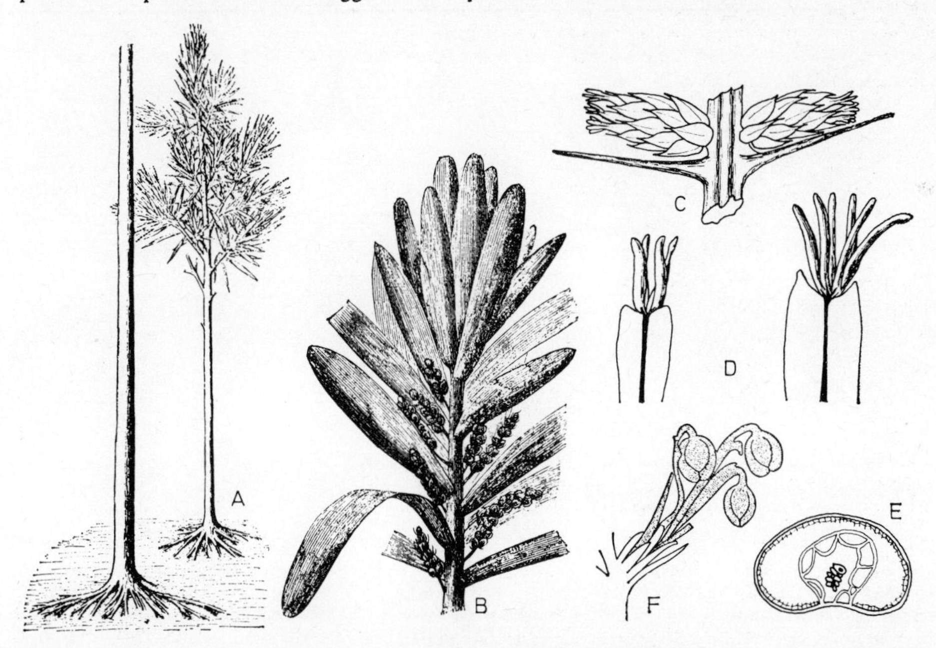

Fig. 630. Cordaitidae (Carboniferous–Permian). **A.** Habit of a species of *Cordaites* (about 10 m high). **B.** Short with lateral buds and inflorescences of *Cordaites laevis* (reduced). **C.** Two ♂ flowers of *Cordaianthus concinnus*. ×2.5 approx. **D.** Stamens with erect pollen sac groups of *C. penjonii*. ×100 approx. **E.** Pollen grain with multinucleate ♂ gametophyte. ×300 approx. **F.** ♀ flower of *C. pseudofluitans* with bract, sterile scales and stalked ovules ('carpels'). ×1.5. **A.** to **B.** After Grand'Eury. **C.** After Delevoryas. **D.** to **F.** After Florin.

B. Subclass: Pinidae (= Coniferae), conifers

The floral structure of the conifers (Fig. 615) is essentially the same as that of the Cordaitidae. It is true that the ♀ flowers are often more greatly reduced and united to form so-called ovuliferous scales; generally these are also united with their bract or subtending leaf. Frequently these bract-ovuliferous scale complexes are aggregated in cones. Fertilization is effected, at least in present-day conifers, by a pollen tube (siphonogamy).

The conifers grow from seedlings with two or many cotyledons into profusely-branched trees, or more rarely shrubs. The stem is usually monopodial, bearing lateral branches of various orders which are arranged in successive tiers; commonly there is a distinct differentiation into long and short shoots. The leaves have basically dichotomously branched, parallel veins. In primitive members they were still dichotomously lobed, but later they became reduced, simple, and linear, needle-shaped or scale-like. Originally spiral, they later became whorled or decussate; although predominantly evergreen, tough-textured and xeromorphic, a few deciduous forms occur. Resin canals are frequent in all parts of the plant (Fig. 130).

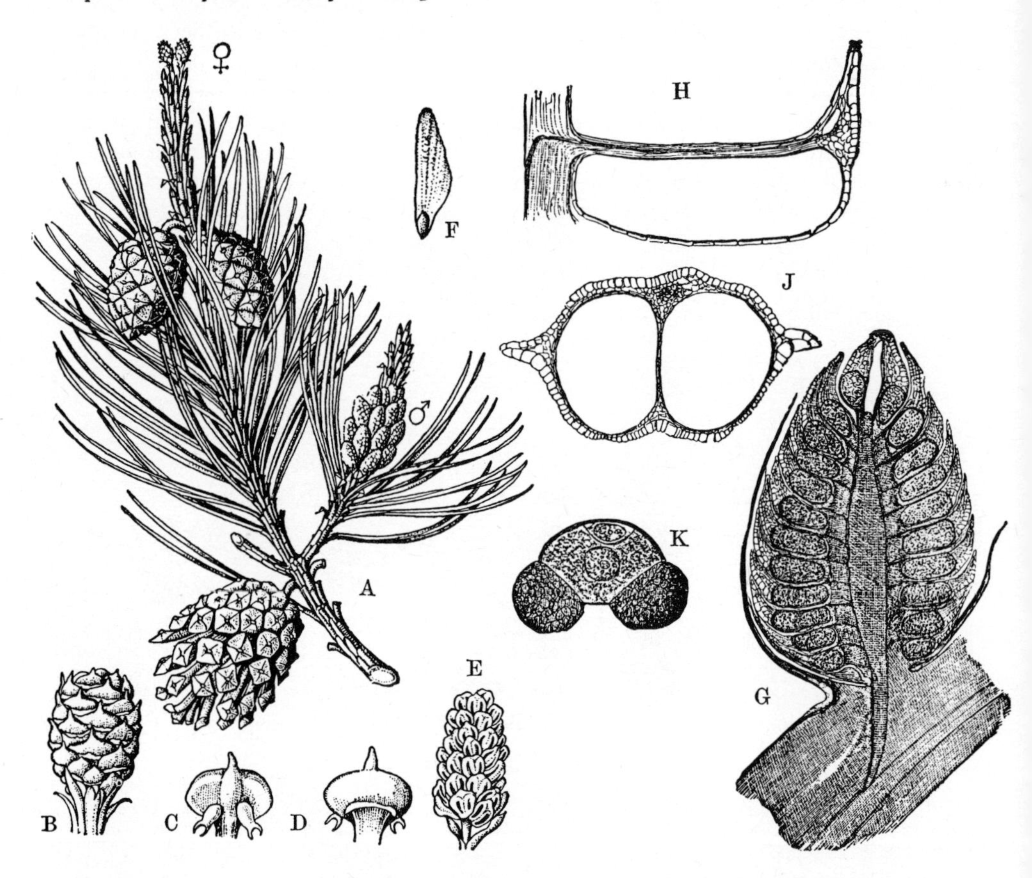

Fig. 631. *Pinus* (**A.** to **F.** and **K.** *P. sylvestris.* **G.** to **J.** *P. mugo*). Flowering and fruiting branch with two-needled short shoots in the axils of abscissed scale leaves (**A.**). ♂ flowers (**E.**, **G.** in L.S.), stamens with two pollen sacs (**H.** in L.S., **J.** in T.S.); vesiculate pollen grain (**K.**). ♀ inflorescence (**B.**), with the bract and ovuliferous scale complex (from above and below, **C.** and **D.**), from which the one-year-old, still green, and two-year-old, mature and dehiscing cones (**A.**) develop, with two winged seeds (**F.**) on the upper side of the ultimately woody scale complex. **A.** to **F.** Somewhat reduced. **B.** to **E.** Enlarged. After Berg and Schmidt, modified. **G.** ×10. **H.** ×20. **J.** ×27. **K.** ×400. After Strasburger.

They are monoecious or dioecious. The cone-like ♂ flowers are generally solitary or in loose aggregations; the floral axis bears dense spirals of numerous dorsiventral stamens with pollen sacs on the underside (Figs. 631E, G, 635B, 636A, etc.). On the other hand, numerous reduced ♀ flowers almost invariably form a cone-like inflorescence. These generally contain numerous spiral, whorled or decussate bracts with ovuliferous scales in their axils. The ovuliferous scales consist of a few ovules (rarely many or only one) which often lie upon a vegetative scale (Fig. 635C–D).

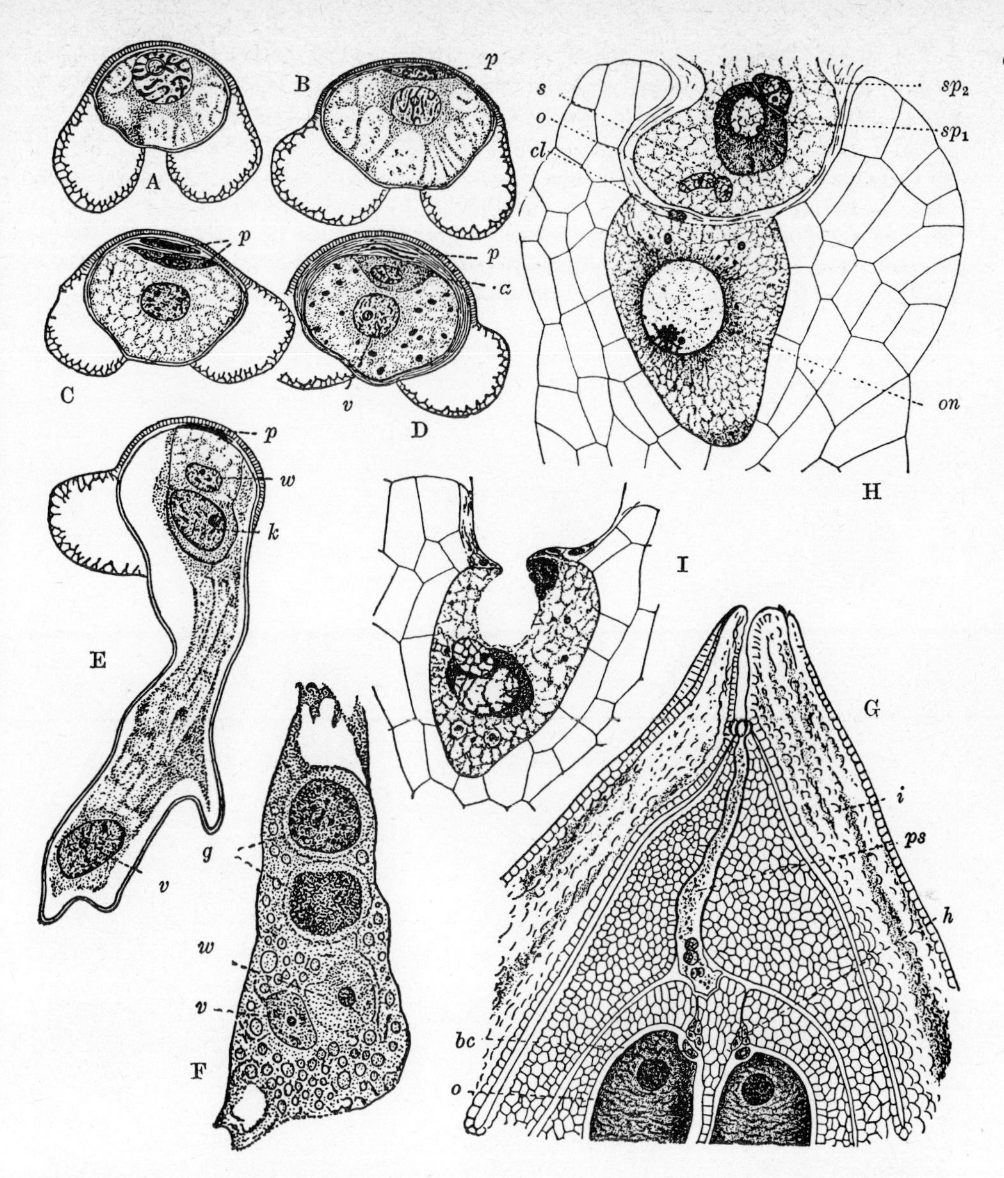

Fig. 632. Pollen germination and fertilization in Pinatae (**A.** to **F.** *Pinus nigra*, and **G.** *P. sylvestris*; Pinaceae; **H.** to **I.** *Torreya taxifolia*; Taxaceae). **A.** to **F.** Development of the ♂ gametophyte in the pollen grain and in the tip of the pollen tube (**F.**); p, prothallial cells; v, nucleus of the vegetative pollen tube cell; a, antheridial cell, from which the stalk cell (w) and spermatogenous cell (k) arise, the latter producing two spermatogenous cells (g). ×500 approx. **G.** Receptive ovule with integument (i), pollen tube (ps) and archegonia with neck- (h), ventral canal- (bc) and egg cells (o). Enlarged. **H.** Pollen tube, with two sperm cells (sp_1, sp_2), pollen tube and stalk cells (o, s), in contact with the egg cell (egg nucleus, on; remains of a neck cell, cl). **I.** Fusion of the egg nucleus with one of the sperm nuclei; the other nucleus degenerating. ×367. **A.** to **F.** After Coulter and Chamberlain. **G.** After Strasburger. **H.** and **I.** After Coulter and Land.

The morphological interpretation of the ♀ cone of the conifers was for long controversial. Frequently the bract-ovuliferous scale complex was regarded as an open carpel and the entire cone as a single flower. It was the classic work of the Swedish palaeobotanist R. Florin on the Voltziales (Fig. 634) which first showed with certainty that the ovuliferous scale corresponds to a short shoot with sterile and fertile scale leaves (ovules). This agrees basically with the ♀ floral structure in the Cordaitidae (Fig. 630F). The coniferous cone is therefore an inflorescence (or infructescence) and not a flower.

Each ovule contains only a single megaspore, the embryo sac cell (Fig. 622). This produces a ♀ gametophyte in the form of a multicellular prothallus (embryo sac) with several archegonia (up to sixty in *Sequoia*!). Each has a large number of neck wall cells, and in the Pinaceae there is also a separate ventral canal cell (Fig. 632G).

The ♂ gametophyte develops inside a pollen grain which is carried by the wind to an ovule where as a rule it germinates at the apex of the nucellus by means of a pollen tube (Fig. 632G). Spermatozoids are not formed; rather the pollen tube conveys the two unaltered sperm cells to the archegonia. Generally fertilization is effected by a single sperm cell: frequently the other one (which is often smaller from the outset) degenerates (Fig. 632F, H–I).

The fertilized egg cell first forms a proembryo, and from this (in various ways in particular families and genera) one or more embryos.

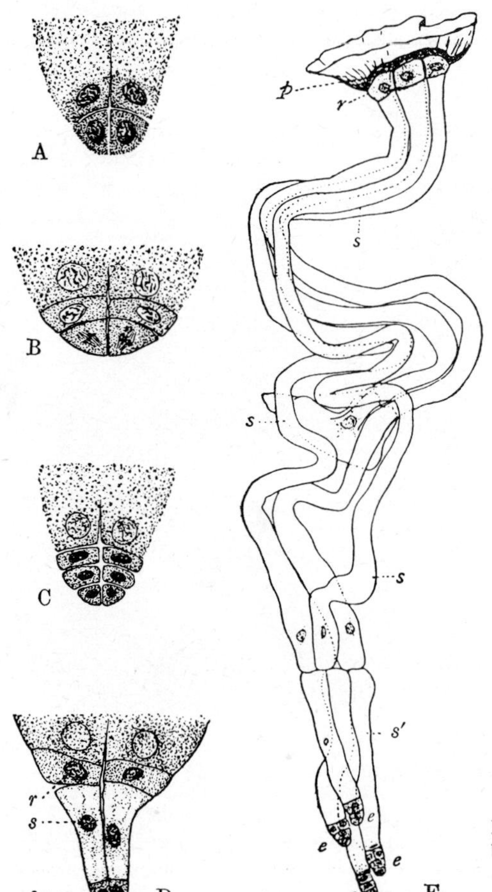

Fig. 633. Formation of embryo in *Pinus* (**A.** to **D.** *Pinus nigra;* **E.** *P. banksiana*). r, rosette; p, basal plate; s, suspensor; s′, secondary suspensor cells; e, embryo. **A.** to **D.** ×100. After Coulter and Chamberlain. **E.** ×80. After Buchholz.

Figure 633 shows this for *Pinus*. The fertilized egg nucleus first divides into four free nuclei which move to the lower end of the egg cell, arranging themselves in a plane. Here they next form two tiers of cells, each with four nuclei, and then, after the formation of cell walls and further divisions in each tier, four tiers each with four cells (A–D). The uppermost tier is not delimited from the remainder of the egg cell, and later disintegrates. The next tier also remains within the original egg cell membrane, but is separated from the uppermost tier by a thick wall, the 'basal plate', and its four cells form a 'rosette'. The cells of the next tier below form the suspensor, which, as the surrounding prothallial cells disintegrate, pushes the lowest tier down into the nutritive prothallial tissue or primary endosperm. Only the cells of this lowest tier take part in embryo-formation, after producing a secondary suspensor as a direct downward continuation of the primary suspensor (E). The four longitudinal series of cells thus produced separate, and each gives rise to one of the four genotypically identical embryos (monozygotic polyembryony). Each embryo, which grows initially by an apical cell with three dividing faces, is thus supported on its own suspensor. Since in *Pinus* several archegonia may be fertilized polyzygotic embryony also occurs. Eventually only one embryo survives, namely the one which is plunged by its suspensor deepest into the prothallus and is therefore presumably the best fed. After a time the remaining embryo develops a variable number (five to eighteen) of cotyledons. Germination is shown in Fig. 626. In other families and genera of the Pinidae the course of development is similar, but usually only one embryo arises from each zygote, no doubt a more primitive condition.

The conifers first appeared in the Upper Carboniferous as the Voltziales which distinctly resembled the cordaites. The cosmopolitan Pinales spread during the Mesozoic, and with six families and some 600 species, some of them dominants in forests, they are today by far the most 'successful' group of gymnosperms. Many produce timber of great economic importance; their resins and ethereal oils are also frequently used (e.g. in medicine).

Order 1: **Voltziales.** These are characterized by the readily recognizable short shoot nature of the ♀ flowers. The sterile scale leaves and the ovules ('carpels') were sometimes still radially organized; they were fused together in a complex of bract and ovuliferous scales (Fig. 634B–D).

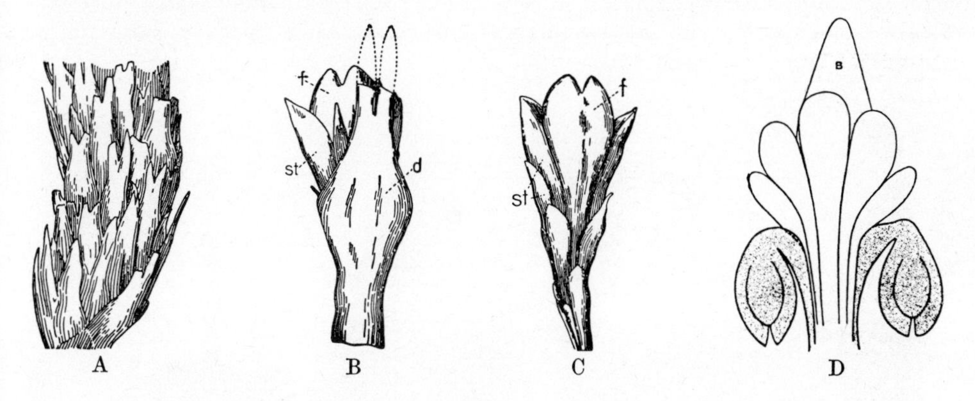

Fig. 634. Voltziales. **A.** to **C.** *Lebachia piniformis* (Lower Permian). **A.** Part of a cone, only the forked bracts visible. ×2. **B.** and **C.** ♀ flower, dorsal and ventral veins. d, bract; st, sterile scales; f, compressed orthotropous ovule with bifid integument. ×5. **D.** *Glyptolepis longibracteata* (Lower Triassic); ♀ flower with bract (B), sterile scales with two anatropous ovules. After Florin, diagrammatic.

The leaves were sometimes forked (Fig. 634A). The secondary wood had 'araucarioid' pits (see below). The arborescent Voltziales played a major role in forming forests from the Upper Carboniferous to the Permian (e.g. *Lebachia* (*Walchia*) and *Ullmannia*), and in the Triassic and Jurassic (*Pseudovoltzia* and *Glyptolepis*); they then became extinct.

The Voltziales are linked in the Mesozoic by forms transitional to the next order, which arose then and has survived to the present.

Order 2: **Pinales.** In these the ♀ flowers consist only of ovuliferous scales more or less extensively united with bracts.

The **Araucariaceae** were widely distributed in the Triassic, even in Europe and Greenland, but are now restricted to the Southern Hemisphere. Their woody cones have numerous one-seeded bract-ovuliferous scale complexes. The tracheids of the secondary xylem have bordered pits arranged in a honeycomb-like pattern (the primitive 'araucarioid' pattern). The species of *Araucaria* are lofty trees with strikingly regular branching and spirally arranged, generally very large needles; *A. excelsa*, the Norfolk Island 'pine' is well-known as a room plant; *A. araucana* is the familiar 'monkey puzzle'. Species of *Agathis* yield hard copal resins.

The **Pinaceae** have spirally arranged, needle-like leaves and woody cones with two seeds to each ovuliferous scale. They include the most important European conifers which have, with the exception of the larch, evergreen and more or less xeromorphic needles (Fig. 179).

In pine, spruce and fir the leaf lives for five to nine years, rarely longer; it is divided into lamina and base. In spruce the lamina forms the 'needle' which abscisses, whilst the leaf base is united to the shoot axis and remains as a 'leaf cushion'.

The genera can be grouped in three subfamilies according to the arrangement of their needles on long or short shoots (Figs. 631, 635–7). Thus in the Abietoideae the needles are borne exclusively on long shoots, e.g. fir (*Abies*, Fig. 635), spruce (*Picea*, Fig. 636), hemlock (*Tsuga*), and Douglas fir (*Pseudotsuga*). In the Laricoideae the needles occur on both long and short shoots, as in the evergreen cedars (*Cedrus*) and the deciduous larches (*Larix*, Fig. 637): in the first year a long shoot bears needles in the axils of which short shoots arise in the second year; these bear tufts of needles and can continue to grow for several years. Finally in the Pinoideae the needles of adult plants are borne exclusively on short shoots, e.g. pine (*Pinus*, Fig. 631). First and second year seedlings do in fact have long shoots with green needles, but later the long shoots bear only brown scale leaves in the axils of which arise (in the same year) short shoots, the contracted axes of which bear first a few membranous bracts and then merely a group of five, three, two or even only one green needle (pp. 624 et seq.).

The axis of a ♂ flower bears a few basal scale-like bracts (as a simple perianth) and then numerous spirally arranged stamens (Figs. 631E, G, 635B). Each stamen has a short stalk, a scale-like upturned tip and two dorsal pollen sacs which open by means of the exothecium (Fig. 631H–J, 636D).

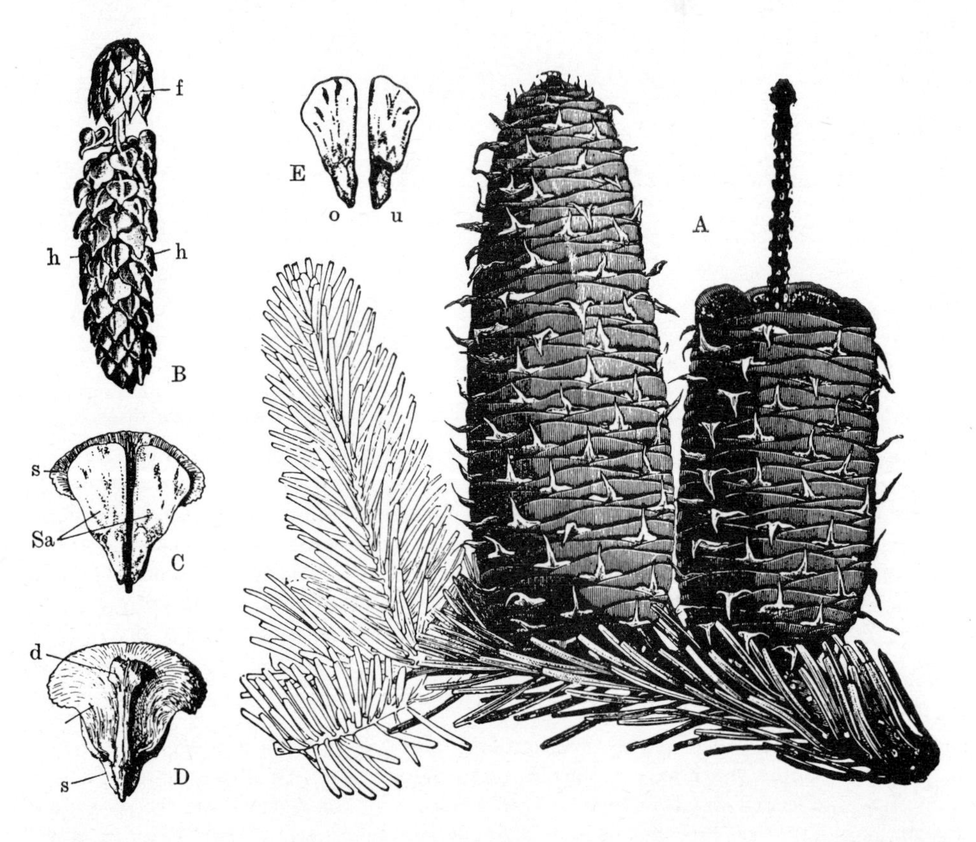

Fig. 635. *Abies* (**A.** *A. nordmanniana*, **B.** to **E.** *A. alba*). **A.** Branch with mature cones, partially disintegrated. Somewhat reduced. **B.** ♂ flower with bracts (f) and stamens (h). ×2 approx. **C.** to **D.** Mature ♀ flower with bract (d), ovuliferous scale (s) and two seeds (Sa, **E.**) seen from the upper (o) and lower (u) sides. Somewhat reduced. **A.** After Berg and Schmidt. **B.** to **D.** After Firbas. **E.** After Eichler.

The ♀ inflorescence bears numerous, spirally arranged, sterile bracts with an ovuliferous scale in the axil of each (Figs. 615, 635D); the bract and ovuliferous scale are more or less united. The ovuliferous scales bear two inverted ovules at the base; each ovule has a single integument. As the inflorescence develops into a cone the ovuliferous scales increase greatly in size, becoming woody 'cone scales'. The bracts may also enlarge so that they project freely from the cone (e.g. *Abies*, Fig. 635A, and *Pseudotsuga*), but generally they remain small and in *Pinus* atrophy completely.

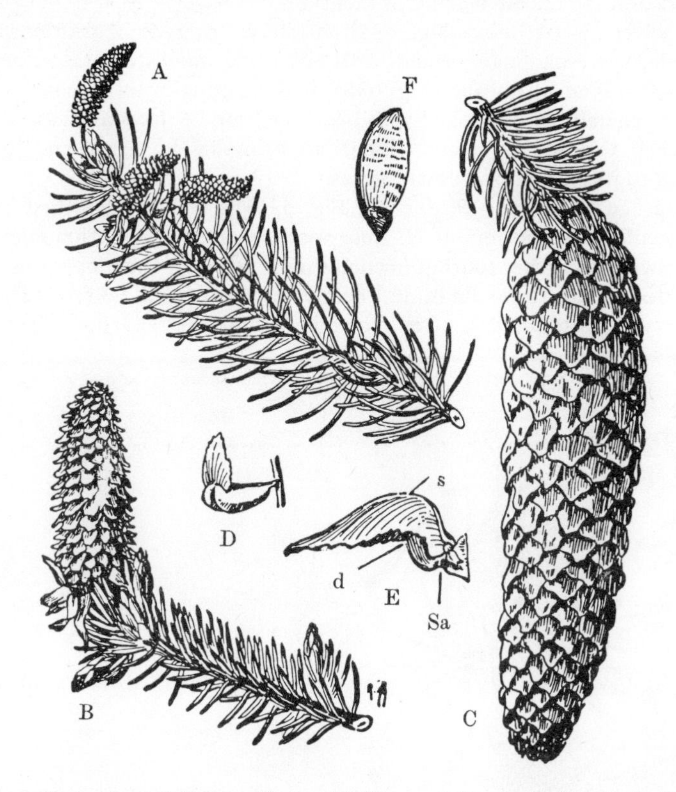

Fig. 636. *Picea abies.* **A.** to **C.** Twigs with ♂ and ♀ flowers and cone. Reduced. **D.** Stamen. **E.** ♀ flower with bract (d), ovuliferous scale (s) and ovules (Sa). Enlarged. **F.** Winged seed. Nat. size. After Karsten.

The relatively large pollen grains of many Pinaceae, e.g. of *Abies, Picea* and *Pinus,* enhance their buoyancy with air bladders (Figs. 632A, 659; vesiculate pollen) formed when an outer layer of exine lifts up as two blisters. When the wind carries the pollen grain to the ovule the megaprothallus is still undeveloped, and indeed the megaspore may not have started to divide. Thus a considerable time elapses between pollination and fertilization, during which the micropyle closes over the pollen grain as it develops inside the ovule. This interval is longest in most species of *Pinus*: flowers pollinated in May develop only in the following spring into small green cones (Fig. 631A) in which are formed on the one hand archegonia and on the other sperm cells in the pollen tube, and fertilization proceeds. Thus a year separates pollination and fertilization and the cones only grow to their full size during the summer following fertilization: the seeds are finally shed early in the subsequent spring. In other European genera pollination and fertilization take place in the same year.

At the time of pollination the ♀ inflorescences are always erect. In *Abies* and *Cedrus* the cones remain upright and when they are ripe the scales fall separately from the axis (Fig. 635A). In other genera, however, the cones become pendulous (Fig. 636C); as they dry out the cone scales open out, releasing the seeds, after which the cone is shed as a whole. The seeds have a membranous wing which aids dispersal (Figs. 631F, 635E, 636F).

The Pinaceae occurred as far back as the Jurassic; today they are almost entirely confined to the Northern Hemisphere, where they are dominant in the coniferous belt of N. America and Eurasia, both in the plains and on the mountains. Together with *Betula* they often form the polar forest limit. Further south they are increasingly restricted to the mountains, but even there they assume dominance only in certain altitudinal zones, especially towards the limits of trees and forests (Fig. 749). It is noticeable that in Europe they avoid the west and south-west where the winters are mild (Fig. 638). The following European genera and species are worthy of mention.

Abies. The silver fir (*A. alba*, Fig. 635), so-called from its light-coloured bark, is recognized by the flat, emarginate needles with two white waxy stomatal stripes below. A mountain tree of C. and S. Europe (Fig. 638), it is exacting in its climatic and edaphic requirements; it generally grows mixed with beech and spruce. Its sensitivity to late frosts (particularly when exposed by clear felling), to browsing by game and to industrial pollution has led to its extinction in many areas.

Picea. The spruce (*P. abies* = *P. excelsa*, Fig. 636) is readily identified by its sharp quadrangular needles. Its continuous N. European–Siberian distribution extends to the Vistula and C. Sweden; further south it becomes more a tree of the mountains (Fig. 636). There it often forms the forest limit, and nearly reaches it in the Arctic. The spruce is economically very important and is often grown in plantations.

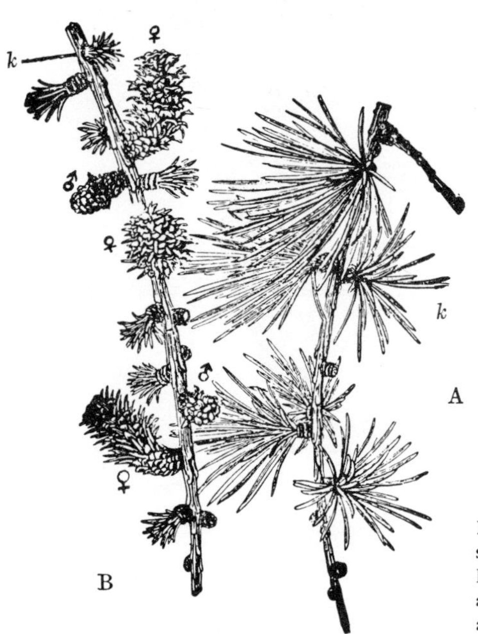

Fig. 637. *Larix decidua.* **A.** Long shoot with short shoots (k) bearing needles in summer. **B.** Long shoots with ♂ flowers, ♀ inflorescences and sprouting short shoots (k) in spring. Nat. size approx. After Willkomm.

Larix. The European larch (*L. decidua*, Fig. 637) is a light-demanding tree especially frequent at the forest limit in the continental climate of the C. Alps. Elsewhere it is found wild only in isolated patches, but it is extensively planted in the lowlands.

Pinus. The Scots fir or pine (*P. sylvestris*, Fig. 631), the only pinaceous tree now native in Britain (Fig. 638), is moderately light demanding, but otherwise remarkably unexacting, tolerating both the hot dry summers of the steppe margins and the winter cold of Siberia; it ascends from the plains to the Alpine forest limit and colonizes dry sandy soils, wet bogs and both calcareous and siliceous substrata. It is therefore most frequent where more demanding trees fail to grow, e.g. on the sandy soils of N.E. Germany. Above the forest limit in the mountains and at lower altitudes on raised bogs several subspecies of the mountain pine (*P. mugo*) are widespread, sometimes as a prostrate shrub (*Legföhre* or *Latsche*), sometimes as an erect tree. Both *P. mugo* and *P. sylvestris* have only two needles on the short shoot whereas there are five in *P. cembra* (Arolla pine) which occurs especially at the forest limit in the Alps and Carpathians, and also in Siberia. Other five-needled pines include the Weymouth or white pine (*P. strobus*) from N. America, frequently planted in Europe but not thriving in Britain, and the bristle pine (*P. aristata*), individuals of which in the White Mountains of California are over 4600 years old. Further species are predominantly Mediterranean, such as the two-needled Austrian or Corsican pine (*P. nigra*) which is native in woods near Vienna and widely planted on dry slopes further north, and the stone or umbrella pine (*P. pinea*). The large seeds of *P. cembra* and *P. pinea* are edible and are dispersed by animals.

Numerous exotic Pinaceae are grown in gardens, but so far only a few are important in European forestry. Besides the Weymouth pine, the Douglas fir (*Pseudotsuga douglasii*, which can reach a height of 127 m!) and the Sitka spruce (*Picea sitchensis*), both from western N. America, are particularly important, as well as the Japanese larch (*L. leptolepis*).

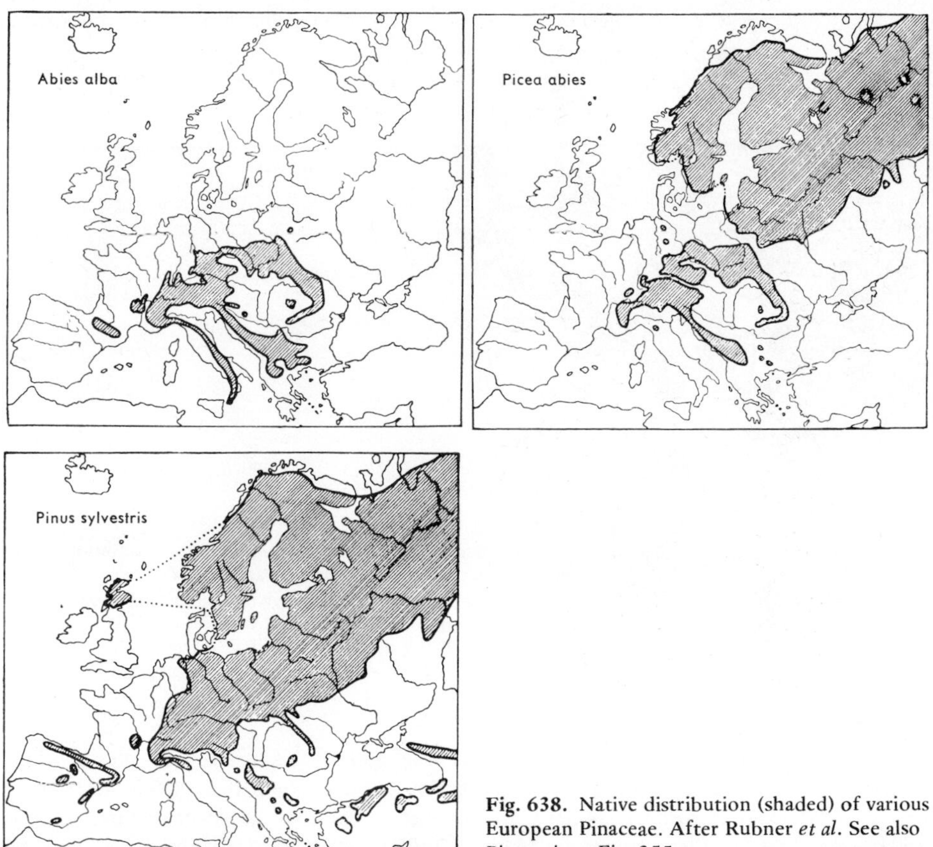

Fig. 638. Native distribution (shaded) of various European Pinaceae. After Rubner *et al.* See also *Pinus nigra*, Fig. 355.

In the next two families the extensively united bract and ovuliferous scales often bear more than two generally erect ovules; air bladders are absent from the pollen. The **Taxodiaceae** have predominantly spirally arranged needles and woody cones. They include, for example, the Californian mammoth trees or redwoods: *Sequoiadendron giganteum* from the Sierra Nevada reaches diameters of over 8 m and an age of over 3000 years. *Sequoia sempervirens* from the coastal range reaches over 100 m in height and grows more quickly; it is a valuable timber tree. Both genera were widespread in the Northern Hemisphere in the Tertiary, along with the umbrella fir (*Sciadopitys*) and water spruce (*Glyptostrobus*), both now confined to E. Asia; the wood of all these is frequent in lignites. *Metasequoia*, first found living in China in 1941, has deciduous short shoots; it was previously known only as a Mesozoic and Tertiary fossil; its remains have been found even in N. America and Spitzbergen. Finally, the swamp cypress (*Taxodium distichum*) forms extensive swamp forests along the north coast of the Gulf of Mexico; it is remarkable for the 'knee-roots' (? breathing roots) which project from the water or mud. The short shoots are shed annually and look like pinnate leaves.

The **Cupressaceae** are cosmopolitan. The scale-like or more rarely needle-like leaves (Fig. 180B) are opposite or in whorls of three. Most genera have woody cones, but the only genus of C. and N. Europe, *Juniperus*, forms fleshy berry-like cones (endozoochory!).

Woody cones occur in, for example, the Mediterranean cypress (*Cupressus sempervirens*) and the genera *Thuja* (arbor vitae) and *Chamaecyparis* which originate from N. America and E. Asia and are often planted in gardens.

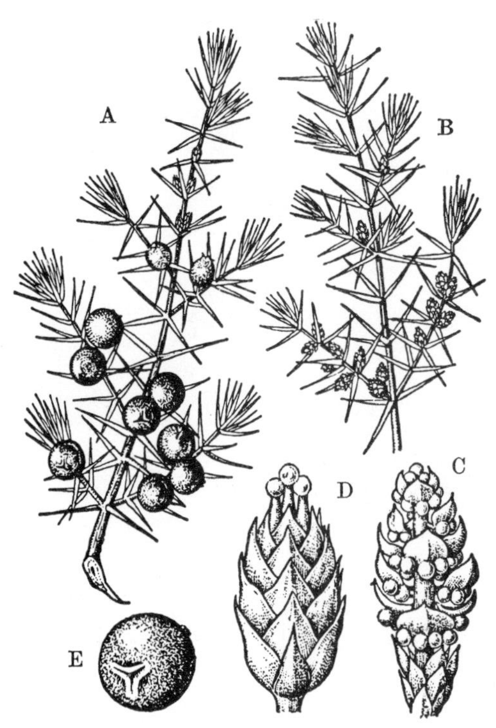

Fig. 639. *Juniperus communis.* **A.** Shoot of a ♀ plant with ♀ flowers and one- and two-year 'berries' (**E.**). **D.** ♀ with pollination droplets. **B.** Shoot of a ♂ plant with ♂ flowers (**C.**). **A.** and **B.** $\times\frac{2}{3}$ approx. **C.** to **E.** Enlarged. **A.** and **B.** After Firbas. **C.** to **E.** After Berg and Schmidt.

Juniperus communis, the dioecious juniper (Fig. 639), bears its sharp needles in whorls of three. The ♂ flowers consist of small axillary shoots which bear a few scale leaves below and several whorls of stamens above, each with three to seven (generally four) pollen sacs. The ♀ inflorescences are likewise axillary short shoots which have numerous scale leaves and terminate in three erect ovules. As in the pine, a year elapses between pollination and fertilization. As the cone develops, the three uppermost scale leaves enlarge, enclosing the maturing seeds and forming a fleshy, globose 'berry' which still shows at its apex the tips of the three scale leaves separated by indented sutures. The common juniper is an undemanding shrub especially characteristic of pastures and dwarf shrub heaths. Above the Alpine forest limit and in the Arctic it is replaced by the prostrate subspecies *nana*.

In the **Podocarpaceae** from the Southern Hemisphere the ♀ inflorescences are much reduced and woody cones are not formed; the ovuliferous scale develops at maturity into a unilateral fleshy covering of the seed.

The **Cephalotaxaceae**, with only one relict genus confined to the Himalayas and E. Asia, are similar in some respects to the Taxidae, but have lateral ♀ flowers in few flowered inflorescences.

C. Subclass: Taxidae

The ♀ flowers have a few pairs of scale leaves at the base and a single, terminal, orthotropous ovule. This arrangement can be derived from the basic ground plan of the Coniferophytina (Fig. 616A) by the loss of all the lateral ovules.

The single family **Taxaceae** includes trees and shrubs with generally spirally arranged needles and is almost entirely confined to the Northern Hemisphere. The only European representative is the yew, *Taxus baccata* (Fig. 640), readily recognized by the acute, flat needles, darker green above than below; they are held in two rows on long shoots. The species is dioecious and the flowers develop in the axils of the leaves. The ♂ flowers have a few scale leaves at the base and a number of peltate stamens above, each with six to eight pollen sacs, reminiscent of the sporangiophores of *Equisetum*.

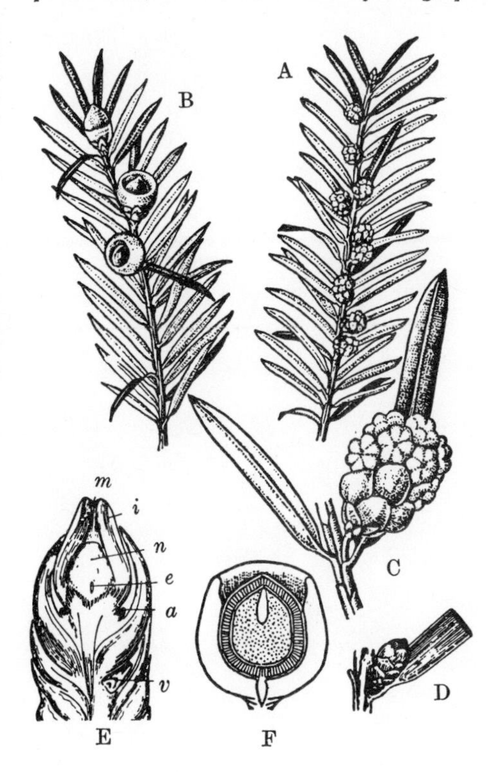

Fig. 640. Taxus baccata. **A.** Flowering ♂ twig. **B.** Fruiting ♀ twig (with one unripe and two ripe seeds). ×¾. **C.** and **D.** ♂ and ♀ flowering shoot, each in the axil of a needle. ×2.5. **E.** L.S. of a ♀ flowering shoot with micropyle (m), integument (i), nucellus (n), embryo sac (e), primordium of the aril (a) and vegetative apex of the primary axillary shoot (v). ×9. **F.** Seed in L.S. with aril, testa, endosperm and embryo. ×2. **A., B.** and **D.** After Firbas. **C.** and **F.** After Wettstein. **E.** After Strasburger.

The ♀ flower arises in the axil of a needle as a small second-order shoot (Fig. 640D, E) bearing only a single orthotropous terminal ovule which secretes from the micropyle a pollination droplet which entraps the pollen grains. The ovule is surrounded at the base by a meristematic ring which can be regarded as an outgrowth of the axis. As the seed matures this outgrowth develops into a red, fleshy, sweet-tasting cup which functions in the dispersal of the seed by birds and is the only part of the plant free from the poisonous alkaloid taxin.

The Taxaceae occurred from the Upper Triassic onwards. In Germany the yew has become rarer in the course of centuries despite its capacity for regeneration from stumps; this may be attributed to its slow growth, the high value placed on its dense, hard wood even in prehistoric times, and its sensitivity to frost. Some of the finest yew woods in Europe occur in the milder climate of the British Isles.

Survey of the evolutionary history of the Coniferophytina

The hypothetical evolutionary history of the Coniferophytina from progymnospermous ancestors (pp. 612 et seq.) is evident in Fig. 628. The Ginkgoatae and Pinatae probably appeared as parallel lines of evolution at the transition from the Upper Devonian to the Lower Carboniferous. Whilst the Ginkgoatae only reached their maximum in the

Mesozoic the differentiation of the Pinatae began in the Upper Palaeophytic with the Cordaitidae. Further developments already set in train in the Carboniferous led to the origins of the Pinidae at the Carboniferous/Permian transition with the Voltziales, and in the Mesozoic to the diversity of the Pinales, and to the appearance of the Taxidae. The Pinales have remained to the present-day as trees of considerable competitive ability, especially in habitats otherwise too extreme for woody plants.

Subdivision 2: **Cycadophytina (Cycadicae), pinnate-leaved gymnosperms**

This second group of gymnosperms is principally characterized by the complex structure of its vegetative and fertile lateral organs: the leaves (trophophylls) have a pinnate basic plan, the stamens (microsporophylls) comprise several groups of pollen sacs (or consist of synangia), and the carpels (megasporophylls) bear several ovules, at least in the primitive forms.

The primitive Cycadophytina branch very little (pachycaulous) but in advanced groups are richly branched (leptocaulous); branching ranges from monopodial to sympodial. The extinct pteridosperms show the evolution of eusteles with an extensive medulla from lobed protosteles. The secondary xylem frequently consists of loosely packed tracheids with scalariform thickening or bordered pits and is often interspersed with broad medullary rays (manoxylous, Fig. 642A). The advanced Gnetatae show anomalous secondary thickening with several cambial rings and the occasional occurrence of vessels with bordered pits. The complex pinnately branched leaves are often reduced to undivided linear or even scale-like forms (Figs. 642B–C, 648–50). The originally dichotomous branching, especially of the terminal pinnae, gradually disappears and correspondingly the venation shows a progression from dichotomous to pinnate and ultimately to reticulate. The originally spiral arrangement of the leaves may become whorled or decussate.

The primitive sporotrophophylls give rise to ♂ and ♀ sporophylls and ultimately by their aggregation on short shoots to flowers. Originally unisexual, they may become hermaphrodite and surrounded with a perianth. Shortening of the floral axis may cause the floral organs to become whorled instead of spirally inserted. Secondarily reduced (or secondarily unisexual) flowers may be aggregated in unisexual or bisexual inflorescences (Gnetatae).

Stamens and carpels were originally branched in three dimensions and, like the leaves, not borne in a plane; vegetative segments were borne as well as numerous groups of pollen sacs or ovules (Figs. 641, 643, 644). The stamens gradually evolved as flattened, pinnately branched, undivided, scale-like or stalk-like structures (Figs. 646–50). The number of pollen sac groups per stamen was reduced, and fusion and the formation of synangia often occurred. Originally the ovules were borne on leaves or leaf segments (phyllosporous); these may have evolved into typically leaf-like carpels (Figs. 644I, 645B) or into cup-shaped or peltate structures (Fig. 644E, G, H); finally, reduction may have led to a single ovule borne directly on the floral axis (secondary stachyospory; Figs. 647A, 649E). In general the ovules are naked, i.e. not enclosed; but see Fig. 644K and p. 633. The inner, first integument is formed by the fusion of enveloping telomes (Fig. 644A–C); generally there is a second outer integument which probably originated from a cupule-like structure (Fig. 644C–D). The pollen grains, ♂ and ♀ gametophytes, fertilization and development of the seed are broadly similar to those of the Coniferophytina, and show similar progressions (p. 615). Rarely no archegonium is formed in the ♀ prothallus (p. 638). Sometimes wind pollination is superseded by animal pollination.

The Cycadophytina apparently go back to the Upper Devonian and are phylogenetically related to the progymnosperms present at that time. Today the group is represented only by a few (some 200) relict species which play a minor role in plant communities ('living fossils').

Class I: Lyginopteridatae (Pteridospermae), seed ferns

This extinct group of fern-like gymnosperms included the progenitors of the Cycadophytina. They had no flowers; instead the groups of pollen sacs or the ovules were borne on definite portions of the generally copiously branched fronds (sporotrophophylls). More rarely they occurred together on their own sporophylls, i.e. stamens or carpels, but these were never aggregated on short shoots of limited growth (i.e. flowers). This agrees closely with the basic ground plant of the Cycadophytina (Fig. 616B).

The pteridosperms developed an extraordinary diversity of forms, apparently as early as the Upper Devonian, but especially in the Carboniferous and Lower Permian; the last representatives survived until the Jurassic and then died out. Since they are mostly only

Fig. 641. *Tetrastichia bupatides* (or a related form; Lower Carboniferous). $\times\frac{1}{5}$ approx. Reconstruction after Andrews.

preserved as fragments, the taxonomy of the group is still chaotic. It was one of the most significant achievements of palaeobotany when in 1904–06 it was demonstrated that stems with secondary thickening, fern-like fronds, groups of microsporangia (?) and in particular certain seeds, all known previously under various names, belonged to the same plant, *Lyginopteris hoeninghausii*. On the other hand we are rather better informed about the extremely important historical differentiation of particular groups of organs.

Primitive characters of the seed ferns are most clearly displayed in the oldest family of the following order.

Order 1: **Lyginopteridales** (Cycadofilicales). These include the important Carboniferous **Lyginopteridaceae**. The thin, scarcely branched stems (about 1 cm in diameter) of *Tetrastichia* (Fig. 641) had a lobed protostele with no medulla; in *Lyginopteris* the thicker stems (about 4 cm in diameter, Fig. 642A) already had a eustele with a central medulla; both had secondary thickening with loosely packed (manoxylous) xylem elements. Tracheids ranged from spiral and scalariform to those with 'araucarioid' bordered pits. The leaves were often pinnate, but sometimes were still dichotomously lobed, often three dimensional and with open dichotomous venation, thus showing telome-like characteristics (Figs. 641, 642B). *Crossotheca* (Fig. 643A, B) probably represented the type of pollen sac group in this family: more or less radially arranged and only slightly united below; these were borne on three-dimensional branches which formed part of a leaf. The ovules of *Lyginopteris* (Fig. 644D) were of the type known as *Lagenostoma*; there were also carried on more or less cylindrical segments of the leaves and also branched in more than one plane. The inner integument was largely united to the nucellus and was surrounded by a lobed, glandular structure, the cupule. A pollen chamber indicated fertilization by spermatozoids. Development of the embryo was very protracted and apparently only took place after the seed was shed (generally leaving the cupule on the parent plant).

The stems of advanced pteridosperms were noteworthy for the progression to polysteles, especially in the Upper Carboniferous/Permian family **Medullosaceae**. The closed stele gave way ever more completely to independent leaf traces forming numerous vascular bundles, and secondary thickening took place around each bundle independently.

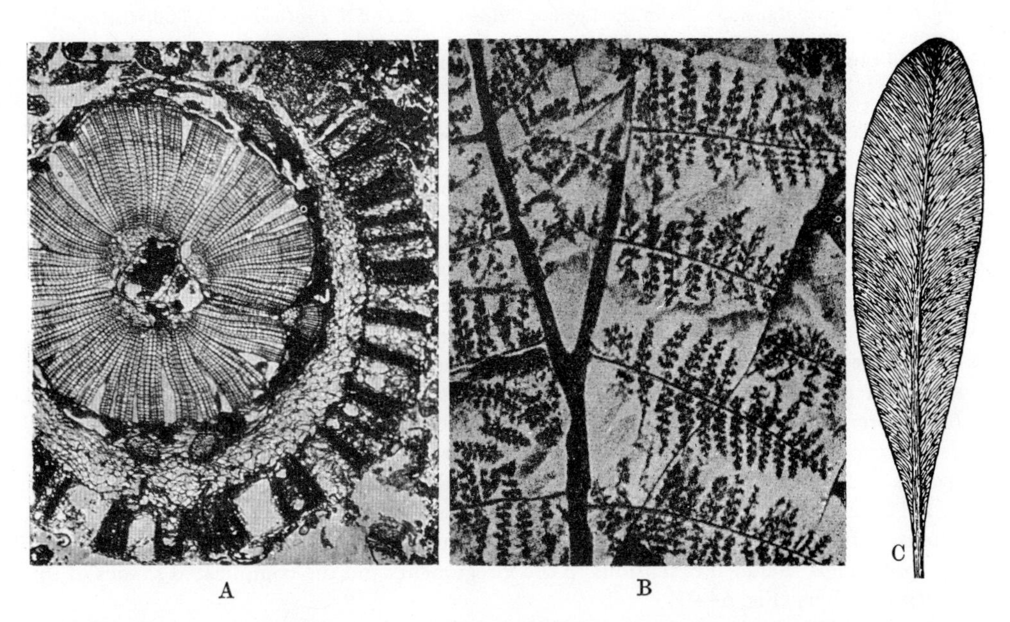

Fig. 642. Vegetative organs of the Lyginopteridatae. **A.** and **B.** *Lyginopteris hoeninghausii* (Upper Carboniferous). **A.** T.S. of stem; from the inside outwards: medulla, ring of secondary xylem (with tracheids and medullary rays), leaf traces (leaves with one strand), inner cortex (with parenchyma), outer cortex with radial anastomosing plates of sclerenchyma. ×3 approx. **B.** Part of a frond with forking midrib. $\times\frac{1}{3}$. **C.** *Glossopteris* (Permian coal measures), entire leaf with reticulate venation. $\times\frac{1}{3}$. **A.** After Scott. **B.** After Potonié. **C.** After Gothan.

Apparently the stem was held together only by the concomitant growth of the intervening parenchyma. Perhaps some of these were lianes.

The leaves of the later pteridosperms showed a progressive tendency towards plane, less divided and finally undivided types, and towards reticulate venation. The tongue-shaped leaves of the **Glossopteridaceae** (Fig. 642C) provide a characteristic example. These are indicator fossils of the very independent Gwondana flora of the Permian coal measures (and Lower Mesozoic) characteristic of the then apparently united landmasses of the Southern Hemisphere (and including India).

The groups of pollen sacs of the more recent pteridosperms show progressive fusion to form synangia and also an increase in numbers of the pollen sacs. The structures such as *Whittleseya* (Fig. 643C, D), *Aulacotheca* (E) and *Potoniea* (G) occur especially in the Medullosaceae. Flat, leaf-like structures only developed from three-dimensional ones relatively late (e.g. *Zeilleria*, Fig. 643F).

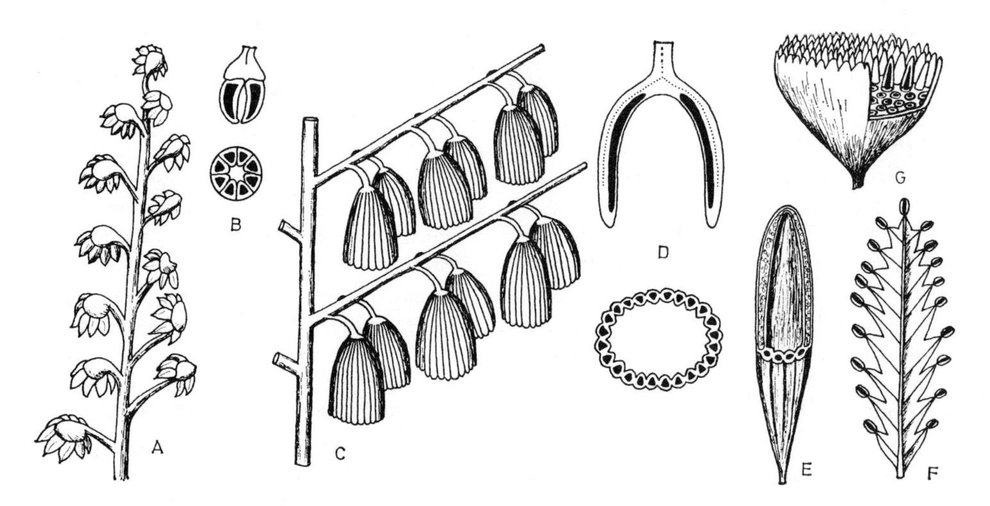

Fig. 643. Microsporangiophores of the Lyginopteridatae. **A.** and **B.** *Crossotheca* (Middle Carboniferous to Lower Permian), general appearance (×1.5 approx.) and L.S. and T.S. of a group of pollen sacs (×3 approx.). **C.** to **D.** The same for *Whittleseya* (Middle Carboniferous; $\times\frac{1}{3}$ and $\frac{2}{3}$ approx. respectively). **E.** and **G.** Groups of pollen sacs of *Aulacotheca* and *Potoniea* (both Middle Carboniferous; ×1½ and ×3 approx. respectively). **F.** Leaf-like structure with numerous groups of pollen sacs in *Zeilleria* (Upper Carboniferous). ×1.5. Sporogenous tissue black. After Hirmer, Remy *et al.*

The progressions shown by the ovules and carpels are particularly noteworthy. Various Lower Carboniferous forms (*Genostoma kidstonii—G. latens—Eurystoma angulare*, Fig. 644A—C) show very strikingly the progressive fusion of enveloping telomes (or sterilized megasporangia) around a central nucellus (fertile megasporangium) to form the first integument. In Upper Carboniferous pteridosperms (e.g. *Lyginopteris, Gnetopsis*, Fig. 644D, H, or in the Medullosaceae) this integument is extensively fused to the nucellus. This process may then be repeated, either round single ovules or groups of ovules, as shown in *Eurystoma* (C) and *Calathospermum* (G), both Lower Carboniferous, and in *Lyginopteris* (D) and *Gnetopsis* (H) (Middle or Upper Carboniferous). In the Lyginopteridaceae cupules surrounding one or more seeds arose in this manner. In the later pteridosperms and their descendants a second, outer integument arose by the failure of the seed to fall from the cupule, by the formation of a single seed inside each cupule, and by the extensive fusion of the cupule with the first, inner integument. Devices which trapped microsporangia included extensions to the integuments and the complicated shapes of the pollen chambers (Fig. 644D).

In the older pteridosperms the branches bearing ovules were more or less radial in T.S., three dimensional and often more or less dichotomous. Sometimes they became modified as fully leaf-like carpels (e.g. *Pecopteris plukenetii*, Upper Carboniferous to Lower Permian, Fig. 644I), but other characteristic developments took place, as for example in the Upper Permian to Triassic **Peltaspermaceae** with peltate structures (Fig. 644E).

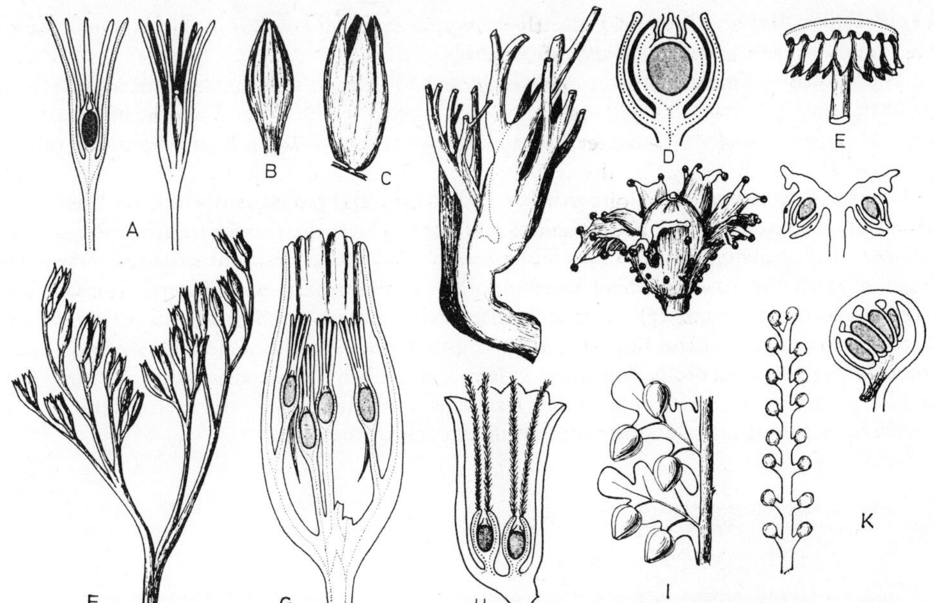

Fig. 644. Megasporangiophores of the Lyginopteridatae. Union of enveloping telomes to form the (first) integument around the ovules of: **A.** *Genomosperma kidstonii* (×1.5). **B.** *G. latens* (×2). **C.** *(left), Eurystoma angulare* (×2.5) (all Lower Carboniferous). Development of the cupule (second integument) from telomes in **C.** *(right), Eurystoma angulare* (Lower Carboniferous; ×2.5). **D.** *Lyginopteris (Lagenostoma) hoeninghausii* (Middle Carboniferous; reconstruction and L.S.; ×2 approx.). **E.** Peltate megasporangiophore of *Peltaspermum rotula* (Triassic; reconstruction and L.S. Nat. size). **F.** Dichotomous branch system, bearing cupules with several seeds in each, of *Stamnostoma huttonense* (Lower Carboniferous; ×½ approx.). Cupules with several ovules of: **G.** *Calathospermum scoticum* (Lower Carboniferous; simplified L.S.; ×¾ approx.) and **H.** *Gnetopsis elliptica* (Upper Carboniferous, with integuments bearing style-like processes; L.S.; ×5 approx.). **I.** Part of the leaf-like megasporangiophore (carpel) of *Pecopteris pluckenetii* (Upper Carboniferous to Lower Permian; somewhat enlarged). **K.** Ovules of *Caytonia* enclosed in pinnate leaf segments (Lias), general appearance, ×½ approx.; and L.S. of a pinna, ×3 approx.; sporogenous tissue or embryo sacs grey. **A.** and **B.** After Andrews. **C.** and **F.** After Long. **D.** After Oliver and Scott. **E.** After Harris, modified. **G.** After Walton. **H.** After Renault and Zeiller. **I.** After Arnold. **K.** After Thomas.

Order 2: **Caytoniales** also showed a very individual development. In the **Caytoniaceae** pinnate leaf segments nearly enclosed several ovules (Fig. 644K). Indeed, the pollen grains, which were provided with air bladders, could apparently be drawn through a small opening up to the micropyle by means of a pollination droplet. This peculiar terminal group of pteridosperms (Upper Triassic to Lower Cretaceous) was thus not truly angiospermous. The leaves of the Caytoniaceae were '*Sagenopteris*' with four palmate lobes; the stamens showed complex branching and bore many synangia with four pollen sacs.

Class II: Cycadatae, cycads

The cycads first occurred in the Triassic; today they are 'living fossils', distinguished from the pteridosperms mainly by the pollen sac groups and ovules being borne on typical microsporophylls (stamens) and megasporophylls (carpels); these are numerous and form simple flowers on shoots with limited growth. Only in *Cycas* do groups of sporophylls and trophophylls alternate on an axis of unlimited growth.

Apart from the extinct Mesozoic order **Nilssoniales** (with very lax ♀ cones) the chief members of the class are the true **Cycadales**. Although numerous in the Mesozoic, there are now only ten genera, each with few species; these show very disjunct distributions in the Tropics and Subtropics: the **Cycadaceae**, with *Cycas* from Madagascar to Polynesia and E. Asia; the **Stangeriaceae**, with *Stangeria* in Africa; and the **Zamiaceae**, with

Lepidozamia, *Macrozamia* and *Bowenia* in Australia, *Encephalartos* in Africa, and *Dioon*, *Microcycas*, *Ceratozamia* and *Zamia* in America.

The cycads resemble palms in habit (Fig. 645A). The stem is massive and unbranched (pachycaulous); it may be quite short or even buried in the ground and bears an apical tuft of large, spirally arranged, frond-like leaves; these are pinnate or more rarely bipinnate.

The leaves continue apical growth for a long time and are initially circinate like ferns. Venation is frequently still dichotomous and open. The formation of transfusion tissue is notable. The growing point alternately produces foliage leaves, and scale leaves which, together with the bases of dead leaves, clothe the older parts of the stem. Stomata are haplocheilous, i.e. guard cells and subsidiary cells develop from different initials. While *Dioon* only forms a single ring of xylem, in other genera additional cambial rings occur towards the outside, producing both xylem and phloem, somewhat as in the Medullosaceae (p. 642). The xylem includes tracheids both with scalariform thickening and bordered pits. Mucilage ducts are present in all parts of the plant.

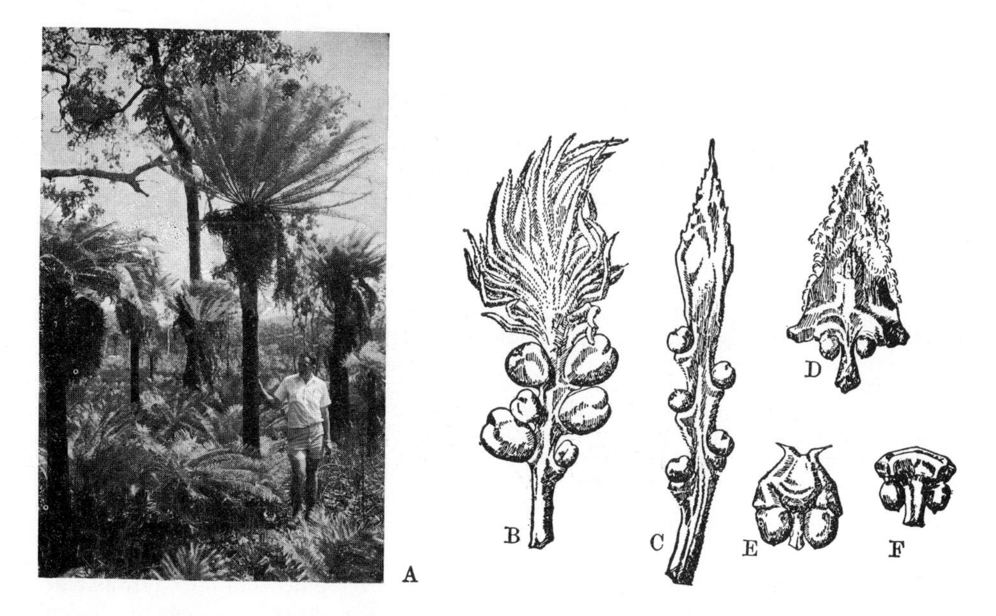

Fig. 645. **Cycadales.** **A.** *Cycas rumphii* in New Guinea, habit. **B.** to **F.** Carpels (megasporophylls) of *Cycas revoluta* (**B.**), *C. circinalis* (**C.**), *Dioon edule* (**D.**), *Ceratozamia mexicana* (**E.**), *Zamia skinneri* (**F.**). **A.** Photo by Ehrendorfer. **B.** and **D.** to **F.** After Firbas. **C.** After Schuster.

The cycads are dioecious. A perianth is never present. The most primitive ♀ 'flower' is that of *Cycas* in which the growing point periodically produces, instead of leaves, a large number of yellowish-brown, densely hairy carpels; that these are homologous with leaves is clearly shown by their markedly pinnate terminal portion (Fig. 645B). They do not become green; instead the basal part bears a number of marginal ovules. The apical growing point is not used up in the formation of 'flowers' and after a time forms further leaves and bracts. The primitive ♀ 'flower' of *Cycas* thus has unlimited growth. The terminal ♀ flowers of all other cycads (and the ♂ flowers of all genera) are formed by a growing point which ceases to function after it has produced a number of carpels or stamens, as in all other seed plants; subsequently the cone is displaced to one side by a new sympodial meristem which continues the growth of the stem. *Macrozamia*, however, has genuinely axillary flowers.

Even within the genus *Cycas* various species show a suppression of the sterile terminal portion of the carpel (Fig. 645B, C) and a reduction in the number of ovules. This tendency is more marked in other genera (D, E), culminating in simple peltate carpels with only two ovules (F); such carpels are borne spirally on an extended axis, their

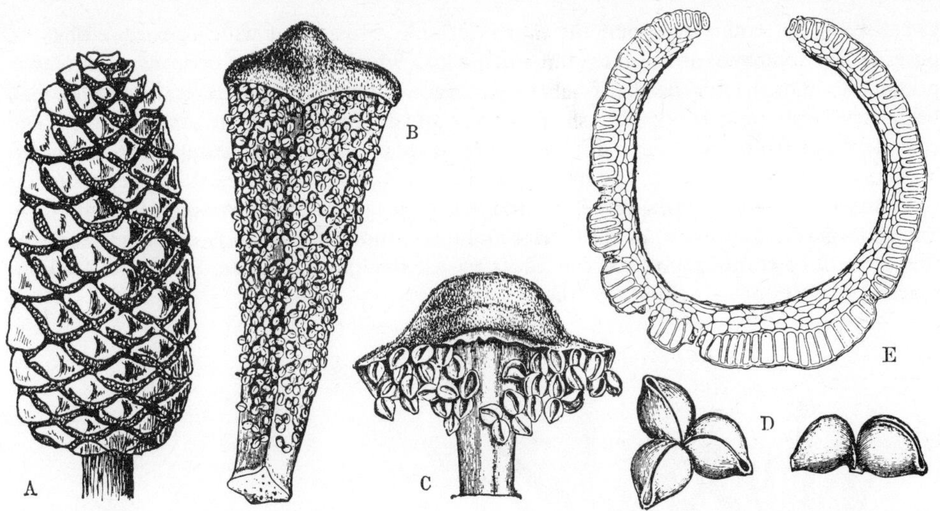

Fig. 646. Cycadales. **A.** ♂ flower of *Encephalartos altensteinii* (reduced). **B.** Stamen of *Cycas circinalis*. ×2 approx. **C.** Stamen of *Zamia integrifolia*. ×6 approx. **D.** Its pollen sac groups. ×20 approx. **E.** T.S. of the wall of a dehisced pollen sac of *Stangeria paradoxa* with exothecium. ×100 approx. **A.** After Troll. **B.** to **D.** After Karsten. **E.** After Goebel.

expanded apices tightly packed together, forming compact cones. At the time of pollination the axis elongates and the scales separate slightly so that the pollen carried by the wind (or in *Encephalartos* by beetles, p. 608) reaches the ovules.

A similar structure occurs in the ♂ flowers of all species (Fig. 646): an elongated axis bears spiral series of stamens having a sterile apical portion and a large number of groups of pollen sacs on the lower surface. The pollen sac opens by means of an exothecium.

The envelope of the ovule is supplied by two vascular systems and has apparently evolved by the fusion of two integuments (Fig. 624A). Below the micropyle the nucellus is hollowed out as a pollen chamber which after a while extends until it reaches an opening in the megaspore wall. The massive megaprothallus contains a variable number of archegonia in a chamber opposite the micropyle (Fig. 624B). Each archegonium has a remarkably large egg cell (up to 6 mm), an evanescent ventral canal cell and generally just two neck cells.

At the time of pollination the ovule secretes through the micropyle a drop of fluid which traps the pollen grains; these have already formed a prothallial cell and the generative cell (Fig. 623B, C). Apparently, as the drop of fluid dries up, the pollen grain is drawn into the pollen chamber which meanwhile closes over towards the outside but becomes continuous inside with the archegonial chamber by the dissolution of nucellar tissue. At this stage the exine of the pollen grain is ruptured and the vegetative cell grows out as a pollen tube which penetrates the tissue of the nucellus (Figs. 623E–H, 624B), apparently functioning here, as in *Ginkgo*, in the nutrition and attachment of the ♂ gametophyte in the pollen chamber and not, as in all other seed plants, the transportation of the ♂ sex cells. As the pollen tube develops the generative cell divides into a stalk cell and a spermatogenous cell which later divides again into the two sperm cells. Each of these forms a freely motile spermatozoid which is liberated by the breakdown of the intine (Fig. 623G, H). In *Microcycas* a large number of spermatozoids is formed (Fig. 623F). These spermatozoids are remarkably large, reaching 0.3 mm in diameter—the largest in the Plant and Animal Kingdoms—and have a spiral band of flagella (Fig. 623J). They are able to swim through the cavity formed by the union of the pollen chamber and archegonial chamber in a fluid apparently discharged from the pollen tubes. One of them enters the egg cell, sheds its plasmatic membrane and the band of flagella, and its nucleus unites with the egg nucleus (zoidiogamy, Fig. 624B). Some months elapse between pollination and fertilization.

The zygote grows rapidly with free nuclear division, forming a proembryo (Fig. 625) in which cells only arise at the lower extremity, and of these only the lowermost few cells give rise to the embryo proper, the upper forming a suspensor which greatly elongates, pushing the embryo down into the prothallus, which now functions as a nutritive, primary (haploid) endosperm. If several archegonia are fertilized, several embryos may begin development, but sooner or later all but one degenerate. At the same time the seed coat develops from the integument; it is fleshy outside (sarcotesta), stony and sclerenchymatous within (sclerotesta). The ovule has thus become a seed. The seedling generally has two cotyledons which remain inside the seed on germination, absorbing nutrients from the endosperm. The cycads have little economic importance: sago is obtained from the starchy pith of some species and the leaves are used as 'palm' on Palm Sunday in Mediterranean countries or in making wreaths.

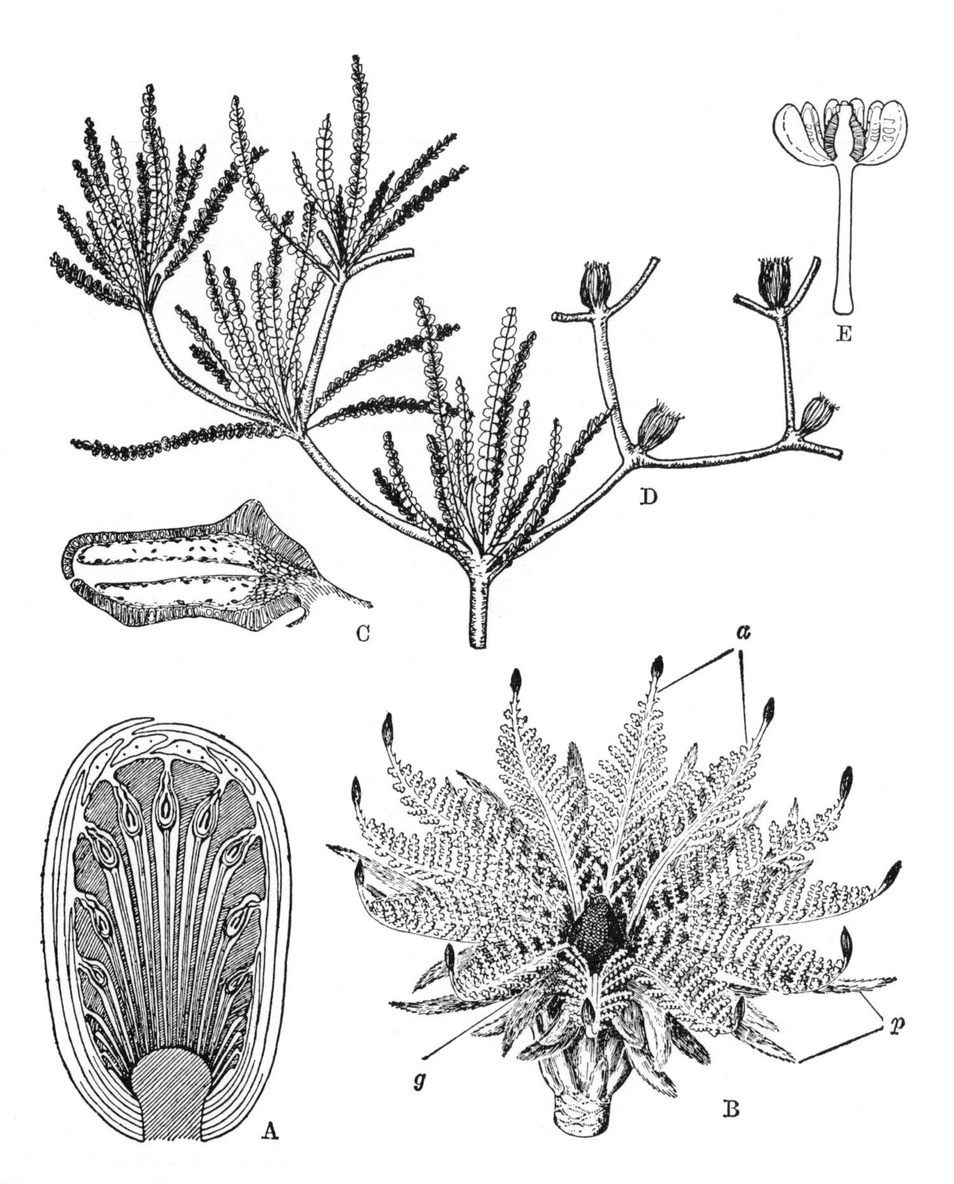

Fig. 647. Bennettitatae. **A.** L.S. of the fruit derived from the ♀ flower of *Bennettites gibsonianus* (Lower Cretaceous) with perianth leaves, interseminal scales and stalked seeds. ×½ approx. **B.** Reconstruction of the hermaphrodite flower of *Cycadeoidea ingens* (Lower Cretaceous) with perianth leaves (p), pinnate stamens (a) and the gynoecium (g). $\times\frac{1}{3}$. **C.** L.S. of a pollen sac group with exothecium of *Cycadeoidea dacotensis*. ×20 approx. **D.** *Wielandiella angustifolia* (Rhaetic), reconstruction with branches, leaves and flowers or fruits. ×¼. **E.** Flower of *Williamsoniella coronata* (Middle Jurassic) with simple stamens and gynaecium. Nat. size approx. **A.** After Solms. **B.** After Wieland, modified. **C.** After Wieland. **D.** After Nathorst. **E.** After Thomas.

Class III: Bennettitatae

This class was extensively represented in the Mesophytic but subsequently became extinct: it differed from the seed ferns and cycads principally in the extremely reduced carpels which consisted of a single stalked ovule borne directly on the floral axis.

The subclass **Bennettitidae** (order **Bennettitales**) is of particular importance in exhibiting for the first time truly hermaphrodite flowers with a perianth; these were probably insect pollinated. The bennettites formed a major part of the land flora from the Upper Triassic to the Lower Cretaceous but then became extinct probably in competition with the angiosperms. For their possible relationships with the Gnetatae see p. 639.

The principal genera were *Williamsonia, Wielandiella, Williamsoniella* and *Cycadeoidea* (Fig. 647). Some were pachycaulous in habit like cycads (e.g. *Williamsonia*) but others were leptocaulous with sympodial branching (e.g. *Wielandiella*, Fig. 647D). Xylem structure was more primitive than that of modern cycads, with simple eusteles and mainly scalariform tracheids. The leaves were similar, pinnate or simple, but had syndetocheilous stomata (i.e. guard cells and subsidiary cells arose from the same initial). Whilst the more primitive *Williamsonia* had unisexual flowers (like Fig. 647A), other genera frequently had hermaphrodite flowers. The flowers were either terminal or lateral. Commonly a perianth of sterile bracts was present. Originally the floral organs were numerous and spirally arranged but became reduced and eventually whorled (Fig. 647B, E). The stamens were sometimes pinnately branched (*Cycadeoidea*, Fig. 647B), sometimes reduced to resemble perigonial leaves (*Williamsoniella*, Fig. 647E). The pollen sacs (with an exothecium, Fig. 647C) formed synangia which were marginal or embedded in the upper surface of the stamens. In the upper, female part (gynaecium) the more or less conically elongated axis bore closely packed, spirally arranged interseminal scales between which were the carpels reduced to stalked ovules. The bennettites were apparently secondarily stachysporous, and the homology between the ♂ and ♀ organs was almost completely obscured. The integument (probably originally double) of the ovule protruded at anthesis beyond the densely packed interseminal scales, thus permitting access to the pollen. Protandry (p. 657), zoophily (? beetles) and fertilization by spermatozoids were highly probable. At maturity the interseminal scales became more or less fleshy and together with the seeds (including the embryo with two cotyledons) apparently formed fruits (Fig. 647A).

A smaller subgroup of the Bennettitatae was the Jurassic subclass (or order) **Pentoxylidae (Pentoxylales)**. These had polystelic stems and tongue-shaped leaves, and the compact ♀ flowers with stachysporous ovules had no interseminal scales.

Class IV: Gnetatae (Chlamydospermae)

The Gnetatae resemble the Bennettitatae in sometimes having hermaphrodite flowers, a perianth, and ovules borne directly on the floral axis. The flowers are, however, extremely reduced with only one or a few stamens and a single ovule. The class comprises only the peculiar genera *Welwitschia, Gnetum* and *Ephedra* as the representatives of the monotypic subclasses (or families) Welwitschiidae (**Welwitschiaceae**), Gnetidae (**Gnetaceae**) and Ephedridae (**Ephedraceae**).

Their habit is very diverse. The secondary xylem includes fibres and vessels (but still with bordered pits) as well as tracheids; the phloem sometimes has companion cells as well as sieve tubes (although these develop from different initials). The opposite (or whorled) leaves are reticulately veined in *Gnetum* but have parallel venation in *Welwitschia* and are reduced to scales in *Ephedra*. The flowers have a definite perianth, are functionally unisexual and are dioecious (or more rarely monoecious); occasionally inflorescences include flowers of both sexes. The stamens usually bear a number of groups of pollen sacs (synangia); the ovules are surrounded by two integuments (or by one as a result of fusion). The gametophytes are more greatly reduced than in other

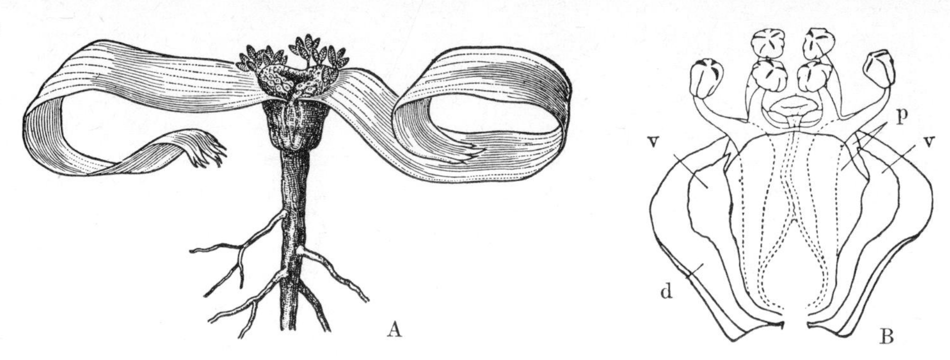

Fig. 648. *Welwitschia bainesii.* **A.** Habit of a young plant with ♀ inflorescences. $\times\frac{1}{20}$ approx. **B.** ♂ flower with bract (d), prophylls (v), perianth leaves (p), united stamens and sterile ovules. ×7. **A.** After Eichler. **B.** After Church.

gymnosperms. Thus the Gnetatae are gymnosperms at a particularly high level of organization, resembling angiosperms in some respects.

The genus *Welwitschia* is represented by the celebrated and fantastic species *W. bainesii* (*W. mirabilis*; Fig. 648A) from the coastal mist deserts of S.W. Africa and Angola. The massive plant has a short tuberous stem developed from the hypocotyl of the seedling and a tap root which penetrates the moister subsoil. The main axis produces, apart from the two transitory cotyledons, only two minute scale leaves and then two broadly strap shaped foliage leaves. These have parallel venation (with anastomoses) and grow from the base throughout the life of the plant to a metre or more in length, dying off at the apex. The flowers are borne in cone-like inflorescences in the axils of bracts. The ♂ flower (Fig. 648B) has an envelope consisting of two prophylls and two perianth segments, six stamens united below and each bearing three fused pollen sacs, and one rudimentary ovule, thus approaching the hermaphrodite state. The ♀ flower consists of two united perianth segments and a single terminal ovule with two integuments. Pollination is apparently effected by insects attracted by pollination droplets secreted from the micropyles of the sterile or fertile ovules.

The species of *Gnetum* are mostly lianes, more rarely trees or shrubs, of Tropical Rain Forests, with elliptical, net-veined leaves (Fig. 649A; the lateral veins, however, develop dichotomously). The unisexual flowers are dioecious or monoecious (cf. Fig. 649B) and are borne in the axils of ring-shaped united pairs of scale leaves in spike-like inflorescences. The ♂ flower consists of one stamen with a perianth, the ♀ of one ovule with two integuments (the inner prolonged as a tube) and a perianth (Fig. 649C–E).

The species of *Ephedra* are Mediterranean switch plants, also occurring in the arid parts of Asia and America. The profusely branched green shoots bear only small, opposite or whorled scale leaves. Most species are dioecious with the flowers solitary, in pairs or in groups at the tips of branches borne in the axils of decussate bracts; or they may be terminal (Fig. 650A, D). The ♂ flower consists of a two-lobed perianth (Fig. 650B, C) and a stamen which is sometimes forked and stalk-like (as a result of the union of several initials); the apex bears extensively united groups of pollen sacs. The ♀ flower has a bilobed perianth and a single ovule surrounded by a tubular integument with a tubular micropyle; the integument develops from two united initials (Fig. 650E). In *E. campylopoda* the ♂ inflorescence also contains sterile ♀ flowers, the ovules of which secrete, like the fertile ones, a sugary pollination droplet; this is attractive to insects (Apidae and Syrphidae) which thereby effect pollination. Very rarely hermaphrodite flowers occur.

The development of the gametophyte in the Gnetatae is characterized by progressive reduction. The ♂ gametophyte has only four cells: pollen tube cell, stalk cell (?) and two sperm cells. The ♀ gametophyte of *Ephedra* is monosporous; archegonia can still be discerned. In *Welwitschia* and *Ephedra* the ♀ prothallus develops from all four meiospores (tetrasporous) and no archegonia are developed. Polyembryony is widespread, sometimes as a result of the formation of several zygotes, sometimes by the division of the proembryos. In all genera the embryos form two cotyledons.

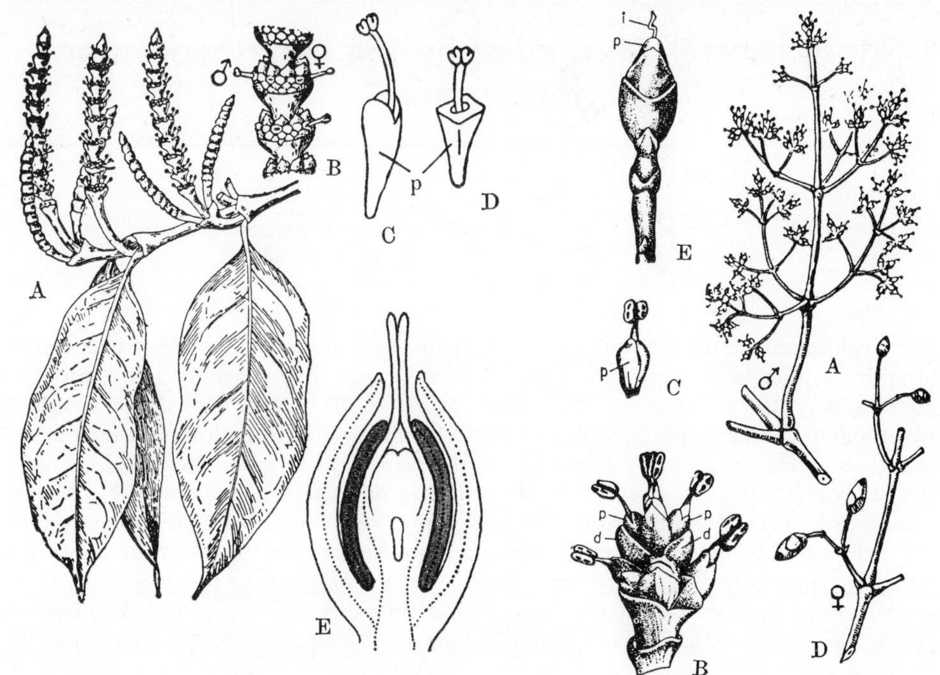

Fig. 649. *Gnetum.* **A.**, **B.** and **E.** *G. gnemon.* **C.** *G. costatum.* **D.** *G. montanum.* **A.** Shoot with ♂ inflorescences. $\times\frac{3}{8}$. **B.** Whorled partial inflorescences, outer with fertile ♂, inner with sterile ♀ flowers. ×1.5. **C.** and **D.** ♂ flowers with perianth (p). **E.** L.S. of ♀ flower with perianth, outer (woody) and inner (elongated) integuments, nucellus and embryo sac. Enlarged. **A.** and **B.** After Karsten and Liebig, modified. **C.** and **D.** After Markgraf. **E.** After Pearson, modified.

Fig. 650. *Ephedra altissima.* **A.** ♂ branch. $\times\frac{2}{3}$. **B.** and **C.** ♂ partial inflorescence and ♂ flower. ×7.5. **D.** and **E.** Branch with unripe seeds ($\times\frac{2}{3}$) and terminal ♀ flower (×2). d, bract; p, perianth; i, elongated, tubular integument. **A.** and **D.** After Karsten. **B.** and **C.** After Stapf. **E.** After Wettstein.

The scanty fossil remains give no clue to the evolutionary history of the Gnetatae. Presumably they represent the extremely ancient and fragmentary remains of a formerly more extensive group. Many of their features suggest that their ancestors were at least similar to the bennettites; a relationship between the Pinatae and *Ephedra* is less likely.

Survey of the evolutionary history of the Cycadophytina

Figure 628 also shows the possible evolutionary relationships of the Cycadophytina. The Lyginopteridaceae (seed ferns) were the fundamental group, clearly related to the progymnosperms (pp. 612 et seq.). These showed a remarkable morphological diversity; their main expansion was in the Palaeophytic, but some (e.g. Peltaspermaceae, Caytoniales) straggled on into the Mesophytic. The Cycadatae and Bennettitatae (with numerous diverse Bennettitidae and the smaller subgroup Pentoxylidae), so important in the Mesophytic, already occurred in the Carboniferous or Permian as evolutionary lines parallel to the seed ferns; they became drastically reduced as the Magnoliophytina (angiosperms) burgeoned forth at the beginning of the Cainophytic (Middle Cretaceous). Whilst the Cycadatae still survive with a few residual forms, the Bennettitatae became extinct. As shown in Fig. 628, there is a distinct possibility that the present-day Gnetatae can be considered as the last but heterogeneous descendants of the Bennettitatae.

Subdivision 3: **Magnoliophytina (Angiospermae), angiosperms**

The angiosperms resemble the gymnospermous Cycadophytina in the possession of primitively pinnate or pinnately veined leaves, several (but generally two) groups of pollen sacs on the stamens and often several ovules on the carpels. They differ from both Cycadophytina and Coniferophytina primarily in the invariable enclosure of their ovules in a closed structure or ovary formed from the carpels, from which they are liberated as ripe seeds only after the ovary, either alone or with other parts of the flower, has become a fruit (Fig. 651). The enclosure of the ovules, the formation of a floral envelope or perianth and the predominance of hermaphrodite flowers are related to the transfer of pollen by animals in all primitive angiosperms. The pollen grains do not gain direct access to the ovules, but are caught on the stigma, a special receptive organ of the carpel. A pollen tube carries the sperm cells from the stigma to the ovules and their embryo sac.

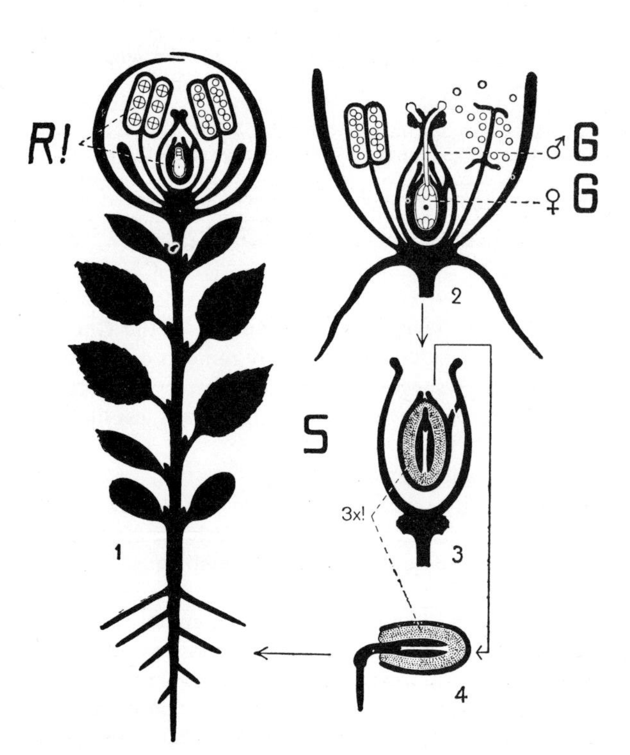

Fig. 651. Life-cycle (diagrammatic) of an angiosperm. Alternation of generations: gametophyte (G), sporophyte (S); nuclear phase: diplophase (solid black), meiosis (R), haplophase, thin lines, and triploid endosperm (3n; dotted). **1.** Whole plant with root, stem, leaves and hermaphrodite flower bud. **2.** Open flower with perianth (sepals and petals), stamens with pollen grains, and carpels (ovary, style, stigma and enclosed ovule; pollination has been effected (pollen tube!) and the egg cell in the embryo sac is just about to be fertilized. **3.** Seed with testa, secondary endosperm and embryo, becoming detached from the one-seeded fruit. **4.** Germinating seed. After Firbas.

Furthermore, the angiospermous gametophytes are more greatly reduced. The pollen grain lacks the prothallial and stalk cells, and spermatozoids are never formed. In the embryo sac within the ovule a multicellular prothallus and differentiated archegonia are also wanting. The ♀ gametophyte is generally reduced to a few cells, one being the egg cell which, after fertilization by one of the two sperm cells, gives rise to the embryo. In addition, the other sperm cell also effects fertilization (double fertilization), leading to the formation of the secondary endosperm. This compression of the development of the gametophytes facilitates, amongst other things, a significant speeding-up of sexual reproduction.

The high degree of differentiation of the tissues of angiosperms is expressed particularly in the development of vessels and sieve tubes with companion cells. Furthermore, their vegetative diversity is distinctly greater than that of the gymnosperms: besides trees and shrubs, they exhibit many herbaceous life forms, including both perennials and annuals.

The angiosperms have dominated the land flora of the earth as the largest group of species since the Middle Cretaceous. Their evolutionary origin is still obscure, but there is much to suggest that they arose within the group of the seed ferns (Lyginopteridatae) in the broad sense.

Vegetative organs. The seedlings of primitive angiosperms have two cotyledons (as in the dicotyledons); the development of a single cotyledon (especially in the monocotyledons) is advanced (Figs. 186, 731A–E). Apical meristems reach the highest levels of differentiation (pp. 89–93, 122–4). Shoot structure has developed from monopodial to partly or wholly sympodial (pp. 132–4). The characteristic eustele of the stems with open bundles and secondary thickening (Figs. 150, 160) is replaced in the monocotyledons by atactosteles with closed bundles without, or with anomalous, secondary thickening (pp. 146–7). Within the dicotyledons the bundle systems show further progressions from those with few to those with many anastomoses, and from collateral to bicollateral bundles.

The angiosperms show similarities to the Cycadophytina in the development of medulla, relatively diffuse secondary xylem and the widespread occurrence of scalariform elements, but both xylem and phloem exhibit much more extensive differentiation. With the exception of a few primitive forms (and some secondarily reduced groups; cf. pp. 680 and 685), vessels and sieve tubes with companion cells (from a common initial; Fig. 115) are uniformly present. The development of vessels from elongated tracheids is perhaps independent but parallel within the dicotyledons and monocotyledons. The vessels exhibit striking evolutionary trends in the progressive breakdown of their transverse walls as a result of the reduction of the scalariform thickenings, and in the progressive shortening and broadening of the vessel segments (Fig. 120 h, i). The sieve tubes show analogous developments in the increase in pore size and shortening and broadening of the elements (Fig. 113). Further progressions are shown in the development of xylem and phloem fibres, the clustering of the xylem parenchyma around the vessels, the spatial separation of supporting and conducting tissues, and the unification and radial elongation of the medullary ray cells. All this no doubt results in a general improvement in conduction and support by the xylem and phloem of woody plants, but in herbaceous types cambial activity and hence secondary thickening are often reduced or fully suppressed.

The primitive ground plan of the leaves of the Magnoliophytina, like those of the Cycadophytina, is basically pinnate (Figs. 173, 174). Of course, the great diversity of form and plasticity makes it difficult to decide whether a compound or simple leaf shape is the more primitive. Venation shows progressive fusion of free lateral veins to give reticulate patterns (open to closed venation). Digitate venation (Fig. 690E) is more advanced than pinnate, as is the parallel venation which predominates amongst the monocotyledons (Figs. 732C, 733H). Trilacunar nodes and three-stranded leaf traces are apparently more primitive than those with more or fewer lacunae or traces (p. 600).

In various groups of angiosperms, but especially the monocotyledons, the primary root is soon suppressed; its functions may be taken over by underground shoots or more commonly by adventitious roots from the stem (allorhizy to secondary homorhizy; Fig. 191).

Taxonomically important characters are also provided by progressive differentiation of secretory and mucilage containing organs, laticifers, stomata, hairs, etc. (cf. pp. 87–121).

In diversity of growth and life forms the angiosperms exceed all other groups of seed plants (pp. 145, 178–201, 600; Fig. 211). It may be assumed that this adaptive radiation arose from lowly, evergreen, pachycaulous trees. From these apparently have arisen in parallel leptocaulous, evergreen or deciduous trees and shrubs, lianes, dwarf shrubs and subshrubs, perennials, and finally herbaceous annuals (but sometimes the woody habit has arisen secondarily). The nature of these and numerous other adaptations arising from the modifications of stems, leaves and roots forms the basis for the evolutionary development of the various large and small related groups of angiosperms.

Flowers. Primitive angiosperms (e.g. Fig. 652A) have relatively large flowers; their floral axis (receptacle) is an elongated cone bearing numerous perianth segments, stamens and carpels in a spiral (acyclic) arrangement; occasionally there are transitions between the floral organs. Advanced groups frequently show progressive diminution in the size (reduction) of the flower and a numerical decrease in the floral members (oligomerization; Fig. 652B–E).

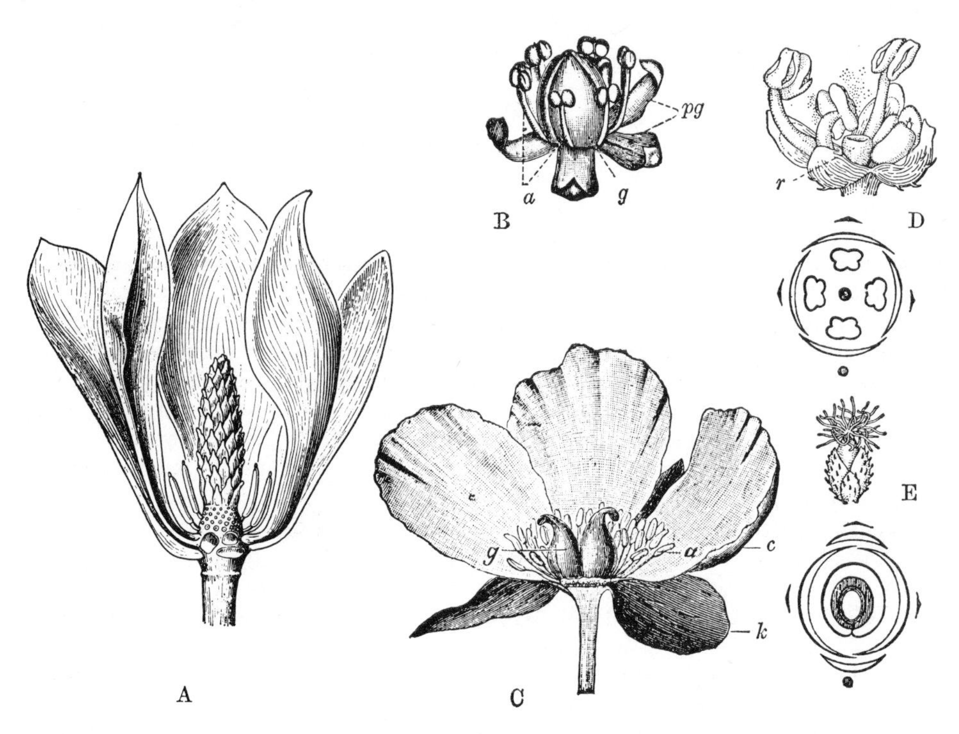

Fig. 652. Flowers of various angiosperms. **A.** *Magnolia*: elongated floral axis with numerous, spirally arranged, free perianth segments, stamens and carpels (partly removed in front). *c.* $\times\frac{1}{3}$. **B.** *Acorus calamus*: short floral axis with floral members in whorls; 3 + 3 greenish perianth segments (pg), 3 + 3 stamens (a), and three united carpels (g). Enlarged. **C.** *Paeonia*: whorled, with five sepals (k), five petals (coloured, c), numerous (secondarily multiplied) stamens (a) and two free carpels (g). $\times\frac{1}{2}$. **D.** to **E.** Flowers of *Urtica dioica*, general views and floral diagrams: main axis, bract, two prophylls, perianth of four inconspicuous segments: **D.** ♂ flower with four stamens and rudimentary ovary (r). **E.** ♀ flower with filamentous stigma, pseudomonomerous ovary and one ovule. **A.** After Zimmermann. **B.** After Eichler. **C.** After Schenck; floral diagrams in **D.** and **E.** after Eichler.

These tendencies in the flower, like the analogous progression from pachycaulous to leptocaulous shoot structure (pp. 600 et seq.) can be explained on the following assumptions: (1) numerous small flowers and fruits rather than fewer larger ones allow more rapid development and minimize the risk of damage; (2) oligomerous rather than polymerous flowers afford more opportunities for spatial and formal integration of the floral members (contrast, for example, Fig. 652A and Fig. 653C).

Along with reduction and oligomerization goes the shortening of the floral axis, leading via spiral/whorled (hemicyclic) insertion of the floral parts to a fully whorled (cyclic) insertion (Figs. 652, 653). Occasionally it is possible to recognize the origin of the whorls from spirals from the developmental sequence or the pattern of overlap (aestivation) of the floral parts, e.g. in the $\frac{2}{5}$ spiral in the whorl of sepals in *Rosa* (Fig. 693C).

Whorled insertion greatly predominates amongst the angiosperms; it facilitates the sharp separation of the floral members in adjacent whorls and the fixing of their numbers

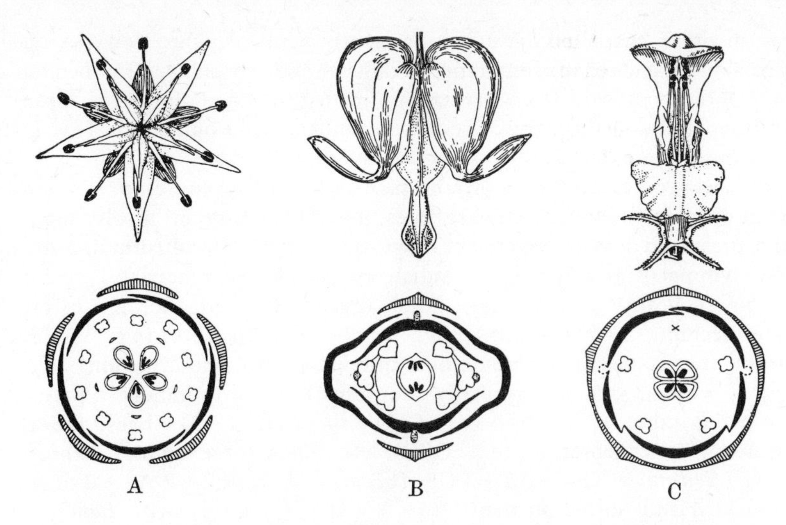

Fig. 653. Floral symmetry and floral diagrams. **A.** *Sedum sexangulare*: polysymmetric (radial). **B.** *Dicentra spectabilis*: disymmetric. **C.** *Lamium album*: monosymmetric (dorsiventral). For floral formulae, see p. 644. Partly after Eichler and Hegi.

(mostly five, but also four, three, and two in dicotyledons, mostly three in monocotyledons). The number of whorls in the flower varies. Pentacyclic flowers with five whorls are particularly common: two perianth whorls (calyx and corolla), two whorls of stamens and one of carpels (e.g. Fig. 653A). In, for example, the sympetalous Asteridae the absence of one whorl of stamens gives rise to tetracyclic flowers (e.g. Fig. 653C). The unisexual flowers of *Urtica* or *Alnus* are only di- or monocyclic (♂ or ♀, Figs. 652D, E, 688). As a rule, the floral parts in successive whorls lie above the gaps between the members in the next whorl below, i.e. they alternate (p. 124). More rarely the members of one whorl lie directly over those below, i.e. they are superposed (p. 647).

The angiosperm flower also shows numerous lines of development with respect to **symmetry** (lateral and mirror-image; cf. p. 76). Apart from (1) the primary asymmetry of acyclic flowers, cyclic arrangements may be (2) polysymmetric (multilateral, radial or actinomorphic) with more than two planes of symmetry (Fig. 653A), (3) disymmetric (bilateral) with two planes of symmetry (Fig. 653B), (4) monosymmetric (dorsiventral or zygomorphic) with only one plane of symmetry (Fig. 653C), or (5) secondarily asymmetric (and cyclic) with no plane of symmetry (e.g. *Canna*, Fig. 736D, pp. 744–5).

The plane passing through the main axis, the floral axis and the subtending bract is called median (cf. Fig. 695); the plane through the floral axis at right angles to this is transversal (cf. Fig. 685B); all other planes are oblique (e.g. Fig. 726B, E). Hence one can distinguish medially, transversely or obliquely monosymmetric (zygomorphic) flowers (cf. also Fig. 140, p. 130). The phylogenetic sequence of these types of symmetry is commonly: $1 \rightarrow 2$, $2 \rightarrow 3$, $2(3) \rightarrow 4$ and 2 or $4 \rightarrow 5$.

Modifications of floral symmetry can often be attributed to the promotion or suppression of particular floral parts, backed up by the occurrence of rudimentary organs (cf. Figs. 336, 653C for the characteristic progression from radial to dorsiventral flowers), of abnormalities (e.g. the radial terminal flowers in *Digitalis*; peloria), or of atavistic mutants (Fig. 340II, B). Whilst disymmetric and secondarily asymmetric flowers are rare, radial and especially dorsiventral flowers are predominant. Dorsiventral floral structure is generally related to the dorsiventrality of animal visitors to the flower: vertically upright or horizontally placed flowers, the development of a landing platform, protective upper lips, etc.

Angiosperm flowers exhibit all possible forms of **distribution of the sexes** (cf. p. 601). The nature of the primitive arrangement was disputed for a long time, but it is now generally accepted that the hermaphrodite flower is more primitive than the unisexual on the following grounds: (1) in almost all unisexual groups, ♀ flowers show rudiments of stamens, ♂ flowers of ovaries (e.g. *Castanea* or *Urtica*, Fig. 652D); (2) groups of angiosperms which are most primitive in other characters have predominantly hermaphrodite flowers (e.g. Magnoliidae); these alone facilitate the simultaneous transfer and acceptance of pollen by animals (p. 608); (3) the transition from hermaphrodite to unisexual flowers is often produced by selection in the change from insect to secondary wind pollination (e.g. in *Acer*, pp. 701–2, or *Fraxinus*, p. 725).

Floral structure is best represented by a ground plan or floral diagram (cf. p. 124 and Fig. 653). Empirical diagrams show the actual realities, theoretical diagrams involve interpretations, e.g. the presumed loss of organs; cf. Figs. 336, 653C. Floral formulae give information about symmetry (⊚ = spiral; $*$ = radial; $+$ or $\dagger$ = disymmetric; $\downarrow$, $\leftarrow$ or $\swarrow$ = zygomorphic; ϟ = cyclically asymmetric), floral organs (P = perigone; K = calyx; C = corolla; A = androecium; G = gynaecium), their number in each whorl (e.g. A5 + 5, two whorls each of five stamens; ∞ = numerous and indefinite), modifications (e.g. $A3^{st}$ = staminodes, 3° = missing, 5^{∞} = secondarily multiplied), union (numbers in parentheses, e.g. C(5) = petals united), insertion (e.g. $G(\underline{5})$ = superior, G–(5) = half-inferior, $G(\overline{5})$ = inferior ovaries), false septation (e.g. $G(\dot{2})$), etc. Thus for example, *Adonis*: $*$/⊚ K5 C6–10 A^{∞} G^{∞}; *Sedum*: $*$ K5 C5 A5 + 5 $G\underline{5}$; *Dicentra*: $+$ K2 C2 + 2 A2 + 2 or (as a result of both divided and united stamens!) $(\frac{1}{2} + 1 + \frac{1}{2}) + (\frac{1}{2} + 1 + \frac{1}{2})$ $G(\underline{2})$; *Lamium*: $\downarrow$ K(5) [C(5) $A1^{\circ}$:4] $G(\underline{\dot{2}})$; *Iris*: $*$ P3 + 3 A3 + 3° G(3); cf. the floral diagrams in Figs. 653, 684S, 734B.

Inflorescences. The individual flowers of angiosperms are commonly grouped in progressively more complex aggregations or inflorescences (pp. 136–8). There is frequently a concomitant reduction in size of the flowers and an increase in their numbers. This tendency seems to have begun with the development of lateral flowers in the leaf axils below a solitary terminal flower: all other types of inflorescence can be derived from such a closed panicle (Fig. 146). These lines of development are of great taxonomic and phylogenetic importance.

The development of foliage leaves is often suppressed in inflorescences (giving rise to inconspicuous bracts and bracteoles or prophylls), and the numerous flowers are often crowded together, thus compensating for the inconspicuousness of the small individual flowers (see, for example, Apiaceae, Fig. 707B); furthermore, for some floral visitors many smaller flowers have a greater attraction than fewer larger ones. A radiating appearance of the inflorescence, optically attractive to visitors, may be enhanced by the increase in size of marginal flowers (e.g. in *Iberis*), by sterile and conspicuous marginal flowers (e.g. *Hydrangea* and *Viburnum opulus*, p. 724), or by the presence of coloured bracts (e.g. *Astrantia* and *Cornus suecica*). Ultimately division of labour between individual flowers and the development of accessory axial or foliar structures give rise to biologically functional units ('blooms', p. 606) analogous to individual flowers, for example, the cyathia of *Euphorbia* (Fig. 704H–K), the capitula of the Dipsacaceae (Fig. 722E) and Asteranae (Fig. 792F, G), or the slippery trap blooms of *Arum* (Fig. 671G). The selective pressure of floral visitors on the origins and adaptations of these compound 'superflowers' or pseudanthia is obvious. These types are characteristic of the terminal phases of evolutionary lines.

Floral axis. In some primitive angiosperms the floral axis or receptacle is still elongate and conical (Fig. 652A), but generally it is abbreviated. Furthermore it may broaden out as a disc or saucer, eventually becoming cup shaped or tubular (Figs. 654, 699J). In the development of such floral cups and tubes (hypanthia) congenital fusion of the bases of calyx, corolla and stamens takes place to such an extent that the demarcation of the various floral regions is next to impossible. The carpels, either free or united into an ovary, are thus sunk in the floral cup, from which they may remain free (as in *Rosa* or *Prunus*, Fig. 693D, F) or with which they may be united (e.g. *Pyrus*, Fig. 693E, or *Conium*, Fig. 707F), whilst the sepals, petals and stamens appear to be raised above the margin of the cup. Corresponding to the progressive fusion of the ovary with the floral axis one can distinguish superior, half-inferior and inferior gynaecia. The terms hypogynous, perigynous and epigynous are frequently used as synonymous, but are best restricted to considering the relative positions of the ovaries and the floral organs other

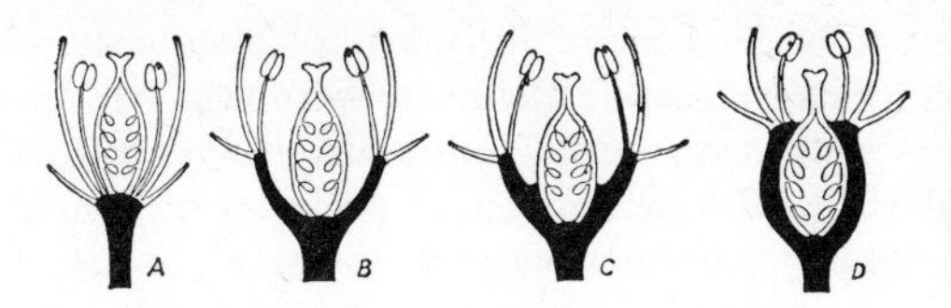

Fig. 654. Position of the ovary (G) relative to the floral axis (black) and to other floral members (white). **A.** G superior, flower hypogynous. **B.** G superior, flower perigynous. **C.** G half inferior, flower perigynous. **D.** G inferior, flower epigynous. After Engler's *Syllabus*.

than the receptacle (cf. Fig. 654).

This immersion of the carpels or ovaries has taken place in all subclasses of the angiosperms and has thus arisen many times in parallel. The significance of this characteristic progression may be that the delicate carpels and ovules become better protected from being damaged or eaten by animal visitors.

Further modifications of the floral axis occur when the internodes secondarily elongate between adjacent floral whorls, raising up the carpels or even the stamens as well; this is seen in the gynophore of the Capparaceae and Brassicaceae (Fig. 710F g) or the androgynophore in, for example, the Passifloraceae (p. 710). Nectar-secreting portions of the floral axis are also of biological significance: these may consist of nectar-glands (e.g. in the Brassicaceae, Fig. 710C) or of more extensive annular structures or discs (e.g. in Rutaceae and Aceraceae, Figs. 700, 710, 703), either between the carpels and stamens (intrastaminal) or between the stamens and corolla (extrastaminal).

Perianth. As in the extinct bennettites and their modern relatives, the Gnetatae, the angiosperms possess a floral envelope or perianth. This is clearly related to pollination by animals: the perianth protects the developing reproductive organs from floral visitors which are attracted in the flowering stage by its coloured members.

The commonest forms of perianth in the angiosperms are (*a*) homoiochlamydeous, with more or less identical perianth segments (perigone segments or tepals) in two or more spiral series or whorls: the multiple perigone, as in *Magnolia* (Fig. 652A) or *Tulipa* (Fig. 733A) with coloured, petaloid tepals, or *Acorus* (Fig. 652B) with inconspicuous greenish tepals; (*b*) heterochlamydeous, with dissimilar perianth segments, namely outer, generally green calyx leaves or sepals and inner, mostly brightly coloured corolla leaves or petals: the 'double' perianth of calyx and corolla, as in *Paeonia* (Fig. 652C); (*c*) haplochlamydeous or monochlamydeous, with only one whorl of perianth segments: the simple perigone, as in *Urtica* (Fig. 652D, E), *Paronychia* (Fig. 717D) or *Beta* (Fig. 717E); and (*d*) apochlamydeous, with perianth lacking, as in *Carpinus* (Fig. 688).

Primitive angiosperm flowers have a homoiochlamydeous perianth with numerous, spirally arranged, free segments, the outer bract-like, the inner progressively more coloured and petaloid (cf. *Magnolia*, Fig. 652A). The development of such a primary perianth from bracts (hypsophylls) is indicated e.g. by *Helleborus* (Fig. 655A–D). But how has the 'double' perianth developed? On the one hand by the differentiation within the multiple perigone (as in certain Magnoliales, and similarly also in the monocotyledons, cf. Fig. 736B, C), but more frequently, it seems, by the transformation of stamens into petals: thus in *Nymphaea* all transitions occur between stamens and petals whilst the four sepals are sharply distinct (Figs. 655E–L, 683D).

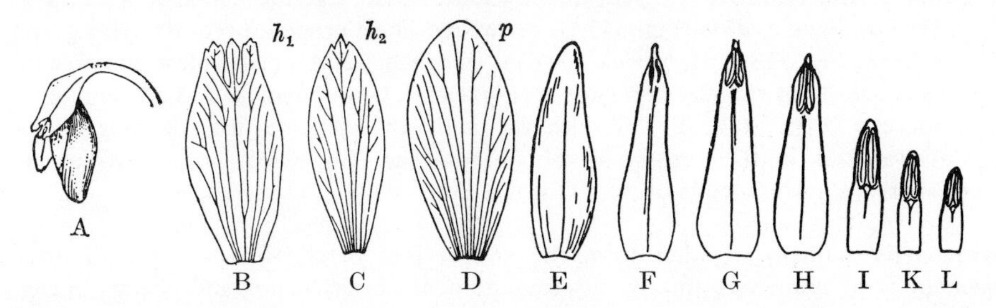

Fig. 655. A. to **D.** Transition from bract (or hypsophyll, h) to perigone segments in the flower bud of *Helleborus niger*. **A.** ×½ approx. **B.** to **D.** Enlarged. **E.** to **L.** Transition from stamens (**L.** to **G.**) to petals (**F.** to **E.**) in *Nymphaea*. After W. Troll, somewhat modified.

In the 'double' (multiplex, 'flore pleno') flowers of many decorative plants the transformation of stamens into petals is an abnormality. In the Ranunculaceae (pp. 682–4, Fig. 684), *Caltha* has a petaloid perigone as well as stamens; *Trollius* also has a petaloid perigone but some of the stamens have developed as inconspicuous nectaries or honey-leaves; in *Ranunculus* the perigone is sepaloid but the honey-leaves are petaloid; finally in *Adonis* there are typical sepals and petals. At least in the dicotyledons the petals have correspondingly generally developed from stamens, like which they are almost invariably supplied with a single vascular bundle. In contrast, sepals can be traced back to perigone segments or bracts (hypsophylls); they are homologous with the basal part of the foliage leaves, like which they are generally supplied with several vascular bundles (cf. Fig. 655). Further research into the origins of the perianth is desirable, particularly concerning the monocotyledonous perigone (? staminal).

Inconspicuous haplo- and apochlamydeous flowers (e.g. *Urtica*, Fig. 652D, E) have arisen in the course of progressive diminution in size and simplification, either more or less directly from polymerous, spirally inserted, homoiochlamydeous perianths (as in the Piperaceae, p. 680, Tetracentracaceae and Trochodendraceae, p. 685 and Fig. 686), or indirectly from heterochlamydeous perianth types (as in the Caryophyllales, Fig. 717D, K–L, or Oleaceae: *Fraxinus*, Fig. 723E, G). Such secondarily simplified flowers are characteristic of wind pollinated groups in which a differentiated perianth is both unnecessary in the absence of animal pollinators and a positive hindrance to the liberation of pollen from stamens and its reception by the stigmas (cf. pp. 663 et seq.).

Conversely, the frequent congenital fusion between the tepals or sepals and especially between the petals is generally closely related to planation (p. 84), to whorled insertion and to specialization for animal pollination. Correspondingly the primitively free arrangement can be contrasted with the advanced united one: choritepalous versus syntepalous (P), chorisepalous versus synsepalous (K), or choripetalous versus sympetalous (C). Examples are *Convallaria* for syntepaly, the Silenoideae (Fig. 717A–B, F) for synsepaly, and most of the Lamianae (cf. *Salvia*, *Sanchezia* or various Scrophulariaceae, Figs. 671A–E, 727) for sympetaly.

Such fusions frequently provide better protection of the reproductive organs against the weather or unwelcome animals, and better spatial coordination and stabilization of the floral organs in relation to their visitors, e.g. in the development of alighting platforms, access to nectar, contact with the anthers and stigmas, etc. Also of obvious floral biological significance are petaloid sepals, as in *Polygala* (Fig. 702I–K) or *Impatiens*, the unequal development of the petals of a corolla (e.g. upper and lower lips, as in Figs. 653C, 671B), the formation of petals with a (generally) nectar-containing spur, as in *Aquilegia* (Fig. 684M, T), *Corydalis* (Fig. 685C), *Viola* (Fig. 709A, B) and *Orchis* (Fig. 735A, B), or of coronas, as in the Silenoideae (Fig. 717A, F) and *Narcissus*, and so forth. Finally, specialized sepals may also function in the dispersal of fruits (e.g. as pappus; Figs. 365, 729N, O).

The taxonomic significance of the insertion and overlapping of the perianth segments (aestivation) should also be pointed out. Starting from the spiral arrangement, for example in the petals or honey-leaves of *Nuphar* (Fig. 683B), we can see the following types of aestivation in whorled perianths: overlapping of the perianth segments (imbricate), either in the primitive $\frac{2}{5}$ arrangement (quincuncial, e.g. the calyx of *Rosa*, Fig. 693C, or the corolla of *Sedum*, Fig. 653A), or twisted like a propellor (contorted, e.g. the corolla in *Malva*, Fig. 713G, H, *Nerium*, or *Gentiana*, Fig. 720A); or cochlear and ascending, as in the corolla of the Caesalpiniaceae (Fig. 695C, D) or cochlear and descending as in the Fabaceae (Fig. 695E, F); or with the perianth segments just touching at the margins (valvate) as in the corolla of the Mimosaceae (Fig. 695A, B), or with gaps between (apert) as in the corolla of *Acer* (Fig. 701A).

Stamens. The stamens together form the androecium of the flower. The primitive arrangement is of numerous spirally inserted stamens (primary polyandry, Fig. 652A). Oligomerization and the change to whorled insertion lead first to several whorls of stamens (e.g. 3 + 3 + 3 + 3 in Fig. 681A), more frequently to two (diplostemony, Fig. 733E), and ultimately to only one (haplostemony, Figs. 653C, 734B). Normally the whorls of stamens laid down from below upwards on the floral axis alternate (p. 124) so that the outer, lower whorl of stamens stands between the petals and above the sepals (episepalous), whilst the inner, upper whorl develops above the petals (epipetalous). This alternation can, however, be upset so that one whorl is absent; thus the single whorl of stamens in the Rhamnales is epipetalous, the outer episepalous whorl being absent (Fig. 703E, F). Furthermore, subsequent growth may displace the epipetalous stamens, which develop later, so that they come to lie outside the earlier, episepalous ones (obdiplostemony, Figs. 653A, 715B).

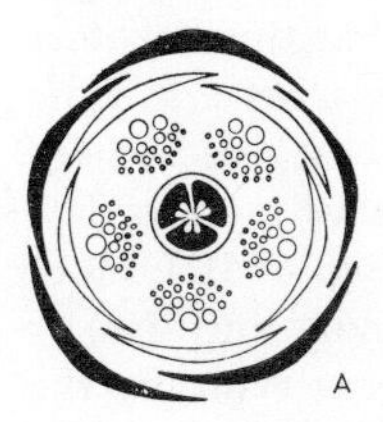

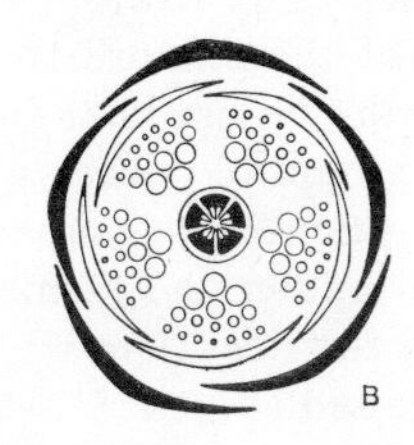

Fig. 656. Secondary increase in the number of stamens from few (five) initials as a result of dédoublement (secondary polyandry). **A.** Centripetal dédoublement in *Melaleuca hypericifolia* (Myrtaceae, Rosidae). **B.** Centrifugal dédoublement in *Hypericum hookeranum* (Hypericaceae, Dilleniidae). Diagrams by Leins.

In the angiosperms the number of stamens is not invariably reduced; not uncommonly it may be increased (secondary polyandry, Fig. 656). As in the formation of accessory buds (p. 130), so enlargement of the meristematic primordia may give rise to groups of stamens in the place of single stamens ('dédoublement'). The originally single primordium may divide during the course of development (as with the five or three bundles of stamens in *Hypericum*, Figs. 656B, 708B), or the primordia may be numerous from the start. Very often study of the course of development may indicate in which direction the accessory stamens develop and where the secondary primordia are active. If the initials develop from the outside inwards towards the apex or centre of the flower, the dédoublement is said to be centripetal, as particularly in the Rosidae (Fig. 656A); development from the inside outwards is centrifugal, as particularly in the Dilleniidae (Fig. 656B); but the initials may also develop laterally or downwards (serial and basipetal dédoublement).

The centripetal or centrifugal development of the androecium is of great taxonomic significance. It is not always certain whether such modifications of development are always linked with secondary rather than with primary polyandry (as for example in the Dilleniidae and Alismatidae). The functional significance of secondary polyandry is also obscure. It is notable, however, that this phenomenon frequently occurs in (? secondary) pollen flowers which offer particularly large amounts of pollen to the floral visitors (p. 659). In any case in the further course of evolution there is a further reduction in the number of stamens even in secondarily polyandrous groups. Whilst dédoublement involves an increase in the number of whole stamens, fission leads to the formation of

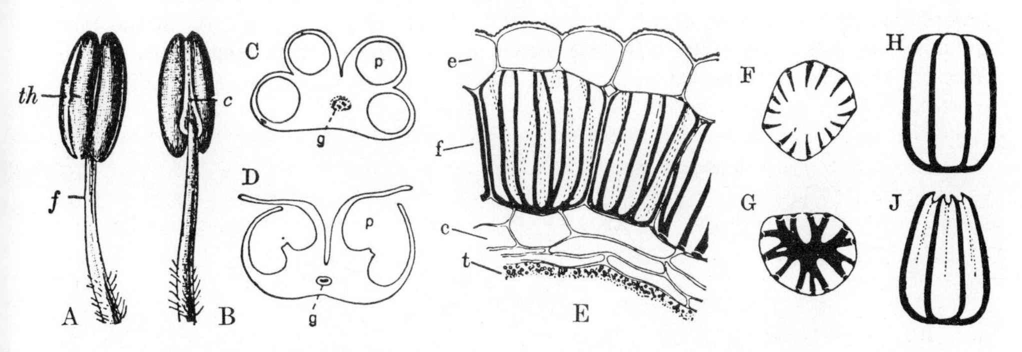

Fig. 657. The angiosperm stamen and its structure. **A.** and **B.** *Hyoscyamus niger.* **C.** and **D.** *Hemerocallis fulva.* **E.** to **G.** *Lilium pyrenaicum.* **A.** and **B.** Whole stamen from in front and behind with filament (f), theca (th) and connective (c). Enlarged. **C.** and **D.** T.S. of anthers with closed and dehisced pollen sacs (p); vascular bundle (g). **E.** T.S. through the anther wall with epidermis (e), fibrous layer (f), intermediate layer (c) and remains of tapetum (t). **F.** and **G.** Individual fibrous cells from above and below. ×150. **H.** and **J.** Diagrams of a fibrous cell before and during shrinking. **A.** and **B.** After A. F. W. Schimper. **C.** and **D.** After Strasburger. **E.** to **J.** After Firbas.

half-stamens (e.g. in Betulaceae, Fig. 688). We may also note the frequent congenital or postgenital fusion of the stamens, either with each other (e.g. Figs. 695E, F; 713G, K, in the region of the filaments; 729E, I, L, in the region of the anthers), or with the perigone or corolla (especially in sympetalous flowers, e.g. Fig. 725B), or with parts of the gynaecium (e.g. Figs. 719H, K; 735A, C). The stamens may also be reduced to sterile staminodes, or be completely missing, or take over new functions, e.g. as nectaries (e.g. Fig. 684I–M) or by becoming petaloid and optically attractive (e.g. Fig. 736B–D). See p. 645 and Fig. 655E–L for the possible derivation of true petals from stamens.

In the development of the angiosperm stamen (Fig. 658A–C) a meristematic transverse zone arises on the ventral (adaxial) side of the primordium so that this becomes peltate (shield shaped). Both the dorsal part of the primordium and the transverse zone now grow upwards, forming on the margins two groups of pollen sacs (thecae), each theca with two pollen sacs. The stamens of the most primitive angiosperms have more or less apical pollen sacs and are scarcely divided into anther and filament (Fig. 663D). More advanced types of stamen bear the pollen sacs at the sides (lateral), ventrally (introrse) or dorsally (extrorse) and may be strongly compressed (Fig. 655G–L); most typically the advanced stamen is divided into a stalk or filament and the true anther with a sterile central portion (connective) and two thecae each with two united pollen sacs. The typical angiosperm stamen is homologous with one microsporophyll with four microsporangia or two bisporangiate synangia. Stamens with only two or more than four pollen sacs are rare.

A transverse section of a young anther (Figs. 618, 657) shows that each pollen sac contains a central pollen-forming archesporium surrounded by a wall of at least four layers; from the outside inwards these are the epidermis, fibrous layer (or endothecium), an evanescent intermediate layer and a tapetum one or two cells thick. The tapetum has a dense cytoplasm and its nuclei usually become polyploid by restitution nucleus formation (p. 36); it serves to nourish the pollen grains. The tapetum may remain as a distinct tissue (secretion tapetum), or the cell walls may break down and the protoplasts, which may fuse together, penetrate between the young pollen grains (plasmodial tapetum).

Except for the epidermis, the wall and archesporium usually originate from a single subepidermal layer of cells, but later development varies (Fig. 618); the differences are taxonomically significant.

The archesporium forms a large number of pollen mother cells (Fig. 618D, E) which after meiosis produce four meiospores each (Figs. 27, 28). These are the uninucleate pollen grains (pollen cells). In dicotyledons wall formation is predominantly simultaneous, in monocotyledons successive. When the pollen grains are mature the fibrous layer causes the dehiscence of the pollen sacs by means of a cohesion mechanism (pp. 379–80).

The cells of the fibrous layers have radial bands which are often stouter and united towards the inside and taper off towards the outside (Fig. 657E–J). As in the fern annulus (Fig. 326), when the cells lose water they can only shrink tangentially (transversely). This sets up tensions which cause the anther wall to rupture, generally by a

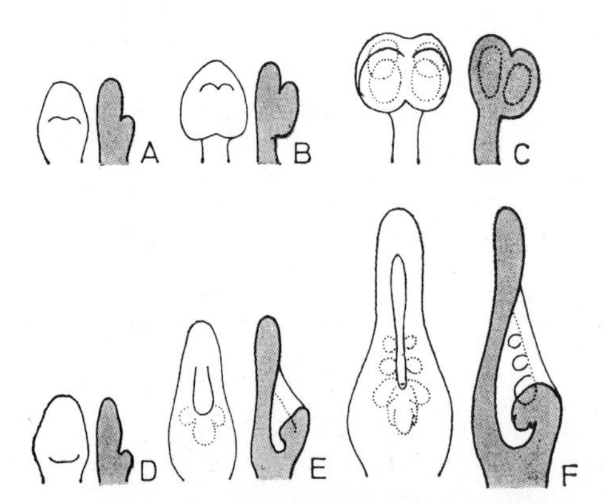

Fig. 658. Ontogeny (diagrammatic) of typical stamens (**A.** to **C.**) and carpels (**D.** to **F.**) in angiosperms, from in front (white) and the side (grey). For further explanation see pp. 648 and 652. Original, partly after Payer, Baum and Leinfellner.

longitudinal split above the septum which separates a pair of pollen sacs; this septum is frequently resorbed (Fig. 657D).

However, some pollen sacs open by the dissolution of tissues at determined places, thus forming pores (Ericaceae, Fig. 715C, D), or the fibrous layer only develops around limited areas which then lift up as valves (Lauraceae, Fig. 681A). Sometimes the thickenings of the fibrous cells are reversed so that on drying the pollen sacs shorten longitudinally (e.g. in the Liliales) or contract so that the pollen is squeezed out of an opening (e.g. Araceae).

Pollen. The fine structure of angiosperm pollen grains is basically similar to that of the gymnosperms (pp. 603 et seq.) but reaches higher levels of differentiation (Fig. 619).

The intine of several layers is surrounded by exine, the innermost layer of which is the compact, two to three layered, partly lamellar nexine, the outer layer of which is called the foot-layer and corresponds in development and staining reactions to the sexine which covers it. The sexine in angiosperms is particularly highly differentiated. In intectate pollen grains it takes the form of rodlets, clubs, spheres, warts or a network inserted upon the nexine. Its columnar elements (columellae, bacula) may, however, be fused together at their distal ends, forming an outer additional layer, the tectum (tectate pollen grains).

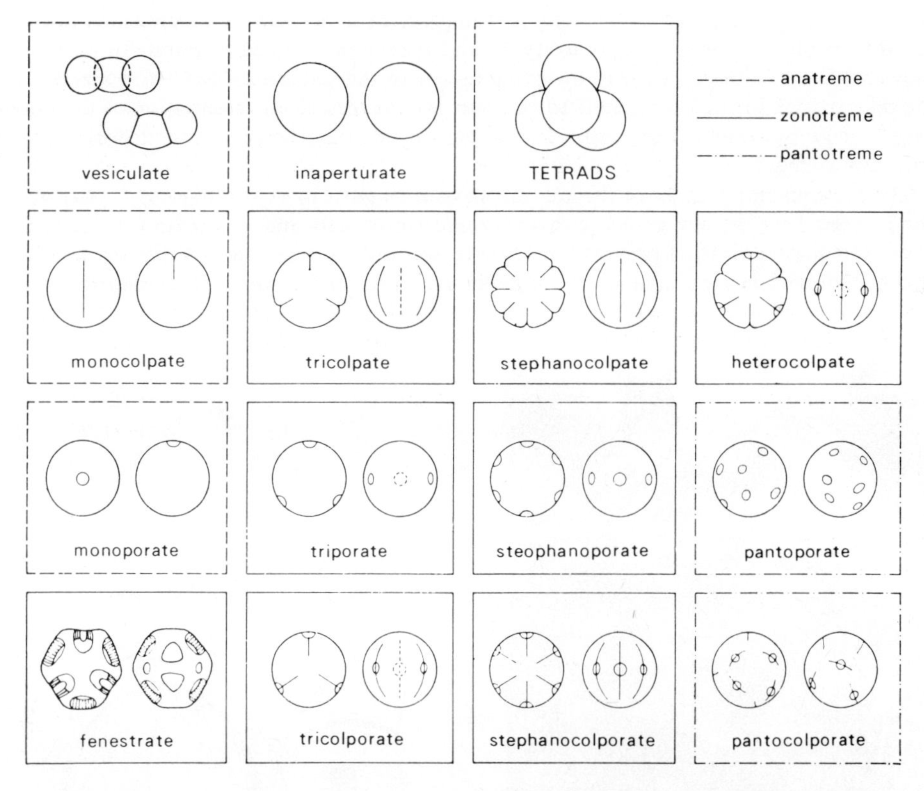

Fig. 659. Tabular survey of some common pollen types in C. European seed plants. In each case left, polar view; right, equatorial view (distal pole above). Monads (single grains): anatreme types: vesiculate (*Abies, Picea, Pinus*), monocolpate (most of the Liliidae), monoporate (Poaceae), inaperturate in otherwise anatreme groups (*Larix, Taxus, Potamogeton*, Cyperaceae). Zonotreme types: tricolpate (Ranunculaceae in part, *Quercus, Acer*, Brassicaceae, *Salix*, Lamiaceae in part), triporate (*Betula, Corylus*, Urticaceae, Onagraceae), tricolporate (*Fagus*, Rosaceae in part, Apiaceae, *Tilia*, Asteraceae), fenestrate (a special case of the tricolporate type, Cichoriaceae), stephanocolpate (Rubiaceae, Lamiaceae in part), stephanoporate (*Alnus, Ulmus*), stephanocolporate (Boraginaceae in part), heterocolpate (*Lythrum, Myosotis*), inaperturate in otherwise zonotreme groups (*Populus, Callitriche*). Pantotreme types: pantoporate (*Juglans*, most of the Caryophyllaceae, Chenopodiaceae, Plantaginaceae), pantocolporate (Polygonaceae in part). Tetrads: in otherwise anatreme groups (Orchidaceae in part, Typhaceae in part), in otherwise zonotreme groups (most of the Ericales). After Faegri and Iversen, and Erdtman; arranged by Teppner.

The tectum may be penetrated by pores of diverse shapes and may itself consist of several layers and be sculptured externally. The cavities beneath the tectum contain oils, waxes, pollen cement, etc.

Studies of development with the electron microscope show that at first a thin layer of fibrillar materials, the primexine, is laid down on the plasmalemma inside the thick callose wall which originated from the pollen mother cell. More compact elements are then laid down which form the columellae by stretching and thickening; these broaden out at both ends to form the tectum and foot-layer and hence the outer exine (ektexine). At the base of the foot-layer is formed the inner exine (endexine) which is at least initially lamellar. The cellulose-containing intine is formed last of all. It is open to doubt whether the specific structure of the sporopollenin-containing exine is determined by the sporophyte (tapetum), by the young pollen grain itself or by the mutual interactions of the pollen grains in a tetrad. So far it appears that the activities of all three components are involved, but control by the young microspore must be very considerable.

A superficial coating of sticky pollen cement (consisting partly of oily substances) is characteristic of the grains of animal pollinated plants (p. 658). It facilitates both the cohesion of the pollen grains and their adhesion to floral visitors; furthermore it prevents the premature liberation of the pollen from the dehisced anther.

In the most primitive angiosperm pollen the apertures (p. 603) are often indistinct or absent. As the germ-pores become more complex, so a number of progressions can be traced from simple colpate to porate types and thence to complex colporate or pororate (pores with a mouth) types (e.g. the progressions in the Cactaceae, Fig. 367; Fig. 659). The diversity of forms from simple to complex is enhanced by differentiation of the pore margin (e.g. by thickenings), the formation of lid-like opercula, the deposition of sculptural elements, etc.

The position and number of the apertures exhibit a great diversity (Figs. 620, 659). The Annonaceae have pollen grains with catatreme (proximal) and anatreme (distal) germ pores: this may indicate a progression (? or regression). A more certain starting point of all other lines of development are those pollen grains which resemble the gymnosperms in

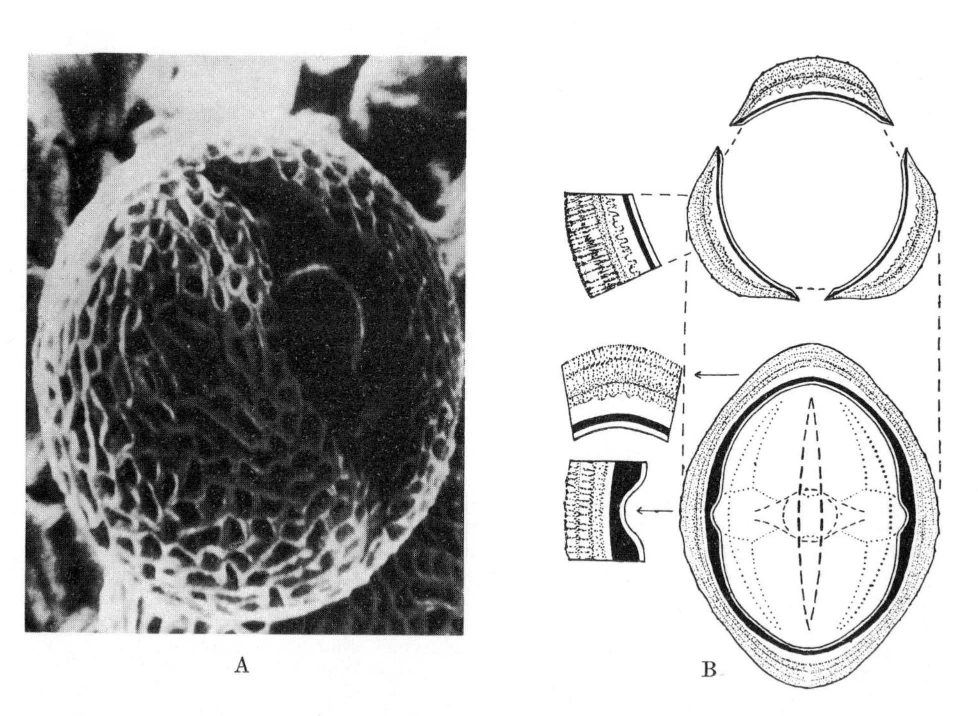

Fig. 660. A. Pollen grain of *Pelargonium* (tricolporate) with coarsely reticulate exine, colpus (furrow) and pore; photograph with Stereoscan electron microscope. ×1000 approx. **B.** Palynogram of the pollen grain of *Centaurea scabiosa* (tricolporate): equatorial view, optical section and details of wall structure. **A.** After Thornhill, Matta and Wood. **B.** After Erdtman.

having one distal colpus (anatreme, monocolpate) as in many Magnoliidae and monocotyledons (Fig. 620, group A). At the same time this group shows various types with more or less transverse colpi (e.g. two colpi around the equator), or with indistinct germination areas (leptomata) or none (analept and atreme respectively; both inaperturate). Within the Magnoliidae one can also see transitions to the pollen forms so characteristic of the advanced dicotyledons, with three longitudinal but equatorial colpi (zonotreme, tricolpate = zono-tricolpate; Fig. 620, group B). The further progression to apertures distributed over the whole surface of the grain (zonotreme → pantotreme) is shown for example by the Cactaceae (Fig. 367), Caryophyllaceae and the genus *Linum*. In this way the number of apertures may be greatly increased (up to forty in the Caryophyllaceae, 100 in Chenopodiaceae), but occasionally is reduced to none (secondarily atreme or inaperturate pollen grains).

The enormous diversity of pollen types can be arranged with the help of an artificial system of classification (NPC-system) based on the number (N), position (P) and character (C) of the apertures. A simplified survey of some frequent C. European pollen types is shown in Fig. 659. There are very many other differences of taxonomic significance involving the symmetry, shape and size of the grains, as well as the fine structure of the exine. Figure 660 shows how these characteristics can be clearly represented by an electron microscope picture or by a so-called palynogram.

Angiosperm pollen grains are not always separate as monads. The daughter cells of one pollen mother cell may remain united in a tetrad and be dispersed as such, as in the Ericales, *Drosera, Epilobium*, Juncaceae, etc. (Figs. 659, 715G). In the Cyperaceae in particular progressive reduction of two or three cells of the tetrad leads to the formation of 'false' monads (pseudomonads). Cases in which only two of the four tetrad cells remain together in pairs are rare (e.g. *Scheuchzeria*). Polyads are formed when the grains from a number of pollen mother cells cohere, forming dispersal units of eight, sixteen or thirty-two pollen grains as in the Mimosaceae. Finally, the entire content of a pollen sac may remain united to form a pollinium, or of two or more sacs to form a pollinarium (in Asclepiadaceae and Orchidaceae).

Carpels. These are homologous with megasporophylls, and together with the ovules borne on them form the gynaecium (gynaeceum, gynoecium) of the angiosperm flower. The carpels thus always enclose the ovules and the Magnoliophytina are 'angiovulate'—but only in advanced groups are they strictly 'angiospermous', i.e. remaining closed around the seeds; cf. pp. 670 et seq. The delicate ovules are thus protected from desiccation and

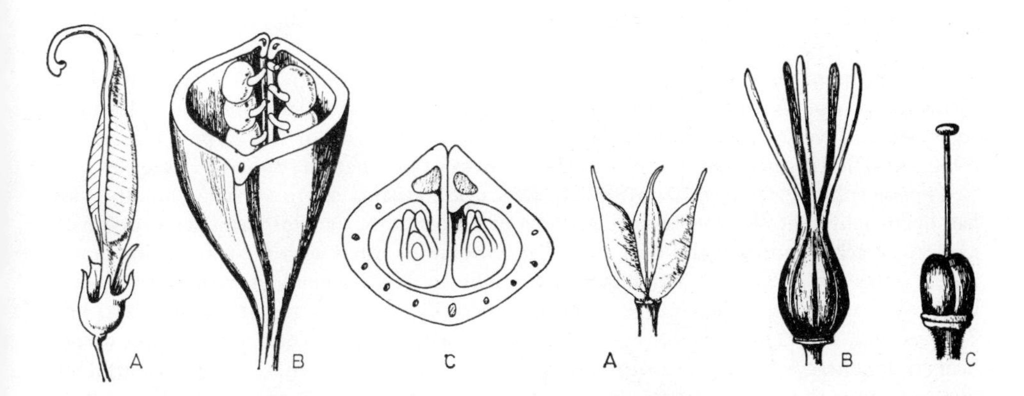

Fig. 661. Structure of carpels. **A.** and **B.** *Colutea arborescens.* **C.** *Delphinium elatum.* **A.** Ventral view of a maturing single, free carpel with closed suture; calyx at the base. ×3 approx. **B.** and **C.** In T.S. with dorsal and two ventral vascular bundles, two placentas, and ovules. ×10 approx. After Troll.

Fig. 662. Progressive fusion of the carpels in the gynaecium. **A.** *Delphinium elatum* (choricarpous). ×2. **B.** *Linum usitatissimum* and **C.** *Nicotiana rustica* (coenocarpous, styles free and fused respectively). Enlarged. **A.** After Troll. **B.** and **C.** After Berg and Schmidt.

the attacks of visiting insects, whilst the direct access of pollen is prevented (for the 'filtration' of pollen of the same individual or of different species cf. pp. 364, 402, 657).

As with the stamens, so in the gynaecium the spiral arrangement of numerous free carpels is primitive in the angiosperms (cf. Fig. 652A), Oligomerization (p. 601) commonly leads to the formation of whorls of free carpels (frequently five, three or two: Figs. 653A, 662A) and ultimately reduction to a single carpel (e.g. Fig. 661A). Secondary increase in the number of carpels is rarer (and then compensates for reduction in the number of ovules per carpel, e.g. in Ranunculaceae, Fig. 684A, B, P, Q). Very commonly fusion of the separate carpels of a choricarpous (apocarpous) gynaecium results in the formation of a coenocarpous one (cf. Figs. 662, 664, 665, p. 653). Both free and more or less united carpels may be fused to the floral axis or receptacle (cf. p. 644).

The development of a free carpel is at first similar to that of a stamen (Figs. 658D–F, 663A–C): from a unifacial stalk-zone there develops ventrally a meristematic transverse zone (cf. Fig. 175) so that as in the stamen there is a peltate stage initially. The margins now grow upwards (more on the dorsal side than the ventral) to form a more or less tubular (utriculate) structure which, however, remains open along the ventral side (ventral suture). Inside the fertile part of the carpel (the ovary) the ovules develop on the placentas. Finally the ventral suture closes by postgenital fusion. A stalk-like terminal portion, the style, often remains sterile; within this the pollen tubes are conducted and nourished. The style receives the pollen grains on a generally papillose or slimy and sticky area, the stigma. Thus a free carpel (Fig. 664A; p. 653) usually exhibits from below upwards a stalk zone, an initially (congenitally) closed tubular zone (ascidiate zone, a), a zone which only closes during development (postgenital; the plicate zone, p), and a stylar zone.

Besides carpels with a well-marked ascidiate zone and an almost circular suture (Fig. 663A–C), there are many in which the ascidiate and stalk-zones are somewhat suppressed whilst the plicate zone and suture are elongated (Fig. 658D–F). Finally epeltate carpels (without a transverse zone meristem) frequently occur in which the stalk and ascidiate zones are completely lacking and the carpel appears folded (conduplicate). Many lines of evidence (p. 677) suggest that phylogenetical development proceeds from ascidiate to conduplicate; nevertheless the contrary view is frequently held.

The placentas bearing the ovules may develop on the inner surface of the carpel (laminal) or close to the margins (submarginal) (Fig. 665A, B). The primitive submarginal placenta was probably O-shaped; from this can be derived U-shaped and then II-shaped (two-ranked) ones, or those which only develop on the transverse zone (median) (Fig. 663B). The development of numerous ovules in each carpel is primitive; reduction to one is advanced (e.g. Fig. 684P, Q). The orientation of the ovules on the placenta varies (e.g. pendulous, horizontal, oblique or erect, with dorsal or ventral raphe; cf. Fig. 684O, Q and p. 604).

The stigmatic zone of primitive carpels is confined to the papillose margins of the suture (Fig. 663C). Subsequently, and often in parallel, a stylar zone and a localized stigma have developed by the elongation of an apical portion of the carpel (Figs. 652E, 691C, 738D). This in general improves the chances of pollination (pp. 657 et seq.).

The morphological interpretation of the carpel remains controversial. Thus the view has been held that the ovules or placentas of all (or some) angiosperms were originally axillary (stachyosporous) and only secondarily became united with the subtending bracts. Against this are the many similarities, especially in development, growth and venation, between carpels and vegetative leaves, especially peltate and tubular ones (e.g. *Tropaeolum*, Fig. 232; *Nepenthes*, Fig. 222G). Furthermore, secondarily leaf-like carpels bear rudimentary ovules more or less marginally. Thus the ovules of angiosperms are phyllosporous (borne on leaves). It has often been stated that carpels have arisen phylogenetically by the 'inrolling' of leaf-like, bifacial structures, very much like the fertile carpels of *Cycas* (Fig. 654B). However, the almost universal occurrence of a peltate or tubular stage of development of the angiosperm carpel (and stamen) and of saucer or cup shaped structures bearing ovules (and pollen sacs) in the pteridosperms (the presumed ancestors of the angiosperms; Figs. 643, 644; pp. 632 et seq., 657) make such a hypothesis, supported neither by fossils nor by ontogeny, quite unnecessary.

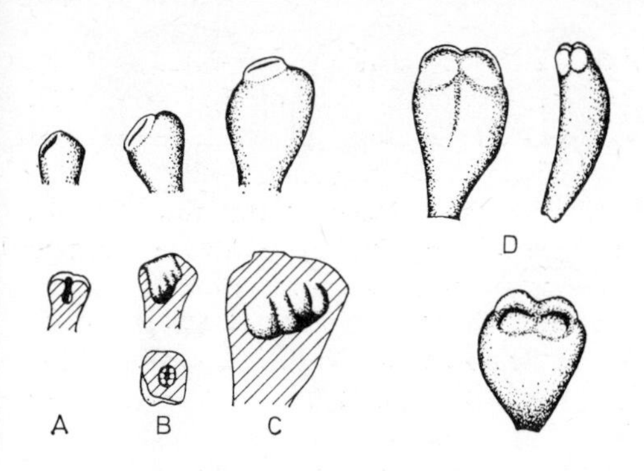

Fig. 663. Primitive, little differentiated, tubular or peltate carpels and stamens in the Winteraceae (*Pseudowintera*). **A.** to **C.** Ontogeny of the carpels with lateral placenta under the ventral opening (general view and in L.S.). **D.** Mature stamens with apical pollen sacs in front, side and oblique views. ×10. After Sampson.

A gynaecium of several fused (coenocarpous) carpels is called in its entirety a pistil, the basal part being the ovary, the sterile apical portion or portions being the styles and stigmas (p. 652). Various transitional forms indicate how free carpels become united (cf. Fig. 662 and Fig. 664A → B → C). This involves the progressive progenital fusion of the stalk, ascidiate and plicate zones; only the apical portions often remain free, but even these may unite to form a single style at the apex of which only the stigmatic lobes indicate the number of fused carpels.

Many coenocarpous carpels show the varying extent of fusion in adjacent zones (Fig. 664C). Thus the basal, ascidiate zone of the carpels shows congenital fusion (synascidiate, sa); above this the plicate zone is wholly (symplicate, sp) or partially (hemisymplicate) fused; whilst the apical zone is often still free (asymplicate, asp); of course these distinctions may be obscured by subsequent postgenital fusions. The four zones may be very variously promoted or perhaps completely suppressed. Placentas and ovules may develop in all zones, but especially in the central part of the ovary.

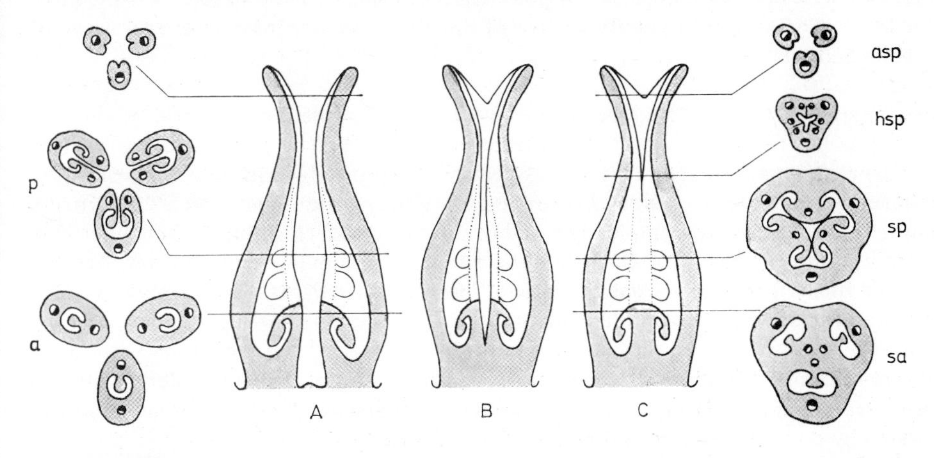

Fig. 664. Diagrams of the structure of choricarpous (**A.**), coenocarpous (**C.**) and intermediate hemisyncarpous (**B.**) gynaecia. L. and T.S. with ascidiate (a), plicate (p), synascidiate (sa), symplicate (sp), hemisymplicate (hsp) and asymplicate (asp) zones. After Leinfellner.

In the primitive coenocarpous gynaecium the ovary is completely divided into chambers (loculi) by 'true' walls or septa and the placentas and ovules are axile. If this septation is present in the greater part of the ovary, we speak of a syncarpous gynaecium (Figs. 665C, D, G, 669E, F, 713A), but if the development of the septa is inhibited a paracarpous ovary with incomplete loculi or none is the result and the placentas are either parietal, with the ovules borne on the walls (Figs. 665E, F, 685C, D, 709B), or central and free with numerous basal ovules, or only a single one (Figs. 665H, I, 716–18). Central placentas apparently arise directly on an upgrowth of the transverse zone of a syncarpous, synascidiate ovary. A survey of the main progressions in the gynaecium is given in Fig. 665.

Reduction of the carpels and of the number of ovules may be accompanied in

Fig. 665. Various types of gynaecium and their hypothetical evolutionary relationships, showing transverse sections of the main fertile zone of the ovule. **A.** Choricarpous, laminal placentation. **B.** Choricarpous, submarginal. **C.** Hemisyncarpous, axile. **D.** and **G.** Syncarpous axile, septa free or congenitally fused. **E.** and **F.** Paracarpous, parietal. **H.** and **I.** Paracarpous, numerous basal ovules or one. Partly after Takhtajan and Engler's *Syllabus*, modified.

coenocarpous gynaecia with the development of sterile loculi (Fig. 722B), or the ovary may appear to be unilocular (pseudomonomerous, e.g. Figs. 652E, 681A) or even completely suppressed (e.g. Fig. 705C). Conversely, ingrowths of tissue may develop as 'false' septa which supplement the loculi (cf. pp. 726, 730 and Fig. 725E). Furthermore, axial tissue may grow around free carpels (Fig. 683B, C) giving rise to false syncarpy. Specialized developments are the septal nectaries between the carpels of many Liliales and the display function of the styles in the Iridaceae (Fig. 734). For the development of inferior from superior ovaries see p. 644.

Whilst with free carpels each stigma must be separately pollinated, the fusion of carpels and styles allows a single act of pollination to bring pollen tubes into each loculus of the ovary. The progressive modifications of the placentas may improve the nutrition of the ovules and seeds.

Ovules (Figs. 667G, 673A, 677A). The ovules are provided with nutrients through vascular bundles from their point of origin on the placenta. The primitive ovule has two integuments (bitegmatic), but only one is present in advanced groups (e.g. in most of the Asteridae) as a result of fusion or suppression (unitegmatic). Very rarely, e.g. in the Loranthaceae, the ovule is completely fused to the ovary wall (Fig. 705C). The innermost layer of the integument is often differentiated as a tapetum-like endothelium. An outgrowth of tissue from the placenta, funicle or integument towards the micropyle is called an obturator; this may function in guiding the pollen tube (Fig. 704M). Reduction in the ovules also involves the nucellus: the progression here leads from a nucellus with many cells (crassinucellate, Fig. 667) to one reduced to little more than an epidermis and an embryo sac (tenuinucellate). This reduction and progressive neoteny of the ovules is facilitated by their inclusion in and protection by the carpels.

As in the archesporium of the pollen sac, so in the nucellus an embryo sac mother cell develops after a few divisions (in advanced tenuinucellate groups this arises direct from a subepidermal cell); sometimes more than one embryo sac mother cell is formed (Figs. 667A–F, 668). This soon becomes conspicuous by its size and cytoplasmic content. In the course of megasporogenesis meiosis gives rise to four haploid meiospores, generally in a linear tetrad. Usually only the lowermost of these becomes an embryo sac. The structure of the wall of this megaspore is much reduced compared with that of the gymnosperms.

Gametophytes. The pollen grains begin to develop as very reduced ♂ gametophytes whilst still in the pollen sac (Fig. 666). The uninucleate pollen cell divides after mitosis very unequally into a large vegetative or pollen tube cell which nearly fills the pollen grain and a small, lens shaped generative or antheridial cell which at first lies against the wall and then moves into the cytoplasm of the pollen tube cell as a spindle shaped structure surrounded by a membrane. After a second mitosis in the pollen grain the generative cell

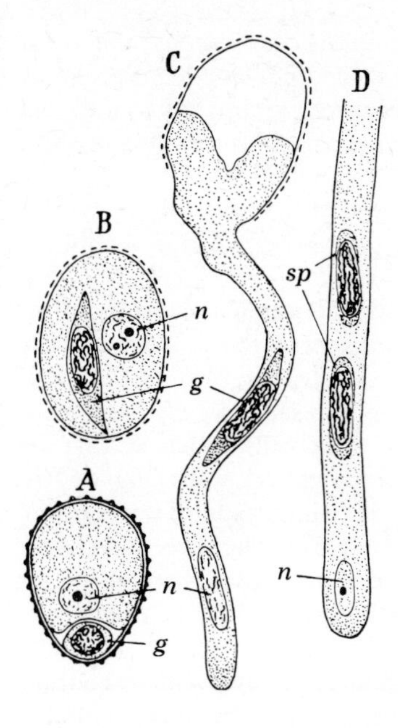

Fig. 666. Development of the ♂ gametophyte (*Lilium martagon*). Vegetative cell (its nucleus, n) and generative cell (g) in the pollen grain (**A.** and **B.**) and pollen tube (**C.**). In the tip of the pollen tube (**D.**) the generative cell has divided into two sperm cells (sp). ×530. After Strasburger, following Guignard; somewhat modified.

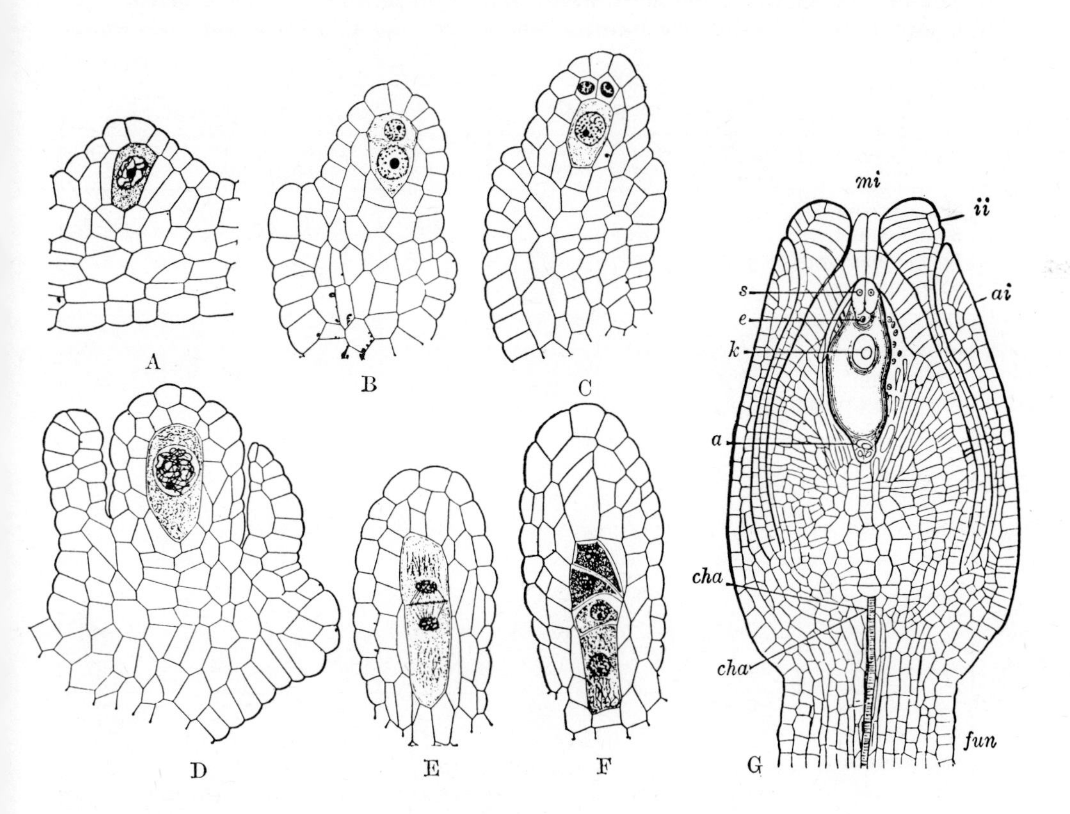

Fig. 667. Development of the ♀ gametophyte. **A.** to **F.** *Hydrilla verticillata*, Hydrocharitaceae. **G.** *Polygonum divaricatum*. In the developing nucellus of the ovule a hypodermal cell (**A.**) differentiates, cutting off a cover cell which again divides (**B., C.**), enlarges to form the embryo sac mother cell (**D.**) and forms after meiosis (**E., F.**) four embryo sac cells of which only the lowermost develops further as an embryo sac. **G.** Mature ovule with micropyle (mi), outer and inner integuments (ai, ii), chalaza (cha) and funicle (fun); the embryo sac contains the synergidae (s), from beneath which the egg cell projects (e), the secondary embryo sac nucleus (k) and the three antipodal cells (a). ×200. **A.** to **F.** After Maheshwari. **G.** After Strasburger.

divides into two sperm cells which still include mitochondria, proplastids and other organelles all enclosed in a thin plasmatic membrane. This second pollen mitosis probably occurs in primitive forms only in the pollen tube, but in advanced groups whilst the grain is still in the pollen sac; consequently at pollination the grain may contain two or three cells.

The ♂ gametophyte of the angiosperms is thus more reduced than that of the gymnosperms; the prothallial and stalk cells are absent and the generative cell is probably the sole vestige of an antheridium (cf. pp. 605 et seq.).

Normal development of the ♀ gametophyte begins with the uninucleate embryo sac (embryo sac cell, megaspore; Fig. 668). During megagametogenesis the primary embryo sac nucleus generally undergoes three successive free nuclear divisions, forming two, four and eventually eight nuclei. At each end of the elongated embryo sac three of these nuclei surround themselves with their own cytoplasm forming three cells which at first are enclosed in membranes; the lower ones subsequently develop rigid cell walls (Fig. 667G). The three upper ones form the egg apparatus, the largest one (which extends deepest into the embryo sac) being the egg cell, the two others the synergidae. The three lower cells are the antipodal cells. The two remaining nuclei, not delimited from the embryo sac cytoplasm, are the polar nuclei; these fuse together either before or after the entrance of the pollen tube to form a diploid, secondary embryo sac nucleus.

There are numerous deviations from this widespread (and probably primitive) normal type of embryo sac development; some of these deviations (which are taxonomically important) are shown in Fig. 668. Whilst in the normal type only one embryo sac cell (megaspore) is involved in the formation of the correspondingly monosporous embryo

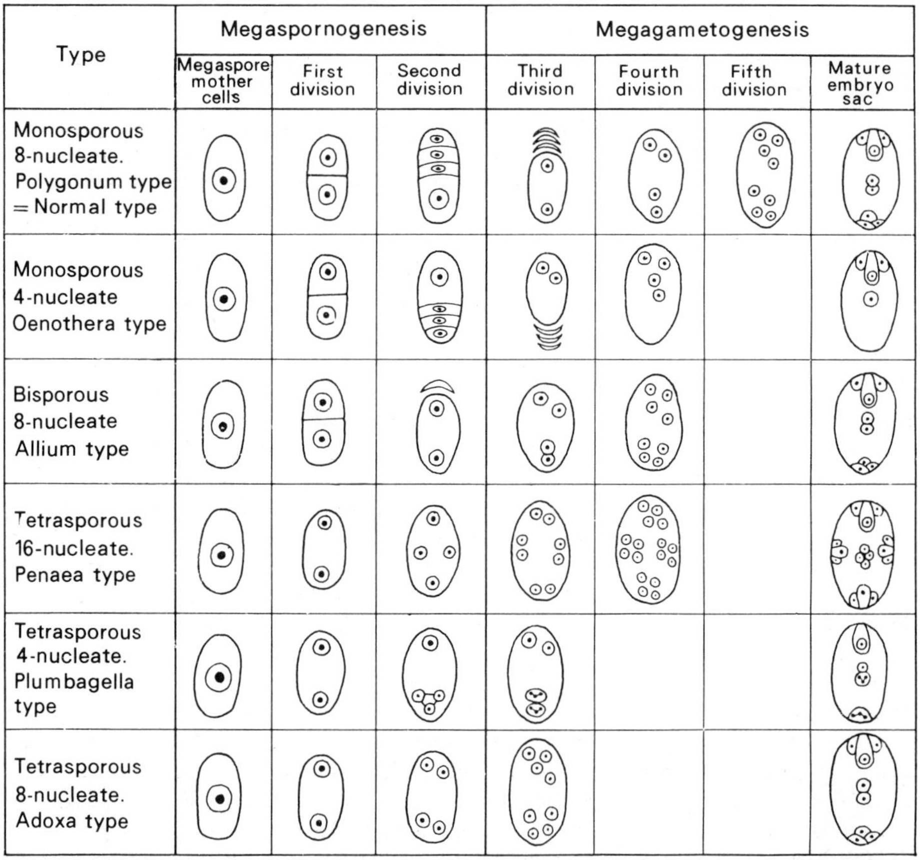

Type	Megaspornogenesis			Megagametogenesis			
	Megaspore mother cells	First division	Second division	Third division	Fourth division	Fifth division	Mature embryo sac
Monosporous 8-nucleate. Polygonum type = Normal type							
Monosporous 4-nucleate Oenothera type							
Bisporous 8-nucleate Allium type							
Tetrasporous 16-nucleate. Penaea type							
Tetrasporous 4-nucleate. Plumbagella type							
Tetrasporous 8-nucleate. Adoxa type							

Fig. 668. Some types of embryo sac development in the angiosperms: meiosis in the diploid embryo sac mother cell (megasporogenesis) and development of the haploid embryo sac cell into the mature embryo sac (megagametogenesis). For further explanation see pp. 656 et seq. After Maheshwari.

sac, in bi- or tetrasporous ones two or four are involved. Other modifications concern the suppression of some of the divisions in megagametogenesis, various arrangements of the cell groups, and cell or nuclear fusions (as in the *Plumbagella*-type). Whilst the mature embryo sac of the *Penaea*-type has sixteen cells or nuclei, the *Oenothera*-type has only four. The antipodal cells are much concerned with nutrition of the embryo sac, but haustoria may also develop from the megaspore, synergidae or antipodal cells.

The ♀ gametophyte of the angiosperms is thus much more reduced than that of the gymnosperms as a result of neoteny (cf. p. 606). This is shown both in the number of cells and nuclei (generally eight but sometimes only four) and in the absence of archegonia. Hence exact homology with the ♀ prothalli of the Coniferophytina and Cycadophytina is somewhat uncertain (e.g. egg apparatus = two neck canal cells + egg cell?).

Pollination. Whilst in the gymnosperms the pollen is conveyed directly through the pollination droplet to the micropyle of the ovule, in angiosperms it is transferred to the sticky, papillate stigma of the carpel (cf. Figs. 651, 673A; pp. 606 et seq., 652). This no doubt gives to sexual reproduction a greater degree of independence of moisture. The development of the stigma is necessitated by the inclusion of the ovules within the carpels; this must be a protective device associated with animal pollination (cf. pp. 652, 662), which is also related to the hermaphrodite flower (pp. 608, 643). All this indicates that the primitive angiosperm was hermaphrodite and zoophilous, giving it considerable advantages over its unisexual and anemophilous gymnosperm progenitors.

The hermaphrodite flower readily lends itself to self-pollination (autogamy) and hence inbreeding (pp. 402, 606). It is therefore not surprising that the angiosperms show numerous floral biological adaptations which favour or enforce cross-pollination (allogamy). In this context the stigma or style play a decisive role as a developmental and physiological 'filter': they generally inhibit the germination of pollen or the growth of the pollen tubes, not only of different species, but in the majority of the angiosperms even of the same plant. This is caused by genetic incompatibility and self-sterility (pp. 364, 402).

The genes responsible for self-sterility were probably already present in the ancestors of the angiosperms. Even in self-compatible plants there is often some filtering effect of the stigma and style in which their own pollen tubes grow more slowly than those of other individuals.

Genetic incompatibility has been reinforced in a number of different lines of angiosperms by heterostyly (p. 403). Heterostylous plants have two or three different lengths of style and positions of the stamens on different individuals, determined by linked genes.

Thus, for example, in many *Primula* species roughly equal numbers of plants are 'pin-eyed' with a long style and stamens inserted at the top of the corolla tube and 'thrum-eyed' with a short style and stamens inserted lower down. Darwin showed that only crossings between the two types result in optimal seed production. The size of the pollen grains corresponds to the size of the stigmatic papillae of the opposite type (Fig. 669). Such 'legitimate' pollination is normally ensured because the floral visitors transfer the pollen from the anthers to those stigmas which are at the same level in the corolla tube. Dimorphic heterostyly is also found in members of the Polygonaceae, Oxalidaceae, Plumbaginaceae, Boraginaceae (Fig. 344A) and Rubiaceae; trimorphic heterostyly, with three types of flower, occurs, for example, in the Lythraceae (*Lythrum salicaria* amongst others).

In dichogamy the stamens and stigmas mature at different times; if the stamens mature first the flower is protandrous (proterandrous); if the stigmas, protogynous (proterogynous). Simultaneous maturation is called homogamy. Hercogamy is the spatial separation of stamens and stigmas. Of course neither dichogamy nor hercogamy precludes the pollination of adjacent flowers, but nevertheless they tend to reduce autogamy and inbreeding in self-compatible plants and promote allogamy.

Protandrous flowers are widespread throughout the angiosperms (Fig. 670); protogyny is much less common (e.g. in *Plantago*). Hercogamy is also common. In *Iris* (Fig. 734C) bumble-bees, for example, bend down the stigmatic lobes and pollinate them with their backs only when they crawl into the partial flower (p. 606); when they crawl out of the

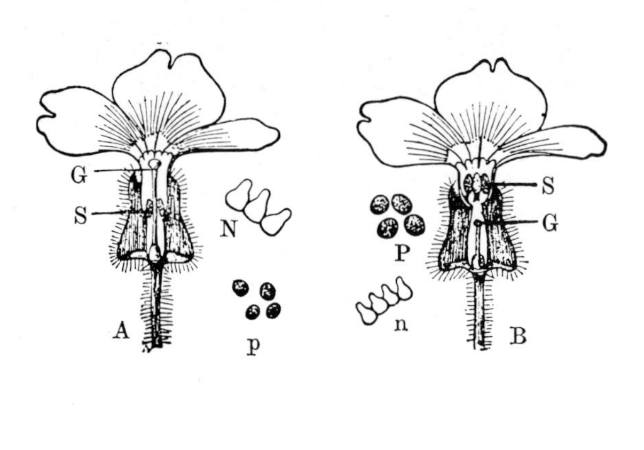

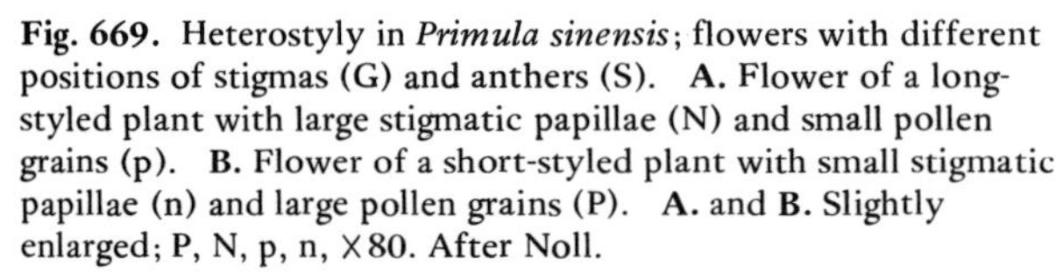

Fig. 669. Heterostyly in *Primula sinensis*; flowers with different positions of stigmas (G) and anthers (S). **A.** Flower of a long-styled plant with large stigmatic papillae (N) and small pollen grains (p). **B.** Flower of a short-styled plant with small stigmatic papillae (n) and large pollen grains (P). **A.** and **B.** Slightly enlarged; P, N, p, n, ×80. After Noll.

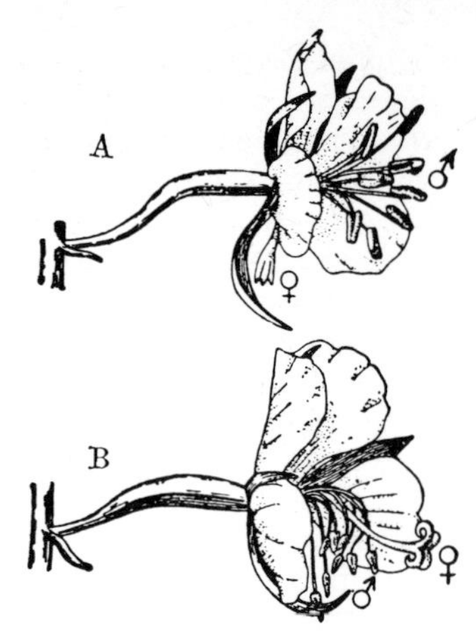

Fig. 670. Protandry in *Epilobium angustifolium*. **A.** Flower in ♂, **B.** in ♀ stage of development. Nat. size. After Clements and Long.

same flower, laden with pollen, the stigmatic lobes are pressed back against the style, preventing self-pollination.

Most angiosperm families include advanced representatives with facultative and eventually obligate self-pollination (autogamy). Pre-requisites for this are the loss of genetic self-incompatibility and the directed transfer of pollen to the stigma in the same flower, for example by the raining down of the pollen, by growth curvatures of the stamens or by closure of the corolla (cf. Figs. 344C, 717H). In cleistogamy self-pollination occurs even in the flower bud and the flowers fail to open.

Autogamy allows even one isolated plant to set seed and reproduce. It is therefore widespread amongst pioneer plants and weeds (p. 403) as well as in the floras of islands (long distance dispersal by single propagules; pp. 762 et seq.). Autogamy frequently affords the only method of sexual reproduction in extreme habitats with few pollinators, for example under arctic, alpine or desert conditions. Obligate self-pollinators generally have inconspicuous flowers without scent or nectar, the size or number of petals and stamens is often reduced, and the quantity of pollen diminished (see, for example, Figs. 344C, 717C, or the size of the anthers in the allogamous rye compared with the autogamous wheat and barley, Fig. 739B, E, G). Allogamy and autogamy commonly form an advantageous balanced system in which, when the flowers open, cross-pollination is facilitated, and then, towards the end of anthesis, self-pollination takes place. In many species of *Viola* and in *Oxalis acetosella* there is a corresponding balance between the formation of normal, open (chasmogamous) flowers and very reduced, bud-like (cleistogamous) ones on the same plant. In *Lamium amplexicaule* there is a tendency towards cleistogamy at the beginning and end of the flowering season.

Secondary unisexuality of the flowers of angiosperms in the form of monoecism, dioecism or polygamy (p. 601) has developed particularly in connection with wind pollination (pp. 663 et seq.), but it also affords a way out of autogamy and inbreeding (as in *Silene dioica* (*Melandrium rubrum*) and *S. alba*).

According to the external agencies which transport the pollen we can distinguish between animal, wind and water pollinated angiosperms.

Animal pollination (zoophily, zoidiophily, zoogamy) pre-supposes that the pollinating animals regularly visit particular flowers and remain in them for a sufficient time; the flower must fulfil the mechanical demands; the pollen and stigma must be touched in a definite sequence; and the pollen must be attached sufficiently firmly to the pollinator for it to be safely transported to the stigma of another flower. Zoophilous flowers (p. 608) therefore provide 'bait' (pollen, nectar, etc.), stimuli (colour, scent, etc.) and adhesive pollen (p. 650).

In the course of evolution angiosperms have undergone very striking developments of luring and stimulating mechanisms and in floral structure, and there was a progressive increase in the number of groups of animals involved in pollination, especially a very wide range of insects and several groups of birds. Starting with casual visits to flowers, very close mutually advantageous ties gradually developed between particular specialized zoophilous flowers and anthophilous animals. Thus increasing precision in the attraction of particular visitors which collected pollen from the stamens and deposited it on the stigmas facilitated more reliable and more economical transfer of the pollen from plant to plant, and hence better setting of seed. While the ratio of pollen grains to ovules in wind pollinated flowers is often some $10^6:1$, in specialized entomophilous flowers (e.g. Orchidaceae) it may be only 1:1. For the specialized floral visitor, competition with other anthophilous animals was reduced and the directed pollination of its own proper food plant was ultimately of benefit to itself. The evolutionary development of zoophilous angiosperms and the anthophilous animals adapted thereto can only be understood as a mutually determined 'co-evolution'. Indeed the interdependence of the partners is so close that neither could exist without the other.

The original attraction of the angiosperm flower was unquestionably food: at first this was the abundant pollen which contains proteins, fats, carbohydrates and vitamins. Such pollen flowers, available to equally primitive insects with biting mouth-parts, are found in the Magnoliidae (e.g. *Anemone*, *Papaver*, *Victoria*), primitive Rosidae (e.g. *Rosa*), Dilleniidae (e.g. *Paeonia*), etc. At a very early stage nutritive tissues were provided for visitors, but more especially a sugary sap (nectar), thus allowing the saving of the 'expensive' materials required in pollen production. An overwhelming majority of present-day angiosperms are indeed nectar flowers. A supply of nectar leads reciprocally to the improvement in the sucking mouth-parts of the floral visitors.

Nectaries are generally developed from a disc on the floral axis (p. 645) or from modified stamens (p. 648) but may also arise from certain regions of the carpels (p. 654), petals or calyx. Originally the nectar was more or less freely exposed and accessible to a variety of visitors with short mouth-parts, e.g. on the carpels of *Magnolia* (Fig. 652A) or on the receptacle of many Rosaceae (Fig. 693); later it became concealed, often in special organs, e.g. stored in the hollow spurs of the flowers of *Viola* (Fig. 709), *Linaria* and *Corydalis*, or in long floral tubes (Figs. 671E, 699J) where it is only accessible to certain animals, e.g. long-tongued lepidoptera.

In some cases the reproductive drive of the animals themselves may be harnessed by plants. This is for example the case in the fig (*Ficus carica*, Fig. 671H–L) in which the familiar pear-shaped hollow inflorescence contains three kinds of flowers: as well as the ♂ there are long and short-styled ♀ ones. The long-styled flowers form seeds; usually the short-styled do not, becoming instead gall-flowers in each of which a gall-wasp (*Blastophaga psenes*) with an ovipositor of the appropriate length deposits an egg, and in which the larva develops. The moth *Tegiticula yuccasella* pollinates the flowers of *Yucca* (Agavaceae) and lays its eggs in the ovaries; the larvae feed on some of the developing seeds. The flowers of the Mediterranean orchidaceous genus *Ophrys* simulate in form, hairiness and scent the females of certain bees or sand wasps; the males attempt to copulate and effect pollination. This case furnishes an example of 'deception' by flowers since the animal derives no return benefit from the visit; trap flowers (e.g. *Arum*, Fig. 671G, p. 660, and the lady's slipper orchid, *Cypripedium*, p. 742) provide other examples.

The means of stimulation by angiosperm flowers are mainly optical and chemical. Very often the two work together, but long and short range actions may be different.

Comprehension of the optical or chemical action of flowers presupposes a sound knowledge of the sensory physiology of the visiting animals which at present exists only for the honey-bee, bumble-bees, the hummingbird hawkmoth, the bee-flies (Bombyliidae) and a few hummingbirds. So far as the colour sense of these and other animals has been investigated, we find, for example, that the honey-bee and bumble-bees cannot see pure red, but can perceive ultraviolet from 400 to 310 nm which is invisible to Man and, amongst the other floral colours, only a yellow waveband from 650 to 520 nm, a blue-violet band from 480 to 400 nm, and white, which is perceived as blue-green. On the other hand the optical perception of birds is more similar to that of Man, and they are particularly sensitive to red. Training experiments with anthophilous insects have shown that various degrees of saturation and brightness, simultaneous brightness and colour contrasts, and the shapes of parts of the flower all participate in the effectiveness of the floral display apparatus. Thus the significance of those floral patterns and spots of colour which have been long regarded as 'honey guides' has now been demonstrated, for example, the orange-yellow lip of the otherwise lemon-yellow flower of *Linaria vulgaris* (Fig. 727I). Sometimes honey guides are only recognized by insects sensitive to ultra-violet (e.g. on the perianth segments of *Caltha palustris*, which to us are uniformly yellow). The movements of whole flowers or parts of flowers may also act as optical stimulants.

The chemical attraction of flowers is mainly effected by perfumes which are emitted from the pollen, the petals or other floral organs. Some of these are unpleasant to Man, e.g. in those Araceae which smell of carrion or dung and are pollinated by coprophilous and saprophilous insects. Whilst optical stimuli induce approach in a straight line, perfumes diffuse irregularly and hence attract in a less certain and more erratic manner. Scent is, however, important over short distances for bees and bumble-bees. Many flowers have besides the optical honey guides, 'scent guides', often in the same part of the flower (e.g. in the corona of *Narcissus*). Development of these stimulating mechanisms goes hand in hand with improvements in the sensory organs of the visitors. Bees and other hymen-optera develop an 'affinity' after a number of productive visits to particular species which they exclusively and intensively exploit over a limited period. This depends both on the stimulating effect of the scent of the pollen and nectar carried into the hive and on the highly developed 'memory' and 'language' (power of communication) of these animals.

Mechanical adaptations ensure that only animals with the appropriate body structure can effect pollination, being guided along a determined route which results in adequate contact with the pollen and stigmas. A certain duration of stay may be enforced, ensuring pollination, or transference of the pollen may be guaranteed by specific lever-, adhesion-, clamp- or discharge-mechanisms.

Thus, for example, the protandrous flowers of *Salvia pratensis* (Fig. 671A–D) are well known for the effective lever-mechanism described by Sprengel (p. 606); each of the two stamens has a greatly elongated connective (c) which forms a lever just under the upper lip, and which is pivoted to the short filament (f). Only the long anterior arm of the lever bears a fertile theca; the shorter, sterile posterior arm is expanded and united with the corresponding arm of the other stamen to form a plate (s) which blocks the way to the nectar. If a bumble-bee now presses this plate, the long anterior ends of the two levers are forced down, pressing the two thecae with their pollen against the insect's back. In older flowers (B) the stigmas take the place formerly occupied by the thecae so that cross-pollination regularly ensues.

A very closely integrated floral mechanism is the slippery-trap inflorescence exempli-fied by *Arum maculatum* (Fig. 671G) and other *Arum* species. In this chemical attraction operates over a distance. The unisexual flowers are borne in monoecious inflorescences on the lower part of a thick spadix which is enveloped by a large light-coloured bract or spathe; the basal part of this forms a chamber with a constricted mouth, whilst the upper part is widely expanded. On the spadix at the bottom of the chamber are the ♀ flowers; above these is a ring of elongated, sterile, bristle-like flowers; then follows a zone of ♂ flowers; on top of these is a further ring of sterile hairs. *Arum nigrum* has been studied in particular detail: above the chamber its spadix is thicker and club-like, and on the morning of the first day after the opening of the spathe gives off both heat and a carrion-like odour attractive to carrion flies and beetles, some of which are already laden with pollen from other inflorescences. When these insects try to alight on and cling to the spadix or the inner surface of the spathe they quickly lose their footing and tumble into the chamber, since their claws and suckers cannot grip the smooth, oily epidermis. Escape is impossible at this stage because both the rings of bristles and the upper part of the chamber wall also have slippery surfaces, and the bristles still bar the exit. The pollen brought in is now dusted on the stigmas. During the following night the ♂ flowers scatter their pollen over the insects in the chamber; meanwhile the emission of scent ceases. Finally, the bristles and spadix wither, so that on the following day the pollen-laden insects can depart, generally only to tumble very soon into another inflorescence. The flowers of *Aristolochia* species also form slippery traps. For further examples of mechanical arrangements see Orchidaceae (p. 742), Asclepiadaceae (p. 723), Fabaceae (p. 696) and Asteranae (p. 731).

The pollen of entomophilous flowers has a copious oily pollen cement and so is transported in clumps. Frequently spines (Fig. 619) or toothed ridges on the surface of the grains facilitate their attachment to the hairs or feathers of animals. A single pollination can therefore effect the fertilization of numerous ovules; accordingly, zoophilous flowers commonly have numerous ovules in each ovary—many thousands in the Orchidaceae, which are pollinated by entire pollinia.

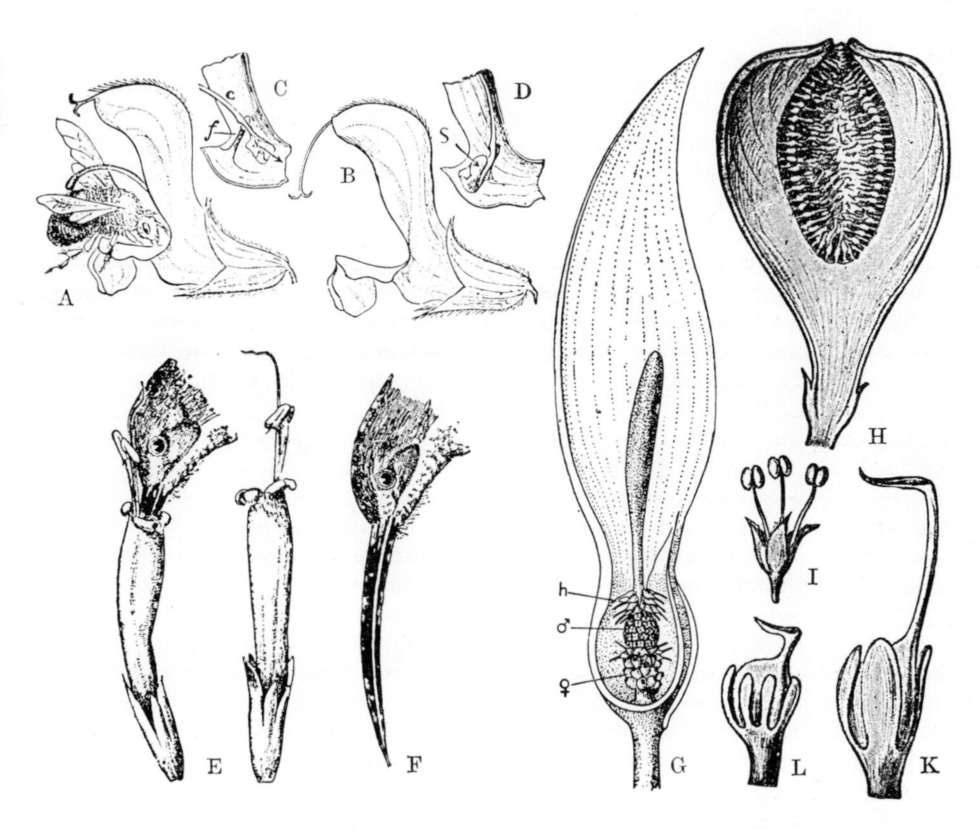

Fig. 671. Animal pollination in various angiosperms. **A.** to **D.** Bumble-bee as visitor of *Salvia pratensis* (violet-blue). Somewhat enlarged. **E.** and **F.** The honey-bird *Arachnothera longirostris* as pollinator of *Sanchezia nobilis* (Acanthaceae, flower yellow, bracts purple). ×¾ approx. **G.** Inflorescence, cut open lengthwise, of *Arum maculatum* (slippery-trap 'flower') with light-green spathe and inconspicuous ♂, ♀ and sterile filamentous (barrier) flowers (h) in the female stage of development. $\times\frac{2}{3}$. **H.** Inflorescence of *Ficus carica* in L.S. (somewhat enlarged) with ♂ (**I.**), fertile, long-styled ♀ (**K.**) and short-styled gall flowers (**L.**) (enlarged). Further explanation on p. 659. **A.** to **D.** After Noll. **E.** and **F.** After Porsch. **G.** After Firbas. **H.** After Karsten. **I.** After Kerner. **K.** and **L.** After Solms-Laubach.

The various zoophilous flowers, whether euanthia, meranthia or pseudanthia (p. 606) can be classified into floral types according to their functional and structural characteristics.

The developmental series of floral types starts with (1) flat discoid and bowl-shaped (acetabuliform) forms, ranging from (*a*) polymerous (e.g. *Magnolia*, Fig. 652A; *Nymphaea*, Fig. 683D) and (*b*) multiradiate (e.g. *Anemone*; *Adonis*; *Matricaria* (pseudanthium), Fig. 729F) to (*c*) pauciradiate (few-rayed) types (e.g. *Berberis*, Fig. 683A; *Cabomba*, Fig. 731F; Rosaceae, Fig. 693; *Acer*, Fig. 701; *Paeonia*, Fig. 652C; *Tilia*, Fig. 713C; *Rheum*, Fig. 718A, B; *Ornithogalum*, Fig. 733C; Apiaceae, Fig. 707B–D; and, remarkably, the pseudanthium of *Euphorbia*, Fig. 704H–K). Further development leads to types with deeper flowers and greater volumes involved in pollination. More or less radial and progressively more elongated are (2) cup- (crateriform) and bell-shaped (campanulate) flowers (e.g. *Hyoscyamus*, Fig. 726C; *Leucojum*, Fig. 733I; *Crocus*, Fig. 734A) and (3) tubular and salver-shaped flowers (e.g. *Silene*, Fig. 717F; *Oenothera*, Fig. 699J; *Primula*, Fig. 669A; *Symphytum*, Fig. 725A, B; *Nicotiana*, Fig. 726G; *Cinchona*, Fig. 721A). Dorsiventral are (4) papilionate (vexillate) flowers (e.g. *Corydalis*; *Pisum*, Fig. 698; *Polygala*, Fig. 702I, K) and (5) ringent (gaping) and labiate flowers (e.g. *Aconitum*, Fig. 684D, E; *Viola*, Fig. 709; Scrophulariaceae, Fig. 727C, D, H–K; *Salvia*, Fig. 671A–D; *Orchis*, Fig. 735A, B; *Curcuma*, Fig. 736B; meranthium: *Iris*, Fig. 734C; pseudanthium: *Mimetes*, Proteaceae). Specialized developments are (6) brush-like and tufted flowers (e.g. *Eugenia*, Fig. 699D; *Eucalyptus*; *Acacia*, Fig. 696A, B and the special case of the pseudanthium of *Salix*, Fig. 711G–L) and (7) trap flowers (e.g. the slippery-trap flowers of *Arum*, Fig. 671G, pp. 660–1, and the clamp-trap flowers of *Asclepias*, Fig. 719I, K), etc. Levels of organization in the evolution of the euanthium of the angiosperms, documented by fossil evidence, correspond to the series of flower types from (1*a*) to (1*c*) and hence to (2) and (3), and from (1*c*) and (2) to (4) and (5). Type (6) belongs to intermediate levels of organization, type (7) and all meranthia and pseudanthia to intermediate or higher levels.

The evolution of the animal pollinators runs parallel to that of the plants. The oldest and most primitive pollinators were apparently the Coleoptera (beetles; since the Permian). Hymenoptera (wasps, ants and particularly the more recent Apidae—bees and bumble-bees) and Diptera (flies, e.g. hoverflies) only appeared later. A third phase of evolution of animal pollinators included the Lepidoptera (especially butterflies, hawk-moths and noctuid moths) and, especially in the Tropics, birds (Trochilidae, humming-birds, in the New World; Meliphagidae, honey-eaters and Nectarinidae, sunbirds, amongst others in the Old World) and also bats. Other animals (e.g. Orthoptera, Hemiptera, Thysanoptera, small mammals) are generally less important.

Many of these animal groups have taken advantage, through their body structure, mouth-parts, behaviour and nutritional demands, of particular features of the flowers visited which they have selectively modified in specific ways: thus modes of pollination have arisen with whole complexes or syndromes of characters.

The validity of this idea is supported by experimental study of the great power of discrimination shown by most pollinators, and by the fact that functionally very similar modes of pollination can arise from quite different floral organs in single flowers (euanthia), in partial flowers (meranthia) and in inflorescences (pseudanthia). In connection with the selective action of the floral visitors on floral structure it should be noted that many flowers are visited by a large number of diverse animals (polyphilous); specialization only led gradually to the development of oligophilous or monophilous flowers with few visitors or only one.

Amongst insect flowers (entomophilous, entomogamous) those visited by beetles must again be placed first (cantharophilous). Beetles are relatively clumsy animals and often do considerable damage with their mouth-parts to floral organs. Correspondingly the mode of pollination of beetle flowers is characterized by readily accessible, robust discoid or saucer-shaped flowers, greenish or white in colour, without honey guides and generally with a strong scent and much pollen as food. Many Magnoliales (e.g. Fig. 652A) are beetle flowers, but so also are some advanced discoid flowers (e.g. the pseudanthia of *Cornus* and *Viburnum*).

Fly flowers (myophilous) are very heterogeneous. They include both small, more or less odourless disc flowers with exposed nectar (e.g. Apiaceae, Fig. 707B–D; *Ruta*, Fig. 700B, C) and carrion-fly flowers (sapromyophilous), which with their green or purple-spotted colours and carrion-like odour resemble the usual habitat of the visitors which are generally taken into service as pollinators of deceptive or trap flowers (e.g. the euanthium of *Aristolochia* and the pseudanthium of *Arum*, Fig. 671G, p. 660).

Bee flowers are both diverse and numerous (melittophilous). Their mode is marked by dorsiventral papilionate, ringent and labiate flowers with landing platforms, frequently yellow, violet or blue colours, light scents, honey guides, and moderately deeply concealed nectar (e.g. *Salvia*, Fig. 671A–D, and the papilionate and ringed flowers above).

Butterfly flowers (psychophilous) are characterized by their erect position, frequently red colour and deeply concealed nectar (e.g. *Dianthus carthusianorum*; *Nicotiana tabacum*, Fig. 726G). In contrast to the butterfly flowers open in the daytime, noctural moth flowers (sphingophilous, phalaenophilous) open in the evening. They have vertically orientated or pendulous, narrowly tubular flowers, white in colour, heavily scented and with deeply concealed nectar (e.g. *Oenothera*, Fig. 699J; *Silene*, Fig. 717F; *Lonicera periclymenum*). The remarkable orchid *Anagraecum sesquipedale* from Madagascar has a spur 32 cm long: it was predicted that the pollinator would be a nocturnal hawkmoth (*Xanthopan morgani* f. *praedicta*) before the insect was actually discovered.

The mode of pollination of bird flowers (ornithophilous, ornithogamous) is in marked contrast to that of insect flowers. Landing platforms are absent since relatively heavy birds must either hover (hummingbirds) or use a rigid perch. The large flowers are frequently of the cup, tube or brush type and the colours or colour contrasts are often bright red, or blue, yellow or even green ('parrot colours'). Scent is absent, corresponding to the poor sense of smell of birds; however, copious but thin nectar, often deeply concealed, is taken up by the bird's tubular or brush-like tongue. The frequency of ornithophily in warmer climates may be related to the enhanced moisture requirements of the birds and the presence of plants in flower throughout the year. The pollen is carried attached to the beak or other parts of the body (Fig. 671E, F). Bird flowers occur in almost all zoophilous Tropical families; amongst those seen in cultivation are *Erythrina* (Fabaceae), *Fuchsia*, *Hibiscus tiliaceus* (Malvaceae), *Tropaeolum majus*, *Salvia splendens* and *Aloe*.

Bat flowers (chiropterophilous) are restricted to the Tropics; in both hemispheres they are visited especially by the long tongued 'flying foxes' and fruit bats. Their adaptations include exposed, robust, generally cup shaped, broadly ringent or brush-like flowers, nocturnal anthesis, sombre colours, a strong odour of fruit or fermentation, and copious nectar or pollen. Examples include *Carnegiea* (Cactaceae), *Adansonia* (Bombacaceae), *Cobaea* (Polemoniaceae), various Bignoniaceae and species of *Musa* and *Agave*.

In general the progressions of modes of pollination run from beetle to fly and hymenoptera flowers, and thence on the one hand to butterfly or moth flowers or to bird and bat flowers on the other, but many other different relationships are known.

Wind pollination (anemophily, anemogamy) involves the production and scattering of large quantities of pollen grains, their rapid and fairly uniform dispersal, their buoyancy for a sufficient period, and the sufficient size and exposure of the stigmas to ensure a high frequency of pollination. Means of enticement and stimulation are absent, the flowers are unisexual and ♂ flowers (or stamens) are much more numerous than ♀ flowers (or ovules), and the pollen is smooth and dry (pp. 606 et seq.).

Secondary wind pollination has arisen many times in the course of evolution in various groups of angiosperms, especially where species occur in pure stands in windy habitats in which insect pollinators are few. In almost all wind pollinated flowers there are vestiges of former hermaphroditism and entomophily.

The syndromes of adaptations shown by secondarily wind pollinated angiosperm flowers resemble in many respects those shown by the primarily wind pollinated gymnosperms (similar selection pressures; cf. p. 606).

Correspondingly the pollen grains lack pollen cement; they are therefore dry, with smooth walls, separate easily and are very buoyant because of their small size. Their production is greatly enhanced by increase in the numbers of ♂ flowers or stamens (cf. Fig. 672, *Poterium*; in *Corylus* there are about $2\frac{1}{2}$ million pollen grains per ovule). Liberation of the pollen is facilitated by flexibility of the filaments (e.g. in grasses, Fig. 783C), of the pedicels (e.g. in *Cannabis*) or of the axis of the inflorescence (as in the dangling ♂ catkins of *Corylus*, *Alnus* and *Quercus*, Figs. 687 and 688); generally the pollen is deposited before it is blown about by the wind. The ♂ flowers of *Urtica* and *Pilea* 'explode' as a result of the elastic tension of the filaments (Fig. 652D).

The styles and stigmas of the ♀ flowers are much enlarged, thus facilitating the reception of pollen. The number of ovules in the ovary is much reduced corresponding to pollination with separate pollen grains. The flowers are exposed, but individually very inconspicuous; a perianth would hinder the transfer of pollen and is thus reduced or

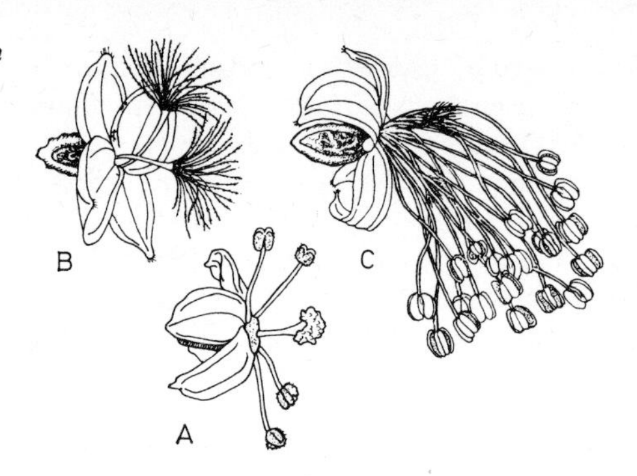

Fig. 672. Transition from entomophily (**A.**) to secondary anemophily (**B.**) in the related genera *Sanguisorba* and *Poterium* (Rosaceae).
A. Hermaphrodite flower of *S. officinalis* with four stamens, verrucose stigma and nectary.
B. and **C.** Unisexual flowers of *P. sanguisorba* with no nectaries, ♀ flower with feathery styles and ♂ with numerous stamens. ×6. After Knoll.

eliminated. Similarly such adaptations to animal pollination as nectaries, scent, and dorsiventrality are usually absent. Unisexuality is common, promoting the unhindered transfer of pollen, but is probably also related to the general reduction of the flowers and the prevention of self-pollination. Pollination is promoted by the early flowering season, often before the leaves form an obstruction (e.g. in oak, alder, elm, poplar, ash). It is clear that wind pollinated angiosperms are dependent on certain environmental factors: they occur almost exclusively in pure stands in savannas, steppes and Arctic–Alpine habitats exposed to wind, or in the tree-layer of subtropical and boreal forests. In humid tropical forests zoophily prevails and anemophilous angiosperms are almost completely absent.

The secondary origin of anemophily is obvious where closely related plants are predominantly zoophilous (e.g. in *Thalictrum*; *Sanguisorba*, Fig. 672; *Acer*; Caryophyllales, Fig. 717K, L; *Fraxinus*, Fig. 723E–G; *Artemisia*, Fig. 730A). Development of the complete anemophilous syndrome can often be traced step by step. Many plants, e.g. *Tilia* and *Calluna* stand at the borderline between entomophily and anemophily since much of their pollen is wind-dispersed. Even in some very large and apparently ancient anemophilous groups there can be no doubt that anemophily is secondary since traces of pollen cement, nectaries, scent and hermaphroditism can be seen, e.g. in the Amentiferae (Hamamelididae), Salicaceae, Euphorbiaceae, Chenopodiaceae, Juncales, Poales and Arecidae. The pollen of *Castanea* has initially much pollen cement and is carried by insects; later it dries out and is wind dispersed. Apparently, however, secondarily anemophilous groups may occasionally revert to entomophily (e.g. *Salix*, Fig. 711G–L; *Euphorbia*, Fig. 704H–K).

Water pollination (hydrophily, hydrogamy) occurs only in a few angiosperms. Rainwater may effect pollination in some erect disc-, bowl- and cup-shaped flowers (e.g. *Ranunculus*, Fig. 684A, B), and rarely cross-pollination may be effected by splashing, but even in aquatic plants hydrophily is by no means universal: the flowers are frequently raised above the water surface and pollinated by wind, as in *Potamogeton*, Fig. 732C. *Callitriche* (p. 731), however, has floating pollen, and in *Vallisneria* and *Elodea* (*Anacharis*) the detached ♀ flowers reach the stigmas which project from the water surface at the time. Submerged pollen carried under water from the ♂ to the ♀ flowers occurs in, for example, *Ceratophyllum* (p. 682), *Najas* and *Zostera* (p. 738; this last with filamentous pollen grains over 2 mm long).

The differentiation of pollination mechanisms related to various external factors and adapted to various pollinating animals give rise to important phylogenetic trends (p. 412). Thus in *Aquilegia* amongst the Ranunculaceae (pp. 431, 683 et seq.; Fig. 369) we find both zoophily (beetle, bee, hawkmoth and bird pollination) and anemophily, as well as obligate self-pollination.

Fertilization. This is initiated by the growth of the pollen tube from the stigma to the ovule (Fig. 673A; cf. also p. 364). At first the pollen tubes generally grow down the style, often on the papillate surface of a hollow stylar canal or in special conducting tissue (in which the middle lamellae are broken down enzymatically); subsequently growth continues across the ovular cavity (often filled with mucilage) to the micropyle of the ovule.

This is porogamy, but in advanced cases (e.g. when the micropyle is blocked) it may be

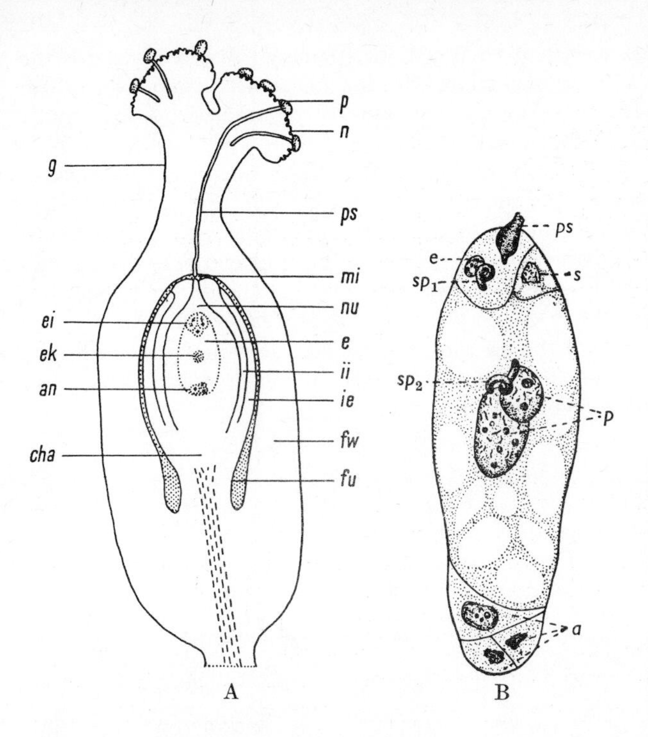

Fig. 673. Pollination and fertilization in angiosperms. **A.** Diagrammatic L.S. of the ovary of *Bilderdykia (Polygonum) convolvulus* with one orthotropous ovule. ×48. fw, ovary wall; g, style; n, stigma; p, pollen grain with pollen tube, ps; ovule with funicle, fu; chalaza, cha; outer and inner integuments, ie, ii; micropyle, mi; and nucellus, nu; embryo sac, e; with egg apparatus, ei; secondary embryo sac nucleus, ek; and antipodal cells, an. ×48. **B.** Embryo sac of *Lilium martagon* during fertilization. Remains of the pollen tube, ps; the two sperm nuclei, sp_1, sp_2; egg cell with egg nucleus, e; one of the two synergidae (s); the two polar nuclei, as yet unfused; and the antipodal cells (a). ×600. After Guignard.

replaced by aporogamy in which the pollen tube may reach the embryo sac by way of the chalaza (chalazogamy).

In the pollen tube the cytoplasm containing the sperm cells and pollen tube nucleus is always confined to the tip since the older parts become empty and are often divided up by callose plugs. Growth rates are very variable, reaching 1–3 mm hr^{-1}, but sometimes very slow indeed as in many Hamamelididae, Cactaceae and Orchidaceae. In these fertilization may take place many weeks or months after pollination; this is often associated with delayed development of the ovule.

On reaching the egg apparatus the pollen tube invariably discharges its contents, not into the egg cell, but into an adjacent synergida, which then dies. Apparently the synergidae are somehow responsible (possibly by secreting enzymes) for breaking down the tip of the pollen tube. The vegetative pollen tube nucleus sooner or later disintegrates, but from the two sperm cells (of which the nuclei are commonly coiled in a spiral and undergo spontaneous amoeboid movements) one nucleus passes to the egg cell, where it fuses with the egg nucleus (Fig. 673B sp_1), whilst the other (sp_2) penetrates further into the embryo sac where it fuses with the secondary embryo sac nucleus or with the two polar nuclei if these have not already united. The angiosperms are thus characterized by a process of double fertilization resulting in a diploid zygotic nucleus and a triploid endosperm nucleus in the embryo sac.

Development of embryo, endosperm and seed. The fertilized egg cell gives rise to the embryo whilst the endosperm nucleus and the residual embryo sac cytoplasm form the nutritive tissue known as the secondary endosperm. The zygote then divides by one or commonly more cross walls forming a short row of cells, the proembryo (Fig. 674A–C). Only the first cell or cells of this series, directed towards the inner part of the embryo sac, form the embryo proper, the growth of which is initiated by the formation of longitudinal walls. The remaining cells form the suspensor which pushes the embryo down into the developing endosperm and probably also conducts nutrients to the embryo. The suspensor often terminates in a cell (hypophysis) which abuts on the embryo and which may, after further divisions, participate in the formation of the root cap and root tip. At first the embryo consists of a spherical structure divided into quadrants and subsequently octants which lies at the end of the suspensor (Fig. 674D–F). This later differentiates into a root primordium (radicle) with root cap directed towards the micropyle and seed leaves (cotyledons) and apical meristem of the shoot (plumule) directed towards the chalaza. Dicotyledons have two lateral cotyledons with the apical meristem between them (Fig. 674G), whilst monocotyledons have only one apparently terminal cotyledon and a laterally displaced apical meristem (Fig. 674H).

The sequence of cell divisions in the proembryo and embryo follows various more or less regular patterns: the commonest are the types seen in the Asteraceae and Onagraceae (or Cruciferae) (Fig. 674A–G). In general, small and straight embryos (e.g. Fig. 682B) belong to a more primitive level of development than large (e.g. Fig. 679B) or curved ones (e.g. Figs. 685E, 710I–L). Embryos may remain few-celled and undifferentiated as a result of reduction; this is often the case with plants with special modes of nutrition such as the mycotrophic Orchidaceae and the parasitic Orobanchaceae which produce a very large number of minutely, scarcely differentiated seeds, of which only a very small fraction has a prospect of reaching favourable habitats.

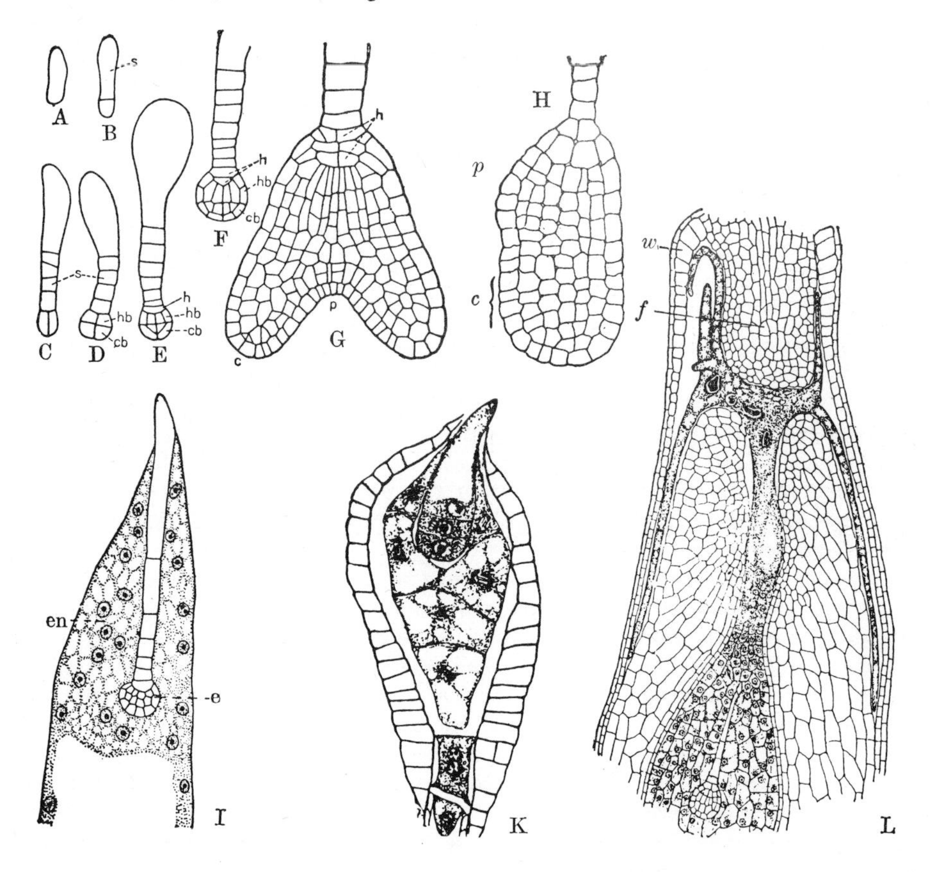

Fig. 674. Development of embryo and secondary endosperm in angiosperms. Zygote (**A.**), suspensor (s), young embryos with hypophysis (h) and the hypocotylar region (hb), cotyledons (cb, c) and apical meristem of the shoot (p). **A.** to **G.** In a dicotyledon (*Capsella bursa-pastoris*) and **H.** in a monocotyledon (*Alisma plantago-aquatica*). ×200 approx. **I.** and **K.** Young embryo (e) with suspensor in the endosperm (free nuclear in **I.** *Lepidium* sp., cellular in **K.** *Ageratum mexicanum*). Enlarged. **L.** L.S. of part of a young seed of *Globularia cordifolia*. A tubular, branched haustorium has grown out of the micropyle from the embryo sac and is appressed partly to the ovary wall (w), partly to the funicle (f). The embryo with its suspensor is also visible in the embryo sac. Enlarged. **A.** to **H.** After Hanstein and Souèges. **I.** After Guignard. **K.** After Dahlgren. **L.** After Billings.

Embryos may develop in some plants without fertilization, so that sexual reproduction is replaced by asexual (apomixis); the seeds are then agamospermous, as in many polyploids of hybrid origin (cf. pp. 403, 421).

In some forms of agamospermy alternation of generations and the formation of an embryo sac may be retained, at least initially: thus a diploid embryo sac cell may arise either as a result of the formation of a restitution nucleus when the embryo sac mother cell undergoes meiosis (diplospory, as in forms of *Taraxacum, Antennaria, Allium,* etc.), or the embryo sac cell may be replaced by a vegetative cell of the nucellus (apospory, e.g. in forms of *Ranunculus, Hieracium* (Fig. 345), *Poa* and other grasses). If the archesporium is multicellular the distinction between diplospory and apospory may be difficult to detect, and both forms of agamospermy may occur simultaneously (as in various Rosaceae: *Potentilla, Rubus, Sorbus, Alchemilla,* Fig. 675B). Development of the egg cell generally takes place spontaneously (parthenogenesis), but development of the endosperm may require previous pollination and fusion of a sperm cell with the secondary embryo sac nucleus (pseudogamy).

Alternation of generations and the formation of an embryo sac may be completely suppressed when embryos develop directly from somatic cells of the ovule (adventitious embryony or nucellar embryony: as in forms of *Citrus, Hosta* (Fig. 675A) and Orchidaceae); like agamospermy this may be spontaneous or induced by pollination. Sometimes adventitious embryos arise alongside sexually produced ones (polyembryony). Adventitious embryony can be regarded as a special case of vegetative propagation.

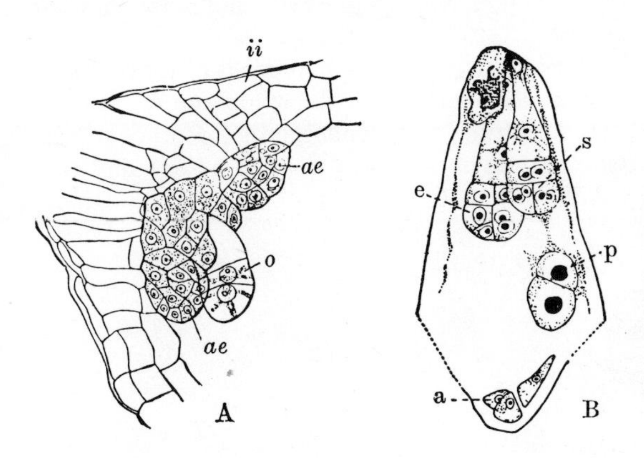

Fig. 675. Asexual development of embryos. **A.** Adventitious embryos (ae) from the apex of the nucellus, and an embryo developing normally from the egg cell (o) in *Hosta albomarginata* (Agavaceae); inner integument, ii, ×120. **B.** Parthenogenetic development of two embryos from the egg cell (e) and a synergida (s) in the unreduced embryo sac of *Alchemilla sericata*; polar nucleus, p; antipodal nucleus, a. ×210. **A.** After Strasburger. **B.** After Murbeck.

The endosperm nucleus generally divides before the first division of the zygote and initiates the formation of the secondary endosperm. At first this nourishes the embryo; later it either forms a storage tissue in the seed, which is used up by the embryo before or after germination, or it may be completely suppressed, in which case the seed is non-endospermous instead of endospermous.

Endosperm formation is generally free nuclear, i.e. the endosperm nucleus divides into a large number (eight to over 2000, depending on the species; Fig. 674I) of nuclei which are generally embedded in a peripheral layer of cytoplasm surrounding a large vacuole formed during the enlargement of the embryo sac after fertilization. Cell walls subsequently form between the free nuclei (cf. p. 44, Fig. 36); eventually the entire embryo sac becomes filled in one way or another by a mass of cells. In some cases (apparently advanced), e.g. in many Asteranae, endosperm formation is cellular, i.e. nuclear division is associated with cell wall formation from the outset (Fig. 674K). Finally in the helobial type (particularly in monocotyledons, e.g. Alismatidae = Helobiae) the upper part of the endosperm shows nuclear division at first whilst the lower part develops cellularly from the beginning.

The formation of embryo and endosperm requires a considerable supply of nutrients. Most commonly the nucellus is displaced and largely or completely used up by the enlarging embryo sac which becomes filled with nutritive tissue. Sometimes tubular absorbing organs or haustoria grow out from the embryo sac (p. 657), or more frequently from the endosperm (Fig. 674L) or suspensor; the haustoria penetrate the surrounding tissues so that the embryo and endosperm behave almost like parasites of the maternal tissues.

In some seeds, e.g. the nutmeg (Fig. 681D–F) and the *Areca* palm, the nucellus, or in other cases, e.g. in the Annonaceae, the integuments, produce lamellae distinguished by their colour and contents which penetrate and branch within the endosperm (ruminate endosperm).

In contrast to the primary endosperm (haploid ♀ prothallus) of the gymnosperms, the secondary (and usually triploid) endosperm of the angiosperms arises after fertilization. This has two advantages: (1) development of the ♀ gametophyte is further abbreviated and accelerated, and (2) the endosperm only develops when it is actually needed for supplying an embryo after fertilization has been achieved (economy!).

As well as the secondary endosperm, the nucellus also can function as a nutritive and storage tissue: such a perisperm is found in addition to the secondary endosperm in Nymphaeaceae, Piperaceae (Fig. 682B) and Zingiberaceae, or by itself in the Caryophyllaceae.

In the seeds of all primitive angiosperms the secondary endosperm (with or without a perisperm) forms an extensive nutritive and storage tissue around the small and little differentiated embryo. Frequently, however, the embryo develops further in the ripening seed whilst still on the parent plant; it may retain a central position (Fig. 676B) or be displaced to one side (Fig. 739L). Related to the rapid mobilization of reserves on germination is the final suppression of the secondary endosperm, whilst the storage function is transferred to the embryo itself, especially to the cotyledons (Fig. 679B). Well-known examples are the Leguminosae, oak (Fig. 687Q), walnut (Fig. 691D) and horse-chestnut with their swollen cotyledons filling the seed.

Very rarely no reserves are stored in the seed (e.g. in the minute seeds of the orchids), but in general starch, protein and aliphatic oils are stored in the cytoplasm (p. 55) and reserve cellulose in the cell wall (p. 64). Accordingly the endosperm (or other storage tissue) is mainly starchy, as in the grasses, fatty as in the coconut, or horny or stony as in many Liliales and some palms, e.g. *Phytelephas* (p. 750).

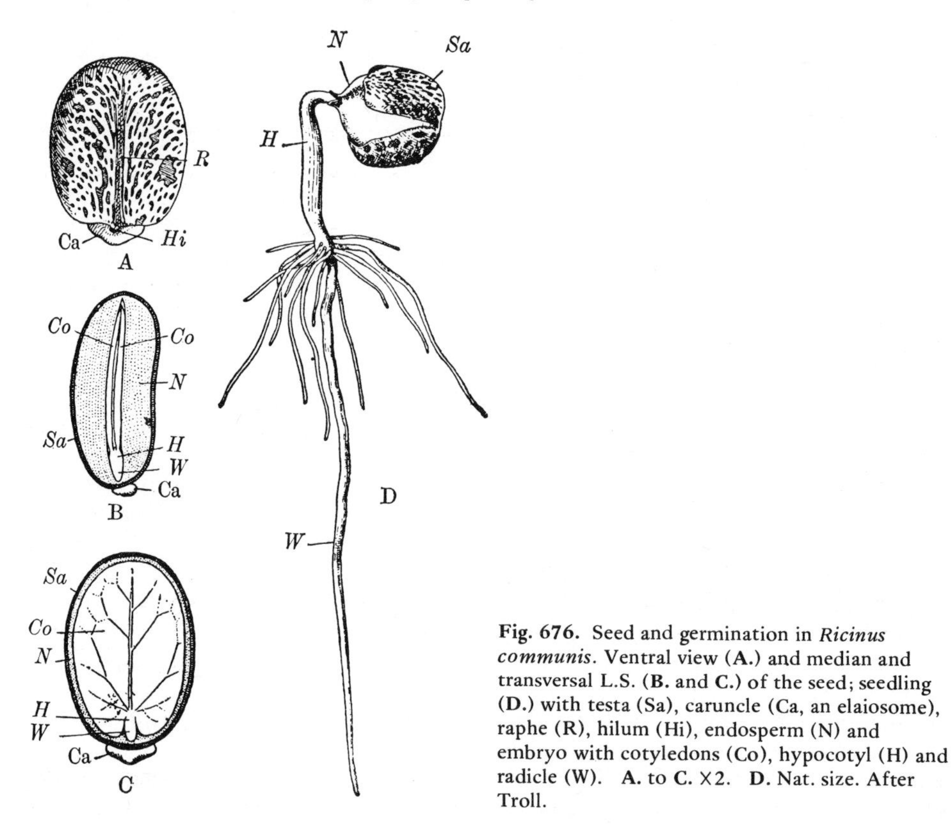

Fig. 676. Seed and germination in *Ricinus communis*. Ventral view (**A.**) and median and transversal L.S. (**B.** and **C.**) of the seed; seedling (**D.**) with testa (Sa), caruncle (Ca, an elaiosome), raphe (R), hilum (Hi), endosperm (N) and embryo with cotyledons (Co), hypocotyl (H) and radicle (W). **A.** to **C.** ×2. **D.** Nat. size. After Troll.

The seed coat (testa) develops from both integuments, or from only one in advanced groups. In the most primitive angiosperms it appears to consist, as in the cycads (pp. 609, 616, 636), of an inner, lignified sclerotesta and an outer, fleshy, mostly brightly coloured (e.g. red) sarcotesta (e.g. in some Winteraceae, Magnoliaceae, *Paeonia*, etc.).

The sarcotesta has apparently given rise to various forms of aril which only partially enclose the seed. Examples are seen in *Euonymus*, or in a much-divided form in *Myristica* (Fig. 681D). In *Nymphaea* the aril forms an air bladder enabling the seed to float (Fig. 677H). Greater reduction of the sarcotesta or aril apparently leads to the formation of a caruncle (beside the micropyle, e.g. in *Ricinus*, Fig. 676A–C) or a strophiole (on the funicle). When such seed appendages are rich in fat, protein or sugar they may play a part as elaiosomes in dispersal by ants (p. 674).

The testa often becomes mucilaginous (e.g. in various Cruciferae, flax, quince, tomato, *Plantago, Juncus*; myxotesta). A dry testa may bear hairs (e.g. *Epilobium*, Fig. 677A–C; cotton, Fig. 713M; *Strophanthus*, Fig. 719G) or wing-like outgrowths (e.g. the gliding seeds of *Zanonia*, Fig. 677D). Commonly, however, there is only a sculptured or more or less smooth sclerotesta (e.g. Fig. 677E). When the seed remains permanently protected by the carpel wall (pp. 670 et seq.), as in the mericarps of the Umbelliferae, the indehiscent fruits of the composites and grasses, or the stone-fruits of the Prunoideae, etc., the testa remains thin and membranous.

Seed size varies enormously, from the double coconut of the Seychelles (Arecaceae: *Lodoicea*) weighing several kilograms and the heavy seeds of the horse-chestnut (*Aesculus*), to the minute, dust-like seeds of the Pyrolaceae, Orobanchaceae and Orchidaceae with weights of only a few micrograms.

Obviously all these progressions, relating to nutrient supply, form of the surface, and size, play an important part as adaptations in the dispersal and germination of seeds (pp. 673 et seq.).

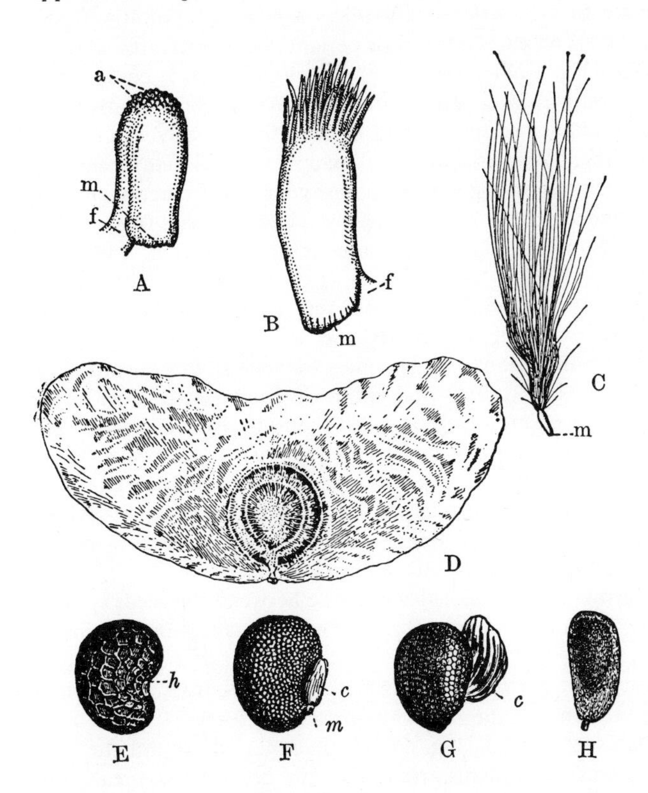

Fig. 677. Seeds and their development. **A.** and **B.** Ovules with funicle (f), micropyle (m) and initials of the seed-hairs (a). ×70. **C.** Ripe seed. ×9. **A.** to **C.** *Epilobium (Chamaenerion) angustifolium*. Seeds of (**D.**) *Zanonia javanica* (Cucurbitaceae), winged, ×½; (**E.**) *Papaver rhoeas* (hilum, h); (**F.**) *Corydalis ochroleuca*, and (**G.**) *Chelidonium majus* with elaiosome (c; micropyle, m); (**H.**) *Nymphaea alba* with sac-shaped aril (enlarged). **A.** to **C.** After Goebel. **D.** After Firbas. **E.** to **H.** After Duchartre.

The fruit. Simultaneously with the formation of seeds from ovules goes the development of the fruit, i.e. the organ arising from the flower, or parts of a flower, or accessory structures or inflorescences, which encloses the seeds until they are ripe and then participates in their dispersal, whether they are merely scattered from the fruit or separated with it from the parent plant. Amongst primitive spermatophytes the functional dispersal unit (diaspore) in the domain of the flower is the naked seed, but in the angiosperms this function was initially hampered by the inclusion of the seed within the carpel. Although this sheltering of the ovules is comprehensible in connection with the biology of pollination (p. 651), it is not surprising that in primitive angiosperms the carpel opens again to expose the seeds, at least at maturity, so that the seeds retain their significance as active agents of dispersal (e.g. Fig. 677). However, the task of dispersal is progressively taken over by fruits formed, firstly, from a single carpel (Figs. 678B, 679D, 694H), then from groups of free, choricarpous carpels as aggregate fruits (with several or many partial fruits or carpidia, e.g. Fig. 694A–G), and finally from coenocarpous gynaecia, with united carpels (pp. 672 et seq.), forming various types of fruit (e.g. Figs. 678C–G, 679A–C, E); at the same time the role of the seed becomes more passive. Originally the ovary alone determined the structure of the fruit, but in advanced cases various foliar and axial organs from the floral and extra-floral regions augment the formation of the fruit. Ultimately entire infructescences (e.g. Fig. 671H) or even whole plants may function as units of dispersal.

The growth of seeds from ovules is closely linked to the increase in size of the ovary and its development as a container for the seeds. Petals, stamens, styles and stigmas usually dry up and fall off as the fruit develops. For further physiological changes, and the occurrence of seedless parthenocarpic fruits (e.g. the cultivated banana and citrus fruits), see p. 313.

The increasingly complex series of units functioning in dispersal is comparable with the series functioning in pollination: sporophyll–(meranthium)–euanthium–pseudanthium (pp. 601, 606, 661 et seq.). Examples are given on p. 669 and pp. 671 et seq. Here we shall briefly indicate the diversity of supplementary structures involved in the fruit: envelopment in axial tissue in a superior ovary (e.g. *Nuphar*, Fig. 683C); fleshy receptacles (e.g. Rosaceae, Fig. 694), or parts of axes or perianths in many inferior fruits (e.g. Apiaceae, Fig. 707E–H; *Iris*, Fig. 678E; *Arctostaphylos*, Fig. 715H, I; comparison with the corresponding superior fruits, e.g. of *Tulipa* or *Atropa* (Fig. 726A) shows that there is no functional distinction between superior and inferior fruits); calyces (e.g. the red 'lanterns' in the winter cherry, *Physalis alkekengi*; pappus in Valerianaceae, Fig. 722D, and composites, Fig. 729N, O); perigone (e.g. fleshy in the infructescence of *Morus*, Fig. 690D, hairy in *Eriophorum*, Fig. 737F, G); prophylls and bracts (e.g. winged in *Carpinus*, Fig. 689F, G, or *Humulus*, Fig. 690H; tubular in *Carex*, Fig. 737M, N; epicalyx in Dipsacaceae, Figs 365, 722F); fruit-stalk (e.g. fleshy in *Anacardium occidentale*); or axial or foliar organs of the infructescence (e.g. cupules of the Fagaceae, Fig. 687A–C, G, O; the fleshy parts of the fig, Fig. 671I, and *Ananas*).

The ovary wall also undergoes modifications as it becomes a fruit wall or pericarp. This is usually differentiated into an exocarp (outer) and endocarp (inner wall), both often only one layer thick, and a multilayered mesocarp between them. In dry fruits (e.g. Figs. 678–9) all the layers of the pericarp are more or less dry and consist of dead cells. There are also various fleshy fruits in which the exocarp, or especially the mesocarp, e.g. in stone fruits with a lignified endocarp (sclerocarpous), is fleshy at maturity; or the whole pericarp becomes fleshy when ripe, the cells remaining alive (as in berries; sarcocarpous); cf. *Olea*, Fig. 723D, or *Saxifraga* and *Ribes*, Fig. 692D, G.

In some fruits the endocarp grows inwards as a fleshy tissue between the seeds (pulpa; e.g. in *Ceratonia*, citrus fruits, p. 700, and the banana). As mentioned, accessory organs of the fruit may also become fleshy.

Dehiscence or opening of the fruit is extremely diverse. Whilst primitive fruits open when ripe, generally by the action of turgor or hygroscopic forces (pp. 376–9; dehiscent fruits; Fig. 678), indehiscent fruits (Fig. 679) remain closed around the seed since these mechanisms are suppressed. Indehiscent fruits may eventually break into portions (fissile fruits), either into partial fruits or mericarps along the sutures of the carpels (schizocarps, Fig. 679C), or across the carpels (as in a lomentum, Fig. 679D). Dry indehiscent fruits which fall as a whole are known as nuts (Fig. 679C), whilst fleshy fruits are often stone-fruits or berries. These progressions are particularly evident when various forms of fruit occur together within a circle of affinity (e.g. within the Ranunculaceae, Figs. 678A, D, 684; Rosaceae, Fig. 694; Brassicaceae, Figs. 679E, 710; or Fabaceae, Figs. 678B, 679D).

Dehiscent fruits open to a varying extent. The most primitive case is probably when the carpels open along their ventral suture (ventricidal, e.g. Fig. 678A) or along the line of union between neighbouring carpels (septicidal, e.g. Fig. 678D; schizocarps divide similarly, Fig. 679C). The formation of a line of dehiscence along the dorsal side of the carpel is more advanced (dorsicidal or loculicidal, Fig. 678C; E, combined with septifragal; poricidal is a special case, e.g. Fig. 678G); so is breakage along the united margins of the carpels (septifragal, e.g. Fig. 678E, combined with dorsicidal). Transverse dehiscence across the entire fruit leads to the circumscissile pyxis (e.g. Fig. 678F) and to the lomentum (e.g. Fig. 679D, E).

Very often the progression from dehiscent to indehiscent fruits is correlated with a reduction in the number of seeds: whilst berries frequently contain many seeds (but stone fruits only one as a rule), partial fruits or the portions of the lomentum and nuts generally contain a single seed, even when the ovule contains several ovaries (cf. Figs. 678, 688, 713B).

The structure of fruits (and seeds) is very closely related to their mode of dispersal and cannot be fully understood without considering the functional ecological relationships. As in pollination, the chief agents are animals (zoochory), wind (anemochory), water (hydrochory) and Man (anthropochory). In addition active dispersal by the plant itself occasionally occurs (autochory, p. 675).

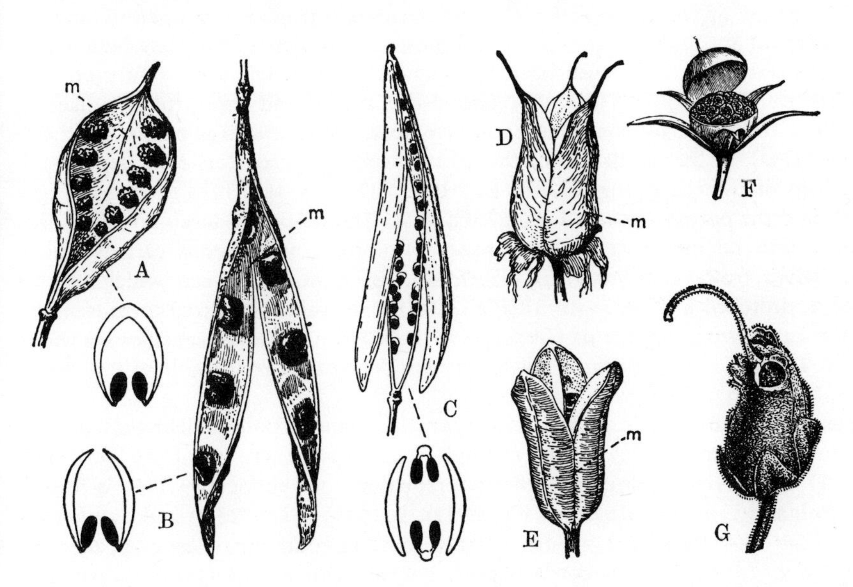

Fig. 678. Dry dehiscent fruits. From a single carpel (unicarpellary): **A.** Follicle (*Consolida regalis*; ×4 approx.); **B.** Legume (*Laburnum anagyroides*; ×1). Coenocarpous fruits: **C.** Siliqua (*Chelidonium majus*; nat. size approx.); **D.** Septicidal capsule (*Hypericum perforatum*; ×3); **E.** Dorsicidal capsule (*Iris sibirica*; ×3); **F.** Pyxis (*Anagallis arvensis*; ×2); **G.** Poricidal capsule (*Antirrhinum majus*; ×¾). m, median (dorsal) line of the carpel. **A.** After Beck-Mannagetta. **B.**, **D.** and **E.** After Firbas. **C.** After Wettstein. **F.** and **G.** After Schimper.

A 'natural' classification of fruits taking all considerations into account is not possible since these organs, which evolved relatively late, are too plastic; as we have seen the many possible methods of division overlap too much. The following is an attempt to take into account both the morphological–anatomical facts and an ecological classification based on the chief means of dispersal (pp. 673 et seq.).

The phylogenetic development of the angiosperm fruit begins, corresponding to the primitive choricarpous structure of the gynaecium, with **(A) choricarpous** fruits. Initially in these a number of adjacent carpels separately function as individual dispersal units, or they are reduced to one per flower; in both cases we have **(I) unicarpellary** fruits, which in primitive cases open ventricidally as (1) follicles (as in many Magnoliidae and the more primitive Dialypetalae, in which several occur in one flower, as in *Paeonia* or *Delphinium*, or reduced to one, as in *Consolida*, Fig. 678A). (2) The legume or pod has both ventral and dorsal dehiscence (several per flower in *Magnolia*, one in the Leguminosae, Fig. 678B). The lomentum, breaking transversely, is a further development of the legume (Fig.

679D). (3) Unicarpellary berries occur in some Annonaceae, in *Actaea* (Ranunculaceae) and the date (with a hard endosperm, Fig. 740E). (4) Unicarpellary drupes occur for example amongst the stone fruits such as the cherry (with lignified endocarp, Fig. 694H). (5) Unicarpellary nuts or nutlets are widespread, e.g. in *Anemone, Ranunculus* Fig. 684B, C) and *Alisma* (Fig. 732B); some of these have functional appendages which may be feathery (e.g. *Clematis, Pulsatilla*) or formed from the recurved style (e.g. species of *Geum*).

Unicarpellary fruits show many transitions to **(II) aggregate fruits** in which more or less numerous choricarpous carpels (each of which forms a carpidium) form a dispersal unit joined together by axial tissue or as a result of post-genital fusion. Thus (1) aggregate follicles occur in *Trollius* (Ranunculaceae) and *Spiraea* (Fig. 694A); (2) aggregate nutlets in *Fragaria* where the reduced carpidia are borne on a fleshy receptacle (Fig. 694C). Blackberries and raspberries (Fig. 694E) can be regarded as (3) aggregate drupelets (drupels); the axis participates in the former but not in the latter. An approach to the coenocarpous condition is seen in the (4) pome or core-fruit in which choricarpous carpels are completely immersed in a fleshy axial cup, as in the medlar with stony carpel walls or the apple with coriaceous (leathery) ones (Fig. 694F, G).

(B) Coenocarpous fruits are derived from choricarpous or hemisyncarpous forms with a cyclic arrangement of the carpels (Fig. 664). **(I) Dehiscent** fruits which open to liberate the seeds (capsules) are relatively primitive and must be considered first. Here belong the great majority of the (1) dry capsules of the angiosperms, which can be further subdivided by the number of loculi (many or one in syncarpous and paracarpous gynaecia, e.g. Figs. 678D, 709C, D), by complete or merely apical dehiscence (as in the 'toothed' capsule, Fig. 717D), or by the nature of the dehiscence (septicidal, dorsicidal, septifragal or combinations of these, e.g. Figs. 678D, E, 704N, 727F, 733H, 735E; the pyxis, Figs. 678F, 726D, and the porose capsule, Figs. 678G, 685C, D, are special developments). The siliqua should also be mentioned: this consists of united, paracarpous carpels which separate as valves from their marginal placentas (Fig. 678C), between which in the Brassicaceae is stretched a 'false septum' (Fig. 710F). A further specialized form is represented by the catapulting capsule of *Geranium* (Fig. 702C). (2) Fleshy capsules are widespread in the Tropics; a native example is *Euonymus*. The explosive capsule of *Impatiens* (Fig. 320, p. 376) also belongs here.

From dehiscent fruits the following five types of coenocarpous indehiscent groups have developed, often in parallel. So, for example, in the two subgroups of the **(II) fleshy** fruits: the (1) coenocarpous drupes, with a sclerenchymatous endocarp which is forced open on germination, include *Juglans* (Fig. 691D, E), *Olea* (Fig. 723D) and *Sambucus*. *Cocos*, a tropical coastal plant, is peculiar in its fibrous, air-containing mesocarp, adapted for floating (Fig. 740C, D); (2) coenocarpous berries with a wholly fleshy pericarp include *Ribes* (Fig. 692G), *Vitis*, *Arctostaphylos* (Fig. 715H, I), *Atropa* (Fig. 726A) and *Convallaria*. Citrus fruits have a fleshy pulpa (p. 700). Gourds, cucumbers, etc. (Cucurbitaceae) have a berry with a hard outer shell (pepo).

The subgroup of coenocarpous **(III) fissile** fruits show quite different lines of development. In the (1) schizocarp the partial fruits or mericarps separate septicidally; these may be several (e.g. in *Malva*, Fig. 713L) or only two (e.g. *Acer*, Fig. 679C); sometimes a central carpophore remains as in the Apiaceae (Fig. 707G). (2) Carpels may also break up transversely or longitudinally in coenocarpous fruits. The lomentose siliqua of some Brassicaceae, which breaks transversely (Fig. 679E), and the fruits of Boraginaceae and Lamiaceae, in which the bicarpellary ovary breaks into four cells or 'nutlets' by both true and false septa (Fig. 725) belong here.

The large group of **(IV) coenocarpous nuts** is also advanced; a nut falls as a unit by means of an abscission layer. Here belong, for example, the winged nuts of *Betula* (Fig. 688N), *Ulmus* (Fig. 690A) and *Fraxinus* (Fig. 723F), the nuts surrounded by a cupule in the Fagaceae (Fig. 687), the nuts with modified bracts and prophylls in *Carpinus* (Fig. 689F, G) and *Humulus* (Fig. 690H), and the nuts with such diverse means of dispersal in the Dipsacaceae which have an epicalyx (Fig. 365, p. 422). The fruit wall and testa are united in the particular cases of the grasses (caryopsis, superior, often enclosed in bracts, Fig. 739) and the composites (achene, inferior, calyx often modified as a pappus, Fig. 729N–Q).

(C) Infructescences are highly advanced dispersal organs at the ends of various lines of development. Examples are the mulberry (Fig. 690D), fig (Fig. 671H) and pineapple with progressively more fleshy perianth segments, floral axes, etc. Multiple or collective fruits also include *Tilia*, with an infructescence of nuts plus a winged prophyll (Fig. 713B, C), and the burdock (*Arctium*) in which the composite capitulum is furnished with hooked bracts (phyllaries) and is dispersed as a unit (Fig. 729G). Finally, entire dried-up plants may function as ball-shaped 'tumbleweeds' or 'steppe witches'; they break off at the base and are rolled about in the wind as diaspores, gradually scattering their fruits (e.g. *Salsola kali*, Chenopodiaceae, and *Eryngium campestre*, Apiaceae).

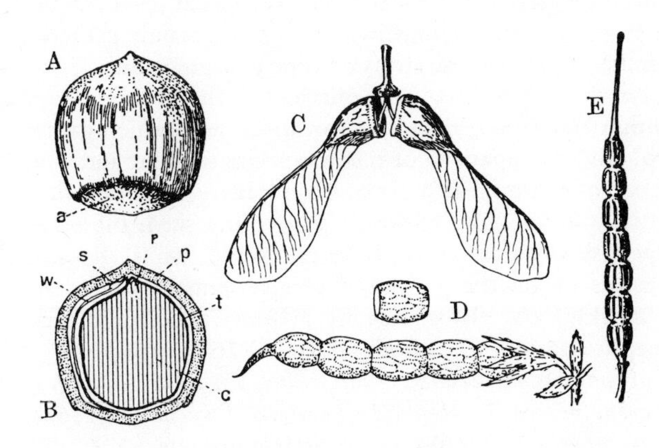

Fig. 679. Dry indehiscent fruits. **A.** and **B.** Nut of *Corylus avellana*; general view and L.S.; a, abscission scar; p, fruit wall; s, abortive ovule; w, vascular bundle to the seed; t, testa; c, cotyledon; r, radicle. Fissile fruits: **C.** Schizocarp (*Acer pseudoplatanus*) with two one-seeded mericarps; **D.** Lomentum (*Ornithopus sativus*; a single carpel with one-seeded segments); **E.** Lomentose siliqua (bilomentum; *Raphanus raphanistrum*; coenocarpous with one-seeded segments). **A.** to **D.** Nat. size. **E.** $\times\frac{2}{3}$. After Firbas.

Dispersal of seeds and fruits. In the following survey of the means of dispersal of seeds and fruits it must be remembered that specialization in this area is often much less advanced than it is in pollination. Very many diaspores are in fact polychorous, i.e. they can be transported in more than one way. Some species are indeed polyspermous or polycarpous, i.e. the same individual produces seeds or fruits with different types of dispersal (e.g. achenes with and without pappus in the heads of various *Leontodon* species); this leads to greater plasticity of dispersal. The structure of diaspores is not always adapted to dispersal over a distance; some plants of specialized habitats have adaptations which prevent dispersal, e.g. anchoring or burying the diaspores as in some desert plants. There is much to be done in the field of seed and fruit biology in the way of exact and experimental study, particularly in the Tropics.

Animal dispersal (zoochory) occurs principally as endozoochory (the diaspores are eaten and then excreted), as myrmecochory (dispersal by ants; only appendages of the diaspores are eaten) and epizoochory (diaspores attached to the surface of animals).

Endozoochory requires the provision of means of attraction (foods such as carbohydrates, proteins, fats and oils, vitamins, organic acids and minerals), means of stimulation (e.g. colour and scent) and protection of the seed against mastication and digestion (sclerotesta, sclerocarp, etc.). Both fleshy and dry seeds or fruits may satisfy these conditions; whilst the former are generally eaten at once by animals, the latter are also suitable for stock-piling. The earliest animals to disperse fruits and seeds were apparently reptiles (saurochory; fossil records!); later followed several groups of birds (ornithochory) and mammals (e.g. primates, rodents, bats). As in the case of pollination, so with endozoochory there is frequently a close relationship between plant and animal which can also be understood as the result of a mutual selective 'improvement' and hence co-evolution.

Fleshy diaspores show characteristic syndromes according to the chief agents of dispersal. Whilst saurochory (for example by certain turtles and lizards) is insignificant at the present-day, ornithochory continues to play an important role: the diaspores generally have bright or contrasting colours (red, yellow, or black and shiny), no odour, moderate or small size, thin skins and in autumn remain attached (winter-persistent). Examples include fleshy seeds (*Magnolia*, *Paeonia*, etc.), unicarpellary fruits (*Prunus avium*, etc.), aggregate fruits (*Fragaria, Rosa, Rubus*, etc.), berries (*Ribes, Vitis, Vaccinium, Paris*), coenocarpous drupes (*Ligustrum, Olea, Sambucus*, etc.) and infructescences (*Morus*, etc.). Mammals are particularly important for endozoochory in the Tropics; since their mouth-parts and senses differ from those of birds the diaspores are generally not so brightly coloured, but have strong odours, are generally large with hard shells, and fall off (eaten on the ground). They include unicarpellary berries and drupes (*Phoenix, Prunus persica*, etc.), aggregate fruits (Rosaceae—Maloideae, etc.), berries with hard or armoured skins (avocado, cocoa, *Citrus*, Cucurbitaceae, persimmon, *Musa*, etc.), and infructescences (*Ficus, Artocarpus*, etc.). Bat-fruits are similar but hang exposed on the trunks or branches (e.g. Sapotaceae).

Dry diaspores include a smaller size-class of seeds and nuts dispersed generally by grain-eating birds, and larger ones (e.g. *Quercus, Fagus, Corylus, Juglans*) mainly gathered and hoarded by rodents (e.g. squirrels); some of these always escape consumption.

Myrmecochory depends on the development of appendages of the fruit or seed (elaiosomes) containing food or other substances attractive to various ants which gather and disperse these diaspores (cf. p. 669). Comparison of related species without and with myrmecochory shows how this mode of dispersal affects the entire plant: thus in *Primula elatior* the long peduncle bears stiffly erect capsules which scatter the seed, the calyx soon becomes dry, the seeds ripen slowly and there is no elaiosome, whilst *Primula vulgaris* has no peduncle, the capsules lie on the ground, the calyx remains green and photosynthetic, the seeds ripen quickly, and elaiosomes are present. Elaiosomes arise from various parts of the seed (e.g. in *Asarum*; *Chelidonium*, Fig. 677G; *Corydalis*, Fig. 677F; species of *Viola*; *Cyclamen purpurascens*; *Melampyrum*; *Allium ursinum*; *Galanthus nivalis*) or on nutlets (e.g. in *Anemone nemorosa, Hepatica, Lamium, Knautia*; Fig. 365). Myrmecochory is common in temperate forests but also occurs in tropical ones.

Epizoochory is effected by numerous arrangements leading to the attachment of diaspores to animals, and their dispersal. The seeds or fruits of many swamp and water plants are minute and are dispersed all over the world in mud attached to wild fowl; this is further facilitated in those diaspores which become sticky and mucilaginous when moist (e.g. *Plantago, Juncus*). Commonly diaspores are attached by glandular hairs (e.g. *Salvia glutinosa*) and especially by barbed hooks. Such attachment devices may arise as hairs or emergences on the carpels (e.g. in *Medicago*, Fabaceae; *Circaea*, Onagraceae, *Galium aparine*), from modified styles (e.g. *Geum urbanum*, Rosaceae), calyces or epicalyces (e.g. Figs. 365, 729P) or bracts (e.g. in the inflorescences of *Arctium*, Fig. 729G, or *Xanthium*, Asteraceae). Whilst these rather slightly built burrs are dispersed mainly in the fur of smaller animals, the particularly robust 'caltrops' become attached to the feet of the larger ungulates which disperse them (e.g. *Tribulus terrestris*, Zygophyllaceae, and many Pedaliaceae).

In special cases animals may also operate catapulting dispersal mechanisms; stiff, spreading stems are pushed back and then catapult the seeds or fruits on the rebound (e.g. various plants with capsules, Lamiaceae and Dipsacaceae, Fig. 365; to some extent, *Arctium*).

In recent geological times Man has become an extremely important agent of dispersal (anthropochory). Many weeds (cf. pp. 403, 772 et seq.) have escaped unintentionally throughout the world, in seed, in wool, and in animal fodder, whilst cultivated plants have been deliberately introduced. In many areas (e.g. in parts of New Zealand and California) anthropochores have overwhelmed the native flora. It is remarkable that the diaspores of arable weeds have become by selection so similar in size and behaviour to those of their respective crops that they can scarcely be separated mechanically (e.g. *Camelina*, Brassicaceae; *Rhinanthus*, Scrophulariaceae; *Bromus*, Poaceae).

Anemochory or **dispersal by wind** may be indirect when the diaspores are shaken out of containers borne on stiff but flexible stems, and the seeds are scattered from capsules (e.g. Figs. 685C, 716B, 735E) or the fruits form heads (e.g. *Bellis*); or direct when the diaspores are just blown away. This latter category includes the minute and dust-like seeds of the Orchidaceae (p. 742) or *Orobanche*; the bladder-like calyx of *Trifolium fragiferum*; hairy seeds (e.g. Figs. 677A–C, 711F, O); feathery styles, e.g. of *Clematis* or *Pulsatilla*, or awns (*Stipa pennata*); perigonial hairs (Fig. 737F, G); pappus hairs (*Ptero-*

cephalus; Fig. 365; or many composites, Fig. 729N, O); winged seeds (Fig. 677D), nuts (Figs. 688N, 723F), mericarps (Fig. 679C), inflorescences (Fig. 713C) or epicalyx (the parachute of *Scabiosa*, Fig. 365); and the 'tumbleweeds' or 'steppe witches' (p. 673).

Hydrochory or **dispersal by water** does occur when diaspores are washed about by rainwater (e.g. when the seeds are washed out of capsules which open only when wet (hydrochastic), as in *Sedum acre* and the Aizoaceae), but more generally when the diaspores float on bodies of water; in this case the seeds or fruits may be unwettable or develop air sacs (e.g. the seeds of *Nymphaea*, Fig. 677H, or the utricles of various species of *Carex*) or specialized flotation tissues (e.g. *Cocos*, Fig. 740C, D, p. 672, and several native swamp and water plants such as *Iris pseudacorus* and *Potamogeton*). The mechanical action of rain is indirect when it catapults the diaspores as in those plants which have fruits arranged like the blades of a turbine on an elastic stem: the impetus of the falling raindrops sets up vibrations so that the seeds are ejected from silicules (e.g. in *Iberis* and *Thlaspi*, both Brassicaceae) or the 'nutlets' from calyces (e.g. in *Prunella* and *Scutellaria*, both Lamiaceae).

Autochory or **self-dispersal**. Whilst many unspecialized diaspores simply fall to the ground (e.g. *Aesculus hippocastanum*), they are actively ejected by autochorous plants. The mechanisms either involve turgor (as in the explosive capsules of *Impatiens*, the squidging action of *Oxalis*, or the ejection of the seeds to over 12 m by the squirting cucumber, *Ecballium*) or hygroscopic movements (e.g. torsion in legumes, Fig. 678B, and *Dictamnus*; the catapulting capsules of *Geranium*; or the squidging action of various species of *Viola*). Finally, some plants either deposit their fruits by growth mechanisms in rock crevices (e.g. *Cymbalaria muralis*) or bury them in the soil (e.g. the peanut, *Arachis hypogaea*, and *Trifolium subterraneum*; drilling or auger fruits in *Erodium* and *Stipa*).

All these biological specializations of seed and fruit are intimately related to environmental factors. This is evident, for example, in our deciduous woodlands in which the low-growing herbs are predominantly myrmecochorous, the taller herbs epizoochorous, the shrubs endozoochorous, and the trees anemochorous, corresponding to the stratification of the principal agents of dispersal (ants, mammals, birds, wind).

Biological specialization of seed and fruit was and is of the greatest importance for the evolutionary development of the angiosperms. The example of the Dipsacaceae has already been considered in some detail (p. 422, Fig. 365).

For the angiosperms as a whole the relevant progressions are not always easy to trace since the preferred means of dispersal may well have changed several times. Nevertheless it appears certain that the endozoochorous fleshy seed is extremely primitive in the angiosperms, whilst anemochory, myrmecochory, epizoochory and eventually hydrochory and autochory, with partly or wholly dry seeds, are advanced. The function of fleshy tissues in the typically endozoochorous seed is taken over secondarily by choricarpous and coenocarpous fruits, by accessory organs, and finally by infructescences. The same is true of structural peculiarities linked with other means of dispersal.

Germination of the seed. The angiosperm is generally similar to the gymnosperm in this respect (p. 610). Epigeal germination is primitive; sometimes the more advanced hypogeal type occurs (p. 163) in which the large cotyledons modified as storage organs remain in the seed and only the epicotyl emerges from the soil (as in *Vicia faba*, *Pisum*, *Phaseolus coccineus*, Fig. 172, *Quercus*, *Juglans*, etc.). The behaviour of many monocotyledons is similar, and the single cotyledon is often modified as an absorbing organ (Fig. 731B–D) which remains in the seed and breaks down the endosperm.

Some seeds show specialized types of germination. Particularly noteworthy are the viviparous seeds of some mangroves (p. 792 et seq.), trees of tropical coastlines (especially Rhizophoraceae). The embryo of the one-seeded fruit germinates whilst still on the parent plant (Fig. 699A–C) and hangs down from the fruit with the radicle and a massive club-shaped hypocotyl which may be 1 m long. When it eventually falls it either anchors itself on the spot or floats away and roots when it reaches land.

Evolution and systematics. Some 226,000 species of living angiosperms are currently known; the total may well lie between 250,000 and 300,000. This huge number is contained in more than 10,000 genera and over 300 families. The subdivision Magnoliophytina is thus by far the largest group of plants. The angiosperms have a remarkable diversity of life-forms (pp. 178–201), they occupy practically every habitat of the biosphere and most terrestrial plant communities are dominated by them (cf. pp. 769–75, 782–93). No other group has anything like the same direct economic importance for man, or the countless numbers of useful and cultivated plants. Despite this, our knowledge of them has many gaps, their systematic arrangement is still controversial, even in broad outline, and their evolutionary origin remains enigmatic.

Fossils (mainly remains of leaves, wood and pollen) indicate clearly that the angiosperms were already represented in the Lower Cretaceous by a range of primitive to moderately advanced families (e.g. from the neighbourhood of the dicotyledonous Magnoliidae, Hamamelididae, Rosidae and Dilleniidae, but also from some monocotyledonous groups); some of these fossils can apparently be classified in still extant genera (e.g. *Magnolia* and *Cercidiphyllum*). Correspondingly the pollen flora of the Lower Cretaceous of monocolpate and gymnosperm-like grain types (comparable with modern Magnolianae) was gradually augmented by tricolpate types, but more advanced pollen types were still lacking. Only a few fossils possibly belonging to the angiosperms (mainly pollen) are known from the Jurassic (? and Triassic). Despite this, the large number of forms known from the Lower Cretaceous leaves no doubt that the angiosperms must go back to the Jurassic and probably to the Triassic (or even earlier). There are certain indications that the oldest angiosperms inhabited (sub)tropical montane regions where conditions unfavourable for fossilization would account for the paucity of early remains.

Quantitative analyses of various Lower Cretaceous floras show that the early angiosperms came from the south, at first in small numbers, to join plant communities of the northern hemisphere dominated by ferns, cycads, bennettites, ancestors of *Ginkgo* and conifers, mainly in montane localities where the angiosperms initially formed small populations. These conditions were apparently favourable for the rapid divergent evolution of the angiosperms (cf. isolation and drift, pp. 404, 407 et seq.). In the Middle Cretaceous they quickly became dominant, generally as extant genera, whilst the bennettites became extinct and the cycads and ginkgos in particular were largely suppressed: thus the 'angiosperm age' (Cainophytic) of geological history was inaugurated; this has lasted up to the present (cf. Fig. 628).

Although palaeobotany gives an impressive picture of the spread of the angiosperms in geological time, it tells us very little about the evolutionary origins of this group. We therefore depend on comparison of living angiosperms with each other and with living and fossil gymnosperms. The first question is whether the Magnoliophytina form a single, natural, evolutionary group (i.e. whether they are monophyletic), or whether they have arisen from diverse gymnospermous ancestors by convergence (i.e. whether they are polyphyletic; cf. p. 426). The following considerations must be taken into account: (1) All Magnoliophytina have in common numerous characters which are not of necessity correlated (e.g. sieve tubes and companion cells from a common initial cell, primitive hermaphrodite flowers with stamens below and carpels above, stamens with two groups of pollen sacs and an endothecium, sac-shaped carpels, ♂ gametophytes with three cells, ♀ gametophytes originally with eight cells comprising egg apparatus, polar nuclei, and antipodal cells, double fertilization, and triploid endosperm). (2) There is a great morphological discontinuity between the angiosperms and all other plants. (3) Within the Magnoliophytina there is nowhere any such unbridgeable discontinuity. These facts, and the improbability of all the similarities mentioned arising by chance convergent evolution, point to the view commonly held today that the Magnoliophytina have probably arisen from a common (if not necessarily uniform) ancestral gymnospermous group.

Which gymnospermous seed plants so far known could have formed this ancestral group? The facts that the Magnoliophytina have complex pinnate leaves, stamens with more than one group of pollen sacs, carpels with several ovules and other characteristics clearly indicate the Cycadophytina, whilst the Coniferophytina can certainly be excluded.

The superficial resemblance of the leaves of *Gnetum* to those of angiosperms, the pollen-bearing structures of *Ephedra*, etc., long ago led to speculation about a phylogenetic relationship between the Gnetatae and the Magnoliophytina. According to this

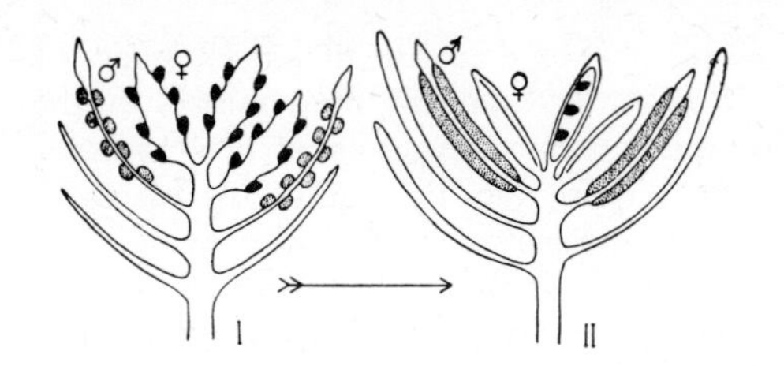

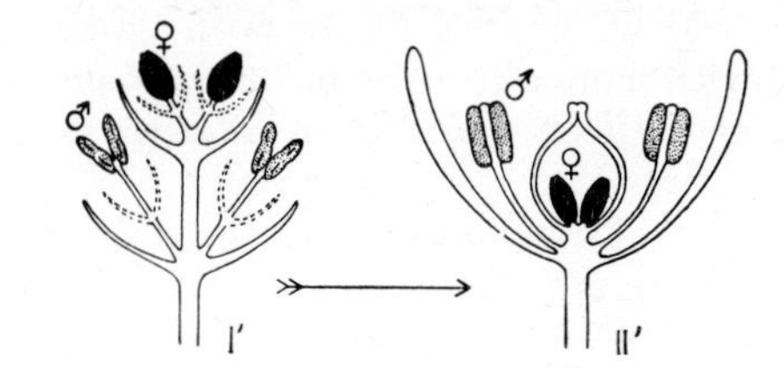

Fig. 680. Hypotheses of the origin of the hermaphrodite angiosperm flower: I, II, 'euanthium theory'; I', II', 'pseudanthium theory'. Pollen sacs dotted, ovules black. I, II, after Arber and Parkin; I', II', after Wettstein's derivation from *Ephedra*.

view the ♂ and ♀ flowers with their bracts of the Gnetatae were transformed into the perigone, stamens, carpels and ovules of the angiosperm flower in a manner reminiscent of the development of the pseudanthium in *Euphorbia* (p. 704). This 'pseudanthium theory' of the origin of the angiosperms (Fig. 680I′–II′) is now completely rejected because of fundamental differences in the Gnetatae (e.g. sieve tubes and companion cells so far as they are present from different initial cells, vessels with bordered pits, quite different ♀ prothalli), because of the difficulty of interpreting the angiosperm flower as an inflorescence, and because of the unlikelihood that a relict group so strongly reduced as the present-day Gnetatae could form the starting point of the apparently very ancient angiosperms.

The hermaphrodite flowers of the Bennettitatae are reminiscent of those of the angiosperms, but their quite different carpels prevent them from being considered as ancestral forms. A more favoured view was that the carpels of angiosperms correspond to 'inrolled' megasporophylls of the type in *Cycas* (Fig. 645B), and the perianth and stamens to the comparable organs of the bennettites (Fig. 647B), all these being united in a hermaphrodite flower. This 'primeval angiosperm' would have consisted, like the true flowers of its descendants, of euanthia. As an attempt to derive the Magnoliophytina directly from the Cycadatae this 'euanthium theory' (Fig. 680I, II) is scarcely considered nowadays because of various anatomical and morphological difficulties and the lack of palaeobotanical evidence. The same goes for attempts to relate them to the Caytoniales (in which the structures containing ovules correspond with leaf pinnae; p. 633) or to the Glossopteridales (the sporotrophophylls of which are insufficiently known; p. 632).

What remains is the widely accepted hypothesis that the Magnoliophytina have descended, not from any of the Cycadophytina groups so far named, but from pteridosperms (Lyginopteridatae) which were the common ancestors of the lot. Any resemblances to the Cycadatae, Bennettitatae and Gnetatae are therefore to be attributed to parallel evolution, to similar 'potential for differentiation' and to their common inheritance from the same pteridosperm ancestors. A 'euanthium theory' of angiosperm origin modified in this way can be based, above all, on certain resemblances between the radial structures bearing groups of pollen sacs in the pteridosperms and the stamens of primitive angiosperms, and between the hollow structures bearing ovules in the pteridosperms and the carpels of angiosperms (Fig. 663). Stamens and carpels have thus in the course of evolution become only to a certain extent leaf-like (and not vice versa). Although fossil intermediates between pteridosperms and angiosperms are quite unknown, there are no insuperable morphological differences between the two groups. The time sequence of the two groups also fits this hypothesis very well: pteridosperms from the (? Upper Devonian) Carboniferous to the Jurassic, the angiosperms from the (?? Triassic, ? Jurassic) Cretaceous to the present-day (cf. Fig. 628).

To which morphological and anatomical properties did the early angiosperms owe their historically documented superiority over their gymnospermous ancestors and relations? (1) Probably hermaphroditism and protection of the ovules in carpels as prerequisites for the replacement of anemophily by the more economically, more efficiently directed zoophily independent of wind (cf. pp. 657 et seq.). (2) Probably the great reduction and neoteny in the flowers and gametophytes as prerequisites for significantly accelerated reproduction (cf. pp. 606, 657, 668). (3) Probably the degree of vegetative plasticity attained by no other group of seed plants, coupled with the development of much more efficient conducting systems in the xylem and phloem, prerequisites for the more effective utilization of areas already occupied and colonization of new habitats. Comparative and experimental studies will be required in the future to test these essentially working hypotheses of the origin of the angiosperms.

How can we make a systematic arrangement of the Magnoliophytina that is reasonably 'natural' and therefore indicates, where possible, degrees of relationship (cf. pp. 425 et seq.)? This requires plants with relatively primitive characters to be placed at the beginning, those with the most advanced ones at the end. As a basis for comparison we shall therefore first try to reconstruct broadly in the light of modern knowledge a 'primitive combination of characters', based on the more important progressions already considered for the Magnoliophytina (pp. 640–75). This runs as follows. Dicotyledonous; small, pachycaulous, more or less sympodially branched (?), evergreen trees with main and lateral roots; leaves pinnate (or at least pinnately veined), spirally inserted, with three leaf traces and trilacunar nodes; eustelic with secondary thickening, secondary xylem and phloem poorly developed, with scalariform tracheids (but no vessels) and narrow sieve tubes (with no companion cells); flowers terminal and solitary, relatively large, hermaphrodite, protandrous; receptacle conical with numerous, free, spirally arranged, little differentiated perianth segments, stamens and carpels, all following the bracts (hypsophylls); outer perianth segments bract-like, the inner more or less coloured; stamens undifferentiated with a cylindrical stalk-zone and two more or less apical pollen sac groups (thecae), each with two pollen sacs, endothecium and secretory tapetum; pollen grains with a proximal (?) germination area or a distal furrow (anatreme and monocolpate), sticky with pollen cement; carpels sac-shaped, the ventral suture progenitally more or less united, the margins papillose and functioning as stigmas, styles absent, placentas more or less superficial or annular, with numerous orthotropous ovules, the latter with two integuments and crassinucellate; ♂ gametophyte with pollen tube cell and generative cell, the latter dividing into two sperm cells only in the pollen tube; ♀ gametophyte with monosporous embryo sac containing egg apparatus (egg cell and two synergidae), two polar nuclei and three antipodal cells; pollination zoophilous, by pollen-eating beetles attracted mainly by floral scent; fertilization porogamous and 'double', giving rise to the zygote, hence a small, straight embryo as well as the triploid, 'secondary' endosperm which arises by free nuclear division; seeds with perisperm in addition; carpels maturing as follicles which soon open, seeds coloured, fleshy with sarcotesta and sclerotesta, dispersed by vertebrates; germination epigeal.

This presumed primitive combination of characters cannot be found in its entirety in any one living member of the Magnoliophytina, but there can be no doubt that some of the members of the Magnoliidae closely approach it; this dicotyledonous subclass is therefore placed at the beginning of the angiosperms. The sequence of the two great classes generally recognized, Magnoliatae (= Dicotyledones) and Liliatae (= Monocotyledones), is similarly determined.

In the past three great groups of dicotyledons were generally recognized: (1) Apetalae (or Monochlamydeae) with no corolla, and either a single, generally inconspicuous perianth or none at all (this included mainly wind-pollinated families, e.g. Fagaceae, Betulaceae, Urticaceae, Euphorbiaceae, Salicaceae, Chenopodiaceae); (2) Dialypetalae with a double perianth, differentiated into calyx and corolla, but with the petals originally free and generally conspicuous (including the more primitive animal-pollinated families, e.g. Ranunculaceae, Rosaceae, Fabaceae, Apiaceae, Brassicaceae, Malvaceae); and (3) Sympetalae, also with a double perianth but with the petals united and conspicuous (including the more specialized animal-pollinated families, e.g. Cucurbitaceae, Ericaceae, Primulaceae, Lamiaceae, Rubiaceae, Asteraceae). The Apetalae and Dialypetalae were often united in the Choripetalae (with petals free, if present).

Although it has been obvious for a long time that the Apetalae, Dialypetalae and Sympetalae do not constitute natural groups, these 'taxa' (p. 433) were retained both by tradition and for want of a better and generally acceptable scheme of classification. Thanks largely to the efforts of A. Takhtajan and A. Cronquist, in recent years the basic principles of an improved classification taking account of genetic relationships have begun to emerge. This more natural system, which is slowly being accepted, forms the basis of the following account.

In this system, the primarily polyandrous and acyclic subclass Magnoliidae (Polycarpicae) is followed directly by a closely related circle of apparently ancient, woody Apetalae, the Hamamelididae (Amentiferae). Then follow three more advanced groups with originally two whorls of stamens, or secondary polyandry, and predominantly cyclic flowers: the Rosidae (Rosiflorae) with centripetal androecia, the Dilleniidae (Cistiflorae) with centrifugal androecia, and the Caryophyllidae (Centrospermae), also with centrifugal androecia but distinguished by their free central placentation. These three subclasses include a few apetalous families (in the Caryophyllidae) and some sympetalous ones ('Pentacyclicae'), especially in the Dilleniidae. The majority of the more advanced Sympetalae (the 'Tetracyclicae') are connected with the Rosidae but are included in a separate subclass, the Asteridae.

The reclassification of the monocotyledons, which are nowadays generally thought to be related to the dicotyledonous Magnoliidae, is less drastic. The primitive acyclic and choricarpous subclass Alismatidae (Helobiae) has to be placed at the beginning of the monocotyledons, followed by the cyclic and predominantly coenocarpous group (treated very broadly here) of the animal and wind-pollinated Liliidae (Liliiflorae), and finally, as a monocotyledonous counterpart of the Amentiferae, the Arecidae (Spadiciflorae) with inflorescences of inconspicuous flowers enclosed in a spathe.

Figure 742 (p. 753) gives a diagrammatic summary of the basic broad groupings of the angiosperms. It should be noted here and in what follows that within the confines of a textbook one can scarcely consider the many existing gaps in knowledge, uncertainties, and divergent opinions concerning the classification of this enormous group. The provisional nature of this taxonomic arrangement, both in broad outline and in detail, must always be kept in mind.

Class I: Magnoliatae (Dicotyledoneae), dicotyledons

The dicotyledons have with few exceptions two lateral seed leaves (cotyledons) in the embryo (Fig. 674A–G). The main root is long-lived in primitive examples (allorhizy; cf. Fig. 191; pp. 177, 641). In a transverse section of the stem the vascular bundles normally appear in a ring (eustele, Fig. 150) and are 'open', that is they can undergo secondary growth in thickness by means of a cambium (pp. 147 et seq., 600). The leaves vary greatly in shape, but usually have a distinct petiole and reticulate venation (Fig. 176); they are often compound; stipules are frequent, leaf sheaths rare (p. 168). The first appendages on axillary shoots are two transverse prophylls (cf. Fig. 688A; p. 131). Flowers with pentamerous or (more rarely) tetramerous whorls including a calyx and corolla predominate (thus K5 C5 A5 + 5 G5 or K4 C4 A4 + 4 G4); but dimery and trimery sometimes occur, or the number of whorls is reduced, or the floral members are borne in spirals. The pollen grains generally develop simultaneously and are often tricolpate (p. 651). The tree form is primitive but generally distributed.

Exceptions to the above characters are important in indicating relationships to the monocotyledons; they are considered on pp. 735 et seq.

Subclass A: Magnoliidae (Polycarpicae)

The great taxonomic interest of the Magnoliidae lies in their many primitive characters (cf. p. 678) and their great diversity, ranging from simple to advanced types which show relationships with the other groups of angiosperms. Thus the Magnoliidae form as it were the 'substructure' or foundations both of the dicotyledons and the monocotyledons (cf. above and Fig. 742). Indeed the great plasticity of this basic group makes it difficult to characterize.

The predominantly choricarpous gynaecium with numerous free carpels (hence Polycarpicae) is highly characteristic, as is the frequently spiral (acyclic) arrangement of the floral parts and their often large and indefinite numbers (polymery). In this respect the widely occurring primary polyandry is particularly important. The perianth is generally robust and often shows no distinction between calyx and corolla.

Life forms include woody plants (a few without vessels), various herbaceous types and even complete parasites. The flowers are mainly hermaphrodite but occasionally unisexual (e.g. in the dioecious bay tree, *Laurus*). The receptacle is often elongate-conical, bearing numerous spirally arranged floral organs (e.g. *Magnolia*, Fig. 652A), but oligomerization of the perianth, carpels and finally even of stamens is often seen and is associated with the development of whorls amongst which dimery and trimery are particularly noteworthy. The members of any of the floral regions may show union. The perianth may even be lacking (e.g. Piperales). The stamens often show no division into filament and anther, and the generally multiovulate carpels may lack a style (Fig. 663). The pollen grains are often still anatreme-monocolpate and have only two cells when the anthers open. The ovules are crassinucellate throughout and have two integuments. As well as pollen flowers visited by beetles there are also many nectar flowers of very diverse pollination types; rarely wind pollination occurs. Unicarpellary fruits, e.g. follicles, predominate. The seeds often still have sarcotesta and sclerotesta; the endosperm is abundant and the embryo is often small. Phytochemically characteristic of the Magnoliidae are alkaloids of the phenylalanine group (particularly benzyl isoquinoline bases, e.g. aporphine).

Woody plants predominate in the first three orders. Their secretory cells contain ethereal oils; the pollen grains are monocolpate. They can be included in the super-order **Magnolianae.**

Notable for many primitive characters is:

Order 1: **Magnoliales.** This includes the evergreen **Winteraceae**, with a disjunct distribution in the Southern Hemisphere, no vessels and with clearly very primitive stamens and carpels (Fig. 663), and the evergreen or deciduous **Magnoliaceae** (Figs. 652A, 745) in the (sub)tropical and warm temperate regions of the Northern Hemisphere, but with a relict distribution. These are woody plants with simple leaves and large flowers, including the familiar genus *Magnolia*, native in S. and E. Asia and N. America, and the N. American tulip tree, *Liriodendron tulipifera*, which is also frequently seen in gardens. Widespread in the Tropics are the **Annonaceae**, with perianth segments frequently in whorls of three, and the **Myristicaceae** with only a single seed in the carpel; the endosperm is ruminate in both families, as seen in the nutmeg (*Myristica fragrans*), a native of the Moluccas (Fig. 681B–F).

Whilst the preceding families have several lacunae at the nodes there is only one in the following. In the **Austrobaileyaceae** the genus *Austrobaileya* is particularly primitive, being the only angiosperm with no companion cells in the phloem (cf. p. 641). The leaves are evergreen, leathery and simple in the tropical-disjunct **Monimiaceae** and the **Lauraceae**, which extends to the Mediterranean region where the best known representative is the bay tree or laurel, *Laurus nobilis*. These, too, have flowers in which most of the whorls are trimerous; the stamens open by valves (Fig. 681A); the ovary is pseudomonomerous and becomes a berry or drupe. Important economic plants are found in the S. and E. Asiatic species of *Cinnamomum*, e.g. the camphor tree (*C. camphora*) from the wood of which camphor is obtained by sublimation, and the cinnamon trees (*C. zeylanicum* in Ceylon and *C. cassia* in S. China), the bark of which contains oil cells and yields cinnamon.

The **Sarcandraceae**, which also lack vessels, connect the Magnoliales to the apetalous:

Order 2: **Piperales.** The **Piperaceae** include tropical woody plants and herbs; the unisexual or hermaphrodite flowers lack a perianth and are borne in spicate or club-shaped axillary inflorescences. The fruit of *Piper*, the most important genus, is a drupe (Fig. 682). An orthotropous ovule gives rise to a seed which contains a massive perisperm as well as the endosperm. *Piper nigrum* is a Malaysian root-climber. Immature dried fruits yield black pepper, ripe ones with the outer parts removed, white pepper.

A very advanced herbaceous line of development from the Magnolianae with generally trimerous but syntepalous flowers and an inferior ovary is:

Order 3: **Aristolochiales** with the single family **Aristolochiaceae**. European representa-

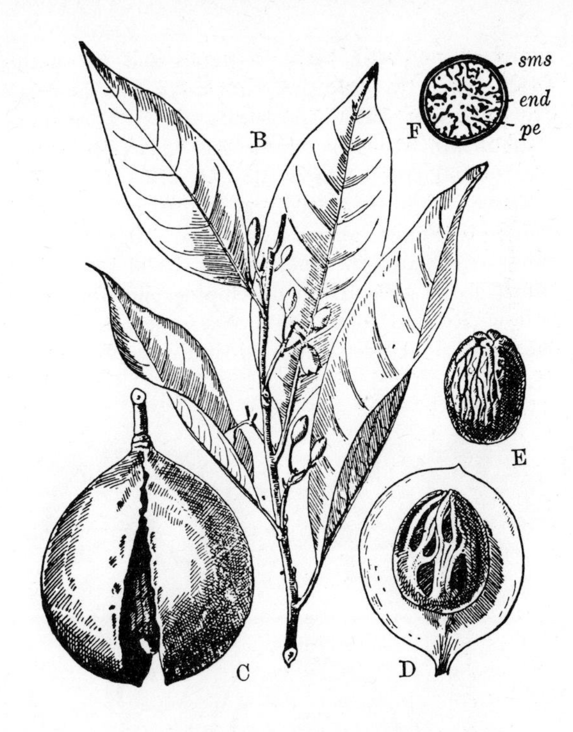

Fig. 681. Magnoliales. **A.** *Cinnamomum zeylanicum.* Perigynous flower cut lengthways, with pseudomonomerous ovary and anthers opening by valves. ×5 approx. **B.** to **F.** *Myristica fragrans*: shoot with ♂ flowers (**B.**), fleshy but dehiscent unicarpellary fruit from outside (**C.**) and opened up (**D.**); a red aril (mace, a spice and drug) surrounds the dark brown seed (**E.**); the seed in T.S. (**F.**) with testa (sms) and ruminate endosperm (end) permeated by branching ingrowths of the nucellus (pe). ×½ approx. **A.** After Baillon. **B.** to **F.** After Karsten.

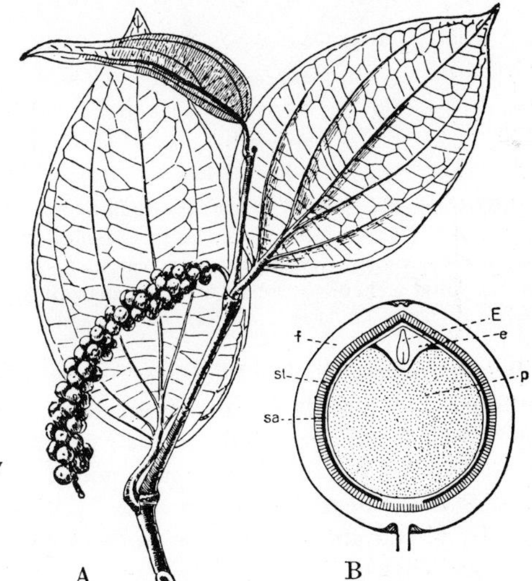

Fig. 682. Piperales. *Piper nigrum.* **A.** Shoot with inflorescence. **B.** Drupe in L.S. with fleshy mesocarp (f), woody endocarp (st), testa (sa), embryo (E), secondary endosperm (e) and perisperm (p). **A.** $\times\frac{1}{3}$. **B.** ×5. **A.** After Karsten. **B.** After Baillon.

tives include asarabacca (*Asarum europaeum*) and birthwort (*Aristolochia clematitis*, with dorsiventral slippery-trap flowers). Very uncertain are the relationships of:

Order 4: **Rafflesiales** (**Rafflesiaceae** and **Hydnoraceae**) which comprises (sub)tropical total parasites (e.g. *Pilostyles*, Fig. 217; *Cytinus* on *Cistus* in the Mediterranean region).

A special place within the Magnoliidae (? a separate superorder) is occupied by:

Order 5: **Nymphaeales** which goes back at least to the Cretaceous. The members are herbaceous aquatics rooted in the mud in shallow water; they have no oil cells, the pollen grains are mostly mono- or tricolpate and placentation is superficial; they show many features of ancestral monocotyledons (cf. pp. 735 et seq.).

The **Nymphaeaceae** have floating leaves which in the celebrated *Victoria amazonica* (*V. regia*) of the Amazon basin reach a diameter of 2 m. The pollen grains are more or less monocolpate. The perianth usually consists of sepals and petals which may be in trimerous whorls (e.g. *Cabomba*, Fig. 731F) or in spirals (as in the native yellow waterlily, *Nuphar luteum*, Fig. 683B). In the white waterlily (*Nymphaea alba*, Fig. 683D) there is a complete gradation between petals and the numerous, spirally inserted stamens (Fig. 655E–L); in this case the green calyx was the original perianth. In *Nuphar*, however, the calyx is bright yellow and the corolla is represented only by inconspicuous honey-leaves (Fig. 683B). Whilst *Cabomba* (Fig. 731F) has choricarpous carpels, in *Nuphar*, *Nymphaea* (Fig. 683D) and *Victoria* they are enclosed in a covering of receptacular tissue which separates from the carpels at maturity in *Nuphar* (Fig. 683C: false coenocarpy). In *Nelumbo* (**Nelumbonaceae**; leaves peltate, raised on long petioles above the water; pollen grains tricolpate) the apex of the floral axis forms an inverted cone, the upper part of which grows around the developing choricarpous carpels so that each seed is sunk in a hollow. Related to the Nymphaeaceae are the rootless submerged aquatic **Ceratophyllaceae** which include the hornwort, *Ceratophyllum*.

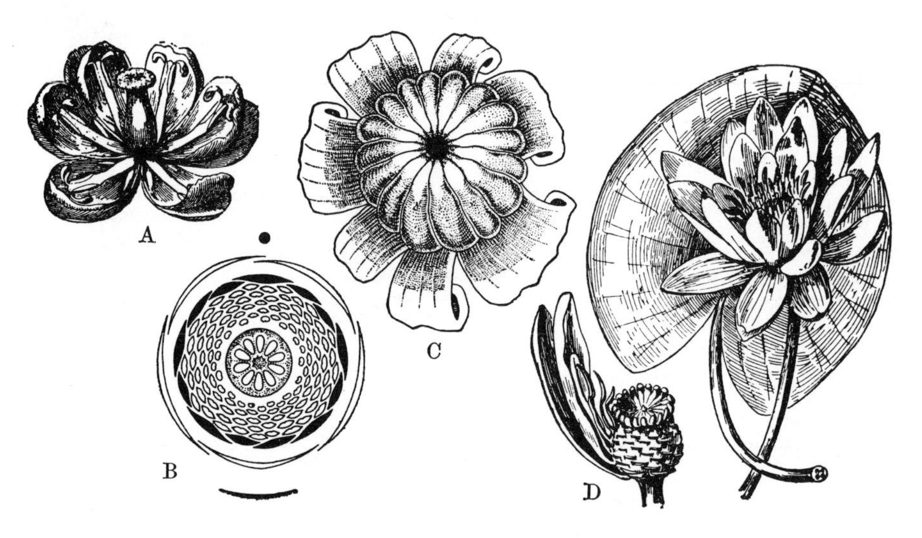

Fig. 683. Nymphaeales (**B.** to **D.**) and Ranunculales, Berberidaceae (**A.**). **A.** *Berberis vulgaris*, flower. ×3. **B.** to **C.** *Nuphar luteum.* **B.** Floral diagram (honey leaves, black; axial tissue, dotted). **C.** Fruit (the axial tissue separates from the free carpels). **D.** *Nymphaea alba.* Floating leaf, flower and ovary with the spiral scars of the petals and stamens which have been removed. ×$\frac{1}{3}$. **A.** After Baillon. **B.** After Eichler. **C.** After Troll. **D.** After Karsten.

Woody intermediates (**Illiciaceae**) with tricolpate or hexacolpate pollen connect the remaining two orders (in the super-order **Ranunculanae**) with the Magnoliales, from which they differ principally in their generally herbaceous habit, frequently compound leaves, lack of oil cells, tricolpate pollen (or related derived forms) and isoquinoline alkaloids of the phenylalanine group.

Order 6: **Ranunculales.** These show essentially the same primitive floral characters as the Magnoliales.

The **Ranunculaceae** are the most important family (cf. the serological relationships, Fig. 371, and the references to floral biology, pp. 411, 664). They are predominantly herbaceous with alternate exstipulate, often dissected leaves (Figs. 182A, 184) and brightly coloured hermaphrodite flowers which have numerous stamens and a choricarpous gynaecium of several or many (rarely one) carpels borne on a convex receptacle (Fig. 684B, G, R–U). The carpels either bear numerous ovules on either side of the ventral suture and develop as many-seeded follicles, or bear a single ovule on the transverse zone and develop into one-seeded indehiscent fruits, especially nutlets (Fig. 684N–Q). Otherwise the flowers show great diversity of form. Although originally radial (actinomorphic), the flowers are sometimes dorsiventral (zygomorphic) as in monkshood (*Aconitum*) and larkspur (*Delphinium, Consolida*); the floral parts are often spirally inserted and numerous, but they may occur in bi-, tri- or pentamerous whorls (Fig. 684R–U). The perianth is sometimes only a simple perigone, e.g. in marsh marigold or kingcup (*Caltha*), windflower or wood anemone (*Anemone nemorosa*) and pasque flower (*Pulsatilla*, Fig. 359), sometimes consisting of calyx and corolla as in the buttercup or

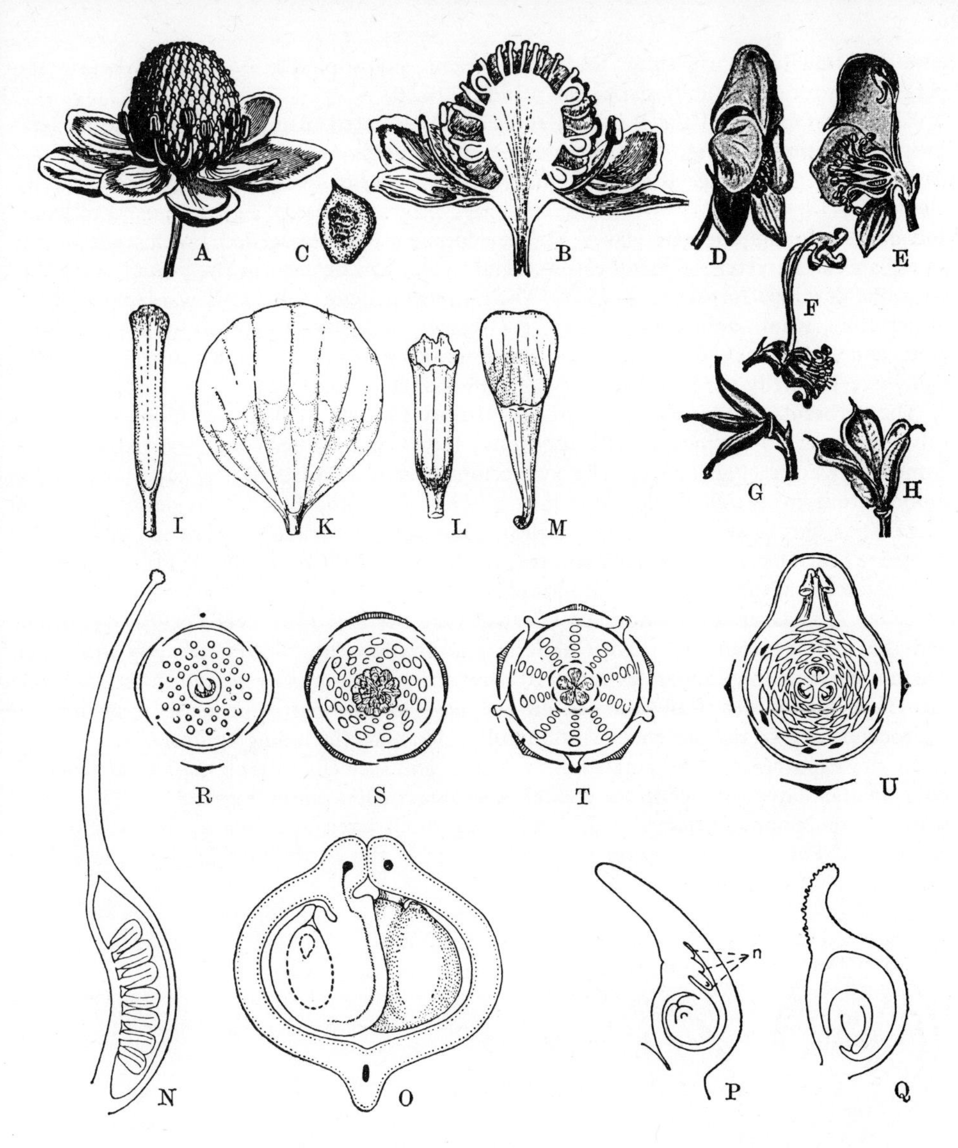

Fig. 684. Ranunculales, Ranunculaceae. **A.** to **C.** *Ranunculus sceleratus.* **A.** Whole flower. **B.** Half flower. **C.** Unicarpellary nutlet. ×4 approx. **D.** to **H.** *Aconitum napellus.* **D.** Oblique anterior view. **E.** Half flower. **F.** After removing the perigone, exposing the two nectaries. **G.** Young and **H.** mature choricarpous gynaecium. $\times\frac{3}{5}$. **I.** to **M.** Honey-leaves. **I.** *Trollius giganteus.* ×2.5. **K.** *Ranunculus auricomus.* ×3. **L.** *Helleborus foetidus.* ×4.5. **M.** *Aquilegia vulgaris.* ×1. **N.** to **Q.** Carpels. **N.** L.S. of *Helleborus orientalis.* ×5. **O.** In T.S. ×18. **P.** *Anemone nemorosa*, in L.S., with abortive ovules, n. ×10. **Q.** *Ranunculus auricomus*, in L.S. ×10. **R.** to **U.** Floral diagrams (nectaries or petals black). **R.** *Cimicifuga racemosa.* **S.** *Adonis aestivalis.* **T.** *Aquilegia vulgaris.* **U.** *Aconitum napellus.* **A.** to **C.** After Baillon. **D.** to **H.** After Karsten. **I.** to **M.**, **N.**, **O.** and **Q.** After Firbas. **P.** After Rassner. **R.** to **U.** After Eichler.

crowfoot (*Ranunculus*, Figs. 182A, 211E, 684A, B) or in pheasant-eye (*Adonis*, Fig. 684S). The perianth segments may show transitions to leaves and bracts (hypsophylls), as in the globe flower (*Trollius*), *Helleborus* (Fig. 655A–D) and to some extent in the winter aconite (*Eranthis*), or to stamens as sometimes in *Hepatica nobilis*. The stamens are often developed as honey-leaves which variously bear the nectar in a cavity or a spur-like outgrowth; sometimes they are inconspicuous, as in *Trollius* or *Helleborus*, sometimes petaloid, as in *Ranunculus* or the columbine (*Aquilegia*; Figs. 369, 684I–M). Thus by comparing various genera one can visualize the development of the perianth from bracts, on the one hand, forming a calyx or perigone, and from stamens or honey-leaves on the other, forming a corolla or in some cases a perigone (cf. p. 645). A calyx-like envelope (involucre) of three undivided bracts (hypsophylls) occurs just below the perigone in, for example, *Hepatica* (cf. also Fig. 182B).

The many species of the Ranunculaceae are distributed mainly in the northern extra-Tropical regions (cf. Fig. 356). Besides perennial herbs (e.g. Figs. 210A, B, 211E–G), annuals also occur (e.g. *Myosurus minimus*, *Ranunculus arvensis*), more rarely woody plants, e.g. *Clematis* (*C. vitalba*, travellers' joy, old man's beard), a genus mostly of lianes and notable for its opposite leaves. The development of a one-seeded, indehiscent fruit is an advanced character; in many cases several ovules can be seen in the young ovary but only one develops further (Fig. 684P). The nutlets ('achenes') often show adaptations for disperal by wind, animals or water, e.g. hairy, elongated styles (*Pulsatilla*), hooked emergences (*Ranunculus arvensis*) and membranous wings or flotation tissue (Fig. 684C). Quite exceptionally the fruit is a berry as in the baneberry (*Actaea*).

The **Berberidaceae** are closely related: herbaceous or woody plants with cyclic flowers: calyx, corolla, sometimes corolla-like honey-leaves and stamens are arranged in di- or trimerous whorls (Fig. 683A). The gynaecium usually consists of one superior pseudo-monomerous ovary which develops into a berry. Only the barberry is native (*Berberis vulgaris*), a shrub well known for its leaf-spines and its irritable filaments, and as the alternate host of the black rust of wheat (pp. 187, 367, 523; Figs. 199, 311).

Much more advanced in floral structure is:

Order 7: **Papaverales.** Predominant features are the dimerous, cyclic calyx and corolla and the coenocarpous ovary with parietal placentation (Fig. 685). Formerly this order was united with the Capparales (pp. 710), but it is now generally placed in the Magnoliidae near the related Ranunculales on the basis of the partly primarily polyandrous androecium and of the presence of isoquinoline alkaloids (including opium).

In the **Papaveraceae** the petals have no spur, and sac-cells (idioblasts) and articulated ducts contain latex (orange in the greater celandine, *Chelidonium*, Figs. 285, 677G, 678C; white in the poppy, *Papaver*, Fig. 211J). The dried latex obtained by incision of the capsules of *Papaver somniferum* (Fig. 685C) yields the alkaloid-containing opium. The

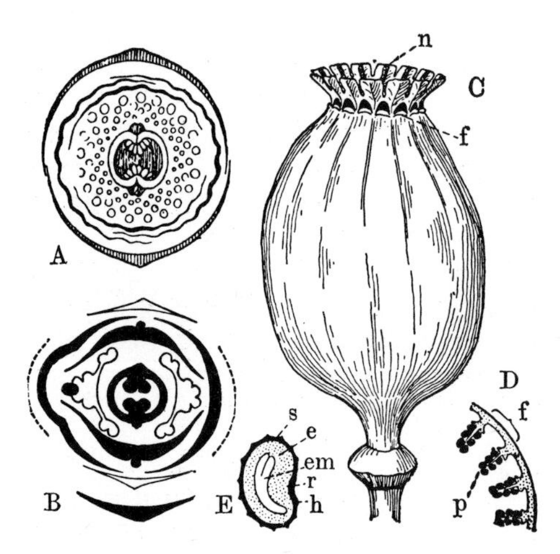

Fig. 685. Papaverales. Floral diagrams of *Glaucium* (**A.**) and *Corydalis cava* (**B.** Inner stamens divided and half of each united to the outer ones: ½ + 1 + ½.) **C.** to **E.** *Papaver somniferum.* **C.** Porose capsule with stigmas (n) and partially broken down fruit wall (f). $\times \frac{1}{2}$. **D.** Part of the T.S. of the fruit with parietal placentas (p). $\times \frac{2}{3}$. **E.** L.S. of seed with testa (s), raphe (r), endosperm (e) and embryo (em). $\times 8$. **A.** and **B.** After Eichler. **C.** to **E.** After Firbas.

calyx soon falls to reveal usually two whorls of petals, numerous stamens and two (or more) carpels united in a superior gynaecium (Fig. 685A: $K2\ C2 + 2\ A\infty\ G(\underline{20-2})$). The seeds have an oily endosperm (Fig. 685E). In the **Fumariaceae** two of the outer petals have a spur so that the flowers are disymmetric (as in the bleeding heart, *Dicentra*, Fig. 653B), or one is spurred and the flower is transversely zygomorphic as in *Corydalis* (Fig. 685B) and *Fumaria* (fumitory). The idioblasts contain no latex. The fruits are either dehiscent (and the seeds often have elaiosomes, Fig. 677F) or are one-seeded nutlets (e.g. *Fumaria*).

Subclass B: Hamamelididae (Amentiflorae)

This includes the majority of those apetalous, wind-pollinated, catkin-bearing plants formerly considered primitive. In contrast to the Magnoliidae they are characterized by greatly reduced, often unisexual flowers with few, cyclic members, and generally coenocarpous ovaries with only one functional ovule; correspondingly the fruits are mainly nuts.

Fig. 686. Trochodendrales, *Trochodendron aralioides.* **A.** Flowering shoot. **B.** Flower (P0 A ∞ G4–11). **C.** L.S. of carpel. **D.** Tricolpate pollen grain. **E.** Young fruit. **F.** Dehiscing fruit. After Takhtajan.

They are almost always woody. In a few cases xylem structure is still very primitive (without vessels) but is generally moderately advanced. Flowers with calyx and corolla occur rarely (Hamamelidaceae); generally there is only a single perigone (homoiochlamydeous) with free segments (probably equivalent to a calyx), or the flowers are naked. Only a few primitive representatives have a number of free carpels of a single carpel, or more than one functional ovule: coenocarpous gynaecia predominate with few or one generally crassinucellate ovules with two or one integuments. Pollen grains are usually zonotreme (mostly tricolpate or triporate) and have little or no pollen cement. With the exception of a few partly insect-pollinated representatives, wind-pollination is the rule. The primitive dehiscent fruit has been replaced almost entirely by the nut. The alkaloids of the phenylalanine group and oil cells so characteristic of the Magnoliales are absent; on the other hand there are phytochemical resemblances to the Rosidae.

The Hamamelididae are recorded as fossils back as far as the Lower Cretaceous. They can be regarded as early derivatives of tropical, woody Magnoliidae (or related primitive members of the Rosidae) in which reduction of the flowers and wind pollination are adaptations to temperate climates and a paucity of insects. Today they stand as an ancient assemblage with little diversity of form, which are not very closely related to each other, and which have evolved for a long time in parallel.

At the beginning can be placed the very isolated:

Order 1: **Trochodendrales** from which vessels are absent. It comprises only two E. Asiatic families each with only one genus and one species: *Tetracentron* and *Trochodendron* (Fig. 686). The polyandrous flowers are to some extent still entomophilous; the few united carpels have a number of seeds; the perianth is single or lacking. Like some other very ancient, very isolated orders or families now restricted to E. Asia (Cercidiphyllaceae, p. 676; Eupteleaceae, etc.; but with vessels in the secondary xylem) the Trochodendrales appear to be closely related to the Magnoliidae.

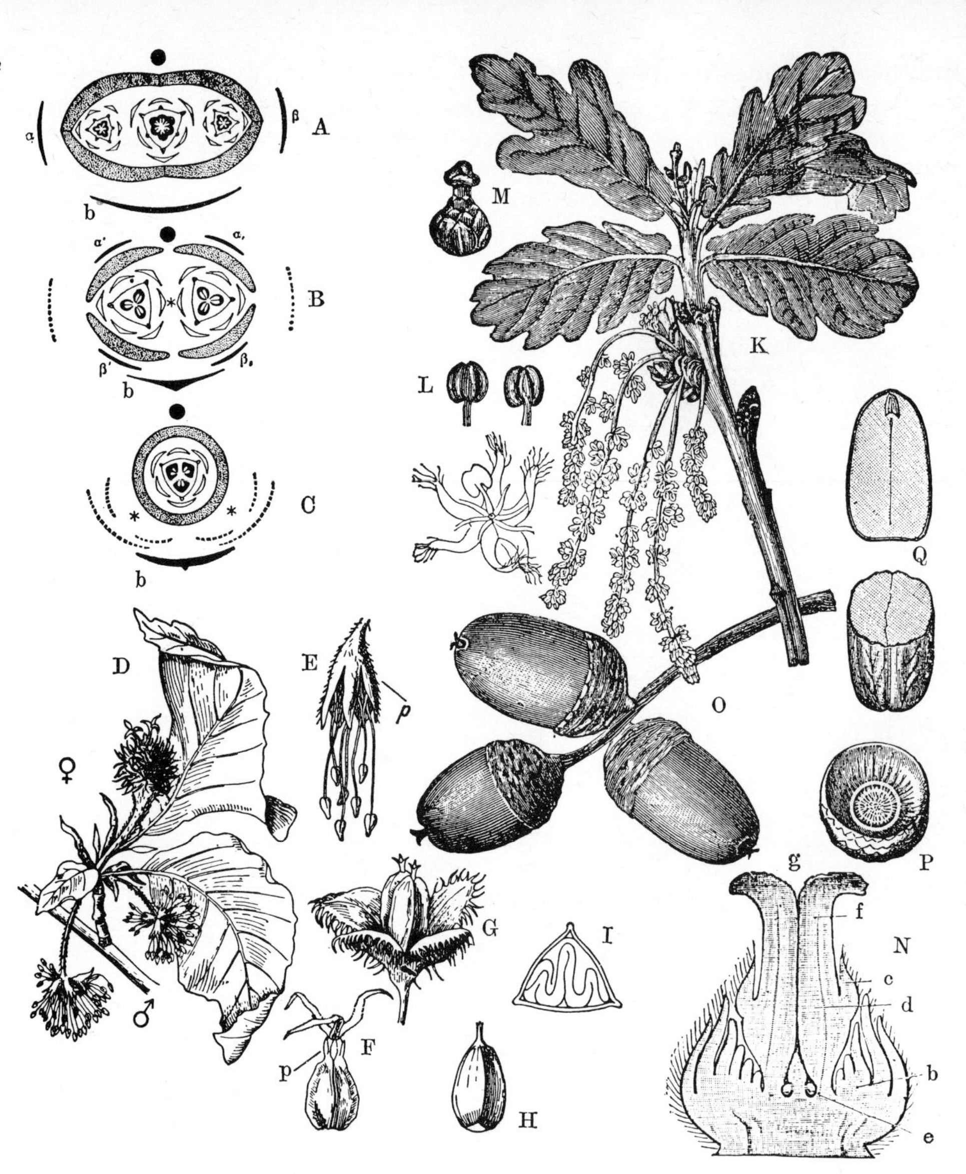
A
B
C
D
E
F
G
H
I
K
L
M
N
O
P
Q
b
f
c
d
e
g
p

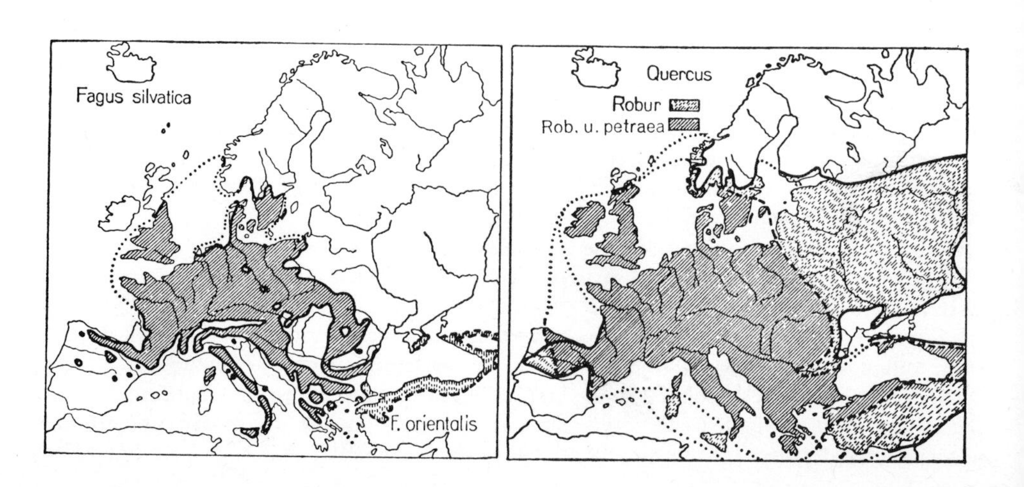
Fagus silvatica
F. orientalis
Quercus
Robur
Rob. u. petraea

All other Hamamelididae have vessels in the secondary xylem. Very primitive is:

Order 2: **Hamamelidales**: some members still have hermaphrodite flowers and a heterochlamydeous perianth. They show many similarities to primitive Rosidae. The coenocarpous **Hamamelidaceae** include *Hamamelis* (witchhazel) and *Liquidambar* (which yields styrax). The **Platanaceae** have unisexual flowers and choricarpous carpels; they include species and hybrids of *Platanus* (plane) commonly planted in streets and parks.

The floral structure is more advanced in the important:

Order 3: **Fagales** which has unisexual but monoecious flowers throughout, a single perianth or none, an inferior coenocarpous ovary and several pendulous anatropous ovules but only one seed without endosperm in the nut. Our most important deciduous forest trees belong here (cf. Fig. 749).

The alternate leaves of the **Fagales** are simple and have stipules which fall early (caducous). The inconspicuous flowers are aggregated in catkins. The flowers are borne in a three (rarely more) flowered dichasium in the axil of a bract, but bracteoles, whole flowers or individual perianth segments or stamens may be suppressed (Figs. 687A–C, 688A, showing various stages of the reduction of an originally several flowered dichasium to a single flower in the axil of a bract). In the ♂ flower the stamens are opposite the perianth segments; the ovary in the ♀ contains two or more ovules but only one develops.

In our native deciduous species the flowers open before or when the leaves unfold. Sometimes, as in hazel and alder, the catkins are fully formed in the previous year so that in the spring they have only to elongate, but the ovules are still little differentiated, developing further only after pollination, so that fertilization is much delayed.

In the **Fagaceae** three or more carpels are present. The ♀ flowers generally have the formula $P3 + 3\ G(\overline{3})$ whilst the ♂ have a varying number of perianth segments and stamens. The fruits are surrounded by a woody cup or cupule derived from axial tissues and covered with scales or spines.

The Spanish or sweet chestnut (*Castanea sativa*, Fig. 687A) is relatively primitive. It is pollinated partly by insects, especially beetles; the ♂ inflorescences are rigid, and the cupule usually contains three edible nuts for the sake of which it was brought from the Mediterranean region to the warmer parts of C. Europe by the Romans.

The beech (*Fagus sylvatica*, Figs. 133E, 687B, D, I; note the entire leaf margin) is always pollinated by wind. Numerous ♂ flowers are borne in (probably dichasial) heads, the ♀ in two flowered dichasia. Two three-angled nuts (the oily beech mast) are borne in each cupule which opens by four valves. As a dominant forest tree its main area of distribution is in C. Europe, especially at moderate altitudes in the mountains, but also in the lowlands where the soil is not too poor, not too dry and well aerated. Its continental frost and drought limit runs from western E. Prussia to the eastern foothills of the Carpathians (Fig. 687).

The ♂ and ♀ dichasia of the oak (*Quercus*, Figs. 133C, 687C, K–Q) have only one flower. The nut (acorn) is thus solitary in the scaly, cup-shaped cupule. Of the native species the pedunculate oak (*Q. robur*, fruit stalked) is distributed over the greater part of Europe from Ireland to the wood-steppes of S. Russia, especially in the lowlands and at the lower altitudes in the mountains; it occurs in a wide range of habitats (Fig. 687). The durmast oak (*Q. petraea*, *Q. sessiliflora*, fruit sessile) has a more restricted distribution. The sub-Mediterranean downy oak (*Q. pubescens*, Fig. 756) is related. The valuable hard timber is used in joinery and building; the bark in tanning. Of the many evergreen species,

◀
Fig. 687. Fagales, Fagaceae. **A.** to **C.** Floral diagrams of the ♀ dichasia of *Castanea* (**A.**), *Fagus* (**B.**) and *Quercus* (**C.**); (bracts and prophylls black, cupules dotted, suppressed flowers*, suppressed bracts and bracteoles - - -; cf. diagram, Fig. 688A, top left). **D.** to **I.** *Fagus sylvatica.* **D.** Flowering shoot. **E.** ♂, and **F.** ♀ flowers with perigone (p). **G.** Cupule with two nuts. **H.** Whole nut. **I.** In section with the folded cotyledons of the embryo. **D.**, **G.** to **H.** Nat. size. **E.**, **F.** and **I.** Enlarged. **K.** to **Q.** *Quercus robur.* **K.** Flowering shoot. **L.** ♂ flower with stamens. **M.** Whole ♀ flower. **N.** In L.S. with stigma (g), style (f), perigone (c), ovary (d), ovules (e) and cupule (b). **O.** Infructescence. **P.** Mature cupule. **Q.** Seed in T.S. and L.S. **L.** to **N.** Enlarged. Maps: natural distribution areas of some European *Fagus* and *Quercus* species. **A.** and **B.** After Eichler. **C.** After Prantl and W. Troll. **D.** to **I.** After Karsten. **K.** to **Q.** After Schimper, and Berg and Schmidt. Distribution maps after Rubner *et al.*

three Mediterranean ones may be mentioned (cf. p. 786): the western Mediterranean cork oak (*Q suber*; cf. p. 104), the kermes oak (*Q. coccifera*) and the holm oak (*Q. ilex*).

The genus *Nothofagus* (cf. pp. 411, 793) occurs in the Southern Hemisphere and the Antarctic.

The **Betulaceae** have only two carpels. Primitive flowers have the formula P2 + 2 A2 + 2 or G($\overline{2}$), but progressive reduction has taken place (Fig. 688A). The stamens are frequently divided (cf. Fig. 688K). In birch (*Betula*) and alder (*Alnus*) the nutlets are borne in the axils of woody scales arising from the fusion of the bract and two bracteoles; in the birch they fall at maturity but in alder they persist in the cone-like infructescence (Fig. 688F, M). In hazel (*Corylus*), hornbeam (*Carpinus*) and the hop hornbeam (*Ostrya*), three genera which are often put into their own family (Corylaceae),

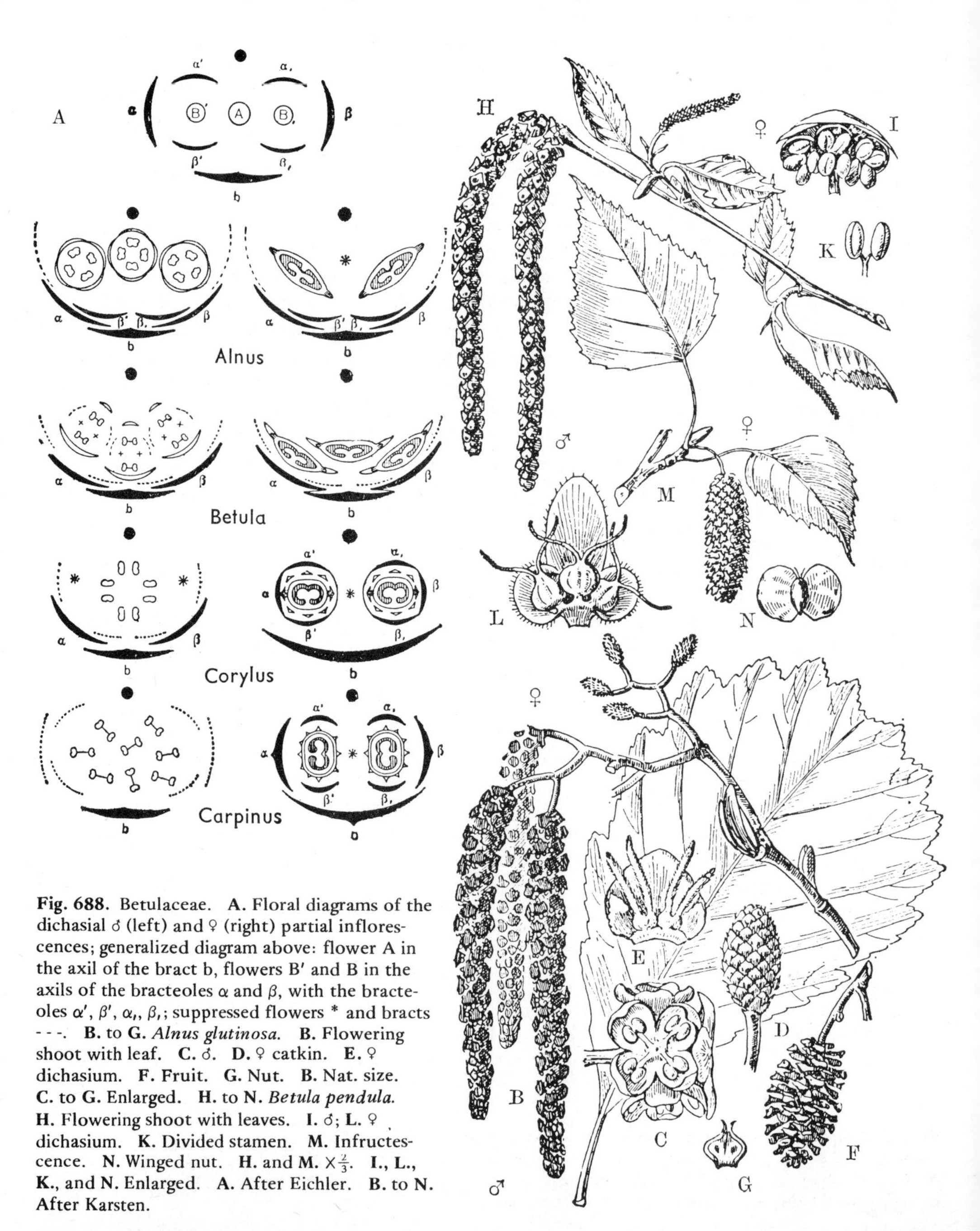

Fig. 688. Betulaceae. **A.** Floral diagrams of the dichasial ♂ (left) and ♀ (right) partial inflorescences; generalized diagram above: flower A in the axil of the bract b, flowers B′ and B in the axils of the bracteoles α and β, with the bracteoles α′, β′, $\alpha_{,}$, $\beta_{,}$; suppressed flowers * and bracts - - -. **B.** to **G.** *Alnus glutinosa.* **B.** Flowering shoot with leaf. **C.** ♂. **D.** ♀ catkin. **E.** ♀ dichasium. **F.** Fruit. **G.** Nut. **B.** Nat. size. **C.** to **G.** Enlarged. **H.** to **N.** *Betula pendula.* **H.** Flowering shoot with leaves. **I.** ♂; **L.** ♀ dichasium. **K.** Divided stamen. **M.** Infructescence. **N.** Winged nut. **H.** and **M.** $\times\frac{2}{3}$. **I.**, **L.**, **K.**, and **N.** Enlarged. **A.** After Eichler. **B.** to **N.** After Karsten.

the nuts are surrounded by an involucre which in each case consists of three united bracts and bracteoles. In the hornbeam the involucre abscisses together with the nut forming an organ of wind dispersal (Fig. 689G).

Species of alder include *Alnus glutinosa* (Fig. 688B–G), widespread throughout Europe and the most important tree of swampy woodlands and river banks in the lowlands; the grey alder (*A. incana*; leaves grey below), a Circumboreal species on river gravels in mountainous districts; and the shrubby *A. viridis*, found near the forest limit in the Alps. The root-nodules of alders contain a nitrogen-fixing actinomycete (cf. pp. 293, 443), hence the grey alder in particular is valuable for afforesting waste lands.

The light-demanding birches (*Betula pendula* = *B. verrucosa*, the silver birch, and *B. pubescens*) are unexacting trees of poor sandy and peaty soils. In northern woods and in the Arctic the round-leaved dwarf birch (*B. nana*) is also important.

The inflorescences of the hazel (*Corylus avellana*, Fig. 145A) are borne on shoots of the previous year and open particularly early in the spring; it occurs in woods and scrub throughout most of Europe. The short ♀ inflorescences are surrounded by bud scales from which the red stigmas project. The heavy nuts are dispersed by the nuthatch, woodpeckers and rodents (Fig. 679A, B); the embryo is very fatty. Species from S.E. Europe are also of commercial value: *C maxima*, the filbert, and *C. colurna*, the Turkish or tree hazel.

The flowers of the hornbeam (*Carpinus betulus*, Fig. 689; note the sharply biserrate leaf margin) are borne only on shoots of the current year (the ♂ often on leafless ones). It occurs throughout C. Europe, being an important tree beyond the limits of the beech and in river-basins since it withstands more continental or hotter climates and moister soils than the beech. In the northern part of the Mediterranean region the similar *Ostrya carpinifolia* is widespread above the *Quercus ilex* zone.

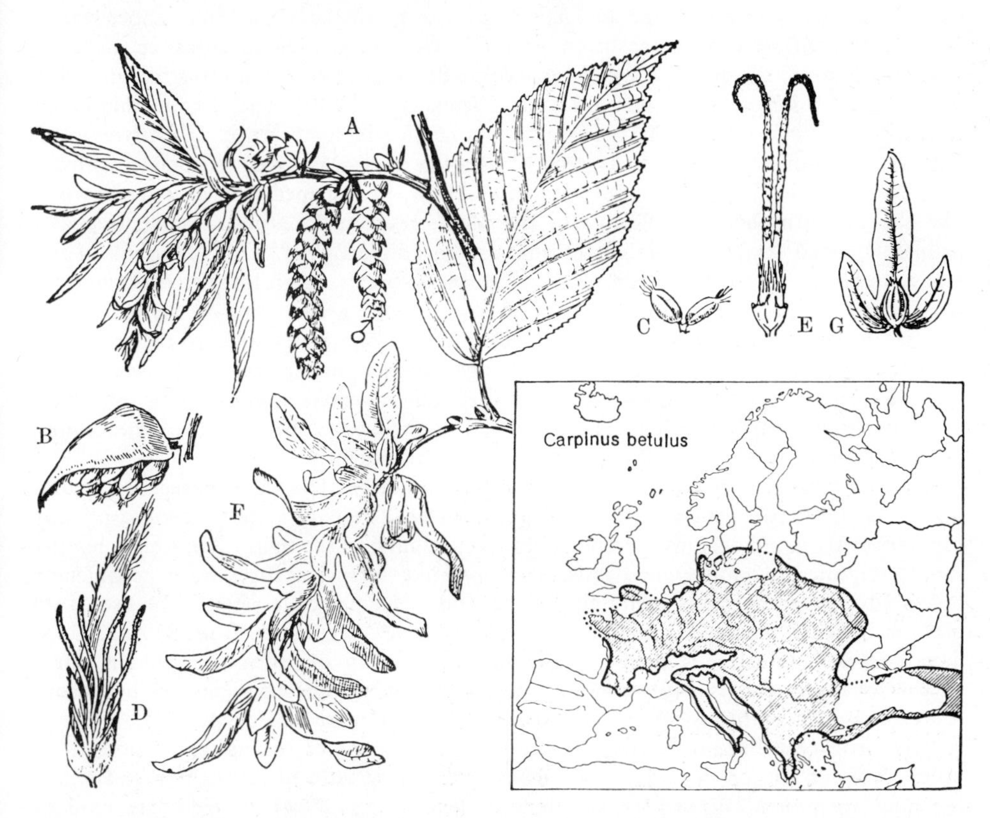

Fig. 689. Betulaceae, *Carpinus betulus.* **A.** Flowering shoot. **B.** ♂; **D.** ♀ dichasium. **C.** Divided stamen. **E.** ♀ flower. **F.** Infructescence. **G.** Nutlet with fused and enlarged bract and bracteoles. **A., F.** to **G.** Nat size approx. **B.** to **E.** Enlarged. Map of natural area of distribution, after Rubner *et al.* **A.** to **D., F.** and **G.** After Karsten. **E.** After Büsgen.

Very isolated and characteristic is:

Order 4: **Casuarinales** (Verticillatae) which comprises only the **Casuarinaceae** and the single genus *Casuarina* (she-oak), drought resistant Australian and Indomalayan trees with switch-like branches and whorls of scale leaves reminiscent of the shoots of *Equisetum*. The flowers are extremely reduced: the ♂ has two perianth segments and a single stamen, the ♀ has a bicarpellary ovary with no perianth. Because of their very simple flowers the Casuarinales have often been regarded as primitive and related to *Ephedra*, and hence have been placed at the start of the angiosperms.

Another circle of affinity, apparently separately derived from Hamamelidales-like ancestors, is:

Order 5: **Urticales.** The ♂ catkins characteristic of the Fagales are absent. The usually superior ovary is derived from two carpels but has a single loculus containing a single ovule; the fruit is a nutlet or drupe.

Woody plants predominate, but herbaceous forms have arisen more than once. The leaves are simple but often lobed and have stipules which are often caducous. The occurrence of commercially important phloem fibres and sometimes of latex and cystoliths (Fig. 55D) should be noted.

The **Ulmaceae** have hermaphrodite flowers; they are woody plants with no latex, represented in C. and N. Europe by the elms (Figs. 144C, 690A–C): the wych elm (*Ulmus glabra* = *U. scabra* = *U. montana*), mainly in montane mixed forests, and the smooth-leaved elm (*U. minor* = *U. carpinifolia* = *U. campestris*) and white elm (*U. laevis* = *U. effusa*) in the lowlands and riverine forests. The leaves are alternate in two rows and markedly asymmetric; the clusters of flowers develop into winged nutlets as the leaves unfold. *Celtis australis* (nettle tree) from S.E. Europe bears drupes and is often planted.

In contrast the woody **Moraceae** have unisexual flowers and latex. The Mexican *Castilloa elastica* and the E. Indian *Ficus elastica* in particular yield India rubber (caoutchouc). The inflorescences are often peculiar. Thus in the monoecious or dioecious mulberry (*Morus*) the perianth segments become fleshy at maturity, uniting the individual fruits of the ♀ inflorescence into the edible 'fruit' (Fig. 690D). The large, edible breadfruit (*Artocarpus*) of Indo-Malaya is also a collective infructescence. In *Dorstenia* the flowers and individual fruits are aggregated on a flat or concave axial receptacle, whilst in *Ficus* (700 species!) they are embedded in a hollow, pear-shaped axial receptacle which like the perianths becomes fleshy. In the Mediterranean *F. carica*, a small tree with palmately lobed leaves, this infructescence is the edible fig (cf. p. 673; Fig. 671H–L). Many species of *Ficus* are woody plants of tropical forests, often lofty trees. One of the most remarkable is the E. Indian banyan (*F. bengalensis*): germinating on the branch of a tree, it first becomes a considerable epiphyte which sends its roots down to the ground; as these develop into columnar trunks, it strangles its supporting tree, and since further roots are sent down from the extensive horizontal branches, a whole grove eventually develops from a single seedling. The undivided or shallowly lobed leaves of the Chinese white mulberry (*Morus alba*) provide the food of the silk-worm.

The last two families are herbaceous and have no latex. The **Cannabaceae** (Cannabinaceae) has only two genera with anatropous ovules. The native hop (*Humulus lupulus*, Fig. 690E–H) is a dioecious, perennial climber, twining to the right, with rough, hooked stems; it grows in moist woodlands. The cone-like infructescences have conspicuous bracts covered with glands containing resins and bitter principles, hence the use of the plant in brewing and medicine. Hemp (*Cannabis sativa*), originating from S. Asia, is also dioecious, but annual; it is mainly cultivated for its 1–2 m long strands of phloem fibres, to a lesser extent for its oily seeds; the dried shoot tips of various forms rich in narcotic resin are smoked as the objectionable intoxicant hashish, marijuana or 'pot'.

The **Urticaceae** have orthotropous ovules. The stamens of the unisexual flowers are inflexed under tension in bud; as the flower opens they suddenly straighten and scatter the powdery pollen (Fig. 652D, E). In some genera (e.g. *Pilea*) the fruits are similarly ejected by staminodes (staminal rudiments). Many Urticaceae, e.g. the nettles (*Urtica*), have stinging hairs (Fig. 107A–C). Fibres are yielded by *Urtica dioica*, but the Asiatic *Boehmeria nivea* (ramie) is more important (pp. 12, 117).

The following two orders appear to be more closely related to each other; they are

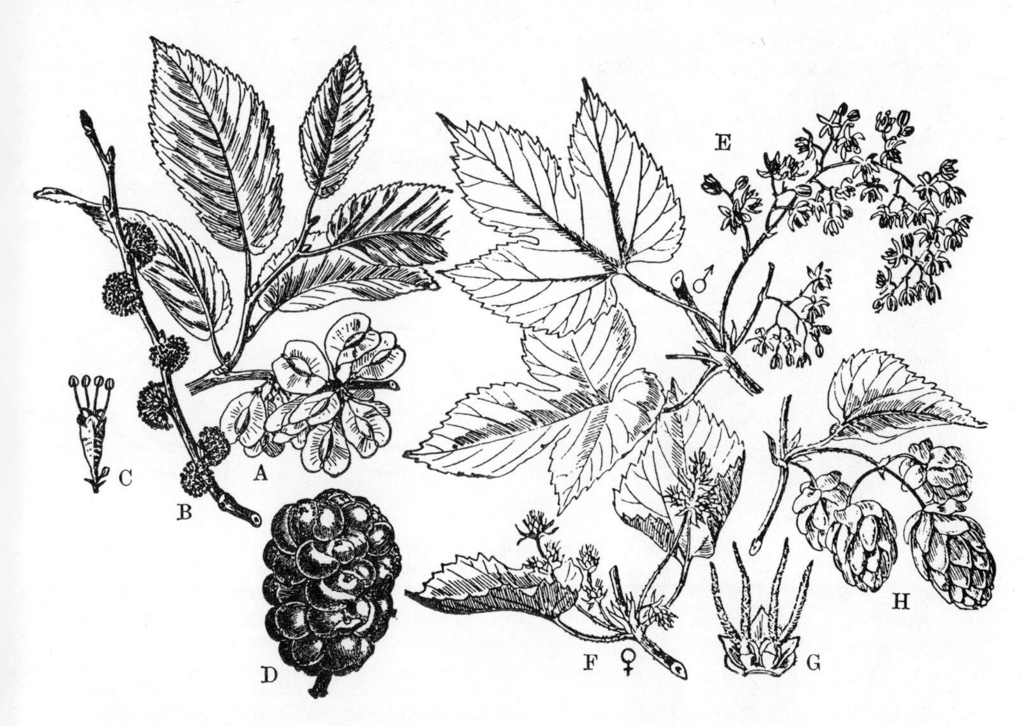

Fig. 690. Urticales. **A.** to **C.** *Ulmus minor.* **A.** Fruiting and **B.** Flowering shoots. ×⅓ approx. **C.** A single flower, enlarged. **D.** *Morus nigra*, infructescence. Nat. size. **E.** to **H.** *Humulus lupulus.* **E.** Flowering ♂, **F.** flowering ♀. **H.** Fruiting shoots. ×½. **G.** ♀ dichasium. Enlarged. **A.** to **C.**, **E.** to **H.** After Karsten. **D.** After Duchartre.

both woody and aromatically scented (glandular hairs, ethereal oils) with one orthotropous ovule in the bicarpellary gynaecium.

Order 6: **Myricales** includes only the **Myricaceae** in which there is no perianth and the leaves are simple, as in the bog myrtle or sweet gale (*Myrica gale*), an aromatic shrub of Atlantic bogs and heaths (p. 785) with actinomycete symbiosis (p. 293).

Order 7: **Juglandales**, sometimes grouped with the Anacardiaceae (p. 700), has a simple perianth and pinnate leaves, and comprises only the **Juglandaceae**, a family of north temperate regions. Commonly planted is the walnut tree (*Juglans regia*, Fig. 691). Numerous ♂ flowers are borne in stout catkins arising from axillary buds on shoots of the previous year, whilst a few ♀ are borne at the tip of the current year's growth. In both kinds of flower three to five perianth segments are fused with a bract and two bracteoles. The fruit is a drupe of which the walnut is a stone; the shell separates on germination into halves along a predetermined line at right angles to the sutures of the two carpels. The oily reserves of the seed are stored in the cotyledons which are much lobed by the ingrowth of numerous incomplete septa. The walnut tree is native in the sub-Mediterranean region; north of the Alps it often suffers frost damage. Like other species of *Juglans* (e.g. the N. American *J. nigra*), the N. American species of *Carya* (hickory, pecan) and the Caucasian wing nut (*Pterocarya fraxinifolia*), the walnut yields a valuable timber.

The following two subclasses, the Rosidae (Rosiflorae) and Dilleniidae (Cistiflorae, p. 709), consisting of groups with free petals (dialypetalous), form the two 'central columns' of the 'middle storeys' of the dicotyledons (Fig. 742). In contrast the Caryophyllidae (p. 717) stand rather as a 'side column'. On the two dialypetalous 'central columns' rests the 'superstructure' or 'attics' of the sympetalous groups: the Asteridae (i.e. the greater part of the Sympetalae tetracyclicae) on the Rosidae, the Ericanae (i.e. Sympetalae pentacyclicae) and the Cucurbitales on the dialypetalous Dilleniidae.

Compared with the more primitive Magnoliidae which form the 'substructure' or 'foundations' of the angiosperms, the 'middle storeys' of the dialypetalous Rosidae and

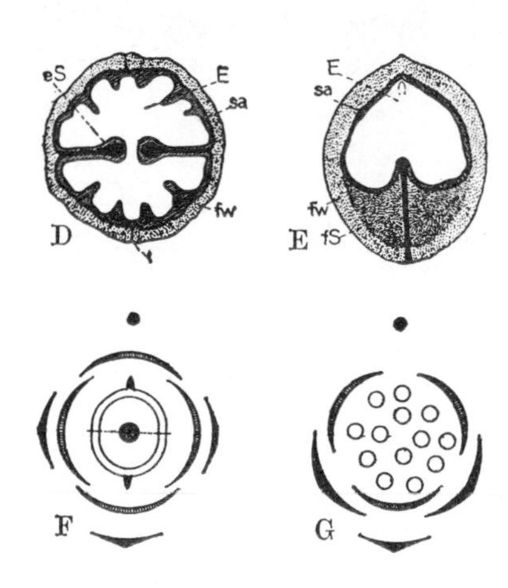

Fig. 691. Juglandales, *Juglans regia*. **A.** Flowering shoot. **B.** ♂ and **C.** ♀ flowers with bract (d), bracteoles (v), perianth segments (p) and stigmas (n). **F.** and **G.** Floral diagrams. **D.** and **E.** Stone ('nut') of the drupe in T.S. and L.S. (median) with stony endocarp ('shell', fw), line of cleavage (t), true septum (eS) and false septum (fS), testa (sa) and embryo (E). **A.** From Hegi. **B.** After Firbas. **C.** After Kirchner. **E.** and **F.** After Eichler, somewhat modified.

Dilleniidae almost entirely lack forms with such primitive characters as absence of vessels in the xylem, as spiral arrangement of the floral organs, primary polyandry, undifferentiated stamens and monocolpate pollen grains. On the other hand the predominance of a cyclic, pentamerous perianth differentiated into calyx and corolla, two whorls of stamens (diplostemony) or secondary polyandry, or finally reduction to one whorl of stamens (haplostemony) is highly characteristic. The essential difference between the Rosidae and Dilleniidae is the sequence of development of the primordia in the secondarily polyandrous androecium: centripetal in the Rosidae, centrifugal in the Dilleniidae (pp. 647, 679, Fig. 656). In diplo- and haplostemonous groups, however, it is not so much single characters as whole groups of morphological relationships which determine the arrangement of the two classes. There are indeed many parallel progressions in the Rosidae and Dilleniidae: choricarpy (free carpels) occurs only in the most primitive groups, otherwise coenocarpy (union of the carpels) predominates; further there is often a trend from superior to inferior ovaries (i.e. hypogyny to perigyny). The perianth repeatedly shows the loss of the corolla (apetaly) and sometimes of the perigone (equivalent to a calyx); on the other hand the petals may become united (sympetaly), leading the way to the above-mentioned 'superstructure' or 'attics' of the dicotyledons. Other characteristic trends are from two- to three-celled pollen, from crassinucellate ovules with two integuments to tenuinucellate ovules with one integument, and from woody plants to herbs.

Previously the Rosidae and Dilleniidae had been thought of as separate lines of development each stemming directly from the Magnoliidae. Recently, however, the following objections have been put forward: (1) There are difficulties in making the distinction between centripetal and centrifugal development in secondarily polyandrous gynaecia (p. 647). (2) The Rosidae and Dilleniidae show many similar characteristics especially in xylem structure and in phytochemistry (e.g. ellagic acid amongst other trihydroxylated compounds, 'iridoid substances' derived from cyclopentanoid monoterpenes) in which they are contrasted to the Magnoliidae. (3) Several groups cannot be assigned with certainty to either of the two subclasses (e.g. Euphorbiales). We can take these observations into account by placing the Rosidae and Dilleniidae (as well as the

Hamamelididae) alongside each other and conceding the possibility of a closer phylogenetic connection between them at the level of their 'basements'; this is indicated in Fig. 742.

Subclass C: Rosidae (Rosiflorae)

In addition to the centrifugal development of the stamens in the secondarily polyandrous representatives and the predominantly cyclic and dialypetalous flowers, the Rosidae are principally distinguished by their tendency to develop cup-shaped or extended disc-like receptacles, the widespread occurrence of central-axile placentation and the frequent reduction in ovule number.

The great diversity of the Rosidae necessitates a grouping of the orders into superorders, even if provisionally. The **Rosanae** take first place, sometimes with carpels still choricarpous or only partly united and with numerous ovules, and frequently with secondarily polyandrous androecia. This superorder is followed by the Myrtanae (p. 698) which are still secondarily polyandrous, have cup-shaped receptacles and coenocarpous, generally inferior gynaecia; the Rutanae (p. 700), with frequently compound leaves, disc-shaped receptacles and superior ovaries; the Celastranae and Proteanae (pp. 702, 706) with a progressive tendency to haplostemony and apetaly; and finally by the Aralianae (p. 706) with haplostemony and inferior ovaries.

At the start of the superorder Rosanae (Orders 1–5) can be placed:

Order 1: **Saxifragales** in which the seeds are still endospermous. Primitive woody representatives with partly more or less choricarpous superior ovaries include the **Cunoniaceae** from the Southern Hemisphere and the **Hydrangeaceae** which includes the mock orange or 'syringa' (*Philadelphus*) and *Hydrangea*. The **Grossulariaceae** are characterized by inferior gynaecia and berries (Fig. 692E–G), especially in the various species of *Ribes*, e.g. the gooseberry (*R. uva-crispa* = *R. grossularia*), the red currant (*R. rubrum*) and the black currant (*R. nigrum*).

In contrast the following families are herbaceous. The **Crassulaceae** are leaf-succulents with many free carpels. These include the familiar native genus *Sedum* (stonecrop, with pentamerous flowers, Fig. 653A), *Sempervivum* (house-leek, with hexamerous or polymerous flowers) and the tropical genera *Kalanchoe* (used in experiments on photoperiodicity) and *Bryophyllum* (adventitious buds, Fig. 223B). The **Saxifragaceae** on the other hand generally only have two more or less united carpels more or less deeply embedded in the receptacle (Fig. 692A–D). Species of *Saxifraga* have various life-forms (especially cushion and rosette plants) which extent in Arctic and Alpine regions to the uttermost climatic limits of vascular plants.

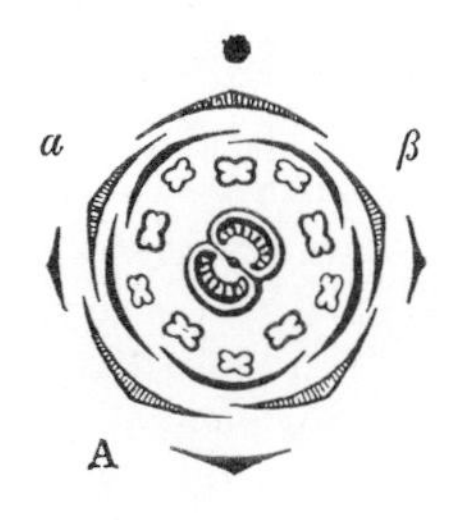

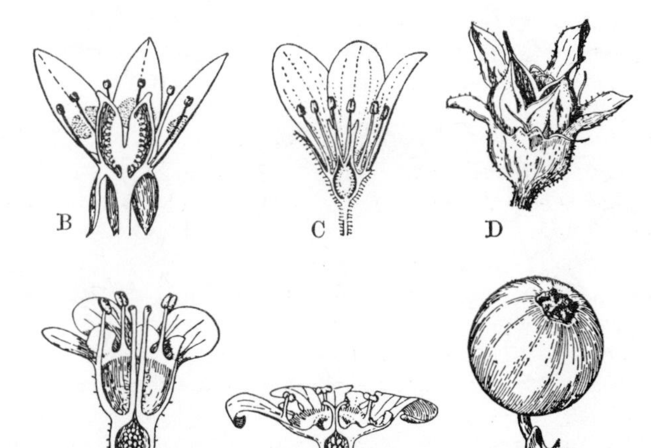

Fig. 692. Saxifragales. **A.** Floral diagram of *Saxifraga granulata* with bracteoles (α, β). **B.** *Saxifraga stellaris*, ×2.5, and **C.** *S. granulata*, ×1.5, half-flowers. **D.** *S. cespitosa*, capsule with calyx. ×3. **E.** *Ribes uva-crispa*, half-flower. ×2.5. **F.** Half-flower, ×3.5, and **G.** berry, ×2, of *R. rubrum*. **A.** After Eichler. **B.** to **G.** After Firbas.

In contrast there is no endosperm in the seeds of:

Order 2: **Rosales.** The family **Rosaceae** is represented in the European flora with a great diversity of forms; it often has a secondarily polyandrous androecium (Fig. 693) and shows in a very striking manner the range of differentiation and progressions in the structure of choricarpous gynaecia and fruits (Figs. 693, 694, cf. p. 672).

In the subfamily Spiraeoideae we still find follicles with many seeds, or aggregate fruits derived therefrom, e.g. in the decorative and commonly planted shrubby species of *Spiraea*. Instead we find in the Rosoideae unicarpellary indehiscent fruits which are usually unicarpellary nutlets as in *Potentilla*. In mountain avens (*Dryas*) and water avens (*Geum*) feathery or hooked stylar appendages function in the dispersal of the partial fruits. Such nutlets may also be united by a receptacle which becomes fleshy, forming an aggregate fruit as in the rose (*Rosa*), where the nutlets are borne inside the flask-shaped receptacle (the hip), or in the strawberry (*Fragaria*), in which the nutlets are borne on the outside of the fleshy conical receptacle. In the raspberry (*Rubus idaeus*) and blackberry (*R. fruticosus* agg.) nutlets are replaced by small drupes (drupels) which are united in an aggregate fruit. Stone fruits may also develop from the falsely syncarpous (p. 654) inferior ovaries of one to five carpels of the Maloideae (core fruits) in which the fleshy tissue is mainly formed by the receptacle. In the hawthorn (*Crataegus*, Fig. 200) and medlar (*Mespilus germanica*) each carpel forms a hard stone inside the fleshy part of the fruit, whilst in the quince (*Cydonia oblonga*), pear (*Pyrus communis*), apple (*Malus sylvestris*) and the genus *Sorbus* (*S. aucuparia*, mountain ash; *S. domestica*, service tree, etc.) each carpel is like a follicle and develops into a parchment-like structure surrounded by relatively few groups of stone cells (pome). Typical unicarpellary drupes are found in the Prunoideae (stone fruits). The single carpel is not united to the concave receptacle and the outer flesh contains a very hard one-seeded stone, as in the sweet cherry (*Prunus avium*), sour cherry (*P. cerasus*), plum and damson (*P. domestica*), peach (*P. persica*), apricot (*P. armeniaca*) and almond (*P. amygdalus*, with a leathery mesocarp). Noteworthy is the occurrence here (as in some Pomoideae) of cyanogenetic glucosides in the seeds.

The Rosaceae include more than 2000 species, mainly in the Northern Hemisphere; *Rosa, Rubus* and *Alchemilla* (lady's mantle) alone have several hundreds of species resulting from polyploidy, hybridization and to some extent agamospermy (cf. pp. 403, 421, 667). The occurrence of anemophily in *Sanguisorba* is noteworthy (Fig. 672).

Economically important are the soft fruits (strawberry, raspberry, blackberry) and the numerous orchard fruits. Wild forms of the apple, pear and sweet cherry are native in C. Europe where they were gathered for food and tended as early as the Neolithic along with

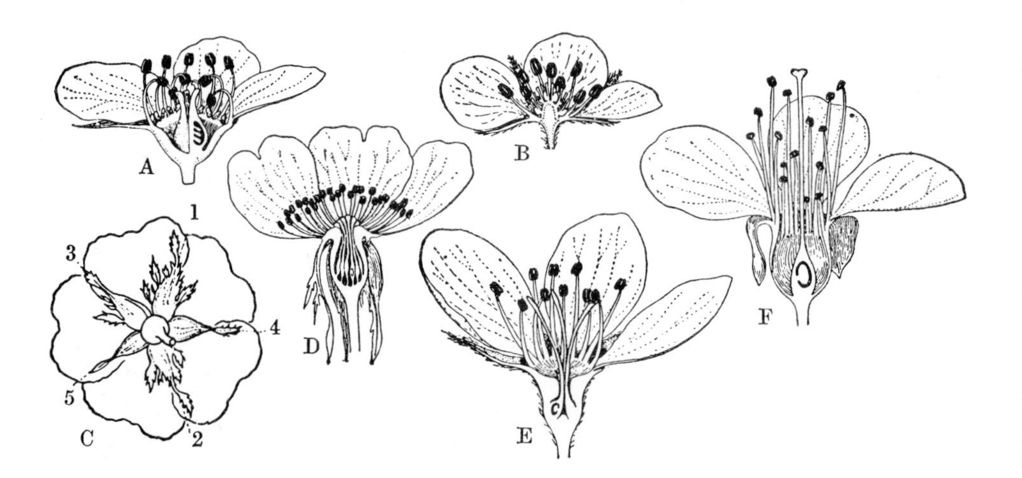

Fig. 693. Rosales, Rosaceae. Half flowers of: **(A.)** *Spiraea lanceolata*; **(B.)** *Fragaria vesca*, ×1.5. **(D.)** *Rosa canina*, ×¾; **(E.)** *Pyrus communis*, ×1.5; **(F.)** *Prunus avium*, ×1.5. **C.** Spiral sequence of the progressively reduced sepals in the quincuncial calyx whorl of *Rosa*. **A., B., D.** to **F.** After Firbas. **C.** After Goebel.

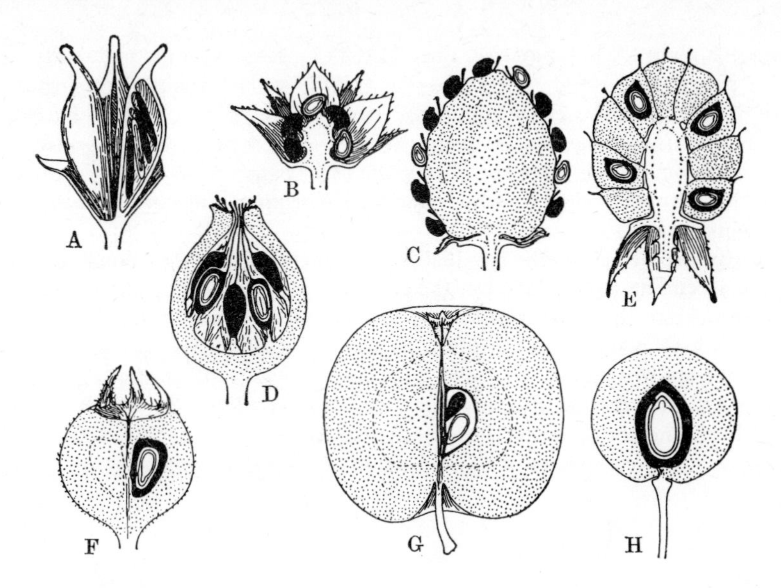

Fig. 694. Rosaceae. Fruits in L.S. (diagrammatic). **A.** *Spiraea*. **B.** *Potentilla*. **C.** *Fragaria*. **D.** *Rosa*. **E.** *Rubus*. **F.** *Mespilus*. **G.** *Malus*. **H.** *Prunus*. Fleshy tissue dotted; vascular bundles, pecked lines; indurated layers of the fruit wall or testa, black. After Firbas.

sloes (*Prunus spinosa*), the bird cherry (*P. padus*), etc. The quince, medlar, almond and sour cherry are native in the Near East where the wild forms of apple, pear and sweet cherry are most numerous, whilst the apricot is native from Turkestan to W. China, the peach only in China. Cultivated forms of all these have come to Europe since Graeco-Roman times.

The legume or pod, a single superior carpel, originally many-seeded and opening both ventricidally and dorsicidally (Fig. 678B) is the chief characteristic of:

Order 3: **Fabales** (Leguminosae), the legumes. These are woody or herbaceous with usually alternate, pinnately compound, stipulate leaves. The tendency for an initially radial flower to develop as a zygomorphic papilionate (butterfly-like) one is pronounced (Figs. 695, 698). The seeds generally lack endosperm.

In most species pulvini enable the leaves to move in response to stimuli (pp. 365, 367 et seq., Figs. 314, 315, 318, 319). The roots bear nodules containing symbiotic nitrogen-fixing species of *Rhizobium* (pp. 285, 293, 443, Fig. 265).

In the **Mimosaceae** the flowers are still radial (actinomorphic) and the stamens are often secondarily increased in number (Fig. 695A, B). They are tropical and subtropical woody plants and herbs, generally with bipinnate, paripinnate leaves and small globular or spike-like inflorescences. The flowers are frequently tetramerous; they are rendered conspicuous by the long coloured filaments of the mostly very numerous stamens (Fig. 696B). The pollen grains often remain clustered together (polyads, p. 651).

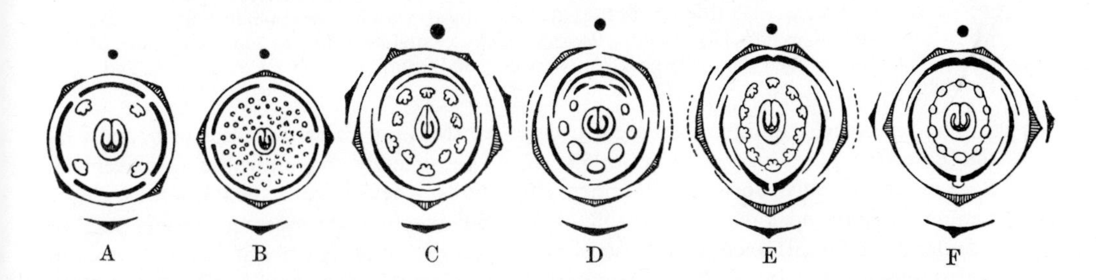

Fig. 695. Fabales. Floral diagrams. Mimosaceae: **A.** *Mimosa pudica*. **B.** *Acacia lophantha*. Caesalpiniaceae: **C.** *Cercis siliquastrum*. **D.** *Cassia caroliniana*. Fabaceae: **E.** *Vicia faba*. **F.** *Laburnum anagyroides*. After Eichler.

The sensitive plant (*Mimosa pudica*, Figs. 314, 315), famous for its irritability, belongs here; it is a pantropical weed. The numerous *Acacia* species, mainly trees, show many features of interest: many, especially those of dry forests in Australia, have leaf-like phyllodes (Fig. 180A); some are ant-plants (Fig. 696C, D); several yield gums from the bark, others tannins. The species of *Entada* are pantropical lianes with a large 'craspedium' or framed pod from which one-seeded segments become detached, leaving the two marginal parts as a frame.

The **Caesalpiniaceae** show the gradual development of the papilionate flower, but aestivation is ascending (Fig. 695C, D): the two lowest petals overlap the two lateral ones which in turn overlap the uppermost one. The stamens are usually free. The members are (sub)tropical woody plants with paripinnate, pinnate or bipinnate leaves. There are two familiar Mediterranean species: the carob or locust tree (*Ceratonia siliqua*), with indehiscent edible pods, and the cauliflorous Judas tree (*Cercis siliquastrum*), often planted in gardens. Others include the N. American *Gleditschia triacanthos* with branching shoot-spines, and tropical species of *Cassia* used in medicine (Fig. 697A).

Fig. 696. Mimosaceae, *Acacia*. **A.** and **B.** *A. catechu*, flowering shoot, ×½, and a single flower, ×5. **C.** and **D.** *A. nicoyensis* from Costa Rica. Twig (reduced) with hollow stipular thorns (d) in which ants live; l, holes bored by the ants; leaves with extra-floral nectaries (n) and food-bodies (Belt's bodies, f) on the pinnules (**D.**); enlarged. **A.** After Berg and Schmidt. **B.** After Baillon. **C.** and **D.** After Noll.

The **Fabaceae** (Papilionaceae) are distinguished from the Caesalpiniaceae principally by the descending aestivation (Fig. 695E, F). The strongly zygomorphic (dorsiventral) flowers are usually borne in racemes (Fig. 698); they have a pentamerous calyx, frequently gamosepalous, and a corolla of five petals: the uppermost (posterior) petal or 'standard' (vexillum) overlaps the two lateral ones or 'wings' (alae) which in turn overlap the two lowest petals which are frequently joined at the lower margin to form the 'keel' (carina). The keel encloses the ten stamens which surround the ovary. All ten stamens are rarely free; generally either all ten or only nine of them are united by their filaments. Hence: ↓ K(5) C5 A(10) or A(9) + 1 $G\underline{1}$.

The imparipinnate leaf is held to be primitive: from this the digitate leaf (*Lupinus*), the trifoliate (*Trifolium*) and finally the simple leaf (representing the terminal leaflet) can be derived. The terminal leaflet or even the upper pinnae may be replaced by tendrils (e.g. in *Vicia, Pisum,* Fig. 213A). Furthermore, photosynthesis may be taken over by the stipules (*Lathyrus aphaca*, Fig. 213B) or by the stem as in many leafless switch-plants and thorn-bushes, e.g. the broom (*Sarothamnus scoparius*), various *Genista* species, and gorse or furze (*Ulex*).

The flowers are pollinated mainly by honey-bees and bumble-bees; they possess various arrangements by which the anthers are pushed or even jerked out, or the pollen is squeezed out, when the insect presses down on the wings or keel which form a landing stage. The pod (Fig. 678B) may be modified as a lomentum, separating into one-seeded segments (Fig. 679D), or even as a one-seeded nut; the seeds are surrounded by an impermeable testa which delays germination (hard seeds). The greatly swollen cotyledons of the embryo contain starch, much protein, and often fats.

Fig. 697. Fabales. **A.** Caesalpiniaceae, *Cassia angustifolia*, flowering shoot and pod. ×½. **B.** Fabaceae, *Astragalus gummifer*, flowering shoot with leaf-thorns. ×½. **A.** After Berg and Schmidt. **B.** After Firbas.

This extremely large family is distributed throughout the world, the woody forms predominating in the Tropics, the herbaceous elsewhere. In general they prefer dry, nitrogen-deficient or calcareous soils; they are therefore prominent in the Eurasian steppes and semi-deserts, where, for example, there are over 2000 species of *Astragalus*, including the hedgehog-like bushes of the section Tragacantha which have characteristic leaf-spines (Fig. 697B). In addition they play a part in various European plant communities.

Economically they are enormously significant. Some are important herbage plants, growing well as nitrogen fixers on nitrogen deficient soils; they are often ploughed in as 'green manure': thus various species of clover (*Trifolium pratense*, *T. hybridum*, *T. repens*, *T. incarnatum*), the alfalfa or lucerne (*Medicaga sativa*), sainfoin (*Onobrychis viciifolia*) and, especially on sandy soils, the seradella (*Ornithopus sativus*) and some originally Mediterranean lupins (*Lupinus angustifolius*, *L. luteus*). The starch and protein rich seeds of others are important as foodstuffs, e.g. the broadbean or horse bean (*Vicia faba*), pea (*Pisum sativum*), chick pea (*Cicer arietinum*) and lentil (*Lens culinaris*), which have been known since the Neolithic, as well as the scarlet runner (*Phaseolus coccineus*) and French or kidney bean (*P. vulgaris*) (Fig. 172) from S. America and grown in gardens; to these the 'sweet lupin' has recently been added, its bitter alkaloids removed by breeding. Oil-plants include the soybean of E. Asia (*Glycine soja* = *Soja hispida*) and the

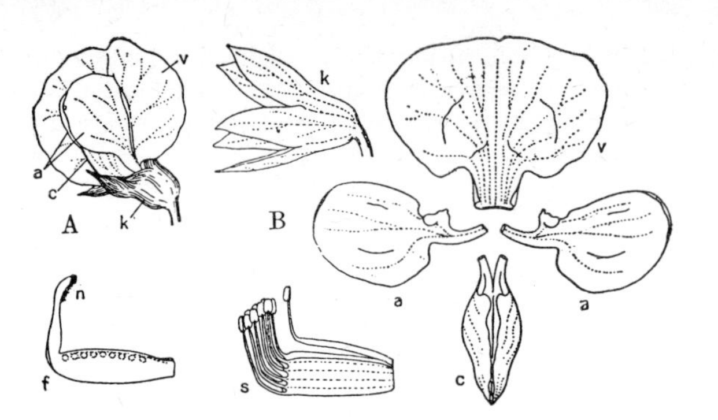

Fig. 698. Fabaceae, *Pisum sativum*. **A.** Whole flower, ×1. **B.** Dissected into: calyx (k), corolla consisting of standard (v), wings (a) and keel (c), stamens (s; 9 + 1) and unicarpellary ovary (f) with stigma (n) and ovules (dotted), ×1.2. After Firbas.

S. American groundnut, or monkeynut or peanut (*Arachis hypogaea*), cultivated in hot countries for its oil and protein rich seeds: in this the carpels are pushed down into the soil, where they ripen, by the growth of a stalk-like elongated zone of the lower part of the ovary. Important trees include the false acacia (*Robinia pseudacacia*) from eastern N. America, used in afforesting waste and dry areas; many others are familiar ornamental plants, e.g. the poisonous golden rain (*Laburnum anagyroides*) from S. Europe and the eastern Asiatic *Wistaria sinensis*. Medicinal plants include the restharrow (*Ononis spinosa*; with leaf-spines) and liquorice (*Glycyrrhiza glabra*), a perennial herb native from the Mediterranean to C. Asia.

Possibly related to the Saxifragales is:

Order 4: **Sarraceniales**, notable for its carnivorous members. Like the Australian **Cephalotaceae** which properly belong to the Saxifragales, the New World **Sarraceniaceae** (with hermaphrodite flowers) and the palaeotropical **Nepenthaceae** (with unisexual flowers) have pitcher-shaped leaves modified as animal traps (Fig. 222G, p. 201). In contrast the **Droseraceae** catch insects by means of sticky tentacles (*Drosera, Drosophyllum*) or irritable leaves which snap together (*Dionaea, Aldrovanda*; cf. pp. 26, 199–201, Fig. 221). The Sarraceniales also show similarities to the Dillenianae (pp. 709 et seq.).

Also related to the Saxifragales is:

Order 5: **Podostemales** (Podostemonales). **Podostemaceae**, the only family, has members with the vegetative parts reduced to thallus-like structures (cf. p. 86); they live in fast-flowing waters in the Tropics.

The following orders (6–8) can be included in the superorder **Myrtanae**. In contrast to the Rosanae the gynaecium is never choricarpous. A cup-shaped or tubular floral axis or receptacle (hypanthium) is diagnostic; the ovary is half-inferior or inferior with central-axile placentation and commonly numerous ovules. The leaves are generally simple, opposite and stipulate; the flowers generally radial (actinomorphic) and often tetramerous. The Myrtanae are apparently related phylogenetically to the Rosanae–Saxifragales.

Order 6: **Myrtales** is characterized by a heterochlamydeous perianth (calyx and corolla), by secondary polyandry (Fig. 656A), coenocarpous carpels with even the styles united, usually by a lack of endosperm, and bicollateral vascular bundles. Some primitive features are shown by the woody **Sonneratiaceae** (half-inferior ovaries: related to Dilleniidae) and **Rhizophoraceae** (still with endosperm and collateral vascular bundles: Fig. 699A–C). These two families include the most important of the tropical mangroves (p. 792), namely *Sonneratia* on the one hand, *Rhizophora, Bruguiera, Kandelia* and *Ceriops* on the other. Stilt-roots, breathing roots (pneumatophores) and vivipary (p. 675) are amongst their adaptations to the peculiar habitat conditions of these coastal communities.

More advanced are the numerous species of the **Myrtaceae** (Fig. 699D–F). These are mostly evergreen (sub)tropical woody plants, distinguished by the invariable presence of lysigenous secretory cavities containing ethereal oils, hence their importance as spices and drugs. The conspicuousness of the flowers is often enhanced by the coloured filaments of the numerous stamens. Of the many species of the tropical genera *Eugenia* and *Sizygium*, the clove (*S. aromaticum* = *E. caryophyllata*), distributed from Ceylon to Borneo, is notable (Fig. 699D, E). In Australia most of the dry forests are dominated by *Eucalyptus* with some 700 species of trees or shrubs; juvenile and adult leaves are often different in shape (cf. p. 181). Pollination is mainly by birds, but also by bats and small marsupials. *E. amygdalina* can reach 150 m in height and 10 m in diameter; it is probably the world's most gigantic tree. Some species, e.g. *E. globulus*, are much planted because of their rapid growth in warmer countries, e.g. in the Mediterranean. In the Mediterranean, too, occurs the only European member of the Myrtaceae, the myrtle (*Myrtus communis*, Fig. 699F) which is occasionally planted further north.

Also woody, but without secretory cavities and related to the Lythraceae (p. 700) are the **Punicaceae** to which the pomegranate (*Punica granatum*) belongs; this originated in the Orient and is often grown for its bright red fleshy seeds and fruits; the red flowers

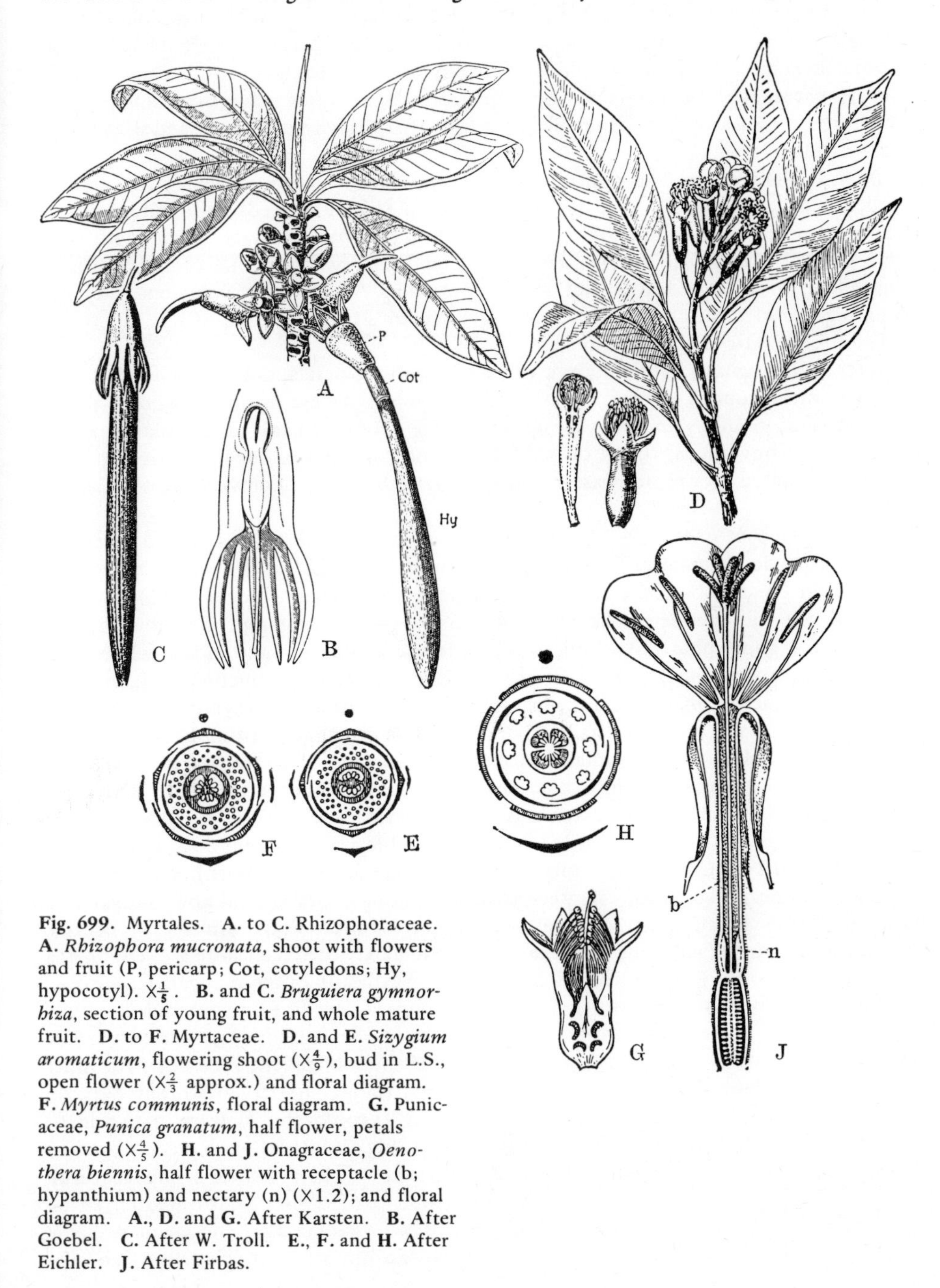

Fig. 699. Myrtales. **A.** to **C.** Rhizophoraceae. **A.** *Rhizophora mucronata*, shoot with flowers and fruit (P, pericarp; Cot, cotyledons; Hy, hypocotyl). $\times\frac{1}{5}$. **B.** and **C.** *Bruguiera gymnorhiza*, section of young fruit, and whole mature fruit. **D.** to **F.** Myrtaceae. **D.** and **E.** *Sizygium aromaticum*, flowering shoot ($\times\frac{4}{9}$), bud in L.S., open flower ($\times\frac{2}{3}$ approx.) and floral diagram. **F.** *Myrtus communis*, floral diagram. **G.** Punicaceae, *Punica granatum*, half flower, petals removed ($\times\frac{4}{5}$). **H.** and **J.** Onagraceae, *Oenothera biennis*, half flower with receptacle (b; hypanthium) and nectary (n) ($\times$1.2); and floral diagram. **A.**, **D.** and **G.** After Karsten. **B.** After Goebel. **C.** After W. Troll. **E.**, **F.** and **H.** After Eichler. **J.** After Firbas.

have the carpels in two or three superposed tiers (Fig. 699G). The numerous **Melastomataceae** are especially characteristic of the S. American Tropics and Subtropics; they lack stipules and the diplostemonous flowers are biologically specialized with 'handles' and other gadgets on the connectives.

In the mainly herbaceous **Onagraceae** the hypanthium is almost always greatly elongated above the inferior ovary (Fig. 699J). The evening primroses (*Oenothera*) originated in America but are now cosmopolitan in waste places; they afford important experimental material for genetic research (p. 418). The bird-pollinated flowers of *Fuchsia* have brightly coloured hypanthia and sepals; they are native mainly in S. and C. America and commonly cultivated. Native members of this family include the genus *Epilobium* (willow herb, Fig. 291A–C) and *Circaea* (enchanter's nightshade; p. 421).

A predominantly herbaceous line of development, usually with only two loculi in the still half-inferior ovary starts with the **Lythraceae**: the purple loosestrife (*Lythrum salicaria*) is notable for its trimorphic heterostyly (p. 657). Transitional to the next order is the **Trapaceae** with the annual floating aquatic *Trapa natans* which is becoming progressively rarer in C. Europe; its drupe-like nuts have sharp calyx-horns furnished with hooks (anchor fruit).

Order 7: **Haloragales**. The position here of this order is questionable: it has collateral bundles, reduced perianths, free styles, indehiscent fruits and seeds with endosperm. It includes, amongst others, the **Haloragaceae** with *Myriophyllum*, water plants with very finely divided leaves, the **Gunneraceae**, woodland and swamp plants of the Southern Hemisphere with large leaves and stems containing *Nostoc* as a symbiont, and the **Hippuridaceae** with the aquatic *Hippuris vulgaris* (mare's-tail, Fig. 132A–C).

Also problematic are the relationships of:

Order 8: **Elaeagnales**. These have a calyx-cup without petals, only one inferior carpel and one orthotropous ovule. The sole family **Elaeagnaceae** includes woody plants rendered silvery and shining by scaly hairs (Fig. 106E, F), as in the native, wind-pollinated sea buckthorn, found on sand dunes and (on the Continent) on river gravels, and in the planted species of oleaster (*Elaeagnus*); both have actinomycete symbionts (cf. p. 293).

A third group of orders (9–19) of the Rosidae is characterized by a discoid extension of the receptacle, frequently forming a nectar-producing 'disc', and a tendency towards reduction in the number of ovules. In the superorder **Rutanae** (orders 9–12) primitive characters are still prevalent: the perianth consists generally of a pentamerous, actinomorphic or zygomorphic calyx and corolla, the stamens are usually in two whorls, the gynaecium is superior and syncarpous (rarely still choricarpous), sometimes still with five carpels and numerous ovules, and the leaves are often compound or divided. The Rutanae may originate from woody Saxifragales, or probably more directly from the Magnoliidae.

Order 9: **Rutales** (= Terebinthales in part) consists mainly of tropical woody plants with secretory cavities containing oils, resins and balsams (hence medicinal and other uses), and with radial flowers often having an intrastaminal disc (Fig. 700D). In the **Rutaceae** lysigenous secretory cavities (sometimes visible as transparent dots on the leaves or fruits: Fig. 129B, C) contain strongly odorous ethereal oils. The most important genus is *Citrus*; the species are small evergreen trees originating in S. Asia and now cultivated in numerous forms in all warm climates, e.g. in the Mediterranean, where they became known through the expeditions of Alexander the Great. Of particular note are: *C. sinensis* (orange, Fig. 700A), *C. aurantium* subsp. *amara* (Seville or bitter orange), *C. maxima* (shaddock), *C. paradisi* (grapefruit), *C. limon* (lemon), *C. medica* (citron) and *C. reticulata* (mandarin). These fruits are berries; the number of carpels is often increased. The edible part consists of juicy emergences arising from the sub-epidermal tissue of the inner side of the carpel wall and projecting into the loculi (pulpa, pp. 102, 670). The warmth-demanding dittany (*Dictamnus albus*) with slightly zygomorphic flowers, native in C. Europe, and the Mediterranean rue (*Ruta graveolens*, Fig. 700B–D), with greenish-yellow flowers, are subshrubs or perennial herbs. Resin canals are characteristic of the **Anacardiaceae** (with *Anacardium occidentale*, the cashew nut, p. 670; *Pistacia*, the Mediterranean species yielding mastic and bearing edible seeds; *Rhus*, yielding dyes and lacquers, some poisonous to touch; *Mangifera indica*, the mango, an important tropical fruit) and the

Fig. 700. Rutales, Rutaceae. **A.** *Citrus sinensis*, flowering shoot. ×½. **B.** to **D.** *Ruta graveolens*, flowering shoot (×½); tetramerous lateral flower and diagram of a pentamerous apical flower with disc (d). **A.** to **C.** After Karsten. **D.** After Eichler.

Burseraceae (*Commiphora*, myrrh; *Boswellia*, frankincense or olibanum; p. 56), whilst bitter substances are characteristic of the **Simaroubaceae** (*Quassia, Simarouba* and *Picrasma* with pharmaceutically important bitter barks and woods; the E. Asiatic tree of heaven, *Ailanthus altissima* is frequently cultivated).

Order 10: **Sapindales** (= Terebinthales in part) also comprises woody plants but without secretory cavities and generally with more or less zygomorphic flowers and an extrastaminal disc (Fig. 701A). Related to the mainly tropical **Sapindaceae** are the **Hippocastanaceae** of the Northern Hemisphere which include the horse-chestnut (*Aesculus hippocastanum*, Fig. 181B), native in the mountains of the Balkans but widely planted. The **Aceraceae** are characterized by the loss of one whorl of stamens and the formation of schizocarps (Figs. 679C, 701A–C); the leaves are opposite and generally palmately lobed (Fig. 181A). This family includes only the genus *Acer*: *A. pseudoplatanus*, the sycamore, is widespread in the mountains of C. Europe, whilst *A. platanoides*, the early flowering Norway maple, and *A. campestre*, the hedge maple, are native at lower altitudes in C. Europe. Noteworthy is the tendency to develop unisexual and sometimes greatly reduced flowers, correlated with the transition from insect to wind pollination (Fig. 701B, C; p. 664); indeed the N. American box elder (*A. negundo*) is dioecious. The pinnate-leaved and haplostemonous **Staphyleaceae** (with the warmth-demanding bladder nut, *Staphylea pinnata*) link the Saxifragales to the Celastrales (pp. 693, 702).

A predominantly herbaceous line of development, generally without secretory cavities or a disc, is exhibited by:

Order 11: **Geraniales** (= Gruinales). The flowers range from actinomorphic to zygomorphic. The androecium is usually obdiplostemonous, more rarely haplostemonous by the loss of the whorl of stamens opposite the petals. The superior ovary often becomes an explosive fruit (sling fruit). A typical floral formula: * or ↓ K5 C5 A5 + 5 G($\underline{5}$) (Fig. 702A, B, D, E, I).

The **Oxalidaceae** still have free styles, several-seeded carpels, two whorls of stamens and actinomorphic flowers (Fig. 702A). The native wood sorrel (*Oxalis acetosella*, Fig.

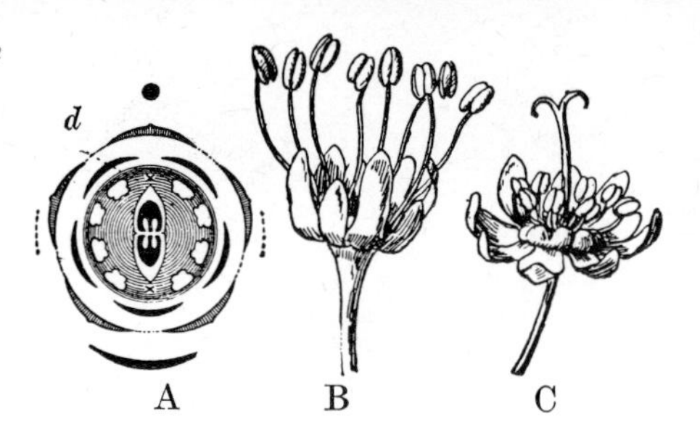

Fig. 701. Sapindales, Aceraceae, *Acer pseudo-platanus*. **A.** Floral diagram (disc, d). **B.** ♂, **C.** ♀ flower with rudimentary stamens. ×2 approx. **A.** After Eichler. **B.** and **C.** After Karsten.

702A) is notable for the movements of its digitate leaflets and the discharge of its seeds from the capsule. Reduction in the androecium (A20 → 5) or in the number of ovules is characteristic of the following two families: the **Linaceae** (Fig. 702E–H) include the annual, narrow-leaved, blue-flowered flax (*Linum usitatissimum*), one of the most ancient of cultivated plants; the phloem fibres of the stem (pp. 12, 116, Fig. 151A) are used in the manufacture of linen (p. 268), whilst the seeds, of which one is borne in each of the ten loculi of the capsule, yield linseed oil. The **Erythroxylaceae** include *Erythroxylum coca* and other S. American species which yield the alkaloid cocaine. The **Zygophyllaceae** (p. 674) occur mainly in deserts and salt steppes and include *Guajacum* from which are obtained various woods and resins also used to some extent in medicine.

United styles and two (or three) whorls of stamens are found in the **Geraniaceae** (Fig. 702B). The fruit is peculiar: the carpels are much elongated and each bears at the base only two ovules, of which only one develops further, whilst the upper sterile part forms a 'beak'. When ripe, the inner parts of the united carpels remain as a central column whilst the outer wall, enclosing one seed at the base, lifts off. In many species of cranesbill (*Geranium*, Fig. 702C) these outer parts of the carpel wall remain attached to the top of the column and the seed is catapulted away, whilst in storksbill (*Erodium*, Fig. 325) they become completely detached as mericarps each containing a seed, and the upper sterile part acts as a hygroscopic awn which drills the seed into the soil. In *Pelargonium*, popular decorative plants from S. and C. Africa, the flowers are dorsiventral with a spur united to the pedicel along its entire length; Fig. 702D). Free spurs are characteristic of two smaller families, the diplostemonous **Tropaeolaceae**, which includes the S. American 'nasturtium', *Tropaeolum majus*, Fig. 232), and the haplostemonous **Balsaminaceae**, with the touch-me-not (*Impatiens*) which discharges the seeds explosively (cf. p. 376, Fig. 320).

Finally, dorsiventral flowers are also characteristic of:

Order 12: **Polygalales**; these are also mostly herbaceous but with simple leaves. The flowers of the **Polygalaceae** (Fig. 702I, K) have two petaloid lateral sepals and a boat-shaped anterior petal furnished with a laciniate appendage; they therefore superficially resemble the butterfly-flowers of the Fabaceae, as in which the stamens (here usually eight) are united in an incomplete ring, open above.

Compared with the Rutanae the superorder **Celastranae** (orders 13–16) is more advanced: the perianth is usually pentamerous or tetramerous, actinomorphic, heterochlamydeous or with the corolla lacking, the stamens are usually in a single whorl, a disc is often present and the gynaecium is syncarpous or secondarily paracarpous, superior or half-inferior, with a progressive reduction in the number of carpels and ovules; the leaves are generally simple and undivided. The mainly woody Celastranae can also be traced back to the Saxifragales.

Hermaphrodite flowers with calyx and corolla and sometimes two but generally only one whorl of stamens (which are then episepalous, i.e. standing in front of the sepals) are found in:

Order 13: **Celastrales.** Here belong, amongst others, the **Aquifoliaceae** (Fig. 703A) with the holly (*Ilex aquifolium*), an evergreen Mediterranean–Atlantic shrub or tree with red drupes, and *I. paraguariensis* from S. America (maté tea); and the **Celastraceae** (Fig. 703B) with the spindle (*Euonymus europaeus*), a native shrub, and other species which have black seeds surrounded by a brilliant orange-red aril.

A parallel group is formed by:

Order 14: **Rhamnales** in which only the epipetalous whorl of stamens remains. The

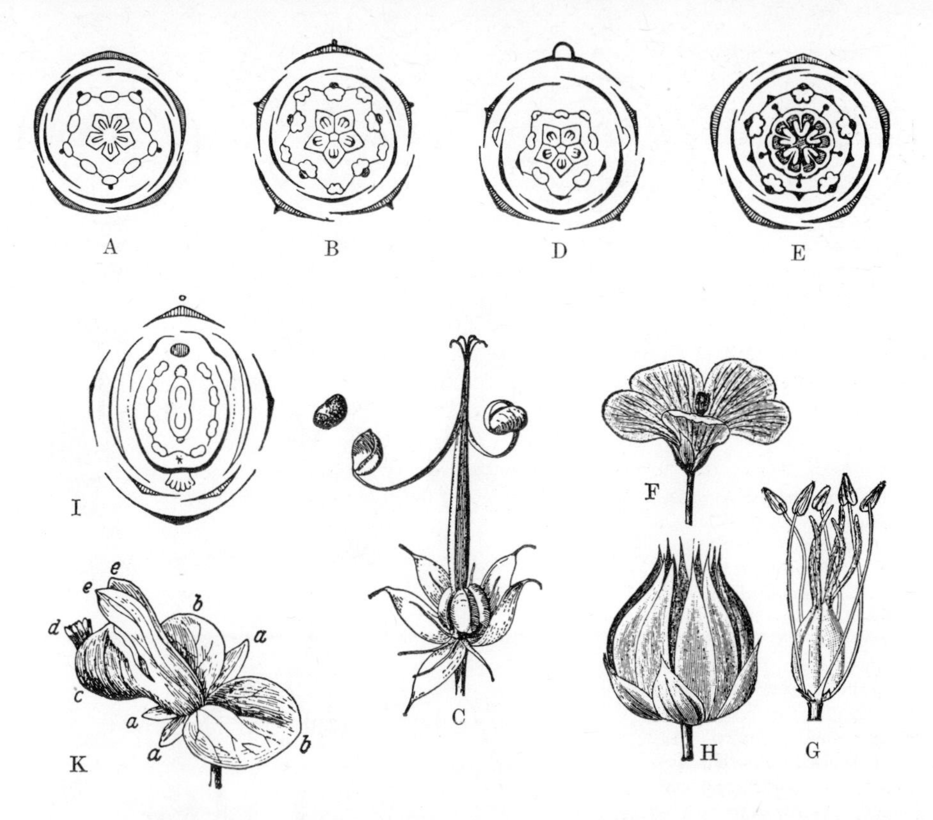

Fig. 702. Geraniales (**A.** to **H.**) and Polygalales (**I.** and **K.**). Floral diagrams: **A.** *Oxalis acetosella.* **B.** *Geranium pratense.* **D.** *Pelargonium zonale.* **E.** *Linum austriacum.* **I.** *Polygala myrtifolia* (formation of disc adaxial to the gynaecium). **C.** *Geranium sanguineum*, fruit ejecting a seed. ×1.6. **F.** to **H.** *Linum usitatissimum*, flower (×1); perianth removed and rudiments of the outer whorl of stamens visible (×3); capsule (×3). **K.** *Polygala seneca*, flower, with (a) greenish and (b) petaloid sepals; (c) anterior petal with (d) appendages, united at the base with (e) the lateral petals. Enlarged. **A., B., D., E.** and **I.** After Eichler. **C.** After Firbas. **F.** and **G.** After Schenck. **K.** After Berg and Schmidt.

Rhamnaceae (Fig. 703C–E) are characterized by a concave receptacle and a half-inferior or inferior ovary, amongst other features. Whilst the flowers of alder buckthorn (*Frangula alnus*), frequent especially in wet woods, are pentamerous and hermaphrodite, those of *Rhamnus*, e.g. the purging buckthorn (*R. catharticus*), are tetramerous and dioecious by the suppression of one or other sex. The fruits are drupaceous with two to four thin-walled stones. The most important member of the **Vitaceae** (Fig. 703F, G) is the grape vine (*Vitis vinifera*), an ancient plant of cultivation, of which numerous varieties are now grown; one of its progenitors was the wild vine (subsp. *sylvestris*) found in riverside forests in the Mediterranean region as well as in the Rhine and Danube valleys. The grape vine is a liane with leaf-opposed tendrils which are interpreted as the ends of the individual members of a sympodial shoot system (cf. p. 134, Fig. 143). The branch system comprises long shoots with short shoots in their axils. The short shoots die back in the autumn as far as a basal axillary bud which grows out in the following year as a long shoot. The inflorescences (panicles) occupy a similar position to the tendrils; the petals are united at the tips and are pushed off together at anthesis (Fig. 703G). The fruit is a four-seeded berry. In some of the species of *Parthenocissus* commonly cultivated as Virginia creeper and Boston ivy the tips of the tendrils are modified as adhesive discs (Fig. 213C).

The two following orders (15, 16) are predominantly monochlamydeous lines of development, possibly from ancestral types resembling the Celastrales: generally loss of the corolla leaves a simple perianth. Furthermore the flowers are unisexual in:

Order 15: **Euphorbiales** (= Tricoccae), which is perhaps related to the Malvanae (pp. 713 et seq.). The superior ovary commonly has three loculi, each containing one pendulous anatropous ovule (rarely two). Close to the Celastrales is the **Buxaceae** with

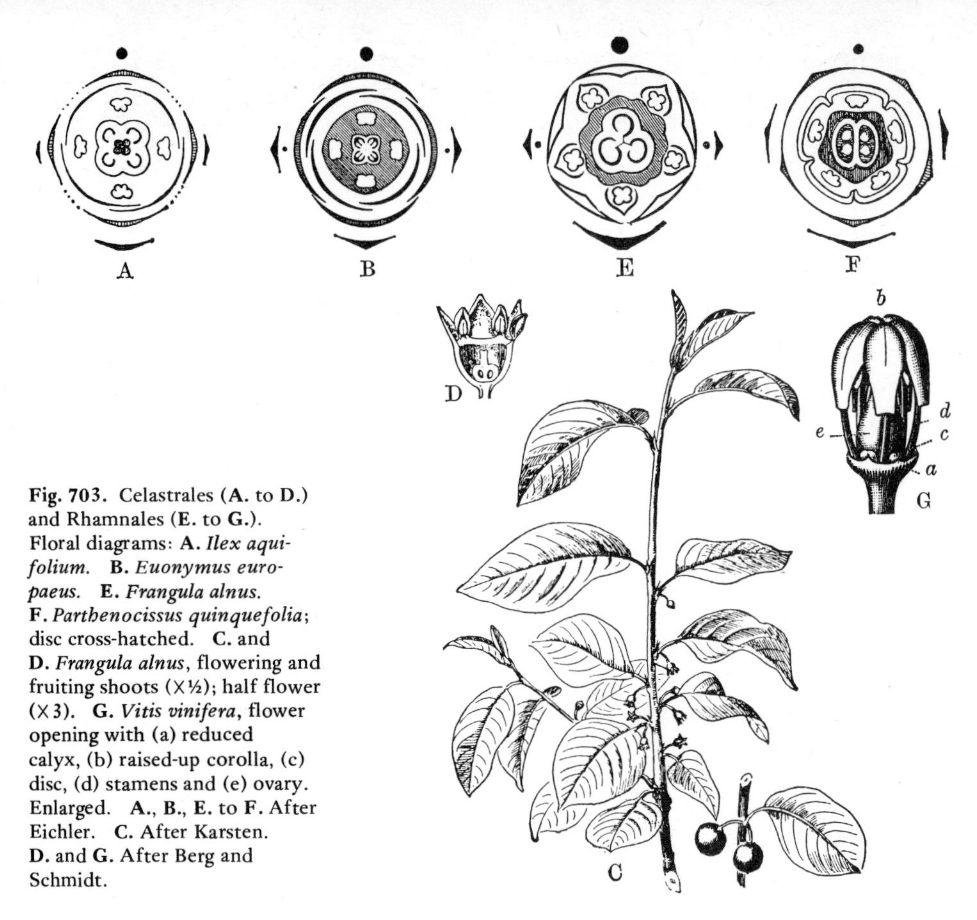

Fig. 703. Celastrales (A. to D.) and Rhamnales (E. to G.). Floral diagrams: A. *Ilex aquifolium.* B. *Euonymus europaeus.* E. *Frangula alnus.* F. *Parthenocissus quinquefolia*; disc cross-hatched. C. and D. *Frangula alnus*, flowering and fruiting shoots (×½); half flower (×3). G. *Vitis vinifera*, flower opening with (a) reduced calyx, (b) raised-up corolla, (c) disc, (d) stamens and (e) ovary. Enlarged. A., B., E. to F. After Eichler. C. After Karsten. D. and G. After Berg and Schmidt.

the Mediterranean–Atlantic, evergreen box (*Buxus sempervirens*). Of particular note is the large family **Euphorbiaceae**, predominantly tropical but also represented in Europe. They are woody or herbaceous with stipulate leaves, or the leaves are suppressed and photosynthesis is taken over by the stems, e.g. in the stem-succulent *Euphorbia* species of the African savannas and semi-deserts; these resemble Cactaceae and provide an excellent example of convergent evolution (Fig. 203B); their leaves are often reduced to a pair of stipular spines (Fig. 704G).

The flowers and inflorescences are also very diverse. The tropical oil plant *Jatropha curcas* (Fig. 704A, B) and many tropical *Croton* species still have both calyx and corolla. Flowers with a single perianth occur, for example, in the European, anemogamous, dioecious species of *Mercurialis* (Fig. 704C, D); in this genus the perianth is trimerous; the ♂ flower has numerous stamens whilst the ♀ has a bi- or tricarpellary ovary as well as three staminodes. The monoecious castor oil plant (*Ricinus communis*, Fig. 704E, F) has similar flowers, but the perianth is usually pentamerous and the stamens are freely branched; it is a tropical African tree with large palmately lobed leaves but in European gardens it is grown only as a vigorous annual herb.

In *Euphorbia* (spurge) the individual flowers are extremely reduced, but they are united in characteristic pseudanthia (cyathia). Each cyathium (Fig. 704H–K) has a single terminal long-stalked but pendulous ♀ flower (in most species without a perianth) surrounded by five groups of stalked and naked ♂ flowers which are apparently arranged in cincinni (cf. p. 137). Each ♂ flower consists merely of a single stamen delimited from the pedicel by a constriction (Fig. 704L). The entire inflorescence is surrounded by five bracts (the subtending bracts of the ♂ partial inflorescences) which generally alternate with elliptical or crescentic nectary-glands. The cyathia themselves are borne in dichasial or pleiochasial compound inflorescences (Fig. 276). That the cyathium (which Linnaeus took for a hermaphrodite flower) is really an inflorescence is shown, *inter alia*, by the articulation of the stamen and its pedicel. In related genera (e.g. *Anthostema*) a simple perianth is still present at the junction (Fig. 704O). The unisexual flowers of the cyathium thus form a pseudanthium, insect-pollinated like the hermaphrodite flower which it resembles. This is clearly related to the development of secondary entomophily from anemophily (cf. p. 664).

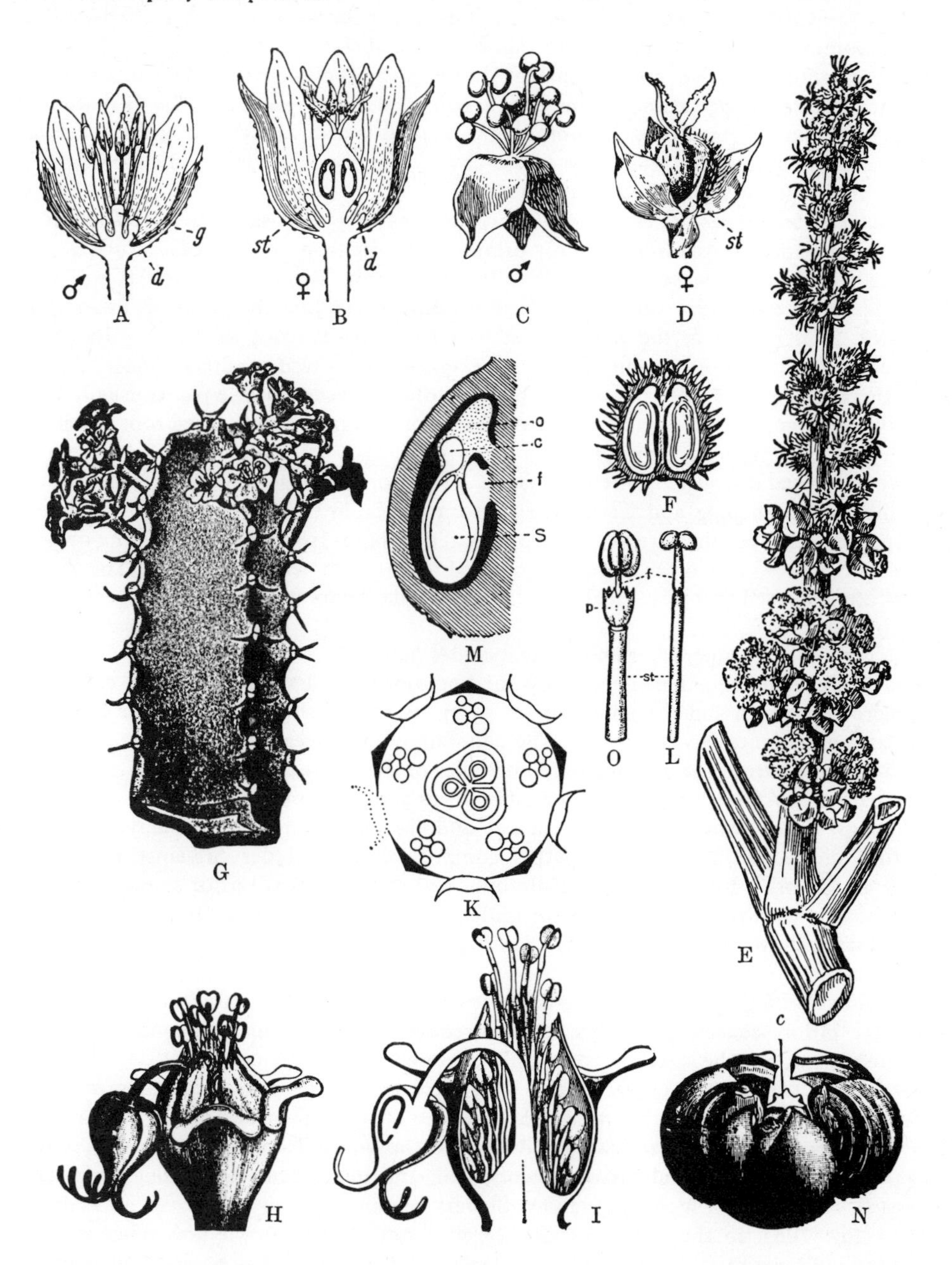

Fig. 704. Euphorbiales, Euphorbiaceae. **A.** and **B.** ♂ and ♀ flowers of *Jatropha curcas*, and of *Mercurialis annua*, (**C.** and **D.**) with (d) scales of the disc, (g) androphore, (st) staminodes. **E.** and **F.** *Ricinus communis*, inflorescence (×½) and immature fruit in L.S. **G.** to **N.** *Euphorbia*. **G.** *E. resinifera*, succulent flowering shoot. ×1. **H.** to **K.** Cyathium, as a whole, in half, and diagram (the gland shown dotted may be absent). **L.** ♂ flower of *E. platyphyllos* with pedicel (st) and filament (f). **M.** Loculus of the ovary (L.S.) of *E. myrsinites* with (S) ovule, (f) funicle, (c) caruncle and (o) obturator (diagrammatic). **N.** Fruit; capsule opening septicidally, dorsicidally and septrifragally, leaving the persistent central columella (c); *E. lathyris*; enlarged. **O.** ♂ flower of *Anthostema senegalense* with perigone (p). Enlarged; cf. L. **A.**, **B.** and **L.** After Pax. **C.** and **D.** After Wettstein, modified. **E.** to **F.** After Karsten. **G.** After Berg and Schmidt. **H.**, **I.**, **N.** and **O.** After Baillon. **K.** After Eichler, modified. **M.** After Schweiger.

Fertilization generally takes place via the 'obturator', an outgrowth from the placenta over the micropyle; this nourishes the pollen tube and guides it towards the ovule (Fig. 704M). The fruit is a capsule, the walls of which completely separate from the central 'columella' and in so doing discharge the seeds (Fig. 704N).

The latex (often poisonous) of many Euphorbiaceae contains caoutchouc; hence they include some of the most important rubber trees, above all *Hevea brasiliensis* (Para rubber). Originally native in the Amazon basin, this is now grown in many tropical countries. Ceará rubber comes from the Brazilian *Manihot glaziovii*. The herbaceous manioc or cassava (*Manihot esculenta* = *M. utilissima*) is also native in tropical America and grown throughout the Tropics for its starchy edible root tubers (tapioca starch).

Also predominantly homoiochlamydeous is:

Order 16: **Santalales.** These show various degrees of parasitism with corresponding modifications and reductions. The most primitive are the tropical **Olacaceae** which stand close to the Celastrales; some are still fully autotrophic, with calyx, corolla and two whorls of stamens. More advanced are the **Santalaceae**, mainly green semiparasites (pp. 199, 292) rooted in the ground; the root haustoria abstract water and salts from their hosts (as in the European species of *Thesium* or bastard toadflax). Most of the **Loranthaceae** (Fig. 705) also have green, photosynthetic leaves, but the plants usually live entirely above ground on woody plants and have correspondingly modified root systems. The deciduous European *Loranthus europaeus* (on Fagaceae) produces 'sinkers' from a massive haustorial disc. The mistletoe (*Viscum album*, p. 199, Figs. 220, 705) is dioecious with dichasial branching and leathery, evergreen leaves; in C. Europe it occurs in three host-specific races confined to silver fir, pine and deciduous trees respectively. The ovules of the Loranthaceae are often completely united to the ovary wall (Fig. 705C). The slimy berries are dispersed by birds. Finally, the tropical **Balanophoraceae** are highly specialized holoparasites.

The very isolated superorder **Proteanae** includes only:

Order 17: **Proteales**, characterized by the homoiochlamydeous but brightly coloured tetramerous perianth and the unicarpellary ovary. A relationship with primitive Rosidae or Celastranae–Santalales is possible. The only family **Proteaceae** contains sclerophyllous trees and shrubs from the Tropics and Subtropics of the Southern Hemisphere, especially Australia and S. Africa: they are pollinated by birds and marsupials.

Amongst these Rosidae which have a disc and are still mainly heterochlamydeous the two final orders (18, 19) which together comprise the superorder **Aralianae** (Umbelliflorae) are greatly advanced, with usually only one episepalous whorl of stamens and an inferior ovary with typically only one pendulous ovule in each loculus. Here too there are close relationships with the woody Saxifragales as well as to the more primitive Celastranae.

The flowers of the Aralianae are small, mostly hermaphrodite and generally borne in umbellate inflorescences. The calyx is often reduced and the ovule frequently has only one integument, but the seed still commonly contains abundant endosperm.

Order 18: **Cornales** is characterized by simple leaves, the general absence of resin canals, a woody habit, and the occasional presence of more than one whorl of stamens. The **Cornaceae** have opposite leaves and include the white-flowered dogwood (*Cornus sanguinea*), a common shrub of open woodland, and the warmth-demanding Cornelian cherry (*Cornus mas*) which has yellow flowers opening before the leaves and edible drupes (Fig. 706). In the Early Tertiary the subfamily Mastixioideae now relict in S.E. Asia was widespread in Europe. The N. American **Garryaceae** has secondarily anemophilous catkins.

Order 19: **Araliales** is characterized by compound or lobed leaves, by schizogenous resin canals with ethereal oils and gum-resins, and by the transition from woody to herbaceous life-forms.

The woody **Araliaceae** are predominantly tropical but include our native ivy (*Hedera helix*), well known as a root climber and for its heterophylly (lobed juvenile or shade leaves, rhomboidal adult or sun leaves on the flowering shoots; pp. 168, 342). Ivy is pollinated in the autumn by flies and wasps and the berries ripen in the following spring.

The family **Apiaceae** (Umbelliferae, umbellifers) is almost entirely herbaceous. The

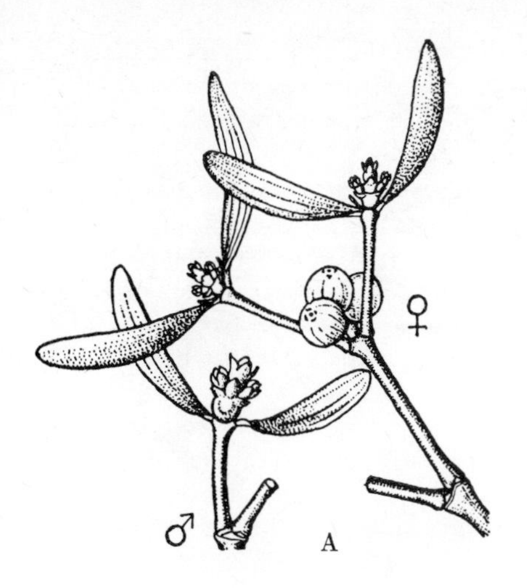

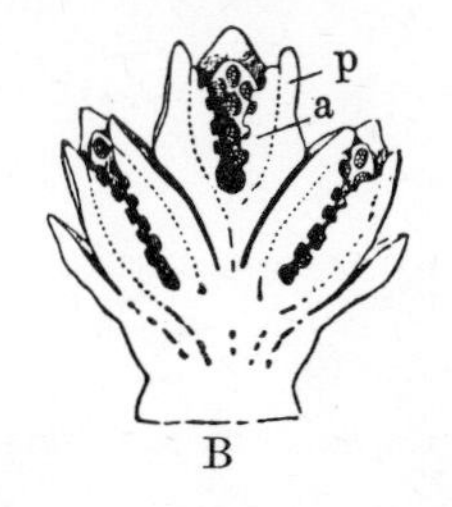

Fig. 705. Santalales, Loranthaceae, *Viscum album.* **A.** Shoots with ♂ and ♀ flowers or fruits. ×½. **B.** ♂ and **C.** ♀ three-flowered dichasia (in L.S.); perianth segments united to the stamens (a) or ovary and ovule (g) respectively: [P4 + A4] and P4 + G($\overline{2}$). ×3 approx. After Firbas.

Fig. 706. Cornales, Cornaceae, *Cornus mas.* **A.** Flowering and **B.** fruiting shoots. ×½. **C.** and **D.** Flower from above and half flower. Enlarged. After Karsten.

habit is very characteristic (Fig. 707B, C): the stems have hollow internodes and bear alternate leaves which are usually divided—often compound—and have broadly sheathing bases. The inflorescence is nearly always a compound umbel (an umbel of partial umbels or umbellules), its bracts forming an involucre and partial involucres. The minute flowers, conspicuous only by being massed together, are white, or less commonly pink or yellow; the typical floral formula is * K5 C5 A5 G($\overline{2}$) (Fig. 707A, D), but the calyx is almost always much reduced. The petals are often incurved at the tips. The styles are borne on top of the ovary on a swollen cushion which functions as a nectar-secreting disc; each of the two loculi contains a single anatropous ovule hanging from the septum (commisural surface; Fig. 707F); a second ovule in each loculus aborts at an early stage. The seed (Fig. 707G, H) contains a small embryo embedded in a copious fat- and protein-rich endosperm, whilst the testa fuses with the fruit wall or pericarp forming a dry schizocarp which splits along the commisural surface into two one-seeded partial fruits or mericarps; at first these hang from a carpophore from which they eventually separate.

The structure of the fruit is extremely characteristic and its details are of great taxonomic importance. The fruit wall (pericarp) of each partial fruit (mericarp) has five main vascular bundles over each of which a primary ridge (two lateral, three dorsal ridges) occurs (Fig. 707H). Between these lie furrows or valleculae in which secondary ridges

Fig. 707. Araliales, Apiaceae. **A.** Floral diagram of *Laser trilobum*. **D.** Flower of *Ammi majus* (×12). **B., C., E.** and **F.** *Conium maculatum*; shoot (×½), umbel, fruit, as a whole and in L.S., with two pendulous ovules (enlarged). **G.** and **H.** Schizocarp of *Carum carvi*, as a whole and in L.S. (×10) and T.S. (×25), with (cp) carpophore, (f) pericarp, (h) primary ridge with vascular bundle, (t) vallecula with vitta below, (s) testa, (e) endosperm and (em) embryo. **A.** After Noll, modified. **B., C.** and **E.** After Karsten. **D.** After Thellung. **F.** After Tschirch and Oesterle. **G.** and **H.** After Berg and Schmidt, somewhat modified.

may be developed. Schizogenous secretory ducts or vittae run under the valleculae and the commisural surface, more rarely under the primary ridges. The distribution of the vittae, the character of the ridges (often winged or spiny), the form of the endosperm, the occurrence of hairs, etc., furnish important characters by which, for example, umbelliferous fruits used as seed or drugs, and any adulterants, can be identified with certainty.

A few umbellifers have undivided leaves, e.g. species of *Bupleurum* and the peltate-leaved marsh pennywort (*Hydrocotyle vulgaris*). In some genera, e.g. *Heracleum* (hogweed), the petals toward the outside of the umbel are enlarged, slightly so near the centre of each umbel or umbellule, more so at the margin, so that the flowers are zygomorphic. In some genera the visual attractiveness of the inflorescence is enhanced by coloured bracts, e.g. by the white involucral bracts of the (simple) umbels of *Astrantia* or the yellow partial involucres of *Bupleurum*. Flies, beetles and other short-tongued insects are the main pollinators; the flowers are almost always protandrous.

More than 3000 species of umbellifers occur in steppes, swamps, meadows and woodlands, mainly in the extratropical regions of the Northern Hemisphere. Gigantic perennial forms, several metres high, occur especially in the steppes of C. Asia (e.g. *Ferula*). In

Europe they are highly characteristic of well-manured meadows (e.g. *Heracleum*). The importance of a large number as condiments, spices and medicinal plants is explained by their high content of ethereal oils; fruits, leaves and roots are all used. Prominent amongst these are the fine-leaved, white-flowered caraway (*Carum carvi*, Fig. 707G, H), found in meadows, the commonly cultivated aniseed (*Pimpinella anisum*) and coriander (*Coriandrum sativum*), and the yellow-flowered dill (*Anethum graveolens*), lovage (*Levisticum officinale*), fennel (*Foeniculum vulgare*) and parsley (*Petroselinum hortense*). Some have swollen edible roots such as the carrot (*Daucus carota*), parsnip (*Pastinaca sativa*) and celeriac (*Apium graveolens*; root plus tuberous hypo- and epicotyl). Poisonous members include the unpleasant smelling hemlock (*Conium maculatum*, with red-spotted stems, Fig. 707B, C, E, F) and the cowbane (*Cicuta virosa*), notable for its chambered rhizome.

Subclass D: Dilleniidae

In the 'middle storeys' of the dicotyledons (cf. pp. 691 et seq.) the Dilleniidae are distinguished from the Rosidae principally by the centrifugal development of the stamens in those groups which are secondarily polyandrous. Compared with the Rosidae the Dilleniidae show a less marked tendency towards the reduction of the androecium to a single whorl of stamens (haplostemony), whilst concave and disc-shaped receptacles are less in evidence. Other characteristics are the syncarpous and especially paracarpous ovaries with numerous ovules and the predominantly simple (not compound) leaves. In contrast to the Rosidae the choripetalous Dilleniidae have often given rise to closely related sympetalous groups. For the rest, various similarities between woody Rosidae–Saxifragales and primitive Dilleniidae suggest their common origins.

The Dilleniidae have also to be divided into superorders. The **Dillenianae** (Cistiflorae, Parietales in a broad sense) have, as well as some primitively choricarpous forms, predominantly syncarpous and especially paracarpous groups with numerous ovules. The syncarpous Malvanae (p. 713) on the other hand are characterized by reduction in the number of ovules. Whilst the Dillenianae and Malvanae are predominantly choripetalous, sympetaly is characteristic of the Ericanae (p. 715).

The Dillenianae (Orders 1–7) start with:

Order 1: **Dilleniales** which has still choricarpous gynaecia, secondarily polyandrous androecia and seeds with a fleshy aril, abundant endosperm and a small embryo. Besides the (sub)tropical and mainly woody **Dilleniaceae**, the subshrubby or herbaceous **Paeoniaceae** of the Northern Hemisphere are of note. The sole genus *Paeonia* (peony, Fig. 652C) is remarkable for a free nuclear stage in the early development of the embryo.

Order 2: **Theales** (= Guttiferae) has superior coenocarpous–syncarpous gynaecia with central-axile placentation and seeds with reduced endosperm and an enlarged embryo; the habit is usually woody. The **Theaceae** have the perianth in part spirally arranged and

Fig. 708. Theales. **A.** *Camellia sinensis*, flowering shoot (×¼), fruit and seed. **B.** *Hypericum maculatum*, floral diagram. **A.** After Karsten. **B.** After Eichler.

include the tea plant (*Camellia* or *Thea sinensis*, Fig. 708A), grown especially in China, Japan and India, and the camellia (*C. japonica*). The **Hypericaceae** have schizogenous secretory cavities (Fig. 129A) and include the native St John's worts (*Hypericum*, Figs. 656B, 708B), whilst the **Dipterocarpaceae** are important palaeotropical forest trees which also yield resin (Fig. 758).

Another coenocarpous–paracarpous line of development with parietal placentation, endosperm, a tendency towards modification of the receptacle, pentamerous calyx and corolla, and hermaphrodite flowers is shown by:

Order 3: **Violales** (Cistales, Parietales in the narrow sense). The first of a group of families with an oily endosperm is the tropical, woody **Flacourtiaceae** with actinomorphic flowers and numerous stamens. A transition to herbaceous forms occurs in the **Violaceae** which have only five stamens and zygomorphic flowers; *Viola* (violet, pansy) has the anterior petal produced as a spur into which processes of the two anterior stamens secrete nectar (Fig. 709). The **Passifloraceae** comprises shoot-tendril climbers (p. 193) with an androgynophore; the (sub)tropical members include the frequently grown passion flower (*Passiflora caerulea*). Sympetaly occurs in the **Caricaceae** which includes the pawpaw (*Carica papaya*), widely grown in the Tropics. A starchy endosperm and choripetalous, actinomorphic flowers are characteristic of the two following families: the **Cistaceae** with the genus *Cistus*, shrubs characteristic of the Mediterranean maquis (macchia; p. 786) with aromatic resins and large, brightly-coloured but quickly falling petals, and the various native species of *Helianthemum* (rock rose) found in dry pastures; and the **Tamaricaceae**, shrubs with scale-leaves (Fig. 195A), including the tamarisk (*Tamarix*) of saline soils, but also planted in gardens.

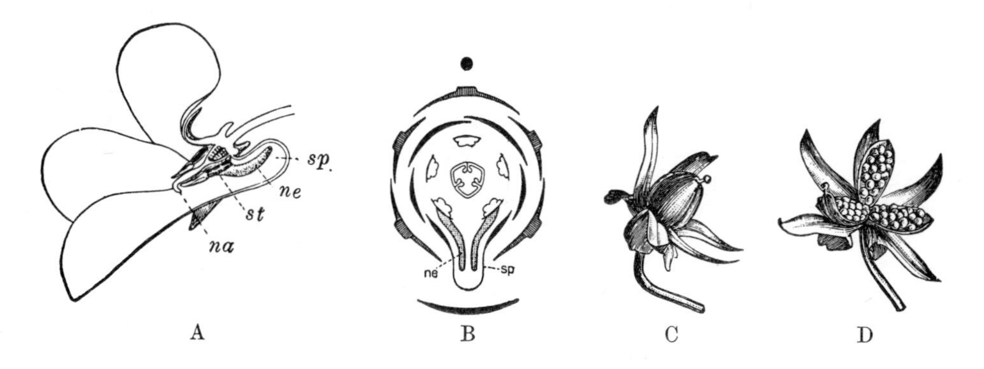

Fig. 709. Violales, *Viola*. **A.** and **B.** *V. odorata*, half flower (×2.3) and floral diagram; sp, spur; st, stamens, some with nectary appendages (ne); na, stigma. **C.** and **D.** *V. tricolor*, dorsicidal capsule before and after dehiscence. **A.** and **B.** After Firbas. **C.** and **D.** After Schimper.

Related to the Violales is:

Order 4: **Capparales** (= Rhoeadales in part). The hermaphrodite flowers are often tetramerous. Also characteristic are the spiral phyllotaxis, gyno- or androgynophores (p. 645), the development of a disc or nectary glands, parietal placentation, seeds without endosperm when ripe, and myrosin cells: when the tissues are damaged these often tubular idioblasts liberate the enzyme myrosinase which breaks down the mustard oil glycosides present in other cells—hence the characteristic pungent taste of many Capparales (capers, mustard, radish).

Formerly the Capparales were often united with the Papaverales (p. 684) in the order Rhoeadales, but their very different cell contents and serological, embryological and palynological findings suggest the present more natural arrangement.

The (sub)tropical, mainly woody family **Capparaceae** (Capparidaceae) can be placed at the beginning of this order. The flower buds of *Capparis spinosa*, a small shrub of rocky places in the Mediterranean, are used as a condiment (capers).

The **Brassicaceae** (Cruciferae, crucifers) are particularly important and easily recognized by their flowers. They are generally herbaceous perennials or annuals with racemose inflorescences lacking bracts, bracteoles and terminal flowers. The disymmetric flowers

(Fig. 710A–C) have four sepals in a whorl, four petals alternating with the sepals, two outer shorter and four longer inner stamens and one superior, often more or less stalked ovary divided by a septum; this ovary consists of two fertile carpels and probably also of two sterile ones. Thus as 'crucifers' the floral formula is K4 C4 A2 : 0 + 4 G($\underline{4}$) or G($\underline{2}$). The fruit is usually a siliqua (or if less than three times as long as broad, a silicule; Fig. 710D–H). When the fruit dehisces the membranous septum remains stretched between the placentas with the seeds still attached at first. The seeds develop from campylotropous ovules and contain a curved oily embryo (Fig. 710I–L).

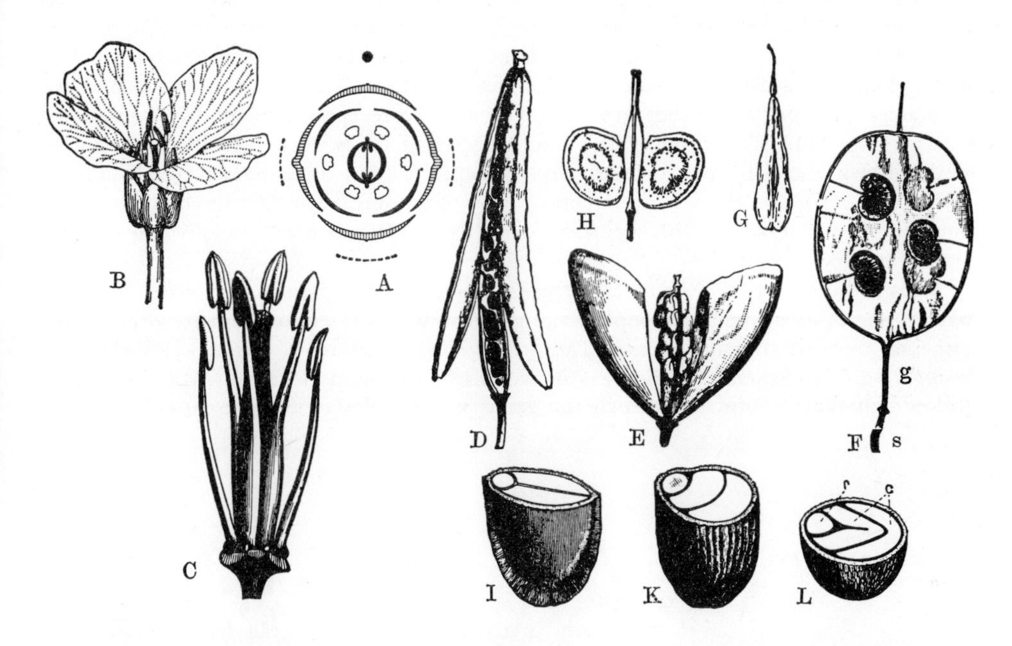

Fig. 710. Capparales, Brassicaceae. **A.** Floral diagram. **B.** and **C.** Flower with (×2) and without perianth (with nectaries at the base; ×4; *Cardamine pratensis*). Fruits of: **D.** *Erysimum cheiri* (siliqua). **E.** *Capsella bursa-pastoris* (silicule). **F.** *Lunaria annua* (silicule, valves removed, hyaline replum exposed; s, pedicel; g, gynophore). **G.** *Isatis tinctoria* (one or two-seeded winged nut). **H.** *Biscutella laevigata* (schizocarp). **I.** to **L.** Seeds cut transversely showing various orientations of the embryo with cotyledons (c), hypocotyl and radicle (r). **I.** *Erysimum cheiri* ('pleurorhizal' ×8); *Alliaria petiolata* ('notorhizal', ×7). **L.** *Brassica nigra* ('orthoplocous', ×9). **A.** After Eichler and Alexander. **B., D., G.** and **H.** After Firbas. **C., E., I.** to **L.** After Baillon.

The floral structure of the Brassicaceae can be understood by reference to the more primitive relationships in the Capparaceae. Earlier attempts to seek homologies in the dimerous flowers of the Papaveraceae have now been abandoned. Important characters for the classification of this large family include the type of fruit (besides the dehiscent siliquae there are such indehiscent types as lomentose siliquae (Fig. 679E), schizocarp-silicules (Fig. 710H) and one- or few-seeded nut-silicules (Fig. 710G)), the orientation of the embryo in the seed (Fig. 710I–L) and the arrangement of the nectaries (most species are insect-pollinated).

The crucifers are mainly found in the extratropical regions of the Northern Hemisphere where in the Arctic and on the highest mountains they reach the extreme limits of vascular plants. The polyploid complex in *Biscutella laevigata* (Fig. 363) is of interest in the history of the flora. Many species have accompanied Man as autogamous weeds of cultivation and as ruderals (e.g. *Capsella bursa-pastoris*, species of *Lepidium* and *Thlaspi*). Economically important are: (1) vegetables and fodder crops such as the many forms of kale and cabbage (*Brassica oleracea*, Figs. 206A, B, 350), the turnip (*B. rapa* subsp. *rapa*), the rutabaga (*B. napus* subsp. *rapifera*), the black and red radishes (*Raphanus sativus*); (2) oil plants and condiments such as rape (*Brassica napus* subsp. *napus*), turnip-rape (*B. rapa* subsp. *oleifera*), black mustard (*B. nigra*) and white mustard (*Sinapis alba*), horseradish

(*Armoracia rusticana*); and (3) many ornamental plants such as the wallflower (*Erysimum* = *Cheiranthus cheiri*), stock (*Matthiola*), candytuft (*Iberis*), etc.

Finally, the **Resedaceae** have weakly dorsiventral flowers. In various species of *Reseda* the carpels are not united above so that the ovules remain visible.

There is a relationship between the Violales (e.g. Flacourtiaceae–Tamaricaceae) and the apetalous:

Order 5: **Salicales**, formerly placed in the Amentiferae. The only family is the **Salicaceae** (Fig. 711) in which the unisexual, dioecious flowers with little or no perianth and the catkin-like inflorescences are characteristic; the paracarpous ovary of two carpels contains numerous seeds with long hairs and no endosperm. The two main genera, *Populus* (poplar) and *Salix* (willow), are trees or shrubs with simple, alternate, stipulate leaves. The catkins usually flower before the leaves unfold. The flowers, borne in the axils of bracts, are greatly reduced: apart from a concave receptacle in the wind pollinated poplars (Fig. 711C, D) or one or two nectar secreting scales in the usually insect pollinated willows, the ♂ flowers consist only of some stamens (in *Populus* several, in *Salix* often only two), and the ♀ flowers consist only of an ovary. The capsules contain a large number of minute seeds with a tuft of hairs (Fig. 711F, O) which usually only remain viable for a few days.

Many willows (e.g. *S. viminalis, S. fragilis, S. alba*) and poplars (e.g. *P. nigra*, the black poplar, and *P. alba*, the silver poplar) can grow on waterlogged soils and are amongst the principal trees of riverside forests. The goat willow or sallow (*S. caprea*) and aspen (*P. tremula*) are widespread as pioneers in woodland gaps and clearings. Many species of willow and their hybrids (in which the genus is particularly rich) are cut every two or

Fig. 711. Salicales. **A.** to **F.** *Populus nigra.* **A.** Flowering ♂ and **B.** fruiting ♀ shoots. ×¾. **C.** ♂ and **D.** ♀ flowers with their subtending bracts. **E.** Fruit. **F.** Seed. Enlarged. **G.** to **O.** *Salix viminalis.* **G.** Flowering ♂ shoot and **I.** ♀ catkin. ×1. **H.** ♂ and **K.** and **L.** ♀ flowers with their subtending bracts. **M.** and **N.** Fruits. **O.** Seed. Enlarged. **A.** to **F.** After Karsten. **G.** to **O.** After Schimper.

three years as pollards; the wands are used in basketwork. Various prostrate 'creeping willows' (e.g. *S. retusa*, *S. herbacea*) are characteristic of high mountains and the Arctic.

The two final orders (6, 7) of the Dillenianae are apparently very advanced lines of development from the Violales, having unisexual flowers, inferior ovaries and a herbaceous habit, but otherwise they are not closely related to each other.

Order 6: **Begoniales** comprises the tropical (often cultivated) **Begoniaceae** notable for their asymmetric leaves (*Begonia*, Fig. 273).

Order 7: **Cucurbitales** is almost always sympetalous with shoot-tendrils (Fig. 309); it was previously placed in the 'Sympetalae', but its only family **Cucurbitaceae** (Fig. 712) is clearly closely related to the Passifloraceae and like most Violales has a crassinucellate ovule with two integuments. The vascular bundles are bicollateral. The tendrils (Fig. 712) are equivalent to leaves borne at the ends of short shoots. The unisexual flowers are monoecious (e.g. *Bryonia alba*) or dioecious (*B. dioica*; for the inheritance of sex see p. 327). In the ♂ flower the stamens usually have a single theca each, but this is frequently curved or S-shaped, and the stamens are united in groups (e.g. 2 + 2 + 1) or all together. The usually tricarpellary, paracarpous inferior ovary has massive incurved placentas and develops into a large, thick-skinned, many-seeded berry (pepo). Familiar representatives include the marrow (*Cucurbita pepo*, with various races yielding vegetables and oil-seeds, native in tropical America), the cucumber (*Cucumis sativus*, native in tropical Asia), the red-fleshed water-melon (*Citrullus lanatus*), the colocynth (*Citrullus colocynthis*, an African–Near Eastern desert plant, rich in bitter substances and purgative; Fig. 712A, B), the calabash or gourd (*Lagenaria*), grown in the Tropics for flasks, etc., and the Mediterranean squirting cucumber (*Ecballium*, p. 675).

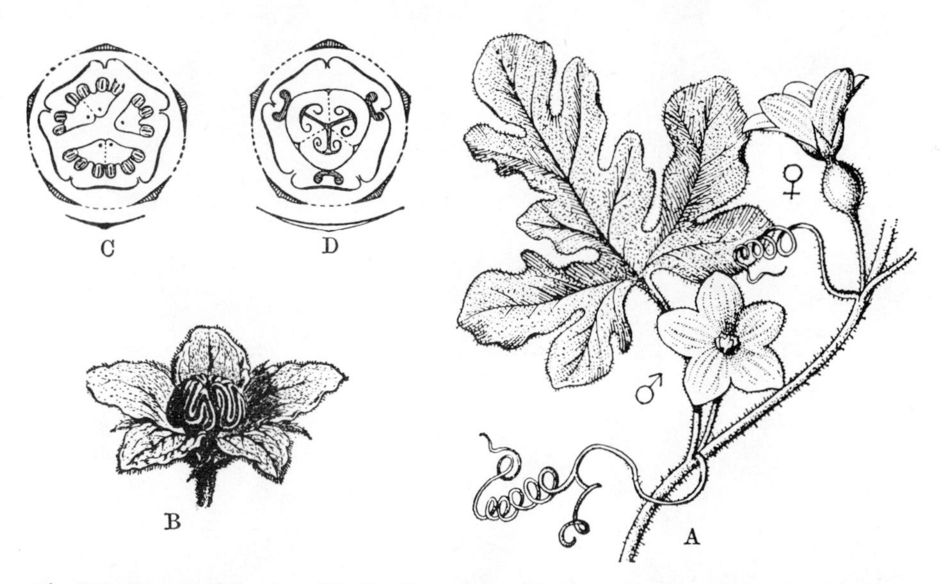

Fig. 712. Cucurbitales. **A.** and **B.** *Citrullus colocynthis*. Flowering shoot (×1 approx.) and ♂ flower (×2¾). **C.** and **D.** Floral diagrams of ♂ and ♀ flowers. **A.** After Firbas. **B.** After Baillon. **C.** and **D.** After Eichler.

The next two orders (8, 9) are characterized above all by their actinomorphic, generally pentamerous hermaphrodite flowers, the syncarpous superior ovaries, a tendency towards reduction in the number of carpels and ovules (which are still crassinucellate with two integuments), and the frequent occurrence of mucilage cells. They can be grouped in the superorder **Malvanae** and probably go back to ancestors similar to the Theales or Violales.

The most important is the generally woody:

Order 8: **Malvales** (Columniferae). The calyx is valvate but the corolla is frequently contorted (convolute) in bud (Fig. 713G, H) and both are free or only slightly united. Of the two whorls of stamens the outer (episepalous) one has a tendency to be suppressed

Fig. 713. Malvales. A. to C. *Tilia*. A. Floral diagram. C. Inflorescence (×1) with peduncle (a) partly united to a wing-like bract (b). B. T.S. of nut with pericarp (p), abortive embryos (f) and one mature seed containing endosperm (e) and embryo (em). ×4. D. to F. *Theobroma cacao*. D. Flowering and fruiting trunks. Greatly reduced. E. Flower and F. androecium with long staminodes. ×2 approx. G. to N. Malvaceae. G. Floral diagram of *Malva* with epicalyx (ak). H. Bud (×1). I. Half flower (×1.5) with K. stamens forming a column and the styles protruding above (×5). L. Schizocarp (×4) of *Malva sylvestris*. M. Flower of *Gossypium herbaceum* (×¾). N. Dehisced capsule showing the hairs of the seeds of *G. vitifolium* (×¾). A. After Eichler. B., M. and N. After Wettstein. C. After Berg and Schmidt. D. to F. After Karsten. G. After Firbas. H. After Schenck. I. to L. After Baillon.

whilst the inner (epipetalous) one is often increased centrifugally. Frequently the filaments are united below to form a tube united to the corolla and surrounding the styles so that the anthers are borne on a column (Fig. 713K). The type is thus $* \ K5\ C5\ A5 - 0 + 5^{\infty}\ G\ (5 - \infty)$.

The mainly tropical woody **Tiliaceae** still has more or less free stamens. The only native representatives are the small-leaved lime (*Tilia cordata*) and the large-leaved lime (*T. platyphyllos*). These are entomophilous trees with dichasial inflorescences partially united to an enlarged wing-like bracteole (Fig. 713C) which later serves to disperse the whole infructescence by wind. The ovary has five loculi, each of which contain two ovules initially, but all but one are suppressed so that the fruit is a one-seeded nut (Fig. 713B). The limes are trees of mixed woodlands on the better soils. Whilst *T. cordata* comes late into leaf and is widespread in Europe, even in the continental lowlands, the earlier flowering *T. platyphyllos* is rather a tree of moderate altitudes, scarcely extending north of the German highlands. Tropical or subtropical representatives include the species of *Corchorus* which yield jute and *Sparmannia africana*, from S. Africa, with irritable stamens.

The two following families are tropical and have the stamens more or less united. The anther still has two thecae in the woody **Bombacaceae**. These include *Ceiba* (kapok, a fibre which cannot be spun is obtained from pericarp hairs) and *Adansonia* (*A. digitata*, the African baobab or monkey bread tree with an enormously thick water-storing trunk, is bat-pollinated). The **Sterculiaceae** are closely related and include cocoa (*Theobroma cacao*, Fig. 713D–F), a native of America cultivated throughout the Tropics; it is a small tree with large simple leaves and cauliflorous flowers pollinated by midges, ants, etc. The large indehiscent fruits contain numerous seeds (cocoa 'beans'); the massive cotyledons contain fat (cocoa butter) and the alkaloid theobromin; after the fat has been partly expressed they yield cocoa powder. The seeds of tropical W. African species of *Cola* contain caffeine (hence used as a stimulant).

The frequently herbaceous **Malvaceae** are the most advanced, having anthers with only one theca. An epicalyx derived from bracts (hypsophylls) is often present (Fig. 713G). The ovaries have three to five or even more (up to fifty) carpels. They may develop as many-seeded capsules, or as a schizocarp dividing into one-seeded mericarps each corresponding to a carpel. The former is the case in the cotton plant (*Gossypium*) which has various shrubby or annual cultivated forms derived by hybridization and polyploidy from tropical–subtropical Asiatic, African and American species (e.g. *G. arboreum* and *G. herbaceum*, 2 *x*; *G. hirsutum*, 4 *x*; Fig. 713M, N); cotton itself consists of unicellular hairs of the testa up to 60 mm long (Fig. 106A, B, p. 101). Cottonseed oil is also of great economic importance, e.g. in the manufacture of margarine. Our native herbaceous mallows (*Malva*, Fig. 713G–L) and the densely hairy species of *Althaea* have schizocarps; *A. officinalis* is the marshmallow, a halophytic, ancient medicinal plant whilst *A. rosea* is the familiar hollyhock.

Basal fusion of the perianth segments in:

Order 9: **Thymelaeales** gives rise to a hypanthium; the corolla is absent. Reduction of the carpels and ovules leads eventually to pseudomonomerous ovaries with one ovule. The **Thymelaeaceae** includes, for example, the poisonous mezereon (*Daphne mezereum*, Fig. 714) in which the purple petaloid calyx opens before the leaves; the berries are sealing-wax red. The systematic position of this order is very uncertain (possibly Rosidae, near the Myrtales).

The final three orders (10–12) of the Dilleniidae can be included in the superorder **Ericanae** (Sympetalae pentacyclicae). As a predominantly sympetalous line of development they can be looked upon as part of the 'superstructure' of the dicotyledons, but they show many close connections with choripetalous Dillenianae (indeed there are occasionally forms with free petals); however, secondary polyandry is almost completely absent. Compared with the main sympetalous groups (Asteridae, p. 721) there is less expression of advanced characters: there are still two whorls of stamens (so that the flowers are pentacyclic), or if only one, the other is indicated by staminodes, or the surviving one is epipetalous; the flowers are predominantly actinomorphic and hermaphrodite with superior ovaries, often of five or four carpels; whilst the ovules vary from

Fig. 714. Thymelaeales. Half flower of *Daphne mezereum.* ×1. After Karsten.

bitegmatic to unitegmatic, and from crassinucellate to tenuinucellate.

Syncarpous gynaecia with five or four carpels and numerous central-axile, tenuinucellate and unitegmatic ovules are found in:

Order 10: **Ericales** (Bicornes). These are mainly shrubs and perennial herbs with simple, often evergreen leaves; they are notably mycotrophic with a tendency towards semi- and holoparasitism. The tetramerous or pentamerous flowers have usually two obdiplostemous whorls of generally free stamens; these often open by pores (adaptation for pollen scattering) and bear two horn-like processes (hence 'Bicornes'; Fig. 715C, D). The pollen grains commonly cohere in tetrads (Fig. 715G). The (sub)tropical **Clethraceae** with free petals and separate pollen grains link this order closely with the Theales. On the other hand tetrads occur in the **Ericaceae**, woody (often suffruticose) plants with generally evergreen, often very small scale- or needle-like leaves well known for their xeromorphy. They are important components or Subarctic and Atlantic dwarf shrub heaths, of raised bogs, of coniferous forests with much raw humus, of the tree-line in the mountains, of the Mediterranean maquis (p. 786), and of the heaths of the Cape of Good Hope (p. 792). Their mycotrophy enables them to colonize soils extremely poor in minerals. Even in the Ericaceae the petals are free in a few cases, e.g. in *Ledum palustre*, but in general they are united. In most genera the ovary is superior and becomes a capsule as in *Rhododendron*, *Andromeda*, *Erica* (e.g. the Atlantic *E. tetralix*, the cross-leaved heath, Fig. 754, and the early-flowering winter heather, *E. carnea*, of the mountains) and *Calluna* (ling or Scots heather); only rarely does it become a berry or, as in the bearberry (*Arctostaphylos uva-ursi*, Fig. 715E–I), a drupe. However, in some genera the ovary is inferior and then the fruit is invariably a berry as in *Vaccinium*, e.g. the whortleberry or bilberry (*V. myrtillus*) and the cowberry (*V. vitis-idaea*).

Progressive dependence on mycotrophy is characteristic of the small family **Pyrolaceae**, with evergreen herbs including the choripetalous species of wintergreen (*Pyrola*) widespread in coniferous forests, and of the **Monotropaceae** which lack chlorophyll (yellow bird's-nest, *Monotropa hypopitys*). In the **Empetraceae** the anthers open by slits;

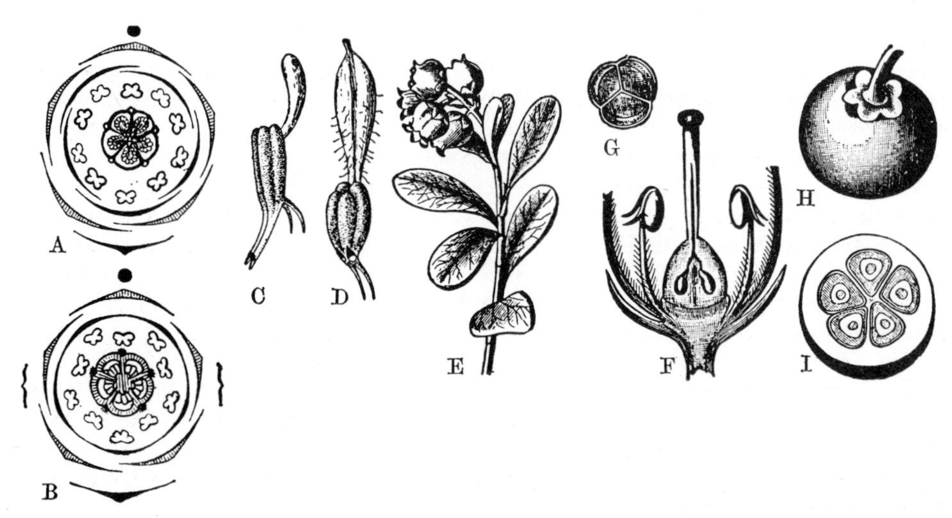

Fig. 715. Ericales. **A.** and **B.** Floral diagrams. **A.** Pyrolaceae (*Pyrola rotundifolia*). **B.** Ericaceae (*Vaccinium vitis-idaea*). **C.** and **D.** Stamens, in their natural orientation, of *Vaccinium myrtillus* and *Andromeda polifolia*. ×10. **E.** to **I.** *Arctostaphylos uva-ursi.* **E.** Flowering shoot. **F.** L.S. of flower. **G.** Pollen tetrad. **H.** and **I.** Drupaceous fruit, as a whole and in T.S., with five stony seeds. **F.** to **I.** Enlarged. **A.** and **B.** After Eichler. **C.** and **D.** After Firbas. **E.** to **I.** After Berg and Schmidt.

the crowberry (*Empetrum nigrum* agg.) belongs here, a prostrate, xeromorphic dwarf shrub with revolute leaves, found on heaths and moors.

The ovules are still sometimes bitegmatic but reduced in number and central-axile, and the stamens are united to the corolla in the (sub)tropical woody

Order 11: **Ebenales.** These include many important economic and medicinal plants, such as species of *Styrax* (**Styracaceae**) yielding benzoin resins, various species of *Diospyros* (**Ebenaceae**), the source of ebony and including *D. kaki*, the originally Asiatic persimmon or date-plum, and the Indo-Malayan species of *Palaquium* and *Payena* (**Sapotaceae**), the latex of which yields gutta percha, used in medicine and the insulation of power cables.

Paracarpous ovaries with central placentation, numerous bitegmatic ovules and united styles, and the frequent absence of the episepalous stamens are characteristic of:

Order 12: **Primulales.** The tropical, woody **Theophrastaceae** and **Myrsinaceae** connect the Ebenales to the temperate and herbaceous **Primulaceae**. Whilst yellow loosestrife (*Lysimachia*) and the scarlet pimpernel (*Anagallis arvensis*, Fig. 678F) have leafy stems, the following genera are rosette plants: *Cyclamen*, a genus centred on the E. Mediterranean with tuberous hypocotyls; *Primula* (primrose, cowslip, oxlip, etc.), a genus found especially in mountains throughout the world, often heterostylous (Fig. 669) and glandular hairy (Fig. 127A); *Soldanella*, in snow patch communities in the European mountains (Fig. 753); and *Androsace*, cushion plants which reach the snow line.

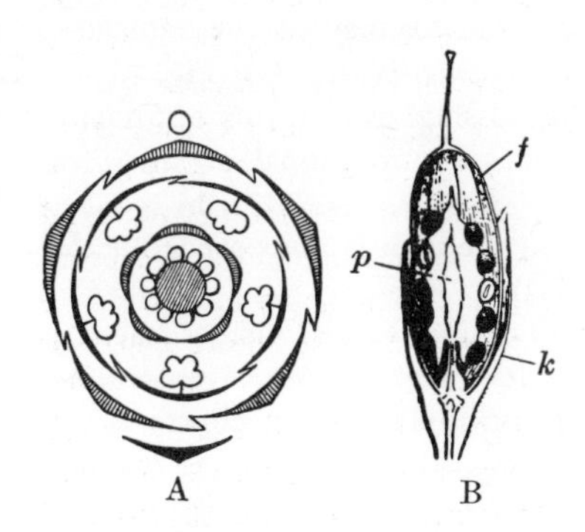

Fig. 716. Primulales, *Primula*. **A.** Floral diagram of *P. vulgaris*. **B.** Nearly ripe fruit of *P. elatior* cut lengthwise with calyx (k), pericarp (f), central placenta (p) and seeds. ×1.5. **A.** After Eichler. **B.** After Firbas.

Subclass E: Caryophyllidae

The last major group in the 'middle storeys' of the dicotyledons is the Caryophyllidae; these lie somewhat off the main line of development. They are similar to the Dilleniidae in that the primitive androecium is diplostemonous and the stamens are often increased in number by centrifugal secondary polyandry, but the tendency towards the reduction of the stamens to one whorl is much more pronounced. The pollen is almost always three-celled and primarily tricolpate but often pantoporate in advanced cases (Fig. 367). Also characteristic are the transformation of gynaecia with several more or less choricarpous or syncarpous carpels to paracarpous gynaecia with a central placenta, and the reduction from numerous ovules to a single basal one. The ovules are often campylotropous, bitegmatic and crassinucellate, and commonly have a perisperm. Further characteristics are the actinomorphic, mostly trimerous or pentamerous, almost always cyclic flowers with a single (rarely spiral) or double perianth, the simple, undivided leaves, the predominantly herbaceous habit and a preference for mineral soils. The Caryophyllidae appear to derive directly from the Magnoliidae and separately from the Rosidae + Dilleniidae.

The occasional occurrence of carpels which are more or less choricarpous and often of numerous ovules leads to:

Order 1: **Caryophyllales** (Centrospermae) being placed first. These are mainly herbaceous plants with tendencies towards succulence and anomalous secondary thickening (supplementary vascular rings). In many families anthocyanins do not occur and are replaced by the chemically quite different, nitrogen-containing betacyanins and betaxanthins (betalains, Fig. 368); these occur nowhere else in the Plant Kingdom. The single perianth (homoiochlamydeous; perigone segments) often becomes double (heterochlamydeous) by the development of free petals homologous with stamens (but conversely the corolla may then be subsequently suppressed). The primitive androecium is apparently two-whorled but it may be greatly increased in number by centrifugal secondary polyandry (dédoublement); in other cases it may be reduced to one whorl. The transition from more or less choricarpous carpels to syncarpous and finally to paracarpous carpels with central placentation, formed by the suppression of the septa between the loculi, is very characteristic; correspondingly the ovules are reduced from many to eventually only one basal one. The endosperm is suppressed in the seed but the curved embryo is generally surrounded by a starchy perisperm.

Anthocyanins are still present in the **Molluginaceae**; this (sub)tropical, partly woody but scarcely succulent family has more or less choricarpous or coenocarpous carpels with numerous seeds; primitive characters include the single perianth, commonly two whorls of stamens and tricolpate pollen.

Fom these can be derived the important, cosmopolitan, almost always herbaceous **Caryophyllaceae**. These have generally narrow, opposite leaves. Dichasial inflorescences are frequent (Fig. 717E). When a corolla is present the hermaphrodite flower generally has the formula: K5 C5 A5 + 5 G($\underline{5}$), as in mouse-ear chickweed (*Cerastium*), corn cockle (*Agrostemma githago*) or catchfly (*Lychnis viscaria*, Fig. 717A). The stamens are obdiplostemonous. The number of carpels is frequently reduced, e.g. G(3) in *Silene* (Fig. 717B, F) and *Stellaria*, G(2) in the pink (*Dianthus*). The stamens may also be reduced to one whorl and may not all be present even then (*Stellaria media*, A5 → 3, Fig. 717C). Further, in many genera and species the corolla may be missing, as in forms of *Stellaria media* or in *Herniaria* and *Paronychia*, Fig. 717D, H). Dioecism is occasional, e.g. in *Silene alba, S. dioica* (= *Melandrium album, M. rubrum*; cf. pp. 328–9, Fig. 283). In reduced flowers the number of ovules is frequently only one, and the fruit is a one-seeded nutlet instead of a capsule (e.g. *Scleranthus, Herniaria*, Fig. 717H).

Classification of the family is based on the possession of free sepals (Alsinoideae, e.g. *Cerastium, Stellaria, Scleranthus*) or united ones (Silenoideae, e.g. *Lychnis, Agrostemma, Silene, Dianthus*), or on the presence of stipules (Paronychioideae = Illecebraceae, e.g. *Spergula, Herniaria*). Some species contain saponins, e.g. the soapwort (*Saponaria officinalis*).

The other families contain betalains (not anthocyanins). The **Phytolaccaceae** are both diverse and primitive in floral morphology and are closely related to the Molluginaceae. They include, for example, *Phytolacca americana* (pokeweed) which yields a red dye. The two following families of succulents with greatly increased numbers of stamens and perianth segments are related to the Phytolaccaceae. The **Aizoaceae** are leaf-succulents with *Mesembryanthemum* and many other large genera found especially in the arid regions of S. Africa; they include plants resembling pebbles embedded in the soil ('living stones'; e.g. *Lithops*, Fig. 201). The numerous petals are derived from the outermost of the many stamens. The fruit is usually a capsule which opens only when moist (hygrochastic).

In contrast, the **Cactaceae** are predominantly American stem-succulents. The columnar or flattened (e.g. *Opuntia*, Fig. 197B), fluted (e.g. *Cereus*, Fig. 203A), or globular and tuberculate (e.g. *Mamillaria*) stems almost always bear leaf-spines, frequently in tufts (areolae) which develop from the leaf primordia of axillary shoots (Figs. 202, 717I). Only *Pereskia* has normal adult leaves, but small scale-like or needle-shaped leaves can be seen in many juvenile stages, e.g. in *Opuntia*. The sessile flowers have many spirally arranged perianth segments, the outer ones sepaloid, the inner petaloid (Fig. 717I), numerous stamens, and a large number of carpels united in an inferior ovary. The fruit is a berry.

The Cactaceae are almost entirely American, occurring mainly in the deserts and semideserts of S.W. United States, Mexico and the Andes; they range from small forms to giants 15 m high (e.g. *Carnegeia gigantea*). Some genera, e.g. *Rhipsalis, Epiphyllum* and *Phyllocactus* (Fig. 139B) are forest epiphytes. The prickly pear (*Opuntia ficus-indicus*) has edible fruit and is widely naturalized in the southern parts of the Mediterranean

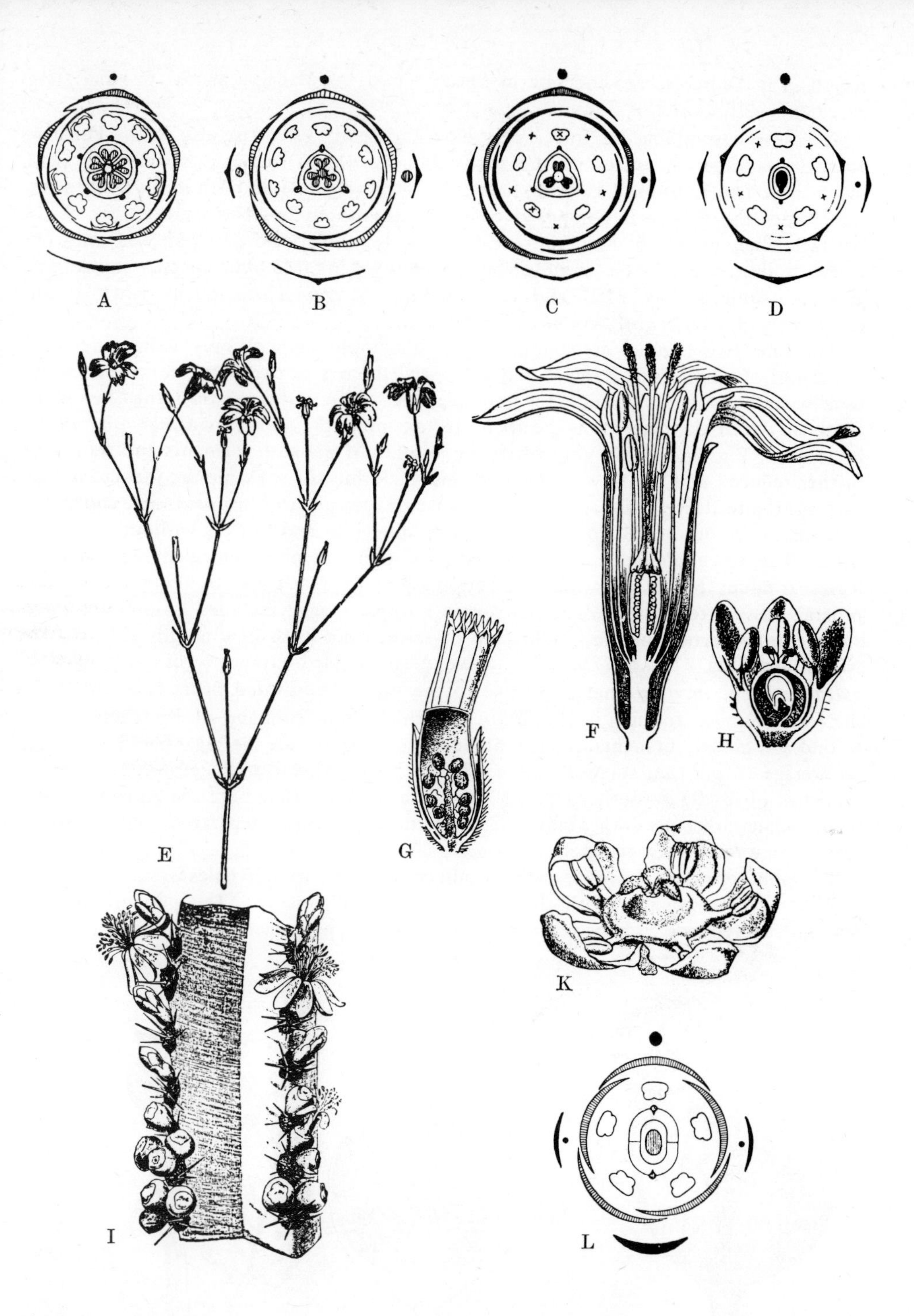

Fig. 717. Caryophyllales. **A.** to **H.** Caryophyllaceae. Floral diagrams of: **A.** *Lychnis viscaria.* **B.** *Silene vulgaris.* **C.** *Stellaria media.* **D.** *Paronychia* sp. **E.** Inflorescence of *Cerastium arvense*: dichasium (cf. Fig. 142B). **F.** and **H.** Flowers of *Silene nutans* and *Herniaria glauca.* **G.** Capsule of *Cerastium holosteoides* (cut away below). Enlarged. **I.** Cactaceae, *Myrtillocactus* (*Cereus*) *geometrizans*, two ribs of a five-ribbed stem with flowers and fruits. ×½. **K.** and **L.** Chenopodiaceae, flower of *Beta trigyna* and floral diagram of *Chenopodium*. **A.** to **D.** and **L.** After Eichler, somewhat modified. **E.** After Duchartre. **F.** to **H.** After Beck-Mannagetta. **I.** After Karsten. **K.** After Baillon.

region. The **Didiereaceae**, endemic in the arid parts of Madagascar, are peculiar stem-succulents with unisexual flowers.

Ancestors resembling Phytolaccaceae have apparently given rise also to the **Portulacaceae** (with capsules; including the native blinks, *Montia*) and the **Basellaceae** (with indehiscent fruits; tropical twiners, some used as vegetables); in both a calyx-like structure under the flower is composed of bracts (hypsophylls). The same is true of the **Nyctaginaceae** in which, however, the perigone is tubular and petaloid, and only one carpel is developed. These include the marvel of Peru (*Mirabilis jalapa*, familiar as a genetical subject, Fig. 279) and the climbing *Bougainvillea*, much grown in the (sub)Tropics for its brightly coloured bracts.

The final two families are predominantly anemophilous with only a single perianth, one whorl of epitepalous stamens and a one-seeded ovary of two (or three) carpels which usually develops into a nutlet. This basic plan of * P5 A5 G($\underline{2}$) occurs in the **Chenopodiaceae** (in which the perianth is green) in, for example, the goosefoot (*Chenopodium*) or beet (*Beta*, Fig. 717K–L). Very often the number of perianth segments and stamens is further reduced and the flowers become unisexual; thus the sea samphire (*Salicornia*) has hermaphrodite flowers with usually only three or four perianth segments and one or two stamens, while orache (*Atriplex*) is frequently dioecious and flowers without a perianth occur. The Chenopodiaceae prefer saline soils; they are often succulent with reduced leaves or none; they have a wide distribution especially in salt deserts, along coasts and as ruderals associated with Man. *Salicornia europaea* agg. includes halophytic stem-succulents important in the building up of saltmarshes in shallow muddy seas, e.g. the North Sea (Fig. 216, p. 771). The native Atlantic–Mediterranean coastal plant *Beta vulgaris* subsp. *maritima* has given rise to the various cultivated beets (Fig. 206C–E); these are biennial, forming in the first year a thick fleshy root and a leafy rosette, in the second a copiously branched panicle of flowers; forms include the sugar beet (containing on average 16 per cent sucrose), spinach-beet, chard, mangold and beetroot. The spinach (*Spinacia olera ea*) is another culinary vegetable. Closely related are the **Amaranthaceae** with membranous perianth segments; these include various decorative and economic plants and weeds of the genus *Amaranthus*.

Probably related to primitive anthocyanin-containing Caryophyllales is:

Order 2: **Polygonales**, with the single family **Polygonaceae**. The perianth remains homoiochlamydeous with two trimerous whorls or one pentamerous one, the stamens are

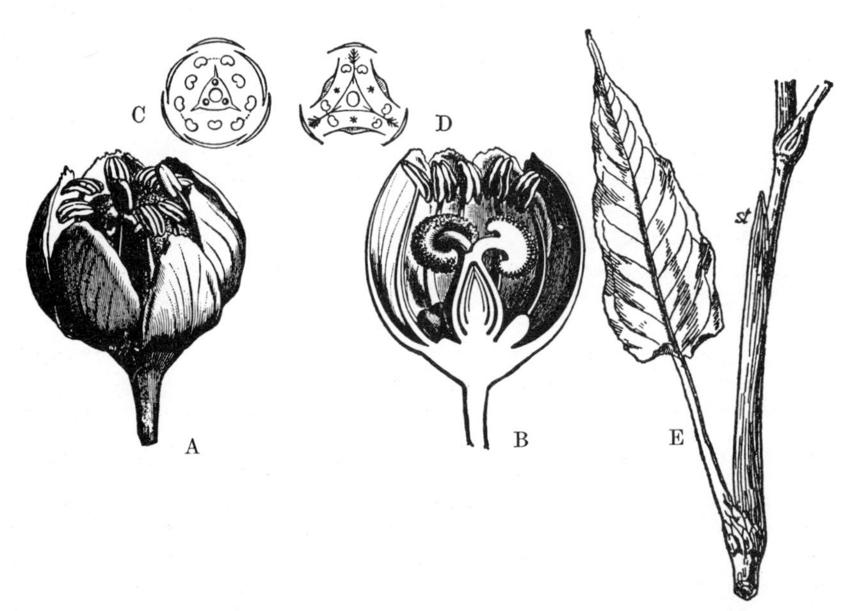

Fig. 718. Polygonales. **A.** and **B.** *Rheum officinale*, whole and half flower. Enlarged. **C.** and **D.** Floral diagrams of *Rheum* and *Rumex*. **E.** Part of a shoot with leaf and ochrea (st) of *Polygonum amplexicaule*. $\times\frac{1}{3}$. **A.** and **B.** After Baillon. **C.** and **D.** After Eichler. **E.** After Karsten.

usually in one or two whorls (rarely three), showing little or no polyandry, and the ovary is unilocular with only one usually orthotropous ovule (Fig. 718A–D); the endosperm is copious. The leaves are alternate and their stipules are united to form a hood or ochrea which initially covers the stem apex, rupturing later and remaining as a tubular membrane around the stem (Fig. 718E). The perianth of the small, hermaphrodite or unisexual flowers is usually inconspicuous (many representatives are anemogamous), more rarely petaloid as in the buckwheat (*Fagopyrum esculentum*, Fig. 235), or in some insect pollinated species of *Polygonum* (knotgrass). In the docks (*Rumex*, Figs. 191A, 718D) the inner perianth persists as an organ of dispersal of the fruit by wind, water or animals. The unilocular ovary consists of three (two to four) united carpels and becomes a nutlet containing a starchy reserve tissue for which buckwheat was often formerly cultivated on poor soils. Species of *Rheum* originating in the mountains of C. and E. Asia are well-known dessert (rhubarb) and medicinal plants.

A heterochlamydeous, pentamerous perianth with calyx and corolla, one epipetalous whorl of stamens, a paracarpous ovary of five carpels with one basal ovule and an endosperm are characteristic of:

Order 3: **Plumbaginales.** These, too, contain anthocyanins and comprise only the **Plumbaginaceae**. These include xerophytes and halophytes of steppes, semi-deserts and coasts, e.g. *Limonium* (including *Statice*; sea lavender) and *Armeria* (sea pink or thrift).

Subclass F: Asteridae (Sympetalae tetracyclicae)

This final subclass of the dicotyledons comprises the most advanced sympetalous orders (apart from the Cucurbitales, p. 713) and forms the greater part of the 'superstructure' of the Magnoliatae (cf. p. 691, Fig. 742). Compared with the likewise sympetalous Ericanae (p. 715), only the whorl of stamens alternating with the petals remains so that the flowers are tetracyclic. Although often actinomorphic (radial), the flowers are predominantly zygomorphic (dorsiventral) and nearly always hermaphrodite and specialized for animal pollination; pseudanthia often occur as the terminal representatives of a series. Secondary polyandry is absent and the stamens (A5 → 4 → 2) are almost always united with the corolla. Compared with the generally pentamerous perianth the number of carpels is usually reduced and commonly there are only two. The syncarpous to paracarpous ovaries frequently become inferior and have central-axile or parietal (or basal) placentation with numerous to one tenuinucellate ovules with only one integument (unitegmatic). A typical floral formula is thus: K(5) C(5) A5 G($\underline{2}$).

The Asteridae form an independent group of related forms which are not directly related to any living choripetalous dicotyledons. However, various morphological, anatomical, phytochemical and serological similarities indicate that they have arisen (probably along various parallel lines) from the general area of the Rosidae.

The orders of the Asteridae can be grouped in two superorders, the **Lamianae** with usually free anthers and the Asteranae (p. 731) with postgenitally more or less united anthers. Within the Lamianae the first group of orders (1–3) with often woody members and actinomorphic flowers shows the closer affinities with the Rosidae whilst the second group (4–6, pp. 725 et seq.) is more advanced and is more often herbaceous and zygomorphic.

Order 1: **Gentianales** (Contortae plus part of the Rubiales) characteristically has actinomorphic pentamerous or tetramerous flowers, often contorted in bud, A5–4, superior or inferior ovaries with two carpels, generally with numerous ovules, free nuclear division of the endosperm, leaves almost invariably simple with an entire margin and opposite, and the common occurrence of bicollateral vascular bundles and of indole alkaloids (tryptophane derivatives). The **Loganiaceae** are particularly primitive; they are (sub)tropical, mostly woody and stipulate, with superior ovaries, but without latex canals. Various poisonous plants belong here, e.g. in the genus *Strychnos* of which many species yield arrow poisons, e.g. the S. American curare; the indole alkaloid strychnine comes from the seeds of the E. Indian *S. nux-vomica* (Fig. 719A–E).

A predominantly herbaceous, exstipulate line of development with noteworthy bitter substances (e.g. gentiopicrine) is seen in the two following families: in the **Gentianaceae**

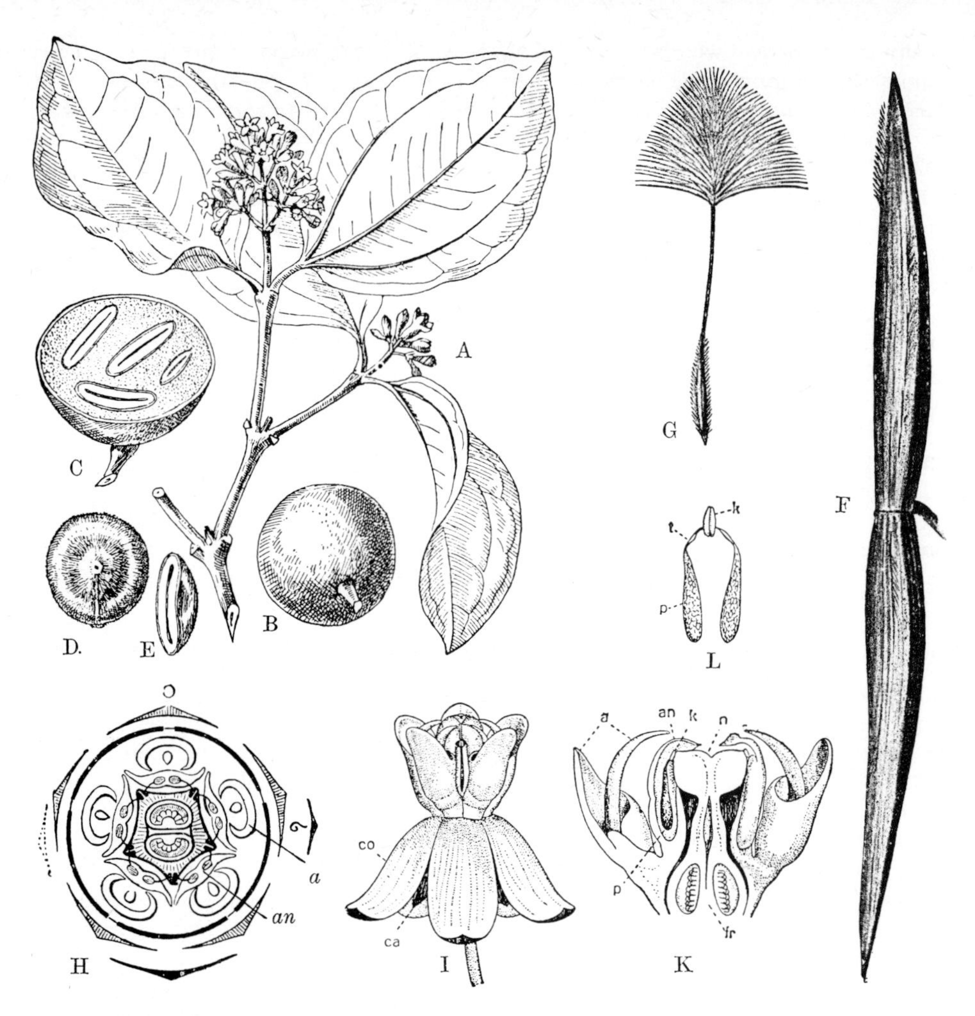

Fig. 719. Gentianales. **A.** to **E.** Loganiaceae. *Strychnos nux-vomica*, flowering shoot, berry and seeds, whole and in section. ×½. **F.** and **G.** *Strophanthus hispidus*, fruit (×½) and seed (×⅔). **H.** to **L.** Asclepiadaceae, *Asclepias syraica*, floral diagram (in the axil of a prophyll, with lateral branch); flower (enlarged) with calyx (ca) and corolla (co); gynostegium (k, in L.S., enlarged) with, opposite the stamens, the appendages of the corona; anthers (a) with pollinia (p) and corpusculum (k) between them; ovary (fr), capitate stigma (n); two pollinia (p) joined by translators (t) and the corpusculum (k). **L.** Enlarged. **A.** to **E.** After Karsten. **F.** and **G.** After Schumann. **H.** After Eichler, somewhat modified. **I.** to **L.** After Engler.

(with simple opposite leaves) the gentians (*Gentiana*, Fig. 720, and *Gentianella*) form the largest genus and ascend high into the mountains of the Northern Hemisphere, whilst in the **Menyanthaceae** (with alternate leaves) are the bogbean (*Menyanthes trifoliata*), a marsh plant with trifoliate leaves, and the fringed waterlily (*Nymphoides peltata*) with small waterlily-like floating leaves.

The next two families, also related to the Loganiaceae, have non-septate latex canals, are rich in alkaloids, often woody and (sub)tropical. Their coenocarpous gynaecia have a marked tendency for the development of the upper, free (asymplicate) zone (only the styles and stigmas are free at the time of flowering); correspondingly the fruits appear to be two lobed and secondarily almost choricarpous (Fig. 719F). In the **Apocynaceae** the anthers are still free and the pollen grains separate. Woody representatives include the Mediterranean oleander (*Nerium oleander*), the African species of *Strophanthus* (Fig. 719F, G containing cardenolides as important cardiac glycosides and arrow poisons), *Rauvolfia* (indole alkaloid reserpine, etc.), and various plants yielding caoutchouc (e.g. the African *Funtumia* and *Landolphia* and the Brazilian *Hancornia*); the native evergreen periwinkle (*Vinca minor*, Fig. 211A) is herbaceous.

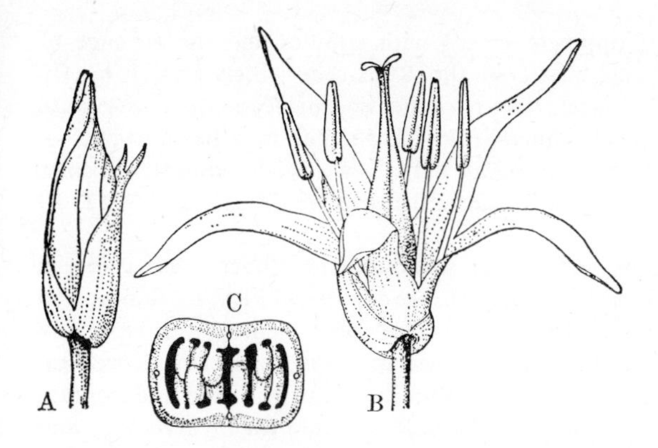

Fig. 720. Gentianaceae, *Gentiana lutea.* **A.** Bud with contorted (convolute) corolla. ×1. **B.** Flower. ×1. **C.** Ovary in T.S. ×3. After Firbas.

In the **Asclepiadaceae** the anthers are united with the stigmatic head to form a 'gynostegium' and the pollen grains generally cohere in pollinia (Fig. 719H–L). Generally two pollinia from neighbouring anthers are connected by bow-shaped 'translators' and a 'corpusculum' formed by the capitate stigma. The proboscis or leg of an insect seeking nectar becomes trapped in a groove of the corpusculum, dragging away the two pollinia and carrying them to another flower. In addition, appendages on the outside of the stamens form a nectar-producing corona. In *Ceropegia* this type of pollination is combined with slippery trap flowers. As well as woody plants the family includes lianes (e.g. *Marsdenia*), epiphytes (e.g. *Dischidia*, Fig. 215), perennial herbs (e.g. the European *Cyananchicum vincetoxicum* and species of *Asclepias*, Fig. 719H–L) and stem-succulents (e.g. the Stapelieae with carrion-fly flowers, found especially in arid regions in Africa, Fig. 203C).

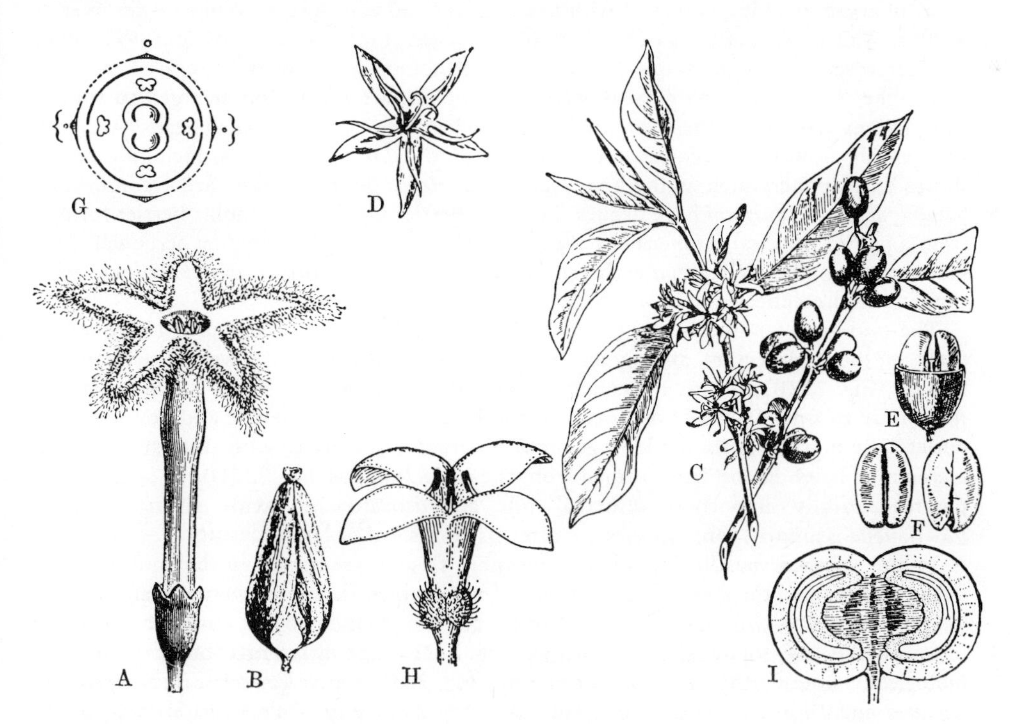

Fig. 721. Rubiaceae. **A.** and **B.** *Cinchona calisaya*, flower (×4), and septicidal capsule opening from below (×1). **C.** to **F.** *Coffea arabica.* **C.** Flowering and fruiting shoots. ×$\frac{3}{8}$. **D.** Flower. **E.** Drupe, fleshy portions partly removed. **F.** Seed without and with parchment-like endocarp. ×¾. **G.** Floral diagram of *Sherardia arvensis.* **H.** Flower of *Galium odoratum* (*Asperula odorata*), woodruff. ×7. **I.** Fleshy schizocarp of *Rubia tinctorum* (in L.S., ×2.7). **A.**, **B.** and **I.** After Baillon. **C.** to **F.** After Karsten. **G.** After Eichler. **H.** After Firbas.

Inferior ovaries, indole alkaloids, opposite leaves with stipules and the absence of bicollateral vascular bundles are characteristic of the **Rubiaceae** which were formerly placed near the Dipsacales but are in fact related to the Loganiaceae. The flowers are usually long and funnel-shaped or salver-shaped (Fig. 721A), but may be shorter (Fig. 721H, D) or even quite flat and rotate (as in many native species of *Galium*). Capsules with numerous ovules are primitive (Fig. 721B); derived forms are drupes (Fig. 721E, F) or schizocarps (Fig. 721G, I).

The Rubiaceae are most numerous as tropical woody plants (over 6000 species). Economically important are the Jesuits' bark trees (*Cinchona*, Fig. 721A, B; quinine and other indole alkaloids as febrifuges) and coffee bushes (*Coffea*; palaeotropical, especially *C. liberica, C. arabica,* Fig. 721C–F, and others amongst the most important of tropical crop plants; the coffee 'beans' consist largely of the endosperm of the seed and contain the purine derivative caffeine which acts like an alkaloid). Notable for their ecology and nutrition are the Indo-Malayan epiphytes *Myrmecodia* and *Hydnophytum* which have tubers inhabited by ants and the tropical species of *Psychotria* and *Pavetta* which have nitrogen-fixing bacteria in small tubercles on the leaves. In the predominantly herbaceous and temperate genus *Galium* (bedstraw) the stipules resemble the leaves and occur with them in whorls of four or more, but axillary shoots are only found in the true leaf axils; alkaloids are absent. *Rubia tinctorum* (Fig. 721I) is related; it was formerly much used as the dye madder.

Order 2: **Dipsacales** (Rubiales in part) is distinguished from the Gentianales by actinomorphic to weakly zygomorphic or asymmetric flowers, the corolla never contorted, inferior ovaries with five to two carpels, often with sterile loculi (compare Figs. 722A and B) and with few ovules, cellular development of the endosperm, compound, lobed or at least crenately toothed leaves without stipules, and no bicollateral bundles or alkaloids. Relationships are shown with the Cornales and Saxifragales. At the beginning come the **Caprifoliaceae**, woody plants still with five stamens and with several to one ovules in each loculus. The elder (*Sambucus*, with pinnate leaves, Figs. 145B, 170) and *Viburnum* (guelder rose, etc., with simple leaves) have actinomorphic flowers aggregated in dense umbel-like thyrses; the marginal flowers of *V. opulus* are sterile and enlarged to form a conspicuous attraction (the garden forms–snowball trees–with globular heads have only such sterile flowers). In contrast the flowers of the *Lonicera* species, including native shrubs and climbers such as the honeysuckle, *Lonicera periclymenum*, are zygomorphic. *Sambucus* and *Viburnum* have drupes, *Lonicera* berries (sometimes double berries formed from a pair of flowers). *Linnaea borealis* is a northern and Alpine creeping perennial. Also herbaceous is the delicate moschatel or townhall clock (*Adoxa moschatellina*) with capitate inflorescences, the only representative of the **Adoxaceae**.

The two final families in this order are mainly herbaceous and show progressive reductions in the androecium and gynoecium. The **Valerianaceae** have weakly asymmetric flowers (Fig. 722B, C) with a generally pentamerous, spurred corolla and stamens ranging from four to one. Of the three loculi of the ovary only one is fertile and develops as a nutlet. The most important native genus is *Valeriana*, a perennial with three stamens and a crown of hairs on the fruit formed from the calyx (pappus, Fig. 722D). *V. officinalis* is pharmaceutically important (ethereal oils and isovalerianic acid: strong odour!). *Valerianella* is annual; some species are eaten as corn salad or lamb's lettuce.

In the **Dipsacaceae** the weakly zygomorphic flowers are united in thyrsoid, capitate inflorescences and the marginal flowers are often enlarged forming rays (pseudanthia, Fig. 722E). The ovary, with one loculus and one seed, is surrounded by an envelope consisting of four bracts (hypsophylls) as an epicalyx (Fig. 722F). The differentiation of the fruits is biologically noteworthy (cf. pp. 422 et seq.; Fig. 365). Native genera include *Scabiosa, Knautia* and *Dipsacus*. The spiny heads of *Dipsacus sativus* (fullers' teasel) are used in raising the nap on woollen fabrics.

A central position between Gentianales and Scrophulariales (and the Celastrales!) is occupied by:

Order 3: **Oleales** (Ligustrales). These are woody and the leaves are usually opposite. The only family **Oleaceae** is characterized by actinomorphic tetramerous flowers, imbricate or valvate vernation of the corolla (rarely petals free), generally only two

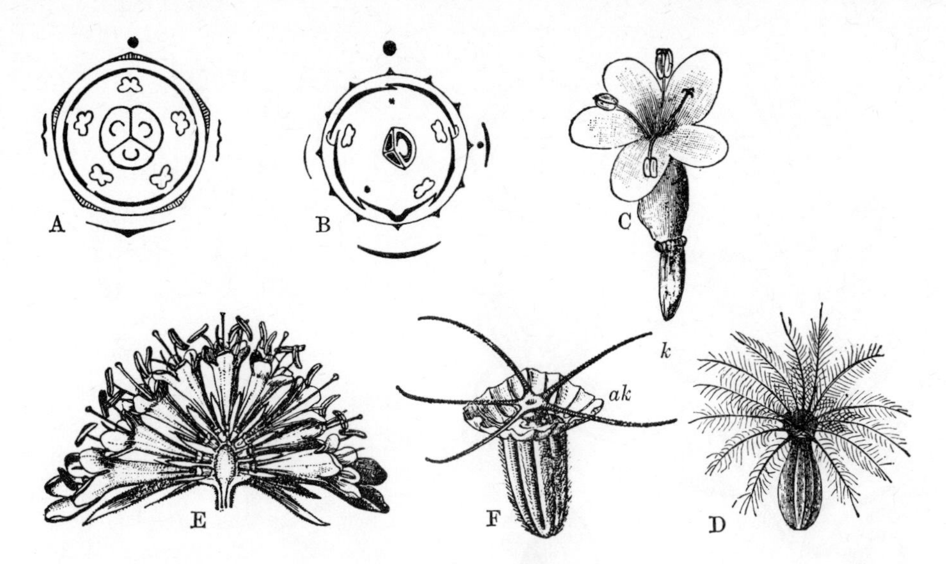

Fig. 722. Dipsacales. **A.** Caprifoliaceae, *Sambucus ebulus*, floral diagram. **B.** to **D.** Valerianaceae, *Valeriana officinalis*, floral diagram, flower (×8) and fruit with pappus (×3). **E.** and **F.** Dipsacaceae, *Scabiosa columbaria*, L.S. of a head and fruit with epicalyx (ak) and calyx (k). Enlarged. **A.** and **B.** After Eichler. **C.** and **D.** After Schenck. **E.** and **F.** From Hegi.

stamens, a superior ovary of two syncarpous carpels, often with only a few ovules, and cellular development of the endosperm; thus * K(4) [C(4) A2] G($\underline{2}$) (Fig. 723A). The fruits are very diverse. The lilac (*Syringa vulgaris*) from S.E. Europe has capsules. The Mediterranean olive (*Olea europaea*, Fig. 723B–D), notable for its simple, silvery-grey leaves, bears drupes with aliphatic oil in the mesocarp and endosperm. The species of ash (*Fraxinus*, with pinnate leaves) have winged one-seeded nuts (ash keys, Fig. 723F). The floral structure of this genus is also of great interest (Fig. 723E, G). The strongly scented, insect pollinated manna ash (*F. ornus*; sub-Mediterranean) is markedly primitive with a deeply divided white corolla and conspicuous panicles of flowers, whilst our native ash (*F. excelsior*, a tree of nutrient-rich soils) is more advanced and wind pollinated: the inconspicuous flowers appear before the leaves and lack both calyx and corolla; unisexual flowers occur as well as hermaphrodite ones. Familiar decorative shrubs are *Jasminum, Forsythia* and *Ligustrum*.

The orders in the second group (orders 4–6) are closely related to each other (and have been placed in the Tubiflorae in the broad sense). Herbaceous plants predominate; alternate and lobed or crenate-dentate leaves are frequent; stipules are absent. The change from actinomorphic to zygomorphic flowers and the concomitant reduction in stamens from five to four and then two are characteristic. Generally the superior ovaries have only two (or three) carpels which are paracarpous or syncarpous with many to few (or even one) ovules. Endosperm development is mostly cellular. The fruits are capsules or berries, or, as a result of the formation of false septa, schizocarps (breaking into 'nutlets' or 'achenes', Fig. 725C–H). There are certain similarities to the Geraniales.

Actinomorphic pentamerous or tetramerous flowers with stamens equal in number to the petals and ovaries with two or three carpels with numerous to few ovules are characteristic of:

Order 4: **Polemoniales** (Tubiflorae in part). The **Polemoniaceae** have ovules directed downwards and capsules usually formed of three carpels. These include the Jacob's ladder (*Polemonium caeruleum*) and the mainly N. American genus *Phlox*, frequently grown in gardens. The **Convolvulaceae** (Fig. 724) have two carpels and capsules with often four seeds; they are mostly twining plants with simple, alternate leaves and funnel-shaped flowers, contorted in bud, such as the bindweed (*Convolvulus arvensis*) or the large-flowered bellbine (*Calystegia sepium*). An important cultivated plant, probably originating in the Neotropics, is the sweet potato (*Ipomaea batatas*) with root tubers rich in starch. Closely related are the almost leafless and more or less chlorophyll-free **Cuscutaceae** with *Cuscuta* (dodder, Fig. 218) parasitic on clover, nettle, willows, etc.

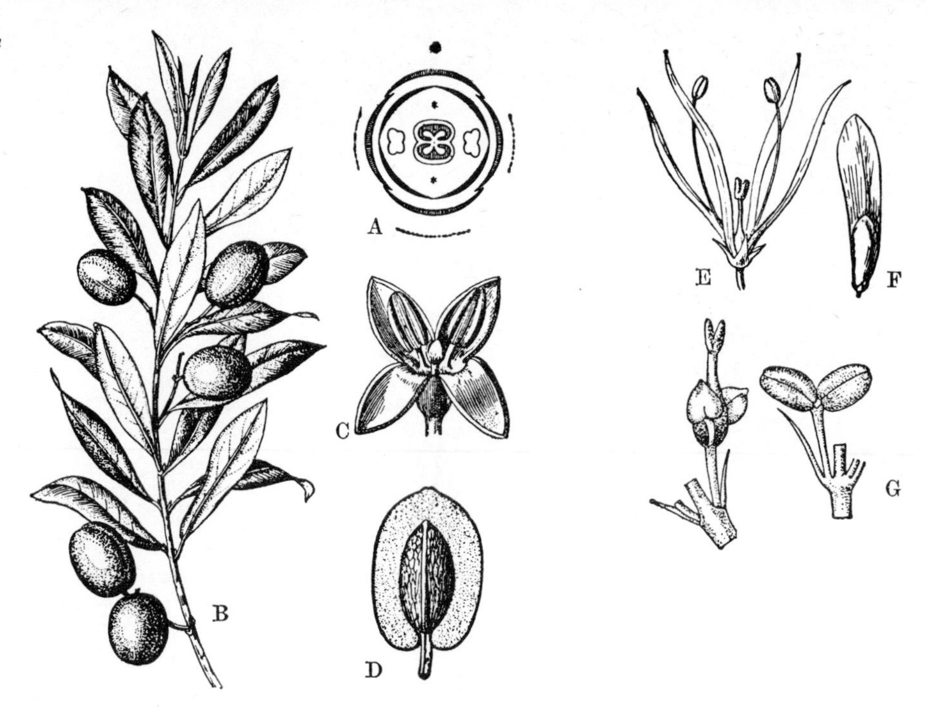

Fig. 723. Oleales. **A.** Floral diagram of *Syringa vulgaris*. **B.** to **D.** *Olea europaea*. **B.** Fruiting shoot (×$\frac{2}{5}$). **C.** Flower (enlarged). **D.** Drupe with the stone exposed (×1). **E.** to **G.** *Fraxinus*. **E.** and **F.** Hermaphrodite flower and winged nut of the entomophilous *F. ornus* (somewhat enlarged). **G.** Hermaphrodite and male flowers of the anemophilous *F. excelsior* (enlarged). **A.** and **B.** After Firbas. **C.** and **D.** From Hegi. **E.** and **F.** After Karsten. **G.** After Hempel and Wilhelm.

Ovules directed upwards are characteristic of the **Hydrophyllaceae**, which have capsules and include the N. American bee-plant *Phacelia*, and the **Boraginaceae** (Fig. 725A–E) which have four one-seeded nutlets in the fruit. These are predominantly herbaceous plants with simple, alternate, mostly hispid leaves. The flowers are usually borne in characteristic double cincinni ('scorpioid cymes', Fig. 725A) and are generally actinomorphic, only rarely, as in viper's bugloss (*Echium*), slightly zygomorphic. The throat of the corolla is frequently folded inwards to form five scales which constrict the entrance to the corolla tube (Fig. 725B). The bicarpellary ovary is divided into four loculi by false septa and develops into four one-seeded nutlets; these differ from the similar nutlets of the Lamiaceae (p. 730) in that the micropyle of the ovule and hence the radicle point upwards (Fig. 725D, H). The floral formula is generally

$$* \ K(5)\ [C(5)\ A5]\ G(\underline{2}) \quad \text{(Fig. 725E).}$$

Lungwort (*Pulmonaria*), forget-me-not (*Myosotis*), comfrey (*Symphytum*), bugloss (*Anchusa*) and borage (*Borago*) are familiar genera.

In contrast to the Polemoniales

Order 5: **Scrophulariales** (Personatae) is distinguished by the more or less zygomorphic flowers, the tendency for the stamens to be reduced from five to four and then to two, by the generally bicarpellary ovaries and capsules or berries with numerous ovules.

The **Solanaceae** which are closely related to the Polemoniales (especially the Convolvulaceae), have generally five stamens, scarcely zygomorphic flowers (but the carpels are oblique; see p. 643) and bicollateral vascular bundles. This family is well-known for the frequent occurrence of alkaloids (especially of the ornithine group, such as the tropine and nicotine types) and consists of both woody and herbaceous plants with alternate leaves. The morphology of the shoot is often difficult to elucidate as a result of fusions and displacements of the axes and leaves (Figs. 141G, 726A). The flowers are

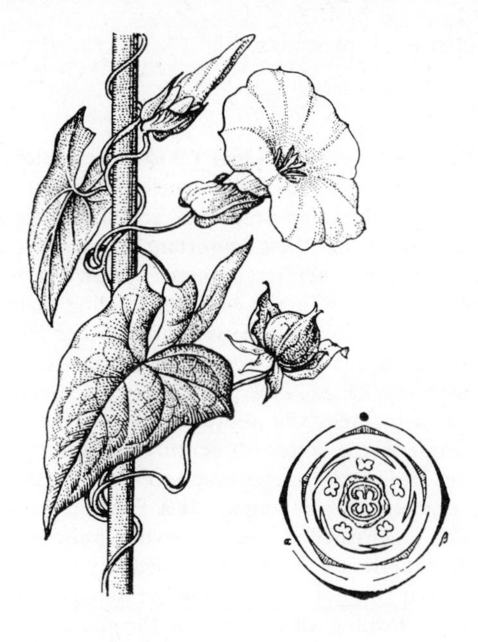

Fig. 724. Polemoniales, Convolvulaceae, *Calystegia sepium*. Flowering and fruiting shoot (×$\frac{1}{3}$) and floral diagram (with prophylls α and β). **A.** After Firbas. **B.** After Eichler.

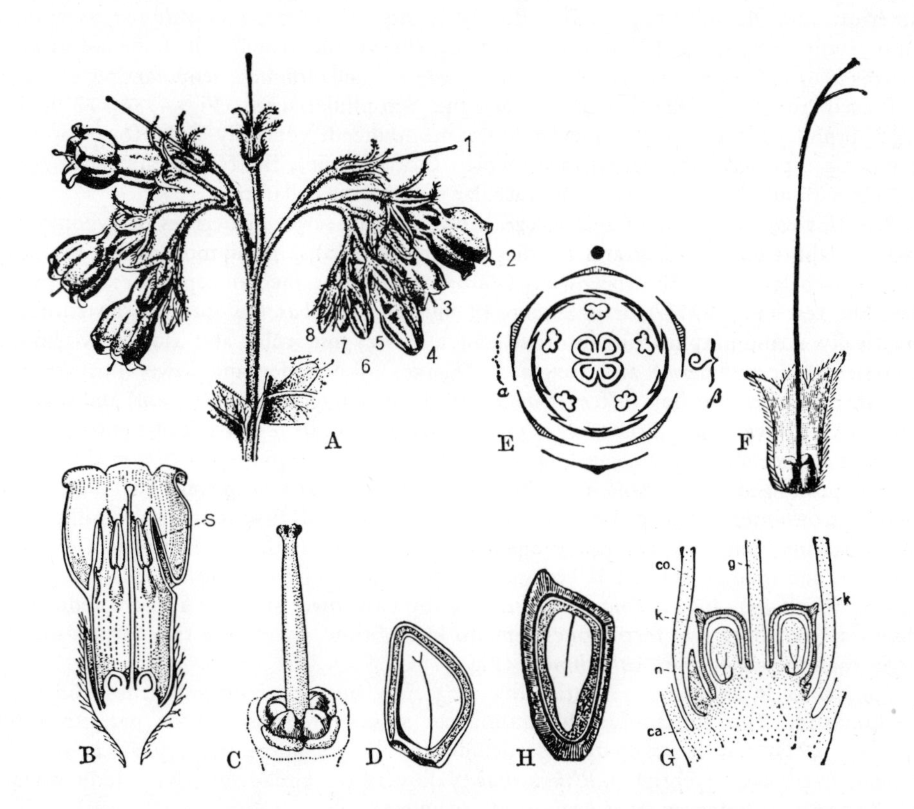

Fig. 725. Polemoniales, Boraginaceae (**A.** to **E.**) and Lamiales, Lamiaceae (**F.** to **H.**). **A.** and **B.** *Symphytum officinale*, double cincinnus (the numbers indicate the successive flowers; ×1 approx.) and half flower with scales (S) in the throat (×3 approx.). **C.** Ovary of *Pulmonaria officinalis* (×6). **D.** Nutlet of *Onosma visianii* (×8). **E.** Floral diagram of *Anchusa officinalis* (a lateral branch indicated in the axil of a prophyll β). **F.** Ovary exposed in the calyx of *Galeopsis segetum* (×2). **G.** L.S. of the base of the flower of *Lamium maculatum* with calyx (ca), corolla (co), nectary (n), developing nutlet with ovule (k) and style (g) (×10). **H.** Ripe nutlet of *Lamium album* in L.S. (enlarged). **A.** and **D.** After Wettstein. **B.** and **H.** After Baillon. **C.** and **G.** After Firbas. **E.** After Eichler. **F.** After Schenck.

often in cincinni. The ovules are borne on massive placentas (Fig. 726H). The floral formula is usually:

* or ↙ K(5) C(5) A5 G(2).

Capsules are found in tobacco (*Nicotiana*). *N. tabacum* (Fig. 726G, H) is allotetraploid and apparently originated in N.W. Argentina from the diploid wild species *N. silvestris* (Fig. 294) and *N. otophora*; it was cultivated in pre-Columbian times in S. and C. America and is now grown in many forms throughout the world. Less important is *N. rustica* (Turkish tobacco) which probably originated in Peru: neither species is known with certainty in the wild (cf. p. 332). Capsules also occur in the popular bedding plant *Petunia* (cf. p. 415) from S. America, and in the poisonous native ruderals henbane (*Hyoscyamus niger*; with a slightly zygomorphic corolla and circumscissile dehiscence, Fig. 726B–D) and thorn apple (*Datura stramonium*, Fig. 726E, F). Berries are characteristic of the very large genus *Solanum* which includes the allotetraploid potato (*S. tuberosum*, Figs. 141G, 207), brought to Europe in the sixteenth century; its ancestral forms are to be sought in the Andes of Peru, Bolivia, N. Argentina and Chile (subsp. *andigerum* and others). *S. melongena*, the egg plant or aubergine, is a Palaeotropical vegetable; *S. dulcamara*, bittersweet or woody nightshade, is a somewhat poisonous scrambler of moist woodlands. The tomato (*Lycopersicum esculentum*) is closely related; it was anciently cultivated in Peru and Mexico but was not widely grown in Europe until the nineteenth century; cultivated forms often exhibit an increase in the number of carpels. Berries are also found in *Capsicum annuum* (chilli, red pepper) from tropical America, and the poisonous native deadly nightshade (*Atropa belladonna*; sympodial shoot complexes, cf. p. 132 and Fig. 726A). The various tropine alkaloids are of great pharmacological importance (hyoscyamine, atropine, belladonnine, scopolamine, etc.).

Closely related to the Solanaceae are the **Scrophulariaceae** (Figs. 336, 727). The zygomorphy of the flowers is even more pronounced; generally only four (or two) stamens are present, the carpels are median (not oblique), the fruit is almost always a capsule with numerous seeds and the vascular bundles are collateral.

The flowers of the Scrophulariaceae show a whole series of degrees of zygomorphy and specialization (Fig. 727 and the diagrams in Fig. 336). Thus in most genera the calyx has five sepals, but in the speedwells (*Veronica* spp.) the median sepal is either smaller than the rest (Fig. 727G) or is absent. In mulleins (*Verbascum* species) the corolla is practically actinomorphic, but in other genera it is zygomorphic and usually two-lipped. In figwort (*Scrophularia*) and foxglove (*Digitalis*) the upper and lower lips are very similar, but generally they differ (*Pedicularis*). In snapdragon (*Antirrhinum*) and toadflax (*Linaria*) the lower lip is inflated as a 'palate' which conceals the throat of the corolla (i.e. masks it, hence Personatae; *persona* = mask). Furthermore, in *Antirrhinum* the corolla tube is prolonged downwards as a short sac, in *Linaria* as a long spur. In *Veronica* the corolla is tetramerous since the two upper petals are united. Five stamens are still present in *Verbascum*, but they are rarely equal and differ in length and hairiness. Generally, however, the median stamen is vestigial (a staminode in *Scrophularia*) or is completely lacking (*Digitalis*). In *Gratiola* and *Veronica* the two lower stamens also are reduced or absent so that only two fertile ones remain. In addition, one of the two carpels may be larger than the other as in *Antirrhinum* (Fig. 678G).

Paulownia is an eastern Asiatic flowering tree, but generally subshrubs and herbs predominate. The subfamily Rhinanthoideae shows progressive root parasitism (cf. pp. 199, 292). Green semiparasites include the perennial genus *Pedicularis* and the annuals *Euphrasia* (eyebright), *Rhinanthus* (yellow rattle) and *Melampyrum* (cow wheat). A white or pink holoparasite is *Lathraea* (toothwort).

Whilst there are no alkaloids in the Scrophulariaceae, the cardiac glycosides of *Digitalis* (particularly of *D. lanata* from the Balkans) are of the greatest pharmacological importance.

Various other families are related to the Scrophulariaceae. The **Globulariaceae** have endospermous seeds, frequently endosperm haustoria (Fig. 674L), and nutlets. The **Orobanchaceae**, holoparasites with scale leaves, are the often strongly host-specific broomrapes (*Orobanche*, Fig. 728). The **Plantaginaceae** are wind pollinated; the reduced

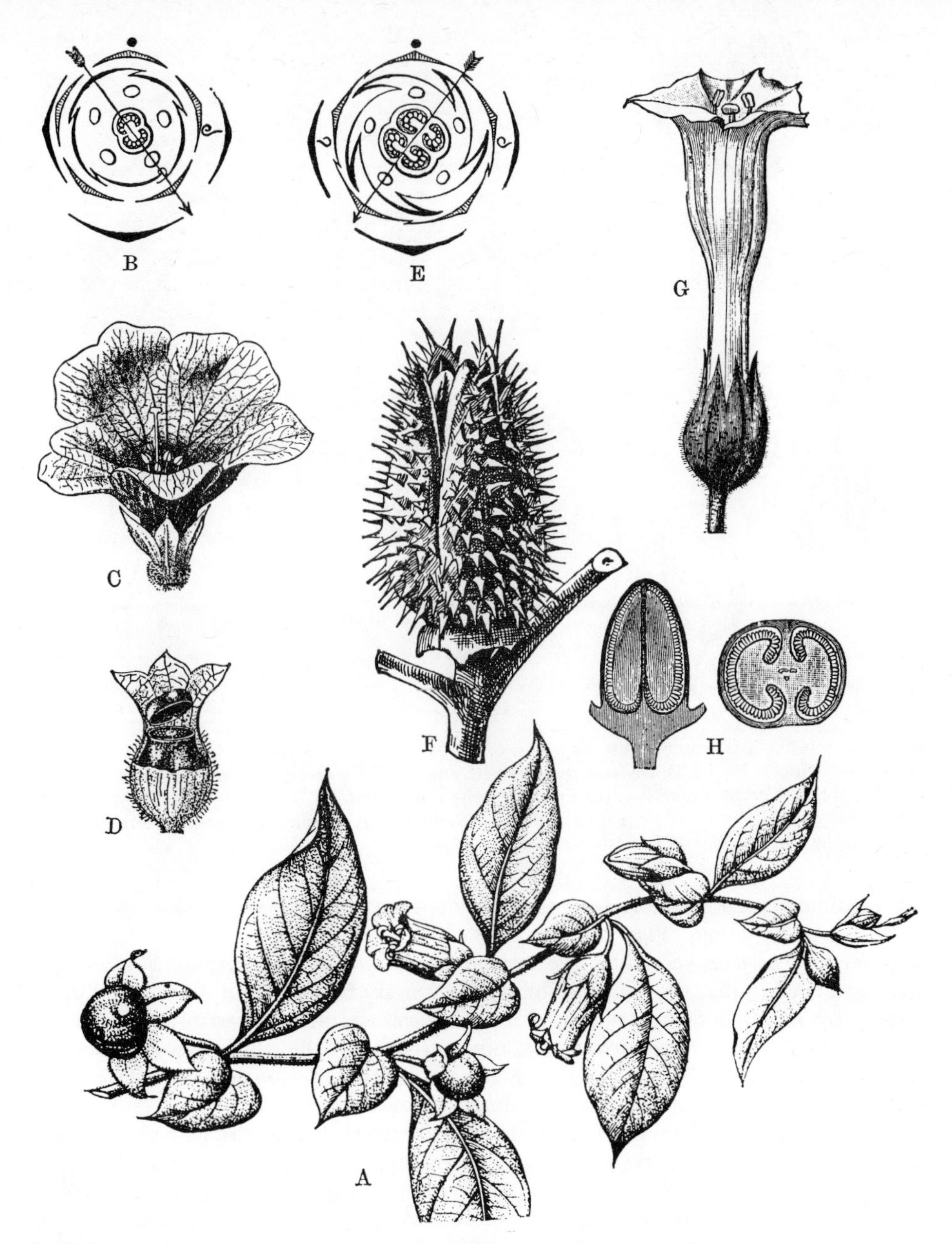

Fig. 726. Scrophulariales, Solanaceae. **A.** *Atropa belladonna*, sympodial shoot system with flowers and berries. ×½. **B.** to **D.** *Hyoscyamus*, floral diagram of *H. albus*, flower and circumscissile capsule of *H. niger* (calyx partly removed; ×1 aporox.). **E.** and **F.** *Datura stramonium*, floral diagram and spiny capsule (×1 approx.). **G.** and **H.** *Nicotiana tabacum*, flower (×1) and young capsule, L.S. and T.S. ×2. **A., F.** to **H.** After Karsten. **B.** and **E.** After Eichler. **C.** and **D.** After Beck-Mannagetta.

tetramerous flowers have membranous petals; they include the protogynous plantains (*Plantago*, Fig. 135D, p. 657) with circumscissile capsules. Without endosperm but with syncarpous carpels are the tropical woody **Bignoniaceae** (garden plants include the Indian bean tree, *Catalpa*, from E. Asia and N. America, and lianes, e.g. *Campsis*) and the generally tropical and herbaceous **Acanthaceae** and **Pedaliaceae** (some with highly specialized fruits, p. 674; the oil plant, *Sesamum indicum*). The ovary is incompletely septate in the **Gesneraceae** (with the 'monophyllous' species of *Streptocarpus*, p. 163; the Mediterranean-montane relict genera *Ramonda* and *Haberlea*; and the well-known decorative African violet, *Saintpaulia ionanthes*, from E. Africa). Finally, the placenta is central in

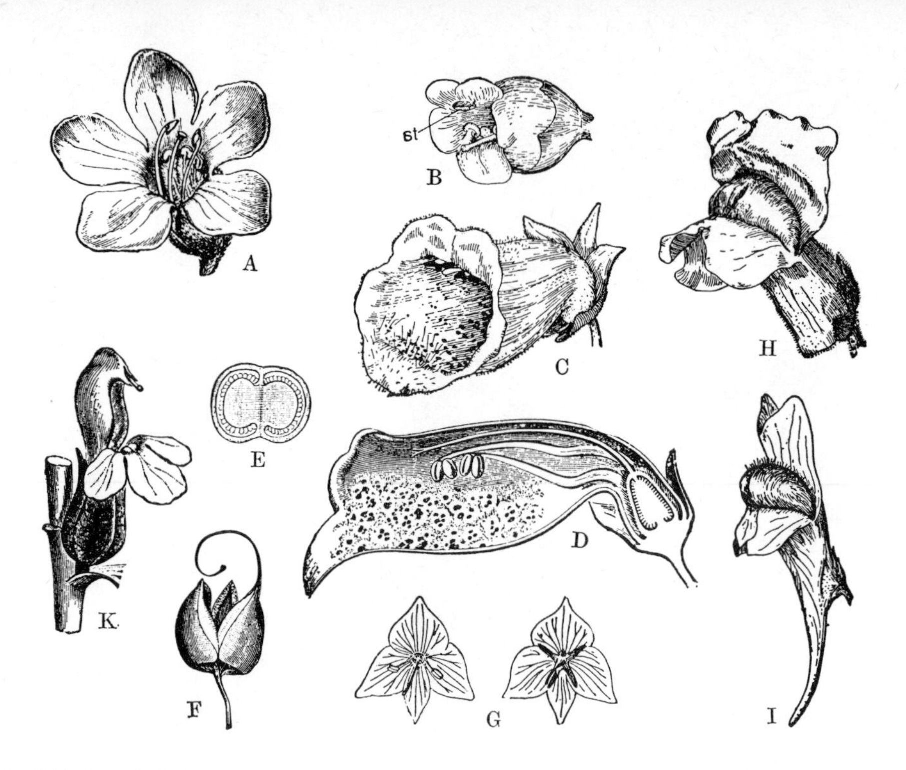

Fig. 727. Scrophulariaceae. Flowers (or fruits) of *Verbascum thapsus* (**A.** ×1.5), *Scrophularia nodosa* (**B.** With staminode, st, ×2.5), *Digitalis purpurea* (**C.** and **D.** Whole and half flower, ×¾ approx. **E.** and **F.** ×1 approx.), *Veronica teucrium* (**G.** From in front and behind, ×1.5), *Antirrhinum majus* (**H.** ×1), *Linaria vulgaris* (**I.** ×1.5), and *Pedicularis palustris* (**K.** ×1.6). **A.**, **D.** and **K.** After Baillon. **B.**, **C.**, **G.** to **I.** After Firbas. **E.** and **F.** After Karsten.

the **Lentibulariaceae**, notable for insectivory (with *Pinguicula* and *Utricularia*, cf. Fig. 127D, E, pp. 199 et seq.; Figs. 222A–F, 328).

Order 6: **Lamiales** resembles the Scrophulariales in its zygomorphic flowers but is more advanced in that the generally bicarpellary ovary has only four downward directed ovules, each of which forms a nutlet, whilst the leaves are almost invariably opposite.

In the predominantly tropical and woody **Verbenaceae** the style arises from the apex of the ovary. Members include the Indo-Malayan teak tree (*Tectona grandis*), *Avicennia* (a mangrove), and the Mediterranean shrub *Vitex agnus-castus*.

In the **Lamiaceae** (Labiatae) the style arises between and at the base of the carpels (Fig. 725G). The members of this very large family of subshrubs and herbs occur especially in hot dry climates (e.g. the Mediterranean region) and are easily recognized vegetatively by their four-angled stems (collenchymatous strands: Figs. 125A, 126H), opposite leaves and aromatic scents (glands with ethereal oils: Fig. 127C). The 'false whorls' of strongly zygomorphic flowers in axillary monochasia or dichasia are also very characteristic (Fig. 653C): a gamosepalous, often two-lipped calyx encloses a corolla with a long tube and an upper lip of two united petals and a lower lip of three. The four stamens (the median one is absent) consist of one longer and one shorter pair; in sage (*Salvia*, Fig. 671A–D) and rosemary (*Rosmarinus*) only the two lower stamens are present or fertile. The superior ovary is already deeply four-lobed at the time of flowering and develops into four nutlets in which the micropyle and radicle are directed downward (Fig. 725F–H). Thus in general:

$$K(5)\ [C(5)\ A4:1^\circ]\ G(\underline{2})$$

Many species are culinary herbs because of their high content of ethereal oils (e.g. marjoram, *Majorana hortensis*; basil, *Ocimum basilicum*; savory, *Satureja hortensis*), or

Fig. 728. Orobanchaceae. The chlorophyll-free, yellowish-brown total parasite *Orobanche minor* on *Trifolium repens*, $\times\frac{2}{5}$; a single flower enlarged. After Karsten.

they are also medicinal plants, e.g. the Mediterranean *Hyssopus officinalis* (hyssop), *Lavandula angustifolia* (lavender), *Rosmarinus officinalis* (rosemary), *Salvia officinalis* (sage), *Thymus vulgaris* (thyme) or the native *Melissa officinalis* (lemon balm), *Mentha* (the numerous kinds of mint, such as the menthol-containing peppermint, *M. piperita*, of hybrid origin from *M. spicata* x *M. aquatica*; p. 421) and *Thymus serpyllum* agg. (wild thyme). Other native genera are, e.g. *Ajuga* (bugle), *Galeopsis* (henbit), *Glechoma* (ground ivy, alehoof), *Lamium* (dead nettle), *Stachys* (woundwort) and *Teucrium* (germander, wood sage).

Here perhaps should be inserted the aquatic **Callitrichaceae** with the water starwort (*Callitriche*) in which the reduced flowers are unisexual and have no perianth, consisting merely of either one stamen or one carpel.

The **Asteranae** (Synandrae, Campanulatae), the other superorder of the Asteridae, are characterized mainly by the postgenitally more or less united anthers (but the filaments remaining free, Fig. 729E, L), by brushes or other stylar devices for collecting the pollen (Fig. 729B–D, I, M), inferior ovaries (often with only one ovule), a marked tendency to form pseudanthia, a generally herbaceous habit, leaves often crenately dentate or lobed, the absence of stipules, and the frequent occurrence of latex canals (Fig. 128) and of inulin in place of starch (p. 53). There are affinities with the Polemoniales and more distantly with the Araliales and Saxifragales.

Order 7: **Campanulales** is relatively primitive, having ovaries which frequently have several carpels (five, three or two; rarely even superior) with numerous ovules, capsules, endosperm, and commonly euanthia. The flowers of the **Campanulaceae** are actinomorphic

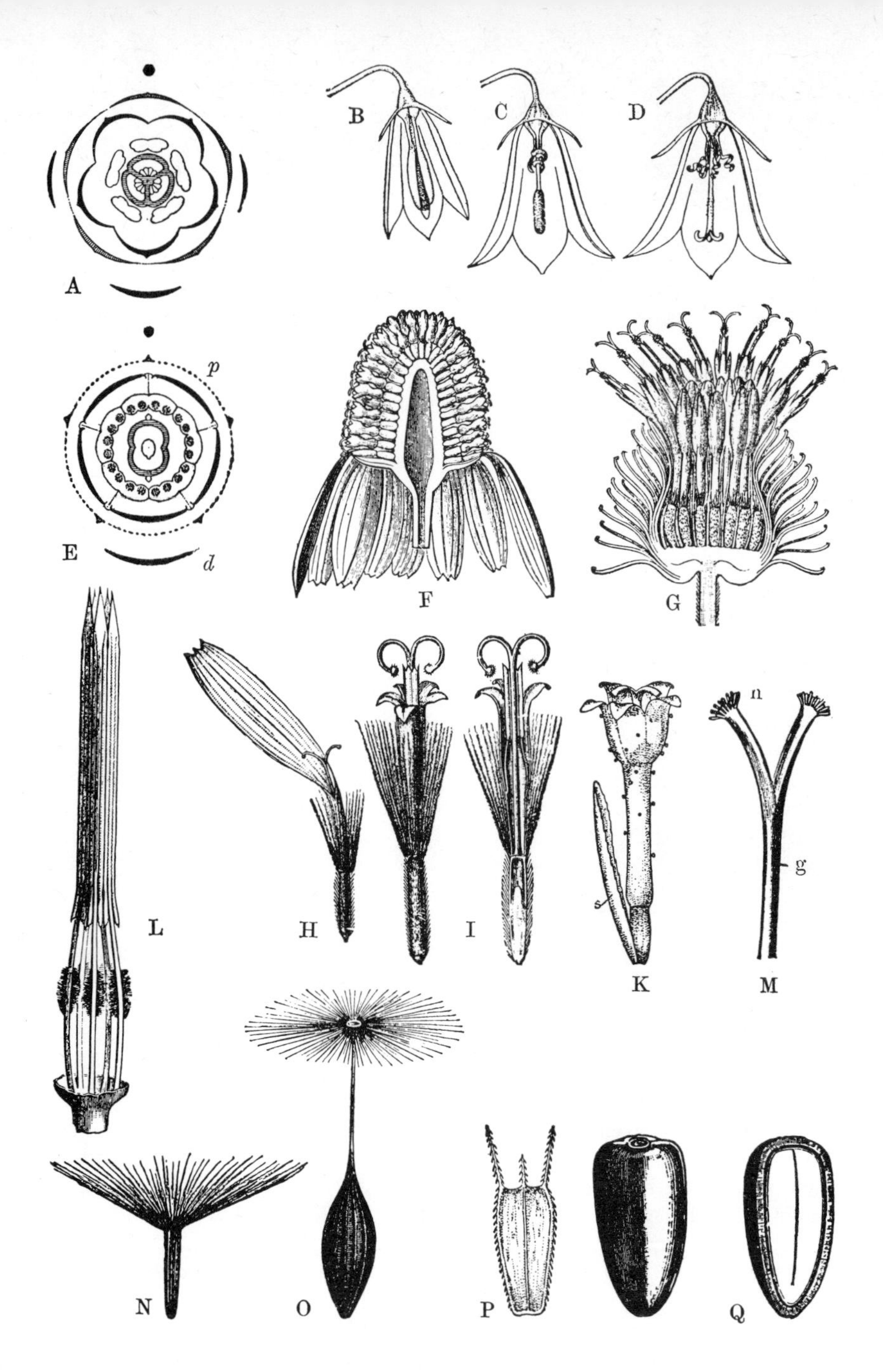

Fig. 729. Campanulales (**A.** to **D.**) and Asterales (**E.** to **Q.**; **N.** and **O.** Cichoriaceae, otherwise Asteraceae). **A.** Floral diagram of *Campanula* sp. **B.** to **D.** Stages of anthesis in *Campanula rotundifolia* (the nearer petals removed): the stamens deposit their pollen onto the style (**B.**) and then shrivel (**C.**); pollen stripped off and the stigmas expanded (**D.**). ×1. **E.** Floral diagram of a disc floret with bract (receptacular scale, d) and pappus (p). **F.** and **G.** Capitula cut lengthwise of *Matricaria chamomilla* (involucral bracts in one row, receptacle hollow, ray (ligulate) florets reflexed, no receptacular scales) and *Arctium lappa* (hooked involucral bracts in several rows, no ligulate flowers, receptacular scales present); enlarged. **H.** to **I.** *Arnica montana.* **K.** *Anthemis nobilis*: ray and disc florets, whole and half flowers; receptacular scale, s; enlarged. **L.** Androecium of *Carduus crispus*; ×10. **M.** Style (g) and stigmas (n) of *Achillea millefolium*; enlarged. Fruits (achenes) of (**N.**) *Hieracium virosum* and (**O.**) *Lactuca virosa*, with hairy pappus; of (**P.**) *Bidens tripartitus* with recurved hooks on the pappus bristles; and of (**Q.**) *Helianthus annuus*, whole and L.S., without pappus. **A.** and **E.** After Eichler, modified. **B.** to **D.** After Clements and Long, somewhat simplified. **F.** to **K.** and **M.** After Berg and Schmidt. **L.**, **N.**, **O.** and **Q.** After Baillon. **P.** After Firbas.

(Fig. 729A–D) and protandrous: the anthers are only loosely united and liberate the pollen before the stigmas are mature on the collecting hairs of the style. The ovary generally has three carpels and develops into a many-seeded capsule which, for example, in bellflower (*Campanula*) opens by pores. In the flowers of rampion (*Phyteuma*) the tips of the corolla remain united at first and the flowers are aggregated in dense inflorescences surrounded by involucral bracts; as in the genus *Jasione* there is a similarity to the capitula of the Asterales. Closely related are the predominantly tropical **Lobeliaceae** with zygomorphic flowers, generally bilocular ovaries and numerous alkaloids (poisonous and medicinal plants). Some species of *Lobelia* are peculiar 'cabbage trees' of the high African mountains; these bear a rosette of very large leaves on a short unbranched stem. *L. dortmanna* is a characteristic Atlantic plant of oligotrophic lakes.

Order 8: **Asterales** (Compositae) in contrast is characterized by the inferior, bicarpellary, unilocular ovary with one ovule, achenes, lack of endosperm and the contraction of the racemose or spicate inflorescences into heads or capitula (pseudanthia) with a calyx-like involucre (Fig. 146P); they are therefore greatly advanced. The inflorescence axis (receptacle) is either elongated and club-shaped or flattened (Fig. 729E–G); sometimes it bears scales, sometimes none (Fig. 729K). The heads either contain tubular florets (disc florets, Fig. 729I, K) which are actinomorphic with the tips of the five petals free, or zygomorphic ligulate florets (ray florets, Fig. 729H, consisting of three or five petals unilaterally united), or both types occur (Fig. 729F; ligulate florets outside, disc florets inside). The calyx is modified as scales, bristles or hairs (as a pappus); or it may be absent; when present it often participates in dispersal of the fruit (Fig. 729N–Q; anemochory: N, O; epizoochory: P, etc.). The filaments of the five epipetalous stamens are free, but the anthers are united by means of their cuticle in a tube into which the pollen is discharged (Fig. 729E, L). Hairs on the outside of the style (Fig. 729I, M) sweep the pollen out of this tube, either by the elongation of the style or by the contraction of the filaments (cf. Fig. 312). Only then do the two lobes of the stigma spread out, exposing their receptive inner surfaces (Fig. 729I, M); the flowers are therefore protandrous. The ovary is bicarpellary but unilocular and contains a single basal anatropous ovule (Fig. 729I). This develops into a nutlet with a more or less united pericarp and testa (achene); the embryo contains protein and oil (Fig. 729Q).

Having some 19,000 species, the Asterales form one of the largest groups of angiosperms and are still undergoing very active evolutionary change (cf. Figs. 338, 342B, 351, 357, 360). They may be divided into closely-related families (or subfamilies) of which the **Asteraceae** (Compositae–Tubuliflorae) have disc florets, with or without ray florets (the rays consisting of three petal apices) and generally also have schizogenous oil cavities (with ethereal oils).

Only disc florets are present in the Eupatorieae, e.g. the native hemp agrimony (*Eupatorium cannabinum*), and in the Cardueae, e.g. the thistles of the genera *Cirsium* (with pinnate pappus hairs) and *Carduus* (simple pappus hairs); in the knapweeds (*Centaurea*; e.g. *C. cyanus*. cornflower) the marginal disc florets are enlarged and sterile, whilst in the burdocks (*Arctium*, Fig. 729G) the hooked involucral bracts function in epizoochorous dispersal (p. 674); useful plants include the Mediterranean globe artichoke (*Cynara scolymus*, a vegetable) and the safflower (*Carthamnus tinctorius*, a dye and oil plant). The other tribes have variously coloured ray florets in addition to the disc florets: alternate leaves and several rows of involucral bracts characterize, for example, the Astereae (including *Bellis* and the genus *Aster*, particularly rich in N. American species); the Inuleae (with the European genera *Inula*, elecampane; *Antennaria*, the dioecious cat's-foot; *Leontopodium*, the various Asiatic and Alpine species of edelweiss with white woolly bracts surrounding a number of heads: *Helichrysum*, the everlasting flowers with scarious coloured involucral bracts; and *Raoulia* from New Zealand, which forms hard cushions); and the Anthemideae (including many medicinal plants and culinary herbs with ethereal oils and sesquiterpene lactones, e.g. the true chamomile, *Matricaria chamomilla* (without receptacular scales, in contrast to the false chamomiles, *Anthemis*); the wild or Roman chamomile, *Ormenis nobilis* (*Chamaemelum nobile*); the yarrow, *Achillea millefolium* agg., Figs. 339, 351, 360; various secondarily anemophilous species of *Artemisia* (Fig. 730A), e.g. the wormwood (*A. absinthium*) and tarragon (*A.*

Fig. 730. Asteraceae. **A.** *Artemisia borealis*, wind pollinated; habit (× ¾ approx.) and capitulum with only disc florets, the outer ones ♀ (enlarged). **B.** *Senecio keniadendron*, 'cabbage trees' in the upper alpine zone in W. Kenya (E. Africa). **A.** After Hegi. **B.** After R. E. Fries.

dracunculus); as well as many decorative plants, e.g. the 'chrysanthemums' in the genus *Tanacetum*. The involucral bracts are in fewer rows in the Senecioneae (including the medicinal coltsfoot, *Tussilago farfara*, and arnica, *Arnica montana*, as well as the large genus *Senecio* which includes annuals, perennial herbs and the 'cabbage trees' of the E. African mountains, Fig. 730B, and the 'candle cactus' *Kleinia*, Fig. 203D, etc.); and in the Calenduleae, with the heterocarpous marigold, *Calendula*. The Heliantheae have opposite leaves; these include the native species of *Bidens*, Fig. 729P; the 'galloping soldiers' escaped from S. America, *Galinsoga*; the Andean 'cabbage trees' *Espeletia*; the widely grown numerous polyploid forms of dahlia, *Dahlia variabilis*, Fig. 209; the important crop plants in the N. American genus *Helianthus* (sunflower), *H. annuus*, oil plants, Fig. 136; Jerusalem artichoke, *H. tuberosus*, with tubers rich in carbohydrates; and the wind-pollinated genera *Ambrosia* and *Xanthium*. Finally, the Helenieae also have opposite leaves and include the decorative genera *Tagetes* and *Gaillardia*.

The **Cichoriaceae** (Compositae–Liguliflorae) have ray florets only (the rays of five petal apices) and latex canals (Fig. 128).

These include the genera *Cichorium* (with *C. intybus*, chicory, a coffee-substitute, and *C. endivia*, endive); *Scorzonera*, sometimes grown as a root vegetable; *Leontodon* (hawk-bits, with pinnate pappus hairs); *Taraxacum* (dandelion) and *Hieracium* (hawkweed), both with simple pappus and extremely rich in species (polyploid hybrids and often agamospermous, cf. pp. 403, 421, Fig. 345); *Crepis* (hawk's-beard); and *Lactuca* (*L. sativa*, cultivated as lettuce; *L. serriola*, well known as a 'compass plant', pp. 181, 350).

Class II: Liliatae (Monocotyledoneae), monocotyledons

The monocotyledons have a single seed-leaf or cotyledon (Fig. 731A–D) which commonly appears to develop at the apex in the embryo (Fig. 674H) with its sheath surrounding the laterally displaced growing point of the stem. Frequently the cotyledon is modified as an organ which transfers nutrients from the endosperm. The primary root is short lived and is soon supplanted by numerous adventitious roots (secondary homorhizy; Fig. 191B, p. 177). The vascular bundles appear scattered in a transverse section of the stem (atactostele, Figs. 153, 155). They have no cambium, i.e. they are closed bundles (Fig. 122), consequently neither shoot nor root has normal secondary thickening. (For the rare, anomalous secondary thickening of certain monocotyledons see Fig. 155 and pp. 147, 739.) Apart from the inflorescences the aerial shoots branch little. The leaves are mostly alternate with a broad insertion on the axis, or a sheathing base; they usually have no petiole. They are generally simple, entire, often linear or elliptical and the venation is almost always 'parallel' (Figs. 732C, 733H, 736A). The axillary branches commonly bear only a single addorsed prophyll (i.e. a bracteole with its back to the main axis) with two veins (cf. Fig. 140II, above; Fig. 732A, p. 131). The flowers are generally composed of trimerous whorls with the formula P3 + 3A 3 + 3 G3 (Fig. 733E). Septal nectaries between the walls of the carpels are very characteristic. Development of the pollen is usually successive, the pollen grains are mostly anatreme or monocolpate, and the endosperm is of the helobial type or generally nuclear. Predominant life-forms are herbaceous marsh and aquatic plants and perennial herbs, especially hemicryptophytes and geophytes perennating underground.

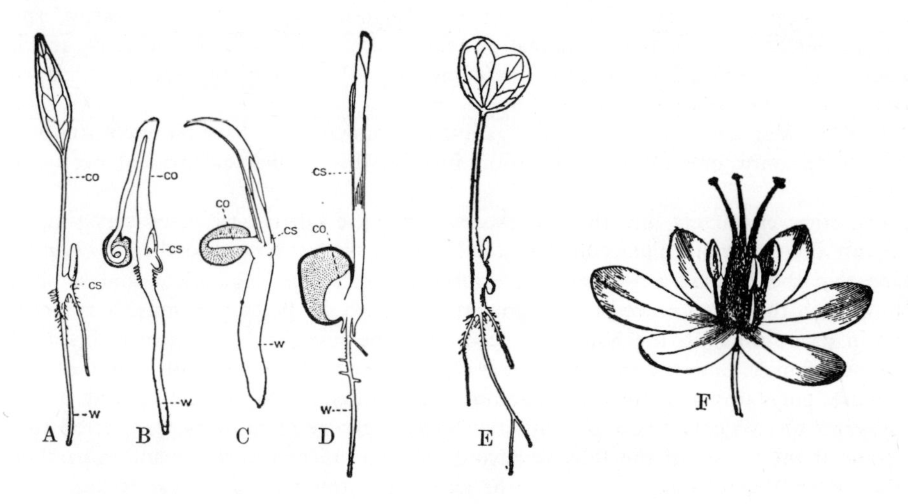

Fig. 731. Seedlings of the Liliatae. **A.** *Paris quadrifolia.* **B.** *Allium cepa.* **C.** *Clivia miniata.* **D.** *Zea mays.* **B.** to **D.** In L.S.; testa or pericarp black, endosperm dotted; cotyledon (co); its sheath (cs); primary root (w). The cotyledons in **C.** to **D.** are partly or wholly modified as absorbing organs. Monocotyledonous characters in dicotyledons: **E.** Seedling of *Ranunculus ficaria* with one cotyledon; **F.** Flower of *Cabomba aquatica* (Nymphaeaceae), P3 + 3, A3, G(3). ×3. **A.** to **E.** After Sachs and Wettstein, modified. **F.** After Baillon.

These and other characters make the Liliatae an easily recognizable and obviously natural class. Nevertheless they are not confined to the Liliatae but occur sporadically in the Magnoliatae, whilst conversely characters of the Magnoliatae occur in the Liliatae. Further consideration shows that the Liliatae can be derived from the Magnoliatae and indeed from the primitive subclass Magnoliidae where these monocotyledonous characters are particularly apparent. The principal evidence for this view is as follows:

1. Plants with one cotyledon may arise from those with two by the congenital union of the two cotyledons (syncotyly) or by the loss of one cotyledon (heterocotyly resulting from progressive anisocotyly). It is debatable in which of these two ways the single seed-leaf of the monocotyledons has arisen, but recent studies suggest that the second is the more likely one. In any case, monocotyledonous plants occur—both as a result of syncotyly and heterocotyly—amongst the dicotyledons (Fig. 731E). Indeed, in *Eranthis* (Ranunculaceae) both syncotyly and heterocotyly can be induced experimentally by the action of growth substances (phenyl boric acid).
2. The vascular bundles in some Magnoliatae are arranged in several rings (Piperales,

Caryophyllidae) or may be scattered (various Magnoliidae, e.g. in the Nymphaeaceae, and *Podophyllum* in the Berberidaceae). In these cases the cambium soon ceases to function. Suppression of cambial activity is also found in herbaceous Magnoliatae, particularly in hydrophytes. Conversely, some Liliatae have the bundles arranged in a ring (e.g. Dioscoreaceae), and here and there traces of a fascicular cambium can be detected.

3. The suppression of the growth of the primary root and its functional replacement by adventitious roots (homorhizy) is not uncommon in the Magnoliatae, particularly in the herbaceous Magnoliidae. This also is related to the loss of secondary thickening, without which the primary root cannot become sufficiently large.

4. Parallel venation of the leaves occurs in some Magnoliatae (e.g. species of *Bupleurum* and *Plantago*), while reticulate venation occurs in Liliatae (e.g. Dioscoreaceae). The monocotyledonous leaf can be regarded as an incompletely developed dicotyledonous leaf. The addorsed prophyll can be equated with two laterally fused prophylls of the Magnoliatae; similar cases occur in the Magnoliidae (Annonaceae, Nymphaeaceae). It can also be regarded as an expression of the tendency towards distichous phyllotaxis in the Liliatae (cf. p. 125, Figs. 132K, 135A–C).

5. Trimerous flowers are characteristic of many Magnoliidae (pp. 680 et seq.). Some indeed have the monocotyledonous floral formula P3 + 3 A3 + 3 G3: in the Nymphaeaceae the genus *Cabomba* is notable for its trimerous flowers (Fig. 731F).

6. Important primitive characters of the Magnoliidae, namely choricarpy, spiral arrangement of the carpels and laminal placentation, also occur in the primitive subclass Alismatidae of the Liliatae which, like some Magnoliidae (Nymphaeaceae), comprises mainly water plants.

7. Many Magnoliidae, for example again the Nymphaeaceae, have monocolpate pollen grains with only one distal germination furrow; this is characteristic of very many Liliatae.

Yet other characters link the two classes. Since the Liliatae are often very simple in structure they were often placed before the Magnoliatae, but nowadays the view is usually taken that the Liliatae arose from ancestors similar to the Magnoliidae and so should follow both these and indeed the Magnoliatae as a whole. Precursors must have been at least in part similar to the Nymphaeaceae since the resemblances between this dicotyledonous family and the Alismatidae are particularly close (anatomy, root and leaf structure, polycarpy and choricarpy, monocolpate pollen, ecology, etc.). Indeed, only the possession of two cotyledons prevents the Nymphaeaceae from being transferred to the Liliatae. If the vessels of the Liliatae have developed independently of and in parallel to those of the Magnoliatae, as is likely, the ancestral forms must also have lacked vessels: and this, too, is found in the Nymphaeaceae.

Finally, the hypothesis that the ancestors of the Liliatae were herbaceous marsh plants illuminates many features and progressions in this class; for example, the reduction in secondary thickening, the suppression of the primary root and the little differentiated leaves. In the evolutionary 'return' of the Liliatae to dry land new types of construction would have arisen, such as secondarily arborescent forms with extreme primary thickening (e.g. palms, Fig. 154A, B) or with new types of secondary thickening (e.g. *Dracaena*, Fig. 155), with stilt roots (e.g. *Pandanus*) or secondarily expanded leaf laminae (e.g. in palms and aroids). However, from these land forms aquatics have arisen once more (e.g. the Lemnaceae from Araceae, cf. p. 752).

Nevertheless, the monocotyledons must have originated very early and developed independently for a long time since their remains occur in the Lower Cretaceous along with the oldest known fossil Magnoliatae. Furthermore the adaptive radiation of the ancestors of the Alismatidae or Dioscoreaceae into Liliidae (including the Commelinidae) and Arecidae (cf. p. 750 and Figs. 628, 742) must have taken place and continued in parallel a very long time ago.

Subclass A: Alismatidae (Helobiae)

Like the Magnoliidae amongst the dicotyledons, the Alismatidae occupy a key position amongst the monocotyledons. As indicated in the introduction (above) the Alismatidae and the Nymphaeales (p. 681) link the two classes and together form the 'substructure' or 'foundations' of the Liliatae (Fig. 742). Correspondingly the Alismatidae show many primitive characters: the gynaecium is choricarpous (or falsely coenocarpous as a result of the participation of the floral axis) and the carpels are sometimes spirally arranged (Fig. 732B); on the other hand the androecium and perianth are cyclic, with three-membered (trimerous) whorls and the occasional polyandry is apparently secondary (dédoublement; Fig. 732A).

The members are exclusively herbaceous aquatic or semi-aquatic plants, with or without very primitive vessels in the roots. The hermaphrodite, actinomorphic flowers often become unisexual or dioecious. Two or more trimerous perianth whorls are present in primitive cases; sometimes these become heterochlamydeous with a distinct calyx and corolla, sometimes they are reduced to one whorl or none. The stamens were probably originally borne in two trimerous whorls; they may be increased by lateral (Fig. 732A) or centrifugal dédoublement, or they may be reduced to one whorl or even to a single stamen. The pollen has usually three cells. The carpels are free and superior, or embedded in the floral axis, with ovules numerous or reduced to one; they exhibit progressive differentiation of the stigma and have laminal or submarginal placentation. The ovules are crassinucellate and bitegmatic; in primitive cases they are numerous, but eventually they are reduced to one. Development of the endosperm is frequently helobial (see p. 667), but the mature seeds generally lack an endosperm and are borne in follicles or unicarpellary nutlets. Reduction in floral structure is clearly closely related to the change from entomophily to anemophily and finally to hydrophily. Like the two other subclasses of the monocotyledons, the Alismatidae occur as fossils as far back as the Cretaceous.

Fig. 732. Alismatidae. **A.** and **B.** Alismatales. **A.** Floral diagram of *Alisma plantago-aquatica*; bracteole or prophyll (v) with two veins and addorsed, stamens showing dédoublement, carpels free with one ovule. **B.** Ripe gynaecium of *Sagittaria sagittifolia* with numerous unicarpellary, spirally arranged nutlets; some of them detached. **C.** and **D.** Najadales. **C.** Flowering shoot of *Potamogeton natans.* X¼. **D.** Flower of *P. perfoliatus* with appendages of the connective of the stamens resembling perigone segments (k). **A.** After Eichler, modified. **B.** After Buchenau. **C.** After Karsten. **D.** From Hegi.

The most primitive combination of characters in the subclass (generally heterochlamydeous perianth, superior gynoecium, etc.) is found in:

Order 1: **Alismatales.** Here the **Butomaceae** exhibit follicles with laminal placentation and monocolpate pollen grains; the native flowering rush (*Butomus umbellatus*) is a member. In contrast the **Alismataceae** have few- or one-seeded nutlets (Fig. 732B) and pollen grains with several apertures. The flowers are hermaphrodite in the water plantains (*Alisma*) with the formula * $P3^k + 3^c$ A 3^2 $G\underline{\infty}$ (Fig. 732A). *Sagittaria* is monoecious and has numerous stamens. In addition to ovate-lanceolate (*Alisma*) or arrow-shaped (*Sagittaria*) aerial leaves, floating leaves and strap-shaped submerged leaves are often present.

Inferior ovaries and commonly unisexual flowers characterize:

Order 2: **Hydrocharitales** with the single family **Hydrocharitaceae**. Frogbit (*Hydrocharis morsus-ranae*) is a floating plant; the water soldier (*Stratiotes aloides*) has large rosettes of leaves which either root or float in still waters; other members are completely submerged such as the Canadian pondweed (*Elodea* or *Anacharis canadensis*; Fig. 96) from N. America, which escaped in Europe in 1836 and has spread extensively by vegetative growth, or the (sub)tropical *Vallisneria spiralis*; the ♀ flowers of this are carried to the surface by a spirally-twisted peduncle, whilst the ♂ flowers become detached under water and rise to the surface where they open and float up to the ♀ flowers. The fruit ripens under water.

The perianth is single or entirely lacking in:

Order 3: **Najadales**. Notable families include the **Scheuchzeriaceae** with the rush-like *Scheuchzeria palustris* on raised bogs: P3 + 3 A3 + 3 G3 − 6 (with many-seeded follicles). The following families are more reduced and have one-seeded nutlets. Hermaphrodite flowers occur in the marsh-dwelling **Juncaginaceae** (including *Triglochin*) and the **Potamogetonaceae** which include the pondweeds (*Potamogeton*), rooting aquatics with or without floating leaves and with tetramerous, wind-pollinated flowers which are borne above the water in spikes; appendages of the connective of the stamens replace the missing perianth (Fig. 732C, D). The eelgrasses of the **Zosteraceae** are completely submerged marine plants with hermaphrodite flowers (of only one stamen and one carpel); the filamentous pollen grains are transported by water; species of *Zostera* occur in the extra-tropical seas, e.g. the North Sea and Baltic, whilst *Posidonia oceanica* is found in the Mediterranean. Finally, unisexual flowers are found in the fresh or brackish water aquatics in the **Zannichelliaceae** (several carpels, e.g. *Zannichellia*, Fig. 193) and the **Najadaceae** (one carpel, e.g. *Najas*).

Subclass B: Liliidae (including Commelinidae)

Compared with the Alismatidae, the Liliidae are more advanced in having generally three more or less united carpels and therefore coenocarpous gynaecia. The floral structure is cyclic throughout and frequently follows the formula P3 + 3 A3 + 3 G(3), in which case it is pentacyclic. The inflorescences are various, but never club-shaped and hardly ever surrounded by a large subtending bract.

The growth forms are predominantly herbaceous although some apparently advanced woody forms do occur (sometimes with anomalous secondary thickening). Diverse adaptations to dry conditions are characteristic. Pollination by animals is frequently associated with the differentiation of calyx and corolla, fusion of sepals and petals, and the development of zygomorphic flowers and inferior ovaries. Frequently nectaries develop from the floral axis as septal nectaries between the walls of the carpels (which are sometimes postgenitally fused). Polyandry scarcely ever occurs, whilst reduction of the stamens from two whorls to one (or even to a single stamen or half of one) is common. As well as capsules, berries are frequent. In wind pollinated groups unisexual flowers, reduction or absence of the perianth and nectaries, reduction in the number of stamens and carpels, reduction of the ovules to one and the development of nutlets are all characteristic. The pollen grains have either two or three cells. Besides crassinucellate and bitegmatic ovules, tenuinucellate and unitegmatic ones also occur. Development of the endosperm is sometimes free nuclear (p. 667), whilst the mature endosperm contains starch, oil, protein or reserve cellulose; sometimes a perisperm is present.

The subclass Liliidae, considered here in a broad sense, is a very large one with ten orders which may be grouped in superorders. The Lilianae (with all perianth segments more or less petaloid; orders 1 and 2) and Bromelianae (with the perianth differentiated into sepaloid and petaloid whorls; orders 3 and 4) frequently have septal nectaries and show a progressive tendency towards zoophily; the Juncanae (with the embryo embedded in the endosperm; orders 5 and 6) and Commelinanae (embryo lateral to the endosperm;

orders 7–10) both lack septal nectaries and incline towards anemophily. The superorder **Lilianae** presents a herbaceous (and often geophytic) line of development, with secondarily woody forms, with a petaloid perianth, actinomorphic or zygomorphic hermaphrodite flowers and an endosperm without starch, or lacking.

Order 1: **Liliales** (Liliiflorae) is characterized particularly by actinomorphic flowers, two whorls of stamens or one, syncarpous ovaries with numerous submarginal ovules and seeds with oil, protein or reserve cellulose (hardly ever starch) in a horny or fleshy (not floury) endosperm. The numerous rhizome, tuberous and bulbous geophytes (with occasional 'cabbage trees') of this order occur especially in arid subtropical regions.

The Liliales form a very diverse and taxonomically difficult basic group which will require further rearrangement in the future. It may well be related to the Alismatidae (cf. for example *Scheuchzeria*, p. 738), but it may be connected via precursors similar to the Dioscoreaceae (pp. 736, 741) directly to the Magnoliidae.

The very large and heterogeneous family **Liliaceae** has still two whorls of stamens and superior ovaries; thus * P3 + 3 A3 + 3 G$(\underline{3})$. Important classificatory characters include vegetative features (e.g. rhizomes → tubers → bulbs; Figs. 205, 208), phytochemistry (e.g. alkaloids), the fruits (e.g. septicidal → loculicidal capsules → berries), and the gametophytes (e.g. normal type → *Allium* type and other forms of embryo sac development, Fig. 668; simultaneous → successive formation of pollen), whilst the progressions from free to united perigones (e.g. in *Convallaria* and *Colchicum*) are, as in other monocotyledons, of lesser systematic significance.

The first group of subfamilies of the Liliaceae (*a–c*) is herbaceous, the leaves have parallel veins and the nectaries still occur on the receptacle or between the only partially united carpels, whilst the styles are free and the ventricidal or more generally septicidal capsules contain seeds with a persistent outer integument (sometimes as a sarcotesta). These comprise: (*a*) the Melanthioideae with rhizomes, including the poisonous *Veratrum album*, a tall perennial herb of mountain meadows, and *Tofieldia*; (*b*) the Wurmbaeoideae with tubers or corms, including for example the meadow saffron or naked ladies, *Colchicum autumnale*, Fig. 733H, which contains the very poisonous alkaloid colchicine; in the autumn the corm produces above ground only flowers with a long floral tube; in the spring the leaves and the developing fruit appear, whilst at the same time a basal internode develops into a new corm; and (*c*) the Lilioideae with bulbs and transitions to loculicidal capsules, including, for example, the tulip, *Tulipa*, Figs. 208, 733A; the yellow star of Bethlehem, *Gagea*; the lilies, *Lilium*; and the fritillaries, *Fritillaria*. The second, partly woody group of families (*d–k*) has septal nectaries, carpels united up to the stylar region and seeds without a sarcotesta. Subfamilies (*d–f*) have berries, and often have rhizomes and leaves with reticulate venation: (*d*) the Asparagoideae include the poisonous, green-flowered tetramerous herb paris, *Paris quadrifolia*, with free styles (Fig. 205A); Solomon's seal, *Polygonatum*, Fig. 205B; may lily, *Maianthemum*; the poisonous lily of the valley, *Convallaria*; and two genera with phylloclades, the native vegetable *Asparagus* and the mainly Mediterranean species of *Ruscus* (butchers' broom, Fig. 197A); (*e*) the Smilacoideae include numerous species of *Smilax*, (sub)tropical lianes with leaf sheaths modified as tendrils; and (*f*) the palaeotropical Dracaenoideae, including the woody *Cordyline* and *Dracaena* with anomalous secondary thickening (Fig. 155), e.g. the dragon tree of the Canary Islands, *Dracaena draco*, which reaches 18 m in height, and the herbaceous fibre producing and decorative species of *Sansevieria*. In contrast loculicidal capsules occur in subfamilies (*g–k*): (*g*) the Asphodeloideae have rhizomes and include the European *Anthericum* and the Mediterranean species of *Asphodelus* with root-tubers; *Gasteria*, Figs. 135A, B, 195B, and the woody, succulent, syntepalous and often bird pollinated species of *Aloe*, Fig. 139A, with resins and bitter substances, especially in S. Africa; (*h*) the isolated New Zealand flax, *Phormium tenax*, a fibre plant, Fig. 125C; and two subfamilies characteristically with bulbs: (*i*) Scilloideae, with racemose inflorescences, e.g. star of Bethlehem, *Ornithogalum*, Fig. 733B–G; the Mediterranean squill, *Urginea maritima*, with cardiac glycosides; the wild squills, *Scilla*; the grape hyacinth, *Muscari*; and the hyacinth, *Hyacinthus*; and (*k*) Allioideae, with falsely umbellate inflorescences, with the genus *Allium*; onion, *A. cepa*; garlic, *A. sativum*; chives, *A. schoenoprasum*, etc.

Fig. 733. Liliales, Liliaceae (**A.** to **H.**) and Amaryllidaceae (**I.**). **A.** Flower of *Tulipa sylvestris*. **B.** to **G.** *Ornithogalum umbellatum*. **B.** Whole plant, reduced. **C.** and **D.** Flower ×1. **E.** Floral diagram. **F.** and **G.** Capsule. **H.** *Colchicum autumnale*, flowering and fruiting. ×$\frac{2}{5}$. **I.** Flower of *Leucojum vernum* with inferior ovary (g). ×1. **A.** After Baillon. **B.** to **D.**, **F.** and **G.** After Schimper. **E.** After Eichler, somewhat modified. **H.** and **I.** After Firbas.

The leaf succulent **Agavaceae** of the arid parts of the New World are closely related to the Liliaceae and have superior to inferior ovaries. The chief genera are *Yucca* and *Agave* (cf. pp. 318, 790, 792). The plant of *Agave* dies after producing an inflorescence; *A. americana* is naturalized in the Mediterranean, *A. sisalana* yields the important fibre sisal. The **Amaryllidaceae** always have inferior ovaries but are related to the Liliaceae–Allioideae and have the floral formula * P3 + 3 A3 + 3 $G(\overline{3})$. These include the familiar European bulbous snowflake (*Leucojum*, Fig. 733I), snowdrop (*Galanthus*) and daffodil (*Narcissus*).

The **Iridaceae** also have inferior ovaries, but the inner whorl of stamens is absent and there is a progressive tendency towards zygomorphy, thus: * or ↓ P3 + 3 A3 + O $G(\overline{3})$ (Fig. 734B). *Crocus* (Fig. 734A) has perennating corms, actinomorphic flowers and a

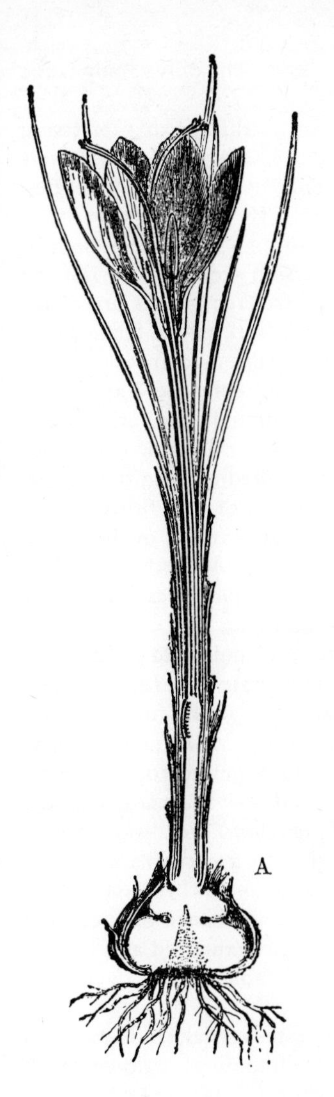

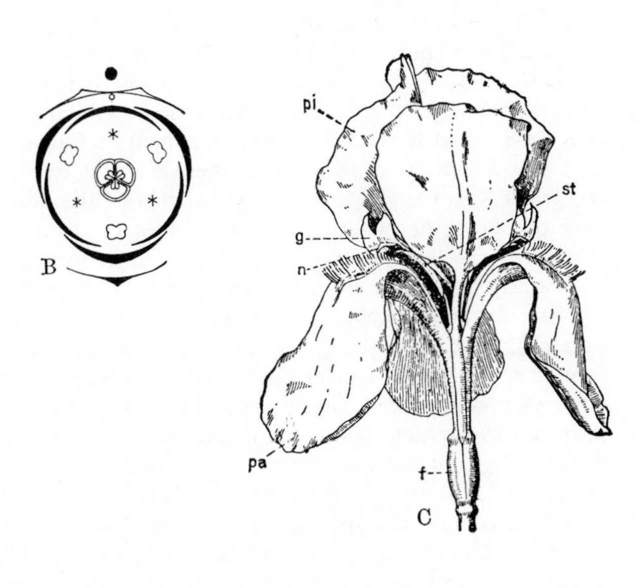

Fig. 734. Iridaceae. **A.** *Crocus sativus*, half of a flowering plant cut lengthwise. **B.** and **C.** *Iris*, floral diagram and flower with outer (pa) and inner (pi) perigone segments, petaloid styles (g) with stigma (n), and ovary (f). ×½. **A.** After Baillon. **B.** After Eichler, modified. **C.** After Firbas.

homoiochlamydeous perianth. The irises or flags (*Iris*, Fig. 734B, C) often have creeping rhizomes and unifacial, sword-like, 'equitant' leaves (Fig. 175E). The inflorescence is a monochasium (rhipidium); the flowers are still actinomorphic but the outer and inner perianth segments are dissimilar. Each flower consists of three partial flowers with a lip, consisting of one outer perianth segment, one petaloid branch of the style (with a triangular stigmatic lobe) and one stamen (meranthium, cf. pp. 606 and 662). Finally, *Gladiolus* has zygomorphic flowers.

The **Dioscoreaceae** are mostly (sub)tropical twining plants which show many dicotyledonous features in their vegetative organs: the cotyledon is not quite terminal, the vascular bundles are arranged in a ring and the leaves are petiolate with reticulate venation. Advanced characters include the mostly unisexual flowers and the inferior ovary. The yam from E. Asia (*Dioscorea batatas*) has edible root tubers; the black bryony (*Tamus communis*), a slender Mediterranean–Atlantic twining plant, is occasional in C. Europe.

The aquatic and semi-aquatic **Pontederiaceae** differ markedly from the Liliales in having a starchy endosperm; the flowers tend towards zygomorphy and syntepaly. This family includes the floating rosette plant *Eichhornia crassipes* (the water hyacinth, a weed of tropical waters).

Order 2: **Orchidales** (Gynandrae, Microspermae) is characterized by zygomorphic flowers, reduction of the androecium to only two stamens or even to one and its fusion with the style or stigma, by the pollen occurring in tetrads in pollinia, by an inferior ovary and by the lack of endosperm in the numerous minute seeds. The only family is the **Orchidaceae** (Fig. 735).

These are humus-loving perennial herbs with endotrophic mycorrhiza, either terrestrial or (especially in the Tropics and Subtropics) epiphytic (Figs. 198, 214, 268). The flowers are usually borne in racemose inflorescences and are generally twisted as they develop through 180° by the torsion of the ovary or pedicel (Fig. 735B). The perigone consists of two whorls and the median (morphologically upper) segment forms a lip or labellum; this is generally highly differentiated and as a result of the inversion (resupination) of the flower forms an underlip which serves as a landing platform; frequently it is prolonged backwards as a spur. Only the two lateral stamens of the inner whorl remain fertile (as in the lady's slipper, *Cypripedium*), or only one persists, namely the median one of the outer whorl (as in *Orchis*, Fig. 735A); the other stamens may be represented as staminodes. The pollen is rarely liberated as separate tetrads; more commonly the entire pollen mass of a pollen sac or of a half anther (theca) is transported as a pollinium. The one or two fertile stamens and the style and stigma of the ovary are united to form the characteristic column or gynostegium which projects from the centre of the flowers. The inferior tricarpellary ovary generally has a single loculus and bears parietal ovules which are scarcely developed at anthesis. Only after pollination do these grow further, developing after fertilization into thousands of minute seeds (with the embryo scarcely differentiated); these are dispersed by the wind almost as easily as spores.

With some 20,000 species the orchids form one of the largest families of plants; they occur especially in the Tropics as epiphytes but also in the temperate zones as terrestrial orchids. Development from seed is in general only possible when a specific fungus infects the seedling and therein forms an endotrophic mycorrhiza (Fig. 268). Subsequently, species with chlorophyll grow autotrophically, but various wholly heterotrophic species with little or no chlorophyll (e.g. the native bird's nest orchid, *Neottia nidus-avis*, the coral-root, *Corallorrhiza*, Fig. 219, and the ghost orchid, *Epipogium*) are always dependent on the mycorrhizal fungus on which they are parasitic to a certain extent (cf. pp. 198 et seq., 294). This mycotrophy is linked with some of the main characteristics of the family, such as the very numerous seeds (because of the small chance of successful germination); the seeds correspondingly minute, and hence undifferentiated and lacking endosperm; and the pollinia which facilitate the fertilization of thousands of ovules as a result of a single act of pollination.

The most primitive subfamily of the Orchidaceae is the terrestrial, Indo-Malayan Apostasioideae with only a few species; *Neuwiedia* has almost actinomorphic flowers with three still fertile stamens, one in the outer, two in the inner whorl (cf. Fig. 735A). Similar precursors have no doubt given rise to the Cypripedioideae (Diandrae) with two inner stamens and the great majority of the orchids in the Orchidoideae (Monandrae) with one outer stamen. The great variety of forms has arisen apparently because of the great diversity of floral specializations. A few examples follow.

In *Orchis militaris* (Fig. 735B–E) and many other native species the lip has a spur (with or without nectar) which opens immediately in front of the gynostegium. The stigmatic surface is a pit filled with tenacious mucilage; the rostellum is an organ which bears the two pollinia. Each pollinium is connected to an adhesive body (viscidium) by a stalk (caudicula). The whole pollination unit is exposed when the anthers open and is called a pollinarium. If an insect which has alighted on the lip attempts to penetrate the flower its head or proboscis comes into contact with the viscidia of the pollinarium which is then withdrawn from the anthers and carried away. Before the insect visits the next flower the caudicles have wilted so that the pollinia move forward and downward so that they come into contact with the stigmatic surface. (The whole process can be readily imitated with the point of a pencil.) In *Ophrys* this type of pollination is linked with the attraction of male bees (Apidae) to flowers which form dummy females (cf. p. 659). In the tropical species of *Catasetum* the pollinaria are ejected onto the pollinating bees as soon as these touch a special tactile process thus releasing certain tissue tensions. In *Cypripedium* and *Stanhopea* legitimate pollination is effected by slippery trap mechanisms. Amongst the many other examples see p. 663 for the 32 cm long spur of *Anagraecum*.

The native terrestrial orchids *Epipactis* and *Cephalanthera* have branching rhizomes. *Orchis, Ophrys, Gymnadenia* and others perennate by means of ovate (Fig. 210C, D) or palmately lobed tubers which arise yearly in the axil of a prophyll through which they subsequently burst (cf. pp. 189–90, Fig. 210C, D); these tubers give rise to the shoots of the following year. The epiphytic orchids show numerous adaptations: the rainwater received intermittently is often stored in thick leathery leaves and in stem tubers (Fig. 214); many have aerial roots (Fig. 111) which may even be photosynthetic (Fig. 198).

The unripe capsules of the Neotropical root climber *Vanilla planifolia* (Fig. 735F) yield vanilla. Many tropical orchids are grown in glasshouses for their unique, colourful and strongly scented flowers (species of *Cattleya, Laelia, Vanda, Dendrobium, Stanhopea,* etc.). The culture of their mycorrhizal fungi has become very important in rearing them.

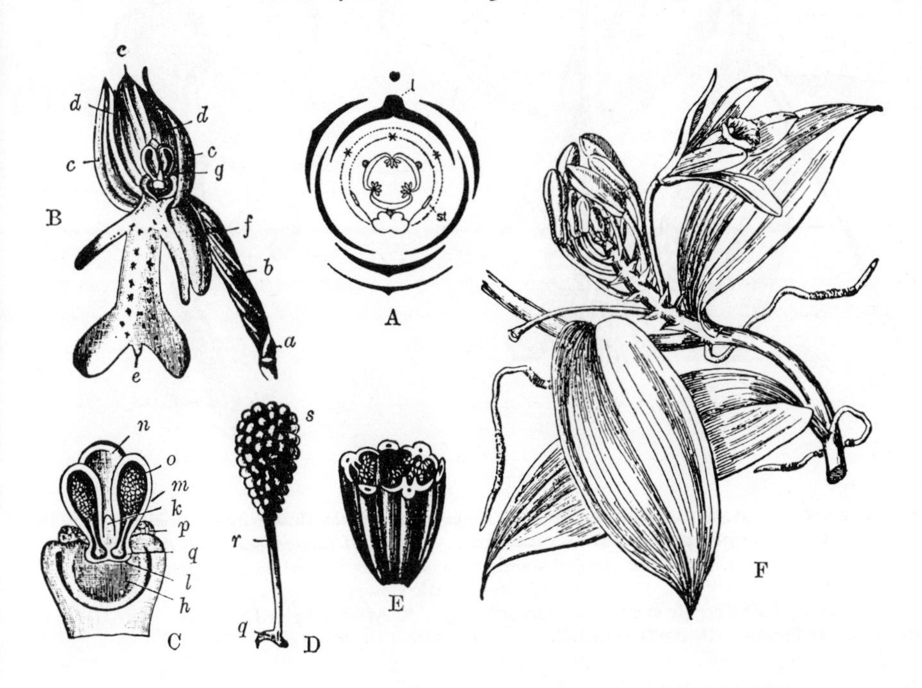

Fig. 735. Orchidales, Orchidaceae. **A.** Floral diagram of the Orchidoideae (of *Orchis*, for example; before resupination); labellum (l); only one of the outer stamens present; in the inner whorl two staminodes (st). **B.** to **E.** *Orchis militaris.* **B.** Flower, resupinate as a result of twisting of the ovary (b); bract (a); outer (c) and inner (d) perigone segments; labellum (e) with spur (f); gynostegium (g). ×2.5 approx. **C.** Gynostegium with stigmatic surface (h), rostellum (l) with process (k), fertile stamen with connective (n), two thecae (m) containing pollinia (o) with caudiculae and viscidia (q); staminodes (p). ×10 approx. **D.** Pollinarium with pollinium (s), caudicula (r) and viscidium (q). ×15 approx. **E.** Capsule cut across. ×8 approx. **F.** *Vanilla planifolia*, flowering shoot with root tendrils. Reduced. **A.** After Eichler, somewhat modified. **D.** to **F.** After Berg and Schmidt.

The next two orders (Bromeliales and Zingiberales) are also zoophilous and may be placed between the Lilianae and Commelinanae in the superorder **Bromelianae**; they have septal nectaries, a perianth differentiated into calyx and corolla, pollen grains with two cells, a generally inferior tricarpellary ovary and a capsule or berry containing numerous seeds with a starchy endosperm.

Order 3: **Bromeliales** comprises the single family **Bromeliaceae**. These are xerophytic or epiphytic rosette plants (rarely 'cabbage trees') from tropical America (Fig. 758). They have narrow leaves and actinomorphic, often bird pollinated flowers with six stamens and an embryo which lies to one side of the endosperm. They take up water by means of peltate hairs (Fig. 112, p. 106). *Tillandsia usneoides* is a reduced epiphyte resembling a lichen. The pineapple (*Ananas sativus*) is a member grown on a large scale for its much esteemed succulent infructescence (p. 673).

Order 4: **Zingiberales** (Scitamineae) on the other hand has zygomorphic or asymmetric

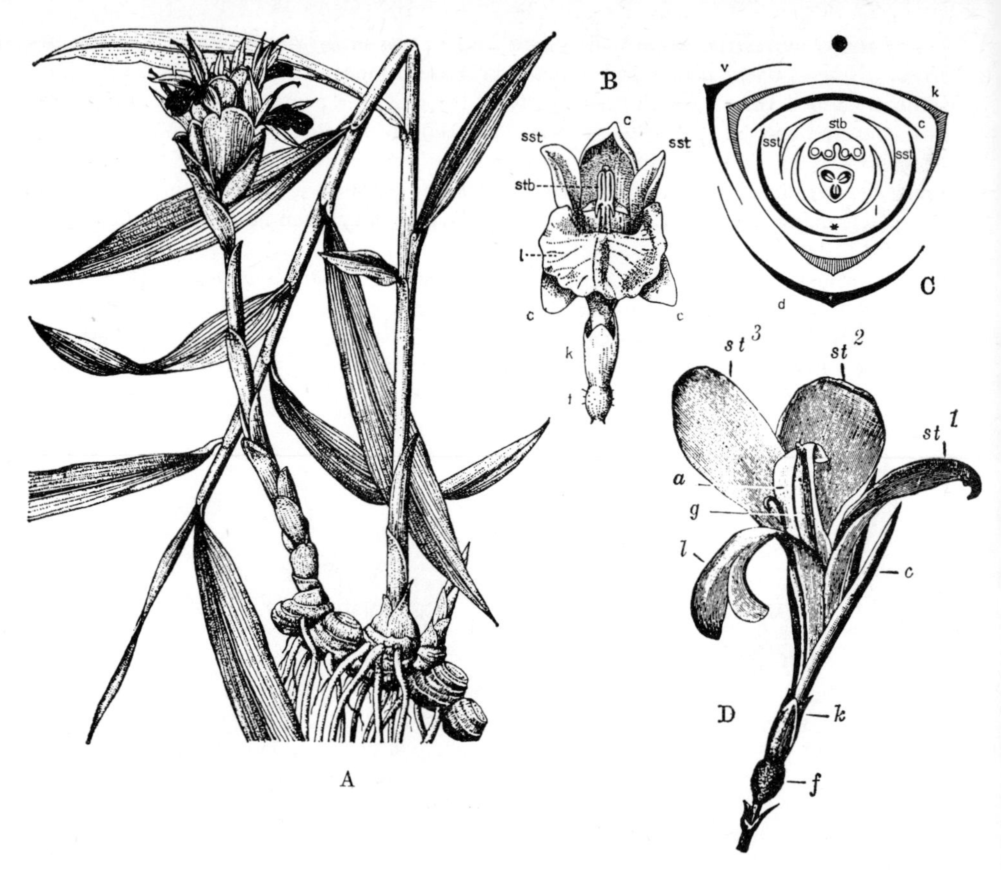

Fig. 736. Zingiberales. **A.** to **C.** Zingiberaceae. **A.** *Zingiber officinale*, flowering plant with rhizomes. $\times\frac{2}{5}$. **B.** Flower of *Curcuma australasica*, and **C.** diagram of *Kaempfera ovalifolia* with bract (d), bracteole (v), calyx (k), corolla (c), lateral staminodes (sst), staminodial labellum (l), the single fertile stamen (stb) and ovary (f). **D.** Cannaceae, asymmetric flower of *Canna iridiflora* with three staminodes (st 1–3), half a fertile stamen (a), and style (g). **A.** After Berg and Schmidt. **B.** After Hooker. **C.** After Eichler. **D.** After Schenck.

flowers with progressive reduction of the stamens or their modification as petaloid staminodes, whilst the seeds contain both endosperm and perisperm. They are tropical, mesophytic, rhizomatous herbs which sometimes have very massive false stems and large entire leaves. The **Musaceae** have a more or less heterochlamydeous perianth and six or, more commonly, only five fertile stamens. They are pollinated by birds or bats. The bananas or plantains are important tropical dessert fruits or sources of flour; they originated in S.E. Asia and are widely grown as diploid and triploid races which are often hybrid in origin and parthenocarpic (seedless; *Musa paradisiaca, M.* x *sapientum* and others). The false stems are formed by the closely overlapping sheaths of the large leaves. The pendulous terminal inflorescence develops the numerous berry-like fruits from double transverse rows of collateral accessory buds. Other members are the fibre plant *Musa textilis* (Manila hemp), the brilliantly coloured, bird pollinated, S. African species of *Strelitzia* and the traveller's palm (*Ravenala*) from Madagascar, with its distichous leaves arranged like a fan. The **Zingiberaceae** also have zygomorphic flowers, but only one fertile stamen (Fig. 736A–C). The two other inner stamens are petaloid and united to form a labellum (contrast the quite different nature of the labellum in the Orchidaceae). This large family occurs principally in tropical S.E. Asia (Fig. 758); it is rich in ethereal oils so that the rhizomes are often used medicinally (e.g. species of *Curcuma, Alpinia* and *Elatteria*) or as spices (especially ginger, *Zingiber officinale*, Fig. 736A). The following two families have asymmetric flowers and only half a fertile stamen (Fig. 736D); its other half, together with the other stamens and the style are petaloid. The **Cannaceae** have capsules with numerous seeds (and include the Neotropical genus *Canna*, a decorative garden plant) whilst the **Marantaceae** have only one to three seeds in the fruit (and leaves with a pulvinus at the base of the lamina; they include various decorative plants and the Neotropical arrowroot, *Maranta arundinaceae*, the rhizomes of which yield starch).

The superorder **Juncanae** (Junciflorae, Cyperales in the broad sense) also comes between the Lilianae and Commelinanae but is anemophilous and correspondingly lacks nectaries. The flowers are more or less actinomorphic, the perianth is basically of two trimerous homoiochlamydeous whorls, inconspicuous and generally scarious (or reduced to hairs or bristles), the pollen is in tetrads or is in 'false' single grains (monads formed directly from the pollen mother cell by the degeneration of three of the meiotic nuclei), the ovary is tri- (or bi-) carpellary, the fruit is a capsule or nutlet, and the embryo is embedded in a starchy (floury) endosperm; the chromosomes are noteworthy in having numerous points of attachment to the spindle (polycentric).

Despite their similarity in habit to the grasses (Poaceae) the Juncanae are readily distinguished vegetatively by their generally solid stems, the absence of swollen nodes and the arrangement of the leaves in three rows (tristichous).

Order 5: **Juncales** with the **Juncaceae** has a floral formula generally like that of the Lilianae: P3 + 3 A3 + 3 G($\underline{3}$) (Fig. 737B); the syncarpous ovaries have numerous (down to three) ovules and develop as capsules. In Europe they are represented by the grass-like woodrushes (*Luzula*) and the rushes (*Juncus*) of which the stems and leaves are often cylindrical (Fig. 175D) and which (corresponding to their wet habitats) contain chambered aerenchyma (p. 95). The flowers are borne in compound inflorescences which sometimes appear lateral since the subtending bract frequently forms a continuation of the main axis.

Order 6: **Cyperales**, comprising only the **Cyperaceae** (sedges), has greater reduction of the flowers, monad pollen, a pseudomonomerous ovary formed from three or two carpels with a single ovule; the fruit is a nutlet. The various genera show stages in progressive reduction. Thus we find for example in *Schoenoplectus, Scirpus, Eleocharis,* etc., hermaphrodite flowers which frequently have six barbed bristles which persist in fruit and aid its dispersal (Fig. 737D, E); in number and position (3 + 3) they correspond to a perigone. The cotton-grasses of fens and raised bogs (*Eriophorum*) also have hermaphrodite flowers but their perigonial bristles are increased in number and form a tassel of white hairs which also function in dispersal of the fruit (Fig. 737F–H). The unisexual flowers of the sedges in the large genus *Carex* (Fig. 737I–P) are even more reduced. Flowers of both sexes almost always occur on the same plant in the axils of bracts in the same or different spikes. The ♂ flowers have three stamens only, the ♀ consist of a two- or three-angled ovary (with a corresponding number of stigmas on a long style) enclosed in a peculiar sac or 'utricle'. Comparison with the Arctic–Alpine genus *Elyna* (Fig. 737Q) shows that the utricle is the modified bracteole or prophyll of the ♀ flower; the bract has a reduced inflorescence in its axil with one ♀ and one ♂ flower, each with its own bracteole.

This large cosmopolitan family occurs mainly in wet places. *Eriophorum vaginatum* and *Trichophorum cespitosum* (Fig. 126F) are important peat forming plants. The papyrus of antiquity consists of compressed longitudinal strips of the pith of the culms of *Cyperus papyrus* from tropical Africa; these culms may exceed a decimetre in diameter.

The final four orders of the Liliidae (7–10) may be included in the superorder **Commelinanae** (Farinosae in the broad sense). Primitive members are still zoophilous but there is a strong tendency towards anemophily. The flowers are actinomorphic (or slightly zygomorphic) and trimerous, nectaries are absent, the perianth is occasionally heterochlamydeous but more generally reduced and inconspicuous, the two-whorled androecium also shows suppression, the pollen has two or three cells; the tri- or bicarpellary ovary is superior and ranges from syncarpous, producing a many-seeded capsule, to pseudomonomerous, producing a one-seeded nutlet; the embryo lies to one side of the starchy endosperm (floury, hence the name Farinosae); also characteristic are the parallel-veined leaves divided into sheath and lamina and the stems with often swollen nodes.

Order 7: **Commelinales** has the most primitive characters: insect pollinated, hermaphrodite flowers producing capsules. This (sub)tropical order includes the

Fig. 737. Juncanae, Juncales (**A.** to **C.**) and Cyperales (**D.** to **Q.**). **A.** to **C.** *Juncus articulatus*, inflorescence, flower and gynaecium. Partly enlarged. **D.** and **E.** Flower of *Schoenoplectus lacustris* (×4) and floral diagram of *Scirpus sylvaticus*. **F.** to **H.** *Eriophorum angustifolium*, infructescence (×1), flower (enlarged), and floral diagram. **I.** to **O.** *Carex*. **I.** Flowering shoot. **K.** ♂, **M.** and **N.** ♀ flowers of *C. riparia*. **L.** and **O.** Floral diagrams. **P.** ♀ flower of *Carex* and **Q.** partial inflorescence with ♂ and ♀ flowers of *Elyna* (diagrammatic) with lateral branch (a) in the axil of a bract and the prophyll (bracteole) modified as a utricle (utr). **A.** to **C.** After Schimper. **D.** After Firbas. **E.**, **H.**, **L.**, **O.**, **P.** and **Q.** After Eichler. **F.** to **G.** After Hoffmann. **I.**, **K.**, **M.** and **N.** After Baillon.

Commelinaceae to which belong the species of *Tradescantia, Zebrina* and *Rhoeo*, frequently grown as house plants, their flowers have a distinct calyx and corolla and are borne in cincinni.

Order 8: **Eriocaulales,** comprising only the (sub)tropical **Eriocaulaceae** has inconspicuous flowers borne in capitate pseudanthia surrounded by an involucre (similar to the composites!).

Order 9: **Restionales** is wind pollinated with generally unisexual, reduced flowers. The **Restionaceae** in S. Africa and Australia occupy to some extent the ecological position of grasses, rushes and sedges.

The end of the anemophilous series within the Commelinanae is formed by the extremely important:

Order 10: **Poales** (Glumiflorae) which includes only the large (some 8000 species) and economically very significant family **Poaceae** (Gramineae), the grasses. The generally hermaphrodite flowers are borne in spikelets and are enclosed in membranous glumes, lemmas and pales (modified bracts, bracteoles and perianth segments); three stamens are usually present and the pollen has three cells; the ovary is pseudomonomerous and contains only a single ovule; at maturity the seed is fused to the carpel wall to form a nutlet or caryopsis.

The grasses are mainly herbaceous plants with cylindrical hollow (Fig. 126G) stems (culms), transversed by diaphragms at the swollen nodes (Fig. 305), and bearing distichous leaves. Each leaf consists of a sheath which surrounds the stem but is usually split on the side opposite its lamina; a long, narrow lamina; and at the junction of sheath and lamina, very often an erect, colourless membrane, the ligule. The form of the stomata should be noted (Fig. 103, p. 99). The walls of the epidermal cells are heavily silicified (as also in the Cyperaceae).

The spikelets are aggregated in compound spikes or panicles (Fig. 739I, A, C, D, F). Each spikelet (Figs. 738A, C; 739B, E, G, H) is surrounded at the base by usually two glumes; above these the lemmas are borne in two rows as the bracts of the individual flowers (Fig. 738A, B). The lemmas are often awned, bearing on the back or at the apex a stiff bristle homologous with the leaf lamina (the lemma itself being homologous with a leaf base). The perianth usually consists of a palea (generally with two keels) and two (rarely three) small scales or lodicules which swell up and force the flower open. The palea and lodicules are apparently homologous with the (partly united) members of the outer and inner perianth whorls (Fig. 738B). Occasionally there are two whorls of stamens (A3 + 3, e.g. in *Oryza*) but generally only the outer one remains. The unilocular ovary apparently consists of three carpels, with three stigmas, but more commonly only two carpels and two stigmas. Each spikelet contains a number of flowers, more rarely (as a result of reduction) only one. The fruit or caryopsis generally develops not from the ovary alone, but with the participation also of the glumes, lemmas or paleae (aids to dispersal and anchoring in the soil). The embryo has one cotyledon modified as the scutellum, a peltate organ of absorption; embryo and scutellum lie to one side of the well developed starchy endosperm (Fig. 739L). The shoot and root apices are both enclosed in sheaths, the coleoptile (? sheath of the cotyledon) and coleorrhiza respectively, which are ruptured on germination (Fig. 731D).

The Poaceae are arranged taxonomically by differences in the structure of the spikelets, the lodicules, the caryopses, the embryos and the seedlings, and in the anatomy of the leaves. In most grasses the mature spikelets disintegrate. The Bambusoideae have primitive flowers (three lodicules and generally 3 + 3 stamens) and the culms are tall and woody; the bamboos occur mainly in the moist (sub)Tropics where they are used both as building materials and as vegetables. The Pooideae (Festucoideae) have several veins in the lemmas and several-flowered spikelets usually in panicles; they include important native genera (used as fodder) such as *Poa* (meadow grass, Figs. 223C, 349), *Festuca* (fescue), *Lolium* (rye-grass) and *Bromus* (brome-grass) with short glumes (Fig. 738C), and *Arrhenatherum* (false oat-grass), *Trisetum* (yellow oat) and *Avena* (oat) with long glumes (Fig. 739H); cereals with spikelets in spikes include *Triticum* (wheat), *Secale* (rye) and *Hordeum* (barley) (Fig. 739A–G). Amongst the native Pooideae with single-flowered spikelets are *Agrostis* (bent) with paniculate inflorescences and *Anthoxanthum* (sweet vernal grass), *Alopecurus* (foxtail) and *Phleum* (timothy) with spike-like panicles. More isolated are the Arundineae with several-flowered spikelets and a hairy spikelet axis: *Phragmites* (reed) dominates in reedswamps (Fig. 747) and *Molinia* (moor-grass) in fen-meadows; and the Stipeae with one-flowered spikelets: the feather and hair grasses (*Stipa*

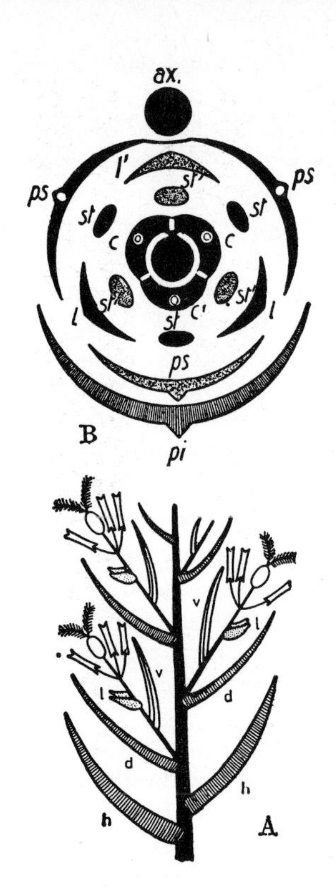

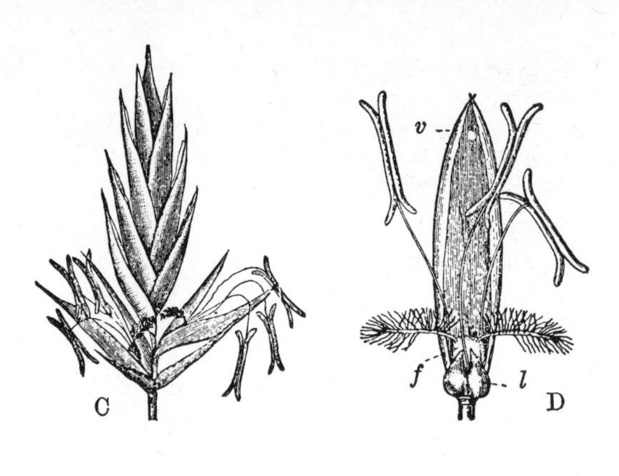

Fig. 738. Poales. **A.** Diagram of a grass spikelet with three flowers, glumes (h), lemmas (d), paleae (v) and lodicules (i). **B.** Theoretical floral diagram of a grass flower (absent floral members dotted grey); axis of spikelet (ax), lemma (pi = bract), palea (ps = outer whorl of perigone), lodicules (l, l′ = inner whorl of perigone), outer (st) and inner (st′) whorls of stamens, carpels (c, c′). **C.** and **D.** *Festuca pratensis.* **C.** Spikelet with two glumes, two open and several unopened flowers. ×3. **D.** A single flower (after removing the lemma) with palea (v), lodicules (l), three stamens, and ovary (f). ×6. **A.** After Firbas. **B.** After Schuster. **C.** and **D.** After Schenck.

pennata and *S. capillata*) with xeromorphic inrolled leaves (Fig. 196B, C) and very long awns on the lemmas; these are characteristic species of the Pontic steppes (p. 000). The tropical Oryzoideae have one-flowered spikelets, but several lemmas; here belongs the rice (*Oryza*, Fig. 739I, K). The lemmas have only one to three veins in the Eragrostoideae, particularly important in the Tropics (including amongst others the Chlorideae); a native representative is *Cynodon* (dog's-tooth or Bermuda grass) with a number of digitate spikes. The usually two-flowered spikelets fall off as a whole in the predominantly tropical Panicoideae in which the spikelets are solitary, as in the various millets (*Panicum, Pennisetum* and *Setaria*), and in the Andropogonoideae in which the spikelets are paired: these include the sugar cane (*Saccharum officinarum*), native in Indo-Malaya, from the pith of which cane sugar is prepared by pressing and concentration; the cereals *Sorghum* (guinea corn) and *Zea* (maize) and the European indicator of dry conditions, *Bothriochloa*.

The grasses are of far-reaching significance in our biosphere. They determine the entire biome in savannas, steppes, meadows and pastures, and have extended their distribution along with these formations since the Upper Cretaceous and especially during the Tertiary. The evolution of major groups of animals, for example the ungulates, is unthinkable without them. For man, the development of civilized societies in the Old and New Worlds has only been possible over the last 7000–9000 years with the inception of cereal culture which allows the rational production of food which can be stored. The cultivated cereals differ from their wild progenitors especially in the increase in number and size of the caryopses (yield!), the suppression of the brittleness of the spikelets and spikelet axes (harvesting!) and to some extent by the increased ease of separation of caryopses (grain) from the chaff.

Important cerals have originated as follows: (A) In W. Asia and the Mediterranean region: (1) wheat (*Triticum*, Figs. 191B, 739C–E; separate mostly three to five flowered spikelets with broad glumes and awned or unawned lemmas). The most important is the hexaploid bread wheat (*T. aestivum* = *T. vulgare*); in warmer countries the tetraploid hard wheat (*T. turgidum* '*durum*') is grown, and occasionally spelt (*T. aestivum* '*spelta*') in S. Germany, whilst the diploid small spelt (*T. monococcum* '*monococcum*') and the tetraploid emmer (*T. turgidum* '*dicoccon*') have practically disappeared from cultivation. (For

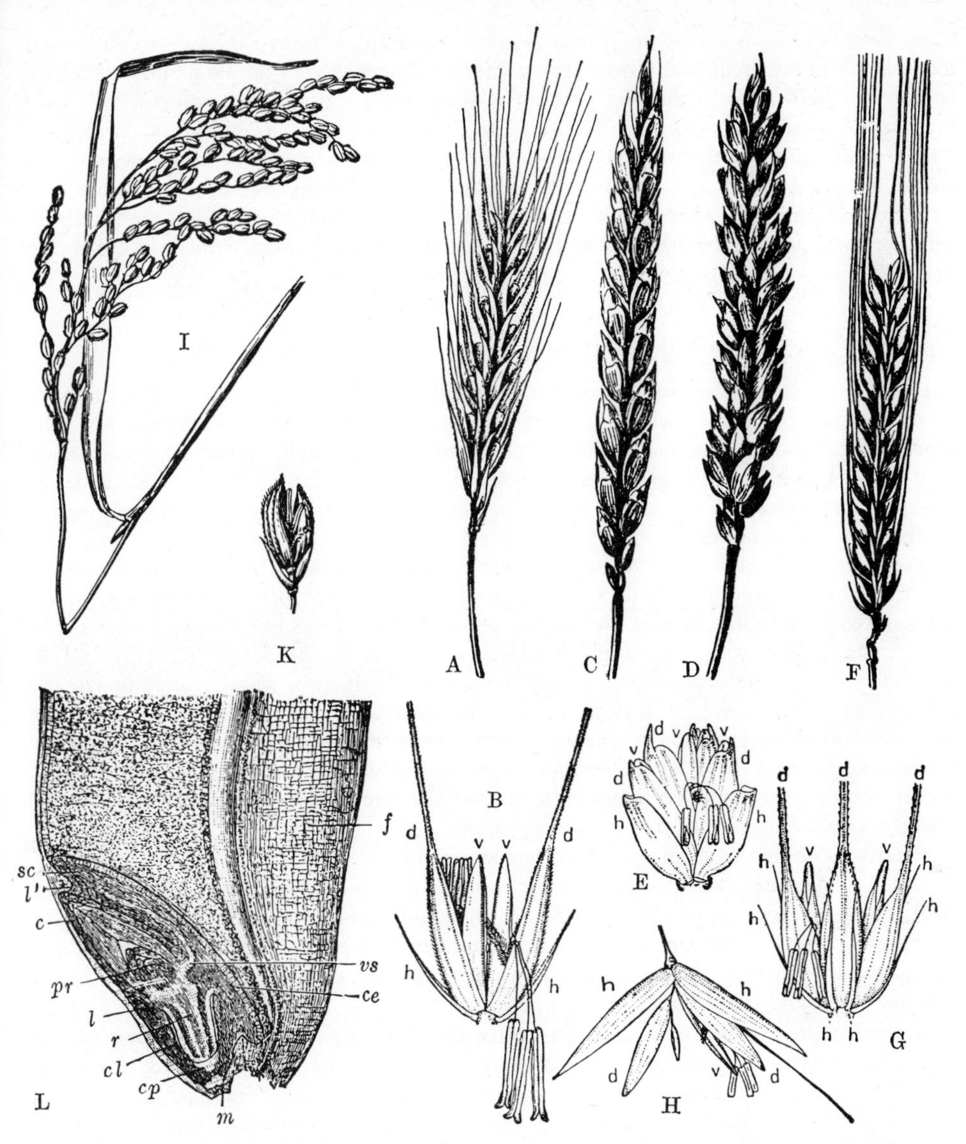

Fig. 739. Poaceae. Cereals. Spikes and spikelets of: **A.** and **B.** Rye, *Secale cereale.* **C.** to **E.** Wheat, *Triticum aestivum*, with **C.** spelt and **D.** and **E.** bread wheat. **F.** and **G.** Barley, *Hordeum vulgare*, with **F.** a two-rowed and **G.** a six-rowed form. **H.** Oat, *Avena sativa.* **I.** to **K.** Rice, *Oryza sativa*; glumes (h), lemmas (d) with their awns only shown in part in **D.** and **G.** paleae (v). **L.** Caryopsis (grain) of wheat, median L.S. through the lower part; one side of the furrow of the grain (f); below left the embryo with scutellum (sc), vascular bundle (vs) and columnar epithelium (ce), ligule (l′), coleoptile (c), apical meristem of the shoot (pr), epiblast (l), coleorrhiza (cl), radicle (r) with root cap (cp) and point of emergence (m). ×14. **A., C., D., F., I.** and **K.** After Karsten. **B., E., G.** and **H.** After Firbas. **L.** After Strasburger.

the origins of cultivated wheats see p. 394 and Fig. 337); (2) barley (*Hordeum vulgare*, Fig. 739F, G; one-flowered spikelets in threes on the axis of the spike; in the six-rowed barleys all spikelets are fertile and awned, in two-rowed barleys only the middle one of the three; cultivated barleys are diploid); (3) rye (*Secale cereale*, Fig. 739A, B; single two-flowered spikelets with narrow glumes and lemmas with long awns; diploid); (4) oat (*Avena*, Fig. 739H; spikelets in panicles; only the hexaploid *Avena sativa* is important). The cultivation of wheat (emmer) and barley goes back to the ninth millenium B.C. in the Near East; even in C. Europe small spelt, emmer, bread wheat and barley were cultivated in the Neolithic. Rye and oats first appeared in the Bronze and Iron Ages; at first they were probably weeds amongst the other cultivated cereals. (B) In S.E. Asia: rice (*Oryza sativa*, Fig. 739I, K: one-flowered spikelets in panicles; diploid) originated from the wild

O. perennis and is the most important tropical–subtropical cereal; it has been cultivated for millenia in S.E. Asia and is now grown in many places (e.g. in the Po plains) especially on flooded fields. (C) In the arid parts of E. Asia, India, Africa, etc.: millets, especially the common millet (*Panicum miliaceum*, Fig. 295), pearl millet (*Pennisetum spicatum*), Italian millet (*Setaria italica*) and guinea corn (*Sorghum bicolor*); often grown as hoed crops. (D) In Mexico and C. America: maize (*Zea mays*; monoecious, ♂ flowers in a terminal panicle, ♀ flowers in compact lateral 'cobs'; diploid); originated from extinct wild forms and by hybridization with wild grasses of the genus *Tripsacum*; cultivated since the sixth century B.C.

Subclass C: Arecidae (Spadiciflorae)

The numerous inconspicuous flowers are generally borne on a club-shaped inflorescence (spadix, hence the name) which is subtended by a conspicuous bract (the spathe). The superior ovaries are occasionally choricarpous but more generally coenocarpous, contain only a few ovules, or one, and invariably form indehiscent fruits. The general tendency towards anemophily is associated with the reduction of the originally hermaphrodite flowers to unisexual ones.

They are often very large woody or herbaceous plants with broad leaves which are undivided (or become divided at a late stage of development), often without the typical parallel venation of the lamina but with a petiole. The flowers are almost always cyclic and generally trimerous, but are otherwise very variable: the perianth is little differentiated (and in some palms is spirally arranged), with two whorls, or reduced; the stamens are often 3 + 3, but occasionally increased and more commonly reduced in number; septal nectaries occur occasionally (in some palms). The indehiscent fruits are berries, drupes or nuts. Their early appearance (palms as early as the Cretaceous) and the occurrence of primitive characters indicate that the Arecidae have developed in parallel with the Liliidae from primitive monocotyledons.

Relatively numerous primitive characters are seen in:

Order 1: **Arecales** (Principes), with the sole family **Arecaceae** (Palmae; the palms, Fig. 740). These are (sub)tropical (Figs. 758, 759), mainly arborescent plants with a slender, unbranched trunk of almost uniform diameter from base to apex and a terminal rosette of enormous, generally long-stalked, pinnate or flabellate (fan-shaped) compound leaves (feather palms and fan palms).

The stem practically reaches its final diameter before extension growth begins (Fig. 154A, B); it is rarely branched (e.g. dichotomous in the African species of *Hyphaene*; sometimes it remains thin and then creeps or scrambles (e.g. the rotang palms, *Calamus*, from which rattan canes are made). The lamina is always undivided at first, but develops as a series of folds (plicate) which subsequently divide up (p. 164). The inflorescences are generally copiously branched and either lateral or terminal; in the latter case the whole plant dies after fruiting, as in the sago palm, *Metroxylon*. The inflorescence is subtended by a rigid spathe (Fig. 740B) and bears hermaphrodite or (generally) unisexual flowers; the plant may then be monoecious or dioecious. The two-whorled perianth is usually not very conspicuous and is more or less homoiochlamydeous. The number of stamens varies but there are always three carpels, but frequently only one of these develops further. The gynaecium is choricarpous (Fig. 740E), or syncarpous; each carpel contains a single ovule. The seed has a copious endosperm which is more or less free of starch. Pollination is by insects (occasionally there is nectar), or by wind.

The various forms of fruit are important both taxonomically and economically. The dioecious date palm (*Phoenix dactylifera*), grown especially in the Saharan oases and eastwards to India, has three free carpels but forms a unicarpellary berry (Fig. 740E); the sugary fleshy pericarp surrounds a seed indurated with reserve cellulose. The dwarf palm (*Chamaerops humilis*) of the S.W. Mediterranean is related; it is the only native European palm. Many other palms have a horny endosperm which in *Phytelephas macrocarpa*, native in tropical America, serves as 'vegetable ivory' (used in the manufacture of buttons, etc.; Fig. 56B); in the Indo-Malayan betel nut palm (*Areca catechu*) it is brown and ruminate; the seeds are borne singly in an initially trilocular ovary and are chewed as a

Fig. 740. Arecales. **A.** to **D.** *Cocos nucifera.* **A.** Twelve-year-old plantation in Java. **B.** Inflorescence (much reduced). **C.** Drupe in L.S. with exocarp, endocarp and mesocarp (ep, me, end), endosperm (esp) and embryo (e) (reduced). **D.** Stone from below with three germination pores (reduced). **E.** *Phoenix dactylifera*, half female flower with choricarpous gynaecium (enlarged). **A.** After Paravicini. **B.** After Karsten. **C.** and **D.** After Wettstein. **E.** After Baillon.

stimulant. The coconut palm (*Cocos nucifera*, Fig. 740A–D) is an important crop plant in all tropical coastal regions; it probably originated in S.E. Asia. The ovary of three syncarpous carpels gives rise to the very large one-seeded drupe. This has a smooth exocarp, a thick, fibrous mesocarp, and a stony endocarp with three germination pores, one of which is perforated on germination by the embryo which lies inside. The outer endosperm is solid and oily (the 'copra' of commerce), the inner is liquid (coconut 'milk'). The mesocarp contains air so that the fruit can float, and yields the fibre known as coir. Also important are the drupes of the oil palms (*Elaeis guineensis* and others) which yield palm oil and palm kernel oil. Finally, the preparation of sago starch from the stems of various palms, especially the Malayan species of *Metroxylon* should be noted.

Order 2: **Cyclanthales** with the Neotropical **Cyclanthaceae** often has bifid leaves and club-shaped inflorescences; it is apparently intermediate between the palms and:

Order 3: **Arales.** These are mainly herbaceous perennials or climbing plants with flat, undivided leaves. The inflorescence is unbranched, forming a spadix subtended by a spathe. The endosperm has copious starch.

The predominantly tropical **Araceae** (aroids) come first; they either perennate with rhizomes or rhizomatous tubers (e.g. *Arum*, Fig. 188) or play a major role in Tropical Rain Forests as large-leaved rosette plants or as epiphytic root-climbing lianes (Fig. 759). The generally broad, undivided cordate or sagittate leaves are sometimes reticulately veined (e.g. in *Arum*). Lobed or even perforated leaves (as in the favourite house-plant *Monstera* or Swiss cheese plant) arise by the death of definite areas of the lamina (p. 164). The flowers vary in structure but are borne, without subtending bracts, on a generally fleshy, club-shaped spadix; the spathe is often brightly coloured. The fruit is usually a berry.

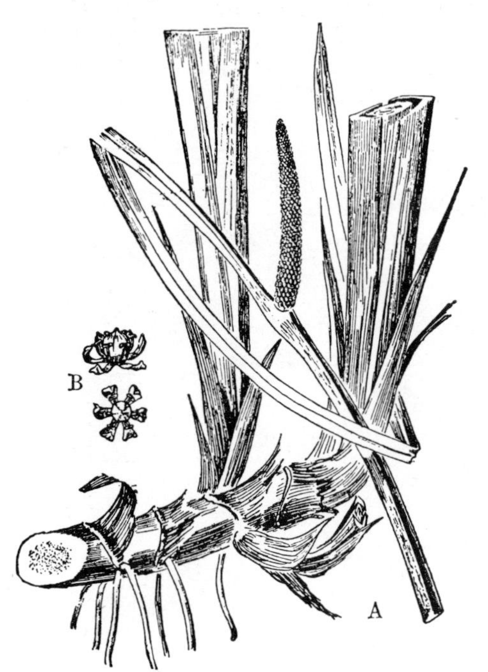

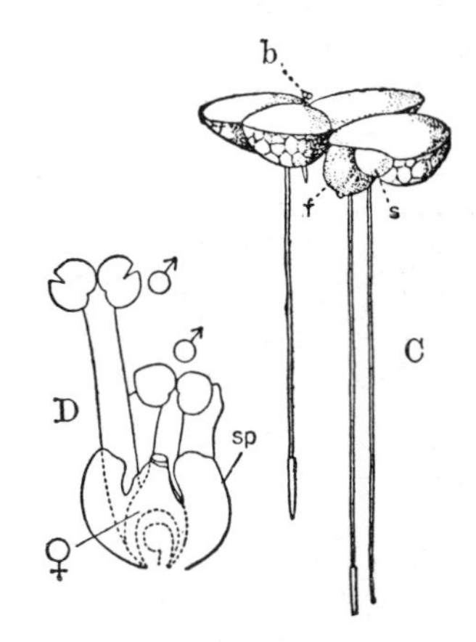

Fig. 741. Arales. **A.** and **B.** *Acorus calamus*, flowering plant (×¼) and single flowers (enlarged). **C.** *Lemna gibba* with young shoot (s), ♂ flower (b) and fruit (f). ×4. **D.** *Lemna trisulca*, inflorescence with spathe (sp), two ♂ and one ♀ flower. Enlarged. **A.** and **B.** After Karsten. **C.** and **D.** After Hegelmaier.

Although few Araceae are native in Europe they demonstrate clearly the progressive reduction of the individual flowers and the increasing specialization of the inflorescences characteristic of the family. Thus the originally E. Asiatic sweet flag (*Acorus calamus*; Figs. 652B, 741A, B; cf. also p. 421) has perfect hermaphrodite flowers with an inconspicuous perigone; thus * P3 + 3 A3 + 3 G($\underline{3}$); the spadix appears to be lateral since the green spathe forms a prolongation of the flattened stem. *Calla palustris*, a plant of swamps has, in contrast, no perigone: * A6–9 G($\underline{3}$), but the spathe is white on the inner side. Finally, the lords-and-ladies, jack-in-the-pulpit or cuckoo pint (*Arum maculatum*), a common plant of moist woodlands, has unisexual flowers consisting either of a few stamens or a single pseudomonomerous ovary: thus, ♂ A3–4, ♀ G1 (Fig. 671G); the inflorescence is concealed within the spathe, the whole forming a slippery trap (p. 660).

Even within the Araceae we find not only a reduction in the individual flowers but also a diminution in their number in the inflorescence, e.g. in the tropical aquatic *Pistia*, with floating leaf rosettes. This development leads ultimately to the **Lemnaceae** (Fig. 741C, D) in which the inflorescence consists of a minute, inconspicuous spathe surrounding a single ♀ flower consisting of a single ovary, and one or two ♂ flowers, each of a single stamen. The duckweeds carpet the surface of still waters and are also very reduced vegetatively (neoteny!), consisting simply of free-floating (in *Lemna trisulca*, submerged) leaf-like segments which rapidly multiply by budding. The larger anterior part of each segment corresponds to a leaf, the posterior part to a reduced stem. Each segment bears several roots in *Spirodela*, only one in *Lemna*, whilst *Wolffia arrhiza*, with a maximum length of 1.5 mm, has no roots and is the smallest of all spermatophytes.

The two final orders have linear leaves; the flowers are always unisexual and usually anemophilous.

Order 4: **Pandanales** comprises Palaeotropical woody plants. The only family **Pandanaceae** includes trees or climbing shrubs which frequently have stilt roots and dichotomous stems. Various species of *Pandanus* are striking tropical shore plants; they are called 'screw-pines' because the sword-shaped spiny leaves are generally borne on the stem in three spiral series. The flowers have no perianth and are borne on a spadix; the ovaries consist of three to many carpels and form dense capitate infructescences.

Order 5: **Typhales** comprises two native families of herbaceous, rhizomatous perennials of reedswamps (Fig. 747), both monoecious. The **Sparganiaceae** (with *Sparganium*, bur-reed) have the ♂ and ♀ flowers in separate globular inflorescences; the membranous perianth has six or three segments. In the closely-related **Typhaceae** (with *Typha*, reedmace) a single continuous axis bears a dense cylinder of ♀ flowers below and ♂ flowers above; a fringe of hairs replaces the perianth. The relationship of this advanced order with the other Arecidae is uncertain; it may belong to the Commelinanae.

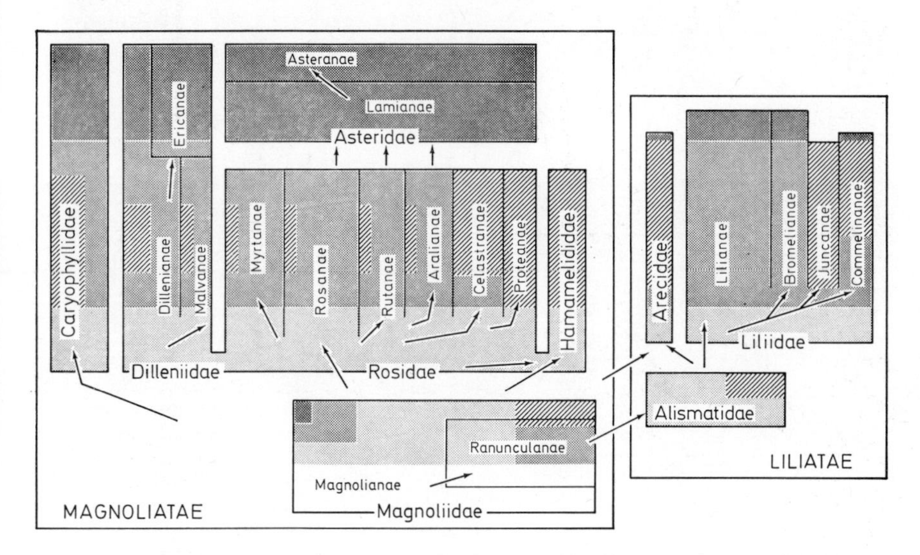

Fig. 742. Scheme of the hypothetical evolutionary relationships between the major taxa (classes, subclasses and superorders) of the Magnoliophytina. The shading indicates some important morphological progressions in the floral region: 1. Perianth spiral (acyclic), free; primary polyandry; choricarpy = white; 2. Perianth cyclic, generally with two whorls, free; polyandry if present, secondary; choricarpy = light grey; 3. Perianth cyclic, reduced to one whorl, or absent; androecium reduced; coenocarpy (generally); often anemophilous = hatched; 4. Perianth cyclic, generally with two whorls, free; androecium secondarily polyandrous, in two whorls or one; coenocarpy = mid-grey; 5. Perianth cyclic, united (sympetalous or syntepalous); androecium of two whorls or one; coenocarpy; often highly specialized for zoophily = dark grey. Further explanation in the text. Original.

The evolution of the Magnoliophytina in retrospect

Figure 628 indicates both the still hypothetical descent of the Magnoliophytina (angiosperms) from the Mesozoic successors of the Lyginopteridatae (seed ferns) and the relatively recent expansion of this subdivision which characterizes the Neophytic era of geological time (Upper Cretaceous to the present). We can best take the primitive dicotyledonous Magnoliidae as model prototypes of the angiosperms (Fig. 742). The monocotyledonous Liliatae have apparently arisen from these by way of precursors resembling Ranunculanae or Nymphaeales. The various groups of dicotyledons or monocotyledons have then undergone further progressive differentiation, to a large extent in parallel (cf. the levels of organization indicated by the shading in the scheme in Fig. 742: hatched = apetalous; white, light grey and mid-grey = dialypetalous; dark grey = sympetalous). Within the Magnoliatae the Magnoliidae are succeeded on the one hand by the Rosidae, Hamamelididae and Dilleniidae, which shows close evolutionary relationships, on the other by the Caryophyllidae. Apart from the Hamamelididae, all these groups have produced sympetalous lines of development; these are particularly far-reaching in the Asteridae which apparently stem from the Rosidae. The diversification of the Liliatae has

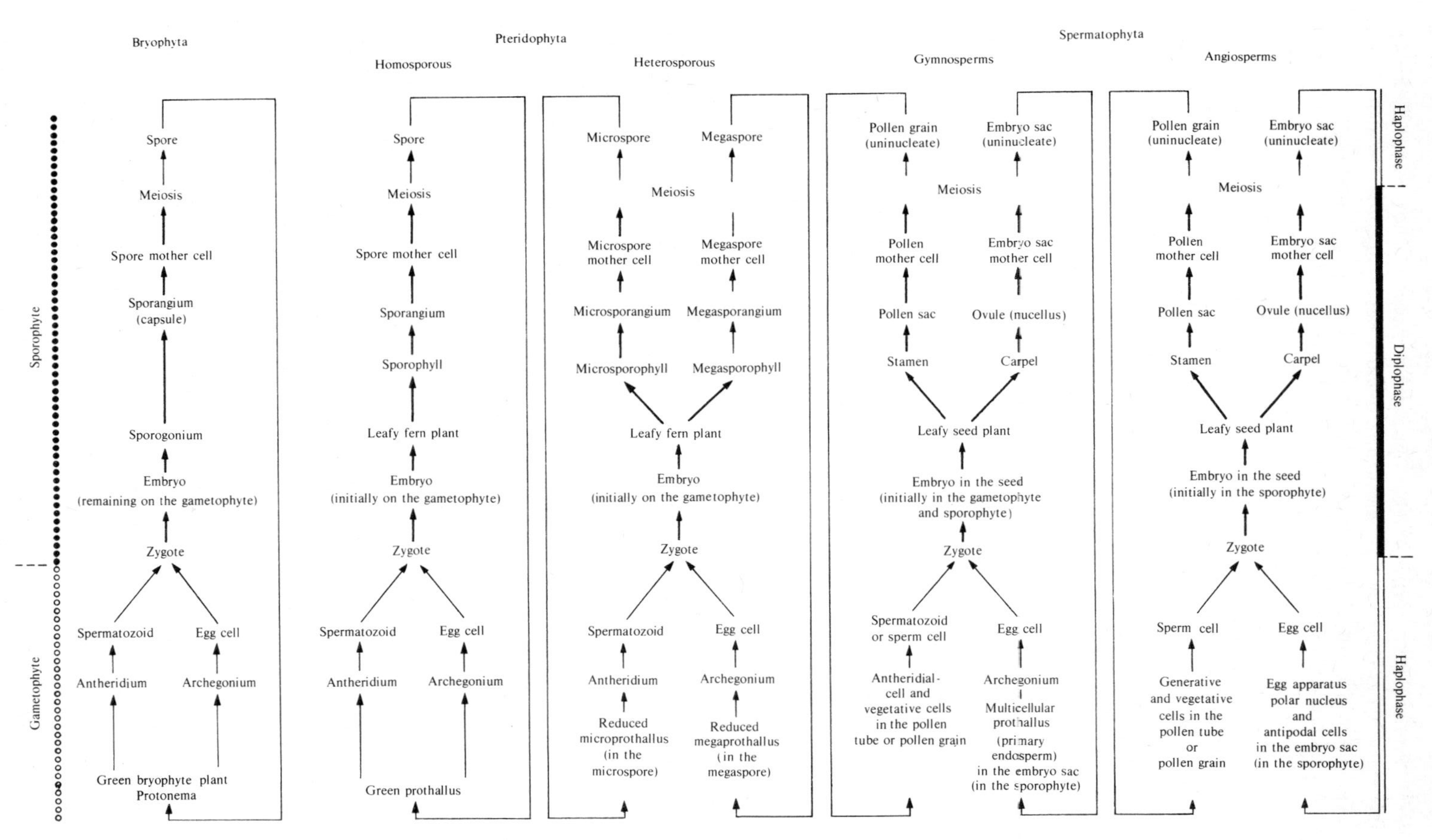
Bryophyta
Pteridophyta
Homosporous
Heterosporous
Spermatophyta
Gymnosperms
Angiosperms
Haplophase
Diplophase
Haplophase
Sporophyte
Gametophyte
Spore
Meiosis
Spore mother cell
Sporangium (capsule)
Sporogonium
Embryo (remaining on the gametophyte)
Zygote
Spermatozoid
Egg cell
Antheridium
Archegonium
Green bryophyte plant
Protonema
Spore
Meiosis
Spore mother cell
Sporangium
Sporophyll
Leafy fern plant
Embryo (initially on the gametophyte)
Zygote
Spermatozoid
Egg cell
Antheridium
Archegonium
Green prothallus
Microspore
Megaspore
Meiosis
Microspore mother cell
Megaspore mother cell
Microsporangium
Megasporangium
Microsporophyll
Megasporophyll
Leafy fern plant
Embryo (initially on the gametophyte)
Zygote
Spermatozoid
Egg cell
Antheridium
Archegonium
Reduced microprothallus (in the microspore)
Reduced megaprothallus (in the megaspore)
Pollen grain (uninucleate)
Embryo sac (uninucleate)
Meiosis
Pollen mother cell
Embryo sac mother cell
Pollen sac
Ovule (nucellus)
Stamen
Carpel
Leafy seed plant
Embryo in the seed (initially in the gametophyte and sporophyte)
Zygote
Spermatozoid or sperm cell
Egg cell
Antheridial-cell and vegetative cells in the pollen tube or pollen grain
Archegonium
Multicellular prothallus (primary endosperm) in the embryo sac (in the sporophyte)
Pollen grain (uninucleate)
Embryo sac (uninucleate)
Meiosis
Pollen mother cell
Embryo sac mother cell
Pollen sac
Ovule (nucellus)
Stamen
Carpel
Leafy seed plant
Embryo in the seed (initially in the sporophyte)
Zygote
Sperm cell
Egg cell
Generative and vegetative cells in the pollen tube or pollen grain
Egg apparatus polar nucleus and antipodal cells in the embryo sac (in the sporophyte)

proceeded similarly: the various groups within the Liliidae and Arecidae must have developed at least in part from precursors resembling the Alismatidae.

The evolution of the Plant Kingdom in retrospect

Finally, some broad features of plant evolution may be summarized with the assistance of Fig. 743. The earliest traces of life on earth are estimated to be 3000 million years old. If the assumption of geophysicists is correct that the primeval atmosphere was devoid of oxygen and reducing (p. 2), these traces must have been of Procaryota similar to bacteria and already organized in cells which obtained the necessary materials and energy from accumulated organic compounds; thus for them we have to postulate primary heterotrophy involving anaerobic fermentation processes. These 'pro-bacteria' must have been preceded in the course of development of living things from a lower molecular level of organization by entirely hypothetical 'Probionta' which probably resembled viruses. The increasing scarcity of readily available organochemical energy sources in the primeval atmosphere necessitated the gradual change towards autotrophy within the early Procaryota: various forms both of anaerobic chemosynthesis and of primitive photosynthesis (e.g. with H_2S as a reactant) probably arose together in this way. The critical step towards typical photosynthesis with the breakdown of water and liberation of oxygen was first taken by 'proto-algae' similar to the Cyanophyceae. As the oxygen content of the primeval atmosphere consequently increased, the conditions were established for the development of the metabolically desirable and more efficient aerobic respiration. The increase in living forms then made possible diverse types of secondary heterotrophy: saprophytism and parasitism. All these fundamental types of differentiation can be postulated by comparing the extant forms of Procaryota.

The next decisive step in evolution was the development of eucaryotic, haplontic organisms closely resembling the flagellates, thus establishing the freely motile unicellular photosynthetic algae as the basic type of the Protobionta. In the Eucaryota, more precise replication by mitosis of the genetic material carried by the chromosomes and the enhanced probability of the recombination of genes by true sexuality and meiosis (with crossing over) facilitated the evolution of organisms with more complex structures and functions. There followed a fanning out of the modes of nutrition of the Protobionta: the algae (Phycophyta) remained autotrophic whilst the fungi (Mycophyta) and animals (Protozoa and subsequently Metazoa = Zoobionta) became secondarily heterotrophic; correspondingly, the algae expanded towards the light whilst the fungi mainly developed within organic substrates.

Whilst algae and fungi have adopted a sedentary mode of existence and grow towards their sources of energy and materials by increasing their surface area ('extroversion'), for animals the determinants are motility, the increase in internal surfaces ('introversion') and the hunt for food. The heterotrophic lines of development often run parallel or even converge: Mycophyta and Protozoa are therefore polyphyletic and consequently entirely conventional taxa. Indeed all the Protobionta exhibit a multidimensional complex of relationships.

The flagellate precursors of the algae already showed a variety of forms of photosynthesis (composition of the pigments, photosynthetic products). The further evolution of the algae proceeded along many parallel lines from naked monads and rhizopodial (amoeboid) types to mainly nonmotile encapsulated forms surrounded by cell walls and mucilage, to unicellular coccoid types, to multicellular filaments (trichoid) and finally to forms dividing in two or three dimensions as complex thallose and plectenchymatous or parenchymatous growth forms. These progressions are related to the change from free-living to sedentary modes of existence in an aqueous medium, and with a progressive increase in size (linked to the struggle for light). At the same time, division of labour between vegetative and sexual reproduction is often associated with the development of alternation of generations, with the transition from isogamy to anisogamy and oogamy (economy of the reproductive processes), and frequently with the promotion of the diplophase at the expense of the haplophase (pp. 42, 402).

The fungi shows lines of development analogous to those in the algae. Both groups only

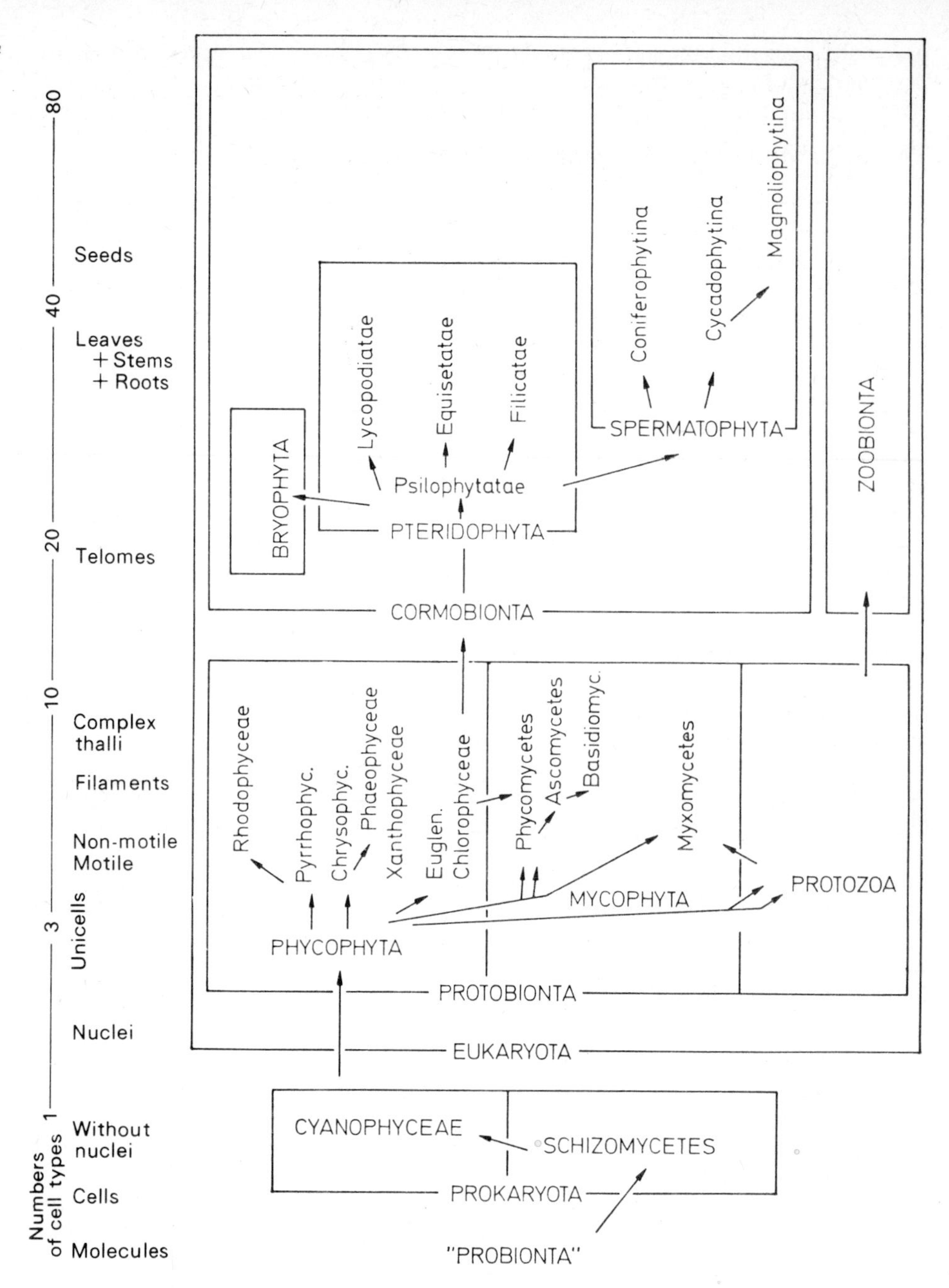

Fig. 743. Scheme of the hypothetical evolutionary relationships between the major taxa of plants (and other organisms). Left: generalizations about the levels of organization reached and the numbers of different cell types present in plants. Further explanation in the text. Original.

advanced to a modest extent onto dry land before the arrival of the true land plants (Cormobionta). In the wake of these the fungi were able to diversify (from the Devonian onwards) as saprophytes and parasites; developments linked with terrestrial existence included gametangiogamy, somatogamy, etc., and the formation and dispersal of drought resistant spores, conidia, etc. In extreme habitats a successful but specialized development was the symbiosis of algae and fungi as lichens.

A further highly significant step in the evolution of the Plant Kingdom (and in the history of the entire biosphere) was the development of the first true land plants. These were the psilophytes (Psilophytatae, cf. Figs. 533, 534) which apparently arose at the junction of water and land in the Upper Silurian (some 400 million years ago) from algae resembling Chlorophyceae which had alternation of generations, complex thalli and apical meristems. They were adapted to terrestrial existence by possessing root hairs, cuticle, epidermis, stomata and vascular bundles, gametangia with a jacket of sterile cells (antheridia and archegonia), and very drought resistant spores. The mosses and liverworts (Bryophyta) developed as an early but not very productive sideline in which the gametophyte in particular was diversified and supported the dependent sporophyte; but in the ferns and related groups (Pteridophyta) the emphasis was placed on the sporophytic diplophase (for this and the following, see the summary on p. 754 and Figs. 614, 628). As size tended to increase yet again in the struggle for light, the telome-like sporophyte generation of the psilophytes underwent division of labour between shoots, with stem and leaves, and roots, whilst the gametophyte remained thallose and became progressively reduced. The various lines of development shown in the lycopods (Lycopodiatae), horsetails (Equisetatae), ferns (Filicatae) and the precursors of the seed plants ('Progymnospermae') differed in the degree of development of their foliar organs, in their spore-bearing structures, and the anatomy of the stems. From initially small herbaceous growth forms all these groups developed, with varying degrees of success, into dominant woody arborescent plants (especially during the Upper Palaeophytic); furthermore, in all of them reduction led to division of labour between ♂ and ♀ gametophytes and (with this division extending back to the sporophyte) to the development of microspory and megaspory (thus: homospory → heterospory). The necessity for the presence of external water for the fertilization of the egg in an archegonium on an independent prothallus was obviated by the development of the seed (retention of the megaprothallus on the mother plant, etc.) at the turn of the Devonian period about 350 million years ago by the Spermatophyta (cf. Fig. 628). After a running-in period in the Carboniferous, these attained a dominant position in the Mesophytic terrestrial vegetation as woody plants (no doubt because of their superior water conducting systems) belonging to the two gymnospermous lines of the Coniferophytina and Cycadophytina. Eventually they were superseded by the angiospermous Magnoliophytina which underwent an unprecedented degree of development and carried the existence of higher plants to the boundaries of the frigid and torrid deserts; these processes started in the Neophytic some 100 million years ago and have continued to the present; they were made possible by the astonishing plasticity and adaptability of angiosperm growth forms, pollination mechanisms, means of seed and fruit dispersal and, in part, by the greatly accelerated development of the flowers, gametophytes, etc.

The evolutionary development of the Plant Kingdom is marked by the advance from lower to higher levels of organization (Fig. 743, left-hand margin): the processes of differentiation were restricted at first to the molecular level and to cell organelles, but subsequently involved cells, tissues, organs and finally complexes of organs. An increase in the number of simple structural units at a lower level of organization frequently gave rise to division of labour, new combinations of units and hence to the development of more complex structural units at the next higher level of organization (cf. p. 424). In general we can trace a general increase in self-regulatory capacity (homeostasis) and hence a greater independence of plants upon the environment. Starting from relatively stable habitats (e.g. the oceans), plants were thus able to invade progressively less stable ones (e.g. semi-deserts) and thus to exploit the opportunities for existence on our planet with greater efficiency and growing productivity.

Part four: **Plant geography**

The aims of plant geography or geobotany are to ascertain the essential features and recurrent patterns of the spatial distribution of plants and to explain their fundamental causes, which lie partly in their ecology, that is to say in the interrelations between their requirements for existence and the conditions actually occurring on the earth, and partly in the history of their evolution and migration.

In general, four branches of geobotany are recognized. The subject of floristic plant geography (areography, chorology) is the distribution of individual taxonomic units; plant sociology is the study of plant communities; these are essentially parts of descriptive plant geography. Ecological plant geography, concerned with the ecological bases of plant distribution, and historical (or genetical) plant geography, concerned with the history of floras and vegetation, are aspects of interpretative plant geography. We shall not adhere rigidly to this division in the limited space available, and only a selection of some of the more important facts and problems will be mentioned, with particular reference to C. and N.W. Europe.

Section one: **Distribution patterns and their causes**

Every species and indeed all taxonomic units of higher or lower rank, i.e. every 'taxon' (p. 391), occupies a certain limited region or area of distribution, since although every plant has the capacity to spread by means of numerous propagules, this spread can occur only (1) if there is an environment with conditions suitable for existence; and (2) if in the course of time the propagules actually reach this habitat, and can develop further, usually in competition with other species. The area of distribution of most species is indeed only a relatively small fraction of the earth's surface; hence the floras of various regions, i.e. their total stock of taxa, are very different.

To understand the pattern of distribution of a species one must first consider the interrelations between its genetically determined requirements for existence and the existing environmental conditions. By environment we mean the totality of all external conditions acting on the plant wherever it grows, and not merely the locality where it happens to be. (The term 'habitat' is often used in English for the *kind* of locality in which the plant occurs, e.g. ponds, woodlands, etc.; of course each habitat provides a distinct environment.)

Environmental conditions or factors are commonly divided into those of the climate and those of the soil (climatic and edaphic factors respectively); furthermore, the influences of animals or of other plants can be regarded as biotic factors. The significance of environmental factors results from their action on individual processes in the living plant (accounts of which are to be found in the physiological and morphological sections of this book). The most important environmental factors are:

1. Light, which varies enormously in intensity, spectral composition, diurnal and seasonal distribution, and according to the proportion of direct and diffuse radiation, to altitude and latitude, to selective absorption in air and water, to shading, etc. It acts on photosynthesis, on germination, extension and other developmental processes (cf. photoperiodicity, pp. 344 et seq.), irritability (e.g. stomatal opening), etc. (cf. pp. 192 et seq.).

2. Temperature, not only the normal, generally regularly varying temperature régimes, with effects on almost all the plant's activities, but also the rare but often particularly important extremes (e.g. frost). Air temperature is closely dependent on the temperature of the solid or liquid surface of the earth, and particularly on soil temperature, which often exerts a direct influence, e.g. on root activity, and on the survival of perennating organs (cf. pp. 185 et seq.).

3. Water, both its supply in the soil and its loss by evaporation (again dependent, *inter alia*, on temperature), with their influence on the water content of land plants. In the soil both the amount and the sustained supply of water actually available to the plant are crucial, so that the physically and chemically (cf. halophytes) determined suction pressures of the soil water are vital. For transpiration the relevant factor is not the absolute humidity, but the saturation deficit; air movements are also significant (cf. pp. 178 et seq.).

4. Carbon dioxide and oxygen concentrations of the air, soil and (for submerged plants) water in relation to photosynthesis and respiration. These concentrations are often dependent on the activities of micro-organisms.

5. Soil-nutrients, in particular their sufficient and sustained supply in available forms. The geological substratum is of overriding importance. Nutrient-rich soils are developed, for example, on basic volcanic rocks (basalt, etc.), most limestones, moraines (particularly the terminal moraines of the Pleistocene glaciations), loess, river alluvium and estuarine

muds. Nutrient-poor soils occur over most granites, coarse schists and sandstones, quartzites and Pleistocene sands and gravels. Rock-weathering and soil-formation are extremely complex processes (cf. p. 231), depending in addition to climate (temperature, precipitation) on the activities of the higher plants themselves (excretion of CO_2, formation of humus, root penetration) and in particular on plant and animal micro-organisms (e.g. nitrification). Eventually the various horizons of the soil undergo impoverishment, modification or enrichment, giving rise to a 'soil profile'. The reaction (hydrogen-ion concentration, pH) of the soil also exerts manifold influences.

6. Mechanical factors, such as high winds, ice- and sand-blasting, avalanches, solifluction, flotation and movement by water, etc.

7. Biotic factors. Animals (as well as wind and water) effect pollination and the dispersal of propagules; many are herbivorous. Plant and animal parasites and symbionts (e.g. mycorrhiza) may also be mentioned, and above all the many and far-reaching ways in which Man exploits and modifies vegetation.

Micro-habitats may show extraordinary diversity over short distances; thus the microclimate enjoyed by a plant may differ very considerably from the general or macroclimate as measured at standard (often widely separated) meteorological stations. For example, in the mountains near Dresden a yearly mean temperature of 23.3°C was found in cushions of the moss *Pohlia nutans* on a rock-face exposed to the south, but only 6.2°C in a patch of the liverwort *Mylia taylori* on a north face only 50 m away. The extreme annual temperature ranges were 66.5°C and 23.0°C respectively. In terms of macroclimate these temperatures correspond to a difference of 40° of latitude! Soil, too, is very heterogeneous: its content of nutrients and water may vary considerably with depth. Furthermore, the action of the environment on biological processes cannot be assessed solely on a basis of generalized physiological knowledge, but must be investigated by direct observation in the natural habitat. In recent years experimental ecology has adapted physiological methods, together with pedological and climatological studies, for the elucidation of the causal relationships of plant communities.

A correlation between plant distribution and particular habitats is evident in the use of such terms as 'water plant', 'rock plant', 'dune plant', etc. So, too, species characteristic of particular soils have long been known, i.e. plants which occur preferentially or exclusively on soils of a particular chemical or physical nature, e.g. halophytes (*Salicornia europaea* agg., Fig. 216, *Aster tripolium*), or calcicole or silicicole (calcifuge) plants. Particularly striking are the so-called vicarious species, i.e. closely related plants which have mutually exclusive distributions, e.g. in the Alps *Rhododendron hirsutum, Achillea atrata* and *Gentiana clusii* on calcareous soils, *R. ferrugineum, A. moschata* and *G. kochiana* on lime-deficient soils. A similar pair in Britain are *Carex lepidocarpa* on calcareous peats and *C. demissa* on non-calcareous wet soils. In such cases the soil reaction (pH) plays an important part (cf. pp. 237, 280).

Such plants can be used as indicators of soil reaction. Thus the wild radish, *Raphanus raphanistrum*, indicates a more or less acid soil reaction, whilst charlock, *Sinapis arvensis*, is found on basic or circumneutral soils. Calcicole weeds include *Ranunculus arvensis, Scandix pecten-veneris* and *Acinos arvensis*; calcifuge, *Rumex acetosella, Scleranthus annuus, Spergula arvensis, Galeopsis segetum, Stachys arvensis,* etc. However, restriction to a given soil type varies with the climate. Some species which occur on any soil in hot, dry regions behave as calcicoles in cool, wet climates where a neutral or alkaline soil reaction is only found on calcareous soils.

In general, of course, edaphic factors only determine the presence or frequency of a species within the limits of its general area of distribution. That these limits are most often climatically determined is shown by their frequent coincidence with definite climatic boundaries, e.g. with the isotherms of the summer or winter months; furthermore, the plants may be observed to suffer damage at these limits in climatically extreme years. Thus, for example, the northern and eastern limits of the holly (*Ilex aquifolium*) coincide roughly with the 0°C January isotherm, but even more closely with a line joining all the easternmost points which have more than 345 days in the year with a temperature maximum above 0° (Fig. 744); this eastern limit of the holly is determined by winter cold. On the other hand, the northern limits of many species are determined by the

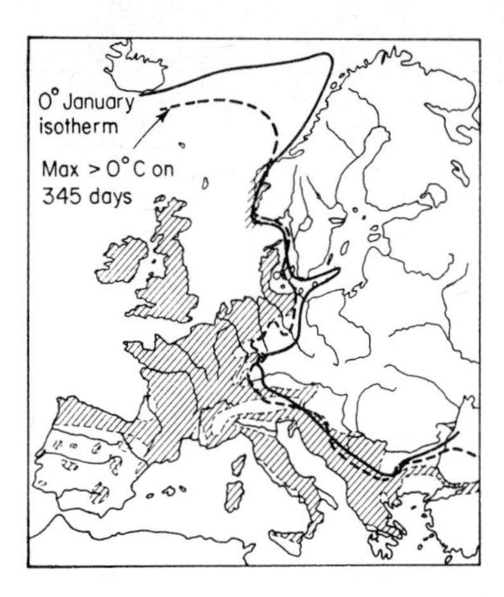

Fig. 744. The area of distribution of *Ilex aquifolium* (shaded) compared with two climatic boundaries. Explanation in the text. After Enquist and Meusel, modified.

coolness of the summer; thus the northern forest limit largely coincides with a line representing a daily mean temperature of 10°C or more on sixty to seventy days in the year.

The agreement between a limit of distribution and a climatic line is, of course, rarely perfect, since the microclimate commonly diverges considerably from the macroclimate. Furthermore, the success of a species and its survival in competition with other species depend on very many factors which may work together or in opposition, their effects thus being enhanced or diminished. (cf. the interaction of climate and edaphic preferences noted above.) In general, the distribution limits of a species practically never occur just where it completely fails to grow in isolation, but rather where, as a result of the progressively unfavourable environment, it begins to succumb in the face of competition. Finally, most species consist of a number of ecotypes, that is to say genetically determined populations (biotypes) which are adapted to different environments and which correspondingly occur in different habitats within the general range of the species (cf. pp. 407 et seq.). Thus there are ecotypes adapted to different lengths of growing season, or with different degrees of sensitivity to late frosts, or with various photoperiodic responses, various nutritional requirements, etc. Individuals of a species from different parts of the distribution area, or from different habitats, may thus behave differently both physiologically and ecologically. Attention to such ecotypes adapted to various climates and soils is important in the selection of plants for use in forestry and agriculture.

It is easy to show that the majority of plants occupy only a fraction of the area in which they are potentially capable of growing. Thus we sometimes encounter plants which, brought by Man to a new area, have escaped and spread rapidly, even at the expense of native plants, and finally have become permanently established, as, for example, the Canadian pondweed, *Elodea canadensis*, has done in Europe since about 1840. The cause of this incomplete exploitation of suitable environments lies partly in the limited effectiveness of the various dispersal mechanisms, and partly in the considerable passage of time which their operation necessitates.

One might suppose that most spores, being so light, would be dispersed throughout the world in a very short time; yet even this is not invariably the case. For almost all seeds and fruits there is a limit, characteristic of each species, to the distance over which dispersal is effective, regardless of whether dispersal is gradual over short distances or whether it leap-frogs over long ones. Frequently a plant comes up against some insurmountable barrier (e.g. a large stretch of water, a high mountain) which prevents it reaching other suitable areas. In other cases insufficient time may have elapsed since a species evolved for its dispersal to have been completed.

Once a species is well established it tends to persist for a very long period, during which it is subjected to the vicissitudes of geological change—the changing distribution of dry land, formation of mountain chains, changing climates, etc. Its present distribution often reflects these events. An area is termed closed or continuous when the various populations are sufficiently close together for dispersal by natural means to take place from one to the other; the greater part of the area of distribution of our forest trees, for example, is closed. On the other hand, the distribution may be fragmented, i.e. disjunct or discontinuous, consisting of separate areas between which dispersal is apparently impossible under present-day conditions. The main area of closed distribution may have one or more widely scattered outposts; or the distribution may consist of two or more closed areas separated by gaps which appear to be unbridgeable. (The greatest distance which a plant can just bridge by natural dispersal is known as its threshold of disjunction.) Such disjunctions or discontinuities may be explained in a number of ways:

1. In exceptional circumstances dispersal might have occurred over enormous distances (e.g. by migratory birds). Although it is difficult to prove this hypothesis, it can be rejected in some instances by considering the distances involved (e.g. the entire width of an ocean) or the means of dispersal (e.g. heavy fruits which cannot be carried by birds or drift in the sea); and in particular when dispersal by Man can be excluded.

2. A species might have arisen independently in the two or more parts of its disjunct distribution, i.e. polytopically. This is not impossible when the presumed ancestor of the species occurs, or did occur, throughout the intervening area, and especially when the species has arisen by hybridization between two wide-ranging parents, or by polyploidy (pp. 36 et seq., pp. 331 et seq., pp. 415 et seq.); but when a systematically isolated taxon is disjunct, polytopic (and therefore polyphyletic) origin is less likely.

3. Disjunct areas may be remnants of an originally closed distribution. In many cases the occurrence of fossil plant material in the intervening areas where the plant does not now grow shows that this is often the correct explanation. Plants which occur in isolated areas separated from their closed area of distribution by unbridgeable gaps are generally interpreted as relicts, i.e. remnants persisting from a time when the closed area of distribution extended as far as these outposts.

In the absence of fossil evidence it is often difficult to decide whether a plant in such an isolated area is truly a relict, rather than a recent immigrant, but relict status is much more probable when a number of species with a similar disjunct distribution occur together in the same locality; cf. many glacial relicts found on bogs, in rocky gorges, etc.

The following types of disjunction are particularly important in the flora of C. and N.W. Europe:

1. European–Asiatic–N. American disjunctions. There are many species of wide distribution in the continents of the North Temperate Zone, but which do not reach into the Arctic. Their regional areas are separated by the breadth of the oceans, and sometimes by large gaps in Asia, e.g. *Humulus lupulus, Hepatica nobilis* (Fig. 356), and genera such as *Fagus, Acer, Aesculus* and *Tilia* with various species in different parts of the world which must have arisen from common ancestors. Even more striking are the distributions of many species or genera which occur, often in restricted areas, on opposite sides of the globe. These are generally absent from Europe, where, however, they grow well if planted. For example, the poison ivy (*Rhus toxicodendron*) occurs in Japan and in Atlantic N. America (and not in Pacific N. America as might have been expected). One of the two species of tulip tree (*Liriodendron*) occurs in eastern N. America, whilst the other is indigenous in E. Asia; the genera *Magnolia* (Fig. 745), *Platanus* and *Morus* have similar distributions. These major disjunctions which are particularly characteristic of forest plants, and to a lesser extent of steppe plants, indicate a formerly closer connection between the areas of distribution of the plants characteristic of a temperate climate in the Northern Hemisphere. This connection must have existed a long time ago, since in the meantime individual species have become extinct over considerable expanses, or new species characteristic of particular areas have evolved from an ancestral form formerly distributed throughout the North Temperate Zone.

2. Arctic–Alpine and Alpine disjunctions. Numerous species of Arctic territories are also found far to the south in high mountain areas, e.g. in the Alpine zone of the Alps, Pyrenees and Carpathians: *Salix reticulata, Dryas octopetala, Loiseleuria procumbens* (Fig. 741), *Eriophorum scheuchzeri, Silene acaulis, Ranunculus glacialis,* etc. Some of these species have isolated stations in the higher uplands of Germany, e.g. *Salix herbacea, Saxifraga oppositifolia, Veronica alpina* and *Gnaphalium supinum*; others occur even in the lowlands, especially on raised bogs (*Betula nana, Carex pauciflora*). Most of these species occur in the mountains of the west and north of Britain, many descending to

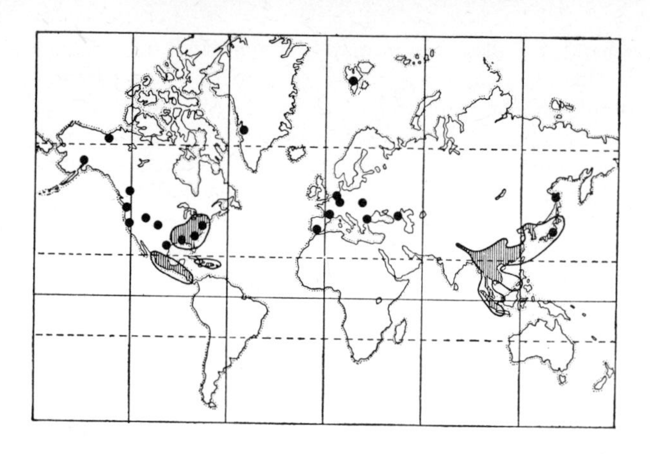

Fig. 745. *Magnolia*: present area of distribution (shaded) and fossil localities (black dots). After Dandy and Good, Berry, etc.

sea-level in the west of Ireland (e.g. *Dryas*) and in the north-west of Scotland. The gaps between stations of these species often exceed 1000 km; the occurrence in the Sudeten Mountains of the otherwise exclusively Arctic species *Saxifraga nivalis* and *Pedicularis sudetica* (the latter about 2000 km from its nearest Arctic stations) is particularly noteworthy. Furthermore, the populations of those species which occur in various European Alpine or highland areas are often widely separated (e.g. many species found in the Pyrenees, Alps and Carpathians, and then often only in restricted areas); or they turn up again in certain highland regions (e.g. the Black Forest, Sudeten Mountains), or in the Alps of S. Europe (e.g. many Alpine species of *Saxifraga, Gentiana, Androsace, Soldanella, Primula, Potentilla,* etc.; cf. Fig. 753). This type of disjunction is explained by assuming that environmental conditions suitable for such Arctic–Alpine and Alpine plants formerly occurred at lower altitudes during a cooler climatic period, when migration in all directions was feasible.

3. Steppe plant disjunctions. Many species of the S. Russian and C. Asiatic–S. Siberian steppes and forest-steppes, and even of some of the Asiatic semideserts (cf. p. 788) extend to a greater or lesser extent north and south of the Carpathians, through Hungary and to some extent along the northern side of the Mediterranean, into the hot, dry 'continental' areas of C. Europe, e.g. into C. Bohemia, the basins of Thuringia and Mainz, the continental valleys of the C. Alps, etc. Some even reach Scandinavia, the south and east of England, France and Spain. In these areas they are associated with many other species demanding high summer temperatures and dry conditions, some of which have their headquarters in C. Europe, others in the Mediterranean. They generally occur together in rocky habitats or xerophytic grasslands with few trees or shrubs (the 'steppe-heaths'); they may also grow in open pine or oak woods, especially on dry, basic, calcareous soils. In these restricted habitats they often occur at remarkable distances from their closed areas of distribution: thus *Onosma arenarium* (Boraginaceae), frequent from Lower Austria eastwards, is also found in a few places several hundreds of kilometres apart in the Mainz basin, in the Valais (Wallis) and in the south of France. Similar European examples include *Adonis vernalis* (Fig. 746), *Astragalus exscapus, Lathyrus pannonicus, Scorzonera purpurea, Linum flavum, Allium strictum, Iris spuria*, etc. In Britain *Silene otites. Medicago falcata* and *Phleum phleoides* are all confined as natives to E. Anglia, and are typical plants of the S. Russian steppes; other British 'steppe plants', e.g. *Hypochaeris maculata* and *Carex humilis*, also exhibit striking disjunctions. The closed areas of many continental species must have extended further west at some time; they appear now in their western outposts as 'xerotherm' relicts.

If a species is indigenous only in a single, naturally restricted area (e.g. a mountain, an island or a particular geographical region), or in a floristically defined and uniform region, it is said to be endemic there. Endemism arises either when a species has evolved in a certain area from which it cannot or has not spread (progressive endemism), or when a formerly much more widespread species has withdrawn to a particular area where competition is less fierce, or where the habitat is otherwise more favourable (conservative or epibiotic endemism).

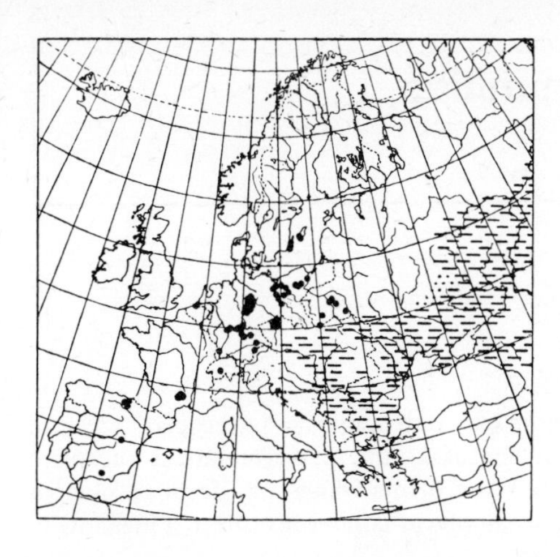

Fig. 746. The area of distribution of *Adonis vernalis* as an example of the disjunction of steppe plants. Closed area: dashes; outposts: dots. After Sterner from Walter.

Progressive endemics which have evolved recently include various species of *Hieracium* of mountainous districts (for example, *H. tubulosum, H. corconticum,* etc., in the Sudeten); other, more ancient ones are found in the Alps. The remarkable wealth of the flora of S.W. Australia is due to progressive endemism of considerable antiquity. On the other hand, the fossil record (cf. p. 777) shows that *Ginkgo biloba* and *Metasequoia glyptostroboides* in E. Asia and *Taxodium distichum* in the south-east of N. America are conservative endemics.

Comparison of the patterns of distribution of a number of species or other taxa (subspecies, genera, etc.) often shows a complete or partial coincidence of their limits, their centres of greatest frequency, etc., so that they can be grouped into geographical elements. The floristic peculiarities of various parts of the world which enable us to distinguish and delimit floristic kingdoms and regions (cf. p. 782) depend on the fact that many species, genera or even families are restricted to these areas, which therefore have characteristic geographical or floristic elements. A summary of the most important elements of the Holartic Kingdom (in so far as they form a significant fraction of the flora of C. and N.W. Europe) is given on p. 784 et seq.

Section two: **Plant communities**

The totality of plants which clothe an area more or less densely is called its plant cover or vegetation, in contrast to its flora (p. 760). The basic unit of vegetation is the plant community, i.e. a regularly recurrent assemblage of plants which may form an obvious and substantial feature of a landscape, e.g. a deciduous or coniferous forest, a meadow or pasture, a reed-swamp, a bog, etc.

Each homogeneous plant community developed in a particular place (i.e. each individual 'stand', e.g. a given piece of woodland) is characterized by the following attributes: (1) by its complement of species (the species list), of which many are absent from or are less important in other, neighbouring stands; (2) by the quantity of each species (number and mass); and by their more or less uniform, homogeneous distribution within the stand. In making an investigation (*Aufnahme, relevé*) of a stand all the species present are recorded, and their amounts and spatial arrangement either measured or at least estimated.

The uniformity of composition of a plant community arises from the following causes:

1. A community includes only those species which can co-exist indefinitely as 'messmates' or commensals in the particular environment of the habitat.

This is particularly significant in open communities in which the individual plants are so widely spaced that they exert no mutual influences, e.g. in the early colonization of a sandy sea-shore, where we find only those species which can tolerate the specialized environment (high salinity, exposure to wind, mobility of the sand, intermittent tidal submersion, etc.). Competition between plants is greatly reduced even in closed communities when the species have different (complementary) growth forms, different seasonal cycles of development or different environmental demands so that they exploit the habitat in various ways, e.g. when their roots penetrate to different depths.

2. Competition further limits the number of species which can continue to grow and reproduce in a particular habitat.

Competition also enhances the sharpness of the boundaries between adjacent stands. Whilst the frequency and vigour of a species growing alone alter only gradually with a gradual change in the environment, a species in competition with another often succumbs abruptly under unfavourable conditions.

3. Many species, especially tall and substantial ones, modify the original environment so that only those species adapted to the new conditions can follow in their wake; these latter are therefore associated with the former which are 'constructive' species (of high 'dynamic value').

Thus, for example, the development of a herb-rich beechwood is restricted, even within the area of distribution of the beech, by the incapacity of this tree to grow on very dry, very poor and acid, and above all on many waterlogged and badly aerated soils; or, for example, in hollows subject to severe late frosts. If, however, a stand of beech has become established in a suitable habitat, it alters all the environmental conditions of its associates. Above all, it creates its own microclimate, since its canopy intercepts a large part of the incoming radiation. The light is quantitatively and qualitatively altered, and its seasonal and diurnal rhythms are modified: the light reaching the woodland floor attains its maximum shortly before the leaves unfold, falling in summer to some $\frac{1}{2}$ per cent of the light outside. The temperature régime inside the wood is more equable than that outside: the surface of the soil, which in the open (when bare of vegetation) may heat up to 70°C, remains in the wood generally cooler than the air temperature. (Only the dry leaf litter in the high light phase of the spring tends to be somewhat warmer.) Evaporation and wind-speed are greatly reduced. In addition, soil development is dependent on the tree cover: the deep roots extract nutrients from the subsoil, returning them to the surface in the annual fall of leaves and twigs (and in a wood not exploited for timber the decay of dead trunks is also important). Furthermore, the litter forms a characteristic raw material (modifiable only within certain limits) for the development of humus. The flora, including the soil microflora, of a beechwood must be adapted to this greatly modified environment. Thus many of the plants largely complete their annual growth cycle during the high light phase before the canopy unfolds its leaves, e.g. *Anemone nemorosa, Arum, Endymion*; others have persistent foliage which utilizes the light phase in the autumn and early spring, e.g. *Hepatica nobilis* (not native in Britain), whilst others make do with the low summer light intensities. Even these subordinate species associated with the beech play an important part in the maintenance of the woodland, since their leaf litter and root activities are also concerned in soil development. Thus, for example, plants such as the bilberry, which form an acid raw humus or 'mor', may in certain circumstances gain a foothold and then spread, so modifying the uppermost layers of the soil that the natural regeneration of the beech, and hence the survival of the community as a forest, may be endangered.

The majority of plant communities are composed similarly of species with various growth forms, often expressed in their stratification or layering (e.g. tree, shrub, herb and bryophyte layers in woodlands), and the predominant growth forms determine the 'physiognomy' characteristic of a community. Competition between the species in a community may also involve the secretion by roots of toxic substances which inhibit the growth of other species (allelopathy: cf. p. 340; cf. the various models of dispersal, p. 675).

A stable plant community thus depends on very complex interactions, more or less in equilibrium, between the plants themselves and the original environment, which then undergoes modification by the vegetation.

The plant communities recognized in an area will, of course, largely coincide with the regional geological, topographic and climatic units, and with the distribution of the local soil types. Hence plant communities and groups of species with similar ecological requirements (ecological groups) are more reliable than single species as indicators of environmental conditions: for example, the forest communities (forest types) mirror the habitat types in a region. In this way the study of vegetation, augmented by assessment of the economic potentialities of various sites, plays an increasingly important practical role.

The description and classification of plant communities involve the following criteria:

1. The basic unit is the plant association. A plant association includes all stands which have virtually the same floristic composition and hence probably exist under very similar environmental conditions.

The names of associations are generally formed by adding the suffix '-etum' to the stem of the name of the diagnostic genus (e.g. Fagetum sylvaticae). The floristic agreement between the stands included in an association and the distinction between this association and others arise from the occurrence of a number of species in all or most of the stands (constancy), and from the tendency of certain species to occur principally only in the community in question (fidelity, exclusiveness), being absent from, or rarer in, other communities (faithful species, index species). Thus a very widespread type of fertile alluvial grassland (the Arrhenatheretum) has numerous constant species (e.g. *Taraxacum officinale, Bellis perennis, Ranunculus acris, Rumex acetosa* and many fodder grasses) as well as the faithful species *Arrhenatherum elatius, Geranium pratense, Heracleum sphondylium, Anthriscus sylvestris, Pastinaca sativa,* etc. Similar floristic criteria can be used to unite associations into higher units (alliances, orders), whilst for the lower units variations in the quantitative relationships can be taken into account. Such units defined by the predominance of certain species in particular strata of the community are termed sociations.

2. Communities can also be classified, regardless of floristic composition, by the predominance in them of particular growth forms, e.g. coniferous forest, evergreen sclerophyllous scrub, dwarf shrub heath, grassland, etc. The most important of such physiognomic units are termed formations.

In so far as the growth forms of plants are adaptations to particular environments, i.e. in so far as they are life forms (cf. pp. 178 et seq., 641; Fig. 211), the structure and physiognomy of formations are also adaptations to environment. Thus, despite their very diverse floristic composition, all the deciduous forests of Eurasia and N. America which lose their leaves in winter reflect the influence of the temperate forest climate with its period of winter dormancy. Within a formation one can further distinguish as a synusia any group of plants with a common life form, and which probably have in consequence similar ecological requirements. Most formations consist of several synusiae, each often constituting a separate stratum or layer.

The concept of the formation is particularly valuable in reviewing the major vegetation types of the world, whilst for more detailed studies of vegetation the consideration of associations is necessary.

Very commonly one plant community is transformed in the course of time into another. Sometimes this depends on the entry of new species into the area in question, but more generally on some modification of the habitat factors. The causes of this change may lie outside the vegetation itself (climatic changes, geological processes, human interference), or they may be due to the alteration of the environment by the vegetation. The development of successive plant communities in the same place is termed succession.

In the case of a body of fresh water, the dead remains of plants and animals accumulate at the bottom, forming peat or organic muds, and in the course of time the water becomes progressively shallower. The distribution of aquatic and riparian vegetation depends very largely on the depth of the water; in consequence, the zoned plant communities replace one another in a centripetal direction and the area of open water diminishes (*Verlandung*). In nutrient-rich (eutrophic) lakes the abundant plankton forms a mud known as 'gyttja'; if copious $CaCO_3$ is deposited a white 'lake chalk' may be formed. The more conspicuous vegetation (the macrophytes) of such a lake in C. Europe

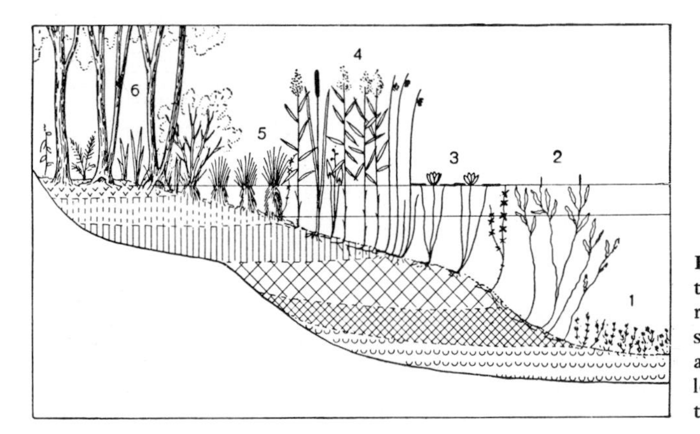

Fig. 747. Diagram of the zonation and stages in succession round a eutrophic lake, with the sediments formed. The upper and lower limits of the water-level are shown; further explanation in the text.

or lowland England (e.g. the Norfolk Broads) undergoes the following succession (or hydrosere), starting at a depth which depends, amongst other things, on the turbidity of the water (Fig. 747): (1) a submerged carpet of charophytes; these, too, form calcareous deposits; (2) a pondweed zone with several species of *Potamogeton*, submerged with only the flowers above water, and with *Myriophyllum spicatum*, *Elodea canadensis*, etc.; (3) a zone of aquatics with floating leaves, e.g. *Nuphar, Nymphaea, Potamogeton natans,* and in very still waters the free-floating *Hydrocharis, Lemna, Stratiotes,* etc.; (4) reed-swamp with dense stands of *Schoenoplectus* (*Scirpus*) *lacustris* (the pioneer), *Phragmites, Typha* and *Sparganium* advancing into the water; (5) a zone of large sedges with tussocks of *Carex paniculata* (Norfolk) and *C. elata* and swards of *C. acutiformis* (Norfolk), on which the pioneer bushes of *Salix cinerea, Frangula alnus* and young alders become established; and (6) an alder wood or 'carr', initially very swampy but becoming progressively drier at the surface as plant remains continue to accumulate; in Norfolk *Alnus* is often mixed with *Fraxinus, Quercus,* etc. Underneath (2) and (3) a mud is formed in which the remains of dead plants progressively increase in size; under (4) a reed-peat, under (5) a sedge-peat, under (6) a brushwood peat. Investigation of such stratified deposits enables one to reconstruct the succession from open water to dry land, a process which may have taken many thousands of years.

The pioneer communities of a succession are intimately affected by the local edaphic conditions and microclimate, but the terminal communities express more clearly the influence of the regional climate; thus whether succession starts from open water or with the colonization of cliffs, rocky slopes, river gravels or dunes, the end-product tends to be forest if the regional climate is suitable. The terminal community largely determined by climate, or the few interrelated terminal communities towards which the natural succession of the vegetation of a climatically uniform region tends, is often called its climax community.

Vegetation can also be divided into zonal, extrazonal and azonal types. Zonal vegetation is the predominant natural type corresponding to the climates of an extended area, e.g. the belts of tundra, coniferous forest, mixed deciduous forest, *Stipa*-steppe and meadow-steppe in E. Europe. Extrazonal communities are those which occur beyond their region of macroclimatically determined predominance, where the local climate or microclimate is similar to their zonal climate; they may therefore appear on moist, shady north-facing slopes far to the south of the zonal belt, or on warm, dry, south-facing slopes far to the north. Plant communities are termed azonal when they occur in ecologically extreme habitats throughout a number of zonal belts (e.g. the semi-aquatic communities of lakes, cliff communities, etc.).

The plant communities of C. and N.W. Europe. A short review of the most important of these follows.

Although rather less than a third of C. Europe is wooded, climatically it is a forest region. Great Britain is one of the least wooded countries in Europe today (less than 5 per cent wooded), yet climatically most of the lowlands are potentially forest areas.

In C. Europe forest has a distinct upper limit as a result of the brevity and coolness of the growing season only in the Alps and Sudeten; in the highlands of the north and west of Britain the forest limit is difficult to determine precisely because of almost universal disforestation. The highest peaks of the Black Forest, the Böhmerwald and the Harz Mountains are very near the forest limit. Scarcely any part of C. Europe is climatically too hot or dry for forests, but a treeless coastal fringe exposed to salt winds, confined in Germany to a very narrow coastal belt and to the islands of the North Sea, is of some significance on the more exposed coasts of the British Isles.

In the west, at low altitudes in C. Europe, and over most of lowland Britain, deciduous woodland generally predominates in the natural state; towards the north-east, at higher altitudes in C. Europe, and in some valleys in Scotland, coniferous forests or mixed forests are found. The period of winter rest enforced by the cool-temperate climate is expressed in the leaf-fall and well-protected buds in the deciduous forests, and by the xeromorphy of the evergreen needles (Fig. 179) in the coniferous ones.

The beechwood has already been described (p. 766) as an example of the ecology of a deciduous forest. Woods of beech, or beech mixed with ash, sycamore, lime, etc., predominate in the western uplands of C. Europe and at moderate elevations in the other highlands and the calcareous Alps. In the plains they occur mainly on the more recent terminal moraines (Fig. 687). Floristically much poorer beechwoods are widespread on well-drained soils in S.E. England, on chalk, limestone and some sands. In C. Europe oak-hornbeam woodlands occur chiefly on the better soils in the lowlands, especially where the more exacting beech is at its limits of distribution or approaching them (N.W. Germany, and in drier continental areas; cf. Figs. 687, 689; they also occur on fertile alluvial soils with a high water table along the great continental rivers, where they commonly succeed the pioneer willows and poplars. In Britain oakwood with hornbeam (usually coppiced) is restricted to sandy loams in S.E. England; more widespread in Britain are *Quercus robur* woodlands with hazel coppice, especially on clays, or *Q. petraea* woodlands on siliceous soils. In C. Europe mixed oak wood rich in species tends to occur on warm dry slopes with thermophilous species such as *Acer monspessulanum*, *Quercus pubescens*, *Sorbus torminalis* and many herbs of southern or eastern origin. The poorest soils often support oak-birch woods, their undergrowth including *Calluna* and other undemanding plants. In the mountains of Europe the oaks and associated species do not ascend as far as the beech. All oak and mixed oak forests are richer in shrubs and herbs than beech forests, since their canopies cast less shade. The oak is a light demanding, the beech a shade bearing tree. The riverine forests of C. Europe are notable for their lianes and twining plants (*Clematis vitalba*, *Vitis vinifera* subsp. *sylvestris*, *Humulus lupulus*, etc.).

Coniferous forests include those of Scots fir (*Pinus sylvestris*), represented in the Scottish highlands by remnants of the Caledonian Forest, where some trees ascend to 600 m. In C. Europe *P. sylvestris* occurs mainly on poor, dry, sandy soils in the plains or hills (Fig. 638). On the poorest soils it is associated only with lichens (*Cladonia*, *Cetraria islandica*, etc.); on slightly better soils *Calluna* and *Vaccinium vitis-idaea*; and where they are still better and moister we find *Vaccinium myrtillus* and numerous mosses, e.g. *Pleurozium schreberi*, *Hylocomium splendens* and *Dicranum scoparium*. The spruce (*Picea abies*) is not native in Britain, and in the lowlands of Europe is frequent only in the north-east; otherwise it is a tree of the upper montane forests, particularly in the Alps, Carpathians and eastern highlands (Fig. 638). These montane spruce forests are rich in *Vaccinium myrtillus*, several grasses (e.g. *Calamagrostis villosa*), ferns (e.g. *Blechnum spicant*), clubmosses and bryophytes. The litter of pine and spruce, and the remains of the dwarf shrubs, commonly form an acid, slowly decaying raw humus or 'mor' in contrast to the less acid or neutral, more rapidly decomposing mild humus or 'mull' generally produced by deciduous trees. (The humus type also depends on the species and the habitat.) The silver fir (*Abies alba*) also enters into the composition of montane forests (Fig. 638), whilst near the timberline, especially in the C. Alps, larch and the Arolla pine (*P. cembra*) also occur. Here the forest terminates in a shrub-belt of *Pinus mugo* and *Alnus viridis*. *Taxus baccata* forms some of the finest yew woods in Europe on the chalk of S. England.

All forests have long been exploited in various ways by Man; hence much attention is paid to their regeneration. In high forest the trees regenerate from seed, whether naturally shed or planted, and they are felled only at an advanced age. The forest is said to be clear-felled when the trees are simultaneously cut down over a large area, whereas selective felling involves only small areas, groups of trees or even single specimens. (There are many such systems of exploitation.) Many deciduous trees are coppiced, i.e. cut back to stools, from which new shoots regenerate; these are cut every ten to thirty years. The bark of oak managed in this way is used in tanning; hazel was formerly coppiced on a large scale in England for hurdles, etc. Such shrubby woodlands treated in this way are called copses. In standards-with-coppice regularly spaced tall trees are left standing over the coppice (e.g. oak standards with hazel coppice, widespread in England). Coppice is less frequent now than it was; today most commercial forests are managed as high forests, particularly the economically more valuable softwoods (pines, spruces, etc.), which are now grown far beyond their natural areas of distribution. These are usually clear-felled, although the sudden exposure of the ground profoundly and often deleteriously affects regeneration and soil development.

Although inconsiderable in area, the naturally treeless parts of C. and N.W. Europe are very diverse in kind.

On cliffs and boulder-slopes we can study the stages in colonization of the rock. The pioneers are crustose lichens (sometimes algae) which dissolve the rock. These are followed by foliose lichens (e.g. species of *Parmelia* and *Gyrophora*) and by mosses forming cushions or mats (e.g. species of *Grimmia*, *Rhacomitrium* and *Hypnum*). In these, and in the rock crevices, the first soil collects in which various higher plants (chomophytes) take root, e.g. ferns of the genus *Asplenium*, tussock-forming grasses, succulent species of *Sedum*, etc. These in turn collect further products of weathering and contribute to the formation of humus, producing an initially very shallow soil which gradually extends over all but the steepest rock faces. Such shallow soils liable to drought are developed particularly on basic or calcareous rocks; these habitats are similar microclimatically, especially on south-facing slopes, to warmer, drier regions, and hence in C. Europe they support steppe-heaths (p. 764) which include many species characteristic of the continental steppes and of the Mediterranean region, e.g. species of *Stipa, Koeleria, Anthericum, Ophrys, Dianthus, Pulsatilla, Astragalus, Linum, Teucrium, Scabiosa,* etc. Some of these steppe plants (e.g. *Veronica spicata, Carex humilis, Hypochaeris maculata*) and Mediterranean species (e.g. *Helianthemum apenninum, Koeleria vallesiana*) are found in less extreme (but still dry) calcareous habitats in England and Wales. At first a more or less open xerophytic grassland with xeromorphic grasses, sedges and herbs, including geophytes and annuals, develops in these habitats; when and where the soil is deeper, shrubs come in (e.g. *Cornus* [*Thelycrania*] *sanguinea, Viburnum lantana*), and in C. Europe the thermophilous mixed oak forest previously mentioned ultimately develops.

The development and colonization of sand-dunes can best be followed on the coast. The foreshore, which is waterlogged, salty and often flooded during winter storms, is first invaded by a community characterized by *Agropyron junceiforme*; only a few other species, annuals or rhizomatous perennials, are present. Wind-blown sand is deposited in the lee of these plants, forming small embryo or primary dunes. Next the salt is leached out of these by rain and they can be colonized by marram, *Ammophila arenaria*. As dune formation proceeds these plants continue to grow upwards through the sand, trapping and stabilizing yet more, so that these yellow or secondary dunes become progressively larger and higher. These yellow dunes also are poor in species, since the sand surface is still mobile between the plants; indeed, storms may break them down again, forming 'blow-outs'. However, if the dunes become less exposed to the wind, by the formation of new dunes on the seaward side for example, they ultimately develop into grey or tertiary dunes completely covered with vegetation. Round the coasts of the North Sea scrub communities with *Hippophae* develop at this stage; here and elsewhere *Salix repens* is frequent, whilst heaths of *Calluna* or *Empetrum* develop where the sand is not too calcareous. On the Baltic coast pine woods eventually develop on the dunes, but if the natural plant cover is destroyed large mobile dunes (e.g. the Culbin Sands in Scotland) may be formed, and dune development begins all over again.

Shallow waters on sheltered coasts also exhibit a classic succession of communities (saltmarshes) where nutrient-rich, sandy or clayey mud is deposited in places uncovered at low tide. In places remaining submerged we find algae (e.g. *Enteromorpha*) and species of eel-grass (*Zostera*). Further landwards, up to the mean high-water mark, the annual glassworts or marsh samphire (*Salicornia europaea* agg., Fig. 216) form dense stands which are particularly effective in causing the deposition of further mud, which can then be colonized by *Puccinellia maritima* and many familiar halophytes (*Aster tripolium, Limonium vulgare, Triglochin maritimum,* etc.). These in turn are succeeded by an initially salty sward of *Festuca rubra* agg.; this may be enclosed by sea walls and grazed, when the salt is gradually leached out. (Under natural conditions the development of woodland may take place.) Artificial methods of speeding up this succession are extremely important in the reclamation of land from the North Sea. On the coasts of N.W. Europe the grass *Spartina townsendii* plays an important part in this process.

The succession from nutrient-rich (eutrophic) standing water to dry land has already been described (p. 768). A generally similar succession takes place in nutrient- and humus-poor (oligotrophic) and nutrient-poor but humus-rich (dystrophic) fresh waters. The pellucid waters, poor in plankton, of many oligotrophic lakes (e.g. in mountainous regions) permit a growth of peculiar bottom-living (benthonic) rosette plants (e.g. species of *Isoetes*); in dystrophic waters a floating mat of *Sphagnum* with sedges, *Menyanthes*, etc., participates in the succession, which may again terminate in the development of swampy woods (carr). Along the great rivers of C. Europe the colonization of gravel, leading eventually to riverine forest (p. 770), is also important.

The natural succession to woodland may often be retarded indefinitely, and indeed existing woodland may be destroyed, by the development of raised bogs. These are found especially in the lowlands of N. Germany, in the foothills of the Alps, and in the montane zone of the highlands and Alps; in Britain in the lowlands of Wales, the west of Scotland and throughout Ireland. A raised bog is one form of 'mire'; a mire is any superficial accumulation of peat, consisting of the persistent remains of mosses and higher plants

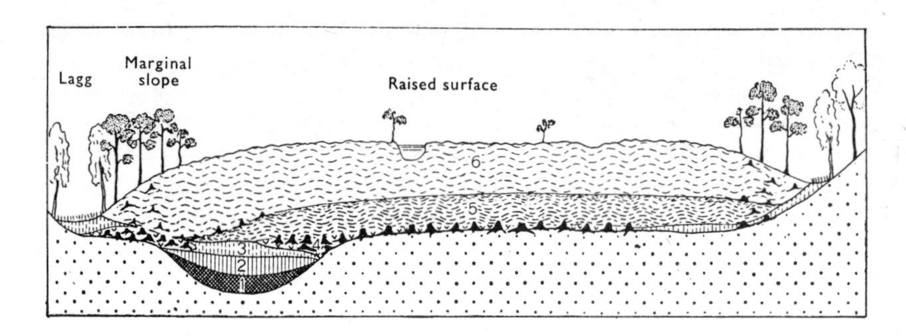

Fig. 748. Diagrammatic section of a C. European raised bog which has developed partly over an infilled lake, partly over former woodland. **1.** Mud. **2.** Reed-peat. **3.** Sedge-peat. **4.** Wood-peat. **5.** Older. **6.** Younger *Sphagnum*-peat. In the centre of the raised surface is a deep pool (the 'eye of the bog'). The mineral substratum is dotted.

which gradually become humified, although retaining their tissue structure indefinitely. Peat can form only where oxygen is largely excluded so that decay is arrested, as for example, near a permanent ground water table, in the succession from open water or in waterlogged mineral soils. More or less nutrient-rich peats of this kind, influenced by eutrophic ground water, are found in fens, in which the peat is neutral or only weakly acid. Such fen peats may support reed-fen, sedge-fen or fen carr communities (e.g. nutrient-rich alder carr). In wetter climates the constantly wet peat surface is colonized by the bog mosses (*Sphagnum* species), the peculiar structure of which (p. 553) enables them to hold water (which becomes extremely acid) like a sponge. If precipitation is copious and the growing season long the individual tussocks of *Sphagnum* coalesce into a continuous carpet, whilst their dead lower parts remain saturated with water; as they grow, so the surface is raised higher and higher. The former vegetation is smothered; tree-roots are killed, so that even forests succumb, and a *Sphagnum* bog with few or no trees is formed. Such extremely nutrient-deficient mires or bogs are fed only by precipitation and atmospheric dust. The continued formation of *Sphagnum* peat elevates the surface of a typical 'raised bog' above its surroundings, and its convex shape is highly characteristic (Fig. 748): the central, more or less level raised surface is surrounded by a steeper marginal slope which is more sharply drained and therefore may support trees, whilst surrounding the whole is a rand or lagg, a marginal fen into which water drains both from the raised bog and its neighbourhood. On the raised surface small hummocks colonized by Ericaceae often alternate with small pools or hollows. Peat is chiefly formed in the infilling hollows and in the young hummocks. When an old hummock ceases to grow upwards it is surrounded by rapidly growing neighbouring hummocks and so becomes a wet hollow; thus in the course of time hummocks and hollows alternate on the same spot. However, in some raised bogs the growth of the *Sphagnum* cover is uniform over the whole surface. The vascular plants of a raised bog have to be adapted to an extremely acid and nutrient-deficient substratum, and be able to grow upwards with the *Sphagnum*. When a hummock reaches a certain height difficulties of water supply appear to be encountered, the more so as the surface of the bog often heats up considerably. Only a few characteristically xeromorphic species can grow under these exacting conditions–various Ericaceae (*Calluna, Vaccinium oxycoccos, V. uliginosum, Andromeda,* etc.), Cyperaceae (*Eriophorum vaginatum, Trichophorum* (*Scirpus*) *cespitosum*) and the various species of *Drosera*. In some districts of C. Europe dwarf *Pinus sylvestris* and *P. mugo* also occur on the raised surface of the bog. Most C. European raised bogs have developed in Post-glacial times on the sites of former forests, but some have grown over infilled lakes and fens (especially in Britain). In the very humid oceanic climate of the west and north of Britain extensive *Sphagnum* bogs can form a continuous cover over the whole landscape except for steep slopes and rock outcrops; these are the so-called 'blanket bogs', which in the Pennines are now commonly dominated by *Eriophorum vaginatum*.

The great majority of our treeless communities have replaced forest under the influence of Man.

These include cultivated areas (arable, gardens) which would rapidly revert to forest if untended by Man. In them are found weed communities in which the various species tolerate interference by cultivation or escape its effects (e.g. as dormant seeds). This, together with the invariable presence of bare soil, and the relatively warm and dry habitats created, explains the high proportion in these communities of annual and biennial species, many of which originated in arid southern and eastern regions (e.g. in the steppes). Closely related are ruderal communities characteristic of nitrogenous soils near human settlement (rubbish tips, waysides, etc.); the typical species, e.g. of *Chenopodium* and *Urtica*, are nutrient-demanding.

The great majority of grasslands and dwarf shrub heaths have also been created by Man. Grasslands are communities with few or no woody plants, consisting predominantly of numerous tussocky and rhizomatous grasses and sedges with various perennial herbs; the soil is moderately to very moist. Natural grasslands occur in the mountains above the forest limit, in a narrow fringe along extremely exposed Atlantic coasts and in saltmarshes (since there are no halophytic woody plants in temperate regions), but the majority of European grasslands have probably arisen by Man's destruction of the original forests and the prevention of the regeneration of trees by regular mowing (meadows) or grazing (pastures). Differences in soil and in methods of exploitation give rise to many different grassland types. The most productive ones (leys, water-meadows, etc.) can be mown two or three times a year and subsequently grazed; they require frequent and heavy applications of manure and a moderately moist soil, so that in the hotter parts of Europe they are found mainly in the valleys. Fen grasslands are generally not manured and are, or were, often used only for litter, e.g. the *Molinia* fens of East Anglia, in which *Carex* species are abundant. Dry soils commonly support pastures with *Festuca ovina* agg., and other somewhat xeromorphic grasses, e.g. *Bromus erectus* and *Brachypodium pinnatum* (especially on calcareous soils). The *Nardus stricta* grasslands of the mountains are also grazed.

Dwarf shrub heaths are treeless communities in which low-growing Ericaceae such as *Calluna* are dominant. Particularly characteristic are the heaths of the oceanic climates of N.W. Germany and much of Britain, in which the predominant *Calluna* is associated with various Atlantic plants such as *Erica tetralix, Genista anglica,* etc. In Britain other species of *Erica*, e.g. *E. cinerea*, and species of *Ulex*, e.g. *U. minor* and *U. gallii*, are also highly characteristic. In parts of N.W. Germany the juniper (resistant to grazing) is the only large shrub. Heath plants grow on poor, generally sandy soils, and since the vegetation itself forms an acid raw humus or mor, and the abundant precipitation tends to leach the soil, a characteristic soil profile (podsol) is developed wherever the drainage is free: beneath the raw humus layer the soil is leached, soluble salts, colloidal aluminium and iron compounds all being washed down or eluviated, leaving virtually pure quartz sand in the A-horizon (bleached sand). Lower down in the rust-coloured or black B-horizon the iron and humus compounds are redeposited (illuviated), often cementing the sand grains together as a 'hard pan' which impedes the penetration of tree roots. Such heaths were very extensive for many centuries (e.g. the Lüneberger Heide and the heaths of East Anglia and the London and Hampshire basins), but most of them are now re-afforested, cultivated or built over. Most heaths apparently originated by the destruction of the former woodland and the prevention of regeneration by grazing, fire (employed to stimulate the growth of new shoots for grazing) and by the paring off of the heather and raw humus for fuel, litter or even manure; hence the soil became progressively poorer. Natural Atlantic heaths occur only in a very narrow coastal belt too exposed for trees or locally where the soil is extremely acid, nutrient deficient and waterlogged. Like most grasslands and weed communities, inland heaths were first created in prehistoric or early historic times.

We conclude this review with an outline of the altitudinal zonation of vegetation in the mountains of C. Europe. (In Scotland and Wales the zonation has been much obscured by disforestation and overgrazing by sheep.) It is well known that the decrease in temperature, the shortening of the growing season. the increase in precipitation and wind strength, the development of snow cover, the modification of the light (with an absolute increase of direct radiation, especially of the shorter wavelengths) and other features of the mountain climate determine the formation of altitudinal zones, of which the following can be seen in the Alps and to some extent in the other mountains of C. Europe (Fig. 749):

1. The Colline (hill) zone. Agricultural land with viticulture predominates. Thermophilous mixed oak woods and steppe-heaths are confined to this zone, which rarely extends above 500–800 m. Pine and oak-hornbeam woods are frequent, and plantations of *Castanea* in the south.

2. The Montane (or lower montane forest) zone with beechwoods and mixed montane forests of beech, silver fir, spruce, sycamore, etc. Cultivation is often still profitable. The upper limit varies between 800 and 1200 m.

3. The Coniferous (or upper montane forest) zone (sometimes included with (4) in the Subalpine zone), in which spruce forest predominates with mountain meadows and pastures in clearings. This zone extends up to the limit of closed forest (the timberline) at 1600–2000 m in the Alps (rarely higher), and at 1300 m in the Sudeten.

4. The Dwarf scrub (or Subalpine) zone with *Pinus mugo, Alnus viridis, Rhododendron hirsutum, R. ferrugineum, Vaccinium myrtillus* and other dwarf shrubs. Although very similar floristically to (3), the low shrubs are protected by a winter snow cover. This zone generally extends some 200–300 m above (3). The tree limit occurs in this zone, often consisting of isolated spruces, which if their crowns reach above the shrub layer are often killed on the weather side by storms and snow-blast, giving characteristic 'flag forms'. In the C. Alps the tree limit is commonly formed by larch and *Pinus cembra*; these ascend higher than the spruce and may form open stands in the dwarf scrub zone.

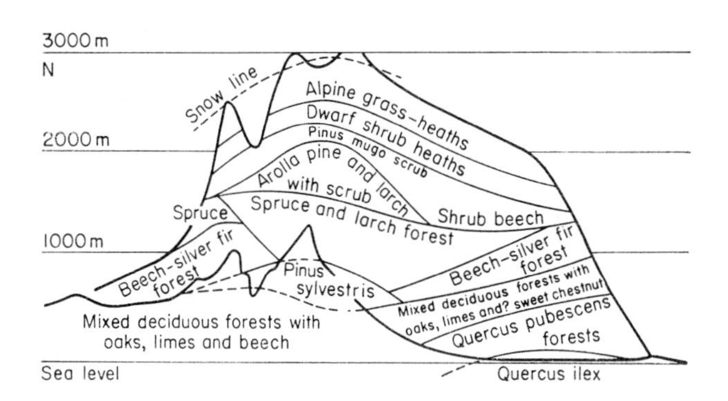

Fig. 749. Diagram of the altitudinal zones of vegetation in the E. Alps along the line Münich–Innsbruck–L. of Garda. The thick lines indicate the altitude of the valleys and passes (below) and the peaks (above). After H. Gams, modified.

5. The Alpine grass-heath (meadow) zone extends up to the snow line, which in the Alps lies between 2450 and 3260 m. (Near Ben Nevis in Scotland the snow line would lie at about 1650 m; Ben Nevis itself reaches 1344 m.) At the lower levels very dwarf shrubs occur, such as the very hardy *Loiseleuria procumbens* (Fig. 751; Ericaceae) with prostrate shoots and small leaves; at high levels turfy communities, often heavily grazed, are found. These become more open and poorer in species as the altitude increases. Like the widespread communities of rocks and screes, these are rich in dwarf perennial herbs with brilliantly coloured flowers, e.g. the genera *Gentiana, Saxifraga* and *Draba* (with many species in Scotland as well), *Androsace, Primula, Phyteuma, Campanula, Minuartia, Senecio, Papaver,* etc. Many of these are limited to the Alpine zone of the high mountains (Alpine species); others also occur in the Arctic (Arctic–Alpine; cf. p. 763). Areas of calcareous and non-calcareous rock exhibit big floristic differences. *Carex curvula* forms extensive turf on acid soils, whilst the grass *Sesleria albicans* (*S. caerulea*) and the sedge *Carex firma* are important on calcareous ones. In moist hollows where the snow lies late (snow patches) very characteristic communities occur, including dwarf willows (especially *Salix herbacea*), species of *Soldanella* and various liverworts and mosses.

6. In the Nival zone above the snow line there are very few flowering plants (e.g. *Ranunculus glacialis*, reaching 4275 m), but there are still many kinds of bryophyte and lichen.

The altitudinal limits of the various zones differ in various parts of a mountain chain, depending, for example, on the aspect (higher levels on south-facing slopes, lower on north-facing; cf. Fig. 749). Comparison of various mountain areas also shows the influence of the massing together of mountains on the character of their climate and vegetation, e.g. on the proportion of coniferous forest. Thus in the more continental climate of the C. Alps, with hotter summers and lower precipitation, deciduous forests are commonly absent from the montane zone, being replaced by larch and Arolla pine. Here too the forest limit is much higher (at up to 2400 m), but its altitude depends as much on the species present as on differences in climate. Conversely, spruce forest is scarcely developed under oceanic climatic conditions in the southern pre-Alps, in the Swiss Jura and the Vosges, where beech and sycamore form the forest limit. In the west of Ireland Arctic–Alpine plants (e.g. *Dryas octopetala*) descend to sea-level in County Clare, where oddly enough they are intimately associated with Mediterranean species such as the orchid *Neotinea intacta*. Many Scottish mountain plants are Arctic–Alpine rather than Alpine (*Cherleria sedoides* is an exception).

Section three: **The history of the flora and vegetation**

The flora and vegetation of the world today have been subject to great geological change in the past, and to understand them fully both their history and their present ecology must be studied. Otherwise we cannot evaluate correctly those processes which take place slowly over considerable periods of time, for example, speciation, climatic changes and migration.

Methods. Certain features of the history of a flora can be deduced from the study of geographical distribution (especially of disjunctions) and of taxonomy (cf. pp. 763 et seq.). but fossil plants from past geological periods give more conclusive evidence. Whilst fossils are preserved only in certain circumstances, particularly in lake deposits, peat and coals derived from these, nevertheless their preservation is often so perfect that extremely detailed anatomical investigation is possible. Even impressions (e.g. in calcareous tufa) and true petrifaction allow fruits, seeds, leaves and wood to be identified with living species, or to be recognized beyond doubt as extinct forms.

The study of fossil pollen (pollen analysis) provides an important palaeontological tool in the elucidation of geologically recent vegetational history, especially forest history, since the pollen grains of many genera and species have characteristic and identifiable shapes (cf. pp. 603 et seq., 649 et seq.) and are well preserved, especially in lake deposits and peats. The pollen grains of many anemogamous forest trees (and even some entomogamous ones) are dispersed in large quantities (in C. Europe several thousand pollen grains are deposited on a square centimetre each year) and become incorporated in accumulating sediments, and the proportion of grains of each kind gives an indication of the quantitative composition of the forests existing in the neighbourhood when a particular stratum was laid down (cf. Fig. 750, in which the occurrence of varying proportions of pollen grains of different species at different times is shown graphically). Comparison of the composition of a forest and its pollen rain at the present-day gives us a check on the validity of our conclusions. Pollen analysis also allows us to follow the history of plants which avoid wet soils and which otherwise are unlikely to be preserved as macro-fossils.

Naturally the determination of the age of any deposit is important. In general, dating depends on geological techniques, which themselves utilize plant remains as index fossils. A fairly reliable absolute chronology of the greater part of the Post-glacial has been worked out by geological methods, e.g. by counting the annual layers or 'varves' in lacustrine clays formed during the retreat of the ice (cf. the table on p. 780). Within recent years these estimates have been checked for organic remains not more than 50,000 years old by C^{14} dating (which depends on the falling content of radioactive carbon of organic material with age). The Post-glacial development of forests underwent very similar trends over wide areas (p. 779); hence the position of a find in a pollen sequence gives a clue to its age. Conversely, a relatively young deposit can be dated if it contains contemporary prehistoric and early historic objects of known age, e.g. the remains of prehistoric settlements in pollen-rich lake deposits (Fig. 750) or the humified remains of timber, cereals, etc., in terrestrial settlement sites. In this way archaeological finds, prehistoric periods and forest history can be closely correlated, and objects dated by pollen analysis of the deposits in which they occurred; hence vegetational history, geology, prehistory and the geography of human settlement share much common ground.

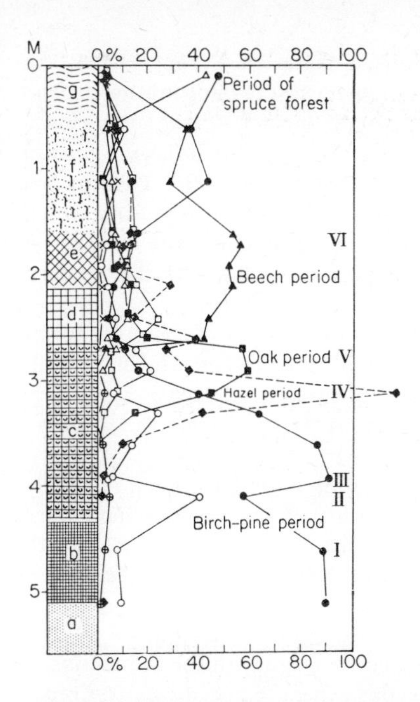

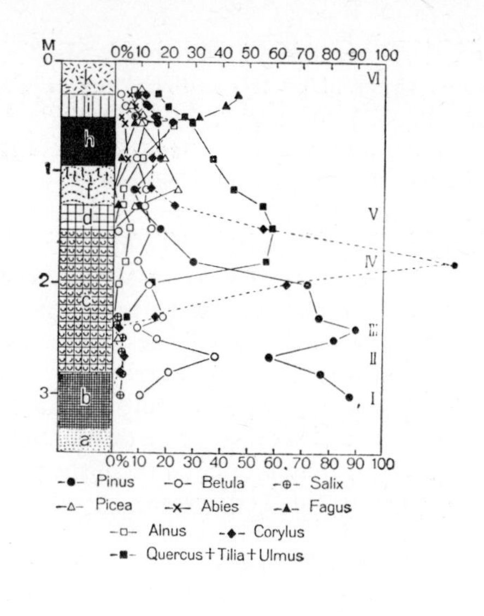

Fig. 750. Two pollen-diagrams from the Federsee district (Upper Swabia). The sequence of deposits is shown to the left of each diagram; the samples investigated are arranged according to their depth below the surface. The abscissae give for each sample the relative frequency of each kind of pollen as a percentage of the total tree pollen in each sample. Successive points for each species are joined. (Hazel, being a shrub, is not included in the total tree pollen; its frequency exceeds 100 per cent when its grains are more numerous than the total tree pollen.) The right-hand profile comes from the site of a late Neolithic settlement which is seen to date from the early beech period. After K. Bertsch, modified. a, glacial clay; b, clay mud; c, calcareous gyttja; d, algal gyttja; e, detritus gyttja; f, *Scheuchzeria* peat; g, raised bog peat; h, horizon of the Neolithic settlement; i, reed peat; k, peaty soil. The Roman numerals indicate horizons of the same age, but do not correspond with the pollen zones recognized by British peat stratigraphers. In Britain beech is never so prominent in pollen diagrams, and spruce did not return as a native tree after the Pleistocene glaciations; otherwise the trends are very similar to those in C. Europe.

The Tertiary Period. Even if we consider only the flora of Europe (and excluding the Mediterranean), we have to go back to the Tertiary. During this period the temperature gradually cooled, reaching a minimum during the Pleistocene ice age. The lower Tertiary of C. Europe and the London basin contains many fossils of tropical–subtropical plants, e.g. palms belonging to the genera *Sabal* and *Nipa*, species of *Cinnamomum* and *Ficus*, etc.; the vegetation must have been very similar to that of the present-day Indo-Malayan Rain Forests. By the Miocene the most thermophilous species were extinct in Europe; instead the lignites contain *Sequoia* species and to a lesser extent *Taxodium*, whilst deciduous trees characteristic of the temperate zone attained predominance. Finally, in the Pliocene conifers and deciduous trees predominated in Europe as at the present-day; indeed, the species were identical or at least very closely related to those now living here or elsewhere in the world, and the greater part of our existing flora was already present towards the end of the Tertiary.

Nevertheless, the Pliocene flora of C. and N.W. Europe was much richer than it is today, including, as it did, many genera now occurring only in other parts of the Holarctic (cf. p. 763), e.g. *Ginkgo, Taxodium, Sequoia, Tsuga, Magnolia, Liriodendron, Sassafras, Liquidambar, Carya,* etc. Similarly, the Tertiary floras of California and Alaska include many plants now confined to Asia. Such genera were therefore widely distributed over the greater part of the North Temperate Zone in the Tertiary, and there must have been an effective migration of species between different parts of the world.

Demonstration of this is provided by the early Tertiary floras of the North Polar lands, especially Greenland and Spitzbergen, but even of the northern part of Ellesmere Island at a latitude of 81° 45′ N, where the present mean annual temperature is −20°C; these fossil floras are very similar to the Miocene and Pliocene floras of Europe and to the modern floras of E. Asia and N. America. They include genera still present in Europe, e.g. *Pinus, Picea, Salix, Populus, Corylus, Betula, Alnus, Quercus, Juglans, Ulmus, Acer, Vitis, Tilia* and *Fraxinus*, and others not now native here, e.g. *Ginkgo, Taxodium, Sequoia, Magnolia* (Fig. 745), *Liriodendron, Sassafras, Platanus, Liquidambar,* etc. Thus in the early Tertiary these Arctic territories had a rich, moderately thermophilous flora which later migrated southwards and which predominated from the upper Oligocene onwards in C. and N. Europe. This Arcto-Tertiary flora is consequently the basis of the modern European flora (excluding the Mediterranean region).

These early Tertiary fossil sites represent a displacement northwards of the polar forest limit in the Northern Hemisphere by 20–30° compared with the present, whilst the limit of the Palmae was 10–15° further north than now. Suggested explanations include greater radiation reaching the earth's surface, displacement of the continents and of ocean currents or a different orientation of the polar axis.

Even before the Tertiary the climate of the Arctic lands must have been relatively warm, since, for example, in the Cretaceous of N. Greenland the remains have been identified not only of *Cinnamomum, Magnolia,* etc., but also of *Artocarpus* (breadfruit), which nowadays occurs wild only in the Indo-Malayan and E. Indian Tropics. The same is true of the fern family Matoniaceae.

The existence of the Arcto-Tertiary flora in the N. Polar regions does not imply that it originated there; indeed, these regions were connected long ago with the mountains of Asia, and taxonomic studies suggest that many herbaceous genera (e.g. *Pedicularis, Crepis, Primula, Saxifraga*) originated in this climatically exceedingly diverse region (E. Tibet, W. China, Altai). The great mountain chains thrown up in the Tertiary have in general played a large part in the evolution of the modern flora.

The Pleistocene ice age. The relatively uniform Arcto-Tertiary flora was driven further southwards by the Glacial climate, and the territories it previously occupied were covered by ice or replaced by treeless periglacial vegetation. In the process the Arcto-Tertiary flora became very impoverished. During Interglacial and Post-glacial times it moved northwards again, recolonizing much of its lost ground, but many species failed to survive these extensive migrations, especially in Europe, where the mountain chains running east and west and the Mediterranean Sea (with its influence on the climate) precluded the displacement of a broad vegetation belt. Hence Europe is today much poorer in species (especially trees and shrubs) than climatically comparable areas in E. Asia and N. America. The Arcto-Tertiary flora and its late Tertiary and Pleistocene fate thus explain many of the Holarctic disjunctions described on pp. 763–5.

During the maximum glaciation the inland ice-sheets extended south to 50° latitude in Europe, reaching the northern slopes of the C. European highlands and attaining a thickness of over 2000 m; they reached 60° in Asia and 37° 30′ in N. America. Further south were extensive mountain glaciers, those of the Alps extending northwards to within 270 km of the northern ice-sheet. The snow-line was depressed by 1200–1300 m compared with the present, whilst the summer temperature of the last glaciation (not the most extreme) is estimated to have been 8–10°C lower in C. Europe. The glacial climate was hostile to vegetation even beyond the areas covered by ice where airborne dust was deposited in a broad belt as loess, whilst as far south as the northern rim of the Mediterranean basin the soil remained frozen to a considerable depth throughout the year (permafrost soil). Since enormous quantities of water were locked up in the ice-sheets, dry land was more extensive than at present; for example, the British Isles and the southern part of the North Sea were both part of the mainland of Europe until some time after the last retreat of the ice.

During the last glaciation C. Europe and the south of England remained treeless except perhaps for stunted woods of birch and pine in exceptionally sheltered areas. The remains of this 'Full Glacial' flora are often found in clayey deposits in lakes formed immediately after the retreat of the ice or in areas which escaped glaciation, e.g. the basin of C. Bohemia, which even today is particularly warm. The index fossil is the Arctic–Alpine *Dryas octopetala*; this gives its name to the '*Dryas* flora', which shows that the unglaciated areas were covered with moss-tundras, dwarf shrub communities, swards of perennial herbs, sedge-fens and impoverished aquatic communities. These included many species which were driven from their original Alpine or Arctic haunts into lower or more southerly sites. When the ice retreated they not only withdrew to their old stations, but some originally Alpine species reached the Arctic, and vice versa; in this way we can explain Arctic–Alpine disjunctions (p. 763). Thermophilous trees and shrubs could survive the successive glaciations only on the northern margin of the Mediterranean or in S.E. Europe.

Index fossils of the *Dryas* flora include species which are now Arctic–Alpine, e.g. *Dryas octopetala, Salix herbacea, Loiseleuria procumbens* (Fig. 751), *Polygonum viviparum* and *Eriophorum scheuchzeri* (only the last of these has not been found living or fossil in Britain); or exclusively Arctic, e.g. *Salix polaris* and *Ranunculus hyperboreus* (both in the British Full Glacial, neither now native); or exclusively Alpine, e.g. *Potentilla aurea, Salix retusa* (neither found in Britain). With these occurred a number of climatically less exacting species, widespread today, e.g. *Menyanthes trifoliata*, or others which have or had relict habitats in the highlands of C. Europe or elsewhere, e.g. *Betula nana* and *Empetrum nigrum* agg.

The glacial fauna of the loess also includes animals of the steppes, and it is very probable that plants of the cold continental steppes could have spread into C. Europe during the Pleistocene, but so far only a very few have been found fossil (e.g. *Crambe tataria*). Nevertheless, a strikingly high percentage of *Artemisia* pollen (p. 734) in glacial deposits suggests a considerable Pleistocene extension of steppe-like plant communities.

The Late and Post-glacial periods. As the climate gradually became warmer (apart from temporary reversions) after the maximum of the last glaciation (about 20,000 years ago), and whilst the ice-sheets largely disappeared during the succeeding 10,000 years of the Late-glacial period, forests and other exacting plant communities extended once more over formerly treeless or ice-covered areas; this continued both during the Late-glacial and the succeeding 10,000 years of the Post-glacial period. Trees with different climatic requirements did not become dominant simultaneously, but successively, so that over large parts of Europe a sequence of forest periods can be distinguished, each period being roughly synchronous. In C. Europe these are as follows (cf. Fig. 750 and the table on p. 780; this is to be read from the bottom upwards):

1. The birch–pine period. As the ice retreated the land remained generally treeless at first, but later birch-woods (especially in the west and north-west) and pine-woods (especially in the east and south-east) developed, giving a landscape similar to that of W. Lapland today. Species such as *Betula nana* and *Empetrum nigrum*, now confined mainly to the mountains and the north, were frequent at low altitudes, so that the climate was probably still quite cold; indeed, after the first invasion of forest, which extended as far as the Baltic during the 'Allerød oscillation', there was a considerable deterioration of the climate just before the onset of the Post-glacial (younger *Dryas* period).

2. The hazel period. The Post-glacial period began with the retreat of the inland ice from the end-moraines in C. Sweden about 8000 B.C.; it apparently became rather warmer and in W. Germany and the highlands especially there was a massive extension of the hazel, followed by trees demanding higher summer temperatures, such as oaks, elms, limes, sycamore and maples, and the ash, with finally the alder and spruce. In consequence, the birch–pine forests were gradually suppressed. The bed of the southern North Sea and of the Baltic was still dry land.

3. The mixed-oak-forest period. In this period the oaks and their associates predominated; as at present, pine-woods were restricted to poor sandy soils. With increasing swampiness of the lowlands extensive alder-woods developed, whilst spruce covered the eastern highlands (westwards to the Harz), the E. Alps and the Carpathians.

4. The beech period. At this time the beech, silver fir and hornbeam became increasingly widespread, suppressing the mixed-oak forests and extending further into the mountains, where mixed forests of beech, silver fir and spruce developed. Meanwhile beechwood predominated at lower altitudes and in the lower and north-western highlands of Germany, whilst in places (e.g. E. Prussia) hornbeam was dominant. During the last

The Late-glacial and Post-glacial Forest History of C. Europe. After Firbas

Estimated date (approximate)		**Period** (after Blytt and Sernander)	**Forest cover in C. Europe**	**Prehistoric culture**	**Development of the Baltic Sea and retreat of the inland ice sheet**
The Present			Managed forests		
				Historical period	Mya-Sea
		Subatlantic	Beech period	La Tène	
800–500 B.C.		Climatic deterioration			Limnaea-Sea
	Post-glacial	**Subboreal**	Transition to beech from the mixed-oak-forest period	Bronze Age	
2500 B.C.				Neolithic Age	
		Atlantic Climatic optimum	Mixed-oak-forest period		Litorina-Sea
5000 B.C.					
		Boreal	Early mixed-oak-forest and hazel period	Mesolithic Age	Ancylus-Lake (Last ice remnants melting)
		Preboreal	Birch and pine period		Yoldia-stage
8000 B.C.					
	Late-glacial	**Younger Dryas**	Period of scattered trees		(Margin of the ice-sheet in C. Sweden)
9000 B.C.				?	
		Allerød	Birch and pine period		Baltic glacial lake
10,000 B.C.				Palaeolithic Age (Magdalénian)	
		Older Dryas	Treeless period		
18,000 B.C.					
		Full Glacial			Margin of the ice-sheet in N. Germany

few millennia the forests have undergone progressive modification by Man.

In England events followed a very similar pattern, but spruce and silver fir (both abundant, for example, in the Clacton Interglacial) never returned as native trees in Post-glacial times. Britain was finally cut off from the Continent by the end of the Boreal period, and climatic considerations apart, we might expect that the formation of the English Channel was a considerable barrier to further plant migration. Beech and to a lesser extent hornbeam increased in S.E. England during the Subatlantic, but they never achieved the predominance attained in C. Europe.

The wood and fruits of thermophilous trees and shrubs have been repeatedly found some hundreds of metres above their present altitudinal limits (e.g. hazelnuts in the Harz at 1000 m) as well as beyond their latitudinal limits in N. Europe, indicating former extensions of the forests both upwards and northwards. Similar extensions of the areas of thermophilous aquatic and semiaquatic plants (e.g. *Phragmites*, species of *Najas*, *Trapa natans* and the cyperaceous *Cladium mariscus*) and of various terrestrial animals and marine molluscs also indicate that these organisms existed, in areas now too cool for them, during the hazel, mixed-oak-forest and early beech periods, which together are spoken of as the Post-glacial warm period, with a climatic optimum in the Atlantic period. A Post-glacial climatic deterioration set in about 800–500 B.C., when the climate became similar to that at the present.

The sequence of forests in the Late- and Post-glacial was probably caused mainly by a roughly corresponding sequence of climates from a glacial tundra climate followed by a cold pine–birch climate to a warm oak climate and back to a rather cooler beech climate.

However, other causes (rates of dispersal of different species, migration routes, the situation of glacial refugia) must also be borne in mind.

Stratigraphical studies have also elucidated the history of succession from open water to fen and raised bog. Raised bogs became frequent only in Post-glacial times; in many places they show an older, highly humified *Sphagnum* peat before the onset of the climatic deterioration and a younger, less-humified peat after (Fig. 748). Such structural features of a bog and of its peats representing stages in succession may give valuable indications of changes in moisture relations (e.g. changes in water level) and their effects on the vegetation.

In addition to those steppe-plants adapted to the cold which could have spread during the Full Glacial and pine period, dispersal of light-demanding and thermophilous steppe-heath plants of southern and eastern provenance was no doubt facilitated, particularly in the hazel and early mixed-oak-forest periods. The extension of increasingly shady woods and finally the climatic deterioration led to their subsequent isolation in scattered relict habitats away from their closed areas of distribution. Many steppe-plant disjunctions (p. 764) can be explained in this way, both in C. Europe and in Britain.

The feeding habits, settlement sites and general economy of prehistoric Man were greatly influenced by these changes in the Quaternary period, especially in Late- and Post-glacial times (cf. the table on p. 780). In S. and C. Germany most Neolithic and Bronze Age settlements largely coincided with the present-day distribution of steppe-heaths; it thus appears that early Man preferred a terrain with thermophilous and xerophytic vegetation, but whether such sites were chosen for their less dense tree cover or because the soils were particularly suitable for the cereals then grown is debatable (steppe-heath theory). Certainly in Britain there was a high density of settlement in the Neolithic and Bronze Age on the light soils of East Anglia and on the chalk of the south of England (cf. Stonehenge).

Interglacial periods. Comparable vegetation sequences can be followed between the Pleistocene glaciations. During the warm Interglacials the vegetation was similar to that of the present-day, but enriched by a few persistent Tertiary species now either extinct in Europe (e.g. the waterlily, *Brasenia schreberi*, still living in E. Asia and N. America) or present only in small relict areas (e.g. *Picea omorika* in the Balkans). Interglacial deposits also show the progressive impoverishment of the vegetation at the onset of each new glaciation; thus the Rinnensdorfer Interglacial (Brandenburg) shows the following forest sequence: (1) older pine, (2) hazel, (3) lime, (4) hornbeam, (5) silver fir, (6) spruce and (7) younger pine. These clearly show an increase in temperature (as well as changes in humidity) to an optimum (? lime period) and then a deterioration. Similar sequences have been demonstrated elsewhere in C. Europe and in England.

Section four: **The floristic kingdoms and regions of the world and their vegetation**

In the course of geological time the plants in different parts of the world have evolved more or less independently, becoming adapted to the conditions prevailing in each area. Consequently, we can distinguish certain major floristic divisions or kingdoms, each subdivided into smaller units (regions, provinces, etc.) and each with a more or less uniform and characteristic flora, i.e. with certain predominant floristic elements and often with characteristic endemics. In so far as vegetation is determined by the available flora, and a given area has its peculiar climate, the floristic peculiarities of an area are also reflected in its vegetation. Moreover, most floristic areas have more or less natural physiographic boundaries.

Six floristic kingdoms are usually distinguished: the Holarctic (or Boreal), the Palaeotropical and Neotropical, and the Australian, S. African and Antarctic Kingdoms. As the vegetation map facing indicates, the kingdoms diminish in area but increase in number from north to south; the greater distances between the continents of the Southern Hemisphere, and probably also their earlier separation, have resulted in a greater degree of independent development of their floras in the past.

The **(I) Holarctic (or Boreal) Kingdom** embraces the entire North Temperate and Arctic zones and is in consequence by far the largest in area. Betulaceae, Salicaceae, Ranunculaceae, Saxifragaceae, Apiaceae, Primulaceae, Campanulaceae and many other families have most of their genera and species in this kingdom. The striking floristic uniformity of this vast area is related to the close affinities between its component floras in the Tertiary.

The core of the Holarctic is formed by the great forest regions of Eurosiberia and northern N. America (see the map), in which the cold winters enforce a period of dormancy whilst the warmth and moisture of the summers permit the almost universal prevalence of forest, which in the natural state would be interrupted only by aquatic communities, bogs and rocky habitats. Beyond the northern forest limit the trees peter out in a belt of scrubby birches and willows transitional to the Arctic tundra. The colder and more continental the climate, the more conifers predominate; the warmer and more oceanic, the more deciduous forests extend. In the north these forest regions are all very similar; to the south they abut on several distinct regions. In Eurasia these are: the small Macaronesian Region (Azores and Cape Verde Islands); the Mediterranean Region, with its sclerophyllous trees and shrubs; the continental Black Sea–C. Asiatic Steppe Region; and the extremely species-rich E. Asiatic Forest Region, which grades imperceptibly into Subtropical Rain Forests. The N. African–Indian deserts are often included here. In N. America, south of the Coniferous Region of N. Alaska and Canada, we can distinguish a Pacific Forest Region, very rich in conifers, and an Atlantic Forest Region, rich in deciduous trees, which passes southwards into Subtropical Rain Forests; between these two forest regions stretch the prairies and deserts. We have space to consider further only a few of the Holarctic floristic regions (cf. the map).

In the **Arctic Tundra Region** the short, cool growing season prevents the growth of trees. Frost-resistant, small-leaved, commonly evergreen dwarf shrubs, e.g. *Juniperus communis* subsp. *nana, Betula nana, Empetrum, Loiseleuria* and *Phyllodoce,* and dwarf willows form dwarf shrub heaths, whilst warmer south-facing slopes support herb-rich communities, but in general the tundras, or cold deserts with permafrost soils, support only swamps rich in Hypnaceae and sedges, or stone-fields colonized mainly by bryophytes and lichens.

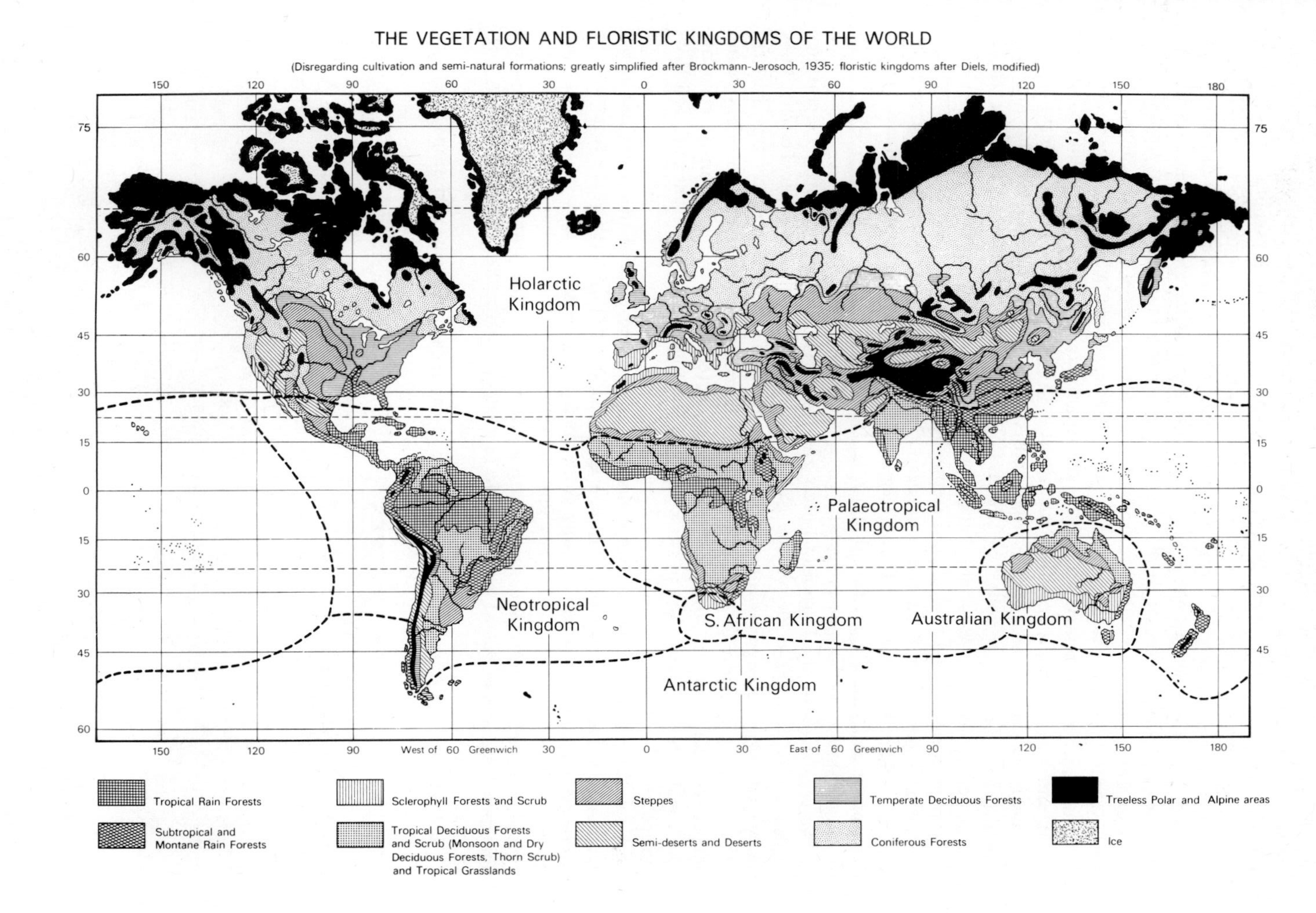
THE VEGETATION AND FLORISTIC KINGDOMS OF THE WORLD
(Disregarding cultivation and semi-natural formations; greatly simplified after Brockmann-Jerosoch, 1935; floristic kingdoms after Diels, modified)
Holarctic Kingdom
Palaeotropical Kingdom
Neotropical Kingdom
S. African Kingdom
Australian Kingdom
Antarctic Kingdom
West of 60 Greenwich
East of 60 Greenwich
Tropical Rain Forests
Subtropical and Montane Rain Forests
Sclerophyll Forests and Scrub
Tropical Deciduous Forests and Scrub (Monsoon and Dry Deciduous Forests, Thorn Scrub) and Tropical Grasslands
Steppes
Semi-deserts and Deserts
Temperate Deciduous Forests
Coniferous Forests
Treeless Polar and Alpine areas
Ice

Central, western and northern Europe belong almost entirely to the great **Eurosiberian Forest Region**, the vegetation of which has already been described on pp. 769–75; in consequence, we shall refer only briefly to its subdivisions here. The evergreen coniferous forests form a belt of increasing width eastwards, where they are subject to the coldest Siberian winters (mean January temperature below −40°C!); they reappear further south at higher altitudes in the mountains. Usually a single species of pine, spruce, fir or larch dominates over a considerable area. A few deciduous trees also occur to a limited extent, especially in natural or artificial gaps in the forest, e.g. birches, alders, poplars or willows. The belt of deciduous forests to the south is very narrow in or even absent from C. Asia, but it becomes wider in W. Europe, reaching the Atlantic coast (from which conifers are usually absent) from Portugal to N. Norway. This deciduous belt can be subdivided: (1) In the west, where the winters are mild, is the Atlantic or *Ilex* Province, including most of Britain and N.W. Germany; conifers are virtually absent; oak and birch forest predominate, but dwarf shrub-heaths with *Calluna* and Atlantic species of *Erica, Genista* and *Ulex* play a major role on poor soils. (2) The C. European or *Fagus sylvatica* Province; this is rich in deciduous trees, especially beech, but pine also occurs in the lowlands, silver fir and spruce in the upland forests, and steppe-heaths occur in dry, warm inland areas. (3) Further east lies the S.E. and E. European *Quercus* Province, as well as the great N. European–Siberian Coniferous Forest Region, which includes most of Scandinavia; in the west spruce and pine are the sole dominants, further east the Siberian larch, the Siberian fir and the spruce (as subspecies *ovata*).

The vegetation of the European Forest and Arctic Tundra Regions suffered enormous displacements during the Pleistocene glaciations. In the north the Arcto-Tertiary flora was wiped out, and large areas were repeatedly recolonized by a relatively few, particularly aggressive species; to the south, however, many species of diverse origin found refugia, especially in the mountains. This is reflected in the various **floristic elements** which make up the basic flora of these regions. The following are relevant to C. and N.W. Europe.

Alpine, Arctic–Alpine and Arctic elements. These are centred mainly in the higher mountains above the forest limit (Alpine species, Fig. 753) or north of the forests in the Arctic (Arctic species). Many occupy both areas (Arctic–Alpine species, Fig. 751) and afford evidence of the former closer connection of the Arctic and Alpine floras during the Pleistocene (cf. p. 763). Species occurring mainly near the forest limits are called Subalpine (e.g. *Pinus mugo*) or Subarctic (e.g. *Betula nana, Rubus chamaemorus*, Fig. 752).

Several hundred Alpine species are found in the high mountains of Europe; these are

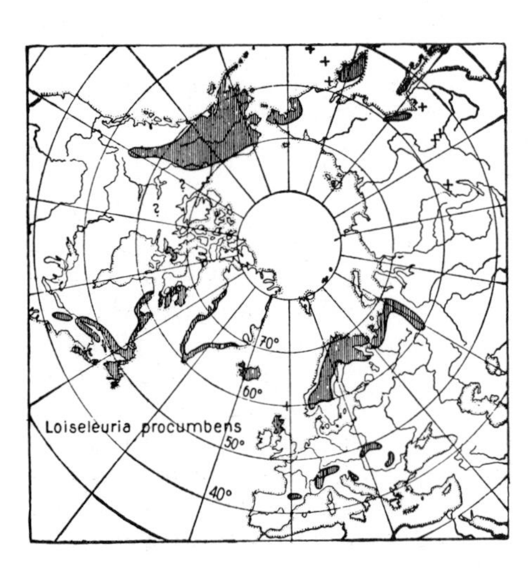

Fig. 751. Area of distribution of *Loiseleuria procumbens* (Arctic–Alpine). Figs. 751–7, after various authors from Meusel, modified.

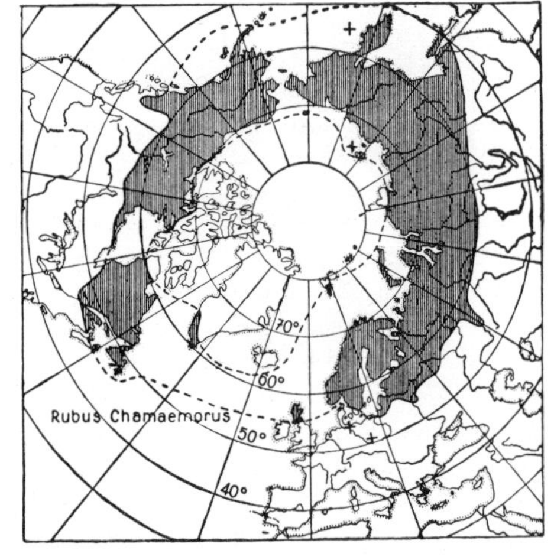

Fig. 752. Area of distribution of *Rubus chamaemorus* (Subarctic).

sometimes related phylogenetically to species or genera of the mountains of C. and E. Asia, e.g. *Rhododendron, Pedicularis, Androsace, Primula* and *Leontopodium*, but more commonly to the flora of the Mediterranean lowlands, e.g. *Sesleria, Crocus, Dianthus, Saponaria, Helianthemum, Globularia,* etc. It seems probable that mountain building in the Tertiary resulted in the enrichment of the European Alpine flora partly from a neighbouring, more thermophilous lowland and montane flora, partly by immigration from elsewhere, e.g. from the geologically older mountains of Asia. The great majority of species of the European mountains, and indeed many genera (e.g. *Soldanella*, Fig. 753) are European–Alpine, i.e. they do not occur outside Europe. Finally, a relatively small number of Arctic–Subarctic species entered C. Europe during the Pleistocene.

The Boreal (Coniferous Forest Belt) element. The centre of distribution of these species is in the great Eurosiberian and N. American Coniferous Forest Regions; further south they occur mainly in the coniferous upper montane forests; more rarely exclusively in the mountains. Many—the more continental—are largely confined to these areas, but others extend beyond their limits, or even tend to occur marginally. As with the other groups, according to whether their area of distribution extends to all the continents of the Holarctic or is more restricted, one can speak of a Circumpolar (or Amphiboreal), a Eurasiatic, a Eurosiberian or a European distribution.

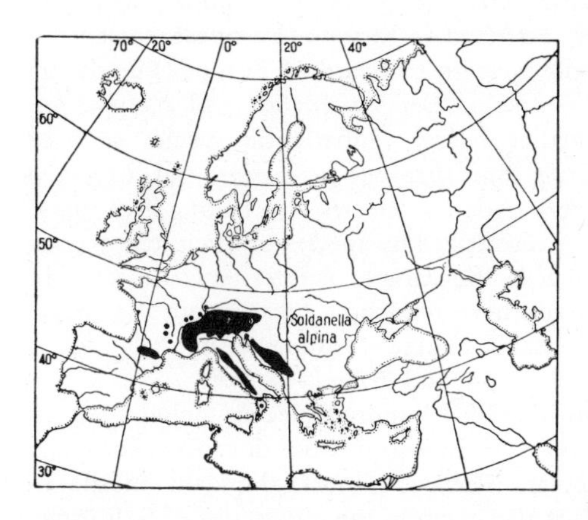

Fig. 753. Area of distribution of *Soldanella alpina* (European–Alpine).

The continental Boreal species include *Alnus incana* (Circumpolar), *Pinus sylvestris* (Eurasiatic, Fig. 638), *Pinus cembra* (Eurosiberian), *Picea abies* (including the closely related subsp. *obovata*, Eurasiatic; *sensu stricto*, European, Fig. 638), *Larix decidua* (European-montane), as well as *Vaccinium vitis-idaea, Linnaea borealis, Ledum palustre*, several species of *Pyrola*, etc. Further examples are *Betula pubescens* and *Polygonum bistorta* (Eurasiatic), *Vaccinium myrtillus* and *Geranium sylvaticum* (Eurosiberian), *Vaccinium uliginosum* and *V. oxycoccos* (Circumpolar–Boreal–Subarctic, Subalpine), *Trientalis europaea* (Circumpolar–Boreal), etc.

The Boreal element comprises relatively few species, but many of them (like the examples quoted) are dominants of coniferous forest, mountain grasslands and bogs.

The temperate (Deciduous Forest Belt) element. This includes numerous species found primarily in the Deciduous Forest Region south of the Coniferous Forest Belt and in the montane mixed forests. However, only a few occur throughout the Holarctic (e.g. *Oxalis acetosella*), and even then generally as different subspecies in different areas (e.g. *Taxus baccata, Anemone nemorosa*). In general, they occupy more restricted areas (e.g. only Europe), but are nevertheless often represented by closely related species in other areas, e.g. E. Asia and Atlantic N. America (e.g. *Hepatica*, Fig. 356); thus *Fagus sylvatica* (Fig. 687) is replaced in N. America by *F. grandifolia*, in E. Asia by *F. sinensis, F. sieboldii* and *F. japonica*. This, of course, reflects the origin of such genera from the much more uniform Arcto-Tertiary flora (p. 778).

This group of European deciduous forest species can be subdivided into the following intergrading subgroups according to the degree of oceanity of the climate:

1. Oceanic species. These are mostly heath, bog or aquatic plants restricted to the

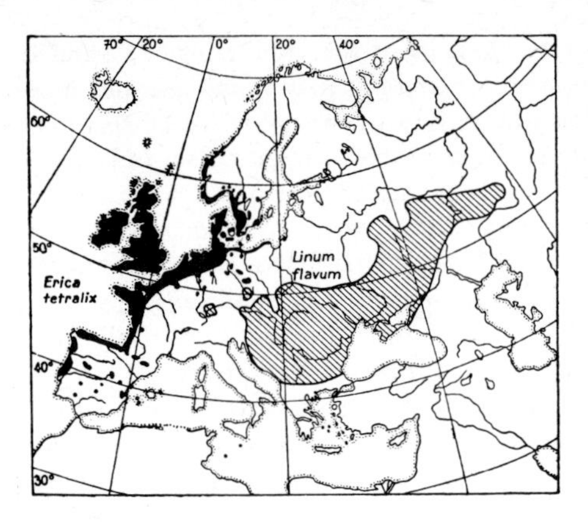

Fig. 754. Areas of distribution of *Erica tetralix* (European–Atlantic) and *Linum flavum* (European–Meridional steppe-plant).

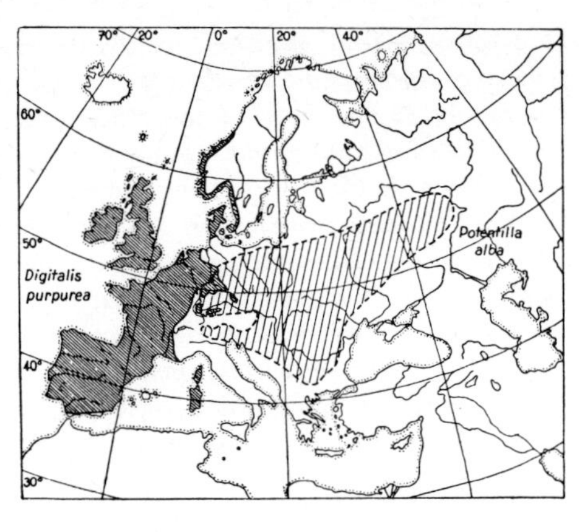

Fig. 755. Areas of distribution of *Digitalis purpurea* (European–Subatlantic) and *Potentilla alba* (Subcontinental deciduous forest plant).

coastal areas either of the entire Holarctic (Amphioceanic species) or of both sides of the Atlantic (Amphiatlantic species); or they are restricted to Europe (European–Atlantic). Besides species found virtually along the whole western seaboard of Europe (Eu-atlantic species), e.g. *Erica tetralix* (Fig. 754), *E. cinerea, Genista anglica* and *Narthecium ossifragum*, or which extend somewhat further inland (Subatlantic species such as *Digitalis purpurea*, Fig. 755, *Galium saxatile, Sarothamnus scoparius* and *Lonicera periclymenum*), the Boreal–Oceanic and Boreal–Atlantic species also belong here; these occur northwards to the Subarctic in regions characterized by species of birch, e.g. *Myrica gale, Cornus* (*Chamaepericlymenum*) *suecica* (Suboceanic), *Isoetes setacea*, etc. In contrast to these are Oceanic species with a southern tendency; these occur also in the Mediterranean area, especially in the mountains. These Mediterranean–Atlantic or –Subatlantic species include *Ilex aquifolium* (Fig. 744) and *Primula vulgaris*.

2. Central European species. These comprise several hundred species, including many characteristic of our deciduous woodlands; most are European ('South-central European species'), fewer are Eurasiatic or Circumpolar. They are most abundant in the C. European deciduous forests and grasslands, but also occur in the mountains of S. Europe, where indeed several genera are best represented (e.g. *Abies, Dentaria*). Besides species of relatively restricted distribution, e.g. *Abies alba* (Fig. 638), *Acer pseudoplatanus, Fagus sylvatica* (Fig. 687), *Quercus petraea, Colchicum autumnale* and *Viola reichenbachiana*, others are more widely distributed, e.g. *Carpinus betulus* (Fig. 689), *Fraxinus excelsior, Quercus robur* (Fig. 689), *Ulmus minor* (= *carpinifolia*) *Alnus glutinosa, Galium odoratum* (= *Asperula odorata*) and *Mercurialis perennis*. All these except the *Abies* and *Acer* are native in Britain. This group grades into the:

3. Subcontinental species, e.g. *Ulmus laevis, Acer tataricum, Euonymus verrucosus, Potentilla alba* (Fig. 755), *Melampyrum nemorosum* and *Digitalis grandiflora*, which are frequent in the warmer E. European mixed-oak forests; none of these named is native in Britain.

The **Mediterranean Region** comprises the sea-board and islands of the Mediterranean. Precipitation is often ample, falling in winter or spring, or further south only in the winter. The mild, relatively frost-free winters are suitable for evergreen trees and shrubs of many different families, whilst the hot, dry summers are responsible for their xeromorphy; they are woody sclerophylls, usually with small, rigid and dull, or even needle-shaped and hairy leaves. In virgin sclerophyll woodlands the holm oak (*Quercus ilex*) plays an important part, with the cork oak (*Q. suber*) in the west and the kermes oak (*Q. coccifera*) in the east, as well as the olive (*Olea europaea*) and the carob tree (*Ceratonia siliqua*), accompanied by various Mediterranean pines, e.g. *P. pinea, P. halepensis* and *P. pinaster* (= *P. maritima*; completely naturalized near Bournemouth). In this ancient centre of civilization the soil has mostly suffered serious erosion for thousands of years, and the original forests have been replaced almost entirely by sclerophyllous scrub or maquis, or by shrubby heaths (garigue, phrygana, etc.) composed of numerous evergreen shrubs (species of *Eriea* and *Cistus*, *Pistacia lentiscus*, *Juniperus oxycedrus*, the strawberry tree *Arbutus unedo*, the spiny shrubs *Ulex* and *Calicotome*, lianes such as *Smilax aspera*, etc., of which the natural habitats were in open woodlands, on shallow soils, in rocky terrain, etc.). This vegetation is rich in colourful and aromatic plants which give the landscape its essential character. Summer drought is particularly evident in the rocky heaths rich in species (e.g. many Labiatae), in the large numbers of annuals (therophytes) and geophytes (many Orchidaceae and Liliaceae, e.g. *Asphodelus*). Amongst cultivated plants the olive is particularly characteristic in the lowlands, and the sweet chestnut (*Castanea sativa*) in the mountains, where because of the higher precipitation the vegetation resembles that of the Eurosiberian Region; nevertheless, it still has characteristic features, e.g. forests of *Cedrus*, Mediterranean species of *Abies*, etc. Also characteristic are the grape, the fig and the various *Citrus* fruits which originated in S. and S.E. Asia.

The Mediterranean flora shows various affinities with that of the Subtropics and Tropics, from which it has evolved independently since the Miocene at the latest. The spatial separation of the various parts of the Mediterranean basin, its many mountainous areas and its escape from the full rigours of the Pleistocene glaciations have led to the development of many endemics, both conservative and progressive.

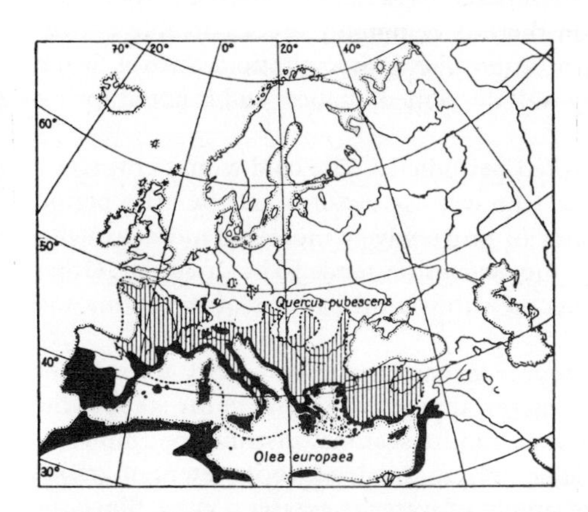

Fig. 756. Areas of distribution of *Olea europaea* (Mediterranean) and *Quercus pubescens* (Submediterranean).

Amongst the floristic elements of the Mediterranean, the characteristic evergreen species, e.g. *Quercus ilex* and *Olea europaea* (Fig. 756), do not extend very far north, although a few reach Ireland (e.g. *Arbutus unedo*) or S.W. England and Wales (e.g. *Rubia peregrina*), but many originally Mediterranean annuals and geophytes (e.g. species of *Allium*, *Ornithogalum*, *Muscari*, *Vicia* and *Crepis*) occur as weeds of cultivation or as ruderals in C. and N.W. Europe. In addition, many Submediterranean species with their headquarters in the hills on the north side of the Mediterranean (Mediterranean–Montane species) also occur in natural plant communities further north, especially in steppe-heaths and thermophilous mixed-oak forests. These include various deciduous trees and shrubs, e.g. *Quercus pubescens* (Fig. 756), *Acer monspessulanum*, *Ostrya carpinifolia*, *Fraxinus ornus*, *Colutea arborescens* and *Cornus mas* (none of these British, although, e.g. *Clematis vitalba* and *Viburnum lantana* have Submediterranean tendencies), and the Austrian or Corsican pine (*Pinus nigra*), as well as many Orchidaceae (many of them in Britain), such as *Himantoglossum hircinum*, *Orchis purpurea* and various species of *Ophrys*. A remarkable group of Mediterranean–Montane species of limestone in S.W. England includes *Helianthemum apenninum*, *Trinia glauca*, *Koeleria vallesiana* and *Lithospermum purpureocaeruleum*, whilst in the south and west of Britain *Tuberaria guttata* is confined to a few localities on acid soils. Mild winters are enjoyed both by Mediterranean and Mediterranean–Atlantic species, whilst Mediterranean–Pontic species such as *Salvia pratensis* and *Odontites lutea* link the E. Mediterranean with the S.E. European steppes.

In C. Europe the Submediterranean species occur mainly in the south-west, where the winters are less severe. Large-scale disjunctions, e.g. the isolated occurrence of *Globularia* and *Fumana* on the Baltic islands of Öland and Götland, indicate that some species formerly had a wider distribution, presumably during the Post-glacial climatic optimum.

Similar sclerophyllous but taxonomically unrelated woody plants are also characteristic of other subtropical areas on the west sides of continental landmasses where the winters are moist and the summers dry (Etesian regions: S.W. Australia, Cape of Good Hope, and the Californian and C. Chilean coasts).

The **Pontic–Central Asiatic Region** is a vast area of semi-desert and steppe with a dry continental climate, extending from the Pontic area on the north side of the Black Sea and the highlands of Asia Minor through C. Asia to W. China; despite great variations of altitude, the area is floristically very uniform.

Steppes are plant communities in which xeromorphic grasses, perennial herbs and dwarf sub-shrubs together with geophytes and annuals, form a more or less closed cover. The steppes of S. Russia include species of *Stipa, Festuca, Koeleria* and *Poa* and numerous xerophytic perennial herbs (e.g. *Artemisia, Centaurea, Salvia*); many of these also occur in the C. European steppe-heaths and a few, e.g. *Artemisia campestris* and *Silene otites*, even extend westwards to East Anglia. In the moister steppes the sward of both tussocky and rhizomatous grasses may reach over 1 m in height, but in drier areas scarcely reaches 1 dm. The root systems are extremely fine and copiously branched, completely exploiting the looser upper layers (generally some decimetres in depth) of the soil, which is blackened by the considerable amounts of mild humus (mull) produced and deposited each year. Below this dark horizon there is commonly a considerable enrichment of $CaCO_3$ at the limit of penetration of atmospheric precipitation. Such a 'black earth' (*Schwarzerde*, chernozem) is frequently well developed on loess and is noted for its fertility and suitability for cereals.

The vegetation of the steppes shows a marked periodicity. The cold winter enforces a period of dormancy. Spring and early summer (the wettest seasons) are the main period of growth, first of annuals and geophytes, later of progressively more and more drought-resistant plants. In high summer and autumn the vegetation tends to be dried out. Other steppe areas, e.g. the N. American prairies and the subtropical pampas of Argentina and Uruguay, are similar.

The conditions under which such grass steppes occur instead of forests are still not completely understood. The decisive climatic factor appears to be that in the steppes the precipitation, as a result of its moderate amount and its seasonal distribution, moistens only the upper layers of the soil and is available only to the dense root system of grasses and herbs, whereas trees require an adequate supply of water at greater depths. Since the depth to which a soil is moistened depends on its texture (e.g. fine-grained soils such as loess retain water in the upper layers), this also is an important factor. Grazing and fire also are important factors hostile to the growth of trees and furthering the development and maintenance of steppes, not only as a result of Man's activities but also in the natural state (e.g. grazing by large mammals, fire caused by lightning).

Forest-steppes occur at the margin of the forest regions (in Russia where the rainfall is between 400 and 500 mm); they are interspersed with woodlands of oak, elm, birch, etc., along the river valleys, in damp hollows and on more permeable soils (e.g. sands); many steppe plants occur in the undergrowth of the more open woodlands. Conversely, as the precipitation decreases the vegetation gradually becomes a semi-desert: the grass thins out and is replaced by progressively more widely spaced dwarf spiny shrubs and herbs (*Artemisia, Tanacetum,* etc.); correspondingly less humus is formed and the soils are brown (burosem), chestnut coloured (kastanosem) or grey (serosem) rather than black (chernosem), whilst saline soils dominated by succulent Chenopodiaceae become increasingly frequent. Only where the soil is again somewhat moister (e.g. in the mountain ravines of C. Asia) are giant herbs found, e.g. rhubarb and the medicinal umbellifers *Ferula* and *Dorema*.

The Pontic–C. Asiatic Region also is rich in endemics, and the enormous numbers of species in certain genera, e.g. *Astragalus, Allium, Salvia, Carduus* and *Cirsium*, are particularly characteristic. The flora is related phylogenetically to the mountain floras of C. Asia, the Near East and the Balkans, from which many species evolved during the progressive extension of the C. Asiatic steppes and semi-deserts in the Tertiary and Pleistocene. Most species thus have a Eurasiatic distribution, although many are confined to Europe.

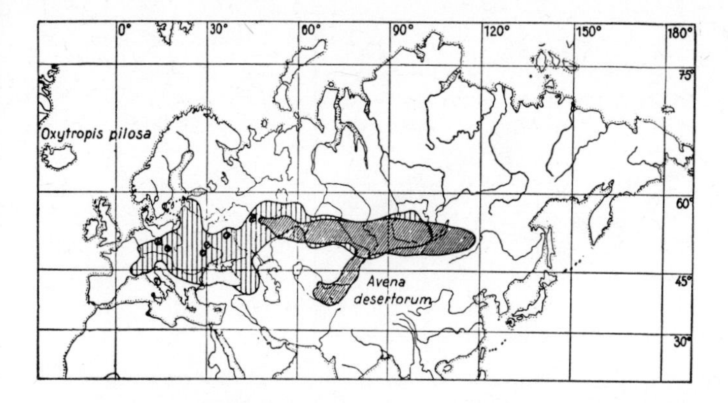

Fig. 757. Distribution of *Oxytropis pilosa* and *Helictotrichon* (= *Avena*) *desertorum* (species of the Pontic–C. Asiatic floristic element).

Some 400 of these Pontic–C. Asiatic (Meridional– and Temperate–Continental) species extend westwards to C. Europe. Those which occur in natural communities, e.g. steppe-heaths, probably entered in the Late-glacial and early Post-glacial (cf. p. 764), but many first invaded C. Europe in the wake of Man as weeds of cultivation or as ruderals. Quite a number of species found in steppe-heaths and steppe-heath woodlands have their headquarters in the forest-steppes, e.g. *Anemone patens, A. sylvestris, Allium strictum, Avena desertorum* (Fig. 757) and *Carex ericetorum* (which reaches the east and north of England). In Siberia these species often occur in intimate association with Continental, Alpine and Arctic–Alpine species such as *Aster alpinus* and *Anemone narcissiflora*. Furthermore, certain European ('Sarmatic') species such as *Gypsophila fastigiata, Dianthus arenarius* and *Astragalus arenarius* also favour the northern areas. Nevertheless, the majority inhabit the meadow-steppes and *Stipa*-steppes to the south, e.g. the Eurasiatic species *Stipa stenophylla, S. capillata, S. joannis, Festuca rupicola, F. valesciaca, Adonis vernalis* (Fig. 746), *Oxytropis pilosa* (Fig. 757), *Carex humilis* (which extends to S. England) and *Artemisia campestris* (found in East Anglia). A European group including *Iris pumila, I. aphylla, Linum flavum* (Fig. 754) and *Aster linosyris* (on limestone on the west coast of England) is often called 'Pontic' or 'Pontic–Pannonian'. Finally, a few species of the C. Asiatic desert steppes are scattered in C. Europe, e.g. *Ephedra distachya* and *Kochia arenaria*, whilst some C. Asiatic halophytes, such as *Artemisia maritima* agg. and *Salsola kali*, reach the sea-coasts of Europe, where they have a second centre of distribution.

Of the remaining Holarctic Regions only the **N. African–Indian Desert Region** will be considered here. The most characteristic feature of desert vegetation is the scanty plant cover; this is related to the enormous difficulty of maintaining a balanced water economy, since the rainfall is not only extremely low on average (generally less than 300 mm and often, as also in S.W. Africa, only 10–20 mm per year, and this with very high daily maximum temperatures) but it also tends to fall in short downpours so that the water quickly runs off the bare and often indurated surface of the soil, whilst in some years rainfall may be nil. These extreme conditions, aggravated by great extremes of temperature, atmospheric dryness, storms, and soil movements, permit the growth of only two main groups of plants: (1) annual or geophytic rain plants which remain dormant for the greater part of the time as seeds or underground organs, very rapidly sprouting, flowering and fruiting after rain, but commonly showing no other adaptations to drought, and (2) the 'true', extremely xeromorphic desert xerophytes, which despite the very small amounts of water available remain above ground throughout the year; these include succulents (e.g. Cactaceae in the less extreme deserts of N. America), but in general are dwarf, usually spiny, sometimes deciduous shrubs, sometimes with photosynthetic stems; any herbs and grasses present have high suction pressures and well-developed root systems which enable them to withdraw water from a large volume of relatively dry soil. In the arid desert climate salts produced by weathering are not leached out and washed away; indeed, they accumulate by evaporation at the surface so that the soils generally have

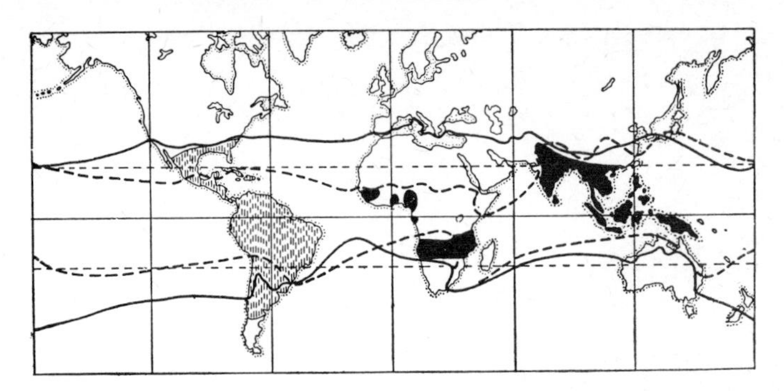

Fig. 758. Areas of distribution of Tropical–Subtropical families. Northern and southern limits of the Arecaceae (solid lines) and Zingiberaceae (broken lines), both Pantropical. Distribution of the Dipterocarpaceae (Palaeotropical), black, and Bromeliaceae (Neotropical), shaded. From Vester.

high osmotic suction pressures and the growth of vascular plants is often completely inhibited. However, certain Chenopodiaceae and Plumbaginaceae extend far into the salt deserts. The superficial accumulation of salts also hinders germination. In general, the ecology of desert plants is very diverse, since there are rocky, gravelly and sandy deserts, plateaux and desert valleys (wadis), etc. Sometimes roots reach an underground water-table, e.g. in the oases of the N. African–Indian Region, where the date palm (*Phoenix dactylifera*) is particularly characteristic.

The tropical floristic kingdoms comprise the Tropics and certain subtropical areas, namely the **(II) Palaeotropical Kingdom** of the Old World and the **(III) Neotropical Kingdom** of the New World; hence both are of enormous extent. A number of exclusively or mainly tropical families predominate in both, e.g. Arecaceae (palms), Araceae, Zingiberaceae, Lauraceae, Myrtaceae, Melastomataceae, Euphorbiaceae, Piperaceae, Moraceae, Asclepiadaceae, etc. (Pantropical elements). Palaeotropical elements include Pandanaceae and Dipterocarpaceae; Neotropical are Cactaceae, Bromeliaceae and Cannaceae, as well as the genera *Agave, Yucca, Fuchsia*, etc. (Fig. 758).

The most characteristic tropical formation is the species-rich Tropical Rain Forest (Fig. 759) developed in those hot tropical climates in which the mean annual temperature usually lies between 24° and 30°C, with only small variations, and in which the mean temperature of the coldest month is never less than 18°C, whilst the uniformly high precipitation reaches 2000–5000 mm per year or more, often falling in showers which may fall regularly at a particular time of the day; prolonged periods without rain never occur. These constantly high humidities and temperatures are responsible for the great luxuriance of the vegetation.

Tropical Rain Forest is extraordinarily rich in evergreen trees of the most diverse families (although conifers are usually absent); this diversity is evident even at a distance in the uneven height and chequered colour of the many-layered canopy. The trunks of the trees are generally tall and straight, but relatively slender, they have buttresses at the base, and the bark is often thin. The crowns are little-branched, with shining, leathery, often strikingly uniform, elliptical leaves which unfold from scarcely protected buds at all times of the year, since the growth rhythms usually present are not geared to any seasonal periodicity of climate. During sunny periods the crown of the tree is subject to considerable internal water deficits, but inside the forest the air remains nearly saturated with water vapour and the CO_2 concentration is higher than normal. The leaves of the herbaceous plants of the undergrowth (e.g. Araceae, Urticaceae, Zingiberales, Melastomataceae, Begoniaceae) are large and relatively soft; sometimes they are variegated or velvety. The light intensities in the forest are very variable, since the canopy is irregular and the leaves of the trees have a high reflectivity. The struggle for light is manifested especially in the wealth of lianes and epiphytes. Lianes include scrambling palms (e.g. *Calamus*), root-climbing Araceae, Piperaceae and species of *Ficus*, and many plants with twining stems or tendrils; these often weave the canopy into a continuous web. Amongst the epiphytes, bryophytes and lichens occur mainly on the surface of leaves, whereas humus- and water-retaining 'nest epiphytes' (e.g. the fern *Asplenium nidus*), many other

Fig. 759. Tropical Rain Forest with numerous palms (Arecaceae), evergreen broad-leaved trees, root climbing Araceae, lianes and epiphytes in central America (Atlantic coast of Costa Rica). (Photograph by E. Ehrendorfer.)

ferns, the Neotropical Bromeliaceae with their leaf 'reservoirs', water-storing Orchidaceae, various species of *Selaginella, Lycopodium* and so forth, are completely at home on the branches and trunks. (See pp. 194–5 for their adaptations). Flowers are rarely conspicuous in Rain Forests, partly because the flowering season is extended over the whole year. On the other hand, the occurrence of cauliflory is noteworthy in many species, i.e. the flowers are borne on the trunk and older branches (as, for example, in *Theobroma*, Fig. 713D).

Where the climate is cooler and misty, e.g. at higher altitudes in the tropical mountains, the forests are less luxuriant, but bryophytes and lichens become prominent as epiphytes and tree ferns are more frequent. Similar modifications occur, especially on the eastern sides of the continents, where Tropical Rain Forests grade into subtropical ones.

However, Rain Forest covers only a fraction of the Tropics. Where pronounced dry and wet seasons alternate during the year most of the trees lose their leaves at the beginning of the dry season, renewing growth at the onset of the rains. Such monsoon forests, leafy in the rainy season, form an ecological counterpart to our deciduous forests, green in summer. The forests dominated by teak (*Tectona grandis*, Verbenaceae, valued for its resistant timber) of the Indo-Malayan Region afford a good example. These forests are still rich in lianes and herbaceous epiphytes, but the influence of the dry season is seen not only in the deciduous foliage but also in the lesser height of the trees, the thicker bark, a greater degree of branching and the smaller leaves.

As the rainfall decreases further and the dry season becomes longer and more severe, the forests become even more open, the trees even smaller, the root systems more extensive, the bark even thicker and the leaves, whether evergreen or deciduous, strongly xeromorphic. The vegetation resembles an open park landscape with closed 'gallery forest' confined to the river valleys.

The undergrowth of these Tropical Dry Forests often consists of a thorny thicket which replaces the forest itself as the rainfall further decreases. These thickets include both evergreen and deciduous species; they are often rich in succulents. Eventually they become dwarfer and more open, grading into the semi-deserts and deserts.

In other types of Dry Forest the undergrowth consists not of thorny shrubs, but of grasses and herbs: such forests grade into tropical grasslands, generally composed of tall, tussock-forming, xeromorphic grasses which cease growth during the dry season. Trees and shrubs commonly occur singly or in groups between the grasses. These woody plants are also markedly xeromorphic; many, such as the African baobab (*Adansonia digitata*), have enormously thickened water-storing trunks; others are small 'umbrella trees' with crowns branching copiously in a horizontal plane, e.g. the well-known umbrella *Acacia* species; but this very striking growth-form is seen in a number of families. Tropical grasslands interspersed with trees, either solitary or in groups, are termed savannas. They are particularly widespread in Africa but are also important in S. America.

As well as climatic factors, others determine the occurrence of these tropical grasslands, especially edaphic factors; thus fine-textured soils highly retentive of water are particularly suitable for the dense root systems of tussock-forming grasses. The flooding of low ground and fire (caused naturally by lightning, or by Man) are also important. It follows that in an area with a uniform climate patches of forest and grassland may alternate; indeed, in the Monsoon Forest belt grassland may completely replace forest. In consequence, the following climatically determined vegetation belts lying between the Evergreen Tropical Rain Forests, on the one hand, and the semi-deserts and deserts, on the other, may be characterized either by the woody plants or the grasses which are present:

1. The Moist Savanna belt with deciduous monsoon forests and Savanna grasslands; the grasses are taller than a man and are interspersed with solitary thornless trees or evergreen gallery forests, e.g. the llanos of the Orinoco and the campos of Brazil.
2. The Dry Savanna belt with open deciduous dry forests (e.g. the miombo woodlands of Africa, rich in Leguminosae) and steppe-like Savanna grasslands, the grasses reaching breast-height.
3. The Thorn Savanna belt with thorn woodland and thorn scrub, in which the woody plants often have green, photosynthetic bark, with succulents (e.g. the caatingas of Brazil, rich in Mimosaceae and Cactaceae) and a knee-high sward of grasses, herbs and succulents.

Of the various tropical plant formations occurring in specialized habitats we shall mention only the Mangroves. These also extend into the Subtropics, but are richest in species in the Indo-Malayan Region. Mangroves consist of coastal, halophytic, evergreen trees and shrubs which colonize sheltered intertidal areas in bays, lagoons or estuaries, growing on saline silts or sandy muds alternately exposed and covered by the tides; they also occur on coral reefs. The species belong to various families, e.g. Rhizophoraceae (*Rhizophora, Bruguiera, Ceriops*), Verbenaceae (*Avicennia*), Sonneratiaceae (*Sonneratia*), etc., but they show many similar adaptations (cf. p. 179 and Fig. 699A–C), namely organs which stabilize the trees in the mud (especially stilt roots), pneumatophores which allow aerobic respiration in the oxygen-deficient soil, and the viviparous seedlings which germinate on and hang down from the parent plant.

Nowadays much of the Tropics is cultivated. Amongst important cultivated plants the following are palaeotropical (or subtropical) in origin: rice, sugar cane, tea, coffee, pepper, cinnamon, ginger, coconut, banana and the *Citrus* fruits. Originally neotropical: maize, potato (montane), groundnut, cocoa, paprika, tomato, pumpkin, quinine, agave, etc. The ancestors of the cotton-plant and the various rubber trees originate from both the Old and New Worlds. All these plants are now cultivated far beyond their native areas.

The remaining floristic kingdoms are much smaller. The **(IV) Australian Kingdom** has mild winters, and the natural vegetation largely consists of sclerophyll forests and woodlands, park-like savannas, dry scrubs and steppe and desert communities (but with hardly any succulents!); all are rich in xerophytes. Subtropical Rain Forests are confined to the north and east coasts. The number of endemics is remarkable—some 8000 out of about 10,000 species. Important genera include *Eucalyptus* (Myrtaceae, with hundreds of evergreen species, ranging from the world's tallest trees to dwarf shrubs) and *Melaleuca* (also Myrtaceae); *Grevillea, Hakea, Banksia* and other Proteaceae; numerous phyllodic species of *Acacia*; most of the species of *Casuarina*; the 'grass trees' of the genus *Xanthorrhoea* (Liliaceae), etc.

The **(V) South African Kingdom** (Cape Region), the smallest of all in area, has a warm temperature climate, dry in summer, wet in winter. Trees are almost completely absent; instead the ground is covered by evergreen, small-leaved, sclerophyllous shrubs and heaths with a large number of geophytes and annuals. The flora is remarkably rich in species, including many Proteaceae and over 450 species of *Erica*; many Liliiflorae, species of

Pelargonium, and especially in the drier marginal areas the succulent species of *Mesembryanthemum* and *Euphorbia* also play an important part.

The **(VI) Antarctic Kingdom** includes southernmost S. America, with its perpetually moist montane forests rich in bryophytes and pteridophytes and dominated by the Antarctic beech *Nothofagus*; both deciduous and evergreen species are present. The kingdom also includes the treeless islands of the Antarctic with their peculiar cushion-plants (e.g. *Azorella*, Apiaceae, Fig. 204A). Sometimes the New Zealand Region is included in this kingdom, which probably represents (as shown by the striking disjunctions shown at the present day) the last remnants of a formerly much richer circumpolar flora which extended further to the south and of which the last traces are discernible in Australia, New Guinea, New Zealand and Tasmania. (*Nothofagus* has been found in Tertiary strata on Seymour Island (64°S, 56°W) in the W. Antarctic.)

Throughout the world mountains profoundly modify the flora and vegetation (cf. Fig. 749). In many respects the sequence of altitudinal zones resembles the sequence of latitudinal zones of vegetation. Where mountains are surrounded by forests these extend upwards only to a certain height. Where mountains rise from a treeless desert, the montane forest zone (generally at a considerable altitude) has lower as well as upper limits. In general, a mountain flora is very different from that of the lowlands, not only because of the peculiar environmental conditions but also because its geographical isolation has led to the independent evolution of the flora; it is often rich in endemics, the survival of which is favoured by the great diversity of habitats at different altitudes.

Appendix

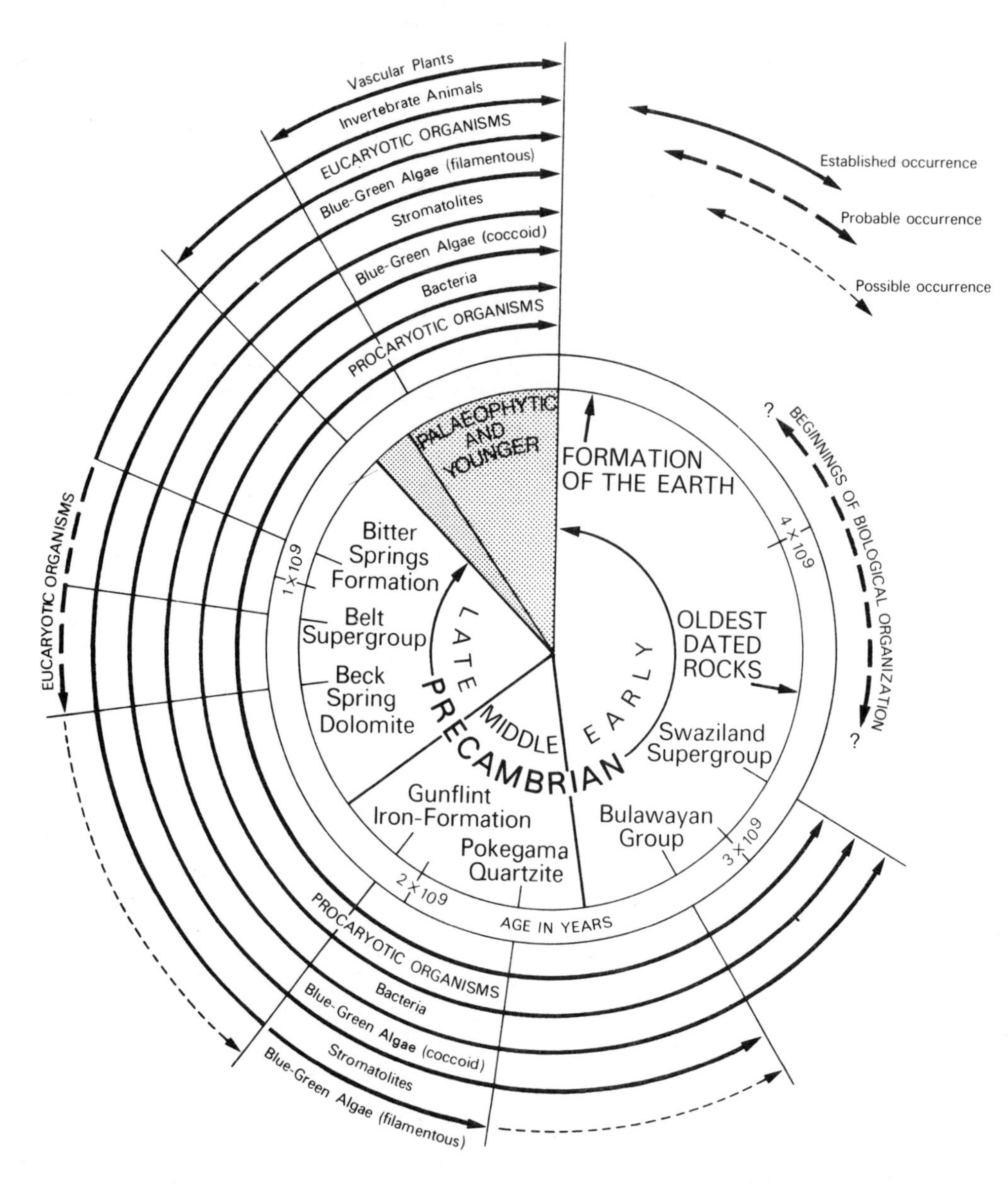

Fig. A1 The fossil record of plants in relation to the age of the Earth. The shaded area represents the Palaeozoic era and younger (see Fig. A2). Note the shortness of the post-Cambrian in relation to the pre-Cambrian history of life. After Schopf (*Biol. Revs.* **45**, 346, 1970), modified.

Era	Period (with age of beginning in years)		
Cainozoic	Quaternary $2 \cdot 5 \times 10^6$	Holocene	Cainophytic
		Pleistocene	
	Tertiary 65×10^6	Pliocene	
		Miocene	
		Oligocene	
		Eocene	
		Palaeocene	
Mesozoic	Cretaceous 136×10^6	Senonian	
		Turonian	
		Cenomanian	
		Albian	Mesophytic
		Aptian	
		Neocomian	
	Jurassic 190×10^6	Upper	
		Middle (Oolite)	
		Lower (Lias)	
	Triassic 225×10^6	Keuper	
		Muschelkalk	
		Bunter	
Palaeozoic	Permian 280×10^6	Zechstein	
		Rotliegendes	Palaeophytic
	Carboniferous 345×10^6	Upper	
		Lower	
	Devonian 395×10^6	Upper	
		Middle	
		Lower	
	Silurian 430×10^6		
	Ordovician 500×10^6		Eophytic
	Cambrian 570×10^6		
Pre-Cambrian	$> 4\,000 \times 10^6$		

Fig. A2 The Geological Table showing the fossil-bearing strata in more detail, the ages of the periods according to McAlester (*The History of Life*, Prentice-Hall Inc., Englewood Cliffs, N.J., 1968). The Eophytic era is considered as beginning about 1800 x 10^6 years ago (Gunflint formation. See Fig. A1).

References

The works listed here cover in greater detail the topics considered, albeit briefly, in the text, and in them will be found further references to the specialized literature, much of it in the form of original papers in the scientific journals. Articles summarizing recent research in the various branches of botanical science appear in the following reviewing periodicals (among others): *Advances in Botanical Research, Advances in Ecological Research, Advances in Genetics, Annual Review of Biochemistry, Annual Review of Plant Physiology, Biological Reviews, Botanical Review, Fortschritte der Botanik, Vistas in Botany*. Collections of papers on particularly important and rapidly developing aspects of biology appear annually in the *Cold Spring Harbor Symposia on Quantitative Biology* and the *Symposia of the Society for Experimental Biology*.

For the sake of brevity, the following abbreviations are used here for the titles of periodicals and handbooks (the abbreviations given in parentheses are those in the *World List of Scientific Periodicals*, 3rd edn., 1952): AB = *Annals of Botany* (*Ann. Bot., Lond.*); BB = *Bibliotheca botanica* (*Bibl. bot., Stuttgart*); BBC = *Beihefte zum Botanischen Zentralblatt* (*Beih. bot. Zbl.*); BDBG = *Bericht der Deutschen Botanischen Gesellschaft* (*Ber. dtsch. bot. Ges.*); BJSy = *Botanische Jahrbücher für Systematik, Pflanzengeschichte und Pflanzengeographie* (*Bot. Jb.*); EB = *Ergebnisse der Biologie* (*Ergebn. Biol.*); FdB = *Fortschritte der Botanik*; HA = *Handbuch der Pflanzenanatomie*, Berlin; HN = *Handwörterbuch d. Naturwiss.*, 2. Aufl. (ed. R. Dittler *et al.*), Jena (1931–35); HV = *Handbuch der Vererbungswissenschaft* (*Handb. Vererbungsw.*); HP = *Handbuch der Pflanzenphysiologie*, Berlin, Göttingen, Heidelberg; JWB = *Jahrbuch für wissenschaftliche Botanik* (*Jb. wiss. Bot.*); NW = *Naturwissenschaften* (*Naturwissenschaften*); PL = *Planta* (*Planta*); PRB = *Progressus rei botanicae* (*Progr. Rei bot.*); ZB = *Zeitschrift für Botanik* (*Z. Bot.*).

Introduction

(1) The living state. M. HARTMANN, *Allgemeine Biologie*, 4 Aufl., Stuttgart, 1953. E. SCHRÖDINGER, *What is Life?* Cambridge, 1948. H. STAUDINGER, *Makromolekulare Chemie und Biologie*, Basel, 1947. W. TROLL, *Das Virusproblem in ontologischer Sicht*, Wiesbaden, 1951. H. F. FREKSA, Die stammgeschichtliche Stellung der Virusarten u. d. Probl. d. Urzeugung, in G. HEBERER, *Die Evolution der Organismen*, 2. Aufl., Stuttgart, 1954. W. WEIDEL, *Virus*, Berl., Götting, Heidelb., 1957. W. M. STANLEY & E. G. VALENS, *Virus and the Nature of Life*, London, 1962. W. WEIDEL, *Virus und Molekularbiologie*, 2. Aufl., Berl., Götting., Heidelb., 1964. G. NASS, *The Molecules of Life*, London, 1970.

(2) Origin of life. A. J. OPARIN, *The Origin of Life on the Earth*, 3rd edn., Edinburgh, 1957. H. LINSER, Das lebende System, *Wissenschaft und Weltbild* **7**, 1954. E. IHNE, Dr HOFFMANN, *Ber. Oberhess. Ges. Nat. Heilk.*, 1893. S. L. MILLER, A production of amino acids under possible primitive earth conditions, *Science* **117**, 528, 1953. C. PONNAMPERUMA, *The Origins of Life*, London, 1972. L. E. ORGEL, *The Origins of Life*, London, 1973. S. R. MILLER & L. E. ORGEL, *The Origins of Life on the Earth*, London, 1974.

Cytology

(3) General works. L. GEITLER, *Grundriss d. Cytologie*, Berlin, 1934. L. W. SHARP, *Fundamentals of Cytology*, New York, London, 1943. E. KÜSTER, *Die Pflanzenzelle*, 3. Aufl., Stuttgart, 1957. A. FREY-WYSSLING, *Macromolecules in Cell Structure*, Cambridge, Mass., 1957. C. P. SWANSON, *Cytology and Cytogenetics*, London, 1958. C. D. DARLINGTON, *The Evolution of Genetic Systems*, 2nd edn., Edinburgh, 1958. J. BRACHET & A. E. MIRSKY, *The Cells*, vols. 1–6, New York, 1959–64. L. PICKEN, *The Organization of Cells and other Organisms*, Oxford, 1960. M. J. D. WHITE, *The Chromosomes*, 5th edn., London, 1961. G. B. WILSON & I. H. MORRISON, *Cytology*, New York, 1961. E. GRUNDMANN, *Allgemeine Cytologie*, Stuttgart, 1964. P. E. PILET, *La Cellule, Structure et Fonctions*, Paris, 1964. J. BONNER & J. E. VARNER, *Plant Biochemistry*, New York & London, 1965. H. METZNER (ed.), *Die Zelle, Struktur und Funktion*, Stuttgart, 1966. P. SITTE (ed.), *Probleme der Biologischen Reduplikation*, Berl., Heidelb., New York, 1966. H. J. BOGEN, *Knaurs Buch der modernen Biologie*, München, 1967. H. BIELKA (ed.), *Molekulare Biologie der Zelle*, Stuttgart, 1969. R. BUVAT, *Die Organisation der Lebendigen*, München, 1969. A. LIMA-DE-FARIA (ed.), *Handbook of Molecular Cytology*, Amsterdam, 1969. A. FREY-WYSSLING, Comparative organography of the cytoplasm, *Protoplasmatologia,* 3G, Wien, 1973.

(4) Significance of cellular organization. L. GEITLER, Normale u. pathol. Anatomie d. Zelle, HP 1, 123, 1955. S. STRUGGER, Das lebende Stoffsystem Protoplasma u. d. Zelle, in *Vom Unbelebten zum Lebendigen, eine Ringvorlesung d. Univ. Münster*, Stuttgart, 1956.

(5) Structure of a typical plant cell. E. STRASBURGER, *Botanisches Praktikum*, 7. Aufl., Jena, 1923. H. SCHNEIDER & ZIMMERMANN, *Bot. Mikrotechnik*, 2. Aufl., Jena, 1922. H. LUNDEGÅRDH, Zelle u. Cytoplasma, HA **1**, 1922. C. R. BIEBL & H. GERM, *Praktikum der Pflanzenanatomie*, Wien, 1950. E. KÜSTER, *Die Pflanzenzelle*, 3. Aufl., Jena, 1956. P. SITTE, *Bau und Funktion der Pflanzenzelle*, Stuttgart, 1965. A. FREY-WYSSLING & K. MÜHLETHALER, *Ultrastructural Plant Cytology*, Amst., London, New York, 1965. F. A. L. CLOWES & B. E. JUNIPER, *Plant Cells*, Oxford, 1968. J. B. PRIDHAM (ed.), *Plant Cell Organelles*, London, 1968. A. S. SPIRIN & L. P. GAVRILOVA, *The Ribosome*, Berl., Heidelb., New York, 1969. R. BUVAT, *Plant Cells*, London, 1970. A. W. ROBARDS, *Electron Microscopy and Plant Ultrastructure*, London, 1970. A. W. ROBARDS, *Dynamic Aspects of Plant Ultrastructure*, London, 1974.

(6) Cells and energids. *See* (4), L. GEITLER.

(7) Cytoplasm. K. PAECH, *Chemie u. Biochemie sek. Pflanzenstoffe*, Berl., Götting., Heidelb., 1950. S. G. WILDMAN, M. COHEN, W. SEIFRITZ & K. STEFFEN, Das Cytoplasma, HP **1**, 243–409, 1955. H. J. BOGEN, Die Koordination der Reaktionssysteme, HP **2**, 607, 1956; Fortschr. i. d. Erforschung d. pflanz. Zelle, *Nat. Rdsch.* **9**, 1956. N. KAMYA, *Protoplasmic Streaming*, Vienna, 1959 (*Protoplasmatologia* **8**).

(8) Submicroscopic structure of cytoplasm. A. FREY-WYSSLING, *Sub-microscopic Morphology of Protoplasm*, 2nd edn., Amsterdam, 1953 (also as *Protoplasmatologia* **2**). A. ENGSTROM & J. B. FINEAN, *Biological Untrastructure*, New York, 1958. R. BUVAT, Electron microscopy of plant protoplasm, *Int. Rev. Cytol.* **14**, 41, 1963. K. E. WOHLFARTH-BOTTERMANN, Grundelemente der Zellstruktur, NW **50**, 237, 1963. K. E. WOHLFARTH-BOTTERMANN & H. KOMNICK, Morphologie des Cytoplasmas, *Fort. Zool.* **17**, 1, 1965. J. L. KAVENAN, *Structure and Function in Biological Membranes*, San Francisco, London, Amst., 1965. A. LIMA-DE-FARIA (ed.), *Handbook of Molecular Cytology*, Amsterdam, 1969. D. BRANTON, Membrane structure, *A. Rev. Plant Phys.* **20**, 209–38, 1969. D. F. H. WALLACH, The plasma membrane, *Heidlb. Science Lib.* **18**, 1972. D. BRUNTON & D. W. DEAMER, Membrane Structure, *Protoplasmatologia* 2E1, Wien, 1972. M. S. BRETSCHER, Membrane structure, *Science* **181**, 622, 1973. H. H.

MOLLENHAUER & D. J. MORRE, Golgi apparatus and plant secretion, *A. Rev. Plant Phys.* **17**, 27, 1966. E. H. NEWCOMB, Plant microtubules, *A. Rev. Plant Phys.* **20**, 253, 1969. M. V. PARTHASARATHY & K. MÜHLETHALER, Cytoplasmic microfilaments in plant cells, *J. Ultrastruct. Res.* **38**, 46, 1972.

(9) Vacuome and cell sap. A. PISEK, H. DRAWERT & P. J. KRAMER, Zellsaft und Vacuolen, HP **1**, 614–67, 1955. P. DANGEARD, Le Vacuome de la Cellule Végétale, *Protoplasmatologia* **3**, Wien, 1956. W. KARRER, *Konstitution und Vorkommen der organischen Pflanzenstoffe*, Basel, 1958. D. HESS, Blütenfarbstoff als Modelle für die Wirkungsweise von Genen, *Umschau* **64**, 758, 1964. W. L. KRETOWITSCH, *Grundzüge der Biochemie der Pflanzen*, Jena, 1965. PH. MATILE & H. MOOR, Vacuolation, PL **80**, 159, 1968.

(10) Nucleus and cell division. G. TISCHLER, Allgemeine Pflanzenkaryologie, HA **2**, 1934, 1951, 2. Aufl. P. F. MILOVIDOW, Physik u. Chemie d. Zellk., *Protoplasma-Monogr., Berl.* **20**, 1949. L. GEITLER, *Schnellmethoden d. Kern- u. Chromosomen-unters.*, 3. Aufl., Berlin, 1949. T. O. CASPERSON, *Cell Growth and Cell Function*, New York, 1950, J. A. SERRA, Der Zellkern, HP **1**, 413–99, 1955. F. OEHLKERS, *Das Leben d. Gewächse*, Bd **1**, s. 14 u. 174, Berl., Götting., Heidelb., 1956. J. HÄMMERLING, Nucleo-cytoplasmic interactions in *Acetabularia* and other cells, *A. Rev. Pl. Physiol.* **14**, 65, 1963. G. TISCHLER, *Die Chromosomenzahlen der Gefässpflanzen Mitteleuropas*, S'Gravenhage, 1950. C. D. DARLINGTON and L. F. LaCOUR, *The Handling of Chromosomes*, 3rd edn., London, 1960. A. BARTHELMESS, Grundlagen der Vererbung, *Hdb. Biol.* **3**, 437, 1963–65. A. KÜHN, *Grundriss der Vererbungslehre*, 4. Aufl., Heidelberg, 1965. K. ESSER & R. KUENEN, *Genetics of Fungi*, Berlin, 1967. R. RIEGER, A. MICHAELIS & M. M. GREEN, *A Glossary of Genetics and Cytogenetics*, 3. Aufl., Berl., Heidelb., New York, 1968. H. BUSCH & K. SMETANA, *The Nucleolus*, New York, 1970. H. BUSCH (ed.), *The Cell Nucleus*, New York, 1974.

(11) Chromosome structure. J. STRAUB, ZB **33**, 65, 1938; NW **31**, 97, 1943. L. GEITLER, Chromosomenbau, *Protoplasma-Monogr.*, Berl. **14**, 1938. F. OEHLKERS, *see* **(10)**. E. HEITZ, Struktur d. Chromosomen u. Chloroplasten, *Nova Acta Leopoldina* NF **17**, 517, 1955. H. MARQUARDT, *Natürl. u. kunstl. Erbänderungen*, Heidelberg, 1957. C. D. DARLINGTON, *Chromosome Botany and the Origins of Cultivated Plants*, London, 1963. E. HARBERS, G. F. DOMAGK & W. MÜLLER, *Die Nukleinsäuren*, Stuttgart, 1964. W. NAGL, Puffing of polytene chromosomes in a plant, NW **56**, 221, 1969. H. G. CALLAN, Replication of DNA in the chromosomes of eukaryotes, *Proc. R. Soc.* B **181**, 19, 1972.

(12) Polyploidy. L. GEITLER, Endomitose u. endomitotische Polyploïdisierung, *Protoplasmatologia* **6**, Wien, 1953. H. MARQUARDT, *see* **(11)**. G. L. STEBBINS, *Variation and Evolution in Plants*, London, 1950; *Chromosomal Evolution in Higher Plants*, London, 1971. C. D. DARLINGTON, *see* **(11)**.

(13) Plastids. P. N. SCHÜRHOFF, Plastiden, HA **1**, 1924. E. HEITZ, PL **26**, 134, 1937. S. GRANICK, Plastid structure, development and inheritance, HP **1**, 507, 1955. E. J. RABINOWITSCH, The chloroplasts and chromoplasts, in *Photosynthesis*, vol. **1**, 355–437, 1945; **2**, 1979–93, 1956. S. STRUGGER & E. PERNER, Beob. z. Frage d. ontogenetischen Entw. d. somatischen Chloroplasten, *Protoplasma* **46**, 711, 1956. K. MÜHLETHALER & A. FREY-WYSSLING, Entwicklung und Struktur der Proplastiden, *J. Biophys. Biochem. Cytol.* **6**, 507, 1959. W. MENKE, Structure and chemistry of plastids, *A. Rev. Pl. Phys.* **13**, 27, 1962. S. GRANICK, The Plastids, in M. LOCKE (ed.), *Cytodifferentiation and Macromolecular Synthesis*, 144, New York & London, 1963. W. MENKE, Feinbau und Entwicklung der Plastiden, BDBG **77**, 340, 1964. K. MÜHLETHALER, H. MOOR & I. W. STARKOWSKI, The ultrastructure of chloroplast lamellae, PL **67**, 305, 1965. D. BRANTON & R. PARK, Subunits in chloroplast lamellae, *J. Ultrastr. Res.* **19**, 283, 1966. W. KREUTZ, Röntgenographische Strukturuntersuchungen in der Photosyntheseforschung, *Umschau* **66**, 806, 1966. I. T. O. KIRK & R. A. E. TILNEY-BASSET, *The Plastids*, San Francisco, 1967. H. MOHR, *Lehrbuch der Pflanzenphysiologie*, Berl., Heidelb., New York, 1969. A. ALLSOPP, Phylogenetic relationships of the Procaryota and the origin of the eucaryotic cell, *New Phytol.* **68**, 591, 1969. P. R. BELL, Are plastids autonomous? *Symp. Soc. exp. Biol.* **24**, 109, 1970.

(14) Mitochondria, etc. K. STEFFEN, Chondriosomen und Mikrosomen (Sphärosomen), HP **1**, 574, 1955. A. L. LEHNINGER, *The Mitochondrion*, New York & Amst., 1965. D. B. ROODYN & D. WILKIE, *The Biogenesis of Mitochondria*, London, 1968.

(15) Flagella and cilia. I. MANTON, Plant cilia and associated organelles, in D. RUDNICK (ed.), *Cellular Differentiation and Growth*, Princeton, 1956. P. R. BELL, Microtubules in relation to flagellogenesis in *Pteridium* spermatozoids, *J. Cell Sci.* **15**, 99, 1974.

(16) Cell wall. W. WERGIN, BDBG **63**, 12, 1950. I. W. BAILEY, *Contributions to Plant Anatomy*, Waltham, 1954. E. TREIBER & R. D. PRESTON, Die Zellwand, HP **1**, 668–751, 1955. E. TREIBER, *Die Chemie der Pflanzenzellwand*, Berlin, 1957. A. FREY-WYSSLING, *Die pflanzliche Zellwand*, Berlin, 1959. P. A. ROELOFSEN, *The Plant Cell Wall*, HA, 2. Aufl., 3, 4, Berlin, 1959. K. MÜHLETHALER, Ultrastructure and formation of plant cell walls, *A. Rev. Plant Phys.* **18**, 1, 1967. D. H. NORTHCOTE, Chemistry of the plant cell wall, *A. Rev. Plant Phys.* **23**, 113, 1972. R. D. PRESTON, *The Physical Biology of Plant Cell Walls*, London, 1974.

The grades of morphological organization

(17) Protophyta. F. E. FRITSCH, *The Structure and Reproduction of the Algae*, Cambridge, 1945, 1948. B. SCHUSSNIG, *Handbuch d. Protophytenkunde* **1** & **2**, Jena, 1953, 1960. A. W. HAUPT, *Plant Morphology*, New York, Toronto, London, 1953. P. R. BELL & C. L. F. WOODCOCK, *The Diversity of Green Plants*, 2nd edn., London, 1971.

(18) Thallophytes. F. OLTMANNS, *Morphologie u. Biologie d. Algen*, 2. Aufl., Jena, 1922, 1923. F. E. FRITSCH, *see* **(17)**. G. M. SMITH, *Manual of Phycology*, Waltham, 1951. B. FOTT, *Algenkunde*, Jena, 1959. E. BÜNNING, Polarität u. inaequale Teilung d. Pflanzlichen Protoplasten, *Protoplasmatologia* **8**, Wien, 1957.

(19) Thallus and cormus. W. TROLL, *Allg. Botanik*, Stuttgart, 1954. O. STOCKER, *Grundriss d. Botanik*, Berl., Götting., Heidelb., 1952; Wasseraufnahme u. Wasserspeicherung b. Thallophyten, HP **3**, 160, 1956; Die Transpiration d. Thallophyten, HP **3**, 313, 1956; Die Wasserleitung d. Thallophyten, HP **3**, 514, 1956. K. GOEBEL, *Organographie* **2**, 3. Aufl., Jena, 1930. W. LORCH, Anatomie der Laubmoose, HA **7**, 1932. TH. HERZOG, Anatomie der Lebermoose, HA **7**, 1925. B. HUBER, Wasserleitung, HP **3**, 511, 1956; Gefässleitung, HP **3**, 541, 1956.

(20) Telome theory. W. ZIMMERMANN, *Geschichte der Pflanzen*, 2 Aufl., Stuttgart, 1969; *Phylogenie der Pflanzen*, 2. Aufl., Stuttgart, 1959. K. R. SPORNE, *The Morphology of Pteridophytes*, London, 1962.

Histology

(21) General works. G. HABERLANDT, *Physiologische Pflanzenanatomie*, 6. Aufl., Leipzig, 1924. W. ROTHERT & L. JOST, Gewebe d. Pflanzen, HN **5**, 1–139, 1934. A. S. FOSTER, *Practical Plant Anatomy*, New York, Toronto, London, 1950. C. R. BIEBL & H. GERM, *see* **(5)**. A. J. EAMES & L. H. MacDANIELS, *An Introduction to Plant Anatomy*, 2nd edn., New York, London, 1951. R. A. POPHAM, *Developmental Plant Anatomy*, Columbus, 1952. K. ESAU, *Plant Anatomy*, New York, London, 1952; *Anatomy of Seed Plants*, New York, 1960. I. W. BAILEY, *see* **(16)**. H. von GUTTENBERG, *Lehrb. d. Allg. Bot.* 4. Aufl., 1955. B. KAUSSMANN, *Pflanzenanatomie*, Jena, 1963. W. BRAUNE, A. LEHMANN & H. TAUBERT, *Pflanzenanatomisches Praktikum*, Jena, 1967. A. FAHN, *Plant Anatomy*, 2nd end., London, 1974.

(22) Meristems. O. SCHUEPP, Meristeme, HA **4**, 1926; *Meristeme, Experientia Suppl.* 11, Basel, 1966. J. HELM, Untersuchungen über die Differenzierung des Sprossscheitelmeristems, PL **15**, 105, 1932. R. KAPLAN, Über die Bildung der Stele aus dem Urmeristem, PL **27**, 224, 1934. E. M. GIFFORD, The shoot apex in angiosperms, *Bot. Rev.* **20**, 8, 1954. H. von GUTTENBERG, Grundzüge der Histogenese höherer Pflanzen, HA **8**, 1960, 1961. F. A. L. CLOWES, *Apical Meristems*, Oxford, 1961. K. ESAU, *see* **(21)**. C. W. WARDLAW, *Morphogenesis in Plants*, 2nd edn., London, 1968.

(23) Meristemoids. E. BÜNNING, *Entwicklungs- u. Bewegungsphysiologie d. Pflanzen*, 3. Aufl., Berl., Götting., Heidelb., 1953; Morphogenesis in plants, *Surv. Biol. Progr.* **2**, 1952; *see also* **(18)**.

(24) Permanent tissues. F. J. MEYER, Assimilationsgewebe, HA **4**, 1923. H. PFEIFFER, Trennungsgewebe, HA **5**, 1928. K. LINSBAUER, Epidermis, HA **4**, 1930. F. NETOLITZKY, Pflanzenhaare, HA **4**, 1932; Speichergewebe, HA **4**, 1935. A. SPERLICH, Exkretionsgewebe, HA **4**, 1939. F. TOBLER, Die mechanischen Elemente, HA **4**, 1939. H. von GUTTENBERG, Physiol. Scheiden, HA **5**, 1943. B. HUBER, Gefässleitung, HP **3**, 541, 1956; *Die Saftströme der Pflanzen*, Berl., Götting., Heidelb., 1956; *Grundzüge der Pflanzenanatomie*, Berl., Götting., Heidelb., 1961. K. ESAU, The Phloem, 2. Aufl., HA **5**, 1969. W. R. PHILIPSON, J. M. WARD & B. G. BUTTERFIELD, *The Vascular Cambium*, London, 1971. B. A. MEYLAN & B. G. BUTTERFIELD, *Three-Dimensional Structure of Wood*, London, 1972.

Morphology and anatomy of the plant body

(25) General works. J. W. von GOETHE, *Versuch die Metamorphose der Pflanzen zu erklären*, 1790. K. GOEBEL, *Organographie*, 3. Aufl., Jena, 1928. W. TROLL, *Vergleichende Morphologie der höheren Pflanzen*, Berlin, 1935–43; *Praktische Einführung i. d. Pflanzenmorphologie*, **1**, 1954 & **2**, 1957, Jena. W. RAUH, *Morphologie der Nutzpflanzen*, 2. Aufl., Heidelberg, 1950. L. GEITLER, *Die Morphologie der Pflanzen*, Berlin, 1945. E. W. SINNOTT, *Plant Morphogenesis*, New York, 1960. C. R. METCALFE & L. CHALK, *Anatomy of the Dicotyledons*, Oxford, 1950; *Anatomy of the Monocotyledons*, Oxford, 1960–61. H. C. BOLD, *Morphology of Plants*, 3rd edn., New York, 1973. D. W. BIERHORST, *Morphology of Vascular Plants*, New York, 1971.

(26) Leaf arrangement and branching of the shoot. G. von ITERSON, *Mathematische und mikroskopischanatomische Untersuchungen über Blattstellungen*, Jena, 1907. M. HIRMER, Zur Kenntnis der Schraubenstellungen im Pflanzenreich, PL **14**, 132, 1931; Neue Untersuchungen auf dem Gebiet der Organstellungen, BDBG **52**, (26), 1934. B. HACCIUS, Untersuchungen über die Deutung der Distichie, *Bot. Arch.* **40**, 58, 1939. W. TROLL, *see* **(25)**. C. W. WARDLAW, *Morphogenesis in Plants*, London, 1952; *Phylogeny and Morphogenesis*, London, 1952. F. O. BOWER, *Size and Form in Plants*, London, 1930. R. SNOW, Problems of phyllotaxis and leaf determination, *Endeavour* **14**, 99–107, 1955. W. TROLL, *Die Infloreszenzen*, Jena, 1964. W. ZIMMERMANN, Die Blutenstände, ihr System und ihre Phylogenie, BDBG **78**, 3, 1965. K. J. DORMER, *Shoot Organization in Vascular Plants*, London, 1972; *see also* AB **39**, 455–524, 1975.

(27) Stem anatomy. K. ESAU, *see* **(21)**. W. TROLL & W. RAUH, Das Erstarkungswachstum der krautigen Dikotylen, *Ber. Heidelb. Akad. Wiss.*, 1950. H. J. BRAUN, *Die Organisation des Stammes von Bäumen und Sträuchern*, Stuttgart, 1963. E. CUTTER, *see* **(21)**.

(28) Stelar theory. W. ZIMMERMANN, *see* **(20)**. K. ESAU, Primary vascular differentiation in plants, *Biol. Rev.* **29**, 46, 1954; *see also* **(21)**.

(29) Leaves. W. TROLL, *Vergleichende Morphologie*, Bd **1**, 3. Teil, Berlin, 1939. F. L. MILTHORPE (ed.), *The Growth of Leaves*, London, 1956. A. ARBER, *Monocotyledons*, Cambridge, 1925; *The Natural Philosophy of Plant Form*, Cambridge, 1950. R. MAKSYMOWYCH, *Analysis of Leaf Development*, Cambridge, 1973.

(30) The root. W. TROLL, *Vergleichende Morphologie*, Bd **1**, 4. Teil, Berlin, 1939. H. von GUTTENBERG, Prim. Bau d. Gymnosp. u. Angiosperm. wurzel, HA **8**, 1940. H. WEBER, *Die Bewurzelungsverhältnisse der Pflanzen*, Freiburg, 1953.

Modifications and adaptations

(31) C. DARWIN, *Insectivorous Plants*, London, 1875. K. GOEBEL, *Pflanzenbiologische Schilderungen*, Marburg, 1889–93. H. SCHENCK, *Beitr. z. Biologie u. Anatomie d. Lianen*, Jena, 1892–93. A. KERNER von MARILAUN (translated by F. W. OLIVER), *The Natural History of Plants*, London, 1894. F. W. NEGER, *Biologie der Pflanzen*, Stuttgart, 1913. A. KERNER von MARILAUN & A. HANSEN, *Pflanzenleben*, 3. Aufl., Leipzig, Wien, 1921. J. W. BEWS, *Life Forms*, London, 1925. H. FITTING, *Ökologische Morphologie der Pflanzen*, Jena, 1926. H. PFEIFFER, Abnormes Dickenwachstum, HA **9**, 1926. N. A. MAXIMOV, *The Plant in Relation to Water*, London, 1929. A. SEYBOLD, EB **6**, 680, 1930. G. KARSTEN, HN **3**, 711. E. GÄUMANN, Parasitismus der Pflanzen, HN **7**, 720–33, 1932. H. BURGEFF, Saprophyten, HN **8**, 713–22, 1933; *Saprophytismus u. Symbiose*, Jena, 1932. C. RAUNKIAER, *Life Forms of Plants*, Oxford, 1934. A. F. W. SCHIMPER & F. C. von FABER, *Pflanzengeographie auf physiologischer Grundlage*, 3. Aufl., Jena, 1935 (first edn. trans. by P. GROOM & I. B. BALFOUR as *Plant Geography upon a Physiological Basis*, Oxford, 1903). B. HUBER, Xerophyten, HN **10**, 702, 1935; *Der Wärmehaushalt der Pflanzen*, Freiburg-München, 1935. F. E. LLOYD, BBC **54**A, 292, 1935; *The Carnivorous Plants*, Waltham, 1942. M. SKENE, *The Biology of Flowering Plants*, London, 1938. M. G. STÅLFELT, Morphologie u. Anatomie d. Blattes als Transpirationsorgan, HP **3**, 324, 1956. TH. SCHMUCKER, Höhere Parasiten, HP **9**, 480, 1959; Saprophytismus b. Kormophyten, HP **9**, 386, 1959. TH. SCHMUCKER & G. LINNEMANN, Carnivorie, HP **9**, 198, 1959. H. WALTER, *Die Vegetation der Erde in ökologischer Betrachtung*, Bd **1**, *Die tropischen und subtropischen Zonen*, Jena, 1962. C. D. SCULTHORPE, *The Biology of Aquatic Vascular Plants*, London, 1967. W. LÖTSCHERT, *Pflanzen an Grenzstandorten*, Stuttgart, 1969.

Reproduction

(32) H. KNIEP, *Die Sexualität der niederen Pflanzen*, Jena, 1928. M. HARTMANN, *Geschlecht und Geschlechtsbestimmung i. Tier- u. Pflanzenreich*, Berlin, 1951; *Die Sexualität*, 2. Aufl., Stuttgart, 1956. F. WIDDER, Grundformen des pflanzlichen Phasenwechsels, *Phyton* **3**, 252, 1951. G. LUCKHAUS, *Fortpflanzung und Nomenklatur im Pflanzen- und Tierreich*, Berl., Hamb., 1965. D. von DENFFER, Ein Vorschlag zur Vereinheitlichung der Sporennomenklatur, BDBG **80**, 371, 1966.

Physiology

(1) **General works.** W. PFEFFER, *Pflanzenphysiologie*, Leipzig, 1897–1904. P. BOYSEN JENSEN, *Die Elemente der Pflanzenphysiologie*, Heidelberg, 1939. O. F. CURTIS & D. G. CLARK, *An Introduction to Plant Physiology*, New York, Toronto, London, 1950. J. BONNER & A. W. GALSTON, *Principles of Plant Physiology*, San Francisco, 1952. F. C. STEWARD, *Plant Physiology*, 6 vols., New York, London, 1959→. M. THOMAS, S. L. RANSON & J. RICHARDSON, *Plant Physiology*, 5th edn., London, 1973. A. W. GALSTON, *The Life of the Green Plant*, Englewood Cliffs, 1961. G. E. FOGG, *Growth of Plants*, London, 1963. W. O. JAMES, *An Introduction to Plant Physiology*, 7th edn., Oxford, 1973. B. S. MEYER, D. B. ANDERSON & R. H. BÖHNING, *Introduction to Plant Physiology*, New York, 1973. W. STILES & E. C. COCKING, *An Introduction to the Principles of Plant Physiology*, 3rd edn., London, 1969. H. MOHR, *Lehrbuch der Pflanzenphysiologie*, Berlin, Heidelb., New York, 1969. P. S. NOBEL, *Biophysical Plant Physiology*, San Francisco, 1974. *See also* HP and *A. Rev. Pl. Physiol.*

Physiology of metabolism

(2) **General works.** J. BONNER, *Plant Biochemistry*, New York, 1950. K. PAECH, *Biochemie und Physiologie der sekundären Pflanzenstoffe*, Berlin, 1950. K. PAECH &

M. V. TRACEY, *Moderne Methoden der Pflanzenanalyse,* 4, Bde., Berlin, 1955–56. F. LEUTHARDT, *Lehrbuch d. physiolog. Chemie*, 14. Aufl., Berlin, 1959. D. D. DAVIES, *Intermediary Metabolism in Plants*, Cambridge, 1961. E. BALDWIN, *Dynamic Aspects of Biochemistry*, 5th edn., Cambridge, 1967. D. D. DAVIES, J. GIOVANELLI & T. AP REES, *Plant Biochemistry*, Oxford, 1964. C. P. WHITTINGHAM, *Chemistry of Plant Processes*, London, 1964. J. BONNER & J. E. VARNER, *Plant Biochemistry*, New York, 1965. H. BIELKA (ed.), *Molekulare Biologie der Zelle*, Jena, 1969. E. E. CONN & P. K. STUMPF, *Outlines of Biochemistry*, 3rd edn., New York, 1972. M. LUCKNER, *Secondary Metabolism in Plants and Animals*, London, 1972. H. STREET & W. COCKBURN, *Plant Metabolism*, 2nd edn., Oxford, 1972.

(3) Plant composition. E. WOLFF, *Aschen-Analysen von landwirthschaftlichen Producten*, Berlin, 1871→. H. H. MAYR & H. KUHN, *Die Zusammensetzung von Kulturpflanzen*, in K. SCHARRER & H. LINSER **(10)**. J. B. HARBORNE, *Phytochemical Methods*, London, 1972.

(4) Imbibition. J. R. KATZ, Quellung, *Erg. d. exakt. Naturw.* **3**, 317, 1924 and **4**, 154, 1925. R. HÖBER, *Physikalische Chemie der Zellen und Gewebe*, Ber, 1947. A. E. ALEXANDER & P. JOHNSON, *Colloid Science*, 2 vols., Oxford, 1949. A. S. CRAFTS, H. B. CURRIER & C. R. STOCKING, *Water in the Physiology of Plants*, Waltham, 1949, H. WALTER, *Grundlagen des Pflanzenlebens*, s. 72–85, Stuttgart, 1950. H. STAUDINGER, *Organische Kolloidchemie*, Berlin, 1950. A. FREY-WYSSLING, *Submicroscopic Morphology of Protoplasm*, Amsterdam, 1953. L. BRAUNER, *Handbuch der Biologie*, 1/6–8, 1958. H. NETTER, *Theoretische Biochemie*, Berlin, 1959. T. T. KOZLOWSKI, *Water Metabolism in Plants*, New York, 1964.

(5) Diffusion and osmosis. W. PFEFFER, *Osmotische Untersuchungen*, 1877. H. DeVRIES, *Eine Methode zur Analyse der Turgorkraft*, JWB **14**, 427, 1884. A. URSPRUNG & G. BLUM, Zur Methode der Saugkraftmessung, BDBG **34**, 525–39, 1916. A. URSPRUNG, Die Messung der osmotischen Zustandsgrössen pflanzlicher Zellen und Gewebe, in E. ABDERHALDEN, *Handbuch der biologischen Arbeitsmethoden*, Abt. 11, Teil 4, H.7, 1109–572, 1938. B. S. MEYER, Wall and turgor pressur, HP **2**, 38–56, 1956. E. STADELMANN, Plasmolyse und Deplasmolyse, HP **2**, 71–115, 1956. C. R. STOCKING, Osmotic pressure or osmotic value, HP **2**, 57–70, 1956. A. S. CRAFTS *et al.*, *see* **(4)**. H. WALTER, *Grundlagen des Pflanzenlebens*, s. 56–72, Stuttgart, 1950. J. DAINTY, Water relations of plant cells, in R. D. PRESTON (ed.), *Advances in Botanical Research* **1**, 1963. H. WALTER & K. KREEB, Die Hydration u. Hydratur des Protoplasmas, *Protoplasmatologia*, **2**C6, Wien, 1970.

(6) Water uptake. H. WALTER, *Die Hydratur der Pflanze*, Jena, 1931. P. J. KRAMER, *Plant and Soil Water Relationships*, New York, 1969. W. H. ARISZ *et al.*, Analysis of the exudation process in tomato plants, *J. Exp. Bot.* **2**, 257–97, 1951. L. D. BAVER, *Soil Physics*, 3rd edn., New York, London, 1961. C. A. BLACK, *Soil–Plant Relationships*, 2nd edn., New York, London, 1968. E. J. RUSSELL, *Soil Conditions and Plant Growth*, 9th edn. (rewritten by E. W. RUSSELL), London, New York, Toronto, 1961. D. HILLEL, *Soil and Water*, New York, 1971. *See also* HP **3**, 1956 (Water relations of plants).

(7) Transpiration: mechanism. N. A. MAXIMOV, *The Plant in Relation to Water*, London, 1929. A. SEYBOLD, EB **5**, 1929 and **6**, 1930. *See also* HP **3**, 1956.

(8) Transpiration: significance. A. F. W. SCHIMPER & F. C. von FABER, *Pflanzengeographie auf physiologischer Grundlage*, 3. Aufl., Jena, 1935. H. WALTER, *Standortslehre*, Stuttgart, 1951; *Grundlagen der Pflanzenverbreitung*, Stuttgart, 1960. D. M. GATES, *Energy Exchange in the Biosphere*, New York, 1962. A. J. RUTTER & F. H. WHITEHEAD (eds.), *The Water Relations of Plants*, Oxford, 1963.

(9) Water conduction. S. STRUGGER, *Der aufsteigende Saftstrom*, NW **31**, 181, 1943. R. D. PRESTON, Movement of water in higher plants, in A. FREY-WYSSLING (ed.), *Deformation and Flow in Biological Systems*, Amsterdam, 1952, B. HUBER, Die Gefässleitung, HP **3**, 541–82, 1956. G. E. BRIGGS, *Movement of Water in Plants*, Oxford, 1967. C. R. HOUSE, *Water Transport in Cells and Tissues*, London, 1974.

(10) Mineral nutrients. H. LUNDEGÅRDH, *Die Nährstoffaufnahme der Pflanzen*, Jena, 1932. K. SCHARRER, *Biochem. d. Spurelemente*, 3. Aufl., Berl., Hamb., 1955. H. BURSTRÖM, Mineralstoffwechsel, *Fort. Bot.* **17**, 1955 and **21**, 1958. E. J. RUSSELL, *see* **(6)**. W. STILES, *Trace Elements in Plants*, 3rd edn., Cambridge, 1961. K. MENGEL, *Ernährung und Stoffwechsel der Pflanze*, **4**. Aufl., Jena, 1972. K. SCHARRER & H. LINSER, *Handbuch der Pflanzenernährung und Düngung*, Wien, New York, 1969. E. EPSTEIN, *Mineral Nutrition of Plants*, New York, 1972. E. J. HEWITT & T. A. SMITH, *Plant Mineral Nutrition*, London, 1974. *See also* HP **4**, 1958 (Mineral nutrition of plants).

(11) Culture techniques. W. SCHROPP, Die Methodik der Wesserkultur, *Handb. d. landwirtsch. Versuchs- u. Untersuchungsmethodik* **8**, Radebeul, Berlin, 1951. E. J. HEWITT, *Sand and Water Culture Methods*, Commonwealth Agricultural Bureaux Publication **22**, 1952.

(12) Uptake of nutrients. G. E. BRIGGS, A. B. HOPE & R. N. ROBERTSON, *Electrolytes and Plant Cells*, Oxford, 1961. J. F. SUTCLIFFE, *Mineral Salts Absorption in Plants*, Oxford, 1962. D. H. JENNINGS, *The Absorption of Solutes by Plant Cells*, Edinburgh, 1963. W. P. ANDERSON (ed.), *Ion Transport in Plants*, London, 1973. A. J. PEEL, *Transport of Nutrients in Plants*, London, 1974.

(13) Photosynthesis. R. WILLSTÄTTER & A. STOLL, *Untersuchungen über die Assimilation der Kohlensäure*, Berlin, 1918. J. FRANCK & W. E. LOOMIS, *Photosynthesis in Plants*, Ames, 1949. E. I. RABINOWITCH, *Photosynthesis and Related Processes*, New York, **1**, 1945, **2.i**, 1951, **2.ii**, 1956. R. HILL & C. P. WHITTINGHAM, *Photosynthesis*, London, 1955. J. A. BASSHAM & M. CALVIN, *The Path of Carbon in Photosynthesis*, Englewood Cliffs, 1957. M. CALVIN & J. A. BASSHAM, *The Photosynthesis of Carbon Compounds*, New York, 1962. O. WARBURG, G. KRIPPAHL & C. JETSCHMANN, Widerlegung der Photolyse des Wassers, *Z. Naturf.* **20B**, 993, 1965. M. GIBBS (ed.), *Structure and Function of Chloroplasts*, Berl., Heidlb., New York, 1971. C. P. WHITTINGHAM, *The Mechanism of Photosynthesis*, London, 1974. R. MARCELLE (ed.), *Environmental and Biological Control of Photosynthesis*, The Hague, 1975. *See also* HP **5**, 1960.

(14) Chemosynthesis. P. W. WILSON, *The Biochemistry of Symbiotic Nitrogen Fixation*, Madison, 1940. P. W. WILSON & R. H. BURRIS, Biological nitrogen fixation, *A. Rev. Microbiol.* **7**, 1953. H. LEES, *Biochemistry of Autotrophic Bacteria*, London, 1955. A. RIPPEL-BALDES, *Grundriss der Mikrobiologie*, 3. Aufl., Berl., Götting., Heidelb., 1955. I. C. GUNSALUS & R. Y. STANIER (eds.), *The Bacteria*, New York, 1960→. K. V. THIMANN, *The Life of Bacteria*, New York, 1963. H. GEST, A. SAN PIETRO & L. P. VERNON, *Bacterial Photosynthesis*, Yellow Springs, 1963. W. D. P. STEWART, *Nitrogen Fixation in Plants*, London, 1966.

(15) Sugar metabolism. K. HEYNS, *Die neueren Ergebnisse der Stärkeforschung*, Braunschweig, 1949. K. H. MEYER, *Natural and Synthetic High Polymers*, 2nd edn., New York, 1950. A. FREY-WYSSLING, *see* **(4)**. M. STACEY & S. A. BARKER, *Carbohydrates of Living Tissues*, London, 1962. F. LOEWUS (ed.), *Biogenesis of Plant Cell Wall Polysaccharides*, New York, 1973.

(16) Enzymes. J. B. SUMNER & G. F. SOMERS, *Chemistry and Methods of Enzymes*, 3rd edn., New York, 1953. T. BERSIN, *Kurzes Lehrbuch der Enzymologie*, 3. Aufl., Leipzig, 1951. K. MYRBÄCK, *Enzymatische Katalyse*, Berlin, 1953. O. HOFFMANN-OSTENHOF, *Enzymologie*, Wien, 1954. R. AMMON & W. DIRSCHERL (eds.), *Fermente, Hormone, Vitamine und die Beziehungen dieser Wirkstoffe zueinander*. Bd. **1**, *Fermente* (ed. R. AMMON & K. MYRBÄCK), Stuttgart, 1959. P. D. BOYER, H. LARDY & K. MYRBÄCK, *The Enzymes*, vols. **1–8**, New York, 1961→. M. DIXON & E. C. WEBB, *Enzymes*, 2nd edn., London, 1964. C. H. WYNN, *The Structure and Function of Enzymes*, London, 1973.

(17) Respiration: general. J. W. FOSTER, *Chemical Activities of Fungi*, New York, 1949. W. O. JAMES, *Plant Respiration*, Oxford, 1953; *Endeavour* **13**, 1954. W. STILES & W. LEACH, *Respiration in Plants*, 4th edn., London, 1960. H. BEEVERS, *Respiratory Metabolism in Plants*, Evanston, 1961. W. O. JAMES, *Cell Respiration*, London, 1971.

M. D. HATCH, C. B. OSMOND & R. O. SLAYTER (eds.), *Photosynthesis and Photorespiration*, New York, 1971. *See also* **(2)**; HP **12**, 1960; *A. Rev. Pl. Physiol.*

(18) Oxidation. H. WIELAND, *On the Mechanism of Oxidation*, New Haven, 1933. O. WARBURG, *Schwermetalle als Wirkungsgruppen*, Berlin, 1948; *Wasserstoffüberträgende Fermente*, Berlin, 1948. A. L. LEHNINGER, Oxidative phosphorylation, *Harvey Lect.*, Ser. **49**, 176, 1953–54.

(19) Energy and dynamics. A. SZENT-GYÖRGI, *Bioenergetics*, New York, 1957. H. A. KREBS & H. L. KORNBERGER, *Energy Transformations in Living Matter*, Berlin, 1957. I. M. KLOTZ, *Some Principles of Energetics in Biochemical Reactions*, New York, 1957. E. BALDWIN, *see* **(2)**. A. L. LEHNINGER, *Bioenergetics*, New York, 1965. *See also* HP **12**, 1960.

(20) Luminescence. H. MOLISCH, *Leuchtende Pflanzen*, Jena, 1912. E. N. HARVEY, *Living Light*, Princeton, 1940.

(21) Bacterial respiration. *See* **(14)**.

(22) Proteins. E. WALDSCHMIDT-LEITZ, *Chemie der Eiweisskörper*, Stuttgart, 1950. H. HELLMANN, *Eiweiss*, Stuttgart, 1952. B. FLASCHENTRÄGER, *Physiologische Chemie*, Bd. **1**, Berlin, 1952. A. MEISTER, *Biochemistry of Amino Acids*, New York, 1957. F. HAUROWITZ, *Biosynthese der Proteine*, NW **46**, 60, 1959. H. TUPPY, *Aminosäure–Sequenzen in Proteinen*, NW **46**, 35, 1959. G. C. WEBSTER, *Nitrogen Metabolism in Plants*, Evanston, 1959. D. O. JORDAN, *The Chemistry of Nucleic Acids*, London, 1960. R. J. C. HARRIS (ed.), *The Biosynthesis of Protein*, New York, 1960. H. S. McKEE, *Nitrogen Metabolism in Plants*, Oxford, 1962. L. BOSCH (ed.), *The Mechanism of Protein Synthesis*, Amsterdam, 1972. L. BEEVERS, *Nitrogen Metabolism in Plants*, London, 1975. A. SMITH, *Protein Synthesis*, London, 1975.

(23) Hydrogen ion concentration. L. MICHAELIS, *Hydrogen Ion Concentration*, Baltimore, 1926. J. SMALL, *Modern Aspects of pH*, London, 1954.

(24) Translocation. F. C. STEWARD, *Plant Physiology*, **2** (Plants in relation to water and solutes), New York, 1960. M. H. ZIMMERMANN, Transport in the phloem, *A. Rev. Pl. Physiol.* **11**, 167–90, 1960. A. L. KURSANOV, Metabolism and the transport of organic substances in the phloem, in R. D. PRESTON, *Advances in Botanical Research*, **1**, London, New York, 1963. M. J. CANNY, *Phloem Translocation*, London, 1973. A. J. PELL, *Transport of Nutrients in Plants*, London, 1974.

(25) Excretion and secretion. A. FREY-WYSSLING, *Die Stoffausscheidung der höheren Pflanzen*, Berlin, 1935. K. PAECH, *see* **(2)**. E. SCHNEPF, Sekretion u. Exkretion bei Pflanzen, *Protoplasmatologia*, **8**, 8, 1969.

(26) Saprophytes. H. BURGEFF, *Saprophytismus und Symbiose*, Jena, 1932. E. G. PRINGSHEIM & H. BURGEFF, Saprophyten, HN **8**, 1933. J. W. FOSTER, *see* **(17)**. V. G. LILLY & H. L. BARNETT, *Physiology of the Fungi*, New York, 1951. L. E. HAWKER, A. H. LINTON, B. F. FOLKES & M. J. CARLILE, *An Introduction to the Biology of Microorganisms*, London, 1960. L. E. HAWKER & A. H. LINTON (eds.), *Micro-organisms*, London, 1971. For **(26)–(29)** *see also* HP **11**, 1959 (Heterotrophy).

(27) Parasites. E. GÄUMANN, Parasitismus der Pflanzen, HN **7**, 720–33, 1932; *Pflanzliche Infektionslehre*, Basel, 1946; *Principles of Plant Infection*, London, 1950.

(28) Symbiosis. R. SCHAEDE, *Die pflanzliche Symbiosen*, 2. Aufl., Jena, 1948. A. RIPPEL-BALDES, *see* **(14)**. E. G. HALLSWORTH (ed.), *Nutrition of Legumes*, London, 1958. S. D. GARRETT, *Soil Fungi and Soil Fertility*, Oxford, 1963. S. M. HENRY, *Symbiosis*, New York, 1966. M. E. HALE, *The Biology of Lichens*, London, 1974.

(29) Mycorrhiza. H. BURGEFF, *Samenkeimung der Orchideen*, Jena, 1936. A. P. KELLEY, *Mycotrophy in Plants*, Waltham, 1950. J. L. HARLEY, *The Biology of Mycorrhiza*, 2nd edn., London, 1969.

(30) Cyclic metabolism. H. LUNDEGÅRDH, *Der Kreislauf der Kohlensäure in der Natur*, Jena, 1924; Kreislauf der Stoffe i.d. org. Welt, HN **5**, 1934. P. BOYSEN JENSEN,

Die Stoffproduktion der Pflanzen, Jena, 1932. E. P. ODUM, *Fundamentals of Ecology*, London, 1953. D. M. GATES, *see* (8). J. D. OVINGTON, Quantitative ecology and the woodland ecosystem concept, *Advances in Ecological Research*, **1**, 1962.

Physiology of growth

(31) General works. H. WINKLER, Entwicklungsmechanik oder Entwicklungsphysiologie der Pflanzen, HN **3**, 620–49, 1933. E. BÜNNING, *Entwicklungs- und Bewegungsphysiologie der Pflanze*, 3. Aufl., Berl., Götting., Heidelb., 1953. P. R. WHITE, *The Cultivation of Plant and Animal Cells*, New York, 1954. A. KÜHN, *Vorlesungen über Entwicklungsphysiologie*, 2. Aufl., Berlin, 1965. A. C. GIESE, *Cell Physiology*, 4th edn., London & Philadelphia, 1973. D. MAZIA & A. TYLER (eds.), *General Physiology of Cell Specialization*, New York, 1963. M. B. WILKINS (ed.), *The Physiology of Plant Growth and Development*, London, 1969. G. C. EVANS, *The Quantitative Analysis of Plant Growth*, 2nd edn., Oxford, 1972.

(32) Virus multiplication. F. C. BAWDEN, *Plant Viruses and Virus Diseases*, 4th edn., Waltham, 1964. W. WEIDEL, Bacteriophagen, FB **17**, 1954. M. H. ADAMS, *Bacteriophages*, New York, 1959. K. M. SMITH, *Plant Viruses*, 4th edn., London, 1968. R. ENGLER & G. SCHRAMM, Die Bildung der infektiösen Ribonucleinsäure während der Vermehrung des Tabakmosaikvirus, *Z. Naturf.* **15B**, 38–45, 1960. W. WEIDEL, *Virus u. Molekularbiologie*, Heidelb., 1963.

(33) Vitamins. W. H. SCHOPFER, *Plants and Vitamins*, Waltham, 1949. Water-soluble vitamins I–III; Fat-soluble vitamins, *A. Rev. Biochem.* **25**, 397–536, 1956. *See also* HP **14**, 1960.

(34) Auxins and growth. G. S. AVERY & E. B. JOHNSON, *Hormones and Horticulture*, New York, London, 1947. A. FREY-WYSSLING, Physiology of cell wall growth, *A. Rev. Pl. Physiol.*, **1**, 169–82, 1950. H. SÖDING, *Die Wuchsstofflehre*, Stuttgart, 1952. H. STERN, The physiology of cell division, *A. Rev. Pl. Physiol.* **7**, 91–114, 1956. F. SCHRADER, *Mitosis: The Movement of Chromosomes in Cell Division*, New York, 1953. R. L. WAIN & F. WIGHTMAN (eds.), *The Chemistry and Mode of Action of Plant Growth Substances*, London, 1956. B. B. STOWE & T. YAMAKI, The history and physiological action of the gibberellins, *A. Rev. Pl. Physiol.* **8**, 181–216, 1957. L. J. AUDUS, *Plant Growth Substances*, 3rd edn., London, 1972. P. F. WAREING & I. D. J. PHILLIPS, *The Control of Growth and Differentiation in Plants*, Oxford, 1970. *See also* HP **14**, 1960.

(35) Polarity. H. VÖCHTING, *Untersuchungen zur experimentale Anatomie und Pathologie*, Bd. **1**, 1908; Bd. **2**, *Die Polarität der Gewächse*, Tübingen, 1918. J. HAMMERLING, Neuere Versuche über Polarität, *Biol. Zentralbl.* **74**, 545, 1955. E. BÜNNING, Polarität und inaequale Teilung, *Protoplasmatologia* **8**, 9a, Wien, 1957. W. HAUPT, Die Entstehung der Polarität, EM **25**, 1962. A. KÜHN, *see* (**31**).

(36) Restitution. H. VÖCHTING, *Über Organbildung im Pflanzenreich, 2. Bde, Bonn*, 1878, 1884, J. LOEB, *Regeneration*, New York, 1924. N. P. KRENKE, *Wundkompensation, Transplantation und Chimären*, Berlin, 1933.

(37) Grafting. H. WINKLER, *Untersuchungen über Propfbastarden*, Jena, 1912. W. NEILSON JONES, *Plant Chimaeras and Graft Hybrids*, London, 1934. P. J. S. CRAMER, Chimaeras, *Bibliographica genet.* **16**, 193–381, 1954. E. SANKEWITSCH, *Die Arbeitsmethoden der Mitschurinschen Pflanzenzüchtung*, Stuttgart, 1950.

(38) Morphogenetic substances. J. HÄMMERLING, Nucleo-cytoplasmic interactions in *Acetabularia* and other cells, *A. Rev. Pl. Physiol.* **14**, 65, 1963.

(39) Gall formation. H. ROSS & H. HEDICKE, *Die Pflanzengallen (Cecidien) Mittel- und Nordeuropas*, 2. Aufl., Jena, 1927. E. KÜSTER, Anatomie der Gallen, HA **5**, 1930; Gallen, HN **4**, Jena, 1934. M. S. MANI, *The Ecology of Plant Galls*, The Hague, 1964.

(40) Ageing and death. H. MOLISCH, *Die Lebensdauer der Pflanzen*, Jena, 1929. B. L. STRECHLER, *Biology of Aging*, Washington, 1960; *Time, Cells and Aging*, London,

1962. H. W. WOOLHOUSE, Longevity and senescence in plants, *Sci. Prog., Oxf.* **61**, 123, 1974.

(41) Genetics. E. ALTENBURG, *Genetics*, London, 1957. C. P. SWANSON, *Cytology and Cytogenetics*, London, 1958. R. SAGER & F. J. RYAN, *Cell Heredity*, New York, 1961. WATKIN WILLIAMS, *Genetical Principles and Plant Breeding*, Oxford, 1964. A. BARTHELMESS, *Hdb. Biol.* **3**, 437, 1963–65. A. KÜHN, *Grundriss der Vererbungslehre*, 4. Aufl., Heidelb., 1965. A. M. SRB, R. D. OWEN & R. S. EDGAR, *General Genetics*, 2nd edn., San Francisco, 1965. A. MÜNTZING, *Genetics*, Stockholm, 1967. K. ESSER & R. KUENEN, *Genetics of Fungi*, Berlin, 1967. R. RIEGER, A. MICHAELIS & M. M. GREEN, *A Glossary of Genetics and Cytogenetics*, 3. Aufl., Berl., Heidelb., New York, 1968. D. HESS, *Biochemische Genetik*, Berl., Heidelb., New York, 1968. B. LEWIN, *The Molecular Basis of Gene Expression*, London, 1970. H. BRESCH, *Klassische u. Molekulare Genetik*, Heidelb., 1970. J. R. S. FINCHAM & P. R. DAY, *Fungal Genetics*, 3rd edn., Oxford, 1971. E. B. FORD, *Ecological Genetics*, 3rd edn., London, 1971. H. L. K. WHITEHOUSE, *The Mechanism of Heredity*, 3rd edn., London, 1973. *See also* HV from 1926.

(42) Genetics. W. JOHANNSEN, *Elemente d. exakt. Erblichkeitslehre*, 3. Aufl., Jena, 1926.

(43) Genetics. G. MENDEL, Experiments in plant hybridization, translation reprinted in E. W. SINNOTT, L. C. DUNN & T. DOBZHANSKY, *Principles of Genetics*, 5th edn., New York, Toronto, London, 1958.

(44) Sex-determination. C. CORRENS, Bestimmung, Vererbung, und Verteilung des Geschlechts bei den höheren Pflanzen, HV **2**, 1928. M. HARTMANN, *Die Sexualität*, 2. Aufl., Stuttgart, 1956.

(45) Mutations. N. W. TIMOFEEFF-RESSOVSKY, The experimental production of mutations, *Biol. Rev.* **9**, 411, 1934; Exp. Mutationsforsch. in der Vererbungsl., *Wiss. Forschungsber.* **42**, 1937. H. STUBBE, *Spontane u. strahleninduz. Mutabilität*, Leipzig, 1937. G. TISCHLER, Polyploidie und Artbildung, NW **30**, 713, 1942; Über die Siedlungsfähigkeit von Polyploiden, *Z. Naturf.* **1**, 157, 1946. J. STRAUB, Cytogenetic, *Fort. Bot.* **17**, 670, 1954. A. HOLLAENDER (ed.), *Radiation Biology*, 3 vols., New York, 1954–56. Cold Spring Harb. Symp. quant. Biol.: **16**, 1951, *Genes and Mutations*; **21**, 1956, *Genetic Mechanisms: Structure and Function*. R. RIEGER, *Die Genom-Mutationen*, Jena, 1963. R. RIEGER & A. MICHAELIS, *Chromosomenmutationen*, Jena, 1967.

(46) Extra-chromosomal inheritance. J. L. JINKS, *Extra-chromosomal Inheritance*, Englewood Cliffs, 1964. D. WILKIE, *The Cytoplasm in Heredity*, London, 1964. R. HAGEMANN, *Plasmatische Vererbung*, Jena, 1964. P. MICHAELIS, Cytoplasmic inheritance in *Epilobium*, *The Nucleus* **8**, 83, 1965; **9**, 1, 1966. W. STUBBE, Die Plastiden als Erbträger, in P. SITTE (ed.), *Probleme der biolog. Reduplikation*, Berl., Heidelb., New York, 1966. F. SCHÖTZ, Extrachromosale Vererbung, BDBG **80**, 523, 1967.

(47) Temperature; vernalization. A. E. MURNEEK & R. O. WHYTE, *Vernalization and Photoperiodism*, Waltham, 1948. G. MELCHERS & A. LANG, Die Physiologie der Blütenbildung, *Biol. Zbl.* **67**, 105–74, 1948. J. LEVITT, Frost, drought and heat resistance, *A. Rev. Pl. Physiol.* **2**, 245–68, 1951.

(48) Allelopathy. H. WALTER, *see* **(8)**. G. GRÜMMER, *Die gegenseitige Beeinflussung höheren Pflanzen: Allelopathie*, Jena, 1955. R. KNAPP, *Experimentelle Soziologie der höheren Pflanzen*, Stuttgart, 1954.

(49) Development and environment. G. KLEBS, *Willkürliche Entwickelungsänderungen bei Pflanzen*, Jena, 1903: *Die Bedingungen der Fortpflanzung bei einigen Algen und Pilzen*, Jena, 1896. L. JOST, Physiologie der Fortpflanzung, HN **4**, 435–50, 1934. G. MELCHERS & A. LANG, *see* **(47)**. J. L. LIVERMAN, The physiology of flowering, *A. Rev. Pl. Physiol.* **6**, 177–210, 1955. E. BÜNNING, Endogenous rhythms in plants, *A. Rev. Pl. Physiol.* **7**, 71–90, 1956. W. CROCKER & L. V. BARTON, *Physiology of Seeds*, Waltham, 1957. F. W. WENT, *The Experimental Control of Plant Growth*, Waltham,

1957. R. B. WITHROW (ed.), *Photoperiodism and Related Phenomena in Plants and Animals*, Washington, 1959. R. van der VEEN & G. MEIJER, *Light and Plant Growth*, Eindhoven, 1959. M. B. ALLEN (ed.), *Comparative Biochemistry of Photoreactive Systems*, New York, 1960. F. B. SALISBURY, *The Flowering Process*, Oxford, 1963. L. T. EVANS (ed.), *Environmental Control of Plant Growth*, New York, London, 1963. E. BÜNNING, *Die physiologische Uhr*, Berl., Heidelb., New York, 1963. H. MOHR, The control of plant growth and development by light, *Biol. Revs.* **39**, 87, 1964. J. LEVITT, *The Response of Plants to Environmental Stress*, New York, 1972.

Movement

(50) General works. P. METZNER, Bewegungen der Pflanzen, HN **1**, 946–61, 1931. L. JOST, Reizerscheinungen der Pflanzen, Allgemeiner Teil, Taxien, HN **8**, 353–76, 1933. E. BÜNNING, *see* **(31)**. *See also* HP **17**, 1959, 1962.

(51) Tropisms. H. FITTING, Reizerscheinungen, Tropismen, HN **8**, 376–421, 1933. L. BRAUNER, Tropisms and nastic movements, *A. Rev. Pl. Physiol.* **5**, 163–82, 1954.

(52) Phototropism. H. Du BUY & E. NUERNBERGK, Phototropismus und Wachstum der Pflanzen, EB **9**, 358–544, 1932; EB **10**, 207–322, 1934. A. R. SCHRANK, Plant tropisms, *A. Rev. Pl. Physiol.* **1**, 59–74, 1950. E. BÜNNING, Die Funktion gelber Pigmente beim Phototropismus, ZB **43**, 167, 1955. L. BRAUNER, Über die Funktion der Spitzenzone beim Phototropismus, ZB **43**, 467, 1955. W. R. BRIGGS, Phototropism in higher plants, in A. C. GIESE (ed.), *Photophysiology* **1**, New York, 1964. *See also* H. SÖDING **(34)**.

(53) Geotropism. F. RAWITSCHER, *Geotropismus der Pflanzen*, Jena, 1932.

(54) Chemotropism. F. G. FISCHER & G. WERNER, Eine Analyse des Chemotropismus einiger Pilze, *Z. physiol. Chem.* **300**, 211–36, 1955. H. ZIEGLER, HP **17** (2), 396, 1962.

(55) Nastic movements. F. RAWITSCHER, Reizerscheinungen, Nastieen, HN **8**, 421–47, 1933.

(56) Autonomic movements. K. GOEBEL, *Entfaltungsbewegungen der Pflanzen*, 2. Aufl., Jena, 1924.

(57) Nyctinastic movements. E. BÜNNING, *see* **(32)** and **(49)**.

(58) Stomata. P. METZNER, *see* **(50)**. M. G. STÅLFELT, Die stomatäre Transpiration und die Physiologie der Spaltöffnungen, HP **3**, 351–426, 1956. O. V. S. HEATH, The water relations of stomatal cells and the mechanism of stomatal movement, in F. C. STEWARD (ed.), *Plant Physiology* **2**, 1959. H. MEIDNER & T. A. MANSFIELD, *Physiology of Stomata*, London, 1968.

(59) Turgor mechanisms. C. T. INGOLD, *Spore Liberation*, Oxford, 1965. H. STRAKA, Nicht durch Reize ausgelöste Bewegungen, HP **17** (2), 716, 1962.

(60) Hygroscopic movements. H. von GUTTENBERG, Die Bewegungsgewebe, HA **5**, I. Abt., 2. Teil, 1926. C. T. INGOLD, *see* **(59)**.

(61) Tactic movements. L. JOST, *see* **(50)**. P. METZNER, *see* **(50)**. A. FREY-WYSSLING, *Die submikroskopische Struktur des Cytoplasmas*, Wien, 1955.

Systematics and evolution

General principles

Theory and processes of evolution: Classic works. J. B. P. A. de LAMARCK, *Philosophie Zoologique*, Paris, 1809. C. DARWIN, *Origin of Species by means of Natural Selection*, London, 1859. E. HAECKEL, *Natürliche Schöpfungsgeschichte*, 1st–10th edns., Berlin, 1868–1902 (trans. by E. LANKESTER as *The History of Creation*, 3rd edn.,

London, 1883). A. KERNER von MARILAUN, *Pflanzenleben*, Leipzig, Vienna, 1888–91 (trans. by F. W. OLIVER as *The Natural History of Plants*, London, 1894–95, 1902). H. de VRIES, *Die Mutationstheorie*, Leipzig, 1901–3 (trans. by J. B. FARMER & A. D. DARBISHIRE as *The Mutation Theory*, London, 1910–11). A. WEISMAN, *The Evolution Theory*, London, 1904.

General works. G. HEBERER (ed.), *Die Evolution der Organismen*, 2nd and 3rd edns., Stuttgart, 1959, 1966→. B. RENSCH, *Neuere Probleme der* Abstammungslehre, 2nd edn., Stuttgart, 1954. G. G. SIMPSON, *Tempo and Mode in Evolution*, New York, 1944. W. ZIMMERMANN, *Vererbung 'erworbener Eigenschaften' unter Auslese*, 2nd edn., Stuttgart, 1969. J. HUXLEY, *Evolution, the Modern Synthesis*, 2nd edn., London, 1963. T. DOBZHANSKY, *Genetics and the Origin of Species*, 3rd edn., New York, 1951. E. MAYR, *Systematics and the Origin of Species*, New York, 1942; *The Species Problem*, Washington, D.C., 1957. G. L. STEBBINS, Jr., *Variation and Evolution in Plants*, New York, 1950; *Processes of Organic Evolution*, 2nd edn., Englewood Cliffs, N.J., 1971; Adaptive radiation and trends of evolution in higher plants, *Evol. Biol.* **1**, 1967; *Chromosomal Evolution in Higher Plants*, London, 1971. J. CLAUSEN, *Stages in the Evolution of Plant Species*, Ithaca, 1951. D. BRIGGS & S. M. WALTERS, *Plant Variation and Evolution*, London, 1969. F. SCHWANITZ, *The Origin of Cultivated Plants*, Cambridge, Mass., 1966.

Variation, recombination. H. STUBBE, *Genetik und Zytologie von* Antirrhinum *L. sect.* Antirrhinum, Jena, 1966. J. CLAUSEN, D. D. KECK & W. M. HIESEY, Experimental studies on the nature of species. II. Plant evolution through amphiploidy and autoploidy with examples from the Madiinae, *Publs Carnegie Instn* **564**, 1945. C. D. DARLINGTON, *Evolution of Genetic Systems*, 2nd edn., Edinburgh, 1958. V. GRANT, The regulation of genetic recombination in plants, *Cold Spring Harb. Symp. quant. Biol.* **23**, 337–63, 1958. J. G. HAWKES (ed.), *Reproductive Biology and Taxonomy of Vascular Plants*, Oxford, 1966. A. GUSTAFSSON, Apomixis in higher plants, *Acta Univ. Lund.* **42–3**, 1946–47.

Population structure, selection, formation of races. S. WRIGHT, *Evolution and the Genetics of Populations*, **1**, Chicago, 1968. R. A. FISHER, *The Genetical Theory of Selection*, New York, 1958. G. TURESSON, The genotypical response of the plant species to the habitat, *Hereditas*, **3**, 211–350, 1922. J. CLAUSEN, D. D. KECK & W. M. HIESEY, Experimental studies on the nature of species, I, III, IV, *Publns Carnegie Instn* **520**, 1940; **581**, 1948; **615**, 1958. J. HESLOP-HARRISON, Forty years of genecology, *Adv. ecol. Res.* **2**, 159–247, 1964.

Chromosomal differentiation. C. D. DARLINGTON, *Chromosome Botany and the Origins of Cultivated Plants*, 3rd edn., London, 1973. I. MANTON, *Problems of Cytology and Evolution in the Pteridophyta*, Cambridge, 1950. E. B. BABCOCK, *The genus* Crepis, **1–2**, Berkeley, Calif., 1947. F. EHRENDORFER, Cytologie, Taxonomie und Evolution bei Samenpflanzen, *Vistas in Botany* (ed. W. B. TURRILL), **4**, 99–186, 1964. C. FAVARGER, Cytologie et distribution des plantes, *Biol. Rev.* **42**, 163–206, 1967. A. A. FEDOROV (ed.), *Chromosome Numbers of Flowering Plants*, Leningrad, 1969; Index to Plant Chromosome Numbers, *Regnum veg.* **50**→, 1967→.

Barriers to crossing. G. TURESSON, Zur Natur und Begrenzung der Arteneinheiten, *Hereditas* **12**, 323–34, 1927. G. L. STEBBINS, The inviability, weakness, and sterility of interspecific hybrids, *Adv. Genet.* **9**, 147–215, 1958.

Hybridization. E. ANDERSON, *Introgressive Hybridization*, New York, London, 1949. G. L. STEBBINS, The role of hybridization in evolution, *Proc. Am. phil. Soc.* **103**, 231–49, 1959. F. EHRENDORFER, Differentiation–hybridization cycles and polyploidy in *Achillea*, *Cold Spring Harb. Symp. quant. Biol.* **24**, 141–52, 1959. R. E. CLELAND, Oenothera, *Cytogenetics and Evolution*, London, 1972. *See also* the references under 'Physiology' to genetics (**41–6**); the general works on systematics and evolutionary history of the Plant Kingdom; and specialized articles in the periodicals *Evolution, Lancaster, Pa., Evolutionary Biology*, etc.

Systematics, taxonomy and nomenclature: General works. W. ROTHMALER, *Allgemeine Taxonomie und Chorologie der Pflanzen*, 2nd edn., Jena, 1955. H. MERX-

MÜLLER, Moderne Probleme der Pflanzensystematik, *Arbeitsgem. Forschung Nordrhein-Westfalen* **183**, 1968. J. HESLOP-HARRISON, *New Concepts in Flowering Plant Taxonomy*, London, 1953, 1960. L. BENSON, *Plant Taxonomy, Methods and Principles*, New York, 1962. P. H. DAVIS & V. H. HEYWOOD, *Principles of Angiosperm Taxonomy*, Edinburgh, 1963. P. H. RAVEN & T. R. MERTENS, Plant systematics, *Biol. Sci. Pamphl.* **23**, Boston, Mass., 1965. V. H. HEYWOOD, *Plant Taxonomy*, London, 1967; (ed.), *Modern Methods in Plant Taxonomy*, London, 1968. P. W. LEENHOUTS, A guide to the practice of herbarium taxonomy, *Regnum veg.* **58**, 1–60, 1968.

Numerical taxonomy. R. R. SOKAL & P. H. A. SNEATH, *Principles of Numerical Taxonomy*, San Francisco, 1963. W. T. WILLIAMS, Numbers, taxonomy and judgement, *Bot. Rev.* **33**, 379–86, 1967. N. JARDINE & R. SIBSON, *Mathematical Taxonomy*, London, 1971.

Nomenclature, terminology. International Code of Botanical Nomenclature, *Regnum veg.* **82**, 1–426, 1972. W. T. STEARN, *Botanical Latin, History, Grammar, Syntax and Vocabulary*, 2nd edn., Newton Abbot, 1973. C. F. WERNER, *Wortelemente lateinisch-griechischer Fachausdrücke in den biologischen Wissenschaften*, 3rd edn., Leipzig, 1968.

The Plant Kingdom

General works. A. ENGLER, *Syllabus der Pflanzenfamilien*, **1–2**, 12th edn., revised by H. MELCHIOR & E. WERDERMANN, Berlin, 1954, 1964. A. ENGLER & K. PRANTL (eds.), *Die natürlichen Pflanzenfamilien*, 1st and 2nd edns., Leipzig, Berlin, 1889–1902, 1923→. R. WETTSTEIN, *Handbuch der systematischen Botanik*, 4th edn., **1–2**, Leipzig, Vienna, 1933, 1935. W. ZIMMERMANN, *Die Phylogenie der Pflanzen*, 2nd edn., Stuttgart, 1959; *Geschichte der Pflanzen*, 2nd edn., Stuttgart, 1969. G. HEBERER, *Die Evolution der Organismen*, 2nd and 3rd edns., Stuttgart, 1954, 1966→. M. C. COULTER & H. J. DITTMER, *Phylogeny and Form in the Plant Kingdom*, Princeton, N.J., 1964. R. F. SCAGEL *et al.*, An Evolutionary Survey of the Plant Kingdom, Belmont, Calif., 1965. H. C. BOLD, *Morphology of Plants*, 3rd edn., New York, 1973. A. CRONQUIST, *Introductory Botany*, New York, 1961. E. J. H. CORNER, *The Life of Plants*, London, 1964. M. CHADEFAUD & L. EMBERGER, *Traité de Botanique Systématique*, **1–2**, Paris, 1960. H. des ABBAYES *et al.*, *Botanique*, Paris, 1963.

Palaeobotany. W. GOTHAN & H. WEYLAND, *Lehrbuch der Paläobotanik*, 2nd edn., Berlin, 1964. M. HIRMER, *Handbuch der Paläobotanik* **1**, Munich, Berlin, 1927. E. BOUREAU, *Traité de Paléobotanique*, Paris, 1964→. L. EMBERGER, *Les Plantes Fossiles*, 2nd edn., Paris, 1968. T. DELEVORYAS, *Morphology and Evolution of Fossil Plants*, New York, 1962. F. SCHAARSCHMIDT, *Paläobotanik*, Mannheim, 1968. H. N. ANDREWS, *Studies in Palaeobotany*, New York, 1961. D. H. SCOTT, *Studies in Fossil Botany*, 3rd edn., London, 1920–23 (reprint, New York, 1962). A. C. SEWARD, *Plant Life through the Ages*, 2nd edn., Cambridge, 1933 (reprint 1941). K. MÄGDEFRAU, *Paläobiologie der Pflanzen*, 4th edn., Jena, Stuttgart, 1968.

Morphology, anatomy and reproduction (see also the references under 'Morphology'). K. GOEBEL, *Organographie der Pflanzen*, **1–3**, 3rd edn., Jena, 1928–33. W. ZIMMERMANN, *Die Telomtheorie*, Stuttgart, 1965. S. CARLQUIST, *Comparative Plant Anatomy*, New York, 1961. F. KNOLL, *Fortpflanzung und Vermehrung der Gewächse* (*Handb. Biol. III/1*), Konstanz, 1963.

Palynology. G. ERDTMAN, *Handbook of Palynology*, Copenhagen, 1969.

Phytochemistry and serology. R. HEGNAUER, *Chemotaxonomie der Pflanzen*, **1**→, Basel, 1962→. R. E. ALSTON & B. L. TURNER, *Biochemical Systematics*, Englewood Cliffs, N.J., 1963. T. SWAIN (ed.), *Comparative Phytochemistry*, London, New York, 1966. J. G. HAWKES (ed.), *Chemotaxonomy and Serotaxonomy*, London, New York, 1968. T. J. MABRY, R. E. ALSTON & V. C. RUNECKLES (eds.), *Recent Advances in Phytochemistry*, **1**, New York, 1968. H. MERXMÜLLER, Chemotaxonomie? BDBG **80**, 1968. J. B. HARBORNE (ed.), *Phytochemical Phylogeny*, London, New York, 1970.

Lower Plants

(1) General works. L. RABENHORST, *Kryptogamenflora von Deutschland, Österreich und der Schweiz*, **1–15**, Leipzig, 1884–1968. R. M. KLEIN & A. CRONQUIST, The evolutionary and taxonomic significance of biochemical ... characters in the thallophytes, *Quart. Rev. Biol.* **42**, 105–296, 1967. G. M. SMITH, *Cryptogamic Botany*, **1–2**, 2nd edn., New York, 1955.

(2) Schizophyta. H. G. SCHLEGEL, *Allgemeine Mikrobiologie*, Stuttgart, 1969. R. Y. STANIER, M. DOUDOROFF & E. A. ADELBERG, *General Microbiology*, 3rd edn., London, 1971. I. C. GUNSALUS & R. Y. STANIER, *The Bacteria*, **1–5**, New York, 1960–64. K. V. THIMANN, *The Life of Bacteria: Their Growth, Metabolism and Relationship*, 2nd edn., New York, 1963. N. A. KRASSILNIKOW, *Diagnostik der Bacterien und Actinomyceten*, Jena, 1959. D. H. BERGEY, *Manual of Determinative Bacteriology*, 7th edn., by R. S. BREED *et al.*, Baltimore, 1957. M. J. R. SALTON, *The Bacterial Cell Wall*, Amsterdam, 1964. J. R. SOKATCH, *Bacterial Physiology and Metabolism*, London, 1969. W. HAYES, *The Genetics of Bacteria and their Viruses*, 2nd edn., Oxford, 1968. S. A. WAKSMAN, *The Actinomycetes*, Waltham, 1950. G. SYKES & F. A. SKINNER (eds.), *Actinomycetales*, London, 1973.

(3) W. WEIDEL, *Virus und Molekularbiologie*, Berlin, 1964. F. C. BAWDEN, *Plant Viruses and Virus Diseases*, 4th edn., New York, 1964. M. KLINKOWSKY, *Pflanzliche Virologie*, 2nd edn., Berlin, 1968. K. M. SMITH, *A Textbook of Plant Virus Diseases*, 3rd edn., London, 1972. M. H. ADAMS, *Bacteriophages*, New York, 1959.

(4) L. GEITLER, *Schizophyceen*, HA **6**, 1, 2nd edn., Berlin, 1960. G. E. FOGG, W. D. P. STEWART, P. FAY & A. E. WALSBY, *The Blue-Green Algae*, London, 1973. N. G. CARR & B. A. WHITTON (eds.), *The Biology of the Blue-Green Algae*, Oxford, 1973.

(5) Phycophyta. F. OLTMANNS, *Morphologie und Biologie der Algen*, **1–3**, 2nd edn., Jena, 1922–23. L. NEWTON, *A Handbook of the British Seaweeds*, London, 1931. F. E. FRITSCH, *The Structure and Reproduction of the Algae*, **1–2**, Cambridge, 1935, 1945. B. SCHUSSNIG, *Handbuch der Protophytenkunde*, **1–2**, Jena, 1953, 1959. G. M. SMITH, *Cryptogamic Botany*, 2nd edn., **1**, New York, 1955. B. FOTT & H. GLENK, *Algenkunde*, 2nd edn., Jena, 1971. V. J. CHAPMAN, *The Algae*, London, 1962. K. GRELL, *Protozoologie*, 2nd edn., Berlin, 1968. R. A. LEWIN (ed.), *Physiology and Biochemistry of the Algae*, New York, 1962. P.-P. GRASSÉ, *Traité de Zoologie*, **1**, Paris, 1952. E. G. PRINGSHEIM, *Farblose Algen*, Stuttgart, 1963; *Pure Cultures of Algae*, Cambridge, 1946. G. HUBER-PESTALOZZI, *Das Phytoplankton des Süsswassers*, Teile 1–7 (*Die Binnengewässer* **16**), Stuttgart, 1938–69. P. BOURELLY, *Les Algues d'Eau Douce*, **1–2**, Paris, 1966, 1968. A. PASCHER, *Die Süsswasserflora Deutschlands, Österreichs und der Schweiz*, **1–15**, Jena, 1913–36. G. M. SMITH, *The Freshwater Algae of the United States*, 2nd edn., New York, 1950. H. KYLIN, *Die Gattungen der Rhodophyceen*, Lund, 1956. J. D. DODGE, *The Fine Structure of Algal Cells*, London, 1973. W. P. D. STEWART (ed.), *Algal Physiology and Biochemistry*, Oxford, 1974.

(6) F. OLTMANNS, **3**, *see* **(5)**. F. GESSNER, *Hydrobotanik*, **1–2**, Berlin, 1955, 1959; *Meer und Strand*, 2nd edn., Berlin, 1957. S. A. SERNOW, *Allgemeine Hydrobiologie*, Berlin, 1958. F. RUTTNER, *Grundriss der Limnologie*, 3rd edn., Berlin, 1962 (trans. by D. G. FREY as *Fundamentals of Limnology*, Toronto, 1963). F. E. ROUND, *The Biology of the Algae*, 2nd edn., London, 1973. H. FRIEDRICH, *Meeresbiologie*, Berlin-Nikolassee, 1955 (trans. by G. VEVERS as *Marine Biology*, London, 1969).

(7) Mycophyta. G. C. AINSWORTH, A. S. SUSSMAN *et al.*, *The Fungi*, **1–4**, New York, 1965–73. C. J. ALEXOPOULOS, *Introductory Mycology*, 2nd edn., New York, 1962. E. A. BESSEY, *Morphology and Taxonomy of Fungi*, Philadelphia, 1950 (reprint 1964). E. GÄUMANN, *Die Pilze*, 2nd edn., Basel, Stuttgart, 1964. J. A. von ARX, *Pilzkunde*, 2nd edn., Lehre, 1968. J. H. BURNETT, *Fundamentals of Mycology*, London, 1968. E. MÜLLER & W. LÖFFLER, *Mykologie*, Stuttgart, 1971. J. WEBSTER, *Introduction to Fungi*, Cambridge, 1970. C. T. INGOLD, *The Biology of Fungi*, London, 1961. K. ESSER & R. KUENEN, *Genetik der Pilze*, Berl., Heidlb., 1965 (translated as *Genetics of*

Fungi by E. STEINER, 1967). J. R. S. FINCHAM & P. R. DAY, *Fungal Genetics*, 3rd edn., Oxford, 1971. G. C. AINSWORTH, *Dictionary of the Fungi*, 6th edn., Kew, 1971. H. KREISEL, *Grundzüge eines natürlichen Systems der Pilze*, Jena, 1969.

(8) J. T. BONNER, *The Cellular Slime Molds*, 2nd edn., Princeton, 1967. A. L. LISTER, *A Monograph of Mycetozoa*, 3rd edn., London, 1925 (reprint 1964). W. D. GRAY & C. J. ALEXOPOULOS, *Biology of the Myxomycetes*, New York, 1968.

(9) F. K. SPARROW, *Aquatic Phycomycetes*, 2nd edn., Ann Arbor, 1960. H. M. FITZPATRICK, *The Lower Fungi (Phycomycetes)*, London, 1930 (reprint 1966). J. S. KARLING, *The Plasmodiophorales*, 2nd edn., New York, 1968.

(10) R. W. G. Dennis, *British Ascomycetes*, Lehre, 1968. E. MICHAEL & B. HENNIG, *Handbuch für Pilzfreunde*, **1–2**, Jena, 1958–60. M. MOSER, *Ascomyceten* (GAMS, *Kleine Kryptogamenflora*, **2a**), Stuttgart, 1963.

(11) J. LODDER (ed.), *The Yeasts*, 2nd edn., Amsterdam, 1970. E. MRAK, *The Life of Yeasts*, Cambridge, Mass., 1966. A. H. ROSE & J. S. HARRISON (eds.), *The Yeasts*, **1–2**, London, 1969–71.

(12) K. B. RAPER & C. THOM, *A Manual of the Penicillia*, London, 1949 (reprint 1968). K. B. RAPER & D. I. FENNELL, *The Genus Aspergillus*, Baltimore, 1965. S. BLUMER, *Echte Mehltaupilze (Erysiphaceae)*, Jena, 1967.

(13) E. MÜLLER & J. A. von ARX, Die Gattungen der Pyrenomyceten, *Beitr. z. Kryptogamenflora der Schweiz*, **11**, 1954–62. H.-W. SCHELOSKE, *Beitrage zur Biologie der Laboulbeniales*, Jena, 1969.

(14) *See* DENNIS **(10)**. MOSER, *see* **(10)**. F. J. SEAVER, *North American Cup Fungi*, **1–2**, New York, 1961.

(15) E. GÄUMANN, *Die Rostpilze Mitteleuropas*, Bern, 1959. S. BLUMER, *Rost- und Brandpilze auf Kulturpflanzen*, Jena, 1963. SORAUER, **III/4**, 1962, *see* **(18)**.

(16) E. MICHAEL & B. HENNIG, **1**, **3**, **4**, **5**, *see* **(10)**. R. SINGER, *The Agaricales in Modern Taxonomy*, Weinheim, 1962. R. KUEHNER & H. ROMAGNESI, *Flore Analytique des Champignons supérieurs*, Paris, 1953. M. MOSER, *Basidiomyceten, II, Röhrlinge und Blätterpilze* (*Agaricales*), 3rd edn. (GAMS, *Kleine Kryptogamenflora* **2.b.2**), 1967. R. WATLING, *Identification of the Larger Fungi*, Amersham, 1973.

(17) H. L. BARNETT, *Illustrated Genera of Fungi Imperfecti*, 2nd edn., Minneapolis, 1960. G. L. BARRON, *The Genera of Hyphomycetes from the Soil*, Baltimore, 1968.

(18) M. KLINKOWSKI, E. MÜHLE & E. REINMUTH, *Krankheiten und Schädlinge gärtnerischer Kulturpflanzen*, **1–3**, Berlin, 1965–68. P. SORAUER, *Handbuch der Pflanzen Krankheiten*, 6th end., **3** (Pilzliche Krankheiten), Berl., Hamb., 1962→. J. C. WALKER, *Plant Pathology*, 3rd edn., London, 1969. W. BRANDENBURGER, *Vademecum zum Sammeln parasitischer Pilze*, Stuttgart, 1963. H. BRAUN & E. RIEHM, *Krankheiten und Schädlinge der Kulturpflanzen und ihre Bekämpfung*, 8th edn., Berl., Hamb., 1957. W. KOTTE, *Krankheiten und Schädlinge im Obstbau*, 3rd edn., Berl., Hamb., 1958. E. GÄUMANN, *Pflanzliche Infektionslehre*, 2nd edn., Basel, 195[illegible]. H. MARTIN, *Die wissenschaftl. Grundlagen des Pflanzenschutzes*, Weinheim, 1967. R. SCHAEDE, *Die pflanzlichen Symbiosen*, 3rd edn., Jena, 1962. S. M. HENRY, *Symbiosis*, **1–2**, London, 1966–67. P. BUCHNER, *Endosymbiose der Tiere mit pflanzlichen Mikroorganismen*, Basel, Stuttgart, 1953. J. G. HORSFALL & A. E. DIMOND (eds.), *Plant Pathology*, **1–2**, London, 1959, 1960. B. E. J. WHEELER, *An Introduction to Plant Diseases*, London, 1969. R. K. S. WOOD, A. BALLIO & A. GRANITI (eds.), *Phytotoxins in Plant Diseases*, London, 1972. J. L. HARLEY, *The Biology of Mycorrhiza*, 2nd edn., London, 1969. H. J. HUDSON, *Fungal Saprophytism*, London, 1972. S. D. GARRETT, *Pathogenic Root-infecting Fungi*, Cambridge, 1970. T. W. JOHNSON & F. K. SPARROW, *Fungi in Oceans and Estuaries*, Weinheim, 1961. H. ZÄHNER, *Biologie der Antibiotika*, Heidelberg, 1965. V. RYPACEK, *Biologie holzzerstörender Pilze*, Jena, 1966. H. KREISEL, *Die phytopathogenen Grosspilze*, Jena, 1961. BAWENDAMM, *Der Hausschwamm*, Stuttgart, 1969.

(19) Lichenes. H. des ABBAYES, *Traité de lichénologie, Encyclopédie Biol.*, **41**, Paris, 1951. M. E. HALE, *Biology of Lichens*, London, 1967. V. AHMADJIAN, *The Lichen Symbiosis*, Waltham, Mass., 1967. U. K. DUNCAN, *Introduction to British Lichens*, Arbroath, 1970. C. F. CULBERSON, *Chemical and Botanical Guide to Lichen Products*, Chapel Hill, N.C., 1969. K. L. ALVIN & K. A. KERSHAW, *The Observer's Book of Lichens*, London, 1963 (reprinted 1966). P. OZENDA, *Lichens, Handbuch der Pflanzenanatomie*, 2nd edn., **VI/9**, Berlin, 1963. J. ANDERS, *Die Laub- und Strauchflechten Mitteleuropas*, Jena, 1928. K. BERTSCH, *Flechtenflora von Nordwestdeutschland*, Stuttgart, 1957. J. POELT, *Bestimmungschlüssel europäischer Flechten*, 2nd edn., Lehre, 1969. H. GAMS, *Flechten (Kleine Kryptogamenflora,* **3**), Stuttgart, 1967. O. GALLØE, *Natural History of Danish Lichens*, **1–9**, Copenhagen, 1927–54. O. KLEMENT, Prodromus der mitteleuropaischen Flechtengesellschaften, *Fedde's Repert., Beih.* **125**, 1955. M. STEINER in *Handbuch der Pflanzenphysiologie* **15/I**, 1965; *Flechtensymposium*, 1969, Stuttgart, 1970.

(20) Bryophyta. D. H. CAMPBELL, *The Structure and Development of Mosses and Ferns*, 3rd edn., New York, 1918. F. VERDOORN, *Manual of Bryology*, The Hague, 1932 (reprint 1967). G. M. SMITH, **2**, *see* **(1)**. N. S. PARIHAR, *An Introduction to Embryophyta*, **1** *Bryophyta*, 5th edn., Allahabad, 1965. E. V. WATSON, *The Structure and Life of Bryophytes*, 3rd edn., London, 1971. K. GOEBEL, *Organographie der Pflanzen*, 3rd edn., **2**, Jena, 1930. TH. HERZOG, *Geographie der Moose*, Jena, 1926. K. BERTSCH, *Moosflora von S.W. Deutschland*, 3rd edn., Stuttgart, 1966. H. GAMS, *Die Moos- und Farnpflanzen* (*Kleine Kryptogamenflora*, **4**), 4th edn., Stuttgart, 1957. R. M. SCHUSTER, *The Hepaticae and Anthocerotae of N. America*, **1–2**, New York, 1966, 1969. E. V. WATSON, *British Mosses and Liverworts*, Cambridge, 1968.

(21) Pteridophyta. K. GOEBEL, *see* **(20)**. D. H. CAMPBELL, *see* **(20)**. F. VERDOORN, *Manual of Pteridology*, The Hague, 1938 (reprint 1967). F. O. BOWER, *Primitive Land Plants*, London, 1935 (reprint 1959). A. J. EAMES, *Morphology of Vascular Plants (Lower Groups)*, New York, 1936. A. S. FOSTER & E. M. GIFFORD, *Comparative Morphology of Vascular Plants*, San Francisco, 1959. K. R. SPORNE, *The Morphology of Pteridophytes*, 3rd edn., London, 1970. N. S. PARIHAR, *An Introduction to Embryophyta,* **2** *Pteridophyta*, 4th edn., Allahabad, 1963. W. TROLL, *Vergleichende Morphologie der höheren Pflanzen*, Berlin, 1937→. I. MANTON, *Problems of Cytology and Evolution in the Pteridophyta*, Cambridge, 1950. H. CHRIST, *Geographie der Farne*, Jena, 1910. K. RASBACH & O. WILMANS, *Die Farnpflanzen Zentraleuropas*, Heidelberg, 1968. For fossil Pteridophyta see under 'Palaeobotany'.

(22) H. NESSEL, *Die Bärlappgewächse (Lycopodiaceae)*, Jena, 1959.

(23) E. B. COPELAND, *Genera Filicum*, Waltham, Mass., 1947. F. O. BOWER, *The Ferns*, **1–3**, Cambridge, 1923–28 (reprint 1964).

Spermatophyta

General works. G. H. M. LAWRENCE, *Taxonomy of Vascular Plants*, New York, 1951. A. CRONQUIST, A. TAKHTAJAN & W. ZIMMERMANN, On the higher taxa of Embryobionta, *Taxon* **15**, 1966. J. C. WILLIS, *A Dictionary of Flowering Plants and Ferns*, 8th edn. (revised by H. K. AIRY SHAW), Cambridge, 1973.

Morphology, reproduction, etc. W. TROLL, *Vergleichende Morphologie der höheren Pflanzen*, Berlin, 1937–43 (reprint 1967). W. TROLL, *Praktische Einführung in die Pflanzenmorphologie,* **1–2**, Jena, 1954, 1957. A. S. FOSTER & E. M. GIFFORD, *Comparative Morphology of Vascular Plants*, San Francisco, 1959. A. D. J. MEEUSE, *Fundamentals of Phytomorphology*, New York, 1966; Fortpflanzung der Samenpflanzen, HN **4**, 1934; Sexualität, Fortpflanzung, Generationswechsel, HP **18**, 1967. F. J. WIDDER, Der Generationswechsel der Spermatophyten, *Aquilo, Ser. Bot.* **6**, 1967. G. ERDTMAN, *Pollen Morphology and Plant Taxonomy; Angiosperms*, Stockholm, 1952 (corrected reprint New York, 1966); *Gymnospermae*, etc., 1957, 1965. K. FAEGRI & J. IVERSEN, *Textbook of Modern Pollen Analysis*, 2nd edn., Oxford, 1964. D. A. JOHANSEN, *Plant Embryology*, Waltham, 1950.

Floras (a selection). T. G. TUTIN *et al.*, *Flora Europaea*, **1** et seq., Cambridge, 1964→. G. HEGI, *Illustrierte Flora von Mitteleuropa*, **1–7**, 1st edn., 1906–31; 2nd edn., 1935→; 3rd edn., 1966→. O. SCHMEIL-FITSCHEN, *Flora von Deutschland*, 81st edn. (W. RAUH & K. SENGHAS), Heidelberg, 1968. W. ROTHMALER, *Exkursionsflora von Deutschland, Gefässpflanzen*, 6th edn., Berlin, 1968, with atlas & supplement. E. OBERDORFER, *Pflanzensoziologische Exkursionsflora*, 3rd edn., Stuttgart, 1970. F. EHRENDORFER *et al.*, *Liste der Gefässpflanzen Mitteleuropas*, Graz, 1967. A. R. CLAPHAM, T. G. TUTIN & E. F. WARBURG, *Flora of the British Isles*, 2nd edn., Cambridge, 1962.

Economic and medicinal plants. W. BAUMEISTER & REICHART, *Lehrbuch der angewandten Botanik*, Stuttgart, 1969. J. C. T. UPHOF, *Dictionary of Economic Plants*, Weinheim, 1959. R. MANSFELD, *Vorläufiges Verzeichnis landwirtschaftlich oder gärtnerisch Kultivierter Pflanzenarten*, Berlin, 1962. A. SPRECHER von BERNEGG, *Tropische und subtropische Weltwirtschaftpflanzen*, **1–5**, Stuttgart, 1st edn., 1929–36, 2nd edn., 1960→. J. BAERNER & J. F. MUELLER, *Die Nutzhölzer der Welt*, **1–4**, Neudamm-Weinheim, 1942–61. G. KRÜSSMANN, *Handbuch der Laubgehölze*, **1–2**, Berlin, 1960–61. PAREYS, *Blumengärtnerei*, 2nd edn., **1–3**, 1958–61. O. GESSNER, *Die Gift- und Arzneipflanzen von Mitteleuropa*, 2nd edn., Berlin, 1956. F. BERGER, *Handbuch der Drogenkunde*, **1–5**, Vienna, 1946–60. For medicinal plants and drugs cf. also the latest editions of the Pharmacopeias of Germany, Austria, Switzerland and Britain. W. J. BEAN, *Trees and Shrubs Hardy in the British Isles*, 8th edn., **1–4**, 1970→. J. W. PURSEGLOVE, *Tropical Crop Plants: Dicotyledons*, **1–2**, London, 1968; *Monocotyledons*, **1–2**, London, 1972. I. H. BURKILL, *A Dictionary of the Economic Products of the Malay Peninsula*, **1–2**, Kuala Lumpur, 1966.

Coniferophytina and Cycadophytina (Gymnospermae). C. J. CHAMBERLAIN, *Gymnosperms*, Chicago, 1935. K. R. SPORNE, *The Morphology of Gymnosperms*, 2nd edn., London, 1974. H. GAUSSEN, *Les Gymnospermes actuelles et fossiles*, **1**→, Toulouse, 1943→. K. SCHNARF, Embryologie der Gymnospermen, HA **2/2**, 1933; Gymnospermensamen, HA **10/1**, 1937. C. B. BECK, The woody, fern-like trees of the Devonian, *Mem. Torrey bot. Club*, **21**, 1964. H. P. BANKS, The Early History of Land Plants, in E. T. DRAKE (ed.), *Evolution and Environment*, New Haven, 1968. R. FLORIN, Die Koniferen des Oberkarbons und des unteren Perms, *Palaeontographica*, Abt. **B85**, 1938–45; Evolution in cordaites and conifers, *Acta Horti Bergiani*, **15/11**, 1951; On the morphology and relationships of the Taxaceae, *Bot Gaz.* **110**, 1948. H. N. ANDREWS, Early seed plants, *Science, N.Y.*, **142**, 1963. T. M. HARRIS, *The Yorkshire Jurassic flora. III. Bennettitales*, London, 1969.

Magnoliophytina (Angiospermae) (see also the references to the Plant Kingdom, and to the Spermatophyta): **General works.** A. GUNDERSEN, *Families of Dicotyledons*, Waltham, 1950. J. HUTCHINSON, *The Families of Flowering Plants*, **1–2**, 2nd edn., Oxford, 1959; *Evolution and Phylogeny of Flowering Plants, Dicotyledons*, London, 1969. H. J. LAM, Reflections on Angiosperm phylogeny, *Proc. K. ned. Akad. Wet., Series C*, **64**, 1961. A. CRONQUIST, *The Evolution and Classification of Flowering Plants*, London, 1968. A. TAKHTAJAN, *Die Evolution der Angiospermen*, Jena, 1959; *Flowering Plants—Origin and Dispersal* (trans. C. JEFFREY), Edinburgh, 1969. T. ECKHARDT, Die natürliche Verwandtschaft bei den Blütenpflanzen, *Umschau* **16**, 1964. O. KIRCHNER, E. LOEW, C. SCHRÖTER, W. WANGERIN & C. SCHMUCKER, *Lebensgeschichte der Blütenpflanzen Mitteleuropas*, Stuttgart, 1908→. G. L. STEBBINS, Natural selection and the differentiation of angiosperm families, *Evolution* **5**, 1951.

General morphology and anatomy. A. J. EAMES, *Morphology of the Angiosperms*, New York, 1961. K. R. SPORNE, *The Morphology of Angiosperms*, London, 1974. C. R. METCALFE & L. CHALK, *Anatomy of the Dicotyledons*, **1–2**, Oxford, 1950. C. R. METCALFE, P. B. TOMLINSON *et al.*, *Anatomy of the Monocotyledons*, **1**→, Oxford, 1960→. I. W. BAILEY, The potentialities and limitations of wood anatomy in the study of phylogeny and classification of angiosperms, *J. Arnold Arbor.*, **38**, 1957.

Floral morphology. A. EICHLER, *Blütendiagramme*, **1–2**, Leipzig, 1875, 1878. W. TROLL, *Organisation und Gestalt im Bereich der Blüte*, Berlin, 1928. J. B. PAYER,

Traité d'Organogénie Comparée de la Fleur, Paris, 1857. R. MELVILLE, A new theory of the Angiosperm flower: I, The gynoecium, *Kew Bull.* **16**, 1962; II, The androecium, **17**, 1963. V. PURI, The role of floral anatomy in the solution of morphological problems, *Bot. Rev.* **17**, 1951. P. HIEPKO, Vergleichend-morphologische und entwicklungsgeschichtliche Untersuchungen über das Perianth bei den Polycarpicae, BJSy **84**, 1965. W. LEINFELLNER, Die petaloiden Staubblätter und ihre Beziehungen zu den Kronblättern, *Österr. Bot. Z.* **101**, 1954. H. BAUM & W. LEINFELLNER, Die ontogenetischen Abänderungen des diplophyllen Grundbaues der Staubblätter, *Österr. Bot. Z.* **100**, 1953. P. LEINS, Das zentripetale und zentrifugale Androecium, BDBG **77**, 1964. G. ECKERT, Entwicklungsgeschichtliche und blütenanatomische Untersuchungen zum Problem der Obdiplostemonie, BJSy **85**, 1966. TH. ECKARDT, Untersuchungen über Morphologie, Entwicklungsgeschichte und systematische Bedeutung des pseudomonomeren Gynoeceums, *Nova Acta Leopold.*, N.F. **5**, 1937; Vergleichende Studie über die Beziehungen zwischen Fruchtblatt, Samenanlage, und Blütenachse bei einigen Angiospermen, *Neue Hefte zur Morphologie* **3**, 1957. W. LEINFELLNER, Der Bauplan des synkarpen Gynözeums, *Österr. Bot. Z.* **97**, 1950.

Gametophytes, embryology. K. SCHNARF, *Vergleichende Morphologie der Angiospermen*, Berlin, 1931. P. MAHESHWARI, *An Introduction to the Embryology of Angiosperms*, New York, 1950; *Recent Advances in the Embryology of Angiosperms*, Delhi, 1963. G. L. DAVIS, *Systematic Embryology of the Angiosperms*, New York, 1966. R. WUNDERLICH, Zur Frage der Phylogenie der Endospermtypen . . ., *Österr. Bot. Z.* **106**, 1959. J. L. BREWBAKER, The distribution and phylogenetic significance of binucleate and trinucleate pollen grains in the angiosperms, *Amer. J. Bot.* **54**, 1967. Å. GUSTAFSON, Apomixis in higher plants, *Acta Univ. Lund* **42–3**, 1946–47. A. NYGREN, Apomixis in the angiosperms, *Bot. Rev.* **20**, 1954. A. RUTISHAUSER, Fortpflanzungsmodus und Meiose apomiktischer Blütenpflanzen, *Protoplasmatalogia* **6F3**, 1967.

Pollination. N. T. ARASU, Self-incompatibility in angiosperms, *Genetica* **39**, 1968. H. G. BAKER, Pollination mechanisms and inbreeders, *Recent Advances in Botany*, 1961. P. KNUTH, *Handbuch der Blütenbiologie*, Leipzig, 1898–**1905**. F. KNOLL, Insekten und Blumen, *Abh. zool.-bot. Ges. Wien* **12**, 1921–26; *Die Biologie der Blüte*, Berlin, 1956. H. KUGLER, *Einführung in die Blütenökologie*, 2nd edn., Stuttgart, 1970. E. WERTH, *Bau und Leben der Blumen*, Stuttgart, 1956. K. FAEGRI & L. van der PIJL, *The Principles of Pollination Ecology*, London, 2nd edn., 1971. E. E. LEPPIK, Morphogenic classification of flower types, *Phytomorphology* **18**, 1969. H. KUGLER, *UV-Male auf Blüten*, BDBG **79**, 1966. K. von FRISCH, *Aus dem Leben der Bienen*, Berlin, 1959; *Tanzsprache und Orientierung der Bienen*, Berlin, 1965. O. PORSCH, Vogelblumenstudien, JWB **63**, 1924; **70**, 1929; *Biol. Gener.* **1–5**, 1926–30; **9–12**, 1933–36. K. A. GRANT & V. GRANT, *Humming-birds and their Flowers*, New York, 1968. S. VOGEL, Chiropterophilie in der neotropischen Flora, *Flora (Abt. B)* **157**, 1969→; *Blütenbiologische Typen als Elemente der Sippengliederung*, Jena, 1954. V. GRANT & K. A. GRANT, *Flower Pollination in the Phlox Family*, New York, 1965. L. van der PIJL & C. H. DODSON, *Orchid Flowers–their Pollination and Evolution*, Coral Gables, 1966. J. B. FREE, *Insect Pollination of Crops*, London, 1970. M. C. F. PROCTOR & P. F. YEO, *The Pollination of Flowers*, London, 1973.

Seeds, Fruits, and their dispersal. F. NETOLITSKY, Angiospermen-Samen, HA **10**, 1926. G. V. BECK, A. PASCHER & F. POHL, Frucht und Same, HN **4**, 1934. W. BROUWER & A. STÄHLIN, *Handbuch der Samenkunde*, Frankfurt-am-Main, 1955. E. S. SMIRNOVA, Die Samenstruktur der Blütenpflanzen, *Biol. Zentralbl.* **84**, 1965. E. ULBRICH, *Biologie der Früchte und Samen*, Berlin, 1928. H. N. RIDLEY, *The Dispersal of Plants throughout the World*, Ashford, 1930. K. STOPP, Karpologische Studien, *Abh. Akad. Wiss. Mainz, math.-naturn. Kl.* **7** & **17**, 1950–51. P. MÜLLER, Verbreitungsbiologie der Blütenpflanzen, *Veröff. Geobot. Inst. Rübel Zurich* **30**, **1955**. L. van der PIJL, *Principles of Dispersal in Higher Plants*, Berlin, 1969. T. T. KOZLOWSKI (ed.), *Seed Biology*, **1–3**, New York, 1972.

Systematics of select groups. K. KUBITSKI, Chemosystematische Betrachtungen zur Grossgliederung der Dicotylen, *Taxon* **18**, 1969. F. EHRENDORFER *et al.*, Chromosome numbers and evolution in primitive angiosperms, *Taxon* **17**, 1968. U. JENSEN, Serologische Beiträge zur Systematik der Ranunculaceae, BJSy **88**, 1968. L. GELIUS, Studien zur Entwicklungsgeschichte an Blüten der Saxifragales sensu lato mit besonderer Berüchsichtigung des Androeceums, BJSy **87**, 1967. H. HUBER, Die Verwandtschaftsverhältnisse der Rosifloren, *Mitt. Bot. München* **5**, 1963; Die Samenmerkmale und Verwandtschaftsverhältnisse der Liliifloren, *Mitt. Bot. München* **8**, 1969. H. MERXMÜLLER & P. LEINS, Die Verwandtschaftsbeziehungen der Kreuzblütler und Mohngewächse, BJSy **86**, 1967. For other references see the periodicals FdB, BJSy, *Taxon* and Index to European Taxonomic Literature, *Regnum vegetabile,* **45**, 1966→.

Plant geography

General works. L. DIELS & F. MATTICK, *Pflanzengeographie*, 5th edn., Sammlung Göschen, 1958. H. WALTER, *Einführung in die Phytologie,* **3**, 2nd edn. & **4**, Stuttgart, 1951–70. J. SCHMITHÜSEN, *Allgemeine Vegetationsgeographie*, 3rd edn., Berlin, 1968. G. SCHMIDT, *Vegetationsgeographie auf ökologisch-soziologischer Grundlage*, Leipzig, 1969. P. DANSEREAU, *Biogeography, an Ecological Perspective*, New York, 1957. H. FREITAG, *Einführung in die Biogeographie von Mitteleuropa . . .*, Stuttgart, 1962.

Distribution patterns and floristics. H. SOLMS-LAUBACH, *Die leitenden Gesichtspunkte einer allgemeinen Pflanzengeographie*, Leipzig, 1905. H. MEUSEL, *Vergleichende Arealkunde*, Berlin, 1943. H. WALTER & H. STRAKA, *Arealkunde, Einführung in die Phytologie,* **3/2**, 2nd edn., Stuttgart, 1970. E. V. WULFF, *An Introduction to Historical Plant Geography*, Waltham, 1943. S. A. CAIN, *Foundations of Plant Geography*, New York, 1944. R. GOOD, *The Geography of Flowering Plants*, 4th edn., London, 1974. W. WANGERIN, Florenelemente und Arealtypen, BBC, Erg.-Bd. **49**, 1932. TH. HERZOG, *Geographie der Moose*, Jena, 1926. E. HANNIG & H. WINKLER, *Pflanzenareale*, Jena, from 1926. E. HULTÉN, *Atlas of the Distribution of Vascular Plants in N.W. Europe*, Stockholm, 1950. K. FAEGRI, *The Distribution of Coast Plants*, Oslo, 1960. F. H. PERRING & S. M. WALTERS, *Atlas of the British Flora*, London, 1962; *Supplement*, 1968. H. MERXMÜLLER, Untersuchungen zur Sippengliederung und Arealbildung in den Alpen, *Jb. Ver. Schutze Alpenpflanzen und -tiere*, 1952. H. MEUSEL *et al.*, *Vergleichende Chorologie der zentraleuropäischen Flora*, Jena, 1965.

Plant communities. J. BRAUN-BLANQUET, *Pflanzensoziologie*, 2nd edn., Vienna, 1964. H. ELLENBERG, *Aufgaben und Methoden der Vegetationskunde*; in H. WALTER, *Einführung in die Phytologie,* **4/1**, Stuttgart, 1956; *Landwirtschaftliche Pflanzensoziologie*, Stuttgart, 1950–54. R. KNAPP, *Experimentelle Soziologie der höheren Pflanzen*, 2nd edn., Stuttgart, 1967. P. GREIG-SMITH, *Quantitative Plant Ecology*, 2nd edn., London, 1964. V. SUKACHEV & N. DYLIS, *Fundamentals of Forest Biogeocoenology*, Edinburgh, 1968. A. GRISEBACH, *Die Vegetation der Erde*, 2nd edn., Leipzig, 1885. E. RÜBEL, *Pflanzengesellschaften der Erde*, Bern–Berlin, 1930. H. WALTER, *Die Vegetation der Erde in ökophysiologischer Betrachtung:* I, *Die tropischen und subtropischen Zonen*, 2nd edn., Jena, 1964; II, *Die gemässigten und arktischen Zonen*, Jena, 1968. C. TROLL, Der asymmetrische Aufbau der Vegetationszonen, *Ber. Geobot. Inst. Rübel Zürich*, 1948. A. ENGLER & O. DRUDE, *Die Vegetation der Erde* (monographs), Leipzig, 1896→. H. WALTER, *Vegetationsbilder*, Jena, 1903→. *Pflanzensoziologie* (regional monographs), 1931→. H. ELLENBERG, *Vegetation Mitteleuropas mit den Alpen*, in H. WALTER, *Einführung in die Phytologie,* **4/2**, Stuttgart, 1963. R. GRADMANN, *Das Pflanzenleben der Schwäbischen Alb*, 4th edn., Stuttgart, 1950. C. SCHROETER, *Das Pflanzenleben der Alpen*, 2nd edn., Zürich, 1926. J. BRAUN-BLANQUET, H. PALLMANN & R. BACH, Pflanzensoziologische und bodenkundliche Untersuchungen im Schweizerischen Nationalpark, *Erg. wiss. Unters. Schweiz. Nationalparks*, Liestal, 1954. H. JENNY-LIPS, *Vegetation der Schweizer Alpen*, Zürich, 1948. R. SCHARFETTER, *Das Pflanzenleben der Ostalpen*, Vienna, 1938. A. G. TANSLEY, *The British Islands and their Vegetation*, Cambridge, 1939 (reprint in 2 vols., 1953); *Britain's*

Green Mantle, 2nd edn., revised by M. C. F. PROCTOR, London, 1968; *Introduction to Plant Ecology*, revised by A. J. WILLIS, London, 1973. H. SJÖRS, *Nordisk Växtgeografi*, Stockholm, 1956. W. SZAFER, *The Vegetation of Poland*, Oxford, 1966. H. WALTER, *Die Vegetation des europäischen Russlands*, Berlin, 1942. E. M. LAVRENKO & V. B. SOCHAVA, *Rastitel'nyj Pokrov C.C.C.R.* [*Vegetation of Russia*, with geobotanical map and key in English] **1–2**, Leningrad, 1956. J. T. CURTIS, *The Vegetation of Wisconsin*, Madison, 1959. M. RIKLI, *Das Pflanzenleben der Mittelmeerländer*, Bern, 1943–46. J. BRAUN-BLANQUET, *Les Groupements Végétaux de la France Méditerranéenne*, Montpellier, 1951. J. BRAUN-BLANQUET & R. TÜXEN, Irische Pflanzengesellschaften in: Die Pflanzenwelt Irlands, *Veröff. Geobot. Inst. Rübel Zürich* **25**, 1952. J. H. BURNETT (ed.), *The Vegetation of Scotland*, Edinburgh, 1964. P. W. RICHARDS, *The Tropical Rain Forest*, Cambridge, 1952. E. BÜNNING, *Der tropische Regenwald*, Berlin, 1956. C. TROLL, Zur Physiognomik der Tropengewächse, *Jb. Ges. Freund. Förd. Univ. Bonn*, 1958; Die tropischen Gebirge, *Bonner Geogr. Abh.* **25**, 1959. R. SCHNELL, *Introduction à la Phytogeographie des Pays Tropicaux*, **1–2**, Paris, 1970–71. H. T. ODUM (ed.), *A Tropical Rain Forest*, Springfield, Virginia, 1970.

Ecological plant geography. H. WALTER, *Einführung in die Phytologie*, **3/1**; *Standortslehre*, 2nd edn., Stuttgart, 1960. W. SCHIMPER, *Pflanzengeographie auf physiologischer Grundlage*, 3rd edn. by F. C. von FABER, Jena, 1935. E. WARMING & P. GRAEBNER, *Lehrbuch der okologischen Pflanzengeographie*, 4th edn., Berlin, 1933. H. LUNDEGÅRDH, *Klima und Boden*, 5th edn., Jena, 1957. J. E. WEAVER & F. E. CLEMENTS, *Plant Ecology*, 2nd edn., New York, 1938. R. F. DAUBENMIRE, *Plants and Environment*, 2nd edn., New York, 1959. S. R. EYRE, *Vegetation and Soils*, 2nd edn., London, 1968. C. RAUNKIAER, *The Life Forms of Plants*, Oxford, 1934. K. RUBNER with F. REINOLD, *Die pflanzengeographischen Grundlagen des Waldbaues*, 4th edn., Radebeul-Berlin, 1953. E. AICHINGER, *Pflanzen als forstliche Standortszeiger*, Vienna, 1967. W. LAATSCH, *Dynamik der mitteleuropäischen Mineralböden*, 3rd edn., Dresden, 1954. W. L. KUBIENA, *Entwicklungslehre des Bodens*, Vienna, 1948; *The Soils of Europe*, London, 1953; Bestimmungsbuch und Systematik der Böden Europas, Stuttgart, 1953. R. GEIGER, *Das Klima der bodennahen Luftschicht*, 4th edn., Braunschweig, 1961. F. RUTTNER, *Grundriss der Limnologie*, 3rd edn., Berlin, 1962 (transl. as *Fundamentals of Limnology*, Toronto, 1963). J. L. MONTEITH, *Principles of Environmental Physics*, London, 1973. E. W. RUSSELL, *Soil Conditions and Plant Growth*, 10th edn., London, 1973.

Historical plant geography (see also the references to Systematics and Evolution; the Plant Kingdom; Palaeobotany; and the references to distribution and floristics under Plant Geography). A. ENGLER, *Versuch einer Entwicklungsgeschichte der Pflanzenwelt*, Leipzig, 1879–83. A. TAKHTAJAN, *Flowering Plants, Origin and Dispersal*, Edinburgh, 1969. R. FLORIN, The distribution of conifer and taxad genera in time and space, *Acta Horti Berg.* **20(4)**, 1963. H. TRALAU, Evolutionary trends in the genus *Ginkgo*, *Lethaia (Oslo)* **1**, 1968. M. HIRMER, Die Forschungsergebnisse auf dem Gebiet der Känophytischen Flora, BJSy **72**, 1942. B. FRENZEL, *Grundzüge der pleistozänen Vegetationsgeschichte Nord-Eurasiens*, Wiesbaden, 1968. H. GODWIN, *The History of the British Flora*, 2nd edn., Cambridge, 1975. F. FIRBAS, *Spät- und nacheiszeitiche Waldgeschichte Mitteleuropas*, Jena, 1949, 1952. P. WOLDSTEDT, *Das Eiszeitalter*, **1–3**, 2nd & 3rd edns., Stuttgart, 1958→. P. FRENZEL, *Die Klimaschwankungen des Eiszeitalters*, Braunschweig, 1967. F. ZEUNER, *Dating the Past*, 3rd edn., London, 1952. M. SCHWARZBACH, *Das Klima der Vorzeit*, 2nd edn., Stuttgart, 1961. G. ERDTMAN, *Handbook of Palynology*, Copenhagen, 1969. K. FAEGRI & J. IVERSEN, *Textbook of Pollen Analysis*, 2nd edn., Oxford, 1964. W. PENNINGTON, *The History of British Vegetation*, London, 1969.

Index

Note: A page reference may relate to the text, to a figure, or to both.